Lecture Notes in Artificial Intelligence 3614

Edited by J. G. Carbonell and J. Siekmann

Subseries of Lecture Notes in Computer Science

Lipo Wang   Yaochu Jin (Eds.)

Fuzzy Systems and Knowledge Discovery

Second International Conference, FSKD 2005
Changsha, China, August 27-29, 2005
Proceedings, Part II

 Springer

Series Editors

Jaime G. Carbonell, Carnegie Mellon University, Pittsburgh, PA, USA
Jörg Siekmann, University of Saarland, Saarbrücken, Germany

Volume Editors

Lipo Wang
Nanyang Technological University
School of Electrical and Electronic Engineering
Block S1, 50 Nanyang Avenue, Singapore 639798
E-mail: elpwang@ntu.edu.sg

Yaochu Jin
Honda Research Institute Europe
Carl-Legien-Str. 30, 63073 Offenbach/Main, Germany
E-mail: yaochu.jin@honda-ri.de

Library of Congress Control Number: 2005930642

CR Subject Classification (1998): I.2, F.4.1, F.1, F.2, G.2, I.2.3, I.4, I.5

ISSN   0302-9743
ISBN-10   3-540-28331-5 Springer Berlin Heidelberg New York
ISBN-13   978-3-540-28331-7 Springer Berlin Heidelberg New York

This work is subject to copyright. All rights are reserved, whether the whole or part of the material is concerned, specifically the rights of translation, reprinting, re-use of illustrations, recitation, broadcasting, reproduction on microfilms or in any other way, and storage in data banks. Duplication of this publication or parts thereof is permitted only under the provisions of the German Copyright Law of September 9, 1965, in its current version, and permission for use must always be obtained from Springer. Violations are liable to prosecution under the German Copyright Law.

Springer is a part of Springer Science+Business Media

springeronline.com

© Springer-Verlag Berlin Heidelberg 2005
Printed in Germany

Typesetting: Camera-ready by author, data conversion by Scientific Publishing Services, Chennai, India
Printed on acid-free paper   SPIN: 11540007   06/3142   5 4 3 2 1 0

Preface

This book and its sister volume, LNAI 3613 and 3614, constitute the proceedings of the Second International Conference on Fuzzy Systems and Knowledge Discovery (FSKD 2005), jointly held with the First International Conference on Natural Computation (ICNC 2005, LNCS 3610, 3611, and 3612) from August 27–29, 2005 in Changsha, Hunan, China. FSKD 2005 successfully attracted 1249 submissions from 32 countries/regions (the joint ICNC-FSKD 2005 received 3136 submissions). After rigorous reviews, 333 high-quality papers, i.e., 206 long papers and 127 short papers, were included in the FSKD 2005 proceedings, representing an acceptance rate of 26.7%.

The ICNC-FSKD 2005 conference featured the most up-to-date research results in computational algorithms inspired from nature, including biological, ecological, and physical systems. It is an exciting and emerging interdisciplinary area in which a wide range of techniques and methods are being studied for dealing with large, complex, and dynamic problems. The joint conferences also promoted cross-fertilization over these exciting and yet closely-related areas, which had a significant impact on the advancement of these important technologies. Specific areas included computation with words, fuzzy computation, granular computation, neural computation, quantum computation, evolutionary computation, DNA computation, chemical computation, information processing in cells and tissues, molecular computation, artificial life, swarm intelligence, ants colony, artificial immune systems, etc., with innovative applications to knowledge discovery, finance, operations research, and more. In addition to the large number of submitted papers, we were blessed with the presence of four renowned keynote speakers and several distinguished panelists.

On behalf of the Organizing Committee, we thank Xiangtan University for sponsorship, and the IEEE Circuits and Systems Society, the IEEE Computational Intelligence Society, and the IEEE Control Systems Society for technical co-sponsorship. We are grateful for the technical cooperation from the International Neural Network Society, the European Neural Network Society, the Chinese Association for Artificial Intelligence, the Japanese Neural Network Society, the International Fuzzy Systems Association, the Asia-Pacific Neural Network Assembly, the Fuzzy Mathematics and Systems Association of China, and the Hunan Computer Federation. We thank the members of the Organizing Committee, the Advisory Board, and the Program Committee for their hard work over the past 18 months. We wish to express our heart-felt appreciation to the keynote and panel speakers, special session organizers, session chairs, reviewers, and student helpers. Our special thanks go to the publisher, Springer, for publishing the FSKD 2005 proceedings as two volumes of the Lecture Notes in Artificial Intelligence series (and the ICNC 2005 proceedings as three volumes of the Lecture Notes in Computer Science series). Finally, we thank all the authors and participants for their great contributions that made this conference possible and all the hard work worthwhile.

August 2005　　　　　　　　　　　　　　　　　　　　　　　　　　　Lipo Wang
　　　　　　　　　　　　　　　　　　　　　　　　　　　　　　　　　Yaochu Jin

Organization

FSKD 2005 was organized by Xiangtan University and technically co-sponsored by the IEEE Circuits and Systems Society, the IEEE Computational Intelligence Society, and the IEEE Control Systems Society, in cooperation with the International Neural Network Society, the European Neural Network Society, the Chinese Association for Artificial Intelligence, the Japanese Neural Network Society, the International Fuzzy Systems Association, the Asia-Pacific Neural Network Assembly, the Fuzzy Mathematics and Systems Association of China, and the Hunan Computer Federation.

Organizing Committee

Honorary Conference Chairs:	Shun-ichi Amari, *Japan*
	Lotfi A. Zadeh, *USA*
General Chair:	He-An Luo, *China*
General Co-chairs:	Lipo Wang , *Singapore*
	Yunqing Huang, *China*
Program Chair:	Yaochu Jin, *Germany*
Local Arrangement Chairs:	Renren Liu, *China*
	Xieping Gao, *China*
Proceedings Chair:	Fen Xiao, *China*
Publicity Chair:	Hepu Deng, *Australia*
Sponsorship/Exhibits Chairs:	Shaoping Ling, *China*
	Geok See Ng, *Singapore*
Webmasters:	Linai Kuang, *China*
	Yanyu Liu, *China*

Advisory Board

Toshio Fukuda, *Japan*
Kunihiko Fukushima, *Japan*
Tom Gedeon, *Australia*
Aike Guo, *China*
Zhenya He, *China*
Janusz Kacprzyk, *Poland*
Nikola Kasabov, *New Zealand*
John A. Keane, *UK*
Soo-Young Lee, *Korea*
Erkki Oja, *Finland*
Nikhil R. Pal, *India*

Witold Pedrycz, *Canada*
Jose C. Principe, *USA*
Harold Szu, *USA*
Shiro Usui, *Japan*
Xindong Wu, *USA*
Lei Xu, *Hong Kong*
Xin Yao, *UK*
Syozo Yasui, *Japan*
Bo Zhang, *China*
Yixin Zhong, *China*
Jacek M. Zurada, *USA*

Program Committee Members

Janos Abonyi, *Hungary*
Jorge Casillas, *Spain*
Pen-Chann Chang, *Taiwan*
Chaochang Chiu, *Taiwan*
Feng Chu, *Singapore*
Oscar Cordon, *Spain*
Honghua Dai, *Australia*
Fernando Gomide, *Brazil*
Saman Halgamuge, *Australia*
Kaoru Hirota, *Japan*
Frank Hoffmann, *Germany*
Jinglu Hu, *Japan*
Weili Hu, *China*
Chongfu Huang, *China*
Eyke Hüllermeier, *Germany*
Hisao Ishibuchi, *Japan*
Frank Klawoon, *Germany*
Naoyuki Kubota, *Japan*
Sam Kwong, *Hong Kong*
Zongmin Ma, *China*

Michael Margaliot, *Israel*
Ralf Mikut, *Germany*
Pabitra Mitra, *India*
Tadahiko Murata, *Japan*
Detlef Nauck, *UK*
Hajime Nobuhara, *Japan*
Andreas Nürnberger, *Germany*
Da Ruan, *Belgium*
Thomas Runkler, *Germany*
Rudy Setiono, *Singapore*
Takao Terano, *Japan*
Kai Ming Ting, *Australia*
Yiyu Yao, *Canada*
Gary Yen, *USA*
Xinghuo Yu, *Australia*
Jun Zhang, *China*
Shichao Zhang, *Australia*
Yanqing Zhang, *USA*
Zhi-Hua Zhou, *China*

Special Sessions Organizers

David Siu-Yeung Cho, *Singapore*
Vlad Dimitrov, *Australia*
Jinwu Gao, *China*
Zheng Guo, *China*
Bob Hodge, *Australia*
Jiman Hong, *Korea*
Jae-Woo Lee, *Korea*
Xia Li, *China*

Zongmin Ma, *China,*
Geok-See Ng, *Singapore*
Shaoqi Rao, *China*
Slobodan Ribari, *Croatia*
Sung Y. Shin, *USA*
Yasufumi Takama, *Japan*
Robert Woog, *Australia*

Reviewers

Nitin V. Afzulpurkar
Davut Akdas
Kürat Ayan
Yasar Becerikli
Dexue Bi
Rong-Fang Bie

Liu Bin
Tao Bo
Hongbin Cai
Yunze Cai
Jian Cao
Chunguang Chang

An-Long Chen
Dewang Chen
Gang Chen
Guangzhu Chen
Jian Chen
Shengyong Chen

Shi-Jay Chen
Xuerong Chen
Yijiang Chen
Zhimei Chen
Zushun Chen
Hongqi Chen
Qimei Chen
Wei Cheng
Xiang Cheng
Tae-Ho Cho
Xun-Xue Cui
Ho Daniel
Hepu Deng
Tingquan Deng
Yong Deng
Zhi-Hong Deng
Mingli Ding
Wei-Long Ding
Fangyan Dong
Jingxin Dong
Lihua Dong
Yihong Dong
Haifeng Du
Weifeng Du
Liu Fang
Zhilin Feng
Li Gang
Chuanhou Gao
Yu Gao
Zhi Geng
O. Nezih Gerek
Rongjie Gu
Chonghui Guo
Gongde Guo
Huawei Guo
Mengshu Guo
Zhongming Han
Bo He
Pilian He
Liu Hong
Kongfa Hu
Qiao Hu
Shiqiang Hu
Zhikun Hu
Zhonghui Hu

Zhonghui Hu
Changchun Hua
Jin Huang
Qian Huang
Yanxin Huang
Yuansheng Huang
Kohei Inoue
Mahdi Jalili-Kharaajoo
Caiyan Jia
Ling-Ling Jiang
Michael Jiang
Xiaoyue Jiang
Yanping Jiang
Yunliang Jiang
Cheng Jin
Hanjun Jin
Hong Jin
Ningde Jin
Xue-Bo Jin
Min-Soo Kim
Sungshin Kim
Taehan Kim
Ibrahim Beklan Kucukdemiral
Rakesh Kumar Arya
Ho Jae Lee
Sang-Hyuk Lee
Sang-Won Lee
Wol Young Lee
Xiuren Lei
Bicheng Li
Chunyan Li
Dequan Li
Dingfang Li
Gang Li
Hongyu Li
Qing Li
Ruqiang Li
Tian-Rui Li
Weigang Li
Yu Li
Zhichao Li
Zhonghua Li
Hongxing Li
Xiaobei Liang

Ling-Zhi Liao
Lei Lin
Caixia Liu
Fei Liu
Guangli Liu
Haowen Liu
Honghai Liu
Jian-Guo Liu
Lanjuan Liu
Peng Liu
Qihe Liu
Sheng Liu
Xiaohua Liu
Xiaojian Liu
Yang Liu
Qiang Luo
Yanbin Luo
Zhi-Jun Lv
Jian Ma
Jixin Ma
Longhua Ma
Ming Ma
Yingcang Ma
Dong Miao
Zhinong Miao
Fan Min
Zhang Min
Zhao Min
Daniel Neagu
Yiu-Kai Ng
Wu-Ming Pan
Jong Sou Park
Yonghong Peng
Punpiti Piamsa-Nga
Heng-Nian Qi
Gao Qiang
Wu Qing
Celia Ghedini Ralha
Wang Rong
Hongyuan Shen
Zhenghao Shi
Jeong-Hoon Shin
Sung Chul Shin
Chonghui Song
Chunyue Song

Guangda Su
Baolin Sun
Changyin Sun
Ling Sun
Zhengxing Sun
Chang-Jie Tang
Shanhu Tang
N K Tiwari
Jiang Ping Wan
Chong-Jun Wang
Danli Wang
Fang Wang
Fei Wang
Houfeng Wang
Hui Wang
Laisheng Wang
Lin Wang
Ling Wang
Shitong Wang
Shu-Bin Wang
Xun Wang
Yong Wang
Zhe Wang
Zhenlei Wang
Zhongjie Wang
Runsheng Wang
Li Wei
Weidong Wen
Xiangjun Wen
Taegkeun Whangbo
Huaiyu Wu
Jiangning Wu
Jiangqin Wu

Jianping Wu
Shunxiang Wu
Xiaojun Wu
Yuying Wu
Changcheng Xiang
Jun Xiao
Xiaoming Xiao
Wei Xie
Gao Xin
Zongyi Xing
Hua Xu
Lijun Xu
Pengfei Xu
Weijun Xu
Xiao Xu
Xinli Xu
Yaoqun Xu
De Xu
Maode Yan
Shaoze Yan
Hai Dong Yang
Jihui Yang
Wei-Min Yang
Yong Yang
Zuyuan Yang
Li Yao
Shengbao Yao
Bin Ye
Guo Yi
Jianwei Yin
Xiang-Gang Yin
Yilong Yin
Deng Yong

Chun-Hai Yu
Haibin Yuan
Jixue Yuan
Weiqi Yuan
Chuanhua Zeng
Wenyi Zeng
Yurong Zeng
Guojun Zhang
Jian Ying Zhang
Junping Zhang
Ling Zhang
Zhi-Zheng Zhang
Yongjin Zhang
Yongkui Zhang
Jun Zhao
Quanming Zhao
Xin Zhao
Yong Zhao
Zhicheng Zhao
Dongjian Zheng
Wenming Zheng
Zhonglong Zheng
Weimin Zhong
Hang Zhou
Hui-Cheng Zhou
Qiang Zhou
Yuanfeng Zhou
Yue Zhou
Daniel Zhu
Hongwei Zhu
Xinglong Zhu

* The term after a name may represent either a country or a region.

Table of Contents – Part II

Dimensionality Reduction

Dimensionality Reduction for Semi-supervised Face Recognition
Weiwei Du, Kohei Inoue, Kiichi Urahama 1

Cross-Document Transliterated Personal Name Coreference Resolution
Houfeng Wang 11

Difference-Similitude Matrix in Text Classification
Xiaochun Huang, Ming Wu, Delin Xia, Puliu Yan 21

A Study on Feature Selection for Toxicity Prediction
Gongde Guo, Daniel Neagu, Mark T.D. Cronin 31

Application of Feature Selection for Unsupervised Learning in Prosecutors' Office
Peng Liu, Jiaxian Zhu, Lanjuan Liu, Yanhong Li, Xuefeng Zhang 35

A Novel Field Learning Algorithm for Dual Imbalance Text Classification
Ling Zhuang, Honghua Dai, Xiaoshu Hang 39

Supervised Learning for Classification
Hongyu Li, Wenbin Chen, I-Fan Shen 49

Feature Selection for Hyperspectral Data Classification Using Double Parallel Feedforward Neural Networks
Mingyi He, Rui Huang 58

Robust Nonlinear Dimension Reduction: A Self-organizing Approach
Yuexian Hou, Liyue Yao, Pilian He 67

An Effective Feature Selection Scheme via Genetic Algorithm Using Mutual Information
Chunkai K. Zhang, Hong Hu 73

Pattern Recognition and Trend Analysis

Pattern Classification Using Rectified Nearest Feature Line Segment
Hao Du, Yan Qiu Chen 81

Palmprint Identification Algorithm Using Hu Invariant Moments
Jin Soo Noh, Kang Hyeon Rhee 91

Generalized Locally Nearest Neighbor Classifiers for Object Classification
Wenming Zheng, Cairong Zou, Li Zhao 95

Nearest Neighbor Classification Using Cam Weighted Distance
Chang Yin Zhou, Yan Qiu Chen 100

A PPM Prediction Model Based on Web Objects' Popularity
Lei Shi, Zhimin Gu, Yunxia Pei, Lin Wei 110

An On-line Sketch Recognition Algorithm for Composite Shape
Zhan Ding, Yin Zhang, Wei Peng, Xiuzi Ye, Huaqiang Hu 120

Axial Representation of Character by Using Wavelet Transform
Xinge You, Bin Fang, Yuan Yan Tang, Luoqing Li, Dan Zhang 130

Representing and Recognizing Scenario Patterns
Jixin Ma, Bin Luo 140

A Hybrid Artificial Intelligent-Based Criteria-Matching with Classification Algorithm
Alex T.H. Sim, Vincent C.S. Lee 150

Auto-generation of Detection Rules with Tree Induction Algorithm
Minsoo Kim, Jae-Hyun Seo, Il-Ahn Cheong, Bong-Nam Noh 160

Hand Gesture Recognition System Using Fuzzy Algorithm and RDBMS for Post PC
Jung-Hyun Kim, Dong-Gyu Kim, Jeong-Hoon Shin, Sang-Won Lee, Kwang-Seok Hong 170

An Ontology-Based Method for Project and Domain Expert Matching
Jiangning Wu, Guangfei Yang 176

Pattern Classification and Recognition of Movement Behavior of Medaka (*Oryzias Latipes*) Using Decision Tree
Sengtai Lee, Jeehoon Kim, Jae-Yeon Baek, Man-Wi Han, Tae-Soo Chon 186

A New Algorithm for Computing the Minimal Enclosing Sphere in
Feature Space
 Chonghui Guo, Mingyu Lu, Jiantao Sun, Yuchang Lu 196

Y-AOI: Y-Means Based Attribute Oriented Induction Identifying Root
Cause for IDSs
 Jungtae Kim, Gunhee Lee, Jung-taek Seo, Eung-ki Park,
 Choon-sik Park, Dong-kyoo Kim 205

New Segmentation Algorithm for Individual Offline Handwritten
Character Segmentation
 K.B.M.R. Batuwita, G.E.M.D.C. Bandara 215

A Method Based on the Continuous Spectrum Analysis for Fingerprint
Image Ridge Distance Estimation
 Xiaosi Zhan, Zhaocai Sun, Yilong Yin, Yayun Chu 230

A Method Based on the Markov Chain Monte Carlo for Fingerprint
Image Segmentation
 Xiaosi Zhan, Zhaocai Sun, Yilong Yin, Yun Chen 240

Unsupervised Speaker Adaptation for Phonetic Transcription Based
Voice Dialing
 Weon-Goo Kim, MinSeok Jang, Chin-Hui Lee 249

A Phase-Field Based Segmentation Algorithm for Jacquard Images
Using Multi-start Fuzzy Optimization Strategy
 Zhilin Feng, Jianwei Yin, Hui Zhang, Jinxiang Dong 255

Dynamic Modeling, Prediction and Analysis of Cytotoxicity on
Microelectronic Sensors
 Biao Huang, James Z. Xing 265

Generalized Fuzzy Morphological Operators
 Tingquan Deng, Yanmei Chen 275

Signature Verification Method Based on the Combination of Shape and
Dynamic Feature
 Yingna Deng, Hong Zhu, Shu Li, Tao Wang 285

Study on the Matching Similarity Measure Method for Image Target
Recognition
 Xiaogang Yang, Dong Miao, Fei Cao, Yongkang Ma 289

3-D Head Pose Estimation for Monocular Image
 Yingjie Pan, Hong Zhu, Ruirui Ji 293

The Speech Recognition Based on the Bark Wavelet Front-End
Processing
  Xueying Zhang, Zhiping Jiao, Zhefeng Zhao 302

An Accurate and Fast Iris Location Method Based on the Features of
Human Eyes
  Weiqi Yuan, Lu Xu, Zhonghua Lin 306

A Hybrid Classifier for Mass Classification with Different Kinds of
Features in Mammography
  Ping Zhang, Kuldeep Kumar, Brijesh Verma 316

Data Mining Methods for Anomaly Detection of HTTP Request
Exploitations
  Xiao-Feng Wang, Jing-Li Zhou, Sheng-Sheng Yu, Long-Zheng Cai .. 320

Exploring Content-Based and Image-Based Features for Nude Image
Detection
  *Shi-lin Wang, Hong Hui, Sheng-hong Li, Hao Zhang, Yong-yu Shi,
  Wen-tao Qu* 324

Collision Recognition and Direction Changes Using Fuzzy Logic for
Small Scale Fish Robots by Acceleration Sensor Data
  Seung Y. Na, Daejung Shin, Jin Y. Kim, Su-Il Choi 329

Fault Diagnosis Approach Based on Qualitative Model of Signed
Directed Graph and Reasoning Rules
  Bingshu Wang, Wenliang Cao, Liangyu Ma, Ji Zhang 339

Visual Tracking Algorithm for Laparoscopic Robot Surgery
  Min-Seok Kim, Jin-Seok Heo, Jung-Ju Lee 344

Toward a Sound Analysis System for Telemedicine
  Cong Phuong Nguyen, Thi Nyoc Yen Pham, Castelli Eric 352

Other Topics in FSKD Methods

Structural Learning of Graphical Models and Its Applications to
Traditional Chinese Medicine
  Ke Deng, Delin Liu, Shan Gao, Zhi Geng 362

Study of Ensemble Strategies in Discovering Linear Causal Models
  Gang Li, Honghua Dai 368

The Entropy of Relations and a New Approach for Decision Tree
Learning
    Dan Hu, HongXing Li 378

Effectively Extracting Rules from Trained Neural Networks Based
on the New Measurement Method of the Classification Power of
Attributes
    Dexian Zhang, Yang Liu, Ziqiang Wang 388

EDTs: Evidential Decision Trees
    Huawei Guo, Wenkang Shi, Feng Du 398

GSMA: A Structural Matching Algorithm for Schema Matching in
Data Warehousing
    Wei Cheng, Yufang Sun 408

A New Algorithm to Get the Correspondences from the Image Sequences
    Zhiquan Feng, Xiangxu Meng, Chenglei Yang 412

An Efficiently Algorithm Based on Itemsets-Lattice and Bitmap Index
for Finding Frequent Itemsets
    Fuzan Chen, Minqiang Li 420

Weighted Fuzzy Queries in Relational Databases
    Ying-Chao Zhang, Yi-Fei Chen, Xiao-ling Ye, Jie-Liang Zheng 430

Study of Multiuser Detection: The Support Vector Machine Approach
    Tao Yang, Bo Hu 442

Robust and Adaptive Backstepping Control for Nonlinear Systems
Using Fuzzy Logic Systems
    Gang Chen, Shuqing Wang, Jianming Zhang 452

Online Mining Dynamic Web News Patterns Using Machine Learn
Methods
    Jian-Wei Liu, Shou-Jian Yu, Jia-Jin Le 462

A New Fuzzy MCDM Method Based on Trapezoidal Fuzzy AHP and
Hierarchical Fuzzy Integral
    Chao Zhang, Cun-bao Ma, Jia-dong Xu 466

Fast Granular Analysis Based on Watershed in Microscopic Mineral
Images
    Danping Zou, Desheng Hu, Qizhen Liu 475

Cost-Sensitive Ensemble of Support Vector Machines for Effective
Detection of Microcalcification in Breast Cancer Diagnosis
  Yonghong Peng, Qian Huang, Ping Jiang, Jianmin Jiang 483

High-Dimensional Shared Nearest Neighbor Clustering Algorithm
  Jian Yin, Xianli Fan, Yiqun Chen, Jiangtao Ren 494

A New Method for Fuzzy Group Decision Making Based on α-Level
Cut and Similarity
  Jibin Lan, Liping He, Zhongxing Wang 503

Modeling Nonlinear Systems: An Approach of Boosted Linguistic Models
  Keun-Chang Kwak, Witold Pedrycz, Myung-Geun Chun 514

Multi-criterion Fuzzy Optimization Approach to Imaging from
Incomplete Projections
  Xin Gao, Shuqian Luo 524

Transductive Knowledge Based Fuzzy Inference System for Personalized
Modeling
  Qun Song, Tianmin Ma, Nikola Kasabov 528

A Sampling-Based Method for Mining Frequent Patterns from
Databases
  Yen-Liang Chen, Chin-Yuan Ho 536

Lagrange Problem in Fuzzy Reversed Posynomial Geometric
Programming
  Bing-yuan Cao 546

Direct Candidates Generation: A Novel Algorithm for Discovering
Complete Share-Frequent Itemsets
  Yu-Chiang Li, Jieh-Shan Yeh, Chin-Chen Chang 551

A Three-Step Preprocessing Algorithm for Minimizing E-Mail
Document's Atypical Characteristics
  Ok-Ran Jeong, Dong-Sub Cho 561

Failure Detection Method Based on Fuzzy Comprehensive Evaluation
for Integrated Navigation System
  Guoliang Liu, Yingchun Zhang, Wenyi Qiang, Zengqi Sun 567

Product Quality Improvement Analysis Using Data Mining: A Case
Study in Ultra-Precision Manufacturing Industry
  Hailiang Huang, Dianliang Wu 577

Two-Tier Based Intrusion Detection System
  Byung-Joo Kim, Il Kon Kim 581

SuffixMiner: Efficiently Mining Frequent Itemsets in Data Streams by
Suffix-Forest
  Lifeng Jia, Chunguang Zhou, Zhe Wang, Xiujuan Xu 592

Improvement of Lee-Kim-Yoo's Remote User Authentication Scheme
Using Smart Cards
  Da-Zhi Sun, Zhen-Fu Cao 596

Mining of Spatial, Textual, Image and Time-Series Data

Grapheme-to-Phoneme Conversion Based on a Fast TBL Algorithm in
Mandarin TTS Systems
  Min Zheng, Qin Shi, Wei Zhang, Lianhong Cai 600

Clarity Ranking for Digital Images
  Shutao Li, Guangsheng Chen 610

Attribute Uncertainty in GIS Data
  Shuliang Wang, Wenzhong Shi, Hanning Yuan, Guoqing Chen 614

Association Classification Based on Sample Weighting
  Jin Zhang, Xiaoyun Chen, Yi Chen, Yunfa Hu 624

Using Fuzzy Logic for Automatic Analysis of Astronomical Pipelines
  Lior Shamir, Robert J. Nemiroff 634

On the On-line Learning Algorithms for EEG Signal Classification in
Brain Computer Interfaces
  Shiliang Sun, Changshui Zhang, Naijiang Lu 638

Automatic Keyphrase Extraction from Chinese News Documents
  Houfeng Wang, Sujian Li, Shiwen Yu 648

A New Model of Document Structure Analysis
  Zhiqi Wang, Yongcheng Wang, Kai Gao 658

Prediction for Silicon Content in Molten Iron Using a Combined
Fuzzy-Associative-Rules Bank
  Shi-Hua Luo, Xiang-Guan Liu, Min Zhao 667

An Investigation into the Use of Delay Coordinate Embedding Technique with MIMO ANFIS for Nonlinear Prediction of Chaotic Signals
Jun Zhang, Weiwei Dai, Muhui Fan, Henry Chung, Zhi Wei, D. Bi 677

Replay Scene Based Sports Video Abstraction
Jian-quan Ouyang, Jin-tao Li, Yong-dong Zhang 689

Mapping Web Usage Patterns to MDP Model and Mining with Reinforcement Learning
Yang Gao, Zongwei Luo, Ning Li 698

Study on Wavelet-Based Fuzzy Multiscale Edge Detection Method
Wen Zhu, Beiping Hou, Zhegen Zhang, Kening Zhou 703

Sense Rank AALesk: A Semantic Solution for Word Sense Disambiguation
Yiqun Chen, Jian Yin 710

Automatic Video Knowledge Mining for Summary Generation Based on Un-supervised Statistical Learning
Jian Ling, Yiqun Lian, Yueting Zhuang 718

A Model for Classification of Topological Relationships Between Two Spatial Objects
Wu Yang, Ya Luo, Ping Guo, HuangFu Tao, Bo He 723

A New Feature of Uniformity of Image Texture Directions Coinciding with the Human Eyes Perception
Xing-Jian He, Yue Zhang, Tat-Ming Lok, Michael R. Lyu 727

Sunspot Time Series Prediction Using Parallel-Structure Fuzzy System
Min-Soo Kim, Chan-Soo Chung 731

A Similarity Computing Algorithm for Volumetric Data Sets
Tao Zhang, Wei Chen, Min Hu, Qunsheng Peng 742

Extraction of Representative Keywords Considering Co-occurrence in Positive Documents
Byeong-Man Kim, Qing Li, KwangHo Lee, Bo-Yeong Kang 752

On the Effective Similarity Measures for the Similarity-Based Pattern Retrieval in Multidimensional Sequence Databases
Seok-Lyong Lee, Ju-Hong Lee, Seok-Ju Chun 762

Crossing the Language Barrier Using Fuzzy Logic
Rowena Chau, Chung-Hsing Yeh 768

New Algorithm Mining Intrusion Patterns
Wu Liu, Jian-Ping Wu, Hai-Xin Duan, Xing Li 774

Dual Filtering Strategy for Chinese Term Extraction
Xiaoming Chen, Xuening Li, Yi Hu, Ruzhan Lu 778

White Blood Cell Segmentation and Classification in Microscopic Bone Marrow Images
Nipon Theera-Umpon 787

KNN Based Evolutionary Techniques for Updating Query Cost Models
Zhining Liao, Hui Wang, David Glass, Gongde Guo 797

A SVM Method for Web Page Categorization Based on Weight Adjustment and Boosting Mechanism
Mingyu Lu, Chonghui Guo, Jiantao Sun, Yuchang Lu 801

Fuzzy Systems in Bioinformatics and Bio-medical Engineering

Feature Selection for Specific Antibody Deficiency Syndrome by Neural Network with Weighted Fuzzy Membership Functions
*Joon S. Lim, Tae W. Ryu, Ho J. Kim,
Sudhir Gupta* 811

Evaluation and Fuzzy Classification of Gene Finding Programs on Human Genome Sequences
Atulya Nagar, Sujita Purushothaman, Hissam Tawfik 821

Application of a Genetic Algorithm — Support Vector Machine Hybrid for Prediction of Clinical Phenotypes Based on Genome-Wide SNP Profiles of Sib Pairs
*Binsheng Gong, Zheng Guo, Jing Li, Guohua Zhu, Sali Lv,
Shaoqi Rao, Xia Li* 830

A New Method for Gene Functional Prediction Based on Homologous Expression Profile
*Sali Lv, Qianghu Wang, Guangmei Zhang, Fengxia Wen,
Zhenzhen Wang, Xia Li* 836

Analysis of Sib-Pair IBD Profiles and Genomic Context for Identification of the Relevant Molecular Signatures for Alcoholism
  Chuanxing Li, Lei Du, Xia Li, Binsheng Gong, Jie Zhang, Shaoqi Rao 845

A Novel Ensemble Decision Tree Approach for Mining Genes Coding Ion Channels for Cardiopathy Subtype
  Jie Zhang, Xia Li, Wei Jiang, Yanqiu Wang, Chuanxing Li, Qiuju Wang, Shaoqi Rao 852

A Permutation-Based Genetic Algorithm for Predicting RNA Secondary Structure — A Practicable Approach
  Yongqiang Zhan, Maozu Guo 861

G Protein Binding Sites Analysis
  Fan Zhang, Zhicheng Liu, Xia Li, Shaoqi Rao 865

A Novel Feature Ensemble Technology to Improve Prediction Performance of Multiple Heterogeneous Phenotypes Based on Microarray Data
  Haiyun Wang, Qingpu Zhang, Yadong Wang, Xia Li, Shaoqi Rao, Zuquan Ding 869

Fuzzy Systems in Expert System and Informatics

Fuzzy Routing in QoS Networks
  Runtong Zhang, Xiaomin Zhu 880

Component Content Soft-Sensor Based on Adaptive Fuzzy System in Rare-Earth Countercurrent Extraction Process
  Hui Yang, Chonghui Song, Chunyan Yang, Tianyou Chai 891

The Fuzzy-Logic-Based Reasoning Mechanism for Product Development Process
  Ying-Kui Gu, Hong-Zhong Huang, Wei-Dong Wu, Chun-Sheng Liu 897

Single Machine Scheduling Problem with Fuzzy Precedence Delays and Fuzzy Processing Times
  Yuan Xie, Jianying Xie, Jun Liu 907

Fuzzy-Based Dynamic Bandwidth Allocation System
  Fang-Yie Leu, Shi-Jie Yan, Wen-Kui Chang 911

Self-localization of a Mobile Robot by Local Map Matching Using Fuzzy Logic
   Jinxia Yu, Zixing Cai, Xiaobing Zou, Zhuohua Duan 921

Navigation of Mobile Robots in Unstructured Environment Using Grid Based Fuzzy Maps
   Özhan Karaman, Hakan Temeltas 925

A Fuzzy Mixed Projects and Securities Portfolio Selection Model
   Yong Fang, K.K. Lai, Shou-Yang Wang 931

Contract Net Protocol Using Fuzzy Case Based Reasoning
   Wunan Wan, Xiaojing Wang, Yang Liu 941

A Fuzzy Approach for Equilibrium Programming with Simulated Annealing Algorithm
   Jie Su, Junpeng Yuan, Qiang Han, Jin Huang 945

Image Processing Application with a TSK Fuzzy Model
   Perfecto Mariño, Vicente Pastoriza, Miguel Santamaría,
   Emilio Martínez 950

A Fuzzy Dead Reckoning Algorithm for Distributed Interactive Applications
   Ling Chen, Gencai Chen 961

Intelligent Automated Negotiation Mechanism Based on Fuzzy Method
   Hong Zhang, Yuhui Qiu 972

Congestion Control in Differentiated Services Networks by Means of Fuzzy Logic
   Morteza Mosavi, Mehdi Galily 976

Fuzzy Systems in Pattern Recognition and Diagnostics

Fault Diagnosis System Based on Rough Set Theory and Support Vector Machine
   Yitian Xu, Laisheng Wang 980

A Fuzzy Framework for Flashover Monitoring
   Chang-Gun Um, Chang-Gi Jung, Byung-Gil Han, Young-Chul Song,
   Doo-Hyun Choi 989

Feature Recognition Technique from 2D Ship Drawings Using Fuzzy
Inference System
  Deok-Eun Kim, Sung-Chul Shin, Soo-Young Kim 994

Transmission Relay Method for Balanced Energy Depletion in Wireless
Sensor Networks Using Fuzzy Logic
  Seung-Beom Baeg, Tae-Ho Cho 998

Validation and Comparison of Microscopic Car-Following Models Using
Beijing Traffic Flow Data
  Dewang Chen, Yueming Yuan, Baiheng Li, Jianping Wu 1008

Apply Fuzzy-Logic-Based Functional-Center Hierarchies as Inference
Engines for Self-learning Manufacture Process Diagnoses
  Yu-Shu Hu, Mohammad Modarres 1012

Fuzzy Spatial Location Model and Its Application in Spatial Query
  Yongjian Yang, Chunling Cao 1022

Segmentation of Multimodality Osteosarcoma MRI with Vectorial
Fuzzy-Connectedness Theory
  Jing Ma, Minglu Li, Yongqiang Zhao 1027

Knowledge Discovery in Bioinformatics and Bio-medical Engineering

A Global Optimization Algorithm for Protein Folds Prediction in 3D
Space
  Xiaoguang Liu, Gang Wang, Jing Liu 1031

Classification Analysis of SAGE Data Using Maximum Entropy Model
  Jin Xin, Rongfang Bie 1037

DNA Sequence Identification by Statistics-Based Models
  Jitimon Keinduangjun, Punpiti Piamsa-nga, Yong Poovorawan ... 1041

A New Method to Mine Gene Regulation Relationship Information
  De Pan, Fei Wang, Jiankui Guo, Jianhua Ding 1051

Knowledge Discovery in Expert System and Informatics

Shot Transition Detection by Compensating for Global and Local
Motions
  Seok-Woo Jang, Gye-Young Kim, Hyung-Il Choi 1061

Hybrid Methods for Stock Index Modeling
    Yuehui Chen, Ajith Abraham, Ju Yang, Bo Yang 1067

Designing an Intelligent Web Information System of Government Based
on Web Mining
    Gye Hang Hong, Jang Hee Lee 1071

Automatic Segmentation and Diagnosis of Breast Lesions Using
Morphology Method Based on Ultrasound
    In-Sung Jung, Devinder Thapa, Gi-Nam Wang 1079

Composition of Web Services Using Ontology with Monotonic
Inheritance
    Changyun Li, Beishui Liao, Aimin Yang, Lijun Liao 1089

Ontology-DTD Matching Algorithm for Efficient XML Query
    Myung Sook Kim, Yong Hae Kong 1093

An Approach to Web Service Discovery Based on the Semantics
    Jing Fan, Bo Ren, Li-Rong Xiong 1103

Non-deterministic Event Correlation Based on C-F Model
    Qiuhua Zheng, Yuntao Qian, Min Yao 1107

Flexible Goal Recognition via Graph Construction and Analysis
    Minghao Yin, Wenxiang Gu, Yinghua Lu 1118

An Implementation for Mapping SBML to BioSPI
    *Zhupeng Dong, Xiaoju Dong, Xian Xu, Yuxi Fu, Zhizhou Zhang,
    Lin He* 1128

Knowledge-Based Faults Diagnosis System for Wastewater
Treatment
    Jang-Hwan Park, Byong-Hee Jun, Myung-Geun Chun 1132

Study on Intelligent Information Integration of Knowledge Portals
    Yongjin Zhang, Hongqi Chen, Jiancang Xie 1136

The Risk Identification and Assessment in E-Business
Development
    Lin Wang, Yurong Zeng 1142

A Novel Wavelet Transform Based on Polar Coordinates for Datamining
Applications
    Seonggoo Kang, Sangjun Lee, Sukho Lee 1150

Impact on the Writing Granularity for Incremental Checkpointing
*Junyoung Heo, Xuefeng Piao, Sangho Yi, Geunyoung Park,
Minkyu Park, Jiman Hong, Yookun Cho* 1154

Using Feedback Cycle for Developing an Adjustable Security Design Metric
Charlie Y. Shim, Jung Y. Kim, Sung Y. Shin, Jiman Hong 1158

w-LLC: Weighted Low-Energy Localized Clustering for Embedded Networked Sensors
Joongheon Kim, Wonjun Lee, Eunkyo Kim, Choonhwa Lee 1162

Energy Efficient Dynamic Cluster Based Clock Synchronization for Wireless Sensor Network
*Md. Mamun-Or-Rashid, Choong Seon Hong,
Jinsung Cho* 1166

An Intelligent Power Management Scheme for Wireless Embedded Systems Using Channel State Feedbacks
Hyukjun Oh, Jiman Hong, Heejune Ahn 1170

Analyze and Guess Type of Piece in the Computer Game Intelligent System
Z.Y. Xia, Y.A. Hu, J. Wang, Y.C. Jiang, X.L. Qin 1174

Large-Scale Ensemble Decision Analysis of Sib-Pair IBD Profiles for Identification of the Relevant Molecular Signatures for Alcoholism
Xia Li, Shaoqi Rao, Wei Zhang, Guo Zheng, Wei Jiang, Lei Du 1184

A Novel Visualization Classifier and Its Applications
Jie Li, Xiang Long Tang, Xia Li.............................. 1190

Active Information Gathering on the Web

Automatic Creation of Links: An Approach Based on Decision Tree
Peng Li, Seiji Yamada 1200

Extraction of Structural Information from the Web
Tsuyoshi Murata... 1204

Blog Search with Keyword Map-Based Relevance Feedback
Yasufumi Takama, Tomoki Kajinami, Akio Matsumura 1208

An One Class Classification Approach to Non-relevance Feedback
Document Retrieval
   Takashi Onoda, Hiroshi Murata, Seiji Yamada 1216

Automated Knowledge Extraction from Internet for a Crisis
Communication Portal
   Ong Sing Goh, Chun Che Fung 1226

Neural and Fuzzy Computation in Cognitive Computer Vision

Probabilistic Principal Surface Classifier
   Kuiyu Chang, Joydeep Ghosh 1236

Probabilistic Based Recursive Model for Face Recognition
   Siu-Yeung Cho, Jia-Jun Wong 1245

Performance Characterization in Computer Vision: The Role of Visual
Cognition Theory
   Aimin Wu, De Xu, Xu Yang, Jianhui Zheng 1255

Generic Solution for Image Object Recognition Based on Vision
Cognition Theory
   Aimin Wu, De Xu, Xu Yang, Jianhui Zheng 1265

Cognition Theory Motivated Image Semantics and Image Language
   Aimin Wu, De Xu, Xu Yang, Jianhui Zheng 1276

Neuro-Fuzzy Inference System to Learn Expert Decision: Between
Performance and Intelligibility
   Laurence Cornez, Manuel Samuelides, Jean-Denis Muller 1281

Fuzzy Patterns in Multi-level of Satisfaction for MCDM Model Using
Modified Smooth S-Curve MF
   Pandian Vasant, A. Bhattacharya, N.N. Barsoum 1294

Author Index 1305

Table of Contents – Part I

Fuzzy Theory and Models

On Fuzzy Inclusion in the Interval-Valued Sense
Jin Han Park, Jong Seo Park, Young Chel Kwun 1

Fuzzy Evaluation Based Multi-objective Reactive Power Optimization in Distribution Networks
Jiachuan Shi, Yutian Liu 11

Note on Interval-Valued Fuzzy Set
Wenyi Zeng, Yu Shi 20

Knowledge Structuring and Evaluation Based on Grey Theory
Chen Huang, Yushun Fan 26

A Propositional Calculus Formal Deductive System $\mathcal{L}^U$ of Universal Logic and Its Completeness
Minxia Luo, Huacan He 31

Entropy and Subsethood for General Interval-Valued Intuitionistic Fuzzy Sets
Xiao-dong Liu, Su-hua Zheng, Feng-lan Xiong 42

The Comparative Study of Logical Operator Set and Its Corresponding General Fuzzy Rough Approximation Operator Set
Suhua Zheng, Xiaodong Liu, Fenglan Xiong 53

Associative Classification Based on Correlation Analysis
Jian Chen, Jian Yin, Jin Huang, Ming Feng 59

Design of Interpretable and Accurate Fuzzy Models from Data
Zong-yi Xing, Yong Zhang, Li-min Jia, Wei-li Hu 69

Generating Extended Fuzzy Basis Function Networks Using Hybrid Algorithm
Bin Ye, Chengzhi Zhu, Chuangxin Guo, Yijia Cao 79

Analysis of Temporal Uncertainty of Trains Converging Based on Fuzzy Time Petri Nets
Yangdong Ye, Juan Wang, Limin Jia 89

Interval Regression Analysis Using Support Vector Machine and
Quantile Regression
  *Changha Hwang, Dug Hun Hong, Eunyoung Na, Hyejung Park,
  Jooyong Shim* 100

An Approach Based on Similarity Measure to Multiple Attribute
Decision Making with Trapezoid Fuzzy Linguistic Variables
  Zeshui Xu 110

Research on Index System and Fuzzy Comprehensive Evaluation
Method for Passenger Satisfaction
  Yuanfeng Zhou, Jianping Wu, Yuanhua Jia 118

Research on Predicting Hydatidiform Mole Canceration Tendency by a
Fuzzy Integral Model
  Yecai Guo, Yi Guo, Wei Rao, Wei Ma 122

Consensus Measures and Adjusting Inconsistency of Linguistic
Preference Relations in Group Decision Making
  Zhi-Ping Fan, Xia Chen 130

Fuzzy Variation Coefficients Programming of Fuzzy Systems and Its
Application
  Xiaobei Liang, Daoli Zhu, Bingyong Tang 140

Weighted Possibilistic Variance of Fuzzy Number and Its Application
in Portfolio Theory
  Xun Wang, Weijun Xu, Weiguo Zhang, Maolin Hu 148

Another Discussion About Optimal Solution to Fuzzy Constraints
Linear Programming
  Yun-feng Tan, Bing-yuan Cao 156

Fuzzy Ultra Filters and Fuzzy G-Filters of MTL-Algebras
  Xiao-hong Zhang, Yong-quan Wang, Yong-lin Liu 160

A Study on Relationship Between Fuzzy Rough Approximation
Operators and Fuzzy Topological Spaces
  Wei-Zhi Wu 167

A Case Retrieval Model Based on Factor-Structure Connection and
λ−Similarity in Fuzzy Case-Based Reasoning
  Dan Meng, Zaiqiang Zhang, Yang Xu 175

A TSK Fuzzy Inference Algorithm for Online Identification
*Kyoungjung Kim, Eun Ju Whang, Chang-Woo Park, Euntai Kim,
Mignon Park* 179

Histogram-Based Generation Method of Membership Function for
Extracting Features of Brain Tissues on MRI Images
*Weibei Dou, Yuan Ren, Yanping Chen, Su Ruan, Daniel Bloyet,
Jean-Marc Constans* 189

Uncertainty Management in Data Mining

On Identity-Discrepancy-Contrary Connection Degree in SPA and Its
Applications
Yunliang Jiang, Yueting Zhuang, Yong Liu, Keqin Zhao 195

A Mathematic Model for Automatic Summarization
Zhiqi Wang, Yongcheng Wang, Kai Gao 199

Reliable Data Selection with Fuzzy Entropy
Sang-Hyuk Lee, Youn-Tae Kim, Seong-Pyo Cheon, Sungshin Kim ... 203

Uncertainty Management and Probabilistic Methods in Data Mining

Optimization of Concept Discovery in Approximate Information
System Based on FCA
Hanjun Jin, Changhua Wei, Xiaorong Wang, Jia Fu 213

Geometrical Probability Covering Algorithm
Junping Zhang, Stan Z. Li, Jue Wang 223

Approximate Reasoning

Extended Fuzzy ALCN and Its Tableau Algorithm
Jianjiang Lu, Baowen Xu, Yanhui Li, Dazhou Kang, Peng Wang ... 232

Type II Topological Logic $\mathbb{C}^2$ and Approximate Reasoning
Yalin Zheng, Changshui Zhang, Yinglong Xia 243

Type-I Topological Logic $\mathbb{C}^1_{\mathcal{T}}$ and Approximate Reasoning
Yalin Zheng, Changshui Zhang, Xin Yao 253

Vagueness and Extensionality
Shunsuke Yatabe, Hiroyuki Inaoka 263

Using Fuzzy Analogical Reasoning to Refine the Query Answers for Relational Databases with Imprecise Information
    Z.M. Ma, Li Yan, Gui Li 267

A Linguistic Truth-Valued Uncertainty Reasoning Model Based on Lattice-Valued Logic
    Shuwei Chen, Yang Xu, Jun Ma 276

Axiomatic Foundation

Fuzzy Programming Model for Lot Sizing Production Planning Problem
    Weizhen Yan, Jianhua Zhao, Zhe Cao 285

Fuzzy Dominance Based on Credibility Distributions
    Jin Peng, Henry M.K. Mok, Wai-Man Tse 295

Fuzzy Chance-Constrained Programming for Capital Budgeting Problem with Fuzzy Decisions
    Jinwu Gao, Jianhua Zhao, Xiaoyu Ji 304

Genetic Algorithms for Dissimilar Shortest Paths Based on Optimal Fuzzy Dissimilar Measure and Applications
    Yinzhen Li, Ruichun He, Linzhong Liu, Yaohuang Guo 312

Convergence Criteria and Convergence Relations for Sequences of Fuzzy Random Variables
    Yan-Kui Liu, Jinwu Gao 321

Hybrid Genetic-SPSA Algorithm Based on Random Fuzzy Simulation for Chance-Constrained Programming
    Yufu Ning, Wansheng Tang, Hui Wang 332

Random Fuzzy Age-Dependent Replacement Policy
    Song Xu, Jiashun Zhang, Ruiqing Zhao 336

A Theorem for Fuzzy Random Alternating Renewal Processes
    Ruiqing Zhao, Wansheng Tang, Guofei Li 340

Three Equilibrium Strategies for Two-Person Zero-Sum Game with Fuzzy Payoffs
    Lin Xu, Ruiqing Zhao, Tingting Shu 350

Fuzzy Classifiers

An Improved Rectangular Decomposition Algorithm for Imprecise and
Uncertain Knowledge Discovery
Jiyoung Song, Younghee Im, Daihee Park 355

XPEV: A Storage Model for Well-Formed XML Documents
Jie Qin, Shu-Mei Zhao, Shu-Qiang Yang, Wen-Hua Dou 360

Fuzzy-Rough Set Based Nearest Neighbor Clustering Classification
Algorithm
Xiangyang Wang, Jie Yang, Xiaolong Teng, Ningsong Peng 370

An Efficient Text Categorization Algorithm Based on Category
Memberships
Zhi-Hong Deng, Shi-Wei Tang, Ming Zhang 374

The Integrated Location Algorithm Based on Fuzzy Identification and
Data Fusion with Signal Decomposition
Zhao Ping, Haoshan Shi 383

A Web Document Classification Approach Based on Fuzzy Association
Concept
Jingsheng Lei, Yaohong Kang, Chunyan Lu, Zhang Yan 388

Optimized Fuzzy Classification Using Genetic Algorithm
Myung Won Kim, Joung Woo Ryu 392

Dynamic Test-Sensitive Decision Trees with Multiple Cost Scales
Zhenxing Qin, Chengqi Zhang, Xuehui Xie, Shichao Zhang 402

Design of T–S Fuzzy Classifier via Linear Matrix Inequality Approach
Moon Hwan Kim, Jin Bae Park, Young Hoon Joo, Ho Jae Lee 406

Design of Fuzzy Rule-Based Classifier: Pruning and Learning
Do Wan Kim, Jin Bae Park, Young Hoon Joo 416

Fuzzy Sets Theory Based Region Merging for Robust Image
Segmentation
Hongwei Zhu, Otman Basir 426

A New Interactive Segmentation Scheme Based on Fuzzy Affinity and
Live-Wire
Huiguang He, Jie Tian, Yao Lin, Ke Lu 436

Fuzzy Clustering

The Fuzzy Mega-cluster: Robustifying FCM by Scaling Down Memberships
Amit Banerjee, Rajesh N. Davé 444

Robust Kernel Fuzzy Clustering
Weiwei Du, Kohei Inoue, Kiichi Urahama 454

Spatial Homogeneity-Based Fuzzy c-Means Algorithm for Image Segmentation
Bo-Yeong Kang, Dae-Won Kim, Qing Li 462

A Novel Fuzzy-Connectedness-Based Incremental Clustering Algorithm for Large Databases
Yihong Dong, Xiaoying Tai, Jieyu Zhao 470

Classification of MPEG VBR Video Data Using Gradient-Based FCM with Divergence Measure
Dong-Chul Park 475

Fuzzy-C-Mean Determines the Principle Component Pairs to Estimate the Degree of Emotion from Facial Expressions
M. Ashraful Amin, Nitin V. Afzulpurkar, Matthew N. Dailey, Vatcharaporn Esichaikul, Dentcho N. Batanov 484

An Improved Clustering Algorithm for Information Granulation
Qinghua Hu, Daren Yu 494

A Novel Segmentation Method for MR Brain Images Based on Fuzzy Connectedness and FCM
Xian Fan, Jie Yang, Lishui Cheng 505

Improved-FCM-Based Readout Segmentation and PRML Detection for Photochromic Optical Disks
Jiqi Jian, Cheng Ma, Huibo Jia 514

Fuzzy Reward Modeling for Run-Time Peer Selection in Peer-to-Peer Networks
Huaxiang Zhang, Xiyu Liu, Peide Liu 523

KFCSA: A Novel Clustering Algorithm for High-Dimension Data
Kan Li, Yushu Liu 531

Fuzzy Database Mining and Information Retrieval

An Improved VSM Based Information Retrieval System and Fuzzy
Query Expansion
*Jiangning Wu, Hiroki Tanioka, Shizhu Wang, Donghua Pan,
Kenichi Yamamoto, Zhongtuo Wang* 537

The Extraction of Image's Salient Points for Image Retrieval
Wenyin Zhang, Jianguo Tang, Chao Li 547

A Sentence-Based Copy Detection Approach for Web Documents
Rajiv Yerra, Yiu-Kai Ng 557

The Research on Query Expansion for Chinese Question Answering
System
Zhengtao Yu, Xiaozhong Fan, Lirong Song, Jianyi Guo 571

Multinomial Approach and Multiple-Bernoulli Approach for
Information Retrieval Based on Language Modeling
Hua Huo, Junqiang Liu, Boqin Feng 580

Adaptive Query Refinement Based on Global and Local Analysis
Chaoyuan Cui, Hanxiong Chen, Kazutaka Furuse, Nobuo Ohbo 584

Information Push-Delivery for User-Centered and Personalized Service
Zhiyun Xin, Jizhong Zhao, Chihong Chi, Jiaguang Sun 594

Mining Association Rules Based on Seed Items and Weights
Chen Xiang, Zhang Yi, Wu Yue 603

An Algorithm of Online Goods Information Extraction with Two-Stage
Working Pattern
Wang Xun, Ling Yun, Yu-lian Fei 609

A Novel Method of Image Retrieval Based on Combination of Semantic
and Visual Features
Ming Li, Tong Wang, Bao-wei Zhang, Bi-Cheng Ye 619

Using Fuzzy Pattern Recognition to Detect Unknown Malicious
Executables Code
Boyun Zhang, Jianping Yin, Jingbo Hao 629

Method of Risk Discernment in Technological Innovation Based on
Path Graph and Variable Weight Fuzzy Synthetic Evaluation
Yuan-sheng Huang, Jian-xun Qi, Jun-hua Zhou 635

Application of Fuzzy Similarity to Prediction of Epileptic Seizures
Using EEG Signals
  Xiaoli Li, Xin Yao 645

A Fuzzy Multicriteria Analysis Approach to the Optimal Use of
Reserved Land for Agriculture
  Hepu Deng, Guifang Yang 653

Fuzzy Comprehensive Evaluation for the Optimal Management of
Responding to Oil Spill
  Xin Liu, Kai W. Wirtz, Susanne Adam 662

Information Fusion

Fuzzy Fusion for Face Recognition
  Xuerong Chen, Zhongliang Jing, Gang Xiao 672

A Group Decision Making Method for Integrating Outcome Preferences
in Hypergame Situations
  Yexin Song, Qian Wang, Zhijun Li 676

A Method Based on IA Operator for Multiple Attribute Group Decision
Making with Uncertain Linguistic Information
  Zeshui Xu 684

A New Prioritized Information Fusion Method for Handling Fuzzy
Information Retrieval Problems
  Won-Sin Hong, Shi-Jay Chen, Li-Hui Wang, Shyi-Ming Chen 694

Multi-context Fusion Based Robust Face Detection in Dynamic
Environments
  Mi Young Nam, Phill Kyu Rhee 698

Unscented Fuzzy Tracking Algorithm for Maneuvering Target
  Shi-qiang Hu, Li-wei Guo, Zhong-liang Jing 708

A Pixel-Level Multisensor Image Fusion Algorithm Based on Fuzzy Logic
  Long Zhao, Baochang Xu, Weilong Tang, Zhe Chen 717

Neuro-Fuzzy Systems

Approximation Bound for Fuzzy-Neural Networks with Bell
Membership Function
  Weimin Ma, Guoqing Chen 721

A Neuro-fuzzy Method of Forecasting the Network Traffic of Accessing Web Server
  Ai-Min Yang, Xing-Min Sun, Chang-Yun Li, Ping Liu 728

A Fuzzy Neural Network System Based on Generalized Class Cover Problem
  Yanxin Huang, Yan Wang, Wengang Zhou, Chunguang Zhou 735

A Self-constructing Compensatory Fuzzy Wavelet Network and Its Applications
  Haibin Yu, Qianjin Guo, Aidong Xu 743

A New Balancing Method for Flexible Rotors Based on Neuro-fuzzy System and Information Fusion
  Shi Liu 757

Recognition of Identifiers from Shipping Container Images Using Fuzzy Binarization and Enhanced Fuzzy Neural Network
  Kwang-Baek Kim 761

Directed Knowledge Discovery Methodology for the Prediction of Ozone Concentration
  Seong-Pyo Cheon, Sungshin Kim 772

Application of Fuzzy Systems in the Car-Following Behaviour Analysis
  Pengjun Zheng, Mike McDonald 782

Fuzzy Control

GA-Based Composite Sliding Mode Fuzzy Control for Double-Pendulum-Type Overhead Crane
  Diantong Liu, Weiping Guo, Jianqiang Yi 792

A Balanced Model Reduction for T-S Fuzzy Systems with Integral Quadratic Constraints
  Seog-Hwan Yoo, Byung-Jae Choi 802

An Integrated Navigation System of NGIMU/ GPS Using a Fuzzy Logic Adaptive Kalman Filter
  Mingli Ding, Qi Wang 812

Method of Fuzzy-PID Control on Vehicle Longitudinal Dynamics System
  Yinong Li, Zheng Ling, Yang Liu, Yanjuan Qiao 822

Design of Fuzzy Controller and Parameter Optimizer for Non-linear System Based on Operator's Knowledge
Hyeon Bae, Sungshin Kim, Yejin Kim 833

A New Pre-processing Method for Multi-channel Echo Cancellation Based on Fuzzy Control
Xiaolu Li, Wang Jie, Shengli Xie 837

Robust Adaptive Fuzzy Control for Uncertain Nonlinear Systems
Chen Gang, Shuqing Wang, Jianming Zhang 841

Intelligent Fuzzy Systems for Aircraft Landing Control
Jih-Gau Juang, Bo-Shian Lin, Kuo-Chih Chin 851

Scheduling Design of Controllers with Fuzzy Deadline
Hong Jin, Hongan Wang, Hui Wang, Danli Wang 861

A Preference Method with Fuzzy Logic in Service Scheduling of Grid Computing
Yanxiang He, Haowen Liu, Weidong Wen, Hui Jin 865

H_∞ Robust Fuzzy Control of Ultra-High Rise / High Speed Elevators with Uncertainty
Hu Qing, Qingding Guo, Dongmei Yu, Xiying Ding 872

A Dual-Mode Fuzzy Model Predictive Control Scheme for Unknown Continuous Nonlinear System
Chonghui Song, Shucheng Yang, Hui yang, Huaguang Zhang, Tianyou Chai 876

Fuzzy Modeling Strategy for Control of Nonlinear Dynamical Systems
Bin Ye, Chengzhi Zhu, Chuangxin Guo, Yijia Cao 882

Intelligent Digital Control for Nonlinear Systems with Multirate Sampling
Do Wan Kim, Jin Bae Park, Young Hoon Joo 886

Feedback Control of Humanoid Robot Locomotion
Xusheng Lei, Jianbo Su 890

Application of Computational Intelligence (Fuzzy Logic, Neural Networks and Evolutionary Programming) to Active Networking Technology
Mehdi Galily, Farzad Habibipour Roudsari, Mohammadreza Sadri ... 900

Fuel-Efficient Maneuvers for Constellation Initialization Using Fuzzy
Logic Control
  Mengfei Yang, Honghua Zhang, Rucai Che, Zengqi Sun   910

Design of Interceptor Guidance Law Using Fuzzy Logic
  Ya-dong Lu, Ming Yang, Zi-cai Wang   922

Relaxed LMIs Observer-Based Controller Design via Improved T-S
Fuzzy Model Structure
  Wei Xie, Huaiyu Wu, Xin Zhao   930

Fuzzy Virtual Coupling Design for High Performance Haptic Display
  D. Bi, J. Zhang, G.L. Wang   942

Linguistic Model for the Controlled Object
  Zhinong Miao, Xiangyu Zhao, Yang Xu   950

Fuzzy Sliding Mode Control for Uncertain Nonlinear Systems
  Shao-Cheng Qu, Yong-Ji Wang   960

Fuzzy Control of Nonlinear Pipeline Systems with Bounds on Output
Peak
  Fei Liu, Jun Chen   969

Grading Fuzzy Sliding Mode Control in AC Servo System
  Hu Qing, Qingding Guo, Dongmei Yu, Xiying Ding   977

A Robust Single Input Adaptive Sliding Mode Fuzzy Logic Controller
for Automotive Active Suspension System
  *Ibrahim B. Kucukdemiral, Seref N. Engin, Vasfi E. Omurlu,
  Galip Cansever*   981

Construction of Fuzzy Models for Dynamic Systems Using
Multi-population Cooperative Particle Swarm Optimizer
  Ben Niu, Yunlong Zhu, Xiaoxian He   987

Human Clustering for a Partner Robot Based on Computational
Intelligence
  Indra Adji Sulistijono, Naoyuki Kubota 1001

Fuzzy Switching Controller for Multiple Model
  Baozhu Jia, Guang Ren, Zhihong Xiu 1011

Generation of Fuzzy Rules and Learning Algorithms for Cooperative
Behavior of Autonomouse Mobile Robots(AMRs)
  Jang-Hyun Kim, Jin-Bae Park, Hyun-Seok Yang, Young-Pil Park ... 1015

UML-Based Design and Fuzzy Control of Automated Vehicles
  Abdelkader El Kamel, Jean-Pierre Bourey 1025

Fuzzy Hardware

Design of an Analog Adaptive Fuzzy Logic Controller
  Zhihao Xu, Dongming Jin, Zhijian Li 1034

VLSI Implementation of a Self-tuning Fuzzy Controller Based on Variable Universe of Discourse
  Weiwei Shan, Dongming Jin, Weiwei Jin, Zhihao Xu 1044

Knowledge Visualization and Exploration

Method to Balance the Communication Among Multi-agents in Real Time Traffic Synchronization
  Li Weigang, Marcos Vinícius Pinheiro Dib,
  Alba Cristina Magalhães de Melo 1053

A Celerity Association Rules Method Based on Data Sort Search
  Zhiwei Huang, Qin Liao 1063

Using Web Services to Create the Collaborative Model for Enterprise Digital Content Portal
  Ruey-Ming Chao, Chin-Wen Yang 1067

Emotion-Based Textile Indexing Using Colors and Texture
  Eun Yi Kim, Soo-jeong Kim, Hyun-jin Koo, Karpjoo Jeong,
  Jee-in Kim 1077

Optimal Space Launcher Design Using a Refined Response Surface Method
  Jae-Woo Lee, Kwon-Su Jeon, Yung-Hwan Byun, Sang-Jin Kim 1081

MEDIC: A MDO-Enabling Distributed Computing Framework
  Shenyi Jin, Kwangsik Kim, Karpjoo Jeong, Jaewoo Lee,
  Jonghwa Kim, Hoyon Hwang, Hae-Gook Suh 1092

Time and Space Efficient Search for Small Alphabets with Suffix Arrays
  Jeong Seop Sim 1102

Optimal Supersonic Air-Launching Rocket Design Using Multidisciplinary System Optimization Approach
  Jae-Woo Lee, Young Chang Choi, Yung-Hwan Byun 1108

Numerical Visualization of Flow Instability in Microchannel
Considering Surface Wettability
*Doyoung Byun, Budiono, Ji Hye Yang, Changjin Lee,
Ki Won Lim* 1113

A Interactive Molecular Modeling System Based on Web Service
Sungjun Park, Bosoon Kim, Jee-In Kim 1117

On the Filter Size of DMM for Passive Scalar in Complex Flow
Yang Na, Dongshin Shin, Seungbae Lee 1127

Visualization Process for Design and Manufacturing of End Mills
Sung-Lim Ko, Trung-Thanh Pham, Yong-Hyun Kim 1133

IP Address Lookup with the Visualizable Biased Segment Tree
Inbok Lee, Jeong-Shik Mun, Sung-Ryul Kim 1137

A Surface Reconstruction Algorithm Using Weighted Alpha Shapes
Si Hyung Park, Seoung Soo Lee, Jong Hwa Kim 1141

Sequential Data Analysis

HYBRID: From Atom-Clusters to Molecule-Clusters
Zhou Bing, Jun-yi Shen, Qin-ke Peng 1151

A Fuzzy Adaptive Filter for State Estimation of Unknown Structural
System and Evaluation for Sound Environment
Akira Ikuta, Hisako Masuike, Yegui Xiao, Mitsuo Ohta 1161

Preventing Meaningless Stock Time Series Pattern Discovery by
Changing Perceptually Important Point Detection
Tak-chung Fu, Fu-lai Chung, Robert Luk, Chak-man Ng 1171

Discovering Frequent Itemsets Using Transaction Identifiers
Duckjin Chai, Heeyoung Choi, Buhyun Hwang 1175

Incremental DFT Based Search Algorithm for Similar Sequence
Quan Zheng, Zhikai Feng, Ming Zhu 1185

Parallel and Distributed Data Mining

Computing High Dimensional MOLAP with Parallel Shell Mini-cubes
Kong-fa Hu, Chen Ling, Shen Jie, Gu Qi, Xiao-li Tang 1192

Sampling Ensembles for Frequent Patterns
    Caiyan Jia, Ruqian Lu 1197

Distributed Data Mining on Clusters with Bayesian Mixture Modeling
    M. Viswanathan, Y.K. Yang, T.K. Whangbo 1207

A Method of Data Classification Based on Parallel Genetic Algorithm
    Yuexiang Shi, Zuqiang Meng, Zixing Cai, B. Benhabib 1217

Rough Sets

Rough Computation Based on Similarity Matrix
    Huang Bing, Guo Ling, He Xin, Xian-zhong Zhou 1223

The Relationship Among Several Knowledge Reduction Approaches
    Keyun Qin, Zheng Pei, Weifeng Du 1232

Rough Approximation of a Preference Relation for Stochastic
Multi-attribute Decision Problems
    Chaoyuan Yue, Shengbao Yao, Peng Zhang, Wanan Cui 1242

Incremental Target Recognition Algorithm Based on Improved
Discernibility Matrix
    Liu Yong, Xu Congfu, Yan Zhiyong, Pan Yunhe 1246

Problems Relating to the Phonetic Encoding of Words in the Creation
of a Phonetic Spelling Recognition Program
    Michael Higgins, Wang Shudong 1256

Diversity Measure for Multiple Classifier Systems
    Qinghua Hu, Daren Yu 1261

A Successive Design Method of Rough Controller Using Extra Excitation
    Geng Wang, Jun Zhao, Jixin Qian 1266

A Soft Sensor Model Based on Rough Set Theory and Its Application
in Estimation of Oxygen Concentration
    Xingsheng Gu, Dazhong Sun 1271

A Divide-and-Conquer Discretization Algorithm
    Fan Min, Lijun Xie, Qihe Liu, Hongbin Cai 1277

A Hybrid Classifier Based on Rough Set Theory and Support Vector
Machines
    Gexiang Zhang, Zhexin Cao, Yajun Gu 1287

A Heuristic Algorithm for Maximum Distribution Reduction
  Xiaobing Pei, YuanZhen Wang 1297

The Minimization of Axiom Sets Characterizing Generalized Fuzzy
Rough Approximation Operators
  Xiao-Ping Yang 1303

The Representation and Resolution of Rough Sets Based on the
Extended Concept Lattice
  Xuegang Hu, Yuhong Zhang, Xinya Wang........................ 1309

Study of Integrate Models of Rough Sets and Grey Systems
  Wu Shunxiang, Liu Sifeng, Li Maoqing 1313

Author Index 1325

Dimensionality Reduction for Semi-supervised Face Recognition

Weiwei Du, Kohei Inoue, and Kiichi Urahama

Kyushu University, Fukuoka-shi, 815-8540, Japan

Abstract. A dimensionality reduction technique is presented for semi-supervised face recognition where image data are mapped into a low dimensional space with a spectral method. A mapping of learning data is generalized to a new datum which is classified in the low dimensional space with the nearest neighbor rule. The same generalization is also devised for regularized regression methods which work in the original space without dimensionality reduction. It is shown with experiments that the spectral mapping method outperforms the regularized regression. A modification scheme for data similarity matrices on the basis of label information and a simple selection rule for data to be labeled are also devised.

1 Introduction

Supervised learning demands manual labeling of all learning data which is laborious, hence semi-supervised learning is useful practically, where a new datum is classified on the basis of learning data only few of which are labeled and many remaining data are unlabeled[1]. There have been presented graph-oriented semi-supervised learning methods where labels are propagated from labeled data to unlabeled ones on the basis of regularization on graphs[2,3]. In these methods, however, classification is tested only for learning data and any generalization scheme of the classification rule to a new datum has not been presented. Furthermore, the methods are tested with examples where classes are well separated such as numeral images. If the classes are complicatedly entangled, erroneous propagation occurs and their classification performance deteriorates.

Additionally, in the test of these methods, labeled data are selected randomly in each class. This selection scheme is, however, not possible in practice because collected data have not been partitioned into each class before their classification. Therefore we can only select data to be labeled randomly from entire learning data, hence it is possible that some classes are given no labeled datum. Thus there remain some practical questions on the previous semi-supervised learning methods[2,3].

In this paper, we present a spectral mapping method for semi-supervised pattern classification where a generalization scheme of a classification rule on learning data to a new datum is incorporated. The same generalization scheme is also devised for the regularized regression methods, and we show that our spectral mapping method outperforms the regularization methods for complexly entangled data such as face images. Additionally, we present a modification scheme of similarity matrices on the basis of label information and a simple selection rule of data to be labeled on the basis of clustering of learning data.

2 Spectral Mapping for Dimensionality Reduction of Data

Similarity data are represented with an undirected graph. In this section, we briefly introduce three representative graph spectral methods for mapping similarity data into a low dimensional space.

Let there be m data whose similarity $s_{ij}(=s_{ji})$ is given by $s_{ij} = e^{-\alpha\|f_i - f_j\|^2}$ where f_i is the feature vector of the i-th datum. Let $S = [s_{ij}]$ and $D = \mathrm{diag}(d_1, ..., d_m); d_i = \sum_{j=1}^{m} s_{ij}$. Let the coordinate of data on the first dimension of a mapped space be $x = [x_1, ..., x_m]^T$.

2.1 Basic Mapping Without Normalization

A method where x is obtained by

$$\min_{x} \sum_{i=1}^{m}\sum_{j=1}^{m} s_{ij}(x_i - x_j)^2 \quad \mathrm{subj.to} \quad \sum_{i=1}^{m} x_i^2 = 1 \quad (1)$$

is the most basic spectral mapping scheme popularly used in the multivariate analysis. Equation (1) is rewritten in the vector form as

$$\min_{x} \; x^T(D-S)x \quad \mathrm{subj.to} \quad x^T x = 1 \quad (2)$$

The solution of this optimization problem is given by

$$\min_{x}\max_{\lambda} \; x^T(D-S)x + \lambda(x^T x - 1) \quad (3)$$

and is the eigenvector of the unnormalized Laplacian $D - S$ with the minimal eigenvalue. Note that this principal eigenvector is $x = [1, ..., 1]^T/\sqrt{m}$ for regular graphs, which gives no information on data structure, hence it is generally discarded. Nevertheless we use it here, hence mapped dimension in this paper is larger by one than that in the conventional multivariate analysis.

This solution x is the coordinate of the first dimension. When we map data into an n-dimensional space, we compute n eigenvectors of $D - S$ with n smallest eigenvalues, then the i-th datum is mapped to the point $[x_{i1}, ..., x_{in}]^T$ where we denote the k-th eigenvector by $x_k = [x_{1k}, ..., x_{mk}]^T$.

2.2 Mapping with Symmetric Normalization

If we modify eq.(1) to

$$\min_x \sum_{i=1}^{m}\sum_{j=1}^{m} s_{ij}\left(\frac{x_i}{\sqrt{d_i}} - \frac{x_j}{\sqrt{d_j}}\right)^2$$
$$\text{subj.to} \quad \sum_{i=1}^{m} x_i^2 = 1 \quad (4)$$

then eq.(2) is modified to

$$\min_x \quad x^T(I - D^{-1/2}SD^{-1/2})x$$
$$\text{subj.to} \quad x^T x = 1 \quad (5)$$

of which solution is the eigenvector of $D^{-1/2}SD^{-1/2}$ with the maximal eigenvalue. The matrix $I - D^{-1/2}SD^{-1/2} = D^{-1/2}(D - S)D^{-1/2}$ is the normalized Laplacian and this mapping has been used in pre-processing for spectral clustering[4].

2.3 Mapping with Asymmetric Normalization

If eq.(1) is modified to

$$\min_x \sum_{i=1}^{m}\sum_{j=1}^{m} s_{ij}(x_i - x_j)^2$$
$$\text{subj.to} \quad \sum_{i=1}^{m} d_i x_i^2 = 1 \quad (6)$$

then eq.(2) is modified to

$$\min_x \quad x^T(D - S)x$$
$$\text{subj.to} \quad x^T D x = 1 \quad (7)$$

of which solution is the principal eigenvector of $D^{-1}S$ which is the transition matrix of a random walk on the graph. This mapping is called the Laplacian eigenmap[5] and has been used for graph drawing[6].

3 Semi-supervised Classification with Spectral Mapping

All of these mappings are nonlinear and their enhancement property of data proximity relationship is superior than linear mappings such as the principal component analysis. Hence, mapping of data into a low dimensional space with these mappings is expected to raise classification rates from that in the original feature space. In addition, low dimensionality speeds up the computation of distances between data. These mappings give, however, coordinates of only learning data, hence cannot be generalized to a new datum in contrast to linear mappings. Re-computation for whole data appended with the new datum is time-consuming. We, therefore, generalize the mapping computed from the learning data to a new datum in a similar way to the approach[7].

3.1 Generalization of Mapping to New Datum

Let us be given m learning data. We firstly compute their coordinates in a low dimensional space. If we map them into n-dimensional space, we compute and save n eigenvectors $x_1, ..., x_n$ and n eigenvalues $\lambda_1, ..., \lambda_n$.

We then compute the coordinate of a new datum as follows. Let the feature vector of the new datum be f. For instance, we consider the basic mapping in section 2.1 where the coordinates of learning data are the eigenvectors of $D - S$. Hence the coordinate of the k-th dimension $x_k = [x_{1k}, ..., x_{mk}]^T$ satisfies $(D - S)x_k = \lambda_k x_k$ which can be written elementwisely as $(d_i - \lambda_k)x_{ik} - \sum_{j=1}^{m} s_{ij} x_{jk} = 0$ from which we get

$$x_{ik} = \frac{1}{d_i - \lambda_k} \sum_{j=1}^{m} s_{ij} x_{jk} \qquad (8)$$

which is the k-th coordinate of the i-th datum whose feature vector is f_i, hence the k-th coordinate of the new datum whose feature vector is f becomes

$$x_k(f) = \frac{1}{d(f) - \lambda_k} \sum_{j=1}^{m} s_j(f) x_{jk} \qquad (9)$$

where $s_j(f) = e^{-\alpha \|f - f_j\|^2}$, $d(f) = \sum_{j=1}^{m} s_j(f)$. In eq.(9), we use λ_k and x_{jk} computed for the learning data. By computing eq.(9) for every dimension, we get the coordinate of the new datum $x(f) = [x_1(f), ..., x_n(f)]^T$. Generalization for the mapping in section 2.2 and that in section 2.3 is similar. Since $s_j(f)$ is the Gaussian kernel, eq.(9) is a form of kernel regression, hence we call this mapping a spectral kernel regression.

3.2 Classification

We classify a new datum by the simple nearest neighbor rule. We compute the coordinate of the new datum with the above spectral kernel regression and classify it to a class to which the nearest labeled learning datum belongs. This method needs the computation of the distance between the new datum and only the labeled learning data in the mapped space whose dimension is lower than the original feature space.

4 Regularized Regression

The above spectral mapping is the method proposed in this paper. In this section, we briefly review the regularized regression methods previously presented for semi-supervised learning[2,3]. In the above spectral mapping, x is the coordinate of data in mapped spaces, while in the regularized regression, this is used as the membership of data in each class, i.e. x_{ik} is the membership of the i-th datum in the k-th class and the datum i is classified to the class $k_* = \arg\max_k \{x_{ik}\}$.

Since the membership of labeled data is known, x_{ik} is given for those data. Hence the normalization condition for $x_k = [x_{1k}, ..., x_{mk}]^T$ in the above spectral mapping methods becomes unnecessary and the Lagrange multiplier λ becomes constant. Thus eq.(3), for instance, reduces to

$$\min_{x_k} \quad x_k^T(D-S)x_k + \lambda x_k^T x_k \qquad (10)$$

which is rewritten elementwisely as

$$\min_{x_k} \quad \sum_{i \notin T} \sum_{j \notin T} s_{ij}(x_{ik} - x_{jk})^2 + \sum_{i \notin T} \sum_{j \in T} s_{ij}(x_{ik} - t_{jk})^2 + \lambda \sum_{i \notin T} x_{ik}^2 \qquad (11)$$

where T is the set of labeled data whose membership t_{jk} is $t_{jk_*} = 1$ if the datum j belongs to the class k_* and the remaining $t_{jk} = 0 (k \neq k_*)$.

In contrast to the spectral mapping methods, computation of x_{ik} is needed for only data $i \notin T$ in this regularized regression. Let x_k be divided as $x_k = [x_{ku}^T, x_{kt}^T]^T$ where x_{ku}^T is the portion of x_k for $i \notin T$ and x_{kt}^T is that for $i \in T$, then x_{kt} is given as $x_{kt} = t_k$ with the membership t_k of the labeled data and the remaining x_{ku} is obtained from eq.(11) as

$$x_{ku} = (D_{uu} - S_{uu} + \lambda I_{uu})^{-1} S_{ut} t_k \qquad (12)$$

where D_{uu} and S_{uu} are diagonal submatrix of D and S for $i \notin T$, S_{ut} is the nondiagonal submatrix in S, and I_{uu} is the diagonal identity submatrix for $i \notin T$.

Similar to the spectral mappings, this regularized regression gives memberships for only learning data. Its generalization to a new datum is also similar to that in the spectral mapping as follows. By partially differentiating eq.(11) with x_{ik} and setting it to zero, we get

$$x_{ik} = \frac{1}{d_i + \lambda} (\sum_{j \notin T} s_{ij} x_{jk} + \sum_{j \in T} s_{ij} t_{jk}) \qquad (13)$$

Note that we can solve this equation for $x_{ik} \in x_{ku}$ with, for instance, an iterative method as was done in [3], of which converged solution coincides with eq.(12). The converged x_{ik} of eq.(13) is the k-th coordinate of the i-th datum whose feature vector is f_i, hence the k-th coordinate of the new datum whose feature vector is f becomes

$$x_k(f) = \frac{1}{d_j(f) + \lambda} [\sum_{j \notin T} s_j(f) x_{jk} + \sum_{j \in T} s_j(f) t_{jk}] \qquad (14)$$

where $s_j(f) = e^{-\alpha \|f - f_j\|^2}$, $d(f) = \sum_{j=1}^{m} s_j(f)$. In eq.(14), we use x_{jk} computed for the learning data by eq.(12), i.e. the converged values of iterations for eq.(13).

This regularized regression is the unnormalized one corresponding the spectral mapping in section 2.1. Regularized regressions corresponding to the normalized mapping in section 2.2 and that in section 2.3 are similarly derived and their derivation is omitted here.

Zhu et al.[2] presented an unnormalized regularized regression method for semi-supervised learning with experiments of classification only of learning data. Zhou et al.[3] proposed a symmetrically normalized regularized regression method without fixation of the memberships of labeled data and tested their classification also only for learning data. Asymmetrically normalized regularization is novel in this paper.

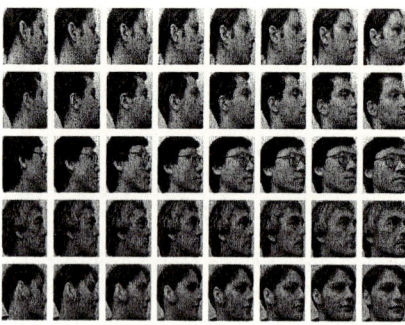

Fig. 1. Face images used in experiments

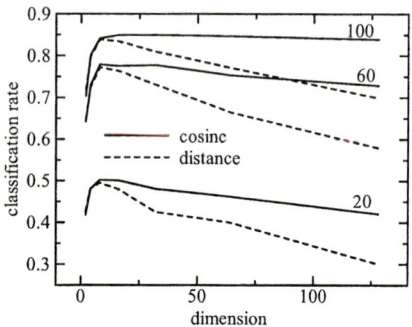

Fig. 2. Classification rates of unnormalized spectral mapping

5 Experiments

We have experimented the above three spectral mapping methods and three regularized regression methods corresponding to them. We used the dataset of "UMIST Face Database" of 20 persons face images of size 112 × 92, some of which are shown in Fig.1. This dataset consists of face images photographed from various viewpoints, hence data distributions of each person are complexly entangled mutually. A subset of 290 images in 575 data were used in learning and the remaining 285 images were used in test. Labeled data were selected randomly from 290 learning data. Feature vectors are the array of pixel grayscales and we set $\alpha = 10^{-6}$ in the similarity $s_{ij} = e^{-\alpha \|f_i - f_j\|^2}$.

5.1 Spectral Mapping Methods

The classification rate of the unnormalized mapping method in section 2.1 is illustrated in Fig.2 where broken lines denote the nearest neighbor classification with the Euclidean distance and the solid lines denote the rates with the cosine measure where attached (20,60,100) are the numbers of labeled data. Next the classification rates of the symmetrically normalized mapping in section 2.2 are

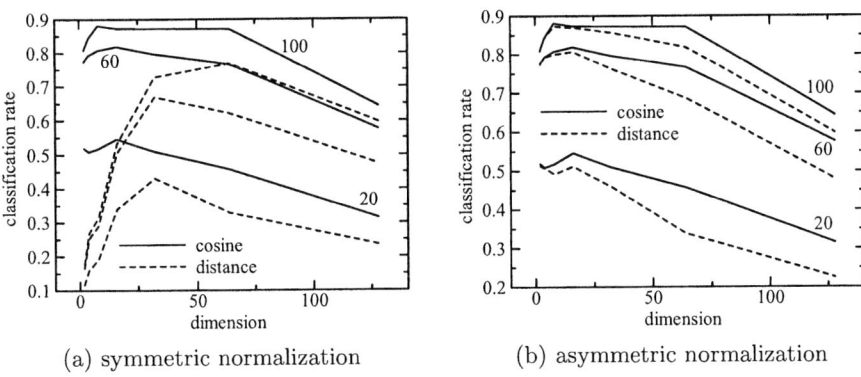

Fig. 3. Classification rates of normalized spectral mapping

shown in Fig.3(a) and those of the asymmetrically normalized mapping in section 2.3 are plotted in Fig.3(b).

The cosine measure outperforms the Euclidean distance in all of these experiments. Note that the cosine is the same in the symmetric normalization (solid lines in Fig.3(a)) and in the asymmetric case (solid lines in Fig.3(b)). Hence we adopt the normalized mapping with cosine measure in the subsequent experiments with the mapped dimension set to 8 around which the classification rates take their maximum.

5.2 Regularized Regression

The classification rates of three regularized regression methods are illustrated in Fig.4 where the dotted line denotes unnormalized regression, the broken line denotes symmetrically normalized regression and the solid lines represents asymmetrically normalized one. The regularization parameter λ was set to 0.01.

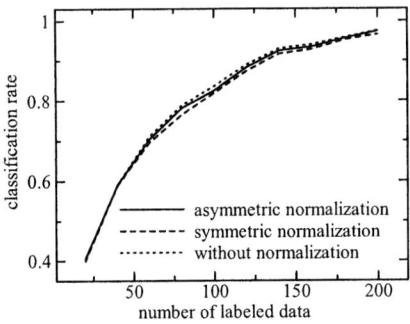

Fig. 4. Classification rates of regularized regression methods

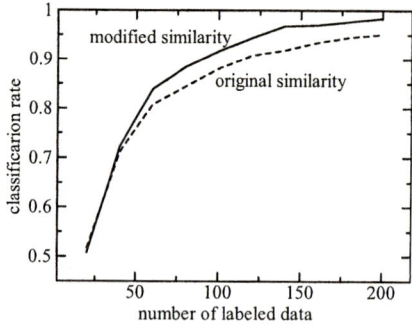

Fig. 5. Effects of similarity modification

6 Modification of Similarity Matrix in Spectral Mapping Method

As is shown above, the spectral mapping method outperforms the regularized regression. The classification rates of the spectral mapping method can be furthermore raised by modifying the similarity matrix $S = [s_{ij}]$. In the methods in section 2, no information on labels was utilized for mapping itself. We enhance the similarity relationship between data by modifying it as follows. For labeled data, if their class coincides then $s_{ij} = 1$, else i.e. if their classes are different then $s_{ij} = 0$, and for the remaining unlabeled data, no modification is done, i.e. $s_{ij} = e^{-\alpha \|f_i - f_j\|^2}$ for them. The effect of this modification is shown in Fig.5 where the broken line denotes the classification rate with the original similarity matrix and the solid line is that after the modification. The improvement of classification by this modification is attributed to the rearrangement of data in the mapped space where data with the same class are mutually attracted while data in different classes become more distant, hence the separation of classes is increased. Note that this modification exerts no effect for the regularized regression because the memberships of labeled data are fixed there.

7 Selection of Data to Be Labeled

In the above experiments, the labeled data are selected randomly. If we can select them appropriately, then the classification rate is expected to increase. Active learning strategies for such data selection are generally complicated. We propose here a simple selection scheme based on the k-means clustering which is consistent with the nearest neighbor rule by exploiting the spectral mapping. We use the spherical k-means clustering[8] which is consistent with the cosine measure. We firstly map entire learning data into a low dimensional space with the spectral mapping by using the original similarity matrix. We next cluster data in the mapped space with the spherical k-means algorithm. We select every prototype of clusters as data to be labeled. The similarity matrix is then modified

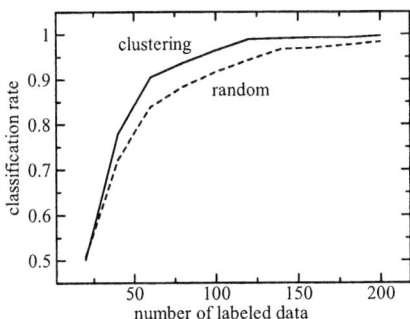

Fig. 6. Effects of labeling data selection with clustering

on the basis of those labeled data and the mapping is re-calculated with this modified similarity matrix. Since prototypes obtained by the spherical k-means depend on initial values, we run it several, e.g. ten, times and adopt the best prototypes with the minimal quantization error of the clustering.

The effect of this selection scheme of labeled data is illustrated in Fig.6 where the broken line shows the case with randomly selected labeled data (i.e. the solid line in Fig.5) and the solid line denotes the case with labeled data selected with clustering.

8 Overall Comparison of Classification Rates

The comparison of classification rates is summarized in Fig.7 where the solid line is the rate of the modified spectral mapping method with labeled data selected by clustering (i.e. the solid line in Fig.6), the broken line denotes the unnormalized regularized regression, and the dotted broken line is the rate of the direct nearest neighbor classification in the original feature space. We also experimented the dimensionality reduction by the principal component analysis (i.e. semi-supervised eigenface method), but its classification rate is lower than any of the lines in Fig.7. Thus linear mapping is useless for complexly structured data such as face images. In every experiment, the labeled data are selected with the scheme in section 7 (hence the broken line in Fig.7 differs from the dotted line in Fig.4).

In Fig.7, the classification rate of the regularized regression method is not so high and almost equal to the simple nearest neighbor classifier. This is due to the data used here in which images of different persons are complexly mixed in the raw feature space, and hence labels are erroneously propagated to different persons in the regularized regression. Dimensionality reduction with the spectral mapping is revealed to be effective for such complicated data. This effectiveness is attributed to the mapping into a low dimensional space where the separation of classes is enhanced than the original feature space. This is confirmed with the observation of Fig.3(a) and Fig.3(b) where the classification rates decrease as

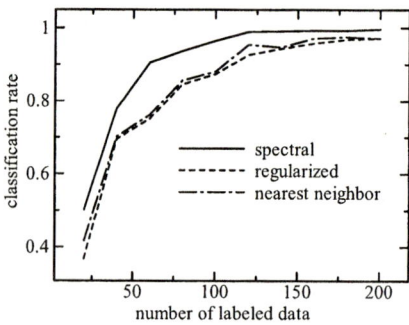

Fig. 7. Overall comparison of classification rates

the mapped dimension increases and eventually approach those in the original feature space. This is also the reason for the poor performance of the regularized regression since it propagates labels on the basis of similarity between data in the original feature space.

9 Conclusion

We have presented a dimensionality reduction method with a spectral mapping for semi-supervised face recognition and have shown its higher classification rates than the regularized regression methods previously proposed for semi-supervised learning. Experiments of other pattern recognition and more elaborated improvement of the spectral mapping method are now being examined.

References

1. Seeger, M.: Learning with labeled and unlabeled data. Tech. Reports, Edinburgh Univ. (2001)
2. Zhu, X., Ghahramani, Z., Lafferty, J.: Semi-supervised learning using Gaussian fields and harmonic functions. Proc. ICML-2003 (2003) 912–919
3. Zhou, D., Bousquet, O., Lal, T. N., Weston, J., Scholkopf, B.: Learning with local and global consistency. Proc. NIPS'03 (2003)
4. Ng, A. Y., Jordan, M. I., Weiss, Y.: On spectral clustering: Analysis and an algorithm. Proc. NIPS'01 (2001) 849–856
5. Belkin, M., Niyogi, P.: Laplacian eigenmaps for dimensionality reduction and data representation. Neural Comp. **15** (2003) 1373–1396
6. Koren, Y.: On spectral graph drawing. Proc. COCOON'03 (2003) 496–508
7. Bengio, Y., Paiement, J.-F., Vincent, P.: Out-of-sample extensions for LLE, Isomap, MDS, eigenmaps, and spectral clustering. Proc. NIPS'03 (2003)
8. Dhillon, I. S., Modha, D. M.: Concept decompositions for large sparse text data using clustering. Mach. Learning **42** (2001) 143–175

Cross-Document Transliterated Personal Name Coreference Resolution[1]

Houfeng Wang

Department of Computer Science and Technology,
School of Electronic Engineering and Computer Science,
Peking University, Beijing, 100871, China
wanghf@pku.edu.cn

Abstract. This paper presents a two-step approach to determining whether a transliterated personal name from different Chinese texts stands for the same referent. A heuristic strategy based on biographical information and "colleague" names is firstly used to form an initial set of coreference chains, and then, a clustering algorithm based Vector Space Model (VSM) is applied to merge chains under the control of a full name consistent constraint. Experimental results show that this approach achieves a good performance.

1 Introduction

Coreference Resolution is the process of determining whether two expressions in natural language refer to the same entity in the world [3]. It is an important subtask in NLP and a key component of application systems such as IE (Information Extraction), IR (Information Retrieval), multi-document summarization, etc. In the past decade, coreference resolution caused increasing concern and great advances have been made. However, such research focused mainly on coreference resolution within a single text, especially those coreferences of distinct expressions (noun and pronoun), under the assumption that the same name always stands for the same individual in the same text.

When an identical name occurs in different texts, however, cases might be different. The same name could refer to different individuals. An internet search for transliterated name 约翰逊 (*Johnson or Johansson*) by Chinese search engine *TianWang* shows 50938 results and this name refers to 4 different individuals in the first 10. With rapid development of Internet, an identical name instances referring to different referents will be more and more common.

Multi-document coreference resolution is very important in that it helps people get more information about an entity from multiple text sources. It is a central tool for information fusion and automatic summarization for multiple documents.

Early in MUC-6 and TIPSER phase III, multi-document coreference resolution was proposed as a potential subtask, but it was not included in final evaluation task because of being considered to be too difficult and too ambitious [1].

This paper presents a two-step approach to multi-document transliterated personal name disambiguation in Chinese. Firstly, a heuristic rule based on biographical

---

[1] Supported by National Natural Science Foundation of China (No.60473138, 60173005).

information and "colleague" names is used to form an initial set of coreference chains. In general, most chains could contain just one element after this processing. Then, a clustering algorithm based on VSM is employed to cluster these chains. A full name consistent constraint remains activated during the process. This paper also discusses feature selection and its weight computing method.

2 Related Work

Little research on multi-document coreference resolution has been documented as compared with coreference resolution in a single text. Almost no work on Chinese multi-document coreference resolution has been published although some typical methods have been developed in English. One of the earliest researches was done by Bagga and Baldwin [1]. They presented a VSM based approach to multi-document coreference resolution. They first form coreference chains within single text and generate a summary for each chain. These summaries then are represented as feature vectors of word bag and are clustered by using the standard VSM. They get a F-measure of 84.6% by testing on 173 New York Times articles. Fleischman and Hovy adopted another approach using both Maximum Entropy model and agglomerative clustering technique [4].

Dozier and Zielund [5] presented a cross-document coreference resolution algorithm based on biographical records created through automatic and manual modes in advance. This algorithm is then used to automatically acquire more information about specific people in legal field from multiple text sources. This resolution algorithm firstly applies IE technique to extract information about each individual whose name is the same as that in biographical records. And then it employs Bayesian network to resolve coreference by computing the probability that a given biographical record matches the same personal name in extracted templates. When the document gives the full personal name and highly stereotypical syntax, the precision and recall are 98% and 95% respectively. Mann and Yarowsky[6] combined IE technique with agglomerative clustering algorithm to carry out the task. IE is used to extract biographical information from texts, such as date of birth, occupation, affiliation and so on. Such information is closely related to a single individual and thus can help to resolve personal name ambiguity with higher accuracy. However, IE itself is a hard task. Mann and Yarowsky get a precision of 88% and recall of 73% with MI (Mutual Information) weight method by testing on real articles. They also get an accuracy of 82.9% testing on pseudonames with MI weight and only proper nouns as features.

Statistics methods are main ways for multi-document coreference resolution at present. Gooi and Allen [7] compared and evaluated three different techniques based on statistics method: Incremental Vector Space, KL-Divergence and agglomerative Vector Space. The evaluation results show that agglomerative vector space clustering algorithm outperforms the incremental vector space disambiguation model and the vector approaches are better than KL divergence.

3 Approach

VSM is thought as an easily implemented model, for it only needs to extract terms (words) as features without involving deep analysis of sentences or discourse in NLP. However, this method is inherently limited. In our approach, we combine it with a heuristic strategy.

3.1 Heuristic Strategy on Biographical Words and Personal Names

Some crucial words or phrases play decisive role in determining whether a personal name refers to a specific individual. For example, those words expressing occupation, affiliation and title etc. For this reason, IE technique can improve the precision of multi-document coreference resolution -- extracted information is from such important words [5][6]. However, as mentioned above, IE technique is still difficult to implement effectively because exact information must be extracted in order to fill slots in a template.

Generally speaking, the words around an interesting personal name are very important although they do not always express exactly biographical information of the person. Especially, they can easily be extracted from texts. Sentences given in Fig.1 are such examples. The two nouns (phrases) immediately preceding name 约翰逊 are obviously clue words. The biographical words will be paid special attention to in our approach. By *biographical word* we mean the two nouns (phrases) immediately preceding an interesting personal name and a noun following it.

```
(1) 美国/ns    著名/a    跨栏/vn    选手/n    阿兰·约翰逊/nr
    American  famous    hurdler              Allen Johnson
(2) 南非/ns    民众/n    送别/vn    艾滋病/n   小/h  斗士/n   约翰逊/nr
    South Africa crowd  saw off   AIDS      little fighter  Johnson
(3) 工程师/n   约翰逊/nr  被/p    凶残/a    处死/v
    engineer   Johnson   was     cruelly   killed
(4) 集团/n    副/b   总裁/n    凯文·约翰逊/nr
    Corp.    vice   president  Kevin Johnson
```

Fig. 1. Examples of two words (phrases) immediately proceeding the personal name 约翰逊

In addition, other personal names in a text are also important. Generally, it is easy to confirm the object to which a personal name refers if his (her) colleague names are mentioned at the same times. We found that when three or more personal names occur in two different texts together, the same name in both texts usually refers to the same individual.

In this approach, we extract biographical words of an interesting personal name from all texts to form the set Bio_set, and personal names except the interesting name to form the set Per_set. For each text, they are presented as Boolean vectors depending on whether they occur in the text, i.e.,

BV_per-x(d): a Per_set Boolean vector of text d in which x is the interesting personal name.
BV_bio-x(d): a Bio_set Boolean vector similar to BV_per-x(d).

We use the following **heuristic strategy** to confirm personal name coreference:

Rule-1:

if $(BV\_per\text{-}x(A) \bullet BV\_per\text{-}x(B)) > v1$,
or $(BV\_per\text{-}x(A) \bullet BV\_per\text{-}x(B) + BV\_bio\text{-}x(A) \bullet BV\_bio\text{-}x(B)) > v2$ (1)
then personal name x in text A and B refers to the same referent

Where the "$\bullet$" stand for operation of inner product. Both v1 and v2 are thresholds.

3.2 VSM Based Clustering Technique

Heuristic strategy can be used to confirm personal name coreference, but not to negate potential coreference even if the inner product is below a pre-defined threshold. In this case, we employ a VSM based agglomerative clustering algorithm to finally determine whether an interesting personal name appearing in different texts refers to the same referent. Cosine formula is used to compute similarity value between two vectors. The clustering algorithm always selects the text-pair with the highest similarity value above a pre-defined threshold and merges them into "a new single text" by simply adding the corresponding vectors together. This means that m*(m-1)/2 comparisons are always needed for m texts. This process repeats until no such text-pair is left. After each step, the number of texts will decrease by one.

3.2.1 Feature Selection

A text is represented as a feature vector with real-value weight. A term, selected as a feature, is either a named entity or one word with tag as follows:

(1) *{n, nr, nt, ns, nx, nz}* for nouns, where, n stands for common noun, nr for person name, nt for organization name, ns for place name, *nx* for non-Chinese character string and nz for other proper nouns [8]
(2) j for Abbreviations (acronym).
(3) vn for verbal noun.

Selected terms have more than two bytes (or one Chinese character) in length. For a named entity, composed of more than one word, the whole and its components all are selected as features. For example, the whole 华盛顿 大学 of organization name [华盛顿/ns 大学/n]nt (Washington University), its components 华盛顿 and 大学 will all be selected as features.

3.2.2 Term Weight

The term frequency is thought as very important in IR. In our approach, we replace term frequency by weighted term frequency, as follows:

$$gfreq(t_{i,j}) = freq(t_{i,j}) * \beta \qquad (2)$$

Where,
$freq(t_{i,j})$ is term frequency of t_i in text j and $gfreq(t_{i,j})$ is its weighted frequency.

Different kinds of terms in a text themselves have different effects on a personal name disambiguation. In order to deal with them differently, we introduce a weighted factor β. It is simply set as follows:

$$\beta = \begin{cases} 2.3, & tag\ of\ term\ t_i = nt; \\ 1.9, & tag\ of\ term\ t_i \in \{nz, nx, j\}; \\ 1.3, & tag\ of\ term\ t_i = ns; \\ 1.0, & tag\ of\ term\ t_i \in \{n, vn\}; \end{cases} \quad (3)$$

Our approach employs the weight strategy used in IR to calculate weight $w_j(t_i)$ of term t_i in text j and the formula is as follow:

$$w_j(t_i) = (1 + \log(gfreq(t_{i,j})))\log(\frac{N}{df(t_i)}) \quad (4)$$

Where, N is the number of documents, and $df(t_i)$ is the document frequency of term t_i, i.e. the number of documents in which t_i occurs.

3.2.3 Feature Filter

Not each feature is significant. Some even can cause noise. These trivial terms should be eliminated from feature set.

We firstly filter out some terms by using frequency, as follows:

Rule-2: A term t_i is eliminated from feature set if

$$\sum_{j=1}^{N} gfreq(t_{i,j}) \leq (\frac{N}{100} + 2) \quad (5)$$

Furthermore, we use variance to measure how well a term is suited to distinguish entities. The variance $V(t_i)$ of term t_i is defined as:

$$V(t_i) = \sum_{j=1}^{N}(w_j(t_i) - \mu_1(t_i))^2, \ where, \ \mu_1(t_i) = \frac{1}{N}\sum_{j=1}^{N}w_j(t_i) \quad (6)$$

The higher variance of a term is, the better it separates entities. Thus, terms with high variance should remain. Rule-3 is given to filter out those terms with the lowest variances:

Rule-3: 2/3 terms with the highest variance is extracted as features after Rule-2 is applied. If the number is still more than 600, only 600 terms with the highest variance remain while the others are filtered out.

Filtering out trivial terms will be helpful to improve effectiveness and efficiency of a VSM based algorithm.

3.3 Full Name Consistent Constraint

A full personal name (e.g. 迈克尔·约翰逊 - Michael Johnson) is composed of firstname (迈克尔 - Michael) and surname (约翰逊 - Johnson). Both firstname and surname are called part-name here. Although a part-name can be more frequently used in both spoken language and a written text, the full name is usually mentioned in the same context. A full personal name contains more information and thus it is more easily disambiguated. For example, we can easily determine that the entity to which 迈克尔·约翰逊 refers is different from that 本·约翰逊 refers to by simply matching the two name strings; however, it is impossible for only using surname 约翰逊. In order to make use of such information, an interesting part-name *PN* in text *X* will be mapped into its full name *FullN* if possible. The mapping is denoted as:

$$FN(PN, X) = \begin{cases} FullN; & \text{if the full name FullN occurs in } X \\ 0; & \text{otherwise} \end{cases} \quad (7)$$

A constraint rule based on full name is given as follows:

Rule-4: if (FN(PN, X) ≠0 ∧ FN(PN, Y) ≠0 ∧ FN(PN, X) ≠ FN(PN, Y)), then, name PN in text X does not refer to an individual for which name PN in Y stands.

If the condition of Rule-4 is true to a person name PN in text-pair (X, Y), it will not be necessary to apply Heuristic strategy or cluster algorithm to this pair.

4 Implementation and Evaluation

4.1 Outline of Algorithm

The outline of our approach is informally described as follows:

Input: Set-of-document D = {d_1, d_2, ..., d_n}, where each document is segmented into Chinese words and tagged with PKU tag set[8].
Steps:
Step1. Extract words from each text d to form Boolean vectors BV_per-x(d) and BV_bio-x(d) respectively; and, at the same time, to build feature vector with real-value weight.
Step2. Filter out some features by using Rule-2 and Rule-3 in section 3.2.3.
Step3. Confirm the interesting personal name coreference by using heuristic strategy in section 3.1.
Step4. Use an agglomerative clustering algorithm to resolve multi-document personal name coreference.
Output: a set of coreference chains.

As mentioned above, the full name consistent constraint (Rule-4) is used before both step3 and step4, and only if its condition is unsatisfied, the two steps will be executed.

4.2 Test Data

In order to test our approach, we downloaded 278 Chinese web pages containing the personal name 约翰逊 at random. These pages are transformed into plain text format, and then are segmented from character string into word sequence and tagged with our integrated tool, in which proper nouns like personal names, place names, organization names and so on, are automatically identified.

In these 278 texts, personal name 约翰逊 refers to 19 different referents. Table 1 gives the simple descriptions of all referents and the number of texts related to each referent.

Table 1. 19 different referents and the number of texts related to each referent

No.	Referent	Number
1	阿兰·约翰逊 (Allen Johnson) : hurdler	23 texts
2	本·约翰逊 (Ben Johnson) : 100m sprinter	22 texts
3	德马尔·约翰逊 (DerMarr Johnson) : NBA player	23 texts
4	菲利浦·约翰逊 (Phillip Johnson) : Dean of American Architects	15 texts
5	格伦·约翰逊 (Glen Johnson) : England Footballer	21 texts
6	林登·约翰逊 (Lyndon Johnson) : former U.S. president	22 texts
7	埃文·约翰逊 (Earvin Johnson) : NBA "Magic"	22 text
8	柯克·约翰逊 / 科克·约翰逊 (Kirk Johnson) : Canadian Boxer	24 text
9	斯宾塞·约翰逊 (Spencer Johnson) : Thinker and author	12 text
10	斯嘉丽·约翰逊 (Scarlett Johansson) : Actress	18 texts
11	迈克尔·约翰逊 (Michael Johnson) : 200m and 400m sprinter	26 texts
12	保罗·约翰逊 (Paul Johnson) : American hostage beheaded in Saudi	22 texts
13	恩科西·约翰逊 (Nkosi Johnson) : South Africa's child with HIV+	8 texts
14	德韦尼·约翰逊 (Dwayne Douglas Johnson) : Actor	4 texts
15	卡尔·约翰逊 (Carl Johnson) : dramatis personae in "Grand Theft Auto: San Andreas"	2 texts
16	塞缪尔·约翰逊 (Samuel Johnson) : lexicographer & litterateur	5 texts
17	凯文·约翰逊 (Kevin Johnson) : vice president of Micorsoft Corp.	7 texts
18	约翰逊 (Johnson): a major of army	1 text
19	玫丽莎·约翰逊 (Melissa Johnson) : Streaker(London student)	1 text

4.3 Scoring Strategy

Bagga and Baldwin discussed two evaluation methods for coreference resolution: MUC and B-CUBED [1][2]. The latter will be used in our evaluation. It is described as follows:

Assume that a system partitions a set of n testing documents D into k subsets (chains), and the referent of the interesting personal name in text d is denoted by $referent_d$. For each document d ($\in$ a subset A), its precision and recall are:

$$precision_d = \frac{number\ of\ documents\ containing\ the\ referent_d\ in\ Subset\ A}{size\ of\ Subset\ A} \quad (8)$$

$$recall_d = \frac{number\ of\ documents\ containing\ the\ referent_d\ in\ Subset\ A}{total\ of\ documents\ containing\ the\ referent_d\ in\ Set\ D} \quad (9)$$

Overall precision and recall value are determined by averaging each document values, i.e.

$$P = \frac{1}{n}\sum_{d \in D} precision_d \quad \text{and} \quad R = \frac{1}{n}\sum_{d \in D} recall_d \quad (10)$$

We also use F measure to evaluate the result, i.e.

$$F = \frac{2 \times P \times R}{P + R} \tag{11}$$

4.4 Evaluation

We independently test heuristic strategy and the agglomerative clustering algorithm on 约翰逊 corpus not using full name consistent constraint (Rule-4).

Threshold $v1$ and $v2$ in formula (1) are set to 2 and 3 respectively. The result by only using heuristic strategy is F measure of 41.12%, recall of 100% and precision of 25.87%.

Fig.2 shows how recall and precision trade off against each other as the similarity threshold (horizontal axis) varies by only using the agglomerative clustering algorithm. In this figure, F-measure reaches the peak value 75.171% at threshold=0.25. Its corresponding precision and recall are 72.374% and 78.193% respectively.

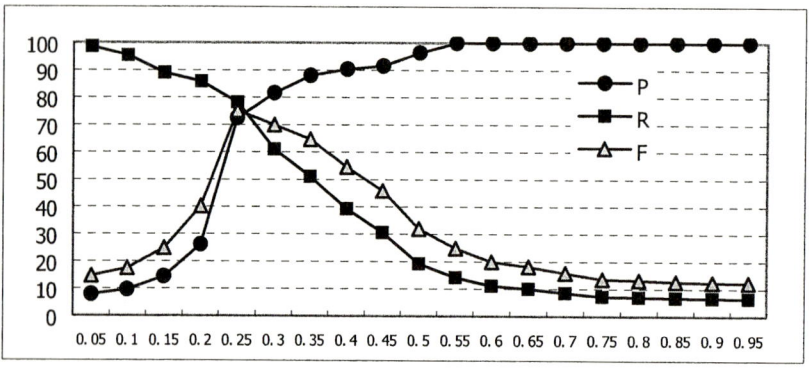

Fig. 2. The results of multi-document personal name coreference on 约翰逊 corpus by only using agglomerative clustering algorithm

From Fig.2, we can see that F measure remains very low values (below 60%) in large regions. Thus, it will be difficult to select a fit threshold where the F measure reaches a high point.

If both heuristic strategy and agglomerative clustering algorithm are used under that the Rule-4 is unused, the performance is obviously improved at most thresholds, although the peak value is almost the same (75.94% vs 75.171%). Especially, the F measure by using "Cluster+Heuristic" falls more gently than that using "only Cluster" as the threshold increases from 0.25 to 0.95. A comparison of F-measures at different thresholds is given in Fig.3.

Fig.3 also indicates that both F measures fall rapidly as the threshold decreases from 0.25 to 0.05. However, if the full name consistent constraint (Rule-4) is activated as described in section 4.1, the performance is encouraging. Fig.4 shows the F measure, precision and recall at different thresholds in this case. Obviously, the F measure almost keeps increasing as threshold decreases and the highest F measure is 86.35% at the minimum threshold point in this figure.

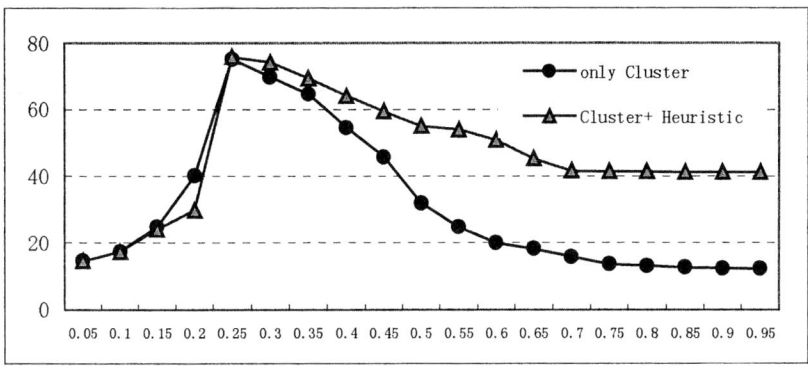

Fig. 3. A result comparison of F measures using only cluster and using cluster+Heuristic

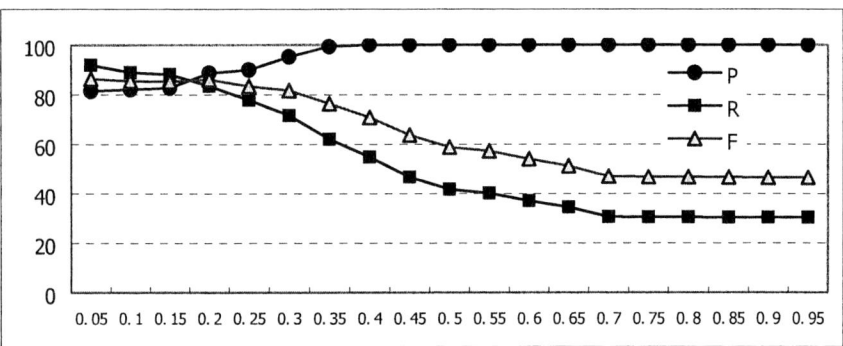

Fig. 4. The results by using Heuristic+Cluster+Full consistent

Fig.4 also shows the highest recall is 91.835% (at threshold=0.05), which is much lower than the value 100% in theory. The reason is that there are two transliterated names 科克·约翰逊 and 柯克·约翰逊 in our test corpus for original name "Kirk Johnson". When the full name consistent constraint is activated, the name (约翰逊) with different full name(such as 科克·约翰逊 and 柯克·约翰逊) will be determined as referring to the different individuals. In fact, the two different full names stand for the same referent and are in the same coreference-chain in our manual reference answer. The evaluation by comparing automatic results with manual ones is thus below 100%. If they are denoted in different chains in manual answer, the test shows the highest recall reaches 100% and the corresponding F measure is about 89.0%.

5 Conclusion

This paper presented a two-step approach to multi-document transliterated personal name coreference resolution in Chinese. The approach achieves a good testing performance.

Clustering technique is a mainly used solution to this problem. However, our test shows that the technique is limited. We introduced a heuristic strategy based Boolean vectors into our approach as a pre-processing step. Both independent test and combinational test with clustering algorithm show that heuristic strategy is useful. It only needs an inner-production operation between two vectors and thus is faster than cosine operation used in clustering method. Fig. 3 shows the performance is obviously better using the strategy in clustering algorithm than that not using it at almost all thresholds. Furthermore, we presented a full name consistent constraint to avoid errors caused by both heuristic strategy and clustering algorithm. It is very important for a transliterated part-name across documents coreference resolution. The test shows that performance is surprisingly improved after this rule is used.

References

1. Bagga Amit and Breck Baldwin. Entity-Based Cross-Document Coreferencing Using the Vector Space Model. Proc. of the 36th Annual Meeting of the ACL and the 17th International Conreference on Computational Linguistics (COLING-ACL)(1998), 79-85
2. Bagga Amit and Biermann Alan. A Methodology for Cross-Document Coreference. In Proc. of the fifth Joint Conference on Information Sciences(JCIS) (2000), pp.207-210
3. Soon Wee meng, Hwee tou Ng and Daniel Chung Yong Lim. A machine Learning Approach to Coreference Resolution of Noun Phrases. Computational Linguistics(Special Issue on Computational Anaphora Resolution), Vol.27, No.4(2001) 521-544
4. Fleischman Michael and Eduard Hovy. Multi-Document Personal Name Resolution. Proc. Reference Resolution Workshop, 42nd Annual Meeting of the ACL(2004), 1-8
5. Dozier Christopher and Thomas Zielund. Cross Document Co-Reference Resolution Application for People in the Legal Domain. Proc. Reference Resolution Workshop, 42nd Annual Meeting of the ACL(2004) 9-16
6. Mann Gideon and David Yarowsky. Unsupervised Personal Name Disambiguation.Proc. of CoNLL, Edmonton,Canada(2003) 33-40
7. Chung Heong Gooi and James Allan. Cross-Document Coreference on a Large Scale Corpus. Proc. of HLT-NAACL2004, Boston(2004) 9-16
8. Yu Shiwen. The Grammatical Knowledge-Base of Contemporary Chinese – A Complete Specification, Tsinghua University Press (China) 2003.

Difference-Similitude Matrix in Text Classification

Xiaochun Huang, Ming Wu, Delin Xia, and Puliu Yan

School of Electronic Information, Wuhan University, Wuhan, 430079, Hubei, China
xiaochun.huang@gmail.com
ming-wu@vip.sina.com, ypl@whu.edu.cn

Abstract. Text classification can greatly improve the performance of information retrieval and information filtering, but high dimensionality of documents baffles the applications of most classification approaches. This paper proposed a Difference-Similitude Matrix (DSM) based method to solve the problem. The method represents a pre-classified collection as an item-document matrix, in which documents in same categories are described with similarities while documents in different categories with differences. Using the DSM reduction algorithm, simpler and more efficient than rough set reduction, we reduced the dimensionality of document space and generated rules for text classification.

1 Introduction

Discovering knowledge from text is an essential and urgent task, because a great deal of visual information of real world is recorded in text documents and the amount of text information continually increases sharply with the growth of internet. How to get expected knowledge quickly and correctly from text becomes a more and more serious problem. Automatic text classification can help to speed up information retrieval and to dig out latent knowledge. Text documents are usually represented in terms of weighted words and described as a *Vector Space Model* (VSM) [1]. Dimensions of document spaces are always too high to deal with directly for many classification algorithms, such as Neural Network classification [2] and K-Nearest Neighbor classification [3] etc. Moreover, many collections of documents only contain a very small vocabulary of words that are really useful for classification. Dimensionality reduction techniques are a successful avenue for solving the problem.

Dimensionality reduction techniques can be divided into two kinds: attribute reduction and sample reduction [4]. As for text classification, that means selecting a small number of keywords to present document content and to describe classification rules, and the rules should be as few as possible. We take keywords as attributes to denote words or phrases that are important for classification. A lot of researches have demonstrated that rough set based methods are good for dimensionality reduction to a great extent [5,6]. As a practical approximating approach, rough sets have been used widely in data analysis, especially in soft computation and knowledge induction. The basic idea of rough set theory is to describe and induct the indiscernible objects using their upper and lower approximations [7]. In the discernibility matrix of rough sets, only differences between objects and indiscernibility relation are represented, while the similarities of them are ignored. In order to make full use of the knowledge that datasets provide, and to decrease the computational complexity, a new theory named *Difference-Similitude Matrix* (**DSM**) was proposed [8], which is quite similar to but easier and more expressive than rough set based methods. Compared with RS methods that use discernibility matrix, DSM-based methods take both differences and similarities of objects into account, and can get good reduced attributes and rules without complex mathmatical operations, such as caculating for every instance the indiscernibility relation. Jiang etc. show their expeimental results in [9] that DSM methods can get a smaller set of more correct rules than discernibility matrix based methods, especially for large scale databases. In this paper, we apply a DSM-based approach [10] to reduce the dimensionality of item-by-document matrix which represents pre-specified collections of document, and generate rules for text classification.

The background of DSM theory and text classification is introduced in section 2. Section 3 describes the DSM-based method for text classification, and gives a simple example to demonstrate the process of reduction and rule generation. The experimental results are discussed in section 4. Section 5 is our conclusion about this paper and the plan of future work.

2 Background

2.1 Basic Theory of DSM

The main cause that makes DSM differentiated from rough sets is the inclusion of similarities among objects. To understand the usage of DSM in our classifier, we introduce the basic knowledge of DSM simply here first. The detailed description can be found in [8].

Suppose *IS* is an information system, and $IS = \langle U, C, D, V, f \rangle$, where U denotes the system object set; C denotes the condition attribute set; D denotes the decision attribute set; $V = \bigcup(V_a : a \in (C \cup D))$ denotes the attribute value set; $f : U \times (C \cup D) \to V$ is the function that specifies the attribute values. As for the objects in information system, we can define a $m \times m$ difference-similitude matrix M_{DS} to represent their attributes and values. The matrix has two types of elements, similarity item m_{ij}^s and difference m_{ij}^d, which can be defined as Eq. (1), where m is the number of condition attributes and n is the number of instances in dataset. Here we modify the definition of m_{ij}^s a little for convenience of computation, that is, using m_{ii}^s to denote the elements on the diagonal instead of zero as before, and we need not to change the related algorithm.

$$m_{ij} = \begin{cases} m_{ij}^s = \begin{cases} \{q \in C : f(q, x_i) = f(q, x_j)\} D(x_i) = D(x_j) \\ \{\phi : \forall (f(q, x_i) \neq f(q, x_j))\} D(x_i) = D(x_j) \end{cases} & i = 1,2,\cdots,m \\ m_{ij}^d = \begin{cases} \{q \in C : f(q, x_i) \neq f(q, x_j)\} D(x_i) \neq D(x_j) \\ \{\phi : \forall (f(q, x_i) = f(q, x_j))\} D(x_i) \neq D(x_j) \end{cases} & j = 1,2,\cdots,m \\ D(x) = 1,2,\cdots,n \end{cases} \quad (1)$$

C_i^b (core basic condition attribute set) and B_i^{op} (best optimal condition attribute set), are essential to define similarity significance, $Sigp(D)$, and difference significance,

$Sigq(D)$, as well as to deduce fundamental lemmas for attribute reduction. Though we use these concepts during reduction, we would not describe them here due to the length limit of the paper. The principle that DSM-based reduction conforms is to get the following things without losing information of the original system after reduction:

- Minimum number of remained attributes to describe rules;
- Minimum number of classification rules.

The principle is quite compatible with the purpose of dimensionality reduction in text classification as we mentioned above. The DSM reduction algorithm is described in section 3.

2.2 Text Classification

The purpose of text classification is to classify documents into different predefined categories. For this purpose, we usually extract keywords from those different categories at first, and count the word frequencies of these keywords in a document to be classified, and compare the results with those of pre-classified documents. Then we put the document into the closest category.

A pre-classified collection and text documents in it can be described as follows: Let $C = \{c_1, \cdots, c_i, \cdots, c_m\}$ represent a set of predefined categories, $D = \{d_1, \cdots, d_j, \cdots, d_n\}$ represent a collection of documents assigned into these categories, where m is the number of categories, n is the number of classified documents in collection. A document is usually described as pairs of keywords and their weights like:

$$d_j = \{<t_1, f_1>, \cdots, <t_k, f_k>, \cdots, <t_p, f_p>\} \qquad (2)$$

where p is the number of different keywords in d_j, t_k is a keyword, f_k is the word frequency of t_k. Then, the collection can be denoted with a matrix composed by keyword-document pairs.

TFIDF [11] is a popular method to represent word frequency, but it doesn't contain position information of words. The importance of words in various blocks of a document differs a lot, for example, keywords in the title and the body of a webpage are often more important than those in link text. In this paper we proposed a weighted TFIDF to represent word frequency:

$$f_k' = \sum_{i=1}^{q} n_{ki} w_i$$
$$f_k = f_k' \cdot \log\left(\frac{N}{N_k}\right) \bigg/ \sqrt{\sum_{l=1}^{M}\left[\log\left(\frac{N}{N_l}\right)\right]^2} \qquad (3)$$

where n_{ki} is the number of t_k in block i of d_j, q is the number of blocks in d_j, w_i is the weight of block i which is user-defined according to specific applications, N is the number of documents in the collection that contain t_k, N_k is the number of t_k in the collection, and M is the number of words that appear at least once in the collection.

Due to the large quantity of documents and the limited performance of classification algorithms, high dimensionality of keywords becomes an obstacle to text classification. Keyword selection is one of the most important steps for dimensionality reduction. After choosing words with word frequency high enough to represent documents, we also apply DSM method to reduce dimension.

Rule-based methods [12,13] and distance-based methods [14,15] are the two most popular approaches for text classification. Rule-based methods use small subsets of keywords as condition attributes of decision rules, which means only part of the keywords need to be examined by rules and the speed of classifying new document is faster than distance-based methods. Therefore, we chose DSM method as our feature selection and rule generation method. The generated rules can be written as:

$$r_{ik} : (h_1 = a_1) \wedge (h_2 = a_2) \wedge \cdots \wedge (h_n = a_n) \Rightarrow d_j \rightarrow C_i \quad (4)$$

where r_{ik} is the kth rule to decide whether d_j belongs to c_i, h_n is the word frequency of remained keyword t_n after DSM reduction and a_n is the corresponding value.

3 DSM Based Text Classification

We presented a DSM-based text classifier in this paper. DSM method in our system plays two roles: dimensionality reduction and classification rule generation.

3.1 Preprocessing

Content of text documents is always unstructured data. Before text classification, we should preprocess them to satisfy the requirements of document representation. The usual tasks of preprocessing are:

- removing html tags or tags in other formats
- word segmentation, stemming and removing stop list
- computing word frequency for each remained word, i.e. keyword, in different categories, and removing keywords with low word frequencies. We call the remained keywords as items in the rest of the paper.

Then we can describe a pre-classified collection with a *Text Information Matrix* (TIM), instance j of which is document d_j. d_j in TIM can be represented in the following form:

$$d_j = \begin{bmatrix} F & \bar{C} \end{bmatrix} = \begin{bmatrix} f_1 & \cdots & f_k & \cdots & f_n & \bar{C} \end{bmatrix} \quad (5)$$

where n is the total number of different items of the collection, and the word frequencies of these items are $f_1, \cdots, f_n$. $\bar{C}$ denotes the category that d_j belongs to.

Since the algorithm DSM method can only deal with nominal value, while item frequency is usually continuous, so we must discretize it before transforming TIM into a difference-similitude matrix. The discretization method we select is equal-width approach. That is, divide value range of f_k into a number of intervals with equal width. By doing this, a large amount of possible item frequencies can be mapped into a small number of intervals so that the dimension of classification rules generated will be reduced correspondently.

3.2 DSM Reduction and Rule Generation

The collection has been transformed into an item-by-document matrix after the preprocessing mentioned above. Considering that documents in the same category share some frequent item sets, while documents of different categories share few [16], we applied DSM-based reduction algorithm to select the most representative items to reduce the dimensionality of attributes. During the procedure, we can achieve the related classification rules at the same time.

The DSM reduction algorithm was proposed based on several lemmas which can be found in [17]. We adjusted the algorithm to satisfy the requirements of text classification as following:

DSM_reduction:
Input: TIM ($m \times (n+1)$) dimensions, where m denotes the number of documents in a collection, n denotes the number of items)
Output: reduced attribute (items) set - RA,
classification rule set - CR
Steps:
(1) Order the objects in TIM by class;
(2) Construct Difference-Similitude matrix M_{DS} ; let $ON(k)$ denotes the array for category k ; let $k=1$, $i=1$, $s=\phi$;
(3) for $k=1 \sim m$:
Record the start index, $is(k)$, the end index, $ie(k)$, of category k;
For $i= is(k)+1 \sim ie(k)$:
a) if $i \in s$
then $r(i)=r(i-1)$; $i=i+1$;
b) Get c_i^b and $sig_{a \in c}(\overline{C})$ (importance of each item); $a = \phi$;
c) If $c_i^b \cap (\forall m_{ij}^d) \neq \phi$
then record the index of document j that satisfies $c_i^b \cap m_{ij}^s \neq \phi$ in s;
else seek_ c_i^a ;
d) $c_i^b \cup c_i^a \to B_i^{op}$; $RA = RA \cup B_i^{op}$
e) $r_k(i): B_i^{op}(v_i) \to \overline{C}_k$; $CR = CR \cup \{r_k(i)\}$
(4) Remove redundant rules in CR.

Function seek_ c_i^a is for computing the complementary attribute set (for details see [16]).

To explain the process, let's illustrate by a simple example. In this example, the collection has seven documents, $d_1, \cdots, d_7$, and is divided into three categories, $\overline{C}_1, \cdots, \overline{C}_3$. After preprocessing, the four items, a, b, c and d, are chosen to represent the collection. The content of item-by-document matrix is shown as a decision table as Table 1.

Then discretize the values of attributes into three ranks using equal-width method. The number of intervals for each attribute is 3. Table 2 is the discretization result.

The corresponding difference-similitude matrix is shown in Table 3. Elements printed in bold type are the similarities of items.

Table 1. Item-by-document matrix

Attr. Doc.	a	b	c	d	$\overline{C}$
d_1	0.084	0	0.033	0.027	1
d_2	0.131	0.013	0.033	0.005	1
d_3	0.022	0	0.027	0.013	2
d_4	0.110	0.027	0.012	0.018	2
d_5	0.196	0.019	0	0.045	3
d_6	0.115	0.033	0.064	0.033	3
d_7	0.233	0.056	0.051	0.036	3

Table 2. Discretized item-by-document matrix

Attr. Doc.	a	b	c	d	$\overline{C}$
d_1	1	0	1	1	1
d_2	1	0	1	0	1
d_3	0	0	1	0	2
d_4	1	1	0	1	2
d_5	1	1	0	2	3
d_6	2	1	2	2	3
d_7	2	2	0	2	3

Table 3. Difference-Similitude matrix

abcd	abc	ad	bc	bcd	abcd	abcd
	abcd	a	bcd	bcd	abcd	abcd
		abcd	∅	abcd	abcd	abcd
			abcd	d	acd	abd
				abcd	bd	cd
					abcd	ad
						abcd

First we can find the core attribute set c_i^b and the complementary attribute set c_i^a, if exists. In this example they are:

$$\{c_i^b\} = \{\phi, \{a\}, \phi, \{d\}, \phi, \phi, \phi\}$$
$$\{c_i^a\} = \{\{ac\}, \{c\}, \{a\}, \phi, \{d\}, \{d\}, \{d\}\}$$

Thus according to $c_i^b \cup c_i^a \to B_i^{op}$; $RA = RA \cup B_i^{op}$, the best optimal attribute set B_i^{op} is:

$$\{B_i^{op}\} = \{\{ac\}, \{ac\}, \{a\}, \{d\}, \{d\}, \{d\}, \{d\}\}$$

then RA for representing the collection is $\{a, c, d\}$. The final generated rule set CR after redundancy removals are:

$r_{1,1}$: $(a=1)$ AND $(c=1)$ → $\overline{C_1}$ (d_1, d_2)

$r_{2,1}$: $(a=0)$ → $\overline{C_2}$ (d_3)

$r_{2,2}$: $(a=1)$ AND $(c=2)$ → $\overline{C_2}$ (d_4)

$r_{3,1}$: $(d=2)$ → $\overline{C_3}$ (d_5, d_6, d_7)

d_j s at the end of each rule are the documents that support the rule.

When a new document comes to be classified, we count and discretize the item frequencies of attributes appearing in B_i^{op}, then use the discretized results as condition attributes to try to match the above four rules. If any successes, then put the document into the category that rule describes.

4 Experiment Evaluation

We used three categories of Reuters-21578 Text Categorization Collection as our dataset to train and test the DSM-based classifier. Reuter-21578 is a collection of documents on Reuter newswire in 1987, the documents of which are classified and indexed manually. It can be freely downloaded from: http://kdd.ics.uci.edu/databases/reuters21578/reuters21578.html. The three categories are listed below in Table 4.

Table 4. Subcategories of Reuter-21578

Dataset	TotalNum.	TrainNum.	TestNum.
acq	2427	614	125
Money-fx	745	532	76
Grain	601	473	54

To evaluate the performance of our method, we chose attribute reduced ratio, R_a, and sample reduced ratio, R_s, as evaluation criteria for reduction, and chose the most commonly used evaluation criteria, precision and recall, for classification:

R_a = (original attributes − remained attributes) / original attributes.
R_s = (original samples − rules generated) / original samples.
Precision = documents classified correctly / all documents classified in a particular category.
Recall = documents classified correctly / all documents that should be classified in a category.

As for the example of section 3.3, values of evaluation criteria are: R_a=100*(4−3)/4%=25%, R_s=100*(7−4)/7%=42.8%.

Using different number of top ranked items results in different values of evaluation criteria. Table 5 shows precision values and recall values of the three categories at the case of 100, 500, 1000 and 2000 items per category.

Table 5. Precision value and recall value of different number of items

Dataset	100		500		1000		2000	
	Pre	Rec	Pre	Rec	Pre	Rec	Pre	Rec
acq	0.882	0.925	0.931	0.982	0.955	0.937	0.904	0.893
Money-fx	0.745	0.973	0.671	0.981	0.689	0.964	0.703	0.976
Grain	0.945	0.960	0.866	0.978	0.913	0.965	0.947	0.913

We noted that increasing items can improve the classification quality, but after the quality is bettered to a certain degree, it will debase the quality instead. We can call the point that quality becomes worse with the increase of items as an upper limit. Generally if contents of documents in the same category consistent with each other or are unambiguous, the upper limit would be much higher.

Category Grain includes all the documents containing word *grain* in Reuters-21578. We counted the attribute reduction R_a and sample reduction R_s for it, and gave the results in Figure 1. These results demonstrate that using DSM-based classifier can

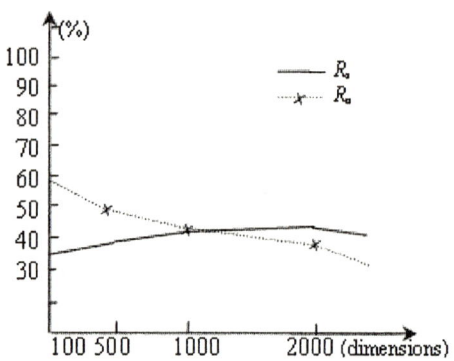

Fig. 1. Performance measure values of Grain

reduce the dimensionalities of samples and rules significantly, and the quality of classification is good. In addition, they also show that too large numbers of items do no good to attribute reduction and rule reduction. We think it might because of impurity added into the process by some words.

5 Conclusions

The experimental results have shown that DSM method can be used in text classification, and can get considerable good results. It reduces the dimensionalities both of samples and attributes of a collection of documents. However, we have only applied DSM-based classifier on the three categories of Reuter-21578 yet, and lots of further work should be done in the future. There is an incremental machine learning algorithm based on DSM method [17], which is able to classify new document, generate and update the rules of classifier dynamically. We will use it to improve the quality of our classifier.

Acknowledgements

This word has been partially funded by the National Natural Science Foundation of China (90204008).

References

1. Salton, G, Wong, A and Yang, C. S: A vector space model for information retrieval. Communications of the ACM, 18(11) (1975) 613-620
2. R. Setiono and H. Liu: Neural network feature selector. IEEE Transactions on Neural Networks, vol.8, no. 3 9 (1997) 645-662
3. Allen L. Barker: Selection of Distance Metrics and Feature Subsets for k-Nearest Neighbor Classifiers (1997)
4. Z. Pawlak: Rough Sets: Theoretical Aspects of Reasoning about Data [M]. Dordrecht: Kluwer Acasemic Publishers (1991)
5. Z. Pawlak: Rough Classification. International Journal of Man-Machine Studies, 20(5) (1984) 469-483
6. Nguyen and Hung Son: Scalable classification method based on rough sets. Proceedings of Rough Sets and Current Trends in Computing (2002) 433-440
7. Z. Pawlak: Rough Sets. Informational Journal of Information and Computer Sciences, vol. 11(5) (1982) 341–356
8. Delin Xia and Puliu Yan: A New Method of Knowledge Reduction for Information System – DSM Approach [R]. Wuhan: Research Report of Wuhan University (2001)
9. Hao Jiang, Puliu Yan, Delin Xia: A New Reduction Algorithm – Difference-Similitude Matrix. Proceedings of the Second International Conference on Machine Learning and Cybernetics, Xi'an, 2-5 (2004) 1533-1537
10. Ming Wu, Delin Xia and Puliu Yan: A New Knowledge Reduction Method Based on Difference-Similitude Set Theory. Proceedings of the Third International Conference on Machine Learning and Cybernetics, Shanghai, vol. 3 (2004) 1413–1418

11. A. Aizawa: The feature quantity: An information theoretic perspective of tfidf-like measures. Proceedings of SIGIR 2000 (2000) 104–111
12. Yixin Chen and James Z. Wang: Support Vector Learning for Fuzzy Rule-Based Classification System. IEEE Transactions on Fuzzy Systems, vol. 11, no. 6 (2003) 716-728
13. Hang Li and Y. Kenji: Text Classification Using ESC-based Stochastic Decision List. Proceedings of the 8th ACM International Conference on Information and Knowledge Management (CIKM'99) (1999) 122-130
14. Han, Eui-Hong and Vipin Kumar: Text Categorization Using Weight Adjusted k-Nearest Neighbor Classification. Technical Report #99-019 (1999)
15. Kamal Nigam, Andrew McCallum, Sebastian Thrun, and Tom Mitchell: Using EM to Classify Text from Labeled and Unlabeled Documents. Technical Report CMU-CS-98-120, School of Computer Science, CMU, Pittsburgh, PA 15213 (1998)
16. Fung, B. C. M., Wang, K. and Ester M: Hierarchical Document Clustering Using Frequent Itemsets. Proceedings of the SIAM International Conference on Data Mining (2003)
17. Jianguo Zhou, Delin Xia and Puliu Yan: Incremental Machine Learning Theorem and Algorithm Based on DSM Method. Proceedings of the Third International Conference on Machine Learning and Cybernetics, Shanghai, vol. 3 (2004) 2202–2207

A Study on Feature Selection for Toxicity Prediction

Gongde Guo[1], Daniel Neagu[1], and Mark T.D. Cronin[2]

[1] Department of Computing, University of Bradford, Bradford,
BD7 1DP, UK
{G.Guo, D.Neagu}@Bradford.ac.uk
[2] School of Pharmacy and Chemistry, Liverpool John Moores University,
L3 3AF, UK
M.T.Cronin@Livjm.ac.uk

Abstract. The increasing amount and complexity of data used in predictive toxicology calls for efficient and effective feature selection methods in data pre-processing for data mining. In this paper, we propose a kNN model-based feature selection method (kNNMFS) aimed at overcoming the weaknesses of ReliefF method. It modifies the ReliefF method by: (1) using a kNN model as the starter selection aimed at choosing a set of more meaningful representatives to replace the original data for feature selection; (2) integration of the Heterogeneous Value Difference Metric to handle heterogeneous applications – those with both ordinal and nominal features; and (3) presenting a simple method of difference function calculation. The performance of kNNMFS was evaluated on a toxicity data set Phenols using a linear regression algorithm. Experimental results indicate that kNNMFS has a significant improvement in the classification accuracy for the trial data set.

1 Introduction

The success of applying machine learning methods to real-world problems depends on many factors. One such factor is the quality of available data. The more the collected data contain irrelevant or redundant information, or contain noisy and unreliable information, the more difficult for any machine learning algorithm to discover or obtain acceptable and practicable results. Feature subset selection is the process of identifying and removing as much of the irrelevant and redundant information as possible. Regardless of whether a learner attempts to select features itself, or ignores the issue, feature selection prior to learning has obvious merits [1]:

1) Reduction of the size of the hypothesis space allows algorithms to operate faster and more effectively.
2) A more compact, easily interpreted representation of the target concept can be obtained.
3) Improvement of classification accuracy can be achieved in some cases.

The aim of this study was to investigate an optimised approach for feature selection, termed kNNMFS (kNN Model-based Feature Selection). This augments the typical feature subset selection algorithm ReliefF [2]. The resulting algorithm was run on different data sets to assess the effect of a reduction of the training data.

2 kNN Model-Based Feature Selection

A kNN model-based feature selection method, *kNNMFS* is proposed in this study. It takes the output of kNNModel [3] as seeds for further feature selection. Given a new instance, kNNMFS finds the nearest representative for each class and then directly uses the inductive information of each representative generated by kNNModel for feature weight calculation. This means the k in ReliefF is varied in our algorithm. Its value depends on the number of instances covered by each nearest representative used for feature weight calculation. The kNNMFS algorithm is described as follows:

Algorithm kNNMFS
Input: the entire training data D and parameter ε.
Output: the vector W of estimations of the qualities of attributes.
1. Set all weights $W[A_i]=0.0$, $i=1,2,...,p$;
2. $M:=kNNModel(D, \varepsilon)$; $m=|M|$;
3. for $j=1$ to m do begin
4.   Select representative $X_j=<Cls(d_j), Sim(d_j), Num(d_j), Rep(d_j), Rep(d_{j1}), Rep(d_{j2})>$ from M
5.   for each class $C \neq Cls(d_j)$ find its nearest miss $M_v(C)$ from M;
6.   for $k=1$ to p do begin
7. $W[A_k]=W[A_k]-(diff(A_k, d_j, d_{j1})+diff(A_k, d_j, d_{j2}))\times \frac{Sim(d_j)}{Num(d_j)})/(2m) +$

$$\sum_{C \neq Cls(d_j)} \left(\frac{P(C)}{1-P(Cls(d_v))} \times (diff(A_k, d_j, d_{v1}(C)) + diff(A_k, d_j, d_{v2}(C)) \times \frac{Sim(d_v)}{Num(d_v)})/(2m) \right)$$

8.   end;
9. end;

Fig. 1. Pseudo code of the kNNMFS algorithm

In the algorithm above, ε is the allowed error rate in each representative; p is the number of attributes in the data set; m is the number of representatives which is obtained from kNNModel(D, ε) and is used for feature selection. Each chosen representative d_j is represented in the form of $<Cls(d_j), Sim(d_j), Num(d_j), Rep(d_j), Rep(d_{j1}), Rep(d_{j2})>$ which respectively represents the class label of d_j; the similarity of d_j to the furthest instance among the instances covered by N_j; the number of instances covered by N_j; a representation of instance d_j; the nearest neighbour and the furthest neighbour covered by N_j. *diff()* uses HVDM [4] as a different function for calculating the difference between two values from an attribute.

Compared to ReliefF, kNNMFS speeds up the feature selection process by focussing on a few selected representatives instead of the whole data set. These representatives are obtained by learning from the original data set. Each of them is an optimal representation of a local data distribution. Using these representatives as seeds for feature selection better reflects the influence of each attribute on different classes, thus giving more accurate weights to attributes. Moreover, a change was made to the original difference function to allow kNNMFS to make use of the generated information in each representative such as $Sim(d_j)$ and $Num(d_j)$ from the created model of kNNModel for the calculation of weights. This modification reduces the computational cost further.

3 Experiments and Evaluation

To evaluate the effectiveness of the newly introduced algorithm kNNMFS, we performed some experiments on a data set of toxicity values for approximately 250 chemicals, all which contained a similar chemical feature, namely a phenolic group [5]. For the prediction of continuous class values, e.g. the toxicity values in the phenols data set, dependent criteria: Correlation Coefficient (CC), Mean Absolute (MAE), Root Mean Squared Error (RMSE), Relative Absolute Error (RAE), and Root Relative Squared Error (RRSE) are chosen to evaluate the goodness of different feature selection algorithms in the experiments. These evaluation measures are used frequently to compare the performance of different feature selection methods.

In this experiment, eight feature selection methods including ReliefF and kNNMFS were performed on the phenols data set to choose a set of optimal subsets based on different evaluation criteria. Besides kNNMFS that was implemented in our own prototype, seven other feature selection methods are implemented in the Weka [6] software package.

The experimental results performed on subsets obtained by different feature selection methods are presented in Table 1. In the experiments, a 10-fold cross validation method was used for evaluation. It is obvious that the proposed kNNMFS method performs better than any other feature selection methods evaluated by the linear regression algorithm on the phenols data set. The performance on the subset after feature selection by kNNMFS using linear regression algorithm is significantly better than those on the original data set and on the subset of the 12 most used features chosen from eight subsets in Table 1. Compared to ReliefF, kNNMFS gives a 3.28% improvement in the correlation coefficient.

Table 1. Performance of linear regression algorithm on different phenols subsets

FSM	NSF	Evaluation Using Linear Regression				
		CC	MAE	RSE	RAE	RRSE
Phenols	173	0.8039	0.3993	0.5427	59.4360%	65.3601%
MostU	12	0.7543	0.4088	0.5454	60.8533%	65.6853%
GR	20	0.7722	0.4083	0.5291	60.7675%	63.7304%
IG	20	0.7662	0.3942	0.5325	58.6724%	63.1352%
Chi	20	0.7570	0.4065	0.5439	60.5101%	65.5146%
ReliefF	20	0.8353	0.3455	0.4568	51.4319%	55.0232%
SVM	20	0.8239	0.3564	0.4697	53.0501%	56.5722%
CS	13	0.7702	0.3982	0.5292	59.2748%	63.7334%
CFS	7	0.8049	0.3681	0.4908	54.7891%	59.1181%
kNNMFS	35	**0.8627**	**0.3150**	**0.4226**	**46.8855%**	**50.8992%**

The meaning of the column titles in Table 1 is as follows: FSM – Feature Selection Method; NSF – Number of Selected Features. The feature selection methods studied include: Phenols – the original phenols data set with 173 features; MostU – the 12 most used features; GR – Gain Ratio feature evaluator; IG – Information Gain ranking filter; Chi – Chi-squared ranking filter; ReliefF – ReliefF feature selection method; SVM- SVM feature evaluator; CS – Consistency Subset evaluator; CFS – Correlation-based Feature Selection; kNNMFS – kNN Model-based feature selection.

4 Conclusions

In this paper we present a novel solution to deal with the shortcomings of ReliefF. To solve the problem of choosing a set of seeds for ReliefF, we modified the original kNNModel method by choosing a few more meaningful representatives from the training set, in addition to some extra information to represent the whole training set, and used it as a starter reference for ReliefF. In the selection of each representative we used the optimal but different k, decided automatically for each data set itself. The representatives obtained can be used directly for feature selection.

Experimental results showed that the performance evaluated by a linear regression algorithm on the subset of the phenol data set by kNNMFS is better than that of using any other feature selection methods. The improvement is significant compared to ReliefF and other feature selection methods. The results obtained using the proposed algorithm for chemical descriptors analysis applied in predictive toxicology is encouraging and show that the method is worthy of further research.

Acknowledgment

This work was supported partly by the EPSRC project PYTHIA – Predictive Toxicology Knowledge representation and Processing Tool based on a Hybrid Intelligent Systems Approach, Grant Reference: GR/T02508/01.

References

1. Hall, M. A.: Correlation-based Feature Selection for Discrete and Numeric Class Machine Learning, In Proc. of ICML'00, the 17th International Conference on Machine Learning (2000) 359 – 366
2. Kononenko, I.: Estimating attributes: Analysis and Extension of Relief. In Proc. of ECML'94, the Seventh European Conference in Machine Learning, Springer-Verlag (1994) 171-182
3. Guo, G., Wang, H., Bell, D. Bi, Y. and Greer, K.: KNN Model-based Approach in Classification, CoopIS/DOA/ODBASE 2003, Springer-Verlag (2003) 986-996
4. Wilson, D.R. and Martinez, T.R.: Improved Heterogeneous Distance Functions, Journal of Artificial Intelligence Research (JAIR), Vol. 6-1 (1997) 1-34
5. Cronin, M.T.D., Aptula, A.O., Duffy, J. C. et al.: Comparative Assessment of Methods to Develop QSARs for the Prediction of the Toxicity of Phenols to Tetrahymena Pyriformis, Chemosphere 49 (2002) 1201-1221
6. Witten, I.H. and Frank, E.: Data Mining: Practical Machine Learning Tools with Java Implementations, Morgan Kaufmann, San Francisco (2000)

Application of Feature Selection for Unsupervised Learning in Prosecutors' Office

Peng Liu, Jiaxian Zhu, Lanjuan Liu, Yanhong Li, and Xuefeng Zhang

School of Information Management and Engineering,
Shanghai University of Finance and Economics,
Shanghai, 200433, P.R. China
{liupeng, zhujiaxian, lljuan, lyhong, xfzhang}@mail.shufe.edu.cn
http://www.shufe.edu.cn/sime/index.htm

Abstract. Feature selection is effective in removing irrelevant data. However, the result of feature selection in unsupervised learning is not as satisfying as that in supervised learning. In this paper, we propose a novel methodology ULAC (Feature Selection for Unsupervised Learning Based on Attribute Correlation Analysis and Clustering Algorithm) to identify important features for unsupervised learning. We also apply ULAC into prosecutors' office to solve the real world application for unsupervised learning.

1 Introduction

The main idea of feature selection is to choose a subset of all variables by eliminating features with little discriminative and predictive information [1]. Feature selection is effective in removing irrelevant data, increasing learning accuracy, and improving result comprehensibility [2].

As we all know, the methods of feature selection for supervised learning perform pretty well for its practice and simplicity. However, as data mining are being applied into more fields, feature selection for unsupervised learning is attracting more and more researchers. Unsupervised learning means learning without a prior knowledge about the classification of samples and learning without a teacher [3].

Data mining has been well developed and applied in the fields of telecom, finance, insurance, etc. Now we are trying to introduce it into a new field--Prosecutors' Office. How to make use of these data efficiently and explore valuable information are essential to the stability and development of people and nation. After preliminary analysis of dataset, in which we found the most important problem is unsupervised learning dataset without any class information. Therefore, the key to the solution of application in prosecutors' office is feature selection for unsupervised learning. In Section 2, we introduce a novel methodology ULAC (Feature Selection for Unsupervised Learning Based on Attribute Correlation Analysis and Clustering Algorithm) in detail. The application of data mining model in prosecutors' office is described in Section 3. Conclusions are given in Section 4 with discussion on future topics interested.

2 ULAC Method

The methods of feature selection for supervised learning can be grouped as filter and wrapper approach [4]. Brodley (2004) introduced wrapper methodology into unsupervised learning and summarized the framework of wrapper approach for unsupervised learning (Fig.1) [5].

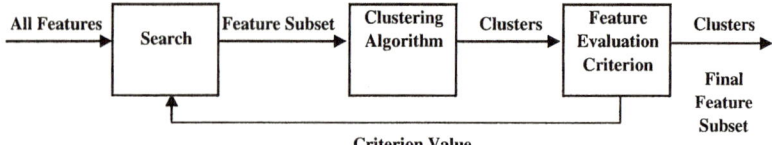

Fig. 1. Wrapper Approach for Unsupervised Learning

Based on the above approach, a heuristic in all the experiments is a novel methodology ULAC (Feature Selection for Unsupervised Learning Based on Attribute Correlation Analysis and Clustering Algorithm) (Fig.2).

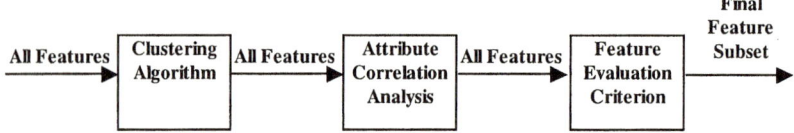

Fig. 2. ULAC for Unsupervised Learning

Our new ULAC methodology removes the step of searching feature subset from the traditional wrapper approach and adds the step of Attribute Correlation Analysis. Attribute Correlation Analysis contributes to remove those weak-relative and irrelative features. Meanwhile, it is unnecessary to circulate between search and feature evaluation criterion to find the final feature subset with the best criterion value. The step of Attribute Correlation Analysis is a very effective and efficient step to rank the importance of features for unsupervised learning.

Firstly, we produce an artificial class feature through the clustering algorithm. The problem of unsupervised learning can be turned into supervised learning. Secondly, in the step of Attribute Correlation Analysis we use artificial class feature to rank features. However, we still didn't know whether the ranked features are real discriminatory to the dataset itself. Finally, we apply Feature Evaluation Criterion to obtain classifier accuracies on our testing samples according to different number of feature subset. The accuracy also can be considered as contribution of feature subset to modeling. In terms of importance order of features the least important feature is removed each time, we can obtain the accuracy of different number of feature subset. Feature subset with the highest accuracy is chosen for modeling. As to the performance of ULAC methodology, experiments on UCI data can prove its efficiency on feature selection for unsupervised learning [6].

3 Application in Prosecutors' Office

The prosecution of dereliction of duty is a very important supervising function in prosecutors' office. One of the most important problems from prosecutors' office is how to identify subject characteristics of criminals. The issue is that dataset is of too much features for unsupervised learning and using the clustering of traditional statistics results in an unsatisfying and unexplainable model. ULAC methodology can deal with above tough problem and reach a satisfying result. We choose dataset about criminals who committed dereliction of duty in China in 2003 as data samples. The original dataset consists of 924 instances and 108 variables without class label. It is obviously difficult for us to use all 108 variables to model. By observing distribution of every variable, 90 unvalued and irrelative variables are removed (Table 1). So ULAC method needs to identify the remained 18 features to improve model efficiency.

Table 1. Irrelative Features Identified by Distribution

	Experience	Missing (>50%)	Feature Value Different Absolutely	Feature Value Unique	Distribution Unbalanced (>97%)	Express Same Character	Total
# Irrelative Features	36	30	3	2	9	10	90

ULAC consists of three main parts. We choose K-Means (KM) and Expectation-Maximization clustering (EM) as Clustering Algorithm. Relief-F, Information Gain and Chi-Squared methods are used as Attribute Correlation Analysis to rank the importance of all features and accurate rate of C4.5 as Feature Evaluation Criterion. Making use of different clustering algorithms and attribute correlation analysis result in different ranks of features. We will remove common unimportant features from different ranks in order to reduce variation by only one result. Result of KM and EM both consists of NO. 2,5,7,9,12,14 variables. Therefore, these six features are considered as unimportant ones and removed (Table 2).

Table 2. Unimportant Features Identified by ULAC

	Unimportant Feature Subset--KM	Unimportant Feature Subset--EM
Relief-F	2, 4, 5, 7, 9, 12, 14, 18	2, 5, 9, 11, 12, 14
Information Gain	2, 4, 5, 7, 9, 12, 14, 18	2, 5, 7, 9, 11, 12, 14
Chi-Square	2, 4, 5, 7, 9, 12, 14, 18	1, 2, 5, 6, 7, 9, 11, 12, 14
Results	**2, 4, 5, 7, 9, 12, 14, 18**	**2, 5, 7, 9, 11, 12, 14**

To prove the efficiency of ULAC, the performance of feature subset should be better than that of all features to modeling. We use two feature subsets before and after ULAC to clustering and compare their accuracies of C4.5 (Table.3). Before ULAC, we have to use 18 variables to clustering and accuracy of KM and EM is 79.65% and 92.53%. However after ULAC, the accuracy is up to 85.50% and 93.61% with 12 variables. So we learn that without reducing accuracy of modeling ULAC can decrease the number of irrelative variables and increase the efficiency and explanation of modeling. By

ULAC, we solve the problem of feature selection in prosecutors' office. Decreasing of irrelative and weak-relative variables improves the efficiency and understandability of data mining model. Reducing the number of irrelevant features drastically reduces the running time of a learning algorithm and yields a more general concept. This helps in getting a better insight into application of model on prosecutors' office.

Table 3. Accuracy of Clustering Before and After ULAC

	Accuracy of Clustering--KM	Accuracy of Clustering--EM
Before ULAC	79.65%	92.53%
After ULAC	85.50%	93.61%

4 Conclusions and Future Work

This paper presents a novel methodology ULAC of feature selection for unsupervised learning. The satisfying performance of application shows that the methodology can identify important features for unsupervised learning and can be used in the practical application. The further work will focus on applying ULAC into more fields.

References

1. Blum, A., Langley, P.: Selection of Relevant Features and Examples in Machine Learning, Artificial Intelligence (1997) 245-271
2. Liu, H., Motoda, H., Yu, L.: Feature selection with selective sampling. Proceedings of the Nineteenth International Conference on Machine Learning (2002)395-402
3. Kohonen, T.: Self-Organizing Maps, Springer, Germany (1997)
4. Kohavi, R., John, G.H.: Wrappers for Feature Subset Selection, Artificial Intelligence (1997)273–324
5. Jennifer, G., Brodley, C.E.: Feature Selection for Unsupervised Learning, Journal of Machine Learning Research (2004) 845-889
6. Zhu, J.X., Liu, P.: Feature Selection for Unsupervised Learning Based on Attribute Correlation Analysis and Clustering Algorithm, Proceedings of IWIIMST05 (2005)

A Novel Field Learning Algorithm for Dual Imbalance Text Classification

Ling Zhuang, Honghua Dai, and Xiaoshu Hang

School of Information Technology, Deakin University,
221 Burwood Highway, VIC 3125, Australia
lzhu@deakin.edu.au hdai@deakin.edu.au xhan@deakin.edu.au

Abstract. Fish-net algorithm is a novel field learning algorithm which derives classification rules by looking at the range of values of each attribute instead of the individual point values. In this paper, we present a Feature Selection Fish-net learning algorithm to solve the Dual Imbalance problem on text classification. Dual imbalance includes the instance imbalance and feature imbalance. The instance imbalance is caused by the unevenly distributed classes and feature imbalance is due to the different document length. The proposed approach consists of two phases: (1) select a feature subset which consists of the features that are more supportive to difficult minority class; (2) construct classification rules based on the original Fish-net algorithm. Our experimental results on Reuters21578 show that the proposed approach achieves better balanced accuracy rate on both majority and minority class than Naive Bayes MultiNomial and SVM.

1 Introduction

Data set imbalance is a commonly encountered problem in text categorization. Given a training set consists of N classes, one of the simplest classification scheme is to build N binary classifier for every individual class. Each classifier will distinguish the instances from one specific topic and all the others. Apparently, in the process of constructing binary classifier, the training set are separated into two sections: the target class, which we will call it minority class; the remaining classes, which we will call it majority class. In this case, whether the classes are evenly distributed in the collection or not, it will easily cause the data set imbalance.

The dimensionality of text data is normally in thousands. Numerous feature selection approaches have been presented in order to eliminate the irrelevant features which can be ignored without degradation in the classifier performance. However, as discussed in [1], most existing methods fail to produce predictive features for difficult class. [1] summarizes the reasons for this as follows:

1. Very few training examples for the class, and/or
2. Lack of good predictive features for that class.

The first situation is the instance imbalance. In text classification, along with the instance imbalance, it will also come with the feature imbalance. Assume that we separate the feature set from the majority and minority classes. Since the majority class has a larger number of documents than the minority one, it is more likely to have a larger vocabulary(feature set) than the minority. We call this **Dual Imbalance** and this is an interesting research issue to be looked into.

The research purpose of our work is to improve the classification accuracy on difficult minority class. We present a feature selection method which extracts features supportive to the minority class. Instead of employing traditional classification algorithms, we build the learning scheme based on the field learning strategy.

2 Related Work

Feature selection on imbalanced text data is a relatively new issue in recent literature. In [1], based on the observations, the authors pointed out that existing feature selection mechanisms tend to focus on features that are useful predictors for easier class, while the features for difficult class are easily ignored. Their solution is to apply round-robin turn to let each class propose features. That is, for each class in the data set, rank all features using a certain feature scoring method, such as IG or CHI, and take the best features suggested from each class in turn. Their experiment on some benchmark data set demonstrated consistent improvement for multi-class SVM and Naive Bayes over basic IG or CHI. In [2], given the size of the feature set l, which is pre-defined, positive feature set of size l_1 and negative feature set of size l_2 are generated by ranking the features according to some feature scoring methods. The combination of the positive and negative features is optimized on test or training set by changing the size ratio l_1/l ranging from 0 to 1. Their results show that feature selection could significantly improve the performance of both Naive bayes and regularized logistic regression on imbalanced data.

3 Preliminaries

We use D, to denote a training document set; m, number of total documents; n, number of total terms. We regard each term as a unique attribute for the documents. The definition of head rope is given as follows [3]:

Definition: Head rope
In an $m \times n$ dimension space Ω, a head rope $h_j (1 \leq j \leq n)$ with respect to attribute j consists of the lower and upper bounds of a point set D_j, where $D_j \subseteq \Omega$ is the set of values of the attribute j occur in the instances in the given instance set.

$$h_j = \{h_{l_j}, h_{u_j}\} = \{min_{1 \leq i \leq m}\{a_{ij}\}, max_{1 \leq i \leq m}\{a_{ij}\}\} \quad (1)$$

Let D^+ be the positive document class and D^- be the negative one; h_j is the positive head rope if h_j is derived from D^+. Otherwise, it is the negative one. Positive and negative head ropes construct the PN head rope pair for an attribute.

The original Fish-Net algorithm [3,4,5] can be summarized as below:

Fish-net Learning Algorithm
Input: A training data set D with a set of class labels $C = \{P, N\}$.
Output: An β-rule which is composed of contribution functions for each attribute, a threshold α and resultant headrope.
1. For each attribute A_j, find out its fields regarding each class.
2. For each attribute A_j, construct its contribution function using its fields.
3. According to the contribution function, work out resultant head rope pair $\langle h^+, h^- \rangle$. For each instance in the training set, we compute the contribution by averaging the contribution values of each attribute. The average contribution of all positive instances compose the positive resultant head rope h^+ and h^- is constructed in the same manner.
4. Determine the threshold α by examining the discovered head rope pair.

The contribution function is used to calculate and measure the contribution of one attribute to the desired class. In [5], the author illustrated six possible relationships between h^+ and h^- as shown in Figure 1.

4 Fish-Net for Text Classification

The original Fish-Net was applied to data set with continuous numeric variables and it is proven to achieve significantly higher prediction accuracy rates than point learning algorithms, such as C4.5. Its training time is linear in both the number of attributes and the number of instances [5]. However, will it still have the high performance on text data? In this section, we will examine the characteristics unbalanced text data has and present our feature selection Fish-net algorithm. Basically, our approach consists of two phases: first, select features supportive to the minorities; second, construct the classification rule based on the original Fish-net.

4.1 Feature Selection on Imbalance Text Data

Table 1 gives a simple example of document-term matrix with two classes. How could we calculate the head rope with 0 values in it? If we take the minimum and maximum value as the lower and upper bound, apparently, a certain number of head ropes will end up beginning with zero. For instance, head rope [0, 3] will be achieved on both classes for *result*. This draws the conclusion that the support of *result* for both classes is similar. Is this the true case? Note that in *cran*, *result* is only contained in one instance while it appears in four instances of *med*. *Result* should have stronger prediction capability for *med* class. Thus, not only we need to consider the value of one attribute, but also we should incorporate its distribution among documents.

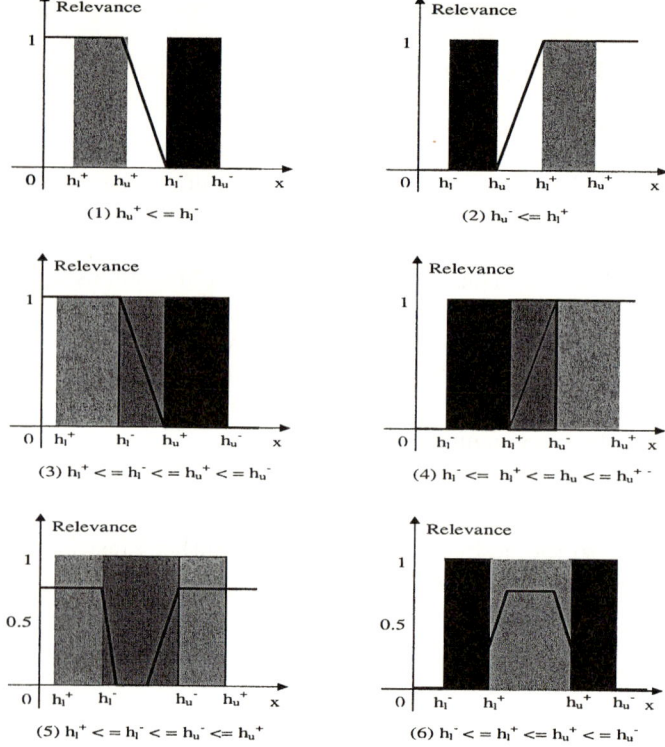

Fig. 1. Six Different Cases for Head Ropes

An alternative way is to calculate the lower bound of the head rope as the average value minus the variance; the upper bound as the average plus the variance. Average indicates the average value of one feature over the entire class and variance indicates how dynamic its distribution in different instances is. However, this approach is not able to detect relevant features in some extreme cases, as shown in the examples below.

Example 1: Suppose that both positive and negative class have 20 instances. Feature A appears in each instance of positive class with frequency 1; in one instance of negative class with frequency 20.

Discussion: *The average value for both positive and negative class is 1. The variance for positive class is 0 while for negative class it is much bigger. Hence, the resulting head rope pair falls in case 6 as in Figure 1. However, this feature is a good predictive feature for positive class from our observation.*

Example 2: The data set is as shown in Table 2.

Discussion: *The average values for both classes are still equal to 1. The resulting head rope pair will either fit in case 5 or 6. Normally features fit in these two cases are regarded as non-informative for both classes and could be discarded.*

Table 1. An Example of Document-feature Data Set

Doc.	flow	form	layer	patient	result	treat
cran.1	1	1	1	0	0	0
cran.2	2	0	1	0	0	0
cran.3	2	1	2	0	3	0
cran.4	2	0	3	0	0	0
cran.5	1	0	2	0	0	0
med.1	0	0	0	8	1	2
med.2	0	1	0	4	3	1
med.3	0	0	0	3	0	2
med.4	0	0	0	6	3	3
med.5	0	1	0	4	0	0
med.6	0	0	0	9	1	1

Table 2. Example 2

P1	P2	P3	P4	P5	P6	P7	P8	P9	P10	N1	N2	N3	N4	N5
0	0	5	1	0	0	3	0	0	1	1	2	1	1	0

It appears in four instances of both positive and negative class. However, in the negative class, those four instances comprise the 80% of the entire class while in the positive class, they only comprise 40%. Apparently, this feature could be more supportive to negative class.

In order to overcome these difficulties, we present a varied calculation of the standard average and variance.

$$Average' = \bar{x}' = \frac{Df}{N} \times \frac{Sum}{N} \quad (2)$$

$$Variance' = \varepsilon' = \frac{\frac{Df}{N} \cdot \sum(x_i - \bar{x})^2 + \frac{\overline{Df}}{N} \cdot \sum(x_i - \bar{x})^2}{N - 1} \quad (3)$$

Df is the number of documents contain a feature f in a single class and $\overline{Df}$ is the number of those does not. If f appears in every document, i.e., $Df = N$, then it turns out to be the normal average and variance. Apparently, Df/N reflects the popularity f is in that class and this value is a tradeoff between the feature distribution and its normal average value. If f does not appear in most instances, even its value in the existing ones are high, the average will still be low. The more frequent f is in the class, the higher weight Df/N will give to the normal average.

The variance calculation is based on the following assumption: the instances are separated into those ones with the feature f and those without. The popularity rate Df/N and $\overline{Df}/N$ give weights on the two sections. If f appears in more than half of the instances, then the first part of variance will dominate the final result, otherwise the second part will.

According to the above discussion, the detailed algorithm is described as follows:

Algorithm1: Range-oriented Feature Selection Algorithm

Input: A pre-processed training document matrix with binary class labels {P,N}. The original feature set is F.
Output: A selected feature subset Fs.
1. For each feature $f \in F$, calculate its average and variance in both positive and negative class according to formula (2) and (3).
2. Work out the head rope pair for each feature $f \in F$:

$$h_j^+ = [h_{l_j}^+, h_{u_j}^+] = [\overline{x_j^+} - \varepsilon_j^+, \overline{x_j^+} + \varepsilon_j^+], h_j^- = [h_{l_j}^-, h_{u_j}^-] = [\overline{x_j^-} - \varepsilon_j^-, \overline{x_j^-} + \varepsilon_j^-]$$

3. For each $f \in F$, find out which case its PN head rope pair fits in.
4. Select those features whose PN head rope pair fits in case 2 as in Figure 1. These comprise the feature subset Fs.

4.2 Classification Rule Construction Based on Fish-Net

The second phase of our algorithm is to construct the classification rule on the training data with the selected features. In this section, we will present the detailed algorithm first, then we will further justify our approach. Let I be instance set: $I = I^+ \bigcup I^-$, where I^+ is the positive instance set and I^- is the negative instance set.

Algorithm 2: Improved Fish-Net Algorithm:
Input: The pre-processed training document matrix with selected feature subset Fs.
Output: An β-rule which is composed of contribution functions for each selected attribute, a threshold α and resultant head rope.
1. For each selected feature $f \in Fs$, find out its fields regarding each class as follows:

$$h_j^+ = [h_{l_j}^+, h_{u_j}^+] = [min_{1 \le i \le m}\{a_{ij}(I_i \in I^+)\}, max_{1 \le i \le m}\{a_{ij}(I_i \in I^+)\}(a_{ij} \ne 0)] \quad (4)$$

The same technique applies to derive the negative head rope $h_j^- = [h_{l_j}^-, h_{u_j}^-]$.
2. For each selected feature $f \in Fs$, construct its contribution function using fields $[h_{l_j}^+, h_{u_j}^+]$ and $[h_{l_j}^-, h_{u_j}^-]$.
3. According to the contribution function, work out resultant head rope pair $\langle h^+, h^- \rangle$. For each instance in the training set, we compute the contribution as follows:

$$Contribution = \frac{Sum}{N} * \frac{N}{N_{total}} \quad (5)$$

where Sum is the sum of contribution values of all attributes in each instance; N is the number of non-zero values the instance has in Fs; N_{total} is the number of features(including non-selected ones) the instance has. The positive resultant head rope h^+ is constructed from all positive instances and h^- is constructed from all negative instances.
4. Determine the threshold α by examining the discovered head rope pair.

The first step of the algorithm is to set up the real head rope pair for each selected feature. We calculate the real fields by ignoring all 0 values and taking the minimum and maximum value as the lower and upper bound of the head rope. The reason for us to do this can be seen from the following case study.

Case Study:
Given a data set with 20 positive instances and 500 negative instances.
Feature A appears in all the positive instances and appears in only 20 negative instances. Feature B appears in every positive instance and does not occur in any negative instance. Assume the frequency is 1.

Discussion: *Both Feature A and B will be selected as supportive for minority class. However, if we only consider Feature B in classification, the positive and negative classes can be separated precisely. If only considering Feature A, although most negative instances are classified correctly, there are still 20 negative instances which could possibly be misclassified.*

In other words, among the selected features, there still exists different levels with respect to classification performance. Step1 and 2 in our algorithm helps to further classify the selected features into six cases.

In Step 3, the contribution value for each instance is calculated. In the original approach, it is obtained by averaging the sum of all contribution values. However, this is not feasible in text data. First of all, the number of features a document includes varies and mostly depends on the document length. This easily causes the feature imbalance problem. If we average the sum of contribution values with the total number of features, we will find the longer documents have higher contribution values and this makes shorter documents difficult to classify.

N/N_{total} is the percentage of features selected for classification in an instance. This adds weight to the average contribution value. The reason for this is by considering this situation: in a feature subset, a longer document could possibly have the same amount of features selected as the short ones. However, for the longer document, it could also have a much larger vocabulary which are not selected and more supportive to the majority class. For the short document, the selected features could already be all the words it has.

5 Experimental Work

5.1 Data Set Description

We use **Reuters-21578** Modified Apte ("ModApte") Split to test our algorithm. The collection contains 9603 documents in the training set and 3299 documents in the test set. We preprocessed the documents using the standard stop word removing, stemming and converted the documents to high-dimensional vectors using TFIDF weighting scheme. We choose 10 most frequent topic categories in the experiments. Table 3 summarizes the details. It lists, for each specific topic, the number of positive documents in the training set(#+Training), the number of positive documents in the test set(#+Test). The total number of

Table 3. Reuters-21578 ModApte Dataset Description

Data set	Earn	Acq	Money-fx	Grain	Crude	Trade	Interest	Ship	Wheat	Corn
#+Training	2866	1632	475	371	330	369	347	197	212	181
#+Test	1083	715	151	127	160	117	131	89	71	56

unique terms, the average number of terms per document are staying the same due to the same preprocessing procedure. In order to reduce the size of the term set, we discarded terms which appear in less than 5 documents. The total number of terms extracted finally is 6362 and the average number of terms per document is 41.

5.2 Evaluation Measurement

Table 4 illustrates the contingency table derived from the classification results for a specific category c_i. Note that True Positive Rate(T.P.R. = TP/(TP+FN)) indicates the percentage of correctly classified positive instances in the actual positive class, and False Positive Rate(F.P.R. = FP/(FP+TN)) indicates the percentage of incorrectly classified negative instances in the actual negative class. They are the major measurements we use in our experimental work. We also employ Receiver Operating Characteristic (ROC) curve analysis to characterize the T.P.R. and F.P.R. Accuracy is measured by Area Under Curve(AUC) which refers to the area under the ROC curve. A classifier which can produce the ROC curve with a very sharp rise from $(0,0)$ and lead to the AUC value close to 1 is regarded as the best.

Table 4. The contingency table for category c_i

C_i	Pos(Standard)	Neg(Standard)
Pos(Classifier)	TP	FP
NegClassifier)	FN	TN

5.3 Experimental Results

The Feature Selection Fish-net learning algorithm is implemented in Java. We compare our approach with Naive Bayes Multinomial implemented in WEKA [6] and SVM in SVMLight [7].

Table 5 reports the classification accuracy on the ten frequent Reuters topics from these three classifiers. The measurement we use is T.P.R. The left column under each classifier is for positive minority class and the right one is for the negative majority class. In general, the classification accuracy of all three learning algorithms on majority class is very high, reaching more than 95% in most cases. But on minority class, the performance varies. For Naive Bayes MultiNomial, the T.P.R. decreases dramatically along with the reduced number of positive instances. On the last five topics, it even has not reached 50%. On each topic's

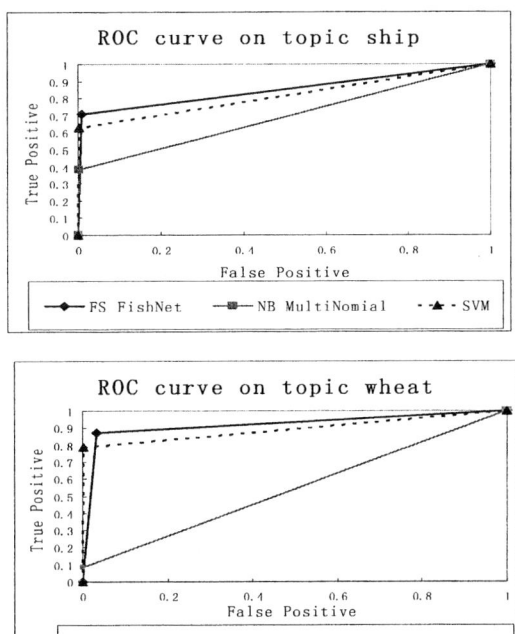

Fig. 2. ROC curves on topic ship and wheat

majority, SVM achieves nearly 100% accuracy rate. However, on minority class, Feature Selection Fish-net achieves better accuracy rate in most cases, especially with small number of positive instances. The accuracy rates of our algorithm on majority and minority are more balanced.

Figure 2 gives the ROC curve obtained on topic *ship* and *wheat* respectively. Apparently, our FS Fish-net performs the best on these three unbalanced text data with larger AUC values.

6 Conclusion

In this paper, we investigate the problem of learning classification rules from dual imbalance text data, which appears to be a common problem in reality. Our approach is designed to improve the classification accuracy on the minority without sacrificing the performance on majority. Our experimental work on the benchmark data set Reuters21578 proves that our approach performs better in achieving balanced accuracy rate than Naive Bayes MultiNomial and SVM.

Our future work will focus on investigating the efficiency issue of the Feature Selection Fish-Net and the possibilities of applying our algorithm to real applications, such as e-mail spam detection, specific target document identification.

Table 5. True Positive Rate on Positive and Negative Class from Feature Selection Fish-Net, Naive Bayes MultiNomial and SVM

Dataset	FS FishNet		NB		SVM	
	P	N	P	N	P	N
earn	0.874	0.99	0.93	0.992	0.977	0.994
acq	0.866	0.968	0.757	0.997	0.922	0.992
money-fx	0.883	0.956	0.419	0.994	0.698	0.99
grain	0.899	0.947	0.57	0.997	0.879	0.999
crude	0.847	0.935	0.635	0.996	0.836	0.993
trade	0.863	0.898	0.331	1	0.735	0.994
interest	0.756	0.974	0.008	0.999	0.573	0.998
ship	0.708	0.992	0.382	0.998	0.629	0.998
wheat	0.873	0.967	0.085	1	0.789	0.998
corn	0.679	0.972	0.089	1	0.839	0.999

References

1. Forman, G.: A pitfall and solution in multi-class feature selection for text classification. In: Proceedings of the 21st International Conference on Machine Learning. (2004)
2. Zheng, Z., Wu, X., Srihari, R.: Feature selection for text categorization on imbalanced data. ACM SIGKDD Explorations Newsletter :Special issue on learning from imbalanced datasets **6** (2004) 80–89
3. Dai, H., Hang, X., Li, G.: Inexact field learning: An approach to induce high quality rules from low quality data. In: Proceedings of 2001 IEEE International Conference on Data Mining. (2001)
4. Ciesielski, V., Dai, H.: Fisherman: a comprehensive discovery, learning and forecasting systems. In: Proceedings of 2nd Singapore International Conference on Intelligent System. (1994) B297(1)–B297(6)
5. Dai, H., Ciesielski, V.: Learning of inexact rules by the fish-net algorithm from low quality data. In: Proceedings of the Eighth Australian Joint Artificial Intelligence Conference. (1994) 108–115
6. Witten, I.H., Frank, E.: Data mining: practical machine learning tools and techniques with Java implementations. Morgan Kaufmann (1999)
7. Joachims, T.: Making large-scale support vector machine learning practical. In B. Scholkopf, C. Burges, A.S., ed.: Advances in Kernel Methods: Support Vector Machines. MIT Press, Cambridge, MA (1998)

Supervised Learning for Classification

Hongyu Li[1], Wenbin Chen[2], and I-Fan Shen[1]

[1] Department of Computer Science and Engineering,
Fudan University, Shanghai, China
{hongyuli, yfshen}@fudan.edu.cn
[2] Department of Mathematics,
Fudan University, Shanghai, China
wbchen@fudan.edu.cn

Abstract. Supervised local tangent space alignment is proposed for data classification in this paper. It is an extension of local tangent space alignment, for short, LTSA, from unsupervised to supervised learning. Supervised LTSA is a supervised dimension reduction method. It make use of the class membership of each data to be trained in the case of multiple classes, to improve the quality of classification. Furthermore we present how to determine the related parameters for classification and apply this method to a number of artificial and realistic data. Experimental results show that supervised LTSA is superior for classification to other popular methods of dimension reduction when combined with simple classifiers such as the k-nearest neighbor classifier.

1 Introduction

In many fields of application, classification is a key step for many tasks, whose aim is to discover unknown relationships and/or patterns from a large set of data. However, original data taken with various capturing devices are usually high-dimensional and most classification methods are more effective in a low-dimensional feature space. As a result, these raw data are in general unsuitable to be directly classified on a high-dimensional space. We expect to first extract some useful features from these data and then classify them only on such a feature space. The procedure of feature extraction is also called dimension reduction. Dimension reduction is an important preprocessing for classification of high dimensional data and acts as an important role to improve the accuracy of classification. Its goal is to obtain compact representations of the original data while reduce unimportant or noisy factors.

Traditional methods for dimension reduction are mainly linear, and include selection of subsets of measurements and linear mappings to lower-dimensional spaces [1,2,3]. In the recent years, a number of techniques have been proposed to perform nonlinear mappings, such as the self-organizing map [4] and generative topographic mapping [5], principal curves and surfaces [6], auto-encoder neural networks [7] and mixtures of linear models [8]. All of these, however, are problematic in application in some way.

Recently, a conceptually simple yet powerful method for nonlinear dimension reduction has been proposed by Zhang and Zha [9]: local tangent space alignment

(LTSA). The basic idea is that the information about the global structure of a nonlinear manifold[1] can be obtained from a careful analysis of the interactions of the overlapping local tangent spaces. LTSA is superior to another popular nonlinear mapping method, locally linear embedding (LLE), proposed by Saul and Roweis [10] since the LTSA method generally discovers better dimensionalities than LLE [11].

Although the authors demonstrate their algorithm on a number of artificial and realistic data sets, there have as yet been few reports of application of LTSA [12]. In this paper we present how class label information can be used in a supervised application of LTSA. Problems one might encounter with supervised LTSA are that there are two important parameters to be set, which greatly influence the results of dimension reduction and classification, and a high computational demand. In this paper, these problems will be addressed. Besides, we apply this method to a number of artificial and realistic data. Experimental results show that the combination of supervised LTSA and simple classifiers has very good performance.

The remainder of the paper is divided into the following parts. Section 2 presents a general framework of LTSA. In section 3 supervised LTSA is brought forward. The determination of related parameters is described and some experimental results are presented in section 4. Finally, section 5 ends with some conclusions.

2 Summary of Local Tangent Space Alignment

In general, images or documents used for recognition can be considered as data points in a high-dimensional space and thus can be represented by their coordinate vectors. To accurately classify these data, further reducing their dimensions is essential.

Let us consider a set of input points with coordinate vectors $X = \{x_i\}_{i=1}^n$ in R^m. Our aim is to obtain a set of output vectors $Y = \{y_i\}_{i=1}^n$ in a d-dimensional space where $d < m$. In this section, we describe local tangent space alignment (LTSA [9]) to achieve this goal. LTSA assumes that all data lie on or close to a nonlinear manifold and the global geometrical structure of this manifold can be learned by analyzing its overlapping local geometrical structure. It makes use of tangent spaces of points as such geometry and preserve them in a lower-dimensional space. By aligning those tangent spaces between the high- and low-dimensional space, much local geometry of the high-dimensional space is preserved in the low-dimensional space. Meanwhile the corresponding low-dimensional coordinates can be discovered in the process of alignment.

As stated above, the whole LTSA algorithm is composed of two main stages: (I) locally fitting tangent space of each data point x_i, and extracting such local geometrical information; (II) aligning those local tangent spaces between the high- and low-dimensional space to find the lower-dimensional coordinates y_i for each x_i

In this paper we will not in detail explain the numerical computation method for LTSA (For details, please refer to [9]). Next we briefly present the implementation of the whole algorithm to deal with discrete data.

Given n m-dimensional points x_i, LTSA produces n d-dimensional coordinates y_i constructed from k local nearest neighbors:

---

[1] Dimension reduction can also been considered as discovering a low-dimensional manifold embedded in a high-dimensional data set.

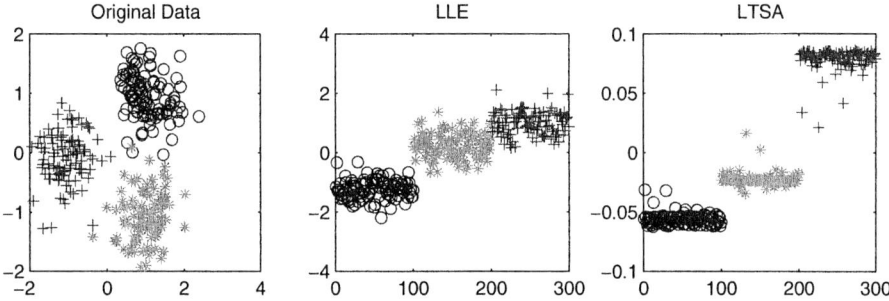

Fig. 1. Comparison between LLE and LTSA. Three Gaussian data sets (left); Global coordinates with LLE (middle); Global coordinates with LTSA (right).

1. Finding k nearest neighbors $X_i = \{x_i^j\}, j = 1, \ldots, k$ for each point x_i.
2. Extracting the local geometrical information by calculating the d largest eigenvectors $g_1, \ldots, g_d$ of the correlation matrix $(X_i - \bar{x}_i e^T)^T (X_i - \bar{x}_i e^T)$. $\bar{x}_i$ represents the average of the neighborhood of x_i, $\bar{x}_i = \frac{1}{k}\sum_j x_i^j$. Set $G_i = [e/\sqrt{k}, g_1, \ldots, g_d]$.
3. Constructing the alignment matrix B by locally summing as follows:

$$B(I_i, I_i) \leftarrow B(I_i, I_i) + I - G_i G_i^T, i = 1, \ldots, n$$

with initial $B = 0$.

4. Computing the $d+1$ smallest eigenvectors of B and picking up the eigenvector matrix $[u_2, \ldots, u_{d+1}]$ corresponding to the 2nd to d+1st smallest eigenvalues, and setting the global coordinates

$$Y = [y_1, \ldots, y_n] = [u_2, \ldots, u_{d+1}]^T.$$

One advantage of LTSA is that it can potentially discover the key degrees of freedom (dimensionalities) of the underlying manifold, which are very beneficial for accurate classification. Fig. 1 is a good example. Locally linear embedding (LLE, [10]) is another famous method for nonlinear dimension reduction, it is based on a similar idea with LTSA that overlapping local geometry can represent global geometry of a manifold. However, LLE do not use local tangent spaces, but local linear combination, as such local geometry. In the left panel of Fig. 1, there are three two-dimensional Gaussian data sets. Simultaneously projecting them into a one-dimensional feature space with the LLE and LTSA methods, we find that the feature space discovered by LTSA is better than the one discovered by LLE because in the former the three Gaussians are clearly separate (plotted in the right panel). In the latter, however, many points belonging to different groups overlap each other (plotted in the middle panel). It is clear that LTSA is more suitable as a preprocessing step for classification.

3 Supervised LTSA

Original LTSA proposed by [9] belongs to unsupervised methods; it does not make use of the class membership of points to be projected. Such methods are mostly intended for data mining and visualization where the number of classes and relationships between elements of different classes are unknown and users want to see the data structure in order to make a decision about what to do next. But they can not usually perform well in the field of classification where the membership information of training samples is known and the center of each class needs to be searched.

Consequently, in this paper we propose a *superevised* LTSA (SLTSA) method for classification. The term implies that membership information is employed to form the neighborhood of each point, that is, nearest neighbors of a given point x_i are chosen only from representatives of the same class as that of x_i. This can be achieved by artificially adding the shift distance between samples belonging to different classes, but leaving them unchanged if samples are from the same class. To select the neighbors of samples, we can define a $n \times n$ distance matrix D where each entry d_{ij} represents the *Euclidean* distance between two samples x_i and x_j. Furthermore, considering the membership information, we can get a variant D' of D,

$$D' = D + \rho \delta \qquad (1)$$

where the shift distance ρ is assigned a relatively very large value in comparison with the distance between any pairs of points, δ is a $n \times n$ matrix whose entries δ_{ij} are 1 if x_i and x_j are in different classes, and 0 otherwise. The shift distance ρ enlarges the distances between a pair of points belong to different classes and promises that nearest neighbors of a sample in a class will always be picked from the same class.

In nature the selection of neighborhood presents an opportunity to incorporate one's priori knowledge such as class membership, time or space order etc. For different purposes, different methods can be devised to find a desirable neighborhood. For example, in video data we can also select nearest neighbors of each frame in terms of the temporal sequence, not the spatial order (Euclidean distance).

Supervised LTSA we propose here is specially for dealing with data sets containing multiple classes and improving the correct rate of classification. Therefore SLTSA is expected to be able to gather all samples belonging to a certain class around its center of class in a low-dimensional space. Thus new arriving samples can be correctly projected

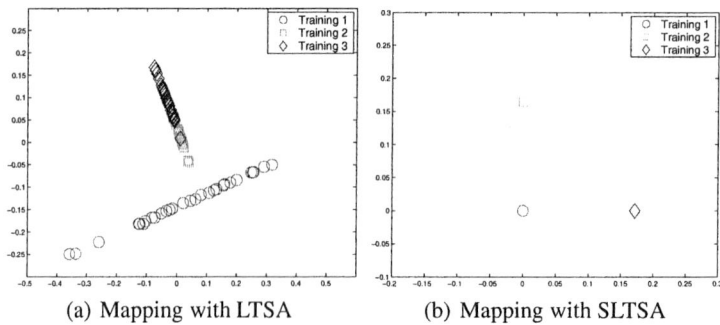

Fig. 2. Mapping **iris** data (m=4, c= 3)into a 2-D feature space with LTSA and SLTSA. The first 100 points are trained as original input samples. (a) Mapping the input samples to a 2-D feature space with LTSA. (b)Mapping the input samples with SLTSA.

around its center of class in the low-dimensional space, further they can be correctly classified. Note that the premise of SLTSA for accurate classification is that the distance between a pair of sample points among different classes is large enough, for within a class the distance is very small. Then new arriving points can be correctly projected by the method of generalization mentioned in [11].

Fig. 2 is an example illustrating that the projection obtained with SLTSA is different from the one obtained with original LTSA. This set of **iris** data [13] are composed of 150 4-D data points belonging to 3 different classes. Here first 100 data points are selected as training samples and mapped from the 4-D input space to a 2-D feature space respectively by SLTSA and LTSA. Fig. 2(a) is the result obtained by LTSA, it sufficiently shows the geometrical structure of **iris** data in the 2-D feature space. But the result is not beneficial for classification since these data distribute very loose and some points belonging to different classes overlap together so that some centers of classes are possibly very close and the boundaries among them can not be accurately determined.

With the SLTSA method, the 100 training samples are mapped to a 2-D feature space where all data in the same class are projected to a point shown in Fig. 2(b). In terms of classification it is excellent if all high-dimensional points belonging to the same class correspond to one point in the low-dimensional feature space. At this time, those centers of classes are distant which is very helpful for determining the boundaries among different classes.

4 SLTSA for Classification

Unlike unsupervised methods, an important application of supervised methods is data classification and they can improve the correct rate of classification. Thus SLTSA is more suitable as a feature extraction step prior to classification compared to LTSA. Before we discuss how to apply SLTSA to implement data classification, we first introduce how to determine the related parameters to this algorithm.

4.1 Determination of Parameters

The previous sections showed that to find a good SLTSA mapping, two parameters will have to be set: the dimensionality d to map to and the number of neighbors k to take into account. Mapping quality is quite sensitive to these parameters. If d is set too high, the mapping will enhance noise; if it is set too low, distinct parts of the data set might be mapped on top of each other. Just like PCA [14], although internal structure of each class is (partially) lost during mapping, class overlap can easily be differentiated in the $c-1$ dimensional space (c is the number of classes). As a result, for accurate classification we will reduce high dimensions the original input data to $c-1$ dimensions in this paper.

The value of the number of neighbors k has influence on the performance of SLTSA. If k is set too small, the mapping will not reflect any global properties; if it is too high, the mapping will lose its nonlinear character and behave like traditional PCA, as the entire data set is seen as local neighborhood. Fig. 3 is an example of mapping the iris data to a 2-D space including 100 known training samples and 50 unknown test samples

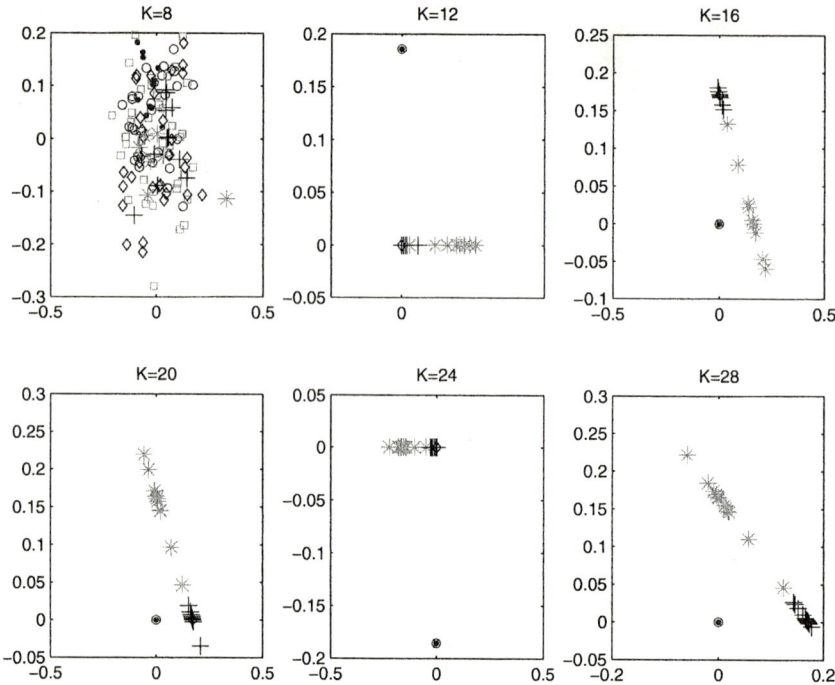

Fig. 3. The influence of the number of neighbors k on the performance of SLTSA. Map the iris data to a 2-D space where k varies from 8 to 28.

with SLTSA where k varies from 8 to 28. It illustrates that when $k = 8$, the test data distribute unorderly in the low-dimensional space. When k increases from 8 to 28, the test data become very orderly. The results of classification are stable over a wide range of k but do break down as k becomes too large. Furthermore the SLTSA algorithm requires that $k > d$ and it must also be less than the number of training samples of each class. Meanwhile, in consideration of the computation cost, k is not more than 50 in general. Ideally, optimal k should make the error rate of classification smallest.

4.2 Experimental Results

To examine the performance of SLTSA in terms of data classification, it was applied to a number of data sets varying in number of samples n, dimensions m and classes c. Most of the sets were obtained from the repository [13] and some are constructed by the authors.

Handwritten Digits. The Binarydigits set consists of 20×16-pixel binary images of pre-processed handwritten digits[2]. Here, we just deal with recognition of three digits: 0, 1 and 2 shown in Fig. 4(a). 90 of the 117 binary images are used as training samples and

---

[2] Download from http://www.cs.toronto.edu/ roweis/data.html.

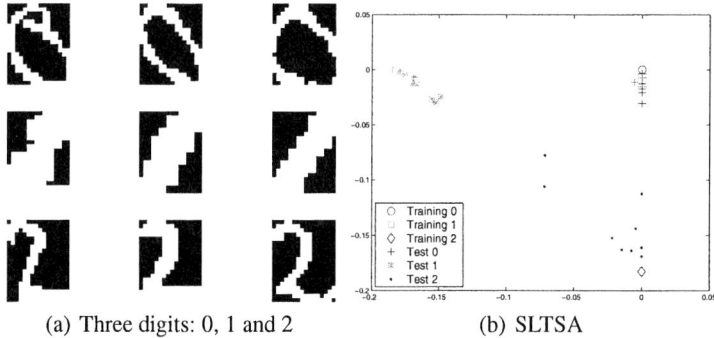

(a) Three digits: 0, 1 and 2 (b) SLTSA

Fig. 4. Recognition of three digits: 0, 1 and 2. (a) Some of these three digits are shown here. They are represented by 20×16 binary images which can be considered as points in a 320-dimensional space. (b) Mapping these binary digits into a 2-dimensional feature space with SLTSA, which provides better clustering information.

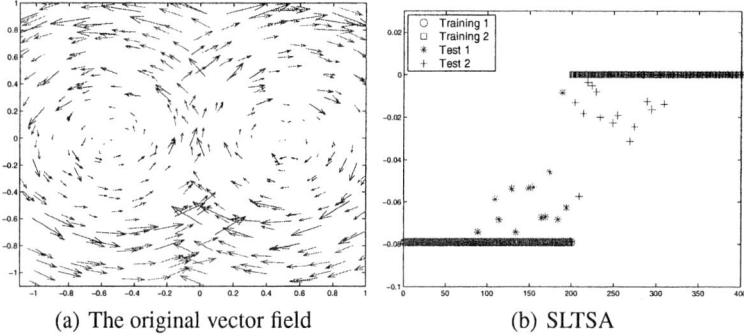

(a) The original vector field (b) SLTSA

Fig. 5. Classification of vector data. (a) A discrete vector field with noise composed of two circular singularities, where 320 training (red) and 80 test (blue) vectors. (b)Mapping the 4-D vector data (including their coordinate information) into 1-D feature space with SLTSA, which provides better classification information.

others as test samples. The results after dimension reduction from 320 to 2 with LTSA and SLTSA are displayed in Fig. 4(b). The feature space obtained with SLTSA provides perfect classification information, each digit in test samples can be rightly assigned to a class. The correct rate of classification of test samples reaches to 100%.

Synthetic Vector Data. Fig. 5 shows an example of classification of vector data. Fig. 5(a) displays a vector field with two circular singularities distributed symmetrically. These vector data are contaminated by noise during sampling. In Fig. 5(a), blue vectors are used as test samples and red ones as training samples. To classify vectors, SLTSA has been considered and Fig. 5(b) shows those test samples in two different clusters are separated satisfactorily where the test error is only 2.5% less than that when directly

Table 1. Error rate (in %) of classification in original input and reduced feature spaces with simple classifiers, *lda* and *k-nn*, here $k = 15$. The symbol '/' means that the current 'column' method can not directly work for the current 'row' data set.

Data Name		PCA	LLE	SLLE	LTSA	SLTSA
iris	k-nn	6	8	2	12	2
($n = 150, m = 4, d = 2, c = 3$)	lda	6	2	2	14	2
wine	k-nn	35	35	66.67	31.67	33.33
($n = 168, m = 13, d = 2, c = 3$)	lda	28.33	68.33	/	35	33.33
glass	k-nn	62.5	47.5	42.5	47.5	42.5
($n = 146, m = 9, d = 1, c = 2$)	lda	80	47.5	47.5	27.5	42.5
optdigits	k-nn	4	23	4	4	4
($n = 600, m = 64, d = 9, c = 10$)	lda	5	23	4	4	4
binarydigits	k-nn	3.7	/	/	22.52	0
($n = 117, m = 320, d = 2, c = 3$)	lda	7.41	/	/	18.52	0
vector field	k-nn	100	50	5	78.75	2.5
($n = 400, m = 4, d = 1, c = 2$)	lda	50	51.25	5	51.25	2.5
scurve&swissroll	k-nn	9.5	10.25	1	6.25	1
($n = 2400, m = 3, d = 1, c = 2$)	lda	19.5	21.25	1	9.25	1

operating on the original vector field or using original LTSA with some simple classifiers (for details, please refer to Table 1).

In our experiments, as a rule a data set was randomly split 10 times into a training set (80%) and a test set (20%). Two popular classifiers were used: *lda*, the linear discriminate analysis classifiers and *k-nn*, the *k*-nearest neighbor classifier. To compare the SLTSA method to more traditional feature extraction techniques, the classifiers were also trained on mapped data with PCA and SLLE [15].

Table 1 presents average error rates of classification on test sets (in %). Mappings were calculated for a range of values of k, the neighborhood size parameter. Only the best result found in the range of values for k is shown. The results confirm that SLTSA generally leads to better classification performance than LTSA and, usually, any other mapping technique. Besides this, there are a number of interesting observations:

1. From the table, it is obvious that in most cases SLTSA is finer than PCA, LLE, SLLE and LTSA. Maybe the reason is that SLTSA can better extract key features or dimensionalities of a high-dimensional input data and accurately discover centers of different classes.
2. The classification results with SLTSA are relatively constant, i.e., the results are independent of the classifier since the within-class coupling and between-class dispersion are very high in the supervised feature space.

5 Conclusions

Local tangent space alignment is proposed for unsupervised dimension reduction. Its key techniques rest with the construction of local tangent spaces to represent local geometries, and their global alignment to obtain the global coordinate system for the underlying manifold. LTSA can potentially detect the key dimensionality of the underlying manifold. However, the selection of the neighboring points proves very crucial to our purpose of application. In terms of classification, we hope that the neighborhood of a given point is constructed by those points belonging to the same class with it. So we take into account the class label information during selecting nearest neighbors and extend the LTSA method to a supervised version, SLTSA.

Furthermore we present how to determine the related parameters for classification and compare SLTSA to other popular linear and nonlinear mapping methods such as PCA, unsupervised and supervised LLE. Our experiments show that in the field of classification SLTSA perform better than other popular methods when combined with such simple classifiers as k-nn classifier.

Acknowledgments

This work was supported by NSFC under contract 60473104 and STCSM under contract 045115013.

References

1. Devijver, P., Kittler, J.: Pattern recognition, a statistical approach. Prentice-Hall, London (1982)
2. Duda, R.O., Hart, P.E., Stork, D.G.: Pattern classification. 2 edn. John Wiley & Sons, New York, NY (2001)
3. Jain, A., Duin, R., Mao, J.: Statistical pattern recognition: a review. IEEE Transactions on Pattern Analysis and Machine Intelligence **22** (2000) 4–37
4. Kohonen, T.: Self-organizing Maps. 3rd edn. Springer-Verlag (2000)
5. Bishop, C., Svensén, M., Williams, C.: Gtm: The generative topographic mapping. Neural Computation **10** (1998) 215–234
6. Hastie, T., Stuetzle, W.: Principal curves. J. Am. Statistical Assoc. **84** (1988)
7. DeMers, D., Cottrell, G.: Non-linear dimensionality reduction. In Giles, C., Hanson, S., Cowan, J., eds.: Advances in Neural Information Processing Systems 5, San Mateo, CA, Morgan Kaufmann (1993) 580–587
8. Tipping, M., Bishop, C.: Mixtures of probabilistic principal component analyzers. Neural Computation **11** (1999) 443–482
9. Zhang, Z., Zha, H.: Principal manifolds and nonlinear dimension reduction via local tangent space alignment. SIAM Journal of Scientific Computing **26** (2004) 313–338
10. Roweis, S., Saul, L.: Nonlinear dimension reduction by locally linear embedding. Science **290** (2000) 2323–2326
11. Li, H., Teng, L., Chen, W., Shen, I.F.: Supervised learning on local tangent space. In Wang, J., Liao, X., Yi, Z., eds.: ISNN 2005, LNCS. Volume 3496. (2005) 546–551
12. Li, H., Chen, W., Shen, I.F.: Supervised local tangent space alignment for classification. In: IJCAI 2005, poster paper. (to appear)
13. Blake, C., Merz, C.: Uci repository of machine learning databases (1998)
14. Turk, M., Pentland, A.: Eigenfaces for recognition. Journal of Cognitive Neuroscience **13** (1991) 71–86
15. de Ridder, D., Duin, R.: Locally linear embedding for classification. Technical Report PH-2002-01, Pattern Recogniion Group, Dept. of Imaging Science and Technology, Delft University of Technology, Delft, The Netherlands (2002)

Feature Selection for Hyperspectral Data Classification Using Double Parallel Feedforward Neural Networks

Mingyi He and Rui Huang

School of Electronics and Information, Northwestern Polytechnical University,
Shaanxi Key Laboratory of Information Acquisition and Processing,
Xi'an, Shaanxi, 710072, P.R. China

Abstract. Double parallel feedforward neural network (DPFNN) based approach is proposed for dimensionality reduction, which is one of very significant problems in multi- and hyperspectral image processing and is of high potential value in lunar and Mars exploration, new earth observation system, and biomedical engineering etc. Instead of using sequential search like most feature selection methods based on neural network (NN), the new approach adopts feature weighting strategy to cut down the computational cost significantly. DPFNN is trained by a mean square error function with regulation terms which can improve the generation performance and classification accuracy. Four experiments are carried out to assesses the performance of DPFNN selector for high-dimensional data classification. The first three experiments with the benchmark data sets are designed to make comparison between DPFNN selector and some NN based selectors. In the fourth experiment, hyperspectral data, that is an airborne visible/infrared imaging spectrometer (AVIRIS) data set, is used to compare DPFNN selector with widely used forward sequential search methods using the Maximum Likelihood classifier (MLC) as criterion. Experiments show the effectiveness of the new feature selection method based on DPFNNs.

1 Instruction

Band selection is a special application of dimensionality reduction in hyperspectral data processing to reduce computational amount and alleviate the Hughes phenomenon [1]. Compared with band extraction, band selection is better in the respect of preserving the physical meanings of original spectral bands, as it tries to identify a subset of original bands for a given task. According to whether to use a predetermined learning algorithms as the criterion function, band selection can also be generally divided into two categories, namely filter model and wrapper model [2]. In literatures, the widely applied separability indices like discriminant analysis, divergence, transformed divergence, Bhattacharyya distance and Jeffreys-Matusita distance [3,4,5,6] are all classifier-independent. Unlike them, wrapper methods directly use the classifier to select features and thus the accuracy level is higher but with cost of expensive computation [7,8]. It is obvious that different learning algorithms may obtain quite different feature sub-sets, even using the same training set.

The Neural Network (NN) based feature selection becomes a promising wrapper method and recently, a few algorithms using Multi-layer Forward Neural Network (MLFNN) have been proposed. Generally, feature selection based on MLFNN can be achieved by adding more or removing less relevant features through some saliency measure to rank features [9, 10, 11, 12] or training NN according to some modified cost function [13, 14]. However, there exist two main problems for various feature selection by using MLFNNs. On one hand, as a learning algorithm, MLFNN has some disadvantages such as local minimal points on the error surface and over-fitting phenomena. On the other hand, since the selection procedure should run many times due to the inherent randomness in NNs, the subset search based on the sequential strategy or feature weighting becomes computationally prohibited. The Radial Basis Function Neural Network (RBFNN) has also been used as the classifier to help select meaningful features which could be generated by feature ranking through a separability-correlation measure [15] or the genetic algorithm [16] with successful application in feature selection [17] and feature extraction [18]. Compared with MLFNN, RBFNN needs to specify the center and width for each Gaussian kernel function. In addition, feature selection using Support Vector Machines (SVM) is also presented recently [19].

In this paper, a DPFNN (Double Parallel Feedforward Neural Networks) selector is proposed. DPFNN [20, 21, 22, 23] presented by the author is composed of MLFNN and SLFNN (Single-Layer Feedforward Neural Network) which parallelly connects each other and thus has the merits of both: (1) Good nonlinear mapping capability; (2) High learning speed for linear-like problem. At first, the SLFNN is used to obtain a near-linear separable solution during a very short training time, and then MLFNN is used to mapping the desired nonlinearity by adjusting the weights connected to the hidden neurons. Thus, DPFNN selection method is put forward with consideration of the high spectral correlation among bands in hyperspectral data. Classification experiments on four real-world data sets are carried out to demonstrate the DPFNN selector's effectiveness.

2 Double Parallel Feedforward Neural Networks

Let $\mathbf{X} = (x_1, x_2, \ldots, x_n)^T$ and $\mathbf{Y} = (y_1, y_2, \ldots, y_m)^T$ be the input and output of DPFNN, respectively. Let $\mathbf{V}$ denote the weights of SLFNN directly from the input layer to the output layer and $\mathbf{W}^{(l)}$ the weights in lth layer of MLFNN. Then, when feeding the ith input $\mathbf{X}_i$, the output signals $\mathbf{Y}_i$ is obtained by

$$\mathbf{Y}_i = f^{(L)}\left(\mathbf{W}^{(L)} f^{(L-1)}\left(\cdots f^{(1)}(\mathbf{W}^{(1)}\mathbf{X}_i)\right) + \mathbf{V}\mathbf{X}_i\right) \qquad (1)$$

where the activation function $f(\cdot)$ is usually a sigmoid function or hyperbolic function. The structure of DPFNN is depicted in Fig.1. From the figure, MLFNN and SLFNN connect parallelly each other. The idea of DPFNN comes from the

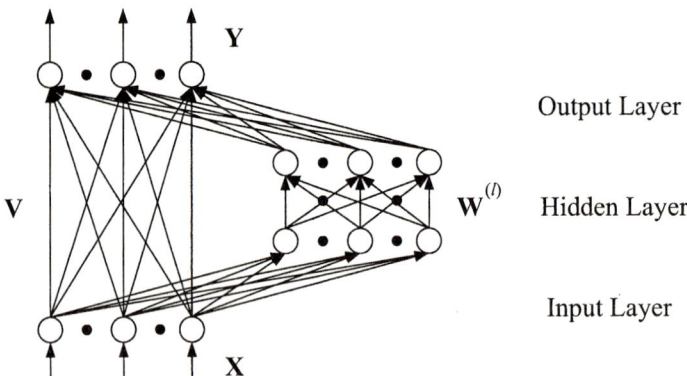

Fig. 1. Structure of DPFNN

fact that not only the indirect information is needed when we are cognising the world, but the intuitional knowledge is indispensable. In MLFNN, the neurons of output layer only can utilize the information from the hidden neurons connected to them and have no way to directly use the input signals. Through a parallel connection between MLFNN and SLFNN, it is possible to improve the learning capacity via the indirect and direct information.

3 The DPFNN Selector

3.1 The Cost Function

Assume there are N samples and the L-layer network structure is n-$h^{(1)}$-$\cdots$-$h^{(L-1)}$-m. The rule for training the network is the well-known mean square error function with regulation terms of weight magnitudes of the output layer neurons, which has been proposed in [22] and shown strong generalization performance in classification of hyperspectral images. The cost function is calculated as follows

$$F(\mathbf{W}, \mathbf{V}) = \frac{\lambda}{mN}\|\mathbf{d} - \mathbf{Y}\|^2 + \frac{1-\lambda}{mh^{(L-1)}}(\|\mathbf{W}\|^2 + \|\mathbf{V}\|^2) \qquad (2)$$

where $\|\cdot\|$ is norm, λ is a constant.

3.2 Feature Selection Strategy

Most of selection strategies based on NNs belong to backwards search, in which features are removed one by one through evaluation. Such search method is prohibitively computational especially when NN selector applied to high-dimensional space like hyperspectral data, because of the time-consuming procedure in training a network and the necessity of many times of running the selection process. Reference [9] summarized a selection approach based on the idea of feature weighting which assign weights to features individually and select those with greater weights. It is obviously faster than the sequential search. However, as feature weighting considers little about the correlation among features, the manipulation of removing redundant features should follow it. In the selection procedure, we firstly remove the irrelevant features via feature weighting method, and then delete the redundant ones according to the correlation among features. A detailed process is listed as follows:

(1) Normalize the features between -1 and 1.
(2) Introduce a Uniform(0,1) noise feature to the original set of features $\mathbf{S}_0$.
(3) Randomly select the training, validate and test sets.
(4) Train the network.
(5) Compute the Tarr's saliency of all features as

$$\tau_i = \alpha 1 \sum_{j=1}^{h^{(1)}} \left(w_{ji}^{(1)} \right)^2 + \alpha 2 \sum_{k=1}^{m} \left(v_{ki}^{(1)} \right)^2 \qquad (3)$$

where τ_i is Tarr's saliency measure for feature i, $w_{ji}^{(1)}$ and $v_{ki}^{(1)}$ are the first layer weights from input node i to hidden node j and output node k, respectively, $\alpha 1$ and $\alpha 2$ are preset constants.

(6) Repeat steps 4 and 5 R times.
(7) Assume the average saliency of noise is normally distributed and find the two-sided percent confidence interval for the mean value of the saliency of noise.
(8) Choose only those features whose average saliency values fall outside this confidence interval as feature subset $\mathbf{S}_1$.
(9) If $\mathbf{S}_0 = \mathbf{S}_1$, goto step (12).
(10) Let $\mathbf{S}_0 \leftarrow \mathbf{S}_1$ and take the subset $\mathbf{S}_0$ with a Uniform (0,1) noise feature as network input.
(11) goto step (4).
(12) Retrain the network with the feature subset $\mathbf{S}_0$ R times.
(13) Compute the average accuracy $\bar{A}_{S_0}$ for the valid set.
(14) Sort the subset $\mathbf{S}_0$ in the descending order of their average saliency and from the second feature, remove the features with higher correlation than a preset threshold to get a smaller subset $\mathbf{S}_1 = \mathbf{S}_0 - \mathbf{S}_2$.
(15) Train the network with the subset $\mathbf{S}_1$ R times
(16) Compute the average accuracy $\bar{A}_{S_1}$ for the valid set and the drop ΔA in the accuracy compared with $\bar{A}_{S_0}$.
(17) If $\Delta A < \Delta A_0$ let $\mathbf{S}_0 \leftarrow \mathbf{S}_1$ and go to Step 14, where ΔA_0 is the acceptable drop in the classification accuracy.
(18) If there is only one element in $\mathbf{S}_2$, $\mathbf{S}_0$ is the final subset and the selection procedure stops.
(19) Add the features in $\mathbf{S}_2$ to $\mathbf{S}_1$ except the feature with the smallest salient value, and goto step (15).

4 Experiments

To assesses the performance of DPFNN selector for high-dimensional data classification, four experiments are carried out. The first three with the benchmark data sets, which are available in www.ics.uci.edu/mlearn/MLRepository.htm, aim to make comparison between DPFNN selector and some NN based selectors. To make the results comparable, we use the same learning and testing conditions and the number of hidden neurons is 12. In the last experiment with an airborne visible/infrared imaging spectrometer (AVIRIS) data set, we compare DPFNN selector with widely used forward sequential search methods in literatures which use the Maximum Likelihood classifier (MLC) as criterion. To accelerate the convergence speed of DPFNN, instead of standard backpropagation algorithm (BP), scaled conjugate gradient algorithm (SCG)is applied [24]. And for DPFNN selector, a 99 percent two-sided confidence interval is constructed for the mean saliency of the injected noise feature. Let $\alpha 1 = \alpha 2 = 0.5$, $\lambda = 0.95$.

4.1 Breast Cancer Data Set

The University of Wisconsin Breast Cancer data set consists of 699 patterns. Amongst them there are 458 benign samples and 241 malignant ones. Each of these patterns consists of nine measurements taken from fine needle aspirates from a patients breast. The data set is randomly divided into training (315 samples), validation (35 samples) and testing (349 samples) sets.

Table 1 shows the classification accuracy of training and testing sets when using four feature selection methods based on NNs, namely the proposed, NNFS [13], SNR [10], Verikas and Bacauskiene's method [14]. Compared with NNFS and SNR, Verikas etc's method and our method perform better. Since the former takes the testing accuracy as the criterion to determine whether to continue removing features, it yields the best results. At the same time, NNFS and Verikas etc's methods select features in a backward fashion through which features are get rid of one by one according to some evaluation function. Obviously, it is a time-consuming procedure compared to feature weighting approach.

Table 1. comparison of classification accuracy for breast cancer data set

	proposed	NNFS	SNR	Verikas etc 's
All features				
Training set	99.92(0.14)	100.0(0.00)	97.66(0.18)	97.93(0.54)
Testing set	95.99(0.91)	93.94(0.94)	96.49(0.15)	96.44(0.31)
Selected features				
# of features	2	2.7(1.02)	1	2(0.0)
Training set	96.20(0.88)	98.05(1.31)	94.03(0.97)	95.69(0.44)
Testing set	94.67(0.97)	94.15(1.00)	92.53(0.77)	95.77(0.41)

4.2 Voting Records Data Set

The data set consists of the voting records of 435 congressmen on 16 major issues in the 98th Congress which has 267 Democrats and 168 Republicans. The votes are categorized into one of the three types of votes: Yea, Nay and Unknown. The task is to predict the correct political party affiliation of each congressman (democrat or republican). The data set is randomly divided into training (197 samples), validation (21 samples) and testing (217 samples) sets.

Table 2 shows the classification accuracy of training and testing sets when using the four feature selection methods based on NNs. From it, we can see the proposed method performs best except that the highest training accuracy of the feature subset is achieved by SNR.

Table 2. comparison of classification accuracy for voting records data set

	proposed	NNFS	SNR	Verikas etc 's
All features				
Training set	100.0(0.00)	100.0(0.00)	99.82(0.22)	99.32(0.13)
Testing set	96.65(0.53)	92.00(0.96)	95.42(0.18)	96.04(0.14)
Selected features				
# of features	1	2.03(0.18)	1	1(0.0)
Training set	95.72(0.37)	95.63(0.43)	96.62(0.30)	95.71(0.25)
Testing set	95.67(0.39)	94.79(1.60)	94.69(0.20)	95.66(0.18)

4.3 Diabetes Diagnosis Data Set

The Pima Indians Diabetes data set consists of 768 samples taken from patients who may show signs of diabetes. Each sample is described by eight features. There are 500 samples from patients who do not have diabetes and 268 ones from patients who are known to have diabetes. The data set is randomly divided into training (345 samples), validation (39 samples) and testing (384 samples) sets.

Table 3 gives the comparison of classification accuracy. For the all features, Verikas etc's method obtains the best testing accuracy at expense of the quite low training accuracy (only 80.35%). For the selected features, Verikas etc's method has the highest training and testing accuracy with two features chosen. Compared with it, the proposed method selects one feature only with 1-2% decrease in accuracies. Therefore, from the whole view, the proposed method performs best.

4.4 Hyperspectral Data Set

The hyperspectral data set used is a segment of one AVIRIS data scene taken of NW Indiana's Indian Pine test site in 1992. It has been reduced from the original 220 bands to 190 bands by discarding the atmospheric water bands and

Table 3. comparison of classification accuracy for diabetes diagnosis data set

	proposed	NNFS	SNR	Verikas etc 's
All features				
Training set	95.24(0.63)	93.59(2.77)	80.35(0.67)	80.64(0.53)
Testing set	74.77(0.97)	71.03(1.74)	75.91(0.34)	77.83(0.30)
Selected features				
# of features	1	2.03(0.18)	1	2(0.0)
Training set	75.67(0.52)	74.02(6.00)	75.53(1.40)	76.83(0.52)
Testing set	74.43(0.64)	74.29(3.25)	73.35(1.16)	76.81(0.45)

low SNR bands. Four classes are selected and each has 50 training samples, 12 validation samples and 88 testing samples. The number of hidden neurons is 20.

Table 4 presents the comparison of classification accuracies. For DPFNN selector, the correlation threshold is set to the square of maximal correlation coefficient in the current subset. Finally, 18 bands are selected by it. For two sequential methods SFS and SFFS [25], the evaluation values of subsets are decided by the minimums of corresponding training accuracy and validation accuracy. From the table, the proposed method obtains the highest overall and detail accuracies except class 4.

Table 4. Comparison of classification accuracy with two sequential search methods when 18 bands selected

	proposed				SFS				SFFS			
All features												
Training set	97.50(0.96)				-				-			
Testing set	92.23(2.24)				-				-			
18 features												
Training set	99.93(0.37)				100.0				100.0			
Testing set	92.48(0.95)				88.35				88.07			
Confusion matrix	79.00	0.23	5.63	3.13	73	0	4	11	70	0	4	14
	1.27	86.70	0	0.03	2	86	0	0	1	86	0	1
	3.47	0.07	82.60	1.87	9	0	71	8	8	1	72	7
	4.67	0.60	5.50	77.23	4	0	3	81	3	1	2	82

5 Conclusion

The paper proposes a new feature selection method based on DPFNNs for high-dimensional data classification. It involves injection of a noise as an irrelevant feature to the network, assignment of weights to every bands with the salient values relative to the noise, preservation of bands with larger salient values and elimination of redundant bands due to spectral correlation. The method based on feature weighting is capable of alleviating the computation load which is heavy for sequential search usually used in NN selectors. Experiment results demonstrate the good performance of the proposed DPFNN selector via comparison with three common NN selectors and two commonly used sequential search methods in literature.

Acknowledgement

This work is supported by the national fundamental research program, Aviation research fund and Ph.D research fund in China.

References

1. Hughes, G.F.: On the mean accuracy of statistical pattern recognizers. IEEE Trans. Inform. Theory **14** (1968)
2. Kohavi, R., John, G.H.: Wrappers for feature subset selection. Artificial Intelligence **97** (1997) 273–324
3. Swain, P., King, R.: Two effective feature selection criteria for multispectral remote sensing. In: Proceedings of the 1st International Joint Conference on Pattern Recognition, Washington, DC (1973) 536–540
4. Kavzoglu, T., Mather, P.M.: The role of feature selection in artificial neural network applications. Int. J. Remote Sensing **23** (2002) 2919–2937
5. Sheffer, D., Ultchin, Y.: Comparison of band selection results using different class separation measures in various day and night conditions. In: Proc. SPIE, Algorithms and Technologies for Multispectral, Hyperspectral, and Ultraspectral Imagery IX. Volume 5093. (2003) 452–461
6. Serpico, S.B., Bruzzone, L.: A new search algorithm for feature selection in hyperspectral remote sensing images. IEEE Trans. Geosci. Remote Sensing **39** (2001) 1360–1367
7. Dash, M., Liu, H.: Feature selection for classification. Intelligent Data Analysis **1** (1997) 131–156
8. Das, S.: Filters, wrappers and a boosting-based hybrid for feature selection. In: Proc. 18th International Conference on Machine Learning. (2001) 74–81
9. Belue, L.M., Bauer, K.W.: Determining input features for multilayer perceptrons. Neurocomputing **7** (1995) 111–121
10. Bauer, K.W., Alsing, S.G., Greene, K.A.: Feature screening using signal-to-noise ratios. Neurocomputing **31** (2000) 29–44
11. De, R.K., Pal, N.R., Pal, S.K.: Feature analysis: neural network and fuzzy set theoretic approaches. Pattern Recognition **30** (1997) 1579–1590
12. Kavzoglu, T., Mather, P.M.: The use of feature selection techniques in the context of artificial neural networks. In: Proceedings of the 26th Annual Conference of the Remote Sensing Society, Leicester, UK, (2000)
13. Setiono, R., Liu, H.: Neural-network feature selector. IEEE Trans. Neural Networks **8** (1997) 654–662
14. Verikas, A., Bacauskiene, M.: Feature selection with neural networks. Pattern Recognition Letters **23** (2002) 1323–1335
15. Fu, X.J., Wang, L.P.: Data dimensionality reduction with application to simplifying rbf network structure and improving classification performance. IEEE Trans. Syst., Man, Cybern. B **33** (2003) 399–409

16. Fu, X.J., Wang, L.P.: A ga-based novel rbf classifier with class-dependent features. In: Proceedings of the 2002 Congress on Evolutionary Computation CEC2002. Volume 2., IEEE Press (2002) 1890–1894
17. Oh, I.S., Lee, J.S., Moon, B.R.: Hybrid genetic algorithms for feature selection. IEEE Trans. Pattern Anal. Machine Intell. **26** (2004) 1424–1437
18. Raymer, M.L., Punch, W.F., Goodman, E.D., Kuhn, L.A., Jain, A.K.: Dimensionality reduction using genetic algorithms. IEEE Trans. Evol. Comput. **4** (2000) 164–171
19. Sindhwani, V., Rakshit, S., Deodhare, D., Erdogmus, D., Principe, J.C., Niyog, P.: Feature selection in mlps and svms based on maximum output information. IEEE Trans. Neural Networks **15** (2000) 937–948
20. He, M.: Theory, application and related problems of double parallel feedforward neural networks. PhD thesis, Xidian University, Xi'an (1993)
21. He, M., Bao, Z.: Neural network and information processing system: limited precision design theory. Northwestern Polytechnical University publisher, Xi'an (1998)
22. He, M., Xia, J.: High dimensional multispectral image fusion: classification by neural network. In: Proc. SPIE, Image Processing and Pattern Recognition in Remote Sensing. Volume 4898. (2003) 36–43
23. He, M.: Multispectral image processing (invited lecture), IEEE Singapore Section, NTU, Singapore (2004)
24. Moller, M.F.: A scaled conjugate gradient algorithm for fast supervised learning. Neural Networks **6** (1993) 525–533
25. Pudil, P., Novovicova, J., Kittler, J.: Floating search methods in feature selection. Pattern Recognit. Lett. **15** (1994) 1119–1125

Robust Nonlinear Dimension Reduction: A Self-organizing Approach*

Yuexian Hou, Liyue Yao, and Pilian He

School of Electronic Information Engineering, Tianjin University, 300072
{yxhou, yaoliyue, plhe}@tju.edu.cn

Abstract. Most NDR algorithms need to solve large-scale eigenvalue problems or some variation of eigenvalue problems, which is of quadratic complexity of time and might be unpractical in case of large-size data sets. Besides, current algorithms are global, which are often sensitive to noise and disturbed by ill-conditioned matrix. In this paper, we propose a novel self-organizing NDR algorithm: SIE. The time complexity of SIE is O(NlogN). The main computing procedure of SIE is local, which improves the robustness of the algorithm remarkably.

1 Introduction

As an important subject of machine learning, dimension reduction (DR) can relevantly imitate essentially formal characters of abstraction ability of human intelligence. Any meaningful understanding depends on high correlation among huge observations [1]. The goal of DR is to make use of correlations to found the compressed representation and abstract description of the universe.

Classical linear dimension reduction algorithms, such as PCA and CMDS is easy to implement and can find real linear subspace structures in higher dimension space. But linear dimension reduction algorithms cannot illuminate nonlinear manifold structure [2]. Joshua B. Tenenbaum *et al* and Sam T. Roweis *et al* proposed two seminal nonlinear dimension reduction (NDR) algorithms: Isomap [2] and LLE [3], respectively. A few sequential researches are inspired by their work, e.g., Laplacian Eigenmap [4] and Hessian Eigenmap [5]. But all algorithms mentioned above suffer some flaws in common. First, these algorithms need to solve large-scale eigenvalue problems or its variations, which are usually of quadratic time complexity and might be unpractical in case of large-size data sets. Second, current algorithms are global, which are often sensitive to noise and disturbed by ill-conditioned matrix. Recently, K. Dimitris proposed a locally self-organizing NDR algorithm SPE [6]. But simulations show that the reconstruction precision of SPE is far inferior than algorithms based on spectral method, e.g., Isomap.

In this paper, we propose SIE (Self-organizing Isometric Embedding) that combines the advantages of global algorithms and self-organizing algorithms. The time com-

---
* The research is supported by natural science fund of Tianjin (granted no 05YFJMJC11700).

plexity of SIE is O($N\log N$). Besides, the main computing procedure of SIE is local, which improves the robustness against noise, circumvents the trouble introduced by ill-conditioned matrix and facilitate distributed processing.

2 Algorithm

We introduce the SIE algorithm first and explain how it works later. Let $\mathbf{P}_1,\mathbf{P}_2,...,\mathbf{P}_n$ be n-point configuration in R^M. We need to embed $\mathbf{P}_1,\mathbf{P}_2,...,\mathbf{P}_n$ in a lower dimension embedding space R^m to form n-point configuration $\mathbf{Q}_1,\mathbf{Q}_2,...,\mathbf{Q}_n$, which preserves the pair-wise manifold distance, e.g., geodesic distance [2], among all points in $\mathbf{P}_1,\mathbf{P}_2,...,\mathbf{P}_n$. Define cost function $S\equiv\Sigma|d_M(i,j)-d_m(i,j)|$, $1\leq i<j\leq n$, where $d_M(i,j)$ is the geodesic distance between point $\mathbf{P}_i$ and $\mathbf{P}_j$, $d_m(i,j)$ is the Euclidian distance between $\mathbf{Q}_i$ and $\mathbf{Q}_j$. The geodesic distance of SIE is similar to that of Isomap, but SIE permits nonsymmetrical geodesic distance, which can depress noise (see the 4th section).

SIE adopt self-organizing principle to optimize S. There are three main computing procedures in SIE. First, select a set of anchor points from input points set, find k-neighbors of all input point and computing the geodesic distance from anchors points to all points by means of a standard algorithm of graph shortest distance [2]. Secondly, embed the anchor points set by mean of a basic self-organizing embedding algorithm (algorithm 2-1). Lastly, embed all non-anchors points according to their geodesic distance relative to anchor points.

Algorithm 2-1: Basic n-point configuration self-organizing embedding algorithm
Algorithm input: The geodesic distance $\{d_M(i,j)\}$ among $\mathbf{P}_1,\mathbf{P}_2,...,\mathbf{P}_n\in R^M$, $1\leq i,j\leq n$, m (embedding dimension), *steps* (the number learning steps), and λ (learning rate).
Algorithm output: n-point configuration $\mathbf{Q}_1,\mathbf{Q}_2,...,\mathbf{Q}_n$ in embedding space R^m.
1) Initialize n-point configuration $\mathbf{Q}_1,\mathbf{Q}_2,...,\mathbf{Q}_n\in R^m$ randomly.
2) Outer cycle $i=0:1:steps$
3) $x=\mod(i,n)+1$
4) $\Delta\mathbf{Q}_x=0$
5) Inner cycle $y=1:1:n$
6) Compute the Euclidean distance $d_m(x,y)$ between $\mathbf{Q}_x$ and $\mathbf{Q}_y$
7) Compute the error of distance $e_{xy}=d_M(x,y)-d_m(x,y)$
8) Compute offset vector $\Delta\mathbf{Q}_x=\Delta\mathbf{Q}_x-\lambda\tanh(e_{xy})(\mathbf{Q}_y-\mathbf{Q}_x)$
9) Inner cycle finishes.
10) $\mathbf{Q}_x=\mathbf{Q}_x+\Delta\mathbf{Q}_x$
11) Outer cycle finishes.
12) Return $\mathbf{Q}_1,\mathbf{Q}_2,...,\mathbf{Q}_n$.

Simulation shows that algorithm 2-1 can effectively reconstruct n-point configuration for small n (n<100). Based on algorithm 2-1, SIE is depicted as follow.

Algorithm 2-2: Self-organizing Isometric Embedding algorithm (SIE)
Algorithm input: The Euclidean distance $\{d_{ME}(i,j)\}$ among $\mathbf{P}_1,\mathbf{P}_2,...,\mathbf{P}_n\in R^M$, $1\leq i, j\leq n$, m(embedding dimension), *steps* (learning step), and λ (learning rate).

Algorithm output: n-point configuration $Q_1, Q_2, ..., Q_n$ in embedding space R^m.

1) Choose n_a (the number of anchor points) such that $n_a \geq m+2$.
2) Select n_a points from $P_1, P_2, ..., P_n$ as anchor points set notated by $\{A_i\}$, $i=1,2,...,n_a$.
3) Compute the geodesic distance $\{d_M(A_i, P_j)\}$, $1 \leq i \leq n_a$, $1 \leq j \leq n$, among all points in $\{A_i\}$ and $P_1, P_2, ..., P_n$.
4) Initialize n-point configuration $Q_1, Q_2, ..., Q_n \in R^m$ randomly.
5) Making use of algorithm 2-1, reconstruct $\{A_i\}$ in R^m to form the embedding of $\{A_i\}$ according to pair-wise geodesic distance among anchor points.
6) Making use of algorithm 2-1, reconstruct all non-anchor points in R^m to form $Q_1, Q_2, ..., Q_n$ according to pair-wise geodesic distance among anchor points and non-anchor points.

Based on local conditions, i.e., geodesic distances from anchor points set to a non-anchor point, SIE can achieve global isometry embedding. It can be demonstrated that local conditions imply global isometry with probability one as long as the size of anchor points set is proper, i.e., $n_a \geq m+2$ [8].

3 Complexity Analysis of the Algorithm

The time cost of SIE lies on two computing procedures. The first procedure computes geodesic distance among anchor points and all points of the sample set. By means of Dijkstra algorithm with binary heap [7], its time cost is $O(n_a E \log(N))$ where n_a is the number of anchor points, E is the number of edges and N is the number of samples. In most cases, we only embed data sets into low-dimensional space. So n_a can be fixed in an interval such as from 10 to 30. SIE adopts k-Neighbors to compute geodesic distance such that $E=kN$. So the time cost of Dijkstra algorithm is $O(N \log N)$. Then SIE reconstructs all points based on their geodesic distance from anchor points. Since n_a is usually far least than N, the time cost of reconstructions of anchor points can be ignored. The reconstructions of non-anchor points are independent with one another. The time cost of reconstructing one point is $O(bn_a)$, where b is a constant. Therefore, the overall computing cost in this procedure is $O(N)$. So the total time complexity of SIE is $O(N \log N)$. Comparing with the algorithms whose reconstructing effect is approximate to SIE, SIE gains a speedup of $O(N/\log N)$.

4 Simulation Results

Define normalized cost $NS \equiv \Sigma |d_M(i,j) - d_m(i,j)| / \Sigma d_M(i,j)$, $1 \leq i < j \leq n$, to evaluate reconstructing quality of NDR algorithms quantitatively. To verify the robustness of SIE against noise, two kinds of noise are considered: noise leading to false free-degrees (notated as FDN) and noise leading to false connectivity (notated as FCN).

FDN are obviously separated from factual manifold and often lead to false free-degrees. For analytical NDR algorithms, such as Isomap, FDN will influence reconstructing manifolds globally and depress reconstructing quality. Following the local self-organizing principle, SIE can effectively reduce the global effect of FDN and gain optimized reconstructing quality. When reconstructing Swiss data set [2] with FDN in 2-dimensional embedding space, the NS of Isomap, LLE, SPE and SIE are 0.0345, 8.049, 0.5726/0.031 and 0.0157/0.001 respectively. Since SPE and SIE are random algorithms, we run 50 random instances respectively and calculate NS's mean and standard deviation (the two numbers separated by "/") among results of runs. FDN points are produced by the following rule: translating data points whose parameters φ locate in interval $(9.1,9,2) \cup (11.1,11.2) \cup (12.9,13.0)$ along with the outer normal of manifold's tangent plane with probability 0.5, the translation scaling is 20. We select optimal parameters configures for the above fours algorithm respectively. In Fig. 1, sub-figures (a) and (b) show two-dimensional projection and three-dimensional manifold of Swiss data with FDN, sub-figures (c)-(f) show the corresponding reconstructing manifold of SIE, Isomap, LLE and SPE respectively.

FCN is the noise leading to false connectivity, which distort the global topology of embedding manifold and depress the embedding quality markedly. FCN is produced by the following rule: every sample points translate along with the normal (inner or outer) of manifold's tangent plane with probability 0.05, the translation scaling is 2. The sub-figure (a) and (b) of Fig. 2 show two-dimensional projection and three-dimensional manifold of Swiss data with FDN. Since Isomap adopt analytical embedding technique and eigenvalue computing is involved, the geodesic distance matrix should be symmetric to insure the availability of CMDS. So the isolated FCN points are easy to destroy the manifold's global topology. The similar problem trouble LLE. Following the self-organizing principle, SIE permits a directional graph representation of sample set and non-symmetric matrix of geodesic distance could be used in SIE. As a result, though the distance of noise points to non-noise points might be finite, the distance of non-noise points to noise points are likely to be infinite. Thus the influence of a few noise points on manifold's global topology can be eliminated effectively. Sub-figures (c)-(f) show the reconstructing result of SIE, Isomap, LLE and SPE respectively. Obviously, the result of SIE is the only qualitatively correct one. Since geodesic distance matrices of Isomap and SIE are different in the case of data set with FCN, we will not compare the NS of their embedding results.

5 Conclusions

SIE may be criticized for problems of parameters selection. Actually, there exist effective rules of parameters identification. The number of anchor points should be great than $m+2$., where m is embedding dimensions. For the sake of reconstructing precision, n_a are often selected from $5m$ to $10m$. Learning rate λ is not a sensitive parameter. Simulations show that fine reconstructing quality can be achieved with λ attenuating from $1/n_a$ to $0.01/n_a$ linearly.

SIE is of $O(N/\log(N))$ speedup compared with analytical algorithms and robust against noise. Therefore it is suited for applications in large-scale data set, such as text analysis, biology information and web data mining etc.

Robust Nonlinear Dimension Reduction: A Self-organizing Approach 71

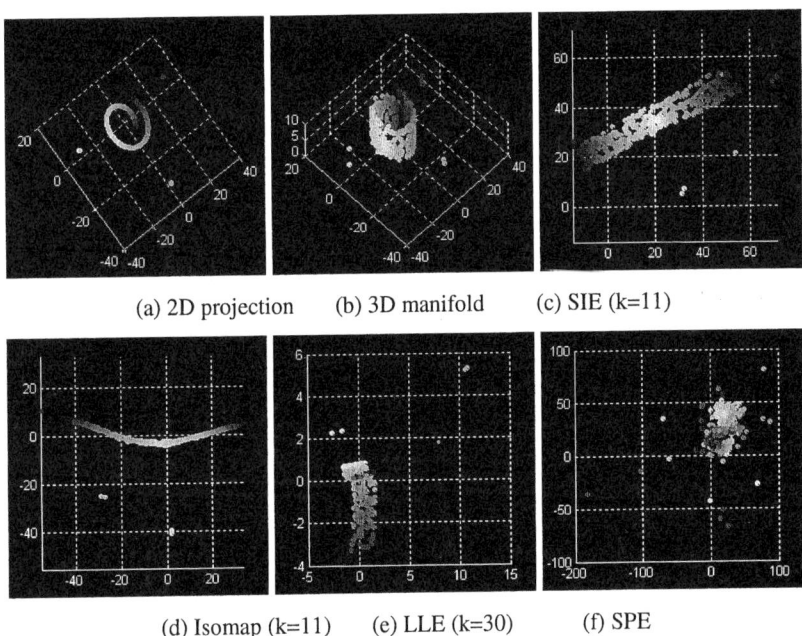

Fig. 1. Swiss data set with FDN

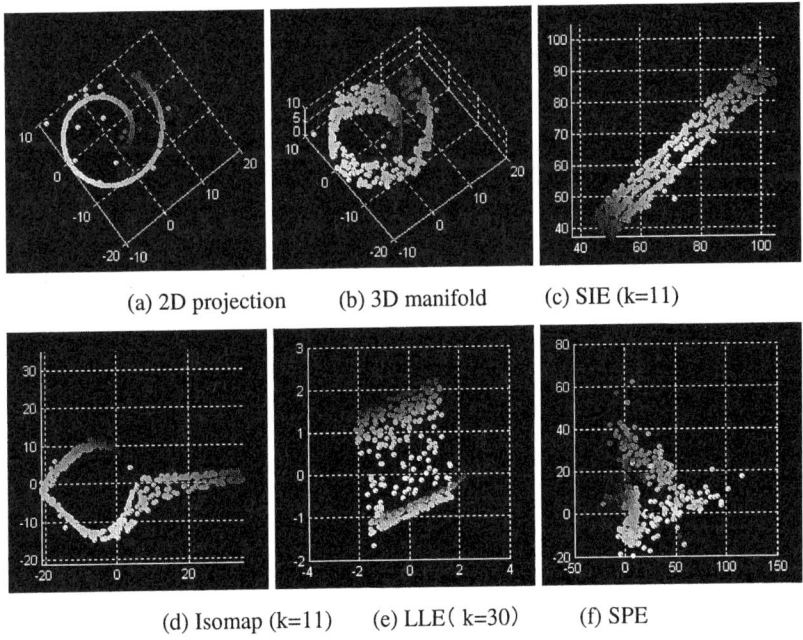

Fig. 2. Swiss data set with FCN

References

1. Eric Baum, *What Is Thought?*, MIT Press, Cambridge, MA, 2004.
2. Joshua B. Tenenbaum et al, Science, Vol. 290, December 22, 2000.
3. Sam T. Roweis et al, Science, Vol. 290, December 22, 2000.
4. M. Belkin et al, NIPS'2001.
5. David L. Donoho et al, PNAS, vol. 100, no. 10, May 13, 2003
6. A self-organizing principle for learning nonlinear manifolds, Dimitris K. Agrafiotis et al, PNAS Early Edition.
7. Cormen, Leiserson, et al., Introduction to Algorithms (2nd edition), MIT press, 2001.
8. Hou Yuexian, Ding Zheng, He Pilian, Self-organizing Isometric Embedding, Computer research and development, 2005, no. 2, (in Chinese).

An Effective Feature Selection Scheme via Genetic Algorithm Using Mutual Information[1]

Chunkai K. Zhang and Hong Hu

Member IEEE,
Department of Mechanical Engineering and Automation, Harbin Institute of Technology,
Shenzhen Graduate School, Shenzhen, China, 518055
ckzhang@hotmail.com

Abstract. In the artificial neural networks (ANNs), feature selection is a well-researched problem, which can improve the network performance and speed up the training of the network. The statistical-based methods and the artificial intelligence-based methods have been widely used to feature selection, and the latter are more attractive. In this paper, using genetic algorithm (GA) combining with mutual information (MI) to evolve a nearoptimal input feature subset for ANNs is proposed, in which mutual information between each input and each output of the data set is employed in mutation in evolutionary process to purposefully guide search direction based on some criterions. By examining the forecasting at the Australian Bureau of Meteorology, the simulation of three different methods of feature selection shows that the proposed method can reduce the dimensionality of inputs, speed up the training of the network and get better performance.

1 Introduction

In the artificial neural networks (ANNs), feature selection is a well-researched problem, aimed at reducing the dimensionality and noise in input set to improve the network performance and speed up the training of the network [1].

Many algorithms for feature selection have been proposed. Conventional methods are based on the statistical tools, such as the partial F-test, correlation coefficient, residual mean square [2,3,4,5] and mutual information (MI) [6,7,8,9]. Although the statistical-based feature selection techniques are widely used, they suffer from many limitations [10]. Firstly, most of them are computationally expensive, because the comparison of all feature subsets is equivalent to a combinatorial problem whose size exponentially increases with the growing number of features. Secondly, the selected feature subset cannot be guaranteed optimal. For example, in the mutual information method, selecting a fixed number of inputs from a ranked list consisting of combinations along with single entries is somewhat problematical, and once a feature is added at an early step, it cannot be removed although it may not constitute the best subset of features in conjunction with the later selected features. Finally, there are a number of parameters that need to be set a priori. For example, the number of features added or removed, the significance level for selecting features and the final feature size.

---

[1] This work is supported by High-tech Industrialization Special Research Project of China

Because the problem of feature selection can be formulated as a search problem to find a nearoptimal input subset, so the artificial intelligence techniques, such as genetic algorithm (GA), is used to selects the optimal subset of features [11,12, 13]. In contrast with the statistical-based methods, the artificial intelligence-based methods are more attractive, as they can find nearoptimal feature subset in lower computational cost and the search process involves no user selectable parameters, such as the final feature size and the signification level etc.. In addition, they have the potential to simultaneous network evolution and feature selection. In most GA-based methods, only the correct recognition rate of a certain neural network is utilized to guide the search direction. In [14], Il-Seok Oh proposed the hybrid GAs for feature selection, which embeds local search operations into the simple GA, but useful information such as statistical information between inputs and outputs in data set don't be added in search process.

In this paper, we proposed a new feature selection scheme for ANNs via genetic algorithm using mutual information. In this method, mutual information (MI) between input and output is employed in mutation in GA to purposefully guide the evolutionary search direction based on some criterions, which can speed up the search process and get better performance. By examining the forecasting at the Australian Bureau of Meteorology [15], the simulation of three different methods of feature selection shows that the proposed method can reduce the dimensionality of inputs, speed up the training of the network and get better performance.

The rest of this paper is organized as follows. Section 2 describes mutual information (MI) and GA. Section 3 the hybrid of GA and MI is used to evolve an optimum input subset for an ANN. Section 4 presents experimental results in a real forecasting problem. The paper is concluded in Section 5.

2 Background

2.1 Definition of Mutual Information

In the information theory founded by Shannon [16], the uncertainty of a random variable C is measured by entropy $H(C)$. For two variables X and C, the conditional entropy $H(C|X)$ measures the uncertainty about C when X is known, and MI, $I(X;C)$, measures the certainty about C that is resolved by X. Apparently, the relation of $H(C)$, $H(C|X)$ and $I(X;C)$ is:

$$H(C) = H(C|X) + I(X;C) \tag{1}$$

or, equivalently,

$$I(X;C) = H(C) - H(C|X),$$

As we know, the goal of training classification model is to reduce the uncertainty about predictions on class labels C for the known observations X as much as possible. In terms of the mutual information, the purpose is just to increase MI $I(X;C)$

as much as possible, and the goal of feature selection is naturally to achieve the higher $I(X;C)$ with the fewer features.

With the entropy defined by Shannon, the prior entropy of class variable C is expressed as

$$H_s(C) = -\sum_{c \in C} P(c) \log P(c) \qquad (2)$$

where $P(c)$ represents the probability of C, while the conditional entropy is $H(C|X)$ is

$$H_s(C|X) = -\int_x p(x)(\sum_{c \in C} p(c|x) \log p(c|x)) dx \qquad (3)$$

The MI between X and C is

$$I_s(X;C) = \sum_{c \in C} \int_x p(c,x) \log \frac{p(c,x)}{P(c)p(x)} dx \qquad (4)$$

Mutual information can, in principle, be calculated exactly if the probability density function of the data is known. Exact calculations have been made for the Gaussian probability density function. However, in most cases the data is not distributed in a fixed pattern and the mutual information has to be estimated. In this study, the mutual information between each input and each output of the data set is estimated using Fraser & Swinney's method [9].

The mutual information of independent variables is zero, but is large between two strongly dependent variables with the maximum possible value depending on the size of the data set. And this assumes that all the inputs are independent and that no output is in fact a complex function of two or more of the input variables.

2.2 Genetic Algorithm

GA is an efficient search method due to its inherent parallelism and powerful capability of searching complex space based on the mechanics of natural selection and population genetics. The method of using GA to select input features in the neural network is straightforward. In GA, every candidate feature is mapped into individual (binary chromosomes) where a bit "1" (gene) denotes the corresponding feature is selected and a bit of "0" (gene) denotes the feature is eliminated. Successive populations are generated using a breeding process that favors fitter individuals. The fitness of an individual is considered a measure of the success of the input vector. Individuals with higher fitness will have a higher probability of contributing to the offspring in the next generation ('Survival of the Fittest').

There are three main operators that can interact to produce the next generation. In replication individual strings are copied directly into the next generation. The higher the fitness value of an individual, the higher the probability that that individual will be copied. New individuals are produced by mating existing individuals. The probability that a string will be chosen as a parent is fitness dependent. A number of crossover points are randomly chosen along the string. A child is produced by copying from one parent until a crossover point is reached, copying then switching to the other parent and repeating this process as often as required. An N bit string can have anything from 1 to N-1 crossover points. Strings produced by either reproduction or crossover may then be mutated. This involves randomly flipping the state of one or more bits. Mutation is needed so new generations are more than just a reorganization of existing genetic material. After a new generation is produced, each individual is evaluated and the process repeated until a satisfactory solution is reached. The procedure of GA for feature selection is expressed as follows:

Procedure of genetic algorithm for feature selection
Initialization
$N \quad \rightarrow \quad$ *Population size*
$P \quad \rightarrow \quad$ *Initial population with N subsets of Y*
$P_c \quad \rightarrow \quad$ *Crossover probability*
$P_m \quad \rightarrow \quad$ *Mutation probability*
$T \quad \rightarrow \quad$ *Maximum number of generations*
$k \quad \rightarrow \quad 0$
Evolution
Evaluation of fitness of P
while ($k < T$ and P does not converge) do
  Breeder Selection
  Crossover with P_c
  Mutation with P_m
  Evaluation of fitness of P Replication
  Dispersal
  $k+1 \rightarrow k$

3 The Proposed Method for Feature Selection

In order to reduce time of calculating MI between single input and output in the whole data set, we randomly select some data from data set with probability 0.5 to construct a data set named *MI* set. Using Fraser & Swinney's method, the mutual information x_i between each candidate input and each output in *MI* set is estimated, which construct a data set $D = \{x_i, i = 1,...,N\}$, x_i represents the mutual information of i th candidate input, and N means there are N candidate inputs.

Then calculate the mathematical statistics of x_i: the mean $\bar{x}$ and standard deviation s_N

$$\bar{x} = \frac{1}{N}\sum_{i=1}^{N} x_i \qquad (5)$$

$$S_N = \sqrt{\frac{1}{N}\sum_{i=1}^{N}(x_i - \bar{x})^2} \qquad (6)$$

And define the three sets which satisfy $D = D_1 \cup D_2 \cup D_3$:

$$D_1 = \{x_i \mid x_i - \bar{x} > \frac{S_N}{2}\},$$

$$D_2 = \{x_i \mid -\frac{S_N}{2} \leq x_i - \bar{x} \leq \frac{S_N}{2}\}$$

$$D_3 = \{x_i \mid x_i - \bar{x} < -\frac{S_N}{2}\}$$

In GA, we use mutual information between each candidate input and each output to guide the mutation based on some criterions, as follows:

$$g_i = \begin{cases} 1 & x_i \in D_1 \\ 0 & x_i \in D_3 \\ rand & x_i \in D_2 \end{cases} \qquad (7)$$

where g_i represents i th gene in a binary chromosome, it means i th candidate input. If the mutual information x_i of i th candidate input belongs to D_1, it means it is a highly correlated input for each output, so include it into input feature subset; if the mutual information x_i of i th candidate input belongs to D_2, it means it is a general correlated input for each output, so randomly include it into input feature subset; If the mutual information x_i of i th candidate input belongs to D_3, it means it is little correlated input for each output, so exclude it from input feature subset.

The procedure of the proposed method for feature selection is same as the procedure of GA for feature selection except the step of "mutation with P_m".

> Mutation with P_m
>
> > If x_i of i th candidate input belongs to D_1, include it into input feature subset;
> >
> > If x_i of i th candidate input belongs to D_2, randomly include it into input feature subset;
> >
> > If x_i of i th candidate input belongs to D_3, exclude it from input feature subset.

4 Experimental Studies

The temperature data for Australia was taken from the TOVS instrument equipped NOAA12 satellite in 1995 [13]. Infrared sounding of 30km horizontal resolution was supplemented with microwave soundings of 150 km horizontal resolution. This data set was used to evaluate the techniques for selecting the input subset. A number of single output networks were developed, each estimating the actual temperature at one of 4 pressure levels (1000, 700, 300 & 150 hPa) given the radiances measured by satellite. These are four of the standard pressure levels (levels 1, 3, 6 and 9) measured by satellite and radiosonde sounders. The input set of TOVS readings to be used by these networks was extracted using each of the three techniques: GA, MI [6] and the proposed method. The appropriate target output temperature was provided by collocated radiosonde measurement.

In MI method, a common input vector length of 8 was used as initial experimentation had proved this to be a suitable value. In GA and the proposed method, $N=50$, $T=60$, $P_c=0.6$ and $P_m=0.02$. And the m-12-1 network uses a learning rate of 0.1 and momentum of 0.8 for 10,000 iterations, where m represents the number of inputs. And the fitness function is defined to be $1/RMSE$, and the root mean square error (*RMSE*) is calculated by

$$RMSE = (\sum (Y - Y_r)^2)^{1/2} \qquad (8)$$

where Y_r is the desired target value, and Y is the output of network.

After selecting an optimal input subset using one of the above techniques, these inputs were assessed by means of an evaluation neural network whose architecture was chosen based on initial experiments. The network used 12 hidden neurons and was trained using fixed parameters to facilitate comparison between the various techniques. It was trained for 2000 passes through the data set using a learning rate of 0.1 and a momentum of 0.8. The network was tested after each pass though the training data with the best result being recorded. The overall performance of this testing network was assumed to reflect the appropriateness of this particular selection of inputs.

The results reported are the mean *RMSE* values obtained from training the ten evaluation networks at each level and should be a reasonable reflection of the inherent worth of the input selection. The results using the full input set (all available inputs) are included in the table for comparison. Mean of *RMSE* (K) derived from all 3 techniques and using all inputs for levels 1,3, 6 & 9 is indicated in Table 1, and selected input subset is indicated in Table 2.

Table 1. Mean of *RMSE* (K) derived from all 3 techniques and using all inputs

	Full	GA	MI	the proposed method
Level 1	2.9	2.6	2.7	2.4
Level 3	2.7	2.9	3.6	2.8
Level 6	2.6	2.5	2.4	2.2
Level 9	3.9	3.6	3.4	3.3

Table 2. Selected input subset for various levels

	GA	MI	the proposed method
Level 1	1, 3, 7, 15, 17, 18, 19, 20, 21	22, 20, 14, 1, 2, 4, 13, 12	3, 8, 14, 20, 22, 4, 2
Level 3	0, 3, 6, 8, 10, 11, 14, 15, 17, 18, 19, 21	4, 21, 17, 15, 20, 3, 1, 9	1, 3, 4, 7, 10, 11, 14, 15, 17, 18, 19, 20, 22
Level 6	3, 6, 8, 14, 18, 21	14, 20, 4, 3, 15, 13, 22, 12	14, 20, 3, 4, 18, 13
Level 9	0, 4, 6, 8, 14, 16, 17, 18, 22	13, 14, 5, 4, 8, 6, 12, 20	4, 5, 6, 8, 12, 14, 20

Table 1 indicates that the proposed method exhibited better performance than the other techniques at all levels. GA was marginally better than MI, outperforming it in levels 1 and 3. Level 3 is interesting in that all three techniques produced networks with worse performance, especially MI. This seems to indicate that the predictive capability at this level is spread more across the inputs – there is less redundancy of information. It should be noted that at this difficult level GA and the proposed method outperformed MI.

Table 2 indicates that although there is considerable similarity between GA and the proposed method there are substantial differences between the inputs selected, and GA and the proposed method selected the different number of inputs for all level, especially in level 3, the number of inputs is large than MI, which explains the reason why the performance of them is better than MI. In contrast with GA, the proposed method can get more little number of inputs without loss of performance, and the content of input subset is the hybrid of that GA and MI. In addition, it was found that there was very little increase in performance after 43 generations for the proposed method, but 56 generations for GA.

5 Conclusion

We proposed an effective feature selection scheme using genetic algorithm (GA) combining with mutual information (MI), in which mutual information between each input and each output of the data set is employed in mutation in evolutionary process to purposefully guide search direction based on some criterions. By examining the forecasting at the Australian Bureau of Meteorology, the simulation of three different methods of feature selection shows that the proposed method can reduce the dimensionality of inputs, speed up the training of the network and get better performance.

References

1. Dash, M., Liu, H.: Feature selection for classification. Intelligent Data Analysis, vol. 1 (1997). 131–156.
2. D.C. Montgomery and E.A. Peck: Introduction to Linear Regression Analysis. John Wiley & Sons, New York (1982).
3. A. Sen and M. Serivastava: Regression Analysis: Theory, Methods, and Applications. Springer-Verlag, New York (1990).

4. Holz, H. J. and Loew, M. H.: Relative feature importance: A classifier-independent approach to feature selection. In: Gelsema, E. S. and Kanal, L. N. (eds.), Pattern Recognition in Practice IV. Amsterdam: Elsevier (1994) 473-487.
5. H. Wang, D. Bell, F. Murtagh: Automatic approach to feature subset selection based on relevance. IEEE Trans. PAMI 21 (3). (1999) 271–277.
6. Belinda Choi, Tim Hendtlass, and Kevin Bluff: A Comparison of Neural Network Input Vector Selection Techniques. LNAI 3029. Springer-Verlag Berlin Heidelberg (2004) 1-10.
7. N. Kwak, C-H. Choi: Input feature selection by mutual information based on parzen window. IEEE Trans. PAMI 24 (12). (2002) 1667–1671.
8. D. Huang, Tommy W.S. Chow: Effective feature selection scheme using mutual information. Neurocomputing vol. 63. (2005) 325 – 343.
9. A.M. Fraser & H.L. Swinney: Independent Coordinates for Strange Attractors from Mutual Information. Physical Review A, Vol. 33/2. (1986) 1134 – 1140.
10. T. Cibas, F.F. Soulie, P. Gallinari and S. Raudys: Variable selection with neural networks. Neurocomputing 12. (1996) 223–248.
11. W. Siedlechi and J. Sklansky: A note on genetic algorithms for large-scale feature selection. Pattern Recognition Letters, 10. (1989) 335–347.
12. C. Emmanouilidis, A. Hunter, J. Macintyre and C. Cox: Selecting features in neurofuzzy modeling by multiobjective genetic algorithms. Artificial Neural Networks, (1999) 749–754.
13. J.H. Yang and V. Honavar: Feature Subset Selection Using a Genetic Algorithm. IEEE Intelligent Systems, vol. 13, no. 2. (1998) 44-49.
14. Oh, I.-S., Lee, J.-S., Moon, B.-R.: Hybrid genetic algorithms for feature selection. IEEE Trans. Pattern Analysis and Machine Intelligence 26. (2004) 1424–1437.
15. J. LeMarshall: An Intercomparison of Temperature and Moisture Fields Derived from TIROS Operational Vertical Sounder Data by Different Retrieval Techniques. Part I: Basic Statistics. Journal of Applied Meteorology, Vol 27. (1988) 1282 – 1293.
16. T.M. Cover, J.A.: Thomas: Elements of Information Theory. Wiley, New York, (1991).

Pattern Classification Using Rectified Nearest Feature Line Segment

Hao Du and Yan Qiu Chen*

Department of Computer Science and Engineering,
School of Information Science and Engineering,
Fudan University, Shanghai 200433, China
chenyq@fudan.edu.cn

Abstract. This paper proposes a new classification method termed Rectified Nearest Feature Line Segment (RNFLS). It overcomes the drawbacks of the original Nearest Feature Line (NFL) classifier and possesses a novel property that centralizes the probability density of the initial sample distribution, which significantly enhances the classification ability. Another remarkable merit is that RNFLS is applicable to complex problems such as two-spirals, which the original NFL cannot deal with properly. Experimental comparisons with NFL, NN(Nearest Neighbor), k-NN and NNL (Nearest Neighbor Line) using artificial and real-world datasets demonstrate that RNFLS offers the best performance.

1 Introduction

Nearest Feature Line (NFL) [1], a newly developed nonparametric pattern classification method, has recently received considerable attention. It attempts to enhance the representational capacity of a sample set of limited size by using the lines passing through each pair of the samples belonging to the same class. Simple yet effective, NFL shows good performance in many applications, including face recognition [1] [2], audio retrieval [3], image classification [4], speaker identification [5] and object recognition [6].

On the other hand, feature lines may produce detrimental effects that lead to increased decision errors. Compared with the well-known Nearest Neighbor (NN) classifier [7], NFL has obvious drawbacks under certain situations that limit its further potential. The authors of [8] pointed out one of the problems – extrapolation inaccuracy, and proposed a solution called Nearest Neighbor Line (NNL). This extrapolation inaccuracy may lead to enormous decision errors in a low dimensional feature space while a simple NN classifier easily reaches a perfect correct classification rate of 100%. Another drawback of NFL is interpolation inaccuracy. Distributions assuming a complex shape (two-spiral problem for example) often fall into this category, where, by the original NFL, the interpolating parts of the feature lines of one class break up the area of another class and severely damage the decision region.

---

* Corresponding author.

In this paper, a new nonparametric classification method, *Rectified Nearest Feature Line Segment* (RNFLS), is proposed that addresses both of the above-mentioned drawbacks and significantly improves the performance of NFL. The original NFL can conceptually be viewed as a two-stage algorithm – building representational subspaces for each class and then performing the nearest distance classification. We focus mainly on the first stage. To overcome extrapolation inaccuracy, *Nearest Feature Line Segment subspace* (NFLS-subspace) is developed. For the interpolation inaccuracy, the "territory" of each sample point and each class is defined, and we obtain *Rectified Nearest Feature Line Segment subspace*(RNFLS-subspace) from NFLS-subspace by eliminating those feature line segments trespassing the territory of other classes. As a result, RNFLS works well for all shapes of sample distribution, which is a significant improvement.

Another remarkable advantage of RNFLS is that it centralizes the probability density of the initial sample distribution. We show, in an experiment, that the decision region created by RNFLS gets closer to the one built by using the optimal Bayesian rule, bringing the correct classification rate higher. Comparisons with NN, k-NN, NFL, NNL using artificial and real-world datasets demonstrate that the proposed RNFLS method offers remarkably superior performance.

2 Background

2.1 The Nearest Feature Line Method

The Nearest Feature Line (NFL) [1] method constructs a feature subspace for each class, consisting of straight lines passing through every pair of the samples belonging to that class. The straight line passing through samples x_i, x_j of the same class, denoted by $\overline{x_i x_j}$, is called a *feature line* of that class. All the feature lines of class ω constitute an *NFL-subspace* to represent class ω, denoted by $S_\omega = \{\overline{x_i^\omega x_j^\omega} | x_i, x_j \in \omega, x_i \neq x_j\}$, which is a subset of the entire feature space.

During classification, a query point q is classified to class ω if q assumes the smallest distance to S_ω than to any other $S_{\omega'}$, $(\omega \neq \omega')$. The distance from q to S_ω is

$$d(q, S_\omega) = \min_{\overline{x_i x_j} \in S_\omega} d(q, \overline{x_i x_j}). \tag{1}$$

$$= \min_{\overline{x_i x_j} \in S_\omega} \|q - p_{ij}\| \tag{2}$$

where p_{ij} is the projection point of q onto line $\overline{x_i x_j}$.

The projection point can be computed by

$$p_{ij} = (1 - \mu)x_i + \mu x_j, \tag{3}$$

where

$$\mu = \frac{(q - x_i).(x_j - x_i)}{(x_j - x_i).(x_j - x_i)}. \tag{4}$$

2.2 Shortcomings

NFL extends the samples of one class by adding the straight lines linking each pair. A good argument for doing this is that it adds extra information to the sample set. The extra information, however, is a double-edged sword. When a straight line of one class trespasses into the territory of another class, it will lead to increased error probability. There are two types of trespassing, causing two types of inaccuracies: extrapolation inaccuracy and interpolation inaccuracy.

Fig.1(a) shows a classification problem in which the *extrapolation inaccuracy* occurs. The query q, surrounded by four "cross" sample points, is in the territory of the "cross" class, leading to the expectation that q should be classified to the "cross" class. But the extrapolating part of feature line $\overline{x_1 x_2}$ makes the distance from q to $\overline{x_1 x_2}$ smaller. Thus, $d(q, \mathcal{S}_{circle}) < d(q, \mathcal{S}_{cross})$, and NFL will assign q the label "circle", not "cross". This is very likely to be a decision error. Similarly, the *interpolation inaccuracy* caused by the interpolating part of a feature line is illustrated in Fig.1(b).

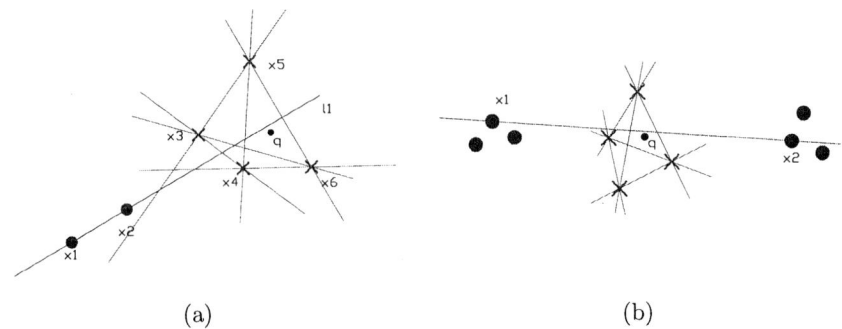

Fig. 1. (a)Extrapolation inaccuracy. (b)Interpolation inaccuracy.

The above inaccuracies are drawbacks that limit the applicability of NFL. In the following section we pursue a more systematic approach in which a new feature subspace for each class is constructed to avoid both drawbacks. The original advantage of NFL that linearly extending the representational capacity of the original samples is retained in our method.

3 Rectified Nearest Feature Line Segment

3.1 Using Feature Line Segments

To avoid extrapolation inaccuracy, we propose to use line segments between pairs of the sample points to construct a *Nearest Feature Line Segment subspace* (NFLS-subspace) instead of the original NFL-subspace to represent each class. Let $X^\omega = \{x_i^\omega | 1 \le i \le N_\omega\}$ be the set of N_ω samples belonging to class ω. The NFLS-subspace ($\widetilde{\mathcal{S}}_\omega$) representing class ω is

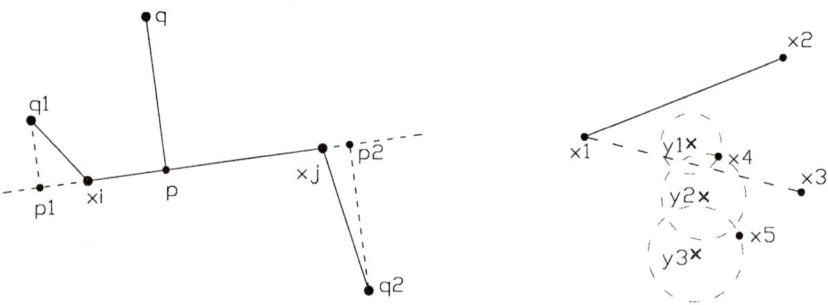

Fig. 2. Distance (solid lines) from feature points to feature line segment $\widetilde{x_i x_j}$

Fig. 3. The territory of "cross"-samples shown in dashed circle

$$\widetilde{S}_\omega = \{\widetilde{x_i^\omega x_j^\omega} | 1 \leq i, j \leq N_\omega\}, \tag{5}$$

where $\widetilde{x_i^\omega x_j^\omega}$ denotes the line segment connecting point x_i^ω and x_j^ω. Note that a degenerative line segment $\widetilde{x_i^\omega x_i^\omega}$ ($1 \leq i \leq N_\omega$), which is a point in the feature space, is also a member of $\widetilde{S}_\omega$.

The distance from a query point q to an NFLS-subspace $\widetilde{S}_\omega$ is defined as

$$d(q, \widetilde{S}_\omega) = \min_{\widetilde{x_i x_j} \in \widetilde{S}_{\omega_k}} d(q, \widetilde{x_i x_j}) \tag{6}$$

where

$$d(q, \widetilde{x_i x_j}) = \min_{y \in \widetilde{x_i x_j}} \|q - y\|. \tag{7}$$

And to calculate $d(q, \widetilde{x_i x_j})$, there are two cases. If $x_i = x_j$, the answer is simply the point to point distance,

$$d(q, \widetilde{x_i x_i}) = \|q - x_i\|. \tag{8}$$

Otherwise, the projection point p of q onto $\overline{x_i x_j}$ is to be located first by using Equ.(3) and Equ.(4). Then, different reference points are chosen to calculate $d(q, \widetilde{x_i x_j})$ according to the position parameter μ. When $0 < \mu < 1$, p is an interpolation point between x_i and x_j, so $d(q, \widetilde{x_i x_j}) = \|q - p\|$. When $\mu < 0$, p is a "backward" extrapolation point on the x_i side, so $d(q, \widetilde{x_i x_j}) = \|q - x_i\|$. When $\mu > 1$, p is a "forward" extrapolation point on the x_j side, so $d(q, \widetilde{x_i x_j}) = \|q - x_j\|$. Fig.2 shows an example.

In the classification stage, a query q is classified to class ω_k when $d(q, \widetilde{S}_{\omega_k})$ is smaller than the distance from q to any other $\widetilde{S}_{\omega_i}$ ($\omega_i \neq \omega_k$).

3.2 Rectifying the Feature Line Segment Subspace

The next step is to rectify the NFLS-subspace to eliminate interpolation inaccuracy. Our motivation is to have the inappropriate line segments removed from the NFLS-subspace $\widetilde{S}_{\omega_k}$ for each class ω_{k_i}. The resulting subspace denoted by $\widetilde{S}^*_{\omega_k}$ is a subset of $\widetilde{S}_{\omega_k}$ termed *Rectified Nearest Feature Line Segment subspace* (RNFLS-subspace).

Territory. We begin with the definitions of two types of territories. One is *sample-territory*, $T_x \in \Re^n$, that is the territory of a sample point x; the other is *class-territory*, $T_\omega \in \Re^n$, that is the territory of class ω. Suppose the sample set X is $\{(x_1, \theta_1), (x_2, \theta_2), ..., (x_m, \theta_m)\}$, which means x_i belongs to class θ_i. The radius r_{x_k} of the sample-territory T_{x_k} is,

$$r_{x_k} = \min_{\forall x_i, \theta_i \neq \theta_k} \|x_i - x_k\|. \tag{9}$$

Thus,
$$T_{x_k} = \{y \in \Re^n \mid \|y - x_k\| < r_{x_k}\}. \tag{10}$$

The class-territory T_{ω_k} is defined to be

$$T_{\omega_k} = \bigcup_{\theta_i = \omega_k} T_{x_i}, \quad (x_i, \theta_i) \in X. \tag{11}$$

In Fig.3, the points denoted by "circle" and "cross" represent the samples from two classes. Each of the "cross"-points (y_1, y_2, y_3) has its own sample-territory as shown by the dashed circle. The union of these sample-territories is T_{cross}. T_{circle} is obtained in a similar way.

Building RNFLS-subspace. For class ω_k, its RNFLS-subspace $\widetilde{S}^*_{\omega_k}$ is built from the NFLS-subspace $\widetilde{S}_{\omega_k}$ by having those line segments trespassing the class-territories of other classes removed. That is

$$\widetilde{S}^*_{\omega_k} = \widetilde{S}_{\omega_k} - \widetilde{U}_{\omega_k}, \tag{12}$$

where '−' is the set difference operator, and

$$\widetilde{U}_{\omega_k} = \{\widetilde{x_i x_j} \mid \exists \omega_y, \omega_k \neq \omega_y \wedge \widetilde{x_i x_j} \in \widetilde{S}^*_{\omega_k} \wedge \widetilde{x_i x_j} \cap T_{\omega_y} \neq \phi\}$$
$$= \{\widetilde{x_i x_j} \mid \exists (x_y, \theta_y) \in X, \widetilde{x_i x_j} \in \widetilde{S}^*_{\omega_k} \wedge \omega_k \neq \theta_y \wedge d(x_y, \widetilde{x_i x_j}) < r_{x_y}\}. \tag{13}$$

Classifying Using RNFLS-subspaces. To perform classification using RNFLS-subspaces is similar to using NFLS-subspaces, since the only difference between an RNFLS-subspace and an NFLS-subspace is $\widetilde{S}^*_{\omega_k} = \widetilde{S}_{\omega_k} - \widetilde{U}_{\omega_k}$, where, except for some removed line segments, $\widetilde{S}^*_{\omega_k}$ is still a set consisting of line segments. The distance measure from a query point to the RNFLS-subspace remains the same.

3.3 Analyzing the Centralization Property

In many real-world pattern recognition problems, samples from one class tend to scatter around a certain center point because of systematic error and random

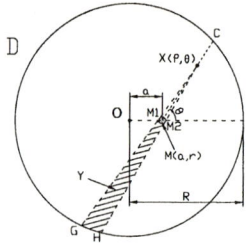

Fig. 4. Calculating N_a^ω for a uniform sample point density on a disk

noise. Gaussian distribution is an example. Two scattered classes may overlap each other, causing decision errors. Compared with the original ideal sample distribution without noise, the NFLS-subspace of each class has an impressive property – distribution centralization, which can be viewed as the converse of scattering. With the help of NFLS-subspace, the distribution overlapping is reduced, and the probability distribution grows closer to the original. And so, we get a higher correct classification rate.

The simplest case to show the centralization property is when the distribution is uniform in a two-dimensional feature space. Suppose that the sample points of class ω are uniformly distributed in a disk D whose radius is R and the center is at O, as shown in Fig.4. For the NFLS-subspace of the class, consider a small region $M(a,r)$ ($a \leq R$), that is a round area with an arbitrarily small radius r and distance a from O. Let N_a^ω be the probability of a randomly selected feature line segment of class ω passing through $M(a,r)$.

Proposition 1. *Given an arbitrarily small r, N_a^ω is decreasing on a.*

Proof. We calculate N_a^ω in a polar coordinate system by choosing the center of $M(a,r)$ as pole and $\overrightarrow{OM}$ as polar axis. For a line segment $\widetilde{XY}$ passing through $M(a,r)$, given one endpoint $X(\rho, \theta)$ in D, the other endpoint Y has to appear in the corresponding $\Box_{M_1M_2HG}$, as shown in Fig.4. Thus we obtain

$$N_a^\omega = \iint_D \frac{1}{\pi R^2} A(\rho, \theta) \rho d\rho d\theta$$

$$= \int_0^{2\pi} \int_0^{|MC|} \frac{1}{\pi R^2} A(\rho, \theta) \rho d\rho d\theta \qquad (14)$$

where $A(\rho, \theta)$ is the probability that the randomly generated endpoint Y appears in $\Box_{M_1M_2HG}$,

$$A(\rho, \theta) = \frac{1}{\pi R^2} \left[\frac{1}{2}(2r + |GH|) \cdot |MG| + o(r) \right] \qquad (15)$$

According to Equ.(14) and (15)

$$N_a^\omega = \frac{2r(R^2 - a^2)}{(\pi R^2)^2} \int_0^{2\pi} \sqrt{R^2 - a^2 \sin^2 \theta} \cdot d\theta + o(r). \qquad (16)$$

Thus, for a fixed r, N_a^ω gets smaller when a gets larger.

Proposition 1 indicates that the distribution of line segments in the NFLS-subspace is denser at the center than at the boundary if the original sample points distribution is under a uniform density. A Gaussian distribution can be viewed as a pile-up of several uniform distribution disks with the same center but different radius. It is conjectured that this centralization property also applies to the Gaussian case, and can be extended to classification problems in which the overlapping is caused by noise scattering of two or more classes under similar distribution but different centers. It reverses the scattering and achieves a substantial improvement.

4 Experiment Results and Discussions

The performance of the RNFLS method is compared with four classifiers - NN, k-NN, NFL and NNL - using two artificial datasets as well as a group of real-world benchmarks widely used to evaluate classifiers. The results on these datasets, representing various distributions and different dimensions, demonstrate that RNFLS possesses remarkably stronger classification ability than the other four methods.

4.1 The Two-Spiral Problem

The two-spiral problem is now included by many authors as one of the benchmarks for evaluation of new classification algorithms. The two-spiral curves in a two-dimensional feature space is described as follows

$$spiral1: \begin{cases} x = k\theta \cos(\theta) \\ y = k\theta \sin(\theta) \end{cases} \qquad spiral2: \begin{cases} x = k\theta \cos(\theta + \pi) \\ y = k\theta \sin(\theta + \pi) \end{cases} \qquad (17)$$

where $\theta \geq \pi/2$ is the parameter. If the probability density of each class is uniform along the corresponding curve, an instance of such distribution is shown in Fig.5(a).

In our experiment, Gaussian noise is added to the samples so that the distribution regions of the two classes may overlap each other, as shown in Fig.5(b). If the prior distribution density were known, according to the optimal Bayesian rule, Fig.5(d) should be the optimal decision region. This, however, can hardly be achieved because the only information we have is from a finite number of sample points.

The original NFL is not a good choice for this classification problem. We may imagine how fragmented the decision region is carved up because of its interpolation and extrapolation inaccuracy. The decision region created by NN rule is shown in Fig.5(e). When it comes to RNFLS, Fig.5(c) is the RNFLS-subspaces and Fig.5(f) is the corresponding decision region. Compared with the decision region created by NN, RNFLS produces a much better one in which the boundary is smoother and some incorrect regions caused by isolated noise points

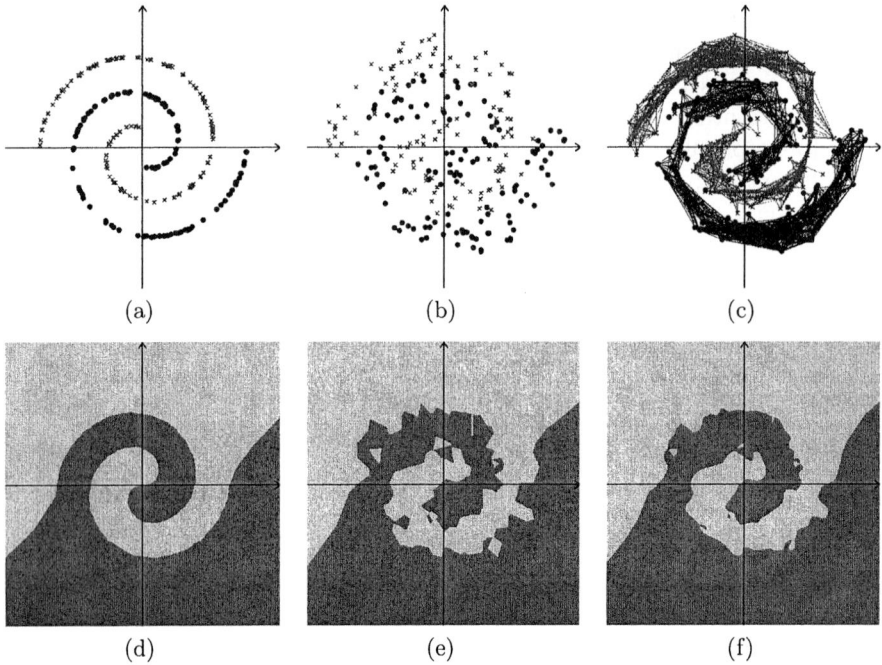

Fig. 5. (a)Two-spiral problem. (b)Two-spiral problem with Gaussian noise. (c)RNFLS subspaces. (d)Bayesian decision region. (e)NN classification result. (f)RNFLS classification result.

is smaller. This significant enhancement can be attributed to the centralization property.

As a concrete test, let $\theta \in [\pi/2, 3\pi]$ and the Gaussian noise is of a variance $\sigma = 1.7$ and an expectation $\mu = 0$. We produce 500 points according to the well-defined distribution, where 250 belong to class ω_1 and the other 250 belong to class ω_2. Then, half of them are randomly chosen to form the sample set and the remaining half constitute the test set. The classifiers, NN, k-NN(k=3), NFL, NNL and RNFLS, are applied to this task for 10 times, and Table 1 shows the results.

Table 1. Performance evaluation on the two-spiral problem using NN, 3-NN, NFL[1], NNL[8] and RNFLS. (CCR: correct classification rate, percentage)

Classifier	CCR (average)	CCR (min)	CCR(max)
NN	83.2	80.4	85.3
k-NN(k=3)	85.3	83.2	87.3
NFL	53.2	49.8	56.7
NNL	72.4	69.0	78.0
RNFLS	**86.1**	**84.0**	**88.2**

4.2 Real-World Classification Problems

We test the RNFLS classifier on a group of real-world datasets as listed in Table 2. All of the datasets are obtained from the U.C. Irvine repository [9]. Since we do not deal with the issue of missing data, instances with missing values are removed. For the fairness of the procedure, attributes of the instances are standardized (normalized) by their means and standard deviations before submitted to the classifiers. The performance in CCR is obtained using the leave-one-out procedure.

Table 2. CCR(%) for NN, 3-NN, NFL, NNL and RNFLS on the real-world datasets

Dataset	#Classes	#Instances	#Attributes	NN	3NN	NFL	NNL	RNFLS
1 iris	3	150	4	94.7	94.7	88.7	94.7	**95.3**
2 housing	6	506	13	70.8	73.0	71.1	67.6	**73.5**
3 pima	2	768	8	70.6	**73.6**	67.1	62.8	73.0
4 wine	3	178	13	95.5	95.5	92.7	78.7	**97.2**
5 bupa	2	345	6	63.2	65.2	63.5	57.4	**66.4**
6 ionosphere	2	351	34	86.3	84.6	85.2	87.2	**94.3**
7 wpbc	2	194	32	72.7	68.6	72.7	54.1	**75.8**
8 wdbc	2	569	30	95.1	96.5	95.3	64.0	**97.2**

It can be seen that RNFLS performs well on both two-category and multi-category classification problems in both low and high dimensional feature spaces. This is encouraging since these datasets represent real-world problems and none of them is specially designed to suit a specific classifier. Since one common characteristic of real-world problems is distribution dispersing caused by noise, the centralization property of RNFLS helps improving the correct classification rate.

5 Conclusions and Future Work

A new classification method RNFLS is developed. It enhances the representational capacity of the original sample points and constitutes a substantial improvement to NFL. It works well independent of the distribution shape and the feature-space dimension. In particular, viewed as the converse of sample scattering, RNFLS is able to centralize the initial distribution of the sample points and offers a higher correct classification rates for common classification problems.

Further investigation into RNFLS seems warranted. In the rectification process it would be helpful to reduce the runtime-complexity, perhaps using some kind of probability algorithms. It may also be helpful to treat the trespassing feature line segments more specifically, for example, finding a way to cut off a part of a trespasser instead of eliminating the whole feature line segments. Also worth more investigation is the centralization property, which might be of great potential.

Acknowledgments

The research work presented in this paper is supported by National Natural Science Foundation of China, project No.60275010; Science and Technology Commission of Shanghai Municipality, project No. 04JC14014; and National Grand Fundamental Research Program of China, project No. 2001CB309401.

References

1. S. Z. Li and J. W. Lu: Face recognition using the nearest feature line method. IEEE Trans. Neural Networks **10** (1999) 439-443
2. J. T. Chien and C. C. Wu: Discriminant waveletfaces and nearest feature classifiers for face recognition. IEEE Trans. Pattern Anal. Machine Intell. **24** (2002) 1644-1649
3. S. Z. Li: Content-based audio classification and retrieval using the nearest feature line method. IEEE Trans. Speech Audio Processing **8** (2000) 619-625
4. S. Z. Li, K. L. Chan and C. L. Wang: Performance evaluation of the nearest feature line method in image classification and retrieval. IEEE Trans. Pattern Anal. Machine Intell. **22** (2000) 1335-1339
5. K. Chen, T. Y. Wu and H. J. Zhang: On the use of nearest feature line for speaker identification. Pattern Recognition Letters **23** (2002) 1735-1746
6. J. H. Chen and C. S. Chen: Object recognition based on image sequences by using inter-feature-line consistencies. Pattern Recognition **37** (2004) 1913-1923
7. T. M. Cover and P. E. Hart: Nearest neighbor pattern classification. IEEE Trans. Inform. Theory **13** (1967) 21-27
8. W. M. Zheng, L. Zhao and C. R. Zou: Locally nearest neighbor classifiers for pattern classification. Pattern Recognition **37** (2004) 1307-1309
9. C. L. Blake and C. J. Merz: UCI repository of machine learning databases. http://www.ics.uci.edu/~mlearn/MLRepository.html (1998) [Online]

Palmprint Identification Algorithm Using Hu Invariant Moments

Jin Soo Noh and Kang Hyeon Rhee*

Dept. of Electronic Eng., Multimedia & Biometrics Lab.,
Chosun University, Gwangju Metropolitan city,
Korea(Daehanminkook)501-759
njinsoo@vlsi.chosun.ac.kr, multimedia@chosun.ac.kr

Abstract. Recently, Biometrics-based personal identification is regarded as an effective method of person's identity with recognition automation and high performance. In this paper, the palmprint recognition method based on Hu invariant moment is proposed. And the low-resolution (75dpi) palmprint image (135×135 Pixel) is used for the small scale database of the effectual palmprint recognition system. The proposed system is consists of two parts: firstly, the palmprint fixed equipment for the acquisition of the correctly palmprint image and secondly, the algorithm of the efficient processing for the palmprint recognition.

1 Introduction

The recent developments in information technology have energized the ongoing studies of personal identification using physiological characteristics of an individual. The physiological characteristics used in personal identification should be only available for each individual and they should have features that are consistent over time. The physiological characteristics satisfying above conditions may include fingerprints, face, iris and retina of the eye, veins on back of the hand, and palmprints. Presently, there are wide variety ongoing studies of Biometrics system which recognize an individual by identifying the specific characteristics for that individual.

There are many fields which objectives are supervising security and personal identification may largely emphasize the importance of using mechanical devices which are scientific methods to employ personal identification. The fingerprint is regarded as the most representative application for a personal identification data which demonstrates unique features for every individual and its consistency in formation throughout the lifetime.

The dermal pattern of a palmprint[1] is completely formulated at birth like the fingerprints and the pattern that is formed would not change over lifetime thus it could be used as the tool for personal identification. However, there are difficult problems associated when applying the palmprints as the identification tool due to the following aspects such as the locations of the specific patterns which are to be classified or or-

---

* Corresponding author.

ganized tend to be widely spread out thus larger amount of processing data than the fingerprints would be needed and the difficulties in obtaining data correctly from the regions through the input device. Moreover, there may be problems affecting the formation or the location of specific patterns generated from the partial damages in the regions of those specific patterns due to the conditions of the hand usage, however principal lines, wrinkles, ridges and minutiae point which are found in palmprint may provide effective means of personal identification[2,3]. Furthermore, the fingerprint tend to have intensive amount of information in a very limited region thus it could be forged partially and people like factory employees may have their fingerprints erased due to the nature of their work. Palmprints may be more secured than fingerprints since they are less likely to be erased or changed and they have a strong tendency against the forgeries due to the wide distributions of detectable locations in identifying the data.

We proposed the palmprint identification algorithm using invariant moments. The palmprint fixed equipment was designed and installed into the scanner to better attaining the specific region of the palmprint accurately. And the following items; the histogram equalization, the smoothing filter and the binarization block[4] were inserted into the palmprint identification process in ways to compensate for those partial damages in palmprint. The Hu invariant moments[5] were applied for identifying the palmprint and the palmprints with low resolution of 75 dpi (dots per inch) were used as the input data in order to minimize the amount of data processing and palmprint identification activation period.

2 The Palmprint Acquisition System

The palmprint acquisition system composed of the palmprint fixed equipment and the palmprint extraction section to acquire the palmprint image that may have little or no changes. In order to minimize the calculation period of palmprint authentication system, the palmprint image with resolution of 75 dpi (dot per inch) was selected by considering the reference [3] and the size of palmprint image was fixed to 135×135 pixel after replicated experiments.

3 The Palmprint Authentication Algorithm

The palmprint authentication algorithm proposed in this paper is composed of 5 systematic steps listed as followings; the histogram equalization, the smoothing filter, the Otsu binarization, the invariant moment and the search algorithm.

The histogram equalization and the smoothing filter allow the accurate distinctions for background regions and wrinkles as well as principle lines of palmprints which are needed in the palmprint identification system. Moreover, they also provide subtle protections against the changes occurring in brightness or the damages in palmprints. The binarization process is the process to transfer the palmprint image into the binary data of '0' and '1' thus allow the calculations of invariant moment. This invariant moment calculation would provide an original value of palmprints for distinguishing each individual. Fig. 1 shows the proposed algorithm of palmprint authentication.

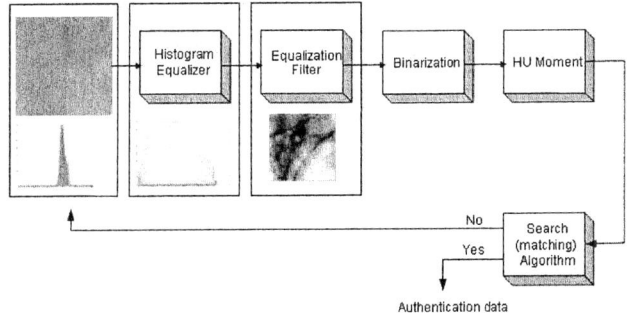

Fig. 1. The Proposed Algorithm of Palmprint Identification

4 Experiment and Result

4.1 Palmprint Database

The performance of proposed algorithm was measured by collecting total 378 units of palmprints from 189 people by using the palmprint acquisition device designed in this paper. The palmprints were provided by 168 males and 21 females between the ages of 21 to 28. The acquired palmprints are in the size of 135×135 pixels and any possible changes that might occur rotationally and positionally were minimized by using the palmprint fixed equipment.

In order to establish the palmprint database, one palmprint was measured 3 times and 5 units of the Hu invariance moment were collected from each palmprint. And then the average of 3 moments was calculated to complete the database.

4.2 The Palmprint Authentication

The matching sequence of the Hu invariance moments stored in palmprint database and those moments of input palmprints sent from the palmprint acquisition device was selected as the palmprint authentication technique. The input palmprint was authenticated as the valid palmprint if the Euclidean distance between the stored and input was below the certain value.

The performance of the proposed algorithm was compared to the authentication system [3] after selecting the most advanced system among the authentication systems based on the hand shape, the finger prints and the palmprint. Table 1 shows the values of FAR and GAR and the marked section would be optimally suggested FAR and GAR in [3].

Table 1. FAR and GAR performance measured using 3rd run authentication

Critical value (coeff.)		FAR(I)[%]		GAR(1-I)[%]
0.001	55	0.038	7	98.1
0.002	71	0.049	7	98.1
0.003	79	0.055	6	98.4
0.004	125	0.087	4	98.9

5 Conclusion

The palmprint authentication algorithm based on the invariance moment was proposed in this paper. This technique largely constituted of the palmprint acquisition system and the palmprint authentication system. The palmprint acquisition system utilized the palmprint fixing device to provide the accuracy in palmprint image acquisition and in order to shorten the time needed for the invariance moment calculations, it composed of various image processing steps such as the histogram equalization, the smoothing filter, and Otsu Binarization. And by allowing the palmprint authentication system to calculate the invariant moment for those palmprint data which passed through the image processing steps, it was installed with the structure to authenticate the palmprint using the algorithm which searched the minimum Eucledean distance with the data stored in the database.

In this paper, FAR, FRR and GAR of proposed algorithm were calculated after acquiring 378 units of palmprint data from 198 students. As shown in Table 1 FAR and GAR measurements of the proposed algorithm after inserting the re-authentication process algorithm contributed 0.038% and 98.1% for FAR and GAR respectively while maintaining the critical value at 0.001. This result suggested that it was improved by 0.002% and 0.1% for FAR and GAR than selecting [3] as the comparative data in this paper.

Acknowledgement. This research is supported by Chosun University, Korea, 2005.

References

1. K. Yamato, Y. Hara, "Fingerprint Identification System by Ravine Thinning," IEICE, Vol. J71-D, No.2, pp. 329–335, 1988.
2. W. Shu and D Zhang, "Automated Personal Identification by Palmprint," Optical Eng., vol. 37m no. 8, pp. 2659–2362, 1998.
3. David Zhang, Wai-Kin Kong, Jane You, Michael Wong, "Online Palmprint Identification," IEEE Trans. Pattern analysis and Machine Intelligence, vol. 25, no. 9, pp. 1041–1050, 2003.
4. N. Otsu, "A threshold selection method from gray level historgam," IEEE SMC-9, no. 1, pp. 62–66, 1979.
5. Rafael C., Digital Image Processing, Addison Wesley, pp. 514–518, 1993.

Generalized Locally Nearest Neighbor Classifiers for Object Classification

Wenming Zheng[1], Cairong Zou[2], and Li Zhao[2]

[1] Research Center for Science of Learning,
Southeast University, Nanjing, Jiangsu, 210096, China
[2] Engineering Research Center of Information Processing and Application,
Southeast University, Nanjing, Jiangsu 210096, China
{wenming_zheng, cairong, zhaoli}@seu.edu.cn

Abstract. In this paper, we extend the locally nearest neighbor classifiers to tackle the nonlinear classification problems via the kernel trick. The better performance is confirmed by the handwritten zip code digits classification experiments on the US Postal Service (USPS) database.

1 Introduction

The Nearest Neighbor (NN) classifier is one of the most popular nonparametric techniques in pattern recognition. It finds the training sample that is closest to the query sample in the training data set. It was shown in literature [4] that the one nearest neighbor classifier has asymptotic error rate at most twice the Bayes error [3]. In practice, however, the performance of the NN classifier is always limited by the available samples of the training data set. To overcome this problem, Zheng et al. [5] proposed the locally nearest neighbor classifiers based on the locally linear embedding (LLE) method [2]. The nearest neighbor line (NNL) method and the nearest neighbor plane (NNP) method [5] are two special cases of this method. However, a major drawback of the locally nearest neighbor classifiers is that they only work well in the case that the training data set is linearly separate, and will fail for nonlinearly separate problems. Thus it is necessary to extend the locally nearest neighbor classifiers to be suitable for the nonlinear case.

Motivated by support vector machine (SVM) [4] and kernel principal component analysis (KPCA) [1], in this paper we propose the generalized locally nearest neighbor classifiers, which are the extensions of the locally nearest neighbor classifiers via the kernel trick. The main idea of this method is to map the input space into a high-dimensional (even infinite dimensional) feature space with linearly separate properties, and then perform the locally nearest neighbor classifiers in the feature space, where the kernel trick is used to make the computation possible.

2 Generalized Locally Nearest Neighbor Classifiers

2.1 Locally Nearest Neighbor Classifiers

Let $\mathbf{X} = \{\mathbf{x}_i^j\}_{i=1,\cdots,c; j=1,\cdots,N_i}$ be an n-dimensional sample set with N elements belonging to c classes, where $\mathbf{x}_i^j$ is the jth sample of the ith class, and N_i is the number of the samples in the ith class. Let $\mathbf{x}$ be the query sample. Then the NNL $\overline{\mathbf{x}_c^{N(1)}\mathbf{x}_c^{N(2)}}$ is defined as [5]

$$\overline{\mathbf{x}_c^{N(1)}\mathbf{x}_c^{N(2)}} = \arg\min_i d(\mathbf{x},\overline{\mathbf{x}_i^{N(1)}\mathbf{x}_i^{N(2)}}) \tag{1}$$

where $\overline{\mathbf{x}_i^{N(1)}\mathbf{x}_i^{N(2)}}$ is defined as the neighbor line (NL) of $\mathbf{x}$ in the ith class sample set, $\mathbf{x}_i^{N(1)}$ and $\mathbf{x}_i^{N(2)}$ are the two nearest neighbors of the query $\mathbf{x}$ in the ith class, and $d(\mathbf{x},\overline{\mathbf{x}_i^{N(1)}\mathbf{x}_i^{N(2)}})$ stands for the distance between $\mathbf{x}$ and the NL $\overline{\mathbf{x}_i^{N(1)}\mathbf{x}_i^{N(2)}}$. Figure 1 illustrates an NL example associated with the query point $\mathbf{x}$, where $p_{N(1)N(2)}^i$ is the projection of the query point $\mathbf{x}$ onto the NL $\overline{\mathbf{x}_i^{N(1)}\mathbf{x}_i^{N(2)}}$.

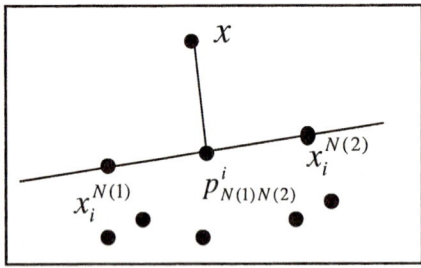

Fig. 1. An NL example, where $\mathbf{x}$ is the query sample, $\mathbf{x}_i^{N(1)}$ and $\mathbf{x}_i^{N(2)}$ are two neighbors of $\mathbf{x}$ in ith class

The NNL classifier can be easily extended to the NNP classifier [5] by adding one neighbor of the query in each class, or even more general form: K-nearest neighbors space (K-NNS) classifier, where K (K>2) nearest neighbors of the query $\mathbf{x}$ in the ith class are used to construct a K-neighbors space (K-NS) $\mathbf{S}_{N(1)N(2)\cdots N(k)}^i$, which is spanned by the K-nearest neighbors under the constraint that the weights sum to one. More specifically, let $\mathbf{x}_i^{N(1)}$ $\mathbf{x}_i^{N(2)}$ $\cdots$ $\mathbf{x}_i^{N(k)}$ be the K nearest neighbors of the query $\mathbf{x}$ in the ith class, then the K-NS in the ith class is defined as

$$\mathbf{S}_{N(1)N(2)\cdots N(k)}^i = \sum_{j=1}^{k} t_i^{(j)} \mathbf{x}_i^{N(j)} \quad \text{s.t.} \quad \sum_{j=1}^{k} t_i^{(j)} = 1 \tag{2}$$

The K-NNS is then defined as the K-NS which has the smallest distance to the query $\mathbf{x}$. In other words, let $\mathbf{S}_{N(1)N(2)\cdots N(k)}^{c^*}$ denote the NNS of $\mathbf{x}$, then we have

$$\mathbf{S}_{N(1)N(2)\cdots N(k)}^{c^*} = \arg\min_i d(\mathbf{x},\mathbf{S}_{N(1)N(2)\cdots N(k)}^i) \tag{3}$$

where $d(\mathbf{x},\mathbf{S}_{N(1)N(2)\cdots N(k)}^i) = \|\mathbf{x}-\mathbf{p}_{N(1)N(2)\cdots N(k)}^i\|$ stands for the distance between $\mathbf{x}$ and $\mathbf{S}_{N(1)N(2)\cdots N(k)}^i$, $\|\cdot\|$ stands for the Euclidean distance, $\mathbf{p}_{N(1)N(2)\cdots N(k)}^i$ is the projection of the query point onto the neighbor space $\mathbf{S}_{N(1)N(2)\cdots N(k)}^i$.

2.2 Generalized Locally Nearest Neighbor Classifiers

Let **X** be mapped from the input space into the reproducing kernel Hilbert space F through a nonlinear mapping Φ:

$$\Phi: \mathbf{X} \to F, \quad \mathbf{x}_i^j \to \Phi(\mathbf{x}_i^j)$$

Let $\Phi(\mathbf{x})$ and $\Phi(\mathbf{y})$ be the two mappings of **x** and **y** in F, then the dot product of these two points in F can be calculated according to the kernel function:

$$k(\mathbf{x}, \mathbf{y}) = <\Phi(\mathbf{x}), \Phi(\mathbf{y})> = (\Phi(\mathbf{y}))^T \Phi(\mathbf{x})$$

where $<\cdot,\cdot>$ stands for the inner product operator [1].

Let $\Phi(\mathbf{x}_i^{N(1)})$ $\Phi(\mathbf{x}_i^{N(2)})$ $\cdots$ $\Phi(\mathbf{x}_i^{N(k)})$ be the K nearest neighbors in F of the query $\Phi(\mathbf{x})$ in the i th class. For the simplicity of notations, we still denote the K-NS in the i th class in F by $\mathbf{S}_{N(1)N(2)\cdots N(k)}^i$, and called it the K-generalized neighbors space (K-GNS) in i th class. Then $\mathbf{S}_{N(1)N(2)\cdots N(k)}^i$ can be expressed by

$$\mathbf{S}_{N(1)N(2)\cdots N(k)}^i = \sum_{j=1}^{k} t_i^{(j)} \Phi(\mathbf{x}_i^{N(j)}) \quad \text{s.t.} \quad \sum_{j=1}^{k} t_i^{(j)} = 1 \tag{4}$$

Let $\mathbf{p}_{N(1)N(2)\cdots N(k)}^i$ be the projection of $\Phi(\mathbf{x})$ onto $\mathbf{S}_{N(1)N(2)\cdots N(k)}^i$, then we have

$$\mathbf{p}_{N(1)N(2)\cdots N(k)}^i = \sum_{j=1}^{k} t_i^{(j)} \Phi(\mathbf{x}_i^{N(j)}) \tag{5}$$

where the coefficients $t_i^{(j)}$ ($j = 1, \cdots, k$) are the ones satisfying the following optimal formula:

$$\min_{t_i^{(j)}, j=1,\cdots,k} \left\| \Phi(\mathbf{x}) - \sum_{j=1}^{k} t_i^{(j)} \Phi(\mathbf{x}_i^{N(j)}) \right\| \tag{6}$$

where $\sum_{j=1}^{k} t_i^{(j)} = 1$.

According to literatures [2, 5], the coefficients $t_i^{(j)}$ can be calculated by

$$t_i^{(j)} = \frac{\sum_k \mathbf{C}_{jk}^{-1}}{\sum_{lm} \mathbf{C}_{lm}^{-1}} \quad (j = 1, \cdots, k) \tag{7}$$

where $\mathbf{C} = (\mathbf{C}_{lm})_{l=1,2,\cdots,k; m=1,2,\cdots,k}$ calculated using the following formula:

$$\begin{aligned}\mathbf{C}_{jk} &= (\Phi(\mathbf{x}) - \Phi(\mathbf{x}_i^{N(j)}))^T (\Phi(\mathbf{x}) - \Phi(\mathbf{x}_i^{N(k)})) \\ &= k(\mathbf{x}, \mathbf{x}) - k(\mathbf{x}, \mathbf{x}_i^{N(k)}) - k(\mathbf{x}_i^{N(j)}, \mathbf{x}) + k(\mathbf{x}_i^{N(j)}, \mathbf{x}_i^{N(k)})\end{aligned} \tag{8}$$

According to the definition of KNNS, the K-generalized nearest neighbors space (K-GNNS) associated with the query sample $\Phi(\mathbf{x})$ is given by

$$\mathbf{S}_{N(1)N(2)\cdots N(k)}^{c^*} = \arg \min_i d(\mathbf{x}, \mathbf{S}_{N(1)N(2)\cdots N(k)}^i) \tag{9}$$

where

$$(d(x, S^i_{N(1)N(2)\cdots N(k)}))^2$$
$$= \left\| \Phi(x) - p^i_{N(1)N(2)\cdots N(k)} \right\|^2 = (\Phi(x) - p^i_{N(1)N(2)\cdots N(k)})^T (\Phi(x) - p^i_{N(1)N(2)\cdots N(k)})$$
$$= (\Phi(x) - \sum_{p=1}^{k} t_i^{(p)} \Phi(x_i^{N(p)}))^T (\Phi(x) - \sum_{q=1}^{k} t_i^{(q)} \Phi(x_i^{N(q)}))$$ (10)
$$= k(x,x) - \sum_{q=1}^{k} t_i^{(q)} k(x, x_i^{N(q)}) - \sum_{p=1}^{k} t_i^{(p)} k(x_i^{N(p)}, x) + \sum_{p=1}^{k}\sum_{q=1}^{k} t_i^{(p)} t_i^{(q)} k(x_i^{N(p)}, x_i^{N(q)})$$

Based on the nearest rule, the query sample Φ(x) is therefore classified into the c^* th class.

3 Experiments

In this experiment, we will perform the handwritten character classification task on the US Postal Service (USPS) zip code digits database to test the performance of the proposed method. The USPS database contains 9298 data points with dimensionality 256, where 7291 points are used as training data and 2007 points as test data [1]. The polynomial kernel defined by $k(\mathbf{x},\mathbf{y}) = (\mathbf{x}^T \mathbf{y})^d$ is used over the experiments, where d is the degree of the polynomial kernel. Table 1 illustrates the experimental results of the test error rates using different degree of the polynomial kernel on the different choice of the nearest neighbors, and Table 2 shows the best classification error rates of several systems on the USPS database, From Table 1, we can see that the generalized locally nearest neighbor classifiers achieves better performance than the linear ones. Especially, when the polynomial kernel with degree three and 11 nearest

Table 1. The experimental results of the test error rates on the USPS handwritten digit database by using KNNS

Number of nearest neighbors	Test error rate for polynomial degree (%)				
	1	2	3	4	5
K = 2	5.08	4.53	4.43	4.58	5.38
K = 3	4.63	4.19	4.14	4.43	4.83
K = 5	4.33	**4.14**	4.24	4.14	4.58
K = 7	4.14	3.89	3.84	4.09	4.58
K = 9	**4.09**	4.09	3.94	**4.04**	**4.43**
K = 11	4.33	3.94	**3.69**	4.09	4.48
K = 13	4.38	4.09	3.84	4.14	4.43

neighbors (K=11) are used, we can obtain the classification error rate as low as 3.69%, whereas the best result obtained by the original linear nearest neighbor classifiers is 4.09% (where K=9 are used). Table 2 also shows that the proposed method is competitive with the nonlinear SVM method (= 4.0%) [6], the KPCA method (= 4.0%) [1], and the convolutional five-layer neural networks method (= 5.0%) [7]. In addition to the better recognition performance, it is also notable that the computational cost of the proposed method is much less than both the KPCA method and the SVM method.

Table 2. Classification error rates of several systems on the USPS database

Methods	K-GNNS	K-NNS	KPCA [1]	SVM [6]	Neural Networks [7]
Error Rates (%)	**3.69**	4.09	4.0	4.0	5.0

4 Conclusion

In this paper, we have presented generalized locally nearest neighbor classifiers via the kernel trick and applied them to the digit character classification task. The experimental results on USPS handwritten zip code digits database have shown that the proposed method can achieve better performance than the original locally linear nearest neighbor classifiers.

References

1. Schölkopf B., Smola A. J., and Müller K.-R.: Nonlinear Component Analysis as a Kernel Eigenvalue Problem. Neural Computation, vol.10, MIT Press (1998) 1299-1319.
2. Roweis S.T., Saul L.K.: Nonlinear dimensionality reduction by locally linear embedding, Science 290 (2000) 2323-2326.
3. Cover T.M. and Hart P.E.: Nearest Neighbor Pattern Classification, IEEE Transaction On Information Theory 13 (1967) 21-27.
4. Vapnik V. N.: The Nature of Statistical Learning Theory. Springer (1995).
5. Zheng W., Zhao L., Zou C.: Locally nearest neighbor classifiers for pattern classification, Pattern Recognition 37 (2004) 1307-1309.
6. Schölkopf B., Burges C., and Vapnik V.: Extracting support data for a given task. In U.M. Fayyad & R. Uthurusamy (Eds.), Proceedings, First Intl. Conference on Knowledge Discovery and Data Mining. Menlo Park, CA: AAAI Press.
7. Le Cun Y., Boser B., Denker J.S., Henderson D., Howard R.E., Hubbard W., and Jackel L.J.: Backpropagation applied to handwritten zip code recognition. Neural Computation, Vol.1, MIT Press (1989) 541-551.

Nearest Neighbor Classification Using Cam Weighted Distance

Chang Yin Zhou and Yan Qiu Chen*

Department of Computer Science and Engineering,
School of Information Science and Engineering, Fudan University,
Shanghai 200433, P.R. China
chenyq@fudan.edu.cn

Abstract. Nearest Neighbor (NN) classification assumes class conditional probabilities to be locally constant, and suffers from bias in high dimensions with a small sample set. In this paper, we propose a novel cam weighted distance to ameliorate the curse of dimensionality. Different from the existing neighbor-based methods, which only analyze a small space emanating from the query sample, the proposed nearest neighbor classification using cam weighted distance (CamNN) optimizes the distance measure based on the analysis of the inter-prototype relationships. Experiments show that CamNN significantly outperforms one nearest neighbor classification (1-NN) and k-nearest neighbor classification (k-NN) in most benchmarks, while its computational complexity is competitive with 1-NN classification.

1 Introduction

In a classification problem, given C pattern classes and N labelled training prototypes, the NN classifier, a simple yet appealing approach, assigns to a query pattern the class label of its nearest neighbor [1]. When the sample size approaches infinity, the error rate of NN classifier converges asymptotically, for all sample distributions, to a value between L^* and $2L^*(1 - L^*)$, where L^* is the Bayes risk [2] [3].

The finite sample size of many real world problems poses a new challenge. The statistics of x_0 may no longer be the same as that of its nearest neighbor. So, Many methods [4] [5] [6] [7] have been proposed to modify the distance metric or measure in order to make the finite sample risk be closer to the asymptotic risk.

The existing methods however tackle this problem only from the aspect of the query point. These methods [4] [5] [6] [7] take advantage of the local information around the query point. These approaches only examine a small local region surrounding the query sample, and the most of the inter-prototype information is neglected.

The proposed CamNN classifier optimizes the distance measure from the aspect of prototypes based on the analysis of the inter-prototype relations. Our motivation comes from the understanding that prototypes are not isolated instances.

---

* Corresponding author.

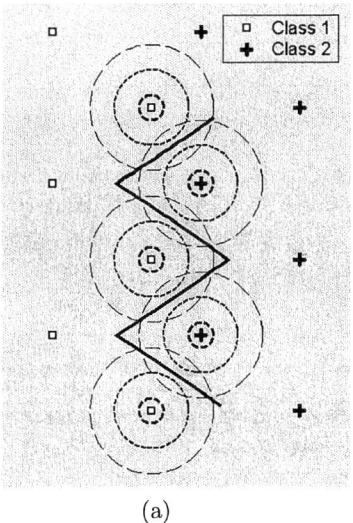 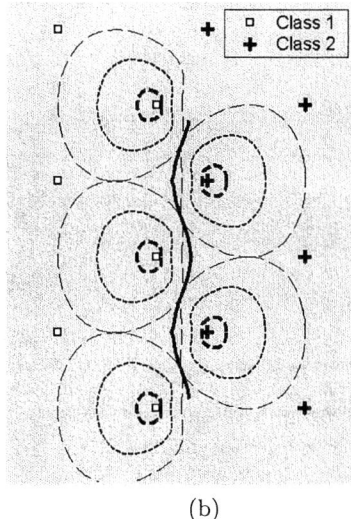

(a) (b)

Fig. 1. Figure (a) shows a traditional 1-NN classification. Figure (b) shows 1-NN classifier with ideal Cam contours. The dash lines are the equi-distance contours around the prototypes. The black solid line in each figure is the corresponding decision boundary.

The nearby prototypes actively affect the confidence level of the information provided by the prototype being considered. So, not only should the distance measure to one prototype vary with orientation, the distance measure to each prototype should also be treated discriminately according to its different surroundings.

From Figure 1, it can be seen clearly that an orientation sensitive distance measure can greatly improve the classification performance. For a traditional NN classification shown in the left, the equi-distance contour is circular because of the isolation assumption. When deformable cam contours are adopted for the prototypes, reflecting the attraction and repulsion they receive from their neighbors, the decision boundary becomes smoother and more desirable as is shown in Figure (b).

Figure 2 presents another common situation, where one prototype of Class 2 falls into an area with many prototypes of Class 1, and shows that it could be more reasonable if each prototype is granted a different but appropriate distance scale. The traditional NN classification treats all the prototypes equally regardless of their surroundings, so that a large region S1 will be decided to belong to Class 2, likely leading to higher error rate. However, the prototypes are not isolated instances, the inter-prototype relationships should not be neglected. Because of the great weakening effects the solitary prototype receives from its oppose neighbors, the distance measure scale of this prototype is diminished and then the region belonging to Class 2 is compressed from S1 to the smaller S2, which should be more desirable and reasonable.

While the idea of optimizing the distance measure from the aspect of prototypes and constructing an orientation sensitive and scale adaptive distance

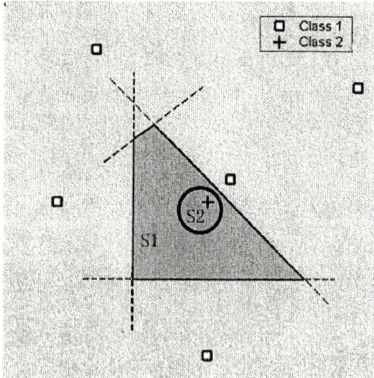

Fig. 2. S1 (containing S2) is the region of Class 2 by the traditional NN classification. S2 is the region of Class 2 by a revised NN classification, who considers the interprototype relations and compresses the distance scale when measuring the distance to the solitary prototype of Class 2.

measure may seem obvious, few proposal along this line could be found in the literature. A literature review is provided in Section 4.

2 Cam Weighted Distance

In a classification problem, each prototype can be regarded as the center of a probability distribution and the similarity to the prototype can be expressed by the corresponding class-conditional probability. In the traditional NN method, the distribution can be a standard normal distribution so that the Euclidean distance is equivalent to the class-conditional probability. However, because of the attraction, repulsion, strengthening effect and weakening effect each prototype receives from its neighbors, the standard normal distributions have actually been greatly distorted.

We construct a simple yet effective transformation $X = (a + b \cdot \frac{Y'\tau}{\|Y\|}) \cdot Y$ to simulate such a distortion, where Y denotes the original distribution and τ is a normalized vector denoting the distortion orientation. We call the eccentric distribution Cam Distribution, if Y subjects to a standard normal distribution. For each prototype representing a Cam distribution, its neighbor prototypes are used to estimate the corresponding distribution parameters a, b, and τ. When a, b, and τ are obtained, an inverse transformation can be performed to eliminate the distortion. Such an inverse transformation will lead to the proposed cam weighted distance.

2.1 Cam Distribution

Definition 1 (Cam Distribution). *Consider a p-dimensional random vector $Y = (Y_1, Y_2, \ldots, Y_p)^T$ that takes a standard p-dimensional normal distribution $N(0, I)$. Let a random vector X be defined by the transformation*

$$X = (a + b \cdot \frac{Y'\tau}{\|Y\|}) \cdot Y, \quad (1)$$

where $a > b \geq 0$ and τ is a normalized vector. Then the distribution of X is called the **Cam distribution**, denoted as $X \sim Cam_p(a, b, \tau)$.

Theorem 1. If a random vector $X \sim Cam_p(a, b, \tau)$, then

$$E(X) = c_1 \cdot b \cdot \tau \quad (2)$$

and

$$E(\|X\|) = c_2 \cdot a, \quad (3)$$

where c_1 and c_2 are constants:

$$\begin{aligned} c_1 &= 2^{1/2}/p \cdot \Gamma(\tfrac{p+1}{2})/\Gamma(\tfrac{p}{2}) \\ c_2 &= 2^{1/2} \cdot \Gamma(\tfrac{p+1}{2})/\Gamma(\tfrac{p}{2}). \end{aligned} \quad (4)$$

2.2 Cam Weighted Distance

As mentioned above, the cam distribution is an eccentric distribution that biases towards a given direction. It is obtained from a standard normal distribution by the transformation $X = Y \cdot (a + b\cos\theta)$. In this model, the Euclidean distance is not suitable to directly describe the similarity, since the assumed normal distribution has been distorted. Instead, we firstly restore the distortion by an inverse transformation $Y = X/(a + b\cos\theta)$, and then measure the distance. This weighted distance redresses the distortion and should be more suitable to describe the similarity.

Definition 2 (Cam Weighted Distance). Assume $x_0 \in \Re^p$ is the center of a Cam Distribution $Cam_p(a, b, \tau)$. The Cam Weighted Distance from a point $x \in \Re^p$ to x_0 is defined to be

$$CamDist(x_0, x) = \|x - x_0\|/(a + b\cos\theta), \quad (5)$$

where θ is the included angle of vectors $x - x_0$ and τ.

Figure 3 shows from left to right three cam distributions $Cam_2(1, 0, [0.8, 0.6])$, $Cam_2(1, 0.4, [0.8, 0.6])$, and $Cam_2(1, 0.8, [0.8, 0.6])$ respectively. By examining the equi-distance contour $CamDist(x_0, x) = d_0$, we can find that the parameter a reflects the overall scale of the distance measure and b reflects the extent of eccentricity in distance measure. When $b = 0$, the contour is circular. As b increases, it looks more like a cam curve. When b approaches to a, the contour becomes a heart curve. In most cases, b is a medium value with respect to a, which represents a cam contour. That is why we call it cam weighted distance.

We should point out that cam weighted distance measure is just a weighted distance, but not a metric, since $CamDist(x_0, x)$ may not equal to $CamDist(x, x_0)$, and $CamDist(x, x_0)$ is even not defined.

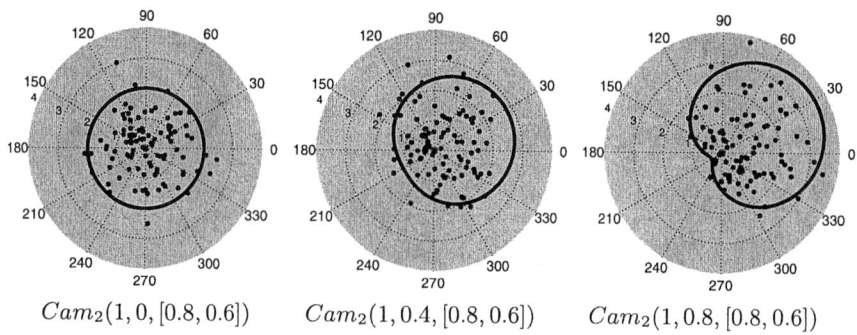

$Cam_2(1, 0, [0.8, 0.6])$ $Cam_2(1, 0.4, [0.8, 0.6])$ $Cam_2(1, 0.8, [0.8, 0.6])$

Fig. 3. Three cam distributions $Cam_2(1, 0, [0.8, 0.6])$, $Cam_2(1, 0.4, [0.8, 0.6])$, $Cam_2(1, 0.8, [0.8, 0.6])$ are shown up respectively, each one with one hundred samples. The samples are marked by black dots. The black solid line in each figure is an equi-distance contour according to the cam weighted distance.

2.3 Parameter Estimation

Parameter estimation has been made simple by Theorem 1. For an arbitrary prototype $x_i \in D$, we assume that it represents a cam distribution and is the origin of this cam distribution. Then, we use its k nearest neighbors $X_i = \{x_{i1}, x_{i2}, \ldots, x_{ik}\}$ to estimate the parameters of the cam distribution, including a_i, b_i and τ_i.

First, we convert X_i to a set of relative vectors $V_i = \{v_{ij} | v_{ij} = x_{ij} - x_i, j = 1, 2, \ldots, k\}$. Then, we use the gravity of mass, $\widehat{G_i}$, and the averaged vector length, $\widehat{L_i}$

$$\begin{cases} \widehat{G_i} = \sum_{j=1}^{k} v_{ij}/k \\ \widehat{L_i} = \sum_{j=1}^{k} \|v_{ij}\|/k \end{cases} \quad (6)$$

to estimate $E(\eta)$ and $E(\|\eta\|)$ respectively. According to Theorem 1, we get an estimation to a_i, b_i, and τ_i:

$$\begin{cases} \widehat{a_i} = \widehat{L_i}/c_2 \\ \widehat{b_i} = \|\widehat{G_i}\|/c_1 \\ \widehat{\tau_i} = \widehat{G_i}/\|\widehat{G_i}\|. \end{cases} \quad (7)$$

The above estimation focuses on a single class situation and assumes all k nearest neighbors of x_i have the same class label as x_i, but in a multiple-class classification problem, for an arbitrary prototype x_i, its k nearest neighbors $X_i = \{x_{i1}, x_{i2}, \ldots, x_{ik}\}$ may come from other opposite classes, so we should not use these neighbor prototypes directly for parameter estimation. A simple skill is employed in our implementation to solve this problem. Assume y_{i0} is the label of x_i and y_{ij} is the label of the neighbors x_{ij}, $j = 0, 1, \ldots, k$. We convert V_i in Equation (6) to W_i, according to

$$w_{ij} = \begin{cases} v_{ij} & \text{if } y_{ij} = y_{i0} \\ -\frac{1}{2} \cdot v_{ij} & \text{if } y_{ij} \neq y_{i0}, \end{cases} \quad (8)$$

where $j = 1, 2, \ldots, k$. Then, Equation (6) is revised to be

$$\begin{cases} \widehat{G_i} = \sum_{j=1}^{k} w_{ij}/k \\ \widehat{L_i} = \sum_{j=1}^{k} \|w_{ij}\|/k. \end{cases} \quad (9)$$

Such a simple transformation not only reserves most of the sample scatter information, but also reflects the relative position of the current class to the nearby opposite classes, so that the orietation information can be reserved.

3   CamNN Classification

The proposed Cam weighted distance can be more suitable for measuring the similarity than the Euclidean distance in many cases, since it exploits the relevant information of the inter-prototype relationships. So, we propose a novel classification method CamNN to improve the neighbor-based classifiers by using the Cam weighted distance.

By the virtue of the simplicity of parameter estimation, the process of CamNN is fairly simple. Its whole process can be divided into two phases. In

Table 1. CamNN Classification Process

Phase 1: Preprocessing
Given a prototype set $D = \{x_i\}$, the corresponding class labels $C = \{y_i\}$ and a parameter k, for each prototype $x_i \in D$,

1) Find its k nearest neighbors $X_i = \{x_{i1}, x_{i2}, \ldots, x_{ik}\}$, $X_i \subset D$
2) Obtain V_i from X_i by $v_{ij} = x_{ij} - x_i$, $j = 1, \ldots, k$
3) Estimate a_i, b_i, τ_i according to Equation (6) and (7)
4) Save a_i, b_i, τ_i to A_i

Phase 2: Classification
For an arbitrary query $q \in \Re^p$,

5) Calculate the cam weighted distance from q to each prototype x_i:
$CamDist(x_i, q) = \|q - x_i\|/(a_i + b_i \cos \theta_i)$,
where θ_i is the included angle of vectors $q - x_i$ and τ_i
6) Find the nearest neighbor $x^* \in D$, which satifies
$CamDist(x^*, q) = \min_{x_i \in D} CamDist(x_i, q)$
7) Return the label y^*, where y^* is the class label of x^*

the preprocessing phase, for each prototype x_i in the training set D, CamNN firstly finds its k nearest prototypes by the Euclidean distance, and then uses these k nearest prototypes to estimate the three cam weighting parameters a_i, b_i and τ_i, according to Equation (6) and (7). After this phase, a parameter matrix A is obtained: $A_i = [a_i, b_i, \tau_i], i = 1, 2, \ldots, \|D\|$, so that we will be able to calculate the cam weighted distance $CamDist(q, x_i)$, from any query point $q \in \Re^p$ to an arbitrary prototype $x_i \in D$, according to Equation (5). In the following classification phase, for any query $q \in \Re^p$, we find the prototype with the shortest cam weighted distance and assign to q the label of this prototype. The detailed steps of this proposed method CamNN are listed in Table 1.

It is remarkable that CamNN is computationally competitive with the traditional NN classification when it significantly outperforms the traditional NN classification (See Section 5). Given a classification problem with M prototypes and N queries, the computational complexity in the preprocessing phase is $O(k * M)$ and the computational complexity in the classification phase is $O(2 * N)$. Compared with k-NN whose complexity is $O(k * N)$ and other sophisticated neighbor-based methods such as [4], [5], [6] and [7], CamNN has great computational advantage in classification.

4 Literature Review

Hastie [7] introduces Discriminate Adaptive NN classification(DANN) metric which combines the advantage of Linear Discriminant (LDA) classifier and NN classifier to ameliorate the curse of dimensionality. For each query, DANN iteratively adjusts its metric while searching for the k nearest neighbors. DANN elongates the distance along the linear discriminate boundary, which is believed to have improved the performance of k-NN.

Friedman [6] integrates tree-structured recursive partitioning techniques and regular k-NN methods, to estimate the local relevance of each query point, and then uses this information to customize the metric measure centered at the query.

Short [4] uses the k nearest neighbors of the query point to construct a direction vector, defines the distance as the multiplication of a vector with this direction vector and then selects the nearest one from k nearest neighbors to classify the query x_0.

From the view point of information retrieval, all of these methods are very different from our proposed CamNN. All these methods [4] [6] [7] [5] take advantage of the local information around the query point. They analyze the measurement space around the query point, and study how the neighbors should be weighted according to their relations with the input point. In contrast, our proposed CamNN analyzes and takes advantage of the inter-prototype relationships. In many cases, the information of the inter-prototype relationships is very important, but is difficult to be obtained from the aspect of the query point.

5 Experimental Evaluation

We perform two sets of experiments to examine the effects of the cam weighted distance on the performance of NN classification. To evaluate the improvement thoroughly, CamNN will also be compared with the k-NN classifier. Especially, to be fair, we always choose the best k for k-NN classification in each experiment.

5.1 Experiments on Two Artificial Problems

First, we perform experiments to check whether CamNN has fulfilled our motivation explained in the introduction. The experiment is performed on the problem shown in Figure 1, and the results of 1-NN, 5-NN and our proposed CamNN are presented in Figure 4. In another experiment, we apply 1-NN, 5-NN and CamNN to classify two classes with independent standard normal distribution

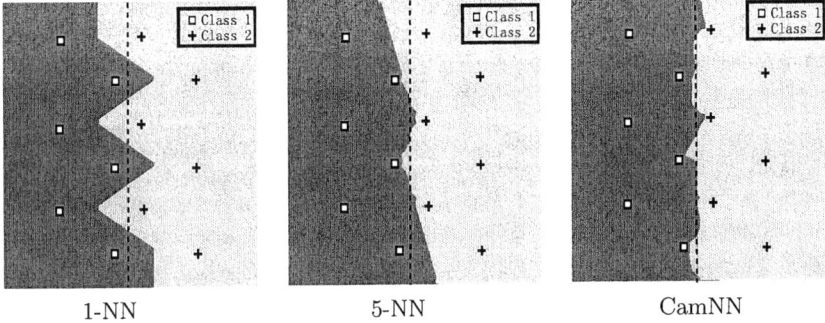

Fig. 4. The results of 1-NN, 5-NN and CamNN(K=5) are shown up respectively from the left to the right. Any points in the left grayed area will be classified to Class 1. It can be seen that the decision boundary of CamNN is more desirable.

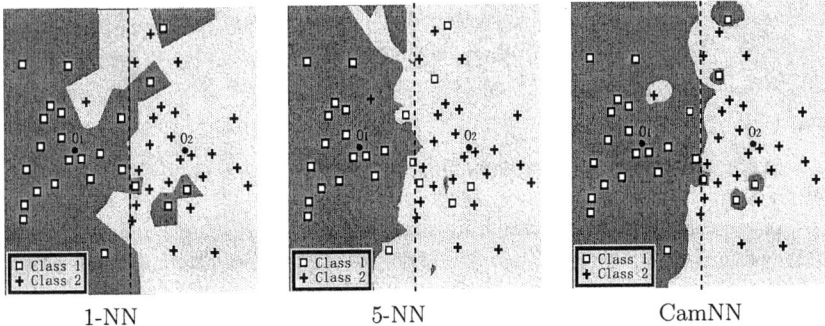

Fig. 5. The marked points are training data coming from two independent standard normal distributions centered at O_1 and O_2 respectively ($\|O_1 - O_2\| = 2$). The classification results of 1-NN, 5-NN and CamNN (k=5) are shown from left to right. Any points in the grayed area will be classified to Class 1.

N(0, I) centered at (-1, 0) and (1, 0) respectively. The classification results are shown in Figure 5.

It can be seen from both Figure 4 and 5 that, the decision boundary of CamNN is smoother and closer to the Bayesian decision boundary than those of 1-NN and 5-NN. CamNN greatly outperforming 1-NN in the experiments shows the great effectiveness of the cam weighted distance in measuring similarity.

5.2 Experiments on UCI Machine Learning Database

UCI machine learning database is a well-known set of benchmarks for machine learning *(http://www.ics.uci.edu/~mlearn/MLRepository.html)*. For the real world datasets in UCI machine learning database, leave-one-out [8] cross-validation is performed to evaluate the performance. The comparison results of 1-NN, k-NN and CamNN on UCI database are given in Table 2.

Table 2. Comparison Results on UCI datasets

Dataset	#C	#Dim	#Samples	1-NN Error Rate(%)	k-NN Error Rate(%)	K	CamNN Error Rate(%)	K
1 auto-MPG	3	7	392	26.7	26.5	7	**24.2**	8
2 balance-Scale	3	4	625	19.7	9.8	7	**8.4**	5
3 bcw	2	9	699	4.9	3.3	7	**3.3**	9
4 wdbc	2	30	569	4.9	**3.2**	9	3.5	5
5 ionosphere	2	33	351	13.4	13.4	1	**6.8**	60
6 iris	3	4	150	5.3	4	7	**3.3**	6
7 pima	2	8	768	29.3	25.8	5	**24.7**	4
8 wine	3	10	178	6.7	4.5	3	**2.8**	7

* The best performer for each dataset is bolded.
* Best k is selected for k-NN classification in each experiment.

Again, CamNN greatly outperforms 1-NN for all data sets and outperforms k-NN for seven of eight data sets. For the remaining one data set, CamNN is only slightly inferior to the k-NN classifier. In particular, it can be observed that CamNN is by far the best performer on 'balance-scale', 'ionosphere' and 'wine'.

6 Summary and Conclusions

This paper presents a novel direction to optimize the distance measure for the neighbor-based classifiers. Our motivation is that the prototypes are not isolated and by analyzing the inter-prototype relationships, we should be able to obtain useful relevant information to optimize the distance measure.

We have also proposed a method CamNN to analyze and take advantage of these inter-prototype relationships. The cam weighted distance, the core of CamNN, has two essential characters, orientational sensitivity and scale adaptivity, which enable it to express the inter-prototype relationships effectively, so that a better classification performance is achieved. The efficacy of our method is validated by the experiments using both artificial and real world data. Moreover, the proposed CamNN is computationally competitive with 1-NN classification.

Acknowledgements

The research work presented in this paper is supported by National Natural Science Foundation of China, project No.60275010; Science and Technology Commission of Shanghai Municipality, project No. 04JC14014; and National Grand Fundamental Research Program of China, project No. 2001CB309401.

References

1. Hart, P., Cover, T.: Nearest neighbor pattern classification. IEEE Transactions on Information Theory. (1967) 13:21–27
2. Devroye, L.: On the inequality of cover and hart in nearest neighbor discrimination. IEEE Transactions on Pattern Analysis and Machine Intelligence. (1981) 3:75–79
3. Wagner, T.: Convergence of the nearest neighbor rule. IEEE Transactions on Information Theory. (1971) 17(5):566–571
4. II Short, R., Fukunaga, K.: The optimal distance measure for nearest neighbor classification. IEEE Transactions on Information Theory. (1981) 27(5):622–627
5. Domeniconi, C., Jing Peng, Gunopulos, D.: Locally adaptive metric nearest-neighbor classification. IEEE Transactions on Pattern Analysis and Machine Intelligence. (2002) 24(9):1281–1285
6. Jerome, H., Friedman: Flexible metric nearest neighbor classication. The Pennsylvania State University CiteSeer Archives. September 24 1999.
7. Hastie, T., Tibshirani, R.: Discriminant adaptive nearest neighbor classification. IEEE Transactions on Pattern Analysis and Machine Intelligence. (1996) 18(6):607–616
8. Hayes, R.R., Fukunaga, K.: Estimation of classifier performance. IEEE Transactions on Pattern Analysis and Machine Intelligence. (1989) 11(10):1087 – 1101

A PPM Prediction Model Based on Web Objects' Popularity

Lei Shi[1,2], Zhimin Gu[1], Yunxia Pei[2], and Lin Wei[2]

[1] Department of Computer Science and Engineering, Beijing Institute of Technology,
Beijing 100081, China
`shilei@zzu.edu.cn, zmgu@x263.net`
[2] College of Information Engineering, Zhengzhou University, Zhengzhou 450052, China
`pyx@hngazk.edu.cn, weilin@shengda.edu.cn`

Abstract. Web prefetching technique is one of the primary solutions used to reduce Web access latency and improve the quality of service. This paper makes use of Zipf's 1st law and Zipf's 2nd law to model the Web objects' popularity, where Zipf's 1st law is employed to model the high frequency Web objects and 2nd law for the low frequency Web objects, and proposes a PPM prediction model based on Web objects' popularity for Web prefetching. A performance evaluation of the model is presented using real server logs. Trace-driven simulation results show that not only the model is easily to be implemented, but also can achieve a high prediction precision at the cost of relative low storage complexity and network traffic.

1 Introduction

Web access latency is one of the main problems leading to low network QoS, which depends on many factors such as network bandwidth, transmission delay, etc. Presently caching and prefetching techniques are the primary solutions used to reduce Web access latency. Web caching technique makes use of the temporal locality principle to cache the most frequently used Web objects near the clients, while prefetching technique is based on the spatial locality principle in order to fetch the most likely Web pages before the users take the action. Web caching has been widely used in different places of Internet. However, approaches that rely solely on caching offer limited performance improvement [1][2] because it is difficult for caching to handle the large number of increasingly diverse network resources. Studies have shown that Web prefetching technique with smoothing traffic can substantially lower Web access latency [3]. Usually, the hit ratio of caching is ranged from 24% to 45%, no more than 50% in many cases, but for prefetching, it can improve the hit ratio to 60% or even more. Web prefetching is becoming more and more important and demanding [1][2][3].

An important task for prefetching is to build a simple and effective prediction model. Prediction by Partial Match (PPM) is a commonly used technique in Web prefetching, where prefetching decisions are made based on historical URLs in a dynamically maintained Markov prediction tree. Existing PPM prediction models, which are proposed by T. Palpanas [4] and J. Pitkow [5] etc., have the common limitations that they take too much storage space for storing millions of historical Web pages and thus corresponding algorithms are always time-consuming and space-consuming.

In the work of this paper, we present a prediction model based on Web objects' popularity. By use of the Zipf's law, we build Web objects' popularity information into the Markov prediction tree; Using Zipf's 1st law that used to describe the high frequency Web objects' popularity, we only prefetch the most popular Web objects of which popularity is bigger than or equal to a threshold value β to control additional network traffic; Based on the Zipf's 2nd law used to depict the low frequency Web objects' popularity, we remove all most unpopular Web objects of which popularity is lower than or equal to a threshold value θ for reducing model size. The experiments have shown that comparing with existing models, not only the model is easily to be implemented, but also can achieve a high prediction precision at the cost of relative low storage complexity and network traffic.

The rest of this paper is organized as follows: Section 2 introduces how to make use of Zipf's law to model the popularity of Web objects. Section 3 describes related work and presents our PPM prediction model for Web prefetching. The experiment results are discussed in Section 4. Section 5 is the summary and conclusions.

2 Modeling Web Objects' Popularity

Many researches show that there are some hot places in the Web objects access distribution [12]. In other words, Web objects include the high frequency objects and the low frequency objects.

Zipf's 1st law has been used to depict the high frequency Web objects' popularity. Reported values of α from recent studies [6] range from 0.75 to 0.85 at Web servers and 0.64 to 0.83 at Web proxies. We model the high frequency Web objects' popularity by modifying Zipf's 1st law as follows:

$$P(i) = C/i^a, \quad \alpha \in [0.5, 1] \qquad (1)$$

where parameter C is a constant, i is the rank of popularity and P(i) is the conditional probability of the Web page ranking i. The Zipf exponent α reflects the degree of popularity skew, while the proportionality constant C represents the number of requests for the most popular Web page (i=1). Let N represent the total number of the high frequency Web pages. Since the sum of all probabilities is equal to 1, then:

$$\sum_{i=1}^{N} P(i) = 1 \qquad (2)$$

Thus C can be calculated as:

$$C = \left(\sum_{i=1}^{N} \frac{1}{i^a} \right)^{-1} \qquad (3)$$

Zipf's law described above is also called the high frequency law, because it is valid for the high frequency objects but invalid for the low frequency objects. Zipf's 2nd law is suitable for the description of the low frequency objects access distribution. We borrow Zipf's 2nd law to describe the low frequency Web objects' popularity, which can be modeled as follows:

$$\frac{I_m}{I_1} = \frac{2}{m(m+1)} \qquad (4)$$

where m is the critical value for Web low frequency district pages, I_1 is the total number of Web pages of which popularity is 1, while I_m for the popularity of m.

From the Zipf's 2nd law, we can estimate that objects occurred just once in the total distinct objects is about 50-60%, objects occurred just twice in the total distinct objects is about 17%, and for the three times objects, about 8.3%, etc. Studies of Web server and Web proxy workloads [6] have shown that usually, the percentage of distinct documents of total number of requests is between 25% and 40%, many documents requested from a server are rarely reused, about 15% to 40% of the unique files accessed from a Web server are accessed only once and for Web proxy access logs, the one-timers can account for more than 50% of the documents. This implies that the low frequency Web objects account for a large percentage of the request documents. The facts conform to the Zipf's 1st law and Zipf's 2nd law approximately. Removing the low frequency Web objects can reduce model size to some degree.

3 Prediction Model

Prediction by Partial Match (PPM) belongs to the context models [7]. The algorithms employing the model in the compression community tend to achieve superior performance. PPM used in Web prefetching describes user's surfing patterns in a dynamically maintained Markov prediction tree. The model that uses *m* preceding Web pages to determine the probability of the next one is called order-m PPM model. An order-m PPM model maintains the Markov prediction tree with height m+1 which corresponds to context of length 0 to m. Each node in a tree represents the access sequence of Web pages that can be found by traversing the tree from the root to that node.

3.1 Existing PPM Models

There are two representative PPM prediction models in Web prefetching. The first PPM prediction model is standard PPM model. Figure 1a shows the prediction tree structure of the standard PPM model for three access sequences of {CDABA}, {ABA}, {ACD}. It uses arbitrary URL for a root node, records every subsequent URL and the number of times the URL occurs in the path from the root node in the tree rooted by the first URL. For example, the notation B/2 indicates that URL B was accessed twice. The advantage of this model is that it is not very complex and is easy to be implemented. However, this model takes up too much space because it records every accessed URL.

The other PPM prediction model is LRS (Longest Repeating Sequences) PPM model, which keeps the longest repeating subsequences and stores only long branches with frequently accessed URL. The method for building LRS PPM model is: building a standard PPM model, then scanning each branch to eliminate non repeating sequences. Figure 1b shows the prediction tree structure of the LRS PPM model for three access sequences of {CDABA}, {ABA}, {ACD}. Relative to the standard PPM model, the LRS PPM model offers a lower storage requirement and higher prediction precision. However, because the prediction tree keeps only a number of frequently accessed branches, so overall prefetching hit rate can be low, further more, there are still many Web pages leading to inaccurate prediction in the LRS PPM model. As a result, taking too much storage space for PPM models is still the key problem.

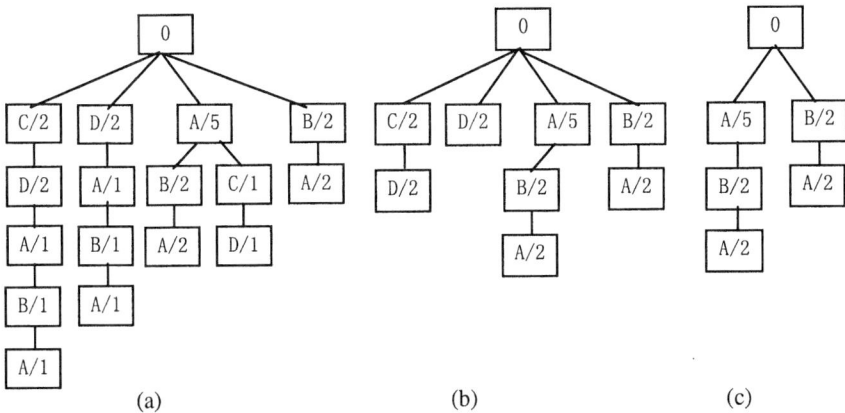

Fig. 1. Three Models (a) Standard PPM Model (b) LRS PPM Model (c) BWOP PPM Model.

3.2 PPM Model Based on Web Objects' Popularity

Our PPM prediction model based on Web objects' popularity consists of two parts: model construction and model prediction.

(1) Model construction

Our model (BWOP PPM Model) is built by the use of Web server log file LF. Firstly initialize variables; Secondly based on Zipf's 2nd law, select the pages of which popularity is bigger than a threshold value θ into Selected_URL; Thirdly transform Selected_URL into a collection of user sessions Session_Set; Fourthly add each page in the Session_Set into model tree T; Finally return model tree T.

BWOP PPM model can be updated as a new page request comes in. In order to facilitate updating the model an additional data structure cur_web[0...m] keeps track of the current context of length j, for $0 \leq j \leq m$. A length-j current context embodies the j last pages of the access sequence. Then, a new page request is added to the model in the following fashion:

①For each current context of length j, check whether any of the child nodes represents the new page request.

②If such a node exists (i.e., the same sequence has been seen before) then set the current context of length j+1 to this node, and increase the number of occurrences of this context. Otherwise, create a new child node, and proceed with the same operations above.

The algorithm for constructing the BWOP model is described as follows:

```
Algorithm ModelConstruction (Train LF, Pop-threshold θ)
Begin
Step1: Tree T=:NULL;
Step2: Read log file LF;
       Count popularity of each URL, Select the pages
       of which popularity is bigger than a threshold
       value θ into Selected_URL;
Step3: Process Selected_URL into Session_Set;
Step4:
For each session S in Session_Set
Begin
   Cur_web[0]:=root node of T;
   Cur_web[1...m]:=NULL;
   For each page in S
     For length j=m down to 0
       If cur_web[j] has child-node C representing a
       new page request R
     Begin
       C.Count++ ;
       Cur_web[j+1]:=node C;
     End
     Else
     Begin
       Create child node C representing a new page re-
       quest R;
       C.Count=:1;
       Cur_web[j+1]:=node C;
     End
End
Step5: Return T
End
```

From three access sequences of {CDABA}, {ABA}, {ACD}, we can easily obtain that the popularity of the page A is the highest, the popularity for the page B is higher and lowest for the page C and D, though the absolute access times for the page C and D is equal to ones for the page B. Figure 1c shows the prediction tree structure of the model for three access sequences of {CDABA}, {ABA}, {ACD}.

(2) Model prediction

Based on the Zipf's 1st law used to depict the high frequency Web objects' popularity, we propose a hybrid prefetch algorithm that uses combination of probability and popularity thresholds by "OR" them. The algorithm for constructing PT is described as follows, where PT is the popularity table used to track the popularity of each requested page before prediction.

```
Algorithm BuildPopularityTable (Selected_URL)
Begin
Step1:
For each URL in Selected_URL
   If URL exists in PT
     Begin
       Increase its popularity;
       Age its popularity according to time difference
       of last access time and this access time;
     End
   Else
     Insert a new URL in PT and initialize its popular-
     ity to 1;
Step2:   Return PT
End
```

By the use of prediction model T and popularity table PT, we can predict user's future page request using the last k requests ($0 \leq k \leq m$) in the current access sequence under the control of popularity threshold POP_TH and probability threshold PRO_TH. Model construction and model prediction are not actually two separate phase because popularity table PT and cur_web are updated as new page requests arrive. The prediction algorithm is as follows.

```
Algorithm HybridPrediction(T, PT, PRO_TH, POP_TH)
Begin
Step1:   Initialize a set P of predicted pages P=:NULL;
Step2:
For length j=k down to 1
 For each child-node C of Cur_web[j]
    Begin
      Fetch the popularity of node C from PT;
      If C.Count/parent. Count>PRO_TH or popularity of
      C>POP_TH
      P:=P+C;
    End
Step3: Remove duplicate pages from P;
Step4: Return P
End
```

4 Experiment Results

We make the trace-driven simulation by the use of two log files. First data comes from NASA [9] that records HTTP requests of two months. Another is from UCB-CS [10] that contains HTTP requests of one week. For all log datum, the experiments take 4/5 of the log as training set and the remaining 1/5 as prediction set. In each test, order-4 PPM models are employed; popularity threshold POP_TH is set to 3.

For the evaluation of our algorithm, the following performance metrics are defined:

(1) Prediction Precision

Precision=$P^+/(P^+ + P^-)$

If users access the predicted pages in a subsequent prediction window, this prediction is considered to be correct, otherwise it is incorrect. In the above equation, P^+ denotes the number of correct predictions and P^- for the number of incorrect predictions.

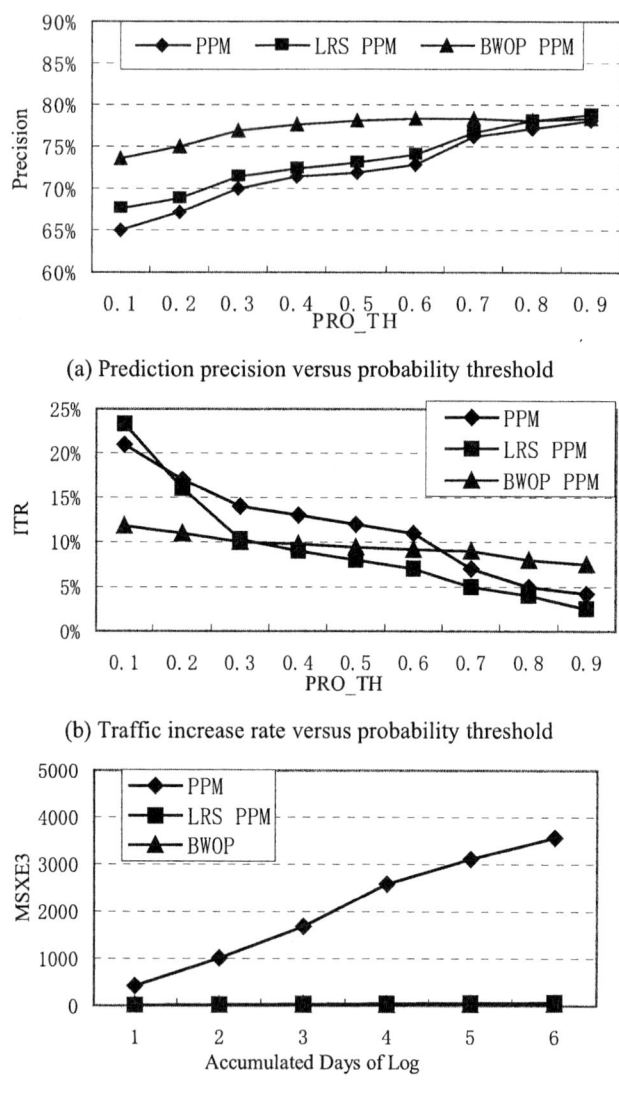

(a) Prediction precision versus probability threshold

(b) Traffic increase rate versus probability threshold

(c) Model size versus log of accumulated days

Fig. 2. Performance comparison among the three PPM models using NASA log

(2) ITR (Traffic Increase Rate)
ITR=(TP-TSP)/TT
where TP stands for the traffic resulting from prefetching, TSP for the traffic resulting from correct prefetching, TT for the traffic required without prefetching.
(3) Model Size (MS)
MS refers to the number of nodes in the model.

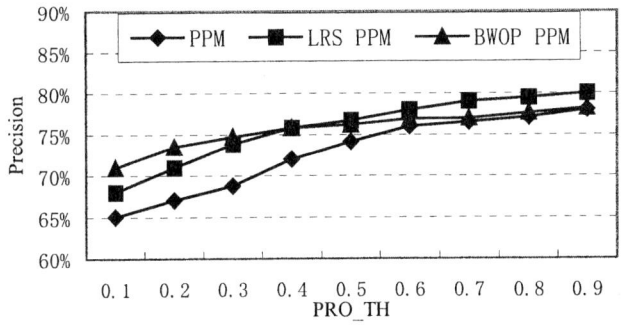

(a) Prediction precision versus probability threshold

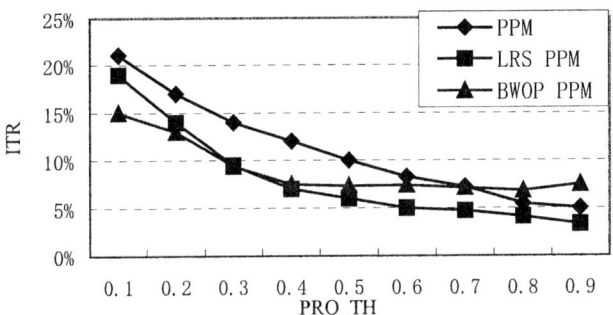

(b) Traffic increase rate versus probability threshold

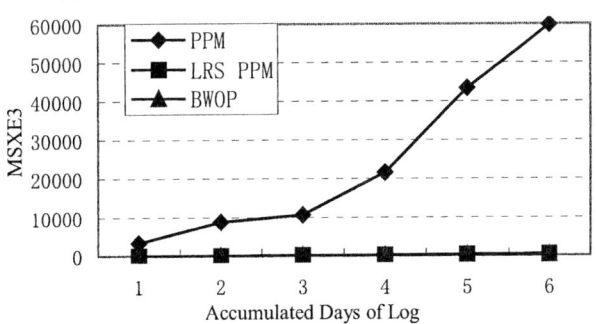

(c) Model size versus log of accumulated days

Fig. 3. Performance comparison among the three PPM models using UCB-CS log

Figure 2 compares the performance metrics among the three PPM models using the NASA [9] log file. Figure 2a shows that the prediction precisions for the three PPM models increase when the probability threshold varies from 0.1 to 0.9. We can find that the prediction precision of BWOP model can achieve a relative good performance and varies more smoothly than the other two models as probability threshold changes. There are several reasons to explain what Figure 2a presents. First, the most unpopular pages leading to incorrect prediction are removed based on Zipf's 2nd law; Second, the principle of each order matching is employed when making prediction; Third, popularity information is considered when prefetching the users' future page requests, accordingly popular pages without passing probability threshold can be prefetched. Figure 2b indicates that traffic increase rates for the three PPM models decreases as the probability threshold varies from 0.1 to 0.9. The traffic increase rate in BWOP model is not the lowest at the probability threshold of 0.4 to 0.6, even highest for the probability threshold of 0.7 to 0.9, however, taking one with another, BWOP model achieves relative low traffic increase rate while keeping the traffic increase rate augment more smoothly with probability threshold varying. Figure 2c displays that the number of nodes stored by each of the three PPM models increase as the number of log accumulated days varies from 1 to 6. For the standard PPM model, the number of nodes dramatically increase with the number of log accumulated days increasing and for LRS PPM model and BWOP PPM model more slowly and keep smaller size, the size of BWOP model is slightly smaller than that of LRS PPM model due to the space optimizations of combination of popularity information with PPM prediction model.

Figure 3 compares the performance metrics among the three PPM models using the UCB-CS [10] log data. Figure 3a shows that the prediction precisions for the three PPM models increase when the probability threshold varies from 0.1 to 0.9. Although the prediction precision of BWOP model is not the highest when probability threshold is bigger than 0.5, it varies more smoothly than the other two models as probability threshold changes. What figure 3b and figure 2b show are similar to that of figure 3c and figure 2c respectively. Figure 3 and figure 2 indicate that the size of our model is relative small, and it achieves the prediction precision of 76%-78% at the cost of traffic increase of 10% when the probability threshold is 0.4.

5 Conclusions

Many prefetching models have been discussed in recent years. How to set up an effective model with high access hit ratio and low cost and complexity is still the goal of the research.

The work of this paper tries to present a simple but effective way of Web prefetching. Zipf's 1st law and Zipf's 2nd law are employed to model the Web objects' popularity and a popularity-based PPM model is presented. The experiment results show that the model has a better tradeoff between prediction precision and model size and traffic increase rate at the probability threshold of 0.3 to 0.4.

Acknowledgements

This work was partially supported by the National Natural Science Foundation of China (Grant No.50207005), the National Fund for Studying Abroad (Grant No.21307D05) and the Research Foundation of Beijing Institute of Technology (Grant No. 0301F18). We would like to thank Dr. Yun Shi of China State Post Bureau and Professor Jun Zou of Tsinghua University for their helpful and constructive comments.

References

1. Lei Shi, Zhimin Gu, Lin Wei, Yun Shi. Popularity-based Selective Markov Model. IEEE/WIC/ACM International Conference on Web Intelligence, Beijing (2004) 504-507.
2. M.K.Thomas, DEL. Darrel, C.M. Jeffrey. Exploring the bounds of Web latency reduction from caching and prefetching. Proceedings of the USENIX Symposium on Internet Technologies and Systems. California: USENIX Association (1997) 13-22.
3. M. Crovella, P. Barford. The network effects of prefetching. Proceedings of the IEEE Conference on Computer and Communications. San Francisco (1998) 1232-1240.
4. T. Palpanas and A. Mendelzon.Web prefetching using partial match prediction. Proceedings of Web Caching Workshop. San Diego, California, March (1999).
5. J. Pitkow and P. Pirolli, Mining Longest Repeating Subsequences to Predict World Wide Web Surfing. Proc. Usenix Technical Conf., Usenix (1999) 139-150.
6. M. Busari, C. Williamson. On the sensitivity of Web proxy cache performance to workload characteristics. IEEE INFOCOM (2001) 1225-1234.
7. J. G. Cleary and I. H.Witten, Data compression using adaptive coding and partial string matching, IEEE Transactions on Communications, Vol. 32, No. 4 (1984) 396-402.
8. L. Breslau, P. Cao, L. Fan, G. Phillips, S. Shenker. Web caching and Zipf-like distributions: evidence and implications. IEEE INFOCOM (1999) 126-134.
9. Lawrence Berkeley National Laboratory, URL: http://ita.ee.lbl.gov/
10. Computer Science Department, University of California, Berkeley, URL: http://www.cs.berkeley.edu/logs/.
11. J.I. Khan and Q. Tao, Partial Prefetch for Faster Surfing in Composite Hypermedia, Proc. Usenix Symp. Internet Technologies and Systems, Usenix (2001) 13-24.
12. A. Mahanti. Web Proxy Workload Characterization and Modeling, M.Sc. Thesis, Department of Computer Science, University of Saskatchewan, September (1999).
13. M. Busari, C. Williamson. On the sensitivity of Web proxy cache performance to workload characteristics. IEEE INFOCOM (2001) 1225-1234.

An On-line Sketch Recognition Algorithm for Composite Shape

Zhan Ding[1], Yin Zhang[1], Wei Peng[1], Xiuzi Ye[1,2], and Huaqiang Hu[1]

[1] College of Computer Science/State Key Lab of CAD&CG,
Zhejiang University, Hangzhou 310027, P.R. China
dingzh@hotmail.com
{weip, xiuzi}@zju.edu.cn
[2] SolidWorks Corporation, 300 Baker Avenue, Concord, MA 01742, USA

Abstract. Existing sketch recognition algorithms are mainly on recognizing single segments or simple geometric objects (such as rectangles) in a stroke. We present in this paper an on-line sketch recognition algorithm for composite shapes. It can recognize single shape segments such as straight line, polygon, circle, circular arc, ellipse, elliptical arc, hyperbola, and parabola curves in a stroke, as well as any composition of these segments in a stroke. Our algorithm first segments the stroke into multi-segments based on a key point detection algorithm. Then we use "combination" fitting method to fit segments in sequence iteratively. The algorithm is already incorporated into a hand sketching based modeling prototype, and experiments show that our algorithm is efficient and well suited for real time on-line applications.

1 Introduction

The de facto method for drawing graphic objects using computer is to use mouse/keyboard with the help of toolbar buttons or menu items. However, this is not the most natural and convenient way for human beings. In order to adapt such systems to users, pen/tablet devices are invented as an important extension of mouse/keyboard for input. Now, they are mainly used for handwriting character input or for replacement of the mouse during directly drawing regular shape graphic objects. The most convenient and natural way for human beings to draw graphics should be to use a pen to draw sketches, just like drawing on a real sheet of paper. This interactive way is also called calligraphic interfaces. Moreover, it is even better to recognize and convert the sketchy curves drawn by the user to their rigid and regular shapes immediately. In this paper, we refer to the approach and process of immediately converting the input sketchy composite curve (contains multi-shapes) in a stroke to a serial rigid and regular geometry shapes as on-line composite shape recognition.

On-line handwriting recognition is very common to many users and its prevalence is increasing. However, very few research works have been done on on-line sketchy composite shape recognition. Ajay et al.[1] have proposed an algorithm of recognizing simple shapes based on filter, but the algorithm is sensitive to orientation and their precondition is somewhat too strict. Fonseca et al. [5,6] have extended Ajay's work by providing more filters and using fuzzy logic. However, because of being based on global area and perimeter calculation, these filters can hardly distinguish ambiguous shapes such as pentagon and hexagon. Sezgin[9] and Shpitalni[10] gives a method of detecting corner point in sketch and a method of curve approximation. Their work is valuable for sketch recognition. Liu et al. [7] have proposed a recognition system Smart Sketchpad. They have supplied three recognition methods: rule based, SVM based and ANN based. And they showed good performance in recognition of single shape and simple objects. Qin et al. [8] proposed a sketch recognition algorithm based on fuzzy logic in their 2D and 3D modeling system. Their algorithm also shows good performance in single shape recognition, however it cannot recognize any sketchy composite shape. Arvo et al. [2] also provide an on-line graphics recognition algorithm. Their approach continuously morphs the sketchy curve to the guessed shape while the user is drawing the curve. However, their main purpose focuses on the user studies of such sketch recognition system. Moreover, their recognition approach only handles two simplest classes of shapes (circles and rectangles) drawn in single strokes. Chetverikov et al. [3] have presented an algorithm:IPAN99 to detect corner point. They have compared IPAN99 with other four corner point detecting algorithms, and it shows the best performance. This algorithm is simple, effective and accurate to find sketch corner-points. Here, we refined IPAN99 as our key points detecting algorithm. Fitzgibbon et al.[4] have compared several kinds of conic approximation methods. Their work gives us some good suggestions in conic fitting.

In this paper, we proposed an on-line sketch recognition algorithm for composite shape. It cannot only recognize the basic sketch shape in a stroke, including line, polygon (open or closed), ellipse, elliptical arc, circle, circular arc, hyperbola, parabola, but also can recognize the sketchy composite shape in a stroke.

2 Architecture of Algorithm

Figure 1 shows the architecture of our on-line sketch recognition algorithm for composite shape.

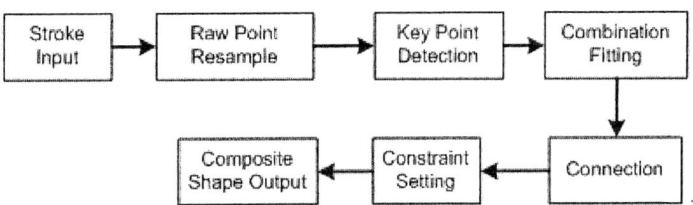

Fig. 1. Architecture of the sketchy composite shape recognition

The input is a stroke drawn by a user. A stroke is a trajectory of the pen movement on a tablet between the time when the pen-tip begins to touch the tablet and the time when the pen-tip is lifted up from the tablet. It is represented in a chain of points. The chain of points is then resampled.

The density of raw sampling points is relative to the sketching speed. When the user sketches slowly or rapidly, the distance between neighbor points can be small or large. For correct and effective recognition, we need to resample raw points. The goal of resampling is to delete redundancy points for reducing calculation when raw points' density is high, and to add more points for reducing recognition error when the density is low.

After the raw points are re-sampled, key points are recognized or added for recognizing polygon and other composite shapes. A key point is defined as a point which contains the geometric feature of the sketch, such as high curvature point, tangency point, corner point and inflexion point. Key points are likely to be the segmenting points which separate the composite sketch into multiple simple segments. Here we take the key points detection algorithm IPAN99 from Chetverikov[3].

After finding the key points, we can construct new sub-chains of points between any two key points. For each such sub-chain, we fit the points in the sub-chain by a straight line or a conic segment (a circle, a circular arc, an ellipse, an elliptical arc, a parabola, and a hyperbola). Since not all key points are the actual segmenting points of composite shapes, e.g., we may get 6 key points for a circle, and get a hexagon instead of a circle, which is really wanted, we must distinguish the actual segmenting points between "false" key points. We combine conic segments if possible. If the fitting error exceeds a predefined tolerance, we fit the sub-chain with a B-spline, and recursively adding knots to the B-spline to achieve the given tolerance.

Since a composite shape in a stroke is recognized in a piecewise fashion, the resultant segments may be disconnected and the users' intended constraints may not be maintained. We will modify the end points of each segment to ensure the segments are connected for sketch in one stroke. We then estimate geometric relationships such as parallel and, perpendicular relationships between neighboring segments within given tolerances, and add the applicable constraints to the segments. We can now output the recognized composite shapes from the sketch in a stroke.

3 Key Algorithms

In the section, we discuss several key algorithms during the recognition process.

3.1 Algorithm for Detecting Key Points

This section deals with detection of high curvature points from a chain of points. It is well known that human beings are very sensitive to high curvature extreme points. Locations of significant changes in curve slope are, in that respect, similar to intensity edges. If these characteristic contour points are identified properly, a shape can be represented in an efficient and compact way with accuracy sufficient in many shape analysis problems.

Here we take a fast and efficient algorithm of detection of key points based on Chetverikov[3]. This algorithm should scan the whole resample chain points two pass, so it is also called two-pass scanning.

Two-pass scanning algorithm defines a corner in a simple and intuitively appealing way, as a location where a triangle of specified size and opening angle could be inscribed in a curve. A curve is represented by a sequence of points in the image plane. The ordered points are densely sampled along the curve. A chain-coded curve can also be handled if converted to a sequence of grid points. In the first pass scanning, the algorithm scans the sequence and selects candidate key points. The second pass scanning is post-processing to remove superfluous candidates.

First pass. In each curve point P the detector tries to inscribe in the curve a variable triangle (P^-, P, P^+) (P^- is the pre point of P, P^+ is the successor point of P) the constrained by a set of simple rules:

$$\begin{cases} d_{min}^2 \leq |p-p^+|^2 \leq d_{max}^2 \\ d_{min}^2 \leq |p-p^-|^2 \leq d_{max}^2 \\ \alpha \leq \alpha_{max} \end{cases} \quad (1)$$

where $|P-P^+|=|a|=a$ is the distance between P^+ and P, $|P-P^-|=|b|=b$ the distance between P^- and P, and $\alpha \in [0, \pi]$ the opening angle of the triangle of (P^-, P, P^+) (see figure 2-a). The latter is computed as $\alpha = \arccos(\frac{a^2+b^2-c^2}{2ab})$.

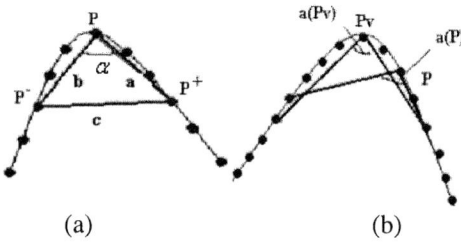

(a) (b)

Fig. 2. Detecting key points in two pass scanning. (a)First pass scanning: determining if P is a candidate point. (b)Second pass scanning: testing P for sharpness non-maxima suppression and remove superfluous candidate point.

Variations of the triangle that satisfy the conditions (1) are called admissible. Search for the admissible variations starts from P outwards and stops if any of the conditions (1) is violated. (That is, a limited number of neighboring points are only considered.) Among the admissible variations, the least opening angle $\alpha_{(P)}$ is selected. If no admissible triangle can be inscribed, P is rejected and no sharpness is assigned.

Second pass. The sharpness based non-maxima suppression procedure is illustrated in figure 2b. A key point detector can respond to the same corner in a few consecutive points. Similarly to edge detection, a post-processing step is needed to select the strongest response by discarding the non-maxima points.

A candidate point P is discarded if it has a sharper valid neighbor P_v: $\alpha_{(P)} > \alpha_{(P_v)}$. In the current implementation, a candidate point P_v is a valid neighbor of P if $|P-P_v|^2 \leq d_{max}^2$. As alternative definitions, one can use $|P-P_v|^2 \leq d_{min}^2$ or the points adjacent to P.

Parameters d_{min}, d_{max} and α_{max} are the controlling parameters of the two pass scanning. d_{min} sets the scale (resolution), with small values responding to fine corners. The upper limit d_{max} is necessary to avoid false sharp triangles formed by distant points in highly varying curves. α_{max} is the angle limit that determines the minimum sharpness accepted as high curvature. In IPAN99, they set $d_{max}=5$, $d_{min}=3$ and $\alpha_{max}=150°$. But in this paper, we should adjust these parameters to adapt for the shape size. So we cannot take the d_{max} and d_{min} as a const value. And we take two relative parameters into account: $k_{d\,min}$ and $k_{d\,max}$. We substitute d_{min}, d_{max} with $k_{d\,min}l$ and $k_{d\,max}l$, l is the sum of the resample chain of points' distance. Meanwhile, for limiting the power of l, we take another two parameters into the algorithm: D_{min} and D_{max}, so the condition(1) has been changed with:

$$\begin{cases} \max^2(k_{d\,min}l, D_{min}) \le |p-p^+|^2 \le \min^2(k_{d\,max}l, D_{max}) \\ \max^2(k_{d\,min}l, D_{min}) \le |p-p^-|^2 \le \min^2(k_{d\,max}l, D_{max}) \\ \alpha \le \alpha_{max} \end{cases} \quad (2)$$

Figure 3 shows three key point detection examples: the thick points are the key points recognized using the two-pass scanning algorithm, and the thin points are the raw re-sampled points.

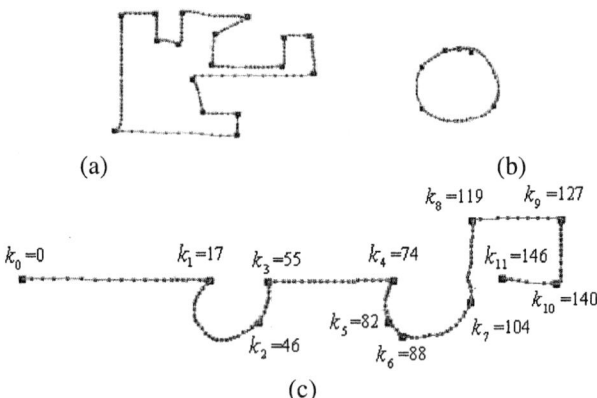

Fig. 3. Samples using two pass scanning. Notice: Both start point and end point are default treated as key points.

3.2 Algorithm for Conic Approximation

For points in each sub-chain between any two key points, we approximate them first with a straight line or a conic segment. Any conic that does not pass origin point can be expressed as follows:

$$f(X,Y) = a_0Y^2 + a_1XY + a_2X^2 + a_3X + a_4Y = 1 \qquad (3)$$

And for the conic which passes the origin, we can translate it away the origin, so formula (3) is still valid. That is to say, we first move the raw points $L\{(x_0, y_0), (x_1, y_1), \ldots, (x_n, y_n)\}$ away from the origin point, then fit the curve as in formula (3) to get the coefficient vector: $a(a_0, a_1, a_2, a_3, a_4)$, and then move the conic segment back to its original place to get the result.

We use the Least Squares' Fitting method to get our coefficient vector:

$$\varphi(a_0, a_1, a_2, a_3, a_4) = \sum_{i=1}^{n}(a_0Y_i^2 + a_1X_iY_i + a_2X_i^2 + a_3X_i + a_4Y_i - 1)^2 \qquad (4)$$

by letting φ to be minimum. The minimization problem can be converted to the following linear equation system:

$$\begin{bmatrix} \sum y_i^4 & \sum x_i y_i^3 & \sum x_i^2 y_i^2 & \sum x_i y_i^2 & \sum y_i^3 \\ \sum x_i y_i^3 & \sum x_i^2 y_i^2 & \sum x_i^3 y_i & \sum x_i^2 y_i & \sum x_i y_i^2 \\ \sum x_i^2 y_i^2 & \sum x_i^3 y_i & \sum x_i^4 & \sum x_i^3 & \sum x_i^2 y_i \\ \sum x_i y_i^2 & \sum x_i^2 y_i & \sum x_i^3 & \sum x_i^2 & \sum x_i y_i \\ \sum y_i^3 & \sum x_i y_i^2 & \sum x_i^2 y_i & \sum x_i y_i & \sum y_i^2 \end{bmatrix} \begin{bmatrix} a_0 \\ a_1 \\ a_2 \\ a_3 \\ a_4 \end{bmatrix} = \begin{bmatrix} \sum y_i^2 \\ \sum x_i y_i \\ \sum x_i^2 \\ \sum x_i \\ \sum y_i \end{bmatrix} \qquad (5)$$

Solve the linear equation system (5), we will obtain the coefficient vector $a(a_0, a_1, a_2, a_3, a_4)$. Then we can use translations and rotations to convert (3) to the following (6):

$$\frac{X^2}{A} + \frac{Y^2}{B} = 1 \qquad (6)$$

Formula (6) can only represent circle (A=B, A>0), ellipse (A>0, B>0) and hyperbola (A*B<0). By moving and rotating, we can get the conic center $C(C_x, C_y)$ and the rotate angle θ as follows:

$$\begin{cases} C_x = \dfrac{a_1 a_4 - 2a_0 a_3}{(4a_0 a_2 - a_1^2)^2}, C_y = \dfrac{a_1 a_3 - 2a_2 a_4}{(4a_0 a_2 - a_1^2)^2} \\ \theta = \dfrac{1}{2} arctg(\dfrac{a_1}{a_2 - a_0}) \end{cases} \qquad (7)$$

The standard formula (6) can not represent parabola and line. We solve this problem according to relation of parameter A, B, the center C and angle θ.

The flow of fitting conic curve lists as follows:

(1) Move the re-sampled chain of points $L\{(x_0, y_0), (x_1, y_1), \ldots, (x_n, y_n)\}$ a distance d along the x-axis.

(2) Least Squares' Fitting of the points, and get the conic in forms of (3) and then transform it to the form in (6), meanwhile, compute the center $C(C_x, C_y)$ and the rotating angle θ. Obtain $k=\sqrt{|\frac{A}{B}|}$, k is the ratio of long axis to short axis in ellipse.

(3) If $k > 50$ or $k < 0.02$, the curve is recognized as a straight line, we get the parameter of the line using the least square method and terminate the recognition algorithm. Otherwise, go to step (4).

(4) else if $|C_x|$ or $|C_y|$ is larger than a predefined (big) number C_{max}, we make an educated guess on $I = \begin{vmatrix} a_0 & \frac{1}{2}a_1 \\ \frac{1}{2}a_1 & a_2 \end{vmatrix}$ as equals zero, and the conic is a parabola. We compute the parabola parameters from the coefficient vector $a(a_0, a_1, a_2, a_3, a_4)$ obtained, and translate the parabola back to the original point position.

(5) else if A>0, B>0 and $|k-1| < k_c$ ($k_c = 0.2$ in our implementation), meaning that the major axis radius and minor axis radius of the ellipse are very close, and the conic is considered as a circle (or circular arc). The radius of the circle is the average of two radii. If A>0, B>0 and $|k-1| > k_c$, the conic is considered as an ellipse (or an elliptical arc). The circle or ellipse or circular/elliptical arc is then transformed back to the original point position. We need to determine whether the conic is closed or not. We calculate the sum of the triangle angles of the adjacency raw points and the center point (positive if clockwise and negative if count-clockwise). If the absolute value of the sum is larger than 2π or near 2π, then we consider the conic is closed.

(6) If A*B<0, then the conic is considered as a hyperbola. We compute the parameters from the coefficient vector $a(a_0, a_1, a_2, a_3, a_4)$ obtained, and translate the hyperbola back to the original point position.

3.3 Algorithm for Combination Fitting

As mentioned above, not all key points will be the real segmenting points of composite shapes. For example, in Figure 3.c, we sketched a composite shape in a stroke, then after detection of key points, we got twelve key points. In fact, key points k_2, k_5 and k_6 are not real segmenting points of the composite shape. If we directly fit resample points which located between k_1, k_2 and k_2, k_3, we got an elliptical arc and a line. In fact we want to fit resample points between k_1 and k_3, and hope to get only one elliptical arc. The same thing will happen in recognizing curves among k_4, k_5, k_6, k_7. So we should distinguish the real segmenting points from key points. Here we take a combination fit method.

First, we define a fit error:

$$s = \sum_{i=1}^{n} \sqrt{d_i^2} \Big/ (L*n) \tag{8}$$

where d_i is the shortest distance between the raw points and the approximation curve. L is the length of the re-sampled curve, and n is the num of re-sampled points.

"Combination fitting" is a method that regroups sub-chains of points from re-sampled chains of points based on key points. The destination of combination fitting is to delete "false" key points and sub-divide the raw curve into several parts. Each part is recognized as a single conic or poly-line. The rule of regrouping is to take r serial key points from key points array and make all points between the r key points into one sub-chain of points. Here our combination method is implemented based on a window (A, W), where A stands for the index of key points array, and W stands for the window's size. A and W are initialized to be 1 and the length of key points array, respectively. Given a re-sampled chain of points R[1~M], where M is the length of the chain. Each element in R is a position in 2D space. Given a key points array K[1~N], where N is the number of the key points. Each element in K is an index of R. We call the following procedure Recognition(L,M,K,N,A,W), and save the recognition result in the result array RecognitionResult. The algorithm of combination fitting lists as follows.

```
Procedure Recognition( PointArray * resamplepoints, // In: resample points array
int pointslength,              // In: length of the points array
KeyPointArray * keypoints,     // In: key points array
int keypointslength,           // In: length of key points array
int a, int w)                  // In: window's position and size
{
//When window's size equals to zero or window moves to the end, exit.
if(w= =0 || a= =N)   return;
//Using the algorithm of conic fit mentioned in section 3.2, fit resample points which locate between key points a and a+w.
shape=FitResamplePoints(resamplepoints[keypoints[a]~ keypoints[a+w-1] ] );
//Using formula (8), calculate the fitting error s
s=CalculateError(resamplepoints[keypoints[a]~ keypoints[a+w-1] ], shape );

if(s<D) {  // D is a given tolerance
   //Combination Fitting succeeded, save the shape to result array
   RecognitionResult.Add(shape);
   //Split the resample chain of points into two sub-chains of points: FRP and SRP, and the length of these two arrays are FRL and SRL.
   FRL=keypoints[a]-1;
   SRL=pointslength-keypoints[a+w-1];
   FRP=resamplepoints[keypoints[1] ~ keypoints[a]];
   SRP=resamplepoints[keypoints[a+w-1] ~ pointslength];
   //Split key points array into two sub key points array: FKP and SKP, the length of array is FKL and SKL.
   FKL=a-1;
   SKL=keypointslength-(a+w-1);
```

```
    FKP=keypoints[1~a];
    SKP=keypoints[(a+w-1) ~ keypointslength];
    if(FRP is not NULL)
         Recognition (FRP,FRL,FKP,FKL,1,FKL);
    if(SRP is not NULL)
         Recognition (SRP,SRL,SKP, SKL,1,SKL);
  }
  else{
       //Fitting failed. Adjust the window's size or position
       if(a+w-1<keypointslength)
           a++;      //move window's position
       else
           w--;         //reduce the window's size
       Recognition( resamplepoints, pointslength, keypoints, keypointslength,a,w);
   }
}
```

4 Experiments and Evaluation

Figure 4 shows some examples for our proposed composite shape recognition algorithm. We use a computer with CPU PIII 800MHZ, RAM 256M.

(a)Hand sketching in a stroke (16 key points). (b) Recognition result of a: Four elliptical arcs (recognition time: 0.332 sec)
(c) Hand sketching in a stroke (25 key points). (d) Recognition result of c: One circular arc, two ellipses and eleven lines (0.554 sec)
(e) Hand sketching in a stroke (51 key points)   (f) Recognition result of e: One ellipse, five elliptical arcs and twenty- five lines (0.742 sec)

Fig. 4. Examples of hand sketching recognitions for composite shapes. Notice: The recognition result drawn by green color has not been processed using constraints setting. It means that no other constraint relationship has been fulfilled.

5 Concluding Remarks

An on-line sketch recognition algorithm for composite shape is presented. It can not only recognize line, polygon, ellipse (elliptical arc), circle (circular arc), parabola, hyperbola in a stroke, but can also recognize complex shapes consisting of these basic shapes. This algorithm has been used in a 3D sketch based concept modeling prototype system. The system provides users a natural, convenient, and efficient way to input geometric shapes by hand sketching.

Acknowledgments

The authors would like to thank the support from the China NSF under grant #60273060, #60473106 and #60333010, China Ministry of Education under grant# 20030335064, China Ministry of Science and Technology under grant# 2003AA4Z3120.

References

1. Ajay A., Van V., Takayuki D. K.: Recognizing Multistroke Geometric Shapes: An Experimental Evaluation. In Proc. of the 6th Annual ACM Symposium on User Interface Software and Technology, 1993: 121-128
2. Arvo J. and Novins K.: Fluid Sketches: Continuous Recognition and Morphing of Simple Hand-Drawn Shapes. In Proc. of the 13th Annual ACM Symposium on User Interface Software and Technology, San Diego, California, November, 2000.
3. Chetverikov D, Szabo Z.: A Simple and Efficient Algorithm for Detection of High Curvature Points in Planar Curves. In Proc of 23rd Workshop of the Austrian Pattern Recognition Group, 1999: 175-184
4. Fitzgibbon A W, Fisher R B.: A buyer's guide to conic fitting. In Proc of British Machine Vision Conference, Birmingam, 1995.
5. Fonseca M J, Jorge J.A.: Experimental evaluation of an on-line scribble recognizer. Pattern Recognition Letters, 22(12), 2001: 1311-1319
6. Fonseca M J, Jorge J A.: Using Fuzzy Logic to Recognize Geometric Shapes Interactively. In Proc. of 9th IEEE Conf. on Fuzzy Systems, Vol. 1, 2000: 291-296
7. Liu W Y, Qian W J, Rong X, Jin X,Y. Smart Sketchpad—An On-line Graphics Recognition System. In: Proceedings of Sixth International Conference on Document Analysis and Recognition (ICDAR2001), Seattle, September 2001: 1050-1054
8. Qin S F, Wright D K, Jordanov I N. From on-line sketching to 2D and 3D geometry: a system based on fuzzy knowledge. Computer-Aided Design. 2000, 32: 851-866
9. Sezgin M. Feature Point Detection and Curve Approximation for Early Processing of Free-Hand Sketches. Master's thesis, Massachusetts Institute of Technology, 2001
10. Shpitalni M, Lipson H. Classification of Sketch Strokes and Corner Detection using Conic Sections and Adaptive Clustering. Trans. of ASME J. of Mechanical Design. Vol. 19, No. 2, 1997: 131-135

Axial Representation of Character by Using Wavelet Transform

Xinge You[1,2], Bin Fang[2,3], Yuan Yan Tang[1,2], Luoqing Li[1], and Dan Zhang[1]

[1] Faculty of Mathematics and Computer Science, Hubei University, P.R. China
{xyou, bfang, yytang}@comp.hkbu.edu.hk
[2] Department of Computer Science, Hong Kong Baptist University
[3] Chongqing University

Abstract. Axial representation plays a significant role in character recognition. The strokes of a character may consist of two regions, i.e. singular and regular regions. Therefore, a method to extract the central axis of a character requires two different processes to compute the axis in theses two different regions. The major problem of most traditional algorithms is that the extracted central axis in the singular region may be distorted by artifacts and branches. To overcome this problem, the wavelet-based amendment processing technique is developed to link the primary axis, so that the central axis in the singular region can be produced. Combining with our previously developed method for computing the primary axis in the regular region, we develop a novel scheme of extracting the central axis of character based on the wavelet transform (WT). Experimental results show that the final axis obtained from the proposed scheme closely resembles the human perceptions. It is applicable to both binary image and gray-level image as well. The axis representation is robust against noise.

1 Introduction

Axis representation of character is especially suitable for describing character since they have natural and inherent axes [4]. From a practical point of view, the skeleton-based axis representation of characters by sets of thin curves rather than by a raster of pixels may consume less storage space and processing time, and be sufficient to describe their shapes for many applications. It has found that this representation is particularly effective in extracting relevant features of the character for optical character recognition [3]. At present, it covers a wide range of applications, such as handwriting recognition, signature verification, etc. Thus, we focus on the skeleton representation of character in this paper.

Generally, the axis of a character image is the locus of the midpoints or the symmetric axis of the character stroke [4]. Different local symmetry analysis techniques may generate different symmetric axes (hereafter we call it skeleton) [4,6,14,16].

To serve the purpose of extracting skeleton axis, the character stroke can be partitioned into regular and singular regions [12]. The singular region corresponds to ends, intersections and turns, and the regular region covers the other parts of the strokes. Thus, the extracting axis process of character based on the symmetry analysis contains two main steps: (1) Computing the primary skeleton in the regular region of the character strokes; (2) Amending the skeleton in the singular region of the character strokes.

To extract the primary axis in the regular region of the character stroke, the key technique is to construct a local symmetry of the contour of the regular region of a stroke, and to compute the symmetric center points, which produce the skeleton. In [14], we developed a new symmetric analysis based on the modulus maxima of the WT to compute successfully the primary skeleton in the regular of the character stroke.

Character strokes contain a variety of intersections and junctions, which belong to the singular region. A major problem with the traditional skeletonization methods is that their results do not always conform to the human perceptions, since they often contain unwanted artifacts such as noisy spurs and spurious short branch in the singular region of character strokes [1,2,4,8]. Although the skeleton is an attractive representation of character, its application to pattern recognition and image processing is limited if the artifacts cannot be removed effectively. To remove all sorts of artifacts from the primary skeleton in their singular region is still another challenging topic. This paper focuses on developing a simple and efficient wavelet-based amendment processing to extract the skeleton points in the singular region of the character strokes. The key of this technique is to position the skeleton points in the junction of the singular region in terms of three types of characteristic points. To measure three types of characteristic points, multiscale-based corner detection technique in [5] is also improved by using a new wavelet function.

The remainder of this paper is organized as follows: For the sake of completeness, Section 2 is dedicated to review a technique of extracting the primary skeletonin in our previous work, some more descriptions are given for clarifying the reasons why construction of new wavelet function is necessary. Amendment processing is detailed in Section 3. Experimental results are presented in Section 4. The conclusions are described in Section 5.

2 Review of Wavelet-Based Primary Skeleton Extraction

To extract the primary skeleton of the regular region of the character stroke, we developed a new symmetric analysis based on the modulus maxima of the WT in our previous works [14]. We also further clarified why a new wavelet function needs to be constructed there [14].

Symmetry analysis of the regular region depends on the detection of a pair of edge curves of the regular region of the stroke. Wavelet-based edge detection is to extract the local maxima of the WT moduli. The details can be found in [9,13].

In'fact, the position of the local maximum of WT modulus is not the exact edge point, a slight shift exists. As illustrated in Fig. 1, the positions of the modulus maxima slightly shift around the exact edge point with respect to scale $s = 1, 2$ and 3, respectively. Actually, the ideal points of the edge are optimized by the trade-off in both localization and bandwidth [9].

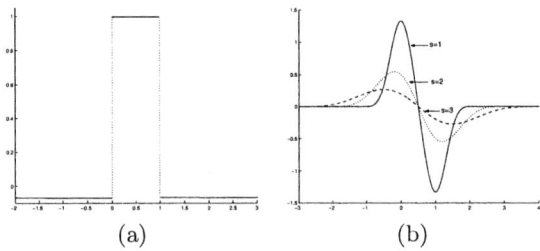

Fig. 1. The spline wavelet function is selected to perform WT on characteristic function. (a) The characteristic function $f(x)$; (b) The output of the WT $W_s f(x)$.

As far as the skeleton extraction is concerned, the desired wavelet not only locates the edge points by the local maxima of the WT moduli, but also adjust properly the position of local maxima around the exact edge points, in order to find easily the center of the regular region of a stroke by a symmetric pair of local maxima. Ideally, we wish to construct such a wavelet that a pair of contour curves extracted by the corresponding WT is located outside the original edges of the shape, and the middle axis (skeleton) of the shape is the symmetrical axis of these contour curves. In addition, the distance between these two contour curves depends strongly on the scale of the WT. Although many wavelet functions have been constructed, the construction of an appropriate wavelet function for the particular application is still a great challenge.

Theoretically, wavelet $\hat{\psi}(0,0) = 0$, which implies that $\psi(x,y)$ is a band-pass filter. It is easy to see that the partial derivatives of a low-pass function can become the candidates of the wavelet functions. The WT $W_s^1 f(x,y) = (f * \psi_s^1)(x,y) = (f * s\frac{\partial}{\partial x}\theta_s)(x,y) = s\frac{\partial}{\partial x}(f * \theta_s)(x,y)$ is the derivative of the smoothed function along the horizontal axis, and the local maxima of the derivative function occur at the positions, where the function has a sharp variation along the horizontal axis. Where $\theta(x,y)$ denotes a real function satisfying: 1. $\theta(x,y)$ fast decreases at infinity; 2. $\theta(x,y)$ is an even function on both x and y. 3. $\hat{\theta}(0,0) = 1$.

In fact, the purpose of smoothing $f(x,y)$ is to remove the noise, not the edge points. For this reason, $\theta(x,y)$ should be localized, such that $f * \theta_s(x,y) \sim f(x,y)$ when $s(s > 0)$ is small enough. This implies that the smoothed signal is almost the same as the original one, when the scale is very small. More details are described in [9].

Although the Gaussian function and quadratic spline would be the good candidates of $\theta(x,y)$, and they have been extensively applied to multi-scale edge detection, they still have some problems to treat the skeleton. Mokhtarian and Mackworth [10] pointed that the representation should satisfy the efficiency and ease for implementation such that it is useful for the practical tasks of the shape recognition in computer vision. Obviously, the Gaussian function does not satisfy such criteria since the computational burden is high at large scales and it is not compactly supported as well. Hence, it is not the best candidate of characterizing the singular point in practice.

The B-spline is a good approximation to the Guassian function [15]. It has been shown that the B-spline wavelet performs better than other wavelets for singularities detection [10,15]. Especially, it is suitable for characterizing Dirac-structure edge [13] and corner detection [5]. Unfortunately, it has been proved that the position of modulus maxima of the WT with respect to the Dirac-structure edge and corner detection is not width-invariant [5,13]. Thus, both the Gaussian and guadratic spline do not satisfy the such extra requirements for skeleton extraction [13,14].

Based on these reasons, the new wavelet, $\psi(x,y)$, which is constructed in our previous work [14], not only has the advantages of the Gaussian function and quadratic spline, but also possesses some desirable properties. These properties are helpful for extracting a pair of contours of a character stroke, further producing the primary skeleton. According to the characteristics of the modulus maxima of the WT with the new wavelet function, we define the wavelet-based local modulus maxima symmetry (WLMMS). For more details about the implementation of the WLMMS, one can refer to [14].

3 Amendment Processing of Singular Region of Stroke

Obviously, the primary skeleton usually seems to be rough with poor visual appearance since the skeleton loci are not continuous and smooth. This is because that the symmetric counterparts of some contour points of the segments of the regular region of a stroke cannot be found due to some factors, such as the image digitization, approximation, noise, computation error, and so on. Therefore, some skeleton points are lost from the primary skeleton. These lost points can be easily amended by using the previously proposed smooth processing to generate a continuous skeleton segment in [14]. In terms of the smooth processing, two types of contour points of the character strokes can be obtained: *stable* and *unstable* contour points, which will be used for the classification of the characteristic points in next section.

A key issue of our amendment processing is to compute the skeletons of the singular region of the stroke. In fact, most existing skeletonization algorithms still perform poorly for extracting these skeleton points in the singular region.

The singular regions of the character strokes can be categorized into five typical patterns, namely, X-pattern, V-pattern, T-pattern, Y-pattern, and K-pattern [16]. In order to compute the skeleton points of the singular regions of the strokes, we develop a wavelet-based amendment processing technique. The key of this technique is to position the junction skeleton points in the singular region by detecting three types of characteristic points: terminal point, divider point, and corner point.

3.1 Detection of Wavelet-Based Characteristic Points

Here, the start or end point of the skeleton segment obtained from smooth processing is defined as terminal point. Obviously, they can be easily obtained from the skeleton segment after smooth processing.

The intersection point between the stable and the unstable contour segments is called as divider point. Accordingly, along the gradient direction of the terminal point the divider points can be detected easily. Thus, the main task of the detection of the characteristic point aims at computing the corner point based on the following WT.

By using the new wavelet function, we improve the multiscale corner detection technique [5] to extract the corner points in the contour curves. According to Lipschitz exponents of the WT, the singularity points are divided into three types: step-structure, roof-structure and dirac-structure [5,9,13]. Although the singularity points correspond to the local maxima of the modulus image after the WT, we can not identify what kind of singularity points from the local maxima, based on the desirable properties of the new wavelet function, which are proved

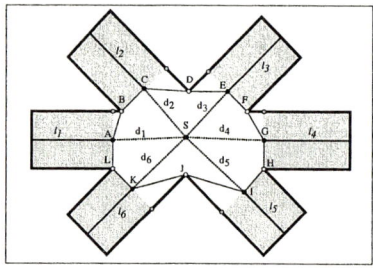

Fig. 2. Illustration of modifying processing in the singular region

in our latest work [14], it is very suitable for detecting the singularity points with dirac-structure. Thus, the new wavelet function instead of the spline function will be used to improve the performance of the detection of the multiscale corner (dirac-structure singularity point of curve) significantly [5].

We denote the contour curve of the character stroke using parametric equation $C(t) = (X(t), Y(t))$, where t is the arc length. The orientation of the curve is defined as [7] $\alpha(t) = tan^{-1}(\frac{dY/dt}{dX/dt})$, where dY/dt and dX/dx stand for the derivatives along t of Y and X respectively. To improve the orientation resolution, the orientation at the point P_i can be defined by simply replacing the above derivative by the first difference. Hence, the orientation at the point P_i is defined as:

$$\alpha(i) = tan^{-1}((Y_{i+q} - Y_{i-q})/(X_{i+q} - X_{i-q})) \qquad (1)$$

where $q > 1$, typically, q is chosen to be 3 [5]. A detailed description of the orientation curve can be found in [11].

We apply the WT to the orientation $\alpha(x)$, the corner points are obtained by computing the local maxima $| W_s\alpha(i) |$. The WT on the orientation $\alpha(x)$ is given as follows:

$$W_s\alpha(x) = \int_{-\infty}^{\infty} \alpha(x-t)\psi_s(t)dt. \qquad (2)$$

Its discrete form is

$$W_s\alpha(i) = \int_{-\infty}^{\infty} \alpha(i-t)\psi_s(t)dt = \sum_{k\in Z}\int_{k-1}^{k}\alpha(k)\psi_s(i-t)dt$$

$$= \sum_{k\in Z}\alpha(k)\int_{\frac{i-k}{s}}^{\infty}\psi(t)dt - \sum_{k\in Z}\alpha(k+1)\int_{\frac{i-k}{s}}^{\infty}\psi(t)dt = \sum_{k\in Z}[\alpha(k)-\alpha(k+1)]\psi_{i-k}^{s}; \quad (3)$$

where $\psi_i^s = \int_{\frac{i}{s}}^{\infty}\psi(t)dt$. Since $\psi(t)$ is an odd function, we have $\psi_{-i}^s = \int_{\frac{-i}{s}}^{\frac{i}{s}}\psi(t)dt + \int_{\frac{i}{s}}^{\infty}\psi(t)dt = \psi_i^s$. Hence, in practice, we simply calculate ψ_i^s for all non-negative integer i.

Obviously, a corner point should be an unstable contour point as well. Based on this principle, we determine the corner points from the candidates, and exclude ineligible candidates. Noticeably, the proposed corner detection method does not need further confirmation process for all candidates using the isolated algorithm in [5].

3.2 Estimation of Junction Skeleton Point

The junction skeleton point can be estimated by two ways: (1) It is the centroid of the polygon, which is composed of the above three types of characteristic points. (2) The position of the junction skeleton point can be calculated by the least square method with the characteristic points.

After obtaining all characteristic points in the singular region of the character strokes, these related characteristic points will be connected to produce a polygon. The junction skeleton point in the singular region can be determined by the centroid of such a polygon or by least square method.

In practice, a polygon is generated as follows: For each terminal point in the skeleton segment, along its gradient (positive and negative) direction, one of two divider points should be found. Along its corresponding unstable contour segment, another divider point (or corner point) can be found. All these characteristic points related to the terminal point are recorded and form a subgroup. When two subgroups share some characteristic points, two subgroups are merged into a bigger subgroup. The same search continues until a new characteristic point is included. Finally, all these subgroup generate a union group, which includes all characteristic points in the singular region, and they can be connected to construct a polygon, as shown in Fig. 2.

The junction point of the skeleton is estimated by the centroid of the produced polygon. Assume $(x_i, y_i), (i = 1, 2, ..., n)$ are the coordinates of characteristic points, which generate the polygon, the coordinate (x, y) of the junction point of the skeleton is the centroid of the polygon, as shown in Fig. 2.

$$\begin{cases} x = \frac{1}{n}\sum_{i=1}^{n} x_i \\ y = \frac{1}{n}\sum_{i=1}^{n} y_i \end{cases}$$

An alternative method to find the junction point of the skeleton is least square method. In this way, the determination of the junction point is by minimizing the cumulative distance from point (x, y) to all terminal points,

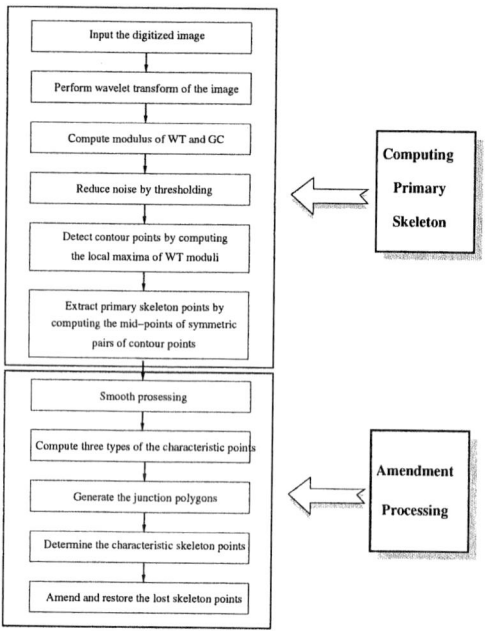

Fig. 3. Flowchart of the proposed algorithm

such as, A, C, E, G, I, and K, in Fig. 2. Let $d_1, d_2, ..., d_n$ be the distances from the junction skeleton point (x, y) to all terminal points $(x_i, y_i), (i = 1, 2, ..., n)$, that is, $d_i(x, y) =: \sqrt{(x - x_i)^2 + (y - y_i)^2}$, then the cumulative distance equals $R(x, y) = \sum_{i=1}^{n} d_i^2(x, y)$, by minimizing $R(x, y)$, (x, y) can be solved by the following equations $\begin{cases} \frac{\partial R(x,y)}{\partial x} = 0 \\ \frac{\partial R(x,y)}{\partial y} = 0. \end{cases}$ The amendment processing is finalized by connecting the junction skeleton point with all terminal points. For example, in Fig. 2, we connect the junction skeleton point S with the terminal points A, C, E, G, I, and K, the lost skeleton loci AS, CS, ES, GS, IS and KS can be obtained.

Thus all skeleton segments are connected together by the junction skeleton points to form the final skeleton.

For the sake of completeness, the complete flowchart of skeleton extraction is summarized in Fig. 3.

4 Experiment Result

We apply the proposed method to English and Chinese characters, which are represented by both the binary images and noise-distorted images. Fig. 4 consists of some Chinese characters, each character is composed of the strokes of different widths. The scale adopted in wavelet transform is $s = 4$, the images of the modulus maxima of the wavelet transform are shown in Fig. 4 (b). The primary skeletons extracted by the proposed algorithm are shown in Fig. 4(c). Apparently, the primary skeleton loci are lost in both the regular and the singular region of the character strokes. These lost loci can be amended by smooth processing and amendment processing respectively, and the resulting skeletons have pleasing visual appearance, which can be found in Figs. 4(d) and closely resemble human perceptions of the underlying shapes.

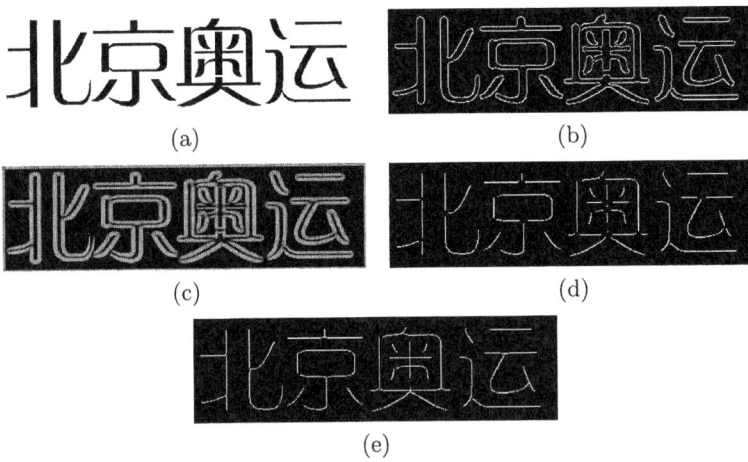

Fig. 4. (a) The original image; (b) The image of the modulus maxima of the wavelet transform with $s = 4$; (c) The primary skeletons extracted by the proposed algorithm; (d) The final skeletons obtained by applying the proposed modifying algorithm; (e) The skeleton obtained from ZSM method; (f) The skeleton obtained from CYM method.

The proposed method is also applied to the noise-contaminated image. However, most of the existing algorithms often fail to process these gray-level images. The patterns are harmed by the the "salt and pepper" noise in Fig. 5(a). The WT with scale $s = 4$ is applied to the noisy image. To remove the noise of the modulus image, the threshold processing is used with threshold $T = 0.35$. The image of the modulus maxima of the wavelet transform after thresholding is shown in Fig. 5(b). Fig. 5(c) shows the primary skeleton extracted by the proposed WLMMS. The final skeleton after the amendment processing is displayed in Figs. 5 (d), which demonstrates that the proposed approach is robust to white noise.

5 Conclusions

We propose a simple and efficient scheme for skeleton extraction. The new scheme contains two main steps: (1) primary skeleton extraction of the regular stroke, and (2) amendment processing of the singular region of the stroke. Extraction of the primary skeleton of the regular stroke depends on the new symmetric analysis of the modulus maxima of the WT. Amendment process is used to compute the skeletons in the singular regions of the strokes. It includes two main steps: detecting wavelet-based characteristic point and estimating junction skeleton point. Its essential operation is to position the junction point of the skeleton from three types of the characteristic points and then connect the junction skeleton point with the terminal points to finalize the skeleton. The proposed skeleletonization

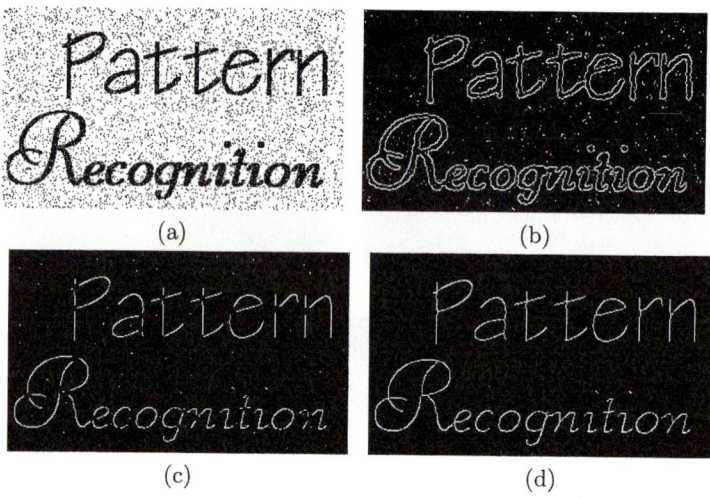

Fig. 5. (a) The original image is contaminated by adding "salt and pepper" noises; (b) The modulus image after threshold processing; (c) The primary skeleton obtained from the proposed approach; (d) The final skeletons after the modifying processing.

scheme has the following features: 1)the extracted skeleton is centered inside the underlying stroke in the mathematical sense while the skeleton obtained by most of the existing algorithms slightly deviates the center; 2) The skeletons in the singular region of character strokes can be extracted accurately, retain sufficient information of the original characters and have pleasing visual appearance; 3) The proposed skeleletonization method is robust against the noise; 4) The algorithm are applicable for both binary and gray-level images. Although it is a promising technique, problems still exist. How to automatically select the scale of the WT according to the width of the character strokes, and how to choose the denoising threshold are currently under investigation.

Acknowledgments

This research was partially supported by a grant (60403011 and 10371033) from National Natural Science Foundation of China and a grant (2003ABA012) and (20045006071-17) from Science &Technology Department, Hubei province and

Wuhan respectively, China. This research was also supported by the grants (RGC and FRG) from Hong Kong Baptist University.

References

1. Chang, H. S., Yan, H.: Analysis of Stroke Structures of Handwritten Chinese Characters. IEEE Trans. Systems, Man, Cybernetics (B). **29** (1999) 47–61
2. Ge, Y., Fitzpatrick, J. M.: On the Generation of Skeletons from Discrete Euclidean Distance Maps. IEEE Trans. Pattern Anal. Mach. Intell. **18** (1996) 1055–1066.
3. Kegl, B., A. Krzyżak.: Piecewise linear skeletonization using principal curves. IEEE Transactions on Pattern Analysis and Machine Intelligence, **24(1)** (2002)59–74.
4. Lam, L., Lee, S. W., Suen, C. Y. Thinning Methodologies - a Comprehensive Survey. IEEE Trans. Pattern Anal. Mach. Intell. **14** (1992) 869–885
5. Lee, J. S., Sun, Y. N., Chen. C. H.: Multiscale Corner Detection by Using Transform. IEEE Trans. Image Processing. **4** (1995) 100–104
6. Leyton, M.: A process-grammar for shape. Artifial Intell. **34** (1988) 213–247
7. Liu, H. C., Srinath, M. D.: Partial Shape Classification Using Contour Matching in Distance Transformation. IEEE Trans. Pattern Anal. Mach. Intell.**12** (1990) 1072–1079
8. Lu, S. W., Xu, H.: false stroke detection and elimination for character recognition. Pattern Recognition Lett. **13** (1992) 745–755
9. Mallat, S. : Wavelet Tour of Signal Processing. Academic Press, San Diego, USA, 1998.
10. Mokhtarian, F., Mackworth, A. K.: A theory of multiscale curvature-based shape representation for planar curves. IEEE Trans. Pattern Anal. Mach. Intell. **14(8)** (1992) 789–805
11. Rattarangsi, A., Chin, R. T.: Scale-based detection of corners of planar curves. IEEE Trans. Pattern Anal. Mach. Intell. **14** (1992) 430–448
12. Simon, J. C. A complemental approach to feature detection. In J. C. Simon, editor, From Pixels to Features. North-Holland, Amsterdam. (1989) 229–236
13. Tang, Y. Y., Yang, L. H., and Liu, J. M.: Characterization of Dirac-Structure Edges with Wavelet Transform. IEEE Trans. Systems, Man, Cybernetics (B). **30(1)** (2000) 93–109
14. Tang, Y. Y. and You,X. G.: Skeletonization of ribbon-like shapes based on a new wavelet function. IEEE Transactions on Pattern Analysis and Machine Intelligence. **25(9)** (2003) 1118–1133
15. Wang, Y. P., Lee, S. L.: Scale-space derived from B-splines. IEEE Trans. Pattern Anal. Mach. Intell. **20(10)** (1998)1040–1050
16. Zou J. J., Yan, H.: Skeletonization of ribbon-like shapes based on regularity and singularity analyses. IEEE Trans. Systems. Man. Cybernetics (B). **31** (2001) 401–407

Representing and Recognizing Scenario Patterns

Jixin Ma[1] and Bin Luo[2]

[1] School of Computing and Mathematical Sciences,
University of Greenwich, U.K.
[2] School of Computer Science,
AnhHui University, P.R. China
j.ma@gre.ac.uk,luobin@ahu.edu.cn

Abstract. This paper presents a formal method for representing and recognizing scenario patterns with rich internal temporal aspects. A scenario is presented as a collection of time-independent fluents, together with the corresponding temporal knowledge that can be relative and/or with absolute values. A graphical representation for temporal scenarios is introduced which supports consistence checking as for the temporal constraints. In terms of such a graphical representation, graph-matching algorithms/methodologies can be directly adopted for recognizing scenario patterns.

1 Introduction

Pattern recognition is a research area aiming at the operation and design of technologies to pick up meaningful patterns in data. In particular, pattern recognition is incredibly important in all automation, information handling and retrieval applications [1]. Pattern recognition starts with classification – in order to recognize something we need to be able to decide which category it belongs to. While pattern classification is about putting a particular instance of a pattern in a category, the goal of pattern matching is to determine how similar a pair of patterns are [2].

Generally speaking, temporal representation and reasoning is essential for many computer-based systems. In particular, an appropriate representation and reasoning for temporal knowledge seems necessary for recognizing scenario patterns that usually involve rich internal temporal aspects, rather than just distinct objects or episodes. Recognizing temporal scenario patterns actually plays an important role in solving problems such as prediction, forecast, explanation, diagnosis, decision, and so on. For instance, an obvious goal of research in weather forecast/diagnosis is to identify major patterns contribute to extreme, short-term events such as major droughts and floods, and patterns of climate variability on decadal and longer time scales, including natural variations and human-induced changes. Without a good understanding of climate phenomena based on past observations the weather expert cannot make good predictions of the future. In fact, to provide correct and accurate forecast, the weather expert needs to know not only the current weather parameters summarized as temperature, air pressure, precipitation amount, wind speed and residual snow/ice amount, etc., but also the weather scenarios over some certain prior periods (for instance: How long did the heat wave last? Was there lightning before or during the rain? Did snow melt then refreeze? And so on).

Most applications require temporal knowledge to be expressed by means of associating time-dependent statements with absolute time values. However, there are some other cases where there may be just some relative temporal knowledge about the time-depended statements, where their precise time characters are not available. For example, we may only know that event A happened before event B, without knowing their precise starting and finishing time. Relative temporal knowledge such as this is typically derived from humans, where absolute times are not always remembered, but relative temporal relationships are, and require less data storage than that presented in the complete description [3]. Over the last 30 decades, many approaches have been proposed to accommodate the characteristics of relative temporal information. For example, as an early attempt at mechanizing part of the understanding of relative temporal relationships within an artificial intelligence content, Kahn and Gorry's time specialist was developed [4], endowed with the capacity to order temporal facts in three major ways: 1. Relating events to dates, 2. Relating events to special reference events, 3. Relating events together into before-after chains. However, the most influential work dealing with incomplete relative temporal information in the field of artificial intelligence is probably that of Allen's interval-based theory of time [5, 6]. In [5], Allen introduces his interval-based temporal logic, where intervals are addressed as primitive time elements, and between time intervals there are 13 possible temporal relations; i.e., Equal, Before, Meets, Overlaps, Starts, Started_by, Duration, Contains, Finishes, Finished_by, Overlapped_by, Met_by and After, which may be formally defined in terms of the single primitive relation "Meets" [7].

The objective of this paper is to introduce a formal method for scenario pattern representation and recognition. As the temporal basis for the formalism, a simple point-based time model is presented in section 2, allowing expression of both absolute time values and relative temporal relations. Section 3 proposes a formal characterization of fluents and temporal scenarios, where a graphical presentation of temporal scenarios is introduced in section 4. Finally, section 5 provides a brief summary and concludes the paper.

2 The Time Model

In this paper, we shall use **R** to denote the set of real numbers, and **T**, the set of time elements. Each time element t is defined as a subset of the set of real numbers **R** and must be in one of the following four forms:

$$(p_1, p_2) = \{p \mid p \in \mathbf{R} \land p_1 < p < p_2\}$$
$$[p_1, p_2) = \{p \mid p \in \mathbf{R} \land p_1 \leq p < p_2\}$$
$$(p_1, p_2] = \{p \mid p \in \mathbf{R} \land p_1 < p \leq p_2\}$$
$$[p_1, p_2] = \{p \mid p \in \mathbf{R} \land p_1 \leq p \leq p_2\}$$

In the above, p_1 and p_2 are real numbers, and we shall call them the left-bound and right-bound of time element t, respectively. The absolute values as for the left and/or right bounds of some time elements might be unknown. In this case, real number variables are used for expressing relative relations to other time elements (see later).

In this paper, if the left-bound and right-bound of time element t are the same, we shall call t a time point, otherwise t is called a time interval. Without confusion, we shall take time element [p, p] as identical to p. Also, if a time element is not specified as open or closed at its left (right) bound, we shall use "<" (or ">") instead of "(" and "[" (or ")" and "]") as for its left (or right) bracket. In addition, we define the duration of a time element t, D(t), as the distance between its left bound and right bound. In other words:

$$t = <p_1, p_2> \Leftrightarrow D(t) = p_2 - p_1$$

Following Allen's terminology [5], we shall use Meets to denote the immediate predecessor order relation over time elements:

$$\begin{aligned}
\text{Meets}(t_1, t_2) \Leftrightarrow \exists p_1, p, p_2 \in \mathbf{R}(&t_1 = (p_1, p) \wedge t_2 = [p, p_2) \\
\vee\ &t_1 = [p_1, p) \wedge t_2 = [p, p_2)) \\
\vee\ &t_1 = (p_1, p) \wedge t_2 = [p, p_2] \\
\vee\ &t_1 = [p_1, p) \wedge t_2 = [p, p_2] \\
\vee\ &t_1 = (p_1, p] \wedge t_2 = (p, p_2) \\
\vee\ &t_1 = [p_1, p] \wedge t_2 = (p, p_2) \\
\vee\ &t_1 = (p_1, p] \wedge t_2 = (p, p_2] \\
\vee\ &t_1 = [p_1, p] \wedge t_2 = (p, p_2])
\end{aligned}$$

It is easy to see that the intuitive meaning of Meets(t_1, t_2) is that, on the one hand, time elements t_1 and t_2 don't overlap each other (i.e., they don't have any part in common, not even a point); on the other hand, there is not any other time element standing between them.

Analogous to the 13 relations introduced by Allen for intervals [5], there are 30 exclusive temporal order relations over time elements including both time points and time intervals, which can be classified into the following 4 groups:

- {Equal, Before, After}
  which relate points to points;
- {Before, After, Meets, Met_by, Starts, During. Finishes}
  which relate points to intervals;
- {Before, After, Meets, Met_by, Started_by, Contains, Finished_by}
  which relate intervals to points;
- {Equal, Before, After, Meets, Met_by, Overlaps, Overlapped_by, Starts, Started_by, During, Contains, Finishes, Finished_by}
  which relate intervals to intervals.

The definition of these derived temporal order relations in terms of the single relation Meets is straightforward. In fact:

Equal(t_1, t_2) $\Leftrightarrow$ $\exists t_3, t_4 \in$ **T**(Meets(t_3, t_1) $\wedge$ Meets(t_3, t_2) $\wedge$ Meets(t_1, t_4) $\wedge$ Meets(t_2, t_4))
Before(t_1, t_2) $\Leftrightarrow$ $\exists t \in$ **T**(Meets(t_1, t) $\wedge$ Meets(t, t_2))
Starts(t_1, t_2) $\Leftrightarrow$ $\exists t_3, t, t_4 \in$ **T**(Meets(t_3, t_1) $\wedge$ Meets(t_3, t_2) $\wedge$ Meets(t_1, t)
$\wedge$ Meets(t, t_4) $\wedge$ Meets(t_2, t_4))
Finishes(t_1, t_2) $\Leftrightarrow$ $\exists t_3, t, t_4 \in$ **T**(Meets(t_3, t) $\wedge$ Meets(t_3, t_2) $\wedge$ Meets(t, t_1)
$\wedge$ Meets(t_1, t_4) $\wedge$ Meets(t_2, t_4))
During(t_1, t_2) $\Leftrightarrow$ $\exists t_3, t_4 \in$ **T**(Meets(t_3, t_1) $\wedge$ Meets(t_1, t_4) $\wedge$ Starts(t_3, t_2) $\wedge$ Finishes(t_4, t_2))
Overlaps(t_1, t_2) $\Leftrightarrow$ $\exists t \in$ **T**(Finishes(t, t_1) $\wedge$ Starts(t, t_2))
After(t_1, t_2) $\Leftrightarrow$ Before((t_2, t_1)
Met-by(t_1, t_2) $\Leftrightarrow$ Meets(t_2, t_1)
Overlapped-by(t_1, t_2) $\Leftrightarrow$ Overlaps(t_2, t_1)
Started-by(t_1, t_2) $\Leftrightarrow$ Starts(t_2, t_1)
Contains(t_1, t_2) $\Leftrightarrow$ During(t_2, t_1)
Finished-by(t_1, t_2) $\Leftrightarrow$ Finishes(t_2, t_1)

In what follows in this paper, we shall use **TR** to denote the set of 13 exclusive temporal order relations:

TR = {Equal, Before, After, Meets, Overlaps, Overlapped-by, Met-by,
Starts, Started-by, During, Contains, Finishes, Finished-by}

It is important to note that the distinction between the assertion that "point p Meets interval t" and the assertion that "point p Starts interval t" is critical: while Starts(p, t) states that point p is the starting part of interval t, Meets(p, t) states that point p is one of the immediate predecessors of interval t but p is not a part of t at all. In other words, Starts(p, t) implies interval t is left-closed at point p, and Meets(p, t) implies interval t is left-open at point p. Similarly, this applies to the distinction between the assertion that "interval t is Finished-by point p" and the assertion that "interval t is Met-by point p", i.e., Finished-by(t, p) implies interval t is right-closed at point p, and Met-by(t, p) implies interval t is right-open at point p.

3 Fluents and Scenarios

A fluent is a statement (or proposition) whose truth-value is dependent on time elements. The set of fluents, **F**, is defined as the minimal set closed under the following rules:

- If $f_1, f_2 \in$ **F** then $f_1 \vee f_2 \in$ **F**
- If $f \in$ **F** then not(f) $\in$ **F**

Here, following Galton's notation [8], the negation of fluent f is denoted as not(f), while the ordinary sentence-negation is symbolized by "¬".

In order to associate a fluent with a time element, we shall use a meta-predicate [6, 9], Holds, to substitute the formula Holds(f, t) for each pair of a fluent f and a time element t, denoting that fluent f holds true over time t.

(H1) $Holds(f_1 \vee f_2, t) \Leftrightarrow Holds(f_1, t) \vee Holds(f_1, t)$
(H2) $Holds(f, <p_1, p_2>)$
$\Leftrightarrow \forall p_3, p_4 \in \mathbf{P}(p_1 \leq p_3 \wedge p_4 \leq p_2 \Rightarrow Holds(f, <p_3, p_4>))$
(H3) $Holds(f, <p_1, p_2>) \wedge Holds(f, <p_2, p_3>) \wedge Meets(<p_1, p_2>, <p_2, p_3>)$
$\Rightarrow Holds(f, <p_1, p_3>)$

It is worth pointing out that the time model and the formulae introduce in the above allows temporal knowledge with absolute values, as well as temporal knowledge expressed in terms of relative relations. For instance, consider the following scenario:

(*) John took 5 minutes to walk to the train station after having his breakfast at home. Then he immediately boarded the train at 7:50am and the train left the station 8:00am. John traveled on the train for 30 minutes and arrived at his office at 8:40. He spent 20 minutes to write a report whilst waiting for his colleague. Then they went to the auction preview.

Here, the scenario consists of the following fluents:

f_1: John had his breakfast at home.
f_2: John walked to the train station.
f_3: John boarded the train.
f_4: The train left the station.
f_5: John traveled on the train.
f_6: John arrived at his office.
f_7: John waited for his colleague.
f_8: John wrote a report.
f_9: John and his colleague went to the auction preview.

together with the corresponding temporal knowledge that can be express as below:

$Holds(f_1, t_1) \wedge Holds(f_2, t_2) \wedge Before(t_1, t_2) \wedge D(t_2) = 5 \wedge Holds(f_3, 7.50) \wedge Meets(t_2, 7.50) \wedge Holds(f_4, 8.00) \wedge Holds(f_5, <8.00, 8.30>) \wedge Holds(f_6, 8.40) \wedge Holds(f_7, t_7) \wedge Meets(8.40, t_7) \wedge Holds(f_8, t_8) \wedge D(t_8) = 20 \wedge During(t_8, t_7) \wedge Holds(f_9, t_9) \wedge Meets(t_7, t_9)$

In this paper, we shall use **S** to denote the set of scenarios, where each scenario s can be expressed in one of the following two equivalent schemas.

3.1 Schema I for Temporal Scenarios

In the first schema, Schema I, a temporal scenario is represented as a quadruple <sf, sh, st, sd>, where sf is a collection of fluents, sh is a collection of Holds formulae, st is the collection of temporal order relations, and sd is a collection of duration knowledge. That is:

s = <sf, sh, st, sd> where
  sf = $\{f_i \mid f_i \in \mathbf{F}, i = 1, ..., m\}$,
  sh = $\{Holds(f_i, t_i) \mid t_i \in \mathbf{T}, 1 \leq i \leq m\}$
  st = $\{relation_{ij}(t_i, t_j) \mid relation_{ij} \in \mathbf{TR}, 1 \leq i, j \leq m\}$
  sd = $\{D(t_i) = r_i \mid t_i \in \mathbf{T}, r_i \in \mathbf{R}, 1 \leq i \leq m\}$

For example, in terms of schema I, scenario (*) as described in the above can be expressed as:

s = <sf, sh, st, sd> where
  sf = $\{f_1, ..., f_9\}$
  sh = $\{Holds(f_1, t_1), Holds(f_2, t_2), Holds(f_3, 7.50),$
    $Holds(f_4, 8.00), Holds(f_5, <8.00, 8.30>), Holds(f_6, 8.40),$
    $Holds(f_7, t_7), Holds(f_8, t_8), Holds(f_9, t_9)\}$
  st = $\{Before(t_1, t_2), Meets(t_2, 7.50), Meets(8.40, t_7), During(t_8, t_7), Meets(t_7, t_9)\}$
  sd = $\{D(t_2) = 5, D(t_8) = 20\}$

N.B. Here, some temporal order relations that can be directly deduced from absolute values may not be explicitly expressed in st. For instance, Before(7.50, 8.00), Before(8.30, 8.40), etc.

3.2 Schema II for Temporal Scenarios

It is interested to note that, the temporal relationships presented in the above schema I are given in the form of a collection of order relations each of which can be any one of those 13 in **TR**, that is, Equal, Before, After, Meets, Overlaps, Overlapped-by, Met-by, Starts, Started-by, During, Contains, Finishes and Finished-by. However, since all these order relations can be derived from the single Meets relation, we shall have another schema, Schema II, which is equivalent to the schema I:

s = <sf, sh, sm, sd> where
  sf = $\{f_i \mid f_i \in \mathbf{F}, i = 1, ..., m\}$,
  sh = $\{Holds(f_i, t_i) \mid t_i \in \mathbf{T}, 1 \leq i \leq m\}$
  sm = $\{Meets(t', t'') \mid t', t'' \in \mathbf{T_s}\}$
  sd = $\{D(t) = r \mid t \in \mathbf{T_s}, r \in \mathbf{R}\}$

Here, $\mathbf{T_s}$ is the minimal subset of **T** closed under the following rules:

- $t_i \in \mathbf{T_s}$, i = 1, ..., m.
- $t \in \mathbf{T_s} \Leftrightarrow \exists t' \in \mathbf{T_s}(Meets(t, t') \vee Meets(t', t))$

For example, as for scenario (*) described in the above,

st = {Before(t_1, t_2), Meets(t_2, 7.50), Meets(8.40, t_7), During(t_8, t_7), Meets(t_7, t_9)}

can be equivalently as:

sm = {Meets(t_1, t_{12}), Meets(t_{12}, t_2), Meets(t_2, 7.50), Meets(8.40, t_7),
Meets($t_{8.40,8}$, t_8), Meets(t_8, $t_{8,9}$), Meets($t_{8,9}$, t_9), Meets(t_7, t_9)}

4 Graphical Representation of Scenarios

In [10], a graphical representation for expressing temporal knowledge in terms of Meets relations and duration knowledge has been introduced by means of a directed and partially weighted graph, where time elements are denoted as arcs of the graph, relation Meets(t_i, t_j) is represented by t_i being in-arc and t_j being out-arc to a common node, and for time elements with known duration, the corresponding arcs are weighted by their durations respectively.

Such a graphical representation can be directly extended to express temporal scenarios presented in Schema II. In fact, for each formula Holds(f, t), we can simply double label the corresponding arc by both f and t. Also, if the duration of a time element is available, it will be expressed as a real number in bracket alongside the corresponding arc, denoting a weight of the arc.

N.B. For those temporal order relations that can be directly deduced from absolute values and are not originally explicitly expressed in st, such as Before(7.50, 8.00), Before(8.30, 8.40), etc., their equivalent expressions in Meets relations may be needed to add to sm. For instance, Meets(7.50, $t_{7.50,8.00}$) and Meets($t_{7.50,8.00}$, 8.00) are needed to express Before(7.50, 8.00); and Meets(8.30, $t_{8.30,8.40}$) and Meets($t_{8.30,8.40}$, 8.40) are needed to express Before(8.30, 8.40).

As an illustration, consider again the scenario (*) as described in section 3. In terms of Schema II, this temporal scenario is expressed as:

s = <sf, sh, sm, sd> where
    sf = {$f_1, ..., f_9$}
    sh = {Holds(f_1, t_1), Holds(f_2, t_2), Holds(f_3, 7.50),
        Holds(f_4, 8.00), Holds(f_5, <8.00, 8.30>), Holds(f_6, 8.40),
        Holds(f_7, <8.40, p_2>), Holds(f_8, t_3), Holds(f_9, t_4)}
    sm = {Meets(t_1, t_{12}), Meets(t_{12}, t_2), Meets(t_2, 7.50), Meets(8.40, t_7),
        Meets($t_{8.40,8}$, t_8), Meets(t_8, $t_{8,9}$), Meets($t_{8,9}$, t_9), Meets(t_7, t_9)}
    sd = {D(t_2) = 5, D(t_8) = 20}

The corresponding graphical representation of temporal scenario s = <sf, sh, st, sd> can be shown in Fig 1. as below:

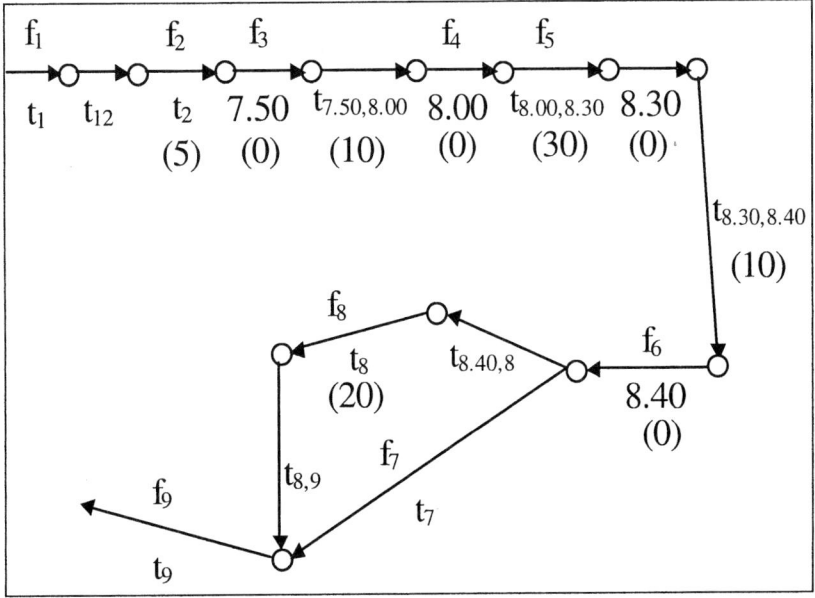

Fig. 1. Graphical representation of temporal scenarios

As an example of consistence checking, consider another temporal scenario s_1 = $<sf_1, sh_1, sm_1, sd_1>$ where

sf_1 = {f_1, f_2, f_3, f_4}
sh_1 = {Holds(f_1, t_1), Holds(f_2, t_2), Holds(f_3, t_4), Holds(f_4, t_8)}
sm_1 = {Meets(t_1,t_2), Meets(t_1,t_3), Meets(t_2,t_5), Meets(t_2,t_6), Meets(t_3,t_4),
  Meets(t_4,t_7), Meets(t_5,t_8), Meets(t_6,t_7), Meets(t_7,t_8)}
sd_1 = {$D(t_2) = 1, D(t_4) = 0.5, D(t_6) = 0, D(t_8) = 0.3$}

The corresponding graphical representation of temporal scenario s_1 can be shown in Fig 2. as below:

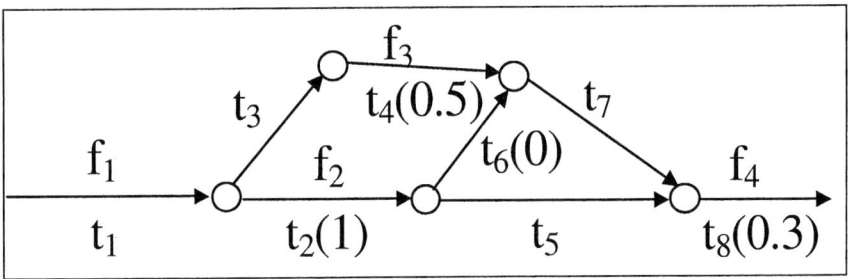

Fig. 2. Graphical representation of temporal scenarios$_1$

In the above graph, there are two simple circuits as shown below in Fig3.:

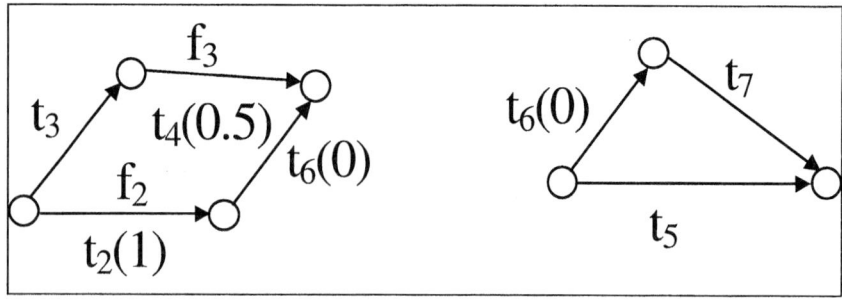

Fig. 3. Two simple circuits

It is easy to see that, to ensure the scenario is temporally consistent, the directed sum of weights in each of these simple circles has to be 0 [10]. This imposes the following two linear constraints:

$Dur(t_2) + Dur(t_6) = Dur(t_3) + Dur(t_4)$
$Dur(t_5) = Dur(t_6) + Dur(t_7)$

We can easily find a solution to the above linear programming, for instance: $Dur(t_3) = 0.5$, $Dur(t_5) = Dur(t_7) = 1.5$ (In fact, the duration assignment to t_5 and t_7 can be any positive real number, provided that $Dur(t_5) = Dur(t_7)$). Therefore, the scenario is temporally consistent. Otherwise, if there is no solution to the corresponding linear programming, it will be temporally inconsistent [10], that is, the scenario will be temporally impossible.

From the above, we can see that a temporal scenario can be actually denoted as a directed (in terms of Meets relations) graph whose arcs might be at most triple-labeled (in terms of fluents and the associated time elements, and the duration knowledge). Therefore, scenario pattern matching can be simply transformed into graph matching. Due to the scope of the paper, we shall not address this issue further.

5 Conclusion

In this paper we have introduced a simple point-based time structure based on which two equivalent schemas for representing temporal scenarios have been proposed. Such schemas allow expression of both absolute time values and relative temporal relations as temporal knowledge associated with scenario patterns. It is shown in this paper that a temporal scenario can be expressed as a directed graph whose arc might be at most triple-labeled. Therefore, it is suggested that scenario pattern recognition and matching can be simply transformed into graph matching.

References

1. Theodoridis, S. and Koutroumbas, K.: *Pattern Recognition*. Second Edition, Academic Press (2003).
2. Tveter, D.: *The Pattern Recognition Basis of Artificial Intelligence*. Wiley-IEEE Computer Society Press (1998).
3. Knight, B. and Ma, J.: A Temporal Database Model Supporting Relative and Absolute Time, *the Computer Journal* 37(7), (1994) 588-597.
4. Kahn, M. and Gorry, A.: Mechanizing Temporal Knowledge, *Artificial Intelligence* 9, (1977) 87-108.
5. Allen, J.: Maintaining knowledge about temporal intervals. *Communications of the ACM* 26 (11), (1983) 832-843.
6. Allen, J.: Towards a General Theory of Action and Time. *Artificial Intelligence* 23, (1984) 123-154.
7. Allen, J. and Hayes, P.: Moments and Points in an Interval-based Temporal-based Logic. *Computational Intelligence* 5, (1989) 225-238.
8. Galton, A.: A Critical Examination of Allen's Theory of Action and Time, *Artificial Intelligence*, **42** (1990) 159-188.
9. Ma, J. and Knight, B.: Reified Temporal logic: An Overview, *Artificial Intelligence Review*, 15 (2001) 189-217.
10. Knight, B. and Ma, J.: A General Temporal Model Supporting Duration Reasoning, *Artificial Intelligence Communication*, Vol.5(2), (1992) 75-84.

A Hybrid Artificial Intelligent-Based Criteria-Matching with Classification Algorithm

Alex T.H. Sim and Vincent C.S. Lee

School of Business Systems, Monash University, PO Box 63B, Clayton,
Wellington Road, Victoria 3800, Australia. Tel: +61 3-99052360
vincent.lee@infotech.monash.edu.au
http://www.bsys.monash.edu.au/index.html

Abstract. Classifying dynamic behavioural based events, for example human behaviour profile, is a non-trivial task. In this paper, we propose an AI-based criteria-matching with classification algorithm which can be used to classify preference based decision outcome. The proposed algorithm is mathematically justified and with more practical benefits than a conventional multivariate discriminant analysis algorithm which is widely used for prediction tasks. Real world (Singapore) diamond dataset test results revealed the practical usefulness of our proposed algorithm to diamond sellers in Singapore.

1 Introduction

Manufacturers and traders of industrial products often need to formulate a useful marketing strategy through market segmentation. Market segmentation involves viewing heterogeneous market into smaller homogeneous groups [1]. By focusing on an easier-to-capture (i.e., minimum difference between marketers' and customers' criteria), a desired and usually also enlarged group of customers should emerge, leading to finite cost and effort saving, and hence improved customer profitability. There are several techniques and algorithms that allow us to identify the customer groups effectively.

In this paper, we demonstrated a hybrid artificial intelligence (a non-statistical) algorithm that can overcome the limitations of discriminant analysis. Our proposed algorithm is called the criteria-matching with classification (CMC). This algorithm has two major steps. In the first step a pre-classification algorithm is used to interactively select a set of suitable explanatory variables while converting the incoming data into "degree of membership". The second step classifies the multidimensional dataset into groups via translation of the multi-dimensional data into single dimension points using neuro-fuzzy hybrid structure.

2 Criteria-Matching with Classification (CMC) Model

We introduce the concept of CMC in this section. Assuming a first party, $X = <x_1, x_2, x_3, \ldots x_n>$ and a second party, $Y = <y_1, y_2, y_3, \ldots y_n>$ are concurrently present in a competitive business environment. The distance between the two parties or objects [3], [4], [5] is usually in a derived form of Minkowski similarity function, $\left[\sum_i^n w_i |X_i - Y_i|^q\right]^{\frac{1}{q}}$. The choice of parameter q produces indexes of different implications. For example, Manhattan distance (i.e., q = '1') is to measure the tangent distance between two parties and Euclidean distance (i.e., q = '2') is to measure the nearest distance between the two objects.

X and Y are explanatory variables defined by several attributes and are of multi dimension. If the variables can be dimensionally reduced into indexes (points), the distance difference between two or more multi-dimensional variables is just the value difference between them. The similarity function is as simple as $\left[\sum_i^n w_i |X_i - Y_i|^1\right]^1$.

Besides, speeding up computation, index differences (i.e., within single dimension) have less statistical properties to be considered for classification.

2.1 Mathematical Justifications on CMC's Functions

CMC can be represented by a six-layer feed-forward weighted structure with min-max learning rules. It is used to first, measure the distance between two multi dimensional variables after converting them into indexes, extract the classification rules and to group instances into known groups. Its weighted structure takes the form of a neural network. This is an important step before index differences between two parties can be later calculated. Mathematically, this is normally performed using normalisation. However, the structure has the extra ability to first, represent the domain using first order Sugeno fuzzy rules before normalising the values into indexes using calculated degree of membership as a denominator. A comparison of indexes between a reference and other points yield final indexes within [0 1] that are easier to discriminate.

Assuming that there are N instances, first party (reference point), X = <x_1, x_2, x_3, ... x_n>, a second party (test point), Y = <y_1, y_2, y_3, ... y_n> and " $'$ ", denotes normalisation, CMC performs the following processes,
For reference point 'X',

Layer 1: T_1: $X \rightarrow X' = \mu_A(X_{i,i+1,...,N}) = \dfrac{X_{i,i+1,...,N}}{\max(X_{i,i+1,...,N})}$,

Layer 2: T_2: $X' \rightarrow kX' + k_0$, (default: set of parameter, $k_0 = 0$)
Layer 3: T_3: $kX' \rightarrow X'$, (due to default number of MF, $k = 1$)
Layer 4: T_4: $X' \rightarrow wX'$, (applying w, selection matrix (0 or 1) for variable used)
Layer 5: T_5: $wX' \rightarrow (kX')(wX')$,

(first $X`$ is a test point and the second X' is a reference point. At this point, they are the same.)

Layer 6: T_6: $(kX')(wX') \rightarrow \dfrac{\sum kX'wX'}{\sum wX'}$,

(denominator X' is a ref. point. Contrast it when test point is 'Y')

For test point 'Y',

Layer 1: T_1: $Y \rightarrow Y' = \mu_A(Y_{i,i+1,\ldots,N}) = \dfrac{Y_{i,i+1,\ldots,N}}{\max(Y_{i,i+1,\ldots,N})}$,

Layer 2: T_2: $Y' \rightarrow kY' + k_0$,

Layer 5: T_5: $kY' \rightarrow (kY')(wX')$

Layer 6: T_6: $(kY')(wX') \rightarrow \dfrac{\sum kY'wX'}{\sum wX'}$

Index, $I = |X - Y| = \left|\dfrac{\sum kX'wX'}{\sum wX'} - \dfrac{\sum kY'wX'}{\sum wX'}\right|$

However, $(X, Y) \in Z$ (Universe of discourse) and $X = \max(Z)$, therefore $Y \leq X$. Hence, index, I is scaled in relate to X' and is a measure of distance similarity and is consisted of single dimensional indexes. $w's$ is the level of contribution for each criterion. We argue that, $\forall X, Y: I \rightarrow G$, or $G = F(X, Y) + \varepsilon$, where ε is the error.

That is, group label, G is in single dimension. Mapping function, F maps multiple dimensional variables X, Y to G after converting them into X' and Y' and scaled them to a reference point X' (i.e., normalized into single dimension for relative differences). CMC, a six layers feed-forward weight structural is used to model each sub functions (i.e. $T_{1,\ldots 6}$) hence the complete function, F. The hybrid structure of CMC is shown in Figure 1.

2.2 Mathematical Justifications on CMC's Learning Rules

Group label consists of a total of n number of groups,
$G \in \{g_1, g_2, \ldots, g_n\}$,
$G \in \{f_1(x,y), f_2(x,y), \ldots, f_c(x,y)\}$, $f_1(x,y) \rightarrow g_1$,

Since $g_1 \in G$ and $i_1 \in I$ are of single dimension,
$f_1(x,y) = [\min(i_1), \max(i_1)]$,
similarly, $G \in \{[\min(i_1), \max(i_1)], \ldots, [\min(i_n), \max(i_n)]\}$
or $G \in \{[\min(I), \max(I)]\}$, where $I = |X - Y|$

Hence, it is possible to predict the group label, G, if the minimum and the maximum value of each converted variables, X' and Y' can be found. We have justified on the use of learning rules in CMC (i.e., min-max learning rules).

CMC learns the border values from each group using the training data. These border values are confirmed through visual inspection to constrain undesired statistical properties (e.g. critical values of skewness and kurtosis) on its distribution. This increases the uniqueness of each 'univariate' group, which is defined by single dimension and scaled primary explanatory variables.

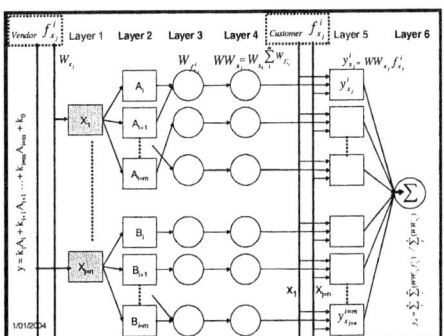 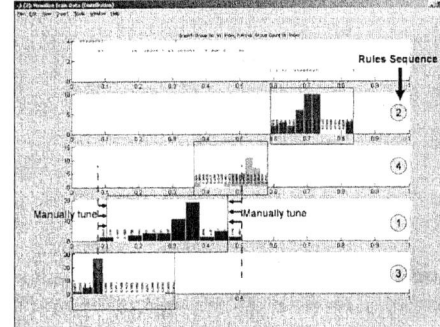

Fig. 1. A six-layer feed-forward weighted structure

Fig. 2. Visual inspection to fine-tune over-spreading phenomena. (Visualise train data distributions of Diamond dataset)

The min-max values between two and three standard deviations do not affect much to the prediction (or classification) accuracy on group labels. From Fig 2, the third sub group has a normal distribution shape with stable but thin and long tails; note that its tail can be constrained in order to minimize its spread. The loss of some points (<4%) in a group is compensated by much points gained from other groups by constraining this learning rule. Visual inspection helps in fine tuning classification rules using simply statistical properties for better rules.

2.3 Mathematical Justifications on CMC's Assumptions on Explanatory Variables

If $(X, Y) \in Z$ (Universe of discourse) and $X = \max(Z)$, $\therefore Y \leq X$
Index, $I = |X - Y|$

$$= |\frac{\sum kX'wX'}{\sum wX'} - \frac{\sum kY'wY'}{\sum wX'}|$$

$= |aX' - bX'|$; where a and b are parameter

The maximum value in each set of parameter, $k, w = 1$ and $X' \rightarrow [0\ 1]$,
$\therefore$ Index, $I \leq |a - b| \rightarrow [0\ 1]$

CMC is to use the single dimensional indexes (that is, the linear distance between single dimensional X' and Y') to classify group labels. Apart from this assumption CMC does not assume a correlation between the explanatory variables, X and Y (i.e., the origin of X' and Y'), i.e. X and Y are reference and test variables representing two different parties. They are interaction free.

In contrast, discriminate function uses the correlation of variables (i.e., assumption on interaction) to calculate the coefficients matrix, w. In adaptive paradigm, neural network approximates a black-box function to represent all possible but unknown sub functions and is used to predict group labels. In conclusion, CMC is unique as it neither needs to consider on the interaction between explanatory variables nor it requires to function approximate an unknown level of function using the underlying variables.

3 A Walk-Through on Diamond Dataset

Chu [6] discussed his work based on a diamond dataset, which resulted in the use of statistical model for a real world diamond price prediction. In this research, Diamond dataset has been formulated as a classification problem. Besides classifying the dataset, it is also demonstrated that CMC has the capabilities of

1) Learning classification rules from each dataset to make predictions about groups.
2) Identifying and quantifying the gap between each customer group's demand and a seller's (plan or existing) supply criteria

Table 1 shows a portion of data instances within a diamond dataset. The commonly used criteria in selecting a diamond (especially in Singapore from where the dataset was obtained) are carat, colour, clarity and the certification body/authority. Therefore, four explanatory variables or criteria define a customer group. Descriptions of customer groups are: Group-1 customer pays "low price" (S$ 0 – S$2000) on diamond, Group-2 customer pays "moderate price" (S$ 2000.1 – S$4000) on diamond, Group-3 customer pays "high price" (S$ 4000.1 – S$8000) on diamond and Group-4 customer pays "excessive price" (S$ 8000.1 – S$16000) on diamond.

Table 1. A portion of instances within a diamond dataset

Carats (Interval)	Colour (Nominal)	Clarity (Ordinal)	Certification Body (Nominal)	Customer Price Group
0.31	E	VS1	GIA	Grp1
0.31	G	VVS2	GIA	Grp1
1.1	H	VS2	GIA	Grp4
0.21	D	VS1	IGI	Grp1
0.58	H	IF	HRD	Grp2
1.01	I	VVS1	HRD	Grp4
1.09	I	VVS2	HRD	Grp4

The criterion carat (weight) is an interval measurement. The criteria of colour and the Certification Body are more towards nominal measurement [7, 8]. Clarity is generally an ordinal measurement; the rarer the stone, the more valuable it will be.

3.1 Pre-classification

CMC uses the concept of fuzzy logic to convert and relate nominal instances to a reference point (i.e. a seller's perspective). Each instance which comes with an ordinal and higher level of measurement will have a degree of membership in a maximum class. Each instance with a nominal measurement will have a degree of membership in an introduced class, "Suitability". A seller, by default, takes the maximum value of a class, [1,1,1,1]; Grp4. Converted instances from Table 1 are shown in Table 2.

Table 2. Degree of membership on a portion of instances within Diamond dataset, it symbolises customers' purchase criteria. A is the "$\mu(MaxClassofCarats)$; Interval". B is the "$\mu(SuitabilityinColour)$; Nominal". C is the "$\mu(MaxofClarity)$; Ordinal". D is the "$\mu(SuitabilityinCertBody)$; Nominal". E is the "Customer Price Group".

A	B	C	D	E
0.28	0.83	0.60	0.67	Grp1
0.28	0.50	0.40	0.67	Grp1
1.00	0.33	0.80	0.67	Grp4
0.19	1.00	0.60	1.00	Grp1
0.53	0.33	1.00	0.33	Grp2
0.92	0.17	0.20	0.33	Grp4
0.99	0.17	0.40	0.33	Grp4

An incremental liked variable selection algorithm called Search Algorithm (SA) augments the use of explanatory variables (i.e., A to D). It selects the most contributing set of variables that distinguish instances into groups. It changes according to the changes in the volume of data that was fed into the model over a period of time. For example, at 10% train data, [0,0,0,1] (Cert. Body) is being used and hence [1,0,0,0] (carat) is found to be best in grouping customers into groups with good accuracy.

3.2 Dimensional Reduction and Classification Using Neuro-fuzzy Structure

A seller and multiple customers' set of criteria are compared for indexes that are both unique and scaled within [0 1] using mathematical equations proposed on section 2.1, earlier. A seller holds a maximum value, set of [1,1,1,1], any indexes that are closer to the max-value within a range [0 1], are therefore nearer to a seller.

3.3 Min-max Classification Learning Rules (Min-max)

Classification rules are mathematically justified by extracting only the min and max value from each group. Four fuzzy classification rules can be extracted. If the index difference between a seller and a customer is represented using the symbol e, any value near to '0' is considered to be very near to a seller's supply criteria. Also, these instances have a linguistic value, "Very Near" (that is, near '0'). Using 50% of diamond dataset, four extracted classification rules are as follows:

Classification Rule 1
  IF e is Far (min-max point: [0.591 0.836])
  THEN a customer belongs to Group-1
Classification Rule 2
  IF e is Medium (min-max point: [0.364 0.582])
  THEN a customer belongs to Group-2   (1)
Classification Rule 3
  IF e is Near (min-max point: [0.082 0.509])
  THEN a customer belongs to Group-3
Classification Rule 4
  IF e is Very Near (min-max point: [0 0.309])
  THEN a customer belongs to Group-4

Above classification rules can be visualised through CMC as shown in Fig.4.

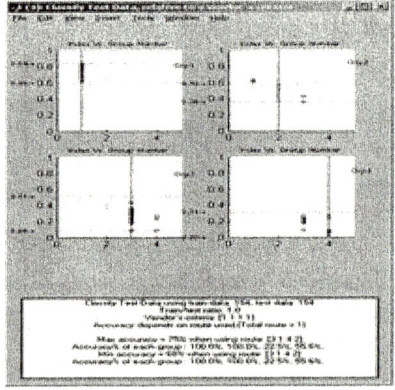

Fig. 3. Classify test data of Diamond dataset at 50% training data. The dots represent all instances that have been converted into indexes. They are associated with their groups. The dotted horizontal lines in each plot represent the min and max value. At 50% training data, some dots appear in other groups. These are misclassified instance. At this stage, without a fine tuning, the accuracy ranges from 68% to 75% depending on the training data distribution.

4 Analysis and Implication on Classification Results

On average, CMC is 71.1% accurate (using leave-one-out test) in determining the individual purchase criteria for diamonds in Singapore. That is, seven out of ten individuals can be grouped and predicted correctly with this model. Chu (2001) asserts that the customer profile in the training dataset is in fact the representative profile of a general population of Singapore. Also the diamond composition mix has relatively remained constant and consistent with that of the training dataset. Experiment shows that a 50% training data yield results much similar to 100% training data using CMC technique.

4.1 Post-Classification: Fine-Tuning Classification Accuracy via Visualisation

In making a good prediction about a customer group, classification rules, values and learning rule's sequence are important. If a different sequence of a learning rule is being used, a different accuracy result might be obtained. At 50% training, it follows a classification rule sequence of 3 → 1 → 4 → 2. This is determined by the total number of instances within a group. That is, a group with more instances should have higher priority.

In order to increase classification accuracy (i.e., based on statistical understanding in section 2.2 and Fig. 2), each training data group is plotted for its distribution and visually inspect for fine tuning opportunity. The border of group-3 is too long and is constrained to free up more instances for Group-2 and Group-4. A revised classification rule at 50% training data is as follows:

Classification Rule 3

IF e is <u>Near</u> (min-max point: [0.082 0.509]) adj. → [0.1032 0.46635]
THEN a customer <u>belongs to Group-3</u>   (2)
Note: $(0.0925+0.1139)/2 = 0.1032$, $(0.4557+0.4770)/2 = 0.46635$

With this post classification process, classification using CMC is completed and its accuracy burst from 71.1% to 92.2% (92.2% using the stringent leave-one-out test or 92.6% using 100-fold cross validation) at remarkable 50% training data. Statistical discriminant needs all 100% training data and achieves 95.5% accuracy. That is, results obtained using CMC only differs by 3 persons at every 100 customers, if compared to discriminant technique.

4.2 CMC Advantages

Although CMC produced results are slightly lower than discriminant analysis but more benefits can be obtained: (1) CMC has fast convergence learning ability and in most test cases [2] 50% of training data is enough for accurate reporting, (2) explanatory variables are ensured to be primary (i.e., mathematically there is no interaction between variables, carat is directly affecting the prediction on customer group), (3) changes in market would cause a change in the uses of explanatory variables (for example at less than 10% training data, certification body is more important than the use of variable carat). Hence, CMC is a visually viable tool to at least diamond sellers. Besides, the relative distance and size of customer groups are made known to a seller in simulating and understanding customer groups in a dynamic market. Lastly, each group's reference point, which is the averaging of all single dimensional variables give the seller a set of referencing point.

4.3 Use of Classification Results

CMC has the default seller supply criteria at [1,1,1,1]. Judging by customer price groups, size and the customers' criteria distance from a seller, the above model (i.e., Fig. 2 to Fig. 5 and classification rules (1) and (2)) can be interpreted as follows:

1) Customer Group-4 (S$8,000.1 – S$16,000) has demand criteria nearest to a seller's existing supply criteria. The difference is valued within [0 0.31]. Since, value '0' is an exact match with a seller's supply criteria, this means

that this seller is currently selling diamonds to this group of customers. A distribution plot on the 50% training data in Fig 2 reveals that the distribution is skewed right, leaving almost nothing on the right. Therefore, a seller can align marketing strategies to only import products that fulfill the need of customers, plotted on the left (that is, with an index near or equal to zero). This decision depends on the seller's business resources and subjective to a seller's projected profit.

2) Group-3 (S$4,000.1 – S$8,000) has criteria differences rather near from an existing seller's supply criteria. A moderate change in supply criteria might attract a larger number of customers. Since Group-3 has a wider customer spread, with low customer demand on both ends of the distribution, a seller would need more resources in order to target this group. It is better to target customers with index 0.3 to 0.4 within this group, if budget is a concern.

3) Group-2 (S$2,000.1 – S$4,000) is rather far to a seller's existing supply criteria. If a seller has the flexibility to make minor change in the supply criteria, this could be a fairly good niche market to concentrate on. The distribution has skewed left which means there are more customers preferring lower price diamond within this group range.

4) Lastly, Group-1 (S$0 – S$2,000) is furthest from a seller's existing business. However, it is perhaps the best target group in a diamond market for a capital constraint seller. First, this group can be identified easily, with not much overlapping with other groups. It has the most number of customers and the cost of the diamond is lower. A seller has the opportunity to sell more diamonds at lower cost to fulfill many customers who can be identified fairly easily.

The soft-decision to formulate correct marketing strategies lies with the seller. CMC, a criteria-matching technique with classification, has shown the customer groups, gap, its size with possible distribution based on training data. A seller is free to change his or her supply reference point in order to study the customer groups around him / her. For example, in order to sell to Group-1 customers, who are looking at a price of less than S$2000, the referencing criteria is 0.26, which equals to ~0.286 carats unit or at a price of ~S$1225.

5 Conclusion

The proposed integrated technique can help a seller to learn and predict customer profiles. CMC technique has advantages over discriminant analysis in providing a prediction on customer group label, the size and their gap in a dynamic market. This helps a resource constrained niche seller to target the desired and most suitable group of customers (for example, the largest), who are the nearest to a seller's existing supply criteria. The proposed technique is also unique as it converges fast, variables are interaction-free, augmenting to changes in data and is a single dimension classification technique. In brief, CMC is a potential tool that is suitable for use in predicting customer groups and showing criteria distances to a novice seller.

References

1. Smith, W.: Product Differentiation and Market Segmentation as Alternative Marketing Strategies. Journal of Marketing, vol. 21 (1956) 3–8
2. Sim, A.T.H.: A Technique Integrating Criteria-matching with Classification for Predicting Customer Groups. Thesis, Monash University (2004) 1–229
3. Zeng, Z. and Yan, H.: Region Matching by Optimal Fuzzy Dissimilarity. Proceedings of the ACM International Conference Proceeding Series: Pan-Sydney workshop on Visualisation. Australian Computer Society Inc., Darlinghurst, Australia (2000) 75–84
4. Zeng, Z. and Yan, H.: Region Matching and Optimal Matching Pair Theorem. Proceedings of International Conference on Computer Graphics. IEEE Computer Society, Washington, DC, USA (2001) 232–239
5. Khan, M., Ding, Q. and Perrizo, W.: k-nearest Neighbor Classification on Spatial Data Streams Using P-trees, Proceedings of the Pacific-Asia Conference on Knowledge Discovery and Data Mining. CiteSeer (2002) 517–518
6. Chu, S.: Pricing the C's of Diamond Stones. Journal of Statistics Education, Vol. 9 (2). On-line (2001)
7. Chard: The Very Highest Quality Diamond Information. Available: *[http://www.24carat.co.uk/diamondscolour.html]* (1 Feb 2004)
8. A.D.H.C.: The HRD Diamond Certificate. Available: *[http://www.diamonds.be/professional/certificate/dia_certi.htm]* (1 Feb 2004)
9. Chu, S.: Diamond Ring Pricing Using Linear Regression. Journal of Statistics Education, Vol. 4 (3). On-line (1996)

Auto-generation of Detection Rules with Tree Induction Algorithm

Minsoo Kim[1], Jae-Hyun Seo[1], Il-Ahn Cheong[2], and Bong-Nam Noh[3]

[1] Dept. of Information Security, Mokpo Nat'l Univ., Mokpo, 534-729, Korea
{phoenix, jhseo}@mokpo.ac.kr
[2] Electronics and Telecommunications Research Institute, Daejeon, 305-700, Korea
qubcia@etri.re.kr
[3] Div. of Electr-Comp. & Inform-Engin., Chonnam Nat'l Univ., Gwangju, 500-757, Korea
bbong@jnu.ac.kr

Abstract. A generation of rule for detecting an attack from enormous network data is very difficult, and this is commonly required an expert's experiences. An auto-generation of detection rules cut down on maintenance or management expenses of intrusion detection systems, but the problem is accuracy for the time being. In this paper, we propose an automatic generation method of detection rules with a tree induction algorithm that is adequate to search special rules based on entropy theory. While we progress the experiment on rule generation and detection with extracted information from network session data, we found a problem in selecting measures. To solve the problem, we present a method of converting the continuous measures into categorical measures and a method of choosing a good measure according to the accuracy of the generated detection rules. As the result, the detection rules for each attack are automatically generated without any help of the experts. Also, the correctness of detection improves according to the selection of network measures.

1 Introduction

A numerous varied attacks have been appeared owing to the open source of attacks and more intelligent attacks. However, IDSs (Intrusion Detection Systems) immediately couldn't bring an apt counter-move for this. The variety of attacks led the market of IDSs to depression after all.

Previous IDSs detect mainly existing attacks by comparing features, and for detecting unknown attacks, an anomaly technique is applied. However the use of the technique is a burden to IDSs because of long learning time and many false-positives. While the common rule-based detection method serves high accuracy and ability in real-time detection, it has a problem that more time is required to make rules because experts have to analyze the attacks and generate the detection rules [1]. This problem can cause heavy damages due to delay of response against new attacks. This is why we begin to study about generation method of rules for detecting attacks without any help of the experts.

We propose an automatic generation method of detection rules with a tree induction algorithm. This algorithm is a classification method that takes a priority measure according to entropy [2]. Using this method, we automatically generate detection rules with extracted information from network session data and verify an accuracy of generated rules. We find a problem in accuracy of generated detection rules because of the continuous measure. So we suggest a method of converting the continuous measures into categorical measures. We generate rules while choosing measures which are good for rule generation and removing measures which give a negative effect. As the result, the detection rules for each attack are automatically generated without any help of the experts. Also, the correctness of detection becomes better according to the selection of network measures.

2 Manuscript Preparation

Recently, the researches for detecting anomalous behaviors and selecting measure on network with packet analysis and artificial intelligent method based on network have been progressed at Florida University and New Mexico University, etc. Florida University generated detection rules for features from training data based on TCP/IP protocol by using LERAD (Learning Rules for Anomaly Detection) that is a learning algorithm generating conditional rules [3]. In result of their experiment against 23 extracted measures from the 1999 DARPA IDS Evaluation dataset, it showed high detection rate. However, in case of a training data including attacks, it showed low detection rate.

New Mexico University detected each attack type using SVM (Support Vector Machine) method that classifies objects having two-class considering information such as session-based information on network packet and field characteristics of protocol header, etc [4]. Their research showed quick training and high accuracy for suggested measures from KDD-CUP 99, but the processing needed much time for collecting information and it was hard to block prior attacks. Because used measures include system logs as well as network packets.

Artificial intelligent or data mining has been used as a method of analyzing an amount of data. Resent researches have progressed in an automatic generation of rules and detection of attacks. The represent of such researches was ARL:UT (the Applied Research Laboratories of University of Teas at Austin) applied decision tree and genetic algorithm [5]. Genetic algorithm in this research was a logical platform to solve some problems occurring in intrusion detection, and decision tree algorithm was used to generate decisive rules. However, the rules have simple measures and could be extracted only after confirming attacks.

3 Automatic Generation of Rules Using Tree Induction Algorithm

3.1 The Selection of Algorithm

The tree induction algorithm, which is one of the classification methods in data mining, presents combinations of existing attributes among measures of a class and converts a classification model into the form of a tree after analyzing the collected data. In case of having many unnecessary attributes for constructing trees, the algorithm can select measures easily because the attributes automatically excepted do not affect

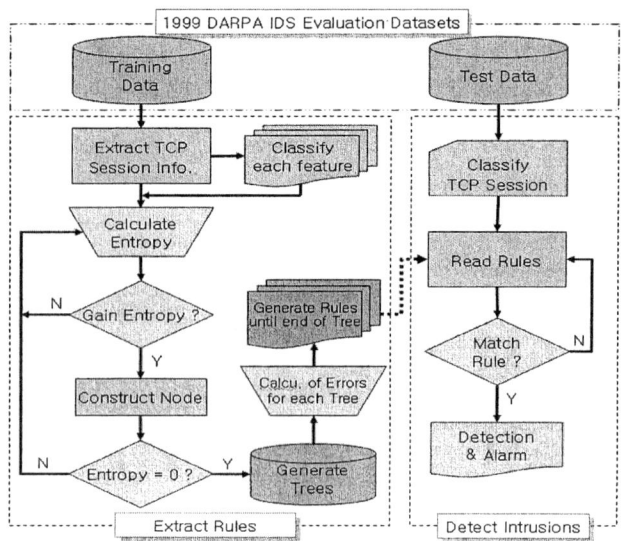

Fig. 1. Model for Automatic Generation of Rules and Detection of Attacks

the classification. The tree structure can be quickly composed by reducing the loads consumed in the step of data conversion, because the algorithm processes continuous data as well as categorical data [6].

The C4.5 algorithm, which is the tree induction algorithm, automatically generates trees by calculating the IGs (Information Gains) according to the reducing values of entropy. The constructed tree shows the superior measure and what it means. If the algorithm is applied to detecting attacks, it is appropriate for the examination of the association between detection rules and attacks because it is easy to find how a combination of the constituted variables affects the target variables which are added to previous combined variables. The algorithm is appropriate for being applied to network data since it can classify data whose variables have the both continuous and categorical attributes.

Figure 1 shows the model that generates rules and detects attacks using tree induction algorithm, and the algorithm is modified from the C4.5 to process network data based on TCP session. In order to generate detection rules, the first step is extracting and selecting information about each session from training data and the C4.5 constructs nodes by calculating IGs according to entropy reduction. The rules are generated by calculating errors of the leaf nodes for each tree. The errors are used for the process of detecting attacks after constructing trees by repeating these processes.

3.2 Generation of Detection Rules According to Selected Measures

3.2.1 Categorical Measures

In order to construct the efficient tree for the categorical measures, the measure which gains the highest IG according to the reduction of the entropy must be constructed for the upper node after calculating the entropy. The equation (1) calculates the entropy for all the classes (where, C: the number of classes, N: the number of total data, N_i: the number of data which is classified to the ith class).

$$Entropy(I) = -\sum_{i=1}^{C}(N_i/N)\log_2(N_i/N) \qquad (1)$$

The amount of the entropy reduction between measures is calculated after constituting each node of the tree. The amount which is calculated through the following *i) ~ iv)* steps is the easiest way to distinguish the classes.

- *i)* Let a_{kj} is the *j*th attribute of the *k*th rule set. According to a_{kj}, divide the population with the *level-1*.
- *ii)* Let n_{kj} is a population which is one of the divided branches and $n_{kj}(i)$ is the # of rules belonging to class *i*. The entropy for the branch belonging to class *i* is calculated by equation (2).

$$Entropy(I, A_k, j) = -\sum_{i=1}^{C}(n_{ki}(i)/n_{kj})\log_2(n_{ki}(i)/n_{kj}) \qquad (2)$$

After calculating the entropies of all branches, total entropy is calculated by equation (3). The total entropy is calculated by multiplying a probability of the rules on each branch and summing the result values.

$$Entropy(I, A_k, j) = \sum_{i=1}^{J}\sum_{i=1}^{C}(n_{kj}/\sum_{j}n_{kj})[-(n_{ki}(i)/n_{kj})\log_2(n_{ki}(i)/n_{kj})] \qquad (3)$$

- *iii)* As in the equation (4), the amount of the entropy reduction is calculated by the amount difference between the total entropy and the entropy of the divided population with *level-1*.

$$\Delta Entropy(A_t) = Entropy(I) - Entropy(A_k) \qquad (4)$$

- *iv)* Let h ($h=1, 2, \ldots, k, h \neq h_0$) is each component of rules. The most upper node is selected by the value whose entropy reduction is the largest. That is, if $\Delta E_h > \Delta E_{h0}$, then the most upper node is constructed by ΔE_h.
- *v)* After deciding the most upper node, a tree, which has *level-1*, is completely constructed. The steps, *i) ~ iv)*, should be progressed iteratively until the entropy of the most upper node becomes 0.

3.2.2 Continuous Measures

A classification with continuous measures should use a threshold that is a criterion of dividing a set of variables in the direction of decreasing largely the entropy. However, it takes much time to search accurate threshold [7].

Calculating entropy for continuous measures needs to decide the limit values [8]. Partially continuous measures as (b) in the Fig. 2 make the trees generate differently according to the decision of limit value and detection performance decrease by the results. Therefore, a method properly dealing with partially continuous measures is required. We experiment with destination ports among continuous measures and evaluate the performance of converted measure.

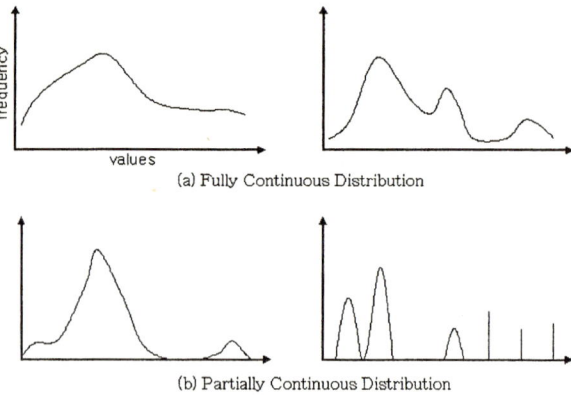

Fig. 2. Distribution types of continuous measures

3.2.3 Change of Selection for Network Measures

The less important measures need the classifying time and have no effect on improving the detection performance. We used the method of selecting the high priority measures among network measures, because the more important measures should be chosen first. The measures (R_G) are reflected next learning because they affect positively in generating of detection rules and have high frequencies. On the other hand, the negative measures (R_B) are excluded. The selection rate of measures having high priority, GRR, is calculated by equation (5) (where, α is 0.01, as value in order to prevent calculating error which divided by 0).

$$GRR(Good\ Rule\ Rate) = \frac{The\ \#\ of\ R_G}{The\ \#\ of\ (R_G + R_B + \alpha)} \quad (5)$$

4 The Result and Analysis of Experiments

4.1 Generation and Verification of Detection Rules

For focusing the TCP information, we selected attacks against only TCP included in DARPA99 dataset. To select network measures, we modified TCP trace [10] and extracted 80 features from TCP session information as you see in Table 1. We used the Week 2 data among DARPA99 dataset as training data. The data consists of 252,296 TCP sessions. It is used to generate detection rules by learning each attack in the data.

Figure 3 shows a part of generated rules. Figure 4 shows the format of a generated rule. It consists of rule number, attack type, attack name respectively and many conditions. Each condition part in a rule is connected by AND operation. As shown below column, the generated rules are converted into the form of detection rule. For example, the detection rules (the rule number is 14) of *satan* attack is below column in Figure 4 as shown with bold string in Figure 3. When the construction of tree is completed, detection rules are decided.

Table 1. The Measures based on TCP Session

Session Information	Measure Name		Type
TCP Connection Information	complete connection		categorical
	FIN/SYN counts		categorical
	Elapsed time		continuous
	total packets		continuous
Host A or Host B Connection Information	total packets	ack packets sent	continuous
	pure acks sent	sack packets sent	continuous
	Dsack packets sent	max sack	continuous
	unique bytes	SYN/FIN packets sent	continuous
	urgent data pkts/bytes	mss requested	continuous
	max/min segment size	avg segment size	continuous
	ttl stream length	missed data	continuous
	truncated data	data transmit time	continuous
	idletime max	Throughput	continuous

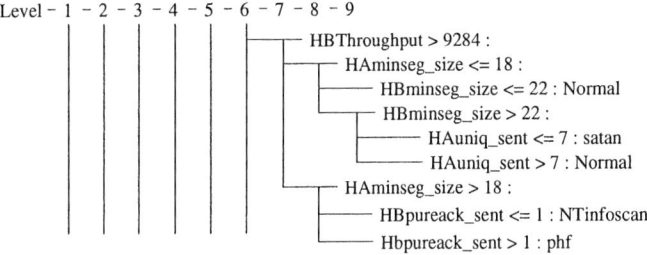

```
Level - 1 - 2 - 3 - 4 - 5 - 6 - 7 - 8 - 9
                    ├── HBThroughput > 9284 :
                    │   ├── HAminseg_size <= 18 :
                    │   │   ├── HBminseg_size <= 22 : Normal
                    │   │   └── HBminseg_size > 22 :
                    │   │       ├── HAuniq_sent <= 7 : satan
                    │   │       └── HAuniq_sent > 7 : Normal
                    │   └── HAminseg_size > 18 :
                    │       ├── HBpureack_sent <= 1 : NTinfoscan
                    │       └── Hbpureack_sent > 1 : phf
```

Fig. 3. A Part of Generated Rules

Real Number	Attack Type	Attack Name	Measure Name	Symbol	Measure Value	...
R#1	dos /	Neptune;	Haack_sent	<=	0;	...
R#14; dos/satan; ...; HBThroughput > 9284; HAminseg_size <= 18; HBminseg_size > 22; HAuniq_sent <= 7 (in case of satan attack)						

Fig. 4. Form and Example of Detection Rules

4.2 The Comparison and Analysis for Experiment to Detect Attacks

To evaluate the detection rate, we used Week 4 data which consists of 191,077 TCP sessions in DARPA99. Figure 5 shows the result of comparing our research with

LERAD for each attack. In the figure 5, we found that unknown attacks as well as attacks for training were detected, and we found the association among attacks having similar property. The reason why *back, land, phf,* and *neptune* attack were not detected is that these attacks were not included in Week 2 data. The LERAD in Florida University is a learning algorithm to generate condition rules for features. When we compare our research with the LERAD in case of *netcat* attack (↓), the former detected it but the latter not. However, in case of the other five attacks (↑), such as *httptunnel* attack, *xsnoop* attack, *guest* attack, etc., the detection rate of our research was totally higher than LERAD by about 10%.

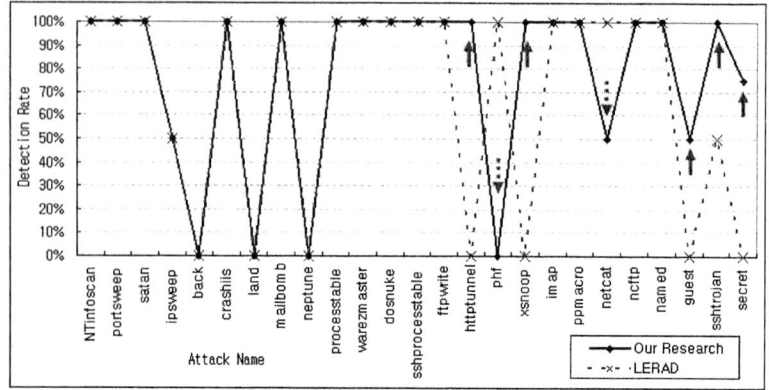

Fig. 5. Comparison of detection rate for Each Attack

4.3 Experiments for Detection Rate and False Alarm Rate

Figure 6 shows the improvement in detection rate and false alarm rate after converting continuous measures into categorical measures. It also shows that the detection rate increased about 20% and the false alarm rate decreased by 15%.

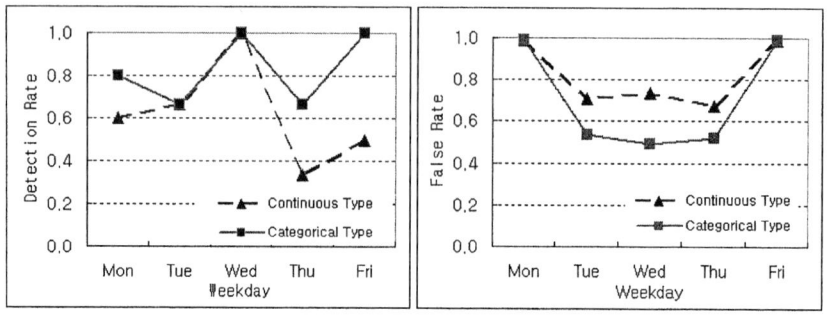

Fig. 6. The change of detection rate and false alarm rate after conversion

Auto-generation of Detection Rules with Tree Induction Algorithm   167

Table 2. The step of selecting the network measures

No	Step0					Step1					Step2					Step3			
	RG	RB	GRR	RST	SLT	RG	RB	GRR	RST	SLT	RG	RB	GRR	RST	SLT	RG	RB	GRR	RST
1	0	0	0.00	I		3	1	0.75	G		11	1	0.92	G		11	1	0.92	G
3	0	0	0.00	I		0	0	0.00	I		7	7	0.50	B	X	0	0	0.00	I
4	7	28	0.20	B	X	0	0	0.00	I	X	0	0	0.00	I	X	0	0	0.00	I
5	10	5	0.67	G		18	10	0.64	G		19	11	0.63	G		19	9	0.68	G
6	6	7	0.46	B	X	0	0	0.00	I	X	0	0	0.00	I	X	0	0	0.00	I
8	0	0	0.00	I		30	20	0.60	G		38	14	0.73	G		38	12	0.76	G
10	3	0	1.00	G		8	4	0.67	G		6	1	0.86	G		6	1	0.86	G
13	0	0	0.00	I		0	0	0.00	I		61	15	0.80	G		61	13	0.82	G
20	37	30	0.55	G		35	39	0.47	B	X	0	0	0.00	I	X	0	0	0.00	I
21	0	0	0.00	I		2	0	1.00	G		10	0	1.00	G		10	0	1.00	G
22	0	0	0.00	I		0	0	0.00	I		5	0	1.00	G		5	0	1.00	G
24	0	0	0.00	I		5	1	0.83	G		55	14	0.80	G		55	12	0.82	G
26	17	11	0.61	G		23	14	0.62	G		34	13	0.72	G		34	11	0.76	G
27	10	5	0.67	G		20	10	0.67	G		34	13	0.72	G		34	11	0.76	G
28	0	0	0.00	I		0	0	0.00	I		2	0	1.00	G		2	0	1.00	G
38	4	0	1.00	G		3	1	0.75	G		3	2	0.60	G		3	2	0.60	G
39	4	1	0.80	G		4	1	0.80	G		0	0	0.00	I	X	0	0	0.00	I
41	19	26	0.42	B	X	0	0	0.00	I	X	0	0	0.00	I	X	0	0	0.00	I
43	0	0	0.00	I		2	3	0.40	B	X	0	0	0.00	I	X	0	0	0.00	I
45	21	18	0.54	G		2	0	1.00	G		13	0	1.00	G		13	0	1.00	G
46	0	0	0.00	I		6	4	0.60	G		9	4	0.69	G		9	4	0.69	G
47	10	5	0.67	G		20	10	0.67	G		22	11	0.67	G		22	9	0.71	G
48	3	3	0.50	B	X	0	0	0.00	I	X	0	0	0.00	I	X	0	0	0.00	I
49	0	0	0.00	I		0	0	0.00	I		6	0	1.00	G		6	0	1.00	G
50	27	26	0.51	G		20	37	0.35	B	X	0	0	0.00	I	X	0	0	0.00	I
51	10	8	0.56	G		7	13	0.35	B	X	0	0	0.00	I	X	0	0	0.00	I
52	0	0	0.00	I		2	0	1.00	G		0	0	0.00	I	X	0	0	0.00	I
53	0	0	0.00	I		0	0	0.00	I		26	12	0.68	G		26	10	0.72	G
54	49	48	0.51	G		49	52	0.49	B	X	0	0	0.00	I	X	0	0	0.00	I
55	11	4	0.73	G		21	9	0.70	G		12	1	0.92	G		12	1	0.92	G
56	7	12	0.37	B	X	0	0	0.00	I	X	0	0	0.00	I	X	0	0	0.00	I
60	4	1	0.80	G		9	3	0.75	G		64	17	0.79	G		64	15	0.81	G
61	3	0	1.00	G		0	0	0.00	I		8	5	0.61	G		8	5	0.61	G
62	0	0	0.00	I		10	3	0.77	G		25	2	0.93	G		25	2	0.93	G
63	0	0	0.00	I		2	0	1.00	G		5	0	1.00	G		5	0	1.00	G
64	0	0	0.00	I		5	2	0.71	G		3	0	1.00	G		10	5	0.67	G
65	12	12	0.50	B	X	0	0	0.00	I	X	0	0	0.00	I	X	0	0	0.00	I
66	6	5	0.54	G		8	17	0.32	B	X	0	0	0.00	I	X	0	0	0.00	I
67	2	0	1.00	G		3	8	0.27	B	X	0	0	0.00	I	X	0	0	0.00	I
68	18	29	0.38	B	X	0	0	0.00	I	X	0	0	0.00	I	X	0	0	0.00	I
75	1	2	0.33	B	X	0	0	0.00	I	X	0	0	0.00	I	X	0	0	0.00	I
76	4	1	0.80	G		46	51	0.47	B	X	0	0	0.00	I	X	0	0	0.00	I
77	46	48	0.49	B	X	0	0	0.00	I	X	0	0	0.00	I	X	0	0	0.00	I
78	0	0	0.00	I		13	26	0.33	B	X	0	0	0.00	I	X	0	0	0.00	I
79	17	11	0.61	G		25	14	0.64	G		32	13	0.71	G		32	11	0.74	G
80	6	0	1.00	G		13	10	0.56	G		14	5	0.74	G		14	5	0.74	G

Removed measure's numbers: 2, 7, 9, 11-12, 14-19, 23, 25, 29-37, 40, 43-44, 57-59, 69-74

Table 2 shows the step of selecting the network measures according to GRR. In this table, 'Step0' means that all network measures were used. If GRR of a measure is the less than threshold (0.5), the measure is removed in measure set, and then the process of generating rules is continued again. Other 'Step#' in Table 2 means the process of selecting measures. Each 'G', 'B' and 'I' in Table 2 means 'Good measure', 'Bad measure' and 'Ignorable measure'. 'B' or 'I', as GRR of the measure was less than the threshold, is marked with 'X' and is not used in the next step. After all, the removed measure number is showed in the last row of Table 2.

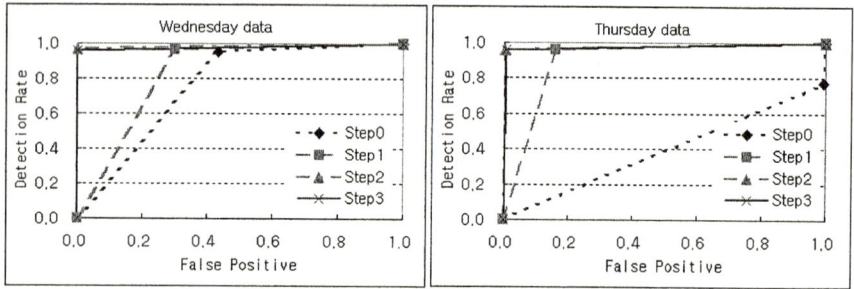

Fig. 7. ROC for Week 4 of the DARPA 99 dataset

Figure 7 shows the ROC for Week 4 data of the DARPA99 dataset. It is noticed that the detection rate and false alarm rate for measures selected by GRR improved while selecting the good measures. In the Step 3, detection rate was the highest and false positive was the lowest.

5 Conclusion

As a tree induction algorithm offers an explanation of classification and prediction for given data, it is easy to understand generated rules and analyze which attributes critically affect each class. Also, this can process both continuous and categorical data so that construct quickly models by reducing time and load to convert data.

In this paper, we studied the method that automatically generates detection rules using tree induction algorithm. A tree induction algorithm classifies the selected measures which are firstly obtained by calculating high information gain according to entropy reduction. This method made detection rules against various attacks were generated by extracting the 80 measures from network session data. These rules can detect attacks effectively and reduce a lot of costs and time because of automatic generation without support of experts.

We found a problem in selection from continuous measures. To solve the problem, we presented the method that is converting the continuous measures into categorical measures. Therefore we verified the accuracy for the generated detection rules through experiments and selected measures according to the accuracy. As the results, detection rate gets increased and false alarm rate gets decreased.

Acknowledgement

This research was supported by the MIC (Ministry of Information and Communication), Korea, under the ITRC (Information Technology Research Center) support program supervised by the IITA (Institute of Information Technology Assessment).

References

1. Mahoney, M., Chan, P.: Learning Nonstationary Models of Normal Network Traffic for Detecting Novel Attacks. Proc. of 8th ACM SIGKDD Int. C. on KDD, 1. (2002) 376-385
2. Ross Quinlan, J.: C4.5: Programs for Machine Learning, Morgan Kaufmann Pub (1993)
3. Mahoney, M., Chan, P.: Learning Models of Network Traffic for Detecting Novel Attacks. Florida Institute of Tech. TR. CS-2002-08, (2002)
4. Mukkamala, S., Sung, A.: Identifying Significant Features for Network Forensic Analysis using Artificial Intelligent Techniques. Int. J. Digit. Evid. 1 (2003)
5. Chris, S., Lyn, P., Sara, M.: An Application of Machine Learning to Network Intrusion Detection. 54th Ann. Computer Security Applications C. 12 (1999)
6. Kamber, H.: Data Mining Concepts and Techniques, Morgan Kaufmann Pub (2001)
7. Yoh-Han P.: Adaptive Pattern Recognition and Neural Networks, Addison-Wesley (1989)
8. Dombi, J., Zsiros, A.: Learning Decision Trees in Continuous Space. Acta Cybernetica 15. (2001) 213-224
9. Das, K.: Attack Development for Intrusion Detection Evaluation, MIT Mast. Th. 6. (2000)
10. Ostermann, S.: TCPtrace. http://www.tcptrace.org. (1994)

Hand Gesture Recognition System Using Fuzzy Algorithm and RDBMS for Post PC

Jung-Hyun Kim, Dong-Gyu Kim, Jeong-Hoon Shin, Sang-Won Lee, and Kwang-Seok Hong

School of Information and Communication Engineering,
Sungkyunkwan University, 300, Chunchun-dong,
Jangan-gu, Suwon, KyungKi-do, 440-746, Korea
{kjh0328, kdgyu13, swlee}@skku.edu, only4you@chol.com,
kshong@skku.ac.kr
http://hci.skku.ac.kr

Abstract. In this paper, we implement hand gesture recognition system using union of fuzzy algorithm and Relational Database Management System (hereafter, RDBMS) module for Post PC (the embedded-ubiquitous environment using blue-tooth module, embedded i.MX21 board and note-book computer for smart gate). The learning and recognition model due to the RDBMS is used with input variable of fuzzy algorithm (fuzzy max-min composition), and recognize user's dynamic gesture through efficient and rational fuzzy reasoning process. The proposed gesture recognition interface consists of three modules: 1) gesture input module that processes motion of dynamic hand to input data, 2) RDBMS module to segment significant gestures from inputted data, and 3) fuzzy max-min recognition module to recognize significant gesture of continuous, dynamic gestures and extensity of recognition. Experimental result shows the average recognition rate of 98.2% for significant dynamic gestures.

1 Introduction

The Post PC refers to a wearable PC that has the ability to process information and networking power. Post PC that integrates sensors and human interface technologies will provide human-centered services and excellent portability and convenience [1]. And also, the haptic technology can analyze user's intention distinctly, such as finding out about whether user gives instruction by some intention, judging and taking suitable response for the user intention [2][3]. In this paper, in order to exploit the advanced technology and react to the changed user requirements / environments appropriately, we enhance the traditional data entry pattern, which is limited to wired communications, up to ubiquitous environment (based on the Post PC platform), using wireless communication interface between data acquisition devices and other modules. The propose gesture recognition system that is consisted of union by fuzzy max-min composition and RDBMS can improve the efficiency in data acquisition, provide user with more convenience, and recognize significant dynamic gesture of user in real-time and anytime-anywhere manner.

2 Concepts of the Post PC

For Post PC, we have two concepts. The first is wearable and the second is small & light for humanization. The wearable computers are full-function PC that is designed to be worn and operated on the body. The Post PC's processors and input devices can be connected to wireless local area networks or other communications systems [1] [4]. The Post PC has two concepts as shown in Fig.1 [5].

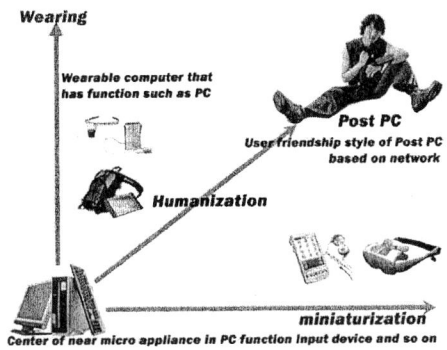

Fig. 1. Concepts of the Post PC

3 Gesture Input Module and Regulation of Hand Gesture

In this paper, we use 5DT Company's 5th Data Glove System (wireless) which is one of the most popular input devices in haptic application field. 5th Data Glove System is basic gesture recognition equipment that can capture the degree of finger stooping using fiber-optic flex sensor and acquires data through this [7]. In order to implement gesture recognition interface, this study choose 19 basis hand gestures through priority "Korean Standard Sign Language Dictionary [6]" analysis. And 4 overlapping gestures are classified as pitch and roll degree. Fig. 2 shows a classification of basis hand gesture used in this study.

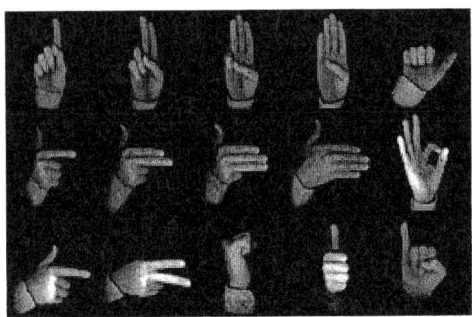

Fig. 2. Classification of basis hand gesture

Two main functions of gesture input module are 1) to input the significant hand gesture data (in analog format) of users, which is generated/delivered from (wireless) 5th Data Glove System, and 2) to execute the calibration control function as a pre-processing step. The captured dynamic gesture data of various users is transmitted to embedded i.MX21 board and server (Oracle 10g RDBMS) through notebook-computer. The gesture data transmitted to server that is used to as a training data for gesture recognition model by Oracle 10g RDBMS's SQL Analytic Function. And, gesture data that is transmitted to embedded i.MX21 is used as the input to fuzzy recognition module for significant gesture recognition. The architecture of gesture input module is shown in Fig. 3.

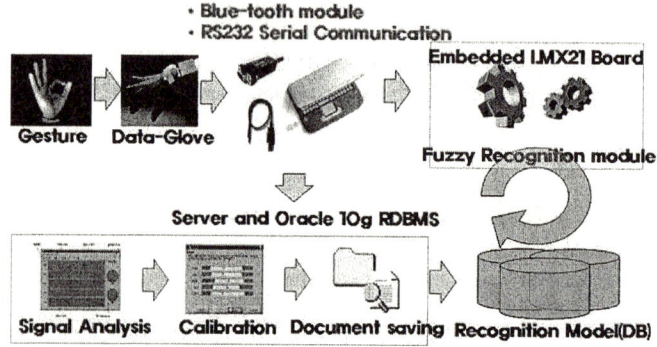

Fig. 3. The architecture of gesture input module

4 Training and Recognition Model Using RDBMS

The RDBMS is used to classify stored gesture document data from gesture input module into valid and invalid gesture record set (that is, status transition record set) and to analyze valid record set efficiently. Fig. 4 shows the process of recognition model construction for validity gesture.

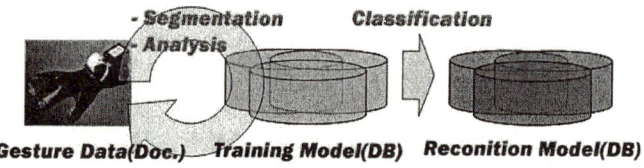

Fig. 4. Process of recognition model construction using RDBMS

The process of implementing training model is to analyze the input data from data glove system and to segment the data into valid gesture record set and status transition record set. A rule to segment valid and invalid gesture record set (changing gesture set) is as following.

Hand Gesture Recognition System Using Fuzzy Algorithm and RDBMS for Post PC

- If the difference between preceding average (preceding 3 and 1) and current row value is over 5, the current value is regarded as transition gesture record.
- If one of five finger's data value is over 5, current value data is also regarded as changing gesture record.

5 Fuzzy Max-min Module for Hand Gesture Recognition

The fuzzy logic is a powerful problem-solving methodology with a myriad of applications in embedded control and information processing. Fuzzy provides a remarkably simple way to draw definite conclusions from vague, ambiguous or imprecise information [8]. The fuzzy logic system consists of the fuzzy set, fuzzy rule base, fuzzy reasoning engine, fuzzifier, and defuzzifier. Because each fuzzy relation is a generalization of relation enemy ordinarily, the fuzzy relations can be composed. The fuzzy max-min CRI for fuzzy relations that is proposed in this paper is defined in Fig. 5. The learning and recognition model by the RDBMS is used with input variable of fuzzy algorithm and recognizes user's dynamic gesture through efficient and rational fuzzy reasoning process. And, we decide to give a weight to each parameter and do fuzzy max-min reasoning through comparison with recognition model constructed using RDBMS module. The proposed membership function of fuzzy set is defined as in the following formula (1).

$$\mu_{tz} = \begin{bmatrix} \frac{1}{(s-p)}(x-s)+1 & p<x\leq s \\ 1 & s<x\leq t \\ -\frac{1}{(q-t)}(x-t)+1 & t<x\leq q \end{bmatrix} \quad (1)$$

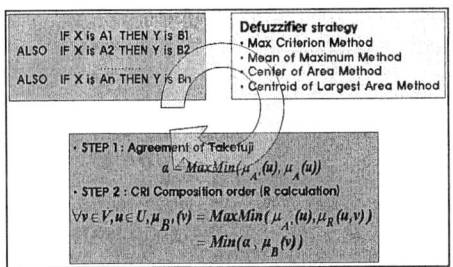

Fig. 5. Fuzzy Max-Min CRI (Direct Method)

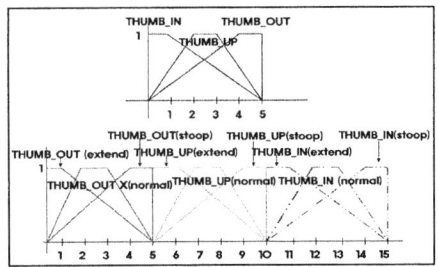

Fig. 6. Fuzzy membership functions

6 Experiments and Results

The overall process of recognition system consists of three major steps. In the first step, the user prescribes user's hand gesture using 5th Data Glove System (wireless), and in the second step, the gesture input module captures user's data and change characteristics of data by parameters, using RDBMS module to segment recognition model. In the last step, it recognizes user's dynamic gesture through a fuzzy reasoning process and a union process in fuzzy min-max recognition module, and can react, based on the recognition result, to user via speech and graphics offer. The process of automatic human gesture recognition is shown in Fig. 7.

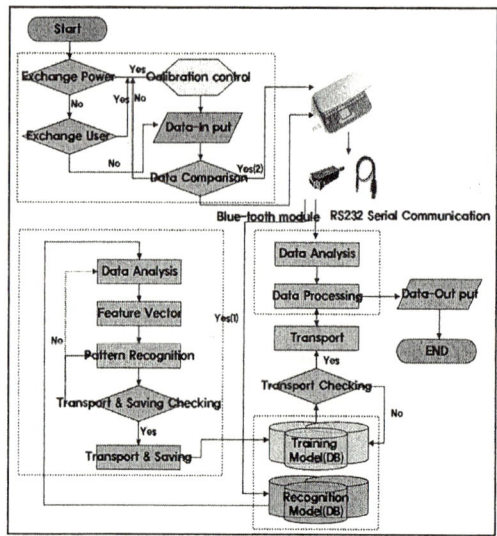

Fig. 7. Flow-chart of the recognition system

Experimental set-up is as follows. The distance between server (Oracle 10g RDBMS) and smart-gate is about radius 10M's ellipse form. As the gesture, we move 5th Data Glove System to prescribed position. For every 20 reagents, we repeat this action 10 times. Experimental result shows the average recognition rate of 98.2% for significantly dynamic gestures. Fig. 8 shows an example of fuzzy value's production and recognition results.

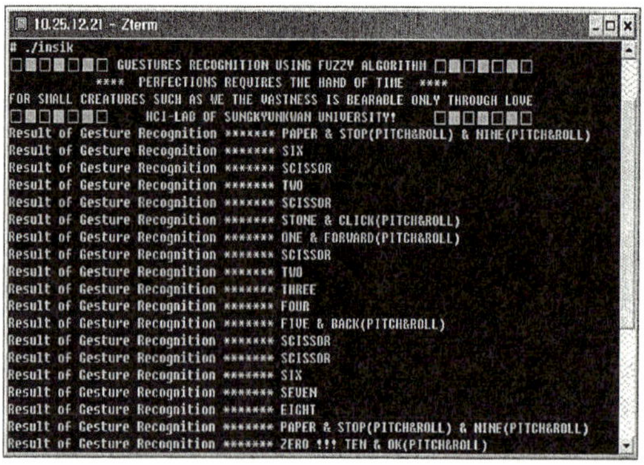

Fig. 8. Fuzzy value's production and results of hand gesture recognition

7 Conclusions

In this paper, we implemented hand gesture recognition system that analyzes user's intention more efficiently and more accurately can recognize 19 significant dynamic hand gestures in real time using fuzzy max-min recognition module in Post PC platform. Also, the hand gesture recognition system by union of fuzzy algorithms (the fuzzy relation's composition) and RDBMS module is the improvement of recognition efficiency and can be more powerful pattern recognition methods in the near future. The result of the experiment for hand gesture recognition was successful and satisfactory. To more flexible and natural hand gesture recognition, artificial intelligence and pattern recognition algorithm such as fuzzy-neural network or fuzzy-control should be applied. And also we must encode human's gesture to five senses fusion for the Post PC.

Acknowledgement

This work was supported by the MIC (The Ministry of Information & Communication) and NGV (Next Generation Vehicle technology).

References

1. http://iita57.iita.re.kr/IITAPortalEn/New/PostPC.htm
2. Jong-Sung Kim, Won Jang, Zeungnam Bien.: A Dynamic Gesture Recognition System for the Korean Sign Language (KSL). Ph.D. Thesis (1996)
3. Balaniuk R, Laugier C.: Haptic interfaces in generic virtual reality systems. In: Proceedings of the IEEE/RSJ International Conference on Intelligent Robots and Systems (2000) 1310–5
4. http://www.itfind.or.kr
5. http://www.etri.re.kr
6. Seung-Guk Kim.: Korean Standard Sign Language Dictionary, Osung publishing company (1995) 107–112
7. http://www.5dt.com/products/pdataglove5.html
8. http://www.aptronix.com/fide/whyfuzzy.htm

An Ontology-Based Method for Project and Domain Expert Matching

Jiangning Wu and Guangfei Yang

Institute of Systems Engineering, Dalian University of Technology, Dalian, 116024, China
jnwu@dlut.edu.cn, gfyang@student.dlut.edu.cn

Abstract. In this paper, we present a novel method to find the right expert who matches a certain project well. The idea behind this method includes building domain ontologies to describe projects and experts and calculating similarities between projects and domain experts for matching. The developed system consists of four main components: ontology building, document formalization, similarity calculation and user interface. First, we utilize Protégé to develop the predetermined domain ontologies in which some related concepts are defined. Then, documents concerning experts and projects are formalized by means of concept trees with weights. This process can be done either automatically or manually. Finally, a new method that integrates node-based and edge-based approach is proposed to measure the semantic similarities between projects and experts with the help of the domain ontologies. The experimental results show that the developed information matching system can reach the satisfied recall and precision.

1 Introduction

The word "ontology" has gained a great popularity within the information retrieval community. There are a number of definitions about the ambiguous term "ontology" [1]. Different definitions have different level of detail and logic [9]. In our work, the ontology is a collection of concepts and their interrelationships, and serves as a conceptualized vocabulary to describe an application domain. So far ontology has been successfully developed to solve different kinds of real problems. In this work, we focus on the project and domain expert matching problem.

For project reviewing problem, it is not easy to choose an appropriate domain expert for a certain project if experts' research areas and the contents of the projects are not known ahead very well. It is also a hard work when the number of projects is much high. So there is a great need for the effective technology that can capture the knowledge involved in both domain experts and projects. Nowadays, the popular strategy for such problem is people choose proper experts for the given projects with the help of querying database through computer information systems. These kinds of systems can only provide keyword-based search. They cannot identify which expert is more suitable than the others. In addition, the effectiveness of keyword-based retrieval system is not good.

To solve the above matching problem effectively, we propose an ontology-based method that presents knowledge related to both experts and projects by using a set of concepts coming from the ontology library. The derived concept tree is much suitable for computing the semantic similarity between each pair of project and expert. Once the semantic similarity is obtained, the matching process can be done by selecting the expert with the highest similarity value who is thought to be the best choice for the given project.

There are two main ways to calculating the semantic similarity for two concepts, they are node-based and edge-based approaches [8]. But these two approaches only discuss the similarity between two individual concepts without considering the similarities between two sets of concepts. However, a text document is usually represented by a set of concepts. When matching documents, it is necessary to compute the semantic similarities between two sets of concepts. Such kind of problem is solved in this paper by using a new integrated method whose main idea comes from the node-based and edge-based approaches. To apply the proposed method to our matching problem, we first select a domain of computer science and engineering, and then develop an ontology-based information matching prototype system. The experimental results show that our system can get better *Recall* and *Precision*.

The advantage of our work lies in incorporating the statistical information with the semantic information embedded in the hierarchical structure of the ontology. The proposed integrated method can give more intelligent supports for projects and domain experts matching.

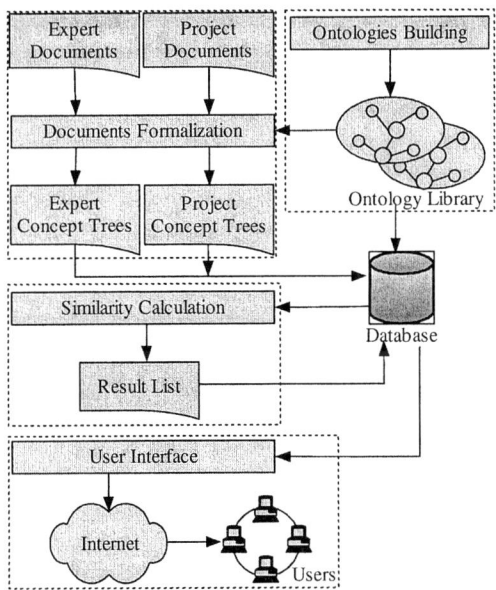

Fig. 1. System architecture

2 System Architecture

Shown as in Figure 1, four main modules consist of our proposed ontology-based information matching system: Ontologies building, Documents formalization,

Similarity calculation and User interface. The brief descriptions about these four parts are given as follows.

Ontology building: We adopt Protégé [11], developed by Stanford University, to build our domain ontologies. The concepts and relations are from the Chinese standard subject classification.

Document formalization: Benefiting from the ontologies that we have built, we can use the concepts to formalize the documents containing information about domain experts and projects.

Similarity calculation: By conducting the proposed integrated method to the concept trees corresponding to projects and domain experts respectively, we can calculate the similarities between them and rank the candidate domain experts afterwards. As a result, the most appropriate domain expert can be obtained.

User interface: This matching system implements the typical client-server paradigm. End users can access and query the system from the Internet, while domain experts or system administrators can manipulate the formalization and ontology building process.

2.1 Ontology Building

In the first step, we use the Chinese standard subject classification to build the ontology prototype with Protégé. Before using the ontology to formalize the documents, we should carry out the disambiguation process to remove the polysemy phenomena. Our system performs this phase by assigning a central concept for its all synonyms.

2.2 Document Formalization and Concept Extraction

We propose two ways to formalizing the documents (See Figure 2): automatically or manually.

In order to automatically process the unstructured documents like html, doc, txt, etc, we should firstly construct a Chinese domain word list for segmentation, which consists of concepts frequently used in the specified domain. When segmenting the Chinese documents into words, MM Method (Maximum Matching Method) is used. The outputs of segmentation are a list of words each of which is with a weight value. This value can be calculated by the frequency of each word.

Below we give an example to show the automatic process. After segmentation, document A is represented by a word list,

$$L_A = \{ w_1, w_2, \ldots, w_i, \ldots, w_n \}, \qquad (1)$$

where w_i represents the ith ($1 \leq i \leq n$) word, n is the number of words in document A.

The corresponding frequencies are

$$F_A = \{ f_1, f_2, \ldots, f_i, \ldots, f_n \}, \qquad (2)$$

where f_i represents the frequency value of the ith ($1 \leq i \leq n$) word.

$$f_k = \text{Max}(f_i), \tag{3}$$

$$f_i' = \frac{f_i}{f_k}, \tag{4}$$

$$F_A' = \{f_1', f_2', \cdots, f_n'\}, \tag{5}$$

where F_A' is the frequency list and we call it the weight vector of the concepts. Therefore document A can be formalized by using the following weight vector.

$$Weight(A) = \{weight_1, weight_2, \ldots, weight_n\} = \{f_1', f_2', \cdots, f_n'\}, \tag{6}$$

where $weight_i$ is the weight of the word w_i.

In manual way, we develop a tool, named Concept Filler, to process the document into words by assigning their weights manually in order to improve the precision of concept extraction (see also Figure 2). The Concept Filler is simply an interface to help the user assign proper concepts and weights by hand. The interface for specifying the concepts by the concept filler is shown in Figure 3. The concepts selected from the ontology hierarchy are finally stored into the database. The corresponding weight value ranging from 0 to 1 is assigned to each concept aiming to distinguish its importance, the bigger value the more importance.

After document formalizing, the remaining task is to calculate the similarities between concepts for matching. In the following section, we will discuss the similarity calculation method in somewhat detail.

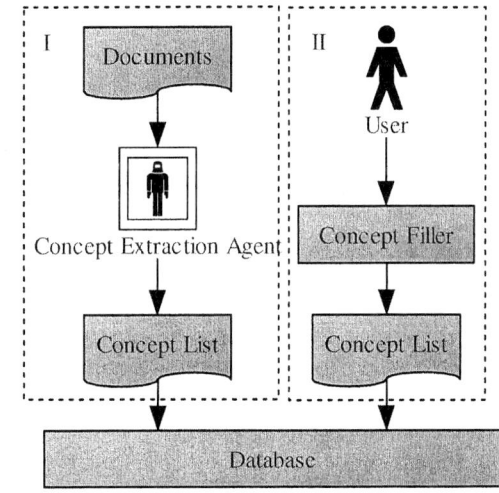

Fig. 2. Document formalization

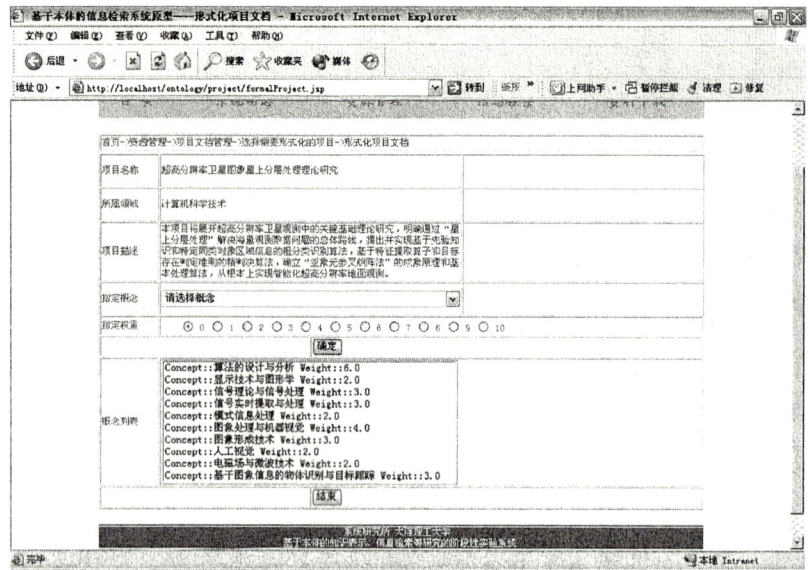

Fig. 3. The interface for specifying concepts by the concept filler

3 Similarity Calculation and Matching Process

As we have known, there are many kinds of interrelations between concepts in ontology construction, such as IS-A relation, Kind-of relation, Part-of relation, Substance-of relation, and so on. Calculating the similarity between concepts based on the complex interrelations is a challenging work. However, there unfortunately are not any methods that can deal with the above problem effectively up to now. From the other hand, IS-A (hyponym / hypernym) relation is the most common concern in ontology presentation. Some similarity calculation methods have been developed based on this simple relation [8]. So in this scenario, only IS-A relation is considered in our research. And we let the other relations be the future research topics.

3.1 Node-Based Approach and Edge-Based Approach

Here we want to discuss two classes of methods for calculating semantic similarities between concepts, they are node-based method and edge-based method [8].

Resnik used information content to measure the similarity [3] [7]. His point is that the more information content two concepts share, the more similarity two concepts have. The similarity of two concepts c_1 and c_2 is quantified as:

$$sim(c_1, c_2) = \max_{c \in Sup(c_1, c_2)} [-\log p(c)], \quad (7)$$

where $Sup(c_1, c_2)$ is the set of concepts whose child concepts contain c_1 and c_2, $p(c)$ is the probability of encountering an instance of concept c, and

$$p(c) = \frac{freq(c)}{N}, \tag{8}$$

where $freq(c)$ is simply the statistical frequency of concept c, and N is the total number of concepts contained in the given document. The frequency of concept c can be computed beforehand. Considering multiple inheritances where many words may have more than one concept sense, similarity calculation should be modified as:

$$sim(w_1, w_2) = \max_{c_1 \in sen(w_1), c_2 \in sen(w_2)} [sim(c_1, c_2)], \tag{9}$$

where $sen(w)$ means the set of possible different concepts denoted by the same word.

Another important method to quantify the similarity is the edge-based approach. Leacock and Chodorow summed up the shortest path length and converted this statistical distance to the similarity measure [4]:

$$sim(w_1, w_2) = -\log[\frac{\min_{c_1 \in sen(w_1), c_2 \in sen(w_2)} len(c_1, c_2)}{2 \times d_{max}}]. \tag{10}$$

where $len(c_1, c_2)$ is the number of edges along the shortest path between concepts c_1 and c_2, and d_{max} is the maximum depth of the ontology hierarchy.

3.2 An Integrated and Improved Approach

From the above section, we can see that there are three main shortcomings in the above two methods. By analyzing problems involved we suggest some improvements accordingly.

1. Both node-based and edge-based methods only simply consider two concepts in the concept tree without expanding to two lists of concepts in different concept trees, whereas the documents are usually formalized into lists of concepts. To solve this problem, we have to propose a new method that can calculate the similarities between two lists of concepts, by which the quantified similarity value can show how similar the documents are.

2. The node-based method does not concern the distance between concepts. In Figure 4 for example, if concepts B_1, C_1 and C_6 have only one sense individually and the same frequency that determines the same information content, we may have the following result according to the node-based method:

$$sim(B_1, C_1) = sim(B_1, C_6). \tag{11}$$

From Figure 4, it is easy to see that concepts B_1 and C_1 are more similar.

3. The edge-based method does not consider the weight of each concept. For example, both concepts C_1 and C_2 in Figure 4 have only one edge with B_1, we may get the same similarity value using the edge-base method:

$$sim(B_1, C_1) = sim(B_1, C_2). \tag{12}$$

But, if C_1 has bigger weight than C_2, C_1 is usually more important and the similarity between B_1 and C_1 should be greater.

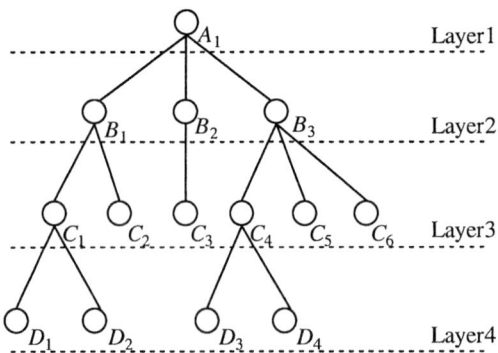

Fig. 4. Concept tree with four hierarchies

To overcome the shortcomings of both node-based and edge-based methods, a new integrated method is proposed in this paper to calculate the similarity between two documents. Before conducting the proposed method, the documents related to projects and domain experts should be formalized first that results in two vectors containing the concepts with their frequencies. Suppose a document $Document(1)$ describing a project, and a document $Document(2)$ describing a domain expert, the formalization results are:

$$Document(1) = \{c_{11}, c_{22}, ..., c_{1m}\}, \quad (13)$$

$$Document(2) = \{c_{21}, c_{22}, ..., c_{2n}\}, \quad (14)$$

with their corresponding frequencies:

$$Weight(1) = \{weight_{11}, weight_{12}, ..., weight_{1m}\}, \quad (15)$$

$$Weight(2) = \{weight_{21}, weight_{22}, ..., weight_{2n}\}. \quad (16)$$

For each pair of concepts (c_{1i}, c_{2j}) in the concept tree, there must be a concept c, for which both c_{1i} and c_{2j} are child concepts, and the path length is minimum. Concept c is the nearest parent concept for both c_{1i} and c_{2j}. The similarity between c_{1i} and c_{2j} can be calculated by

$$sim(c_{1i}, c_{2j}) = -\log(\frac{weight_{1i}}{len(c, c_{1i})+1} + \frac{weight_{2j}}{len(c, c_{2j})+1}), \quad (17)$$

where $len(c, c_{1i})$ is the path length between c and c_{1i}. Considering multiple senses of the concepts, we improve the calculation equation as:

$$sim(C_{1i}, C_{2j}) = \max_{c_{1i} \in sen(C_{1i}), c_{2j} \in sen(C_{2j})} [sim(c_{1i}, c_{2j})], \quad (18)$$

where C_{1i} is the sense representation. We calculate the maximum similarity value among all candidate concepts:

$$SIM = \max[sim(C_{1i}, C_{2j})] . \qquad (19)$$

Then, the similarity between two documents can be calculated by the following formula:

$$sim(Document(1), Document(2)) = \frac{\sum_{i=1}^{m}\sum_{j=1}^{n}\frac{sim(C_{1i}, C_{2j})}{SIM}}{m \times n} . \qquad (20)$$

4 Experimentation and Evaluation

We carry out a series of experiments to compare and evaluate edge-based approach, node-based approach and our integrated approach. Normally we use two measures precision and recall to evaluate an information retrieval system. In our research, we also use these two measures to verify the efficiency of the developed information matching system. Let R be the set of relevant documents, and A be the answer set of documents. The precision and recall are defined as follows respectively:

$$Precision = \frac{|A \cap R|}{|A|} \times 100\%, \qquad (21)$$

$$Recall = \frac{|A \cap R|}{|R|} \times 100\% . \qquad (22)$$

In the experiment, we collect around 800 domain experts (including professors, engineers, researchers, etc) and over 500 projects within the domain of computer science and engineering. Table 1 shows the different precision and recall results using three different methods with different number of projects. Also the comparison charts are given in Figures 5 and 6 respectively.

Table 1. Precision and recall comparison. E-based denotes edge-based approach, N-based denotes node-based approach, Integrated denotes integrated approach.

	Projects	Precision			Recall		
		E-based	N-based	Integrated	E-based	N-based	Integrated
1	100	20.65%	25.99%	30.71%	30.56%	32.28%	39.04%
2	200	22.32%	25.85%	28.93%	31.00%	33.98%	34.73%
3	300	27.55%	19.32%	32.79%	23.56%	30.46%	42.92%
4	400	20.38%	27.61%	31.59%	30.87%	35.43%	32.96%
5	500	23.40%	23.44%	29.63%	33.70%	43.75%	49.74%

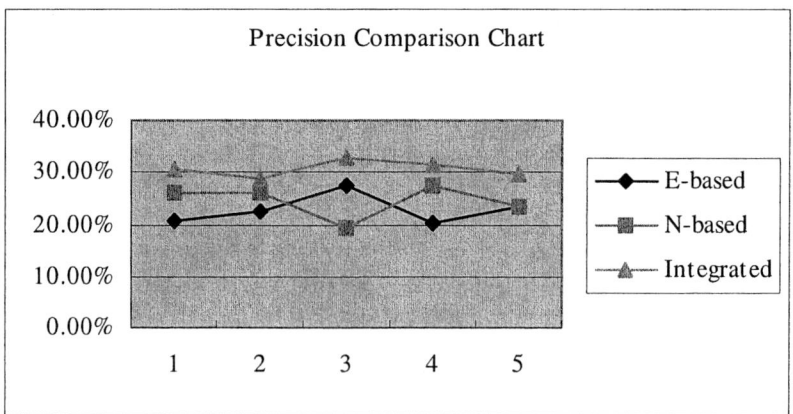

Fig. 5. Precision comparison

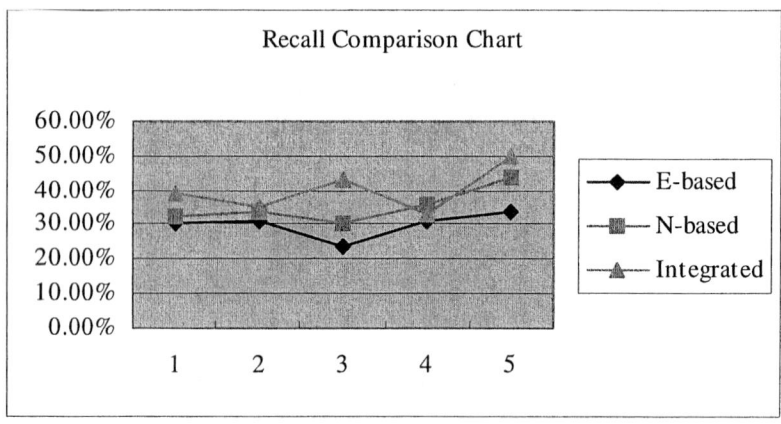

Fig. 6. Recall comparison

5 Conclusions and Future Works

In this paper, we present an ontology-based method to match projects and domain experts. The prototype system we developed contains four modules: Ontology building, Document formalization, Similarity calculation and User interface. Specifically, we discuss node-based and edge-based approaches to computing the semantic similarity, and propose an integrated and improved approach to calculating the semantic similarity between two documents. The experimental results show that our information matching system can reach better recall and precision.

As mentioned previously, only the simplest relation "IS-A relation" is considered in our study. When dealing with the more complex ontology whose concepts are restricted by logic or axiom, our method is not powerful enough to describe the real semantic meaning by merely considering the hierarchical structure. So the future work should focus on the other kinds of relations that are used in ontology construction. In other words, it will be an exciting and challenging work for us to compute the semantic similarity upon various relations in the future.

References

1. Guarino, N., Giaretta, P.: Ontologies and knowledge bases: Towards a terminological clarification. In N.J.I. Mars, editor, Towards Very Large Know ledge Bases. IOS Press (1995)
2. Heftin, J., Hendler, J.: Searching the Web with SHOE. In Artificial Intelligence for Web Search. Papers from the AAAI Workshop. 01. AAAI Press (2000) 35–40
3. Resnick, P.: Using information content to evaluate semantic similarity in a taxonomy. In Proceedings of the Fourteenth International Joint Conference on Artificial Intelligence (IJCAI-95) (1995) 448–453
4. Leacock, C., Chodorow, M.: Filling in a sparse training space for word sense identification. ms (1994)
5. Salton, G., Wong, A.: A Vector Space Model for Automatic Indexing. Communications of the ACM 18 (1975) 613-620
6. Guarino, N., Masolo, C., Vetere, G.: OntoSeek: Content-Based Access to the Web. IEEE Intelligent Systems, 14(3) (1999) 70–80
7. Resnik, P.: Semantic similarity in a taxonomy: An information-based measure and its application to problems of ambiguity in natural language. Journal of Artificial Intelligence Research 11 (1999) 95-130
8. Jiang, J., Conrath, D.: Semantic similarity based on corpus statistics and lexical taxonomy, In Proc. of Int'l Conf. on Research on Computational Linguistics, Taiwan (1997)
9. Gruber, T.: Toward principles for the design of ontologies used for knowledge sharing. Technical Report KSL-93-4, Knowledge Systems Laboratory, Stanford University. Communications of the ACM 37(7) (1993) 48–53
10. Decker, S., Erdmann, M., Fensel, D., Studer, R.: Ontobroker: Ontology based access to distributed and semi-structured information. In R. Meersman, editor, Semantic Issues in Multimedia Systems. Proceedings of DS-8. Kluwer Academic Publisher (1999) 351–369
11. Protégé: http://protege.stanford.edu

Pattern Classification and Recognition of Movement Behavior of Medaka (*Oryzias Latipes*) Using Decision Tree

Sengtai Lee[1], Jeehoon Kim[2], Jae-Yeon Baek[2], Man-Wi Han[2], and Tae-Soo Chon[3]

[1] School of Electrical Engineering, Pusan National University,
Jangjeon-dong, Geumjeong-gu, 609-735 Busan, Korea
youandi@pusan.ac.kr
[2] Korea Minjok Leadership Academy,
Sosa-ri, Anheung-myeon, Heongseong-gun, Gangwon-do, 225-823, Korea
{fantasy002, mrswoolf}@hanmail.net, manwihan@chol.com
[3] Division of Biological Sciences, Pusan National University,
Jangjeon-dong, Geumjeong-gu, 609-735 Busan, Korea
tschon@pusan.ac.kr

Abstract. Behavioral sequences of the medaka (*Oryzias latipes*) were continuously investigated through an automatic image recognition system in increasing temperature from 25°C to 35°C. The observation of behavior through the movement tracking program showed many patterns of the medaka. After much observation, behavioral patterns could be divided into basically 4 patterns: active-smooth, active-shaking, inactive-smooth, and inactive-shaking. The "smooth" and "shaking" patterns were shown as normal movement behavior, while the "smooth" pattern was more frequently observed in increasing temperature (35°C) than the "shaking" pattern. Each pattern was classified using a devised decision tree after the feature choice. It provides a natural way to incorporate prior knowledge from human experts in fish behavior and contains the information in a logical expression tree. The main focus of this study was to determine whether the decision tree could be useful in interpreting and classifying behavior patterns of the medaka.

1 Introduction

Ecological data are very complex, unbalanced, and contain missing values. Relationships between variables may be strongly nonlinear and involve high-order interactions. The commonly used exploratory and statistical modeling techniques often fail to find meaningful ecological patterns from such data [1], [2], [3]. The behavioral or ecological monitoring of water quality is important regarding bio-monitoring and risk assessment [4], [5]. An adaptive computational method was utilized to analyze behavioral data in this study. Decision tree is modern statistical techniques ideally suited for both exploring and modeling such data. It is constructed by repeatedly splitting the data, defined by a simple rule based on a single explanatory variable.

The observation of the movement tracks of small sized animals has been separately initiated in the field of search behavior in chemical ecology [6] and computational behavior [7], [8]. For searching behavior, the servometer and other tools were used for investigating the continuous movement tracks of insects, including cockroaches, in characterizing the effects of wind [9], pheromone [10], [11], relative humidity [12], and sucrose feeding [13].

These computational methods convey useful mathematical information regarding similarities present in data of the movement tracks, for instance, correlation coefficients or fractal dimensions. By these methods, however, the parameters are obtained through mathematical transformations of the movement data, and information is in a general highly condensed state. These methods are usually not interpretable for uniquely and directly characterizing the actual shape of the movement tracks.

In this paper, we utilized the decision tree for the classification of response behaviors and attempted to explain the shapes of the movement tracks through feature extraction in increasing temperature. Realizing there is a limit to observing with the naked eye, computational methods were used to conduct our research more effectively. First, statistical analysis in total moving distance, average speed, and in sectional domination was conducted as a feature extraction. Furthermore, we devised a new analysis method for pattern isolation based on a decision tree to differentiate the patterns we thought were distinctive. This research can help the biosensor field in detecting defects in fish, or in finding out chemical toxicants that exist in the water by observing specific behavior patterns of fish.

2 Experiment for Data Acquisition

The specimens of medaka (*Oryzias latipes*) used in our experiment were obtained from the Toxicology Research Center, Korea Research Institute of Chemical Technology (KRICT; Taejon, Korea). Only the specimens six to twelve months from birth were used. The medaka is about 4cm in length and lives for about 1-2 years.

Before the experiment, they were maintained in a glass tank and were reared with an artificial dry diet (Tetramin™, Tetra Werke, Germany) under the light regime of Light 10: Dark 14 at a temperature of $25 \pm 0.5°C$. The water in the tank was continually oxygenated prior to experimenting.

A day before experimentation, the medaka was put into the observation tank and was given approximately twelve hours to adjust. The specimen was kept in a temperature of 25°C and was given sufficient amount of oxygen during these twelve hours prior to the experiment. The specimens used were male and about 4cm in length. In order to achieve image processing and pattern recognition effectively, stable conditions were maintained in the monitoring system. Disturbances to observation tanks and changes in experimental conditions were minimized. Aeration, water exchange and food were not provided to test specimens during the observation period and the light regime was kept consistent.

The aquarium used was based on the experiments on fish behavior in KRICT. The observed aquarium size was 40cm×20cm×10cm. The temperature was adjusted by using the circulator. The heated water from the circulator flows into the tank and then flows back into the circulator. The rate of water flow was 16m/s. The analog data captured by the camera set in front of aquarium were digitized by using the video overlay board every 0.25 seconds and were sent to the image recognition system to locate the target in spatial time domains. The spatial position of the medaka was recorded in two-dimensional x, y coordinate values.

After giving the experimenting specimen approximately twelve hours to adjust to the observation aquarium, the experiment was started. The experiment was started at about 8:00~8:30 AM every day. The initial temperature was 25°C. After 2 hours, the setting temperature was increased to 35°C using the water circulatory system. Depending on the day and external temperature, the time it took for the aquarium to elevate to 35°C varied. 90 minutes for the temperature of 25°C and 35°C, 30~60 minutes for the transition period were used as data. Each data from a movement pattern had an interval of one minute, and were overlapped every 30 seconds for analysis.

3 Feature Extraction Process

In this paper, the movement patterns of the medaka were classified into shaking and smooth patterns as shown in Fig. 1. The behavior of the medaka in a one minute period of time was used to classify them into 5 patterns: active-smooth, active-shaking, inactive-smooth, inactive-shaking, and not determined in each case. "Not determined" are patterns that were not classified into any one of these four categories. By the observation of an expert in fish behavior to initiate pattern isolation, the features were observed and the following three feature variables could be defined: high-speed ratio, FFT (Fast Fourier transformation) to angle transition, and projection to x- and y-axes. Fig. 2 shows the schematic diagram in one minute of the movement analysis for the process of extracting three distinctive characteristics from the data we acquired and classifying 5 patterns based on this information. It is possible that for medaka treated with sub-lethal chemicals, there might be patterns that cannot be classified. However, for these cases, a new analysis can be done to add new patterns and update the decision tree.

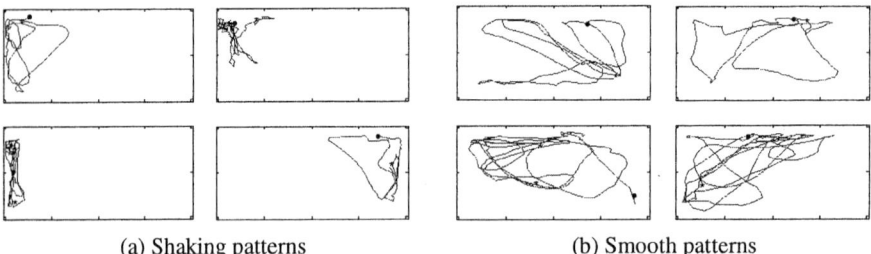

(a) Shaking patterns (b) Smooth patterns

Fig. 1. Examples of the shaking and smooth patterns in one minute (•: start, *: end)

In order to know the activeness of a medaka, speed information was used to define high-speed ratio. Speed of the medaka shows that it is an active movement or inactive movement. The formula of speed is as the following:

$$S = \sqrt{(x_{n+1} - x_n)^2 + (y_{n+1} - y_n)^2} \quad n = 1, 2, 3, \cdots \quad (1)$$

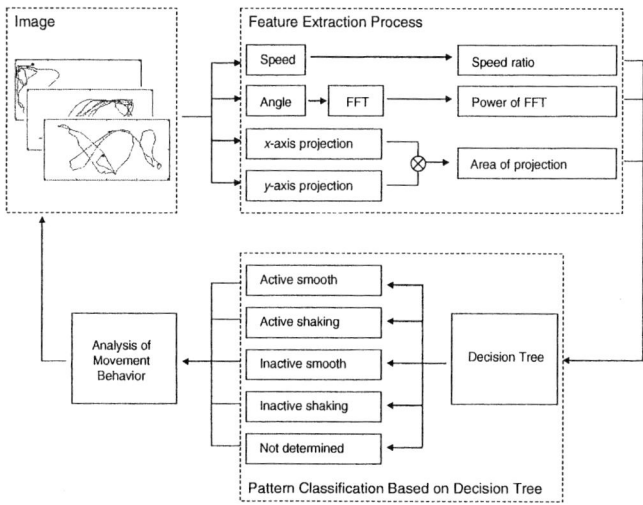

Fig. 2. Schematic diagram for automatic pattern isolation

Where, x_n and y_n are the position values of the medaka in a sampled time. The ratio that exceeded the calculated average speed of the overall 7 data sets, 21mm/sec, was used as the first feature variable. High-speed ratio is calculated as the following equation.

$$S_{ratio} = \frac{\text{Number of samples above A2}}{\text{Number of samples in one minute}} \times 100(\%) \tag{2}$$

The change of direction in the movement track was observed to consider movement of medaka. The change of direction is represented as an angle transition to classify the movement behavior of medaka. Angle transition between two sampled times denoted as H is calculated in the following equation. Where x_n and y_n shows the coordinate value for the x and y axes.

$$H = \arctan\left(\frac{y_{n+1} - y_n}{x_{n+1} - x_n}\right), \quad n = 1, 2, \ldots \tag{3}$$

Fourier transformation is used to transform signals in the time domain to signals in the frequency domain [20]. We apply the Fast Fourier transform (FFT) to the signal of angle transition to calculate energy. The FFT for a given discrete signal $x[n]$ is calculated through the following equation:

$$X[k] = \sum_{n=0}^{N-1} x[n] \cdot e^{-j(2\pi k n / N)}, \quad k = 0, 1, \cdots, N-1. \tag{4}$$

After applying the FFT to angle transition, the power of FFT (PF) is calculated in the following equation for the amplitudes above a median.

$$P = \sqrt{\sum_{i=1}^{k} x_i^2} \tag{5}$$

Where x_i is the amplitudes above a median. We use all sets to find median in experiments. We are used to FFT power because of the calculation in qualified angle transition. The PF is employed as a second feature variable for pattern isolation. Fig. 3 shows a program that was devised by Matlab environment to analyze the data that was acquitted consecutively.

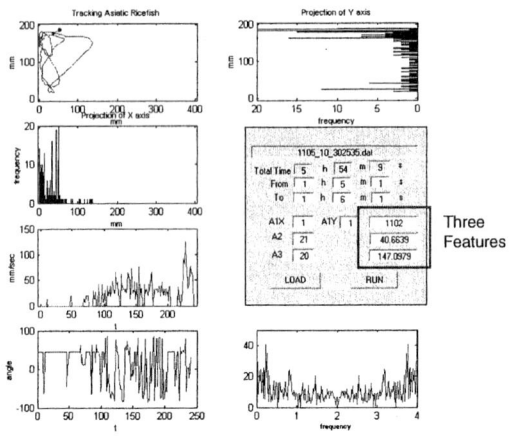

Fig. 3. Movement tracking program

4 Pattern Classification Based on Decision Tree

4.1 Decision Tree

A decision tree is a graph of decisions and their possible consequences, used to create a plan to reach a goal. Decision trees are constructed in order to help make decisions. It has interpretability in its own tree structure. Such interpretability has manifestations which can easily interpret the decision for any particular test pattern as the conjunction of decisions along the path to its corresponding leaf node. Another manifestation can occasionally get clear interpretations of the categories themselves, by creating logical descriptions using conjunctions and disjunctions [3], [19].

Many people related to artificial intelligence research has developed a number of algorithms that automatically construct decision tree out of a given number of cases, e.g. CART [1], ID3 [14], [15], C4.5 [16], [17], [18]. The C4.5 algorithm, the successor and refinement of ID3, is the most popular in a series of "classification" tree methods. In it, real-valued variables are treated the same as in CART.

A decision tree consists of nodes(N) and queries(T). The fundamental principle underlying tree creation is that of simplicity. We prefer decisions that lead to a simple, compact tree with few nodes. During the process of building the decision tree, we seek a property query T at each node N that makes the data reaching the immediate descendent nodes as "pure" as possible. It turns out to be more convenient to define the impurity, rather than the purity of a node. Several different mathematical measures of impurity have been proposed, i.e. entropy impurity (or occasionally information impurity), variance impurity, *Gini* impurity, misclassification impurity in equation (6), (7), (8), (9), respectively.

$$i(N) = -\sum_j P(\omega_j) \log_2 P(\omega_j) \tag{6}$$

Where $i(N)$ denote the impurity of a node and $P(w_i)$ is the fraction of patterns at node N that are in category w_j.

$$i(N) = P(\omega_1)P(\omega_2) \tag{7}$$

$$i(N) = \sum_{i \neq j} P(\omega_i) P(\omega_j) = 1 - \sum_j P^2(\omega_j) \tag{8}$$

$$i(N) = 1 - \max_j P(\omega_j) \tag{9}$$

All of them have basically the same behaviors. By the well-known properties of entropy, if all the patterns are of the same category, the entropy impurity is 0. A variance impurity is particularly useful in the two-category case. A generalization of the variance impurity, applicable to two or more categories, is the *Gini* impurity in equation (8). This is just the expected error rate at node N if the category label is selected randomly from the class distribution present at N. The misclassification impurity measures the minimum probability that a training pattern would be misclassified at N. Of the impurity measures typically considered, this measure is the most strongly peaked at equal probabilities.

In order to drop in impurity, we used the equation (10)

$$\Delta i(N) = i(N) - P_L i(N_L) - (1 - P_L) i(N_R) \tag{10}$$

Where N_L and N_R are the left and right descendent nodes, $i(N_L)$ and $i(N_R)$ are their impurities, and P_L is the fraction of patterns at node N that will go to N_L when property query T is used. Then the "best" query value s is the choice for T that maximizes $\Delta i(T)$.

If we continue to grow the tree fully until each leaf node corresponds to the lowest impurity, then the data have been typically overfitted. Conversely, if splitting is stopped too early, then the error on the training data is not sufficiently low and hence performance may suffer. To search sufficient splitting value, we used cross-validation (hold-out method). In validation, the tree is trained using a subset of the data with the remaining kept as a validation set.

4.2 Implementation of Decision Tree

We analyzed movement tracks of the medaka using Matlab6.1. The decision tree is employed and programmed to express the classification in the form of a tree and as a set of *IF-THEN* rules.

In order to classify the patterns into active smooth, active shaking, inactive smooth, and inactive shaking divided by experts in fish behavior, the following features were used: high speed ratio (HSR), power of FFT (PF), and area of projection product (APP). These 3 features were used as input variables to decision tree. The training data for the decision tree consisted of 30 data in each patterns. The decision tree gives a rough picture of the relative importance of the features influencing movement tracks of the medaka. We continue splitting nodes in successive layers until the error on the validation data is minimized. The principal alternative approach to stopped splitting is pruning. Fig. 4 shows the decision tree applied to evaluated pruning. This benefit of this logic is that the reduced rule set may give improved interpretability.

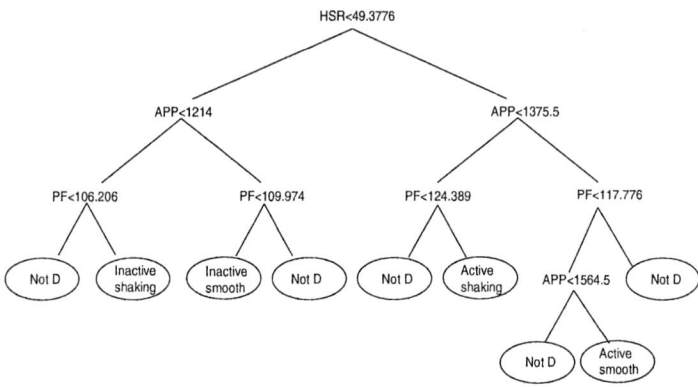

Fig. 4. The decision logic for pattern classification generated by decision tree applied to pruning. (HSR: high-speed ratio, APP: area of projection product, PF: power of FFT).

5 Behavior Analysis and Discussion

5.1 Analysis of Movement Behavior

Results were calculated for the decision logic for 90 minutes at a temperature of 25°C. This was the same for the temperature at 35°C. Also, a time period of 30~60 minutes was calculated for the transition period, in which the temperature was raised from 25°C to 35°C. The total number of specimens used in the experiment was 7. The recognition is calculated by 4 patterns over 5 patterns that includes "not determined." "Smooth" means that "active smooth" patterns and "inactive smooth" patterns appeared in the decision tree logic. "Shaking" means that "active shaking" patterns and "inactive shaking" patterns appeared in the decision tree logic. "Not determined" means that neither "smooth" nor "shaking" appeared in the decision tree logic.

Fig. 5 shows the ratio of smooth and shaking patterns. Each specimen is represented by 3 bar graphs. The first bar graph shows the ratio of smooth and shaking patterns in 25°C, the second bar graph shows the ratio in the temperature elevation period from 25°C to 35°C, and the third bar graph shows the ratio in 35°C. Most specimens showed an increase in smooth patterns detected by the decision tree logic in the temperature elevation period from 25°C to 35°C. In the temperature elevation period (25~35) there were significantly more smooth patterns compared to 25°C but compared to 35°C, 3sets (b, c, g) showed more smooth patterns, 2 sets showed less smooth patterns and the remaining 2 sets showed little difference.

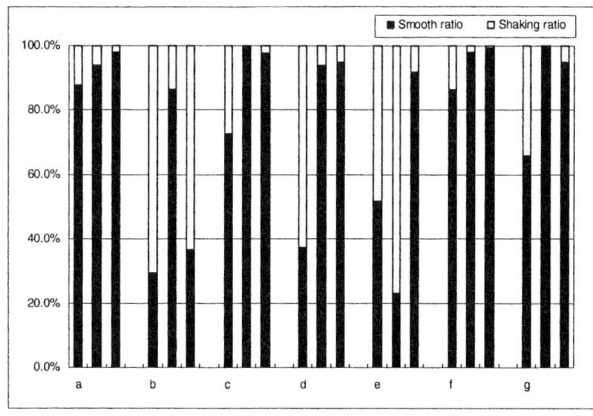

Fig. 5. Smooth ratio vs shaking ratio

5.2 Discussion

This study demonstrated that behavioral differences of animals in response to an insecticide could be detected by a decision tree with 3 features of behavior. One difficulty of conducting this type of monitoring study is the necessity of handling a large amount of data. This produced a gigantic amount of data. The automatic pattern recognition system solved this time-consuming problem in detecting response behaviors. Besides time consumption in recognition, objectivity in judgments for classification has been another problem for manual recording. The application of machine intelligence to behavioral data has the advantage of classifying the movement patterns on a more objective basis. In this regard the pattern recognition by a decision tree was demonstrated as an alternative for detecting the movement tracks of animals.

Another problem that arouses from this experiment is that biological specimens such as the medaka show too many different types of movement patterns. This makes selecting certain characteristics for a certain pattern difficult. This is why so many artificial systems such as neural networks and fuzzy are being used [21], [22]. However, although neural networks are sufficiently able to differentiate patterns, it is impossible to interpret exactly how much a certain pattern the specimen shows.

The results revealed that after differentiating smooth and shaking patterns through a decision tree, temperature increase caused the smooth ratio to increase. This can be seen as a pattern that appears in response to a new environment, such as temperature change, and is a process of adaptation. Shaking patterns show a lot of angle change and can be seen as a pattern right before adaptation, and it can be said that it appears the most frequently. Speed ratio of the medaka shows whether it is an active movement or inactive movement as shown in Fig. 4. Also, the area of projection product interprets smooth or shaking pattern. Power of FFT distinguishes specific patterns from unknown patterns.

Biologically, results showed that variable such as smooth ratio vs shaking ratio distinguished in the constant temperatures of 25°C and 35°C in Fig. 5. There was a significant increase in those variables in the transition period from 25°C to 35°C. It can be inferred from these results that the activity did increase as the temperature began to rise. Although this is a short period of time it may be seen as a case of fast acclimation to temperature change by the medaka.

6 Conclusions

The complex movement data were used to construct a decision tree with 3 features that could represent movement tracks of medaka: speed ratio, power of FFT, and x- and y-axes projection product. As new input data were given to the decision logic, it was possible to recognize the changes of pattern in increasing temperature. It is possible that for medaka treated with sub-lethal chemicals, there might be patterns that cannot be classified. However, in these cases, a new analysis can be done to add new patterns and update the decision tree. The results of the decision tree revealed that medaka was interpretable in different temperature as speed, angle, area of projection to x- and y-axes. If this is applied to more sets of data, it is thought that more distinctive and accurate methods of differentiating the behavior patterns can be created. Also, this research in differentiating patterns may help in the field of research for the special characteristics of living organisms. This research can help the biosensor field in detecting defects in fish, or in finding out chemical toxicants that exist in the water by observing specific behavior patterns of fish.

References

1. Breiman, L., J. H. Friedman, R. A. Olshen, and C.G.Stone.: Classification and Regression Trees, Wadsworth International Group, Belmont, California. USA (1984)
2. Ripley, B. D.: Pattern recognition and neural networks. Cambridge University Press, Cambridge, UK (1996)
3. Richard, O. D., Peter, E.H., David, G.S.: Pattern Classification 2nd edn. Wiley Interscience, USA (2001)
4. Dutta, H., Marcelino, J., Richmonds, Ch.: Brain acetylcholinesterase activity and optomotor behavior in bluefills, Lepomis macrochirus exposed to different concentrations of diazinon. Arch. Intern. Physiol. Biochim. Biophys., 100(5) (1993) 331-334
5. Lemly, A. D., Smith, R. J.: A behavioral assay for assessing effects of pollutants of fish chemoreception. Ecotoxicology and Enviornmental Safety 11(2) (1986) 210-218
6. Bell, W. J.: Searching behavior patterns in insects. Annual Review of Entomology 35, (1990) 447-467
7. Alt, W., Hoffman, G. (Eds): Biological Motion. Lecture notes in Biomathematics 89. Springer-Verlag, Berlin (1989)
8. Tourtellot, M. K., Collins, R. D., Bell, W. J.: the problem of movelength and turn definition in analysis of orientation data. Journal of Theoretical Biology 150 (1991) 287-297
9. Bell, W. J., Kramer, E.: Search and anemotactic orientation of cockroach. Journal of Insect Physiology 25 (1975) 631-640
10. Bell, W. J., Kramer, E.: Sex pheromone-stimulated orientation of the American cockroach on a servosphere apparatus. Journal of Chemical Ecology 6 (1980) 287-295
11. Bell, W. J., Tobin, R. T.: Orientation to sex pheromone in the American cockroach: analysis of chemo-orientation mechanisms. Journal of Insect Physiology 27 (1981) 501-508

12. Sorensen, K. A., Bell, W. J.: Orientation responses of an isopod to temporal changes in relative humidity simulation of a "humid patch" in a "dry habitat," Journal of Insect Physiology 32 (1986) 51-57
13. White, J., Tobin, T. R., Bell, W. J., 1984. Local search in the house fly Musca domestica after feeding on sucrose. Journal of Insect Physiology 30 (1984) 477-487
14. Quinlan, J. R.: Discovering rules by induction from large collections of examples. In: Micjie, E. (Ed.), Expert Systems in the Micro-Electornic Age, Edinburgh University Press, Edinburgh. (1979) 168-201
15. Quinlan, J. R.: Induction of decision trees. Machine Learning, 1(1) (1986) 81-106
16. Quinlan, J. R.: C4.5: programs for machine Learning. Morgan Kaufmann, San Francisco, CA, (1993)
17. Quinlan, J. R.: Improved use of continuous attributes in C4.5, Journal of Artificial Intelligence, 4. (1996) 77-90
18. Quinlan, J. R., Ronald L.: Rivest.Inferring decision trees using the minimum description length principle, Information and Computation. 80(3) (1989) 227-248
19. Tom M. Mitchell.: Machine Learning. McGraw-Hill, New York. (1997)
20. Kreyszig, Erwin: Advanced Engineering Mathematics, 8th Ed, Wiley. (1999)
21. Chon, T.-S., Park, Y. S., Moon, K. H., Cha, E. Y.: Patternizing communities by using an artificial neural network, Ecological Modeling 90 (1996) 69-78
22. I. S. Kwak, T. S. Chon, H. M. Kang, N. I. Chung, J. S. Kim, S. C. Koh, S. K. Lee, Y. S. Kim.: Pattern recognition of the movement tracks of medaka (*Oryzias latipes*) in response to sub-lethal treatments of an insecticide by using artificial neural networks. Environmental Pollution, 120 (2002) 671-681

A New Algorithm for Computing the Minimal Enclosing Sphere in Feature Space

Chonghui Guo[1], Mingyu Lu[2], Jiantao Sun[2], and Yuchang Lu[2]

[1] Department of Applied Mathematics,
Dalian University of Technology, Dalian 116024, China
guochonghui@tsinghua.org.cn
[2] Department of Computer Science, Tsinghua University, Beijing 100084, China
{my-lu,lyc}@tsinghua.edu.cn, sjt@mails.tsinghua.edu.cn

Abstract. The problem of computing the minimal enclosing sphere (MES) of a set of points in the high dimensional kernel-induced feature space is considered. In this paper we develop an entropy-based algorithm that is suitable for any Mercer kernel mapping. The proposed algorithm is based on maximum entropy principle and it is very simple to implement. The convergence of the novel algorithm is analyzed and the validity of this algorithm is confirmed by preliminary numerical results.

1 Introduction

Given a set of l points in the n-dimensional Euclidean space, the minimal enclosing sphere (MES) problem is to find the sphere of minimum radius that encloses all given points. The MES problem arises in various application areas, such as pattern recognition and location analysis, and it is itself of interest as a classical problem in computational geometry. More description of these applications can be found in [1,2,3]. Many algorithms have been developed for the problem. Megiddo [4] presented a deterministic $O(l)$ algorithm for the case where $n \leq 3$. Welzl [5] developed a simple randomized algorithm with expected linear time in l for the case where n is small, and Gärtner described a C++ implementation thereof in [6]. The existing algorithms, however, are not efficient for solving problems in high-dimensional space.

In pattern recognition, the MES problem arises in VC-dimension estimation [3,7] and support vector clustering [8,9]. In this case, the data points are mapped by means of a inner product kernel to a high (or even infinite) dimension feature space, and we search for the MES that encloses the image of the data in kernel-induced feature space. This is in contrast to the cases arisen in other application areas where the dimension number of space is small. Solving the MES problem in feature space leads to a quadratic programming (QP) problem with non-negative constraints and one normality condition. This type of optimization problem can be solved via an off-the-shelf QP package to compute the solution. However, the MES problem possesses features that set it apart from general QPs, most notably the simplicity of the constraints. In fact, the MES

QP problem can be solved by the SMO algorithm originally proposed for support vector machines training [10]. But some modifications are required to adapt it to the MES QP. The goal of this paper is to present a simple and efficient algorithm which takes advantage of these features. We derive an entropy-based algorithm for the considered problem by means of Lagrangian duality and the Jaynes' maximum entropy principle. The research is motivated by the use of information entropy and maximum entropy formalism in the solution of nonlinear programming problems [11,12].

The article is organized as follows. In the next section, we formulate the MES problem in feature space as a convex quadratic programming and discuss some properties of the MES problem. We propose an entropy-based algorithm for the considered problem in section 3 and analyze the convergence of the algorithm in section 4. In section 5, we present some preliminary numerical results for the proposed algorithm. We give conclusions in section 6.

2 Problem Formulation

Let $X = \{x_1, x_2, ..., x_l\}$ be a set of l input vectors in Euclidean space R^n. Using a nonlinear transformation Φ from the input space R^n to the feature space H. We wish to compute the radius r of the smallest sphere in feature space H which encloses the mapped training data. We formulate the problem as follows:

$$\begin{cases} \min r^2 \\ s.t. \ \|\Phi(x_i) - c\|^2 \leq r^2, \quad i = 1, 2, ..., l, \end{cases} \quad (1)$$

where $\|\cdot\|$ is the Euclidean norm and $c \in H$ is the position vector of the center of the sphere. For very high-dimensional feature spaces it is not practical to solve the problem in the present form. However, this problem can be translated into a dual form that is solvable. To solve this problem, we introduce the Lagrangian

$$L(r, c, \lambda) = r^2 - \sum_{i=1}^{l} \lambda_i (r^2 - \|\Phi(x_i) - c\|^2), \quad (2)$$

where $\lambda_i \geq 0, i = 1, 2, ..., l$ are Lagrange multipliers.

Differentiating $L(r, c, \lambda)$ with respect to r and c, and setting the results equal to zero, lead to

$$\frac{\partial L}{\partial r} = 2r - 2r \sum_{i=1}^{l} \lambda_i = 0, \quad (3)$$

and

$$\frac{\partial L}{\partial c} = 2c - 2 \sum_{i=1}^{l} \lambda_i \Phi(x_i) = 0. \quad (4)$$

From (3) and (4), we can derive

$$\sum_{i=1}^{l} \lambda_i = 1, \quad (5)$$

and
$$\mathbf{c} = \sum_{i=1}^{l} \lambda_i \Phi(\mathbf{x}_i). \tag{6}$$

By substituting (5) and (6) into the Lagrangian (2) and taking into account the non-negativity of the Lagrange multipliers, we can obtain the Wolfe dual

$$\begin{cases} \max L(\lambda) = \sum_{i=1}^{l} \lambda_i \Phi^2(\mathbf{x}_i) - \sum_{i=1}^{l} \sum_{j=1}^{l} \lambda_i \lambda_j \Phi(\mathbf{x}_i) \cdot \Phi(\mathbf{x}_j) \\ s.t. \ \sum_{i=1}^{l} \lambda_i = 1 \\ \lambda_i \geq 0, \quad i = 1, 2, ..., l. \end{cases} \tag{7}$$

We follow the support vector machines method and represent the dot products $\Phi(\mathbf{x_i}) \cdot \Phi(\mathbf{x_j})$ by an appropriate Mercer kernel $K(\mathbf{x_i}, \mathbf{x_j})$. The optimization problem (7) is now written as

$$\begin{cases} \max L(\lambda) = \sum_{i=1}^{l} \lambda_i K(\mathbf{x}_i, \mathbf{x}_i) - \sum_{i=1}^{l} \sum_{j=1}^{l} \lambda_i \lambda_j K(\mathbf{x}_i, \mathbf{x}_j) \\ s.t. \ \sum_{i=1}^{l} \lambda_i = 1 \\ \lambda_i \geq 0, \quad i = 1, 2, ..., l. \end{cases} \tag{8}$$

Note that the optimal solution λ^*, r^* and $\mathbf{c}^*$ must satisfy the KKT complementary conditions

$$\lambda_i^*((r^*)^2 - \|\Phi(\mathbf{x}_i) - \mathbf{c}^*\|^2) = 0, \quad i = 1, 2, ..., l. \tag{9}$$

The KKT complementary conditions state that any Lagrangian multiplier λ_i is nonzero only if the corresponding data point $\Phi(\mathbf{x}_i)$ in feature space satisfies the constraint of optimization problem (1) with equality. Equation (9) implies that the data point $\Phi(\mathbf{x}_i)$ for which the Lagrangian multiplier λ_i is nonzero lies on the surface of the sphere in feature space. Such a point will be referred to as a support vector.

At each point $\mathbf{x}$, we define the distance of its image in feature space from the center of the sphere:
$$r^2(\mathbf{x}) = \|\Phi(\mathbf{x}) - \mathbf{c}\|^2. \tag{10}$$

In view of (6) and the definition of the kernel, we have

$$r^2(\mathbf{x}) = K(\mathbf{x}, \mathbf{x}) - 2\sum_{i=1}^{l} \lambda_i K(\mathbf{x}_i, \mathbf{x}) + \sum_{i=1}^{l} \sum_{j=1}^{l} \lambda_i \lambda_j K(\mathbf{x}_i, \mathbf{x}_j). \tag{11}$$

The radius of the MES in feature space is

$$r^* = \{r(\mathbf{x}_i) \mid \mathbf{x}_i \text{ is a support vector}\}. \tag{12}$$

In view of equation (12), support vectors lie on the MES surface and the other points lie inside the MES in feature space.

The following proposition gives an alterative approach to compute the radius of the MES in feature space.

Proposition 1. *Suppose that λ^* is an optimal solution to optimization problem (7) or (8), r^* is the radius of the MES in feature space, then*

$$(r^*)^2 = L(\lambda^*). \tag{13}$$

Proof. From the KKT complementary conditions (9), we have

$$\sum_{i=1}^{l} \lambda_i^*((r^*)^2 - \|\Phi(\mathbf{x}_i) - \mathbf{c}^*\|^2) = 0, \tag{14}$$

It follows from (14) that

$$\sum_{i=1}^{l} \lambda_i^*(r^*)^2 = \sum_{i=1}^{l} \lambda_i^* \|\Phi(\mathbf{x}_i) - \mathbf{c}^*\|^2, \tag{15}$$

By (5) and (6), we have

$$(r^*)^2 = \sum_{i=1}^{l} \lambda_i^* \Phi^2(\mathbf{x}_i) - \sum_{i=1}^{l}\sum_{j=1}^{l} \lambda_i^* \lambda_j^* \Phi(\mathbf{x}_i) \cdot \Phi(\mathbf{x}_j) = L(\lambda^*). \tag{16}$$

This completes the proof of Proposition 1.

3 Entropy-Based Approach

The last section has formulated a QP (8) for computing the MES problem in kernel-induced feature space. From the constraints of optimization, we know that the dual variables go into the range $[0, 1]$ and sum to one, so they meet the definition of probability. Our approach to the solution of (8) is based on a probabilistic interpretation that the center $\mathbf{c}$ of the sphere represents the mean vector of all points $\Phi(\mathbf{x}_i), i = 1, 2, ..., l$ and the Lagrangian multiplier λ_i represents the probability that $\mathbf{x}_i$ is support vector. Using this probabilistic point of view, we may consider searching for the MES as a procedure of probability assignments which should follow the Jaynes' maximum entropy principle. Thus instead of QP problem (8), we construct a composite maximization problem

$$\begin{cases} \max L_p(\lambda) = L(\lambda) + H(\lambda)/p \\ \text{s.t.} \sum_{i=1}^{l} \lambda_i = 1 \\ \lambda_i \geq 0, \quad i = 1, 2, ..., l, \end{cases} \tag{17}$$

where p is a non-negative parameter, and

$$H(\lambda) = -\sum_{i=1}^{l} \lambda_i \ln \lambda_i. \tag{18}$$

From information theory perspectives, $H(\lambda)$ represents an information entropy of the multipliers $\lambda_i, i = 1, 2, ..., l$. The additional term $H(\lambda)/p$ is commensurate with the application of an extra criterion of maximizing the multipliers entropy to the original MES QP problem (8). It is intuitively obvious that the entropy term on the solution of (17) will diminish as p approaches infinity.

To solve this problem we introduce the Lagrangian

$$L_p(\lambda, \alpha) = L(\lambda) + H(\lambda)/p + \alpha(\sum_{i=1}^{l} \lambda_i - 1), \tag{19}$$

where α is Lagrange multiplier. Setting to zero the derivative of $L_p(\lambda, \alpha)$ with respect to λ and α, respectively, leads to

$$\frac{\partial L}{\partial \lambda_i} - \frac{1}{p}(1 + \ln \lambda_i) + \alpha = 0, \quad i = 1, 2, ..., l, \tag{20}$$

and the normality condition (5). Solving (20) for $\lambda_i, i = 1, 2, ..., l$ gives

$$\lambda_i = \exp(p(\frac{\partial L}{\partial \lambda_i} + \alpha) - 1), \quad i = 1, 2, ..., l. \tag{21}$$

Substituting λ from (21) into (5) we obtain

$$\exp(p\alpha - 1) \sum_{i=1}^{l} \exp(p \frac{\partial L}{\partial \lambda_i}) = 1. \tag{22}$$

The term $\exp(p\alpha - 1)$ may be eliminated between (21) and (22) to give

$$\lambda_i = \exp(p \frac{\partial L}{\partial \lambda_i}) / \sum_{i=1}^{l} \exp(p \frac{\partial L}{\partial \lambda_i}), \quad i = 1, 2, ..., l, \tag{23}$$

By optimization problem (8), we have

$$L_{\lambda_i}(\lambda) \stackrel{\text{def}}{=} \frac{\partial L}{\partial \lambda_i} = K(\mathbf{x}_i, \mathbf{x}_i) - 2\sum_{j=1}^{l} \lambda_j K(\mathbf{x}_i, \mathbf{x}_j), \quad i = 1, 2, ..., l. \tag{24}$$

Thus we obtain the iterative formula

$$\lambda_i^{(k+1)} \stackrel{\text{def}}{=} g_i(\lambda^{(k)}) = \frac{\exp(p^{(k)} L_{\lambda_i}(\lambda^{(k)}))}{\sum_{i=1}^{l} \exp(p^{(k)} L_{\lambda_i}(\lambda^{(k)}))}, \quad i = 1, 2, ..., l. \tag{25}$$

Based on formulas (23)-(25), we obtain the entropy-based iterative algorithm for the solution of optimization problem (8) as follows:

Step 1. Let $p^{(0)} = 0$, from (23) we get $\lambda_i^{(0)} = 1/l, i = 1, 2, ..., l$; Let $\Delta p \in (0, +\infty)$ and set $k := 0$;

Step 2. Based on formulas (24) and (25), computing $\lambda_i^{(k+1)}, i = 1, 2, ..., l$; Let $p^{(k+1)} = p^{(k)} + \Delta p$;

Step 3. If stop criterion satisfied, then stop; Otherwise, we set $k := k+1$ then return to step 2.

In short, we start with rough estimates of Lagrange multipliers, calculate improved estimates by iterative formula (25), and repeat until some convergence criterion is met. The new algorithm for MES problem is similar to the generalized expectation-maximization algorithm.

4 Convergence Analysis

In order to analyze the convergence of the iterative formula (25), the following definition and lemma will be useful [13].

Definition 1. *A function G from $D \subset R^n$ into R^n has a fixed point at $x^* \in R^n$ if $G(x^*) = x^*$.*

Lemma 1. *Let $D = \{x \in R^n \mid a \leq x \leq b\}$ for some constants vector a and b. Suppose G is a continuous function from $D \subset R^n$ into R^n with the property that $G(x) \in D$ whenever $x \in D$. Then G has a fixed point in D. Suppose further that G is continuously differentiable in the interior of D and that its Jacobian matrix J_G can be continuously extended to all of D such that the spectral radius $\rho(J_G) < 1$. Then the sequence $x^{(k+1)} = G(x^{(k)}), k = 0, 1, 2, ...$, converges for each $x^{(0)} \in D$ to the unique fixed point $x^* \in D$.*

Denote $G(\lambda) = (g_1(\lambda), g_2(\lambda), ..., g_l(\lambda))^T$. For a fixed p, the Jacobian matrix of iterative formula (25) is

$$J_G = (\nabla g_1(\lambda), \nabla g_2(\lambda), ..., \nabla g_l(\lambda))^T, \qquad (26)$$

where

$$\nabla g_i(\lambda) = pg_i(\lambda) \sum_{j=1}^{l} g_j(\lambda))(\nabla L_{\lambda_i}(\lambda) - \nabla L_{\lambda_j}(\lambda)), \quad i = 1, 2, ..., l. \qquad (27)$$

From Lemma 1, we can obtain the following proposition.

Proposition 2. *For a fixed p, the iterative formula (25) has a fixed point in $[0,1]^l$. If the spectral radius $\rho(J_G) < 1$, then the sequence generated by (25) converges for each $\lambda^{(0)} \in [0,1]^l$ with $\sum_{i=1}^l \lambda_i^{(0)} = 1$ to the unique fixed point $\lambda_p^* \in [0,1]^l$.*

Proposition 2 provides a sufficient condition for iterative formula (25) to converge. Note that the convergence properties depend on the concrete Mercer kernel functions.

Proposition 3. *Suppose that λ_p^* is an optimal solution to optimization problem(17) with non-negative parameter p, then $L(\lambda_p^*)$ is increasing with respect to p and $H(\lambda_p^*)$ is decreasing with respect to p.*

Proof. For $p_2 \geq p_1 > 0$, suppose that $\lambda_{p_2}^*$ and $\lambda_{p_1}^*$ are optimal solutions to optimization problem (17) with non-negative parameter p_2 and p_1, respectively. Then we have

$$L(\lambda_{p_2}^*) + H(\lambda_{p_2}^*)/p_1 \leq L(\lambda_{p_1}^*) + H(\lambda_{p_1}^*)/p_1 \qquad (28)$$

and

$$L(\lambda_{p_1}^*) + H(\lambda_{p_1}^*)/p_2 \leq L(\lambda_{p_2}^*) + H(\lambda_{p_2}^*)/p_2 \qquad (29)$$

From (28) and (29), we have

$$H(\lambda_{p_2}^*) \leq H(\lambda_{p_1}^*), L(\lambda_{p_2}^*) \geq L(\lambda_{p_1}^*) \qquad (30)$$

This completes the proof of Proposition 3.

5 Numerical Examples

In this section, we present some preliminary numerical results for the proposed algorithm. The results were computed on a 500 MHz PC running Matlab, Version 6.1.

In the first example (Fig. 1) we illustrate the entropy-based iterative algorithm for computing MES of 50 simulated data points in the plane. In this case, we use the linear kernel $K(x_i, x_j) = x_i \cdot x_j$. The parameters are chosen to be $p^{(0)} = 0$ and $\Delta p = 0.2$. The initial solution is selected to be $\lambda_i^{(0)} = 1/50$. Fig. 1 shows some of the proposed algorithm iterations on the simulated data. Fig. 1(a) shows the simulated data indicated by dots and the initial solution (i.e. $k = 0$) for the MES problem. Fig. 1(b), (c) and (d) show the first, third and fifth iteration, respectively. We can see that Fig. 1(d) gives a solution that is nearly the exact solution. The support vector is depicted by squares in Fig. 1(d). The Lagrange multiplier λ_i represents the probability that x_i is a support vector, so clearly points located far from the MES center have the largest support values.

In the second example we test the proposed algorithm on the iris data, which is a standard benchmark in the pattern recognition literature, and can be obtained from [14]. The data set contains 150 instances each composed of four measurements of an iris flower. In this case, we use the Gaussian kernel $K(x_i, x_j) = \exp(-q\|x_i - x_j\|^2)$. The parameters are chosen to be $q = 10$, $p^{(0)} = 0$ and Δp is a control parameter. The initial solution is selected to be $\lambda_i^{(0)} = 1/150$. We compare the entropy-based algorithm to the Matlab QP(quadprog) algorithm applied to optimization problem (8) for the second example. To compare the computational efficiency, we run our algorithm until the approximating square of radius is within the tolerance 0.01.

In Table 1, the comparison of CPU time and number of iteration between the entropy-based algorithm and the Matlab QP(quadprog) algorithm on the

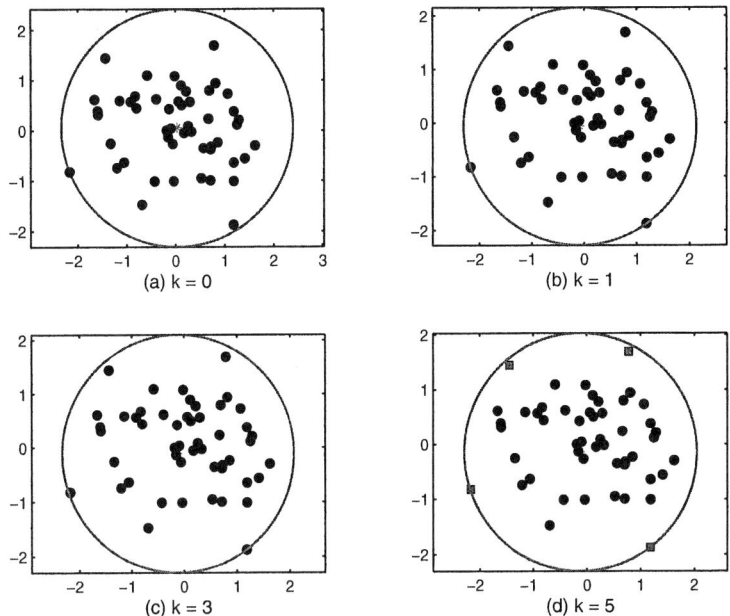

Fig. 1. Successive iterations of the proposed algorithm on the 50 simulated data points

Table 1. Numerical results for the entropy-based algorithm on the iris data

Item	Entropy-based algorithm			QP in Matlab
	$\Delta p = 0.2$	$\Delta p = 0.1$	$\Delta p = 0.05$	
CPU time(seconds)	0.1210	0.1910	0.3600	7.1710
Number of iterations	9	17	33	44

iris data is reported. From the numerical results on the iris data reported in Table 1, it is obvious that the entropy-based algorithm is more efficient than the Matlab QP algorithm for the MES problem in feature space. The performance of the proposed algorithm, however, is sensitive to the selection of the control parameter Δp. The obvious question to ask is how the appropriate value of Δp is chosen. Although no complete answer to this question is known for the entropy algorithm, the adaptive feedback adjustment rule in [15] can be used to obtain more robust and effective algorithms, which remains as a direction for future research.

6 Conclusions

We have considered the problem of computing the MES of a set of points in the high dimensional kernel-induced feature space. Based on maximum entropy principle, we proposed a novel iterative algorithm for computing the minimal enclosing sphere in feature space. Due to non-negative constraints and the normality condition, we considered searching for the MES as a procedure of probability assignments, and the solution follows from solving a set of entropy optimization problems, instead of quadratic programming. The new algorithm is extremely simple to code. The convergence of the novel algorithm is analyzed and the validity of this algorithm is confirmed by preliminary numerical results.

Acknowledgements

This research work was supported in part by the National Basic Research Program of China (G1998030414) and the National Natural Science Foundation of China (60473115, 60403021).

References

1. Berg M.: Computational geometry: algorithms and application, Springer, Newyork, 1997.
2. Elzinga D. J., Hearn D. W.: The minimum covering sphere problem, Management Science, **19**(1972), 96-104.
3. Schölkopf B., Burges C., Vapnik V.: Extracting support data for a given task, In Fayyad U. M. and Uthurusamy R., editors, First International Conference on Knowledge Discovery & Data Mining, Proceedings, 252-257. AAAI Press, 1995.
4. Megiddo N.: Linear-time algorithms for linear programming in R^3 and related problems, SIAM J. Comput., **12**(1983), 759-776.
5. Welzl E.: Smallest enclosing disks (balls and ellipses), In Maurer H., ed., New Results and New Trends in Computer Science, LNCS 555, 359–370. Springer-Verlag, 1991.
6. Gärtner B.: Fast and robust smallest enclosing balls, In Nesetril J., Editor, Algorithms - ESA'99, 7th Annual European Symposium, Proceedings, LNCS 1643, 325-338, Springer-Verlag, 1999.
7. Vapnik V.: Statistical learning theory, John wiley & Sons, New York, 1998.
8. Ben-Hur A., Horn D., Siegelmann H. T., Vapnik V.: Support vector clustering, Journal of Machine Learning Research, **2**(2001), 125-137.
9. Horn D.: Clustering via Hibert space, Physica A, **302**(2001), 70-79.
10. Platt J.: Fast training of support vector machines using sequential minimal optimization, In Scholkopf B., Burges C. J. C. and Smola A. J., Editors. Advances in Kernel Methods — Support Vector Learning, 185-208, MIT Press, 1999.
11. Li X. S.: An information entropy for optimization problems, Chinese Journal of Operations Research, **8**(1)(1989), 759-776.
12. Templeman A. B., Li X. S.: An maximum entropy approach to constrained nonlinear programming, Engineering Optimization, **12**(1987), 191-205.
13. Kress R.: Numerical analysis, Springer-Verlag, New York, 1998.
14. Blake C. L., Merz C. J.: UCI repository of machine learning databases, avaible at http://www.ics.uci.edu/ mlearn/MLRepository.html, 1998.
15. Polak E., Royset J. O., Womersley R. S.: Algorithms for adaptive smoothing for finite minimax problem, Journal of Optimization Theory and Applications, **119**(3)(2003), 459-484.

Y-AOI: Y-Means Based Attribute Oriented Induction Identifying Root Cause for IDSs*

Jungtae Kim[1], Gunhee Lee[1], Jung-taek Seo[2], Eung-ki Park[2], Choon-sik Park[2], and Dong-kyoo Kim[1]

[1] Graduate School of Information Communication, Ajou University, Suwon, Korea
{coolpeace, icezzoco, dkkim}@ajou.ac.kr
[2] National Security Research Institute, Hwaam-dong, Yuseong-gu, Daejeon, Korea
{seojt, ekpark, csp}@etri.re.kr

Abstract. The attribute oriented induction (AOI) is a kind of aggregation method. By generalizing the attributes of the alert, it creates several clusters that includes a set of alerts having similar or the same cause. However, if the attributes are excessively abstracted, the administrator does not identify the root cause of the attack. In addition, deciding time interval of clustering and deciding *min_size* are one of the most critical problems. In this paper, we describe about the over-generalization problem because of the unbalanced generalization hierarchy and discuss the solution of the problem. We also discuss problem to decide time interval and meaningful *min_size*, and propose reasonable method to solve these problems.

1 Introduction

Recently, various attacks using system vulnerability are considered as a serious threat to e-business over the Internet. To control those threats properly, the most system administrators employ Intrusion Detection System (IDS). If the system detects the attacks, it generates a number of alerts. Periodically or continuously the administrator analyzes those alerts, and he/she searches for the cause of the alerts. According to the cause, he/she responds against the threat in order to remove it. However, it is difficult and burden work since there are extremely many alerts on a system. Moreover, the IDS reports extremely many alerts as compared with the number of the real attack [1], [2]. To control this situation, there are some researches on the efficient alerts handling method such as aggregation and correlation [3]. The attribute oriented induction (AOI) method is a kind of the aggregation method [4]. Since it uses all the alert attributes as the criteria, it properly identifies and removes the most predominant root causes.

The classical AOI method has some issues to be solved such as deciding time interval of clustering and over-generalization problem. Usually, this method clusters the alerts using discrete time interval such as hour, a day, or a month.

---
* This study is supported by the National Security Research Institute in Korea and the Brain Korea 21 project in 2004.

However, if the clustering time interval is unreasonable, the classic AOI method creates unreliable result. If the discrete time interval is too long (i.e., a month), the alerts caused by different attacks can be merged into a cluster. On the other hand, if the discrete time interval is too short (i.e., an hour), the alerts caused by an attack are divided into different clusters. For example, if the DDoS attack occurs from 2005-11-26 11:45:00 to 2005-11-27 01:11:11, and if the clustering time interval is an hour or a day, the alerts caused by the same DDoS attack are divided into different clusters. When system administrators analyze clustering result, he/she get into trouble since these unreliable clustering results. Therefore, clustering time interval of AOI algorithm should be decided more reasonably. In addition, classical AOI method has over-generalization problem confuses the administrator due to excessively abstracted attribute value. Although there is an attempt to solve it by K. Julisch, it still has the problem because of the unbalanced generalization hierarchy [5], [6].

In this paper, we will discuss about these problems of the classical AOI algorithm and propose Y-AOI method to cope with these problems. The proposed method adopts Y-means algorithm to solve clustering interval problem and enhances the classical AOI algorithm to solve the over-generalization problem of it. Experimental result shows that the proposed method properly handles these problems and generates more meaningful clusters.

This paper is organized as follows. We describe the Y-means algorithm and discuss the over-generalization problem of the classical AOI algorithm in section 2. In section 3, we describe proposed algorithm in detail. This is followed by the experimental result of the proposed method in section 4. Section 5 concludes.

2 Related Works

2.1 Y-Means Algorithm

The Y-means algorithm is proposed as anomaly detection method [7]. It is based on the K-means algorithm and other related clustering algorithm. It overcomes two disadvantages of the K-means; number of clusters dependency and degeneracy. Number of clusters dependency is that the value of k is very critical to the clustering result. Obtaining the optimal k for a given data set is an NP-hard problem [8]. Degeneracy means that the clustering may end with some empty clusters. This is not what the system administrator expects since the classes of the empty clusters are meaningless for the classification. To overcome these problems of the K-means algorithm, the Y-means algorithm automatically partitions a data set into a reasonable number of clusters so as to classify the instances into *normal* clusters and *abnormal* clusters.

In the Y-means algorithm, the first step is to partition the normalized data into k cluster. The number of cluster, k, is an initialized integer between 1 and n, where n is the total number of instances. The second step is to find whether there are any empty clusters. If they exist, they are removed and new clusters are created to replace these empty clusters; and then instances will be reassigned to existing clusters. This processing is iterated until there is no empty cluster. Subsequently, the outliers of cluster will be removed to form a new cluster, in which instances are more similar to each other; and overlapped adjacent cluster will be merged into a new cluster. In this way, the value of k will be determined automatically by splitting or merging clusters.

2.2 AOI Algorithm

Attribute Oriented Induction is operated on relational database tables and repeatedly replaces attribute values by more generalized values. The more generalized values are taken from user defined generalization hierarchy [4]. By the generalization, previous distinct alerts become identical, and then it can be merged into single one. In this way, huge relational tables can be condensed into short and highly comprehensible summary tables.

However, the classic AOI algorithm has over-generalization problem that is important detail can be lost. K. Julisch modified classic AOI algorithm to prevent this problem. It abandons the generalization threshold d_i. Instead, it search alerts $a \in T$ that have a count bigger than min_size (i.e. $a[count] > min\_size$) where $min\_size \in N$ is a user defined constant.

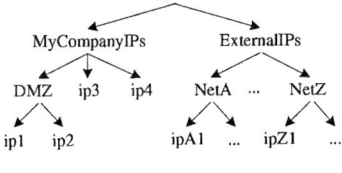

Src-IP	Dst-IP	Count
ipA1	ip1	1
ipJ1	ip1	1
ipK1	ip4	1
ipZ1	ip4	1
ip3	ipA1	1
ip3	ipZ1	1

a) Alert table

b) Generalization hierarchy for IP address

Fig. 1. Sample alert table and generalization hierarchy

Table 1. Clustering result from classic AOI

Algorithm	Src-IP	Dest-IP	Count
Classic AOI algorithm	ANY-IP	ExternalIPs	26
	ExternalIPs	ANY-IP	26
K. Julisch's algorithm	ip3	ExternalIPs	26
	ExternalIPs	ANY-IP	26

Fig. 1 represents an example generalization hierarchy and alert table having the attributes *Src-IP* (the Source IP) and *Dest-IP* (the Destination IP). Table 1 shows the clustering result of the classic AOI algorithm and K. Julisch's algorithm. Both algorithms clustered with example alerts and generalization hierarchy in Fig. 1. The classical AOI algorithm use 2 as threshold d_i and K.

Julisch's algorithm use 10 as *min_size*. In the result of classic AOI, the *Src-IP* of first record is represented with *ANY-IP*. On the other hand, in the result of K. Julisch's, it is represented with ip3 that is more specific and informative.

Even though K. Julisch's algorithm improve over-generalization of the traditional AOI, it does not properly handle the problem owing to the configuration of generalization hierarchy. Note that, for the node *MyCompanyIPs* in the Fig. 1-(b), the depths of sub trees are not equal. The depth of its left sub-tree DMZ^1 is 2 and another two (*ip3* and *ip4*) are 1. We call this kind of node unbalanced node. The unbalanced generalization hierarchy is a tree that has more than one unbalanced node.

In the result of K. Julisch's algorithm, the *Dest-IP* of the second record is represented with *ANY-IP*. However, it is more reasonable and meaningful that *Dest-IP* is represented with *MyCompanyIPs*. If *Dest-IP* is abstracted to *ANY-IP*, system administrator can't identify whether target of attack is home-network or external-network. It is another over-generalization problem, caused by the unbalanced node *MyCompanyIPs*. The solution of this problem will be explained in the section 3.2.

3 The Y-AOI Method

The proposed method consists of two steps; time based clustering and attributes oriented induction. In the first step, using Y-means algorithm, the alerts are divided into several clusters based on its occurrence time. In the second step, using enhanced AOI algorithm, the clusters generated in the first step are inducted into short and highly comprehensible summary tables. Detail of the each step will be explained in the next two sections.

3.1 Time Based Clustering

We adopt the Y-means algorithm to flexibly decide clustering time interval of the AOI algorithm. In the proposed method, the alerts are divided into several groups according to the density of the alerts in the specific time interval. If the density is high in the specific interval such as interval B and D of the Fig. 2-(b), the alerts make a dense alert group. On the other hand, if the density is low in the specific interval such as interval A and C of the Fig. 2-(b), it makes a sparse alert group. Fig. 2 shows the comparison between grouping result of discrete time interval and flexible time interval. As represented in the figure, using the Y-mean method, we can flexibly divide time interval based on the alert occurrence time and its density.

The first advantage of flexible time interval is that it generates more meaningful cluster. In the real situation, most of system administrators cluster alerts based on discrete time interval such as an hour, a day, or a month. However, if

---

[1] In this paper, when we refer a tree, we will use the name of its root node with the term *sub-tree*.

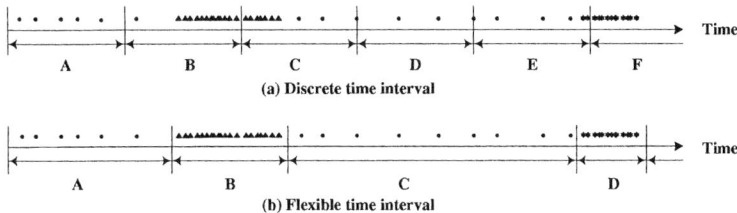

Fig. 2. Alert occurrence according to time

we uses discrete time interval, it is possible that the alerts caused by an attack are divided into different clusters since attack does not occur in the fixed time interval. For example, in the Fig. 2-(a), the alerts caused by the same attack are divided into B and C due to discrete time interval. Furthermore, in some cases, it cannot be a cluster because the alerts caused by an attack are divided into several parts. In the previous example, if the number of alerts caused by the same root cause in the time interval B is 50 and in the time interval C is 60 where the min_size is 100, it does not become a cluster since the number of alerts of each part are smaller than min_size. On the other hand, if the time interval is too long, different alerts can be inducted a cluster. For example, if the discrete time interval is a day and two DDoS attacks occurred in 10:30:00 and 21:00:00, these alerts generated by two DDoS attacks are inducted a cluster since these are contained in a same time interval. Even though these are the same kind alerts caused by DDoS attack, these are generated by different attacks. Thus, these alerts should be divided into different clusters. If we use flexible time interval, we can prevent these problems. In the proposed method, the alerts caused by the same root cause are contained within a same time interval. In addition, the alerts within a same time interval are strongly related. Therefore, using flexible time interval, we can generate more meaningful cluster.

3.2 Enhanced Attribute Oriented Induction Algorithm

As we mentioned in section 2, classical AOI algorithm has over-generalization problems. To solve the problem, we propose enhanced AOI algorithm that improve two factors of classical AOI algorithm. First, proposed algorithm improves deciding min_size mechanism. The classical AOI algorithm used static min_size. However, determining min_size is a critical issue of the AOI algorithm. If the min_size is two high, many real attacks can be ignored since the number of alerts caused by it is lower than the min_size. On the other hand, if the min_size is too low, the AOI algorithm creates too many unnecessary clusters. Thus, in the proposed method, we use adaptive min_size. According to the number of alerts in the specific time interval, the min_size is changed; if the number is high, the min_size is increased and if the number is low, the min_size is reduced.

Second, to improve over-generalization problem we improve the classical AOI algorithm. Fig. 3 shows the proposed algorithm to prevent over-generalization.

To handle the unbalanced generalization hierarchy, the algorithm employs a new attribute *hold count* in generalization hierarchy. This value means the number of the generalization until all the lower level nodes of the current node are generalized to the current node. If the levels of sub trees are the same or the number of sub tree is less one or zero, the *hold count* is initialized to zero. Otherwise, it is initialized to maximum value among the differences between the level of current node and the one of leaf nodes belonged to its sub tree. For example, in the Fig. 1-(b), since the level of the sub-trees for the *ExternalIPs* are equal, the *hold count* is set to zero. For the *MyCompanyIPs*, on the other hand, the levels of sub-tree are not equal, thus the *hold count* is set to two, which is the difference between level of *MyCompanyIPs* and *ip1*. Thus in our proposed algorithm, each node of generalization hierarchy has two attributes that are *generalization attribute value* (e.g., *ExternalIPs*, *MyCompanyIPs*) and *hold count*. During the generalization procedure, the algorithm checks the *hold count*. According to the values the algorithm determines whether to generalize the attribute or not.

```
Input: An alert clustering problem (L, min size, H_i, ..., H_n)
Output: A heuristic solution for (L, min size, H_i, ..., H_n)
1:  T := Store log L in table T
2:  H_ik[V] := Value of node k in generalization hierarchy H_i
3:  H_ik[HC] := hold count of node k in generalization hierarchy H_i
4:  for all alerts a in T do a[count] := 1           // initialize counts
5:  while all a ∈ T : a[count] < min size do {
6:      Use heuristics to select an attribute A_p, i ∈ {1..... n}
7:      for all alerts a in T do
8:          for all attribute A_i of alert a do
9:              if a[A_i] = H_ik[V] and H_ik[HC] = 0    // if hold count is zero
10:                 a[A_i] := father of a[A_i] in H_i   // generalize attribute A_i
11:             if H_ik[HC] > 0
12:                 H_ik[HC] := H_ik[HC] - 1            // decrease hold count
13:     while identical alerts a, a' exist do          // merge identical alerts
14:         Set a[count] := a[count] + a'[count] and delete a' from T
    }
```

Fig. 3. Enhanced alert clustering algorithm

In more detail, the algorithm starts with the alert log L and repeatedly generalize the alerts in L. Generalizing alerts is done by choosing an attribute A_i of the alert and replacing the attribute values of all alerts in by their parents in generalization hierarchy of H_i. At this point, the *hold count* of a node k in H_i that is matched with the attribute value $a[A_i]$ is must be zero. This process continues until an alert has been found to which at least min size of the original alerts can be generalize.

The algorithm considers *hold count* during the generalization step (line 6 12 in the Fig. 3). When a node is to be generalized in the hierarchy, it first needs to check the *hold count* of the node. If the *hold count* of the node is not zero, then it implies that there are several records that should be generalized to the node. Therefore, it waits until no such node left. In other words, it should wait until the generalization level of all sub trees is to the current node.

For example, in the Fig. 1-(b), *ip1* and *ip2* is generalized to *DMZ* and *ip3* and *ip4* is generalized to *MyCompanyIPs* at the first generalizations. At the second generalization, there exist records which attribute value belongs to the sub-tree of *MyCompanyIPs* and it is not generalized to *MyCompanyIPs* in the alert table. Thus, the records having *MyCompanyIPs* are waiting until *DMZ* is generalized to *MyCompanyIPs*. It is the reason why we modify previous algorithm like described above is that to prevent the rapid generalization to the upper level of the node when the levels of sub-trees are different.

At every generalization step, the *hold count* is decreasing by one where it is bigger than zero. As stated above *hold count* represents the number of the generalization until all the lower level nodes of the current node are generalized to the current node. Therefore, all generalization steps, it is decreased by one. Table 2 shows the result of the clustering using the proposed algorithm with the data in Fig. 1. In the result, generalization of the *Dest-IP* of the second record is stopped at *MyCompanyIPs*.

Table 2. Clustering result from the proposed algorithm

Src-IP	Dest-IP	Count
ip3	ExternalIPs	26
ExternalIPs	MyCompanyIPs	26

4 Experimental Results

To measure effectiveness of the proposed method, we used DARPA1998 data sets [9]. The simulation network is configured as shown in Fig. 4-(a). Based on the router, the network is divided into two parts; inside network and outside network. The inside network consists of 7 sub-networks from *subnetA* to *subnetG*, while the outside network represents theInternet.

As a network sensor, we used Snort. For the experiment, we choose a data set among several data sets in the DARPA1998 data sets. Selected data set contains various attacks done in a day. Using the data set, the Snort generates alerts and these are logged into mySQL database. In the experiment, we used five attributes of alerts for the clustering such as alert type, source IP, destination IP, source port and destination port. Based on those attributes, the proposed method created an initial set of the alerts from the raw alerts generated by the sensor. For the attribute generalization, we used the generalization hierarchy as shown in Fig. 4-(b) for the IP and the one in Fig. 4-(c) for the port number. Seven sub-networks from *subnetA* to *subnetG* are abstracted to inside host. However, for more informative alert cluster, we didn't generalize the type of the raw alerts. Since the alert type is very critical information, if it is abstracted, it is difficult that we find out what kind of attack occurred.

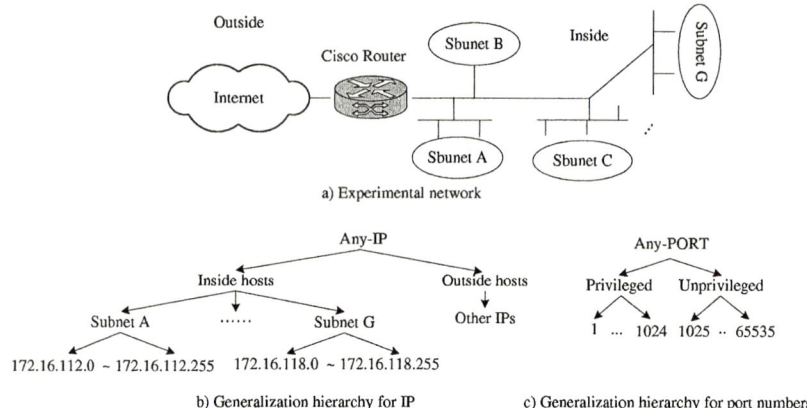

Fig. 4. Experimental network and generalization hierarchy

Fig. 5 shows the number of alerts according to time. As showed in the graph, the alerts are drastically increased in three points; from 21:20 to 21:30, from 00:20 to 01:20, and from 02:10 to 03:20. Analyzing the raw alerts, we found out that the experimental dataset contain three actual attacks and these attacks occurred in three times; from 1998-07-09 21:27:03 to 1998-07-09 21:27:27, from 1998-07-10 00:17:48 to 1998-07-10 01:27:35, and from 1998-07-10 02:12:35 to 1998-07-10 03:15:07. These points are exactly matched with three points. As shown in Fig. 5, at that points, the number of alerts increases drastically. The almost of the other alerts were false positive.

Table 3 shows that the result of the clustering using the proposed method and Table 4 shows that the result using the classical AOI algorithm. The proposed method di-vides the alerts caused by two different attacks into two groups (2nd 4th and 5th 7th), while the classical AOI method merges these alerts into a cluster. Even though two attacks are the same kind attack, if these occurred in different time and originated from different source, these are considered as a different attack. In addition, the gen-eralization level of the proposed method is lower than the classical AOI method since proposed method used various min_size according to the number of alerts within the each time interval.

When we used only classical AOI method, the false positive is reported as a cluster; the 5th cluster in the table 4 is not informative since it is too much generalized. As mentioned above in the section 3.1, most of false positives are generated continually and regularly. Thus, if clustering time interval is unreasonable, a little number of false positives that are continually generated can be a cluster. Therefore, for more informa-tive cluster, clustering method should not

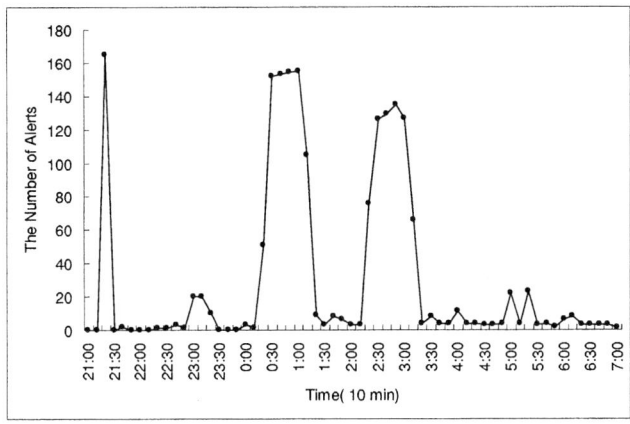

Fig. 5. Experimental network and generalization hierarchy

Table 3. Experimental result of the proposed algorithm

Alert Type (Alert Occurence Time)	Src-IP	Dest-IP	Src-Port	Dest-Port	Count
ICMP PING NMAP (21:27:03 - 21:27:27)	205.231.28.163	Subnet B	Unknown	Unknown	165
SNMP request tcp (00:17:48 - 01:27:35)	230.1.10.20	172.16.112.50	Unprivileged	161	242
SNMP trap tcp (00:17:48 - 01:27:35)	230.1.10.20	172.16.112.50	Unprivileged	162	250
SNMP AgentX/tcp request (00:17:48 - 01:27:35)	230.1.10.20	172.16.112.50	Unprivileged	705	246
SNMP request tcp (02:12:35 - 03:15:07)	210.107.195.50	172.16.112.50	Unprivileged	161	200
SNMP trap tcp (02:12:35 - 03:15:07)	210.107.195.50	172.16.112.50	Unprivileged	162	200
SNMP AgentX/tcp request (02:12:35 - 03:15:07)	210.107.195.50	172.16.112.50	Unprivileged	705	200

create these uninformative clusters. The proposed method ignores the time interval if the alert density is too low. Thus, the result of the proposed method did not create false positive alert cluster.

5 Conclusion

In this paper, we have made the following contributions. First, we set up a scheme to flexibly decide AOI clustering interval. Second, we proposed an algorithm that handles over-generalization problem of AOI algorithm owing to unbalanced generalization hierarchy and proposed adaptive *min_size*. From the

Table 4. Experimental result of the proposed algorithm

Alert Type	Src-IP	Dest-IP	Src-Port	Dest-Port	Count
ICMP PING NMAP	205.231.28.163	Subnet B	Unknown	Unknown	165
SNMP request tcp	Outside hosts	172.16.112.50	Unprivileged	161	442
SNMP trap tcp	Outside hosts	172.16.112.50	Unprivileged	162	450
SNMP AgentX/tcp reqeust	Outside hosts	172.16.112.50	Unprivileged	705	446
SCAN FIN	Outside hosts	Inside hosts	Unprivileged	Privileged	117

experimental results, the proposed method generates more informative cluster than previous algorithm. Therefore, using the proposed method, system administrator identifies the root causes of the alert more easily and more effectively. In the future work, we will try to evaluate the proposed method in more various situations and enhance the proposed method for real time situation.

References

1. Valdes, A., Skinner, K.: Probabilistic Alert Correlation, Proceedings of Recent Advances in Intrusion Detection, LNCS 2212 (2001) 54-68
2. Axelsson, S.: The Base-Rate Fallacy and the Difficulty of Intrusion Detection, ACM Transactions on Information and System Security, Vol. 3, No. 3 (2000) 186-205
3. Debar, H., Wespi, A.: Aggregation and Correlation of Intrusion-Detection Alerts, Proceedings of Recent Advances in Intrusion Detection, LNCS 2212 (2001) 85-103
4. Han, J., Cai, Y.: Data-Driven Discovery of Quantitative Rules in Relational Databases. IEEE Transactions on Knowledge and Data Engineering, Vol. 5, No. 1 (1993) 29-40
5. Julisch, K.: Clustering intrusion detection alarms to support root cause analysis, ACM Transactions on Information and System Security, Vol. 6, No. 4 (2002) 443-471
6. Julisch, K.: Mining Intrusion Detection Alarms for Actionable Knowledge, Proceedings of the eighth ACM SIGKDD international conference on Knowledge discovery and data mining (2002) 366-375
7. Guan, Y., Ali, A.: Y-MEANS: A Clustering Method for Intrusion Detection, Canadian Conference on Electrical and Computer Engineering, (2003) 1083-1086
8. Hansen, P., Mladenovic, N.: J-means: a new local search heuristic for minimum sum-of-squares clustering, Pattern Recognition, Vol. 34, No. 2 (2002) 405-413
9. DARPA data set, http://www.ll.mit.edu/IST/ideval/index.html

New Segmentation Algorithm for Individual Offline Handwritten Character Segmentation

K.B.M.R. Batuwita[1] and G.E.M.D.C. Bandara[2]

[1] Department of Statistics and Computer Science, Faculty of Science,
University of Peradeniya,
Peradeniya, SriLanka
[2] Department of Production Engineering, Faculty of Engineering,
University of Peradeniya,
Peradeniya, SriLanka
rukshanbatuwita@yahoo.com, dcb@pdn.ac.lk

Abstract. Handwritten character recognition has been an intensive research for last decade. A handwritten character recognition fuzzy system with an automatically generated rule base possesses the features of flexibility, efficiency and online adaptability. A major requirement of such a fuzzy system for either online or offline handwritten character recognition is, the segmentation of individual characters into meaningful segments. Then these segments can be used for the calculation of fuzzy features and the recognition process. This paper describes a new segmentation algorithm for offline handwritten character segmentation, which segments the individual handwritten character skeletons into meaningful segments. Therefore, this algorithm is a good candidate for an offline handwritten character recognition fuzzy system.

1 Introduction

Handwritten character recognition is one of the benchmark problems of Artificial Intelligence research. Recognizing characters is a problem that at first seems simple, but is an extremely difficult task to program a computer to do it. Automated character recognition is of vital importance in many industries, such as banking and shipping. Many methodologies have been developed for handwritten character recognition, such as connective learning based methods like neural networks [4], statistical methods like the Hidden Markov model [7], Fuzzy Logic [3][6] and Hybrid methods [1].

Fuzzy Logic plays a major role in the area of both online and offline handwritten character recognition since, a fuzzy system with an automatically generated rule base can posses the following features [6].

1. Flexibility
2. Efficiency
3. Online adaptability.

A block diagram of a fuzzy system which can be used for offline handwritten character recognition is depicted in Figure 1.

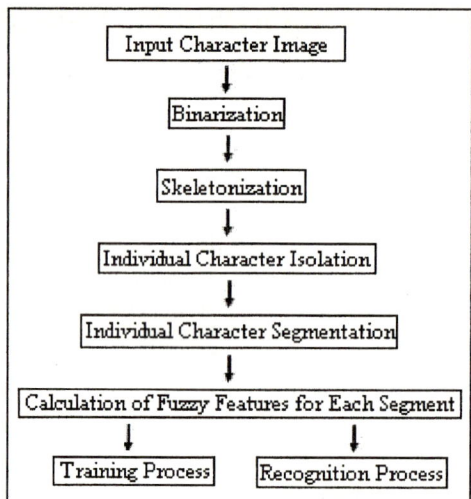

Fig. 1. Block diagram of a fuzzy system for offline handwritten character recognition

Compared to the method which was proposed in [2][8] for online handwritten character recognition, the method given by Figure 1 has three additional steps. They are the preprocessing phases of the image, namely, binarization, skeletonization and individual character isolation. Those are the steps that must be performed prior to the segmentation process. Since every scanned black and white image is a collection of pixels having intensities from 0% to 100%, the image should undergo a binarization process before it is subjected to the skeletonization. In the binarization process the image is processed pixelwise and each pixel is set to either black or white according to a particular threshold value. The binarized image is then subjected to the process of skeletonization, which involves getting the skeletons of the characters in the image. The individual character isolation is the process of isolating each character skeleton into a separate *skeleton area* that is defined in section 3.

One of the major requirements of a fuzzy system for either offline (as depicted in Figure 1) or online [2][8] character recognition is, the individual character segmentation, which involves the segmentation of individual character patterns into meaningful segments. Online character segmentation is benefited from a time-ordered sequence of data and pen up and down movements [2][8], but offline character segmentation is not. Therefore, it is a quite tedious task to develop an algorithm to do the segmentation of offline handwritten characters. If we could develop an algorithm for offline character segmentation, a system which is depicted in Figure 1 could be implemented using the same methods used under online character recognition [2][8] for the calculation of fuzzy features, recognition and training processes. In the "Fuzzy Recognition of Chinese Characters" [9], the dominant point method was used for the segmentation, but that method does not preserve the meaningful arc segments like "C Like", "D Like" etc. The meaningful segments are described in section 2.

This paper describes a new segmentation algorithm developed for offline handwritten character segmentation, which can be used with the system depicted in Figure1. The proposed segmentation algorithm traverses through the paths in an offline character skeleton, and segments it into a set of meaningful segments. The segmentation is done based on the abrupt change in the *written direction* of the character skeleton. The term, *written direction* is defined in section 7.

This paper is organized as follows. The major requirements that should be satisfied by the outputs of the algorithm are described in section 2. In section 3, the requirements which should be satisfied by the input of the algorithm are described. Section 4 outlines the main routing of the algorithm. The identification of major starter points is described in section 5 and section 6 outlines the traversal routing through the skeleton paths. Section 7 discusses how to determine the segmentation, while section 8 discusses how to remove the segments resulted due to the noises in the image. Section 9 presents some experimental results, and the paper concludes with some conclusions and future work outlines, which are stated in section 10.

2 Requirements of the Output

The output of the proposed segmentation algorithm should be a set of segments, which a character skeleton is broken into. Once a skeleton is properly segmented, each of the resulted segments can be used to calculate a set of fuzzy features (MHP, MVP, etc.) as described in [2].

In order to calculate these fuzzy features more accurately, each resulted segment should be a meaningful segment. That is, each resulted segment should be a meaningful straight line or a meaningful arc as described in Table 1, but not an arbitrary segment. As an example, the handwritten character skeleton 'B' (in Figure 2) should be segmented into two "Positive Slanted" lines and two "D Like" arcs.

Table 1. Types of meaningful straight lines and meningful arcs that a character skeleton should be broken into

Meaningful Straight Lines		Meaningful Arcs	
Horizontal Line	————	C like	C
Vertical Line	|	D like	⊃
Positive Slanted	\	U like	∪
Negative Slanted	/	A like	∩
		O like	O

The segmentation algorithm should be able to segment all the handwritten character skeletons in a particular character set, into meaningful segments for it to be useful with the character recognition system proposed in Figure1. The task of segmenting every character in a particular character set into a set of meaningful segments gets harder since different individuals write the same character in many different ways.

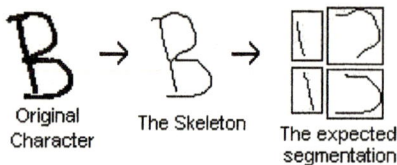

Fig. 2. The expected segmentation of character skeleton 'B'

3 Requirements of the Input

Definition 1: A *skeleton area* is a rectangular area in the image, which contains a character skeleton to be segmented. A *skeleton area* can be represented, as depicted in Figure 3.

The input for this segmentation algorithm is the *skeleton area* of a character skeleton. In order to get the expected segmentation results, the input character skeleton should be in one pixel thickness and it should not contain any spurious branches. It has been observed that the skeletonization algorithm proposed in [5] gave skeletonized results, which satisfies the above two conditions in most of the times.

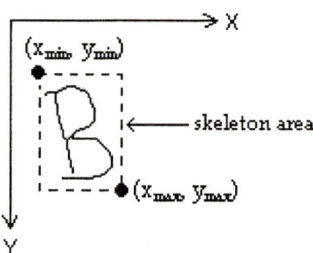

Fig. 3. The representation of the *skeleton area* of character skeleton 'B'

4 The Algorithm

This section outlines the main routine of the algorithm. Apart from that some definitions are presented here. These definitions will be used through out this paper to describe the algorithm in detail.

Definition 2: A *starter point* is a pixel point on the character skeleton with which the traversal through the skeleton could be started. *Starter points* are in twofold; *major starter points* and *minor starter points*

Definition 3: A *major starter point* is a *starter point*, which is identified before starting the traversal through the skeleton. The identification of *major starter points* is described in Section 5.

Definition 4: A *minor starter point* is a *starter point*, which is identified during the traversal through the skeleton. The identification of *minor starter points* is described in Section 6.2.

Two types of major data structures are used in this algorithm, namely, the *Point*, which holds the X and Y coordinate values of a pixel point and the *Segment*, which is a *Point* array.

The main routing of the segmentation algorithm can be described as follows;

Function Segmentation_algo (skeleton_area) returns all_Segments.
Begin
  major_starters = empty // *a queue of Point to store major starter points.*
  m_Segments = empty // *an array of Segment, to store the segments identified*
                          by starting traversal with a particular major starter point.
  all_Segments = empty // *an array of Segment, to store all the identified*
                          segments.

  major_starters = find all major starter points (skeleton_area) // *(Section 5)*

  for each of the major start points in major_starters do
    m_Segments = traverse(major_starter_point) // *(Section 6)*
    add all the segments in the m_Segment to all_Segments
  end for.

  noice_removal(all_Segments). // *(Section 8)*

  return all_Segments
End

5 Identification of Major Starter Points

The algorithm starts on finding all the *major starter points* in the given *skeleton area*. In order to find the *major starter points*, two techniques can be used. In both of these techniques, the pixels in the given *skeleton area* are processed rowwise.

5.1 Technique 1

In Technique 1, all the pixel points in the *skeleton area*, those having only one neighboring pixel are taken as *major starter points*. It is assumed here that the skeleton is in one pixel thickness. As an example, the skeleton pattern 'B' depicted in Figure 4(i) has three *major starter points* namely 'a', 'b' and 'c', which can be identified using Technique 1.

5.2 Technique 2

The *major starter points* of some skeleton patterns (in the case of character 'O' and number zero) cannot be obtained by the Technique 1, since all the points in such a one pixel thickness, closed 'O' like curve (Figure 4(ii)) would have at least two neighboring pixels. In such a case, the first pixel which is found on the character skeleton is taken as its *major starter point*. This one and only *major starter point* would be the highest most pixel point in the skeleton (the pixel point 'a' in Figure 4(ii)), since the *skeleton area* is processed rowwise.

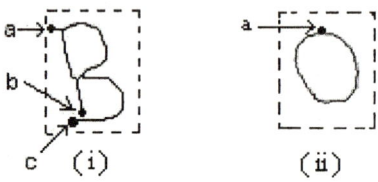

Fig. 4. *Major starter point* of character skeletons 'B' and 'O'

6 Traversal Through the Character Skeleton

Definition 5: The *current traversal direction* is the direction from the current pixel to the next pixel to be visited during the traversal. The determination of *current traversal direction* is described in Section 6.3.

After finding all the *major starter points*, the algorithm starts traversing through the character skeleton, starting from the *major starter point* which was found first. While this traversal, the segments are identified in the traversal path. Moreover, the *minor starter points* are also identified at each junction of the skeleton and they are queued to a different queue, which is hereafter referred to as *minor_starters*. (The identification of *minor starter points* is discussed in section 6.2). Once the traversal reaches an end point, which a pixel point there is no neighboring pixel to visit next, the focus is shifted to the identified *minor starter points* in the *minor_starters* queue. Then the algorithm starts traversing the unvisited paths of the skeleton by starting with each *minor starter point* in the *minor_starters* queue. While in these traversals, the algorithm also segments the path being visited into meaningful segments.

The above mentioned process is continued with all the unvisited *major starter points* in the *major_starters* queue, until all the unvisited paths in the *skeleton area* are visited. The risk of visiting the same pixel (hence the same path) more than once during traversals is eliminated by memorizing all the visited pixel points and only visiting the unvisited pixels in the later traversals. Therefore it is guaranteed that a path in the skeleton is visited only once and hence the same segment is not identified twice.

The traversal routine is as follows.

```
Function traverse(major_starter_point) return segments.
Begin
  minor_starters = empty // Point queue, to store minor starter points.
  current_segment = empty // Segment, to store points of the current segment
  current_direction = empty // String, to hold the current traversal direction.
  current_point = empty // Point, refers to the current pixel point.
  next_point = empty // Point, refers to the next pixel point to be visited.
  neighbors = empty // Point array, to hold all the unvisited neighboring points of
                         the current point.
  segments = empty // Segment array, to hold the identified segments.

  minor_starters.enque(major_starter_point)

    while(there are more points in the minor_starters queue OR current_point is
            nonempty) do

      if(current_point = empty) then
        current_point = minor_starters.deque()
        Initialize the current_segment
        if(current_point is unvisited) then
          current_segment.add(current_point)
          mark the current_point as visited
        end if
      end if


      if(unvisited eight adjacent neighbors of current_point exist) then

        neighbors = get all unvisited eight adjacent neighbors of current_point.

        if(# of points in current_segment > 1) then
          if(there is an unvisited neighbor in the current_direction) then
            next_point = get that neighbor in the current_direction.
            enque all other unvisited neighbors into the minor_starters
                    queue. // Finding minor starter points (Section 6.2)
            current_segment.add(next_point)
            mark next_point as visited
            current_point = next_point
          else // if there is no unvisited neighbor in the current_direction.
              I.e. the traversal direction changes.
            tmp_segment = get next 5 pixels in the path. (Section 6.1)
            if(IsAbruptChange(current_segment, tmp_segment)) then
                                                // Section 7.
              // start a new segment
              segments.add(current_segment)
```

                    current_segment = tmp_segment
                else
                    // the traversal can continue with the same segment.
                    add all the points in the tmp_segment to current_segment
                end if
              end if
            end if
          else // number of points in the current_segment is 1
            next_point = choose any neighbor of the current_point // (Section 6.5)
            current_segment.add(next_point)
            mark next_point as visited
            current_direction = get the current traversal direction // (Section 6.3)
            current_point = next_point
          end if

        else // if there are no unvisited neighbors to visit
          segments.add(current_segment)
          current_point = empty
        end if

    end while

    return segments
End

6.1 Getting Next 5 Pixel Points in the Path

The intention of getting next 5 pixel points into a separate data structure refereed to as *tmp_segment*, is to find the new *written direction* as described in Section 7. Getting next 5 pixel points in the path is carried out as follows.

    tmp_segment = empty // array of Point.
    while(there are more unvisited adjacent neighbors of current_point exist and
          the size of tmp_segment is < 5) do
      neighbors = get all unvisited eight adjacent neighbors of current_point
      if(there is an unvisited neighbor in the current_direction)
            next_point = get that neighbor in the current_direction.
      else
        if (there is an unvisited neighbor in the *closest traversal direction* to the
              current_direction) then
            next_point = get the neighbor in the closest direction to the
                                    current_direction   // (Section 6.4)
        else
            next_point = get a neighbor in any direction // (Section 6.5)
        end if

```
              current_direction = get the new traversal direction // (Section 6.3)
           end if.
           enque other neighbors into the minor_starters queue. //Finding minor
                                                     starter points (Section 6.2)
        mark next_point as visited
        tmp_segment.add(next_point)
        current_point = next_point
    end while
```

6.2 Identification of Minor Starter Points

Let us consider the traversal through the character skeleton 'B' in Figure 5. The traversal starts with the *major starter point* 'a' and continues to the junction 'J_1'. At that junction, the current pixel point has two unvisited neighbors. Since there is a neighboring pixel in to the *current traversal direction*, the algorithm chooses that pixel (n_1) between the two neighboring pixels as the next pixel point to visit. It is clear that the other neighboring pixel (n_2) is a starter point of another path in the skeleton. Therefore the point n_2 is identified as a *minor starter point* and inserted into the *minor_starters* queue for later consideration. In every junction in the skeleton, there may be one or more *minor starter points* identified.

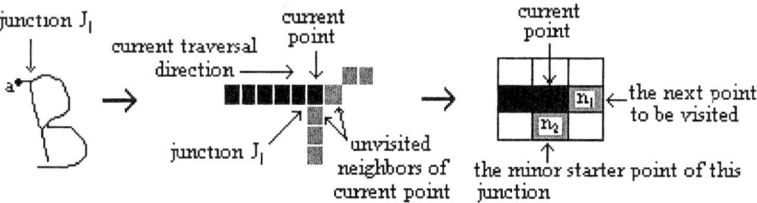

Fig. 5. The identification of *minor starter points* (Already visited pixels are depicted in black color and the unvisited pixels are depicted in ash color)

6.3 Determination of Current Traversal Direction

Let us consider all the eight adjacent neighbors of current pixel (i, j) in Figure 6. The *current traversal direction* can be defined as described in Table 2, according to the neighboring pixel, which is chosen as the next pixel to be visited.

(i-1, j-1)	(i, j-1)	(i+1, j-1)
(i-1, j)	(i, j)	(i+1, j)
(i-1, j+1)	(i, j+1)	(i+1, j+1)

Fig. 6. Eight adjacent neighbors of the current pixel (i,j)

Table 2. Determination of the *current traversal direction*

Neighboring pixel which is chosen as the next pixel point to be visited	Current traversal direction
(i, j-1)	U (UP)
(i+1, j)	R (LEFT)
(I, j+1)	D (DOWN)
(i-1, j)	L (RIGHT)
(i+1, j-1)	RU (RIGHT_UP)
(i+1, j+1)	RD (RIGHT_DOWN)
(i-1, j+1)	LD (LEFT_DOWN)
(i-1, j-1)	LU (LEFT_UP)

6.4 Determination of Closest Traversal Direction

When the *current traversal direction* is given, the *closest traversal directions* to the *current traversal direction* can be determined as described in Table 3.

Table 3. Determination of the *closest traversal direction* to the *current traversal direction*

Current traversal direction	Closest traversal directions
U	RU, LU
R	RU, RD
D	RD, LD
L	LD, LU
RU	U, R
RD	R, D
LD	D, L
LU	L, U

In Table 3, for each *current traversal direction*, two *closest traversal directions* are mentioned. Therefore, to find a neighbor in the *closest traversal direction* to the *current traversal direction*, the algorithm first checks the *closest traversal direction* which would be found first, when considering the directions clockwise starting from the direction U (Fig. 7). As an example, if the *current traversal direction* is U, the algorithm first checks whether there is an unvisited neighbor in the direction RU. If it fails to find such a neighbor, then it checks whether there is an unvisited neighbor in the direction LU.

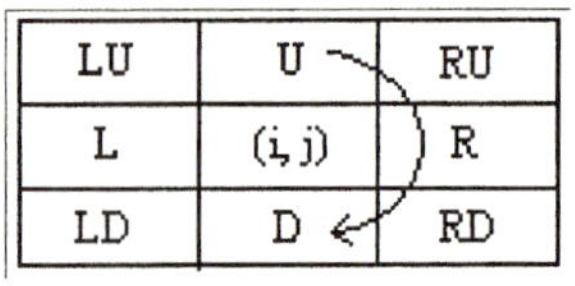

LU	U	RU
L	(i, j)	R
LD	D	RD

Fig. 7. Consideration of the directions in clockwise, staring from the direction UP from the current pixel point (i, j)

6.5 Choosing of an Unvisited Neighbor in Any Direction

To choose an unvisited neighbor, which is in any direction from the current point, as the next point to be visited, the unvisited neighbors are processed starting with the direction U and then in clockwise (Figure 7). That is, first, the algorithm checks whether there is an unvisited neighbor in the direction U to the current point (i, j). If it fails to find such a neighbor then it checks an unvisited neighbor in the RU direction to the current point and so on. Therefore the directions are considered in the order; U, RU, R, RD, D, LD, L, and finally LU.

7 The Segmentation

Definition 6: The w*ritten direction* is the direction of a particular sequence of pixels to which they were written.

The segmentation decision is based on the abrupt change in the *written direction*. According to this algorithm, as long as the *current traversal direction* remains unchanged (if it can find an unvisited neighboring pixel in the *current traversal direction*), the algorithm considers the path, which is being visited belongs to the same segment. If the *current traversal direction* changes, then the algorithm goes and checks for an abrupt change in *written direction*. The calculation of *written direction* of a sequence of pixels is discussed in section 7.1.

In this work, it was found that the *written direction* of a skeleton path could be changed after a sequence of 5 pixels. Therefore, to find out the previous written direction, the *written direction* of last 5 pixel points (which are stored in *current_segment*) in the path is considered. To find out the new written direction, the *written direction* of next 5 pixels in the path (which are stored in *tmp_segment*) is considered. The determination of abrupt change in *written direction* can be carried out as follows.

```
Function IsAbruptChange(current_segment, tmp_segment) returns Boolean.
Begin
  if(the size of tmp_segment < 5 OR size of current_segment < 5 )
      return false
  else
      prev_written_d = written direction of the last 5 pixel points in the
                                                       current_segment.
      new_written_d = written direction of the tmp_segment.
      difference = |prevs_written_d – new_written_d|
      if (difference > 315)
        if(prev_written_d > 315 OR curr_written_d > 315)
            difference = 360 – difference
        end if
      end if

      if (difference > threshold angle)
          return true
```

```
    else
        return false
    end if
```

End.

If there is an abrupt change in *written direction*, the corresponding change in *current traversal direction* is considered as a starting point of a new segment. Otherwise the traversal is continued with the same segment.

The suitable value for the *threshold angle* may vary according to the character set which is dealt with. We have used uppercase English handwritten characters to test the algorithm with various *threshold angles* and the results are depicted in Section 9.

7.1 Calculation of Written Direction

The *written direction* of a given sequence of pixel points equals to the angle between the x-axis and the straight line connecting the start and end points of the pixel sequence. Consider the sequence of points in Figure 8, where the start point is (x_s, y_s) and the end point is (x_e, y_e).

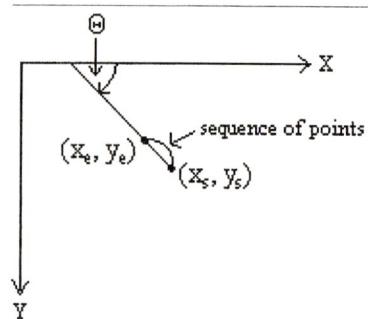

Fig. 8. The calculation of the *written direction*

After calculating the angle Θ (Equation 1), the *written direction* can be calculated as described in Table 4.

$$\Theta = \tan^{-1}(\,|(\,y_s - y_e)/(\,x_s - x_e)|\,) \tag{1}$$

Table 4. Calculation of the *written direction*

	written direction
$(x_s < x_e)$ and $(y_s <= y_e)$	Θ
$(x_s > x_e)$ and $(y_s <= y_e)$	$180 - \Theta$
$(x_s > x_e)$ and $(y_s > y_e)$	$180 + \Theta$
$(x_s < x_e)$ and $(y_s > y_e)$	$360 - \Theta$
$(x_s = x_e)$ and $(y_s <= y_e)$	90
$(x_s = x_e)$ and $(y_s > y_e)$	270

8 Noise Removal

In the process of noise removal, the segments resulted due to the noises in the *skeleton area* are removed. The average size of the character skeletons considered in this research is 30*30 pixels. Accordingly, the minimum size of the segment was taken as 5 pixel points. Therefore under the noise removal, the segments which contain less than 5 pixel points are discarded from the set of resulted segments.

9 Experimental Results

The algorithm was tested with the skeletons obtained by the skeletonization algorithm [5], of handwritten upper case English characters, under different *threshold angles*. Figure 9 shows the input character skeletons to the algorithm and their expected segmentations.

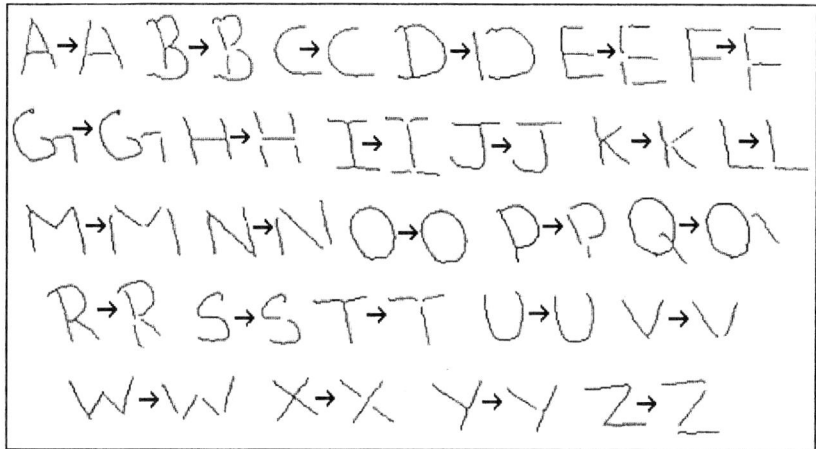

Fig. 9. Input character skeletons and their expected segmentations

By comparing the expected segmentations and the resulted segmentations, the following problems could be identified (Table 5). When the *threshold angle* was equal 40 degrees, the algorithm resulted over-segmentation of characters 'B','G','J','O','Q' and 'S'. When the *threshold angle* was lesser than 40 degrees, more over-segmentations were observed.

When the *threshold angle* was 60 degrees, the algorithm resulted under-segmentation of characters 'B', 'H' and 'L' and when it was larger than 60 degrees, more under-segmentations were resulted. When the threshold angle was equal to an angle between 45 and 60 degrees (ex. 45 degrees), the problem of under-segmentation was eliminated, but still there were some over-segmentation of the characters 'B' 'G' 'J' and 'S'. Except for these problems, other characters were segmented as expected.

Table 5. Problems identified in the segmentation under different threshold angles. The segments are numbered acoording to the order in which they were identified

Threshold angle	Problems in Segmentation
40 degrees	*(hand-drawn character segmentation examples)*
60 degrees	*(hand-drawn character segmentation examples)*
45 degrees	*(hand-drawn character segmentation examples)*

10 Conclusion

When comparing the resulted segments with original character skeletons, the problems of information lost or identification of the same segment twice were not found. Therefore it can be concluded that the above algorithm traverses through each and every path of the character skeleton only once, as it was supposed to do, whether the character skeleton is connected or not. The only problem was with the choice of the *threshold angle*.

According to the observed results, it can be concluded that it is not possible to fix a single value as the value for the *threshold angle*, since different values have given the expected results for the different characters. As an example, under the *threshold angle* equaled to 60 degrees, the character patterns 'J', 'G' and 'S' were segmented as expected, but the character 'H' and 'L' suffered from the under-segmentation. On the other hand, under the *threshold angle* equaled to 45 degrees, the character pattern 'H' and 'L' were segmented as expected, but characters 'J', 'G' and 'S' were over-segmented.

As future works, two methods could be suggested to overcome the above problem. One of them would be to use the *threshold angle* as 45 degrees (which seems to be reasonable since it has given meaningful segments by eliminating the problem of under-segmentation) and use a method to combine the over-segmented segments. The other solution would be to use different *threshold angles* for different situations. If the different situations those requiring different threshold angles could be identified separately, then these different threshold angles could be used for different characters to segment them into meaningful segments. One way to do it may be the use of characteristics of the pixels in the *current_segment* and the *tmp_segment* to differentiate the situations.

References

1. Alessandro, L., Koerich, Yann Leydier: A Hybrid Large Vocabulary Word Recognition System using Neural Networks with Hidden Markov Models, IWFHR (2002)
2. Bandara, G.E.M.D.C., Pathirana, S.D., Ranawana, R.M.: Use of Fuzzy Feature Descriptions to Recognize Handwritten Alphanumeric Characters. In: Proceedings of 1st Conference on Fuzzy Systems and Knowledge Discovery, Singapore (2002) 269-274
3. Bandara, G.E.M.D.C., Ranawana, R.M., Pathirana, S.D.: Use of Fuzzy Feature Descriptions to Recognize Alphanumeric Characters. In: Halgamuge, S., Wang, L. (eds.): Classification and Clustering for Knowledge Discovery. Springer-Verlag (2004) 205-230
4. Guyon, I.: Applications of Neural Networks to Character Recognition. Character and Handwriting Recognition. In: Wang, P.S.P. (eds.): Expanding frontiers, World Scientific (1991) 353-382
5. Lei Huang, Genxun Wan, Changping Liu: An Improved Parallel Thinning Algorithm. In: Proceedings of Seventh International Conference on Document Analysis and Recognition (2003) 780-783
6. Malaviya, A., Peters, L.: Handwriting Recognition with Fuzzy Linguistic Rules. In: Proceedings of Third European Congress on Intelligent Techniques and Soft Computing, Aachen, (1995) 1430-1434
7. Park, Lee: Off line recognition of large-set handwritten characters with multiple hidden markov models. Int. J. Pattern recognition, Vol. 20, no 2. (1996) 231-244
8. Romesh Ranawana, Vasile Palade, Bandara, G.E.M.D.C.: An Efficient Fuzzy Method for Handwritten Character Recognition. Lecture Notes in Computer Science, Vol.3214. Springer-Verlag GmbH (2004) 698-707
9. Singh, S., Amin, A.: Fuzzy Recognition of Chinese Characters. In: Proceedings of Irish *Machine Vision and Image Processing Conference , Dublin, (1999) 219-227*

A Method Based on the Continuous Spectrum Analysis for Fingerprint Image Ridge Distance Estimation*

Xiaosi Zhan[1], Zhaocai Sun[2], Yilong Yin[2], and Yayun Chu[1]

[1] Computer Department, Fuyang Normal College,
236032 Fuyang, P.R. China
{xiaoszhan@263.net, Chuyayun.fync@126.com}
[2] School of Computer Science & Technology, Shandong University,
250100, Jinan, P.R. China
{sunnykiller@126.com, ylyin@sdu.edu.cn}

Abstract. As one kind of image having strong texture character, ridge distance is the important attribute of fingerprint image. It is important to estimate the ridge distance correctly for improving the performance of the automatic fingerprint identification system. The traditional Fourier transform spectral analysis method had the worse redundancy degree in estimating the ridge distance because it was based on the two-dimension discrete Fourier spectrum. The paper introduces the sampling theorem into the fingerprint image ridge distance estimation method, transforms the discrete spectrum into two-dimension continuous spectrum and obtains the ridge distance on the frequency field. The experimental results indicate that the ridge distance obtained from this method is more accurate and has improved the rate of accuracy of the automatic fingerprint identification system to a certain extent.

1 Introduction

Fingerprint identification is the most popular biometric technology, which has drawn a substantial attention recently and the research of the recognition technology of automatic fingerprint becomes a research focus too [1,2]. In order to improve the precision of the automatic fingerprint recognition system and extent the range of application, it is the key to realize effective enhancement to the fingerprint image with the low quality. Forefathers have already researched on the fingerprint image enhancement method deeply and put forward some methods [3,4,5]. These methods can all realize to enhance the fingerprint images with low quality to a certain extent. In the most fingerprint image enhancement algorithms, the ridge distance (or called the ridge frequency) is one important parameter. It is important to accurately estimate the ridge distance for improving the performance of the automatic fingerprint identification system.

---

* Supported by the National Natural Science Foundation of China under Grant No. 06403010, Shandong Province Science Foundation of China under Grant No.Z2004G05 and Anhui Province Education Department Science Foundation of China under Grant No.2005KJ089.

Generally, fingerprint ridge distance (also call ridge mode cycle) is defined as the distance between two adjacent ridges, which can be seen as the distance from the center of one ridge to the center of other adjacent ridge or the pieces of ridge width and valley width in close proximity [6]. Here, ridge and valley mean the dark color and light color ridge in the fingerprint image respectively. The definitions of the ridge, valley and ridge distance are showed as follows Fig.1. The larger the ridge distance is, indicate the sparser here the ridge is. On the contrary, indicate the more intensive here the ridge is. The ridge distance is determined by the following two factors: the fingerprint inherent structure and the image resolution.

Up to now, the fingerprint ridge distance estimation methods can be sum up in two kinds: the ridge distance estimation methods based on the frequency field and the methods based on spatial field.

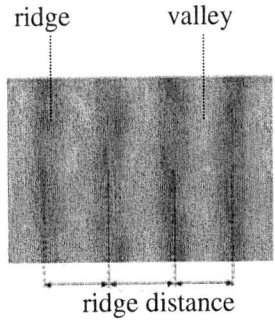

Fig. 1. Definitions of ridge, valley and ridge distance

L. Hong proposed the direction window method to estimate the ridge frequency [3]. This method estimated the ridge frequency reliably when the contrast of the fingerprint image was good and the ridge orientation in the direction window was consistent. But when the noise interference was serious or the ridge direction was not totally unanimous in the direction window, the performance of this method would be influenced seriously. Z. M. Kovace-Vajna brought out two kinds of ridge distance estimation methods: the geometry approach and the spectral analysis approach, which are both based on the block fingerprint image to estimate the ridge distance [7].

O'Gorman and Nickerson used the ridge distance as a key parameter in the design of filters. It is one statistics mean value about the ridge distance [8]. Lin and Dubes attempted to count ridge number in one fingerprint image automatically and assumed the ridge distance is a constant value on the whole fingerprint image [9]. D. Douglas estimated the average distances of the all ridges on the whole fingerprint image [10]. Mario and Maltoni did mathematical characterization of the local frequency of sinusoidal signals and developed a 2Dmodel in order to approximate the ridge-line patterns in his method for ridge-line density estimation in digital images [11].

In addition, Y. Yin proposed the ridge distance estimation method based on regional level. The method divided the fingerprint image into several regions according to the consistency of the orientation information on the whole fingerprint image and calculates the ridge distance to every region respectively [12]. Y. Chen proposed two kinds of methods to estimate the ridge distance: the spectral analysis approach and the statistical window approach. The spectral analysis approach had the following restricted conditions: the definition about the radial distributing function and the method to obtain the value of r' accurately. The statistical window approach was relatively harsh for quality of the fingerprint image [13].

This paper puts forward the ridge distance estimation method based on continuous spectrum in the frequency field. Firstly, the method transforms the fingerprint image expressed in the spatial field into the spectrum expressed in the frequency field adopting two-dimensional discrete Fourier transform. Secondly, the method expands the image on this basis and transforms the two-dimensional discrete frequency spectrum into continuous frequency using the sampling theorem. Lastly, the method calculates two relatively symmetrical extremum points relative to the peak value in this continues frequency spectrum map.

2 The Spectral Analysis Method of the Fingerprint Image

Spectral analysis method that transforms the representation of fingerprint images from spatial field to frequency field is a typical method of signal processing in frequency field. It is a traditional method for ridge distance estimation in fingerprint images. Generally, if $g(x, y)$ is the gray-scale value of the pixel (x, y) in an $N \times N$ image, the DFT of the $g(x, y)$ is defined as follows:

$$G(u,v) = \frac{1}{N^2} \sum_{i=1}^{N} \sum_{j=1}^{N} g(x, y) e^{2\pi j / N <(x,y)(u,v)>} \quad (1)$$

Where j is the imaginary unit, $u, v \in \{1, \cdots N\}$ and $<(x, y)(u, v)> = xu + yv$ is the vector pot product. Obviously $G(u,v)$ is complex. Then the periodic characteristics of the point (u, v) can be represented by the magnitude of $G(u,v)$. In order to illustrate accurately the viewpoint, the paper firstly analyzes the Fourier transform to the regular texture image.

2.1 Spectral Analysis of the Texture Image

The regular texture image has its peculiar good attribute. It is very meaningful to study and utilize the attribute of the texture image fully. Fig.2 (b) shows the spectrum after carrying the Fourier transform to the regular texture image (Fig.2 (a)). It has reflected the frequency information and the orientation information of the ridge in the regular image. The distance among the light spots is in direct proportion to ridge frequency and in inverse proportion to ridge distance. The direction of the line connecting two light spots is vertical with the ridge orientation and parallel with the direction of the normal line of the ridge.

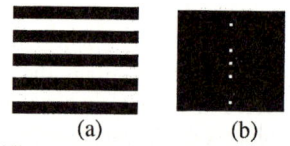

Fig. 2. The regular texture image and the corresponding spectrum ((a) is the regular texture image and (b) is the corresponding spectrum.)

Fingerprint image is one regular texture image. The kind of character should be reflected after we carried the Fourier transform to the fingerprint image. Fig.3 (b) shows the spectrum after carrying the Fourier transform to one block fingerprint image (Fig.3 (a)). Because of the influence of the noise and irregular ridge etc., some light spots are dispersed. For the model region in the fingerprint image, it is especially like this (as Fig.3 (c), (d) show). For the background region, there is not the clear light point (as Fig.3 (e), (f) show). We can obtain the ridge distance of each region in the fingerprint image by calculating the distance between the two light spots too. But, the position of the two light spots in the spectral image is not very obvious, now it is the key how to obtain the distance between two light spots accurately.

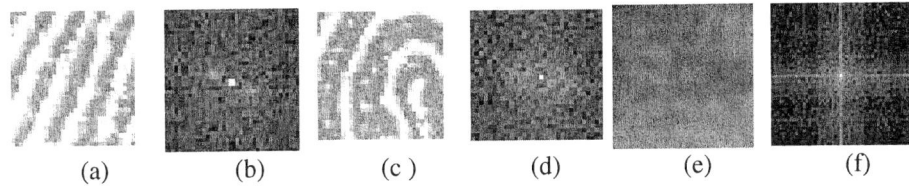

Fig. 3. The fingerprint region, the model region and the background region and the corresponding spectrum image (a), (c) and (e) are the fingerprint region, the model region and the background region; (b), (d) and (f) are the corresponding spectrum.)

2.2 Calculate the Ridge Distance Based on the Discrete Spectrum

Presently, the spectral analysis methods are based on the discrete spectrum mostly, such as the ridge distance estimation proposed in [13] by Y. Chen. Suppose $g(x,y)$ is the gray-scale value of the pixels with coordinates x, y ={0, ... ,N-1} in a $N \times N$ image, $g(x,y)$ is the gray value of the pixel (x, y) in the image, we can obtain the discrete spectrum $G(u,v)$ of the image in carrying on two-dimensional fast Fourier transform. A fingerprint image and its FFT at block level with window size of 32×32 are shown in Fig. 4.

After obtaining the spectrum, the next work is to search for the two light points that are symmetrical with the central lightest point. But it is a hard work. In order to illustrate the issue accurately and effectively, we choose two blocks image (showed in the Fig. 2 (a) and (e), one is the valid fingerprint region and the other is the background

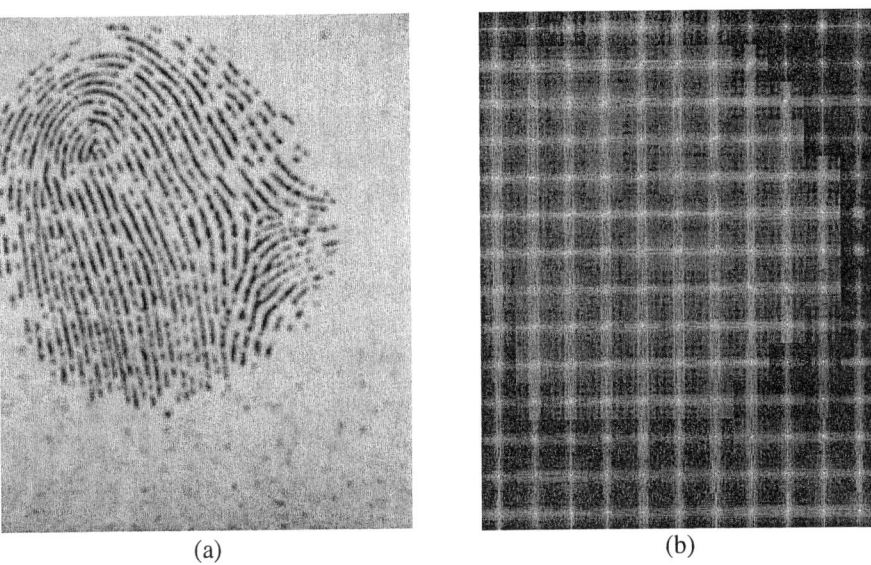

Fig. 4. A fingerprint image and the corresponding spectrum at block level with two-dimension discrete fast Fourier transform ((a) is a fingerprint image (b) is corresponding transform result at block level.)

region in one fingerprint image) and observe them in the three-dimension spatial (shows as Fig.5). The x-axis and the y-axis are the x-coordinate and the y-coordinate of the spectrum image respectively and the z-axis is the energy in the corresponding locality.

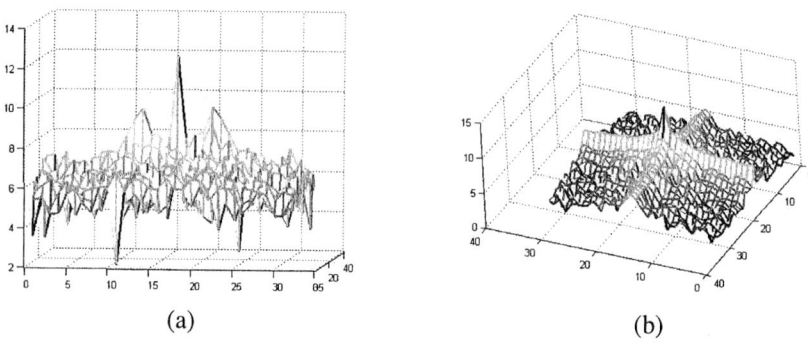

Fig. 5. The sketch map about the three-dimension spectrum ((a) and (b) are the corresponding spectrum of the Fig.3 (b) and (f) respectively.)

In the Fig.5, there are two sub-peak points corresponding to two sub-light points in the Fig.3 in the spectrum of the valid fingerprint image region if we observe the spectrum from the ridge direction. But there aren't the obvious sub-peak points if we observe the spectrum of the background region. From the Fig.5, we can make out that the discrete spectrum is very fuzzy. Perhaps in the area of light spots, that is to say two or three values are relatively high. It is difficult to obtain the position of the light point correctly and quickly. At the same time, the real local extremum is possibly not the discrete value.

For obtaining the right ridge distance value, Y Yin et al. define the radial distribute function in [6] as follows:

$$Q(r) = \frac{1}{\#C_r} \sum_{(u,v) \in C_r} |G_{(u,v)}| \qquad (2)$$

Where C_r is the set of the whole pixels that satisfy the function $\sqrt{u^2 + v^2} = r$, $\#C_r$ is the number of the element of the set C_r. Then define the $Q(r)$ as the distribution intensity of the signal with the cycle N/r in the $N \times N$ image. The value of r corresponding to the peak of $Q(r)$ can be defined as the cycle number of the dominant signal in the $N \times N$ image. Search for the value of r_0 that enables the value of $Q(r_0)$ is the local maximum. At this moment, the ridge distance of the block image can be estimated with $d = N/r_0$. Because the ridge distance of fingerprint image is among 4-12 pieces of pixel, some minute errors in calculating the value of r_0 will cause heavy influences in estimating the ridge distance.

In the course of analyzing the character of the discrete spectrum, we find that the discrete spectrum can be transformed in the continuous spectrum by the sampling theorem. This inspires us to acquire the ridge distance from the continuous spectrum.

3 Ridge Distance Estimation Method Based on the Continues Spectrum

The precision does not meet the requirement if we carry through the discrete Fourier transform. At the same time, the speed can't meet the requirement of the real time disposal if we make the continuous Fourier transform. For these reasons, we make fast discrete Fourier transform firstly, then adopt the two-dimension sampling theorem to transform the discrete spectrum into the continuous spectrum. Lastly, we obtain the position where the energy maximum appears in the normal direction of the ridge. The sampling theorem plays a great role while dealing with the signal. It is also suit to the two-dimension image processing. It was the sample theorem of the two-dimension signal that was adopted here, described it as follows:

3.1 Analysis of the Continuous Spectrum

Suppose the Fourier transform function $F(s_1, s_2)$ about the function f (x1, x2) of L2 (R2) is tight-supported set (namely that the function F is equal to zero except the boundary region D and the boundary region D can be defined as the rectangle region $\{(s_1, s_2) | |s_1| \leq \Omega \text{ and } |s_2| \leq \Omega\}$ in the paper. Here, we firstly assume $\Omega = \pi$ in order to simplify the function. Then the Fourier transform function about the function $f(x_1, x_2)$ can be denoted as follows:

$$F(s_1, s_2) = \sum_{n_1} \sum_{n_2} C_{n_1, n_2} e^{-jn_1 s_1 - jn_2 s_2} \tag{3}$$

Here, the C_{n_1, n_2} is defined as follows:

$$C_{n_1,n_2} = \frac{1}{(2\pi)^2} \int_{-\infty}^{+\infty}\int_{-\infty}^{+\infty} ds_1 ds_2 e^{jn_1 s_1 + jn_2 s_2} F(s_1, s_2) = \frac{1}{2\pi} f(n_1, n_2) \tag{4}$$

Then, we can acquire the following function as:

$$\begin{aligned} f(x_1, x_2) &= \frac{1}{2\pi} \int_{-\infty}^{+\infty}\int_{-\infty}^{+\infty} ds_1 ds_2 e^{jx_1 s_1 + jx_2 s_2} F(s_1, s_2) \\ &= \frac{1}{2\pi} \sum_{n_1}\sum_{n_2} C_{n_1,n_2} \int_{-\pi}^{\pi}\int_{-\pi}^{\pi} ds_1 ds_2 e^{jx_1 s_1 + jx_2 s_2} e^{-jn_1 s_1 - jn_2 s_2} \\ &= \sum_{n_1}\sum_{n_2} C_{n_1,n_2} \frac{\sin \pi(x_1 - n_1)}{\pi(x_1 - n_1)} \frac{\sin \pi(x_2 - n_2)}{\pi(x_2 - n_2)} \end{aligned} \tag{5}$$

In this way, the discrete signal $C_{n1, n2}$ can be recovered for the continuous signal $f(x_1, x_2)$ through the sampling theorem. Then, the discrete frequency spectrum of each block fingerprint image can be recovered for the continuous frequency spectrum. We can try to obtain accurately the local extreme value (that is the "light spot" position we care about) in random small step in the continuous frequency spectrum. Thus we can calculate the ridge distance accurately. The following Fig.6 is the continuous frequency spectrum recovered from the Fig.5 in the step length as 0.1.

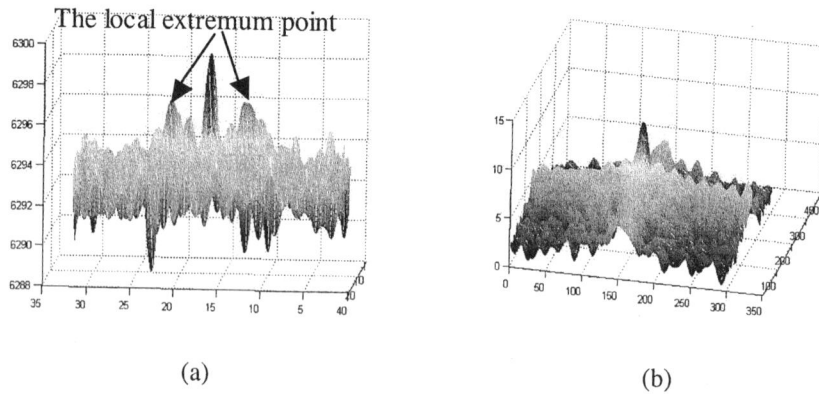

(a)   (b)

Fig. 6. The sketch map about the three-dimension continuous spectrum recovered from the Fig. 5 in the step as 0.1 ((a) and (b) are the corresponding spectrum of the Fig.5 (a) and (b) respectively.)

From the Fig.6, we can find that the local peak value is obvious in the continuous frequency spectrum image. But it is a long period course that we search the continuous spectrum image that is recovered from one $N \times N$ point matrix in a small step for the local peak value. Thus we need to search the continuous spectrum purposefully.

3.2 Ridge Distance Estimation Method Based on the Continuous Spectrum

After observing the Fig.5 and the Fig.6 carefully, we can find that the two light points, which are the local extreme point we care about, always appear in certain region. The direction between them must be vertical with the direction of the ridge and the radius is in relative ranges too. From the relation between the ridge distance and the ridge frequency we have the following function: $r = N/d$. Generally, the ridge distance is from 4 pixels to 12 pixels based on our experience. Thus, the radius that the extreme point appears is in N/12-N/4 by experience. Then we carry on more accurate search on this range. At present the orientation from the Rao [15] method proposed by Lin-Hong is satisfied the requirement.

Suppose that the ridge orientation is θ in the Fig.3 (a), the normal orientation of the ridge is $\theta + \pi/2$. We can obtain the position of the local extreme point in the continuous spectrum (as Fig.6 shows) if we search the region, which is confirmed by the radius: N/12-N/4 and the direction: $\theta + \pi/2$, in the step length as 0.01. As Fig.7 shows, the local extreme points are 11.03, the corresponding radius is 4.71, and the ridge distance of the image is 32/4.71=6.79.

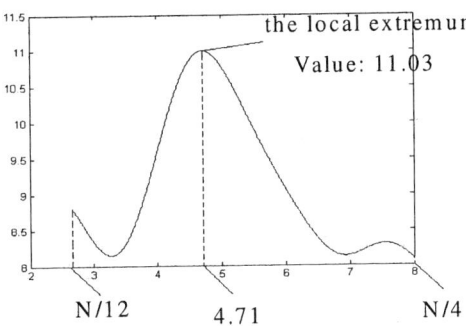

Fig. 7. The cutaway view of the three-dimension continuous spectrum in the normal orientation

3.3 The Steps of Realization

Step 1: Divide the fingerprint image into non-overlap block with the size $N \times N$, the N is equal to 32 generally.

Step 2: To each block fingerprint image $g(i,j)$, carry on two-dimension fast Fourier transform and get the corresponding discrete spectrum $G(u,v)$.

Step 3: To each discrete spectrum $G(u,v)$, apply the sampling theorem to get the continuous spectral function G(x,y).

Step 4: Adopt Rao [15] method to obtain the ridge orientation θ.

Step 5: Search the region confirmed by the radius N/12-N/4 and the direction $\theta+\pi/2$ in a small step length L for finding the radius r corresponding the local extreme point. Generally, the value of L is 0.01.

Step 6: If don't find the local extreme point then think that the ridge distance of the fingerprint image region can't be obtained. Else estimate the fingerprint image ridge distance from d =N/r.

4 Experimental Result Analysis and Conclusion

For illuminating the performance of the ridge distance estimation method proposed in the paper, we choose DB_B in the public data–BVC2004, which is considered the most hard fingerprint image database to process in common and disposed it. The following Fig.8 is the processing results of the two fingerprints from the DB_B in the public data_BVC2004.

From the following Fig.8 we can find that the results are very reasonable and can represent the real ridge distance. Except these strong-noised regions, the ridge distance can be obtained accurately. At the same time, the value of ridge distance is a decimal fraction that has higher precession than the integer in the representation. Contrast the accurate value that obtained from the manual calculation, the result is very exact.

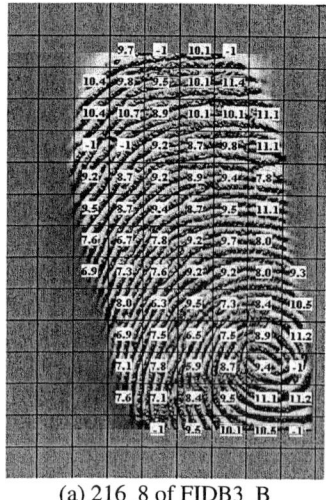 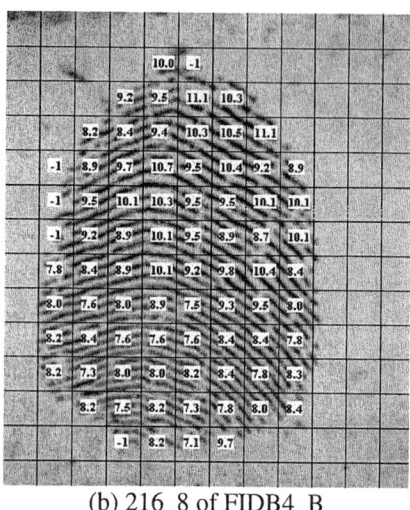

(a) 216_8 of FIDB3_B (b) 216_8 of FIDB4_B

Fig. 8. The ridge distance estimation results of two representative fingerprint images being chosen from the DB_B in the public data–BVC2004 ((a), (b) are the No. 216_8 fingerprint image of FIDB3_B and FIDB4_B. In the results, no value represents the background region, -1 represents these blocks where the ridge distance can not be obtained accurately, the other values is the ridge distance corresponding to the block fingerprint image.)

In order to evaluate the performance of our method, we adopt the performance evaluation method that was put forward by Y. Yin in [6]. There, DER, EA and TC are the three indexes of the performance evaluation method. Where DER indicates the robustness of a method for ridge distance estimation in fingerprint images, EA is the degree of deviation between the estimation result and the actual value of the ridge distance and TC is the time needed for handing a fingerprint image.

To evaluate the performance of the methods, we use 30 typical images (10 good quality, 10 fair quality, 10 poor quality) selected from NJU fingerprint database (1200 live-scan images; 10 per individual) to estimate ridge distance with traditional spectral analysis method, statistical window method and our method, respectively. Here, we choose the NJU database because that the other two methods is tested based on the NJU database and the paper will use the correlative results of the other two methods. Sizes of block images are all 32×32 and the value of ridge distance of each block image is manually measured. The work is performed using the computer with PIV 2.0G, 256M RAM.

From the table 1 we can find that the traditional spectral analysis method has the lowest DER value and EA value with the middle TC value. For spectral analysis method, the biggest problem is how to acquire the value of r' accurately and reliably. The statistical method has the middle DER value and the EA value with the lowest TC value. The average performance of statistical window method is superior to that of the spectral analysis method. But the obvious disadvantage of statistical window method is that it doesn't perform well in regions where there is acute variation of ridge directions. Our method has the higher DER value and the EA value with the higher TC value. It shows that our method has higher performance except the processing time. The method can obtain the ridge distance of most regions in a fingerprint image except the pattern region and the strong-disturbed region because the sub-peak is not obvious in these regions. In addition, the processing time of our method is more that the other two methods because our method is based on the two-dimension continuous spectrum.

Our future research will focus on looking for some better method which can transform the spatial fingerprint image into two-dimension continuous frequency spectrum and making certain the more appropriate step length in order to find the two sub-peak points faster and accurately. In addition, we can fulfill other fingerprint image processing in the two-dimension continuous spectrum, such as fingerprint image segmentation processing, fingerprint image enhancement processing and ridge orientation extraction processing etc.

Table 1. Performance of three methods

Method	DER	EA	TC
Traditional spectral analysis	44.7%	84%	0.42 second
Statistical window	63.8%	93%	0.31 second
Our method	94.6%	95%	0.63 second

References

[1] Hong, L., Jain, A.K., Bolle, R.: Identity Authentication Using Fingerprints. Proceedings of FirstInternational Conference on Audio and Video-Based Biometric Person Authentication, Switzerland (1997) 103-110
[2] Yin, L., Ning, X., Zhang. X.: Development and Application of Automatic Fingerprint Identification Technology. Journal of Nanjing University (Natural Science), (2002) 29-35
[3] Hong, L., Wan, Y., Jain. A. K.: Fingerprint Image Enhancement: Algorithm and Performance Evaluation. IEEE Trans. on Pattern Analysis and Machine Intelligence (1998) 777-789
[4] Sherlock, D., Monro D. M., Millard, K.: Fingerprint Enhancement by Directional Fourier Filter, IEEE Proc. Vis. Image Signal Processing (1994) 87-94
[5] Sherstinsky, A., Picard, R. W.: Restoration and Enhancement of Fingerprint Images Using M-Lattice: A Novel Non-Linear Dynamical System. Proc. 12th, ICPR-B, Jerusalem (1994:) 195-200
[6] Yin, Y., Tian, J., Yang, X.: Ridge Distance Estimation in Fingerprint Images: Algorithm and Performance Evaluation. EURASIP Journal on Applied Signal Processing (2004) 495-502
[7] Kovacs-Vajna, Z. M., Rovatti, R., Frazzoni, M.: Fingerprint Ridge Distance Computation Methodologies. Pattern Recognition (2000) 69-80
[8] O'Gorman, L., Neckerson, J. V.: An Approach to Fingerprint Filter Design. Pattern Recognition, (1989) 29-38
[9] Lin W. C., Dubes, R. C.: A Review of Ridge Counting in Dermatoglyphics. Pattern Recognition, (1983) 1-8
[10] Douglas Hung, D. C., Enhancement Feature Purification of Fingerprint Images, Pattern Recognition (1993) 1661-1671
[11] Maio, D., Maltoni, D.: Ridge-Line Density Estimation in Digital Images. Proceedings of 14th International Conference on Pattern Recognition, Brisbane, Australia (1998) 534-538
[12] Yin, Y., Wang, Y., Yu, F.: A Method Based on Region Level for Ridge Distance Estimation, Chinese Computer Science (2003) 201-208
[13] Chen, Y., Yin, Y., Zhang X.: A Method Based on Statistics Window for Ridge Distance Estimation, Journal of image and graphics, China (2003) 266-270

A Method Based on the Markov Chain Monte Carlo for Fingerprint Image Segmentation*

Xiaosi Zhan[1], Zhaocai Sun[2], Yilong Yin[2], and Yun Chen[1]

[1] Computer Department, Fuyang Normal College,
236032 Fuyang, P.R. China
{xiaoszhan@263.net, Chenyun.fync@163.com}
[2] School of Computer Science & Technology, Shandong University,
250100, Jinan, P.R. China
{sunnykiller@126.com, ylyin@sdu.edu.cn}

Abstract. As one key step of the automatic fingerprint identification system (AFIS), fingerprint image segmentation can decrease the affection of the noises in the background region and handing time of the subsequence algorithms and improve the performance of the AFIS. Markov Chain Monte Carlo (MCMC) method has been applied to medicine image segmentation for decade years. This paper introduces the MCMC method into fingerprint image segmentation and brings forward the fingerprint image segmentation algorithm based on MCMC. Firstly, it generates a random sequence of closed curves as Markov Chain, which is regarded as the boundary between the fingerprint image region and the background image region and uses the boundary curve probability density function (BCPDF) as the index of convergence. Then, it is simulated by Monte Carlo method with BCPDF as parameter, which is converged to the maximum. Lastly, the closed curve whose BCPDF value is maximal is regarded as the ideal boundary curve. The experimental results indicate that the method is robust to the low-quality finger images.

1 Introduction

In recent years, biometric identification represented by fingerprint has been a research focus [1-5]. Automatic Fingerprint Identification System (AFIS) mainly includes image segmentation, image enhancement, minutiae extraction, fingerprint matching, etc. As one of the key problems in fingerprint image processing, the purpose of fingerprint image segmentation is to separate the valid fingerprint image region, with which all the posterior algorithms of AFIS will deal, from background image region. So, it is important for improving the performance of the AFIS to segment the valid fingerprint region from the background region.

There are a lot of literatures that focus on the technology of the fingerprint image segmentation. The present methods of fingerprint image segmentation can be summed up two specials: one is based on block-level [2,3], the other is based on pixel-level

---

* Supported by the National Natural Science Foundation of China under Grant No. 06403010, Shandong Province Science Foundation of China under Grant No.Z2004G05 and Anhui Province Education Department Science Foundation of China under Grant No.2005KJ089.

[4,5]. Both designed the algorithms according to the statistical character (e.g. Variance, Mean) of the gray fingerprint image. J. X. Chen used linear classifier to classify the blocks into foreground and background [6]. The resolution of this method only reached the block level, which made it difficult to judge the reliability of features extracted from the border. The other method is based on pixel level. A. M. Bazen and S. H. Gerez used linear classifier to classify the pixels into foreground and background through the analysis of the pixel features [7]. For different databases, this method provided different segmentation surface. In addition, the performance of linear classifier had certain limits, so it was difficult to reach the ideal segmentation results. Y.L Yin used the model of quadratic curve surface to carry out the fingerprint image segmentation, which regarded the gray variance, the gray mean and the orientation coherence as the parameters of the model [8].

Over the past 40 years, Markov Chain Monte Carlo (MCMC) method had penetrated many subjects, such as statistical physics, seismology, chemistry, biometrics and protein folding, as a general engine for inference and optimization [9]. In computer vision, S. C. Zhu did many works but not fingerprint image [10-12]. Y. L He regarded the fingerprint image as Markov Random Field and carried out the fingerprint image segmentation successfully [13]. But it can generated the boundary curve only where the edge contrast between fingerprint and background is stronger, and it is unsatisfied when the interference of background is stronger.

The paper introduces the MCMC method into fingerprint image segmentation and puts forward the fingerprint image segmentation algorithm based on MCMC. The paper regards the closed curve as research object and generates a random sequence of closed curves as the Markov Chain firstly. Suppose one of these closed curves is the real boundary curve between the fingerprint image region and the background image region. Secondly, the paper defines the boundary curve probability density function (BCPDF) that is regarded as the index of convergence. Following, it is simulated by Monte Carlo method with BCPDF as parameter, which is converged to the maximum. Lastly, the closed curve whose BCPDF value is maximal is regarded as the ideal boundary curve. The experimental results indicate that the method can segment the valid fingerprint image region from the background quickly, exactly and reliably. The more crucial is that the algorithm is robust to the low-quality finger images.

2 Algorithm of Fingerprint Boundary Curve Markov Chain Monte Carlo

2.1 The Model of Boundary Curve Probability

Suppose that one fingerprint image will be segmented well if we can find one closed curve that can separate the valid fingerprint image region from background image region. In the paper, we call this closed curve as boundary curve. As Fig.1 showed, for the four curves A, B, C, D, the curve A separates the fingerprint image region from the background image region successfully and decrease the disturbance caused by the noise (the remainder image region and the peeling image region in the fingerprint image) at the same time. But the curves B, C, D can't accomplish it. Hence, the curve A can be regarded as the right boundary curve in this fingerprint image and the curves B, C, D are all not the correct boundary curve. So, the process of fingerprint image segmentation is to looking for the boundary curve. If we can calculate the probability that a closed curve is the boundary curve of the fingerprint image according to the gray level of the fingerprint image, the closed curve with the biggest probability can be regarded as the boundary curve, for example, curve A as fig.1 showed. Obviously, such boundary curve probability density function (BCPDF) is required to satisfy the following conditions:

(1) The value of BCPDF of the closed curve in background image region (e.g. curve B) is less than that of the boundary curve (e.g. Curve A).

(2) The value of BCPDF of the closed curve within fingerprint image region (e.g. curve C) is less than that of the boundary curve (e.g. Curve A).

(3) The value of BCPDF of the closed curve that has crossed fingerprint image region and background image region (e.g. curve D) is less than that of the boundary curve (e.g. Curve A).

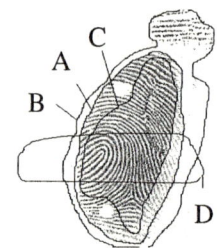

Fig. 1. Fingerprint image and the boundary curve

If we define inward ring and outward ring of a closed curve as fig.2 showed, compared to B, C, D, the boundary curve like as curve A is required:

(1) The outward ring of the boundary curve is in background image region exactly. (\\\ denotes the outward ring as fig.2 showed).

(2) The inward ring of the boundary curve is within fingerprint image region exactly. (/// denotes the outward ring as fig.2 showed).

In this paper, we define the outward background probability density function of a closed curve Γ as $Pout(\Gamma)$ and the inward fingerprint probability density function as $Pin(\Gamma)$. Then, the value of $Pout(\Gamma)$ is the probability that the outward ring is in the background region and the value of $Pin(\Gamma)$ is the probability that the inward ring is in the fingerprint image region. So, if we denote BCPDF of Γ as $PL(\Gamma)$, then, we have:

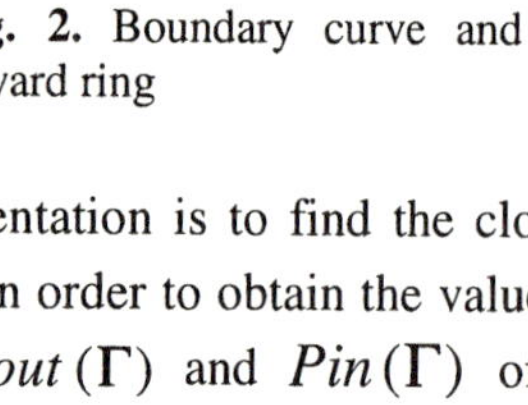

Fig. 2. Boundary curve and the inward ring

$$PL(\Gamma) = Pin(\Gamma)Pout(\Gamma)$$

The following key issue of fingerprint image segmentation is to find the closed curve Γ whose BCPDF value $PL(\Gamma)$ is the maximal. In order to obtain the value of $PL(\Gamma)$ of Γ, we should calculate the values of $Pout(\Gamma)$ and $Pin(\Gamma)$ of Γ firstly. And, the following task is to calculate the outward background probability $Pout(\Gamma)$ and the inward fingerprint probability $Pin(\Gamma)$.

2.2 Calculation for the Values of *Pout* (Γ) and *Pin* (Γ)

Generally, fingerprint image can be segmented into two kinds of regions as background image region and fingerprint image region, which also can be labeled as ridge region and valley region. After studying the fingerprint image carefully, we find that the gray levels of pixels in ridge region are very close, and so as valley region and background region, as fig.3(a) showed. Moreover, gray levels of pixels in valley region or in background region are so close. Hence, the result is that the gray level

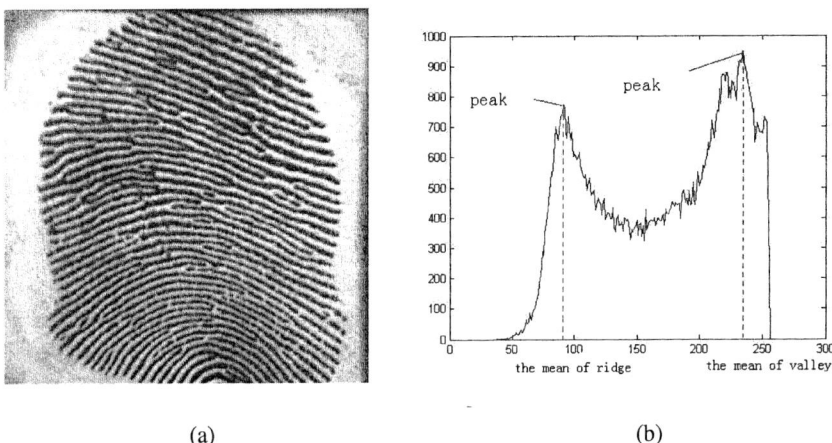

(a)　　　　　　　　　　　　　　　(b)

Fig. 3. The fingerprint image and the gray histogram corresponding to the fingerprint image: (a) is the original fingerprint image;(b) is the corresponding gray histogram.

would gather into two domains and there are two peaks in the histogram of fingerprint image, as fig.3 (b) showed. The gray level where pixels in background region or in valley region gathered is called as the mean of valley, so as the mean of ridge. Then, it can be considered that pixels in ridge region obey the normal distribution with the mean of ridge as the form:

$$p(i(x,y) \mid ridge) = \frac{1}{\sqrt{2\pi}\sigma} e^{-\frac{(g_m - \mu_l)^2}{2\sigma^2}} \quad (1)$$

where g_m denotes the gray level of the pixel i(x,y), μ_l denotes the mean of ridge, σ^2 denotes variance.

At the same time, pixels in background region or in valley region obey the normal distribution with the mean of valley as the form

$$p(i(x,y) \mid valley) = p(i(x,y) \mid back) = \frac{1}{\sqrt{2\pi}\sigma} e^{-\frac{(g_m - \mu_h)^2}{2\sigma^2}} \quad (2)$$

where μ_h denotes the mean of valley.

To a closed curve Γ, if the outward ring of Γ is in the background image region then every pixel is in the background image region and obeys the normal distribution with the mean of valley. In whole, the probability the outward ring of Γ is in the background image region completely can be written as the form:

$$Pout\ (\Gamma) = \prod_{m=1}^{k} \frac{1}{\sqrt{2\pi}\sigma} e^{-\frac{(g_m - \mu_h)^2}{2\sigma^2}} \tag{3}$$

where k denotes the sum of pixels in the outward ring, g_m denotes gray level of the pixel i(x,y), μ_h denotes the mean of valley.

But to fingerprint image region, pixels in it are either in ridge region or in valley region. The gray distribution of valley region is same to that of background region. Hence, it is pixels in ridge region that we judge the inward ring of Γ being in the fingerprint image region according to. We can see, ridge line and valley line are always appear by turn in the fingerprint image region. Hence, it can be considered that the sum of the pixels in ridge region is equal to that of pixels in valley region approximately. In other words, the sum of pixels in ridge region is equal to half of the sum of all pixels in fingerprint image region approximately. So, the inward fingerprint probability can be written as the form:

$$Pin\ (\Gamma) = (1 - |1 - \frac{2k}{N}|) \prod_{m=1}^{k} \frac{1}{\sqrt{2\pi}\sigma} e^{-\frac{(g_m - \mu_l)^2}{2\sigma^2}} \tag{4}$$

where N denotes the sum of all pixels in inward ring of Γ, k denotes the sum of pixels in ridge region, g_m denotes the gray level of the pixel in ridge region, μ_l denotes the mean gray value of ridge. The left coefficient has guaranteed that the value of $PL\ (\Gamma)$ is the biggest only if the sum of pixels in ridge region is half of the sum of all pixels in fingerprint image region, which is also the peculiarity of fingerprint image.

Now, we can calculate BCPDF $PL\ (\Gamma)$ of any closed curve Γ in fingerprint image, through calculating the outward background probability $Pout\ (\Gamma)$ and the inward fingerprint probability $Pin\ (\Gamma)$.

$$PL\ (\Gamma) = Pin\ (\Gamma) Pout\ (\Gamma) \tag{5}$$

In fingerprint image, the closed curve with the biggest value of BCPDF is the optimum solution of the fingerprint image segmentation. So, it is required to find the closed curve with the biggest value of BCPDF. A simple thought is, looking for all closed curves in fingerprint image, and finding one with the biggest BCPDF value. But, it is impossible to look for all the closed curves. Hence, there must be some approximate methods to do it. Markov Chain Monte Carlo (MCMC) will solve this kind of problem well. Generally, it required two steps with MCMC: (1) generating Markov Chain according to the needs of problem; (2) solving it with Monte Carlo and looking for the approximate answer.

2.3 Markov Chain of Boundary Curve in Fingerprint Image

Supposing the sequence $\{\Gamma_1, \Gamma_2, \cdots \Gamma_n\}$ of the boundary curve in fingerprint image is a random Markov Chain, it can be known by the property of Markov Chain that $P(\Gamma_{i+1} | \Gamma_i) = P(\Gamma_{i+1} | \Gamma_1, \Gamma_2, \cdots, \Gamma_i)$. In other words, the next state of the boundary curve Γ_{i+1} depends only on the current state Γ_i, but not the history state $\{\Gamma_1, \Gamma_2, \cdots, \Gamma_{i-1}\}$. There are two basic requirements for designing Markov chain dynamics. Firstly, it should be ergodic and aperiodic. Given any two closed curve Γ, Γ', the Markov chain can travel from Γ to Γ' in finite steps. Secondly, it should observe the stationary equation. This requirement is often replaced by a stronger condition: the so-called detailed balance equations. Brownian motion is a common stochastic process. So we design Markov chain by Brownian motion method.

Step 1: Supposing Γ_i is a set of the point x_i^k, every point x_i^k does Brownian motion, $x_{i+1}^k = B(x_i^k)$.

Step 2: Connect points x_{i+1}^k in turn. $\Gamma_{i+1}^0 = \{x_{i+1}^1, x_{i+1}^2, \ldots, x_{i+1}^k, \cdots, x_{i+1}^{n_i}\}$.

Step 3: Make up the collected curve and get rid of repeated loops, as fig.4 showed, $\Gamma_{i+1} = m(\Gamma_{i+1}^0)$.

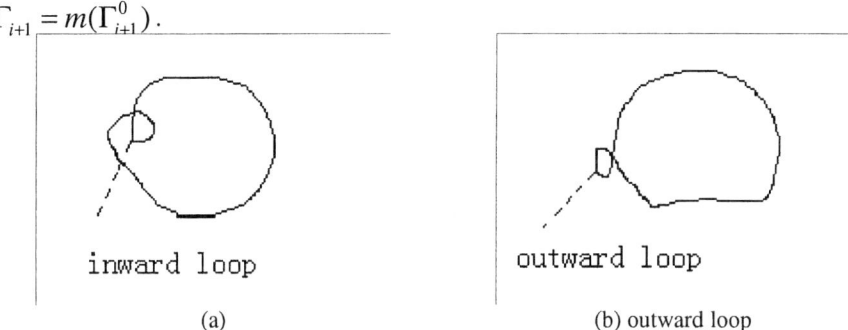

(a)   (b) outward loop

Fig. 4. Two kinds of loops need to get rid of: (a) and (b) are the inward loop and the outward loop to get rid of respectively.

2.4 Monte Carlo Simulation

To the curve Markov chain, as 2.3 discussed, we should introduce some optimize solutions to obtain the value of BCPDF cause of the calculation speed. Monte Carlo is such method of simulation that can converge at the boundary curve quickly. The key of Monte Carlo method is the selection of the kernel. Here we apply the Metropolis-Hastings scheme which has the following function:

$$P(\Gamma_{i+1} | \Gamma_i) = \min\left\{1, \frac{PL(\Gamma_{i+1})}{PL(\Gamma_i)}\right\} \qquad (6)$$

To any state Γ_i and the next state Γ_{i+1} of Markov chain, we calculate the shell and decide if transform or not based on the value of the transform probability $P(\Gamma_{i+1} | \Gamma_i)$. Now, we can summarize that we do it by MCMC method in the following steps.

Step 1: Generate a potential Markov chain as 2.3 illustrated.

Step 2: Supposing the current state is Γ_i and the next state is Γ_{i+1} in Markov chain as step1, calculate the shell $P(\Gamma_{i+1} | \Gamma_i)$ as formula (6).

Step 3: Generate a random variance u with uniform distribution at domain [0,1].

Step 4: If $P(\Gamma_{i+1} | \Gamma_i) \geq u$, Markov chain go to the next state Γ_{i+1} and go to step 2.

Step 5: If $P(\Gamma_{i+1} | \Gamma_i) < u$, Markov chain refuse to transform, $\Gamma_{i+1} = \Gamma_i$, and go to step2.

Step 6: If the number of the continuous repeated transforms is more than 50, stop the process and consider the current answer is the optimum answer.

3 Experiment Results and Conclusion

To examine the effect of the algorithm proposed in the paper, we chose DB_B in the public data–FVC2004, which is considered the most hard fingerprint image database to segmented in common and disposed it. Fig.5 is the representative fingerprint images and the fingerprint image segmented results. Fig.5 (a) and (b) are the two fingerprint image that have been noised by the peeled region and the prior remainder region respectively.

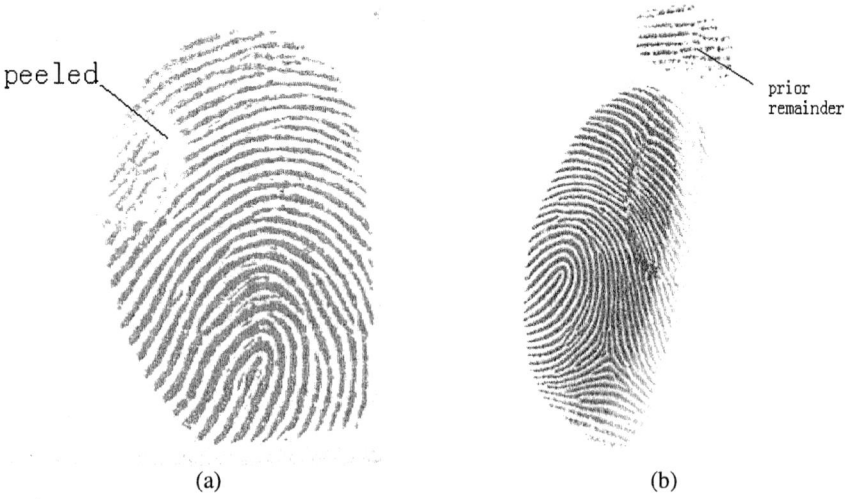

Fig. 5. Some fingerprint images and the segmentation results by our method: (a) and (b) are the peeled fingerprint image and the prior remainder fingerprint image respectively; (c) and (d) are the image segmentation results of (a) and (b) respectively.

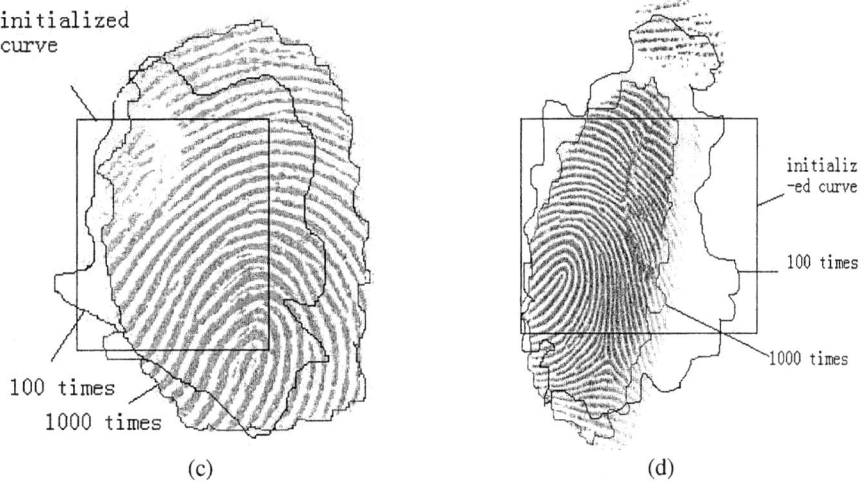

Fig. 5. (*Continued*)

And the fig.5 (b) has some noise region where the gray level is lower than other high-quality fingerprint image region. Fig.5(c) and (d) are the segmentation results with the algorithm in the paper. The initialize curve is the rectangle region curve.

In order to illuminate the effect to the performance of AFIS about our method more accurately and concretely, the paper carries through the systemic test based on the DB_B in the public data–FVC2004 and gets the ROC curve of the performance. Here, suppose other algorithms are the same except the fingerprint image segmentation algorithm for validate the performance of our method effectively. The ROC curves of performance about the AFIS with our method or not are showed as the following Fig.6.

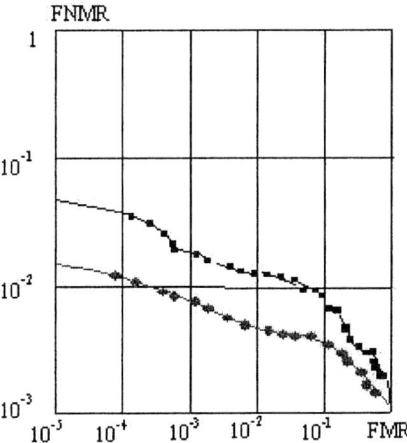

Fig. 6. The ROC curves about the performance with our method or not (where, the FNMR index is the False Not Match Rate and the FMR index is the False Match Rate, the black curve with the black rectangle is the ROC curve adopting the fingerprint segmentation method based on the statistical attribute and the other is the ROC curve adopting our method)

From the Fig.6, we can obtain the following conclusion that the method in the paper is effective and accurate for segmenting the valid fingerprint image region from the background region or the strong noise region. The result of Fig.6 shows that the performance of the AFIS adopting our method is higher than the performance of the AFIS adopting the method based on the statistical attribute (i.e. the gray mean value and the gray variance value) for fingerprint image segmentation.

The paper brings forward the fingerprint image segmentation algorithm based on MCMC method. It takes the closed curve in fingerprint image as research object, randomly generates Markov chain of closed curves, then, it is simulated by Monte Carlo method to convergent to the boundary curve whose boundary curve probability function is the biggest.

The experimental results indicate that the algorithm can segment the valid fingerprint image region from the background quickly, exactly and reliably. At the same time, the algorithm is robust to the fingerprint image with stronger disturbance of background, especially to the peeled fingerprint and the prior remainder fingerprint crucially.

References

1. Jain, A. K., Uludag, U., Hsu, R.L.: Hiding a Face in a Fingerprint Image. Proc. ICPR, Quebec City (2002) 756-759
2. Zhan, X.S.: Research on Several key issues related to AFIS Based on verification mode. Ph.D Dissertation, Najing University (2003)
3. Jain, A.K., Hong, L., Bolle, R.: On-line fingerprint verification. IEEE Transactions on Pattern Analysis and Machine Intelligence (1997) 302-314
4. Mehtre, B.M., Murthy, N.N., Kapoor, S., Chatterjee, B.: Segmentation of fingerprint images using the directional images. Pattern Recognition (1987) 429-435
5. Mehtre, B.M., Chatterjee, B.: Segmentation of fingerprint images-a composite method. Pattern Recognition (1995) 1657-1672
6. Xinjian Chen, Jie Tian, Jianggang Cheng, Xin Yang, Segmentation of Fingerprint Images Using Linear Classifier. EURASIP Journal on Applied Signal Processing (2004) 480-494.
7. Bazen, A.M. and S.H. Gerez. Segmentation of Fingerprint Images, Proc. ProRISC2000, 12th Annual Workshop on Circuits, Systems and Signal Processing, Veldhoven, The Netherlands, Nov 29-30. 2001.
8. Yin,Y.L, Yang, X.K., Chen, X., Wang, H.Y.: Method Based on Quadric Surface Model for Fingerprint Image Segmentation, Defense and Security, Proceedings of SPIE (2004) 417-324
9. Green, P.J.: Reversible Jump Markov Chain Monte Carlo Computation and Bayesian Model Determination. Biometrika (1995) 711-732
10. 10.Zhu, S.C., Zhang, R., Tu, Z.W.: Integrating Bottom- Up/ Top- Down for Object Recognition by Data Driven Markov Chain Monte Carlo . Proc. IEEE Conference on Computer Vision and Pattern Recognition USA : Hilton Head Island (2000) 738 -745
11. Tu, Z.W., Zhu, S.C., Shum, H.Y.: Image Segmentation by Data Driven Markov Chain Monte Carlo. Proc, ICCV 2001. Eighth IEEE International Conference on Computer Vision. Canada : Vancouver (2001) 131 - 138.
12. Tu, Z.W., Zhu, S.C.: Parsing Images into Region and Curve Processes [EB/OL]. http://www.stat.ucla.edu/•ztu/ DDMCMC/curves/ region - curve.htm (2002)
13. He, Y.L., Tian, J., Zhang, X.P.: Fingerprint Segmentation Method Based on Markov Random Field. Proceedings of the 4th China Graph Conference (2002) 149-156

Unsupervised Speaker Adaptation for Phonetic Transcription Based Voice Dialing

Weon-Goo Kim[1], MinSeok Jang[2], and Chin-Hui Lee[3]

[1] School of Electronic and Information Eng.,
Kunsan National Univ., Kunsan, Chonbuk 573-701, Korea,
Biometrics Engineering Research Center
wgkim@kunsan.ac.kr

[2] Dept. of Computer Information Science, Kunsan National Univ.,
Kunsan, Chonbuk, 573-701, Korea
msjang@kunsan.ac.kr

[3] School of Electrical and Computer Eng., Georgia Institute of Technology,
Atlanta, Georgia, 30332, USA
chl@ece.gatech.edu

Abstract. Since the speaker independent phoneme HMM based voice dialing system uses only the phoneme transcription of the input sentence, the storage space could be reduced greatly. However, the performance of the system is worse than that of the speaker dependent system due to the phoneme recognition errors generated when the speaker independent models are used. In order to solve this problem, a new method that jointly estimates the transformation vectors (bias) and transcriptions for the speaker adaptation is presented. The biases and transcriptions are estimated iteratively from the training data of each user with maximum likelihood approach to the stochastic matching using speaker independent phoneme models. Experimental result shows that the proposed method is superior to the conventional method using transcriptions only.

1 Introduction

Voice dialing, in which a spoken word or phrase can be used to dial a phone number, has been successfully deployed in telephone and cellular networks and is gaining user acceptance. As the popularity of this service increases, the requirements of data storage and access for performing speech recognition become crucial. To overcome this problem, methods that use the speaker-specific phonetic templates or phonetic base-form for each label were presented [1,2,3,4]. The advantage of these approaches are that the only information that need to be stored for each speaker is the phonetic string associated with each speaker's words or phrase from speech in terms of speaker independent (SI) sub-word acoustic units (phones), resulting in substantial data reduction. However, this method has two drawbacks. The first one is that a large amount of phoneme recognition errors are generated using SI phoneme HMMs. The other is that the performance of the system is worse than that of the speaker dependent (SD) system since SI models are used.

In this paper, a new method that jointly estimates the transformation function (bias) and transcriptions for the speaker adaptation from training utterances is presented to improve the performance of the personal voice dialing system using SI phoneme HMMs. In training process, the biases and transcriptions are estimated iteratively from the training data of each user with maximum likelihood approach to the stochastic matching using SI phoneme models and then saved for recognition. In recognition process, after SI phoneme models are transformed using speaker specific bias vectors, input sentence is recognized. Experimental result shows that the proposed method is superior to the conventional method using transcriptions only and the storage space for bias vectors for each user is small.

2 Speaker Adaptation of Phonetic Transcription Based Voice Dialing System

A new method that jointly estimates the transformation function (bias) and transcriptions for the speaker adaptation from training utterances is presented to improve the performance of the personal voice dialing system using SI phoneme HMMs. The concept of the proposed system is shown in Fig. 1. There are two kinds of sessions, enrollment and test. In an enrollment session, a user need to repeat a name for two or three times, then input a telephone number associated with the name. The joint estimation stage consists of two pass, recognizer and adaptation. The first pass is standard decoding which produce the phoneme transcription as well as the state segmentation necessary for the estimation of SI phoneme model biases. During the second pass, model biases are estimated using stochastic matching method to generate the new set of HMMs. This process is iterated several times until achieving convergence and then final transcriptions and biases of each user are saved into the database. In a test session, the user just needs to utter the name. For telephone, the identity can be obtained from caller ID or from user's input. After SI phoneme HMMs are adapted using bias vectors, decoder recognize the input speech with transcriptions and adapted SI phoneme HMMs.

The maximum likelihood technique to the stochastic matching for speaker adaptation can be applied as follows [5,6]. The mean μ_y of each mixture component in adapted SI HMMs are derived by adding the bias μ_b to the mean μ_x of the corresponding mixture components in SI HMMs , i.e.,

$$\mu_y = \mu_x + \mu_b . \qquad (1)$$

The transformation vector or bias $\mu_b = \{\mu_1, \mu_2, \ldots, \mu_D\}$ can be obtained by maximum likelihood estimate, such that

$$\mu_{b_i} = \frac{\sum_{t=1}^{T} \sum_{n=1}^{N} \sum_{m=1}^{M} \gamma_t(n,m) \frac{y_{t,i} - \mu_{m,n,i}}{\sigma_{n,m,i}^2}}{\sum_{t=1}^{T} \sum_{n=1}^{N} \sum_{m=1}^{M} \frac{\gamma_t(n,m)}{\sigma_{n,m,i}^2}}, \quad i = 1, \ldots, D \qquad (2)$$

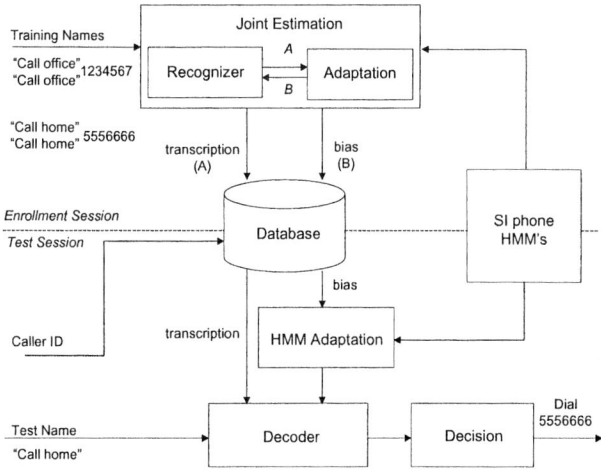

Fig. 1. Performance of the voice dialing system using speaker adaptation algorithm (the number of transformation vector: 1, 2, 3, 9, 14, 42, 180)

and
$$\gamma_t(n,m) = \begin{cases} \frac{w_{n,m}N[y_t;\mu_{n,m},C_{n,m}]}{\sum_{j=1}^{M}w_{n,m}N[y_t;\mu_{n,m},C_{n,m}]}, & \text{if } s=n \\ 0, & \text{otherwise} \end{cases} \quad (3)$$

where N and M are the number of state and mixtures of HMM, D is the dimension of the feature vector, $\mu_{n,m}$, $C_{mn,m}$ are the mean and variance vector corresponding to mixture m in state n, $w_{n,m}$ is the probability of mixture m in state n, N is the normal distribution and s is the state sequence corresponding to the input utterance.

3 Experimental Results

3.1 Database and System Setup

The experimental database consists of 10 speakers, 5 males and 5 females speaking 15 name entries repeated 13 times by each speaker over a period of several weeks [7]. The database evaluation is on a worst-case situation where all the names are "Call ", e.g. "Call office", "Call home", "Call mom", etc. This means that about a half of the contents are the same. Many names are very short in about 1 second, which makes the recognition even more difficult. This database was collected over the telephone network using digital telephone interface. The input speech was sampled at 6.67kHz and saved as 8bit μ-law PCM format. In training, three utterances of each name recorded in one session were used to train a name model. In testing, ten utterances from each speaker collected from 5 different sessions were used to evaluate the recognition performance.

The sampled input speech was pre-emphasized using a first-order filter with a coefficient of 0.97, and the analysis frames were 30-ms width with 20-ms overlap.

A 39-dimensional feature vector was extracted based on tenth order LPC analysis. The feature corresponds to a 12-dimensional cepstrum vector, 12-dimensional delta-cepstrum vector, 12-dimensional delta-delta-cepstrum vector, a normalized log energy, a delta log energy and a delta-delta log energy.

The SI phoneme models were trained using the database collected over the telephone network. SI phoneme HMMs consist of 41 phoneme HMMs and a silence HMM. The phoneme HMM models are left-to-right HMMs and each phoneme HMM have 3 or 5 states consisting of 10 continuous Gaussian mixture components and a silence model has one state consisting of 256 continuous Gaussian mixture components.

3.2 Performance of the Speaker Adaptation

We first conduct baseline experiments to study the effect of speaker adaptation. The base line system is SD voice dialing system using speaker-specific phonetic templates (transcription) and SI phoneme HMMs. Although this kind of system can in substantially reduce storage space, a large amount of phoneme recognition errors are generated using SI phoneme HMMs, especially at the front and end of the input utterance. One way to reduce the phoneme recognition errors is to use the end point detection method. Voice dialing experiment was conducted with and without processing the end point detection method. As expected, using end point detection method reduces percentage word error rates from 4.2% to 3.8% since some of phoneme recognition errors are eliminated.

In order to evaluate the performance of proposed system, transformation vector (bias) and transcriptions for the speaker adaptation are estimated iteratively by using the stochastic matching method. Fig. 2 shows the percentage word error rates and convergence speed of the algorithm according to the number of transformation vector. When iteration number is zero, the recognition error rate is the performance of baseline system. Multiple biases are used according to the phonetic classes. The number of transformation vector is selected according to the acoustic phonetic class, i.e. 1, 2, 3, 9, 14, 42, 180. When a bias is used, all of phonemes have a common transformation vector. When two biases, one is for all of phoneme models and the other is for a silence model. When three biases, the phonetic classes consist of vowels, consonants and a silence. When nine biases, the phonetic classes consist of silence, vowel, diphthongs, semivowels, stops, etc. when 14 biases, the phonetic classes are silence, front vowel, mid vowel, back vowel, diphthongs, liquids, glides, voiced stops, unvoiced stops, etc. When 42 biases, the phonetic classes are all phonemes and a silence. When 180 biases, it is equal to the number of states of all phoneme HMMs. As seen in Figure 2, the performance of proposed system did not enhanced when the number of transformation vector is fewer than 9. However, error reduction could be achieved when the number of transformation vector is more than 14. When 42 transformation vectors are used, the best performance (2.3% word error rate) could be achieved. It corresponds to about a 40% reduction in word error rate as compared to the performance of the baseline system.

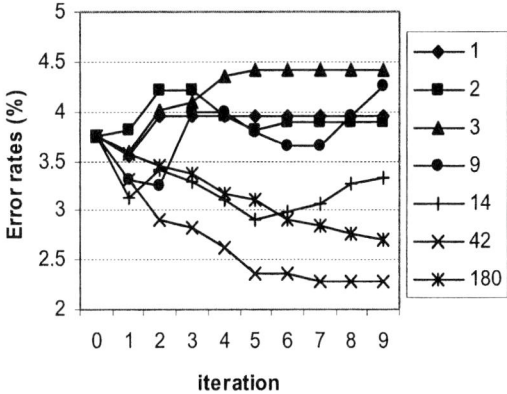

Fig. 2. Performance of the voice dialing system using speaker adaptation algorithm (the number of transformation vector: 1, 2, 3, 9, 14, 42, 180)

Table 1 shows the performance comparison of the proposed method with conventional methods. The error rate of baseline system is 3.8%. The error rate of system adapted with the transformation vector only is 3.3%. This shows that adaptation with the transformation vector only could reduce the error rate. The error rate of proposed system in which the transformation vectors and the transcriptions for the speaker adaptation are estimated iteratively is 2.3%. Finally, the error rate of SD HMMs is 1.8%. Although this is the lowest, large amount of storage space are required.

Table 1. Performance comparison of the proposed speaker adaptation method with conventional ones

Type of system	Baseline	Baseline with bias	Proposed system	SD system
Error rate(%)	3.8	3.3	2.3	1.8

4 Conclusion

A new method that jointly estimates the transformation vectors and the transcriptions for the speaker adaptation is presented in order to improves the performance of the personal voice dialing system in which SI phoneme HMMs are used. The biases and transcriptions are estimated iteratively from the training data of each user with maximum likelihood approach to the stochastic matching using SI phoneme models. Experimental result shows that the performance of proposed system corresponds to about a 40% reduction in word error rate, when compared to the performance of the baseline system.

Acknowledgement

This work was supported by the Korea Science and Engineering Foundation (KOSEF) through the Biometrics Engineering Research Center (BERC) at Yonsei university.

References

1. Jain, N., Cole, R. Barnard, E.: Creating Speaker-Specific Phonetic Templates with a Speaker-Independent Phonetic Recognizer: Implications for Voice Dialing. Proc. of ICASSP'96 (1996) 881–884
2. Fontaine, V., Bourlard, H.: Speaker-Dependent Speech Recognition Based on Phone-Like Units Models-Application to Voice Dialing. Proc. of ICASSP'97 (1997) 1527–1530
3. Ramabhadran, B., Bahl, L.R., deSouza, P.V., Padmanabhan, M.: Acoustic-Only Based Automatic Phonetic Baseform Generation. Proc. of ICASSP'98 (1998) 2275-2278
4. Shozakai, M.: Speech Interface for Car Applications. Proc. of ICASSP'99 (1999) 1386–1389
5. Zavaliagkos, G., Schwartz, R., Makhoul, J.: Batch, Incremental and Instantaneous Adaptation Techniques for Speech Recognition. Proc. of ICASSP'95 (1995) 676–679
6. Sankar, A., Lee, C.H.: A Maximum-Likelihood Approach to Stochastic Matching for Robust Speech Recognition. IEEE Trans. on Speech and Audio Processing, Vol. 4. (1996) 190–202
7. Sukkar, R.A., Lee, C.H.: Vocabulary independent discriminative utterance verification for non-keyword rejection in subword based speech recognition. IEEE Trans. Speech and Au-dio Processing, Vol. 4. (1996) 420–429

A Phase-Field Based Segmentation Algorithm for Jacquard Images Using Multi-start Fuzzy Optimization Strategy

Zhilin Feng[1,2], Jianwei Yin[1], Hui Zhang[1,2], and Jinxiang Dong[1]

[1] State Key Laboratory of CAD & CG, Zhejiang University, Hangzhou, 310027, P.R. China
[2] College of Zhijiang, Zhejiang University of Technology, Hangzhou, 310024, P.R. China

Abstract. Phase field model has been well acknowledged as an important method for image segmentation. This paper discussed the problem of jacquard image segmentation by approaching the phase field paradigm from a numerical approximation perspective. For fuzzy theory provides flexible and efficient techniques for dealing with conflicting optimization probelms, a novel fuzzy optimization algorithm for numerical solving of the model was proposed. To achieve global minimum of the model, a multi-start fuzzy strategy which combined a local minimization procedure with genetic algorithm was enforced. As the local minimization procedure does not guarantee optimality of search process, several random starting points need to be generated and used as input into global search process. In order to construct powerful search procedure by guidance of global exploration, genetic algorithm was applied to scatter the set of quasi-local mimizers into global positions. Experimental results show that the proposed algorithm is feasible, and reaches obvious effects in terms of jacquard image segmentation.

1 Introduction

Jacquard image segmentation heavily influences the performance of jacquard-pattern identification system. An accurate extraction of pattern features from jacquard images promises reliability for jacquard fabric CAD [1]. Most of pattern extraction algorithms have difficulty in capturing complex structure of visual features, such as complex contours of a jacquard pattern. In deed, poor image quality, low contrast, and the complex nature of the shape and appearance of some jacquard patterns may lead to poor accuracy or robustness of existing image segmentation algorithms.

Many algorithms that deal with image processing using phase field models have been presented in the literatures [2,3,4]. The range of applications of phase field models in image processing includes noise removal, image segmentation and shape optimization problems. Feng *et al.* [2] introduced the Allen-Cahn equation in phase transition theory to remove noise from jacquard images. For nonlocal Allen-Cahn equation can generate an area-preserving motion by mean curvature flow, it can perfectly preserve shapes of the fabric texture while in the process of denoising. Benes *et al.* [3] presented an algorithm of image segmentation based on the level set solution of phase field equation. The approach can be understood as a regularization of the level set motion by means of curvature, where a special forcing term is imposed to enforce the initial level set closely surrounding the curves of patterns with different shapes. The phase field method is also a robust and rigorous framework for topology optimization problems. Burger et al. [4] introduced a relaxation scheme for shape optimization based on a phase field model. This approach is similar in spirit to the diffused-interface approximations of sharp-interface models in phases transition, such as Allen-Cahn and Cahn-Hilliard equation.

However, from a numerical point of view, it is not easy to compute a minimizer for the non-convex functional of the phase field. Barrett et al. [5] presented a fully practical finite element approximation of a phase field system with a degenerate mobility and a logarithmic free energy. Provatas et al. [6] developed a large-scale parallel adaptive mesh code for solving phase-field type equations in two and three dimensions. Han et al. [7] implemented a numerical solving of the phase field model with the highly efficient Chebyshev spectral method. For the minimization of phase field model can be defined as an optimization process, we consider the model as an unconstrained optimization problem, and apply the fuzzy set theory to solve it. Since the introduction of the theory of fuzzy logic by Zadeh [8], many different applications of this fuzzy theory have been developed [9,10,11]. The earliest studies of fuzzy techniques are due to Zadeh, who first developed a suitable framework for handling uncertainties. His work was extended by Rosenfeld to image segmentation and pattern recognition systems [9]. Devi and Sarma [10] presented an adaptive scheme based on a fuzzy gradient descent approach for training a classification system. The system uses the classification results for already classified data and a training process to classify each new datum. Rao [11] applied fuzzy concepts to a static finite element analysis, and proposed a methodology using the fuzzy finite element method for vibration analysis of imprecisely defined systems. In this paper, we propose a fuzzy-based optimization algorithm to solve the phase field model on piecewise linear finite element spaces. The proposed algorithm involves two coordinate steps: refining and reorganizing an adaptive triangular mesh to characterize the essential contour structure of a pattern, and minimizing the discrete phase field model by multi-start fuzzy optimization strategy.

The multi-start strategy for fuzzy optimization method is a global optimization strategy which combines a local minimization algorithm with random sampling [12]. Each randomly sampled point is a starting point from which one seeks a local minimizer via a local minimization algorithm. At the end, the local minimizer with lowest function value is taken as an approximation to the global minimizer. The adaptive multi-start strategy proposed by Boese et al.[13] consists of two steps:

(1) Generate a set of random starting points and call iterative algorithm on each starting point, thus determining a set of (local minimum) solutions;

(2) Construct adaptive starting points from the best local minimum solutions found so far, and run the iterative algorithm on these points to yield corresponding new solutions.

Multi-start strategies have been successfully applied to both nonlinear global optimization and combinatorial problems. By combining genetic algorithm with a local search strategy in step (1), Burger et al. [4] showed that the improved strategy was able to solve the shape optimization problem.

2 Phase Field Model

The goal of image segmentation is to extract meaningful objects from an input image. Image segmentation, with wide recognized significance, is one of the most difficult low-level image analysis tasks, as well as the bottle-neck of the development of image processing technology. Recently, the phase field methodology has achieved considerable importance in modeling and numerically simulating a range of phase transitions and complex pattern structures that occur during image recovery [14]. The main advantage of the phase field methods is that the location of the different medium interface is given implicitly by the phase field, which greatly simplifies the handling of merging interfaces. Traditionally, phase transitions have been expressed by free boundary problems, where the interface between regions of different phases is represented by a sharp surface of zero thickness. When the transition width approaches zero, the phase field model with diffuse-interface becomes identical to a sharp-interface level set formulation, and it can also be reduced properly to the classical sharp-interface model.

Let $\Omega \subset R^2$ be a bounded open set and $g \in L^\infty(\Omega)$ represent the original image intensity. The function g has discontinuities that represent the contours of objects in the image. Let $u = u(x,t) \in R$ be a image field, which stands for the state of the image system at the position $x \in \Omega$ and the time $t \geq 0$, and K be the set of discontinuity points of u. Here, we assume that the image field has two stable states corresponding to $u = +1$ and $u = -1$. In this case, the phase-field energy of the image system is often given as follows:

$$E_\varepsilon(u, K) = \int_{\Omega \setminus K} \left(\frac{\varepsilon}{2} (\nabla u(x))^2 + \frac{1}{\varepsilon} F(u(x)) \right) dx \tag{1}$$

where $F(u) = (u^2 - 1)^2 / 4$ is a double well potential with wells at -1 and $+1$ (i.e., a nonnegative function vanishing only at -1 and $+1$). Here, the value $u(x)$ can be interpreted as a phase field (or order parameter), which is related to the structure of the pixel in such a way that $u(x) = +1$ corresponds to one of the two phases and $u(x) = -1$ corresponds to the other. The set of discontinuity points of u parameterizes the interface between the two phases in the corresponding configuration and is denoted by a closed set K. The first term in $E_\varepsilon(u,K)$ represents the interfacial energy, which penalizes high derivatives of the function u and therefore constrains the number of interfaces separating the two phases. The second term represents the double-well potential that tries to force the values of image pixels into one phase or the other. The small parameter ε links to the width of transition layer, which controls a ratio of the contribution from the interfacial energy and the double-well potential.

Heuristically, we expect solutions to Eq. (1) to be smooth and close to the image g at places $x \notin K$, and K constitutes edges of the image. To show existence of solutions to Eq. (1), a weak formulation was proposed by De Giorgi et al. [15] by setting $K = S_u$ (the jumps set of u) and minimizing only over $u \in SBV$, the space of functions of bounded variation. We recall some definitions and properties concerning functions with bounded variation.

Definition 1. Let $u \in L^1(\Omega; R^2)$. We say that u is a function with bounded variation in Ω, and we write $u \in BV(\Omega; R^2)$, if the distributional derivative Du of u is a vector-valued measure on Ω with finite total variation.

Definition 2. Let $u \in L^1(\Omega; R^2)$. We denote by S_u the complement of the Lebesgue set of u, i.e., $x \notin S_u$ if and only if $\lim_{\rho \to 0} \rho^{-n} \int_{B_\rho(x)} |u(y) - z| \, dy = 0$ for some $z \in R^2$, where $B_\rho(x) = \{y \in R^2 : |y - x| < \rho\}$.

Definition 3. Let $u \in BV(\Omega)$. We define three measures $D^a u$, $D^j u$ and $D^c u$ as follows. By the Radon-Nikodym theorem we set $Du = D^a u + D^s u$ where $D^a u$ is the absolutely continuous part of Du, $D^s u$ is the singular part of Du. $D^s u$ can be further decomposed into the part supported on S_u (the jump part $D^j u$) and the rest (the Cantor part $D^c u$): $D^j u = D^s u|_{S_u}$ and $D^c u = D^s u|_{\Omega \setminus S_u}$. Thus, we can then write $Du = D^a u + D^j u + D^c u$.

Definition 4. Let $u \in BV(\Omega)$. We say that u is a special function of bounded variation, and we write $u \in SBV(\Omega)$, if $D^c u = 0$.

De Giorgi et al. [15] gave the weak formulation of the original problem (1) as follows:

$$E(u, K) = E(u, S_u) \tag{2}$$

They also proved that minimizers of the weak problem (2) are minimizers of the original problem (1). However, from a numerical point of view, it is not easy to compute a minimizer for Eq. (2), due to the term $H^1(S_u)$, and to the fact that this functional is not lower-semicontinuous with respect to S_u. It is natural to try to approximate Eq. (2) by simpler functionals defined on *SBV* spaces. Ambrosio and Tortorelli [16] showed that Eq. (2) can be approximated by a sequence of elliptic functionals which are numerically more tractable. The approximation takes place in the sense of the Γ-convergence.

To approximate and compute solutions to Eq. (2), the most popular and successful approach is to use the theory of Γ-convergence. This theory, introduced by De Giorgi et al. [15], is designed to approximate a variational problem by a sequence of regularized variational problems which can be solved numerically by finite difference/finite element methods.

Let $\Omega = (0,1) \times (0,1)$, let $T_\varepsilon(\Omega)$ be the triangulations and let ε denote the greatest length of the edges in the triangulations. Moreover let $V_\varepsilon(\Omega)$ be the finite element space of piecewise affine functions on the mesh $T_\varepsilon(\Omega)$ and let $\{T_{\varepsilon_j}\}$ be a sequence of triangulations with $\varepsilon_j \to 0$.

Modica [17] proved that

Theorem 1. Let $BVC(\Omega) = \{\psi \in BV(\Omega) : \psi(\Omega) \subset \{-1,+1\}\}$, and let $W : R \to [0,+\infty)$ be a continuous function such that $\{z \in R : W(z) = 0\} = \{-1,+1\}$, and $c_1(|z|^\gamma - 1) \le W(z) \le c_2(|z|^\gamma + 1)$ for every $z \in R$, with $\gamma \ge 2$.

Then, the discrete functionals

$$E_\varepsilon(u,T) = \begin{cases} \int_{\Omega \setminus K} \left(\frac{\varepsilon}{2}(\nabla u_T(x))^2 + \frac{1}{\varepsilon}F(u_T(x)) \right) dx, & u \in V_\varepsilon(\Omega), T \in T_\varepsilon(\Omega) \\ +\infty & \text{otherwise} \end{cases} \quad (3)$$

Γ-converge as $\varepsilon \to 0$ to the functional $E(u) = c_0 \int_\Omega \Phi(u)\, dx$ for every Lipschitz set Ω and every function $u \in L^1_{loc}(R^2)$, where $c_0 = \int_{-1}^{1} \sqrt{F(u)}\, du$, and

$$\Phi(u) = \begin{cases} H^1(S_u) & \text{if } u \in BVC(\Omega) \\ +\infty & \text{otherwise} \end{cases} \quad (4)$$

3 Multi-start Fuzzy Optimization Strategy for Numerical Solving

In order to arrive at the joint minimum (u,T) of Eq. (3), we propose a fuzzy optimization algorithm to implement the numerical solving of Eq. (3). The main idea of the proposed algorithm is as follows: alternating back and forth between minimizing Eq. (3) holding T constant and adapting triangulation T holding u constant. The intuitive idea behind the algorithm is that if the triangulation T were known and held constantly, it would be straightforward to calculate the variable u, and if the variable u were known, it would be straightforward to calculate the triangulation T. The segmentation algorithm is summarized as follows:

Step 1. Initialize iteration index: $j \leftarrow 0$.
Step 2. Set initial ε_j and u_j.
Step 3. Generate the adapted triangulation T_{ε_j} by the mesh adaptation algorithm, according to u_j.
Step 4. Minimize $E_{\varepsilon_j}(u_j)$ on the triangulation T_{ε_j} by the multi-start fuzzy optimization strategy.
Step 5. Update the current index: $j \leftarrow j+1$.
Step 6. Generate a new ε_j.
Step 7. If $|\varepsilon_j - \varepsilon_{j-1}| > \mu$, return to Step 3. Otherwise, goto Step 8.
Step 8. Stop.

In the above algorithm, a scheme for the mesh adaptation is first enforced to refine and reorganize a triangular mesh to characterize the essential contour structure. Then, the multi-start fuzzy optimization strategy is applied to find the absolute minimum of the discrete version of the model at each iteration.

For the minimization of the functional $F_\varepsilon(u)$ can be defined as an optimization process, we consider the minimization as an optimization problem. In the area of optimization, there are two kinds of optimization methods: global optimizers and local optimizers. Global optimizers are useful when the search space is likely to have many minima. Local optimizers can quickly find the exact local optimum of a small region of the search space, but they are typically poor global searchers.

Fuzzy theory provides flexible and efficient techniques for handling different and conflicting optimization criteria of the search process. Fuzzy Optimization Method (FOM) is a modified version of the steepest descent method (SDM), in which searching direction vector at nth step is constructed by use of convex conjugation between $(n-1)$th searching direction vector and nth searching direction vector used in SDM [18, 19]. The coefficient of convex conjugation is computed by use of stochastic fuzzy estimation based on data resulting from $(n-1)$th and $(n-2)$th searching direction vectors. Note that FOM was originally invented as a local minimizer search algorithm.

A variety of deterministic and stochastic search methods for finding global optimization solutions have been developed in the past two decades [12, 13]. Both of single-start search and multi-start search are widely used stochastic search algorithms. They use a local search method to obtain the exact local optima and a random sampling process for global search. The advantages of these method are simplicity and low sampling overhead. Both of them can be developed even if deep mathematical properties of the problem domain are not at hand, and still can provide reasonably good solutions [12].

The single-start search starts from an initial solution x and repeats replacing x with a better solution in its neighborhood N until no better solution is found in N. For local procedures do not guarantee optimality of the search process, in practice, several random starting points may be generated and used as input into the local search technique. One way to solve this problem is to restart the search from a new solution once a seach region has been extensively explored. Multi-start strategy can be used to guide the construction of new solutions in a long term horizon of such search process. The multi-start search applies the single-start search to a number of randomly generated initial solutions and outputs the best solution found during the entire search [13]. The multi-start algorithm consists of two nested loops. At each iteration of the external loop we simply generate a random feasible solution, which is improved by the operations performed in the internal loop. The most common criteria used to stop the algorithm is to give a global time limit, or fix the number of iterations of the external loop. The main idea of the multi-start algorithm is as follows:

Step 1. Initialize iteration index: $i \leftarrow 0$.
Step 2. If stopping conditon is satisfied, Stop. Otherwise, Goto Step3.
Step 3. Generation process. Construct solution x_i
Step 4. Search process. Apply a search method to improve x_i. Let x_i' be the solution obtained, if x_i' improves the best, Update the best. Otherwise, Goto Step 5.
Step 5. Update the current index: $i \leftarrow i+1$, return to Step 2.

In the above algorithm, Step 1 is enforced to construct a solution x_i at iteration i and Step 2 is devoted to improving this solution to obtain a new solution x_i'. The algorithm returns the best solution identified during the whole search. The procedure which is applied to generate the random feasible solutions is an implementation of genetic algorithm.

The genetic algorithm proposed by Holland is an adaptive procedure that searches for good solutions by using a collection of search points known as a population in order to maximize some desirable criterion [20]. Basically it consists of making a population of solutions evolve by mutation and reproduction processes. The fittest solutions of the population shall survive and perpetuate their information, while the worse ones will be replaced. After a large number of generations, it is expected that the final population would be composed of highly adapted individuals, or in an optimization application, high-quality solutions of the problem at hand.

The genetic algorithm is used to scatter the set of quasi-local minimizers, i.e., restarting points to the next down-hill procedure, into much higher positions and to let the set of quasi-local minimizers escape completely from the bad region. The three fundamental functions in GAs play an important role in the following way: the selection operation in GAs aids the complete climb up to the summits of the target manifold or cross the ridge. The cross over and mutation operations contribute to the optimal rearrangement of the set of restarting initial points. Let us define operators F, M and R as follows.

1) F : Algorithm due to Fuzzy Optimization Method. This procedure is a down-hill process on the cost manifold.
2) M : Mountain crossing algorithm. This procedure is a up-hill process on the cost manifold.
3) R : Rearrangement algorithm by GAs. In this procedure, starting points for the next down-hill process are rearranged by use of GAs.

Thus, the completed algorithm for multi-start fuzzy optimization strategy is summarized as follows:

Step 1 Generate an initial population W^0 of $E_\varepsilon(u)$ (the set of searchers).

Step 2 Evaluate W^0, and compute $U^n := FW^n$ (the set of local minimizers obtained).

Step 3 Compute $V^n := MU^n$ (the set of quasi-local maximizers obtained).

Step 4 Compute $W^n := RV^n$ (the set of rearranged searchers).

Step 5 Increase generation number $n := n+1$ and repeat Steps from 2 to 4 until the generation number n is beyond the present one.

4 Experiments

In order to demonstrate the effciency of the proposed method, some experiments were carried out on different types of images. Firstly the four synthetic image shown in Figure 1 was processed. Four synthetic images with different shapes were given in Figure 1(a)-(d), respectively. After 8 mesh adaptation processes, the final foreground meshes of Figure 1(a)-(d) were shown in Figure 1(e)-(h).

The order of accuracy for the proposed segmentation algorithm was measured by comparing the length and area error of the four shapes in Figure 1(a)-(d) between the analytical solution and numerical solution. Detailed accuracy measurements was provided in Table 1, which indicated how the length and area error depend on the width of interface ε. As the width ε decreased, the adaptive triangular meshes would approach closer and closer to the boundary of the two shapes in Figure 1(a)-(b) respectively.

Table 1. Number of length and area error changes

		Width of interface (ε)						
		0.004	0.002	0.001	0.0005	0.00025	0.000125	0.0000625
Fig.1a	Length	0.0583	0.0435	0.0384	0.0296	0.0172	0.0115	0.0092
	Area	0.0735	0.0657	0.0593	0.0471	0.0398	0.0316	0.0280
Fig.1b	Length	0.0462	0.0419	0.0392	0.0318	0.0278	0.0196	0.0127
	Area	0.0696	0.0552	0.0519	0.0383	0.0312	0.0271	0.0102
Fig.1c	Length	0.0519	0.0478	0.0365	0.0318	0.0259	0.0211	0.0188
	Area	0.0744	0.0603	0.0598	0.0471	0.0399	0.0311	0.0273
Fig.1d	Length	0.0871	0.0711	0.0659	0.0582	0.0479	0.0332	0.0211
	Area	0.0928	0.0855	0.0792	0.0623	0.0545	0.0426	0.0321

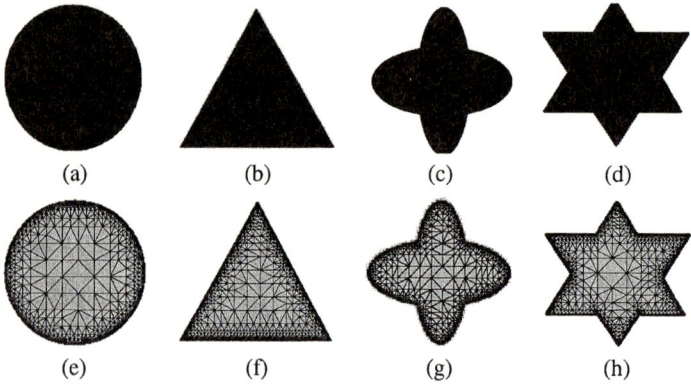

Fig. 1. Segmentation of synthetic images. (a)-(d): Original images. (e)-(h): Final foreground meshes.

Then a second experiment was carried out on several real jacquard images. Figure 2 illustrates the segmentation results of three jacquard images using our algorithm. Figure 2(a)-(c) gives original jacquard images. Figure 2(d)-(f) and Figure 2(g)-(i) show nodes of Delaunay triangulation and the structure meshes of Figure 2(a)-(c). The segmented edge sets of Figure 2(a)-(c) are shown in Figure 2(j)-(l).

The following experiments are designed for comparing efficiency of the proposed algorithm with a numerical solving method for phase field model, i.e., Benes's algorithm [3]. Table 2 shows comparisons of the iteration times(ITs) and the average iteration time(AIT) between the two algorithms. We can see that the two algorithms spend the same iteration time to accomplish the segmentation process, but the proposed algorithm consumes much less time than Benes's algorithm at each iteration.

Table 2. Computational time comparison between the Benes's algorithm and the proposed algorithm

	Benes's algorithm		Proposed algorithm	
	ITs	AIT	ITs	AIT
Fig.2 (a)	10	0.348	10	0.319
Fig.2 (b)	20	0.446	20	0.412
Fig.2 (c)	30	0.552	30	0.510

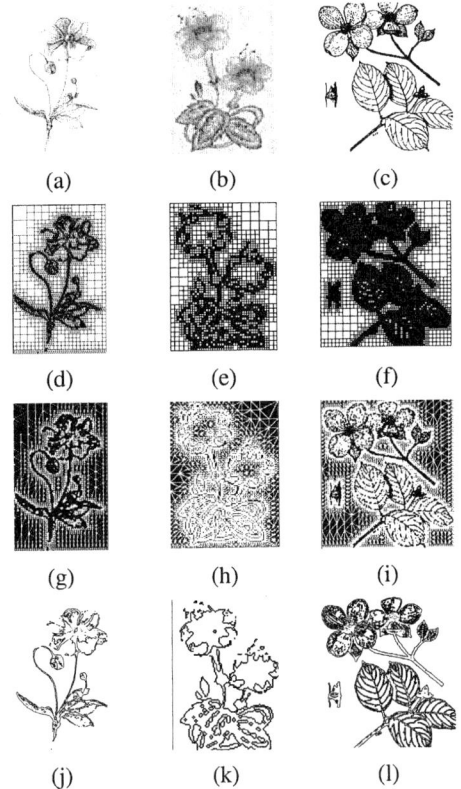

Fig. 2. Segmentation results of three jacquard images. (a)-(c) are original images, (d)-(f) are nodes of Delaunay triangulation, (g)-(i) are the final foreground meshes, (j)-(l) are the segmented edge sets.

5 Conclusions

In this paper, we have proposed and analyzed a phase-field based segmentation algorithm for jacquard images. For solving the corresponding minimization problem of the phase field model, a global optimization method is presented by deploying the multi-start searching strategy. The proposed segmentation algorithm is applied to segment synthetic and jacquard images, and shows its capability of accurate segmentation.

References

1. Sari, S.H., Goddard, J.S.: Vision system for on-loom fabric inspection. IEEE Transaction on Industry Applications, Vol. 35, (1999) 1252-1259
2. Feng, Z.L., Yin, J.W., Chen G., Dong, J.X.: Research on jacquard fabrics image denoising using Allen-Cahn level set model. Journal of Zhejiang University (Engineering Science). Vol. 39 (2005) 185-189
3. Benes, M., Chalupecky, V., Mikula, K.: Geometrical image segmentation by the Allen–Cahn equation. Applied Numerical Mathematics, Vol. 51 (2004) 187-205
4. Burger, M., Capasso, V.: Mathematical modelling and simulation of non-isothermal crystallization of polymers. Mathematical Models and Methods in Applied Sciences, Vol. 6 (2001) 1029-1053
5. Barrett, J.W., Nurnberg, R., Styles, V.: Finite Element Approximation of a Phase Field Model for Void Electromigration. SIAM Journal on Numerical Analysis, Vol. 42 (2004) 738-772
6. Provatas, N., Goldenfeld, N., Dantzig, J.: Adaptive Mesh Refinement Computation of Solidification Microstructures using Dynamic Data Structures. Journal of Computational Physics, Vol. 148 (1999) 265-290
7. Han, B.C., Van der Ven, A., Morgan, D., Ceder, G.: Electrochemical modeling of intercalation processes with phase field models, Electrochemical Acta, Vol. 49 (2004) 4691-4699
8. Zadeh, L.A.,: Fuzzy sets, Information and Control, Vol. 8 (1965) 338-353
9. Rosenfeld, A.: The Fuzzy Geometry of Image Subsets. Pattern Recognition Letters, Vol.2 (1984) 311-317
10. Devi, B.B., Sarma, V.V.S.: A fuzzy approximation scheme for sequential learning in pattern recognition. IEEE Transaction on Systems, Man, and Cybernetics, Vol. 16 (1986) 668-679
11. Rao, S.S., Sawyer, J.P.,: Fuzzy Finite Element Approach for the Analysis of Imprecisely Defined Systems, AIAA Journal, Vol. 33 (1995) 2364-2370
12. Hagen, L.W., Kahng, A.B.: Combining problem reduction and adaptive multi-start: a new technique for superior iterative partitioning, IEEE Transaction on CAD, Vol. 16 (1997), 709-717
13. Boese, K., Kahng, A., Muddu, S.: A new adaptive multi-start technique for combinatorial Global Optimizations. Operations Research Letters, Vol. 16 (1994) 101-113
14. Capuzzo, D., Finzi, V., *et al.*,: Area-preserving curve-shortening flows: From phase separation to image processing. Interfaces and Free Boundaries, Vol. 31 (2002) 325–343
15. De Giorgi, E., Carriero, M., *et al.*,: Existence theorem for a minimum problem with free discontinuity set. Archive for Rational Mechanics and Analysis, Vol. 11 (1990) 291-322
16. Ambrosio, L., Tortorelli, V.M.,: Approximation of functionals depending on jumps by elliptic functionals via Γ-convergence, Communications on Pure and Applied Mathematics, Vol. 43 (1990) 999-1036
17. Modica, L.,: The gradient theory of phase transitions and the minimal interface criterion, Archive for Rational Mechanics and Analysis, Vol. 98 (1987) 123-142
18. Fedrizzi, M., Karcprzyk, J., Verdagay, J.L: A survey of fuzzy optimization and mathematical programming. Interactive Fuzzy Optimization, Lecture Notes in Economics and Mathematical Systems, (1991) 15-28
19. Kawarada, H., Suito, H.: Fuzzy Optimization Method, Computational Science for the 21st Century. John Wiley & Sons, 1997.
20. Cheol W.L., Yung, S.C.: Construction of fuzzy systems using least-squares method and genetic algorithm. Fuzzy Sets and Systems, Vol. 137 (2003) 297-323

Dynamic Modeling, Prediction and Analysis of Cytotoxicity on Microelectronic Sensors

Biao Huang[1] and James Z. Xing[2]

[1] Department of Chemical and Materials Engineering,
University of Alberta, Edmonton, Alberta, Canada T6G 2G6
[2] Department of Laboratory Medicine and Pathology,
University of Alberta, Edmonton, Alberta, Canada T6G 2S2

Abstract. This paper is concerned with dynamic modeling, prediction and analysis of cell cytotoxicity. A real-time cell electronic sensing (RT-CES) system has been used for label-free, dynamic measurements of cell responses to toxicant. Cells were grown onto the surfaces of the microelectronic sensors. Changes in cell number expressed as cell index (CI) have been recorded on-line as time series. The CI data are used for dynamic modeling in this paper. The developed models are verified using data that do not participate in the modeling. Optimal multi-step ahead predictions are calculated and compared with the actual CI. A new framework for dynamic cytotoxicity system analysis is established. Through the analysis of the system impulse response, we have observed that there are considerably similarities between the impulse response curves and the raw dynamic data, but there are also some striking differences between the two, particularly in terms of the initial and final cell killing effects. It is shown that dynamic modeling has great potential in modeling cell dynamics in the presence of toxicant and predicting the response of the cells.

1 Introduction

Time series expression experiments are an increasingly popular method for studying a wide range of biological systems. Time series analysis of biological system may be classified into a hierarchy of four analysis levels: experimental design, data analysis, pattern recognition and networks[1]. This paper will focus on the data analysis aspect.

Cytotoxicity experiments performed by Xing et al.(2005)[6] will be considered in this paper. When exposing to toxic compounds, cells undergo physiological and pathological changes, including morphological dynamics, an increase or decrease in cell adherence to the extracellular matrix, cell cycle arrest, apoptosis due to DNA damage, and necrosis[6]. Such cellular changes are dynamic and depend largely on cell types, the nature of a chemical compound, compound concentration, and compound exposure duration. In addition, certain cellular changes, such as morphological dynamics and adhesive changes, which may not lead to ultimate cell death, are transient and occur only at early or late stages of toxicant exposure. Modeling such diverse information sets is difficult at the analytical level. An experimental approach has to be resorted to.

Dynamic information of cells adds an additional dimension and insight into explanation of cell behavior such as cytotoxicity comparing with conventional fixed point observations method such as cell viability. Dynamic responses are typically described by dynamic models. These models can take many different forms, such as state space model, transfer function, neural net, fuzzy logic etc. To obtain the dynamic model, one usually has to perform dynamic experiments, determine model structure, estimate model parameters, and validate the model[5]. This complete procedure is known as system identification in systems and control literature[3].

To performance system identification, the dynamic experiment is the first step and also the most important step to ensure the quality of the modeling. The dynamic monitoring is difficult to achieve in most of conventional cell based assays because they need chemical or radiation indicators that may kill or disturb target cells. For dynamic detection of a broad range of physiological and pathological responses to toxic agents in living cells, Xing et al (2005)[6] has developed an automatic, real-time cell electronic sensing (RT-CES) system that is used for cell-based assay.

The Cell Index data obtained from [6] will be analyzed in this paper. Dynamic models will be built and verified. A new system analysis framework for the dynamic cytotoxicity will be established. It will be shown in this paper that the dynamic model can play an important role in analyzing intrinsic cell behavior and predict the trajectory of its progress (growth or death) over considerable time horizon. The remainder of this paper is organized as follows: Section 2 starts from discussion of dynamic model expression, followed by introduction of time series modeling and optimal multi-step prediction expression. A novel analysis framework for the dynamic cytotoxicity is discussed in the same section. The experiment setup is discussed in Section 3. In Section 4, modeling, prediction and analysis of cytotoxicity is reported, followed by concluding remarks in Section 5.

2 Dynamic Systems

2.1 Dynamic Representation of Systems

To perform analytical operations such as differentiation and integration, numerical approximations must be utilized. One way of converting continuous-time models to discrete-time form is to use finite difference techniques[4]. In general, a differential equation,

$$\frac{dy(t)}{dt} = f(y, u) \qquad (1)$$

where y is the output variable and u is the input variable, can be numerically integrated (although with some error) by introducing a finite difference approximation for the derivative. For example, the first-order, backward difference approximation to the derivative at time $t = n\Delta t$ is

$$\frac{dy(t)}{dt} \approx \frac{y_n - y_{n-1}}{\Delta t} \qquad (2)$$

where Δt is the integration interval that is specified by the user, y_n is the value of $y(t)$ at $t = n\Delta t$ and y_{n-1} denotes the value at the previous sampling instant $t = (n-1)\Delta t$. Substituting (2) into (1) and evaluating function $f(y, u)$ at the previous values of y and u (i.e., y_{n-1} and u_{n-1}) gives

$$\frac{y_n - y_{n-1}}{\Delta t} \approx f(y_{n-1}, u_{n-1}) \qquad (3)$$

or

$$y_n = y_{n-1} + \Delta t f(y_{n-1}, u_{n-1}) \qquad (4)$$

Equation (4) is a first-order difference equation that can be used to predict the value of y at time step n based on information at the previous time step $(n-1)$, namely, y_{n-1} and $f(y_{n-1}, u_{n-1})$. This type of expression is called a recurrence relation. It can be used to numerically integrate Equation (1) by calculating y_n for $n = 0, 1, 2, \cdots$ starting from known initial conditions, $y(0)$ and $u(0)$. In general, the resulting numerical solution, $\{y_n, n = 1, 2, 3, \cdots\}$ becomes more accurate and approaches the correct solution y_n as Δt decreases. There exist other methods that can exactly convert a differential equation into a difference equation under certain assumption on the input [4].

The backshift operator q^{-1} is an operator which moves a signal one step back, i.e. $q^{-1}y_n = y_{n-1}$. Similarly, $q^{-2}y_n = q^{-1}y_{n-1} = y_{n-2}$ and $qy_n = y_{n+1}$. It is convenient to use the backshift operator to write a difference equation. For example, a difference equation

$$y_n = 1.5 y_{n-1} - 0.5 y_{n-2} + 0.5 u_{n-1} \qquad (5)$$

can be represented as

$$y_n = 1.5 q^{-1} y_n - 0.5 q^{-2} y_n + 0.5 q^{-1} u_n$$

This can be written as

$$\frac{y_n}{u_n} = \frac{0.5 q^{-1}}{1 - 1.5 q^{-1} + 0.5 q^{-2}} \qquad (6)$$

This equation is also regarded as a discrete transfer function. In general, a discrete transfer function can be written as

$$\frac{y_n}{u_n} = G_p(q^{-1})$$

where, in this paper, u_n represents a signal that can be manipulated.

2.2 Optimal Predictions

In general, a time series y_n may be expressed as

$$y_n = H(q^{-1}) e_n = \frac{D(q^{-1})}{C(q^{-1})} e_n$$

where

$$D(q^{-1}) = 1 + d_1 q^{-1} + \ldots + d_q q^{-m}$$
$$C(q^{-1}) = 1 + c_1 q^{-1} + \ldots + c_p q^{-p}$$

For a general dynamic model with exogenous input u_n

$$y_n = G_p(q^{-1})u_n + H(q^{-1})e_n$$

where u_n can be manipulated and e_n is disturbance that can not be manipulated, the optimal one-step prediction is given by [3]:

$$\hat{y}_{n|n-1} = H^{-1}(q^{-1})G_p(q^{-1})u_n + [I - H^{-1}(q^{-1})]y_n$$

As an example, consider the following model:

$$y_n = \frac{bq^{-1}}{1+aq^{-1}}u_n + \frac{1+cq^{-1}}{1+aq^{-1}}e_n$$

The optimal prediction of y_n is given by

$$\hat{y}_{n|n-1} = \frac{bq^{-1}}{1+cq^{-1}}u_n + \frac{(c-a)q^{-1}}{1+cq^{-1}}y_n$$

The k-step ahead predictor is given by [3,2]

$$\hat{y}_{n|n-k} = W_k(q^{-1})G(q^{-1})u_n + (1 - W_k(q^{-1}))y_n \qquad (7)$$

where

$$W_k(q^{-1}) = F_k(q^{-1})H(q^{-1})^{-1} \qquad (8)$$

and

$$F_k(q^{-1}) = \sum_{i=0}^{k-1} g(i)q^{-i} \qquad (9)$$

where $g(i)$ is the impulse response coefficient of $H(q^{-1})$.

2.3 System Analysis

Once the system model is obtained, one will be able to analyze the properties of the system. Typical examples of the system properties include system poles, stability, damping coefficients, rate of response, settling time, steady state behavior, correlation/autocorrelation, power spectrum, system bandwidth, etc. In this paper, we will consider a specific property of the system, namely impulse response.

Let the time series model be written as

$$y_n = H(q^{-1})e_n \qquad (10)$$

where e_n is white noise. The prediction of y_n from this equation should match closely with the actual data with the error of e_n. The output y_n is therefore driven by the white noise e_n. This is the stochastic interpretation of Eq.(10). On the other hand, Eq.(10) may also be interpreted in the deterministic manner. Let's assume that $e_n = 1$ at time $n = 0$; otherwise $e_n = 0$. This may be interpreted as a shock on the system at time zero. The response of the system to this shock is the response without further shock and/or noise. Thus this response represents a purified system behavior, and is also known as impulse response in systems and control literature. By visualizing the impulse response trajectory, one can obtain a variety of information about the system, such as stability, response speed, steady state behavior, settling time etc. Thus impulse response curve is a powerful tool for the interpretation and prediction of the system dynamics. The further interpretation of impulse response will be elaborated in the application section.

3 Experiment Setup

The experimental procedure has been described in Xing et al.(2005)[6]. A parameter termed CI is derived to represent cell status based on the measured electrical impedance. The frequency dependent electrode impedance (resistance) without or with cells present in the wells is represented as $R_b(f)$ and $R_{cell}(f)$, respectively. The CI is calculated by

$$CI = \max_{i=1,...,N} [\frac{R_{cell}(f_i)}{R_b(f_i)} - 1]$$

where N is the number of the frequency points at which the impedance is measured. Several features of the CI can be derived: (i) Under the same physiological conditions, if more cells attach onto the electrodes, the larger impedance value leading to a larger CI value will be detected. If no cells are present on the electrodes or if the cells are not well-attached onto the electrodes, $R_{cell(f)}$ is the same as $R_b(f)$, leading to CI = 0. (ii) A large $R_{cell}(f)$ value leads to a larger CI. Thus, CI is a quantitative measure of the number of cells attached to the sensors. (iii) For the same number of cells attached to the sensors, changes in cell status will lead to a change in CI.

Three environmental toxicants, As(III), mercury(II) chloride, and chromium (VI), were used for cytotoxicity assessment on the 16 sensor device. Three cell lines, namely, NIH 3T3, BALB/c 3T3, and CHO-K1, were tested. The starting cell numbers were 10000, 10000, and 15000 cells per sensor wells, respectively, for NIH 3T3, BALB/c 3T3, and CHO-K1 cells. The cell growth on the sensor device was monitored in real time by the RT-CES system. When the CI values reached a range between 1.0 and 1.2, the cells were then exposed to either As(III), mercury(II) chloride, or chromium(VI) at different concentrations. After compound addition, the cell responses to the compounds were continuously and automatically monitored every hour by the RT-CES system.

4 Dynamic Modeling, Prediction and Analysis

4.1 Dynamic Growth with Toxicity

To test whether the RT-CES system can detect dynamic cell responses to different toxicants, the NIH 3T3 cells were treated with As(III), mercury(II) chloride, or chromium(VI) at different concentrations. After treatment, the CI was recorded in real time every hour for up to 24 h. Notably, the dynamic CI patterns of the NIH 3T3 cells in response to three toxicants are distinct.

Fig 1 shows dynamic cytotoxic response to different doses of As(III): 1.25, 4.06, 6.21, 9.20, 13.58, 20.01, and 29.64 μM, respectively. In the figure, higher dose toxicant yields smaller CI at the end of the curve as expected. It appears

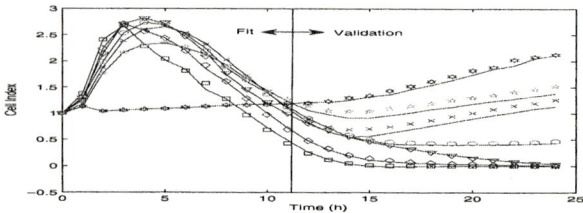

Fig. 1. Dynamic model of cytotoxicity due to As(III) toxicant (comparison based 1 on hr prediction). First 12 hr data for modeling and remaining 12 hr data for validation.

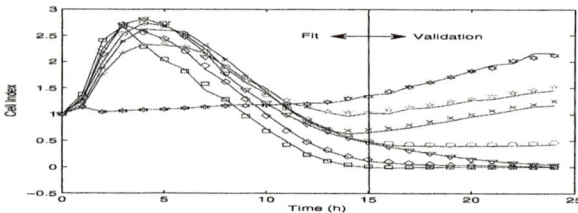

Fig. 2. Dynamic model of cytotoxicity due to As(III) toxicant (comparison based on 1 hr prediction). First 15 hr data for modeling and remaining 9 hr data for validation.

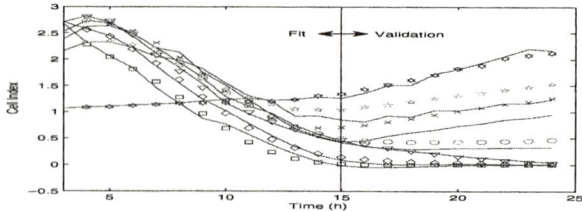

Fig. 3. Dynamic model of cytotoxicity due to As(III) toxicant (3 hr ahead prediction). First 15 hr data for modeling and remaining 9 hr data for validation.

that As(III) actually facilitates the growth of the cells at the beginning. The cell killing effects are only observed after the initial sharp growth. For modeling purpose, the first 12 hr data are used to fit dynamic data and the remaining 12 hr data for validation. The one step ahead predictions are shown in the figure. With reasonable fits overall, one can see that better models and predictions are obtained for larger dose of AS(III).

Fig 2 shows the result when the first 15 hr data are used for modeling and remaining 9 hr data for validation instead. Better fits are observed. To see comparison based on 3 hr ahead prediction, Fig. 3 shows the results in both model fitting and validation sections. The capability of the fitted dynamic models in prediction of longer term is observed from this figure.

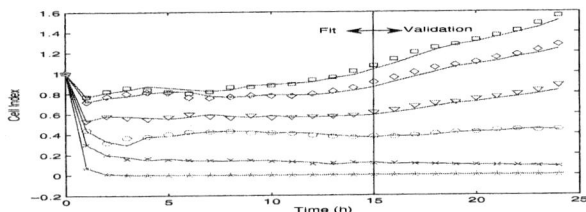

Fig. 4. Dynamic model of cytotoxicity due to mercury toxicant (comparison based on 1 hr prediction). First 15 hr data for modeling and remaining 9 hr data for validation.

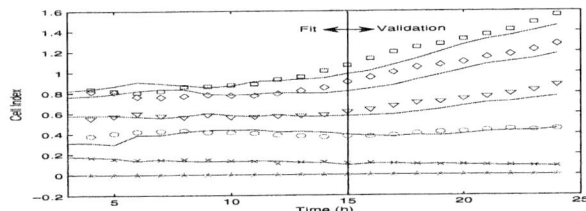

Fig. 5. Dynamic model of cytotoxicity due to mercury toxicant (based on 3 hr prediction). First 15 hr data for modeling and remaining 9 hr data for validation.

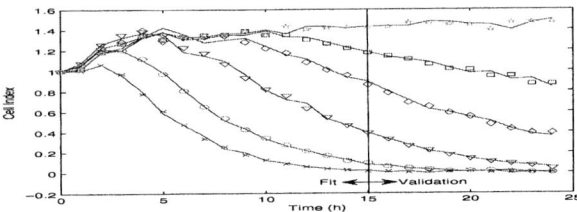

Fig. 6. Dynamic model of cytotoxicity due to chromium toxicant (comparison based on 1 hr prediction). First 15 hr data for modeling and remaining 9 hr data for validation.

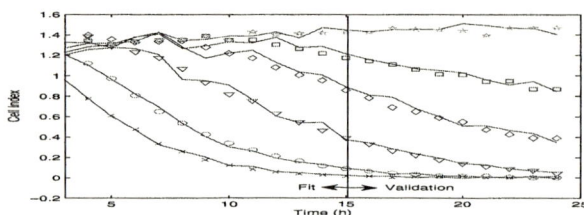

Fig. 7. Dynamic model of cytotoxicity due to chromium toxicant (based on 3 hr prediction). First 15 hr data for modeling and remaining 9 hr data for validation.

The two figures of dynamic cytotoxic response to different doses of mercury are shown in Fig. 4 and 5. The injected mercury consists of the following doses: 10.43, 15.2, 22.35, 32.8, 48.3, and 71 μM. Comparing to As(III), mercury seems to have quick toxic effect at beginning but poorer longer term effect. Better dynamic models and predictions have also obtained, indicating better dynamic models have been achieved.

The following doses of chromium were injected: 0.62, 0.91, 1.97, 2.89, 4.25, and 5.78 μM. The two counterpart figures are shown Fig 6 and 7. It is observed that chromium has slower toxic effect on the cells than Mercury at the beginning. The longer term toxic effect is similar to that of AS(III) but the initial growth is not as sharp as As(III). However, the dynamic models appear to be the best ones.

4.2 Dynamic Cytotoxicity Analysis

The modeling and prediction presented in the previous sections are more or less targeting at matching the data. The intrinsic effects of the toxicant can not be observed from the raw data due to the effect of disturbances/noise. The impulse response curve represents a purified dynamic behavior of the cell in the presence of toxicant.

The impulse responses of the cells to the three toxicants are shown in Fig 8, 9, 10, respectively. For the AS(III) toxicant, by comparing Fig 8 with Fig 2, one can clearly see a smoother response curve has been obtained. Most importantly, all doses of As(III) but 1.25 μM seem to give a stable response, i.e. they possess

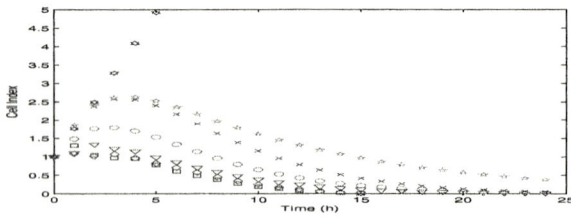

Fig. 8. Impulse response of dynamic model due to As(III) toxicant

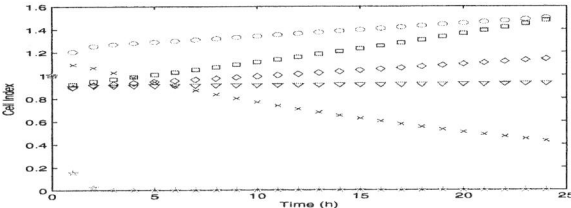

Fig. 9. Impulse response of dynamic model due to mercury toxicant

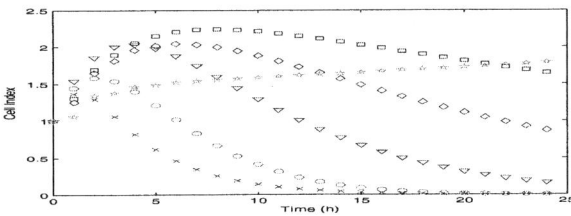

Fig. 10. Impulse response of dynamic model due to chromium toxicant

the property of completely killing the cell at the end, while according to Fig 2, about half of the doses do not have the complete cell killing effect. This result challenges the conclusion that may be drawn from raw data such as the one shown in Fig 2. Is this disagreement due to the disturbance or due to some other reasons? This question posts an interesting topic for further work. Other than this, other dynamic behaviors of the cells are quite similar between the two figures, i.e. they all have fast initial growth of the cells.

For the mercury toxicant, by comparing Fig 9 with Fig 4, one can see that both have good agreement in terms of final cell killing. The dynamic responses are overall similar except for one. However, the initial toxicant effects are quite different between the two figures. The fast initial cell killing effect observed from the raw data in Fig 4 is not consistent with the one observed from the impulse response, challenging the fast initial toxicant effect of mercury.

For the chromium toxicant, by comparing Fig 10 with Fig 6, one can surprisingly find similar results except for one.

To summarize the analysis, we have found that impulse response curves provide an additional insight into the cytotoxicity. There are considerably similarities between the impulse response curves and the raw dynamic data, but there are also some striking differences between the two, particularly in terms of the initial and final cell killing effects. More work is needed to study the impulse response curve as an important tool to explain cytotoxicity.

5 Conclusion

In this paper, we have considered dynamic modeling, prediction and analysis of cytotoxicity. The CI data from a real-time cell electronic sensing (RT-CES) system have been used for dynamic modeling. The developed models are verified using data that do not participate in the modeling. Optimal multi-step ahead predictions are calculated and compared with the actual CI. The dynamic behavior of cytotoxicity has been studied under a novel framework, namely impulse response of the system. It has been shown that dynamic time series models have great potential in modeling cell dynamics in the presence of toxicant and predicting the response of the cells. Through the analysis of the system impulse response, we have observed that there are considerably similarities between the impulse response curves and the raw dynamic data, but there are also some striking differences between the two, particularly in terms of the initial and final cell killing effects. The results in this paper presented the initial step in our effort towards dynamic modeling, prediction and analysis of cell cytotoxicity. Work is in progress towards dynamic experiment design, nonlinear modeling, pattern recognition and control of cytotoxicity.

References

1. Z. Bar-Joseph. Analyzing time series gene expression data. *Bioinformatics*, 20(16):2493–2503, 2004.
2. B. Huang, A. Malhotra, and E.C. Tamayo. Model predictive control relevant identification and validation. *Chemical Engineering Sciences*, 58:2389–2401, 2003.
3. L. Ljung. *System Identification*. Prentice-Hall, 2nd edition, 1999.
4. D.E. Seborg, T.F. Edgar, and D.A. Mellichamp. *Process Dynamics and Control*. John Wiley & Sons, 1989.
5. T. Soderstrom and P. Stoica. *System Identification*. Prentice Hall International, UK, 1989.
6. J.Z. Xing, L. Zhu, J.A. Jackson, S. Gabos, X.J. Sun, X. Wang, and X. Xu. Dynamic monitoring of cytotoxicity on microelectronic sensors. *Chem. Res. Toxicol.*, 18:154–161, 2005.

Generalized Fuzzy Morphological Operators

Tingquan Deng[1,2] and Yanmei Chen[1]

[1] Department of Mathematics, Harbin Institute of Technology,
Harbin, 150001 P.R. China
{Tq_Deng, Chen.yanmei}@163.com
[2] Department of Automation, Tsinghua University, Beijing, 100084 P.R. China

Abstract. The adjunction in lattice theory is an important technique in lattice-based mathematical morphology and fuzzy logical operators are indispensable implements in fuzzy morphology. This paper introduces a set-valued mapping that is compatible with the infimum in a complete lattice and with a conjunction in fuzzy logic. According to the generalized operator, a concept of a fuzzy adjunction is developed to generate fuzzy morphological dilation and erosion. Fundamental properties of the generalized fuzzy morphological operators have been investigated.

1 Introduction

A seminal book [5] by G. Matheron in 1975 laid down the foundations of a novel technique for shape analysis known as mathematical morphology. Mathematical morphology was enriched and subsequently popularized by the highly inspiring publications [8,10,11] in image analysis for binary images, grey-scale images and multi-valued images. In the study of mathematical morphology on complete lattices, the concept of an adjunction shows an important link between a dilation and an erosion, and implies many interesting algebraic properties of basic morphological operators. With the adjunction theory, T.-Q. Deng and H. Heijmans studied a broad class of grey-scale morphology [4] based on fuzzy logic.

On the other hand, many researchers interpreted grey-scale images as fuzzy sets and established frameworks of fuzzy mathematical morphology, where the frequently used models are devoted to D. Sinha and E. Dougherty [9], I. Bloch and H. Maitre [1], and M. Nachtegael and E. Kerre [6]. V. Chatzis and I. Pitas [2] also studied generalized fuzzy morphology and its applications by means of introducing a new inclusion indicator of fuzzy sets. In fuzzy morphology and grey-scale morphology based on fuzzy logic, a t-norm, a t-conorm, and the duality principle in fuzzy logic are indispensable implements in the construction of fuzzy morphological operators.

It is well known that a t-norm, or generally, a conjunction in fuzzy logic plays the role similar to that of the infimum in a complete lattice. This paper introduces a set-valued operator by means of integrating the concepts of the conjunction and the infimum, and fuzzify the concept of the adjunction [4]. According to the fuzzy adjunction, generalized fuzzy morphological operators are able to be derived and many interesting properties of such the operators will be investigated.

2 Extended Fuzzy Logical Operators

2.1 Fuzzy Logical Operators

Suppose that **L** is a complete lattice with its least element **0** and greatest element **1**. The prototype of such a lattice is the unit interval $[0,1]$ or a complete chain.

Definition 1. *An operator $C : \mathbf{L} \times \mathbf{L} \to \mathbf{L}$ is called a fuzzy conjunction if it is non-decreasing in both arguments satisfying $C(\mathbf{0},\mathbf{1}) = C(\mathbf{1},\mathbf{0}) = \mathbf{0}$ and $C(\mathbf{1},\mathbf{1}) = \mathbf{1}$. A fuzzy implication $I : \mathbf{L} \times \mathbf{L} \to \mathbf{L}$ is an operator being non-increasing in its first argument, non-decreasing in its second, and satisfying $I(\mathbf{0},\mathbf{0}) = I(\mathbf{1},\mathbf{1}) = \mathbf{1}$ and $I(\mathbf{1},\mathbf{0}) = \mathbf{0}$. A t-norm C is a commutative and associative fuzzy conjunction satisfying the boundary condition $C(\mathbf{1},s) = s$ for any $s \in \mathbf{L}$.*

For short, a fuzzy conjunction is called a conjunction and a fuzzy implication is called an implication if there are no confusions happened. If C is a conjunction and I is an implication on **L**, then $C(s,\mathbf{0}) = C(\mathbf{0},s) = \mathbf{0}$ and $I(\mathbf{0},s) = I(s,\mathbf{1}) = \mathbf{1}$ for all $s \in \mathbf{L}$.

Definition 2. *An implication I and a conjunction C on **L** are said to form an adjunction if*

$$C(a,s) \leq t \iff s \leq I(a,t) \tag{1}$$

for each $a \in \mathbf{L}$ and for all $s,t \in \mathbf{L}$.

When an implication I and a conjunction C form an adjunction, the pair (I,C) is called an adjunction and I is called the adjunctional implication of C. If C is a t-norm, its adjunctional implication is usually called the adjoint implication or R-implication in fuzzy logic. Many examples of such conjunctions and implications have been presented and the following proposition has been proved in [4].

Proposition 1. *For an implication I and a conjunction C on **L**, if (I,C) forms an adjunction, I and C can be reconstructed from one to the other by*

$$I(a,s) = \vee\{r \in \mathbf{L} \mid C(a,r) \leq s\} \tag{2}$$

and

$$C(a,s) = \wedge\{r \in \mathbf{L} \mid s \leq I(a,r)\}. \tag{3}$$

From the reconstruction property, one knows that $C(\mathbf{1},s) = s$ if and only if $I(\mathbf{1},s) = s$ for all $s \in \mathbf{L}$ provided that (I,C) is an adjunction.

2.2 Fuzzifications of Fuzzy Set Operations

Suppose that E is a nonempty set, called a universe of discourse, and that $\mathcal{F}_\mathbf{L}(E)$ denotes the family of all L-fuzzy sets (fuzzy sets, in short) from E to $\mathbf{L}$. Several definitions of the inclusion degree of one fuzzy set $G \in \mathcal{F}_\mathbf{L}(E)$ into another $F \in \mathcal{F}_\mathbf{L}(E)$ have been, respectively, introduced in [2,3,4], which can be summarized as follows.

$$|G \leq F| = \wedge_{x \in E} I(G(x), F(x)) = \wedge_{x \in E} H(x), \qquad (4)$$

where I is an implication on $\mathbf{L}$.

It is well known [7] that any t-norm in fuzzy logic is less than or equals to the *Gödel-Brouwer* conjunction (Zadeh t-norm) $C(s,t) = \min(s,t)$, and the infimum operation $\wedge$ in lattice theory, to some extent, can be considered as the generalization of this conjunction; in which sense, one may say that $\wedge$ is compatible with the *Gödel-Brouwer* conjunction. From the viewpoint of a mapping, the infimum in $[0,1]$ is generated when the number of the arguments of the *Gödel-Brouwer* conjunction is augmented from two to multitude, even to infinity.

It is true that every implication is closely connected with a conjunction by means of the adjunction relation or the duality principle. To use independently the infimum and the implication or its corresponding conjunction in (4) is unnatural. It is interesting to extend the infimum to a general operator INF, also devoted by $\sqcap$, which is compatible with a conjunction C and with the infimum $\wedge$. From which, the degree of a fuzzy set G being included in F is denoted by

$$\lfloor G \leq F \rfloor = \text{INF}(H) = \sqcap_{x \in E} H(x). \qquad (5)$$

To proceed the definition of INF, a concept, called the sup-generating family of a complete lattice is reviewed.

Definition 3. *A nonempty subset S of a complete lattice $\mathbf{L}$ is called a sup-generating family of $\mathbf{L}$ if for every $l \in \mathbf{L}$, $l = \vee\{s \mid s \in S, s \leq l\}$.*

For example, $\{\{x\} \mid x \in E\}$ is a sup-generating family of $\mathcal{P}(E)$, the power set of E; $S = \{x_\lambda \mid x \in E, \lambda \in \mathbf{L} \setminus \{\mathbf{0}\}\}$ is a sup-generating family of $\mathcal{F}_\mathbf{L}(E)$, where $x_\lambda(y) = \lambda$ if $y = x$; $\mathbf{0}$, otherwise, is a fuzzy point in E.

3 Generalized Fuzzy Logical Operators

Let $\mathbf{L}_E$ be a complete sublattice of $\mathcal{F}_\mathbf{L}(E)$ with a sup-generating family S, i.e., $S \subseteq \mathbf{L}_E$, $\mathbf{L}_E$ is a sublattice of $\mathcal{F}_\mathbf{L}(E)$ and is complete ($\mathbf{L}_E$ may be the whole universe $\mathcal{F}_\mathbf{L}(E)$), the following definition provides an axiomatic characterization of an operator INF from $\mathbf{L}_E$ to $\mathbf{L}$ that is compatible with a conjunction C in fuzzy logic and with the infimum $\wedge$ in a complete lattice.

Definition 4. *Let C be a conjunction on $\mathbf{L}$. INF or $\sqcap$, an extended operator of C, is a continuous mapping from $\mathbf{L}_E$ to $\mathbf{L}$ satisfying the following axioms:*

(1) $H \equiv 1 \iff \text{INF}(H) = 1$;
(2) $\text{INF}(H) \le \inf(H) = \wedge H$;
(3) $H_1 \le H_2 \implies \text{INF}(H_1) \le \text{INF}(H_2)$, where $H_1 \le H_2$ means that $H_1(x) \le H_2(x)$ for all $x \in E$;
(4) $\text{INF}(\bar{r}) \le r$, where $\bar{r}(x) \equiv r \in \mathbf{L}$ for every $r \in \mathbf{L}$ and for any $x \in E$;
(5) If $H(x) = \begin{cases} t_1, & x = x_1, \\ t_2, & x = x_2, \\ 1, & x \ne x_1, x_2, \end{cases}$ then $\text{INF}(H) = C(t_1, t_2)$.

Based on the axiomatic definition of INF, the following proposition holds.

Proposition 2. *(1) Translation invariance.* $\text{INF}(F_x) = \text{INF}(F)$ *for arbitrary* $F \in \mathbf{L}_E$ *and* $x \in E$ *if* E *is a translation invariant additive group with '+' operation, where* $F_x(y) = F(y - x)$, $y - x = y + x^{-1}$, *and* x^{-1} *is the inverse of* x.
(2) Homothetic invariance. If C *is an upper semi-continuous conjunction in its second argument satisfying* $C(C(s,t), C(s,r)) = C(s, C(t,r))$ *and* $C(s, 1) = s$ *for every* $s, t, r \in \mathbf{L}$, *then* $\text{INF}(k \circ F) = C(k, \text{INF}(F))$, *where* $(k \circ F)(x) = C(k, F)(x) = C(k, F(x))$, $x \in E$, $k \in \mathbf{L}$.
(3) Scalar multiplication. $\text{INF}(\lambda F) = \text{INF}(F)$, *where* $(\lambda F)(x) = F(x/\lambda)$, $x \in E$ *and* $\lambda \in \mathbf{L} \setminus \{\mathbf{0}\}$ *if* E *is a linear space.*

Some illustrations of the generalized operator INF and its related conjunction are presented as follows.
(1) When the complete lattice $\mathbf{L}$ is a finite point set, every associative conjunction C on $\mathbf{L}$ can be extended to INF.
(2) In $[0, 1]$, the common inf-operation is a degenerate INF operator and the related conjunction C is the *Gödel-Brouwer* conjunction $\min$.
(3) Let $\mathbf{L}$ be a complete lattice, for every $A \in \mathbf{L}$, let $\text{INF}(A) = \wedge A$, then INF is a generalized operator and the related conjunction is $C(s, t) = s \wedge t$.
(4) Let $E = \overline{R} = R \cup \{-\infty, +\infty\} = [-\infty, +\infty]$ and $\mathbf{L} = [0, 1]$, for every $F \in \mathbf{L}_E$, define

$$\text{INF}(F) = \min(\inf(F), \Pi_{x \in E} Sh(F(x)/2)), \qquad (6)$$

where $Sh(x) = (-x \ln x - (1-x) \ln(1-x))/\ln 2$ is the usual *Shannon* function. The related conjunction is

$$C(s, t) = \min(s, t, Sh(s/2) Sh(t/2)). \qquad (7)$$

(5) Let $\theta : [0, 1] \to [0, +\infty)$ be a decreasing continuous mapping satisfying $\theta(1) = 0$ and $\lim_{x \to 0+} \theta(x) = +\infty$. $\theta(s) = \cot(\pi s/2)$ for example, is such a function. By convention, take notations that $\infty + M = M + \infty = \infty$ for arbitrary $M \in [0, +\infty)$, and reinforce the definition of θ at point 0 by $\theta(0) = +\infty$.

Let $\mathbf{L} = [0, 1]$, if E is a continuous space, say the n dimension Euclidean space $\overline{R}^n$ ($n \ge 1$) for instance, for every $F \in \mathbf{L}_E$, let

$$\text{INF}(F) = \theta^{-1}(\int_E \theta(F(x)) \, d\sigma), \qquad (8)$$

then INF is a generalized operator of the conjunction $C(s,t) = \theta^{-1}(\theta(s) + \theta(t))$; meanwhile, the adjunctional implication of C is $I(s,t) = \begin{cases} 1, & s \leq t, \\ \theta^{-1}(\theta(t) - \theta(s)), & s > t. \end{cases}$

If E is discrete, e.g., $E = \overline{Z}^n = (Z \cup \{-\infty, +\infty\})^n$, let

$$\text{INF}(F) = \theta^{-1}(\Sigma_{x=-\infty}^{+\infty} \theta(F(x))), \tag{9}$$

then INF is also a generalized operator of the same conjunction $C(s,t) = \theta^{-1}(\theta(s) + \theta(t))$.

In both cases, one can deduce that for all $F, G \in \mathbf{L}_E$,

$$\lfloor F \subseteq G \rfloor = 1 \iff F \subseteq G. \tag{10}$$

Especially, $\lfloor F \subseteq F \rfloor = \sqcap_{x \in E} I(F(x), F(x)) = 1$. Furthermore, if $C \leq min$, define $(F \cap G)(x) = C(F(x), G(x))$, $x \in E$, then

$$\lfloor F \cap G \subseteq F \rfloor = \sqcap_{x \in E} I(C(F(x), G(x)), F(x)) = 1. \tag{11}$$

In general, the following statements hold.

Proposition 3. *Let (I, C) be an adjunction on $\mathbf{L}$ satisfying $C(s, 1) = s$, then $\lfloor F \leq F \rfloor = 1$ for any $F \in \mathbf{L}_E$. Furthermore, for all $F, G \in \mathbf{L}_E$, $\lfloor F \leq G \rfloor = 1 \iff F \leq G$.*

Proof. Let $t \in \mathbf{L}$, then $I(t,t) = \vee\{r \in \mathbf{L} \mid C(t,r) \leq t\} = \vee\{r \in \mathbf{L} \mid C(t,r) \leq C(t,1) = t\} = 1$. Hence, for any $F \in \mathbf{L}_E$, $\lfloor F \leq F \rfloor = \sqcap_{x \in E} I(F(x), F(x)) = 1$.

To prove the second assertion.

$\Leftarrow$: If $s \leq t$, $I(s,t) = \vee\{r \in \mathbf{L} \mid C(s,r) \leq t\} = \vee\{r \in \mathbf{L} \mid C(s,r) \leq C(s,1) \leq t\} = 1$. Thus, if $F, G \in \mathbf{L}_E$ and $F \leq G$, then for any $x \in E$, $I(F(x), G(x)) = 1$. Hence, $\lfloor F \leq G \rfloor = \sqcap_{x \in E} I(F(x), G(x)) = 1$.

$\Rightarrow$: Let $F, G \in \mathbf{L}_E$ and $\lfloor F \leq G \rfloor = 1$, then $I(F(x), G(x)) = 1$ for any $x \in E$. So, for any $x \in E$, $F(x) = C(F(x), 1) = C(F(x), I(F(x), G(x))) = C(F(x), \vee\{r \in \mathbf{L} \mid C(F(x), r) \leq G(x)\}) = \vee_{r \in \mathbf{L}, C(F(x),r) \leq G(x)} C(F(x), r) \leq G(x)$.

Definition 5. *Let ν be a negation and C be a conjunction on $\mathbf{L}$, and let $\overline{C}(s,t) = \nu(C(\nu(s), \nu(t)))$, $s, t \in \mathbf{L}$, then $\overline{C}$ is called a (fuzzy) disjunction on $\mathbf{L}$. If INF is an extended operator of C, let $SUP = INF^*$, or explicitly, $SUP(H) = \nu INF(\nu H)$, then SUP, denoted by $\sqcup$, is called the extension of the disjunction $\overline{C}$.*

Proposition 4. *Let (I, C) be an adjunction on $\mathbf{L}$, then for any $s \in \mathbf{L}$, any index set T and any family $\{p_t\}_{t \in T} \subseteq \mathbf{L}$,*

(1) $C(s, \sqcup_{t \in T} p_t) = \sqcup_{t \in T} C(s, p_t)$ and $I(s, \sqcap_{t \in T} p_t) = \sqcap_{t \in T} I(s, p_t)$;
(2) $C(\sqcup_{t \in T} p_t, s) = \sqcup_{t \in T} C(p_t, s) \iff I(\sqcup_{t \in T} p_t, s) = \sqcap_{t \in T} I(p_t, s)$.

Proof. (1) It is sufficient to prove the first equality. By means of the adjunction relation of (I,C) and the definition of $\sqcup$, note that $\sup \leq \sqcup$ and $r \leq \text{SUP}(\bar{\mathbf{r}})$ for every $r \in \mathbf{L}$,

$$\begin{aligned}
C(s, \sqcup_{t\in T} p_t) \leq r &\iff \sqcup_{t\in T} p_t \leq I(s,r) \\
&\iff \forall t \in T, p_t \leq I(s,r) \\
&\iff \forall t \in T, C(s,p_t) \leq r \\
&\iff \sqcup_{t\in T} C(s,p_t) \leq r,
\end{aligned}$$

which means that $C(s, \sqcup_{t\in T} p_t) = \sqcup_{t\in T} C(s,p_t)$.

(2) $\Rightarrow$: Taking $r \in \mathbf{L}$, then

$$\begin{aligned}
r \leq I(\sqcup_{t\in T} p_t, s) &\iff C(\sqcup_{t\in T} p_t, r) \leq s \\
&\iff \sqcup_{t\in T} C(p_t, r) \leq s \\
&\iff \forall t \in T, C(p_t, r) \leq s \\
&\iff \forall t \in T, r \leq I(p_t, s) \\
&\iff r \leq \sqcap_{t\in T} I(p_t, s).
\end{aligned}$$

Therefore, $I(\sqcup_{t\in T} p_t, s) = \sqcap_{t\in T} I(p_t, s)$.

$\Leftarrow$: It can be proved in the same manner.

In view of Proposition 4, if C is commutative in addition, it is trivial that $C(\sqcup_{t\in T} p_t, s) = \sqcup_{t\in T} C(p_t, s)$.

4 Generalized Fuzzy Morphological Operators

Definition 6. *Let A and B be two arbitrary nonempty sets, and $H : A \times B \to \mathbf{L}$ be an arbitrary mapping, if $\sqcap$ is an extended operator of a conjunction C on $\mathbf{L}$ satisfying*

$$\sqcap_{x\in A} \sqcap_{y\in B} H(x,y) = \sqcap_{y\in B} \sqcap_{x\in A} H(x,y), \tag{12}$$

then $\sqcap$ is called commutative and associative on $\mathbf{L}$.

If $\sqcap$ is commutative and associative on $\mathbf{L}$, then for any negation ν, $\sqcup$ is commutative and associative on $\mathbf{L}$, either.

Let E_1 and E_2 be two nonempty sets, $\mathbf{L}$ be a complete lattice, and let $\mathbf{L}_{E_1}$ and $\mathbf{L}_{E_2}$ denote, respectively, the complete sublattices of $\mathcal{F}_\mathbf{L}(E_1)$ and $\mathcal{F}_\mathbf{L}(E_2)$ with their own sup-generating families.

Definition 7. *An operator $\mathcal{E} : \mathbf{L}_{E_1} \to \mathbf{L}_{E_2}$ is called a generalized fuzzy erosion from $\mathbf{L}_{E_1}$ to $\mathbf{L}_{E_2}$ if*

$$\mathcal{E}(\sqcap_{u\in E_1} F_u) = \sqcap_{u\in E_1} \mathcal{E}(F_u) \tag{13}$$

for any family $\{F_u\}_{u\in E_1} \subseteq \mathbf{L}_{E_1}$; an operator $\mathcal{D} : \mathbf{L}_{E_2} \to \mathbf{L}_{E_1}$ is called a generalized fuzzy dilation from $\mathbf{L}_{E_2}$ to $\mathbf{L}_{E_1}$ if

$$\mathcal{D}(\sqcup_{v\in E_2} G_v) = \sqcup_{v\in E_2} \mathcal{D}(G_v) \tag{14}$$

for any family $\{G_v\}_{v\in E_2} \in \mathbf{L}_{E_2}$.

Proposition 5. *Let $\sqcap$ be a commutative and associative operator on $\mathbf{L}$, if for an arbitrary index set T and for each $t \in T$, $\mathcal{E}_t$ is a generalized fuzzy erosion from $\mathbf{L}_{E_1}$ to $\mathbf{L}_{E_2}$ and $\mathcal{D}_t$ is a generalized fuzzy dilation from $\mathbf{L}_{E_2}$ to $\mathbf{L}_{E_1}$, then $\sqcap_{t \in T} \mathcal{E}_t$ is also a generalized fuzzy erosion, while $\sqcup_{t \in T} \mathcal{D}_t$ is a generalized fuzzy dilation.*

Proof. If for each $t \in T$, $\mathcal{E}_t$ is a generalized fuzzy erosion, let $\{F_u\}_{u \in E_1} \subseteq \mathbf{L}_{E_1}$, then $(\sqcap_{t \in T} \mathcal{E}_t)(\sqcap_{u \in E_1} F_u) = \sqcap_{t \in T} \sqcap_{u \in E_1} \mathcal{E}_t(F_u) = \sqcap_{u \in E_1} \sqcap_{t \in T} \mathcal{E}_t(F_u) = \sqcap_{u \in E_1} (\sqcap_{t \in T} \mathcal{E}_t)(F_u)$, which implies that $\sqcap_{t \in T} \mathcal{E}_t$ is a generalized fuzzy erosion from $\mathbf{L}_{E_1}$ to $\mathbf{L}_{E_2}$.

That $\sqcup_{t \in T} \mathcal{D}_t$ is a generalized fuzzy dilation can be proved in the same way.

Definition 8. *Let $\mathcal{E}$ be a mapping from $\mathbf{L}_{E_1}$ to $\mathbf{L}_{E_2}$, and $\mathcal{D}$ be a mapping from $\mathbf{L}_{E_2}$ to $\mathbf{L}_{E_1}$, if*

$$\lfloor \mathcal{D}(G) \leq F \rfloor = \lfloor G \leq \mathcal{E}(F) \rfloor \tag{15}$$

for every $F \in \mathbf{L}_{E_1}$ and $G \in \mathbf{L}_{E_2}$, then the pair $(\mathcal{E}, \mathcal{D})$ is called a fuzzy adjunction between $\mathbf{L}_{E_1}$ and $\mathbf{L}_{E_2}$. If $E_1 = E_2$, $(\mathcal{E}, \mathcal{D})$ is called a fuzzy adjunction on $\mathbf{L}_{E_1}$.

Theorem 1. *Let (I, C) be an adjunction on $\mathbf{L}$ satisfying $C(s, 1) = s$ for every $s \in \mathbf{L}$. If $\sqcap$ is a commutative and associative extended operator of C, and the mappings $\mathcal{E} : \mathbf{L}_{E_1} \to \mathbf{L}_{E_2}$ and $\mathcal{D} : \mathbf{L}_{E_2} \to \mathbf{L}_{E_1}$ form a fuzzy adjunction, then*

(1) $\mathcal{D}\mathcal{E}(F) \leq F$ and $G \leq \mathcal{E}\mathcal{D}(G)$ for any $F \in \mathbf{L}_{E_1}$ and any $G \in \mathbf{L}_{E_2}$;
(2) $\mathcal{E}\mathcal{D}\mathcal{E} = \mathcal{E}$ and $\mathcal{D}\mathcal{E}\mathcal{D} = \mathcal{D}$;
(3) $\mathcal{E}$ *is a generalized fuzzy erosion, and $\mathcal{D}$ is a generalized fuzzy dilation.*

Proof. (1) In $\lfloor \mathcal{D}(G) \leq F \rfloor = \lfloor G \leq \mathcal{E}(F) \rfloor$, taking $G = \mathcal{E}(F)$, one will have that $\lfloor \mathcal{D}\mathcal{E}(F) \leq F \rfloor = \lfloor \mathcal{E}(F) \leq \mathcal{E}(F) \rfloor = 1$. In the same way, taking $F = \mathcal{D}(G)$ in the above equality yields $\lfloor G \leq \mathcal{E}\mathcal{D}(G) \rfloor = \lfloor \mathcal{D}(G) \leq \mathcal{D}(G) \rfloor = 1$. Therefore, it is straightforward from Proposition 3 that $\mathcal{D}\mathcal{E}(F) \leq F$ and $G \leq \mathcal{E}\mathcal{D}(G)$.

(2) From the statements in the proof of (1), one will have that $\lfloor \mathcal{E}(F) \leq \mathcal{E}\mathcal{D}\mathcal{E}(F) \rfloor = 1$ and that $\lfloor \mathcal{D}\mathcal{E}\mathcal{D}(G) \leq \mathcal{D}(G) \rfloor = 1$. Therefore, $\mathcal{E}(F) \leq \mathcal{E}\mathcal{D}\mathcal{E}(F)$ and $\mathcal{D}\mathcal{E}\mathcal{D}(G) \leq \mathcal{D}(G)$.

Let $F, G \in \mathbf{L}_{E_1}$ and $F \leq G$, then $1 = \lfloor \mathcal{D}\mathcal{E}(F) \leq F \rfloor \leq \lfloor \mathcal{D}\mathcal{E}(F) \leq G \rfloor$. Therefore, $\lfloor \mathcal{E}(F) \leq \mathcal{E}(G) \rfloor = \lfloor \mathcal{D}\mathcal{E}(F) \leq G \rfloor = 1$, and so $\mathcal{E}$ is monotone. Similarly, $\mathcal{D}$ is also monotone. From which and the consequences in (1), for arbitrary $F \in \mathbf{L}_{E_1}$ and $G \in \mathbf{L}_{E_2}$, $\mathcal{E}\mathcal{D}\mathcal{E}(F) \leq \mathcal{E}(F)$ and $\mathcal{D}(G) \leq \mathcal{D}\mathcal{E}\mathcal{D}(G)$.

(3) For any family $\{F_u\}_{u \in E_1} \subseteq \mathbf{L}_{E_1}$, and $G \in \mathbf{L}_{E_2}$,

$$\begin{aligned}
\lfloor G \leq \mathcal{E}(\sqcap_{u \in E_1} F_u) \rfloor &= \lfloor \mathcal{D}(G) \leq \sqcap_{u \in E_1} F_u \rfloor \\
&= \sqcap_{x \in E_1} I(\mathcal{D}(G)(x), \sqcap_{u \in E_1} F_u(x)) \\
&= \sqcap_{x \in E_1} \sqcap_{u \in E_1} I(\mathcal{D}(G)(x), F_u(x)) \\
&= \sqcap_{u \in E_1} \sqcap_{x \in E_1} I(\mathcal{D}(G)(x), F_u(x)) \\
&= \sqcap_{u \in E_1} \lfloor \mathcal{D}(G) \leq F_u \rfloor \\
&= \sqcap_{u \in E_1} \lfloor G \leq \mathcal{E}(F_u) \rfloor
\end{aligned}$$

$$= \sqcap_{u \in E_1} \sqcap_{y \in E_2} I(G(y), \mathcal{E}(F_u)(y))$$
$$= \sqcap_{y \in E_2} \sqcap_{u \in E_1} I(G(y), \mathcal{E}(F_u)(y))$$
$$= \sqcap_{y \in E_2} I(G(y), \sqcap_{u \in E_1} \mathcal{E}(F_u)(y))$$
$$= \lfloor G \leq \sqcap_{u \in E_1} \mathcal{E}(F_u) \rfloor.$$

In view of Proposition 3, taking $G = \mathcal{E}(\sqcap_{u \in E_1} F_u)$ and $G = \sqcap_{u \in E_1} \mathcal{E}(F_u)$, respectively, in the above equalities, yields

$$\mathcal{E}(\sqcap_{u \in E_1} F_u) = \sqcap_{u \in E_1} \mathcal{E}(F_u).$$

The assertion that $\mathcal{D}$ is a generalized fuzzy dilation can be proved in the same way.

In particular, for $y \in E_2$ and $\lambda \in \mathbf{L}$, let $\mathcal{D}(y_\lambda)$ denote the generalized fuzzy dilation of fuzzy point $y_\lambda \in \mathbf{L}_{E_2}$, then

$$\mathcal{D}(G) = \sqcup_{\lambda \leq G(y), y \in E_2} \mathcal{D}(y_\lambda) \tag{16}$$

is the generalized fuzzy dilation of $G \in \mathbf{L}_{E_2}$.

From Theorem 1, the following proposition is clear.

Proposition 6. $\alpha = \mathcal{DE}$ *is a generalized fuzzy morphological algebraic opening (increasing, ani-extensive and idempotent) operator, while* $\beta = \mathcal{ED}$ *is a generalized fuzzy morphological algebraic closing (increasing, extensive and idempotent) operator.*

Morphological openings and closings play important roles in the theory and applications of mathematical morphology, especially, in the construction of morphological filters, granulometries and connected operators for image analysis and segmentation.

In the sequel of this section, assume that $E_1 = E_2 = E$, and that E is a translation invariant additive domain.

Theorem 2. *Let* (I, C) *be an adjunction, and* $\sqcap$ *be a commutative and associative extended operator of* C *on* $\mathbf{L}$, *then*

$$\varepsilon_G(F)(x) = \sqcap_{y \in E} I(G(y - x), F(y)) \tag{17}$$

is a fuzzy erosion of $F \in \mathbf{L}_E$ *by* $G \in \mathbf{L}_E$ *(G here is called a structuring element), while*

$$\delta_G(F)(x) = \sqcup_{y \in E} C(G(x - y), F(y)) \tag{18}$$

is a fuzzy dilation of $F \in \mathbf{L}_E$ *by* $G \in \mathbf{L}_E$, $x \in E$.

Proof. Let $\{F_u\}_{u \in E} \subseteq \mathbf{L}_E$, then for any $G \in \mathbf{L}_E$ and any $x \in E$, $\varepsilon_G(\sqcap_{u \in E} F_u)(x)$ $= \sqcap_{y \in E} I(G(y - x), \sqcap_{u \in E} F_u(y)) = \sqcap_{y \in E} \sqcap_{u \in E} I(G(y - x), F_u(y)) = \sqcap_{u \in E} \sqcap_{y \in E} I(G(y - x), F_u(y)) = \sqcap_{u \in E} \varepsilon_G(F_u)(x)$. Therefore, $\varepsilon_G(F)$ is a fuzzy erosion.

The second statement can be proved similarly.

In this theorem, if the operators $\sqcap$ and $\sqcup$ degenerate to the usual infimum and supremum operators, $\wedge$ and $\vee$, in a complete lattice, respectively, the presented generalized fuzzy morphological erosion and fuzzy morphological dilation will be consonant with the existing grey-scale morphological erosion and dilation [4].

In particular, if $\mathbf{L} = [0,1]$, $\sqcap = \wedge = \inf$, and $\sqcup = \vee = \sup$, then the generalized fuzzy morphological operators will reduce to the traditional fuzzy morphological operators.

Theorem 3. *Let I be an implication, and C be a commutative and associative conjunction on $\mathbf{L}$, if $\sqcap$ is a commutative and associative extended operator of C, then (I, C) is an adjunction on $\mathbf{L}$ if and only if $(\varepsilon_G, \delta_G)$ is a fuzzy adjunction for any $G \in \mathbf{L}_E$.*

Proof. $\Rightarrow$: For all $F, G, H \in \mathbf{L}_E$,

$$\lfloor F \leq \varepsilon_G(H) \rfloor = \sqcap_{x \in E} I(F(x), \varepsilon_G(H)(x))$$
$$= \sqcap_{x \in E} I(F(x), \sqcap_{y \in E} I(G(y-x), H(y)))$$
$$= \sqcap_{x \in E} \sqcap_{y \in E} I(F(x), I(G(y-x), H(y)))$$
$$= \sqcap_{x \in E} \sqcap_{y \in E} I(C(G(y-x), F(x)), H(y))$$
$$= \sqcap_{y \in E} \sqcap_{x \in E} I(C(G(y-x), F(x)), H(y))$$
$$= \sqcap_{y \in E} I(\sqcup_{x \in E} C(G(y-x), F(x)), H(y))$$
$$= \sqcap_{y \in E} I(\delta_G(F)(y), H(y))$$
$$= \lfloor \delta_G(F) \leq H \rfloor.$$

$\Leftarrow$: For every $a, s, t \in \mathbf{L}$, define constant functions $G \equiv a$, $F \equiv s$ and $H \equiv t$, then for any $x \in E$, $\delta_G(F)(x) = \sqcup_{y \in E} C(G(x-y), F(y)) = C(a,s)$ and $\varepsilon_G(H)(x) = \sqcap_{y \in E} I(G(y-x), H(y)) = I(a,t)$. By using the fuzzy adjunction of $(\varepsilon_G, \delta_G)$, $\lfloor C(a,s) \leq t \rfloor = \lfloor s \leq I(a,t) \rfloor$. Thus $C(a,s) \leq t \iff s \leq I(a,t)$.

Besides the above statements, generalized fuzzy morphological operators share the following algebraic properties, whose proofs are omitted.

Proposition 7. *Let C be a conjunction satisfying $C(\mathbf{1}, s) = s$ for every $s \in \mathbf{L}$, then for any $G \in \mathbf{L}_E$ with $G(o) = \mathbf{1}$, δ_G is extensive, and ε_G is anti-extensive. Furthermore, $\varepsilon_G \leq \alpha_G \leq \mathrm{id} \leq \beta_G \leq \delta_G$, and $\varepsilon_G(\bar{\mathbf{r}}) = \alpha_G(\bar{\mathbf{r}}) = \beta_G(\bar{\mathbf{r}}) = \delta_G(\bar{\mathbf{r}}) = \bar{\mathbf{r}}$ for any $r \in \mathbf{L}$.*

Proposition 8. *All of the fuzzy morphological dilation δ_G, erosion ε_G, closing $\beta_G = \varepsilon_G \delta_G$, and opening $\alpha_G = \delta_G \varepsilon_G$ are translation invariant for any $G \in \mathbf{L}_E$.*

Proposition 9. *If $G \in \mathcal{P}(E)$, then for any $F \in \mathbf{L}_E$,*

$$\delta_G(F)(x) = \sqcup_{y \in \check{G}_x} F(y) \quad \text{and} \quad \varepsilon_G(F)(x) = \sqcap_{y \in G_x} F(y), \tag{19}$$

where $\check{G} = \{-a \mid a \in G\}$ is the reflection of G.

Proposition 10. *For $\lambda \in \mathbf{L}$ and $x \in E$, let $G = x_\lambda$, then for any $F \in \mathbf{L}_E$, $\delta_{x_\lambda}(F)(y) = C(\lambda, F(y-x))$ and $\varepsilon_{x_\lambda}(F)(y) = I(\lambda, F(x+y))$. If $C(\mathbf{1}, s) = s$ in addition, then $\alpha_{x_1}(F) = \beta_{x_1}(F) = F$, meaning that every $F \in \mathbf{L}_E$ is both $x_1 = \{x\}$ open and $x_1 = \{x\}$ close for any $x \in E$.*

5 Conclusions

This paper extends a fuzzy conjunction in fuzzy logic to a general operator that is compatible with the infimum in a complete lattice. In virtue of this operator, the concept of a fuzzy adjunction is developed to determine a pair of generalized fuzzy morphological dilation and erosion. A new framework of generalized fuzzy morphology by means of such extensions is therefore formed. The algebraic properties of the generalized fuzzy morphological operators show that the definition of grey-scale morphology based on fuzzy logic and the often used models of fuzzy morphology fit well in with the new framework of fuzzy morphology.

Acknowledgement

This research was supported in part by the China Postdoctoral Science Foundation under Grant 2004035334 and the Multi-discipline Scientific Research Foundation of Harbin Inst. Tech., HIT.MD2001.24.

References

1. Bloch, I., Maître, H.: Fuzzy mathematical morphologies: a comparative study. Pattern Recognition **28** (1995) 1341-1387
2. Chatzis, V., Pitas, I.: A generalized fuzzy mathematical morphology and its application in robust 2-D and 3-D object. IEEE Transactions on Image Processing **9** (2000) 1798-1810
3. Cornelis, C., Van der Donck, C., Kerre, E.: Sinha – Dougherty approach to the fuzzifcation of set inclusion revisited. Fuzzy Sets and Systems **134** (2003) 283-295
4. Deng, T.-Q., Heijmans, H.: Grey-scale morphology based on fuzzy logic. Journal of Mathematical Imaging and Vision **16** (2002) 155-171
5. Matheron, G.: Random Sets and Integral Geometry. John Wiley & Sons, New York (1975)
6. Nachtegael, M., Kerre, E.: Connections between binary, grey-scale and fuzzy mathematical morphology. Fuzzy Sets and Systems **124** (2001) 73-85
7. Nguyen, H., Walker, E.: A First Course in Fuzzy Logic. 2nd edn. Chapman & Hall/CRC, Boca Raton, Florida (1999)
8. Serra, J.: Image Analysis and Mathematical Morphology. Academic Press, London (1982)
9. Sinha, D., Dougherty, E.: A general axiomatic theory of intrinsically fuzzy mathematical morphologies. IEEE Transactions on Fuzzy Systems **3** (1995) 389-403
10. Soille, P.: Morphological Image Analysis. 2nd edn. Springer-Verlag, Heidelberg (2003)
11. Sternberg, S.: Grayscale morphology. Comupter Vision, Graphics and Image Processing **35** (1986) 333-355

Signature Verification Method Based on the Combination of Shape and Dynamic Feature

Yingna Deng, Hong Zhu, Shu Li, and Tao Wang

Department of Automation and Information Engineering,
Xi'an University of Technology, 710048 Xi'an China
Tel: +86-029-82312168
dengyingna@126.com

Abstract. A method based on multi-feature and multi-stage verification is proposed in this paper. The direction and texture features obtained by Gabor filter are used for the first-stage verification. Then the sham dynamic features representing the writing force are used for the second-stage verification. Considering the number of samples for training is limited, the support vector machine (SVM) is selected as the classifier. The experimental results show that this method can not only increase the accuracy, but also satisfy the requirement of speed in real-time system as well.

1 Introduction

For signature verification, the key issue is to extract effective features, which can represent the invariability between various genuine signatures. These features can be divided into two categories: shape features, which represent shape information, and sham dynamic features, which represent the writing force. The common features selected in many works, such as moment[1], the ratio of signature height to width, center of gravity, outline, ratio of pixel to area and so on, cannot reflect the signature shape veritably. Stroke analysis[2] is also commonly used to extract shape features. Because the connection state of adjacent strokes is variable, the verification accuracy may not be satisfied.

This paper proposes a method that uses the directional texture information obtained by Gabor filters as shape features, combining with the sham dynamic features, to make a decision through multi-stage verification.

2 Features Extraction

2.1 Direction Feature Extraction

Chinese characters generally consist of horizontal, vertical, and slant stroke segments. A person's writing style is usually consistent; the direction of each stroke also displays certain consistence. So the directional information of signatures can reflect the characteristic of one's signatures.

Directional information enhancement. In this paper, the directional information is extracted by Gabor filter. Gabor filter is generally formed as following in space domain:

$$G(x, y, f, \theta) = \exp\left\{\frac{-1}{2}\left[\frac{x'^2}{\delta_{x'}^2} + \frac{y'^2}{\delta_{y'}^2}\right]\right\} \cos(2\pi f x')$$

$$x' = x\sin\theta + y\cos\theta$$

$$y' = x\cos\theta - y\sin\theta$$

(1)

where f is the frequency of the sinusoidal plane wave along the direction θ from the x-axis, and $\delta_{x'}$, $\delta_{y'}$ are the variances of the Gaussian model along x', y' axis respectively, which are both constant.

Here define the filter frequency f as a constant and set the direction θ as four different values with respect to x-axis, which are $0°$, $45°$, $90°$ and $135°$. Fig.1 shows the filtered results.

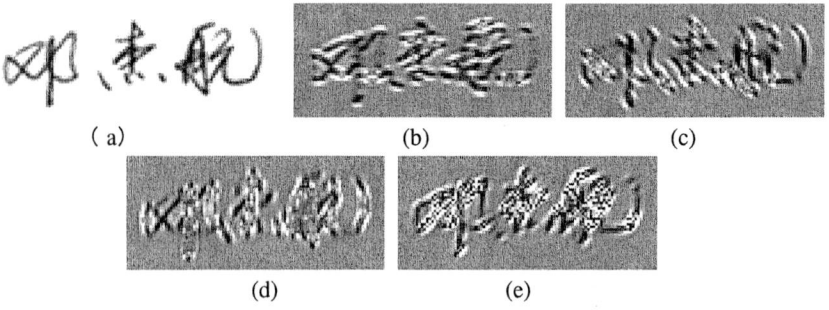

Fig. 1. Examples of filtered images. (a) is the original gray image.(b) is the filtered image with $\theta = 0°$, (c) is the filtered image with $\theta = 45°$, (d) is the filtered image with $\theta = 90°$ and (e) is the filtered image with $\theta = 135°$.

2.2 Sham Dynamic Feature Extraction

This paper deals with the signatures that have already been written on paper, so the dynamic information cannot be obtained. However, the writing force and habit of a certain person could be reflected by his handwriting, therefore, this paper extracts the sham dynamic information for verification.

Low gray level feature extraction. The region with low gray level is defined as low gray level region. Supposing the signature is written on white paper with dark pen, the low gray level region reflects the areas with high pressure, such as weight strokes, pause strokes and stroke crossings. The low gray level image is extracted from a gray image by a threshold. The threshold in this paper is defined as follows:

$$LPR = S_{min} + 0.25 \times (S_{max} - S_{min}) \qquad (2)$$

where S_{min}, S_{max} is the maximal and minimal pixel value separately.

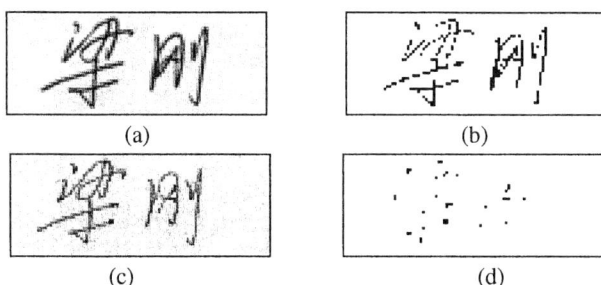

Fig. 2. examples of low gray level region images. (a) is an original gray image, (b) is the low gray level region of (a),(c) is a traced forgery of (a), (d) is the low gray level region of (c).

Gray scale distribution feature extraction. Statistical histogram, which represents the gray scale distribution, can reflect a person's writing characteristics. This feature vector all together with the low gray level feature, compose the sham dynamic feature vector.

3 Signature Verification

Because the number of samples for classifier to learn is limited, this paper uses SVM to perform verification, and takes the radial basis function as its kernel function, as follows:

$$K(x, x_i) = \exp\left\{ -\frac{|x - x_i|^2}{256\delta^2} \right\} \qquad (3)$$

where x is the input vector, δ is the variance. Thus, the classifier is designed; support vectors $x_1, x_2, \ldots x_s$ and the weight w_i can be obtained through training. The output decision maker y is as follows, where b is the threshold.

$$y = \text{sgn}(\sum_{i=1}^{s} w_i K(x_i, x) + b) \qquad (4)$$

This paper carries out a two-stage verification to make decision. For an input image, extract its direction features firstly. If it is justified as forgery, the final output decision is forgery; if it is justified as genuine, then extract the sham dynamic features, to carry out the second stage verification, and the decision of the second stage is taken as the final decision.

4   Experimental Result Analysis

For each person, we have collected 40 genuine and 80 counterfeit signature samples, which were captured as digital images by scanner at 300 dpi. 20 genuine signatures and 20 forgeries are used for training; the last 20 genuine and 60 counterfeit signatures are used for test.

In this paper, the genuine signatures are written by 10 different people with the same pen, so there are 800 test samples in all. The average accuracy rate is above 94 %. Further more, it is effective for traced forgeries. Tab.1 shows the comparison of different methods.

Table 1. Accuracy rate comparison of different methods

Algorithm	Method 1[3]	Method 2[4]	Method 3[5]	Proposed Method
Accuracy rate	78.75%	75%	90.9%	94.1%

References

1. Hwei-Jen Lin, Fu-Wen Yang and Shwu-Huey Yen, Off-Line Verification for Chinese Signatures, International Jounal of Computer Processing of Oriental Language, Vol.14,No.1 (2001) 17-28
2. B. Fang, Y.Y. Wang, C.H. Leung and K.W. Tse, Offline Signature Verification by the Analysis of Cursive Strokes, International Journal of Pattern Recognition and Artificial Intelligence, Vol.15,No.4 (2001) 659-673
3. Wang Guang-song, Off-line Chinese Signature Verification, Master's thesis, Hua Qiao University (2004)
4. Yie Yan, The Investigation of Off-Line Signature Verification Based on Sham Dynamic Feature Extraction, Master's thesis, Shan Dong University(2001)
5. Katsuhiko Ueda, Investigation of Off-Line Japanese Signature Verification Using a Pattern Matching. ICDAR (2003)

Study on the Matching Similarity Measure Method for Image Target Recognition

Xiaogang Yang, Dong Miao, Fei Cao, and Yongkang Ma

303 Branch, Xi'an Research Inst. of High-tech Hongqing Town 710025 China
{doctoryxg,caofeimm}@163.com

Abstract. The matching similarity measures that can be used in image target recognition are surveyed and a novel similarity measure is proposed in this paper. Two basic factors that affect the image matching performance and the merits and faults of two common types of image matching algorithm are firstly analyzed. Then, based on the systematic study of similarity measures, image matching projection similarity measure is defined by simplify the classical normalized correlation measure. An example of the application of the proposed matching similarity measure in image target recognition and position is given at the end of this paper; the experimental results show its feasibility and effectivity.

1 Introduction

Recent years, the rapid development of image processing and computer version technology has led image target recognition to a very hot study topic in the area of pattern recognition, especially in military target recognition [1, 2, 3, 4]. The performance of image target recognition algorithm plays a decisive effect on the recognition accuracy. Nowadays, image matching algorithm has be a popular method in image target recognition [4, 5, 6]. The similarity measurement method is a very important factor, which directly affects the accuracy and efficiency of the image matching algorithm [7, 8]. In this paper, based on the analysis of the principle of image matching algorithm, we propose a novel matching similarity measure method.

2 Analysis on Image Matching Algorithm

There are two element factors in the image matching based target recognition algorithm. One is measure information, which means how to describe the image information efficiently by using its grayscale or image feature. The other is measure criterion, means which similarity measure is adopted in the algorithm.

Matching algorithm can be divided into two classes according to different image measure information that matching algorithm used. One is matching algorithm based on image grayscale, such as AD, MAD and NProd algorithm and so on [5, 6]. These algorithms have a strong anti-noise ability and high matching probability and precision when the images for recognition have little grayscale and geometric distortions. The main fault of these algorithms is their matching speed is relatively slow. The second class is matching algorithm based on image feature. Such as image edge feature, image invariant moment, image texture, image corner and point feature and wavelet feature etc. [5, 6, 7, 8]. This class algorithm has a strong anti-grayscale and geometric distortions ability, but weak to noise disturbance.

The aim of image matching is to determine the deference of two images about one target, which may be acquired by two different sensors at different time. Usually, the known image information is called reference image, while the image to be recognized is called real-measurement image or target image. For the differences of acquisition means between the two images, there always a lot of grayscale distortions or other distortions exist, thus it is impossible to find a reference image which is absolutely same to the target image. So, we use similarity measures to compare the target image and every reference image. According to the geometric relationship between the two image information vectors in the Euclidean space [5], the similarity measures can be divided into two classes [7, 8, 9].

a. Distance similarity measures: 1) Absolute Difference (AD), 2) Mean Absolute Difference (MAD), 3) Square Difference (SD), 4) Mean Square Difference (MSD).

b. Correlation similarity measures: 1) Product correlation (Prod), 2) Normalized Product correlation (NProd).

3 Projection Similarity Measure

As to the NProd measure method [5], we know that during the matching and comparing process, all NProd coefficients have the invariant item $\|\vec{y}\|_2$, so we can omit it in the implement of our matching algorithm. We simplified the NProd measure method and got a new similarity measure:

$$R_{Proj} = \frac{\vec{x} \cdot \vec{y}}{\|\vec{x}\|_2} = \frac{\sum_{i=0}^{n-1}\sum_{j=0}^{n-1} x_{i,j} y_{i,j}}{(\sum_{i=0}^{n-1}\sum_{j=0}^{n-1} x_{i,j}^2)^{1/2}} \quad (1)$$

Where $\vec{x}$ and $\vec{y}$ denote the reference image and target image information vector respectively, $x_{i,j}$ and $y_{i,j}$ denote their element value, n represents their size, R_{Proj} denotes the simplified NProd coefficient. The geometric meaning of (1) is the projection of vector $\vec{y}$ on vector $\vec{x}$. The bigger the projection value is, the more similar the two image are (the maximum value is $\|\vec{y}\|_2$). So, we define the simplified NProd coefficient R_{Proj} as projection similarity measure. By the magnitude of R_{Proj} value, we can judge the similarity between the target image and the reference image, and then recognize the target image and make a corresponding decision.

4 Application Example and Analysis

To demonstrate the feasibility and effectivity of the proposed similarity measure, we take a satellite photo of certain airport as an airplane target recognition and detection example. Figure 1 gives the known airplane reference image. Figure 2 shows the

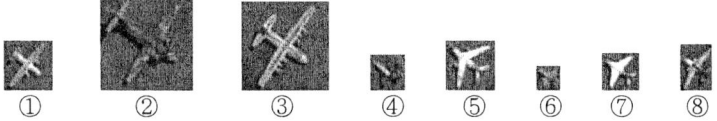

Fig. 1. The reference information images

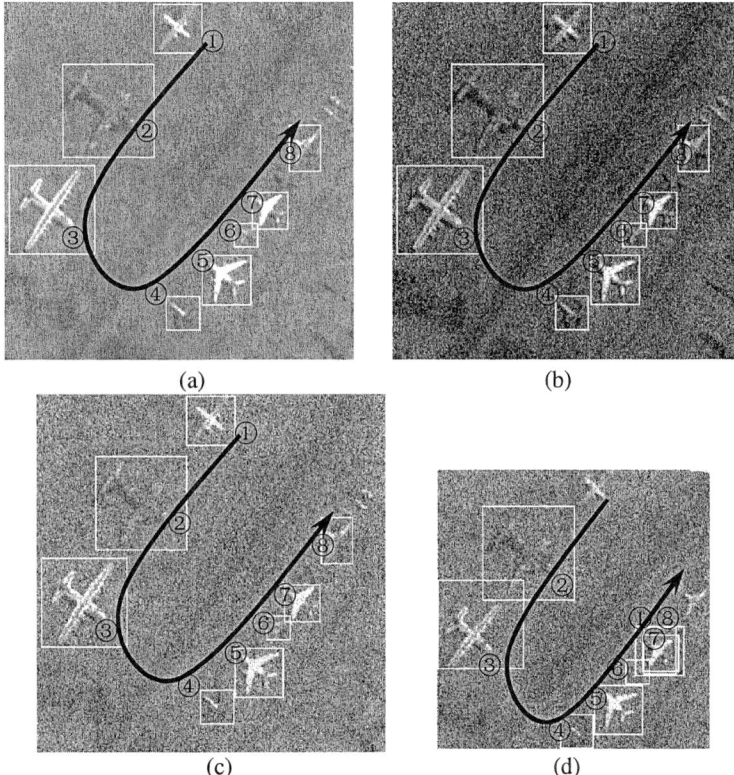

Fig. 2. The recognition and position results under different disturbance

recognition and position results when there have grayscale distortion, noise disturbance and geometric affine distortion [9] in the airport image.

Where Fig.2(a) represents the recognize results of target image with grayscale distortion; Fig.2(b) represents noise disturbance; Fig.2(c) represents grayscale distortion and noise disturbance; Fig.2(d) represents grayscale, affine distortion and noise disturbance.

The principle of the experiment is: based on the image grayscale information, by using the proposed projection measure to implement recognition and position of the airplane image target under different disturbance. The experiment process is similar to scene matching position algorithm [5]. From the experimental results we can see that the proposed similarity measure can accomplish image target recognition efficiently. When we use image grayscale as the measure information in our matching recognition, the projection measure has a good restrain ability to noise disturbance and grayscale distortion, but be sensitive to geometric distortion just like the NProd measure. To improve the anti-geometric distortion ability of the measure, we can use image feature as the measure information, such as image invariant moment, image corner and point feature and so on.

5 Conclusion

In this paper, we defined a novel similarity measure. For that the measure is simplified from the classical NProd measure, it inherits the merit of NProd measure, and reduce the calculation capacity, it has a practical significance in the image matching application area. While similarity measure is only one important factor of image matching algorithm, the selection of measure information, which can describe the image content efficiently, is also a very prospective way to improve the recognition accuracy. The research of practical image feature for image target searching, recognition, classification, position and tracking is our further work.

References

1. Yang, l.: Study on the approaches of two dimension target detection and reorganization. Huazhong Univ.of Sci.& Tech, (1997)1-15
2. Liu,F.H., Ren,S.H., and Zhu,G.M.: Real-Time Image Target Recognition System. Journal of Beijing Institute of Technology,17(6),(1997)717-722
3. Dai,J.Z., Leng,H.Y.: New algorithm study and application for target recognition. Infrared and laser engineering,30(2),(2001)108-111.
4. Matthias O. Franz, Bernhard: Scene-based homing by image matching. Biol. Cybern. 79, (1998)191-202
5. Shen,Z.K.: The Technology of Image Matching. National University of Defense Technology Press, (1987)217-229
6. Zou,L.J., Liu,J.H.: Application of model-matching technology on target recognition in rosette scanning sub-image. Optical Technique,27(2),(2001)166-169
7. Remco C. Veltkamp: Shape Matching: Similarity Measures and Algorithms. Dept. Computing Science, Utrecht University, (2001)
8. B. G"unsel and A. M. Tekalp: Shape similarity matching for query-by-example. Pattern Recognition, 31(7),(1998)931–944
9. M. Hagedoorn, M. Overmars, and R. C. Veltkamp, A new visibility partition for affine pattern matching, In Discrete Geometry for Computer Imagery: 9th International Conference Proceedings DGCI 2000, LNCS 1953, Springer, (2000) 358–370

3-D Head Pose Estimation for Monocular Image

Yingjie Pan, Hong Zhu, and Ruirui Ji

Department of Automation and Information Engineering,
Xi'an University of technology 710048, Xi'an, China   Tel: +86-029-82312168
pan_yingjie11@yahoo.com.cn   zhuhong@xaut.edu.cn
hezirui423@163.com

Abstract. 3-D head pose estimation plays an important role in many applications, such as face recognition, 3-D reconstruction and so on. But it is very difficult to estimate 3-D head pose only from a single monocular image directly without other auxiliary information. This paper proposes a new human face pose estimation algorithm using a single image based on pinhole imaging theory and 3-D head rotation model. The key of this algorithm is to obtain 3-D head pose information based on the relations of projections and the positions changing of seven facial points. Experiments show the proposed method has good performance in both accuracy and robustness for human face pose estimation using only a single monocular image.

1 Introduction

Human face pose estimation is a key problem in many practical applications of computer vision and image processing. Some research achievements have been made on this problem, however, most of these solutions need additional information such as front face view, stereo vision, 3D head model and so on. Considering that it is too hard and too expensive to get these additional information in practical application system, a novel algorithm to analyze human face pose directly from a single image should be designed.

This paper proposes a new human face pose estimation approach from a single face image under arbitrary pose using face geometry characteristic based on pinhole imaging theory[1,2] and 3-D head rotation model.

2 Human Face Pose Estimation

Human face pose estimation is to calculate the rotation angles of the head with respect to front face along the X, Y, and Z-axis respectively. As human head is a non-rigid object in three-dimensional space, the face picture presents us only two-dimensional information, it can be considered from the imaging theory that face image is formed by human head project onto X-Y plane along Z-axis. So the image plane coincides with the X-Y plane. We can use α、β、γ (pitch, yaw and roll, respectively) to denote the three rotation angles of the head about the X, Y and Z-axis respectively.

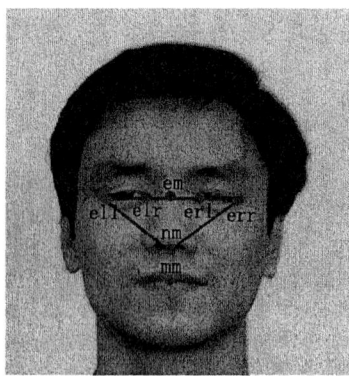

Fig. 1. Facial points

From anthropometry, as Fig. 1 shows, human face is basically bilateral symmetry, two eyes are the same both in size and shape, and the four eye corners are approximately collinear in three-dimension. While γ=0, i.e. the rotation angle around Z-axis is zero, this line is parallel to X-Y plane, and also parallel to X-axis of the face image. Furthermore, in order to describe human head position, the symmetric points (em, nm and mm shown in Fig.1) of eyes, mouth and nose also must be given, as shown in Fig. 1. According to the position change of the seven points listed above, we can calculate the pose of human face directly.

Since human face pose can be described as three rotation angles of the head about the X, Y and Z-axis respectively in three-dimension, we can get human face pose from these angles.

2.1 Roll (The Rotation Angle of Z-Axis, γ)

As Fig.2 shows, roll is the rotation angle of Z-axis, and roll (γ) can be straightforward recovered by the eye corners in the image.

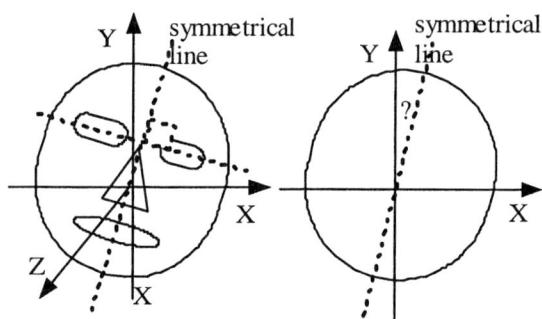

Fig. 2. The rotation angle of Z-axis γ

$$\gamma = (\gamma_1 + \gamma_2 + \gamma_3)/3 \tag{1}$$

$$\gamma_1 = \arctan(\frac{y_{ell} - y_{elr}}{x_{ell} - x_{elr}}) \tag{2}$$

$$\gamma_2 = \arctan(\frac{y_{erl} - y_{err}}{x_{erl} - x_{err}}) \tag{3}$$

$$\gamma_3 = \arctan(\frac{(y_{ell} + y_{elr})/2 - (y_{erl} + y_{err})/2}{(x_{ell} + x_{elr})/2 - (x_{erl} + x_{err})/2}) \tag{4}$$

Set (x_{ell}, y_{ell}), (x_{elr}, y_{elr}), (x_{erl}, y_{erl}), and (x_{err}, y_{err}) denote left corner of left eye, right corner of left eye, left corner of right eye, right corner of right eye respectively.

2.2 Yaw (The Rotation Angle of Y-Axis β)

As Fig.3 shows, yaw is the rotation angle of Y-axis. From anthropometry, it can be known that the four eye corners are collinear and the two eyes have the same size, that is: ‖ Ell−Elr ‖=‖ Erl−Err ‖.

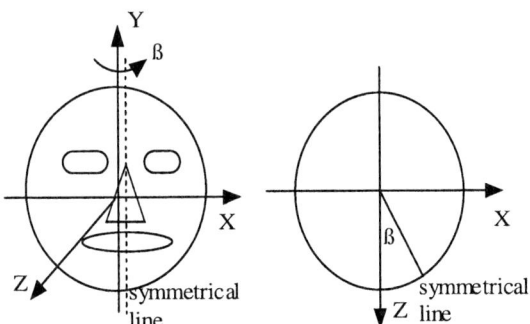

Fig. 3. The rotation angle of Y-axis β

Based on pinhole imaging theory, Fig.4 shows the principal of calculating the Yaw-β. From the projective invariance of the cross-ratios, the following can be derived:

$$I_1 = \frac{(x_{erl} - x_{err})(x_{ell} - x_{elr})}{(x_{ell} - x_{erl})(x_{elr} - x_{err})} = \frac{D^2}{(2D_1 + D)^2} \tag{5}$$

$$D_1 = DQ/2 \tag{6}$$

$$Q = \frac{1}{\sqrt{I_1}} - 1 \tag{7}$$

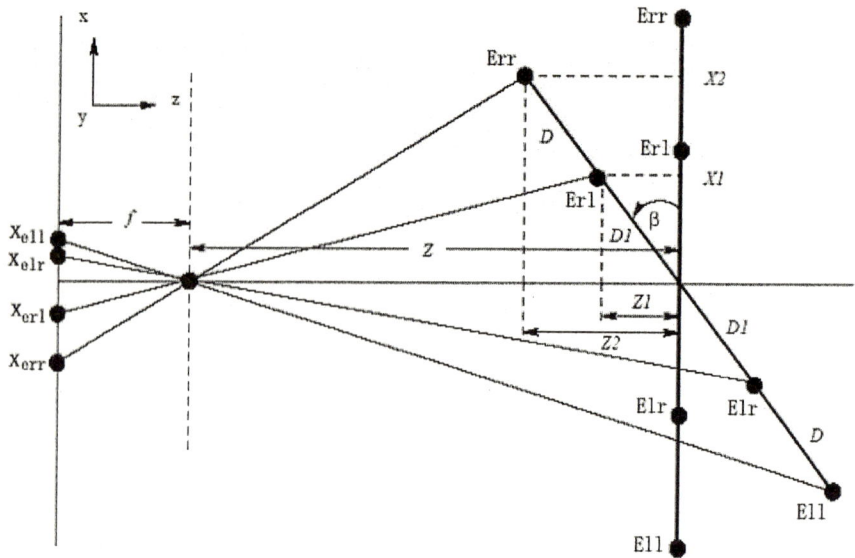

Fig. 4. The principle for calculating β

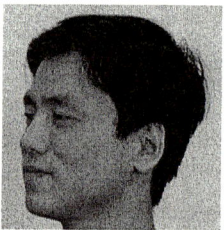

Fig. 5. Head rotation

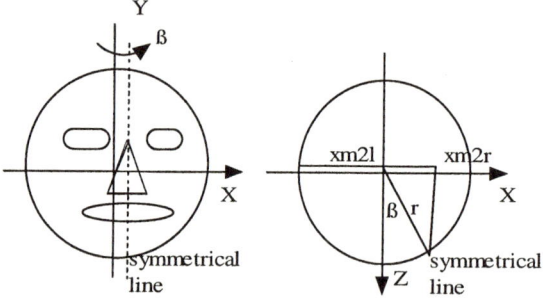

Fig. 6. 3-D head rotation model

$$M = \frac{-1}{2+Q} \tag{8}$$

$$x_{elr} = \frac{fX_1}{Z+Z_1} = \frac{fD_1 \cos\beta_0}{Z + D_1 \sin\beta_0} \tag{9}$$

$$x_{ell} = \frac{fX_2}{Z+Z_2} = \frac{f(D+D_1)\cos\beta_0}{Z + (D+D_1)\sin\beta_0} \tag{10}$$

$$x_{erl} = \frac{fX_1}{Z-Z_1} = \frac{fD_1 \cos\beta_0}{Z - D_1 \sin\beta_0} \tag{11}$$

$$x_{err} = \frac{fX_2}{Z-Z_2} = \frac{f(D+D_1)\cos\beta_0}{Z - (D+D_1)\sin\beta_0} \tag{12}$$

$$Z = Z_1(\frac{fX_1}{x_{elr}Z_1} - 1) = Z_1 S \tag{13}$$

S can be got from equation (14).

$$\frac{x_{ell} - x_{elr}}{x_{erl} - x_{err}} = -\frac{(S-1)(S-(1+2/Q))}{(S+1)(S+(1+2/Q))} \tag{14}$$

$$u = \frac{(x_{ell} - x_{elr})(x_{erl} - x_{err})M(x_{elr} - x_{erl}) - M^2(x_{ell} - x_{err})(x_{elr} - x_{erl})^2}{(x_{ell} - x_{elr})(M(x_{elr} - x_{erl}) - (x_{erl} - x_{err}))} \tag{15}$$

Where f is the focus of the camera, D, D_1, Z, Z_1, Z_2, X_1, X_2 are distance parameters shown in Fig. 4. From formulation (5)-(15), the formulation to calculate β_0 can be got:

$$\beta_0 = \arctan\frac{f}{(S-1)u} \tag{16}$$

However, this method is influenced easily and badly by the point positions, especially great deviation will occur when $\beta>45^0$. For example, as Fig. 5 shows, β_0 calculated by this method is 40.53^0, the actual yaw is 60^0, and the error is 19.47^0. To solve this problem, 3-D head rotation model is introduced. Human head is approximately regarded as a round cylinder, because human face is bilateral symmetry, the symmetric line of face will be coincident with the center line of the cylinder in front human face view; and when human head rotates around Y-axis, there will be offset between the symmetric line and the center line, shown as Fig.6. With this offset, the rotation angle around Y-axis can be calculated.

At first, calculating the rotation angle around Y-axis at eyes, nose and mouth respectively according to the formulation followed:

$$\delta x = [\frac{lx}{2} - \min(xm2l, xm2r)] \tag{17}$$

$$\beta_x = acr\sin(\frac{\delta x}{lx/2}) \tag{18}$$

Where lx could be le, ln and lm, which correspond to the face width at eyes, nose and mouth points respectively; xm2l and xm2r could be em2l, em2r, nm2l, nm2r, mm2l and mm2r respectively, which is the distance from the symmetric points at eyes, nose and mouth to the left edge and right edge of face respectively.

Considering practical applications, the rotation angle around Y-axis could be set in the range of [-90°,+90°]. In the situation of heavy incline, such as β is near -90° or +90°, one eye is covered partly or totally, the symmetric points are near the face contour, the value of min(xm2l,xm2r) is near 0, and β_x is near ±90°.

Then get the mean value of yaws at eyes, nose and mouth according to the formulation as follows:

$$\beta_1 = (\beta_e + \beta_n + \beta_m)/3 \tag{19}$$

Finally, yaw is determined with different formulations in different cases.

$$\beta = \begin{cases} (\beta_0 + \beta_1)/2 & |\beta_1| < 45^0 \\ \beta_1 & 45^0 \leq |\beta_1| \leq 90^0 \end{cases} \tag{20}$$

From 45° to 90°, the nonlinearity is more obvious, the value get from formulation (19) is more exact than that from formulation (16) and more close to the practical value, and the error is less than 5°. So we only use formulation (19) here to be the final answer for this kind situation. How to solve the nonlinearity should be studied further, such as combining these two cases using different weight coefficients.

2.3 Pitch (The Rotation Angle of X-Axis α)

As Fig. 7 shows, pitch is the rotation angle of X-axis.

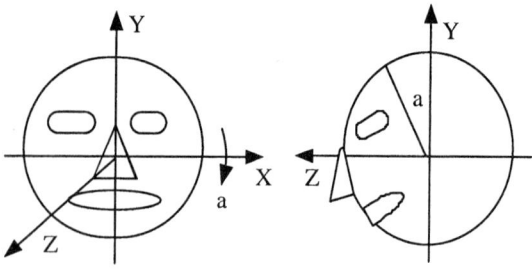

Fig. 7. The rotation angle of Y-axis α

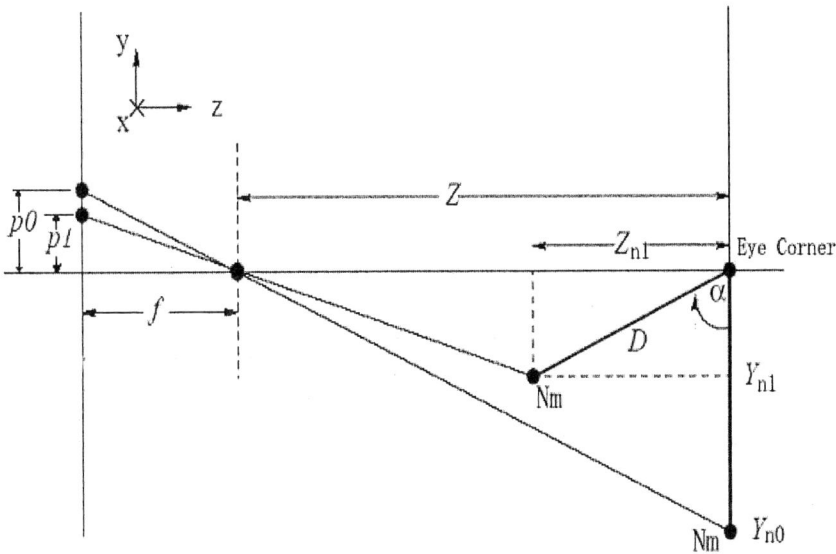

Fig. 8. The principle for calculating α

Supposing p0 denotes the projected length of nose when it is in a front face image, p1 denotes the observed length of nose at the unknown pitch, (Xn0, Yn0, Zn0) and (Xn1,Yn1, Zn1) denote the 3-D coordinates of the nose in these two cases respectively, Set D2 denotes the 3-D vertical distance between the eyes and the tip of the nose. From the projection shown in Fig. 8, the following can be known:

$$\frac{f}{Z} = \frac{p_0}{Yn0} = \frac{p_0}{D_2} \qquad (21)$$

$$\frac{f}{Z - Zn1} = \frac{p_1}{Yn1} = \frac{p_1}{D_2 \cos \alpha} \qquad (22)$$

From formulation (21) and (22), the pitch could be calculated using the following formulation:

$$\alpha = \arcsin E \qquad (23)$$

Where:

$$E = \frac{f}{p_0(p_1^2 + f^2)}[p_1^2 \pm \sqrt{p_0^2 p_1^2 - f^2 p_1^2 + f^2 p_0^2}] \qquad (24)$$

$$p_0 = w \frac{Ln}{Le} \qquad (25)$$

Let Ln denote the average length of the nasal bridge, and Le denote the average length of the eye fissure. The value of Ln/Le could be referenced suitable for north China people from reference [2].

W denotes the eye fissure observed in front facial image. When human head rotates around X-axis, this length will change, if pitch is little, the longer eye fissure observed of the two eyes is used to substitute the eye fissure in front facial image, otherwise using the longer eye fissure divided by cos(β); when pitch is close to 90^0, there will be great error, it can be computed as follows:

$$w = \begin{cases} \max(w_r, w_l) & \|\beta\| < 45^0 \\ \max(w_r, w_l)/\cos\beta & \|\beta\| > 45^0 \end{cases} \quad (26)$$

Where w_l, w_r denote the left eye fissure and right eye fissure respectively.

3 Experimental Results

Considering practical applications, the degree of these angles is limited in $[-90^0, 90^0]$. The experimental camera is nikon-950, with the focus 30 mm. Fig. 9 shows various human face poses and the degrees estimated, the results are very close to the real pose.

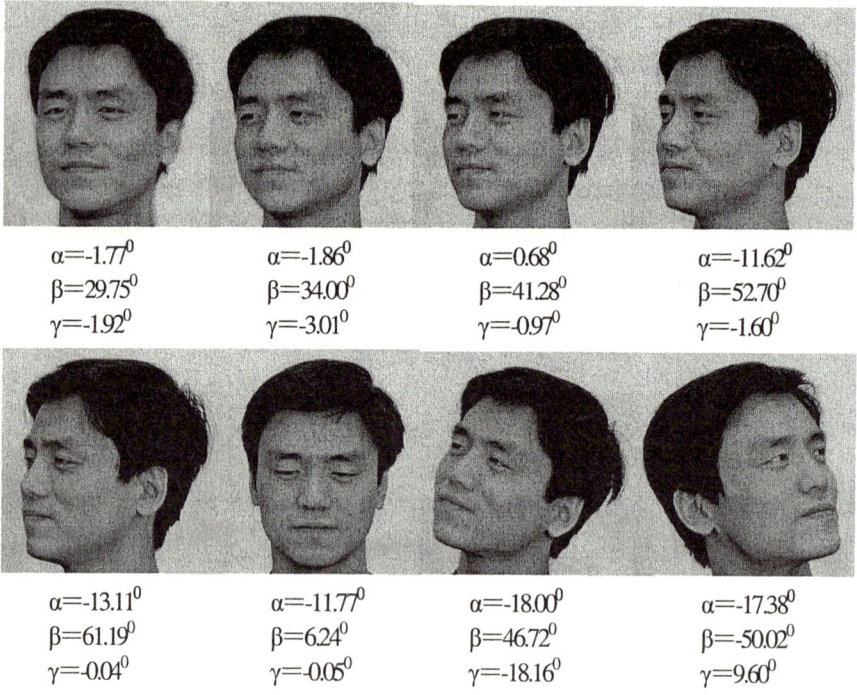

Fig. 9. The result of pose estimation

The average error on three rotations for 25 different pose images through three times experiments is $(\Delta\alpha, \Delta\beta, \Delta\gamma) = (5.12^o, 4.78^o, 1.22^o)$, and when $45^o < \beta < 90^o$, $\Delta\beta$ is far more less than the results done by [2] which $\Delta\beta$ is usually more than 10°. The errors during the course of experiment are mainly brought from the following reasons:

1. Image error that is induced by the location error of feature points.
2. Model error: because of the difference between individual face and facial model which is based on the statistical information and just expresses the characteristic of face approximately. This kind of errors can be reduced by the Expectation-Maximization (EM) algorithm which is proposed to update the facial model parameters according to individual face characteristic.

It should be noted that the experimental results are accurate and very close to real pose, especially the rotation angles around Y-axis and Z-axis. And this method is effective and practical for monocular image.

4 Conclusions

This paper presents a new approach to estimate human face pose quickly and exactly using only seven facial points based on pinhole imaging theory and 3-D head rotation model from monocular image, and it is especially effective for roll and yaw. This method can successfully solve the problem of lack of additional information in estimating human face pose from monocular image.

References

1. T. Horprasert, Y. Yacoob, L. S. Davis, "Computing 3-D head orientation from a monocular image sequence", Proceedings of the 2nd International Conference on Automatic Face and Gesture Recognition (1996)
2. Huaming Li, Mingquan Zhou, and Guohua Geng, "Facial Pose Estimation Based on the Mongolian Race's Feature Characteristic from a Monocular Image", S.Z. Li et al. (Eds.): Sinobiometrics 2004 LNCS 3338
3. T. Cootes, G. Edwards, and C. Taylor. Active appearance models. IEEE PAMI, 23(6), 2001.

The Speech Recognition Based on the Bark Wavelet Front-End Processing

Xueying Zhang, Zhiping Jiao, and Zhefeng Zhao

College of Information Engineering,
Taiyuan University of Technology, 030024 Taiyuan, Shanxi, P.R. China
zhangxy@tyut.edu.cn

Abstract. The paper uses Bark wavelet filter instead of the FIR filter as front-end processor of speech recognition system. Bark wavelet divides frequency band based on critical band and its bandwidths are equal in Bark domain. By selecting suitable parameters, Bark wavelet can overcome the disadvantage of dyadic wavelet and M-band wavelet dividing frequency band based on octave. The paper gave the concept and parameter setting method of Bark wavelet. For signals that are filtered by Bark wavelet, ZCPA features with noise-robust are extracted and used in speech recognition. And recognition network uses HMM. The results show the recognition rates of the system in noise environments are improved.

1 Introduction

Wavelet transform has much better characteristics than Fourier transform. Especially, its multiresolution property has been widely used in image and speech processing. However, most of wavelet transforms, such as dyadic wavelets, wavelet packets transform , their frequency band divisions are based on octave. Such frequency band divisions can not meet the critical frequency band division requirement of speech recognition systems. This paper uses a new Bark wavelet to meet above demand. It is used in front-end processing as filterbank instead of original FIR filters for improving the system performance. ZCPA (Zero-Crossings with Peak Amplitudes) features [1] are used as recognition features and recognition network uses HMM. The experiment results show that Bark wavelet filters are superior to common FIR filters in robust speech recognition system.

2 Bark Wavelet Theory

2.1 The Relation of Bark Frequency and Linear Frequency

The audio frequency band from 20Hz to 16kHz can be divided into 24 bands. The perception of the human auditory system to speech frequency is a nonlinear mapping relation with actual frequency. Traunmular presents the relation of the linear frequency and Bark frequency [2].

$$b = 13\arctan(0.76f) + 3.5\arctan(f/7.5)^2 \tag{1}$$

Where b is Bark frequency, and f is the linear frequency.

2.2 Bark Wavelet

The basic thought of constructing Bark wavelet is as followings.

Firstly, because of the same importance of the time and frequency information in the speech analysis, wavelet mother function selected should satisfy the demand that product of time and bandwidth is least. Secondly, for being consistent with conception of the frequency group, mother wavelet should have the equal bandwidth in the Bark domain and is the unit bandwidth, namely 1 Bark. Thus, the wavelet function is selected as $W(b) = e^{-c_1 b^2}$ in the Bark domain. The constant c_1 is 4ln2. It is easy to prove that Bark wavelet transform can be reconstructed perfectly [2].

Supposing the speech signals analyzed is $s(t)$ and its linear frequency bandwidth satisfies $|f| \in [f_1, f_2]$, and corresponding Bark frequency bandwidth is $[b_1, b_2]$. Thus wavelet function in Bark domain can be defined as:

$$W_k(b) = W(b - b_1 - k\Delta b), k = 0, 1, 2 \ldots K - 1 \tag{2}$$

Where, k is the scale parameter and $\Delta b = \frac{(b_2 - b_1)}{K-1}$ is step length. Thus (2) becomes

$$W_k(b) = e^{-4\ln 2(b - b_1 - k\Delta b)^2} = 2^{-4(b - b_1 - k\Delta b)^2}, k = 0, 1, 2 \ldots K - 1 \tag{3}$$

Then by substituting (1) into (3), the Bark wavelet function in the linear frequency can be described as:

$$W_k(f) = c_2 \cdot 2^{-4[13\arctan(0.76f) + 3.5\arctan(f/7.5)^2 - (b_1 + k\Delta b)]^2} \tag{4}$$

Where, c_2 is the normalization factor and can be obtained by (5):

$$c_2 \sum_{k=0}^{K-1} W_k(b) = 1, \quad 0 < b_1 \leq b \leq b_2 \tag{5}$$

In the frequency domain Bark wavelet transform can be expressed as:

$$s_k(t) = \int_{-\infty}^{\infty} S(f) \cdot W_k(f) \cdot e^{i2\pi f} df \tag{6}$$

Where, $S(f)$ is FFT of signal $s(t)$.

3 The Design of the Bark Wavelet Filter

16 Bark wavelet filter parameters are confirmed according to the auditory frequency bandwidth. Selecting K is 24. Reference [3] showed that by selecting 3 Bark units as 1 filter bandwidth, the recognition rates are best when SNR is higher than 20dB. So we can get the parameters of 16 Bark wavelet filters listed in Table 1. Fig. 1 is the frequency response of the 16 Bark wavelet filters. Fig. 2 is the scheme of the Bark wavelet filters used in the front-end processing. $W(f)$ is the Bark wavelet function and $\hat{s}(t)$ is the signal filtered by Bark wavelet filter.

Table 1. The parameter values of 16 Bark wavelet filters

b1	1	2	3	4	5	6	7	8	9	10	11	12	13	14	15	16
b2	4	5	6	7	8	9	10	11	12	13	14	15	16	17	18	19
k	7	5	2	1	0	1	1	4	4	6	7	9	10	10	10	10

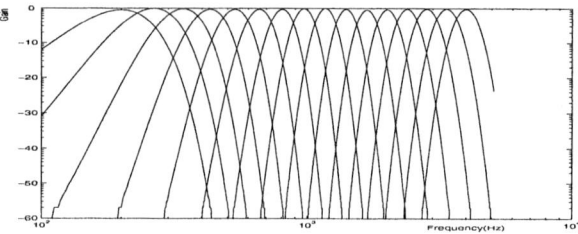

Fig. 1. The frequency response of 16 Bark wavelet filters

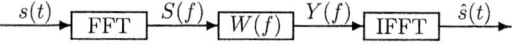

Fig. 2. The scheme of the Bark wavelet filters used in the front-end processing

4 Experiment Results and Conclusions

Speech data files of 10 to 50 words from 16 persons were used in the experiments. For each word, each person speaks 3 times. Speech data of 9 persons were used to train the system, and the other speech data of 7 persons were used to recognize. The sample frequency is 11.025kHz. Firstly we use the Bark wavelet filters instead of the FIR filters to processing the speech data. Secondly, the 1024 dimension ZCPA features are extracted. And then these features are trained and recognized by HMM model. Table 2 is the results based on the various SNRs.

Table 2. The recognition results at various SNRs (%)

	SNR	15dB	20dB	25dB	30dB	Clean
10	ZCPA	85.71	84.76	86.19	85.71	89.05
words	WZCPA	84.00	87.14	90.00	90.48	90.00
20	ZCPA	76.60	81.20	82.38	81.67	85.71
words	WZCPA	77.14	83.57	87.56	84.29	88.20
30	ZCPA	77.14	81.90	83.17	82.86	83.49
words	WZCPA	78.42	83.31	85.71	85.56	86.98
40	ZCPA	76.55	78.26	81.31	82.62	82.98
words	WZCPA	77.50	81.48	84.76	85.00	87.14
50	ZCPA	72.10	74.48	80.09	78.95	81.71
words	WZCPA	73.14	78.20	83.71	85.52	85.24

Where, the ZCPA presents the results with FIR filter and the WZCPA presents Bark wavelet filter. The results show that Bark wavelet filter is superior to FIR filter as front-end processor in speech recognition system.

Acknowledgements

The project is sponsored by the Scientific Research Foundation for the Returned Overseas Chinese Scholars, State Education Ministry of China ([2004] No.176), Natural Science Foundation of China (No.60472094), Shanxi Province Natural Science Foundation (No.20051039), Shanxi Province Scientific Research Foundation for Returned Overseas Chinese Scholars (No.200224) and Shanxi Province Scientific Research Foundation for University Young Scholars ([2004] No.13), Shanxi Province Scientific Technology Research and Development Foundation.(No.200312). The authors gratefully acknowledge them.

References

1. D.S. Kim, S.Y. Lee, M.K. Rhee: Auditory processing of speech signal for robust speech recognition in real-world noisy environments. IEEE Transactions on Speech and Audio Processing, 1999, Vol.7, No.1, 55–68
2. Q. Fu, K.C. Yi: Bark wavelet transform of speech and its application in speech recognition. Journal of Electronics, 2000, Oct, Vol.28, No.10, 102–105.
3. Boiana Gajic, K.Paliwal Kudldip: Robust speech recognition using feature based on zero-crossings with peak amplitudes. ICASSP, 2003, Vol.1, 64–67.

An Accurate and Fast Iris Location Method Based on the Features of Human Eyes*

Weiqi Yuan, Lu Xu, and Zhonghua Lin

Computer Vision Group, Shenyang University of Technology,
Shenyang 110023, China
{yuan60, xulu0330, lzhyhf}@126.com

Abstract. In this paper, we proposed an accurate and fast iris location method based on the features of human eyes. Firstly, according to the gray features of pupil, find a point inside the pupil using a gray value summing operator. Next, starting from this point, find three points on the iris inner boundary using a boundary detection template designed by ourselves, and then calculate the circle parameters of iris inner boundary according to the principle that three points which are not on the same line can define a circle. Finally, find other three points on the iris outer boundary utilizing the similar method and obtain the circle parameters. A large number of experiments on the CASIA iris image database demonstrated that the location results of proposed method are more accurate than any other classical methods, such as Daugman's algorithm and Hough transforms, and the location speed is very fast.

1 Introduction

In recent years, with the great development of communication and information technologies, automated personal identification is becoming more and more important for many fields such as access control, e-commerce, network security, bank automatic teller machine, and so on. However, all of the traditional methods for personal identification (keys, passwords, ID cards, etc.) can be lost, forgot or replicated easily, so they are very inconvenient and unsafe. Therefore, a new method for personal identification named biometric identification has been attracting more and more attention. Biometric identification is the science of identifying a person based on his (her) physiological or behavioral characteristics [1]. Nowadays, there are several kinds of biometric identifications in the world, such as fingerprint recognition, palm print recognition, face recognition, voice recognition, iris recognition, and so on [2].

The human iris, an annular part between the black pupil and the white sclera, has an extraordinary structure and provides many interlacing minute characteristics such as freckles, coronas, stripes, furrows, crypts and so on [3-5]. Compared with other biometric features, iris has many desirable properties as follows [3-8]: (1) Uniqueness. The textures of iris are unique to each subject all over the world. (2) Stability. The

---

* This subject is supported by the National Natural Science Foundation of China under Grant No.60472088.

textures of iris are essentially stable and reliable throughout ones' life. (3) Noninvasiveness. The iris is an internal organ but it is externally visible, so the iris recognition systems can be noninvasive to their users. All of these advantages make iris recognition a wonderful biometric identification technology.

With the increasing interests in iris recognition, more and more researchers gave their attention into this field [3-22]. In an iris recognition system, the iris location is one of the most important steps, and it costs nearly more than a half of recognition time [22]. Furthermore, the iris localization is still very important for latter processing such as iris normalization, iris features extraction and patterns matching. Therefore, the iris localization is crucial for the performance of an iris recognition system, including not only the location accuracy but also the location speed. There are two classical iris location methods, one is the Daugman's algorithm [3-5], and another is the edge detection combining with Hough transforms [16-18].

In this paper, we proposed an accurate and fast iris location method based on the features of human eyes. It doesn't need to find all of the points of iris inner and outer boundaries, so its location speed is very fast. Meanwhile, the circle parameters of iris boundary are calculated according to the classical geometry principle that three points which are not on the same line can define a circle, so the proposed method can not be affected by the random noises in the irrelative regions, and its location accuracy is still very high. Furthermore, the images used in the proposed method are all gray level images, so this method is suitable for any images from different situations such as different skin-color people and different cameras.

2 Finding a Point Inside the Pupil

The first step of proposed method is finding a point inside the pupil accurately. Then this point can be considered as the searching center in the next location step, and its gray value can serve as the reference gray value of pupil.

2.1 The Features of Pupil and the Gray Value Summing Operator

Fig.1 is an eye image chosen from the database random. From many observations we found, in the human eye image, the pupil region is the blackest one. That means the gray value of pupil is a minimum. Furthermore, the gray distribution of pupil is relatively even. Thus, we can find a point inside the pupil according to these features.

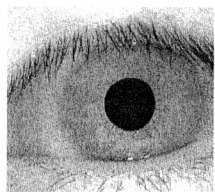

Fig. 1. An eye image

Let $I(x, y)$ be an input human eye image and $f(i, j)$ be the gray value of point (i, j), here $i = 1, 2, ...,$ Image Width and $j = 1, 2, ...,$ Image Height. Meanwhile, another problem we must highlight is that in the whole paper, the x in coordinate (x, y) means the column number in an eye image and the y means the row number.

Then we design a gray value summing operator (GVSO, Fig.2). It has $n*n$ pixels and its center is (x_o, y_o). There are some designing principles as follows: (1) Because the shape of pupil is nearly a circle and the gray distribution is relatively even, in order to adapt these two features, we let the GVSO be a square. (2) In eye images there is not only the pupil but also the iris, the sclera and the eyelid, and the area of pupil is not very large, so the size of GVSO should not be too large. Too large an n will make the point we found outside the pupil. (3) Commonly, there are many eyelashes in the image and the gray values of them are very approximately with that of pupil, and the area of eyelashes region is usually random, so too small an n will make the point we found located in the eyelash region.

$$\begin{bmatrix} f(x_o-\frac{n-1}{2}, y_o-\frac{n-1}{2}) & \cdots & f(x_o, y_o-\frac{n-1}{2}) & \cdots & f(x_o+\frac{n-1}{2}, y_o-\frac{n-1}{2}) \\ \vdots & \ddots & \vdots & \ddots & \vdots \\ f(x_o-\frac{n-1}{2}, y_o) & \cdots & f(x_o, y_o) & \cdots & f(x_o+\frac{n-1}{2}, y_o) \\ \vdots & \ddots & \vdots & \ddots & \vdots \\ f(x_o-\frac{n-1}{2}, y_o+\frac{n-1}{2}) & \cdots & f(x_o, y_o+\frac{n-1}{2}) & \cdots & f(x_o+\frac{n-1}{2}, y_o+\frac{n-1}{2}) \end{bmatrix}$$

Fig. 2. The gray value summing operator (GVSO)

In the eye image, only inside the pupil there are areas with the minimum gray and the even distribution. Thus, we can search in the whole image using the GVSO, when we find a GVSO with a minimum value, it is the most reasonable one and its center (x_o, y_o) must be a point inside the pupil. For the GVSO, we can calculate its value using the formula (1), where $S(x_o, y_o)$ is the value of GVSO which has a center point (x_o, y_o), and n is the size of GVSO.

$$S(x_o, y_o) = \sum_{i=x_o-\frac{n-1}{2}}^{x_o+\frac{n-1}{2}} \sum_{j=y_o-\frac{n-1}{2}}^{y_o+\frac{n-1}{2}} f(i, j) \tag{1}$$

2.2 Steps of Finding a Point Inside the Pupil

According to the features of pupil, the proposed method utilizes four steps to find a point inside the pupil:

(1) Determine the size of GVSO (n). A great number of experiments demonstrated that when $n =$ (Image Width * Image Height)$^{1/4}$ the GVSO is the most reasonable one.
(2) Determine the searching range. In order to avoid errors of pixel lost and data overflow, the column (x) searching range is $[(n - 1)/2,$ Image Width $- (n + 1)/2]$, and the row (y) searching range is $[(n - 1)/2,$ Image Height $- (n + 1)/2]$.
(3) For each pixel (x, y) in the searching ranges, consider (x, y) as the center point of GVSO and calculate the value of $S(x, y)$, then we can get an S set of $\{S_1(x, y), S_2(x, y), S_3(x, y), ...\}$.

(4) Find the minimum value $S_{min}(x, y)$ from the S set, then the point *(x, y)* is the one we need, here we call it $P_o(x_o, y_o)$.

3 Finding the Iris Boundary

In the former section we have found a point inside the pupil, next we will start from this point to find the iris inner and outer boundaries respectively, and then we can separate the pure iris region in the eye image accurately. Because the pupil and the iris are both nearly circles, here we suppose they are two exact circles. The essential principles of finding the iris boundary are as follows: Firstly, find three points on the iris inner and outer boundaries respectively using a boundary detection template designed by us; then, calculate the parameters of two boundary circles according to the principle that three points which are not on the same line can define a circle.

3.1 The Features of Boundary and the Boundary Detection Template

From so many image observations and data analysis we found, in eye images, the gray value of pupil is a minimum and that of sclera is a maximum, but the gray value of iris is between them. Thus, there is a gray suddenly changed region between the iris and the pupil, and in this region the edge intensity of each pixel is very large. If the edge intensity of a pixel in this region is a maximum, this pixel must be a boundary point. The proposed method must detect the maximum of edge intensity, however, the results that we obtained using the classical edge detection operators are very unsatisfactory. There are two reasons about this condition. One is that in the image there are some random noises produced by the eyelashes and the veins in the iris region. Another reason is that all of these operators have only one determining value (edge intensity). So when these operators meet with the noises, they could not find the boundary points exactly. Therefore, in order to find the boundary points and calculate the boundary circle parameters accurately, we must design a boundary detection template by ourselves. The proposed method design a template which includes two determining values and is hard to be affect by the random noises.

This kind of boundary detection template (BDT) is shown in Fig.3. Its initial direction is horizontal, then its width is *m* pixels and its height is *n* pixels, here *m* and *n* are both odd number, and $(m-1)/2 > n$. Let the center of BDT be $O(x, y)$, a horizontal line (*long axis*) through *O* divide it into upper part and lower part, and a vertical line (*short axis*) through *O* divide it into left part and right part, thus the template direction is along the long axis. Furthermore, we stipulate the template direction must be the same as the detecting direction, it means when we detect the left and right boundary points the direction of BDT is horizontal, but when we detect the upper boundary point the direction of BDT is vertical.

Here we give some explanations to four problems: (1) Let *m* and *n* be odd number. Because, when considering the present pixel as the center of BDT, it ensures the two rectangle regions beside the short axis have the same number of pixels. (2) Set $(m-1)/2 > n$. It makes the effect made by the gray value of those irrelative pixels to the calculate results as small as possible and makes the BDT can not be affect by the noises. (3) Let the BDT has direction. It is in order to introduce the operation method of BDT more clearly and readers can understand it easily. (4) Stipulate the direction of BDT must be the same as the detecting direction. It ensures the determining values are same when detecting the horizontal and vertical directions.

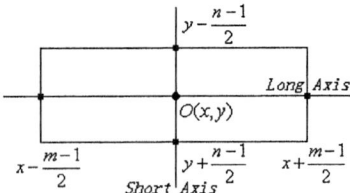

Fig. 3. The boundary detection template (BDT)

The first determining value of BDT is still the edge intensity. It is because whichever kind of templates or operators, in order to detect the boundary points it must calculate the edge intensity of each pixel firstly, then decide whether the pixel is a boundary point or not according to its edge intensity. For each pixel in the searching range, we consider it as the center of BDT and calculate its edge intensity using formula (2) (dif^1 and dif^2 are used for the horizontal and vertical directions respectively). The determining basis of edge intensity is whether it is a maximum.

$$\begin{cases} dif^1(x,y) = \left| \sum_{i=x-\frac{m-1}{2}}^{x-1} \sum_{j=y-\frac{n-1}{2}}^{y+\frac{n-1}{2}} f(i,j) - \sum_{p=x+1}^{x+\frac{m-1}{2}} \sum_{q=y-\frac{n-1}{2}}^{y+\frac{n-1}{2}} f(p,q) \right| \\ dif^2(x,y) = \left| \sum_{i=x-\frac{n-1}{2}}^{x+\frac{n-1}{2}} \sum_{j=y-\frac{m-1}{2}}^{y-1} f(i,j) - \sum_{p=x-\frac{n-1}{2}}^{x+\frac{n-1}{2}} \sum_{q=y+1}^{y+\frac{m-1}{2}} f(p,q) \right| \end{cases} \quad (2)$$

Then the second determining value of BDT can be described that it is the average gray value of all pixels in the half template which is close to the center of pupil (or iris). The reason is that, for an inner (outer) boundary point, if we consider it as the center of

$$\begin{cases} avg^1(x,y) = [\sum_{i=x+1}^{x+\frac{m-1}{2}} \sum_{j=y-\frac{n-1}{2}}^{y+\frac{n-1}{2}} f(i,j)]/(n \times \frac{m-1}{2}) \\ avg^2(x,y) = [\sum_{i=x-\frac{m-1}{2}}^{x-1} \sum_{j=y-\frac{n-1}{2}}^{y+\frac{n-1}{2}} f(i,j)]/(n \times \frac{m-1}{2}) \\ avg^3(x,y) = [\sum_{i=x-\frac{n-1}{2}}^{x+\frac{n-1}{2}} \sum_{j=y+1}^{y+\frac{m-1}{2}} f(i,j)]/(n \times \frac{m-1}{2}) \end{cases} \quad (3)$$

BDT, all pixels in the half template which is close to the center of pupil (iris) must be inside the pupil (iris), so the average value of this half template must be very approximately with the gray value of pupil (iris). Therefore, this kind of feature can also be used to determine a boundary point. For each present pixel, we can calculate the second determining value using the formula (3) (avg^1, avg^2 and avg^3 are used for leftwards, rightwards and upwards respectively). The basis of second determining value is whether it is less than the reference gray value of pupil or iris.

3.2 Finding the Iris Inner Boundary

In the eye image, starting from the point $P_o(x_o, y_o)$ which has been found in section 2, when we detect pixels using the boundary detection template on the leftwards horizontal, rightwards horizontal and upwards vertical directions respectively, we can find three points on the iris inner boundary. Let the center of the iris inner boundary be $P_P(x_P, y_P)$, and the three boundary points be $P_L(x_L, y_L)$, $P_R(x_R, y_R)$ and $P_D(x_D, y_D)$, we can calculate the parameters of inner circle precisely according to the principle that three points which are not on the same line can define a circle. Fig.4 shows the relationships among these points.

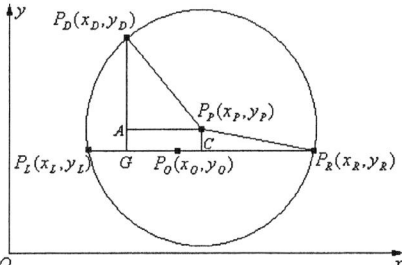

Fig. 4. Three points on the iris inner boundary.

The steps of finding the iris inner boundary are as follows:

(1) Determine the searching range. In our experiments, on the horizontal direction, we detect boundary points along the line $y = y_o$, in the leftwards searching range $[(n-1)/2, x_o - 1]$ we can find the left boundary point P_L, and in the rightwards searching range $[x_o + 1, \text{Image Width} - (n+1)/2]$ the right outer boundary point P_R can be found. Then we detect the upper boundary point along the vertical line $x = x_L + 20$ upwards and the third outer boundary point P_D can be found in the range $[(n-1)/2, y_o - 1]$. So many experiments demonstrated that these searching ranges are all very accurate and effective.

(2) Determine the reference gray value of pupil. Having done many experiments we found, when we supposed the gray value of point $P_o(x_o, y_o)$ is v_o, the value $v = v_o + (v_o)^{1/2}$ is the most reasonable reference value of pupil.

(3) Using the BDT detect the first iris inner boundary point in the leftwards searching range. For each present pixel (x, y), considering it as the center of BDT, calculate its edge intensity $dif_i(x, y)$ and get a set $\{dif_1(x, y), dif_2(x, y), dif_3(x, y), ...\}$. Then put this dif set into the sequence from the maximum to the minimum and get a new set. Then find the first dif which is less than the

reference gray value of pupil (v) in this new set, and this point must be the left boundary point P_L.

(4) Do the same operations as step (3) in the rightwards and upwards searching ranges respectively, then we can get the other two boundary points P_R and P_D.

(5) According to the position relationships shown in Fig.4, calculate the center coordinate of the outer circle (x_P, y_P) and its radius R_P. Here $x_P = (x_R + x_L)/2$, $y_P = [20(x_R - x_L) - 400 + y_R^2 - y_D^2] / [2(y_R - y_D)]$, and $R_P = [(x_P - x_D)^2 + (y_P - y_D)^2]^{1/2}$.

3.3 Finding the Iris Outer Boundary

Commonly, the iris outer and inner boundaries are not concentric circles, and this situation is very serious for some people. So more serious errors would be made if we consider the center point of inner boundary as the center of outer boundary, and the outer circle parameters would be very inaccurate. Alike the feature of inner boundary, the outer boundary is also an approximate circle, so the method of locating the inner boundary can still be used here. However, we must do some modifications because of the particularity of iris region in the eye image.

In the eye image, some areas of iris, especially the outer boundary, is covered by eyelids and eyelashes frequently. In details, the upper and lower areas of the outer boundary commonly can not be detected because of eyelids' coverage. This situation can be clearly seen from Fig.1. However, this situation won't happen for the left and right of outer boundary. Therefore, the outer boundary points can not be found if we start from the center of inner boundary and search them vertically upwards. Done many observations and measurements on human eye images, we found that, after we have found the left and right boundary points utilizing the steps introduced in section 3.2, if we detect the point along the vertical line which is *15* pixels far from the right outer boundary point, we could find a point on the outer boundary accurately.

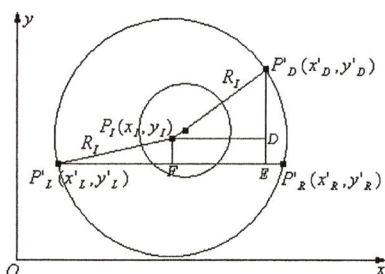

Fig. 5. Three points on the iris outer boundary

The modified method of finding the iris inner boundary includes following steps:

(1) Determine the searching range. In proposed method, on the horizontal directions, along the line $y = y_p + 15$, detect the outer boundary points leftwards and rightwards respectively. The leftwards searching range equals to $[(n-1)/2, x_p - R_p - 1]$ in which the left boundary point $P'_L(x'_L, y'_L)$ can be detected, and the right outer boundary point $P'_R(x'_R, y'_R)$ can be detected in the rightwards searching

range which equals to $[x_p + R_p + 1, Width - (n + 1) / 2]$. Then detect along the vertical line $x = x'_R - 15$ upwards, the third outer boundary point $P'_D(x'_D, y'_D)$ can be found in the range $[(n - 1) / 2, y_p + 15]$.

(2) Detect the reference gray value of the iris. From a lot of experiments we found that, in a small region outside the inner boundary, its average gray value can be used to represent the gray value of the entire iris region. Thus, two 9*9 square regions have been chosen in proposed method. Their locations are 5 pixels far from the outside of inner boundary. Then calculate the average gray value v' of all pixels in these two regions, and then we consider the value $v' + (v')^{1/3}$ as the reference gray value of iris. Experimental results show that this value is very correct and effective.

(3) Using the BDT detect points on the iris outer boundary in the horizontal leftwards range, horizontal rightwards rage, and vertical upwards range respectively. Choose the points which have the highest boundary intensity and whose second determinant value is less than the reference gray value of iris. Ten these points are just located on the iris outer boundary.

(4) According to the position relationships shown in Fig.5, calculate the center coordinate of the outer circle (x_I, y_I) and its radius R_I. Here $x_I = (x'_R + x'_L) / 2$, $y_I = [15(x'_R - x'_L) - 225 + y'^2_L - y'^2_D] / [2(y'_L - y'_D)]$, and $R_I = [(x_I - x'_D)^2 + (y_I - y'_D)^2]^{1/2}$.

4 Experimental Results

Our experiments are performed in Visual C++ (version 6.0) on a PC with Pentium 1.7G Hz processor and 256M DRAM. Our iris images are all supplied by CASIA iris images database [23], which includes 80 people's 108 different eyes, there are 7 images for each eye with a resolution of 320*280 pixels. In our experiments, the iris location algorithm proposed in this paper has been used for each image in the database. Fig.6 are

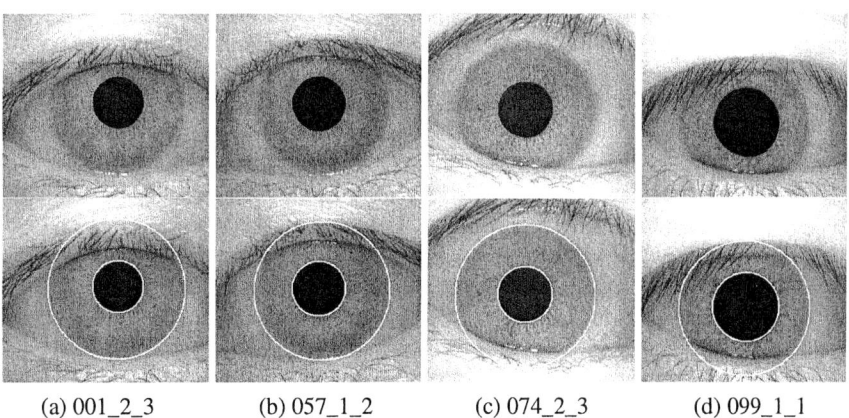

(a) 001_2_3   (b) 057_1_2   (c) 074_2_3   (d) 099_1_1

Fig. 6. Experimental results of four random images

four iris images chosen from CASIA database random, each column shows the original iris image and processed one, and their corresponding numbers in the CASIA are shown below the image.

The experimental data are as follows: 756 iris images have been located in our experiments, it cost 2.5 minutes in total, that is to say each image cost 0.198s in average, and the location accuracy is 99.34%. Meanwhile, we have also done some repetition experiments using the Daugman's algorithm and the edge detection combining with Hough Transforms. In order to be convenient for comparing, we put these data in Table 1.

Table 1. The comparation with other algorithms

Method	Accuracy	Mean time
Daugman	98.4%	5.98s
Hough transforms	95.3%	5.62s
Proposed	99.34%	0.198s

5 Conclusions

In this paper, we proposed an accurate and fast iris location method based on the features of human eyes. We have performed a large number of experiments on the CASIA iris database which has 756 iris images. The location accuracy is 99.34%, and the location speed is less than 0.2 seconds per image. Experimental data demonstrated that the location results of proposed method are more accurate than any other classical methods, such as Daugman's algorithm and Hough transforms, and its location speed is very fast. In our future works, there are two fields to be improved: (1) In order to extract iris region more efficiently without the eyelids coverage, we should do some curve fittings in the contours of upper and lower eyelids. (2) And we should eliminate eyelashes which located inside the iris region in order to avoid the eyelashes' influence for extracting the iris features.

Acknowledgments

The authors would like to thank the Institute of Automation in Chinese Academy of Sciences, they supplied the CASIA iris image database (ver1.0) to us, and this database makes our work complete very smoothly.

References

1. Zhang, M., Pan, Q., Zhang, H., Zhang, S.: Multibiometrics Identification Techniques. Information and Control, Vol. 31, No. 6 (2002) 524-528
2. Jain, A. K., Bolle, R. M., Pankanti, S., Eds: Biometrics: Personal Identification in a Networked Society. Norwell, MA: Kluwer (1999)

3. Daugman, J. G.: High Confidence Visual Recognition of Persons by a Test of Statistical Independence. IEEE Transactions on Pattern Analysis and Machine Intelligence, Vol. 15, No. 1 (1993) 1148-1161
4. Daugman, J. G.: Statistical Richness of Visual Phase Information: Update on Recognizing Persons by Iris Patterns. International Journal of Computer Vision, Vol. 45, No. 1 (2001) 25-38
5. Daugman, J. G.: How Iris Recognition Works. IEEE Transactions on Circuits and Systems for Video Technology, Vol. 14, No. 1 (2004) 21-30
6. Wildes, R. P.: Iris Recognition: An Emerging Biometric Technology. Proceedings of the IEEE, Vol. 85, No. 9 (1997) 1348-1363
7. Wildes, R. P., Asmuth, J., et al: A Machine-vision System for Iris Recognition. Machine Vision and Applications, Vol. 9 (1996) 1-8
8. Boles, W. W., Boashah, B.: A Human Identification Technique Using Images of the Iris and Wavelet Transform. IEEE Transactions on Signal Processing, Vol. 46 (1998) 1185-1188
9. Ma, L., Wang, Y., Tan, T.: Iris Recognition Based on Multichannel Gabor Filters. Proceedings of the Fifth Asian Conference on Computer Vision, Vol. I (2002) 279-283
10. Ma, L., Tan, T., Wang, Y., Zhang, D.: Personal Identification Based on Iris Texture Analysis. IEEE Transactions on Pattern Analysis and Machine Intelligence, Vol. 25, No. 12 (2003) 1519-1533
11. Ma, L., Tan, T., Wang, Y., Zhang, D.: Efficient Iris Recognition by Characterizing Key Local Variations. IEEE Transactions on Image Processing, Vol. 13, No. 6 (2004) 739-750
12. Lim, S., Lee, K., Byeon, O., Kim, T.: Efficient Iris Recognition through Improvement of Feature Vector and Classifier. ETRI Journal, Vol. 23, No. 2 (2001) 61-70
13. Sanchez-Avila, C., Sanchez-Reillo, R.: Iris-based Biometric Recognition Using Wavelet Transforms. IEEE Aerospace and Electronic Systems Magazine (2002) 3-6
14. Kwanghyuk, B., Seungin, N., Kim, J.: Iris Feature Extraction Using Independent Component Analysis. AVBPA 2003, LNCS 2688 (2003) 838-844
15. Wang, C., Ye, H.: Investigation of Iris Identification Algorithm. Journal of Guizhou University of Technology (Natural Science Edition), Vol. 29, No. 3 (2000) 48-52
16. Chen, L., Ye, H.: A New Iris Identification Algorithm. Journal of Test and Measurement Technology, Vol. 14, No. 4 (2000) 211-216
17. Hu, Z., Wang, C., Yu, L.: Iris Location Using Improved Randomized Hough Transform. Chinese Journal of Scientific Instrument, Vol. 24, No. 5 (2003) 477-479
18. Yu, X.: The Study of Iris-orientation Algorithm. Journal of Tianjin University of Science and Technology, Vol. 19, No. 3 (2004) 49-51
19. Huang, X., Liu, H.: Research on Iris Location Technique Based on Gray Gradient. Journal of Kunming University of Science and Technology, Vol. 26, No. 6 (2001) 32-34
20. Wang, C., Hu, Z., Lian, Q.: An Iris Location Algorithm. Journal of Computer-aided Design and Computer Graphics, Vol. 14, No. 10 (2002) 950-952
21. Wang, Y., Zhu, Y., Tan, T.: Biometrics Personal Identification Based on Iris Pattern. Acta Automatica Sinica, Vol. 28, No. 1 (2002) 1-10
22. Ye, X., Zhuang, Z., Zhang, Y.: A New and Fast Algorithm of Iris Location. Computer Engineering and Applications, Vol. 30 (2003) 54-56
23. Institute of Automation, Chinese Academy of Sciences, CASIA Iris Image Database (ver 1.0), http://www.sinobiometrics.com

A Hybrid Classifier for Mass Classification with Different Kinds of Features in Mammography

Ping Zhang[1], Kuldeep Kumar[1], and Brijesh Verma[2]

[1] Faculty of Information Technology Bond University, Gold Coast, QLD 4229, Australia
{pzhang, kkumar}@staff.bond.edu.au
[2] Faculty of Informatics & Comm. Central Queensland University,
Rockhampton, QLD 4702, Australia
b.verma@cqu.edu.au

Abstract. This paper proposes a hybrid system which combines computer extracted features and human interpreted features from the mammogram, with the statistical classifier's output as another kind of features in conjunction with a genetic neural network classifier. The hybrid system produced better results than the single statistical classifier and neural network. The highest classification rate reached 91.3%. The area value under the ROC curve is 0.962. The results indicated that the mixed features contribute greatly for the classification of mass patterns into benign and malignant.

1 Introduction

Radiologists find the calcifications and masses from the mammogram, and classify them as benign and malignant usually by further biopsy test. Digital mammography brought the possibility of developing and using computer aided diagnosis (CAD) system for the classification of benign and malignant patterns. The most important factor which directly affects the CAD system classification result is a feature extraction process. Researchers spend a lot of time in attempt to find a group of features that will aid them in improving the classification rate for malignant and benign cancer. Different classifiers such as wavelet based techniques, statistical classifiers and neural networks have been used for mass classification. It is reported that neural networks achieve relatively better classification rate [1, 2], but statistical methods provide more stable systems [3].

In this paper, we propose a hybrid system combining the Logistic Regression (LR) and Discriminant Analysis (DA) based statistical classifiers with Neural Network (NN), in conjunction with Genetic Algorithm (GA) for feature selection. We concentrated on classifying the mass suspicious areas as benign and malignant. We also combined the statistical calculated features with the human interpreted features such as patient age, mass shape, etc.. The hybrid system showed better performance than using a single classifier.

2 Research Methodology of Proposed Hybrid System

A limitation of ordinary linear statistical models is the requirement that the dependent variable is numerical rather than categorical (eg. benign or malignant). A range of techniques have been developed for analysing data with categorical dependent variables, including DA and LR.

NNs have been used as effective classifiers in pattern recognition. The results of the NN based classification were compared with those obtained using multivariate Baye's classifiers, and the K-nearest neighbor classifier [1]. The NN yielded good results for classification of "difficult-to-diagnose" microcalcifications into benign and malignant categories using the selected image structure features.

GAs have been applied to a wide variety of problems, including search problems, optimization problems. It has been reported efficient for feature selection working with linear and non linear classifiers [4].

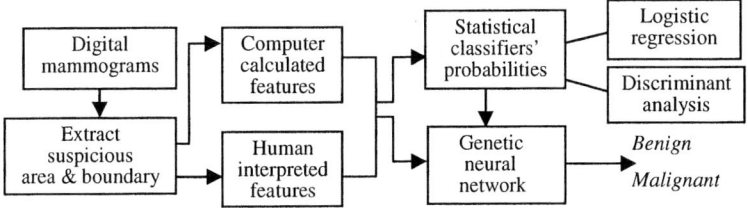

Fig. 1. Proposed LRDA-GNN classifier

The proposed hybrid system combines the two statistical classifiers with NN for classification. We also mixed the features which are statistically calculated from mammogram, with human interpreted features in the system. In the proposed hybrid system (we can call it LRDA-GNN), the membership probability numerical values from LR and DA, rather than the value 0 or 1, are used as the second order features combing with original features to feed the NN for benign and malignant classification. It can be described as Fig 1. A GA was developed for feature selection based on the NN classifier (refer to [5]).

Digital Database for Screening Mammography (DDSM) from University of South Florida is used for our experiments. The establishment of DDSM makes the possibility of comparing results from different research groups [6]. In DDSM, the outlines for the suspicious regions are derived from markings made on the film by at least two experienced radiologists. It also supplies some BI-RADS® (Breast Imaging Reporting and Data System, which contains a guide to standardized mammographic reporting) features and patient ages, which are described as human interpreted features in this paper. The mammograms used in this study are all scanned on the HOWTEK scanner at the full 43.5 micron per pixel spatial resolution.

14 statistical features related to the mammograms are calculated from every extracted suspicious area, which are called computer calculated features. They are: number of pixels, average histogram, average grey level, average boundary grey level, difference, contrast, energy, modified energy, entropy, modified entropy, standard deviation, modified standard deviation, skew, and modified skew. The calculations refer to [5].

The 6 human interpreted features related to the suspicious area are: patient age, breast density, assessment, mass shape, mass margin and subtlety. Some of these features are described with text in the database. We simply numerated them, so that all the features can be represented by different numbers and can be used for further classification. For example, we use "1" represents round and use "2" represents oval for the shape feature.

3 Experimental Results and Analysis

A total of 199 mass suspicious areas were extracted from the digital mammograms taken from DDSM database for experiments. It includes 100 malignant cases and 99 benign cases. A few different splits of the dataset for training and testing were used for the experiments with the different classifiers which are LR, DA, NN, LRDA-NN (the hybrid system using NN without GA for feature selection) and LRDA-GNN for comparison. The results with using 3 splits of the database are shown in Table 1.

Table 1. Classification rates reached with different classifiers

Classifier	Training and testing classification rate(train/test)		
	Split1(130/69)	Split2 (140/59)	Split3 (150/49)
Logistic regression	93.1/87.0	93.1/89.8	92.0/87.8
Discriminant analysis	90.8/85.5	81.4/71.2	84.0/65.3
NN	85.4/89.9	85.7/91.5	81.3/87.8
LRDA-NN	93.1/91.3	92.1/91.5	92.0/91.8
LRDA-GNN	93.1/92.8	92.1/91.5	92.0/93.9

In split1, we randomly selected 130 mass cases for training and 69 cases for testing. Split 2 and split 3 use 140 and 150 cases for training respectively, the rest of whole 199 cases for testing. NN reached a bit higher classification rate for testing set when the classification rate for training is relatively lower than it was with logistic regression model. The disadvantage of NN is it is not so stable as statistical methods. There are many different trained NN models which can reach similar classification rate for testing. More experiments were needed to find the better NN model. From the results we can see the hybrid system reached higher testing classification rate than any other classifiers, with the best classification rates for training as well. Another satisfying thing is that the overall classification rate for the training set is higher and more stable than it is with the NN using 20 original features, as we observed.

We can also see that LRDA-GNN which involved GA for feature selection based on LRDA-NN, improved the classification rate very little than LRDA-NN produced. However in our feature selection result, we found only one feature was constantly selected in the feature sets which reached the higher classification rates. It is the membership probability of the output from the LR model. This indicates that the GA is effective for feature selection. Observing the selected feature sets, although there are a few features were selected a bit more frequently than others we could not really confirm the particular significant features except the probability from LR.

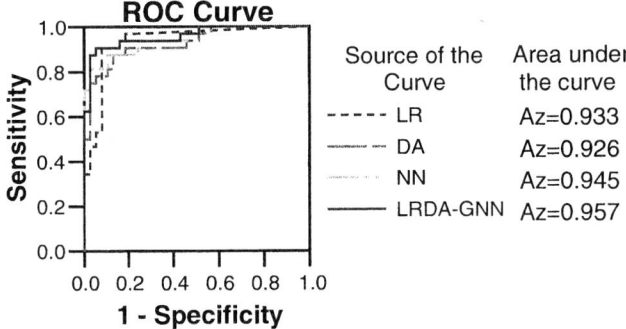

Fig. 2. The ROC curves of 4 different models

Fig. 2 shows the ROC curves produced by different classifiers for further validation of the performances. They are created with the testing result from split1 data set. We can see LRDA-GNN produced the best ROC curve as well.

4 Conclusion and Discussion

In the hybrid system, the second order features which are the output probabilities from the statistical models, showed its great contribution for the final NN classification. The hybrid of the different features shows the potential for the mass diagnosis of benign and malignant. Although GA is effective for feature selection, it didn't show big improvement of the classification rate when working with the current feature set which includes the probability features. Further combination of statistical models with NN will be considered in our future research.

References

1. Chitre, Y., Dhawan, A.P. and Moskowitz, M., Artificial Neural Network Based Classification of Mammographic Microcalcifications Using Image Structure Features, International Journal of Pattern Recognition and Artificial Intelligence, 1993. 7(6): p. 1377-1401.
2. Wood, K.S., Comparative evaluation of pattern recognition techniques for detection of microcalcifications in mammography, International Journal of Pattern Recognition and Artificial Intelligence, 1993. p. 80-85.
3. Sarle, W.S., Neural Networks and Statistical Models, Proceedings of the 19th Annual SAS Users Group International Conference, Cary, NC: SAS Institute, 1994. April: p. 1538-1550.
4. Ripley, B.D., Pattern Recognition & Neural Networks, 1996: Cambridge Press.
5. Zhang, P., Verma, B.K and Kumar, K., Neural vs Statistical Classifier in Conjunction with Genetic Algorithm Feature Selection in Digital Mammography, Proceedings of IEEE Congress on Evolutionary Computation, 2003, p.1206-1213.
6. Heath, M., Bowyer, K.W. and Kopans, D., Current status of the Digital Database for Screening Mammography. Digital Mammography, 1998: p. 457-460.

Data Mining Methods for Anomaly Detection of HTTP Request Exploitations

Xiao-Feng Wang, Jing-Li Zhou, Sheng-Sheng Yu, and Long-Zheng Cai

Department of Computer Science and Technology,
Huazhong University of Science and Technology, Wuhan 430074, China
xfwang@wtwh.com.cn

Abstract. HTTP request exploitations take substantial portion of network-based attacks. This paper presents a novel anomaly detection framework, which uses data mining technologies to build four independent detection models. In the training phase, these models mine specialty of every web program using web server log files as data source, and in the detection phase, each model takes the HTTP requests upon detection as input and calculates at least one anomalous probability as output. All the four models totally generate eight anomalous probabilities, which are weighted and summed up to produce a final probability, and this probability is used to decide whether the request is malicious or not. Experiments prove that our detection framework achieves close to perfect detection rate under very few false positives.

1 Introduction

Web servers and Web-based programs are suffering sharply increasing amounts of web-based attacks. Generally, attacks against either web servers or web-based applications must be carried out through forging specially formatted HTTP requests to exploit potential web-based vulnerabilities, so we name this kind of attacks as HTTP request exploitations.

We provide a novel detection framework dedicated to detecting HTTP request exploitations. The framework is composed of four detection models. Each of the four detection models contains at lease one classifier. The classifiers calculate the anomalous probabilities for HTTP requests upon detection to decide whether they are anomalous or not. Eight classifiers of the framework present eight independent anomalous probabilities for a HTTP request. We assign a weight for each probability and calculate the weighted average as the final output.

2 Methodology

Web log file is composed of a large amount of entries that record basic information of HTTP requests and responses. We extract the source IP and URL of each entry and translate them into serials of name-value pairs. Let $avp=(a, v)$ represent a name-value pair and in this paper we call the name-value pair an attribute which comprises name

$$\underbrace{program = search}_{attribute\ 1}\ \underbrace{ip = 211.67.27.194}_{attribute2}\ \underbrace{hl = zh\text{-}CN}_{attribute3}\ \underbrace{q = computer + science}_{attribute4}$$

Fig. 1. A four-element attribute list is derived from an entry of a web server log file

a and attached value v. Figure 1 shows the example of an entry from a web server log file (ignoring some parts). Attribute list is a set of attributes derived from a web log entry. So a web log file can be mapped into a set of attribute lists. Let *PROB* be the output probability set derived from classifiers and let *WEIGHT* be related weight set. The final anomalous probability is calculated according to Equation 1. The *WEIGHT*, as a relatively isolated subject, is not introduced in this paper.

211.67.27.194 - [Time] "GET search?hl=zh-CN&q=computer+science" 200 2122

$$Final\ Anomalous\ probability = \sum_{\substack{p_m \in PROB \\ w_m \in WEIGHT}} p_m * w_m \quad (1)$$

3 Detection Models

The framework is composed of four independent detection models: attribute relationship, fragment combination, length distribution and character distribution. The corresponding principles for these models are introduced in this section.

3.1 Attribute Relationship

General speaking, there must be some fixed inter-relationships between the attribute lists derived from web log entries. For example, suppose that a web form page contains a hidden tag with a given value. Once a user submits this web form to a specified web program (say, a CGI program dealing with the web form), the given value will be presented in the submitted request as a query attribute instance. Apparently, the specified web program only accepts the given value attached to the hidden tag. This fact is described as an enumeration rule. If malicious users tamper with the given value attached to the hidden tag, the rule is broken. Further more, a HTTP request missing necessary query attribute breaks integrality rule and a HTTP request containing faked query attribute breaks validity rule. Interdependence rule prescribes the interdependence of two attribute instances. Suppose that a web program dealing with an order form accepts two arguments: *id* and *discount*. id is client's identity and discount is client's shopping discount ratio. If *Jason* enjoys 10% discount, an interdependence rule will be discovered between two attribute instances: (*id, Jason*) and (*discount,* 10%). If *Jason* tampers with his discount to 50%, this interdependence rule is broken. A set of attribute lists derived from a given web log file is used to mine the four types of rules. According to the respective kinds of rules, four classifiers calculate four anomalous probabilities: $P_{integrality}$, $P_{validity}$, $P_{enumeration}$ and $P_{interdependence}$. The first three are Boolean values and the last one is continuous value between 0 and 1.

3.2 Fragment Combination

Fragment combination model is based on the fact that most malicious URLs are prone to containing multiple fragments that are assembled according to some specified patterns. For example, most XSS attack URLs contain two signatures "*<script>*" and "*</script>*" which comprises four fragments: "<", ">", "/", "script". None of these fragments is enough to predicate an attack but assemblage of them is quite sure to alarm a XSS attack. Therefore, it is reasonable to conclude that an assemblage of two fragments with high occurring frequency in attack URLs is prone to malicious. Fragment combination model mines such assemblages as many as possible and each assemblage is corresponding to a fragment-assembling rule. Under the condition that the fragment-assembling rule library is setup, an anomalous probability $P_{fragment}$ is calculated as the result of this model. The more fragment pairs match in the specified rule library, the more anomalous probability is approaching 1, i.e. the URL upon detection is more assured to malicious.

3.3 Length Distribution

Benign attribute instances are either script-generated tokens or some printable characters inputted by harmless users, which are short strings of fixed-size or fluctuating in a relatively small range. However, many malicious attribute instances containing attacking code exceed several hundreds bytes easily, which surpass normal range sharply. Length distribution model calculates the mean μ and variance σ^2 for the lengths of the sample attribute values. Then, P_{length}, the anomalous probability of a given attribute value, can be assessed through Chebyshev's inequality, as shown in Equation 2.

$$p(|x-\mu|>=|l-\mu|) <= p(l) = 1/(1+(l-\mu)^2/\sigma^2), \ P_{length} = 1 - p(l) \qquad (2)$$

3.4 Character Distribution

Character distributions of normal attribute value strings always obey some statistical rules. For example, attribute *name* mostly accepts alphanumeric codes as its value, while attribute *age* only accepts Arabic numerals. However, malicious strings are prone to deviating these rules. Kruegel et al. [1] provided a character distribution model focusing on the repeated degree of ASCII characters. First, determine the times of occurrence for each character in the string. Second, divide the occurrence time of each character by the length of the string to generate occurrence probability. Then, sort the occurrence probabilities in descending order. At last, combine those probabilities by aggregating probability values into six segments and get a mean probability vector ($\mu_1, \mu_2, \mu_3, \mu_4, \mu_5, \mu_6$). This model fails to pay attention to the character itself. For example, the string "Ann" and "../" have the same probability vectors. We provide another method to categorize characters. All 256 ASCII characters are aggregated into six segments: capital letters, small letters, Arabic numerals, suspicious punctuations, generic punctuations and other characters. Therefore, a different probability vector can be determined. χ^2-test [2] is used to determine if the observed probability vector derived from the HTTP request upon detection differs significantly from the expected probability vector. Because of the difference in categorizing the characters, the character distribution model provides two anomalous probabilities: $P_{character1}$ and $P_{character2}$.

4 Evaluation

As shown in Table 1, there are totally 57615 records in test set, among which 402 records are marked as anomalous and the total false positive rate is 0.698%. There are 117 records in negative test set, 114 of which are detected. Attribute relationship model behaved very well in detecting form-tampering exploitations. Fragment combination model detected most of XSS and SQL Injections. Length distribution model is sensitive to some attacks containing long strings. Character distribution is an all-powerful model, especially in detecting buffer overflowing and Internet worms.

Table 1. Results of detecting the records in the test set and negative test set

Detection models	Test set	Negative test set 117					
	57615	XSS 15	SQL Injection 15	Buffer overflow 10	Form tamper 27	Worm 5	Others 45
Attribution							
-integrity	41	0	0	0	2	0	0
-validity	35	1	0	0	8	0	4
-enumeration	112	1	0	0	18	0	3
-interdependence	57	0	0	0	11	0	8
Fragment	57	14	11	2	0	1	17
Length	57	7	4	8	0	4	6
Character							
- distribution 1	57	5	8	9	4	5	9
- distribution 2	54	2	3	10	2	5	7
Total detected	402	15	15	10	26	5	43
False positive	0.698%	—	—	—	—	—	—
Detection rate	—	100%	100%	100%	96.3%	100%	95.6%

References

1. Kruegel, C.: Anomaly Detection of Web-based Attacks. CCS'03, October 27–31, 2003, Washington, DC, USA
2. Billingsley, P.: Probability and Measure. Wiley-Interscience, 3 edition, April 1995
3. Mitchell, T.: Machine Learning. McGraw Hill, 1997
4. CGISecurity. The Cross Site Scripting FAQ

Exploring Content-Based and Image-Based Features for Nude Image Detection*

Shi-lin Wang[1], Hong Hui[1], Sheng-hong Li[2], Hao Zhang[2],
Yong-yu Shi[1], and Wen-tao Qu[2]

[1] School of Information Security,
[2] Dept. of Electronic Engineering,
Shanghai Jiaotong Univ., Shanghai, China
{wsl, huih, shli}@sjtu.edu.cn

Abstract. This paper introduces some widely used techniques related to nude image detection. By analyzing the merits and drawbacks of these techniques, a new nude image detection method is proposed. The proposed approach consists of two parts: the content-based approach, which aims to detect the nude image by analyzing whether it contains large mass of skin region, and the image-based approach, which extracts the color and spatial information of the image using the color histogram vector and color coherence vector, and makes classification based on the CHV and CCVs of the training samples. From the experimental results, our algorithm can achieve a classification accuracy of 85% with less than 10% false detection rate.

1 Introduction

With the rapid development of multimedia technology, images become popular on the internet as they are more intuitional and vivid compared to texts. However, among them, there are many nude images which may induce upset for users and do harm to the children. Therefore, an effective method to detect and filter the pornographic images on the world-wide web becomes necessary.

In view of this, many researchers have proposed various methods to detect the pornographic images in recent years. For example, D.A. Forsyth and M.M. Fleck proposed an automatic system [1] to detect whether there are naked people present in an image. However, the object is recognized by their algorithm only if a suitable group can be built. Therefore, the identification of the object depends greatly on non-geometrical cues (e.g. color) and on the interrelations between parts. Recently, Duan et al. [2] put forward a hierarchical method for nude image filtering. In their method, a skin color model is adopted to detect the skin region in the image and images with certain quantities of skin pixels are regarded as nude image candidates. Then the

---

* This paper was supported by National 863 Hi-Tech project fund (Contract no. 2003AA162160) and the National Nature Science Foundation of China (Grant no. 60402019).

SVM and the nearest neighbor methods are used to further verify whether the candidates are nude or benign. Experimental results show that 85% nude images could be filtered accurately.

2 Details of the Proposed Algorithm

By analyzing the merits and drawbacks of the traditional techniques, we proposed a new method to detect nude images, which makes use of both the content-based and image-based features and can be divided into the following three procedures:

(1) Face detection →color filtering →texture analysis →skin region analysis
(2) Color histogram vectors extraction →PCA →SVM classification
(3) Color coherence vectors extraction →PCA →SVM classification

(1) is a content-based method, which aims to detect the nude images based on the distribution of the extracted skin region. Generally speaking, there are large bulks of skin region in the nude images and thus the distribution of the skin region could be a good measurement to differentiate suspicious nude images from benign ones. (2) and (3) are image-based methods, which detect the nude images by analyzing the fundamental color features, the color histogram vector (CHV) and the color coherence vector (CCV). These methods are based on the fact that the color distributions of the nude and benign images are different to some extent. The final decision is drawn by analyzing the classification results of all the three methods mentioned above.

2.1 The Content-Based Approach

The face detection algorithm proposed by Paul Viola and Michael Jones [3] is adopted for its efficiency and relatively high accuracy. The face detection process is performed first to detect such close-face images and mark them as non-nude. After the face detection, the skin pixel "candidates" are first detected by a color filtering process. From the experiments, it is observed that the color distribution between the skin pixels and background ones can be better differentiated in HSI color space than that in the RGB color space. And the proper color thresholds of the skin pixels can be set as $H \in [0,1.6] \cup [5.6, 2\pi]$, $I > 100$, $0.1 < S < 0.88$ and $R > 240$ empirically. In order to reduce the noise influence, a median filtering process is applied to smooth the result.

It is known that the skin regions usually have a specific luminance texture. Analyzing the texture information could further differentiate the skin regions against the background ones. The Gabor filter [4] is employed to describe the texture of a certain image block. Then a training database is built, which contains a great number of skin and non-skin blocks and the Gaussian Mixture Models (GMM) method is employed to model the texture features of both the skin and non-skin blocks in the training database. Finally, a certain block is classified as a skin block if its texture features can better fit the skin model than the non-skin one, and vice versa. After color filtering and texture analysis, the skin region is extracted. The final decision of the content-based approach is made by analyzing the total skin area percentage and the largest connected skin area percentage.

2.2 The Image-Based Approach

The histogram of the image indicates the color distribution of all the pixels. A color histogram H is a vector $<h_1, h_2,..., h_n>$, where h_j stands for the number of pixels whose color information is in the jth bucket. In our approach, each channel is evenly divided into 8 segments and there are 512 bins in total. Hence the color histogram vector is a vector of 512 dimensions.

However, the color histograms lack of spatial information and images with very different appearances may have similar color histograms. To enhance the performance, we adopt another fundamental image feature, the color coherence vector, which takes both the color information and the spatial coherence of the color similar pixels into account and thus the CCV feature provides further differentiation than the feature using the color information only. Similar to the CHV calculation, each color channel is divided into 8 segments and the RGB color space is divided into 512 bins accordingly. The color coherent vector is calculated by analyzing the coherent and incoherent regions of each color bin. The overall dimensions of the CCV equal to 512*2=1024.

From the above analysis, the CHV and CCV are of a high dimension and contain much redundancy, which may lead to confusion for the classifier. Hence the Principle Component Analysis (PCA) is adopted to reduce the dimension of the feature vectors before input to SVM. From the experimental results, it is observed that the first 100 components can express over 97% feature variations.

3 Experimental Results

In order to test the performance of our nude image detection algorithm, a database containing about 9,000 images collected from the internet is built. The total number of nude images is about 3,000 and the total number of non-nude images is 6,000 or so. Half of the non-nude images contain people with various clothing and postures and the other half contains animals, cars, building, scenery, etc.

In order to select proper parameters for the Gabor filter and the Gaussian Mixture Model, we have evaluated various settings of 3 kinds of scales, 5 kinds of rotations, 2

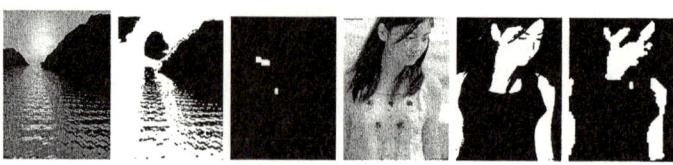

Fig. 1. Original images, skin detection results by color filtering and texture analysis

kinds of block sizes and 4 kinds of number of mixtures. From the evaluation, the setting with 2 scales, 4 rotations and 16*16 image block of the Gabor filter and 3 Gaussian mixtures is of the minimum error rate among all the settings investigated.

Fig. 1 illustrates the importance of the texture analysis procedure in skin detection. For a scenery image, the second image in Fig. 1 shows the corresponding skin detection result by color filtering and it is observed that many pixels whose color information is similar to the skin color are classified as skin pixels. After texture analysis, such erroneous skin pixels are removed because their texture features are different from those of the skin region. In comparison, for an image containing large skin bulks, the right-most image in Fig.1 shows the skin detection result by the color filtering and texture analysis. From the figures, it can be observed that most of the "genuine" skin pixels detected by the color filtering process remain unchanged by the texture analysis.

After the skin region is obtained, the percentage of the total skin area and the percentage of the largest skin bulk are analyzed to differentiate the nude images from the benign ones. An FAR of 0.1 (10%) and an FRR of 0.19 (19%) can be achieved by the content-based approach.

For the image-based approaches, i.e., SVM classification based on the CHV and CCV, different size of the training database is selected and their classification accuracy is evaluated by the equal error rate (EER). EER refers to the error ratio when FAR equals to FRR. The experimental result shows that EER of 15.9% (for CHV) and 16.5% (for CCV) can be achieved when only 3% of all the samples in the database are used for training SVM, which shows good generalization of our approach. It should be noted that the training samples are randomly selected from the database and not included in the test set.

With the classification results of all the three approaches, i.e. content-based approach, image-based approaches (SVM on CHV and SVM on CCV), the final classification result is derived by majority voting among all these three approaches. As a result, 85% nude images can be detected with less than 10% false detection rate.

4 Conclusions

In this paper, we proposed a nude image detection method which is composed of content-based and image-based classification approaches. In the content-based approach, color filtering and texture analysis is used to detect the skin region in the image. And the image is classified as a nude image if it contains large skin bulks. In the image-based approach, the color histogram and coherence vector are extracted to represent the color and spatial information of the image. Then SVM is employed to classify the images into two groups, nude and non-nude, based on the CHV and CCVs of the training samples. An accuracy of 85% can be achieved with less than 10% false detection rate, which demonstrates our method can detect the nude images effectively.

References

1. D.A. Forsyth and M.M. Fleck: Identifying Nude Pictures. Proceedings of IEEE Workshop on Applications of Computer Vision (1996) 103-108.
2. Duan Lijuan, Cui Guoqing, Gao Wen and Zhang Hongming: A Hierarchical Method for Nude Image Filtering. Journal of Computer-Aided Design and Computer Graphics, vol. 14, no. 5, 2002, 404-409.
3. Paul Viola and Michael Joes: Rapid Object Detection using a Boosted Cascade of Simple Features. Proceedings of the Fourth ACM International Conference on Multimedia (1996) 65-73.
4. B.S. Manjunath and W.Y. Ma: Texture Features for Browsing and Retrieval of Image Data. IEEE Transactions on Pattern Analysis and Machine Intelligence, vol. 18, no. 8, 1996, 837-842.

Collision Recognition and Direction Changes Using Fuzzy Logic for Small Scale Fish Robots by Acceleration Sensor Data

Seung Y. Na[1], Daejung Shin[2], Jin Y. Kim[1], and Su-Il Choi[1]

[1] Dept. of Electronics and Computer Engineering, Chonnam National University,
300 Yongbong-dong, Buk-gu, Gwangju, South Korea 500-757
{syna, beyondi, sichoi}@chonnam.ac.kr
[2] Graduate School, Chonnam National University,
300 Yongbong-dong, Buk-gu, Gwangju, South Korea 500-757
djshin71ha@hotmail.com

Abstract. For natural and smooth movement of small scale fish robots, collision detection and direction changes are important. Typical obstacles are walls, rocks, water plants and other nearby robots for a group of small scale fish robots and submersibles that have been constructed in our lab. Two of 2-axes acceleration sensors are employed to measure the three components of collision angles, collision magnitudes, and the angles of robot propulsion. These data are integrated using fuzzy logic to calculate the amount of propulsion direction changes. Because caudal fin provides the main propulsion for a fish robot, there is a periodic swinging noise at the head of a robot. This noise provides a random acceleration effect on the measured acceleration data at the collision instant. We propose an algorithm based on fuzzy logic which shows that the MEMS-type accelerometers are very effective to provide information for direction changes.

1 Introduction

A group of small scale fish robots and submersibles have been constructed in our lab. They have been tested for path design, collision avoidance, maneuverability, posture maintenance, and communication in a tank of 120 x 120 x 180 cm dimension. Depth control using strain gauges, acceleration sensors, illumination, and the control of motors for fins or screws are processed based on the MSP430F149 by TI. Also, stereo images by two cameras are processed by a controller card based on an i386EX processor. User commands, sensor data and images are transmitted by Bluetooth modules of class 1 between robots and a host notebook PC while fish robots are operated within 10 cm depth. An RF module using an operating frequency of about 173MHz from Radiometrix is used when the depth is larger than 10 cm. They are operated in autonomous and manual modes in calm water. Manual operations are by remote control commands in various Bluetooth protocol ranges depending on antenna configurations.

Walls, rocks, water plants and other nearby robots are selected obstacles in the tank for experiments. Sonar sensors are not used to make the robot structure simple and compact. All circuits, sensors and processor cards are contained in a box of 9 x 7 x 4 cm dimension except motors, fins/screws and external covers. Also, image processing results captured by the camera lens at the head are used to avoid collisions. But the results are useful only when the obstacles are far enough for the images processing calculation time to detect them. Otherwise, acceleration sensors are used to detect collision immediately after it happens.

Two of 2-axes acceleration sensors ADXL202JE are employed to measure the three components of collision angles, collision magnitudes, and the angles of robot propulsion. These data are integrated to calculate the amount of propulsion direction change. Since the main propulsion for a fish robot is obtained by the caudal fin, which moves on both sides on the horizontal plane by a servo-motor, there is a continuous swinging action at both tips, head and tail, of the fish robot. This swinging movement results in a certain kind of an unwanted periodic acceleration signal. Acceleration data at the head of the fish robot are measured continuously at 200 Hz rate. Instantaneous values give the information such as collision, touches, or swinging of its body. The average values of a certain interval of time span represent the tilt of fish robot body.

Proper direction changes of the propulsion are necessary to avoid successive collisions once a collision is detected. The angle of collision incident upon an obstacle is the fundamental value to determine a direction change needed to design a following path. Since acceleration is measured in pursuit of finding the collision angle, disturbance from the caudal fin motor should be reduced. To do this, the angle of the caudal fin at the instant of the collision is calculated through a swinging model of the fin. Since a particular swinging frequency of the fin is about 4.2 Hz, while the sampling frequency of the acceleration data is 200 Hz, it may be assumed that the swing is sinusoidal with unknown delay with respect to the motor input signals. The time delay is calculated numerically so that the proposed algorithm produces acceleration data as close as possible for a set of experiments with the given angles of 45, 60, 75 and 90 degrees. Finally, the optimal incident angle of a collision is obtained by compensating the measured acceleration data with the swinging disturbance at the instant of the collision.

2 Components of a Fish Robot

In this presentation, a fuzzy logic algorithm is proposed to estimate the incident collision angles using only two of 2-axes acceleration sensor chips when there is a considerable amount of noise from the propulsion motors. Although a simple small scale fish robot is considered, there are many components necessary to do path design, collision avoidance, maneuverability control and posture maintenance. Two servo motors are used at the caudal fin: one for propulsion and the other for horizontal direction control. Other two servo motors are used at pectoral fins; each one at a side for propulsion and vertical direction control. A strain gauge is used to measure water pressure for depth control. MEMS-type acceleration sensors are installed to measure force changes as well as the posture of the fish body. Illumination by two LED's is for the camera images when additional light is necessary. Stereo images by two cameras are processed by a controller card based on an i386EX processor. All signals other than images are processed based on the MSP430F149 by TI. Also, user commands, sensor data and images are transmitted between robots and a host notebook PC either by Bluetooth modules of class 1 or by an RF module depending on the operation depth.

2.1 Microcontroller MSP430F149

MSP430F149 is adopted for a microcontroller of the small scale structure with many components to be interfaced since it requires a single 3.3V source while two 16 bit timers, eight channels of 12 bit A/D converters, two USART ports are provided internally. Using these merits, the data from a pressure sensor and two acceleration sensors, control signals to four motors are processed every 5ms. Also the sensor data can be transmitted to a host PC using communication modules.

2.2 Acceleration Sensors

One of the most important sensor data for satisfying maneuverability and control of a fish robot is about the posture of its body. The tilt data of three axes are necessary for the analysis of its body posture. Acceleration data are measured continuously at 200 Hz rate by the sensors located at the head of a fish robot. Two of 2-axes MEMS-type acceleration sensors ADXL202JE are used to measure the three components of force changes due to collision, touches, propulsion or disturbances applied on a fish robot's body.

The sensor is an 8-pin IC chip containing a MEMS type sensor inside, measuring acceleration in the range of ±2 G. Either digital or analog type output signals can be interfaced to read the data. The same 3.3 V single operating voltage as for MSP430 is applied to produce an output voltage of 0.0 V for –2 G, 3.3/2 V for 0 G, and 3.3 V for 2 G.

Instantaneous acceleration values give the force information on its body, and the average values of a certain interval of time span represent the tilt of the fish robot body. Since the collision of the fish robot while it is swimming forward is a major concern, the sensors are located at the tip of its head. The output values have relatively small magnitudes since the system is operated in the water. Therefore precision instrumentation amplifiers such as MAX419x are necessary in practical circuit designs.

3 Collision Detection

The most urgent problem in maneuverability and control of the fish robots is collision avoidance. At least, when collision is detected the propulsion direction should be changed properly so that swimming trajectory is modified and successive collision does not occur. Collision incident angle at the instant of collision is the key data upon which desired direction changes of the forward swimming can be designed.

Forward propulsion is provided mainly by the caudal fin motor and partly by the pectoral fin vertical plane respectively for propulsion, there appear swinging disturbance movements on both planes.

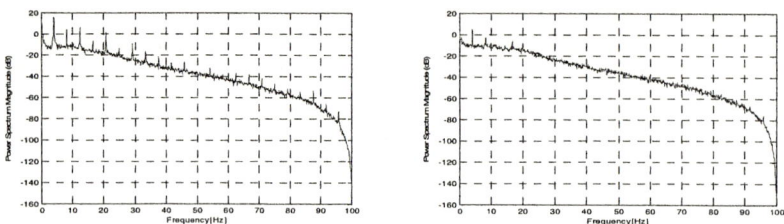

Fig. 1. x and y-axis power spectral density of acceleration data from a fish robot

Although fish robots swim in a three dimensional space freely and therefore collision occurs in the same space, the description of collision detection and the propulsion direction changes hereafter are restricted on the horizontal plane. The same argument can be applied on the vertical plane symmetrically. Since acceleration of the body is monitored to detect the collision and its angle against obstacles, disturbance effects from the caudal fin motor should be reduced.

First, the coordinate system in the x-y plane of the fish robot trajectory is defined y-axis as the heading direction and x-axis as the lateral direction. Figure 1 represents typical power spectral density of the x and y-axis acceleration data when the fish robot is moving forward using caudal and pectoral fins both in calm water without collision. A noticeable characteristic feature from the Figure 1 is the resonance frequency of about 4.2Hz and its harmonics. These harmonics are due to the caudal fin motor since only the horizontal disturbances are considered. Thus its effect is more apparent at x-axis than at y-axis. But the acceleration changes due to collision are more vivid at y-axis than at x-axis. Therefore it is not difficult to detect the collision occurrence by observing y-axis acceleration changes.

4 Estimation of Collision Angles

Acceleration measurement data are considered as noise of swinging action due to the caudal fin motor until a collision is detected. When a collision is detected, proper direction changes of the propulsion are necessary to avoid successive collisions and to modify its trajectory. The angle of a collision incident upon an obstacle is the fundamental value to determine a direction change needed to design a following path. The collision angles are symmetrical for y-axis; therefore, both of x and y-axis acceleration components should be used together to estimate collision angles.

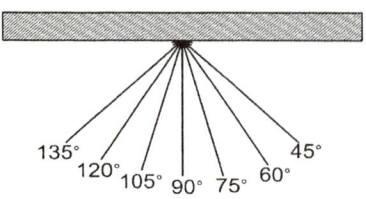

Fig. 2. Collision incident angles

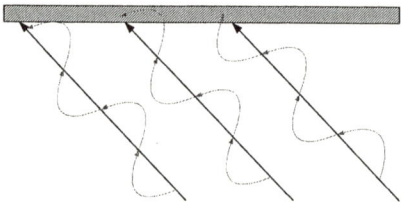

Fig. 3. Disturbance due to caudal fin

As expected, x-axis acceleration component is much influenced by the swinging action due to the caudal fin motor. This phenomenon is exemplified in the Figure 3. It shows such different faulty measured angles, as about -20°, 90°, and 110° respectively, for the same incident angle of 45°. Since acceleration is measured to find the collision angle, disturbance from the caudal fin motor should be reduced as much as possible.

At the beginning, to reject high frequency noise components, we assume the 20Hz low pass filtered data of the measured acceleration as the raw data. The basic approach is to obtain the oscillating components by passing the raw data through the fourth order Butterworth band pass filter of 3-5Hz. Then the time shift due to computation is optimized, that is, the mean squared errors of the compensated acceleration data are calculated. It is obvious to select the sampling time as the necessary time shift when the mean squared error is minimum.

The overall configuration of the collision angles estimation system is represented as in Figure 4. A spectral subtraction method is used for common mode disturbance rejection.

The raw data of acceleration, noise compensated acceleration data by the spectral subtraction method, and the input commands to the caudal fin motor are shown in Figure 5. Motor input signals are represented using "•" at +1 when it is to the right, and at -1 when it is to the left. The time period is 5 $msec$, which is the same for MSP430 processor. The motor shift commands are 24 sampling times which is 0.12 sec for either way. This corresponds to the harmonics of about 4.2 Hz in Figure 1.

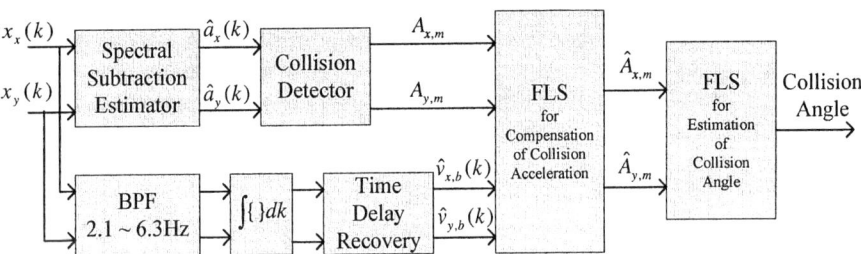

Fig. 4. Block diagram of collision angle estimation system

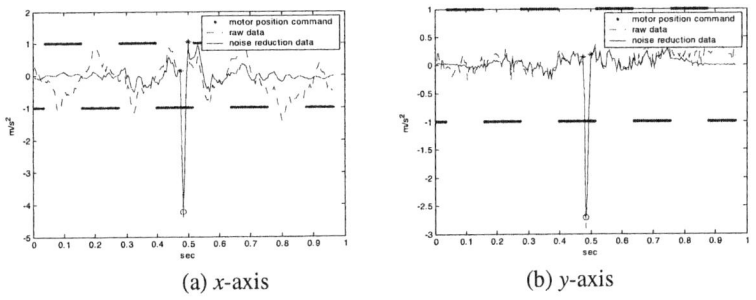

(a) x-axis　　　　　　　　　　(b) y-axis

Fig. 5. Comparison of raw and compensated acceleration signals

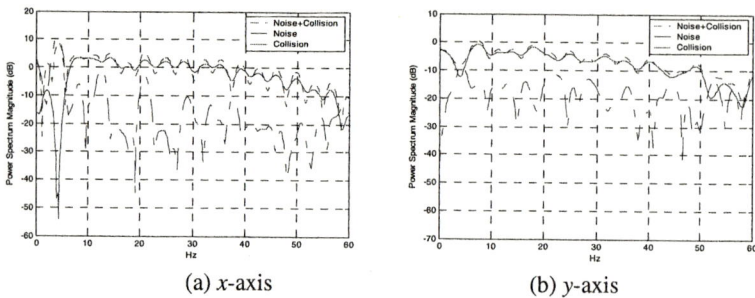

(a) *x*-axis   (b) *y*-axis

Fig. 6. Comparison of raw and compensated acceleration signals in frequency domain

Thirty experiments of the swimming forward and collision for each incident angle of 45°, 60°, 75°, 90°, 105°, 120°, and 135° were carried out for reference acceleration data. There are considerable disturbances from the caudal fin motor as described in Figure 5. Thus the measured acceleration data without compensation are quite dispersed. Measured values are overlapped for different values of angles. Therefore these references can not be used directly to estimate the incident angles.

4.1 Application of Fuzzy Logic to Estimate Collision Angles

Since the main propulsion for a fish robot is provided by the caudal fin which moves on both sides on the horizontal plane by a servo-motor, there is a continuous swinging action at both head and tail of the fish robot. This swinging movement as in Figure 3 results in the dispersion of the measured acceleration data. We propose a collision angles estimation system of Figure 4 based on fuzzy logic algorithm. The reasons for the fuzzy logic are the vagueness of instantaneous noisy acceleration at the instant of collision, and nonlinear effect of waves on the movement of a fish robot. The fuzzy logic system is to improve the estimation of incident angles upon collision using the information of the maximum changes of the measured acceleration data $A_{x,m}$ and $A_{y,m}$. Gaussian membership functions are used for input membership functions.

$$F(x) = e^{-\frac{(x-c)^2}{2\sigma^2}} \qquad (1)$$

The ith fuzzy rule for TSK inference is given as

$$R^i : \text{IF } x_1 \text{ is } F_1^i \text{ and } \ldots x_k \text{ is } F_k^i \ldots, \text{ Then } z^i = c_0^i + c_1^i x_1 + \ldots + c_K^i x_K \qquad (2)$$

where x_k is fuzzy input, z^i is fuzzy output, and F_k^i is a fuzzy membership function.
     $i = 1, 2, \ldots, I$, $k = 1, 2, \ldots, K$
     I : number of fuzzy rules,
     K : number of fuzzy inputs,
       c^i : linear value of fuzzy output

Fuzzy output is given by

$$z^* = \frac{\sum_{i=1}^{I}\prod_{k=1}^{K} F_k^i(x_k) z^i}{\sum_{i=1}^{I}\prod_{k=1}^{K} F_k^i(x_k)} \tag{3}$$

$A_{x,m}$: maximum changes of the measured acceleration data of x-axis
$A_{y,m}$: maximum changes of the measured acceleration data of y-axis

- Compensation of $A_{x,m}$: $A_{x,m}$, $A_{y,m}$, $\hat{v}_{x,b}$ are used as fuzzy input variables
- Compensation of $A_{y,m}$: $A_{x,m}$, $A_{y,i}$, $\hat{v}_{y,b}$ are used as fuzzy input variables

○ Sampling frequency: 200Hz
○ Bandwidth of accelerometer sensors: 50Hz
○ Oscillation frequency: 4.2Hz
○ Number of samples for FFT: 48 samples × 4 periods = 192 samples
○ Data for training: for each angle 15 experiments × 7 = 105 data sets
○ Data for tests: for each angle 15 experiments × 7 = 105 data sets

4.2 Estimation Results of Incident Collision Angles

The collision angles estimation system represented as in Figure 4 was applied for the experiment results with above parameters. Table 1 and Figure 7 are the results for raw data set. Thus the measured acceleration data without compensation are quite dispersed. Measured values are overlapped for different values of angles. Therefore these references can not be used directly to estimate the incident angles.

Table 1. Changes of x-axis and y-axis acceleration

Value	Angle	45°	60°	75°	90°	105°	120°	135°
training	mean	-3.4410	-2.7889	-1.2821	0.1713	0.8972	2.0963	2.5669
	std	0.5461	0.6028	0.7702	1.0390	0.7426	0.6981	1.1319
test	mean	-2.0215	-3.2258	-3.7742	-3.8525	-3.1328	-2.4062	-1.1727
	std	0.2865	0.2510	0.4000	0.3196	0.4981	0.3818	0.5381

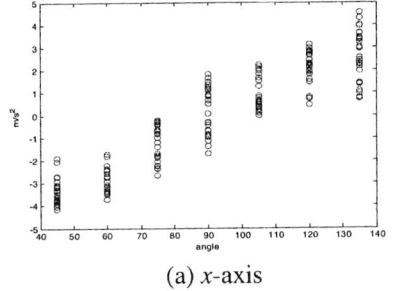
(a) x-axis

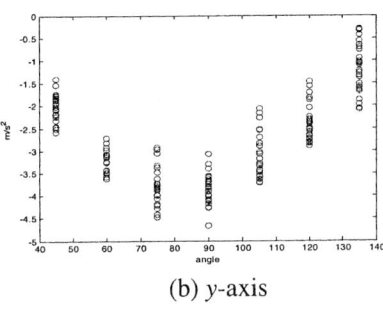
(b) y-axis

Fig. 7. Changes of x-axis and y-axis acceleration

Table 2. Modified changes of x-axis and y-axis acceleration for training data

Value	Angle	45°	60°	75°	90°	105°	120°	135°
training	mean	-3.8590	-3.1856	-1.7624	0.0561	1.5938	2.8044	3.8615
training	std	0.1317	0.3938	0.3833	0.7366	0.3825	0.3249	0.1275
test	mean	-2.3457	-3.2456	-3.8344	-4.2144	-3.6060	-2.8006	-1.7288
test	std	0.1444	0.2952	0.1902	0.1755	0.2886	0.1967	0.1344

Table 3. Modified changes of x-axis and y-axis acceleration for test data

Value	Angle	45°	60°	75°	90°	105°	120°	135°
training	mean	-4.0931	-3.3284	-1.6020	-0.0758	1.5805	2.8653	3.4005
training	std	0.3824	0.4914	0.7619	0.8521	0.7736	0.3057	0.2989
test	mean	-2.4348	-3.3216	-3.8737	-4.1426	-3.6822	-2.8814	-1.9435
test	std	0.2394	0.1510	0.2093	0.2116	0.3259	0.1749	0.3145

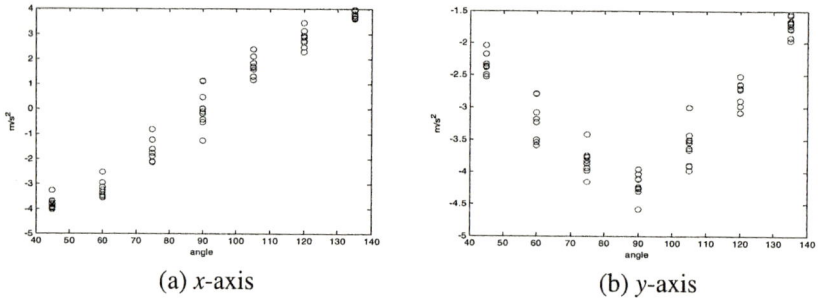

(a) x-axis　　　　　　　　　　(b) y-axis

Fig. 8. Modified changes of x-axis and y-axis acceleration for training data

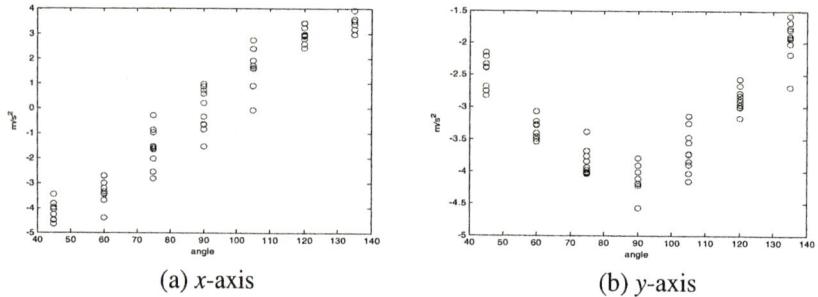

(a) x-axis　　　　　　　　　　(b) y-axis

Fig. 9. Modified changes of x-axis and y-axis acceleration for test data

Table 4. Estimstion errors

Value	Angle	45°	60°	75°	90°	105°	120°	135°
training	mean	-0.0016	-0.0970	-0.0970	0.1811	0.0296	-0.0425	0.0273
	std	0.1420	0.5768	1.2037	1.2498	0.6422	0.5419	0.4911
test	mean	-0.0974	-0.0658	-0.9772	-0.2006	1.2621	-0.0596	0.1383
	std	0.6477	1.4456	3.5524	4.1838	3.1797	1.5464	0.9086

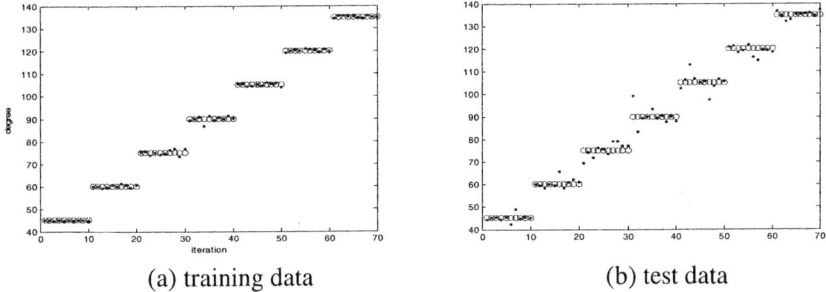

(a) training data (b) test data

Fig. 10. Estimation errors of the fuzzy logic system

5 Conclusions

The angle of a collision incident upon an obstacle is the fundamental value for a fish robot to obtain a direction change needed for better swimming trajectory or to avoid successive collisions. But there are significant amount of disturbances due to a caudal fin motor. This effect gives random acceleration deviations on the measured acceleration data at the collision.

We propose a collision angles estimation system based on fuzzy logic algorithm considering nonlinear effects of caudal fin movements to the head of the fish robot. The system produced improved estimation.

Acknowledgments

This work was supported by the RRC-HECS, CNU under grant R12-1998-007003-0.

References

1. Sfakiotakis, M., Lane, D. M., Davies, J. C.: Review of fish swimming modes for aquatic locomotion. IEEE Journal of Oceanic Engineering, Vol. 24, No. 2,(1999) 237-252
2. Yang, E., Ikeda, T., Mita, T.: Nonlinear Tracking Control of a Nonholonomic Fish Robot in Chained Form. SICE Annual Conference, Hukui University, Japan. (2003) 1260-1265

3. Kim, E. J., Youm, Y.: Design and Dynamic Analysis of Fish Robot: PoTuna. Proceedings of the 2004 IEEE International Conference on Robotics & Automation, New Orleans, LA (2004) 4887-4892
4. Kato, N.: Control Performance in the Horizontal Plane of a Fish Robot with Mechanical Pectoral Fins. IEEE Journal of Oceanic Engineering, Vol. 25, No. 1, (2000) 121- 129
5. Shin, D. J., Na, S. Y., Kim, J. Y.: Reduction of Disturbance Effects and Sustained Oscillation Using Multi-Sensor Fusion in a Flexible Link System. IECON04, Busan, Korea. (2004)
6. Shin, D. J., Na, S. Y., Na, C. H., Kee, C. D.: Robust Dynamic Weight Measurement Using Acceleration Data. GSPx2004, Santa Clara (2004)
7. Jang, J. R.: ANFIS: Adaptive-network-based fuzzy inference system. IEEE Trans. Syst., Man, Cybern., 23 (1993) 665¨C685
8. Frayman, Y., Wang, L.P.: Data mining using dynamically constructed recurrent fuzzy neural networks. Proc. 2nd Pacific-Asia Conference on Knowledge Discovery and Data Mining, LNCS Vol. 1394 (1998) 122-131
9. Wai, R.-J., Chen, P.-C.: Intelligent tracking control for robot manipulator including actuator dynamics via TSK-type fuzzy neural network. IEEE Trans. Fuzzy Systems 12 (2004) 552-560
10. Kiguchi, K., Tanaka, T., Fukuda, T.: Neuro-fuzzy control of a robotic exoskeleton with EMG signals. IEEE Trans. Fuzzy Systems 12 (2004) 481-490

Fault Diagnosis Approach Based on Qualitative Model of Signed Directed Graph and Reasoning Rules

Bingshu Wang, Wenliang Cao, Liangyu Ma, and Ji Zhang

School of Control Science & Engineering, North China Electric Power University,
Baoding 071003, China
caowenliang2002@163.com

Abstract. Signed Directed Graph (SDG) is a fault diagnosis method based on qualitative model and cause and effect analysis, first, it establishes the SDG of the systems and components, simplifies these SDGs corresponding to the fault patterns diagnosed, SDGs are described the many rules forms for shortening the calculating time, then expands the diagnosing rule with expert knowledge to construct the diagnosing rule bank of the system. Second, transforming the quantitative values of the system's variables into qualitative values, the fault patterns can be primary diagnosed. And then the patterns that can not be distinguished are diagnosed by using Fuzzy knowledge to form a qualitative and quantitative model. The case studies show the improved method is valid.

1 Introduction

Signed Directed Graph (SDG) [1] is a very convenient tool to provide a visual description on the industrial system structure by using graph to show the cause and effect relation between system variables [1]. Applying the description system SDG-based and causal analysis, it can identify the root cause by using the information saved on the SDG to search for the possible fault source of disturbance. In a SDG, the value of each node represents the qualitative state of a given process variable: normal (0), high (+) or low (-). Each arc represents the influence of a variable (cause node) on another variable (effect node). A direct relation (both nodes deviate in the same direction) is shown by (+) gain, the inverse by (-) and (0) shows no relation. In other words, in a SDG the gain G_{ij} corresponding to the arc from node x_i and node x_j is a qualitative indicator of the direct effect on x_j due to a perturbation on x_i. Considering the process mathematical model, generally rewritten in the following form:

$$dx_i/dt = f_i(x_1, x_2,..., x_n) \qquad (1)$$

A general expression for is given by:

$$G_{ij} = Sign(\partial x_j/\partial x_i) \qquad (2)$$

---
[1] This paper is supported by the Teacher Fund of Doctor's degree of North China Electric Power University (20041209).

Where Sign function is used to obtain qualitative value, the relationship between sign of $\partial x_j / \partial x_i$ is the state of the arc between x_i and x_j in the SDG model.

A root node is any node in the SDG that has at least one consistent arc connecting it to an effect node and no consistent arc connecting it to a cause node. An arc is said to be consistent if Sign (cause)* Sign (arc)* Sign (effect)=(+). Consistent arcs are the unique paths that can propagate the fault information.

In actual diagnosis process, each fault source will has large amount of according fault patterns, Even though someone shortens the diagnosis time by improving algorithm, the time requirements still can't be satisfied in practical application. So we will introduce a method of transforming the SDG to a series of rules, which can greatly increase the diagnosis speed [1].

(1) In order to conveniently examine if one arc is a consistent arc, we will adopt logic sign to describe, the "P" and "m" are defined as the logic function form of the arc sign "+"and "-". The logic relation of SDG is shown in Table 1. in addition, $A.lt.0 \leftrightarrow A$ is lower than 0; $A.eq.0 \leftrightarrow A$ is equal to 0; $A.ht.0 \leftrightarrow A$ is higher than 0。

Table 1. Logic Relation of SDG

Types	+ $A \to B \Leftrightarrow (pAB)$			− $A \to B \Leftrightarrow (mAB)$		
A \ B	+	0	−	+	0	−
+	T	F	T	F	F	T
0	F	T	F	F	T	F
−	F	F	T	T	F	F

(2) To select one root node, and delete the arcs that point the root node and the node that the root node cannot reach from SDG.

(3) To delete the unmeasurable nodes and to form a graph that composed of measurable nodes according to the sign of arc.

(4) As for each measurable node "n_i", the following rule need to added:
$\cdots and[(*k_1 n_i)or(*k_2 n_i)\cdots or(*k_j n_i)]$

(5) As for positive feedback loop, the following rule need to added:
$\cdots and[(*B_1)or(*B_2)\cdots or(*B_i)]$

In which k_j and B_i are the input nodes of n_i, * is the logic function form $p(+)$ or $m(-)$ from k_j or B_i to n_i.

2 Fault Diagnosis Based on SDG Model

The Thermal Exchange Equipment (TEE)[2] in industry process will be taken as an example to elaborate the reasoning rule and fault diagnosis of SDG in detail; Fig.1 is the corresponding SDG model of TEE. In the case, five fault patterns are taken into consideration, such as the blockage of in V_1; the jam in F_{11}; the increase in T_1; the blockage of in V; the jam in F_{011}; they are marked as P1, P2, P3, P4 and P5 separately.

In actual application, owing to the cause of technology and economy, some nodes in Fig.1 are unmeasurable, such as the node V_1, F_{11} and F_{011}, in addition, the node L_S and F_S are set point. According to the reasoning rules above, for example, the sub SDG corresponding to P1 is shown in Fig.2, the corresponding diagnosis rule is [Rule]. By the same way, we can get the diagnosis rule bank.

[Rule]: $IF(T_0.eq.0) and (T_1.eq.0) and (F_1.lt.0)$ $and(mTT_C) and (pT_C F_C)$
$and[(pF_1L) or (mF_2L)] and (mLL_C) and (mL_C V) and (pVF_2)$
$and(mF_C V_0) and (pV_0 F_{01}) and [(pLT) or (pT_{01}T)] and [(mF_{01}T_{01}) or (pTT_{01})]$

The positive feedback loop between T and T_{01} :

$and IF[(pLT) or (mF_{01}T_{01})]$

THEN V_1 is the possible fault source.

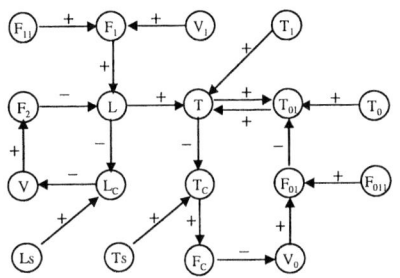 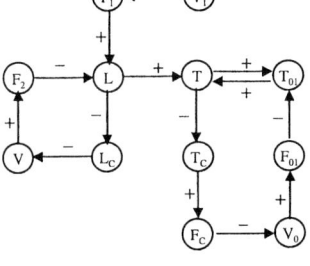

Fig. 1. SDG of TTE   **Fig. 2.** Sub SDG of P1

By using the rules above, the five fault patterns are tested by the instant measurable samples. In Table 2, the case studies show the P3, P4, P5 is unique determined based on the examination of consistent arc. While the fault P1 and P2 can't be distinguished, this is because they are same in qualitative characteristics, that is, SDG only applies qualitative knowledge in the process of establishment without considering other deep quantitative information.

Table 2. Cases Study

Cases	F_{01}	F_1	F_2	F_C	L	L_C	V_0	V	T	T_{01}	T_C	T_1	T_0	Results
1	−	−	−	+	−	+	−	−	−	−	+	0	0	P1
2	−	−	−	+	−	+	−	−	−	−	+	0	0	P2
3	+	0	0	−	0	0	+	0	+	+	−	+	0	P3
4	+	0	−	−	+	−	+	−	+	+	−	0	0	P4
5	−	0	0	−	0	0	+	0	+	+	−	0	0	P5

In order to effective distinguish some fault patterns that are same in qualitative characteristics but different in quantitative values, we should consider to combine the quantitative knowledge to Improve SDG (ISDG) diagnosis resolution, that is, to add the quantitative information replaces the arc sign in SDG with the Gain (rate of changing of influencing (cause node) and influenced variables (effect node)), and by using the fuzzy knowledge the fault sources are determined by calculating the membership grade of the patterns need be diagnosed to the given fault patterns [3][4].

(1) The construction and the scale of application of membership function.

a. The membership function from the root fault node to the one directly connecting with it:

If the arc sign is positive, then   if the arc sign is negative, then

$$\mu(g) = \begin{cases} 1 & g > 0 \\ 0 & g \leq 0 \end{cases} \quad (3) \qquad \mu(g) = \begin{cases} 0 & g \geq 0 \\ 1 & g < 0 \end{cases} \quad (4)$$

b. The membership function between other nodes is shown as following:

$$\mu(g) = \begin{cases} 0 & a \geq 0.3 \\ 1.5 - 5a & 0.1 \leq a \leq 0.3 \\ 1 & -0.1 \leq a \leq 0.1 \\ 1.5 + 5a & -0.3 \leq a \leq -0.1 \\ 0 & a \leq -0.3 \end{cases} \quad (5)$$

In which g^* is the Gain in given fault pattern, g is the Gain of corresponding nodes in fault pattern need to be diagnosed, and assumed the relative Gain $a = (g - g^*)/g^*$.

It should be mentioned that the framework of the membership function is determined by the accuracy of quantitative information and non-linear process of system, in the ISDG, the used quantitative knowledge is the influence from cause node to effect node through different propagate route, so it muse be a tree structure.

(2) As an example of P1 and P2 in Fig.1, the corresponding fault simulating parameters of "P1"and"P2", are shown in Table 3 (the normal value is 100%).

Table 3. Fault simulating parameters

Simulating Parameters		F_0	L	L_C	V_1	F_1	F_C	V_0	P
Fault patterns	P1	0.910	1.096	1.105	0.895	0.882	0.915	1.122	0.903
	P2	0.946	1.053	1.065	0.945	0.932	0.955	1.073	0.931

By following the procedures mentioned above, for example, as for the P1, the gains are calculated between arcs are the following, such as $\Delta L/\Delta F_0 = 1.0$, $\Delta V_0/\Delta F_C = 1.43$ and so on. The sign of arc in sub SDG corresponding to the P1 are replaced by its gains, and then the $ISDG_1$ is shown in Fig.3, in the same way, the corresponding $ISDG_2$ of P2 is shown in Fig.4.

(3) According to Largest membership grade model of Fuzzy mathematics, the formula of calculating the membership grade of fault pattern need be diagnosed to every given fault pattern is following:

$$\mu_i = \min[\mu(g_{i1}), \mu(g_{i2}), \cdots, \mu(g_{ij})] \quad (6)$$

In which i $(i = 1, 2, \cdots, n)$ is given fault pattern, j is the arc of given fault pattern, g_{ij} is the Gain of corresponding arc j in i fault pattern.

To calculate μ_i, and select the fault type i corresponding to the largest μ_i as the diagnosis result. For example, as for the ISDG$_1$, the formula of calculating the membership grade of P1 to every given fault pattern is following:

$$\mu = \min[\mu(\Delta F_0), \mu(\Delta L/\Delta F_0), \mu(\Delta L_C/\Delta L), \mu(\Delta V_1/\Delta L_C),$$
$$\mu(\Delta F_1/\Delta V_1), \mu(\Delta F_C/\Delta F_0), \mu(\Delta V_0/\Delta F_C), \mu(\Delta P/\Delta V_0)]$$

In the same way, the formula of calculating the membership grade of P2 to every given fault pattern can be acquired.

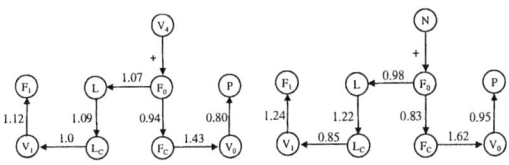

Fig. 3. ISDG$_1$ of P1   **Fig. 4.** ISDG$_2$ of P2

Table 4. The membership grade

Simulating Parameters		P1	P2
Fault patterns	P1	1.00	0.62
	P2	0.56	1.00

In Table.4, it can be seen that the membership grade of these two faults need to diagnosed to the given faults that are the same with them is largest, whose value is 1.0. However, to other given faults is smaller. So the P1 and P2 can be effective distinguished, which shows the ISDG is successful.

Contrasting the diagnosis result with [2], the ISDG model has better resolution.

3 Conclusions

(1) Fault diagnosis method based SDG rules can shorten greatly the time consumption, and it has good completeness, fine resolution and detailed explanation.
(2) The patterns that can not be distinguished are diagnosed by using Fuzzy knowledge to form a qualitative and quantitative model.

References

1. M.A. Kramer and J.B.L. Palowitch. A rule-based approach to fault diagnosis using the signed directed graph. *AIChE. J*, 33:1067-1078,1987
2. Wang Xiaochu. Application of Multivariable Statistical Method on Fault Diagnosis. M.S. Thesis. Zhejiang University. 2003
3. E. E. Tarifa and N.J. Scenna. Fault diagnosis, direct graphs, and fuzzy. Computer & Chemical Engineering, 21:649-654,1997.
4. Weidong Huang, Kechang Wang. Liquid Rocket Engine Fault Diagnosis Based on the Integration of Qualitative and Quantitative Knowledge, Journal of Aerospace Power, 11(3): 281-284,1996.

Visual Tracking Algorithm for Laparoscopic Robot Surgery

Min-Seok Kim, Jin-Seok Heo, and Jung-Ju Lee

[1] Mechanical Engineering Department, Korea Advanced Institute of Science and Technology,
373-1, Guseong-dong Yuseong-Gu, Daejeon, Korea
kidsnkins@samsung.co.kr, {dandyheo3070, leejungju}@kaist.ac.kr

Abstract. In this paper, we present a new real-time visual servoing unit for laparoscopic surgery. This unit can automatically control a laparoscope manipulator through visual tracking of the laparoscopic surgical tool. For the tracking, we present a two-stage adaptive CONDENSATION (conditional density propagation) algorithm to detect the accurate position of the surgical tool tip from a surgical image sequence in real-time. This algorithm can be adaptable to abrupt changes of illumination. The experimental results show that the proposed visual tracking algorithm is highly robust.

1 Introduction

Laparoscopic surgery is minimally invasive surgery (MIS), a new kind of surgery which is becoming increasingly common. In this method, the surgical operation is performed with the help of a laparoscope and several long, thin, rigid instruments through small incisions [1]. In recent years, laparoscope manipulators that involve the use of a camera assistant, for example, AESOP, are being used more widely in laparoscopic surgery. It is a bothersome task, however, for the surgeon to control the laparoscope manipulator manually or with his voice. Because the most important purpose of the laparoscope manipulator is aiming at surgical site with laparoscope, there has been some research on automatic control of laparoscope. To control the laparoscope automatically, the controller must have position information of the tool tip in surgery. Among the methods that obtain the tool tip position in surgery, the visual tracking is an efficient one. The previous methods [2]-[5] cannot avoid the loss of information because the methods use a threshold method in recognizing the tracking target feature. In addition, the methods have no adaptability to changes in the illumination of the surgical environment, so it is risky to apply them to real laparoscopic surgery.

In this paper, we present a new visual tracking algorithm, two-stage adaptive CONDENSATION based on the CONDENSATION algorithm [6]-[10]. The two-stage adaptive CONDENSATION algorithm possesses two advantages: one is adaptability to illumination change and the other is the ability of real-time visual tracking.

2 Visual Tracking Algorithm

We suggest a new visual tracking algorithm, two-stage adaptive CONDENSATION. The new algorithm based on the two-stage verification process in the perception of the tracking target and on the CONDENSATION algorithm has adaptability to the illumination change. In addition, it has high reliability for visual tracking because of the evaluation of color and shape features of the tracking target.

2.1 CONDENSATION Algorithm

The CONDENSATION algorithm is a sampling based algorithm. It uses stochastic propagation of conditional density, which is a probability distribution function constructed by weights of sampling positions in an image. There are three types of probability distribution functions in the algorithm: One is the prior state density, which has information about the tracking target position at a prior phase; Another is the posterior state density, which has information about the position at the posterior phase; And, the other is the process state density which couples the prior state and the posterior state. We take the sampling positions of the tracking target for state x and take the color feature of the surgical tool for observation z of CONDENSATION for the purpose of a simpler structure and faster speed in computation.

2.2 Adaptive Color Model

The adaptive color model [11] modified CONDENSATION algorithm has adaptability to the illumination change. As the model cannot be adaptable to the abrupt illumination change, we cannot use the adaptive color model directly for our case. This means that there are some difficulties in the case of abrupt changes of illumination. Therefore, we present a new adaptive color model for the CONDENSATION algorithm. The new model can be adaptable to the abrupt changes of illumination and the gradual changes of illumination. The model is based on a method in which the color features of the tracking target are updated at each time step. There are two color features updated in the model. One is the color feature of the prior time step and the other is the predefined color feature of the normal state and the abnormal state, as can be seen in table 1.

Table 1. Color Feature of the Surgical Tool

		Environment	Normal	Abnormal
Rod part	Hue	Median	0.9822	0.1592
		Std.	0.0137	0.0137
	Value	Median	0.4471	0.1843
		Std.	0.1589	0.0170
Tip part	Hue	Median	0.9822	0.1250
		Std.	0.0137	0.0225
	R	Median	0.8627	0.3686
		Std.	0.0226	0.0213

2.3 Two-Stage Adaptive CONDENSATION

Although the adaptive color model is adaptable to illumination change, it has a tendency to fail in tracking when the tracking target is hidden by obstacles whose color feature is similar to that of the tracking target. In the medical field, safety is of utmost importance. As the overlapping of surgical tools has often happened in laparoscopic surgery, the visual tracking algorithm cannot detect the visual target. Therefore, we suggest a new visual tracking algorithm, a two-stage adaptive CONDENSATION algorithm, which inherits the advantages of the CONDENSATION and the adaptive color model. Moreover, this algorithm has a robust tracking ability compared with the conventional visual tracking algorithm in several possible surgical situations. The CONDENSATION algorithm with the modified adaptive color model is applied to the each part of laparoscopic surgical tool. Therefore, the new algorithm performs a two-stage verification in perception of the tracking target. It is important that the new algorithm performs not only the verification procedure of the CONDENSATION algorithm at each stage but also the strict verification procedure at the second stage by using the results of the first stage. In the first stage, we apply the CONDENSATION modified by the new adaptive color model to the rod part. Then, we can select the highly weighted sampling positions, as can be seen in (1). This is same procedure of the select phase in the CONDENSATION algorithm.

$$s_1'^{(n)} <= s_1^{(n)} \tag{1}$$

where $s_1'^{(n)}$ is the highly weighted n sampling positions at the first stage.

$$S \equiv \sum_{i=1}^{n} \frac{1}{\sigma_i^2}, \quad S_x \equiv \sum_{i=1}^{n} \frac{s_{1x}'^{(i)}}{\sigma_i^2}, \quad S_y \equiv \sum_{i=1}^{n} \frac{s_{1y}'^{(i)}}{\sigma_i^2}$$

$$t_i = \frac{1}{\sigma_i}(x_i - \frac{s_{1x}'^{(i)}}{S}) \quad S_{tt} = \sum_{i=1}^{n} t_i^2 \tag{2}$$

$$b = \frac{1}{S_{tt}} \sum_{i=1}^{n} \frac{t_i s_{1y}'^{(i)}}{\sigma_i^2} \quad a = \frac{S_y - S_x b}{S}$$

where σ_i is the weighting factor of i-th selected position and $s_{1x}'^{(i)}$ and $s_{1y}'^{(i)}$ are i-th sampling x and y directional position at the first stage.

By using WLSF (weighted least square fitting), we can obtain a line that is the center line of the surgical tool such as

$$y(x) = y(x:a,b) = a + bx \tag{3}$$

We can also calculate the width of the surgical tool by using distances between the center line and the selected positions, as can be seen in (4).

$$\text{distance} = \max(\frac{|b \times s_{1x}'^{(n)} - s_{1y}'^{(n)} + a|}{\sqrt{b^2 + 1}}) \tag{4}$$

$$\text{width} = 2 \times \text{distance}$$

Finally, we can identify the left most position among the selected positions as the left end position of the rod part.

$$(p_x, p_y) = \text{leftmost}(s_{1x}'^{(n)}, s_{1y}'^{(n)}) \tag{5}$$

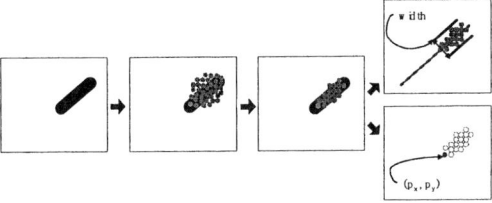

Fig. 1. Schematic diagram of the first stage

In the second stage, we apply the other modified CONDENSATION algorithm to the tip part using the results of the first stage. We can consider a rectangle one side of which is the width of the rod part and the other side of which is an undefined length. We set the length of the rectangle as a third of distance between the leftmost position of the rod part and intersection point of the center line and the image boundary because the size of the rectangle made of that length is large enough for the tool's tip. Then, we can obtain four vertexes of the rectangle, as can be seen (6).

$$\text{length} = \frac{1}{3} \times \sqrt{(0-p_x)^2 + (y(0)-p_y)^2} \tag{6}$$

The optimal distribution of the sampling positions for the second stage tracker is normal distribution along the center line and a perpendicular line of the center line, because the center point of the tip part is located most frequently in the center of the rectangle. It is a bothersome task and a time consuming procedure, therefore, we adapt the master element method to compensate for the computational time. The master element method is a linear transform method commonly used in FEM (finite element method). Using this method, we can create sampling positions in the master element along the normal distribution once for the whole image sequence at the initial time. Then, we can transform the positions in the master element to the inside of a rectangle in the image at each time step. The schematic diagram of the method is shown in Fig. 2.

Therefore, we don't need to calculate all the sampling positions but just calculate four edge positions of the rectangle in the image at each time step. The shape functions, which construct the master element, are shown in (7).

$$
\begin{aligned}
shape\_func_0(x,y) &= \frac{1}{4}(1-x)(1-y) \\
shape\_func_1(x,y) &= \frac{1}{4}(1+x)(1-y) \\
shape\_func_2(x,y) &= \frac{1}{4}(1+x)(1+y) \\
shape\_func_3(x,y) &= \frac{1}{4}(1-x)(1+y)
\end{aligned}
\tag{7}
$$

where *shape_func_i* has value 1 at the point of i-th node and value 0 at the point of other nodes in the master element. Those nodes are four vertexes in the master element. The linear transform is performed by (8).

$$s_{2x}^{(n)}, s_{2y}^{(n)} \equiv \sum_{i=0}^{3} (y_i \times shape\_func_i(R_x^{(n)}, R_y^{(n)})) \tag{8}$$

where (R_x, R_y) is a random position along normal distribution in the master element that is composed of four nodes $(1,1), (-1,1), (-1,-1), (1,-1)$. Once the sampling positions for the second stage are determined, we can calculate the weights of the positions and select highly weighted positions. Then, we can know the left most position of the selected positions as the end position of the tip part.

$$s_2^{'(n)} <= s_2^{(n)}$$
$$(T_x, T_y) = \text{leftmost}(s_{2x}^{'(n)}, s_{2y}^{'(n)}) \qquad (9)$$

where (T_x, T_y) is the end position. The procedure of the second stage is explained in Fig. 3.

Fig. 4 shows the whole procedure of the two-stage adaptive CONDENSATION algorithm. First, we sample random positions in an input image to identify the rod part for the first stage. Then, we sample other random positions transformed from the master element in the restricted region to identify the tip part for the second stage. Finally, we can get the end position of the tip part in the laparoscopic image.

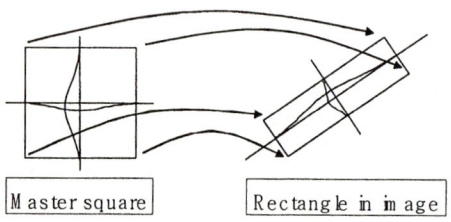

Fig. 2. Schematic diagram of linear transform

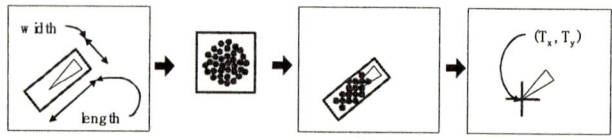

Fig. 3. Schematic diagram of the second stage

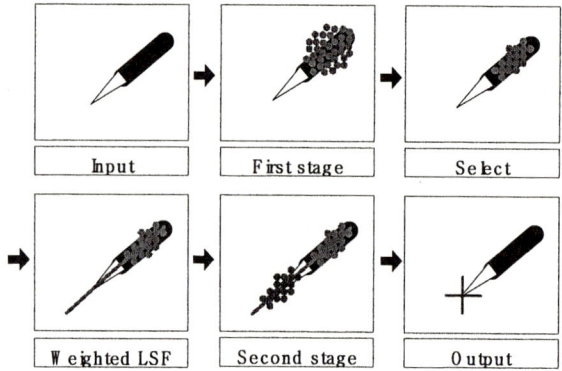

Fig. 4. Two stage adaptive CONDENSATION

3 Experimental Result

3.1 Adaptability to Illumination Change

We tested the tracker that includes the two-stage adaptive CONDENSATION algorithm for illumination change from the normal state to the abnormal state. From the images of Fig. 5 show the good tracking ability of the tracker in the abnormal state. Therefore, we can say that the tracker that includes the two-stage adaptive CONDENSATION algorithm is adaptable of the abrupt illumination change.

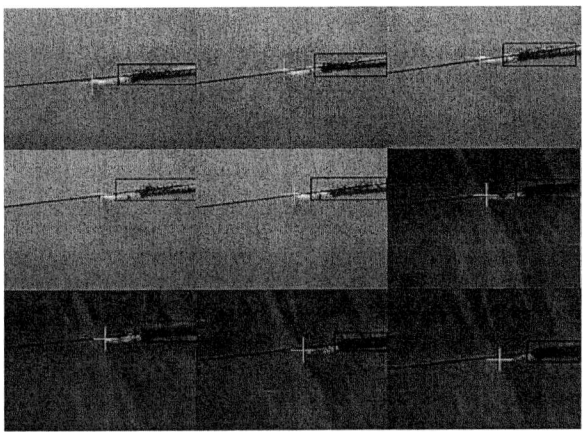

Fig. 5. Image sequence of illumination adaptability

3.2 Intersection and Change of Surgical Tools

We tested the visual tracker in two surgical situations, intersection and change of the surgical tools, which frequently happen in real laparoscopic surgery. Fig. 6 shows the result of the performance of the visual tracker under the condition of an overlapping of the surgical tools. In Fig. 6, the black spots on the rod part of the surgical tool are the sampling positions for the first stage of the algorithm and the rectangle can be generated by the information of the width and the center line. The rectangle is the sampling region for the second stage, and the cross mark represents the end position of the surgical tool identified by the tracker. The visual tracker takes the left tool as the tracking target when the tip part of the left surgical tool is located in the rectangle. However, the tracker does not track the left tool once the left tool is out of the rectangle. This is the reason that the rectangle is constructed by the first stage's result of the tracker for the right tool. Although the tracking target is overlapped by the left surgical forceps, the tracker can detect the end position of the surgical tool. Fig. 7 shows the result of a tool change. We change the left surgical tool from a pair of grasping forceps to a pair of scissors. We set a surgical situation in which a surgeon grips the grasping forceps with the right hand and he or she exchanges the grasping forceps for the laparoscopic scissors. Therefore, the tracking target changes from the grasping forceps to the scissors for the tracker. There are three rectangles in Fig. 7, the largest one represents the safety zone and the other rectangles represent the sampling region for each stage of the visual tracker. The sampling rectangle for the rod part simply represents the limit of the sampling positions and the sampling rectangle for the tip part is made up of the results of the first stage. The line is the center line of the surgical tool in Fig. 7.

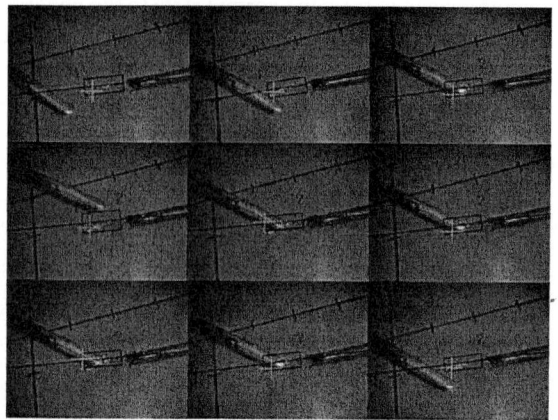

Fig. 6. Image sequence of overlapping of tools

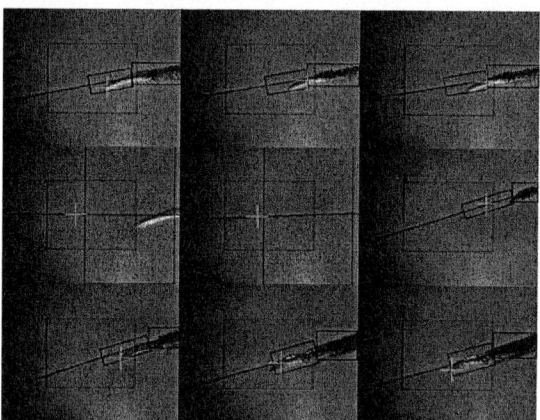

Fig. 7. Image sequence of tool change

When the tracking target disappears in the surgical image, the tracker samples the positions in the right half of the whole image. From those images, we can see that the rectangle for the tip part disappears because that the tracker recognizes that there is no surgical tool and then set the width of the surgical tool as zero. After the visual tracker percepts a new tracking target (scissors), the tracker can detect the new tracking target continuously. From the experimental results, we can know that the tracker using the two-stage adaptive CONDENSATION algorithm is robust to the varied surgical situations and can detect the end position of surgical tools very well.

4 Summary

In this paper, we present a new visual tracking algorithm, the two-stage adaptive CONDENSATION algorithm, which has real-time and robust tracking ability. Also, we can find the optimal color for feature of the laparoscopic surgical tool for the new algorithm. The tracker, which includes the new algorithm, is adaptable to abrupt and gradual illumination changes and robust to the varied surgical situations (change and overlapping of laparoscopic surgical tools). Finally, we can say that our newly developed real-time visual tracker, which includes the two-stage adaptive CONDENSATION, has sufficient performance to be a real time visual servoing system for laparoscopic surgery.

Acknowledgment

We gratefully acknowledge the financial support of the Korea Science and Engineering Foundation (HWRS-ERC).

References

1. Maurice E. Arregui, Robert J. Fotzgibbons, Jr., Namir Katkhouida, J. Barry McKernan, Harry Reich, "Principles of Laparoscopic surgery", Springer-Verlag, 1995.
2. Omote, K., Feussner, H., Ungeheuer, A., Arbter, K., Guo-Qing Wei, "Self-guided robotic camera control for laparoscopic surgery compared with human camera control", The American journal of surgery, Vol. 177, pp. 321-324, April, 1999.
3. Casals, A., Amat, J., Laporte, E., "Automatic guidance of an assistant robot in laparoscopic surgery", IEEE International Conf. on Robotics and Automation, Vol. 1, pp.895–900, 1996.
4. 4. Cheolwhan Lee, Yuan-Fang Wang, Uecker, D.R., Yulun Wang, "Image analysis for automated tracking in robot-assisted endoscopic surgery", Proceedings of the 12th IAPR International Conference , Vol. 1, pp. 88–92, 1994.
5. Guo-Qing Wei, Arbter, K., Hirzinger, G., "Real-time visual servoing for laparoscopic surgery. Controlling robot motion with color image segmentation", IEEE Eng. in Med. and Bio. Magazine, Vol. 16, Issue 1, pp. 40–45, 1997.
6. M. Isard, Andrew Blake, "CONDENSATION - Conditional Density Propagation for Visual Tracking", Int. J. Computer Vision, 29(1), pp.5-28, 1998.
7. Michael Isard, Andrew Blake, "Contour tracking by stochastic propagation of conditional density", In Proc. European Conf. Computer Vision, pp.343-356, 1996.
8. David Reynard, Andrew Wilderberg, Andrew Blake, John Marchant, "Learning dynamics of complex motions from image sequences", Proc. European Conf. Computer Vision, pp. 357-368, 1996.
9. Andrew Blake, Michael Isard, "Active contours", Springer, 1998.
10. Rafael C. Gonzalez, Richard E. Woods, "Digital image processing", Second edition, Prentice Hall, 2001.
11. Gi-jeong Jang, In-so Kweon, "Robust Object Tracking Using an Adaptive Color Model", IEEE International Conf. on Robotics and Automation, pp.1677-1682, 2001

Toward a Sound Analysis System for Telemedicine

Cong Phuong Nguyen, Thi Ngoc Yen Pham, and Castelli Eric

International Research Center MICA, 1 Daicoviet, Hanoi, Vietnam
{Cong-Phuong.Nguyen, Ngoc-Yen.Pham, Eric.Castelli}@mica.edu.vn

Abstract. Our work is within the framework of studying a sound analysis system in a telemedicine project. The task of this system is to detect situations of distress in a patient's room basing on sound analysis. In this paper we present our studies on the constructions of a speech/non-speech discriminator and of a speech/scream-groan discriminator. The first discriminator's task is to distinguish speech signal from non speech signal in a room such as sounds of broken glass, door shutting, chair falling, water in toilette, etc. The second one's task is to detect sounds of scream-groan from speech signal. Results show that these discriminators are applicable to our sound analysis system.

1 Introduction

The system on which we work is developed for the surveillance of elderly, convalescent persons or pregnant women [1]. Its main goal is to detect serious accidents such as falls or faintness at any place in the apartment. Firstly most people do not like to be supervised by a (or some) camera all the day while the presence of microphone can be acceptable. Secondly the supervision field of a microphone is larger that that of a

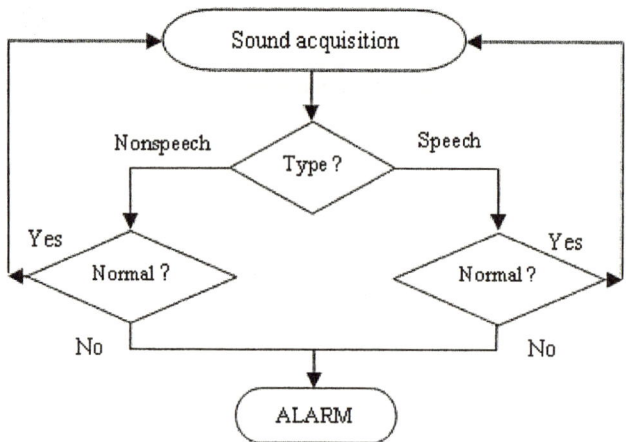

Fig. 1. Sound analysis system. The middle diamond is the speech/non-speech discriminator. The left one is the classifier of sounds of everyday life. The right one is the speech/scream-groan discriminator.

camera. Thirdly, sound processing is much less time consuming than image processing, hence a real time processing solution can be easier to develop. Thus, the originality of our approach consists in replacing the video camera by a system of multichannel sound acquisition. The system analyzes in real time the sound environment of the apartment and detects the abnormal sounds (falls of objects or patient, scream, groan), that could indicate a distress situation in the habitat. This sound analysis system is illustrated in Fig. 1.

Previous works [1] (within this project) already studied two problems of this system which are sound acquisition (the oval in Fig. 1) and sound classification (the left diamond in Fig. 1). Then this paper presents our studies concerning speech/non-speech discriminator (the middle diamond) and speech/scream-groan discriminator (the right diamond).

This paper is presented as follows. Sect. 2 concerns the problems of a speech/non-speech discriminator. In Sect. 3 we show evaluations on speech/scream-groan discrimination. The conclusion is in Sect. 4.

2 Speech/Non-speech Discriminator

Speech/non-speech discrimination has been researched for a long time and widely applied in many multimedia applications, such as automatic speech recognition, speech/music discrimination, audio classification and segmentation, broadcasting transaction. In our works, we restrict ourselves to the problem of distinguishing signals of speech from sounds coming from non speech sources in a room, such as chairs dragged on floor, falls of chairs, typing keyboard, toggle switch, clank of glasses and cups, broken glasses, shutting/opening door, water in toilet-room, and telephone. For convenience, these non speech sounds are referred to as "sound" in the following paragraphs. Because the discrimination algorithm is intended to be implemented in a compact real-time system, its constraints are time consume and size of model.

2.1 Corpus

The corpus for this evaluation consists of speech signals and sounds of everyday life; they are sampled at 16 kHz and quantized at 16 bit [2]. Categories of this corpus were mentioned in paragraphs of introduction. Speech signals (in French) are 2352 seconds extracted from the BREF80 corpus [3]. The total duration of sounds of everyday life is 2325 seconds. From those clean corpora, we created noisy signals whose signal-noise ratio (SNR) varies from 0 to 40dB (additive and convolutive). At last, we have 287 minutes of sound and 283 minutes of speech.

2.2 Feature Set

A set of features is the first problem of a discriminator (the other is the classification model presented in 2.4). In [4] a set of bandwidth, energy and pitch was used for speech/music discrimination. We can also find more feature in larger sets, such as spectral rolloff, spectral centroid, spectral flux, zero-crossing rate (ZCR) [5], [6], Mel frequency cepstral coefficients (MFCC) [6], [9], [11], linear spectral pairs (LSP) [7], [10], perceptual linear predictive (PLP) [8], [11], band energy ratio (BER) [12]. There are some uses of derivatives of feature [6], [8], [9], [10]. Among many features that can be found in the literature, we have chosen those that give different values for signals of speech and signals of sounds in a room. The set of features intended to be used in our evaluations consists of:

- zero crossing rate,
- energy,
- maximal spectral peak frequency,
- spectral centroid,
- spectral rolloff,
- band energy ratio,
- bandwidth,
- Mel frequency cepstral coefficients,
- linear spectral pairs,
- perceptual linear predictive.

These features are extracted from 16-millisecond segments. Some parameters of them (such as percentage of rolloff, order of MFCC, etc.) are selected basing on the Fisher discriminant ratio (FDR). A feature is considered as more discriminative than another one if its FDR value is higher. A primary set of 54 features (representing a 16-ms segment) is established from them. Such a high-dimensional vector is very time consuming for training and estimating models of classification. Besides, it could make the trained model become very complicated. And a model with a high-dimensional input needs a very large training corpus. Therefore we try to reduce the size of feature space for computational purposes by applying the Principal Component Analysis (PCA) method. By choosing a threshold of 90% of the total variability, we get the new 16-dimensional vector mapped (basing on the first principal component) from the original space of 54 features. The selection of the threshold 90% is based on the total variability curve. From now on (in Sect. 2.3 and Sect. 2.4), each 16-ms segment is represented by a 16-dimensional vector.

2.3 Classification Models

After having selected the most discriminant features, in order to find the best way to combine them into our discrimination system, we evaluate different classification models. In the literature, many experiments with many classification models have been performed. We can find Quadratic Gaussian Classifier (QGC) [7], k–Nearest Neighbour (kNN) [5], [7], [10], Gaussian Mixture Model (GMM) [5], [6], [8], [12], Artificial Neural Network (ANN) [11], Decision Tree (DT) [5], Support Vector Machine (SVM) [11] and Fuzzy Inference System (FIS) [13]. GMM seems to be the most widely used model.

In the method described in [13], using of a hierarchical fuzzy tree, no training is needed and the classification can be performed very quickly. For the input of the FIS system, appropriate features are extracted using a simple nearest neighbour classifier and a sequential forward selection method. The membership functions are chosen using Gaussian membership functions. But authors said by themselves that the use of a neural network to train parameters to obtain better classification could be done.

A kNN classifier determines the category of a sample by calculating the distances from the sample to the k nearest points in a multidimensional space. It is simple and it does not need any training process, however, this method is very time consuming because it has to calculate the distances between the test point and every points of the training space. On the other hand, this training space itself is also the model of kNN, making the model become a very big size one. For these reasons we do not estimate kNN model in further experiments.

Parameter(s) of a classification model probably affect(s) its operation, so we have to find those that give the most effective classification. In our evaluation, performances of different types of each classification model are tested (using 10-fold cross validation) to find the most suitable parameters (the optimal configuration) for each model. These evaluations are carried out based on the corpus that was used on the evaluations on discriminative features.

Selected parameters of classification models are summed up in Table 1. QGC is not listed here because it has no parameter. We do not evaluate FIS using the grid partitioning because for a FIS with 16 inputs, each with (for instance) two membership functions, the grid partitioning leads to 65536 (= 2^16) rules, which is too large for a practical training. These six selected models are used in the period of discriminating the whole corpus, except the polynomial SVM (this case will be explained later in Sect. 2.4).

Table 1. Parameters of classification models for speech/non-speech discrimination

Model	Parameter(s)
GMM	65 Gaussians
SVM	Polynomial kernel of degree 2
DT	Twoing criterion
FIS	Scattering partition
ANN	45 tan-sigmoid hidden neurons, quasi Newton training algorithm

2.4 Selection of Model

In order to evaluate models we carry out two evaluations: on 16-ms segments and on 1-second segments. The final purpose is to find a model that can discriminate 1-s segments. This duration is selected because we see that such a length is short enough for a segment of speech or non-speech signal.

At first, signals of corpus are segmented into 16-ms segments. These segments are arranged randomly for a 10-fold cross validation. When selected models (in Table 1) are estimated with the new corpus, the polynomial SVM showed a not very high performance: its discrimination precision is below 50%. This can be because of the new size of the training corpus. Therefore we take another configuration, a Gaussian SVM. Details of performances of this SVM and the four other models (QGC, GMM, ANN and DT) are presented in Table 2. It lists results of discrimination of 16-ms segments of different SNR. The discrimination ratio is defined as the number of correctly discriminated segments divided by the number of tested segments. The false alarm (FA) ratio is calculated by dividing the number of false alarm (defined to occur when a non-speech segment is discriminated as a speech one) by the number of non-speech segments. The missed detection (MD) ratio is calculated by dividing the number of missed detection (defined to occur when a speech segment is not detected) by the number of speech segments.

Table 2. Performances of discrimination on 16-ms segments. c = convolutive; FA = false alarm ratio; MD = missed detection ratio; TC = time consume for discriminating a 16-ms segment; Size = size of model.

Performance		QGC	GMM	ANN	SVM	DT	FIS
Discrimination ratio (%)	Clean	88.19	97.29	97.38	94.64	94.69	93.42
	40dB	80.56	95.74	96.60	92.57	91.56	80.94
	30dB	80.75	92.94	93.86	87.42	87.88	80.68
	20dB	70.59	89.03	90.89	82.78	82.36	75.76
	10dB	81.63	79.27	83.88	76.31	75.11	73.20
	0dB	73.08	71.26	78.51	70.72	70.85	73.19
	-40dB, c	74.44	97.43	96.97	94.23	93.04	90.17
	Total	78.12	89.51	91.53	86.33	85.59	80.82
FA (%)		40.45	12.35	4.90	13.73	15.76	25.15
MD (%)		3,97	8.81	13.35	13.62	13.18	13.26
TC (ms)		0.17	0.43	0.057	22.17	0.39	0.41
Size (kB)		5	359	115	2756	6743	10

Here we can see that in most case (except QGC) the precision decreases when the SNR decreases. But as regards convolutively noised signals, SNR is not an effective factor. QGC has the lowest MD but on the other hand its FA is highest, i.e. nearly most of signals are classified as of one class. It can be supposed that this problem is too nonlinear for QGC. Among the five remaining models, ANN has the best performances (the highest discrimination ratios, the least time consuming, the smallest size, the lowest FA and an acceptable MD), and it is selected to be applied to our speech/non-speech discriminator.

An one-second segment is determined a speech one if the number of detected speech 16-ms segments (which varies from 1 to 62 segments) exceeds a certain threshold. It is clear that when this value increases, FA increases and MD decreases, and vice versa. The optimal number would be the intersection between the received operating curve (ROC) and the diagonal (connecting (0, 0) to (100, 100)) because in that case we get a compromise between getting the lowest possible FA and the lowest possible MD.

The above results are extracted from evaluations on separate signals, which were recorded in soundproof studios. In the next phase of estimation, we apply those trained models to real signals. These signals are recorded in a normal room (non soundproof), hence they are audio sequences composed of different sounds (speech and sounds in a room environment, such as of typing keyboard, shutting window, dragging chair, etc) and naturally mixed. Besides, the distance between audio sources and micros is not fixed. Their total duration is 230 seconds. Applying to this corpus, the ANN-based speech/non-speech discriminator has an FA of 3.70% and an MD of 3.41%.

2.5 Feature Reduction

This system is intended to run in real time so we tried to improve its discriminating speed by reducing the size of the feature set. PCA is a well known method for reducing the dimensionality. It extracts a low-dimensional (16) vector from the original high-dimensional (54) vector. Therefore we still have to extract the original feature set which consists of ten features in our problem. In order to reduce vector extraction time, we try to reduce the number of features. The sixteen largest coefficients of the first principal component correspond to coefficients of MFCC, PLP and BER. It means that they contribute the largest proportions to the total variability, and so we hoped that we can use these three features instead of the ten features mentioned in Sect. 2.2 as elements of a new feature set. MFCC of 16 coefficients, PLP of 12 and BER of 4 constitute a 32-dimensional vector. Once again, applying PCA and choosing a threshold of 90% of the total variability, this vector can be mapped to a 14-dimensional space. The selection of the threshold is based on examining the total variability curve of the 32-dimensional space.

Once again, ANN is chosen as classification model of the reduced-feature discriminator. Experiments showed that the performances of the 3-features discriminator are not far different from those of the 10-features discriminator, specifically the new false alarm is 3.75% and the new missed detection is 3.86%. It is also noted that these results are obtained with a network of 35 hidden neurons (instead of 45).

This discriminator does not function well when we keep on rejecting feature. It is reasonable because the three largest coefficients of the first principal component are three coefficients of MFCC, PLP and BER respectively. In short, we choose MFCC, PLP and BER for our ANN-based discriminator.

3 Speech/Scream-Groan Discriminator

In our opinion, when the patient falls into a situation of distress, he may cry, scream, groan, cough or gasp. In other words those types of sound are probably an indication of such a case. From now on for convenience they are referred to as scream-groan sound. Then when a scream-groan sound is detected, the system has to give alarm. This section presents the module which can detect scream-groan sound from speech signal. In other words this is a speech/scream-groan discriminator. The need of distinguishing scream-groan from speech is discussed in Sect. 3.2. This discriminator is intended to be implemented in a compact real-time, so its constraints are time consume and size of model.

3.1 Corpus

Building a corpus of scream-groan signals which represent situations of distress is rather difficult. Recording this type of signal in hospital is nearly impossible due to the violation of privacy. Recording them in a studio is feasible, but situations of distress are hard to be simulated by speakers, making a not true corpus. For these reasons, we collected scream-groan signals from DVD films. The DVD format is selected because of its high audio quality. These signals come from people who are in indoor dangerous situations (threatened to be killed, about to die, in pain and ache). These people are actors so they are professional in simulating these circumstances and hence the collected signals are expected to be true. From 40 DVD films, we collect a corpus of 114 seconds. It consists of 63 signals of groan, 13 signals of cough and 21 signals of gasp.

3.2 Non-speech, Speech and Scream-Groan

From Fig. 1, a question arises: according to the speech/non-speech discriminator in this system, scream-groan belongs to which class, speech or non-speech? If scream-groan belongs to non-speech class then the problem of speech/scream-groan discrimination becomes nonsense. In order to examine the class of scream, we applied the speech/non-speech discriminator to a corpus of 114 seconds of scream-groan, 151 seconds of speech [3] and 253 seconds of everyday life sounds [2]. From those clean corpora, we created noisy signals whose signal-noise ratio (SNR) varies from 10 to 40dB. They were segmented into segments of 1 second and then fed to the discriminator presented in Sect. 2.5. Our experiments showed that 96.2 % of signals of scream and groan belong to the class of speech and that justifies the need of a second discriminator.

3.3 Feature Set and Classification Model

There are some works on cry/scream detection; most of them are based on detecting impulsive sound. For instance, [14] detects the baby cry by detecting "a fairy loud sound followed by a relative quiet as the infant inhales", and [15] detects audio level. In our opinion, until now there is arguably no works on speech/scream-groan discrimination.

Like the discriminator in Sect. 2, this discriminator also has two problems: a feature set and a classification model. The set of features intended to be used in our evaluations consists of:

- zero crossing rate,
- pitch,
- spectral centroid,
- spectral rolloff,
- peak of power spectrum,
- bandwidth.

This set is empirically selected. These features are extracted from 16-millisecond segments. In other words, each 16-ms segment is represented by a 16-dimensional vector.

In order to find the most appropriate classification model for the discriminator, we examine GMM, ANN, DT and SVM. Performances of the four models (and the most appropriate parameter(s) of each model) tested by a 10-fold cross validation are presented in Table 3. The false alarm (FA) ratio is calculated by dividing the number of false alarm (defined to occur when a speech segment is discriminated as a scream-groan one) by the number of scream-groan segments. The missed detection (MD) ratio is calculated by dividing the number of missed detection (defined to occur when a scream-groan segment is not detected) by the number of speech segments.

Table 3. Performances of models on speech/scream-groan discrimination of 16-ms segments

Model, parameter(s)	Performance		
	MD (%)	FA (%)	TC (ms)
GMM, 96 Gaussians	10.61	15.87	0.21
DT, deviance	9.87	11.87	0.02
SVM, Gaussian kernel	19.02	8.10	30.60
ANN, 15 hidden log sigmoid neurons, 1 linear output, Levenberg-Marquardt training function	9.68	12.84	0.008

SVM has the lowest FA but its MD is the highest, and those make it unavailable for our purpose. Among the four models, DT and ANN have the best performances. ANN is the least time consume but DT has more balanced MD and FA. Our purpose is to discriminate class of one-second segments because we see that such a length is short enough for a segment of speech or scream-groan signal. An one-second segment is determined a scream-groan one if the number of detected groan 16-ms segments (which varies from 1 to 62 segments) exceeds a certain threshold. Using ROCs once again, we choose a threshold of 41 for both DT and ANN. Values of FA and MD of the two models are presented in Table 4. There is no considerable difference between MD of DT and that of ANN. In the other hand, FA of DT is less than a half of that of ANN. Though ANN has a superiority of time consume (see Table 3), but 0.02 ms, that value of DT, is an acceptably small one. In short, we selected the decision tree for our speech/scream-groan discriminator. This discriminator can obtain an FA of 0.43% and an MD of 1.24%.

Table 4. Performance on 1-s segments of the speech/scream-groan discriminator

	Performance	
	MD (%)	FA (%)
DT	1.24	0.43
ANN	1.15	1.10

In an endeavour of reducing the size of feature set, we use PCA once more. Experiments show that pitch, peak of power spectrum and spectral rolloff explain 92% of sum of the total variability. So we presume that this trio plays the most important roles in discriminating speech and scream-groan. Once again, a 10-fold cross validation on the new feature vector is applied in order to evaluate the four models. The most satisfactory results are 12.87% and 17.50% (MD and FA respectively) and belong to a Gaussian kernel SVM. They are much worse that those of DT using the original feature set. In this case, when the dimension of a feature set is not high (6 in this case), the effect of PCA is not evident. Besides worse performance on discrimination, time consume of this SVM is rather high (24.55 ms). These results make us maintain the original feature set.

4 Conclusion

We have presented in this paper a speech/non-speech discriminator and a speech/scream-groan discriminator. They are parts of a sound analysis system within the framework of a habitat telemonitoring system. This system's task is to detect automatically situations of distress in a patient's room. The first discriminator utilises a set of three features and a neuron network. It can achieve an FA of 3.75% and an MD of 3.86%. The second one is based on a set of six features and a decision tree. It's FA and MD is 0.43 and 1.24 respectively. Evaluations are simulated on a PC Windows 2000 2.4 GHz 1G RAM. We think that the combination of the two discriminators can be useful in other audio applications, such as audio information retrieval or a preprocessing module in speech recognition. In the future these two discriminators will be integrated into the sound analysis system. This integration is intended to be an ontology-based system. An audio ontology can be built and in this ontology, audio objects can be described by feature sets of the two discriminators.

References

1. Istrate, D., Vacher, M., Castelli, E., Besacier, L., Sérignat, J.F.: Distress Situation Identification though Sound Processing. An Application to Medical Telemonitoring. European Conference on Computational Biology, Paris (2003)
2. Équipe GEOD Dan Istrate: C.-I. Base de données. Sons de la vie courante (2001)
3. Gauvain, J.L., Lamel, L.F., Eskenazi, M.: Design Considerations and Text Selection for BREF, a large French read-speech corpus. International Conference on Spoken Language Processing 1990, Kobe Japan (1990)
4. Saunders, J.: Real-Time Discrimination of Broadcast Speech/Music. International Conference on Acoustics, Speech and Signal Processing (1996)
5. Scheirer, E., Slaney, M.: Construction and Evaluation of a Robust Multifeature Music/Speech Discriminator. International Conference on Acoustics, Speech and Signal Processing (1997)
6. Carey, M.J., Parris, E.S., Lloyd-Thomas, H.: A Comparison of Features for Speech, Music Discrimination. International Conference on Acoustics, Speech and Signal Processing, Phoenix AZ (1999)
7. El-Maleh, K., Samouelian, A., Kabal, P.: Frame-Level Noise Classification in Mobile Environments. International Conference on Acoustics, Speech and Signal Processing, Phoenix AZ (1999)
8. Wegmann, S., Zhan, P., Gillick, L.: Progress in Broadcast News Transcription at Dragon Systems. International Conference on Acoustics, Speech and Signal Processing, Vol. I, Phoenix AZ (1999) 33–36
9. Moreno, P.J., Rifkin, R.: Using the Fisher Kernel Method for Web Audio Classification. International Conference on Acoustics, Speech and Signal Processing, Vol. 4 (2000) 2417–2420

10. Lu, L., Zhang, H.-J., Jiang, H.: Content Analysis for Audio Classification and Segmentation. IEEE Transaction on speech and audio processing, Vol. 10, No. 7 (2002)
11. Ajmera, J., McCowan, I., Bourlard, H.: Speech/Music Segmentation Using Entropy and Dynamism Features in a HMM Classification Framework. Speech Communication 40 (2003) 351–363
12. McKinney, M.F, Breebaart, J.: Features for Audio and Music Classification. 4th International Conference on Music Information, Maryland USA (2003)
13. Liu, M., Wang, C., Wang, L.P.: Content-Based Audio Classification and Retrieval Using a Fuzzy Logic System: Towards Multimedia Search Engines. Soft Computing 6 (2002) 357 – 364
14. Schmandt, C., Vallejo, G.: "Listenin" to Domestic Environments from Remote Locations. International Conference on Auditory Display Boston USA (2003)
15. http://www.homeguardion.com/

Structural Learning of Graphical Models and Its Applications to Traditional Chinese Medicine

Ke Deng[1], Delin Liu[2], Shan Gao[3], and Zhi Geng[1]

[1] School of Mathematical Sciences, Peking University, Beijing, China
[2] China Academy of Traditional Chinese Medicine, Beijing, China
[3] Peking Union Medical College Hospital, Beijing, China

Abstract. Bayesian networks and undirected graphical models are often used to cope with uncertainty for complex systems with a large number of variables. They can be applied to discover causal relationships and associations between variables. In this paper, we present heuristic algorithms for structural learning of undirected graphical models from observed data. These algorithms are applied to traditional Chinese medicine.

1 Introduction

Graphical models such as undirected graphs, directed acyclic graphs (DAG) and Bayesian networks have been applied widely to many fields, such as data mining, pattern recognition, artificial intelligence and causal discovery [1,2,3,5,6]. Graphical models can be used to cope with uncertainty for a large system with a great number of variables. Structural learning of graphical models from data is NP hard in the number of variables. There are two main kinds of structural learning methods. One is constraint-based learning and the other is score-based learning. Both of them have drawbacks: the former needs a large size of observed data that is not practical in many applications, and the latter needs to search a huge number of models.

In this paper, we discuss structural learning of graphical models and application of graphical models to traditional Chinese medicine. We propose an approach in which constraint-based and score-based methods are combined together. In many applications, the graphical model may be sparse, that is, each variable associates directly with a few of other variables. For example, for a particular application, domain experts may know that each variable associates directly with at most k variables. In such cases, structural learning of graphical models can be simplified by constraining the maximum number k of neighbors for each variable. On the other hand, if the sampling size is small, the maximum number of neighbors must be constrained to an appropriate number such that statistical inference is efficient. For a given maximum number of neighbors, heuristic algorithms are described in this paper for structural learning of undirected graphical models. With a few of modifications, the algorithms can also be used for directed graphical models.

In Section 2, we give notation and definitions of graphical models. In Section 3, two heuristic algorithms for structural learning of undirected graphical models are described. Their applications in traditional Chinese medicine are illustrated in section 4.

2 Notation and Definitions

A graph is a pair $G = (V, E)$ where V is a finite set of *nodes* (also called vertices) and E is a subset $V \times V$ of ordered pairs of distinct nodes, called the set of *edges*. An edge is directed pointing from i to j if $<i, j> \in E$. A directed edge is also called *a arrows*. If $<i, j> \in E$ and $<j, i> \in E$, an edge between nodes i and j is undirected, denoted by (i, j) and depicted by a line in the graph. A graph is *undirected* if it contains only undirected edges. In this paper, we concentrate only on undirected graphs. In an undirected graph, the *neighbor set* of a node i is defined as a set of nodes that have one edge connecting i in G, noted by $\mathcal{N}_i$.

Let $X = (X_1, \ldots, X_p)$ be a p-dimensional vector of random variables. Each variable X_i in X is depicted by a node i in G. An *undirected graphical model* is then a family of probability distributions P_G which has the Markov property over the undirected graph G [4], that is, variables X_i and X_j are conditionally independent given other variables (denoted by $X_i \perp X_j \mid X_{V \setminus \{i,j\}}$) if $(i, j) \notin E$. This property is called the pairwise Markov property.

For constructing an undirected graph from data, we use the mutual information to measure independence between variables. The mutual information for independence between X and Y is defined as

$$I(X, Y) = \int f(x, y) \log \frac{f(x, y)}{f(x)f(y)} dxdy.$$

It equals 0 if and only if X and Y are independent. The conditional mutual information for independence between X and Y given Z is defined as

$$I(X, Y | Z) = \int f(x, y, z) \log \frac{f(x, y|z)}{f(x|z)f(y|z)} dxdydz,$$

which equals 0 if and only if X and Y are independent conditional on Z. Since $I(X, Y|Z) = I(X, Y \cup Z) - I(X, Z)$, the conditional mutual information can be calculated from the mutual information. Now we list some properties to be used in algorithms that we present in Section 3.

Properties.

1. If $X \perp Y | Z$, then $I(X, Z) \geq I(X, Y)$,
2. If $X \perp Y | Z$, then $I(X, Z) = I(X, Y \cup Z)$,
3. $X_i \perp X_{V \setminus (N_i \cup \{i\})} | X_{N_i}$, and
4. $I(X_i, X_{N_i}) = \max_{Y \subseteq V} I(X_i, Y)$.

3 Algorithms for Model Learning

The following algorithms are constructed on the basis of the property 4, which implies that the neighbor set of a variable has the largest mutual information.

Thus for a given variable X_i, its neighbor set can be found by searching for a variable set Z which maximizes the information $I(X_i, Z)$ for $Z \subseteq V$. In fact, the maximum number of neighbors may be set to a constant k by domain experts. To prevent from overfitting a model, we use Akaike Information Criterion (AIC) as a score to penalize complex graphical models and to prune the variable set:

$$AIC(\mathcal{N}_i = Z) = -2 \sum_{j=1}^{n} \log f(x_{1j}, \cdots, x_{pj}; \widehat{\theta}) + 2N$$

$$= (-2 \sum_{j=1}^{n} \log f(x_{ij}; \widehat{\theta}_1 | x_{2j}, \cdots, x_{(i-1)j}, x_{(i+1)j}, \cdots, x_{pj}) + 2N_1)$$

$$+ (-2 \sum_{j=1}^{n} \log f(x_{1j}, \cdots, x_{(i-1)j}, x_{(i+1)j}, \cdots, x_{pj}; \widehat{\theta}_2) + 2N_2)$$

$$= (-2 \sum_{j=1}^{n} \log f(x_{ij}; \widehat{\theta}_1 | z_j) + 2N_1)$$

$$+ (-2 \sum_{j=1}^{n} \log f(x_{1j}, \cdots, x_{(i-1)j}, x_{(i+1)j}, \cdots, x_{pj}; \widehat{\theta}_2) + 2N_2)$$

$$= S(X_i, Z) + T(X_{V \setminus \{i\}})$$

where x_{ij} and z_j are the jth observations of X_i and Z respectively; $\widehat{\theta}$, $\widehat{\theta}_1$ and $\widehat{\theta}_2$ are the MLEs of parameters from likelihood functions $\prod_{j=1}^{n} f(x_{1j}, \cdots, x_{pj}; \theta)$, $\prod_{j=1}^{n} f(x_{ij}; \theta_1 | z_j)$ and $\prod_{j=1}^{n} f(x_{1j}, \cdots, x_{(i-1)j}, x_{(i+1)j}, \cdots, x_{pj}; \theta_2)$ respectively; and $N = |\theta|$, $N_1 = |\theta_1|$ and $N_2 = |\theta_2|$ are the number of parameters in these functions respectively. Since $T(X_{V \setminus \{i\}})$ is constant, for selecting the neighbor set Z of a given X_i, we can use $S(X_i, Z)$ in stead of $AIC(\mathcal{N}_i = Z)$ as the score for comparing different neighbor sets.

Below we describe two algorithms for constructing an undirected graph with minimizing the AIC score.

Information-Based Global Optimization (IBGO) : For each $x \in V$, repeat the following steps:

Step 1. Let $\mathcal{Z}_x = \{Z \subseteq V \setminus \{x\}, |Z| \leq k\}$, and calculate the Score $S(x, Z)$, for all $Z \in \mathcal{Z}_x$;
Step 2. Find the set $Z_x \in \mathcal{Z}_x$ to minimize $S(x, Z)$ for all $Z \in \mathcal{Z}_x$;
Step 3. Define Z_x as the final neighbor set of x, $\mathcal{N}_x$.
This algorithm is time consuming. For a graphical model with n variables and a given maximum number k of neighbors, $n \times C_n^k$ scans are required. So it is recommended only for small n and k. In situations where n or k is large, the following algorithm is more efficient.

Information-Based Local Optimization (IBLO) : Instead of searching all subsets of $V \setminus \{x\}$, IBLO develops a stepwise way to find the neighbor set of x. Let $N^{(t)}(x)$ denote the neighbor set of x in the t th iteration. For each $x \in V$, the following two steps are iterated with an initial value $N^{(0)}(x) = \emptyset$:

Step 1. Find $z^* \in V \setminus (\{x\} \cup N^{(t)}(x))$ to minimize the Score $S(x, \{z\} \cup N^{(t)}(x))$ for $z \in V \setminus (\{x\} \cup N^{(t)}(x))$;
Step 2. If $S(x, \{z^*\} \cup N^{(t)}(x)) \leq S(x, N^{(t)}(x))$, set $N^{(t+1)} = N^{(t)}(x) \cup \{z^*\}$; Otherwise, stop and define $N^{(t)}(x)$ as the final neighbor set of x, $\mathcal{N}_x$.

Bootstrap method is applied to our algorithms to assess the robust of every edges in a graphical model. Resampling the observed data set and then repeating one of the above algorithms, we can get a set of graphs, and count the appearances of each edge. Choosing a threshold, we can delete unstable edges.

The undirected graph obtained from the above algorithms can be treated as the moral graph of a directed graph. Starting with this moral graph, we can orient directions of edges using the decomposing method proposed in [2].

4 Applications to Traditional Chinese Medicine

In this section, we show two applications of graphical models to traditional Chinese medicine. One is for diagnosis of apoplexy patients, and the other is for prescription of herb materials.

Example 1. 303 apoplexy patients were recorded in three hospitals for two years. For each patient, detailed information of 150 variables is recorded in the documents, including basic personal backgrounds (sex, age, etc.), symptoms (paralysis, anaesthesia, temperature, blood pressure, etc.) and electronic imaging records (CT, MRI, etc.). We selected 45 of them for our analysis, as shown in Table 1.

Table 1. Variables involved in the analysis

NO.	variable name	NO.	variable name	NO.	variable name	NO.	variable name
1	sex	13	ra-hypoal	25	dysphagia	37	sphygmus-3
2	CI	14	rl-hypoal	26	vomiting	38	smoking
3	TIA	15	la-hypoal	27	vertigo	39	drinking
4	DM	16	ll-hypoal	28	aphasia	40	time
5	IHD	17	r-dystaxia	29	dysarthria	41	type
6	HL	18	l-dystaxia	30	headache	42	type-course
7	Rap	19	coma	31	dizziness	43	brain-L
8	Rlp	20	pupil	32	insomnia	44	brain-R
9	Lap	21	staring	33	hypertension	45	brain-B
10	Llp	22	psychosis	34	fever		
11	Rp	23	hemianopia	35	sphygmus-1		
12	Lp	24	water-d	36	sphygmus-2		

Using the algorithm IBLO with 50 resamplings and 0.8 as the threshold, we get the undirected graph shown in Figure 1. This graph depicts relationships among these variables. Most of the relationships match doctors' experiences and knowledge. For example, the pattern M1 (M2) describes the relationship among

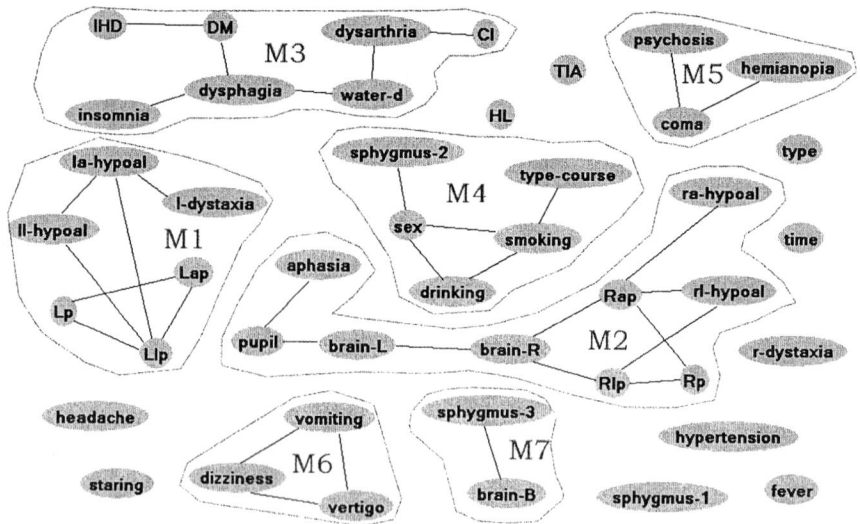

Fig. 1. The graphical model for symptoms of apoplexy

the apoplexy symptoms of left-side (right-side) body; the pattern M3 connects the medical history with the impediments in swallowing, sleeping and talking very well; the connections of sex, smoking, drinking and sphygmus are catched by the pattern M4.

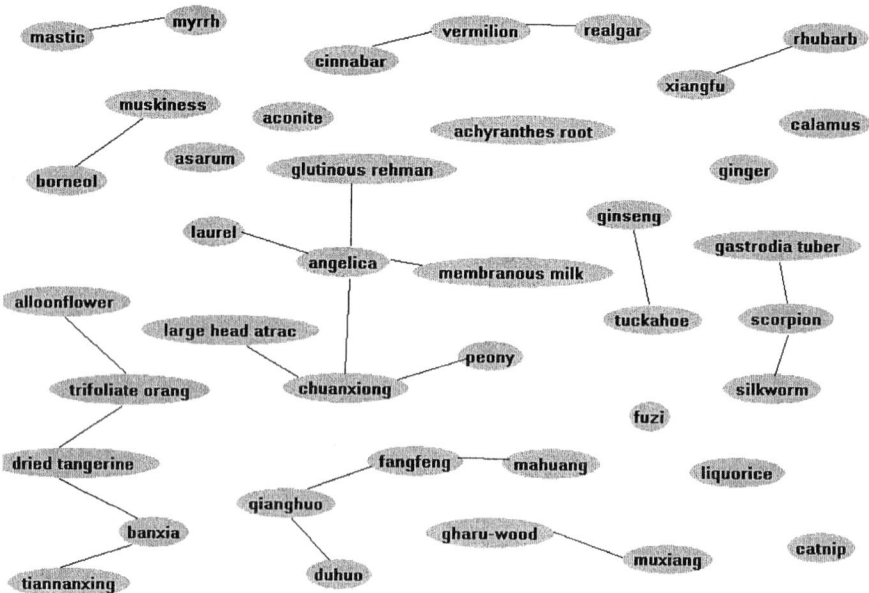

Fig. 2. The graphical model for Chinese medicinal herbs

In this way, for a disease that doctors have few of experiences and knowledge, relationships between backgrounds and symptoms and those between various symptoms may be found by constructing a graphical model from clinical observed data.

Example 2. From books about historic prescriptions of Chinese medicine, we collected 554 prescriptions for apoplexy patients. Hundreds of medicinal herbs appear in those prescriptions. We chose 40 herbs that are the most important. Applying the algorithm IBLO with 100 resamplings and the threshold 0.6, we obtain the graph shown in Figure 2. From the graph, we can see which herbs are frequently used together.

Acknowledgements

This research was supported by NSFC and NBRP 2003CB715900.

References

1. R. G. Cowell, A. P. David, S. L. Lauritzen, D. J. Spiegelhalter, Probabilistic Networks and Expert Systems, Springer Publications, New York, 1999.
2. Z. Geng, C. Wang, and Q. Zhao, Decompsition of search for v-structures in DAGs, J. Multivar. Analy. To appear. (2004)
3. D. Heckerman, A tutorial on learning with Bayesian networks, Learning in Graphical Models, M. I. Jordan (Ed.), 301–354, Kluwer Academic Pub., Netherlands, 1998.
4. S. L. Lauritzen, Graphical models, Oxford, England: Oxford University Press, 1996.
5. J. Pearl, Causality, Cambridge University Press, Cambridge, 2000.
6. P. Spirtes, C. Glymour, R. Scheines, Causation, Prediction and Search, 2nd ed. MIT Press, Cambridge, 2000.
7. T. Verma, J. Pearl, Equivalence and synthesis of causal models, Uncertainty in Artificial Intelligence, Vol. 6, P. Bonissone, M. Henrion, L. N. Kanal and J. F. Lemmer (Eds.), Elsevier, Amsterdam, 1990, pp. 255–268.

Study of Ensemble Strategies in Discovering Linear Causal Models

Gang Li* and Honghua Dai

School of Information Technology, Deakin University,
221 Burwood Highway, Vic 3125, Australia
gang.li@deakin.edu.au hdai@deakin.edu.au

Abstract. Determining the causal structure of a domain is frequently a key task in the area of Data Mining and Knowledge Discovery. This paper introduces ensemble learning into linear causal model discovery, then examines several algorithms based on different ensemble strategies including Bagging, Adaboost and GASEN. Experimental results show that (1) Ensemble discovery algorithm can achieve an improved result compared with individual causal discovery algorithm in terms of accuracy; (2) Among all examined ensemble discovery algorithms, BWV algorithm which uses a simple Bagging strategy works excellently compared to other more sophisticated ensemble strategies; (3) Ensemble method can also improve the stability of parameter estimation. In addition, Ensemble discovery algorithm is amenable to parallel and distributed processing, which is important for data mining in large data sets.

1 Introduction

A class of limited *Graphical Model*, usually referred as *Linear Causal Models*, is widely used in social sciences [1]. In this kind of models, effect variables are strictly linear functions of exogenous variables. Although this is a significant limitation, its adoption allows for a comparatively easy environment in which to develop causal discovery algorithms [2]:

In 1996, Wallace et al. successfully introduced the *Minimum Message Length* (MML) criterion [3] into the discovery of *Linear Causal Models*. After that, a series of work has been done in the reliability and the efficiency issues of the MML-based causal discovery algorithm [4, 5, 6]. To further enhance the accuracy of causal discovery, this paper introduces *Ensemble learning* into causal discovery, and three ensemble strategies are considered for the task of discovering linear causal models, and both discovery accuracy and efficiency are compared to evaluate which method is with the best performance.

The rest of this paper is organized as follows. In Section 2 we briefly introduce MML-based discovery algorithm for linear causal models. In Section 3 we describe four different ensemble algorithms for causal discovery. In Section 5 experimental results of different causal discovery algorithms are compared and analyzed. Finally, we conclude this paper in Section 6.

---

* Corresponding Author.

2 MML Learning of Linear Causal Models

Linear Causal Model is a *Directed Graphical Model* in which every variable involved is a continuous variable. Informally speaking, it consists two parts: *Structure*, which qualitatively describes the relation among different variables; and *Parameters*, which quantitatively describe the relation.

The *Structure* of a *Linear Causal Model* is a directed acyclic graph (DAG) in which each node represents a variable, and a directed edge from V_i to V_j represents that V_i is a parent of V_j. The local relationship between each variable V_i and its parents is captured by a linear function (1):

$$V_i = \sum_{k=1}^{K_i} \alpha_k \times Pa_k(V_i) + R_i \tag{1}$$

Where K_i is the number of parents for node V_i, $\{\alpha_1, \ldots, \alpha_{K_i}\}$ are path coefficients, and R_i is the *Gaussian* noise, i.e. $R_i \sim N(0, \sigma_i^2)$. The set of local *Parameter* θ_i for a node V_i with parents is then $\{\sigma_i^2, \alpha_1, \ldots, \alpha_{K_i}\}$. On the other hand, for a node V_i without any parent, we assume it as a random sample from a *Gaussian* distribution, $V_i \sim N(\mu_i, \sigma_i^2)$, where μ_i is the expect value of node V_i, so the local *Parameter* at node V_i is $\{\mu_i, \sigma_i^2\}$.

Generally speaking, the task of *Linear Causal Model* discovery is: to induce the graph structured knowledge which best summarizes the given training data.

2.1 MML-Based Structure Discovery

For the structure discovery of *Linear Causal Models*, there are two key issues in MML-based method: *Measuring* the MML cost of models and *Searching* through the space of all possible models.

According to the MML criterion [3], the total message length can be approximated using formula (2):

$$\begin{aligned} L &= L(S) + L(\Theta_S) + L(D|S, \Theta_S) \\ &= L(S) + \sum_{i=1}^{n} (L(\theta_i) + L(D_i|\theta_i)) \end{aligned} \tag{2}$$

Where n is the number of nodes, θ_i is the local parameters at node V_i, and D_i is the data set confined to node V_i. $L(S)$ is the encoding length of model structure, while $L(\theta_i)$ is the encoding length for the local parameters at variable V_i, and $L(D_i|\theta_i)$ is the encoding length for the data set confined to variable V_i assuming the model. The detailed encoding scheme can be found in [6].

As for the searching, previous works [7] showed that the *Message Length based Greedy Search* (MLGS) algorithm converges much faster than some more sophisticated search methods while still keeps the accuracy of discovered results: Starting from a seeding graph or a null graph, the MLGS algorithm runs through each pair of nodes attempting to add an edge if there is none or to delete or to reverse it if there already is one. Such adding, deleting or reversing is done only if such change results in a decrease of the total message length. If the new structure is better, it is kept and then another change will be attempted. This process continues until no better structure can be found within a given number of search steps, or the whole structure space has been exploited.

2.2 MML-Based Parameter Estimation

Given a linear causal model structure, its parameters can also be estimated by the Minimum Message Length (MML) based method [2, 6]. For a node without parents, by minimizing the total encoding length, its parameters can be estimated by equations (3):

$$\mu_i = \frac{\sum_{t=1}^{T} v_{it}}{T} \tag{3a}$$

$$\sigma_i^2 = \frac{\sum_{t=1}^{T} (v_{it} - \mu_i)^2}{T - 1} \tag{3b}$$

Where T is the sample size, v_{it} is the value of variable V_i in the t-th instance. For a node with K parents, the MML estimations of $\{\alpha_0, \ldots, \alpha_K\}$ are the same as the estimates by least squares estimation, and the estimation for σ_i^2 is

$$\sigma_i^2 = \frac{\sum_{t=1}^{T}(v_{it} - \sum_k \alpha_k Pa_{i\ kt}^S)^2}{T - K} \tag{4}$$

3 Ensemble Structure Discovery

Ensemble learning is a machine learning paradigm where several individual learning algorithms are trained for the same task, then results of these single algorithms are integrated to get a final result. As for the ensemble structure discovery of linear causal models, there are three key issues needed to be dealt with:

Base Learner is an individual learning process and the building block of ensemble learning algorithm. Normally, it needs to be computationally efficient while learning accuracy may be not so perfect.

Ensemble Strategy is the most important issue in ensemble learning. It decides how to generate data sets from the original data set, how to carry out individual learning using base learner, and how to get the weights which will be used later.

Integration The result from individual learning process is usually a set of individual models, together with weights of each model. Integration will decide how to use these individual models and their corresponding weights to produce a final model.

In this section, we considered structure discovery algorithms based on different ensemble strategies including Bagging, GASEN and Adaboost respectively.

3.1 The Base Learner

The *Base Learner* is the building block of ensemble learning. As for the discovery of *Linear Causal Model*, we use the MML-CI II discovery system which is based on the MLGS algorithm [5, 6]. Although MML-CI II has no mechanism to avoid the local minimum, it can converge very fast to a quite favorite result.

3.2 Three Ensemble Strategies

Bagging. Bagging (Bootstrap Aggregating) is proposed by Breiman [8]. It employs bootstrap sampling [9] to generate several training sets from the original data set, and then induce an individual model from each generated training set, finally these models are integrated to produce the final result.

Based on different seeding methods for base learners, we examine two ensemble causal discovery algorithms using Bagging.

BWV. For a given original data set D, we generate e data sets with the same sample size as of the data set D, here e is the ensemble size. Then, MML-CI II is applied to each data set. In BWV, each base learner starts with a null graph.

BXWV. This algorithm is similar to the BWV, except that in BWV all the base learners start with null graphs, here in BXWV, each base learner will be started with a seeding graph. The algorithm processes as follows: firstly, the bootstrap sampling is applied to the original data set D, and get e data sets $\{D_1, D_2, \cdots, D_e\}$ all with the same sample size as D. then, from each data set D_i, we use MML-CI II to induce one graph S_i as a seeding graph, and then for each sub data set D_i, the base learner will start from every seeding graph S_i, and totally BXWV will generate $e \times e$ individual models.

The weights of individual models calculated from BWV and BXWV are calculated from the message length using a min-max normalization followed by a normalization, as shown in

$$\omega_t^0 = \frac{Max - Min}{L(M_t) - Min + 1}$$
$$\omega_t = \omega_t^0 / \sum \omega_t^0 \qquad (5)$$

Where Max and Min are the maximum and minimum message length of these individual models, respectively, and $L(M_t)$ is the message length of the model M_t. Here ω_t is the normalized weight for the individual model M_t.

GASEN. GASEN (Genetic Algorithm based Selective ENsemble), is a ensemble strategy based on the recognition that ensemble an appropriate subset of individual models may be superior to ensemble all the individual models in some cases, and in addition, the weights are not based on message lengths, but generated from a genetic algorithm.

As described in Zhou et al. [10], the problem of finding out the appropriate subset of individual models can be transformed to an optimization problem where genetic algorithm can be utilized. After inducing a number of individual models, GASEN assigns a random weight to each of the individual models. Then these weights are evolved so that they can characterize the goodness of the individual models in joining the ensemble. In each generation of the evolution, the weights are normalized so that they can be compared with the pre-set threshold λ, here a simple normalization scheme is used, as shown in (6).

$$\omega_t = \omega_t^0 / \sum_{t=1}^{N} \omega_t^0 \tag{6}$$

Where ω_t represents the normalized weight, and ω_t^0 represents the evolved weights of the individual models. Finally GASEN selects these models with weight higher than a pre-set threshold λ to produce a finally model.

In order to evaluate the goodness of different weight vectors, a validation data set D_V bootstrap sampled from the original data set D is used. Let $\boldsymbol{\omega}$ be a weight vector, $M(\boldsymbol{\omega})$ be the model combined using the weight vector $\boldsymbol{\omega}$, and $L_M^{D_V}$ be the encoding message length of the model $M(\boldsymbol{\omega})$ and the data set D_V. Then $L_M^{D_V}$ can be used to express the goodness of $\boldsymbol{\omega}$.

Adaboost. Adaboost (Adaptive boosting) is proposed by Freund and Schapire [11]. It sequentially generates a series of individual models, where the training instances what are weakly described by the previous model will play more important role in the learning of later models, finally all these individual models are combined in which the weights are determined by the Adaboost algorithm itself.

Considering that a *Linear Casual Model* (and all *Graphical Models*) can represent a joint probabilistic distribution on the domain, causal discovery can be formulated as the problem of *Density Estimation*. Many algorithms are available for estimating these densities from training data set. Recently, some work has introduced an Adaboost approach to density estimation [12]. In our experiments Adaboost is used to sequentially discover a set of *Linear Causal Models* from the original data set. The training distribution is updated using the methods adopted in [12]. Finally, those individual models are integrated using the weights determined by the Adaboost algorithm.

3.3 Integration

Integration is the last step for all ensemble causal discovery algorithms. It uses the set of models induced by individual learners together with their corresponding weights to get a final linear causal model. In our experiments, this is carried out as follows: first, for each pair of nodes i and j, we compare the weights of edge $i \rightarrow j$, $i \leftarrow j$ and no edge between i and j, if the weight of $i \rightarrow j$ is the highest, then there will be an edge $i \rightarrow j$ between i and j, or if the weight of

$i \leftarrow j$ is the highest, then there will be an edge $i \leftarrow j$ between i and j, or else if the weight of no edge has the highest weight, there will be no edge between i and j. Then, check the structure of the model, if there is cycle in the model, then remove the edge with the minimal weight within the cycle, until the structure is a valid DAG. Finally, the parameters of the *Linear Causal Model* can be obtained by an MML-based approach [6].

Algorithm 1 Integration Algorithm

Input: models Ms, weights ωs, node number n
Output: a final model M
   $M.struct \Leftarrow nullgraph$
   for $i = 1$ to n **do**
     for $j = i + 1$ to n **do**
       $\omega_{i \to j} \Leftarrow \sum_{(i \to j) \in M_t} \omega_t$
       $\omega_{i \leftarrow j} \Leftarrow \sum_{(i \leftarrow j) \in M_t} \omega_t$
       $\omega_{no\_edge(i,j)} \Leftarrow \sum_{(no\_edge) \in M_t} \omega_t$
       $M.struct+ = $ the edge with maximum weight
     end for
   end for
   while $M.struct$ has cycle **do**
     remove the edge with minimum weight in a cycle
   end while
   $M.\Theta \Leftarrow param\_estimate(M.struct, D)$

---

4 Ensemble Parameter Estimation

After the structure to be discovered, the parameters of this model can also be estimated using ensemble methods. Three issues involved can be designed like this:

Base Learner. the MML estimation algorithms [6] is selected as the base learner for ensemble parameter estimation.

Ensemble Strategy. When the original training data set contains no missing values, we adopt the Bagging to generate an ensemble. Given an ensemble size e and a training data set D consisting of T instances, the algorithm generates $e - 1$ bootstraps samples with each being created by uniformly sampling T instances from D with replacement, then it learns one set of parameters from each bootstrap sample, another set of parameters is estimated from the original training data set. Therefore, if given the graphical model structure and the original data set D, we can finally get e different sets of parameter estimation.

Integration. For a linear causal model, let e be the ensemble size, and let $\hat{\mu}_i^t, \hat{\alpha}_k^t, \hat{\sigma}_i^t$ be some estimation of involved coefficients and variation from the t-th base learner, the final estimation of these parameters are the average of the e estimates:

$$\mu_i = \frac{\sum_{t=1}^{e} \hat{\mu}_i^{\,t}}{e} \tag{7a}$$

$$\alpha_k = \frac{\sum_{t=1}^{e} \hat{\alpha}_k^{\,t}}{e} \tag{7b}$$

$$\sigma_i = \frac{\sum_{t=1}^{e} \hat{\sigma}_i^{\,t}}{e} \tag{7c}$$

5 Empirical Results and Analysis

In this section, we report the empirical results of discovering linear causal models using different algorithms. Intuitively, if a causal discovery algorithm is working perfectly, it should reproduce exactly the model used to generate the data. In practice, sampling errors will result in deviations from the original model, but algorithm which can reproduce a model structure similar to the original structure should be considered to be better than those do not.

Seven benchmark data sets in related literature [2, 5, 6] were used in this experiment, and they are: *Fiji, Evans, Blau, Rodgers, Case9, Case10* and *Case12*.

5.1 Discovery Algorithms

For structure discovery of linear causal model, one single algorithm (MML-CI II [5, 6]) and four ensemble discovery algorithms (BWV, BXWV, GASEN and Adaboost) were compared.

The experiment was done in Matlab with Bayes-Net Toolbox [13]. The ensemble sizes selected for ensemble algorithms are all set to 20. The genetic algorithm employed by GASEN is realized with the GAOT toolbox developed by Houck et al. [14]. The validation set used by GASEN is the original data set, and the preset threshold λ is set to 0.05, which is the reciprocal of the ensemble size.

For the ensemble parameter learning, we assume the model structure as known, and use MML estimation and its ensemble version to estimate the local parameters of each model, then compare these two parameter estimation methods in terms of the average error: $AE = \frac{1}{|E|} \sum_{k \in E} |\alpha_k - \hat{\alpha_k}|$, where $|E|$ is the number of edges, while α_k and $\hat{\alpha_k}$ are the known and estimated path coefficient of the k-th edge, respectively. It is clear that the less the average error, the better the estimated result.

5.2 Result Analysis

For each structure discovery algorithm, we perform 10 runs on each data set and then record the most frequently appeared result by each algorithm, and for each parameter estimation algorithm, the average results from 10 runs by each algorithm are recorded.

For the structure discovery, out of seven data sets, there are four data sets (*Fiji, Case9, Case10* and *Case12*) for which all the above five algorithms converged on same results, and especially for the late three data sets, every algorithm can find the original model. So here we focus our report on the results on four data sets, that is, *Fiji*, *Evans*, *Blau* and *Rodgers*.

Table 1 gives the number of incorrect edges for the results by different algorithms. *The Number of Incorrect Edges* is decomposed into a triple-tuple: $[m, a, r]$, in which m is the number of missing edges, a is the number of added edges, r is the number of reversed edges.

Table 2 gives the the encoding length for each result. Table 3 gives the average running time of different algorithms on seven data sets. The time given is averaged from 10 runs. Here the time is measured in *seconds*. Table 4 gives the results for parameter estimation from the ensemble method and the original MML estimation.

Table 1. Performance Comparison of Algorithms

Alg	Fiji	Evans	Blau	Rodgers
MML-CI II	[2,0,1]	[2,1,3]	[0,1,2]	[2,1,2]
BWV	[2,0,1]	[1,0,2]	[0,0,1]	[0,0,0]
BXWV	[2,0,1]	[2,1,3]	[0,0,1]	[0,0,0]
GASEN	[2,0,1]	[1,0,2]	[0,0,1]	[0,0,0]
Adaboost	[2,0,1]	[2,1,3]	[0,1,2]	[2,1,2]

Table 2. Encoding Length Comparison of Results

Alg	Fiji	Evans	Blau	Rodgers
MML-CI II	5488.977	5472.475	7357.062	9362.275
BWV	5488.977	5472.421	7353.185	9352.063
BXWV	5488.977	5472.475	7353.185	9352.063
GASEN	5488.977	5472.421	7353.185	9352.063
Adaboost	5488.977	5472.475	7357.062	9362.275

Table 3. Time cost comparison of Algorithms

Alg	MML-CI II	BWV	BXWV	GASEN	Adaboost
Fiji	0.094	1.843	44.236	23.141	1.922
Evans	0.172	4.594	69.877	37.156	5.057
Blau	0.360	7.781	93.910	88.062	8.427
Rodgers	0.782	17.406	209.632	153.438	19.395
Case9	1.390	26.189	330.713	214.255	24.733
Case10	1.657	38.391	751.227	429.015	36.110
Case12	2.750	61.466	1007.719	694.779	63.219

From these results, we can see that,

1. In terms of learning accuracy, all these ensemble algorithms are better than or at least equal to the base learner — MML-CI II, and BWV and GASEN achieved the most accurate results among all the compared algorithms.

Table 4. Parameter Estimation: Average Errors

Model	by MML	by Ensemble MML
Fiji	0.025	0.021
Evans	0.026	0.024
Blau	0.022	0.016
Case9	0.016	0.015

2. BWV and GASEN got the same results for each data sets, and Adaboost got the same results as MML-CI II. BXWV is similar to BWV and GASEN except for data set *Evans*.
3. In terms of time efficiency, BWV and Adaboost are faster than the other two ensemble algorithms.
4. For parameter learning, ensemble method can improve the results of MML estimation methods.

Moreover, from the comparison of BWV and BXWV, we can see that seeding graph does not improve the results. In addition, the comparison of BWV and GASEN indicates that the weight calculated from MML can achieve the same result as the weight calculated from the optimal algorithm GASEN. However, if considering computation cost, BWV is more efficient than GASEN.

It is interesting to note that although Adaboost is usually better than Bagging for classification problems, it has no advantage over Bagging for causal discovery, at least for those generated data sets. After a close examination of the weight generated from Adaboost, we find that the weight of the first individual model overwhelms the weights of the other models, this is because the first individual model can usually approximate the unknown density very closely, and the further steps of Adaboost will only slightly update the approximation and keep a tiny weight.

6 Conclusion

Achieving highest accuracy is always one of the essential goal of almost all the research done in the area of machine learning. In this paper, we examined different ensemble approaches that aims to improve the learning accuracy of linear causal models. Four structure discovery algorithms based on different ensemble strategies including Bagging, GASEN and Adaboost are considered, and one ensemble parameter estimation is described, our experimental results reveals that:

1. Among all the examined structure discovery algorithms, BWV is the best choice if considering both accuracy and the time complexity.
2. For parameter estimation, ensemble method can also improve the results of MML estimation.

In addition, ensemble algorithms are also amenable to parallel processing, which is an important characteristics for data mining in large data sets.

Acknowledgement

The first author would like to acknowledge the valuable suggestion from Professor Zhi-Hua Zhou (Nanjing University).

References

1. Bollen, K.: Structural Equations with Latent Variables. Wiley, New York (1989)
2. Wallace, C., Korb, K., Dai, H.: Causal discovery via MML. In: Proceedings of the 13th International Conference on Machine Learning (ICML'96). (1996) 516–524
3. Wallace, C., Boulton, D.: An information measure for classification. Computer Journal **11** (1968) 185–194
4. Dai, H., Korb, K., Wallace, C., Wu, X.: A study of causal discovery with small samples and weak links. In: Proceedings of the 15th International Joint Conference On Artificial Intelligence IJCAI'97, Morgan Kaufmann Publishers, Inc. (1997) 1304–1309
5. Dai, H., Li, G.: An improved approach for the discovery of causal models via MML. In: Proceedings of The 6th Pacific-Asia Conference on Knowledge Discovery and Data Mining (PAKDD-2002), Taiwan (2002) 304–315
6. Li, G., Dai, H., Tu, Y.: Linear causal model discovery using MML criterion. In: Proceedings of 2002 IEEE International Conference on Data Mining, Maebashi City, Japan, IEEE Computer Society (2002) 274–281
7. Dai, H., Li, G., Tu, Y.: An empirical study of encoding schemes and search strategies in discovering causal networks. In: Proceedings of 13th European Conference on Machine Learning (Machine Learning: ECML 2002), Helsinki, Finland, Springer (2002) 48–59
8. Breiman, L.: Bagging predictors. Machine Learning **24** (1996) 123–140
9. Efron, B., Tibshirani, R.: An introduction to the bootstrap. Chapman and Hall, New York (1993)
10. Zhou, Z.H., Wu, J., Tang, W.: Enselbling neural networks: Many could be better than all. Artificial Intelligence **137** (2002) 993–1001
11. Freund, Y., Schapire, R.: A decision-theoretic generalization of on-line learning and an application to boosting. In: Proceedings of EuroCOLT-94, Barcelona, Spain (1995) 23–37
12. Thollard, F., Sebban, M., Philippe, E.: Boosting density function estimators. In: Proceedings of 13th European Conference on Machine Learning (Machine Learning: ECML 2002), Helsinki, Finland, Springer (2002) 431–443
13. Murphy, K.: The Bayes Net Toolbox for Matlab. Computer Science and Statistics **33** (2001) 331–351
14. Houck, C., Joines, J., Kay, M.: A genetic algorithm for function optimization: a matlab implementation. Technical Reports NCSU-IE-TR-95-09, North Carolina State University, Raleigh, NC (1995)

The Entropy of Relations and a New Approach for Decision Tree Learning[*]

Dan Hu and HongXing Li

Beijing Normal University,
Department of Mathematics, 100875 Beijing, China
hufengdd@163.com

Abstract. The formula for scaling how much information in relations on the finite universe is proposed, which is called the entropy of relation R and denoted by $H(R)$. Based on the concept of $H(R)$, the entropy of predicates and the information of propositions are measured. We can use these measures to evaluate predicates and choose the most appropriate predicate for some given cartesian set. At last, $H(R)$ is used to induce decision tree. The experiment show that the new induction algorithm denoted by IDIR do better than ID3 on the aspects of nodes and test time.

1 Introduction

As we all know, relation is an important and basic concept in a lot of aspects including knowledge expression. For two given sets, we often deal with the relations between them. After the relations are given, it should be said that we acquire some information of the connection between them. But how much information have we got? How to compare the information of two relations on the same cartesian set by the way of quantity? Furthermore, through the view of probability and entropy, we can scale the information of the proposition with one variable[2]. But when the proposition have two or more variable, how shall we scale the information of it? For providing tentative answer to these problems, we will discuss the entropy of relations and its application. The rest of this paper is organized as follows. In section 2, we present the definition of $H(R)$, discuss some basic properties of $H(R)$ and indicate the connections of $H(R)$ with classical information theory. In section 3, the applications of $H(R)$ are put forward. We use $H(R)$ as a tool to define the entropy of predicate and the information of proposition. In section 4, a new induction algorithm named IDIR is presented. Through the experimental results on comparing IDIR with ID3, We can find the advantage IDIR. In section 5, we conclude.

---

[*] Supported by The Nature Science Foundation of China (Grant No.60474023), Research Fund for Doctoral Program of Higher Education (20020027013), Science Technology Key Project Fund of Ministry of Education (03184) and Major State Basic Research Development Program of China (2002CB312200).

2 The Entropy of Relations

2.1 The Entropy of Binary Relations

let X and Y be nonempty sets. A relation R from X to Y is a subset of $X \times Y$. if $A \subseteq X$, then
$$R(A) = \{y \in Y | \, xRy \text{ for some } x \text{ in } A\}.$$

Definition 2.1. let U be the finite universe. $X, Y \in \wp(U)$, R is a relation from X to Y.

When $R(X) = Y$, $R^{-1}(Y) = X$, the entropy of R is denoted by $H(R)$ and defined as follows:

$$H(R) = \frac{m}{m+n} H(R \downarrow X) + \frac{n}{m+n} H(R \downarrow Y).$$

$$H(R \downarrow X) = -\sum_{i=1}^{m} \frac{|R(x_i)|}{\sum_{i=1}^{m} |R(x_i)|} \log \frac{|R(x_i)|}{n}.$$

$$H(R \downarrow Y) = -\sum_{j=1}^{n} \frac{|R^{-1}(y_j)|}{\sum_{j=1}^{n} |R^{-1}(y_j)|} \log \frac{|R^{-1}(y_j)|}{m}.$$

When $R(X) \neq Y$ or $R^{-1}(Y) \neq X$, let

$$R'(x_i) = \begin{cases} R(x_i) & x_i \in R^{-1}(Y) \\ Y & x_i \bar{\in} R^{-1}(Y) \end{cases}$$

$$(R'')^{-1}(y_j) = \begin{cases} R^{-1}(y_j) & y_j \in R(X) \\ X & y_j \bar{\in} R(X) \end{cases}$$

The information of R is defined as follows:

$$H(R) = \frac{m}{m+n} H(R' \downarrow X) + \frac{n}{m+n} H(R'' \downarrow Y).$$

Where $H(R \downarrow X)$ is the entropy of R restricted on X, and $H(R \downarrow Y)$ is the entropy of R restricted on Y. $|\cdot|$ is the cardinal number of a set, $m = |X|, n = |Y|$. The base of logarithm is 2, and $0 \log 0 = 0$. The unit of entropy is "bit". All of these are provisions of this paper if we do not have another explanation.

Proposition 2.1. $\forall R \in \wp(X \times Y)$,

$$H(R \downarrow X) = \log n - H(Y|X);$$

$$H(R \downarrow Y) = \log m - H(X|Y).$$

Composition "$\circ$" is a very important operator. $\forall R \in \wp(X \times Y), \forall S \in \wp(Y \times Z)$,

$$S \circ R = \{(x,z) | \exists y \in Y, (x,y) \in R, (y,z) \in S\}.$$

Through the following proposition, we can see the action of composition in the information process.

Proposition 2.2. $\forall R \in \wp(X \times Y), \forall S \in \wp(Y \times Z)$, $T = S \circ R$, $|T| = t$. (i). If R is a function from X to Y, then

$$H(T \downarrow X) = I(S \downarrow Y).$$

If R^{-1} is a function from Y to X, then

$$H(T \downarrow Z) = (\log m - \log n) + H(S \downarrow Y) \leq H(S \downarrow Z).$$

when R is a bijective function,

$$H(T \downarrow Z) = H(S \downarrow Z),$$

and at this time,

$$H(T) = H(S).$$

(ii). If S is a function from Y to Z, then

$$H(T \downarrow Z) = H(R \downarrow Y).$$

If S^{-1} is a function from Z to Y,

$$H(T \downarrow X) = (\log t - \log n) + H(R \downarrow X) \leq H(R \downarrow X).$$

when S is a bijective function,

$$H(T \downarrow X) = H(R \downarrow X),$$

and at this time,

$$H(T) = H(R).$$

(iii). If R, S are all bijective relation, then

$$H(T) = H(R) = H(S).$$

Proposition 2.3. $\forall R \in \wp(X \times Y)$, let $\triangle = |R|$,

$$H(R) = \frac{m}{m+n}\log n + \frac{n}{m+n}\log m - \frac{m}{m+n}\log \triangle + \frac{n}{m+n}I(X;Y) - \frac{n-m}{m+n}H(X);$$

$$H(R) = \frac{m}{m+n}\log n + \frac{n}{m+n}\log m - \frac{n}{m+n}\log \triangle + \frac{m}{m+n}I(X;Y) - \frac{m-n}{m+n}H(Y).$$

When $m = n$,

$$H(R) = \log m - \frac{1}{2}\log \triangle + \frac{1}{2}I(X;Y).$$

From proposition 2.3, the connection of $H(R)$ with classical information theory can be found again except Proposition 2.1.

An equivalence relation R on X is a relation from X to X that is reflexive, symmetric and transitive. For each $x \in X$, the set $R(x)$ is the R-equivalence class of x and is denoted by $[x]_R$. In [3],Kuriyama has put forward the entropy of partition.

Definition 2.2.[3] Suppose $\xi = \{A_1, A_2, \cdots, A_n\}$ is the finite partition of $X = \{x_1, x_2, \cdots, x_m\}$, the entropy of ξ is defined as follows:

$$H(\xi) = -\sum_{i=1}^{n} p(A_i) \log p(A_i) = -\sum_{i=1}^{n} \frac{|A_i|}{m} \log \frac{|A_i|}{m}.$$

Because partition can be get by equivalence relation without other information, we should discuss the connection of our definition of the entropy of relation with the definition of the entropy of partition.

Proposition 2.4. $X = \{x_1, x_2, \cdots, x_m\}, \forall R \in \wp(X \times X)$, if R is an equivalence relation, and $\xi = \{A_1, A_2, \cdots, A_n\}$ is the partition induced by R, then

$$H(R) \le H(\xi).$$

When $\forall x_i \ne x_j, [x_i]_R = [x_j]_R, H(R) = H(\xi)$.

The result $H(R) \le H(\xi)$ maybe will arouse doubt of some readers, because the $H(R)$ as though should equal with $H(\xi)$ for the reason of we can induce R by ξ and induce ξ by R on the other hand without appending other information. But as a matter of fact, we can find their difference in each semantic description. $H(R)$ measures the average information of $x_i R y_i$ while $H(\xi)$ measures the average information of x_i in $[x_i]_R$. $I(R)$ only describe the result that x_i is related to one element by R, but $H(\xi)$ describe the fact that x_i is related to all elements in $[x_i]_R$ by R. They are obviously different, and $H(R) \le H(\xi)$ is the really description of fact.

2.2 The Entropy of n-Ary Relations

After the entropy of binary relations on finite universe is discussed, we will give the definition of the entropy of n-ary relation on finite universe, which is the natural generalization of the entropy of binary relations.

Definition 2.3. let U be the finite universe. $X_i \in \wp(U), i \in I = \{1, \cdots, n\}$, $X_i = \{x_{i1}, x_{i2}, \cdots, x_{im}\}$, then R is a n-nary relation on $\prod_{i=1}^{n} X_i$.
When $\forall i_0 \in I, \bigcap_{j=1}^{m} R(x_{i_0 j}) \ne \emptyset$, the entropy of R is defined as follows:

$$H(R) = \sum_{i=1}^{n} \frac{|X_i|}{\sum_{i=1}^{n} |X_i|} H(R \downarrow X_i).$$

$$H(R \downarrow X_i) = -\sum_{j=1}^{m} \frac{|R(x_{ij})|}{\sum_{j=1}^{m} |R(x_{ij})|} \log \frac{|R(x_{ij})|}{\prod_{k \in I \setminus \{i\}} |X_k|}.$$

When $\exists x_{i_0 j} \in X_{i_0}, R(x_{i_0 j}) = \emptyset$ let

$$R^{i_0}(x_{i_0 j}) = \begin{cases} R(x_{i_0 j}) & R(x_{i_0 j}) \ne \emptyset \\ \prod_{i \in I \setminus \{i_0\}} X_i & R(x_{i_0 j}) = \emptyset. \end{cases}$$

The entropy of R is defined as follows:
$$H(R \downarrow X_{i_0}) = H(R^{io} \downarrow X_{i_0}),$$
$$H(R) = \sum_{i=1}^{n} \frac{|X_i|}{\sum_{i=1}^{n}|X_i|} H(R \downarrow X_i).$$

It's obviously to see that when $n = 2$, definition 2.3 is just the definition 2.1. The properties of the information of binary relation can be easily generalized to n-ary relation, so we will not state those at here any more.

Proposition 2.5. R is a n-ary relation on $\prod_{i=1}^{n} X_i$, let $Y_i = \prod_{k \in I \setminus \{i\}} X_k$, $R_i = X_i \times Y_i$, R_i is the relation from X_i to Y_i Then
$$H(R) \leq \sum_{i=1}^{n} H(R_i).$$

Definition 2.4[4]. R is a n-ary relation on $\prod_{i=1}^{n} X_i$, let
$$q = \{X_{i_1}, X_{i_2}, \cdots, X_{i_k}\} \subseteq \{X_1, X_2, \cdots, X_n\},$$
the projection of R on q is denoted by $P_q(R)$ and defined as
$$P_q(R) = \{(x_{i_1}, x_{i_2}, \cdots, x_{i_k}) | (x_1, x_2, \cdots, x_{i_1}, x_{i_2}, \cdots, x_{i_k}, \cdots, x_n) \in R\}.$$

Definition 2.5[4]. R is a n-ary relation on $\prod_{i=1}^{n} X_i$, $X_1, X_2, \cdots, X_n$ is non-interactional under R if and only if R is separable
$$R = P_{X_1}(R) \times P_{X_2}(R) \times \cdots \times P_{X_n}(R).$$

Proposition 2.6. R is a n-ary relation on $\prod_{i=1}^{n} X_i$, $X_1, X_2, \cdots, X_n$ is non-interactional under R if and only if $H(R) = 0$.

Proposition 2.7. R is a n-ary relation on $\prod_{i=1}^{n} X_i$,
$$q = \{X_{i_1}, X_{i_2}, \cdots, X_{i_k}\} \subseteq \{X_1, X_2, \cdots, X_n\}, \quad k \geq 2,$$
$q' = \{X_1, X_2, \cdots, X_n\} - q$, then
$$H(R) \leq \frac{\sum_{i \in q}|X_i|}{\sum_{i=1}^{n}|X_i|}(H(P_q(R)) + \sum_{s \in q'} \log |X_s|) + \frac{\sum_{i \in q'}|X_i|}{\sum_{i=1}^{n}|X_i|}(H(P_{q'}(R)) + \sum_{s \in q} \log |X_s|).$$

If $\forall X_i, |X_i| = N$, then
$$H(R) \leq \frac{k}{n} H(P_q(R)) + (1 - \frac{k}{n}) H(P_{q'}(R)) + \frac{2k(n-k)}{n} \log N.$$

Proposition 2.8. R is a n-ary relation on $\prod_{i=1}^{n} X_i$, let
$$Q_k = \{q = \{X_{s_1}, X_{s_2}, \cdots, X_{s_k}\} | \{s_1, s_2, \cdots, s_k\} \subset \{1, 2, \cdots, n\}\}$$

then
$$H(R) \leq \sum_{q \in Q_k} \frac{\sum_{i \in q} |X_i|}{C_{n-1}^{k-1} \sum_{i=1}^{n} |X_i|} (H(P_q(R)) + \sum_{s \in q'} \log |X_s|).$$

If $\forall X_i, |X_i| = N$, then
$$C_n^k H(R) \leq k(n-k)\log N + \sum_{q \in Q_k} H(P_q(R)).$$

From proposition 2.7 and 2.8, the connections between $H(R)$ and $H(P_q(R))$ can be find. Those results can actually show the connections between the entropy of whole and the entropy of parts.

3 The Entropy of Predicate and the Information of Proposition

A predicate $P(x_1, x_2, \cdots, x_n)$ is the word to describe property set of elements $\{(x_1, x_2, \cdots, x_n) | P(x_1, x_2, \cdots, x_n)\}$. Every element in the set is an object for which the statement $P(x_1, x_2, \cdots, x_n)$ is true. Obviously,
$$\{(x_1, x_2, \cdots, x_n) | P(x_1, x_2, \cdots, x_n)\}$$
is a subset of $\prod_{i=1}^{n} X_i$, and we can see $\{(x_1, x_2, \cdots, x_n) | P(x_1, x_2, \cdots, x_n)\}$ as a relation on $\prod_{i=1}^{n} X_i$. That is, predicate $P(x_1, x_2, \cdots, x_n)$ on the universe U is corresponding to a relation $R_{P(x_1, x_2, \cdots, x_n)}$ on $\prod_{i=1}^{n} X_i$. Now, we can expediently use $H(R_{P(x_1, x_2, \cdots, x_n)})$ to measure the entropy of predicate $P(x_1, x_2, \cdots, x_n)$.

Definition 3.1. $Q(x_1, x_2, \cdots, x_n)$ is a predicate on $\prod_{i=1}^{n} X_i$,
$$R_Q = \{(x_1, x_2, \cdots, x_n) | Q(x_1, x_2, \cdots, x_n)\}$$
is the relation $Q(x_1, x_2, \cdots, x_n)$ corresponding with. The entropy of predicate **Q** is
$$H(\mathbf{Q}) = H(R_Q).$$
The information of a proposition "$Q(x_1, x_2, \cdots, x_n)$" is
$$I_p(Q(x_1, \cdots, x_n)) = \begin{cases} 0 & Q(x_1, \cdots, x_n) \text{ is false} \\ -\sum_{i=1}^{n} I_p(Q(x_1, \cdots, x_n) \downarrow x_i) & Q(x_1, \cdots, x_n) \text{ is true} \end{cases}$$

where
$$I_p(Q(x_1, \cdots, x_n) \downarrow x_i) = \frac{|R_Q(x_i)|}{\sum_{i=1}^{n} |R_Q(x_i)|} \log \frac{R_Q(x_i)}{\prod_{k \in I \setminus \{i\}} |X_k|}, x_i \in X_i, I = \{1, \cdots, n\}.$$

Example 3.1. Let $X = \{a_1, a_2, a_3, a_4\}$, $Y = \{b_1, b_2, b_3\}$. $Q(x, y)$ is the proposition "x is the friend of y",
$$R_Q = \{(a_1, b_1), (a_1, b_2), (a_3, b_3), (a_4, b_1), (a_2, b_3)\}.$$

The entropy, that is the average information, which given by predicate **Q** is

$$H(\mathbf{Q}) = -\sum_{i=1}^{4} \frac{|R_Q(x_i)|}{\sum_{i=1}^{4}|R_Q(x_i)|} \log \frac{|R_Q(x_i)|}{3} - \sum_{j=1}^{3} \frac{|R_Q^{-1}(y_j)|}{\sum_{j=1}^{3}|R_Q^{-1}(y_j)|} \log \frac{|R_Q^{-1}(y_j)|}{4}$$

$$= 1.14 \ (bit).$$

The information of propositions "a_2 is the friend of b_2", "a_2 is the friend of b_3", and "a_1 is the friend of b_1" respectively is

$$I_p(Q(a_2, b_2)) = 0(bit), \quad I_p(Q(a_2, b_3)) = 1.195(bit), \quad I_p(Q(a_1, b_1)) = 0.792(bit).$$

Given a Cartesian set $\prod_{i=1}^{n} X_i$, we may have many predicates on it, and every predicate describe a relation among these X_i. At this time, we can use the entropy of Predicates to evaluate and choose the most appropriate predicate for $\prod_{i=1}^{n} X_i$.

Example 3.2. $X = \{1, 4, 9, 10\}$, $Y = \{1, 2, 3, 12\}$.

$$Q(x, y): x \text{ is the square of } y,$$

$$T(x, y): x \text{ is the bigger than } y.$$

It's easy to get
$$R_Q(x, y) = \{(1, 1), (4, 2), (9, 3)\},$$
$$R_T(x, y) = \{(4, 1), (4, 2), (4, 3), (9, 1), (9, 2), (9, 3), (10, 1), (10, 2), (10, 3)\}.$$
$$H(\mathbf{Q}) = H(R_Q(x, y)) = 0.86 \ (bit),$$
$$H(\mathbf{T}) = H(R_T(x, y)) = 0.29 \ (bit).$$
$$H(\mathbf{Q}) \gg H(\mathbf{T}).$$

So, "x is the square of y" is more appropriate to describe the relation between X and Y.

4 Building Decision Trees Based on the Information of Relation

As we all know, the construction of decision trees is centered on the selection algorithm of an attribute that generates a partition of the subset of the training data set. Over the years, several techniques for choosing the splitting attribute have been proposed including the entropy gain and the gain ratio [10], the Gini index, the Kolmogorov-Smirnov metric, or a metric derived from Shannon entropy [11-12]. In all of these techniques, Quinlan's ID3 algorithm(1979) is the one that received special attention.

This article presents an algorithm for the induction of decision trees based on the concept of entropy of relations. We named it **IDIR** (Induction of Decision trees By the average information of Relation). The following is the algorithm **IDIR** for building a decision tree on the given training examples.

Algorithm IBIR :
functionDecision-Tree-Learning (*example, attributes, default*) **returns** a decision tree
  inputs: *examples*, set of examples
       attribute, set of attributes
       default, default value for the goal predicate
if *example* is empty **then return** *default*
else if all *examples* have the same classification **then return** the classification
else if *attributes* is empty **then return** Majority-value(*examples*)
else
     best←Choose-Attribute(attributes,examples)
     [function Choose-Attribute(attributes,examples)
       for each C_i *attributes* **do**
         Build a *relation* R_i between the attribute and decision-classification based on *examples*
           Modify the *relation* R_i (* delete the repeated elements in R_i *)
           Compute $H(R_i)$ (the information of R_i)
       end
       Choose the attribute which has the maximal value of $H(R_i)$]
     tree← a new decision tree with root test *best*
       for each value v_j of *best* **do**
         examples$_j$ ←{elements of *examples* with *best*= v_j}
         subtree←Decision-Tree-Learning (*example$_j$*, *attributes-best*,Majority-value(*examples*))
           add a branch to *tree* with label v_j and subtree *subtree*
       end
       return *tree*

The function Choose-Attribute(attributes,examples) in the algorithm IDIR is the kernel of the algorithm, and the process of computing $H(R_i)$ is used to determine the correlation degree of C_i with D. The comparison of the correlation degree of all C_i with D can be defined as follows.

Definition 4.1. Let $S =< U, A = C \cup D, V, f >$ be the database. The element in decision table S can be described as a element in the relation from $\prod_{i=1}^{n} V_{C_i}$ to V_D, and the database can be shown as

$$R_S = \{(c_{1j_1}, \cdots, c_{ij_i}, \cdots, c_{nj_n}, d_j) | c_{ij_i} \in V_{C_i}, d_j \in V_D\}.$$

To a selected condition attribute C_i, we can get a relation from V_{C_i} to V_D, which can be denoted as $R_S|_{C_i}$ and read as the relation restrict on C_i,

$$R_S|_{C_i} = \{(c_{ij_i}, d_j) | c_{ij_i} \in V_{C_i}, d_j \in V_D\}.$$

The elements in $R_S|_{C_i}$ are consistent with the corresponding elements in R_S. Suppose C_{t1} and C_{t2} are two different attributes, C_{t1}, $C_{t2} \in C$. The degree of correlation about C_{t1} with D is more big than C_{t2} with D, if

$$H(R_S|_{C_1}) > H(R_S|_{C_2}).$$

As we all know, learning decision tree is a NP-complete problem. In IDIR, we use $H(R_i)$ to substitute information gain to choose the best attribute. This method do good to simplify the complexity of algorithm. When we compute $H(R_i)$,

$$H(R_i) = -\sum_{k=1}^{m} \frac{|R_i(x_k)|}{\sum_{k=1}^{m}|R_i(x_k)|} \log \frac{|R_i(x_k)|}{n} - \sum_{j=1}^{n} \frac{|R_i^{-1}(y_j)|}{\sum_{j=1}^{n}|R_i^{-1}(y_j)|} \log \frac{|R_i^{-1}(y_j)|}{m}$$

we need know the data $R_i(x_k)$ and $R_i^{-1}(y_j)$. The information complexity is $m+n$, and time to choose the attribute is linear[60].

If we use ID3 to induce the decision tree, we need to compute

$$\frac{Gain(A_i)}{IV(A_i)} = \frac{\sum_{k=1}^{m}\sum_{j=1}^{n} P(a_k d_j) \log \frac{P(a_k d_j)}{P(a_k)P(d_j)}}{-\sum_{k=1}^{m} P(a_k) \log P(a_k)}.$$

The information complexity is $mn + m + n$.

If we use De Mantaras's formula of distance[13] to determine the attribute, we need to compute

$$d(C_i,\ D,\ U) = H(C_i/D) + H(D/C_i)$$
$$= -\sum_{k=1}^{m}\sum_{j=1}^{n} P(a_k d_j) \log \frac{P(a_k d_j)}{P(a_k)} - \sum_{k=1}^{m}\sum_{j=1}^{n} P(a_k d_j) \log \frac{P(a_k d_j)}{P(d_j)}.$$

The information complexity is also $mn + m + n$.

It's obviously that along with the induction of decision tree, IDIR has advantage to reduce the complexity of learning. In fact, we find the learning time of IDIR is less than ID3 by test.

The public domain fingerprint database, DB3 Set A in FVC 2002, is used in the experiment. After transform the images to data, we use the data to inducing decision tree which can distinguish the image "good" and "bad". ID3 is compared with IDIR at the aspects of the number of nodes and the time of training. The results of the experiment are shown in figure 1 and figure 2 respectively.

5 Conclusion

Thus, for a given Cartesian set, we have the tool to scale the entropy of the relations on it. $H(R)$ is more big, the relation R is more suit for the given objects(Cartesian set). For the propositions with two or more variable, the information of them can also be scaled by the definition of the information of proposition.

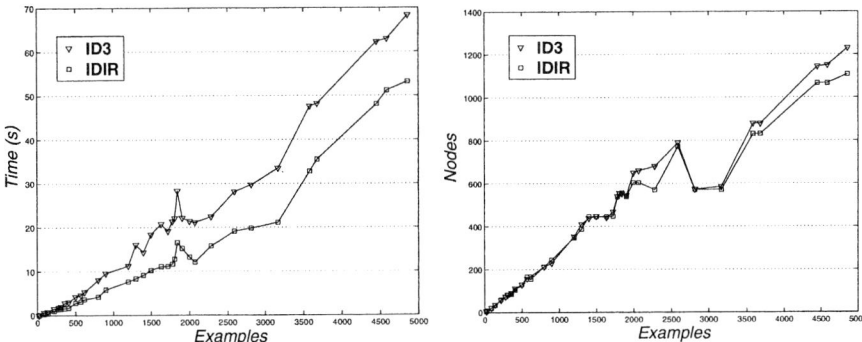

Fig. 1. Compare ID3 with IDIR on the training time

Fig. 2. Compare ID3 with IDIR on the number of nodes

Furthermore, we can evaluate the information of predicate and choose the most appropriate predicate for $\prod_{i=1}^{n} X_i$. Relation, Proposition, Predicate are important concepts in knowledge expression. we discuss them in the way of quantity and get many good results. Some of these must would be well used in artificial intelligence. At last, we use the concept–the entropy of relations to introduce a new algorithm for decision tree learning. Through the comparison of IDIR with ID3, we find the fact that our new algorithm IDIR has advantage to reduce the complexity of learning, which can be measured by test time, and have more simpler tree than ID3.

References

1. Zhu Xuelong, Fundamentals of applied information theory, Tsinghua University press, Beijing(2001)
2. Nu cheng uang, Generalized information theory, University of scince and technology of China press, Hefei(1993)
3. Ken Kuriyama, Entropy of a finite partition of fuzzy sets,J. Math. Anal. Appl.94 (1983)38-43.
4. L.A.Zadeh, The concept of a linguistic variable and its applicaiton to approximate reasoning I, Information Science, vol.8,3(1975),199-251.
5. Quinlan, J.R. C4.5: Programs for Machine Learning. Morgan Kaufmann, San Mateo, CA (1993)
6. De MSantaras, R.L., A distance-based attribute selection measure for decision tree induction, Machine Learning 6(1991)81-92.
7. Simovici,D.A., Jaroszewicz,S., Generalized conditional entropy and decision trees, In: Proceedings of EGC France(2003)369-380
8. D. Dumitrescu, Fuzzy measure and the entropy of fuzzy partition, J. Math. Anal. Appl. 176(1993)359-373.
9. J.Marichal and M.Roubens, Entropy of discrete fuzzy measures,International Journal of uncertainty, fuzziness and knowledge-based systems, vol.8,6(2000)625-640.
10. Bernard Kolman, Robert C.Busby, Sharon Ross, Discrete mathematical structures 3^{rd}ed., Prentice Hall,USA,(1996)

Effectively Extracting Rules from Trained Neural Networks Based on the New Measurement Method of the Classification Power of Attributes

Dexian Zhang[1], Yang Liu[2], and Ziqiang Wang[1]

[1] School of Information Science and Engineering,
Henan University of Technology, Zheng Zhou 450052, China
zdx@haut.edu.cn
[2] Computer College, Northwestern Polytecnical University,
Xi'an 710072, P.R.C(China)

Abstract. The major problems of currently used approaches for extracting symbolic rules from trained neural networks are analyzed. The lack of efficient heuristic information is the fundamental reason that causes the low effectiveness of currently used approaches. In this paper, a new measurement method of the classification power of attributes on the basis of differential information of the trained neural networks is proposed, which is suitable for both continuous attributes and discrete attributes. Based on this new measurement method, a new approach for rule extraction from trained neural networks and classification problems with continuous attributes is proposed. The performance of the new approach is demonstrated by several computing cases. The experimental results prove that the approach proposed can improve the validity of the extracted rules remarkably compared with other rule extracting approaches, especially for the complicated classification problems.

1 Introduction

The existing approaches for extracting the classification rules can be roughly classified into two categories: data drive approaches and model drive approaches. The main characteristic of the data drive approaches is to extract the symbolic rules completely based on the treatment with the sample data. The most representative approach is the ID3 algorithm and corresponding C4.5 system proposed by J.R.Quinalan. This approach has the clear and simple theory and strong ability of rules extraction, which is appropriate to deal with the problems with large amount of samples. But it still has many problems such as too much depending on the quantity and distribution of samples, excessively sensitive to the noise, difficult to process continuous attributes effectively and etc. The main characteristic of the model drive approaches is to build a model at first utilizing the samples, and then extract rules based on the relation between inputs and outputs described by the model. Theoretically, these rule extraction approaches can overcome the shortcomings of data drive approaches mentioned above. The representative approaches are rules extraction approaches based on neural networks. In recent years, neural network techniques have made considerable progress in the improvement of training efficiency and generalization. Therefore, the model drive approaches based on neural networks will be the promising ones for rules extraction.

Existing methods of rules extraction based on neural networks can roughly be divided into two kinds, one kind is the methods based on the structure analysis of single networks [1-3] or network ensembles [4], whose idea is to map the trained neural network architecture to corresponding rules by a search process, and representative examples are Subset algorithm proposed by FU [1], and MOFN method proposed by Towell etc [2]. The other kind is the methods based on the relation conversion of inputs and outputs [5-8], which directly generates the corresponding rules according to the relation between its input units and output units, neglecting the concrete architecture of the neural networks. The representative examples are as follows, similar weight method put forward by Sestito [5], extracting rules based on the learning method put forward by Craven and Shavlik etc [6]. The fundamental problem of rules extraction based on neural networks is how to divide the network input space rationally based on the characteristics of network outputs, and how to express the relation of the inputs and corresponding network outputs in a simple and easily understandable way. Though the methods mentioned above have certain effectiveness for rules extraction, there are still some problems, such as low efficiency and validity, and difficulty in dealing with continuous inputs etc. The fundamental reason causing these problems is that there is the lack of effective heuristic information for the search process of the input space division [9], thus causing the blindness of the search process and the irrationality of the expression and division of the input space. So exploring the effective heuristic information in the search process of the input space division is the key to improve the effectiveness of rules extraction from neural networks [10].

The position and shape characteristics of the classification hypersurface play a very important role in the division of the input space. In ID3 algorithm, the heuristic information for constructing decision tree is the information amount for classification provided by every attributes, which reflects the proportion changes of classes with the change of attribute values, and also reflects the shape changes of classification hypersurface with the change of attribute value. But the reflection to the shape changes of classification hypersurface is obviously indirect and excessively dependent on the sample distribution and amount, which cause many problems such as preferring to select the attributes with many values in the process of decision tree construction. According to the analysis above, the effective heuristic information in the search process of the input space division can be extracted from the position and shape characteristics of classification hypersurface.

On the spatial hypersurface composed of the trained neural network's inputs and output, the sudden changes occurs in the near area of the classification hypersurface, so the partial derivative distribution of the network's output to inputs will describe the position and shape characteristics of the classification hypersurface. Based on the analysis of the relations among the position and shape characteristics of classification hypersurface, the partial derivative distribution of the trained neural network's output to the network's inputs and the decision power of attributes to classifications, this paper mainly studies on the measurement method of the classification power of attributes on the basis of differential information of the trained neural networks and develops new approach for the rule extraction.

2 Measurement of the Classification Power of Attributes

2.1 Quantification and Normalization of Attribute and Classification Values

Assuming the attribute vector of classification problems is $x = [X_1, X_2, \ldots, X_m]$, where m is the number of attributes, and the corresponding classification type is $C(X)$, then the sample of classification problems can be represented as $<X, C(X)>$. For the purpose of processing and comparing, quantification and normalization of the attribute values and classification values will be done as follows.

Quantification is performed for the values of discrete attributes and classification types. In this paper, values of discrete attributes and classification types are quantified as integer in some order, i.e. $0, 1, 2, 3, \cdots$.

Normalization is performed to adjust the range of neural network's inputs and output. For a given attribute value space Ω, assuming the classification type of sample point X is $C(X)$, utilizing the following linear transformation

$$x = bX + b_0 \qquad (1)$$

to map the attribute value X of sample to the input x of neural network input units, making every factors of x in the same range of $[\Delta, -\Delta]$, where $b = (b_{ij})$ is a transformation coefficient matrix and $b_0 = (b_{0i})$ is a transformation vector, $b_{ij} = \begin{cases} \frac{2\Delta}{MaxX_i - MinX_i} & : j = i \\ 0 & : otherwise \end{cases}$, $b_{0i} = \Delta - a_{ii}MaxX_i$. Utilizing the following linear transformation

$$F(X) = \alpha C(X) + \alpha_0 \qquad (2)$$

to map the classification type $C(X)$ of sample X to the desired output $F(X)$ of neural network's output unit, making $F(X)$ in the range of $[\delta, 1 - \delta]$, where δ is the range adjustment coefficient, $0 < \delta < 0.5, \alpha = \frac{(1-2\delta)}{MaxC(X) - MinC(X)}$ and $\alpha_0 = 1 - \delta - a_1 MaxC(X)$ are the proportion transformation coefficient and translation transformation coefficient respectively.

For the given sample set$<x, F(x)>$, the mapping relation $c(x) = Net(x)$ can be established through network training, where $Net(x)$ is the mapping relation between neural network's input x and output $c(x)$. From Equation (2), the following equation can be obtained.

$$C(X) = \frac{c(x)}{\alpha_1} - \alpha_0 \qquad (3)$$

When the activation function of neural unit is a logarithm function or hyperbolic tangent function etc., any order derivatives of network output $c(x)$ to each network input x_k exist.

2.2 Relations Between Distribution of $\frac{\partial c(x)}{\partial x_k}$ and Classification Hypersurface

The key problems in solving the classification rule extraction are how to divide and represent the attribute value space properly according to the position and shape characteristics of classification hypersurface. For convenience of explanation, next we will discuss the relations between the distribution of partial derivatives of trained neural network output $c(x)$ to network input x and the position and shape of classification hypersurface by 2-dimension classi-fication problems. Assuming the attribute value distribution of a 2-dimension classification problem is shown as Fig.1, where the transition area T is the change area of classification types, i.e. the adjacent area of classification hypersurface, the whole area U is the investigating area of the attribute value distribution.

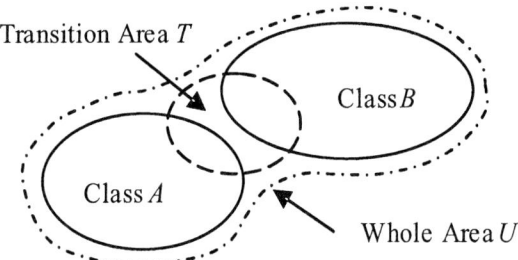

Fig. 1. Distribution of attribute values

Definition 1. For a given small area Γ, $\Gamma \subset U$, local square mean value $\overline{D_{k,\Gamma}(x)}$ of the partial derivative of trained neural network's output $c(x)$ to input x_k is defined as follows.

$$\overline{D_{k,\Gamma}(x)} = \frac{\int_{\Gamma} (\frac{\partial c(x)}{\partial x_k})^2 dx}{\int_{\Gamma} dx} \qquad (4)$$

For the given small area Γ, if $\Gamma \subset T$, $\overline{D_{k,\Gamma}(x)} \geq 0$, while if $\Gamma \subset U - T$, $\overline{D_{k,\Gamma}(x)} \longrightarrow 0$. Therefore, if $\sum_k \overline{D_{k,\Gamma}(x)}$ is big enough, Γ must be in the transition area T, i.e. in the adjacent area of classification hypersurface. The bigger the $\sum_k \overline{D_{k,\Gamma}(x)}$ is, the closer to the classification hypersurface the Γ is.

Meanwhile, when $\Gamma \subset T$, value of $\overline{D_{k,\Gamma}(x)}$ represents the mean perpendicular degree between k attribute axis and classification hypersurface in area Γ or its adjacent area. If $\overline{D_{j,\Gamma}(x)} = Max\overline{D_{k,\Gamma}(x)}$, it means that the mean perpendicular degree between j attribute axis and classification hypersurface is the highest. Therefore the distribution of the partial derivatives of trained neural network's output to inputs can describes the position and shape characteristics of classification hypersurface.

2.3 Measurement Method of the Classification Power of Attributes

For a 2-dimension classification problem, assuming the shape of classification hypersurface in the given area Γ is as shown in Fig.2, in which the level axis is attribute x_1 and the perpendicular axis is attribute x_2. In the cases (a) and (b), the classification power of attribute x_1 is stronger, so area should be divided via attribute x_1. In the case (c), attribute x_1 and attribute x_2 have the equal classification powers. Based on the discussion above, for a given attribute value space Ω, the classification power of each attribute depends on the mean perpendicular degree between each attribute axis and classification hypersurface in space Ω or its adjacent space. The higher is the mean perpendicular degree, the stronger is the classification power.

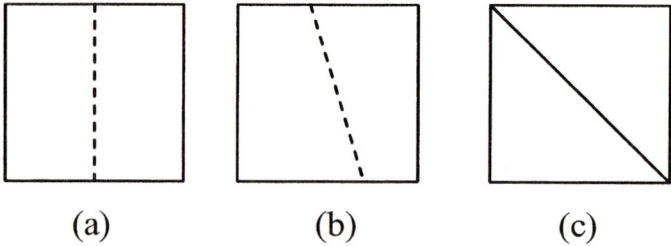

Fig. 2. Typical shapes of classification hypersurface

According to the discussion in section 2.2, the value of $\overline{D_{k,\Omega}(x)}$ represents the mean perpendicular degree between axis k and classification hypersurface in space Ω or its adjacent space. In the case (a) and (b), there must be $\overline{D_{1,\Omega}(x)} \gg \overline{D_{2,\Omega}(x)}$, while in the case (c), there must be $\overline{D_{1,\Omega}(x)} \approx \overline{D_{2,\Omega}(x)}$. So for a given attribute value space Ω, $\Omega \subset U$, the bigger is $\overline{D_{k,\Omega}(x)}$, the stronger is the classification power of attribute k in space Ω. If $\overline{D_{k,\Omega}(x)}$ is close to zero, it means that there is no impact on the classification type changes as the attribute value changes in space Ω. Therefore in space Ω, the distribution of $\frac{\partial c(x)}{\partial x_k}$ will determine the classification power of attributes. The measurements of classification power of attributes can be computed via the distribution of $\frac{\partial c(x)}{\partial x_k}$.

In order to eliminate the influence of noise, the partial derivative distribution function and attribute influence space are introduced here.

Definition 2. For a given attribute value space Ω, the distribution function of the partial derivative of trained neural network's output $c(x)$ to its input x_k is defined as follows.

$$\phi_k(x) = \begin{cases} (\frac{\partial c(x)}{\partial x_k})^2 : (\frac{\partial c(x)}{\partial x_k})^2 \geq \eta \overline{D_{k,\Omega}(x)} \\ 0 : \quad otherwise \end{cases} \quad (5)$$

where η is an adjustment coefficient.

Definition 3. For a given attribute value space Ω, influence space V_k of the attribute x_k is defined as follows.

$$V_k = \{x | \phi_k(x) \neq 0\} \quad (6)$$

Definition 4. For a given attribute value space Ω, the measurement of classification power of attribute x_k is defined as follows.

$$JP(x_k) = \frac{\int_{V_k} \phi_k(x) dx}{\int_{V_k} dx} \quad (7)$$

From equation (5) and (7), the $JP(x_k)$ will become bigger as the adjustment coefficient η increases. So η have two functions: (1) The preference for attribute selection can be adjusted by properly adjusting η. For example, through properly decreasing the η of continuous attributes and increasing the η of discrete attributes, there will be a preference for selecting discrete attributes. (2) The influence of partial derivative noises can be reduced via properly adjusting η.

The measurement $JP(x_k)$ of classification power represents the influence degree of attributes to classification. So in the process of rules extraction, the value $JP(x_k)$ is the important instruction information for selecting attributes and dividing attribute value space. The new method of measuring the classification power of attributes proposed in this paper has many advantages compared with the measurement method of ID3 algorithm based on information entropy. It is suitable for measuring the classification power of continuous attributes and discrete attributes. Because of the strong prediction ability of trained neural networks, it can also overcome the shortcomings of the measurement method of ID3 algorithm such as excess dependency on the amount and distribution of samples, too much sensitive to the noise etc.

The typical classification problem of weather type for playing golf is employed to demonstrate the performance of the new measurement method. The computing results are shown as table 1.The attributes and their values are as follows: Outlook has the value of sunny, overcast and rain, quantified as 0, 1, 2. Temperature has the value of $64 \sim 83$. Humidity has the value of $65 \sim 96$. Windy has the value of true, false, quantified as 0, 1. The size of the training sample set is 14. The construction of the neural network is 4-4-1.

From table 1, in the whole space the measurement value of classification power of attribute outlook is the biggest, so attribute outlook should be selected as the root node of the decision tree, and the attribute value space should be

Table 1. Computing Results of Measurement Value $JP(x_k)$

Attributes	Whole Area	Outlook=0	Outlook=2
Outlook	1.012	-	-
Temperature	0.070	0.021	0.148
Humidity	0.135	0.928	0.041
Windy	0.072	0.157	0.327

divided by its values. While in the subspace of outlook= rain, measurement value of classification power of attribute windy is the biggest. Thus according to this information, the optimal decision tree and corresponding classification rules can be generated.

3 Rules Extraction Method

The rules extraction process from trained neural networks proposed in this paper is quite similar to the process of constructing a decision tree, which includes data pretreatment, neural network training and rules extraction etc. The proposed algorithm for rules extraction from trained neural networks is described as follows in detail.

Step 1. According to the given training sample set of the trained neural networks, generate the attribute value space Ω, generate a queue R for finished rules and a queue U for unfinished rules, choose the interval number of attributes and the allowed number of errors.

Step 2. Pick attribute x_k with the biggest value of $JP(x)$ computed by equation (7) as the extending attribute from all the present attributes. Divide the value range of attribute x_k into intervals according to the chosen interval number. Then for each interval, pick attribute x_j with the biggest $JP(x)$ as the extending attribute for each interval. Merge the pairs of adjacent intervals with the same extending attribute and classification type of the largest proportion. A rule is generated for each merged interval. If the classification error of the generated rule is less than the allowed error, sent it into the queue R, otherwise sent it into the queue U.

Step 3. If $U \neq \emptyset$, the extraction process terminates, otherwise go to Step 4.

Step 4. Pick an unfinished rule from the queue U according to a certain order, and perform division and mergence. A rule is generated for each merged interval. If the classification error of the generated rule is less than the allowed error, then sent it into the queue R, otherwise sent it into the queue U. Go to Step 3.

4 Experiment and Analysis

The spiral problem [9] and balance scale (balance for short), congressional voting records(voting for short), hepatitis, iris plant(iris for short), statlog australian credit approval(credit-a for short) in UCI data sets [11] are employed as computing cases, shown in table 2. The attribute value distribution of spiral problem

Table 2. Computing Cases

	Spiral	Balance	Voting	Hepatitis	Iris	Credit-A
Total Samples	168	625	232	80	150	690
Training Samples	84	157	78	53	50	173
Testing Samples	84	468	154	27	100	517
Classification Numbers	2	3	2	2	3	2
Total Attributes	2	4	16	19	4	15
Discrete Attributes	0	0	16	13	0	9
Continuous Attributes	2	4	0	6	4	6

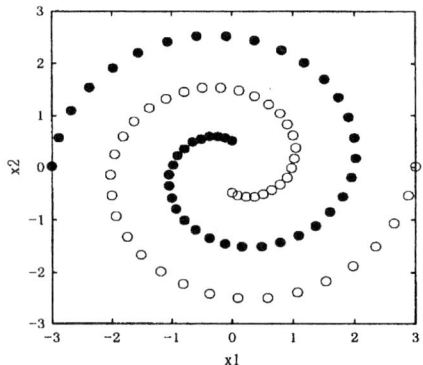

Fig. 3. Samples distribution of spiral problem

Table 3. Experimental results Comparison between New Approach(NA) and C4.5R

	#Rules(NA: C4.5R)	Err.Train(NA: C4.5R)	Err.Test(NA: C4.5R)
Spiral	8: 3	1.2%: 38.1%	0%: 40.5%
Balance	13: 20	15.2%: 10.8%	16.4%: 18.2%
Voting	3: 4	2.5%: 2.6%	2.5%: 3.2%
Hepatitis	5: 5	0%: 3.8%	7.4%: 29.6%
Iris	4: 4	0%: 0%	3%: 10%
Credit-A	4: 3	13.9%: 13.9%	13.9%: 14.9%

is shown as Fig.3, in which solid points are of Class C0, empty points are of Class C1.

Since no other approaches extracting rules from trained neural networks are available, we include a popular rule learning approach i.e.C4.5R for comparison. The experimental results are tabulated in Table 3. For spiral problem, the rules set extracted by the new approach are shown in Table 4. Table 3 shows that the rules extraction results of the new approach are obviously better than that of C4.5R, especially for spiral problem. For the case of spiral problem, C4.5R is difficult to extract effective rules, but the new approach has so impressive results that are beyond our anticipation. This means that the new approach(NA)

Table 4. Rules Set of Spiral Problem (New Approach Proposed)

R1 $x_0 \geq 2.4 \longrightarrow C1$
R2 $x_0[-2.24, -1.34) \wedge x_1 < 1.73 \longrightarrow C1$
R3 $x_0[-1.34, 0) \wedge x_1 < -1.8 \longrightarrow C1$
R4 $x_0[-1.34, 0) \wedge x_1[0.78, 2.26) \longrightarrow C1$
R5 $x_0[0, 1.34) \wedge x_1 < -1.89 \longrightarrow C1$
R6 $x_0[0, 1.34) \wedge x_1[-0.78, 1.8) \longrightarrow C1$
R7 $x_0[1.34, 2.24) \wedge x_1 < -1.73 \longrightarrow C1$
R8 $Default \longrightarrow C0$

proposed can improve the validity of the extracted rules for complicated classification problems remarkably. Moreover, the generalization ability of those rules extracted by the new approach is also better than that of rules extracted by the C4.5R.

5 Conclusions

In this paper, based on the analysis of the relations among the characteristics of position and shape of classification hypersurface, the partial derivative distribution of the trained neural network's output to the network's inputs, a new measurement method based on the differential information of the trained neural networks is proposed, which is suitable for both continuous attributes and discrete attributes, and can overcome the shortcomings of the measurement method based on information entropy. On the basis of this new measurement method, a new approach for rules extraction from trained neural networks is presented, which is also suitable for classification problems with continuous attributes. The performance of the new approach is demonstrated by several typical examples, the computing results prove that the new approach can improve the validity of the extracted rules remarkably compared with other rule extracting approaches, especially for complicated classification problems.

References

1. Fu,L.: Rule generation from neural networks. IEEE Trans. Systems Man. Cybernet. 24(1994)1114-1124
2. Towell,G.G.,Shavlik,J.W.:Extracting refined rules from knowledge-based neural networks.Machine Learning.13(1993):71–101
3. Lu,H.J.,Setiono,R.,Liu.H.: NeuroRule: a connectionist approach to data mining. In Proceedings of 21th International Conference on Very Large Data Bases,Zurich, Switzerland (1995)478–489
4. Zhou, Z.H.,Jiang,Y.Chen,S.F.:Extracting symbolic rules from trained neural network ensembles. AI Communications.16(2003):3–15
5. Sestito,S.,Dillon T.:Knowledge acquisition of conjunctive rules using multilayered neural networks. International Journal of Intelligent Systems.8(1993)779–805

6. Craven,M.W.,Shavlik,J.W.:Using sampling and queries to extract rules from trained neural networks. In Proceedings of the 11th International Conference on Machine Learning, New Brunswick, NJ,USA(1994)37–45
7. Maire,F.:Rule-extraction by backpropagation of polyhedra. Neural Networks.12(1999)717–725
8. Setiono,R.,Leow,W.K.: On mapping decision trees and neural networks.Knowledge Based Systems.12(1999)95–99
9. Kamarthi,S.V.,Pittner S.: Accelerating neural network training using weight extrapolation. Neural Networks.12(1999)1285-1299
10. Fu,X.J.,Wang,L.P.:Data dimensionality reduction with application to simplifying RBF network structure and improving classification performance.IEEE Trans. System, Man, Cybern, Part B-Cybernetics. 33(2003)399-409
11. Blake,C.,Keogh,E.,Merz,C.J.:UCI repository of machine learning databases [http://www.ics.uci.edu/ meearn/ML Repository.htm]. Department of Information and Computer Science, University of California,Irvine,CA,USA(1998)

EDTs: Evidential Decision Trees

Huawei Guo, Wenkang Shi, and Feng Du

School of Electronics and Information Technology, Shanghai Jiao Tong University,
Shanghai, 200030, P.R. China
{hwguo, wkshi, dfengu}@sjtu.edu.cn

Abstract. In uncertain environment, this paper investigates the induction of decision trees based on D-S evidence theory. This framework allows us to handle the case where the test attributes and decision attribute of training instances are all represented by belief functions. A novel attribute selection measure is introduced. We also propose a new evidential combination rule to combine the classification results with different matching coefficients.

1 Introduction

Knowledge acquisition has been universally considered as the bottleneck in knowledge engineering [1]. Decision Trees (DTs) are useful in building knowledge-based systems, however real data is often pervaded with uncertainty [2]. So some generations of DTs have been proposed such as probabilistic decision trees (PDTs) [3], fuzzy decision trees (FDTs) [4] and belief decision trees (BDTs) [5], [6].

Belief decision trees, proposed by Elouedi [5] and Denoeux et al [6], integrate the advantages of evidence theory and decision tree algorithms. However, it is noted that their work assumes only the values of decision attribute are represented by belief functions. However, uncertainty on the test attribute values of the case in the training set is not considered. In fact, the training data may be more complicated and these two kinds of uncertainty may be simultaneously exist, especially when the attribute values are provided by several experts. Therefore, inspired by the publication [5], [6], the objective of this paper is to generalize BDTs to evidential decision trees (EDTs) where the decision attribute and test attributes of training set are all represented by belief functions. This kind of decision tree has not been reported so far.

2 Evidence Theory

Evidence theory was originally introduced by Dempster [7], and later developed by Shafer [8] and Smets [9] et al.

Let Θ be the frame of discernment of event X. The power set of Θ, denoted as $P(\Theta)$, is the set containing all the possible subsets of Θ. The subsets containing only one element are called singleton sets.

Definition 1. A basic belief assignment (BBA) is a mapping $m: 2^\Theta \to [0, 1]$ satisfying $\sum_{A \subseteq \Theta} m(A) = 1$. Any set $A \subseteq \Theta$ with $m(A) > 0$ is called focal element. A BBA m following $m(\phi) = 0$ is said to be normal [8]. But it may be relaxed if one accepts the open-world assumption: the quantity $m(\phi)$ is then represented as the belief assigned to unknown hypothesis that might not lie in Θ [9].

Let m_1 and m_2 be two BBAs on the same frame of discernment Θ, which are induced from distinct pieces of evidence. These BBAs can be combined by conjunctive combination rule, denoted by ⓝ [10]

$$(m_1 \text{ⓝ} m_2)(A) = \sum_{B, C \subseteq \Theta; B \cap C = A} m_1(B) \cdot m_2(C). \tag{1}$$

The conjunctive rule can be seen as an un-normalized Dempster's rule of combination [8]. The latter is based on closed world assumption.

More generally, if all bodies of evidence are labeled by different values of reliability, how to combine them [11]? One alternative way is to discount them by Dempster's discounting rule [8] and then combine them. Unfortunately, the discounting rule assigns the discounted belief to total ignorance, which increases the uncertainty of classification. Therefore, we propose a new combination rule to combine evidence with reliability in Section 4.5.

In order to make decision, the transformation of belief function to probability function may be needed. This transformation can be achieved by pignistic transformation proposed by Smets [10].

$$BetP(A) = \sum_{B \subseteq \Theta} \frac{|A \cap B|}{|B|} \cdot \frac{m(B)}{1 - m(\phi)} \qquad \forall A \subseteq \Theta \tag{2}$$

3 Basics of Decision Trees

In a decision tree, a sample is represented by a set of features expressed with some linguist terms. Their basic ideas are the same: partition the sample space in a data-driven manner and represent the partition as a tree. DTs are mostly based on the approach of top down induction of decision tree (TDIDT) by successive partitioning of the training set [3]. Its main steps can be described as

1. Place the initial training set on the root.
2. Optimal attribute selection: Select the discriminatory attribute from the set of un-used attributes.
3. Partitioning strategy: Create new son-nodes according to the values of the selected attribute, i.e., divide the current training set into training subsets.
4. When the stopping criterion is satisfied, the current training subset will be declared as a leaf.

5. Inference: A new instance is classified by traversing down the proper branches of the tree until its corresponding leafs have been arrived at.

4 Evidential Decision Trees

Consider a set of training instances $S = \{e_1, e_2,...,e_N\}$. Let $\Lambda = \{A^{(1)}, A^{(2)},..., A^{(n)}\}$ be a set of evidential test attributes where each attribute $A^{(j)}$ is represented by a belief function on the set of possible terms, i.e., the frame of discernment $\Theta_j = \{\theta_1^j, \theta_2^j, ..., \theta_{|\Theta_j|}^j\}$. All the subsets of Θ_j, denoted as $a_i^j (i = 1, 2,..., 2^{|\Theta_j|})$, are the elements of power set $P(\Theta_j)$. Let D be the decision attribute whose possible values compose the frame of discernment Θ_d. To illustrate these notations, we consider an example shown in Table 1 which describes a small training set.

Table 1. Training set S with evidential representation (The term 'ordinary' is denoted as Y)

Ability {good, ordinary, bad}	Appearance {good, ordinary}	Property {much, moderate, little}	Decision attribute D {C_1, C_2, C_3}
G:0.8; B:0.2	G:0.6; Y: 0.4	Mu: 0.9; Mo:0.1	C_1:0.8; $C_1 \cup C_2$:0.2
G:0.3; Θ_1:0.7	G:0.4; Y: 0.6	Mu: 0.8; Mo:0.1;L:0.1	C_2:0.6;Θ_d:0.4;
G:0.6; Y:0.4	Y:0.5; Θ_2: 0.5	Mu: 0.7; Mo:0.3	C_1:0.8; Θ_d:0.2
G:0.5; Y:0.2; B:0.3	G:0.3; Y: 0.7	Mo:0.2; L:0.8	C_2:0.7; $C_2 \cup C_3$:0.3;
G:0.2; Y:0.8;	G:0.8; Y: 0.2	Mu: 0.8; Θ_3:0.2	C_1:0.2;C_2:0.55;Θ_d: 0.25
G:0.2; Y:0.1;B:0.7	G:0.1; Y: 0.9	Mo:0.3; L:0.7	C_2: 0.55; C_3:0.45
B:0.8; Θ_1:0.2	G:0.3; Y: 0.7	Mu: 0.3; Mo:0.7	C_3: 0.85; Θ_d:0.15
B:0.9; Θ_1:0.1	Y: 0.7; Θ_2:0.3	Mu: 0.4; Mo:0.6	C_3:0.6;$C_1 \cup C_3$:0.2; Θ_d:0.2
Θ_1=1	Y:0.2; Θ_2: 0.8	Mo: 0.7; L:0.3	$C_1 \cup C_2$:0.3; Θ_d:0.7;

One decision attribute and three test attributes are:

D and $\Theta_d = \{C_1, C_2, C_3\}$,

$A^{(1)} = \{Ability\}$, $\Theta_1 = \{G, Y, B\}$, $P(\Theta_1) = \{\phi, G, Y, B, G \cup Y,..., G \cup Y \cup B\}$,

$A^{(2)} = \{Appearance\}$, $\Theta_2 = \{G, Y\}$, $P(\Theta_2) = \{\phi, G, Y, G \cup Y\}$,

$A^{(3)} = \{Property\}$, $\Theta_3 = \{Mu, Mo, L\}$ and

$P(\Theta_3) = \{\phi, Mu, Mo, L, Mu \cup Mo,..., Mu \cup Mo \cup L\}$.

Instance e_p's BBA on the attribute $A^{(j)}$ is denoted as $m\{e_p\}(\Theta_j)$. For example, in Table 1, instance e_3's BBA on attribute $A^{(2)} = \{Appearance\}$ is

$m\{e_3\}(\Theta_2)$: $m\{e_3\}(\theta_2^2) = m\{e_3\}(Y) = 0.5$ and

$m\{e_3\}(\theta_1^2 \cup \theta_2^2) = m\{e_3\}(G \cup Y) = 0.5$.

The elements a^2 of $P(\Theta_2)$ are $\phi, G, Y, G \cup Y$.

In the following, we shall describe the major steps to build an evidential decision tree and inference procedure for new instances.

4.1 The Test Attribute Selection Measure

In BDTs [5], the authors adapt the gain ratio proposed by Quinlan [3] to the uncertain context with symbolic test attributes. In our context, the test attributes are represented by BBAs, which are continuous. It is not easy to ensure discretization and any implementation will present error. Therefore, we decide to give up the traditional approach and propose a global attribute selection measure. The more discriminatory the test attribute, the closer the similarity between the instances' decision class to the similarity between this test attribute of the instances. For any two instances e_p and e_q ($p,q=1,2,...,N$), the similarity between their decision attributes is denoted as $S_D(e_p, e_q)$. And the similarity between a test attributes $A^{(j)}$ is denoted as $S_{A^{(j)}}(e_p, e_q)$ ($j=1,2,...,n$). We define *mapping dissimilarity* or difference between two kinds of similarities of attribute $A^{(j)}$ as the attribute selection measure

$$Map\_Dis(D, A^{(j)}) = \sum_{p,q=1, p<q}^{p,q=N} \left| S_D(e_p, e_q) - S_{A^{(j)}}(e_p, e_q) \right| \qquad (3)$$

$\forall A^{(j)} \in \Lambda$, we choose attribute $A^{(j)} = \arg\min\{Map\_Dis(D, A^{(1)}), ..., Map\_Dis(D, A^{(n)})\}$ as current test attribute. This idea is inspired by the Sammon's mapping [12], which is a non-linear mapping that maps a set of input points on a plane trying to preserve the relative distance between the input points approximately [13]. Now the key turns to measure the similarity between BBAs. The concepts of similarity relationship and distance are linked in an inverse way.

4.1.1 Evidence Distance
A principled distance metric between two BBAs is given by Jousselme et al [14]. This distance respects all the properties expected of a distance and is an appropriate measure of the dissimilarity between two BBAs [14].

Definition 2. Let m_1 and m_2 be two BBAs defined on the n-element frame of discernment Θ. The distance metric between them is

$$d_{BBA}(m_1, m_2) = \sqrt{\tfrac{1}{2}(\vec{m_1} - \vec{m_2})D(\vec{m_1} - \vec{m_2})'}, \qquad (4)$$

where D is a matrix with size of $2^n \times 2^n$, elements of which are of the form $D(A,B) = \frac{|A \cap B|}{|A \cup B|}$, $A, B \subseteq \Theta$. And we have $0 \le d_{BBA}(m_1, m_2) \le 1$.

4.1.2 A New Attribute Selection Measure
A natural choice for the similarity measure between m_1 and m_2 derived from evidence distance is

$$S(m_1, m_2) = 1 - d_{BBA}(m_1, m_2) \qquad (5)$$

Substituting relation (5) into relation (3), one can compute the *mapping dissimilarity* to select the test attribute. Assume the mapping dissimilarity of attribute $A^{(j)}$ is $Map\_Dis(D, A^{(j)})$, we define the discriminatory capacity of $A^{(j)}$ as

$$Dis\_Cap(A^j) = \frac{\min_{k=1,2,\ldots,n}\{Map\_Dis(D, A^{(k)})\}}{Map\_Dis(D, A^{(j)})\}} \in [0,1], \quad j = 1,2,\ldots,n \qquad (6)$$

If we have $\min_{k=1,2,\ldots,n}\{Map\_Dis(D, A^{(k)})\} = 0$, similar relations are easily defined as (6). The discriminatory capacity of the best attribute is equal to 1 and others decrease in inverse proportion to their values of mapping dissimilarity.

4.2 Partitioning Strategy

Since we deal with evidential test attribute values that are continuous, we must consider how to create edges from one attribute according to the belief structure. It is natural to create an edge for each subset of the framework of discernment. For test attribute $A^{(j)}$, each branch is created by the subset of the set Θ_j and the maximum number of branches is $(2^{|\Theta_j|} - 1)$. We refer to the idea of defining the parameter, *significant level* α, to intercept the training data [4]. At node N with attribute $A^{(j)}$, the instance e_p will follows the branch labeled by a_i^j ($i = 1,2,3,\ldots,2^{|\Theta_j|}$) of $P(\Theta_j)$ when $\max\{m\{e_p\}(\Theta_j)\} = m\{e_p\}(a_i^j)$ and $m\{e_p\}(\alpha_i^j) \geq \alpha$. If $m\{e_p\}(\alpha_i^j) < \alpha$ and the secondary value of $m\{e_p\}(\Theta_j)$ is $m\{e_p\}(\alpha_l^j)$, the instance e_p is thrown away or follow the branch labeled by $\{\alpha_i^j \cup \alpha_l^j\}$. The above procedure is iterated until we get $(m\{e_p\}(a_i^j) + m\{e_p\}(a_l^j) + \ldots) \geq \alpha$. If worse, the attribute BBA $m\{e_p\}(\Theta_j)$ is a vacuous belief function, the instance provides little information regarding attribute $A^{(j)}$ and should be thrown away or follow the branch labeled by Θ_j.

In EDTs, the test attribute will not classify data in crisp way as classical DTs, attribute values will be allowed to have some overlaps. The parameter α, *significant level*, controls the degree of such overlap to build DTs with better classification performance [15]. We recommend to take $\alpha \geq 0.5$. If the cardinality of Θ_j is large, this may increase the size of the tree and we could transform belief function to quasi-Bayesian belief structure with cardinality of $(|\Theta_j|+1)$ by using pignistic transformation. Thus, the instance with $\max\{m\{e_p\}(\Theta_j)\} < \alpha$ ($p = 1,2,\ldots,N$) follows the branch labeled by Θ_j, otherwise the instance follow the branch labeled by θ_i^j ($i = 1,2,3,\ldots,|\Theta_j|$) of Θ_j when $\max\{m\{e_p\}(\Theta_j)\} = m\{e_p\}(\theta_i^j)$ and $m\{e_p\}(\theta_i^j) \geq \alpha$.

4.3 Stopping Criteria

Stopping criteria determine whether or not a training set should be further divided. In classical and fuzzy cases, a node is usually regarded as a leaf if the relative frequency of one class at the node is equal to or greater than a given threshold. In current case, these strategies are proposed as stopping criteria:

1. If the treated node includes only one instance.
2. If there is no further attribute to test.
3. If the treated node includes instances that will follow the same branch based on the un-used attributes.
4. If the value of the average similarity of the decision attribute among the instances attached to one node N is more than or equal to a given threshold β. This average similarity is defined as the truth level of the leaf.

$$Tru\_Lev(N) = 1 - \frac{1}{L} \sum_{\substack{e_i \in N, e_j \in N \\ i,j=1, i<j}}^{|N|} d_{BPA}(m\{e_i\}(\Theta_d) - m\{e_j\}(\Theta_d)), \qquad (7)$$

where $L = \dfrac{|N| \cdot (|N|-1)}{2}$ and $|N|$ is the number of leaves at node N. We recommend to take $\beta \geq 0.8$. In an extreme case, the node includes instances having the same BBA on class attribute will be declared as a leaf.

4.4 Structure of Leaves

In classical DTs, the leaf is labeled by a crisp decision class. In BDTs [5], it is labeled by a BBA defined on the set of decision classes. Different from them, we define the leaf's label as follows:

For the objects attached to the leaf LN, the leaf's label would be represented by a (n+1)-tuple. The first element is a BBA equal to average of the different BBAs of the objects for the decision attribute. The last n elements are BBAs equal to the average of the different BBAs of the objects for the n test attributes respectively. Thus we have a (n+1)-tuple

$$\left(m\{LN\}(\Theta_d),\ m\{LN\}(\Theta_1), m\{LN\}(\Theta_2),\ldots, m\{LN\}(\Theta_j),\ldots, m\{LN\}(\Theta_n) \right), \qquad (8)$$

where $m\{LN\}(\Theta_d) = \dfrac{1}{|LN|} \cdot \sum_{e_i \in LN} m\{e_i\}(\Theta_d)$ and

$$m\{LN\}(\Theta_j) = \frac{1}{|LN|} \cdot \sum_{e_i \in LN} m\{e_i\}(\Theta_j) \qquad j = 1,2,3,\ldots,n.$$

and $|LN|$ denotes the number of instances attached to leaf node LN.

4.5 Classification Procedure

A new instance to classify is characterized by a set of combination values where each one corresponds to a test attribute. At the root node RN with attribute

$A^{(k)}$ ($k \in \{1,2,...,n\}$), the new instance e_0 respectively follows the branches labeled by focal elements of $m\{e_0\}\{\Theta_j\}$. Then at other nodes with $A^{(j)}$ ($j=1,2,...,n, j \neq k$), the instance only follow the branches labeled by a_i^j if $\max(m\{e_0\}(\Theta_j)) = (m\{e_0\}(a_i^j))$. If the branch labeled by a_i^j is empty and the secondary value of $m\{e_p\}(\Theta_j)$ is $m\{e_p\}(\alpha_i^j)$, the instance follow the branches labeled by a_i^j. The process is iterated until it reaches one non-empty branch. Down from the top to the leaf, the corresponding attribute's discriminatory capacity decreases. We temporarily take into account all the matching branches from root node and the major matching branches from non-root nodes. In fact, it is not necessary to apply all the paths to the new instance in order to get the classification result [4].

Thus we get several reached leaves of e_0. How to combine them? As mentioned in Section 4.4, each leaf is represented by a (n+1)-tuple. Maybe we can directly select the first BBAs of one leaf's label as the classification result and then combine all the BBAs. In fact, the matching coefficient of test attributes between the new instance and each reached leaf is different. The higher the matching coefficient, the closer the classification of the new instance to the decision attribute of corresponding leaf. The reached leaves can be regarded all evidence for pattern classification of e_0. The matching coefficient is also regarded as reliability of corresponding evidence. As described in Section 2, Dempster's discounting rule is not suitable. So we propose a new combination rule to combine the BBAs with different matching coefficients.

4.5.1 A New Combination Rule

Assume that the number of the reached leaves is n_{e_0}. As we know, different attribute has different discriminatory capacity. We choose discriminatory capacity of one attribute as the matching coefficient, thus the matching coefficient between e_0 and its reached leaf e_i is defined as

$$R(e_0, e_i) = 1 - \frac{1}{K} \sum_{j=1}^{n} w_j \cdot d_{BPA}(m\{e_0\}(\Theta_j), m\{e_i\}(\Theta_j)) \quad (9)$$

where $w_j = Dis\_Cap(A^j)$ and $K = \sum_{j=1}^{n} Dis\_Cap(A^j)$ is the normalized factor.

Then the proposed rule can be summarized as follows.

(1) For n_{e_0} leaves, the matching coefficient or the reliability of leaf e_i associated with e_0 can be obtained as R_i ($i=1,2,...,n_{e_0}$) by using (9).

(2) Then we combine all the BBAs on the decision attribute. For consistent belief, the conjunctive rule is utilized. For conflicting part, we divide the total conflict into some partial conflict. Then the partial conflict is respectively redistributed considering reliability among the focal elements involving in the conflict. The new combination rule is

$$m(C) = \sum_{C_i \cap C_j \ldots \cap C_w = C} m_1(C_i) \cdot m_2(C_j) \ldots \cdot m_{n_{e0}}(C_w) + \sum \tau(C), \quad C \subseteq \Theta_d, \qquad (10)$$

Where $\tau(C)$ is the part of belief redistributed to hypothesis C by all partial conflict. For example, for partial conflict $m_1(C) \times m_2(X) \ldots \times m_{n_{e0}}(Y)$, the belief redistributed to A is

$$\tau_{1,2}(C) = \frac{R_1 \cdot m_1(C)}{R_1 \cdot m_1(C) + \ldots + R_{n_{e0}} \cdot m_{n_{e0}}(Y)} \cdot [m_1(C) \times \ldots \times m_{n_{e0}}(Y)], \qquad (11)$$

where $C \cap X \ldots \cap Y = \phi$.

4.5.2 Classification

After combining the BBAs of all reached leaves on the decision attribute, the classification result for new instance is obtained. Applying the pignistic transformation on the result, one can get the probability function that the new instance belongs to the possible terms of the decision attribute.

5 Illustrative Example

Let's illustrate our method by the training set shown in Table 1.

5.1 Building a New Evidential Decision Tree

The main steps are described as follows (α=0.5, β=0.8, n=9, N=3)

(1) At root node RN, we apply mapping dissimilarity to select the best attribute based on all the instances in S. Thus we have
Map_Dis (*D, Ability*) = 6.87,
Map_Dis (*D, Appearance*) =11.47 and
Map_Dis (*D, Property*) = 8.64.
Since none of the stopping criteria is satisfied, we choose the *Ability* attribute as the root of the decision tree. Therefore, branches are created below root *Ability* for its possible values respectively $(G, Y, B, G \cup Y, G \cup B, Y \cup B, \Theta_1)$. Instance e_p (p=1,2,...,9) respectively follows the branch labeled by a_i^1 when $\max\{m\{e_p\}(\Theta_1)\} = m\{e_p\}(a_i^1)$ and $m\{e_p\}(a_i^1) \geq \alpha$ (i=1,2,3,..., 8(= 2^{Θ_1})). Delete the empty branches and thus we have four branches: G (e_1, e_3, e_4), Y (e_5), B (e_6, e_7, e_8) and Θ_1 (e_2, e_9).

According to (6), the discriminatory capacities of the attributes are respectively
Dis_Cap(*Ability*) =1,
Dis_Cap(*Appearance*) =0.6 and
Dis_Cap(*Property*) =0.8.
(2) At node $N_1(e_1, e_3, e_4)$, none of the stopping criteria is satisfied and we have
Map_Dis(*D, Appearance*) =1.45,

Map_Dis(D, Property) =0.17.

Thus we choose the *Property* attribute as one node of the decision tree and have two branches: *much* (e_1, e_3), *little* (e_4).

(3) At node N_2 (e_1, e_3), the truth level of this node is $Tru\_Lev(N_2)$ =0.88 >β, so this node is declared as a leaf. The leaf's label is equal to the average of the BBAs of these two instances relative to the attributes.

(4) Applying the above procedure to left nodes, we will construct an original tree. Deleting all the empty branches, we get an evidential decision tree as shown in Fig 1.

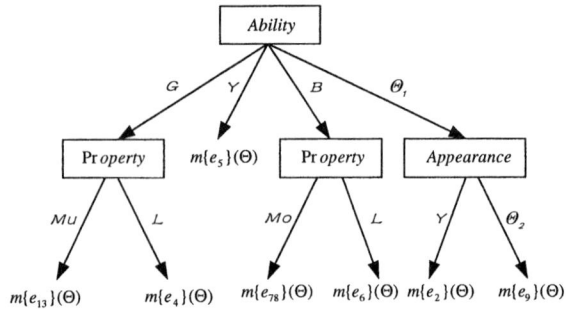

Fig. 1. Evidential decision tree for Table 1: $\Theta=\{\Theta_d, \Theta_1, \Theta_2, \Theta_3\}$

5.2 Classification

We assume the instance e_0 to classify is characterized as

$m\{e_0\}(Ability)=(G: 0.8; Y: 0.2)$,
$m\{e_0\}(Appearance)=(Y: 0.6; \Theta_2: 0.4)$ and
$m\{e_0\}(Property)=(Mu: 0.7; Mo: 0.3)$.

According to Section 4.2, e_0 reaches the leaf e_{13} and e_5. According to (9), the matching coefficient related to e_{13} and e_5 is respectively, 0.87 and 0.51. According to (10), the classification BBA is

$m\{e_0\}\{\phi, C_1, C_2, C_3, C_1C_2, C_1C_3, C_2C_3, C_1C_2C_3\} = \{0, 0.7, 0.24, 0, 0.03, 0, 0, 0.03\}$

Applying pignistic transformation (2), the probability distribution on the classes is
$P\{e_0\}\{C_1, C_2, C_3\} = \{0.73, 0.26, 0.01\}$.

6 Conclusions

In this paper, we propose new decision trees what we have called EDTs based on evidence theory. For the first time we consider the general case where the knowledge about the test attributes and decision attribute are all represented by belief functions.

To handle this case, a new attribute selection measure is proposed. The matching coefficient between the new instance to classify and its reached leaves is defined based on evidence theory. Then we propose a new combination rule to combine these results. Deriving the *production rule* from EDTs is our future research direction.

Acknowledgements

This work was supported National Defence Key Laboratory No. 51476040103JW13.

References

1. Z.-H. Zhou, Y. Jiang: NeC4.5: neural ensemble based C4.5, IEEE Trans. Knowledge and Data Engineering, 16(2004) 770-773
2. Enric Hern'andez, Jordi Recasens: A general framework for induction of decision trees under uncertainty. In: J. Lawry, J. Shanahan, A. Ralescu (eds.): Modelling with Words. Lecture Notes in Artificial Intelligence, Vol. 2873, Springer-Verlag, Berlin Heidelberg New York (2003) 26-43
3. Quinlan, J. R.: Decision trees as probabilistic classifier, Proceedings of the Fourth international Machine Learning (1987) 31-37
4. Y. Yuan, M. J. Shaw: Induction of fuzzy decision trees, Fuzzy sets Syst., 69(1995) 125-139.
5. Z. Elouedi, K. Mellouli, P. Smets: Belief Decision Trees: Theoretical Foundations, Int. J. Approx. Reas., 28(2001) 91-124
6. T. Denoeux, M. Bjanger: Induction of decision trees from partially classified data using belief functions, Proceedings of SMC'2000, 2923-2928, Nashville, USA, 2000
7. Dempster, A.P.: Upper and lower probabilities induced by a multi-valued mapping, Annals of Math. Stat., 38(1967) 325-339
8. G.Shafer: A Mathematical Theory of Evidence, Princeton, NJ: Princeton Univ. Press, 1976
9. P.Smets, R. Kennes: The transferable belief model, Artificial Intelligence, 66(1994) 191-243
10. F. Delmotte, P. Smets: Target identification based on the transferable belief model interpretation of Dempster–Shafer model, IEEE Trans. Syst., Man, Cybern., 34(2004)457-470
11. Y. Deng, W.K. Shi, Z.F. Zhu: Efficient combination approach of conflict evidence, J. Infrared Millim. W., 23(1)(2004) 27-32
12. John W. Sammon, Jr.: A nonlinear mapping for data structure analysis. IEEE Trans. Computers, C-18(1969) 401-409
13. T. Denoeux, M. Masson: EVCLUS: Evidential clustering of proximity data. IEEE Trans. Syst., Man, Cybern., B, 34(2004)95-109
14. Jousselme Anne-Laure, Grenier Dominic, Bossé Éloi: A new distance between two bodies of evidence, Inform. Fusion, 2(2001) 91-101
15. X. Z. Wang, M. H. Zhao, D. H. Wang: Selection of parameters in building fuzzy decision trees, Proceedings of 16th Australian Conference on AI, Perth, Australia, (2003)282-292

GSMA: A Structural Matching Algorithm for Schema Matching in Data Warehousing

Wei Cheng and Yufang Sun

Institute of Software, Chinese Academy of Sciences,
Beijing 100080, P.R. China
{Chengwei,yufang}@ios.cn

Abstract. Schema matching, the problem of finding semantic correspondences between elements of source and warehouse schemas, plays a key role in data warehousing. Currently, schema mapping is largely determined manually by domain experts, thus a labor-intensive process. In this paper, we propose a structural matching algorithm based on semantic similarity propagation. Experimental results on several real-world domains are presented, and show that the algorithm discovers semantic mappings with a high degree of accuracy.

1 Introduction

Schema matching, the problem of finding semantic correspondences between elements of two schemas [1], plays a key role in data warehousing, wherein, data from source format are required to be transformed into the format of data warehouse data. In a typical scenario: the source schemas and the warehouse schemas are created independently, and the warehouse designer determines the mappings between them manually. Thus it is a time-consuming and labor-intensive process. Hence (semi-) automated schema matching approaches are needed.

Lots of research on (semi-)automated schema matching has been done [1,2,3,5,6,7,8]. Matching algorithms used in these prototypes include linguistic-based matching, instance-based matching, and structural-based matching, etc. In this paper, we proposed a structural-based schema matching algorithm GSMA to compute semantic similarity based on similarity propagation. The rest of this paper is organized as follows. Some related work is discussed in Section 2. Details of GSMA algorithm are described in Section 3. Section 4 presents experimental results. Section 5 summarizes the contributions and suggests some future research work.

2 Related Work

A survey of existing techniques and tools for automatic schema matching is given in [9], it proposed orthogonal taxonomy of schema matching. Here, four schema match prototypes are reviewed, namely, Cupid [6], LSD [1], SemInt [5] and SKAT [8].

The LSD system uses machine learning techniques to perform 1:1 matching [1]. It trains four kinds of base learners to assign tags of elements of a mediated schema to data instance of a source schema, the predictions of base learners are weighted and combined by a meta-learner. SemInt is an instance-based matching prototype [5]. It derived matching criteria from instance data values. Then a match signature is determined. Signatures can be clustered according to their Euclidean distance. A neural network is trained on cluster centers to output an attribute category, and thus to determine a list of match candidates. Cupid is a schema-based matching approach [6]. It first computes the similarity between elements of two schemas using linguistic matching. Then, it computes the similarity of schema elements based on their similarity of the contexts or vicinities. Finally, a weighted similarity of the two kinds of similarities is computed. The SKAT prototype is a rule-based schema matching system [8]. In SKAT, match or mismatch relationships of schema elements are expressed by rules defined in first-order logic, and methods are defined by users to derive new matches. Based on that, new generated matches are accepted or rejected.

3 Generic Structural Matching Algorithm

In our prototype, we developed a GSMA (Generic Structural Matching Algorithm) algorithm to automate schema matching based on similarity propagation. The similarity propagation idea is borrowed from the Similarity Flooding algorithm [7]. The structural matching algorithm is based on the intuition: two elements in two object models are similar if they are derived or inherited by similar elements. The prototype operates in three phases. In the first phase, called preprocess phase, schemas are converted into directed graphs, with nodes representing elements of a schema, edges representing relationships between nodes. Some necessary preprocess work are done in this phase: the prototype exploits the whole matching space, creates all theoretic match pairs, and removes constraint-conflict match pairs. The second phase, called linguistic matching, we use an algorithm based on a thesaurus WordNet [10] to compute the semantic distance between two words, then the semantic similarity can be obtained via a transform. The third phase is the structural matching of schema elements with GSMA algorithm.

In the GSMA algorithm, given two graphs A and B, the initial similarity between node pairs is defined as below:

$$Sim_0 = \begin{cases} 0/initialMap & (a \neq b) \\ 1 & (a = b) \end{cases} \quad (1)$$

where a and b are two nodes of A and B.
The similarity Sim_k ($k \geq 1$) of node pair (a, b) after k iterations is defined as:

$$Sim_k = wSim_{k-1} + \frac{w_I}{|I(a)||I(b)|}\sum_{i=1}^{|I(a)|}\sum_{j=1}^{|I(b)|} Sim_{k-1}(I_i(a), I_j(b)) + \frac{w_O}{|O(a)||O(b)|}\sum_{i=1}^{|O(a)|}\sum_{j=1}^{|O(b)|} Sim_{k-1}(O_i(a), O_j(b)) \quad (2)$$

where $I(n)$ and $O(n)$ represent the set of all in_neighbor and out_neighbor nodes of node n, respectively. w, w_I, w_O are the weight values, and $w + w_I + w_O = 1$. Below is the GSMA algorithm.

Input: Directed graph A, B and initial similarity matrix between nodes of A and B
Output: Stable similarity matrix between nodes of A and B after propagations.
Step 1: Initialize the similarity matrix following equation (1).

Step 2: Begin the loop until the similarity get stable and convergent, the convergence rule is:
$$|Sim_k(a,b) - Sim_{k-1}(a,b)| < \varepsilon. \qquad (3)$$
(1) Computing the similarity between pair of nodes (a, b).
  i. Computing the mean of similarities $Sim_kI(a, b)$ between the in_neighbors nodes of (a, b).
  ii. Computing the mean of similarities $Sim_kO(a, b)$ between the out_neighbors nodes of (a, b).
  iii. Computing the similarity between (a, b):
$$Sim_k(a,b) = w \times Sim_{k-1}(a,b) + w_I Sim_{k-1}I(a,b) + w_O Sim_{k-1}O(a,b) \qquad (4)$$
(2) Revising the similarity of (a, b).
  i. Computing the standard deviation $StDev$ of the similarity vectors between node a and any node in B.
  ii. Revising the similarity of (a, b):
$$Sim_k(a,b) = Sim_k(a,b) * (1 + (Sim_k(a,b) - \overline{Sim}) / StDev) \qquad (5)$$

Step 3: Normalize the similarity matrix.
Step 4: Output the similarity matrix $Sim(A,B)$ of A and B.

From equation (2) we know that on each iteration k, the similarity value of (a, b) is updated. In every iteration, each row of the similarity matrix is revised by the standard deviation of that row. When the difference of two similarity values in two sequential iterations is less than a threshold ε, the iteration is terminated.

4 Experimental Evaluation

We have implemented the linguistic and structural matching algorithm in our prototype system and have evaluated them on several real-world domains [4]. The experimental environment is a PIII 667 PC with 256M memory and Redhat Linux 9.0, the develop tool is Anjuta 1.2.1. To evaluate the quality of the matching operations, we use the same match measures used in [2]. Given the number of matches N determined manually, the number of correct matches C selected by our approach and the number of incorrect matches W selected by our approach, three quality measures are computed: Recall = C/N, Precision = $C/(C+W)$, F-Measure = 2/(1/Recall+1/Precision). We defined 8 match tasks, each matching two different schemas. The experimental results are shown in Figure 1.

From experiment, we know that when use the linguistic matcher, the system get an average *F-measure* value of 68%. When the structural matcher with the GSMA algorithm is added, the average *F-measure* value is increased to 90%. This shows the GSMA algorithm can highly improve the matching accuracy. The experimental results also show that the revising operation by standard deviation makes the GSMA algorithm have strong ability to differentiate and cluster nodes, i.e. it make similarity values between nodes that are actually similar far more greater than similarity values between nodes that are actually unsimilar.

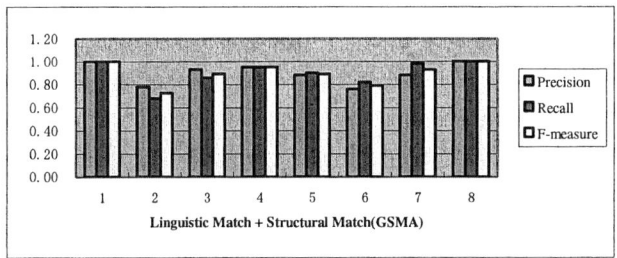

Fig. 1. Measure values of match tasks using linguistic&structural match

5 Conclusion and Future Work

In this paper, we described an efficient schema matching algorithm for the well-known problem of schema matching. The experimental results show that our algorithm works well on several domains. Although experiments show promising results, we do not believe our approach is a robust solution, since it needs the inclusion of other approaches. And the complex schema matching problem, i.e. $n:m$ ($n, m > 1$) matches, should be studied deeply.

References

1. Doan, A., P. Domingos, A. Halevy: Reconciling Schemas of Disparate Data Sources: A Machine-Learning approach. SIGMOD 2001
2. Do, H. and E.Rahm: COMA – A System for flexible combination of schema matching approaches, VLDB 2002
3. Embley,D.W. et al. Multifaceted Exploitation of Metadata for attribute Match Discovery in information Integration. WIIW 2001
4. http://anhai.cs.uiuc.edu/archive/summary.type.html
5. Li W, Clifton: SemInt: A Tool for Identifying Attribute Correspondences in Heterogeneous database Using Neural Network, Data & Knowledge Engineering, 2001
6. Madhavant,J., P.A.Bernstein, E.Rahm: Generic Schema Matching with Cupid. VLDB 2001
7. Melnik S, Garcia-Molina H, Rahm E: Similarity Flooding: A versatile graph matching Algorithm, ICDE 2002.
8. Mitra P, Wiederhold G, Jannink J: Semiautomatic integration of knowledge sources, FUSION 99
9. Rahm, E., P.A. Bernstein. A survey of approaches to automatic schema matching. The VLDB Journal, 10(4):334 ~ 350, 2001
10. WordNet: http://www.cogsci.princeton.edu/~wn/

A New Algorithm to Get the Correspondences from the Image Sequences

Zhiquan Feng[1,2], Xiangxu Meng[1], and Chenglei Yang[1]

[1] School of Computer Science and Technology,
Shandong University,
SanDa South Road 27, Jinan, Shandong, P.R. China, 250100
[2] School of Information and Science Engineering,
Jinan University,
Jiwei Road 106, Jinan, Shandong, P.R. China, 250022

Abstract. Acquisition of the correspondences from the image sequences with high-efficiency and high-precision is a fundamental and key problem in computer virtual techniques and human-computer interaction systems. In this paper, we start from image topological structures, aiming firstly finding the correspondences between topological structures of the image sequences. Then, we further attain accurate correspondences between feature points by adopting local search methods. In order to speed up the search process for desired data, a grid technique is introduced and some new concepts, such as SQVSBS, and related theories are put forward. The specific characteristics of the algorithm are: (1) using the Top-to-Bottom strategy, from rough estimates to accurateness, from local to global; (2) getting better time complexity, $O(Max(f^2,A))$, which is better than that that given in the references [11] and [12],here, f is the number of feature grids; (3) avoiding wrong matches deriving from local optical solutions; and (4) Focusing on the internal topological relationships between feature griddings upon which the characteristic of a feature gridding is based.

1 Introduction

To a point in the 3D space, a image sequences can be obtained according to different internal parameters of a camera, and the two points in different images from a same 3D point is refered as a correspondence[1].

Obtaining correspondences is one of the most important problems, and is one of the most difficult problems as well[2], so it is very important to research on this problem. In mosaic for panoramic image, getting correspondences from the image sequences is the key to image alignment and color smoothing in intersection area. The another application of getting correspondences is related to Virtual Reality(VR) systems. Creating and rendering 3D models are one of the kernel problems in VR systems. As long as the correspondences are identified, automatic camera calibration can be realized and corresponding 3D points can be calculated. In [5,7], correlations of areas or features of pixels are used to drive search correspondences, which search speed can be improved taking into consideration characteristics of image sequences.

For example, the search can be made along an epipolar[7]. The issue of this method is that the drive function is not easy to be identified, and so the precision of the algorithms is not stable. Furthermore, the precondition of using epipolar is that some initial correspondences(at least 5) must be identified in advance, which needs the search in the global scope. Generally, rough correspondences can be identified by hand first of all, and then more precision parameters of cameras can be obtained step by step. In despite of many new improved algorithms[9,10], the following issues are still open:(1) fundamental relationships between feature points needed to be further studied, and internal characteristics of features need to be further dredged up;(2) the known algorithms are limited to local optimal solutions, and (3) algorithms are restricted to the processes point by point, which limits improvement of time complexity and become a bottleneck of computational cost.

A new algorithm to get the correspondences from image sequences is put forward in this paper. This paper is organized as follows: section 2 presents pre-process, section 3 presents the way to obtain the topological relationship between the feature griddings, section 4 presents the method to obtain the correspondences between the feature points, section 5 presents some of our experimental results and section 6 is the conclusions of the paper.

2 Initialization

2.1 The Fast Algorithm of Obtaining Cove-Edge-Griddings

First of all, the image is logically divided by the M(M>0) horizontal lines and N(N>0) vertical lines, consequently, is partitioned into (M+1)×(N+1) small rectangle, each of which is refered as a gridding. Then, the image is scanned along four lings of each gridding by some fixed order, interesting information is gathered together, including the series of different grey values and the number different grey values (which is also called the degree of the gridding in the text), which are called gridding-attributes. The coordinates of top-left-corner are called the coordinates of the griddings(called coordinate for short). Those griddings with the numbers of different grey values more than 1 in the gridding-attributes are called cove-edge-griddings. It is clear that ,a cove-edge-gridding covers at least an edge of the image.

A cover-edge-gridding is illustrated in figure 1 and an instance of a cover-edge-gridding is shown in figure 2. Clearly, there may be those objects which are too small in size to intersect with any gridding, which are called escaped-objects. The algorithm aims at obtain the cover-edge-gridding

Step0: Remove noise and convert the source image into the grey image

Step1: Divide the image by rectangle (M+1)×(N+1)

Step2: Find gridding-attributes of each griddings

Step3: Reserve the cover-edge-griddings

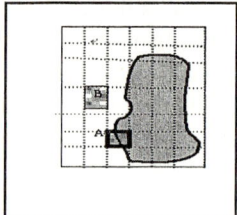  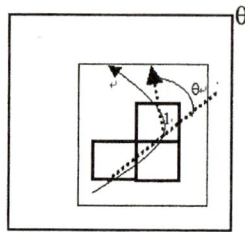

Fig. 1. The gridding A is a cover-edge-gridding and the gridding B is not

Fig. 2. A real example of cover-edge-gridding

Fig. 3. How to find the feature gridding

2.2. Extract the Feature Griddings

Suppose the edges of the image is regarded as the approximation of a set of lines. If one of the following conditions is satisfied, the gridding is called feature gridding: (1) the gridding covers an end of the lines;(2) the degree of the gridding is more than 3. In the first case, the line direction between the point entering into the griddings and the point leaving out of the griddings is found, which is called the go-forward-direction of the gridding. If the angle between the go-forward-direction of the current gridding and the last gridding, θ, is more than the threshold, λ, the current gridding should be a feature gridding.

2.3 Obtain the Adjacent Matrix of Feature Gridding

The adjacent matrix of feature gridding includes desired information for the next stage, which :can be obtained using width-first algorithm with the time complexity $O(N+e)$.Taken into consideration the time of obtaining the griddings, feature griddings and the adjacent matrix of feature gridding, the total time complexity is $O(f+N+e)$, among which f,N,e stand for the total number of griddings, the number of feature griddings and the number of all edges----suppose exists an edge between the arbitrary two feature griddings.

3 Obtain the Topological Relationship Between the Feature Griddings

3.1 Basic Concepts

Define 1. scalar quantity value of sequence $b=(r_1,r_2, ...,r_k)$ based on sequence $a=(y_1,y_2, ...,y_k)$ (SQVSBS)

Suppose $b=(b_1,b_2, ...,b_k)$ is a base series with linearity independence, the value

$$\tau_b(a) = \tau_b(y_1, y_2, \ldots, y_k) = a \bullet b = \sum_{i=1}^{k} y_i r_i$$

is called scalar quantity value of sequence $b=(r_1, r_2, \ldots, r_k)$ based on sequence $a=(y_1, y_2, \ldots, y_k)$, among which, $\bullet$ express dot product operator.

For example, let the base series be $(1, 1/2, 1/4)$, and $b = (10, 15, 20)$, then the SQVSBS of b is $1*10 + 1/2*15 + 1/4*20 = 22.5$.

Suppose, in the following text, that the capital letters express feature griddings, small letters express the related instance value of their corresponding capital letters, and the letters with subscripts express adjacent vertexes of the letters. For example, the letter x stands for the SQVSBS of the vertex ,X, and X_i express one of a adjacent vertex of the vertex X, x_i express SQVSBS of the vertex X_i.

Define 2. Projection on X with the base b of SQVSBS

Let Y is a feature gridding, $Y_1, Y_2, \ldots, Y_k$ is a queue compose of the all adjacent feature griddings of Y, and $Y_1 = X$, we call

$\tau_b(X, Y) = \tau_b(y_1, y_2, \ldots, y_k)$

Projection on X with the base b of SQVSBS, noted as $\tau_b(X, Y)$.

Define 3. SQVSBS of X with field r and base b

$\tau_{b(r)}(X) = \tau_b(\tau_{b(r-1)}(X_1), \tau_{b(r-1)}(X_2), \ldots, \tau_{b(r-1)}(X_k))$, $b = 0, 1, 2, \ldots$; $\tau_{b(0)}(X)$ is the degree of X, among which, $X_1, X_2, \ldots X_k$ is the all adjacent feature griddings, $b(r)$ is the basic sequence in the field r, $\tau_{b(r-1)}(X_1), \tau_{b(r-1)}(X_2), \ldots, \tau_{b(r-1)}(X_k)$ is a series with monotone series. We call $\tau_{b(r)}(X)$ is SQVSBS of X with field r and base b.

3.2 The Algorithm Description

Step1: Initalization. For all adjacent feature gridding Y of the Graph G1 and G2, set $Y(0)$ with the degree value the feature gridding Y, and $r=1, t=1$.

Step2: For any feature gridding X, let the degree isχ, and

    Step2.1: $u_i = \tau_{t_{r-1}}(X, X_i)$

    Step2.2: order the series $\{u_i (1 \leq i \leq \chi)\}$, obtaining $u_{s1}, u_{s2}, \ldots, u_{s\chi}$, among which, $s_1, s_2, \ldots, s_\chi$ is a arrange of the series $1, 2, \ldots, \chi$

    Step2.3: compute $\tau_{b(r)}(X)$:

$$\tau_{b(r)}(X) = \tau_b(u_{s1}, u_{s2}, \ldots, u_{s\chi})$$

Step3: Exist noise feature griddings ?If exist, remove the noise griddings,and goto step1 else goto the next step

Step4: Are the $X^{(r)}$ of all the feature griddings different each other? If yes, goto stwp6 else goto the next srep

Step5: $t_{r+1} \leftarrow t_r+1$, $r \leftarrow r+1$, and then goto Step2

Step6: sort the left feature griddings, obtaining the topological series of G_1.

In our experiment, the base series is set as $\{1,t,t^2,...t^k,...\}$.

After the feature griddings of the two images are obtained, some griddings have probably no the corresponding griddings in the another image, which are called noise feature griddings. Because the noise griddings probably diffused out to other feature griddings before they are moved off, so the noise feature griddings should be removed, and their relations to other griddings are removed as well.

4 Obtain the Correspondences Between the Feature Points

The algorithm is based on the topological relationships among feature griddings, so, holds globally the matching process and may reduces error matching. On the other hand, a giddings is not equal to a pixel because of errors, so obtaining further the correspondences on the pixel scale is necessary. On way to is to reduce the dimension of the griddings, but the number of the griddings will be multipled and so, the time complexity will be greatly increased. Another way is to search the accurate correspondences between the corresponding feature griddings obtained in the former step.

5 Experimental Results

Experiment 1

G1 and G2 the figure 4 are the two input images. Only one loop is needed to obtain the correspondences between the G1 and G2, and the computational results are listed in the table 1.

The correspondences between G1 and G2 are {(A1,A2),(B1,B2),(C1,C2),(D1,D2), (E1,E2),(F1,F2)}. In this example, we set t=0.5. By the way, three times are needed to find the correspondences if bipartite graph matching is used.

Experiment 2

The figures 5 are get from the paper [4] with the dimension 217X166 and the grey level is 256.

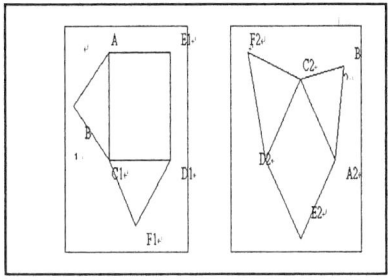

Fig. 4. The topological graph G1 and G2

Table 1. The computational results based on the Fig 4

		R	A(t)	B(t)	C(t)	D(t)	E(t)	F(t)
G1		t=0	3	2	4	3	2	2
		t=1	9.04	7.5	9	8.78125	7.75	7.6875
G2		t=0	3	2	4	3	2	2
		t=1	9.04	7.5	9	8.78125	7.75	7.6875

Fig. 5. The original image sequences from [4]

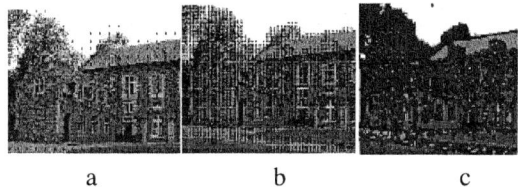

Fig. 6. The feature griddings of the left image in Fig.5. with different dimensions. a,b,c are the the feature griddings with dimension 20X20, 50X50 and 100X100 respectively.

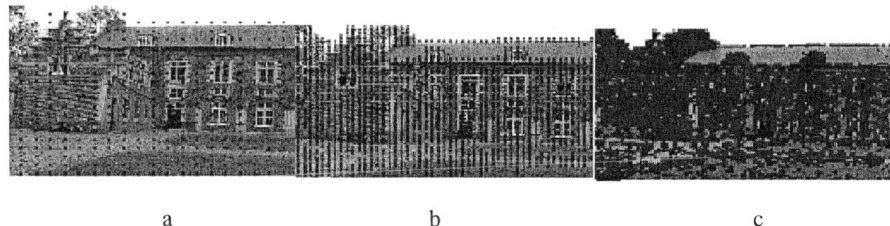

Fig. 7. The feature griddings of the right image in Fig.5. with different dimensions. a,b,c are the feature griddings with dimension 20X20, 50X50 and 100X100 respectively.

The feature griddings of the figure 5 are shown the figure 6 and 7. In our experiment, $\lambda=30°$. The experiment shown that the way the dimension of cover-edge-griddings affects the precisions of the silhouettes. Our algorithm is robust than the robust algorithm to point noise, shown in the figure 8.

Experiment 3

The original input images sequences are given in the figure 9, the cover-edge-griddings are shown in Fig10 with the dimension 45X45. set $\theta=45$, the feature griddings are shown in the figure 11. Part of the corresponding griddings are shown in the figure 12, where X_i correspondences $Y_i (i=1,2,3)$.

Fig. 8. The result after processed by Robert operator to a in Fig 5

Fig. 9. The Original Experimental Image Sequences

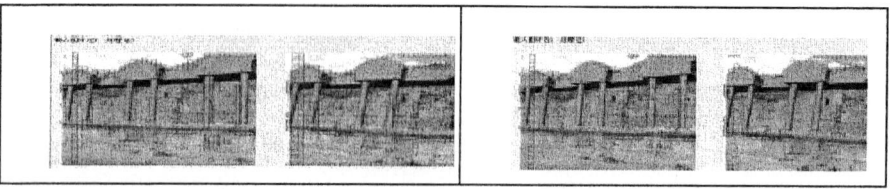

Fig. 10. The Cover-Edge-Griddings of the Fig. 9

Fig. 11. The Feature Griddings of the Fig. 9

Fig. 12. Art of the Correspondences from the sequences in the Fig. 9.

6 Conclusions

This paper aims at obtaining the correspondences in the images taken from different view points with the better time complexity and higher precision, we firstly obtain the global correspondences between the topological structures of the images, then adopt local searching strategy to obtain the accurate correspondences between the feature points. The research has the following characteristics: (1)The design is a process of from coarseness to delicateness;(2)The algorithm has the time complexity of $O(Max(f^2,A))$ which is better compared with the current best one ,$O(|V|(|E| + |V|\log|E|))$;(3) the algorithm is robust to geometrical deformations and noise disturbances characteristic of points and little regions.

References

[1] Barius Burschka, Dana Cobzas,et.al .Recent Methods for Image-based Modeling and Rending. IEEE Virtual Reality 2003 Tutorial 1:55~66
[2] R. Koch, M. Pollefeys, L. Van Gool, *Realistic surface reconstruction of 3D scenes from uncalibrated image sequences*, Journal Visualization and Computer Animation, Vol. 11, pp. 115-127, 2000.
[3] M. Pollefeys, M. Vergauwen, L. Van Gool, *Automatic 3D modeling from image sequences*, invited presentation, International Archive of Photogrammetry and Remote Sensing, Vol. XXXIII, Part B5, pp. 619-626, 2000
[4] M. Pollefeys, R. Koch, M. Vergauwen, B. Deknuydt, L. Van Gool *three-dimensional scene reconstruction from images,* proceedings SPIE Electronic Imaging, Three-Dimensional Image Capture and Applications III, SPIE Proceedings series Vol. 3958, pp.215-226, 2000
[5] Cai Yong,,Liu xuehui,Wu enhua. Virtual Reality System Environment Image-Based Readering. Journal of Software.1997,8(10):721-728
[6] Thaddeus Berier.etc..Feature-Based Image Metamorphosis is. Computer Graphics. July 1992:35-42
[7] M. Pollefeys, R. Koch and L. Van Gool. *Self-Calibration and Metric Reconstruction in spite of Varying and Unknown Internal Camera Parameters*, International Journal of Computer Vision, 32(1), 7-25, 1999:13~15
[8] Yang Ruiyuan Qiu Jianxiong. Survey of Image-Based Modeling and Rendering Journal of Computer-Aided Design &Computer Graphics,2002,14(2):186
[9] D.Huber.M.Hebert.3D Modeling Using a Statistical Sensor Model and Stochastic Search。Proceedings of the IEEE Conference on Computer Vision and Pattern Recognition (CVPR), June, 2003, pp. 858-865
[10] D.Huber,O.Carmichael, and M.Hebert. 3D map reconstruction from range data. Proceedings of the IEEE International Conference on Robotics and Automation (ICRA '00), Vol. 1, April, 2000, pp. 891 - 897
[11] Y. Cheng, V. Wu, R.Collins, A. Hanson, and E. Riseman, "Maximum-Weight Bipartite Matching Technique and Its Application in Image Feature Matching," *SPIE Conference on Visual Communication and Image Processing*, 1996: 1358-1379
[12] Barnard S T,Thompson W B. 1980. Disparity analysis of images. IEEE PAMI-2:333-340
[13] D.B. Goldgof,H. Lee and T.S. Huang, "Matching and motion estimation of three-dimensional point and lines sets using eigenstructure without coresppondences", Pattern Recognition, Vol.25,No.3,pp.271-286,1992
[14] Chia-Hoang Lee and Anupam Joshi, "Correspondence problem in image sequence analysis", Pattern Recognition, Vol.26,No.1,pp47-61,1993
[15] Zhu ZhongTao,Zhang Bo,Zhang ZaiXing.The quasi-invariants of curves under the topological deformation. Chinese J.computers, 1999, 22(9): 897–902.
[16] Zhao Nanyuan.The Topology Properties Algorithm of Image.Journal of Tsinghua University
[17] Guo Lianqi,Li Qingfen.A study of the Graphic Recognition Method by Using the Topological Relation.Journal of Harbin Engineering University, 1998, 9(4):38~43.

An Efficiently Algorithm Based on Itemsets-Lattice and Bitmap Index for Finding Frequent Itemsets[1]

Fuzan Chen and Minqiang Li

School of Management
Tianjin University, China (Tianjin), 300072
{fzchen, mqli}@tju.edu.cn

Abstract. Frequent itemsets play an essential role in many data mining tasks that try to find interesting patterns from databases. A new algorithm based on the lattice theory and bitmap index for mining frequent itemsets is proposed in this paper. Firstly, the algorithm converts the origin transaction database to an itemsets-lattice (which is a directed graph) in the preprocessing, where each itemset vertex has a label to represent its support. So we can change the complicated task of mining frequent itessets in the database to a simpler one of searching vertexes in the lattice, which can speeds up greatly the mining process. Secondly, Support counting in the association rules mining requires a great I/O and computing cost. A bitmap index technique to speed up the counting process is employed in this paper. Saving the intact bitmap usually has a big space requirement. Each bit vector is partitioned into some blocks, and hence every bit block is encoded as a shorter symbol. Therefore the original bitmap is impacted efficiently. At the end experimental and analytical results are presented.

1 Overview

Data mining has recently attracted considerable attention from database practitioners and researchers because of its applicability in many areas such as decision support, market strategy and financial forecasting. One of the most important research topics in data mining is the discovery of association rules in large databases of sales transactions [1,2]. The aim of association rule mining is to identify relationships between items in databases. It has been applied in applications such as market-basket analysis for supermarket, linkage analysis for website, sequence-pattern in bioinformatics etc.

1.1 Related Work

There are many association rule algorithms developed to reduce the scans of transaction database and to shrink number of candidates. These algorithms might as well be classified into two categories: 1) candidate-generation-and-test approach(such as Apriori).The performance of these algorithms degrades incredibly because they perform as many passes over the database as the length of the longest frequent pattern. This incurs high I/O overload for iteratively scanning large database[1,2].

---

[1] This research is supported by National Science Foundation (No. 70171002).

2) pattern-growth approach [11,1,2]. It is difficult to use the FP-Tree in an interactive mining system. However the changing of support threshold may lead to repetition of the whole mining process for the FP-Tree. Besides, the FP-Tree is not suitable for incremental mining when new datasets are inserted into the database.

There remain many challenging issues for the association rule algorithms. The procedure scans the databases multiple times, which make it very hard to be scalable for large data sets. The algorithms generates and tests the huge number of itemsets to make sure that all the frequent itemsets are generated, which results in a complete set of association rules. Many novel ideas have been proposed to deal with these issues. To reduce the search space, some algorithms just mine all the closed frequent itemsets, instead of all frequent itemsets. A set of non-redundant association rules, which can be mined from the closed frequent itemsets, is generated. These rules can be used to infer all the association rules[12]. Other than the traditional horizontal database format for mining, a number of vertical mining algorithms have been proposed recently such as CHARM, VIPER, etc. In vertical database each item is associated with its corresponding transaction id (called tidlist). Mining algorithms using the vertical format have shown to be very effective and usually outperform horizontal approaches because frequent pattern can be counted via tidlist/granule/bitmap intersections in the vertical approach[6,7,8,9,10].

1.2 Outline

Based on the recent advance in these areas, we propose an itemsets-lattice and bitmap index based approach. In this paper we present a new framework, which inherits the advantages of the above methods. The remaining of this paper is organized as follows. In section 2 we give a detailed description of the itemsets-lattice. The definitions and theories about itemsets-lattice, and constructing and decomposition strategies are introduced. In section 3 we analyze the problem of itemsets support counting, which is based bitmap index technology. Some feasible optimizations and experimental results are presented in section 4 and 5 respectively.

We consider an example transaction database D shown in Table1, and assume that the defined minimum support threshold is $minsup$=0.3.There are five different items, i.e., $I = \{A, B, C, D, E\}$, and the frequent itemsets are those that are contained at least in one of $ABCE$, ACD, BCD, CDE.

Table 1. A Transaction Database D

TID	Itemsets
1	ABCE
2	CDE
3	ABCE
4	ACDE
5	ABCDE
6	BCD
7	DE
8	BCE
9	ABCD
10	BE

2 Itemsets-Lattice

In this section, we begin with some basic definition and notions, and then discuss how to use these techniques to represent itemsets and organize the itemsets in an itemsets-lattice structure. Considering the limitation on main memory, we deal with the problem of decomposition in addition.

2.1 Partial Order and Itemsets-Lattice

The lattice theory could be referred in [3,15]. We only introduce some concepts and theorems to be adopted in this paper.

Let $(P, \subseteq)$ be an ordered set, and let A be a subset of P. The $J \in P$ is an **upper bound** of A if $J \subseteq X$ for all $X \in A$, furthermore J is the **least upper bound** of A if there is no element $Y \in P$ that $J \neq Y$ and $U \subseteq Y$. Similarly, an element $M \in P$ is a **lower bound** of A if $M \subseteq X$ for all $x \in A$, M is the **greatest lower bound** of A if there is no element $Y \in P$ that $M \neq Y$ and $Y \subseteq L$. Typically, the least upper bound is called the **join** (or top element) of A, and is denoted as $\vee A$ (or $\overline{A}$); the greatest lower bound is called the **meet** (or bottom element) of A, and is denoted as $\wedge A$ (or $\underline{A}$). For $A = \{X, Y\}$, we also write $X \cup Y$ for the join, and $X \cap Y$ for the meet. Let M be the meet of P and $X \in P$, L be the meet of P, if there is no $y \in P$ that $M \subset Y \subset X$, then we called X is a atom of P. The set consisting of all of the atoms of P is denoted $A(P)$.

Definition 1. An ordered set $(L, \subseteq)$ is a *lattice*, if for any two elements $X, Y \in L$ the join $X \vee Y = X \cup Y$ and meet $X \wedge Y = X \cap Y$ always exist. L is called a *join semi-lattice* if only the join exists. L is called a *meet semi-lattice* if only the meet exists. L is a *complete lattice* if $\wedge S$ and $\vee S$ exist for all $S \subseteq L$. Any finite lattice is complete.

It is deduced that the power set $P(I)$ on the items I derived in the transaction database D is a lattice, and also a complete one, $P(I)$ is called as power-set-lattice. For this $P(I)$ the join is I, the meet is ϕ, and each atom is $\{i | i \in I\}$.

Let I be the itemset derived in transaction database D, $X, Y \subseteq I$ and $X \neq Y$, if Y can be obtained from X by adding a single item, then X is said to be **adjacent** to Y. We call X is a parent of Y, and Y is a child of X. Thus, and itemset may possibly have more than on parent and more than one child. For each itemset X and each $x_k \in X$, itemset $X - \{i_k\}$ is a parent of X.

Definition 2. Let $I = \{i_1, i_2, \cdots, i_m\}$ be the itemset derived in transaction database D, the **itemsets-lattice** L on the ordered powerset $P = (P(I), \subseteq)$ of I be a DAG which each vertex is an itemset $X \subseteq I$. L is constructed as follow: For each primary itemset X, construct a graph with a vertex $v(X)$, each vertex has a label corresponding to the value of its support, denoted as $S(X)$. For any pair of vertices corresponding to itemsets X and Y, if and only if X is a parent of Y, a directed edge exists from $v(X)$ to $v(Y)$, denoted as $E(X, Y)$. The vertex $v(X)$ is said to be the head of the edge $E(X, Y)$, and the vertex $v(Y)$ is said to be the tail of the edge $E(X, Y)$.

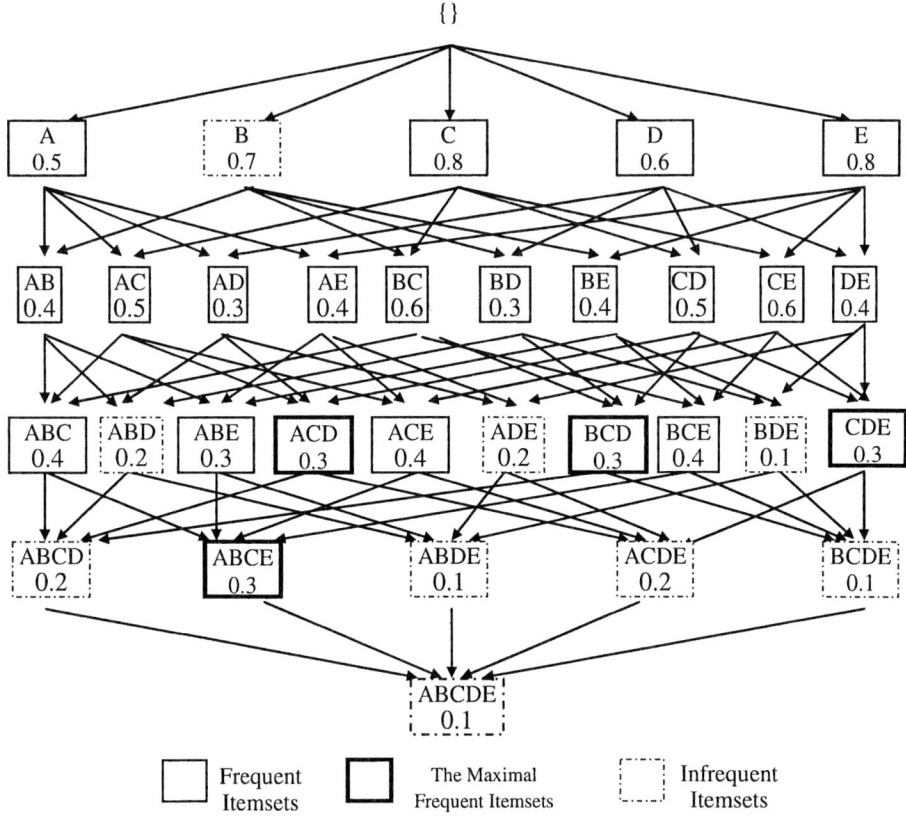

Fig. 1. The itemsets-lattice $L=P(I)$

Thus, the search space of all itemsets can be represented by an itemsets-lattice, with the empty itemset at the top and the set containing all items at the bottom. Itemsets-lattice $L=P(I)$ deduced by the database shown in Table1 is illustrated in Figure 1.

Apparently, the fact there is a directed path from vertex $v(X)$ to vertex $v(Z)$ in an itemsets-lattice L, implies $X \subset Z$. Especially, we call X is an ancestor of Z, and Z is a descendent of X.

2.2 Itemsets-Lattice Decomposition

If we had enough main memory, we could enumerate all the frequent itemsets by traversing the itemsets-lattice. With only a limited amount of main memory in practice, we need to decompose the original lattice into some smaller pieces such that each one can be solved independently in main memory.

Definition 3. Let L be an itemsets-lattice, and $S \subset L$. For each vertex $A, B \in S$, if $A \vee B = A \cup B \in S$ and $A \wedge B = A \cap B \in S$, S is said to be a sub-lattice of L.

Definition 4. Let f be a function on ordered itemset X, which selects the first k items from X, i.e. $f(X,k) = X[1:k]$. If $X \equiv_k Y \Leftrightarrow f(X,k) = f(Y,k)$ for all $X, Y \in L$, then the binary relation $\equiv_k$ is called an equivalence relation on X.

Theorem 1. In itemsets-lattice L, each equivalent class $[X]_k$ induced by equivalence relation $\equiv_k$ is a complete sub-lattice of L.

Proof: $\forall A, B \in [X]_k$, i.e., A and B have the same sub-itemset X. $A \cup B \supseteq X$ implies $A \cup B \in [X]_k$, and $A \cap B \supseteq X$ implies $A \cap B \in [X]_k$. That is to say there are always $A \wedge B = A \cap B \in S$ and $A \vee B = A \cup B \in S$. Then S is a sub-lattice of L following definition 4. Let $S = \{A_1, A_2, \cdots, A_n\}$, meet of S is $\bigcap_{i=1}^{n} A_i = X$, and the join of S is $\bigcup_{i=1}^{n} A_i$. The result follows from definition 3.

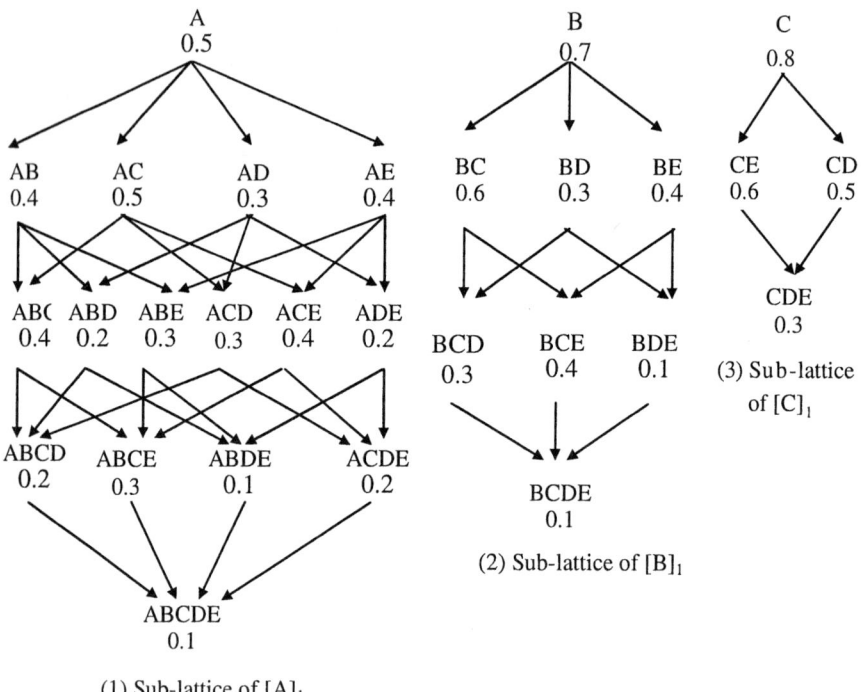

Fig. 2. Sub-lattices of $L=P(I)$ induced by $\equiv_1$

We can generate five independent sub-lattices by applying $\equiv_1$ to itemsets-lattice $L=P(I)$ in Figure1. The sub-lattice deduced by $[A]_1$, $[B]_1$ and $[C]_1$ respectively is shown in Figure2, $[D]_1$ contains two itemsets D and DE, and $[E]_1$ contains only one itemset E.

If a $\equiv_1$ sub-lattice is still too large for the main memory, it can be recursively decomposed until all of its sub-lattices can be solved independently in main memory.

3 Supports Counting

Aiming at efficient calculation of the supports counting, we employ a new bitmap-based technology to optimize both I/O and CPU time. The new bitmap index require less disk space and memory than conventional methods.

The bitmap technique was proposed in the 1960's, and has been used by a variety of products. A typical context is the modern relational DBMS (i.e. Oracle 9i), which implement bitmap indexes for accelerating join and aggregate computation. Bitmap also has been applied in association mining[6,7,8,9,10]. Distinctively, the bitmap proposed is not the same as the one adopted in DBMS. Instead of using the initial bitmaps, we manipulate the encoded ones, which needs fewer space for the disk and main memory.

The key idea of the approach is to use a bitmap index to determine which transactions contain which itemsets. Each transaction has one unique offset position in the bitmap. A bit vector $X.Bit = (b_1, b_2, \cdots, b_n)$ is associated to each itemset X. In $X.Bit$ the ith bit b_i is set to 1 if the transaction i contains the itemset X, and otherwise b_i is set to 0. The ordered collection of these bit vectors composes the bitmap. In a bitmap, a line represents a transaction, while a column corresponds to a given k-itemset. The bitmap for 1-itemsets is just a classical bitmap index implemented in Oracle. Combining the process of performing a logical operation (AND/OR/NOT) on a serious of bitmaps is very efficient, particularly compared with performing similar processes on lists of transaction. The operations between binary bits are not only implemented with software algorithms, but also some appropriative hardware Boolean calculation units.

As an example, bit vectors associated with 1-itemsets in the transaction database D shown in table1 are such:

$\{A\}.Bit = (1011100010)$,
$\{B\}.Bit = (1010110111)$,
$\{C\}.Bit = (1111110110)$,
$\{D\}.Bit = (0101111010)$,
$\{E\}.Bit = (1111101101)$.

The collected bitmap is shown in table2.

During the supports counting the algorithm intensively manipulates bit vectors, and even store the intermediate ones, which requires a lot of disk

Table 2. Bitmap of 1-itemsets

1	2	3	4	5
1	1	1	0	1
0	0	1	1	1
1	1	1	0	1
1	0	1	1	1
1	1	1	1	1
0	1	1	1	0
0	0	0	1	1
0	1	1	0	1
1	1	1	1	0
0	1	0	0	1

space and main memory. So, it is necessary to compress the bitmap in advance. A new block strategy is proposed to encode and decode the bitmap, which is similar to the pagination technology in operating systems. In this approach, every bit vector is partitioned into fractions, called blocks, that can be encoded respectively, so that a bitmap is divided into granules. Each block should have an appropriate size, if the size is too small, the impact is not remarkable; otherwise the encoding is not straightforward. In order to take full advantage of Logical Calculation Units, the block size should be an exponential to 2. Each block is represented as $(p:W)$, p is the number of the block, and W is the block bit vector. Let l be the block size, m the number of transactions in database D. In this way each bit vector $B = (b_1, b_2 \cdots, b_i, \cdots, b_m)$ can be partitioned into $p = \text{INT}(m/l)$ blocks. The kth block vector $W_k = (w_1, w_2, \cdots, w_j, \cdots, w_l)$, $j = \text{MOD}(i,l) = i - l*(k-1)$.

For example, when we initialize the block size as 16, every 16 bits forms a block. Thus the total number of blocks is the total number of transactions divided by 16. If a database contains 10K transactions, each bit vector B consists of $p = \text{INT}(10,000/16) = 625$ blocks. If the 19th bit in B $b_{i=19} = 1$, we can figure out that the 3rd bit $w_{j=3} = 1$ in the $k = 2$ block. If the transaction set $\{1, 3, 4, 9, 11, 18, 20\}$ supports itemset X, and $\{4, 11, 18, 24, 31\}$ supports itemset Y respectively, dividing by 16 yields $X.Bit = \{(1:1011\ 0000\ 1010\ 0000), (2:0101\ 0000\ 0000\ 0000)\}$ and $Y.Bit = \{(1:0001\ 0000\ 0010\ 0000), (2:0100\ 0001\ 0000\ 0010)\}$. The intersection of two bit vectors is accomplished by the logical operation AND. Thus the corresponding bit vector of their join set $X \cup Y$ is $XY.Bit = \{(1:0001\ 0000\ 0010\ 0000), (2:0100\ 0000\ 0000\ 0000)\}$. That is to say transaction $4 + 16^{1-1} = 4, 11 + 16^{1-1} = 11, 2 + 16^{2-1} = 18$ support both itemset X and Y.

Encoding each block as a shorter code can reduce the space demanding. The obligatory principle is that each block can be represented uniquely. The conversion between binary, octal, decimal and hexadecimal can be implemented conveniently, hereby every block can be represented a binary, octal, decimal or hexadecimal code. We use a hexadecimal code, i.e. every four bits in a block encode a hexadecimal code. For instance the encoded hexadecimal vectors associated with previous itemsets X and Y are $X.Bit = \{(1:\text{C0A0}), (2:5000)\}$ and $Y.Bit = \{(1:1020), (2:4102)\}$, and the intersection of their bit vectors is $XY.Bit = \{(1:1020), (2:4000)\}$.

For the sake of efficient counting the number of 1 in a bit vector, figuring out the support of an itemset, we previously store the binary block in a bit array $Bit[1 \cdots l]$, and the hexadecimal blocks in an array $ABit[1 \cdots p]$. The value in $ABit[i]$ is the hexadecimal code of the ith block. Implementation of this support counting algorithm follows.

Algorithm 1. Itemsets Support Counting
Algorithm $Countsupport(X_1, X_2)$
Begin
  $Support=0$;
  For $i=1$ and $i \leq p$ do
    If $X_1.ABit[i] \neq 0$ and $X_2.ABit[i] \neq 0$ then
      For $j=1$ and $j \leq l$ do
        $X.Bit[j] = X_1.Bit[j] \& X_2.Bit[j]$;
        $Support+ = X.Bit[j]$;
        $j++$;
      endfor
      $X.ABit[i] = X_1.ABit[i] \& X_2.ABit[i]$;
    Else
      $X.ABit[i] = 0$;
    endif;
    $i++$;
end;

4 Algorithm Analysis and Optimization

The performance of these candidate-generation-and-test algorithms degrades incredibly because these algorithms perform as many passes over the database as the length of the longest frequent pattern. This incurs high I/O overhead for scanning large disk-resident database many times. Furthermore, this kind of approach finds frequent itemsets according to the given minimum support. When the support changes, algorithm has to be performed on the entire database again, each itemset is also regenerated and its support is also recounting. Thus the previous computed result cannot to be reused is still a side issue. We convert the original transaction database to an itemsets-lattice in a preprocessing, where each itemset vertex has a support label. Since then when the given support changes, we only need traverse the itemsets-lattice straightly to finding frequent itemsets, in spite of manipulating the transaction database complicatedly. Such itemsets-lattice can be mined for many times without numbers. Hence, the preprocessing cost can be contributed, and the more mining task performs, the little average effort is.

The search space of all itemsets contains exactly $2^{|I|}$ different itemsets. If itemset I is large enough, then the naive approach to generate and count the supports of all itemsets over the database can not be achieved within a reasonable period of time. Instead, we could generate only those itemsets that occur at least once in the transaction database. More specifically, we generate all subsets of all transactions in the database. Of course, for large transactions, this number could still be too large. Therefore, as an optimization, we could generate only those subsets of at most a given maximum size. This technique would pay off for sparse transaction databases. For very large or dense databases, an alternative method is to generate subsets of at least a given support threshold. In such a manner those overlong itemsets would be pruned.

This technique can decrease both space and compute effort consumedly. The itemsets-lattice constructing is a long-time spending process. Buffer can decrease the disk I/O, and multiprocessor can parallelize the processing, these ways should accelerate the speed.

The itemset has an important property: if an itemset is infrequent, all of its supper set must be infrequent. In other words all the infrequent itemsets compose a meet semi-lattice. There has no frequent itemset in the sub-lattice which bottom element is infrequent. Thus, we only need to traverse the sub-lattice which bottom element is frequent. Applying this property can minimize the I/O costs when enumerating frequent itemsets.

5 Experimental and Analytical Results

We run the simulation on PC: Intel P4 2.4G CPU, 512MB main memory and Windows 2000 Server. The synthetic data set are generated using a method provided by KDD Research Group in IBM Almaden Recearch Center, referring to Http://www.almaden.ibm.com/cs/quest/syndata#AssocSynData.

Figure 3 shows the response time variation with average transaction size during lattice constructing. The data sets are Tx.I3.D100K and Tx.I3.D100K, the transaction size increases 1 every time. This shows that response time of lattice constructing and the average transaction size is an exponential relation. The computational effort and the scale of constructed lattice are more sensitive to the average transaction size, rather than to the number of transactions. We also find that the computational effort mushrooms when some transactions have very great length, even though the average transaction size is small.

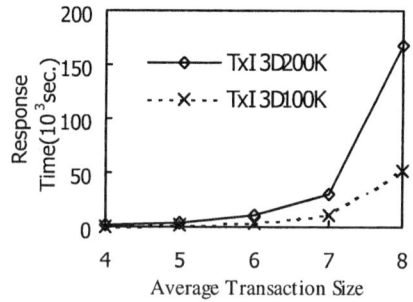

Fig. 3. Response time variation with average transaction size during lattice constructing

Figure 4 shows the response time variation with the number of frequent itemsets enumerated by applying Breadth-First search strategy to itemsets-lattice. The data sets areT5.I3.D100K and T5.I3.D200K. This shows that response time of frequent itemsets enumerating is more sensitive to the number of frequent itemsets searched, rather than to the average transaction size and number of transactions in the data set.

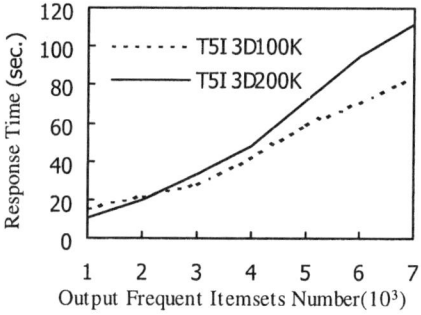

Fig. 4. Response time variation with number of frequent itemsets searched

6 Conclusion

In this paper a new efficient approach for mining frequent itemsets is proposed based on the lattice theory and bitmap index. These two kinds of technique have been fully researched and successfully used in a variety of fields for last decades. By dint of an itemsets-lattice structure, the complicated task of mining frequent itemsets in the database is simplified to vertices searching in the lattice, which can speed up greatly the frequent itemsets mining process. Considering the problem that there may not have enough main memory to process the entire lattice, we discuss the solution for lattice recursive decomposition.

We use an improved compacting bitmaps database format. We count the support of itemset by means of binary bit vectors intersections, which minimizes the I/O and computing cost. To reduce the disk and main memory space demanding, we break the bitmap down into some less blocks, which can be encoded as a shorter code. The blocks of bitmaps are fairly adaptable. Hence the additional space decreases rapidly.

References

1. Qiankun Zhao, Association Rule Mining:A Survey, Technical Report, CAIS, Nanyang Technological University, Singapore , 2003, No. 2003116
2. Jochen Hipp , Algorithms for Association Rule Mining : A General Survey and Comparison, ACM SIGKDD, 2000,Vol.2 Iss.1 pp.58-65
3. J. B. Nation, Notes on Lattice Theory,http://www.Hawaii.edu
4. K.K. Loo, etc., A Lattice-based approach for I/O efficient association rule mining, Information Systems, 27(2002) pp.41-74
5. Huaiguo Fu, Partitioning large data to scale up lattice-based algorithm,ICTAI'03,2003
6. T.Y.Lin, Xiaohua Hu, and Eric Louie, A Fast Association Rule Algorithm Based On Bitmap and Granular Computing , The Proceedings of the IEEE International Conference Fuzzy Systems, 2003, pp.678-683
7. Xiaohua Hu, T.Y.Lin, Eric Louie, Bitmap Techniques for Optimizing Decision Support Queries and Association. Rule Algorithms, ,IDEAS 2003
8. Mikolaj Morzy, Hierarchical Bitmap Index An Efficient and Scalable Indexing Technique for Set-Valued Attributes , ADBIS 2003, LNCS 2798, 2003, pp. 236–252
9. J. Ayres, J. Flannick, J. Gehrke, and T. Yiu, Sequential Pattern Mining Using a Bitmap Representation, Proc. of the 8th Int. Conf. on Knowledge Discovery and Data Mining, 2002, pp. 429-435
10. G. Gardarin, P. Pucheral, and F. Wu, Bitmap Based Algorithms for Mining Association Rules," Proc. of the 14th Bases de Donnes Avancees, 1998, pp. 157-176
11. J. Han, J. Pei, Y. Yin, and R. Mao, Mining Frequent Patterns without Candidate Generation, *Proc. 2000 ACM-SIGMOD Int. Conf. on Management of Data (SIGMOD'00)*, Dallas, TX, May 2000
12. Zaki Mohammed J.,Mining Non-Redundant Association Rules, Data Mining and Knowledge Discovery,3(2004),pp.223-248
13. Haixun Wang,Demand-driven frequent itemset mining using pattern structures, Knowledge and Information Systems ,2004,
14. Shinichi Morishita, Traversing Itemset Lattices with Statistical Metric Pruning, Symposium on Principles of Database Systems, 2000, pp. 226-236
15. Hu Changliu, The basic of Lattice Theory, Henan University publishing company, 1990

Weighted Fuzzy Queries in Relational Databases [1]

Ying-Chao Zhang, Yi-Fei Chen, Xiao-ling Ye, and Jie-Liang Zheng

Dept. of Information and Communication,
Nanjing University of Information Sciences & Technology, Nanjing 210044
yc.nim@163.com, ch_yi_f@etang.com, xl.nim@163.com,
juckyz@sina.com

Abstract. In most of fuzzy querying systems, items in query conditions are assumed to carry equal importance, although it might not actually to be the case. Issues for expressing users' preferences to query conditions are presented here. Relative importance between query items is introduced. Standard SQL is extended to express weighted fuzzy queries, such as sub-queries, multi-table queries; set-oriented queries etc. which make the existing relational database systems more flexible and more intelligent to users.

1 Introduction

In the field of databases, many efforts have been made to enable fuzzy queries in relational databases. Most researches of this kind are based on fuzzy set theory in which the retrieval conditions of SQL queries are described by fuzzy terms represented by fuzzy numbers [1][2][3][5][6]. The extensions of nesting operators, set-oriented operators, and also selection operators are applied to sets of tuples [12][13][14]. The presented fuzzy querying systems allow users to construct their queries conveniently, which make the relational database systems become more flexible and more intelligent to user [9].

But all these methods cannot deal with users' preferences to queries conditions. For example, 'find the houses with cheap rent and moderate size'. A user may feel that the rent is more important than the size. A house with cheap rent, even the size is not so moderate (a little lager or smaller), may be more acceptable to him than a house with moderate size and a little more expensive rent. Few methods can resolve this kind of queries in relational databases, so it is necessary to induce the concept of "importance" in fuzzy queries.

There are numerous researches on "importance" in the field of information retrieval and fuzzy decision-making [8][10]. Some literatures discuss weights in fuzzy queries [4][9]. The method to describe the weights of query terms in conjunction is proposed in [4]. There are few researches introducing weights in multi-table query, nested query, set-oriented query, etc.

The aim of this paper is to present a method to weigh the (fuzzy) terms in fuzzy querying conditions. First of all, we briefly set the base concept and characteristic of

---

[1] This work is supported in part by the Science and technology developing plan of Jiangsu province, under Grant (BR2004012).

fuzzy queries. Then we give the definition of weight and the way to evaluate matching degrees of tuples in different kinds of weighted queries. The effect of weight and threshold to matching degree is discussed. Finally, we summarize and point out to the future.

2 Fuzzy Queries

Fuzzy querying is in the sense of not asking if an element simply satisfies a query or not, but more asking to which extent it satisfies the given condition. Fuzzy queries allow fuzzy terms in WHERE clause in SQL.

2.1 Extension of Simple Queries

SELECT A1,..., An FROM R WHERE fc WITH α (Q1)

In (Q1), R is a relation; A_i is an attribute in R; fc is a fuzzy condition. The result of (Q1) is a fuzzy relation. Elements in it match fc with different degrees, which are named matching degrees MD:

$$MD(t) = \mu_{fc}(t) . \tag{1}$$

MD is a value in interval [0, 1]. fc can imply fuzzy and Boolean basic conditions at the same time. α is a threshold value. The function of clause ' WITH α ' is to discard tuples whose matching degrees are below α.

As we all know, tuples in the result are unique if the prime key is in the list of SELECT clause in SQL. Otherwise, the result is a multi-set. The keyword "distinct" is used to eliminate the duplicates. In the case of fuzzy queries, the extended select block can return duplicate if several tuples have the same value on attributes and eventually different matching degrees [12]. If "distinct" is used in fuzzy queries, only the tuple with the highest matching degree is retained in this paper.

2.2 Extension of Multi-table Queries

Multi-table queries select the tuples belonging to the Cartesian product of the specific relations which are joined by some matching attributes. The general form of fuzzy multi-table queries is:

SELECT R.A, S.B FROM R,S
WHERE fc1(R) AND fc2(S) AND R.C θ S.D (Q2)

θ can be a comparison operator, such as =,<,>,>=,<= or some fuzzy comparison operators, such as >>,<<, $\approx$,etc.. So we have[12]:

$$MD(t) = \min(\mu_{fc1}(x), \mu_{fc2}(y), \mu_\theta(x.C, y.D)) \; x \in R, y \in S . \tag{2}$$

2.3 Extension of Set-Oriented Operators

In SQL99, the available set operators are UNION, INTERSECTION, EXCEPT. UNION is the only one necessary since INTERSECTION and EXCEPT can be stated by means of other constructors. The general form is:
(SELECT A1,...,An FROM R WHERE fc1)

UNION/INTERSECT/EXCEPT (Q3)

(SELECT A1,...,An FROM S WHERE fc2)   WITH α

2.4 Extension of Nested Queries

SQL supports nested queries. Sub-query is nested in the expression SELECT-FROM-WHERE. Operators such as IN, EXIXTS, ALL, ANY are used in nested queries [12].

SELECT * FROM R WHERE fc1 AND A IN
(SELECT B FROM S WHERE fc2) (Q4)

SELECT * FROM R WHERE fc1 AND EXISTS
(SELECT * FROM S WHERE fc2 AND R.A=S.B) (Q5)

SELECT * FROM R WHERE fc1 AND
A= ANY(SELECT B FROM S WHERE fc2) (Q6)

Expression (Q4)(Q5)(Q6) are equivalent.

SELECT R.* FROM R, S WHERE fc1 AND fc2 AND R.A=S.B (Q7)

If the result of (Q7) is a set, (Q4)(Q5)(Q6)(Q7) are equivalent. If we change "=" with θ (fuzzy or crisp comparison operators), the equivalence between (Q5)(Q6)(Q7) is still kept. We also have equivalence between the expressions as follows [12]:

SELECT * FROM R WHERE A θ ALL (SELECT B FROM S WHERE fc) (Q8)

SELECT * FROM R WHERE NOT EXISTS
(SELECT B FROM S WHERE fc AND NOT(A θ B)) (Q9)

3 Weighted Queries in Relational Databases

A query logically connected (AND or OR operators) to other queries is represented by a fuzzy set **A**. The relative importance can be taken into account by weighing **A**. It is necessary to define differently, A_w for disjunctive queries and A^w for conjunctive queries, where w is a value in [0,1]. In both cases, $w = 0$ corresponds to the non-impact of a query, and $w = 1$ to the full impact [10].

Let **A** be a (fuzzy) term with weight w in the unit interval and **t** be a tuple. The membership degree of **t**'s attribute a to **A** is defined as [10]:

$$\mu_{A_w}(t.a) = \min(w, \mu_A(t.a)) . \quad (3)$$

$$\mu_{A^w}(t.a) = \max(1-w, \mu_A(t.a)) . \quad (4)$$

3.1 Importance in Simple Queries

SELECT A FROM R
WHERE $A_1 \theta_1 a_1$ AND/OR…AND/OR $A_k \theta_k a_k$ (Q10)
WEIGHT A_1 is w_1; … A_k is w_k WITH α

A_i is an attribute in relation R, w_i is the weight of A_i, θ_i is a fuzzy comparison operator. "$A_i \theta_i a_i$" may be replaced with "$A_i = F_i$", where F_i is a fuzzy predicate.

As for the weighted conjunctive queries, matching degree is evaluated by:

$$MD(t) = MIN(\max(1-w_1, \mu_1), \ldots, \max(1-w_k, \mu_k)) . \quad (5)$$

The weighted disjunctive queries:

$$MD(t) = MAX(\min(w_1, \mu_1), \ldots, \min(w_k, \mu_k)) . \quad (6)$$

Table 1. Relation Person and membership degrees of fuzzy predicates 'young' and 'tall'

ID	Name	Age	Young	Height	Tall	MD
R1	Jam	20	1	173	0.257	0.257
R2	John	31	0.410	165	0.138	0.138
R3	Amy	22	1	185	0.8	0.8
R4	Tom	40	0.1	175	0.308	0.308
R5	Jack	25	1	183	0.671	0.671
R6	Henry	28	0.735	177	0.372	0.372
R7	Tommy	39	0.113	170	0.2	0.2
R8	Sandy	33	0.281	181	0.552	0.552
R9	Ann	26	0.961	172	0.236	0.236
R10	Bob	31	0.410	183	0.671	0.671

For example:

SELECT * FROM Person WHERE age = young AND height = tall
WEIGHT age is 0.4; height is 0.8 WITH 0.5 (Q11)

Table 1 gives the relation Person, in which the column named "young" (resp "tall") is the membership degree of age (resp height) and the last column is the eventually matching degree computed by formula (5). According to table1, the matching degrees of R3, R5, R8, R10 are greater than or equal to 0.5. So the result of (Q11) is as follows:

Table 2. The result of (Q11)

ID	Name	Age	Height	MD
R3	Amy	22	185	0.8
R5	Jack	25	183	0.671
R10	Bob	31	183	0.6
R8	Sandy	33	181	0.552

If there is no weight in (Q11), we can deduce that the tuples satisfying the query are R3, R5. In (Q11), "height=tall" is assigned a greater weight than "age=young", so the impact of height to the query is greater than that of age. R8 and R10 satisfy (Q11), although they do not if there is no weight.

3.2 Derivation Principle of Weighted Conjunctive Queries

According to (5) iff $MD(t) \geq \alpha$, the tuple t satisfies (Q10). Iff $\forall i, \max(1-w_i, \mu_i) \geq \alpha$, namely $\forall i\ 1-w_i \geq \alpha$ or $\mu_i \geq \alpha$ then $MD(t) \geq \alpha$. The weights are assigned by users, so w_i is known before matching degree is computed.

- If $1-w_i \geq \alpha$, then $\max(1-w_i, \mu_i) \geq \alpha$ without reference to μ_i.
- When $1-w_i < \alpha$, only if $\mu_i \geq \alpha$ we have $\max(1-w_i, \mu_i) \geq \alpha$. If a tuple satisfies (Q10), its attribute A_i's membership degree μ_i is greater or equal to α. So A_i must be in the α-cut ($[a_i^{(\alpha)}, b_i^{(\alpha)}]$) of related fuzzy predicate.

Let us consider (Q11) again, the weight of age is 0.4 and 1-0.4 is greater than 0.5, so we don't have to care for the membership degree of "young". The weight of height is 0.8 and 1-0.8 is lower than 0.5, so we compute the α-cut of fuzzy predicate "tall" (the membership function of "tall" see [1]). (Q11) can be transformed to:

SELECT * FROM Person WHERE height >=180
WEIGHT age is 0.4; height is 0.8   WITH 0.5

R3, R5, R8 and R10 satisfy the condition "height >=180". So we just need to compute these four tuples' the membership degrees to fuzzy predicate "tall". We evaluate their matching degrees according to formula (5).

3.3 Derivation Principle of Weighted Disjunctive Queries

If $\exists i,\ \min(w_i, \mu_i) \geq \alpha$, namely $\exists i, \mu_i \geq \alpha$ and $w_i \geq \alpha$, then $MAX(\min(w_1, \mu_1), \ldots, \min(w_k, \mu_k)) \geq \alpha$, where w_i is assigned by user.

If $w_i < \alpha$, then $\min(w_i, \mu_i) < \alpha$. μ_i has no effect on the eventually result, so the fuzzy term related with μ_i in query can be neglected.

3.4 Importance in Weighted Multi-table Queries

Let us consider a database composed of the relations [12]:

Emp (#emp, name, #dep, age, salary)
Dep (#dep, budget, size, city)

The query: 'find the employee with middling salary who works in the department at SHANGHAI with a high budget' can be expressed as (Q12). There are three conditions in this query: 'working in SHANGHAI'; 'salary is middling'; 'the department has a high budget'. Let us give them different weights:

SELECT #emp, name FROM Emp, Dep

WHERE city="SHANGHAI" AND salary=middling AND budget =high   (Q12)
    AND Emp.#dep=Dep.#dep

WEIGHT city is 0.7; salary is 0.9; budget is 0.4   WITH 0.5

#dep is the prime key of Dep and foreign key of Emp. Emp and Dep is joined by "Emp.#dep=Dep.#dep". The join is crisp, so the weight of "Emp.#dep=Dep.#dep" is 1(it is default). We get the relation REmp by computing the α-cut of fuzzy set "middling salary" and relation RDep by computing the α-cut of fuzzy set 'high budget'. Then we join REmp and RDep by "Emp.#dep=Dep.#dep"

SELECT #emp, name FROM Emp, Dep

WHERE city="SHANGHAI" AND (salary>=40000 AND salary<=70000)   (Q13)
    AND budget >=500000 AND Emp.dep=Dep.dep

WEIGHT city is 0.7 , salary is 0.9 ,budget is 0.4   WITH 0.5

According to the derivation principles in weighted conjunctive queries, 1-0.4 is greater than 0.5, so we can ignore the condition" budget >=500000" and just consider the other two conditions. The tuples satisfy these conditions are in table 3. Then we evaluate the matching degrees of tuples in table3 and order them by descending sequence on matching degree. It is not necessary to establish join on prime key and foreign key. Any attribute from two relations may appear in the join expression if they have comparable data type. Join operators may be crisp (=, <, >,etc.) or fuzzy ($\approx, \ll$,etc).The general form is:

SELECT R.A,S.B FROM R,S

WHERE fc1(R) AND fc2(S) AND R.CθS.D   (Q14)

θ is a (fuzzy) comparison operator .If the weights of three conditions in WHERE clause are w_1, w_2, w_3, we have

$$MD(t) = \min(\max(1-w_1, \mu_{fc1}(x)), \max(1-w_2 \mu_{fc2}(y)), \max(1-w_3, \mu_\theta(x.C, y.D))) \quad (7)$$

Table 3. Tuples satisfying 'city="SHANGHAI" AND (salary>=40000 AND salary<=70000)'

#emp	#dep	age	salary
S5	D4	28	50000
S6	D1	29	65000
S18	D1	45	65000
S19	D4	47	64000

Table 4. The result of (Q12)

#emp	#dep	age	salary	MD
S5	D4	28	50000	1
S19	D4	47	64000	0.8
S6	D1	29	65000	0.6
S18	D1	45	65000	0.6

If θ is a fuzzy join operator, even we compute the α-cut of $\mu_\theta(x.C, y.D)$, (Q14) cannot be transformed to standard SQL query, because C and D are attributes on R and S respectively. So we defuzzy fc1(R) (resp.fc2(S)) with its α-cut, then find the tuples satisfying fc1 (resp.fc2) and get two relations Res_fc1 and Res_fc2. R.CθS.D is used to join Res_fc1 and Res_fc2. At last we have eventually matching degree according formula (7).

The query 'find the employee nd the department, his salary is about 1/100 of the budget of the department' can by expressed by:

SELECT Emp.#emp, Dep.dep FROM Emp, Dep   (Q15)

WHERE age=young AND salary ≈ budget/100

WEIGHT age is 0.7; salary is 0.8 WITH 0.5

In this query, "≈" is the join operator. Let $\mu_\approx(x, y) = \dfrac{\min(x, y)}{\max(x, y)}$ [12].

#emp	#dep	age	salary
S1	D1	20	23000
S4	D3	27	41000
S5	D4	28	50000
S6	D1	29	65000
S13	D2	24	35000
S14	D1	22	28000
S15	D2	26	40000
S20	D3	30	56000

salary ≈ budget/100

#dep	budget	size	city
D1	3500000	12	Shanghai
D2	6000000	30	Nanjing
D3	4000000	32	Beijing
D4	8000000	42	Shanghai
D5	5000000	28	Guangzhou

Fig. 1. Fuzzy join between Emp and Dep

Firstly, we defuzzy the condition "age=young".

SELECT #emp FROM Emp WHERE age<=30   (Q16)

The result of (Q16) is named Res_Emp. Secondly we join Res_Emp and Dep. (See Figure1). Thirdly, we compute matching degrees according formula (7) (see Table 5). Lastly, the tuples whose matching degrees are below threshold are discarded. If θ is a fuzzy join operator, we cannot use α-cut to defuzzy join expression. So the efficiency of query is low if there are numerous tuples.

Table 5. The result of (Q15)

Emp.#emp	Dep.#dep	$\mu_{\approx}$	Young	MD
S1	D1	0.66	1	0.66
S1	D2	0.38	1	0.38
S1	D3	0.58	1	0.58
S1	D4	0.29	1	0.29
S1	D5	0.46	1	0.46
S4	D1	0.85	0.8	0.8
...	...	...	...	
S4	D5	0.82	0.8	0.8
...	...	...	...	
S20	D1	0.625	0.63	0.625
...	...	...	...	
S20	D5	0.89	0.5	0.5

3.5 Importance in Set-Oriented Queries

If we introduce weight in Set-oriented queries, the general form is:

(SELECT A1,...,An FROM R WHERE $fc_{11}, fc_{12},...$

WEIGHT fc_{11} is w_{11}, fc_{12} is w_{12},..... WITH α)

UNION/INTERSECT/EXCEPT   (Q17)

(SELECT A1,...,An FROM S WHERE $fc_{21}, fc_{22},...$

WEIGHT fc_{21} is w_{21}, fc_{22} is w_{22} .... WITH α)

The evaluation in sub-queries is the same as before. When operator "UNION" is used, let us assume the result of former sub-query is Rf and matching degree of a tuple in Rf is μ_{rf_i}, as the same we get μ_{sf_i}. The eventual result is $Rf \cup Sf$, so the matching degree in the result is computed by $\max(\mu_{rf_i}, \mu_{sf_i})$.

The query 'find a department in which the old employee has high salary, or the department has middling size and middling budget':

(SELECT dep FROM Emp WHERE age="old" AND salary="high"
WEIGHT age is 0.6 ; salary is 0.9 WITH 0.5) UNION
(SELECT dep FROM Dmp WHERE budget="high" AND size="middle"
WITHGT budget is 0.5; size is 0.8 WITH 0.5) (Q18)

We may get: Rf-D3 (0.8), Sf-D3(1.0), D5(0.5), Rf ∪ Sf -D3(1.0), D5(0.5).

Queries with INTERSECT/EXCEPT can be also expressed by nested queries with operators EXISTS /NOT EXISTS.

3.6 Importance in Nested Queries

The nested queries with operators IN, ANY, ALL and set-oriented queries with operators INTERSECT, EXCEPT can be stated by means of EXISTS. So we discuss the weighted sub-queries with operators EXISTS.

SELECT * FROM R WHERE <u>fc1</u> AND EXISTS
 (SELECT * FROM S WHERE <u>fc2</u> AND <u>R.A θ S.B</u>) (Q19)

The underlined conditions have weight, the computation of (Q19) is :

$$MD(t) = \min(\max(1-w_1, \mu_{fc1}(x)), \max(1-w_2\mu_{fc2}(y)), \max(1-w_3, \mu_\theta(x.A, y.B))) \quad (8)$$

For example 'find a young employee who work in a department with high budget'

SELECT * FROM Emp
WHERE age='young' AND EXISTS (SELECT * FROM Dep WHERE budget="high" AND Emp.dep=Dep.dep) (Q20)
WEIGHT age is 0.5; budget is 0.8 WITH 0.5

The weight of age is 0.5, so need not to consider age in the WHERE clause.

SELECT #emp FROM Emp
WHERE EXISTS (SELECT * FROM Dep WHERE budget>=5000000 AND Emp.dep=Dep.dep) (Q21)
WEIGHT age is 0.5; budget is 0.8 WITH 0.5

4 Weight, Threshold and Matching Degree

According to Section 3.2, if $\forall i, 1-w_i \geq \alpha$ in weighted conjunction, then all tuples in the relation R will satisfy Q(10). $MD(t)$ will be greater than or equal α no matter what is the membership degrees of t. Let us adjust the weight and threshold in (Q12):

SELECT * FROM Person WHERE age = 'young' AND height = 'tall' (Q22)
WEIGHT age is 0.4; height is 0.3 WITH 0.5

$$MIN(\max(1-w_1, \mu_{young}), \max(1-w_2, \mu_{high\_salary})) = MIN(\max(1-0.4, \mu_{young}),$$
$$\max(1-0.3, \mu_{high\_salary})) = MIN(\max(0.6, \mu_{young}), \max(0.7, \mu_{high\_salary})). \quad (9)$$

Even values of μ_{young} and μ_{tall} are unknown, it is confirmed that formula (9) is greater than 0.5. So all the tuples on relation person satisfy (Q22).

In formula (6), $MD(t)$ will be no less than $1-\max_i w_i$. So if $\alpha < 1-\max_i w_i$, all tuples in relation will satisfy the query. We can increase the weight or decrease the threshold until $\exists i \ \alpha \geq 1-\max_i w_i$ to avoid this problem. In disjunctive queries, $MD(t)$ will not be greater than $\max_i w_i$. If $\alpha > \max_i w_i$, there is no tuples satisfy the query. We can increase threshold or weight to avoid null set. e.g.

SELECT #emp FROM Emp WHERE age ='young' AND salary ='high' (Q10)
WEIGHT age is 0.6; budget is 0.8 WITH 0.9

$$MAX(\min(w_1, \mu_{young}), \min(w_2, \mu_{tall})) = MAX(\min(0.6, \mu_{young}), \min(0.7, \mu_{tall})) \leq 0.7 < 0.9. \quad (10)$$

5 Conclusion

A new method is proposed to weigh fuzzy queries in this paper. We extend SQL to deal with kinds of weighted fuzzy queries, not only simple queries but also mulit-table queries, sub-queries and set-oriented queries, etc. which makes the relational databases queries more flexible and intelligent. In the future we will study other kind of weighted queries, such as aggregation queries and other more efficient methods to defuzzy fuzzy querying conditions.

References

1. Zhou Hong, Xu Xiao-Ling, Wang Le-yu: The design of database query tool based on the fuzzy algorithm. Application research of computers, Vol. 5. (2001) 15-17
2. Zhu Rong: The research of query technique based on the fuzzy theory. Application research of computers, Vol. 5. (2003) 8-29
3. Huang Bo: The representation and implementation of fuzzy query based on extended SQL. Journal of Wuhan Technical university of Surveying and Mapping, Vol. 21.(1996) 86-89
4. Shyi-Ming Chen, Yun-Shyang Lin: A new method for fuzzy query processing in relational database system. Cybernetics and Systems, Vol. 33. (2002) 447-482
5. Yoshikane Takahashi: A fuzzy query language for relational database. IEEE transaction on systems, man, and cybernetics, Vol. 21. (1991)1576-1579
6. Shyi-Ming Chen, Woei-Tzy Jong: Fuzzy query translation for relational database. IEEE transaction on systems, man, and cybernetics-part B: Cybernetics, Vol. 27. (1997) 714-721
7. L.A.Zadeh: A Fuzzy-Set-Theoretic Interpretation of Linguistic Hedges. Journal of Cybernetics, Vol. 2. (1972) 4-34

8. Paul, B.Kantor: The Logic of Weighted Queries. IEEE transaction on systems, man, and cybernetics, Vol. SMC-11. (1981) 816-821
9. Janusz Kacprzyk, Slawomir Zadrozny, Andrzej Ziolkowski: Fquery III +:A "Human_Consistent" database querying system based on fuzzy logic with linguistic quantifiers. Information Sciences, Vol. 14. (1989) 443-453
10. Elie Sanchez: Importance in Knowledge Systems. Information Systems, Vol. 14.(1989) 455-464
11. L.A.Zadeh: The concept of a linguistic variable and its application to approximate reasoning, Part I II.Information Science, Vol. 8. (1975) 199-249,301-357
12. Patrick Bosc, Olivier Pivert: SQLf: A relational Database Language for Fuzzy Querying. IEEE Transactions on Fuzzy Systems, Vol. 3. (1995) 1-17
13. Patrick Bosc: On the permittivity of the division of fuzzy relations. Soft Computing, Vol. 2.(1998)35-47
14. Patrick Bosc, Cédric Legrand, Olivier Pivert: About Fuzzy Query Processing——The Example of the Division. 1999 IEEE International Fuzzy Systems Conference Proceedings II, (1999) 592-597

Appendix: Relation Emp and Dep, and Membership Functions May Be Used

Table 6. Relation Dep

#dep	budget	size	city
D1	3500000	12	Shanghai
D2	6000000	30	Nanjing
D3	4000000	32	Beijing
D4	8000000	42	Shanghai
D5	5000000	28	Guangzhou

Table 7. Relation Emp

#emp	#dep	age	salary	#emp	#dep	age	salary
S1	D1	20	23000	S12	D1	38	80000
S2	D3	31	36000	S13	D2	24	35000
S3	D5	34	53000	S14	D1	22	28000
S4	D3	27	41000	S15	D2	26	40000
S5	D4	28	50000	S16	D3	35	75000
S6	D1	29	65000	S17	D3	39	82000
S7	D5	41	68000	S18	D1	45	65000
S8	D2	44	57000	S19	D4	47	64000
S9	D3	53	70000	S20	D3	30	56000
S10	D5	50	69000	...	...	...	...
S11	D4	36	71000	...	...	...	...

$$\mu_{high\_budget}(x) = \begin{cases} 1 & x \geq 6000000 \\ \dfrac{x - 4000000}{2000000} & 4000000 < x < 6000000 \end{cases}$$

$$\mu_{young}(x) = \begin{cases} 1 & x \leq 25 \\ \dfrac{35 - x}{10} & 25 < x < 35 \end{cases}$$

$$\mu_{high\_sal}(x) = \begin{cases} 1 & x \geq 80000 \\ \dfrac{x - 60000}{20000} & 60000 < x < 80000 \end{cases}$$

$$\mu_{old}(x) = \begin{cases} 1 & x \geq 55 \\ \dfrac{x - 45}{10} & 45 < x < 55 \end{cases}$$

$$\mu_{mid\_sal}(x) = \begin{cases} \dfrac{x - 30000}{20000} & 30000 < x < 50000 \\ 1 & 50000 \leq x \leq 60000 \\ \dfrac{80000 - x}{20000} & 60000 < x < 80000 \end{cases}$$

$$\mu_{middle\_size}(x) = \begin{cases} \dfrac{x - 15}{15} & 15 < x < 30 \\ 1 & 30 \leq x \leq 40 \\ \dfrac{55 - x}{15} & 40 < x < 55 \end{cases}$$

Study of Multiuser Detection: The Support Vector Machine Approach

Tao Yang and Bo Hu

Department of Electronics Engineering, Fudan University, Shanghai, China
taoyang@fudan.edu.cn

Abstract. In this paper, a support vector machine (SVM) multi-user receiver based on competition learning (CL) strategy is proposed. The new algorithm adopts a heuristic approach to extend standard SVM algorithm for multiuser classification problem, and also a clustering analysis is applied to reduce the total amount of computation. In implementation of multi-user receiver, an asymptotical iterative algorithm is used to guide the learning of the input sample pattern. The digital result shows that the new multi-user detector scheme has a relatively good performance comparing with the conventional MMSE detector especially under the heavy interference environment.

1 Introduction

Support vector machine (SVM) was originally designed for binary classification problem, training a SVM amounts to solving a quadratic programming (QP) problem [1] [2]. How to effectively extend it for multiclass classification is still an ongoing research issue. Currently there are two types of approaches for multiclass SVM. One is by constructing and combining several binary classifiers while the other is by directly considering all data in one optimization formulation. The performance comparison between the two methods shows that the former has excellent performance both in training time and classification accuracy rate [3]. Still, the SVM solution need to complete K(K-1)/2 (K is the total classes number) quadratic programming to obtain optimal classification, which pose a great computation burden for a real problem, so how to construct a reduced training model, so as to simplify the sample training process is the focus of application.

Code-division multiple-access (CDMA) technology constitutes an attractive multiuser scheme that allows users to transmit at the same carrier frequency in an uncoordinated manner. However, this creates multiuser interference (MUI) which, if not controlled, can seriously degrade the quality of reception. Mutually orthogonal spreading codes for different users can provide an inherent immunity to MUI in the case of synchronous systems. Unfortunately, multipath distortions are often encountered in CDMA system and will reduce this inherent immunity to MUI. Actually most of multiuser detection (MUD) schemes involve in matrix decomposition or inversion, which is computation expensively and are not suitable for on-line implementation [4]. In this paper, the MUD based on SVM approach at base station is proposed, in this scheme, a cluster splitting and competition learning approach is used to construct the training and test model, which open a new perspective to MUD. The reminder of this paper is organized as follows. Section II presents the multi-class SVM training model based on clustering analysis. Section III gives the formulation of MUD on SVM. Competition learning (CL) scheme description is given in section IV. Section V gives the simulation and analysis. Finally section VI concludes this paper.

2 The Training Model of Multi-class SVM

Consider total K classes in training set, namely $C_1, C_2, ... C_K$, treat each class of K as a cluster and computer the center of these clusters as c_i, $i = 1, 2, ... K$. For any two classes C_i and C_j ($i \neq j$), implement the SVM training in two inverse stages [5]. In first stage, find a way to narrow the scope of training data by splitting the current cluster into sub-cluster, till the minimal training set is found. In second stage, starting the SVM training from inner layer to outer layer till margin error is entirely avoided. A detailed implementation description is given in the following part.

Stage I: Cluster splitting

Initialization:

Select class C_i and C_j (namely cluster c^i and c^j) as the one we will implement SVM training, c_i and c_j represent the cluster center of c^i and c^j, choose the pair (x^+, x^-) that is closest/next closest [1] to each other, set the origin point as $c_0 = (x^+ + x^-)/2$, based on which a normal vector $c_0^\perp$ ($c_0^\perp \perp x^+ x^-$) is obtained, then pass through c_i and c_j respectively to form two hyperplanes h_{c_i} and h_{c_j}, which satisfy $h_{c_i} \parallel h_{c_j} \parallel c_0^\perp$, this two hyperplanes separate cluster c^i and c^j into two sub-clusters respectively (see Fig.1), initialize splitting counter $s = 0$, $k = 0$.

Splitting implementation:

Repeat

    Step 1: calculate the centre of new cluster c^{ik} and c^{jk}, which lie inside of hyperplane $h_{c_{ik}}$ and $h_{c_{jk}}$, take the new centers as $c_{i(k+1)}$ and $c_{j(k+1)}$

    Step 2: if $c_{i(k+1)} = x^+$ or $c_{j(k+1)} = x^-$, then go to end

    Step 3: split current cluster into two subclusters through hyperplane $h_{c_{i(k+1)}}$ and $h_{c_{j(k+1)}}$, $h_{c_{i(k+1)}} \parallel h_{c_{j(k+1)}} \parallel c_0^\perp$, $k = k+1$, $s = s+1$

    Step 4: go to step 1

End

---

[1] Once this pair doesn't give the optimal hyperplane to SVM, then re-choose the pair from neighbor training data.

Stage II: SVM training

Initialization:

Take the split cluster sequence as $\{c^s, c^{s-1}, ..., c^0\}$, where $c^s = c^{is} \bigcup c^{js}$ and the rest may be deduced by analogy, initial training value $t = s$, $f(\cdot)$ is the hyperplane separating function

Step 1: train the SVM on c^t, $c^t = \bigcup_{k=s}^{t} c^k$

Step 2: if $y_k f(x_k) \geq Cf(x^+) \quad \forall x_k \in C_i \bigcup C_j$

then train the SVM on c^i, $c^i = c^t - c^{it}$

if $y_k f(x_k) \geq Cf(x^+)$

then set the training data in c^i as SVs and go to end

else train the SVM on c^j, $c^j = c^t - c^{jt}$

if $y_k f(x_k) \geq Cf(x^+)$

then set the training data in c^j as SVs and go to end.

else set the training data in c^t as SVs and go to end

Step 3: $t = t - 1$, go to step 1

End

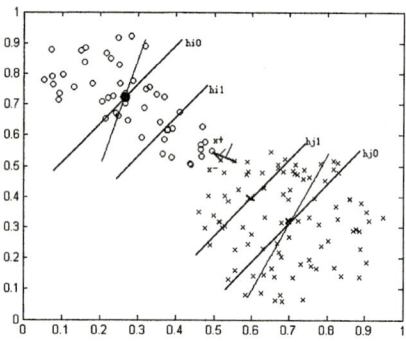

Fig. 1. Illustration of clustering splitting

3 Formulation of MUD on SVM

In our training model, a target vector (TV) and an online learning vector—asymptotic property vector (APV) is introduced to guide the learning of the heavy-interfered user [6]. With the "help" of these two vectors, each user can randomly locate a TV as the one it will shift to, an online updating algorithm through APV will ensure the convergence of user vector to the given TV.

Consider K users in system, denoted as $u_1, u_2, ..., u_K$. Let $\vec{V_i}$ represent the TV (modulated signal by an orthogonal signature code sequence without MAI) and $\vec{A_i}$ denote the APV for user u_i, $n_{\vec{V_i}}$ is an initialization factor used to decide to what extent a user belongs to a given TV. Once a user is classified by the SVCs and the discriminant result indicate that this user belongs to $\vec{V_i}$, then $n_{\vec{V_i}}$ increased with one, else $n_{\vec{V_i}}$ keep unchanged. When a user is classified by all SVCs pair, the index of $n_{\vec{V_i}}$ give the weighting factor of a user winning a $\vec{V_i}$. Consider a worst situation, some users have the same weighting factor in winning a $\vec{V_i}$, then $\vec{V_i}$ is assigned to all of them[2]. Lastly let $|\vec{\mu v}|$ denote the Euclidean distance from $\vec{\mu}$ to $\vec{v}$, i.e., $\|\vec{\mu}-\vec{v}\|$, where $\|.\|$ is the Euclidean norm. Consider the user has the following expression [7]

$$u = s + x \qquad (1)$$

where s is signature waveform of the interested user and is a constant item, x is an interference item. According to de-correlation principle, the de-correlation output with respect to $\vec{V_i}$ and $\vec{V_j}$ is

$$\begin{aligned} d_1 &= u\vec{V_i} = s\vec{V_i} + x\vec{V_i} \\ d_2 &= u\vec{V_j} = s\vec{V_j} + x\vec{V_j} \end{aligned} \qquad (2)$$

obviously, if $s = \vec{V_i}$, then d_1 is the interested de-correlation output. When we acquired a minimization of $x\vec{V_i}$, the optimal detection result is obtained. d_1 is not adopted due to approximately orthogonality between s and $\vec{V_j}$. It's known from distribution of input feature space that the location deviation between interested vector (namely $\vec{V_i}$) and initial point of u is an indirect reflection of interference extent. when iteration algorithm begin from $u(0)$ to $u(k)$ after k iterations, $u(k)$ shift to $\vec{V_i}$ or $\vec{V_j}$ (this paper assume $\vec{V_i}$ is the interested user), then $u(k) \xrightarrow{app} \vec{V_i}$, $x\vec{V_i} \xrightarrow{app} 0$, which is showed in Euclidean distance as $\min x\vec{V_i} \triangleq \max |x\vec{V_i}|$, so $|x\vec{V_i}|$ show a monotonic increase property. Yet for $|x\vec{V_j}|$, assume $u(0)$ shift to $\vec{V_j}$ after k iterations, then $u(k) = \vec{V_i} + x(k) = \vec{V_j}$, from which we get $x(k) = \vec{V_j} - \vec{V_i} \neq 0$, obviously $|x(k)\vec{V_j}| < |x(k)\vec{V_i}|$, this conclusion shows that the result of iteration learning make $u(0)$ shift to $\vec{V_i}$, not to $\vec{V_j}$. Reconsider the iteration direction at the starting point, we take it in two cases as $s = \vec{V_i}$ and $s = \vec{V_j}$, the corresponding initial value for $x(0)$ is $x_i(0) = u(0) - \vec{V_i}$ and $x_j(0) = u(0) - \vec{V_j}$, it's clear $x_i(0)$ and $x_j(0)$ is not a constant under different test pattern, as is just the same for $|x_i(0)\vec{V_i}|$ and $|x_j(0)\vec{V_j}|$, so the iteration process may be represented as an asymptotic process with monotonic or zigzag trajectory.

---

[2] Such a case is analyzed in CL scheme in next part.

In this section we will detail the updating of $\vec{A}$ and u to ensure the algorithm implementation. Assume current iteration move to nth step, iteration direction point to $\vec{V_i}$, we have

$$\overrightarrow{A(n+1)} = \overrightarrow{A(n)} + \delta_n(\overrightarrow{A(n)}, \vec{V_i}, \vec{V_j}) \Theta(\overrightarrow{A(n)}, \vec{V_i}, \vec{V_j}) \qquad (3)$$

where $\delta_n(\cdot)$ is a step factor. $\delta_n(\overrightarrow{A(n)}, \vec{V_i}, \vec{V_j}) = \begin{cases} 1 & if \quad \min(|\overrightarrow{A(n)}\vec{V_i}|, |\overrightarrow{A(n)}\vec{V_j}|) > 1 \\ \min(|\overrightarrow{A(n)}\vec{V_i}|, |\overrightarrow{A(n)}\vec{V_j}|) & otherwise \end{cases}$.

$\Theta(\cdot)$ is a direction vector, defined as

$$\Theta(\overrightarrow{A(n)}, \vec{V_i}, \vec{V_j}) = \begin{cases} \dfrac{\overrightarrow{A(n)}^\perp}{\|\overrightarrow{A(n)}^\perp\|} & \begin{cases} if \quad |x(n+1)\vec{V_i}| > |x(n+1)\vec{V_j}| \text{ and } \overrightarrow{A(n)} \to \vec{V_j} \\ if \quad |x(n+1)\vec{V_i}| < |x(n+1)\vec{V_j}| \text{ and } \overrightarrow{A(n)} \to \vec{V_i} \end{cases} \\ \dfrac{\overrightarrow{A(n)}}{\|\overrightarrow{A(n)}\|} & \begin{cases} if \quad |x(n+1)\vec{V_i}| > |x(n+1)\vec{V_j}| \text{ and } \overrightarrow{A(n)} \to \vec{V_i} \\ if \quad |x(n+1)\vec{V_i}| < |x(n+1)\vec{V_j}| \text{ and } \overrightarrow{A(n)} \to \vec{V_j} \end{cases} \end{cases}$$

here $\overrightarrow{A(n)}^\perp$ represent the nearer direction from current location to $\vec{V_i}$ and $\vec{V_j}$, the initial value $\overrightarrow{A(0)} = \dfrac{x(0)\vec{V_i}}{|x(0)\vec{V_i}|}$. Through such an updating scheme, $\vec{A}$ always point to a direction in which the interference decreases.[3] So we say $\vec{A}$ has the asymptotic property with respect to its associated input user vector.

The updating of item $u(n)$, due to a constant signature waveform s, is actually an updating of $x(n)$, we have

$$x(n+1) = x(n) + \overrightarrow{A(n)} \qquad (4)$$

because $x = u - s$, so $x(0) = u(0) - s$, from the analysis above we know that the initial choice of s doesn't affect the updating of $x(n)$, thereby we get $x(n) \xrightarrow[n \to k]{} 0$, where k is the total steps to complete the iteration.

4 Competition Learning Scheme Description

Assume N among K users need to implement an iteration learning process to attain the given $\vec{V}$, each user is assigned two $\vec{V}$ and the learning process is trigged at the same time, all users form a chain structure (See Fig.2). Assume $\vec{V_1}$ is the first one captured by

---

[3] In initial stage, $x(0)$ can be set to point to $\vec{V_i}$ or $\vec{V_j}$, and finally stay at the given $\vec{V}$ after iterations.

u_N, then remove part I from the chain, so the links between these parts vanish accordingly. The next most possible $\vec{V}$ will be captured is $\vec{V_N}$ by u_{N-1} (otherwise $\vec{V_N}$ will be "left", such a case actually existed but with little probability, we need to reassign $(\vec{V}, u)$ pair among the left unclassified users), so we turn around to the learning of u_{N-1} and u_1, once one of u_{N-1} and u_1 find its corresponding $\vec{V}$, u_{N-2} and u_2 follows, till the last u finished such a search process.

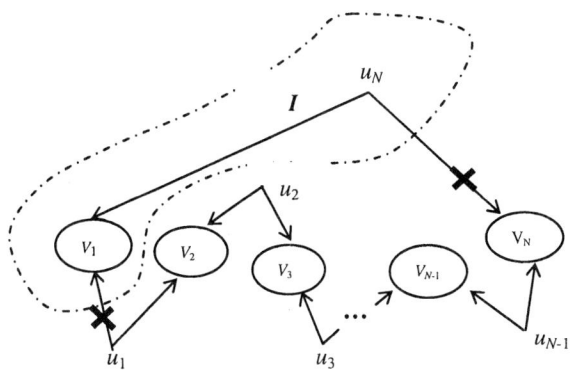

Fig. 2. Representation of CL in chain structure

5 Simulation and Analysis

5.1 Clustering Splitting Simulation

In this section we present experimental results on two relevant problems, the validity of SVCs based on clustering splitting and efficiency performance of MUD based on SVCs. In the first problem, our goal is to compare the difference of training size and training time under different training set. Consider there is four users, u_1, u_2, u_3 and u_4 need to be detected (that means four classes need to be classified), total training set $N = 100$, $\sum_i u_i = N$, u_i represent the training size for user i, the total number of SVCs is $k(k-1)/2$ ($k = 4$), that is $(u_1, u_2) \rightarrow SVC_1$, $(u_1, u_3) \rightarrow SVC_2, \ldots, (u_3, u_4) \rightarrow SVC_6$. For convenience of comparison, we use standard QP algorithm for reference, the training result is given in Table. 1

The data in Table1 shows that clustering splitting scheme has a great advantage in reduce the training size and training time, here we give a quantization expression as $E = \sum_i s_i \times t_i \times e_i$, where s_i, t_i and e_i represent training size, training time and classification error rate for user i, respectively. Which can be used to evaluate the classification performance. From E we know that overall classification efficiency is improved in clustering splitting scheme although the structure itself still needs to be investigated for the further step.

Table 1. A comparison between clustering splitting and standard QP algorithm

	Standard QP algorithm (u_1,u_2) (u_1,u_3) (u_1,u_4) (u_2,u_3) (u_2,u_4) (u_3,u_4)	Clustering splitting (u_1,u_2) (u_1,u_3) (u_1,u_4) (u_2,u_3) (u_2,u_4) (u_3,u_4)
Number of SVs	10 8 13 9 7 8	10 7 11 9 8 7
Training size	58 45 53 47 55 42	32 25 30 26 30 19
Training time (ms)	130 115 137 97 148 123	114 90 95 106 129 103
Classification Error rate	0.10 0.09 0.13 0.14 0.09 0.11	0.12 0.10 0.13 0.16 0.10 0.12
E	4102.8	2096.06

5.2 Experiment of MUD Based on SVC

This simulation example is used to investigate the proposed SVM MUD based on CL scheme and compares its performance with linear MMSE and optimal MUD. It is worth pointing out again that the linear MMSE MUD and the optimal MUD are designed based on the complete knowledge of the system while the SVM MUD is trained using a block of the noisy received signal samples. The detection process is assumed as follows: for any user when $\vec{V}$ is found, this detection is thought successful, else this detection is thought failed and need to be re-arranged. From previous analysis we know that when a certain $\vec{V}$ is "left" in iteration process, we can always re-detect this vector through a reassignment of $(u, \vec{V})$ pair, for sake of convenient comparison, we take each reassignment as a detection error (which is equivalently a bit information detection error), so we can describe detection performance approximately in bit error rate (BER), accordingly the signal component, a signal with an addition of interference item under multiuser environment, is denoted by the signal noise rate(SNR), thus the BER-SNR relation is obtained through previous CL scheme.

Here we first consider a four-user system with eight chips per bit. The orthogonal code sequence of the four users were (-1, -1, +1, +1, +1, +1, -1, -1), (-1, +1, -1, -1, +1, +1, -1, +1), (-1, -1, +1, -1, -1, +1, +1, +1) and (+1, -1, +1, -1, +1, -1, -1, +1), respectively, and the transfer function of the channel impulse response (CIR) is [8]

$$H(z) = 0.3 + 0.7z^{-1} + 0.3z^{-2} \tag{5}$$

the training set for each user is constructed as follows: each user has 50 training samples modulated by bit information +1 or −1 randomly and also a white noise sequence is added to the modulated signal, the received signal is just the source of training set for SVM, among which two training samples modulated by +1 and −1 without noise as a class reference, namely the target vector analyzed in previous section.

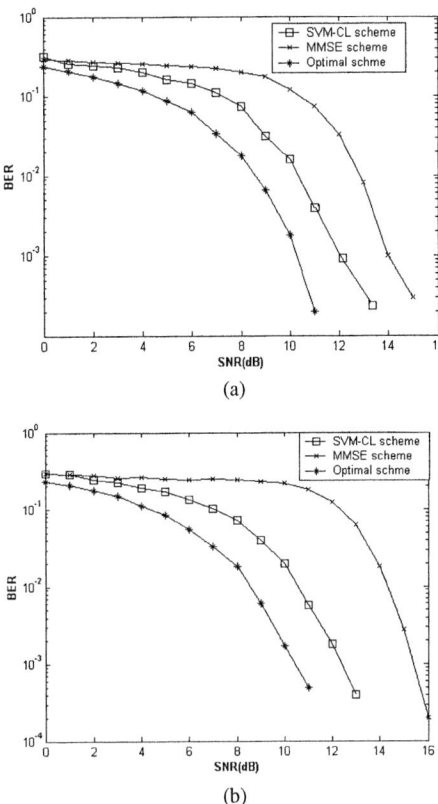

Fig. 3. The detection comparison between three schemes under two MAI environments: (a) light MAI and (b) heavy MAI

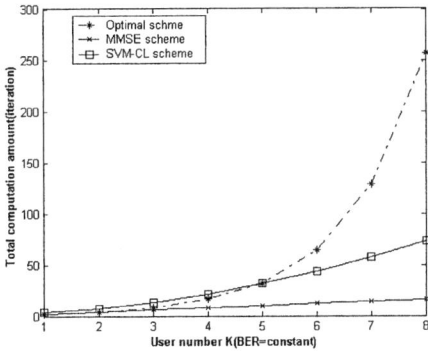

Fig. 4. Computation complexity *vs.* the number of users

Then the test data for online implementation of CL scheme is generated by modulating information bit +1 or −1 in an imperfect orthogonal code sequence – (-1, -1, +1, +1, +1, +1, -1, +1), (-1, +1, -1, +1, +1, +1, +1, +1), (-1, -1, +1, +1, -1, +1, +1, +1) and (+1, -1, +1, -1, +1, +1, -1, +1), the modulated signals is transmitted through the same channel, the received signal can be expressed as [9]

$$y = RAb + n \qquad (6)$$

here R is unitary cross-correlation matrix, A represent signal amplitude. y is the input data for SVCs. also assume these four users have equal signal power, that is, all users have an equal SNR. Fig. 3 gives the Monte Carlo simulation result of 10000 bit information sign in Matlab 6.1 simulation environment, which give performance comparisons between three different detection schemes. Fig.4 gives the comparison result of total computation amount between optimal, MMSE and CL detection scheme.

The simulation shows that under same SNR, different MAI cause a different BER, Fig.3 show us that when MAI increase, the detection performance in MMSE scheme become deteriorated, while SVM-CL scheme get close to optimal detection performance. Fig.4 indicate that the computation complexity of CL scheme lies between optimal and MMSE due to SVM implementation is just a $K(K-1)/2$ linear combination of SVC, as a result, it has a polynomial complexity with user number.

6 Conclusions

This paper propose a novel MUD scheme based on multi-class classification of SVM, this scheme include a problem solution in two aspects, an efficient SVC training model and an effective detection algorithm. The former is constructed through one-against-one mode, a clustering splitting algorithm is used instead of standard QP solution to reduce the training size and training time. The latter has the former as a base, a CL scheme is introduced to form an adaptive mechanism under a heavy MAI environment, the relevant analysis shows that an asymptotic iterative learning ensure each user to find its target vector, the digital result also indicate that in case of a heavy MAI environment, this detection algorithm outperform the conventional MMSE algorithm.

References

1. Vapnik, V. :Statistical Learning Theory. New York: Weiley (1998)
2. Wang, L.P. :Support Vector Machines: Theory and Application. Springer, Berlin Heidelberg New York (2005)
3. Hsu, C.W., Lin, C.J.: A Comparison of Methods for Multiclass Support Vector Machines. IEEE Trans. Neural Networks, Vol. 13 (2002) 415-425
4. Buzzi, S., Lops, M., Poor, H.V.: Code-aided Interference Suppression for DS/CDMA Overlay Systems. Proceedings of the IEEE, Vol. 90, No. 3 (2002) 394-435
5. Chen, J., Chen, C.: Reducing SVM Classification Time Using Multiple Mirror Classifiers. IEEE Trans. Systems, Man and Cybernetics, Part B, Vol.34, No.2 (2004) 1173-1183

6. Friedman, J.: Another Approach to Polychotomous Classification. Department of Statistics. Stanford University, Stanford, CA. (1996)
7. Mucchi, L., Morosi, S., Del Re, E. etc: A New Algorithm for Blind Adaptive Multiuser Detection in Frequency Selective Multipath Fading Channel. IEEE Trans. Wireless Commun., Vol.3, No. 1 (2004) 235-247
8. Chen, S., Hanzo, L.: Block-adaptive Kernel-based CDMA Multiuser Detection. IEEE International Conference on Communication. Vol.2 (2002) 682-686
9. Zhang, X., Bo, Z.: Communication Signal Processing. Press of National Industry of Defense, Beijing (2000)

Robust and Adaptive Backstepping Control for Nonlinear Systems Using Fuzzy Logic Systems

Gang Chen, Shuqing Wang, and Jianming Zhang

National Key Laboratory of Industrial Control Technology,
Institute of Advanced Process Control, Zhejiang University, Hangzhou, 310027, P.R. China
gchen@iipc.zju.edu.cn

Abstract. In this note, a robust adaptive tracking control problem is discussed for a class of affine nonlinear systems in the strict-feedback form with unknown nonlinearities. A unified and systematic procedure is developed to derive a novel robust adaptive fuzzy controller. Compared with most results reported in the literature, the proposed control algorithm has several advantages: 1)the controller singularity problem is avoided perfectly;2)the online computation burden is kept to minimum;3)exponential tracking to the reference trajectory up to an ultimately bounded error is achieved;4)the controllers are particularly suitable for parallel processing and hardware implementation.

1 Introduction

During the last decade, the adaptive control of nonlinear systems has undergone rapid development after the introduction of adaptive backstepping design methodology [1]. For a large class of state feedback linearizable nonlinear systems, the design scheme provides a systematic framework for the design of tracking and regulation strategies. One of the problems with backstepping approaches is that certain functions must be linear in the unknown parameters and some tedious analysis is needed to determine a regression matrix.

In recent years, the study of nonlinear control using universal function approximators has received much attention [2-9]. Usually, a neural network(NN) or a fuzzy logic system(FLS) is used to approximate the unknown functions of the systems .By combining the backstepping design scheme with NN or FLS, the linearity-in-the-parameter assumption of unknown function and the determination of regression matrices are removed. Although significant progress has been made, most results reported in the literature suffer from at least one of the following drawbacks.

1) Because the derivatives of the virtual controllers are included in NN or FLS, the structure of the NN or FLS is very complex and hard to realize. 2) Most controllers only guarantee the output tracking error asymptotically converges to a ball-type error residual set. The size of the residual set increases when the size of the uncertainties increases. Therefore, these controllers are not robust to unmodeled dynamics and external disturbances. 3) In order to avoid the controller singularity problem, the control gain functions are often assumed to be known. However, this assumption can

not be satisfied in many cases. The techniques based on projection and integral-type Lyapunov function can be employed to avoid controller singularity problem, but these methods are difficult to use in practice. 4) Some algorithms require the information of the compact set to which the optimal parameter vector of the universal approximator belongs. 5) Because a lot of parameters are to be updated online, the computation burden is very high.

In this note, by using the advantage of FLS, a novel algorithm is proposed to solve these problems. The new design method can easily incorporate the linguistic information into the controller design. The controllers are highly structural and particularly suitable for parallel processing and hardware implementation.

2 Problem Formulation

2.1 System Description

Consider the following nonlinear system

$$\dot{x}_i = f_i(\bar{x}_i) + g_i(\bar{x}_i)x_{i+1}, \quad 1 \leq i \leq n-1$$
$$\dot{x}_n = f_n(\bar{x}_n) + g_n(\bar{x}_n)u, \quad (1)$$
$$y = x_1,$$

where $\bar{x}_i = (x_1, \cdots, x_i)^T \in R^i, i = 1, \cdots, n$, are state variables, $u \in R$ and $y \in R$ are system input and output, respectively. The control objective in this note is to design a fuzzy control law for system (1) to ensure the plant output y to track desired trajectory y_r within an ultimately bounded error, while maintaining all the signals bounded. Also the ultimate error bounds can be made arbitrarily small by choosing appropriate controller parameters.

We make the following assumptions as commonly being done in the literature.

Assumption 1. The desired output trajectory y_r and its derivatives up to n th order are known and bounded.

Assumption 2. The signs of $g_i(\cdot)$ are known. There exist constants $g_{i1} \geq g_{i0} > 0$ such that $g_{i1} \geq |g_i(\cdot)| \geq g_{i0}, \forall \bar{x}_n \in \Omega \subset R^n$. Without losing generality, we shall assume $g_{i1} \geq g_i(\cdot) \geq g_{i0} > 0, \forall \bar{x}_n \in \Omega \subset R^n$.

Assumption 3. There exists constant $g_{id} > 0$ such that $|\dot{g}_i(\cdot)| \leq g_{id}, \forall \bar{x}_n \in \Omega \subset R^n$.

2.2 The FLS

The basic configuration of the FLS includes a fuzzy rule base, which consists of a collection of fuzzy IF-THEN rules in the following form:

$R^{(\ell)}$: IF x_1 is F_1^ℓ, and $\cdots$, and x_n is F_n^ℓ, THEN $y_\ell = w_\ell$, where F_i^ℓ are fuzzy sets, $x = (x_1, \cdots, x_n)^T \in X \subset R^n$ is an input linguistic vector, w_ℓ is the fuzzy singleton in the

ℓ th rule. Let M be the number of the fuzzy IF-THEN rules. The output of the FLS with center average defuzzifier, product inference, and singleton fuzzifier can be expressed as

$$y(x) = \sum_{\ell=1}^{M} v^{\ell} w_{\ell} \Big/ \sum_{\ell=1}^{M} v^{\ell}, \tag{2}$$

where $\mu_{F_i^{\ell}}(x_i)$ is the membership function value of the fuzzy variable x_i and $v^{\ell} = \prod_{i=1}^{n} \mu_{F_i^{\ell}}(x_i)$ is the true value of the ℓ th implication. Then, equation (2) can be rewritten as

$$y(x) = \theta^{T} \psi(x), \tag{3}$$

where $\theta^{T} = (w_1, \cdots, w_M)$ is an adjustable parameter vector and $\psi^{T}(x) = (\psi^1(x), \cdots \psi^M(x))$ is the fuzzy basic function vector defined as $\psi^{\ell}(x) = v^{\ell} \Big/ \sum_{\ell=1}^{M} v^{\ell}$.

In the rest of the note, $\theta \in R^M$ is referred to as the parameter vector of the FLS, and $\psi : x \to R^M$ the fuzzy basic functions. Based on the universal approximation theorem, the FLS (3) is universal approximator in the sense that given any real continuous function $f : R^n \to R$ in a compact set $X \subset R^n$, and any $k > 0$ there exists a FLS (3) such that $\sup_{x \in X} |y(x) - f(x)| < k$.

According to this result, the function $f(x)$ can be expressed as

$$f(x) = \theta^{*T} \psi(x) + \Delta f(x), \quad \forall \; x \in X \subset R^n \tag{4}$$

where the reconstruction error $\Delta f(x)$ satisfies $\sup_{x \in X} |\Delta f(x)| < k$ and θ^* is the optimal parameter vector $\theta^* = \arg \min_{\theta \in R^M} \left\{ \sup_{x \in X} |\theta^{T} \psi(x) - f(x)| \right\}$.

3 Controller Design

The design procedure consists of n steps. At each step i, the virtual controller α_i will be developed by employing an appropriate Lyapunov function V_i. For clarity, the step 1 is described in detail. The design procedure is given below.

Step 1. Define $z_1 = x_1 - y_r$, $z_2 = x_2 - \alpha_1$, where α_1 is the first virtual controller. Referring to the first equation in (1), we have

$$\dot{z}_1 = f_1(x_1) + g_1(x_1) x_2 - \dot{y}_r. \tag{5}$$

We treat x_2 as its virtual control input. Equation (5) can be rewritten as

$$\dot{z}_1 = g_1(x_1)\left(g_1^{-1}(x_1)f_1(x_1) + x_2 - g_1^{-1}(x_1)\dot{y}_r\right). \tag{6}$$

Since the functions $g_1^{-1}f_1$ and g_1^{-1} are unknown, we use two FLSs $\theta_1^T\varphi_1(x_1)$ and $\delta_1^T\psi_1(x_1)$ to approximate them, respectively. Based on a priori knowledge, the premise parts of the FLSs as well as the nominal vectors $\bar{\theta}_1$ and $\bar{\delta}_1$ are designed first and are fixed. Thus, there exist positive constants ρ_{11} and ρ_{12} such that $\|\theta_1^* - \bar{\theta}_1\| \leq \rho_{11}$, $\|\delta_1^* - \bar{\delta}_1\| \leq \rho_{12}$, where θ_1^* and δ_1^* are the optimal parameter vectors. Define $\tilde{\rho}_{11} = \rho_{11} - \hat{\rho}_{11}(t)$, $\tilde{\rho}_{12} = \rho_{12} - \hat{\rho}_{12}(t)$, where $\hat{\rho}_{11}$ and $\hat{\rho}_{12}$ denote the estimates of ρ_{11} and ρ_{12}, respectively. The FLS reconstruction error $d_1 = \left(g_1^{-1}f_1 - \theta_1^{*T}\varphi_1\right) - \left(g_1^{-1} - \delta_1^{*T}\psi_1\right)\dot{y}_r$ is bounded, i.e., there exists a constant $\varepsilon_{11} > 0$ such that $|d_1| < \varepsilon_{11}$. Throughout the note, we introduce $\theta_i^T\varphi_i(\cdot)$ and $\delta_i^T\psi_i(\cdot)$ as FLSs and define their reconstruction error as $d_i = \left(g_i^{-1}f_i - \theta_i^{*T}\varphi_i\right) - \left(g_i^{-1} - \delta_i^{*T}\psi_i\right)\dot{\alpha}_{i-1}$, where $i = 2, \cdots, n$, θ_i^* and δ_i^* denote the optimal vectors. We assume that there exists positive constant ε_{i1} such that $|d_i| < \varepsilon_{i1}$.

It follows from (6) that

$$\dot{z}_1 = g_1\left(\theta_1^{*T}\varphi_1 - \delta_1^{*T}\psi_1\dot{y}_r + z_2 + \alpha_1 + d_1\right). \tag{7}$$

We choose the first virtual controller as

$$\alpha_1 = -\bar{\theta}_1^T\varphi_1 + \bar{\delta}_1^T\psi_1\dot{y}_r - k_1 z_1 - \hat{\rho}_{11}\|\varphi_1\|\tanh\left(\frac{\|\varphi_1\|z_1}{\varepsilon_{12}}\right) - \hat{\rho}_{12}\|\psi_1\|\dot{y}_r \tanh\left(\frac{\|\psi_1\|\dot{y}_r z_1}{\varepsilon_{13}}\right) \tag{8}$$

where $k_1 > 0$, $\varepsilon_{12} > 0$, and $\varepsilon_{13} > 0$ are constants.

Consider the following Lyapunov function

$$V_1 = \frac{1}{2g_1(x_1)}z_1^2 + \frac{1}{2r_{11}}\tilde{\rho}_{11}^2 + \frac{1}{2r_{12}}\tilde{\rho}_{12}^2 \tag{9}$$

where $r_{11} > 0$ and $r_{12} > 0$ are constants.

From (7) and (8), the time derivative of V_1 is

$$\dot{V}_1 = \frac{z_1\dot{z}_1}{g_1(x_1)} - \frac{\dot{g}_1(x_1)}{2g_1^2(x_1)}z_1^2 - \frac{1}{r_{11}}\tilde{\rho}_{11}\dot{\hat{\rho}}_{11} - \frac{1}{r_{12}}\tilde{\rho}_{12}\dot{\hat{\rho}}_{12}$$

$$\leq \tilde{\rho}_{11}\left(\|\varphi_1\|z_1 \tanh\left(\frac{\|\varphi_1\|z_1}{\varepsilon_{12}}\right) - \frac{1}{r_{11}}\dot{\hat{\rho}}_{11}\right) + \tilde{\rho}_{12}\left(\|\psi_1\|\dot{y}_r z_1 \tanh\left(\frac{\|\psi_1\|\dot{y}_r z_1}{\varepsilon_{13}}\right) - \frac{1}{r_{12}}\dot{\hat{\rho}}_{12}\right) \tag{10}$$

$$-\left(k_1 + \frac{\dot{g}_1}{2g_1^2}\right)z_1^2 + z_1 z_2 + z_1 d_1 + \rho_{11}\varepsilon_{12} + \rho_{12}\varepsilon_{13}.$$

Choosing the following adaptive laws

$$\dot{\hat{\rho}}_{11} = r_{11}\left(\|\varphi_1\|z_1 \tanh\left(\frac{\|\varphi_1\|z_1}{\varepsilon_{12}}\right) - \sigma_{11}\hat{\rho}_{11}\right),$$

$$\dot{\hat{\rho}}_{12} = r_{12}\left(\|\psi_1\|\dot{y}_r z_1 \tanh\left(\frac{\|\psi_1\|\dot{y}_r z_1}{\varepsilon_{13}}\right) - \sigma_{12}\hat{\rho}_{12}\right) \quad (11)$$

where $\sigma_{11} > 0$ and $\sigma_{12} > 0$ are constants.

Let $k_1 = k_{11} + k_{12}$, where $k_{12} > 0$, k_{11} is chosen such that $k_{10} = \left(k_{11} - \frac{g_{1d}}{2g_{10}^2}\right) > 0$. By completing the squares, we have

$$\tilde{\rho}_{11}\hat{\rho}_{11} = \tilde{\rho}_{11}(\rho_{11} - \tilde{\rho}_{11}) \le \frac{\rho_{11}^2}{2} - \frac{\tilde{\rho}_{11}^2}{2},$$

$$\tilde{\rho}_{12}\hat{\rho}_{12} = \tilde{\rho}_{12}(\rho_{12} - \tilde{\rho}_{12}) \le \frac{\rho_{12}^2}{2} - \frac{\tilde{\rho}_{12}^2}{2}, \quad (12)$$

$$-k_{12}z_1^2 + z_1 d_1 \le -k_{12}z_1^2 + \varepsilon_{11}|z_1| \le \frac{\varepsilon_{11}^2}{4k_{12}}.$$

Substituting (11) into (10) and using (12), we have

$$\dot{V}_1 \le z_1 z_2 - k_{10}z_1^2 - \frac{\sigma_{11}}{2}\tilde{\rho}_{11}^2 - \frac{\sigma_{12}}{2}\tilde{\rho}_{12}^2 + \frac{\sigma_{11}}{2}\rho_{11}^2 + \frac{\sigma_{12}}{2}\rho_{12}^2 + \rho_{11}\varepsilon_{12} + \rho_{12}\varepsilon_{13} + \frac{\varepsilon_{11}^2}{4k_{12}}. \quad (13)$$

Step i $(2 \le i < n-1)$. Define $z_{i+1} = x_{i+1} - \alpha_i$. Differentiating $z_i = x_i - \alpha_{i-1}$ yields

$$\begin{aligned}\dot{z}_i &= g_i(\bar{x}_i)\left(g_i^{-1}(\bar{x}_i)f_i(\bar{x}_i) + x_{i+1} - g_i^{-1}(\bar{x}_i)\dot{\alpha}_{i-1}\right) \\ &= g_i(\bar{x}_i)\left(\theta_i^{*T}\varphi_i(\bar{x}_i) - \delta_i^{*T}\psi_i(\bar{x}_i)\dot{\alpha}_{i-1} + z_{i+1} + \alpha_i + d_i\right).\end{aligned} \quad (14)$$

The virtual controller α_i is designed as

$$\alpha_i = -\bar{\theta}_i^T\varphi_i + \bar{\delta}_i^T\psi_i\dot{\alpha}_{i-1} - k_i z_i - z_{i-1} - \hat{\rho}_{i1}\|\varphi_i\|\tanh\left(\frac{\|\varphi_i\|z_i}{\varepsilon_{i2}}\right)$$

$$- \hat{\rho}_{i2}\|\psi_i\|\dot{\alpha}_{i-1}\tanh\left(\frac{\|\psi_i\|\dot{\alpha}_{i-1}z_i}{\varepsilon_{i3}}\right). \quad (15)$$

Consider the Lyapunov function

$$V_i = V_{i-1} + \frac{1}{2g_i(\bar{x}_i)}z_i^2 + \frac{1}{2r_{i1}}\tilde{\rho}_{i1}^2 + \frac{1}{2r_{i2}}\tilde{\rho}_{i2}^2 \quad (16)$$

and the adaptive laws

$$\dot{\hat{p}}_{i1} = r_{i1}\left(\|\varphi_i\|z_i \tanh\left(\frac{\|\varphi_i\|z_i}{\varepsilon_{i2}}\right) - \sigma_{i1}\hat{p}_{i1}\right),$$

$$\dot{\hat{p}}_{i2} = r_{i2}\left(\|\psi_i\|\dot{\alpha}_{i-1}z_i \tanh\left(\frac{\|\psi_i\|\dot{\alpha}_{i-1}z_i}{\varepsilon_{i3}}\right) - \sigma_{i2}\hat{p}_{i2}\right)$$

(17)

where $r_{i1} > 0$, $r_{i2} > 0$, $\sigma_{i1} > 0$, and $\sigma_{i2} > 0$ are constants. Let $k_i = k_{i1} + k_{i2}$, where $k_{i2} > 0$, k_{i1} is chosen such that $k_{i0} = \left(k_{i1} - \frac{g_{id}}{2g_{i0}^2}\right) > 0$.

From (14)-(17), the time derivative of V_i gives

$$\dot{V}_i \leq z_i z_{i+1} - \sum_{\ell=1}^{i} k_{\ell 0} z_\ell^2 - \sum_{\ell=1}^{i}\left(\frac{\sigma_{\ell 1}}{2}\tilde{p}_{\ell 1}^2 + \frac{\sigma_{\ell 2}}{2}\tilde{p}_{\ell 2}^2\right) + \sum_{\ell=1}^{i}\left(\frac{\sigma_{\ell 1}}{2}p_{\ell 1}^2 + \frac{\sigma_{\ell 2}}{2}p_{\ell 2}^2\right)$$
$$+ \sum_{\ell=1}^{i}(p_{\ell 1}\varepsilon_{\ell 2} + p_{\ell 2}\varepsilon_{\ell 3}) + \sum_{\ell=1}^{i}\frac{\varepsilon_{\ell 1}^2}{4k_{\ell 2}}.$$

(18)

Step n. In the final step, we will get the actual controller. Differentiating $z_n = x_n - \alpha_{n-1}$ yields

$$\dot{z}_n = g_n(\bar{x}_n)\left(\theta_n^{*T}\varphi_n(\bar{x}_n) - \delta_n^{*T}\psi_n(\bar{x}_n)\dot{\alpha}_{n-1} + u + d_n\right).$$

(19)

The controller u is designed as

$$u = -\bar{\theta}_n^T \varphi_n + \bar{\delta}_n^T \psi_n \dot{\alpha}_{n-1} - k_n z_n - z_{n-1} - \hat{p}_{n1}\|\varphi_n\|\tanh\left(\frac{\|\varphi_n\|z_n}{\varepsilon_{n2}}\right)$$
$$- \hat{p}_{n2}\|\psi_n\|\dot{\alpha}_{n-1}\tanh\left(\frac{\|\psi_n\|\dot{\alpha}_{n-1}z_n}{\varepsilon_{n3}}\right).$$

(20)

Consider the Lyapunov function

$$V_n = V_{n-1} + \frac{1}{2g_n(\bar{x}_n)}z_n^2 + \frac{1}{2r_{n1}}\tilde{p}_{n1}^2 + \frac{1}{2r_{n2}}\tilde{p}_{n2}^2$$

(21)

and the parameter adaptation laws

$$\dot{\hat{p}}_{n1} = r_{n1}\left(\|\varphi_n\|z_n \tanh\left(\frac{\|\varphi_n\|z_n}{\varepsilon_{n2}}\right) - \sigma_{n1}\hat{p}_{n1}\right),$$

$$\dot{\hat{p}}_{n2} = r_{n2}\left(\|\psi_n\|\dot{\alpha}_{n-1}z_n \tanh\left(\frac{\|\psi_n\|\dot{\alpha}_{n-1}z_n}{\varepsilon_{n3}}\right) - \sigma_{n2}\hat{p}_{n2}\right)$$

(22)

where $r_{n1} > 0$, $r_{n2} > 0$, $\sigma_{n1} > 0$, and $\sigma_{n2} > 0$ are constants. Let $k_n = k_{n1} + k_{n2}$, where $k_{n2} > 0$, k_{n1} is chosen such that $k_{n0} = \left(k_{n1} - \dfrac{g_{nd}}{2g_{n0}^2}\right) > 0$. From (19)-(22), the time derivative of V_n gives

$$\dot{V}_n \leq -\sum_{\ell=1}^{n} k_{\ell 0} z_\ell^2 - \sum_{\ell=1}^{n}\left(\frac{\sigma_{\ell 1}}{2}\tilde{\rho}_{\ell 1}^2 + \frac{\sigma_{\ell 2}}{2}\tilde{\rho}_{\ell 2}^2\right) + \sum_{\ell=1}^{n}\left(\frac{\sigma_{\ell 1}}{2}\rho_{\ell 1}^2 + \frac{\sigma_{\ell 2}}{2}\rho_{\ell 2}^2\right) \\ + \sum_{\ell=1}^{n}(\rho_{\ell 1}\varepsilon_{\ell 2} + \rho_{\ell 2}\varepsilon_{\ell 3}) + \sum_{\ell=1}^{n}\frac{\varepsilon_{\ell 1}^2}{4k_{\ell 2}}. \tag{23}$$

4 Stability and Performance Analysis

Let

$$\varepsilon = \sum_{\ell=1}^{n}\left(\frac{\sigma_{\ell 1}}{2}\rho_{\ell 1}^2 + \frac{\sigma_{\ell 2}}{2}\rho_{\ell 2}^2\right) + \sum_{\ell=1}^{n}(\rho_{\ell 1}\varepsilon_{\ell 2} + \rho_{\ell 2}\varepsilon_{\ell 3}) + \sum_{\ell=1}^{n}\frac{\varepsilon_{\ell 1}^2}{4k_{\ell 2}}, \tag{24}$$

$$c = \min\{2g_{10}k_{10},\cdots,2g_{n0}k_{n0},\sigma_{11}r_{11},\cdots,\sigma_{n1}r_{n1},\sigma_{12}r_{12},\cdots,\sigma_{n2}r_{n2}\}.$$

It follows from (23) that

$$\dot{V}_n \leq -cV_n + \varepsilon, \ \forall \ t \geq 0. \tag{25}$$

Theorem 1. Consider the closed-loop system consisting of system (1) satisfying Assumptions 2-3, controller (20), the parameter updating laws (11), (17), (22). For bounded initial conditions, 1) all the signals in the closed-loop system are bounded; 2) the tracking error exponentially converges to an arbitrarily small neighborhood around zero by an appropriate choice of the design parameters.

Proof.
1) Let $k_{\min} = \min\{k_{10},\cdots,k_{n0}\}$, $\sigma_{\min 1} = \min\{\sigma_{11},\cdots,\sigma_{n1}\}$, $\sigma_{\min 2} = \{\sigma_{12},\cdots,\sigma_{n2}\}$, $z = (z_1,\cdots,z_n)^T$, $\tilde{\rho}_1 = (\tilde{\rho}_{11},\cdots,\tilde{\rho}_{n1})^T$, $\tilde{\rho}_2 = (\tilde{\rho}_{12},\cdots,\tilde{\rho}_{n2})^T$.

From (23), we obtain

$$\dot{V}_n \leq -k_{\min}\|z\|^2 - \frac{\sigma_{\min 1}\|\tilde{\rho}_1\|^2}{2} - \frac{\sigma_{\min 2}\|\tilde{\rho}_2\|^2}{2} + \varepsilon. \tag{26}$$

Therefore, the derivative of global Lyapunov function V_n is negative whenever

$$z \in \Omega_1 = \left\{z \Big| \|z\| > \sqrt{\frac{\varepsilon}{k_{\min}}}\right\}, \tag{27}$$

or

$$\tilde{p}_1 \in \Omega_2 = \left\{ \tilde{p}_1 \Big| \|\tilde{p}_1\| > \sqrt{\frac{2\varepsilon}{\sigma_{\min 1}}} \right\}, \quad (28)$$

or

$$\tilde{p}_2 \in \Omega_3 = \left\{ \tilde{p}_2 \Big| \|\tilde{p}_2\| > \sqrt{\frac{2\varepsilon}{\sigma_{\min 2}}} \right\}. \quad (29)$$

These demonstrate the uniformly ultimately boundedness of z, $\tilde{p}_1$, and $\tilde{p}_2$. Since $z_1 = x_1 - y_r$, we know that x_1 is bounded. From the definition of the first virtual controller, we can get that α_1 is bounded. Noting that $z_2 = x_2 - \alpha_1$, we further get that x_2 is bounded. Recursively using this analysis method, we conclude that x_i, $i = 3, \cdots, n$, and α_j, $j = 2, \cdots, n-1$, are bounded. Thus, the control input u is bounded. From (28) and (29), we know that the estimates of the parameter vector bounds are bounded. Thus, we conclude that all the signals in the closed-loop system are bounded.

2) Equation (25) implies

$$V_n(t) \leq \left\{ V_n(0) - \frac{\varepsilon}{c} \right\} e^{-ct} + \frac{\varepsilon}{c}, \quad \forall \, t \geq 0. \quad (30)$$

From (30), we have

$$\sum_{i=1}^{n} \frac{1}{2g_i(\bar{x}_i)} z_i^2 \leq \left\{ V_n(0) - \frac{\varepsilon}{c} \right\} e^{-ct} + \frac{\varepsilon}{c}, \quad \forall \, t \geq 0. \quad (31)$$

Noting that $g_{i1} \geq g_i(\cdot)$. Let $g_{\max} = \max_{1 \leq i \leq n} \{g_{i1}\}$. Then, we have

$$\frac{1}{2g_{\max}} \sum_{i=1}^{n} z_i^2 \leq \left\{ V_n(0) - \frac{\varepsilon}{c} \right\} e^{-ct} + \frac{\varepsilon}{c}, \quad (32)$$

that is

$$|z_1|^2 \leq 2g_{\max} \left\{ V_n(0) - \frac{\varepsilon}{c} \right\} e^{-ct} + 2g_{\max} \frac{\varepsilon}{c} \to 2g_{\max} \frac{\varepsilon}{c}. \quad (33)$$

Let $X_s \subset R$ be the ball centered at the origin with radius $\max\left\{\sqrt{2g_{\max} V_n(0)}, \sqrt{2g_{\max} \varepsilon/c}\right\}$. For any $z_1(0) \in X_s$, the tracking error $z_1(t)$ tends to a ball centered at the origin with radius $\sqrt{2g_{\max} \varepsilon/c}$, which can be made arbitrarily small by adjusting control gains k_i, the parameters in the robustness terms $\varepsilon_{i2}, \varepsilon_{i3}$, and the parameters in the adaptive laws $r_{i1}, r_{i2}, \sigma_{i1}, \sigma_{i2}$.

Remark 1. No matter how many rules are used in the FLS, our algorithm only requires $2n$ parameters to be updated online, where n denotes the number of the state variables in the designed system. The online computation burden is reduced dramatically. In order to illustrate this idea, we give an example here, i.e., let us assume the input vector is three-dimensional. If we are given three fuzzy sets for every term of the input vector, then 27 parameters are required to be updated online for the conventional adaptive fuzzy controllers. However, only one parameter is needed to be updated in our algorithm.

Remark 2. The algorithm can easily incorporate a priori information of the plant into the controller design. Based on the priori knowledge, we can first design the nominal fuzzy controller. In control engineering, the fuzzy control is very useful when the processes are too complex for analysis using conventional techniques, and have available qualitative knowledge from domain experts for the controller design.

Remark 3. The exponential convergence of the tracking error to an arbitrarily small neighborhood of the origin is achieved. In most existing results based on universal approximators, only the asymptotical convergence property is guaranteed [2-5], [7].

Remark 4. The algorithm is more suitable for practical implementation. For most results reported in the literature [2], [5], [7], the derivatives of the virtual controllers are included in NN or FLS. Thus, the NN or FLS is difficult to realize. From the practical application point of view, we propose a novel design scheme. Based on this method, the controllers and the parameter adaptive laws are highly structural. Such a property is particularly suitable for parallel processing and hardware implementation in the practical applications.

Remark 5. By using a special design technique, the controller singularity problem is avoided.

5 Conclusions

By combining backstepping technique with FLS, a novel control design scheme is developed for a class of nonlinear systems. The design method can easily incorporate a priori information about the system through if-then rules into the controller design. The main feature of the algorithm is the adaptive mechanism with minimal adaptive parameters, that is, no matter how many states and how many rules are used in the FLS, only $2n$ parameters are needed to be updated online. The computation burden is reduced dramatically. The controllers are highly structural and are convenient to realize in control engineering. Furthermore, under the proposed control law, all the signals in the closed-loop system are guaranteed to be semi-globally uniformly ultimately bounded and the tracking error is proved to exponentially converge to a small neighborhood of the origin.

References

1. Krstic, M., Kanellakopoulos, I., Kokotovic, P.V.: Nonlinear and Adaptive Control Design. Wiley, New York (1995)
2. Lee, H., Tomizuka, M.: Robust Adaptive Control Using a Universal Approximator for SISO Nonlinear Systems. IEEE Trans. Fuzzy Syst. 8 (2000) 95-106
3. Koo, T.J.: Stable Model Reference Adaptive Fuzzy Control of a Class of Nonlinear Systems. IEEE Trans. Fuzzy Syst. 9 (2001) 624-636
4. Seshagiri, S., Khalil, H.K.: Output Feedback Control of Nonlinear Systems Using RBF Neural Networks. IEEE Trans. Neural Networks. 11 (2000) 69-79
5. Jagannathan, S., Lewis, F.L.: Robust Backstepping Control of a Class of Nonlinear Systems Using Fuzzy Logic. Inform. Sci. 123 (2000) 223-240
6. Daniel, V.D., Tang, Y.: Adaptive Robust Fuzzy Control of Nonlinear Systems. IEEE Trans. Syst. Man, Cybern. 34 (2004) 1596-1601
7. Yang, Y.S., Feng, G., Ren, J.: A Combined Backstepping and Small-Gain Approach to Robust Adaptive Fuzzy Control for Strict-Feedback Nonlinear Systems. IEEE Trans. Syst. Man, Cybern. 34 (2004) 406-420
8. Chang, Y.C.: Adaptive Fuzzy-Based Tracking Control for Nonlinear SISO Systems via VSS H_∞ Appraoches. IEEE Trans. Fuzzy Syst. 9 (2001) 278-292
9. Li, Y., Qiang, S., Zhuang, X., Kaynak, O.: Robust and Adaptive Backstepping Control for Nonlinear Systems Using RBF Neural Networks. IEEE Trans. Neural Networks. 15 (2004) 693-701

Online Mining Dynamic Web News Patterns Using Machine Learn Methods

Jian-Wei Liu, Shou-Jian Yu, and Jia-Jin Le

College of Information Science & Technology,
Donghua University
liujw@mail.dhu.edu.cn

Abstract. Given the popularity of Web news services, we focus our attention on mining hierarchical patterns from Web news stream data. To address this problem, we propose a novel algorithm, i.e., FARTMAP (fast ARTMAP). We devise a new match and activation function which both simple for computation and understanding. The novelty of the proposed algorithm is the ability to identify meaningful news patterns while reducing the amount of computations by maintaining cluster structure incrementally. Experimental results demonstrate that the proposed clustering algorithm produces high-quality patterns discovery while fulfill a reasonable run time.

1 Introduction

On most Web pages, vast amounts of useful knowledge are embedded into text. Given such large sizes of text datasets, mining tools, which organize the text datasets into structured knowledge, would enhance efficient document access. Given that the Web has become a vehicle for the distribution of information, many news organizations are providing newswire services through the Internet. Given this popularity of the Web news services, we have focused our attention on mining patterns from news streams.

In this paper, we propose a novel algorithm, which we called FARTMAP (fast ARTMAP). We devise a new match and activation function which both simple for computation and understanding. The novelty of the proposed algorithm is the ability to identify meaningful news patterns while reducing the amount of computations by maintaining cluster structure incrementally. Experimental results demonstrate that the proposed clustering algorithm produces high-quality patterns discovery while fulfill a reasonable run time.

2 Proposed Learning Algorithms

Fuzzy ART is a clustering algorithm that operates on vectors with analog-valued elements [1, 2, and 3]. These algorithms have been successfully applied to numerous tasks, including speech recognition [4], handwritten character recognition [5], and target recognition from radar range profiles [6].

After parsing and document "cleaning" and feature extraction and vector space construction, a document is represented as a vector in an n-dimensional vector space.

In our algorithm, a document d_i is represented in the following form: $d_i = (t_i, id_i)$. Where id_i is the document identifier which can be used to retrieve document d_i and t_i is the feature vector of the document: $t_i = (t_{i1}, t_{i2}, \cdots, t_{in})$. Here n is the number of extracted features, and t_{ij} is the weight of the j-th feature, where $j \in \{1, 2, \cdots, n\}$.

Definition 1. cluster C_k: Given M documents in a cluster: $\{d_1, d_2, \cdots, d_M\}$, the Documents Cluster C_k is defined as a triple: $C_k = (N_k, id_k, T_k)$. Where M_k the number of documents in the cluster is C_k, ID_k is the set of the document identifiers of the documents in the cluster, T_k is the feature vector of the document cluster.

Definition 2. $Sim(d_i, d_j)$ Distance [7]: To measure closeness between two documents, we use the Cosine metric. The cosine of the angles between two n-dimensional document vectors (d_i and d_j) is defined by

$$sim(d_i, d_j) = \sum_{i=1}^{n} t_i \cdot t_j \Big/ \|t_i\|_2 \cdot \|t_j\|_2 \qquad (1)$$

We calculate z_j using the concept vector proposed by Inderjit S. Dhillon et al.[8]. Suppose we are given M document vectors $(t_1, t_2, \cdots, t_M)$. Let all neuron weights z_j (j=1...N), which denote a partitioning of the document vectors into k disjoint clusters $C_1, \cdots, C_N$ such that

$$\bigcup_{j=1}^{N} C_j = \{t_1, \cdots t_M\} \quad C_j \cap C_l = \phi \quad if \ j \neq l \qquad (2)$$

For each $1 < j < N$, the mean vector or the centroid of the document vectors contained in the cluster C_j is

$$\frac{1}{n_j} \sum_{x \in C_j} x \qquad (3)$$

Where n_j is the number of document vectors in C_j. The corresponding concept vector of mean vector writes as

$$m_j = \frac{1}{n_j} \sum_{x \in C_j} x \Big/ \left\| \frac{1}{n_j} \sum_{x \in C_j} x \right\| \qquad (4)$$

The concept vector m_j has the following important property. For any unit vector $z \in R^d$, we have from the Cauchy-Schwarz inequality that

$$\sum_{t_i \in C_j} t_i^T z \leq \sum_{t_i \in C_j} t_i^T m_j \qquad (5)$$

Thus, the concept vector may be thought of as the vector that is closest in cosine similarity (in an average sense) to all the document vectors in the cluster c_j.

In our proposed approach, at first, let

$$z_j = m_j \qquad (6)$$

Then we have used the below metric distance for activation functions:

$$T_j(t_i) = t_i \cdot z_j \qquad (7)$$

We define match function as

$$M_j = T_j(t_i) / \max(t^T{}_i t_i, z^T{}_j z_j) \qquad (8)$$

FARTMAP-Algorithm ($\{t_{i,1}, t_{i,2}, \cdots, t_{i,j}\}, \overline{p}, \beta$)

1. $z_0 \leftarrow \{z_{l,1}, z_{l,2}, \cdots, z_{l,q}\}$
2. templates $\leftarrow \{z_0\}$
3. for each $t_{i,k} \in \{t_{i,1}, t_{i,2}, \cdots, t_{i,j}\}$
4. do $T^l{}_{max} \leftarrow 0$ and $z^l{}_{max} \leftarrow none$
5.   for each $z_{l,r} \in$ templates
6.   do if $M_j = T_l(t_{i,k}) / \max(t^T{}_{i,k} t_{i,k}, z^T{}_{l,r} z_{l,r}) \geq \overline{p}$
         and $T(t_{i,k}, z_{l,r}, \beta_l) = t_{i,k}, z_{l,r} > T_{max}, k \neq r$
8.   then
$$T_{max} \leftarrow T(t_{i,k}, z_{l,r}, \beta_l)$$
$$z^l{}_{max} \leftarrow z_{l,r}$$
9.   if $z^l{}_{max} \neq z_0$ and $classlabel(t_{i,k}) = classlabel(w^l{}_{max})$
10.   then $z^l{}_{max} \leftarrow z^l{}_{max} \cdot t_{i,k} / \|z^l{}_{max} \cdot t_{i,k}\|$
11.   else templates $\leftarrow$ templates $\cup \left\{\dfrac{t_{i,k}}{\|t_{i,k}\|}\right\}$
12. return templates

Fig. 1. proposed FARTMAP inline learning algorithms

Fig. 1. shows our proposed algorithm. We called the proposed algorithm as FARTMAP (fast ARTMAP).

3 Conclusions

This work presents a FARTMAP algorithm that works online. We devise a new match and activation function which both simple for computation and understanding. The comprehensive experiments based on several publicly available real data sets shows that significant performance improvement is achieved and produce high-quality clusters in comparison to the previous methods. The algorithm also exhibited linear speedup when the number of news documents is increased.

References

1. G.A. Carpenter, S. Grossberg, N. Markuzon, J.H. Reynolds, D.B. Rosen.: Fuzzy ARTMAP: A Neural Network Architecture for Incremental Supervised Learning of Analog Multidimensional Maps. IEEE Transactions on Neural Networks, (1992)698-713.
2. G.A. Carpenter and W. Ross.: ART-EMAP: A Neural Network Architecture for Learning and Prediction by Evidence Accumulation. IEEE Transactions on Neural Networks, (1995)805-818.
3. G.A. Carpenter, B.L. Milenova, and B.W. Noeske.: Distributed ARTMAP: A Neural Network for Fast Distributed Supervised Learning. Neural Networks. (1998)793-813.
4. Young, D. P.: A Fuzzy ART MAP Neural Network Speech Recognition System Based on Formant Ratios. Master's thesis, Royal Military College of Canada, Kingston, Ontario (1995).
5. Markuzon, N.: Handwritten Digit Recognition Using Fuzzy ARTMAP Network. In World Congress on Neural Networks – San Diego Hillsdale, NJ: Lawrence Erlbaum Associates (1994)117-122.
6. Rubin, M. A.: Issues in AutomaticTarget Recognition from Radar Range Profiles using Fuzzy ARTMAP. In The 1995 World Congress on Neural Networks. Mahwah, NJ: Lawrence Erlbaum Associates (1995) 197-202.
7. G. Salton, and M.J. McGill.: Introduction to Modern Information Retrieval. McGraw-Hill, New York, (1983).
8. Inderjit S., Dhillon Dharmendra, S. Modha. : Concept Decompositions for Large Sparse Text Data Using Clustering.. (2001) 143-175.

A New Fuzzy MCDM Method Based on Trapezoidal Fuzzy AHP and Hierarchical Fuzzy Integral

Chao Zhang[1,2], Cun-bao Ma[1], and Jia-dong Xu[2]

[1] School of Aeronautics, Northwestern Polytechnical University, Xi'an, 710072, P.R. China
{zc85377}@126.com
[2] School of Electrical & Information,
Northwestern Polytechnical University, Xi'an, 710072, P.R. China

Abstract. Fuzzy Multiple Criteria Decision Making (MCDM) has been widely used in evaluating and ranking weapon systems characterized by fuzzy assessments with respect to multiple criteria. However, most criteria have interdependent or interactive characteristic so weapon system cannot be evaluated by conventional evaluation methods. In this paper, a new method based on trapezoidal fuzzy AHP and fuzzy Integral is proposed. The ratings of criteria performance are described by linguistic terms expressed in trapezoidal fuzzy numbers. The weights of the criteria are obtained by trapezoidal fuzzy AHP. And the hierarchical fuzzy integral model is proposed based on λ-fuzzy measure and Sugeno integral to determine the synthesis evaluation of weapon system. Finally, an example of evaluating the best main battle tank is given. The results demonstrate the engineering practicability and effectiveness of this method.

1 Introduction

Weapon system is a large and complex system. Effective evaluation and analysis of weapon systems not only reduce the current cost, but also give our military effective fighting machines. In general, many influence levels and factors must be considered in the process of weapon system evaluation. These weapon systems' performance evaluation and analysis problems are Multiple Criteria Decision-Making (MCDM) problems [1]. Several traditional methods have been proposed to help the researchers evaluation and optimal design weapon system [2,3,4].

The existing methodologies for weapon system evaluation are divided into single-criteria cost/benefit analysis, multiple criteria scoring models and ranking methods, and subjective committee evaluation methods [4]. But, there are have some shortcoming:

(1) Most prior methods are mainly used in nearly crisp decision applications.

(2) The subjective evaluation and preference of decision makers is usually approximated using the linear combination mathematical model.

(3) These methods are assume that the criteria involved are non-interactive and independent, hence, their weighted effects are viewed as additive type.

To overcome these problems, a more realistic fuzzy MCDM method for evaluating weapon system based on trapezoidal fuzzy AHP and hierarchical fuzzy integral is proposed.

2 Fuzzy Set Theory

The fuzzy set theory was introduced by Zadeh to deal with problems in which a source of vagueness is involved. A fuzzy set can be defined mathematically by assigning to each possible individual in the universe of discourse a value representing its grade of membership in the fuzzy set.

2.1 Trapezoidal Fuzzy Number

A fuzzy number is a special fuzzy set $F = \{x \in R | \mu_F(x)\}$, where x takes its values on the real line $R_1 : -\infty < x < +\infty$ and $\mu_F(x)$ is a continuous mapping from R_1 to the close interval [0,1]. A trapezoidal fuzzy number can be denoted as $\tilde{A}=(a_1,a_2,a_3,a_4)$, $a_1 \leq a_2 \leq a_3 \leq a_4$. Its membership function $\mu_{\tilde{A}}(x): R \rightarrow [0,1]$ is equal to

$$\mu_{\tilde{A}}(x) = \begin{cases} 0 & x < a_1 \\ (x-a_1)/(a_2-a_1) & a_1 \leq x \leq a_2 \\ 1 & a_2 \leq x \leq a_3 \\ (x-a_4)/(a_3-a_4) & a_3 \leq x \leq a_4 \\ 0 & x > a_4 \end{cases} \quad (1)$$

Trapezoidal fuzzy numbers are appropriate for quantifying the vague information about most decision problems, and numerous publications on the development of trapezoidal fuzzy number theory have appeared in the academic and professional journals. The primary reason for using trapezoidal fuzzy numbers can be stated as their intuitive and computational-efficient representation.

According to the characteristic of trapezoidal fuzzy numbers and the extension principle, the arithmetic operations laws of two trapezoidal fuzzy number can be see in [3,4,5].

According to [5], for a trapezoidal fuzzy number $\tilde{A}=(a_1,a_2,a_3,a_4)$, its defuzzification value is defined to be $c = (a_1 + a_2 + a_3 + a_4)/4$.

2.2 Linguistic Variable

A linguistic variable is defined as a variable whose values are not numbers, but words or sentences in natural or artificial language [6,7]. The concept of a linguistic variable

Table 1. Linguistic variable for the ratings

Opinion	Value
Very Poor (VP)	(0,0,0.1,0.2)
Poor (P)	(0.1,0.2,0.2,0.3)
Medium Poor (MP)	(0.2,0.3,0.4,0.5)
Fair (F)	(0.4,0.5,0.5,0.6)
Medium Good (MG)	(0.5,0.6,0.7,0.8)
Good (G)	(0.7,0.8,0.8,0.9)
Very Good (VG)	(0.8,0.9,1.0,1.0)

appears as a useful means for providing approximate characterization of phenomena that are too complex or ill-defined to described in conventional quantitative terms. In this paper, the evaluating opinions are described by linguistic terms expressed in trapezoidal fuzzy numbers are shown in Table 1.

2.3 The Overall Valuation of the Fuzzy Judgment

The overall valuation of fuzzy judgment copes with the fact that every respondent differently toward every criterion. The subsequent valuation of the linguistic variable certainly varies among individuals. We integrate the overall fuzzy judgment by Eq.(2).

$$E_{ij} = \tfrac{1}{m}(\times)[E_{ij}^1(+)E_{ij}^2(+)\cdots(+)E_{ij}^m] \qquad (2)$$

where E_{ij} is the overall average performance valuation of entity i under criterion j over m assessors.

3 Trapezoidal Fuzzy AHP

When the evaluating hierarchy structure is built, the trapezoidal fuzzy AHP method in terms of ratio scale is employed to proceed with relative importance of pairwise comparison among every criterion and calculate the trapezoidal fuzzy weights of the evaluating criteria [8,9].

3.1 Trapezoidal Fuzzy Judgment Matrix Construction

According to AHP model, there are k layer in the model and n_k factors on the kth layer, k and n_k are positive integers, and $k \geq 2$. Pairwise comparison between criteria A_h^{k-1} on layer k-1,(h=1,2,...,n^{k-1}), and the corresponding index n_k on the kth is carried through. Then the fuzzy judgment matrix $A = (a_{ij})_{n \times n}$ will be obtained. For each value of i and j, $a_{ij} = (l_{ij}, m_{ij}, n_{ij}, s_{ij})$ representing the relative importance of index i to index j are all trapezoidal fuzzy numbers.

3.2 The Consistency Test

When AHP method is used, it is indispensable to perform the consistency test on the trapezoidal fuzzy judgment matrix in order to ensure the veracity of ranking.

Definition 1 (Consistency). For fuzzy judgment matrix $A = (a_{ij})_{n \times n}$, if $a_{ij} \approx \tfrac{a_{ik}}{a_{jk}}$ ($i,j,k=1,2,\cdots,n$), then A is a fuzzy judgment matrix with consistency.

Theorem 1. For a fuzzy judgment matrix $\tilde{A} = (\tilde{a}_{ij})_{n \times n}$, $\tilde{a}_{ij}$ is a fuzzy number. If $a_{ij} \in (\ker \tilde{a}_{ij} = \{x | \tilde{a}_{ij}(x) = 1\}, i, j = 1,2,\cdots,n)$ exists and $A = (a_{ij})_{n \times n}$ is the fuzzy judgment matrix with consistency, then $\tilde{A}$ is a fuzzy judgment matrix with consistency.

3.3 The Trapezoidal Fuzzy Weights of the Evaluating Criteria

Before the calculation of the trapezoidal fuzzy weights, it is necessary to make assumptions: A is a fuzzy judgment matrix of $m \times m$ order. Element of which is represented as a_{ij}, composed by the results of pairwise comparison between factors, A_1, $A_2,...,A_m$ with one certain criterion on one certain layer, a_{ij}, the element of matrix A, is described as trapezoidal fuzzy number and both $a_{ji} = (a_{ij})^{-1}$ for all $i, j \in \{1,2,\cdots,m\}$ and $a_{ii} = (1,1,1,1)$ for all $i \in \{1,2,\cdots,m\}$ are tenable. To define $\alpha_i = (\prod_{j=1}^{m} l_{ij})^{\frac{1}{m}}$, $\alpha = \sum_{i=1}^{m} \alpha_i$, analogously: $\beta_i = (\prod_{j=1}^{m} m_{ij})^{\frac{1}{m}}$, $\beta = \sum_{i=1}^{m} \beta_i$; $\gamma_i = (\prod_{j=1}^{m} n_{ij})^{\frac{1}{m}}$, $\gamma = \sum_{i=1}^{m} \gamma_i$; $\delta_i = (\prod_{j=1}^{m} s_{ij})^{\frac{1}{m}}$, $\delta = \sum_{i=1}^{m} \delta_i$. With assumptions above, ω_i, the weight of factor A_i, is $(\frac{\alpha_i}{\delta}, \frac{\beta_i}{\gamma}, \frac{\gamma_i}{\beta}, \frac{\delta_i}{\alpha})$, $i \in \{1,2,\cdots,m\}$.

The consistency of the trapezoidal fuzzy judgment matrix is to be tested first. To apply feedback control on it until it matches the consistency, ω_{ih}^{k}, the fuzzy weight of A_i^k on layer k to A_h^{k-1} is worked out, and $\omega_{ih}^{k} = (\frac{\alpha_i}{\delta}, \frac{\beta_i}{\gamma}, \frac{\gamma_i}{\beta}, \frac{\delta_i}{\alpha})$, ($i=1,2,\cdots,n_k$), then $\omega_h^k = (\omega_{1h}^k, \omega_{2h}^k, \omega_{3h}^k, \omega_{4h}^k)$ is the fuzzy weights of all the factors on the kth to factors A_h^{k-1} on layer k-1.

4 The Hierarchical Fuzzy Integral Model

Due to some inherent interdependent or interactive characteristics among the criteria or weapon system [3,4], the non-interactive or independent assumption is not realistic. We proposed the hierarchical fuzzy integral to analyze and solve the interactive and interdependent criteria problem.

4.1 λ – Fuzzy Measure

Definition 3 (λ–fuzzy measure [6]). A λ–fuzzy measure g_λ is a fuzzy measure with the following property.
$\forall A, B \in \beta(X)$ and $A \cap B = \phi$, then

$$g_\lambda(A \cup B) = g_\lambda(A) + g_\lambda(B) + \lambda g_\lambda(A) g_\lambda(B) \quad \text{for } -1 < \lambda < \infty \quad (3)$$

According to the definition of g_λ, the finite set $X = \{x_1, x_2, \cdots, x_n\}$ mapping to function g_λ can be written using fuzzy density $g_i = g_\lambda(\{x_i\})$ as

$$g_\lambda(\{x_1, x_2, \cdots, x_n\}) = \sum_{i=1}^{n} g_i + \lambda \sum_{i1=1}^{n-1} \sum_{i2=i1+1}^{n} g_{i1} g_{i2} + \cdots + \lambda^{n-1} g_1 g_2 \cdots g_n = \frac{1}{\lambda} \left| \prod_{i=1}^{n}(1 + \lambda g_i) - 1 \right| \quad (4)$$

for $-1 < \lambda < \infty$.

Because $g_\lambda(X)=1$, the λ value of λ-fuzzy measure g_λ can be calculated by

$$\lambda+1=\prod_{i=1}^{n}(1+\lambda g_i) \qquad (5)$$

4.2 Fuzzy Integral

Definition 4 (Fuzzy integral). Let g be a fuzzy measure on X and h be a measurable function from X to [0,1]. Assuming that $h(x_1) \geq h(x_2) \geq \cdots \geq h(x_n)$, then the Sugeno fuzzy integral is defined as follows:

$$e(h) = \int_X h dg = \max_{i=1}^{n}(\min(h(x_i), g_\lambda(H_i))) \qquad (6)$$

where $H_1=\{x_1\}, H_2=\{x_1,x_2\}, \cdots, H_n=\{x_1,x_2,\cdots,x_n\}=X$.

Since criteria interact and affect each other in real world, but the fuzzy integral model does not need to assume non-interactive or independency of one criterion from another, it can be employed to conduct non-additive operations for these interdependent or interactive aspects, criteria and sub-criteria. The fuzzy integral of f with respect to g gives the overall evaluation of an alternative.

4.3 The Hierarchical Fuzzy Integral Model

According to AHP structure, the hierarchical fuzzy integral model is constructed as shown in Fig.1. If we see each circle as a node, we can use the evaluation values f and the grades of importance g on the lower-level objects/elements calculated on integral's Eq.(6) to obtain the upper-level's objects/elements evaluation values. For example, $f_1^{11}, f_2^{11}, \cdots, f_{s1}^{11}$ are the evaluation values of the bottom-level's objects/elements;

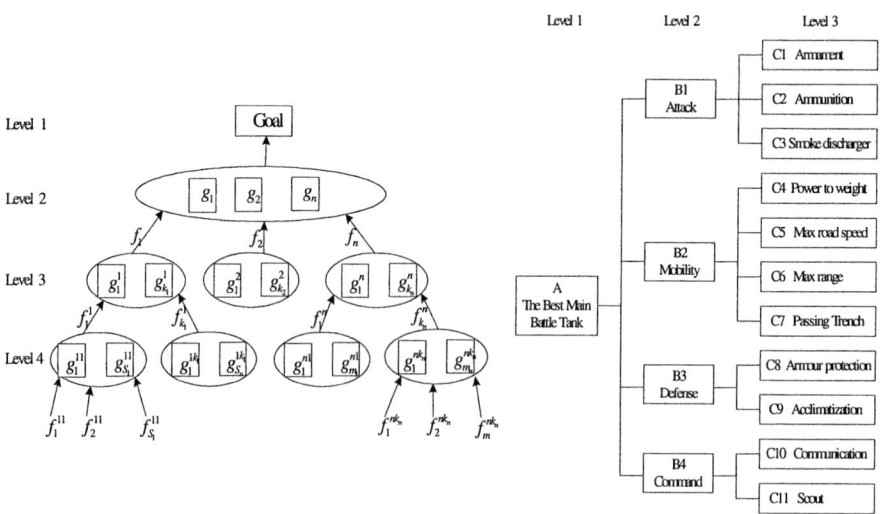

Fig. 1. The hierarchical fuzzy integral model **Fig. 2.** The hierarchy structure of main battle tank

and $g_1^{11}, g_2^{11}, \cdots, g_{s1}^{11}$ are the grades of importance. By using integral's Eq.(6) to compute the subtotal evaluation values of the first node on Level 4, we can get the result f_1^1. The other subtotal evaluation values also can be calculated in the same way. Then all results are $(f_1^1, f_2^1, \cdots, f_{k1}^1), \ldots, (f_1^n, f_2^n, \cdots, f_{k_n}^n)$, respectively. Likewise, there are n nodes on Level 3. Here $f_1^1, f_2^1, \cdots, f_{k1}^1$ are the evaluation values, $g_1^1, g_2^1, \cdots, g_{k1}^1$ are the grades of importance, and the result is f_1. The other subtotal evaluation values are $f_2, \cdots f_n$. Finally, there is only one node on Level 2. Here $f_1, f_2, \cdots, f_n$ are the evaluation values and $g_1, g_2, \cdots, g_n$ are the grades of importance. By using Eq.(6) to compute the overall evaluation value, we get the final result on Level 1.

5 Numerical Example

In this section, we construct an example for evaluating military unmanned combat air vehicle to illustrate our proposed method. Since World War II, the advanced nations have devoted themselves developing unmanned air vehicles for military use [4]. To date, the unmanned air vehicle equipped with velocity, fire power and flexibility, is suitable for complex duty in modern threats from combat air weapons. We select M1A1 (USA), Challenger 2 (UK) and Leopard (Germany) main battle tank as our evaluating entities. On the basis of the comprehensive analysis and the evaluation indices of the main battle tank, the hierarchy structure is constructed as in Fig.2.

5.1 The Weight of Criteria

According to the definition in section 3.2, the judgment matrix (only 1-2 layer as an example) based on trapezoidal fuzzy AHP is shown in Table 2.

Table 2. The judgment matrix

A	B1	B2	B3	B4
B1	(1,1,1,1)	(1/7,1/6,1/6,1/5)	(1/5,1/4,1/4,1/3)	(1/3,1/2,1/2,1)
B2	(5,6,6,7)	(1,1,1,1)	(1/6,1/5,1/5,1/4)	(1/4,1/3,1/3,1/2)
B3	(3,4,4,5)	(4,5,5,6)	(1,1,1,1)	(1/5,1/5,1/5,1/5)
B4	(1,2,2,3)	(2,3,3,4)	(5,5,5,5)	(1,1,1,1)

The consistency test is to be performed. From Table 2, we know

$$B = \begin{bmatrix} 1 & 1/6 & 1/4 & 1/2 \\ 6 & 1 & 1/5 & 1/3 \\ 4 & 5 & 1 & 1/5 \\ 2 & 3 & 5 & 1 \end{bmatrix}$$

And because B is the judgment matrix with consistency, according to criterion, it's obvious that the trapezoidal fuzzy judgment matrix *A-B* is the fuzzy judgment matrix with consistency. So are the trapezoidal fuzzy judgment matrixes *B1-C*, *B2-C*, *B3-C* and *B4-C*.

From the method expressed in section 3 and after the defuzzification and normalization processing, the sequencing of the trapezoidal fuzzy weights of the main battle tank can be obtained as shown in Table 3.

Table 3. The weight of criteria

	B1	B2	B3	B4
	0.1912	0.3437	0.4447	0.1603
C1	0.1396	0	0	0
C2	0.5279	0	0	0
C3	0.3325	0	0	0
C4	0	0.1019	0	0
C5	0	0.1923	0	0
C6	0	0.2356	0	0
C7	0	0.4702	0	0
C8	0	0	0.5457	0
C9	0	0	0.4702	0
C10	0	0	0	0.5936
C11	0	0	0	0.4064

Table 4. The rating of M1A1

	E1	E2	E3	C
C1	G	G	MG	0.6995
C2	VG	G	VG	0.8833
C3	G	MG	G	0.7500
C4	MG	G	G	0.7500
C5	VG	G	G	0.8416
C6	G	MG	G	0.7500
C7	G	G	MG	0.7500
C8	MG	G	MG	0.6995
C9	G	MG	MG	0.6995
C10	G	G	G	0.8000
C11	VG	G	MG	0.7912

5.2 The Rating of Alternative

A committee of three decision-makers E_1, E_2 and E_3 has been formed to select the best main battle tank. The linguistic rating variables and their transform value of fuzzy rating are shown as Table 4, 5 and 6.

Table 5. The rating of Challenger 2

	E1	E2	E3	C
C1	G	MG	G	0.7500
C2	MG	MG	MG	0.6500
C3	MG	G	MG	0.6995
C4	F	MG	F	0.5500
C5	MG	F	F	0.5500
C6	G	MG	MG	0.6995
C7	MG	F	F	0.5500
C8	G	G	G	0.8000
C9	F	F	MP	0.4500
C10	MG	G	G	0.7500
C11	MG	MG	G	0.6995

Table 6. The rating of Leopard

	E1	E2	E3	C
C1	MG	G	G	0.7500
C2	MG	MG	G	0.6995
C3	VG	VG	G	0.8833
C4	G	G	G	0.8000
C5	G	G	MG	0.7500
C6	G	G	G	0.8000
C7	F	MG	MG	0.6000
C8	F	F	F	0.5000
C9	G	MG	MG	0.6995
C10	G	MG	G	0.7500
C11	MG	G	MG	0.6995

5.3 Fuzzy Integral

The procedure of employing a hierarchical structure of fuzzy integral for the multi-criteria comprehensive assessment, as stated in section 4 above, is applied to analyze the evaluation. The calculating procedures for this problem are shown as Table 7, 8,9,10,11. And form Table 11, we can see that the M1A1 is the best main battle tank.

Table 7. The integral result of attack ability

Criteria	g^i	M1A1	Challenger	Leopard
C1	0.1396	0.6995	0.7500	0.7500
C2	0.5279	0.8833	0.6500	0.6995
C3	0.3325	0.7500	0.6995	0.8833
B1	—	0.6995	0.6500	0.6995

Table 8. The integral result of mobility ability

Criteria	g^i	M1A1	Challenger	Leopard
C4	0.1019	0.7500	0.5500	0.8000
C5	0.1923	0.8416	0.5500	0.7500
C6	0.2356	0.7500	0.6995	0.8000
C7	0.4702	0.7500	0.5500	0.6000
B2	—	0.5721	0.5500	0.6000

Table 9. The integral result of defense ability

Criteria	g^i	M1A1	Challenger	Leopard
C8	0.5457	0.6995	0.8000	0.5000
C9	0.4543	0.6995	0.4500	0.6995
B3	—	0.6995	0.5457	0.500

Table 10. The integral result of command ability

Criteria	g^i	M1A1	Challenger	Leopard
C10	0.5936	0.8000	0.7500	0.7500
C11	0.4064	0.7912	0.6995	0.6995
B4	—	0.7912	0.6995	0.6995

Table 11. The integral result of tank

Criteria	g^i	M1A1	Challenger	Leopard
B1	0.1912	0.6995	0.6500	0.6995
B2	0.3437	0.5721	0.5500	0.6000
B3	0.4447	0.6995	0.5457	0.5000
B4	0.1603	0.7912	0.6995	0.6995
A	—	0.6995	0.5500	0.6000
Ranking	—	1	3	2

6 Conclusion

In this paper, a new general and easy fuzzy group multiple criteria decision-making method based on trapezoidal fuzzy AHP and hierarchical fuzzy integral has been proposed, and construct a practical example of evaluating the best main battle tank to illustrate our proposed method. The results demonstrate the engineering practicability and effectiveness of this method.

References

1. Chen, S.J., Hwang, C.L.: Fuzzy Multiple Attribute Decision-Making: Methods and Applications. Springer-Verlag Berlin Heidelberg (1992)
2. Cheng, C.H. and Mon, D.L.: Evaluating Weapon System by Analytical Hierarchy Process Based on Fuzzy Scales. Fuzzy Sets and Systems, Vol.63. (1994) 1-10
3. Cheng, C.H.: Evaluating Weapon Systems using Ranking Fuzzy Numbers. Fuzzy Sets and Systems, Vol.107. (1999) 25-35
4. Cheng, C.H. and Lin, Y.: Evaluating The Best Main Battle Tank using Fuzzy Decision Theory With Linguistic Criteria Evaluation. European Journal of Operational Research, Vol.142. (2002) 174-176
5. Chen, C.T.: A Decision Model for Information System Project Selection. 2002 IEEE International Engineering Management Conference, Aug 18-20. IEMC (2002) 585 – 589
6. Chen, Y.W. and Tzeng, G.H.: Using Fuzzy Integral for Evaluating Subjectively Perceived Travel Costs in a Traffic Assignment Model. European Journal of Operational Research, Vol.130. (2001) 653-664
7. Liu, Y.X., Li, X. and Zhuang, Z.W.: Decision-Level Information Fusion for Target Recognition Based on Choquet Fuzzy Integral. Journal of Electronics and Information Technology, Vol.25. (2003) 695-699
8. Wang, L., Liu, J.Q.: Trapezoidal Fuzzy AHP and its Application to Optimal Selection of Satellite System Schemes. Journal of Harbin Institute of Technology, Vol.34. (2002) 315-319
9. Wu, X.Q., Pu, Fang., Shao, S.H.: Trapezoidal Fuzzy AHP for The Comprehensive Evaluation of Highway Network Programming Schemes in Yangtze River Delta. Proceedings of the 5th World Congress on Intelligent Control and Automation, Hangzhou, P.R.China, June 15-19. WCICA (2004) 5232-5236

Fast Granular Analysis Based on Watershed in Microscopic Mineral Images

Danping Zou[1], Desheng Hu[2], and Qizhen Liu[3]

[1] Computer Science Department, Fudan University, Shanghai 200433, China
dpzou@yeah.net
[2] Baosteel Technology Center, Shanghai 201900, China
[3] Computer Science Department, Fudan University, Shanghai 200433, China
qzliu@fudan.edu.cn

Abstract. The process detecting and measuring granule named granular analysis is very important in mineral analysis. Classical methods for granular analysis are based on Matheron's sieving method. However, these methods are not adequate for mineral microscope image analysis. First, it is not an easy job to choose proper element structure for sieving process. Second, these traditional methods cannot exactly locate the position of each grain in the image. Third, the running cost of these methods on PC is too high to implement an online application. This paper proposes a granular analysis model based on improved watershed which is called varying-ladder watershed. The improved watershed overcomes the over-segment problem by adjusting the ladder height among successive steps and quickly segments the whole image into regions of different textures. Experiments show that using the proposed method gets accurate and detailed results and gains high computational performance....

1 Introduction

Granule is a kind of texture in image processing which exists in many kinds of mineral images. In some conditions, quantity and dimension of granules contained in minerals usually determine the quality of minerals. So it is important for mineral analysis to detect and measure the granules in mineral images. However, the current way of granular detection and measurement depends on technologist's observation and estimation, which leads to a low efficiency and an unstable result as different people has different standards. So that developing automatic method for granular analysis is an urgent request. Using digital image processing method implemented on common PC is a good choice for automatic analyzing. Metheron[1] has proposed a sieving process theory based on mathematic morphology which has been successfully used to get granular information in binary image. The sieving process applies a family of mathematic morphologic openings with nonnegative and size-increasing element structure to the image so that the volume of the image decreases, and finally it yields a size distribution which is called pattern spectrum [2]. The moments extracted from different orders of the pattern spectrum used as texture characteristics are adopted in texture classification [3] [4]. Whereas, these characteristics obtained from morphologic openings do not contain any information about positions so that it may classify two textures with similar size distributions but very different space distributions into the same class. A concept of size distribution called spatial size distribution (SSD) has been introduced in [5]. SSD contains not only the size information but also the spatial information of granules. Using this distribution can highly improve the ability of classification. However, those methods based on Matheron's sieving process are not adequate for mineral analysis for several reasons. First, it is not an easy job to choose a proper element structure to implement the sieving process. Second, these methods could only produce size distributions, while most granular analysis requires more information, such as position or contour information of each granule. Third, the sieving process has high computational complexity because of lots of opening operations. Since the technique of image segmentation has been developing rapidly in recent years and many segmentation methods with high performance have been proposed, we can resort to the segmentation methods to do granular analysis.

Watershed transform [6] has been widely used in image segmentation in recent years, but applying watershed transform usually leads to over-segmentation, so some improved methods have been proposed. These improved methods can be categorized into two kinds. One is called marked-based watershed [8] which imposes the markers on corresponding images, either original image or gradient image, to suppress the disturbance of noises which cause over-segmentation. The other is based on region merging approach which merges similar regions into one region so as to overcome over-segment problems [9]. Although the existing methods aforementioned could avoid over-segmentation remarkably, their high computing complexity makes them not meet the performance requirement of the mineral-analysis applications. Therefore, in this paper, we propose a varying-ladder watershed method which not only solves over-segment problem efficiently but also obtains good results. Furthermore, we describe a new model of granular analysis for chromatic mineral images.

The rest of this paper is organized as follows. Section 2 reviews the definition of watershed by immersion, discusses the over-segment problem and shows how the varying-ladder watershed solves this problem. Section 3 presents a new application model for our granular analysis in mineral images. The last section gives some experiment results using the proposed method.

2 Watershed Transform

2.1 Definition by Immersion

Watershed is originally a concept of topography, which is the joint line between waters coming from different basins. In image processing, gray-scale image is considered as topographic relief where elevation relates to gray level of the image. Such a representation is useful in image study, since it gives a more direct view for better appreciation. Therefore, many concepts of topography can be easily applied to image processing and watershed is just one of them. However, extracting watershed line from digital pictures is far from easy until Vincent proposed an efficient algorithm based on immersion simulations in [6]. He considers grayscale image as a topographic surface with minima pierced, and immerses the surface into water, then the waters from different minima will fill up basins, finally the joint yields watershed lines. Fig. 1 shows this.

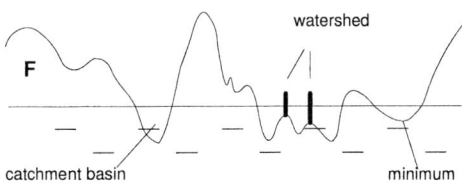

Fig. 1. Image F is immersed in water

Let I be the gray scale image under study, denote $h_{\min}$ the lowest gray level of I and similarly $h_{\max}$ the highest, T_h is the subset with a water level below h and $X(h)$ stands for the watershed subset with a level h, then we get the definition by immersion.

$$\begin{cases} X(h_{\min}) = T(h_{\min}) \\ \forall h \in [h_{\min}, h_{\max} - 1], X(h+1) = \min(h+1) \cup IZ_{T(h+1)}(X(h)) \end{cases} \quad (1)$$

As can be seen, it is a recursion. At each step the watershed subset X(h+1) is the union of the newly appeared minima at level h+1 and the influent zone of the former watershed subset. You can find more exact explanations about the definition in [6]. Although there are other definitions and algorithms for watershed [7], this paper only focuses on the definition by immersion because our method is based on concepts contained in this definition.

2.2 Over-segmentation and Varying-Ladder Watershed

As noises exist in digital pictures under study inevitably, so a lot of minima are added when applying the watershed algorithm, which will cause over-segmentation. As illustrated by Fig. 2, the left is the ideal image which is not corrupted by noises. Applying watershed transform to this image, we can acquire one watershed and two basins. However, the right in Fig. 2 is the same image corrupted by noises; therefore these noises lead to additional minima when applying watershed transform so that causes over segmentation. So how to avoid these additional minima is the key point of the proposed method in the paper.

Let's consider the discrete situations, because most applications are discrete. Denote the difference between two successive gray levels in watershed transform. In definition (1), the value of is always fixed and equal to the gray level interval of gray-scale image usually with value 1. Denote D the homogeneous region. As illustrated by Fig. 3, the homogeneous region D is corrupted by noise so that

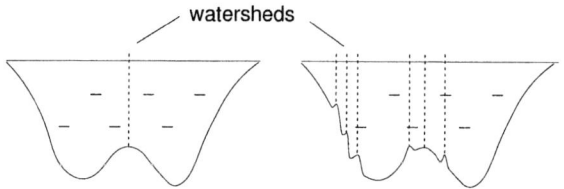

Fig. 2. Over segmentation

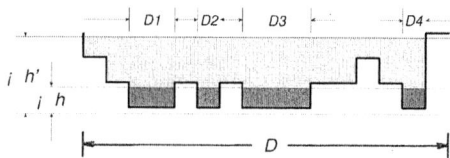

Fig. 3. Varying-ladder watershed

there are some small minima existed in the region. If is less than the peak value in region D as Fig. 3 shows, the area will be over segmented into more than one region. However, if we adjust to which is greater than the peak value of this area, we'll get the expected result and over segmentation is avoided. So properly adjusting the value of according to requirement could efficiently suppress the disturbance of noises, and the over-segment problem comes to solution.

This is the improved watershed which is called varying-ladder watershed in this paper. The *ladder* here means the difference Δh between the successive gray levels in watershed transform. The varying-ladder watershed is defined as

$$\begin{cases} X(h_0) = T(h_0) \\ X(h_{i+1}) = \min(h_{i+1}) \cup IZ_{T(h_{i+1})}(X(h_i)) \\ \Delta h(i) = h_{i+1} - h_i (i = 0...N-1) \end{cases} \quad (2)$$

The influent zones in (2) are different from those in definition (1). In definition by immersion, pixels in the same influent zone have the same value of gray level, whereas in (2) the gray values of pixels are in the same interval with depth of Δh. We can see that in (2) the number of recursive steps is less than that of original definition, so this improved watershed algorithm could both solve the over-segment problem and enhance the running performance, which is the advantage that the existing methods do not have. Either the marker-extracting or region-emerging method is far more complex than varying-ladder watershed, so the varying-ladder watershed is very practical in the applications which have to maintain performance with speed limitations.

3   A Model of Granular Analysis Based on Watershed

As mentioned in introduction, most existing methods are based on Matheron's sieving process, which continuously applies the morphological openings to the image with size-increasing structure elements, and finally yields a curve named pattern spectrum. The moments extracted from the spectrum can be used as features in classification. This kind of method is definitely effective for classification of those images, each of which contains only one texture. However, if there is more than one texture in each image, it becomes unavailable because we should segment the image first. Moreover, these methods can not locate the position of granules while a lot of applications require more information such as the accurate contours of the granules. To solve the problems mentioned above, a model of granular analysis for digital mineral pictures is presented in Fig. 4. The first step is preprocessing, which does some preparative work for following

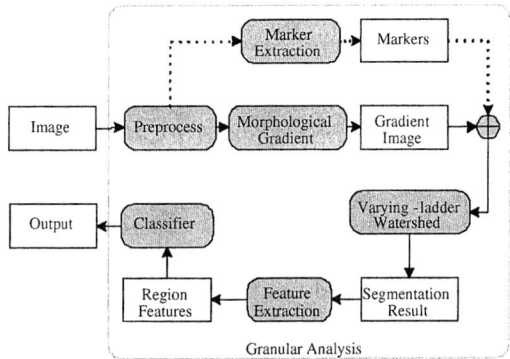

Fig. 4. Model for granular analysis. The flow charted by dot line is optional.

process such as noise cleaning, contrast adjustment etc. Furthermore, in most cases the input image is chromatic, so it must be converted to gray-scale image first. The next step is to calculate the gradient image using morphological gradient defined as

$$grad_s(I) = \delta_s(I) - \varepsilon_s(I) \qquad (3)$$

where I, δ and ε denote the original image, dilation and erosion respectively, s is the structure element. So the morphological gradient under structure element s represents the difference between dilation and erosion image. However, the gradient defined above produces low gradient value at blur edges, we can use the multi-scale gradient [10] instead, which is described as

$$MG(I) = \frac{1}{n}\sum_{i=1}^{n}\{\varepsilon_{g_{i-1}}[grad_{g_i}(I)]\} \qquad (4)$$

where $MG(I)$ represents the multi-scale gradient of image I, g_i is the size-increasing structure element. Having obtained the gradient image, the varying-ladder watershed could generate segmentation regions where each region represents a single granule. The following steps are pattern recognition steps - feature extraction and classification steps. We can choose features such as color, area, perimeter, average gradient of the region, and then put the feature vectors into the classifier. Finally, we'll get the output including not only the segmentation and classification but also the statistical information about the granules such as size and shape distribution.

4 Experiment

Using the proposed granular analyzing model based on the varying-ladder watershed, we have done a lot of experiments on microscope mineral images. The results are well-pleasing because of the high accuracy and efficiency.

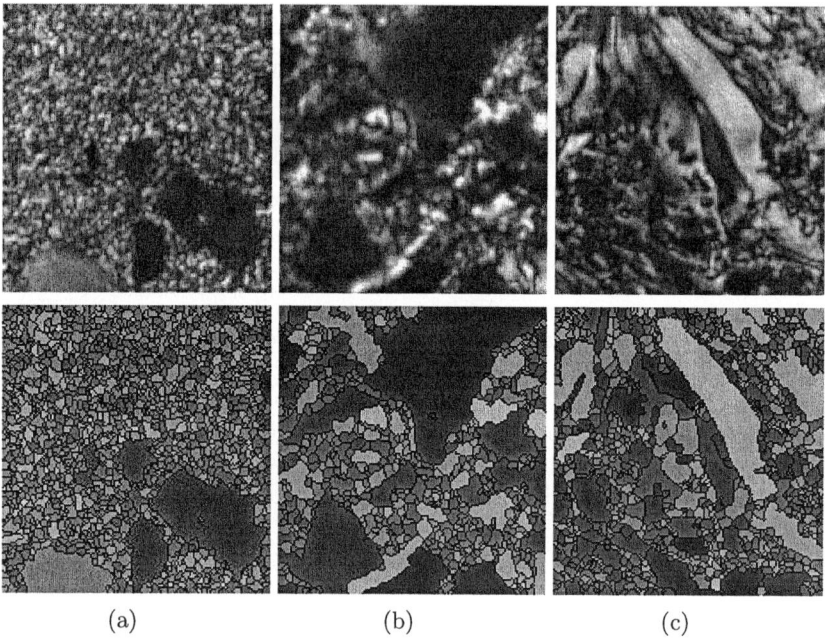

(a)   (b)   (c)

Fig. 5. Part of the results from mineral analysis experiments. The above images are the original images, and the below ones are the segmentation results where each region is filled with average color.

Fig. 5. shows some segment results from our mineral analyzing experiment which automatically segments thousands of digital mineral pictures with a size of 800 × 800 pixels into small granular regions. Because of the size limitation, only the small parts of original pictures are shown in the paper. We can see that the segment results are close to the results acquired from observation. With the segments results, we can both categorize them into different classes according to

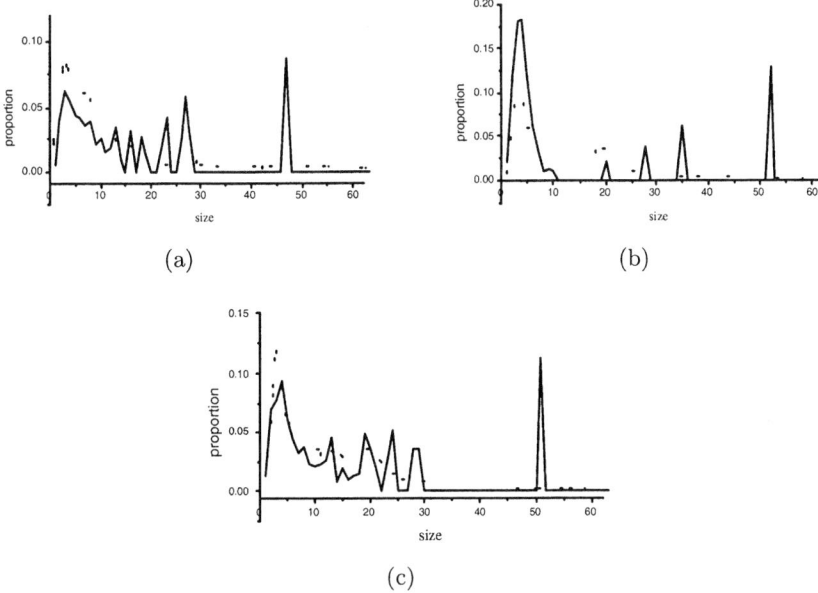

Fig. 6. The Size distributions of samples in Fig. 5

size, shape, color and other region features, and also get statistical information about these granular regions such as size distribution etc.

The pictures in Fig. 6 are the size distributions of the samples in Fig. 5. Solid lines are acquired from the segmentation results with our granular analysis method, while the dot lines are calculated in Matheron's sieving way. The element structure is a disk. As can be seen, the results of sieving method lose the distribution information of large regions, while the results acquired from the method proposed in this paper describe both the large and small regions correctly. Moreover, the computational performance of the proposed way to extract size distribution is much higher than that of the sieving method. The comparative computing costs of the two methods in our experiments are illustrated by Table 1.

Table 1. Time consumptions which are acquired on p4-2.4 PC

Image	Sieving method	Varying ladder
(a)	131,047	219
(b)	135,500	219
(c)	158,406	219

5 Conclusion

Granular analysis is an important task for mineral analysis because the size and quantity of granules contained in the mineral samples decides the quality of the mineral. In the image processing field, there is a classic way for granular analysis based on Matheron's sieving process. However, existing sieving methods are inefficient and unable to locate the position of granules. Since the watershed transform has been successfully used in image segmentation, we adopt it in our granular analysis. However, the watershed transforms may cause over-segmentation when applying them directly to images. There are some existing methods which could overcome this problem, but they all have high computational complexity. The paper proposes an improved watershed called varying-ladder watershed based on the Vincent's immersion definition. The improved method can not only overcome the over-segment problem but also highly enhance the computing performance. Then the paper proposes a new model of granular analysis based on watershed. The model is verified to be efficient and has high performance for granular analysis through large amount of experiments.

References

1. Matheron, G.:Random Sets and Integral Geometry. John Wiley & Sons, London (1975)
2. Maragos, P.: Pattern Spectrum and Multiscale Shape Representation. IEEE Trans, Pattern Analysis and Machine Intelligence, vol.11 (1989) 701-715
3. Dougherty, E. R. Newell, J. Pelz, J.: Morphological Texture-based Maximumlilelihood Pixel Classification Based on Local Granulometric Moments. Pattern Recognition, vol.25 (1992) 1181-1198
4. Ayala, G.; Diaz, E.; Domingo, J.; Epifanio, I.: Moments of Size Distributions Applied to Texture Classification. IEEE Trans, Proceedings of the 3rd International Symposium on Image and Signal Processing and Analysis, vol. 1 (2003) 18-20
5. Ayala, G.; Domingo, J.: Spatial Size Distributions: Applications to Shape and Texture Analysis. IEEE Trans, Pattern Analysis and Machine Intelligence, vol. 23 (2001)
6. Vincent, L.; Soille, P.: Watersheds in Digital Space: An Efficient Algorithm Based on Immersion Simulations. IEEE Trans, Pattern Analysis and Machine Intelligence, vol. 13 (1991)
7. Roerdink, J.; Meijster, A.: The Watershed Transform: Definitions, Algorithms and Parallelization strategies. Fundamenta Informaticae, Vol.41, IOS Press (2001) 187-228
8. Gao, H.; Siu, W.; Hou, C.: Improved Techniques for Automatic Image Segmentation. IEEE Trans, Circuits and Systems for Video Technology, Vol.11 (2001)
9. Haris, K.; Efstratiadis, SN.; Maglaveras, N.; Katsaggelos, AK.: Hybrid Image Segmentation Using Watersheds and Fast Region Merging Image Processing. IEEE Trans, Image Processing, vol.7 (1998) 1684-1699
10. Wang, D.: A Mutiscale Gradient Algorithm for Image Segmentation Using Watersheds. Pattern Recognition, vol.30 (1997) 2043-2052

Cost-Sensitive Ensemble of Support Vector Machines for Effective Detection of Microcalcification in Breast Cancer Diagnosis

Yonghong Peng[1], Qian Huang[2], Ping Jiang[1], and Jianmin Jiang[1]

[1] School of Informatics, University of Bradford, West Yorkshire, BD7 1DP UK.
 {Y.H.Peng,P.Jiang,J.Jiang1}@Bradford.ac.uk
[2] School of Electronic and Information Eng., South China Univ. of Tech. 510640, China
 eeqhuang@scut.edu.cn

Abstract. This paper presents a new approach for the cost-sensitive classification problems based on the Boosting ensemble of support vector machines (SVMs). Different from conventional Boosting ensemble learning methods that adjust the distribution of training instances for minimizing the misclassification rate, the presented approach adjusts the training data distribution so as to minimize the expected cost of classification. This approach has been applied successfully in Microcalcification (MC) detection which is a typical cost-sensitive classification problem in breast cancer diagnosis. Its performance is evaluated by means of Receiver Operating Characteristics (ROC) curves and the expected costs of classification. Experimental results have consistently confirmed that the ROC of the SVM ensemble classifier is very close to the curve enveloping the base classifier ROC curves. This characteristic illustrates that the SVM ensemble is able to always improve the performance of the classification. Furthermore, the experimental results demonstrate that the method presented is able to not only increase the area under the ROC curve (AUC) but also minimize the expected classification cost.

1 Introduction

Most traditional machine learning algorithms for classification have been developed to minimize the misclassification rate. The classifiers generated by these methods are not optimal for many real-world classification problems that having different cost for different classification errors and inherent with imbalanced distribution of sample classes. In medical diagnosis a false positive might lead to unnecessary treatment and unnecessary worry, while a false negative lead to postponed treatment, failure to treat or even death. It is thus expected that a learning algorithm can reflect the cost factors and construct classifiers to work robustly in such a cost-sensitive environment.

Much research has been performed to investigate the effectiveness of ensemble machine learning to construct optimal combination of multiple diverse classifiers for unseen data samples accurately [1]. It has been demonstrated that machine learning ensembles can often perform better than the associated base classifier in terms of accuracy. Boosting is one of the widely-used ensemble machine learning methods, in which a common learning algorithm is used to generate multiple classifiers with re-sampled training instances in order for maximizing the classification accuracy. In addition, several researchers have recently demonstrated that the ensemble machine learning offers a promising solution for cost-sensitive problems (e.g. [2],[3],[4] and [18],[19]). For example, in [4], Ting investigated four different versions of cost-sensitive Boosting algorithms for the ensemble of decision tree C4.5.

A few researchers explored the effectiveness of Support vector machine (SVM) ensembles [5],[6],[7],[8]. Kim et al. [5], [6] reported that improvements over single SVMs have been obtained by SVM ensembles, and Valentini et al [8] showed that an ensemble of SVMs is theoretically promising and suggested two approaches for developing SVM ensembles. Different from the traditional Boosting method that adjusts the distribution of training instances for minimizing the classification error rate, the approach presented intends to adjust the distribution of training instances to minimize the expected cost. The effectiveness of the presented SVM ensemble is evaluated by a Microcalcification detection that is a typical cost-sensitive problem, in which the cost of false positive is significantly different from that of a false negative.

Microcalcifications are tiny calcium deposits that appear as small bright spots in a mammogram, which have become an important indicator for breast cancer diagnosis. MCs are usually difficult to detect, especially in the early stage of development, because of the complex surrounding of breast tissue, their variation in shape, orientation, brightness and size (typically, 0.05-1mm). Manual searching for MCs from mammography is labour intensive, time consuming, and demands great concentration. During the past decades, many image-processing algorithms and data mining techniques have been proposed for the detection of MCs [9]. A two-stage neural network [10], and a hierarchical neural network [11] approaches were proposed. The use of SVM [12] and association rules classification [13],[14] have been also investigated recently. In [12], an SVM is trained through supervised learning to classify each location in the image as MC-present or MC-absent, and a method called 'successive enhancement-learning' was developed to select iteratively the most representative MC-absent examples for training.

2 MC Detection Based on SVM Classification

The process of MC detection is to scan each pixel in a mammogram and determine whether a pixel is located in an MC-present region or not. It can be treated as a two-class classification problem, classifying each pixel into MC-present or MC-absent [12]. For classification, one pixel is characterized by its surrounding pixels with an m by n window (m, n are odd integers, $m=n=9$ is used in this study), i.e. a pixel is represented by a feature vector $x \in R^d$, $d= m \times n$. Pixels located in MC-absent region and MC-present region are denoted by the class label $y=-1$ or $y=1$ respectively.

Unlike most of the modeling methods attempting to minimize an objective function (such as the mean square error) for the whole training instances, SVM finds the hyperplanes that produce the largest separation between the decision function values for the instances located at the borderline between two classes. Let

$S = \{(x_i, y_i), i = 1,2,...,N\}$ denotes a given training set where N is the number of training instances. The target of MC detection is to construct a decision function $f(x)$: $R^d \rightarrow R$, such that for each x_i, the function yields $f(x_i)>0$ for $y_i=+1$, and $f(x_i)<0$ for $y_i=-1$. A SVM employs a linear function $f(x) = W^T x + b$ or a nonlinear function $f(x) = W^T \phi(x) + b$, where $\phi(x)$ is a nonlinear transform function. The $f(x)$ is determined by minimizing $J(w,\xi) = \frac{1}{2}\|W\|^2 + C\sum_{i=1}^{l}\xi_i$ subject to $y_i(W^T x_i + b) \geq 1 - \xi_i$ (linear) or $y_i(W^T \phi(x_i) + b) \geq 1 - \xi_i$ (nonlinear), where $C > 0$ is a regularization parameter and $\xi_i \geq 0$ ($i=1,2,....l$) are slack parameters. The function $K(u,v) = \phi^T(u)\phi(v)$ is usually called the kernel. Three typical kernel functions used in SVM classification are the Linear, Polynomial and Gaussian functions.

3 Cost-Sensitive Ensemble of SVMs

3.1 Boosting Ensemble of Machine Learning

One important discovery during the past years is that ensemble of multiple classifiers can usually produce better accuracy than the individual base classifiers [1]. Boosting has been shown to be an effective method of producing base classifier and improving the classification accuracy [15],[16]. It uses a common learning algorithm to induce multiple individual classifiers sequentially, during which the distribution of training instances is adjusted according to the contribution of classification error. In the end, a vector of weights is produced to reflect the importance of each trained base classifier.

The distribution of training instances is adjusted in terms of their contribution to misclassification, i.e. higher sampling probability is assigned to the instances that were misclassified and less probability to instances classified correctly. These adjustments enable the machine learner to select different examples to learn in each trial and so lead to diverse classifiers. Finally, the individual classifiers are combined to form a composite classifier by weighting the outputs of base classifiers. The AdaBoost, one version of Boosting widely used to generate multiple individual classifiers and their combination [15], is summarized in Fig.1.

3.2 Cost-Sensitive Learning for Classification

Cost-sensitive learning aims at developing an optimal classifier when different misclassification errors incur different costs. For a two class classification problem, the cost matrix has the structure shown in Table 1.

The cost coefficient $C(i,j)=c_{ij}$ denotes the cost predicting an instance as class j when its true class is i. The costs of a false positive and false negative are denoted

Table 1. Cost matrix for two-class classification

	Actual Positive	Actual Negative
Predicted Positive	$C(+1,+1)=c_{11}$	$C(-1,+1)=c_{01}$
Predicted Negative	$C(+1,-1)=c_{10}$	$C(-1,-1)=c_{00}$

respectively by c_{01} and c_{10}. It is reasonable to assume that $c_{01}>c_{00}$ and $c_{10}>c_{11}$ and $c_{11}=c_{00}=0$ [17]. In a cost-sensitive classification, the optimal prediction for an instance x is the class i that is associated with a minimal cost of classification, i.e.

$$\arg\min_i \left(c(i,x) = \sum_j p(j|x)C(j,i) \right) \quad (6)$$

where $p(j|x)$ is the probability of x being class j. In a two-class classification problem, the optimal prediction is positive (+1) if and only if $p(+1|x)c_{01} \le p(-1|x)c_{10}$, i.e. the cost of false positive is less or equal to the cost of false negative. If let $p = p(+1|x)$, a optimal decision is p^* which fulfills $(1-p^*)c_{01} = p^*c_{10}$, namely $p^* = c_{10}/(c_{10}+c_{01})$. Based on the estimated p^*, a cost-sensitive classification can be achieved by

If $p(+1|x) \ge p^*$, it is classified as positive;

If $p(+1|x) < p^*$, it is classified as negative. $\quad (7)$

Training. *Inputs:* 1) $S=\{(x_i, y_i)|x_i \in R^d, y_i \in \{-1,+1\}, i=1\sim N\}$, where N is the number of training instances; 2) Number of desired base classifiers (M). *Outputs:* 1) M base classifiers; 2) The associated weights α_t.

Step 1: The instance distribution is initialized as $D_1(i) = \dfrac{1}{N} \quad (1)$

Step 2: For $t=1\sim M$, repeat the following steps
- Train the t-th classifier, C_t, using the data sampled based on the distribution D_t.
- Calculate the classification error rate of classifier C_t in classifying the current training dataset: $\varepsilon_t = \dfrac{\|\{i \mid C_t(x_i) \ne y_i\}\|}{N} \quad (2)$
- Set $\alpha_t = \dfrac{1}{2}\ln\left(\dfrac{1-\varepsilon_t}{\varepsilon_t}\right) \quad (3)$

    which will be used as the weights in the final ensemble of base classifiers, and also used for updating the distribution of training instances in the next step.
- Update the distribution for training instances:

$$D_{t+1}(i) = D_t(i) \times \begin{cases} e^{-\alpha_t} \text{ if } C_t(x_i) = y_i \\ e^{\alpha_t} \text{ if } C_t(x_i) \ne y_i \end{cases} \text{ and } D_{t+1}(i) = \dfrac{D_{t+1}(i)}{|D_{t+1}|} \quad (4)$$

Classification. Given a new example x, each classifier C_i predicts one class $c_t = C_t(x) \in \{-1,+1\}$, and a final classifier C_{en} is to predicate the class of x by voting the majority class $C_{en}(x) = \text{sign}\left(\sum_{t=1}^M \alpha_t C_t(x)\right). \quad (5)$

Fig. 1. AdaBoost Ensemble of Machine Learning

3.3 Boosting SVM to Construct Cost-Sensitive Classifier

The AdaBoost method provides an effective method to implement the ensemble machine learning. It can be modified to reflect the different cost factors associated to different types of classification error and thus to produce a cost-sensitive classifier. Several researchers have reported the success of using the Boosting algorithm to achieve a cost-sensitive learning [4],[18],[19].

The traditional Adaboost algorithm employs Eqs. (2-4) to adjust the distribution of samples. The proposed new cost-sensitive Boosting method concentrates on learn from the classification costs, as shown in Fig.2, different from the classical Boosting algorithm and the existing cost-sensitive Boosting algorithm (e.g. [4],[18],[19]):

1) The initial distribution of instances is adjusted according to the cost efficient.

$$d_0(i) = \begin{cases} c_{10} \times n_- & \text{if } x_i \in \text{positive} \\ c_{01} \times n_+ & \text{if } x_i \in \text{negative} \end{cases} \text{ and } D_i(i) = \frac{d_0(i)}{\sum_{i=1}^{N} d_0(i)} \quad (8)$$

where n_+ and n_- are the number of positive and negative classes respectively.

2) The cost coefficients c_{01} and c_{10} are used to update the data distribution such that the instances producing higher misclassification cost are assigned with higher sampling probabilities, by replacing the error rate in the Eq. (3) with an relative cost of misclassification calculated by $\gamma_t = \varepsilon_t / \varepsilon_{max}$ where ε_t and ε_{max} are the cost of classification and the maximum cost estimated by:

$$\varepsilon_t = c_{01} \times FP_w + c_{10} \times FN_w \text{ and } \varepsilon_{max} = c_{10} n_{w+} + c_{01} n_{w-} \quad (9)$$

where FP_w and FN_w are the weighted numbers of false positives and false negatives respectively $FP_w = \sum_{i=1}^{M} w_t(i) \delta_-(x_i, y_i)$, $FN_w = \sum_{i=1}^{M} w_t(i) \delta_+(x_i, y_i)$, and $n_{w+} = \sum_{y_i=+1} w_t(i)$ and $n_{w-} = \sum_{y_i=-1} w_t(i)$ are the weighted number of true positives and negatives. Here $\delta_+(x,y)$ and $\delta_-(x,y)$ are $\delta_+(x,y) = \begin{cases} 0 & \text{if } x=y, y=+1 \\ 1 & \text{if } x \neq y, y=+1 \end{cases}$ and $\delta_-(x,y) = \begin{cases} 0 & \text{if } x=y, y=-1 \\ 1 & \text{if } x \neq y, y=-1 \end{cases}$.

3) For the adjustment of training instance distribution, indicated by Eq.(12), two important modifications have been made compared to the existing methods. One is the use of the factor $1 + \log_2(c_{10}/\min(c_{10}, c_{01}))$. This is based on two considerations: (i) it is able to return back as an AdaBoost by setting equal c_{01} and c_{10}, and (ii) the distribution can be adjusted more smoothly by using the logarithm function. The other modification is that, when an instance of rare class (e.g. $y_i=+1$) has been classified correctly, its sampling probability will not be reduced immediately but kept to be the same as in previously. This reflects the significance of rare instances.

$$D_{t+1}(i) = D_t(i) \times \begin{cases} 1 & \text{if } c_t(x_i) = y_i, y_i = +1 \\ e^{-\alpha_t} & \text{if } c_t(x_i) = y_i, y_i = -1 \\ \left(1 + \log_2\left(\frac{c_{10}}{\min(c_{10}, c_{01})}\right)\right) e^{\alpha_t} & \text{if } c_t(x_i) \neq y_i, y_i = +1 \\ \left(1 + \log_2\left(\frac{c_{10}}{\min(c_{10}, c_{01})}\right)\right) e^{\alpha_t} & \text{if } c_t(x_i) \neq y_i, y_i = +1 \end{cases} \quad (12)$$

4) In classification, the expected cost is calculated to generate a final classification.

> **Training.** *Inputs:* 1) S={(x_i, y_i)| $x_i \in R^d$, $y_i \in \{-1,+1\}$, i=1~N}, where N is the number of training instances. 2) the cost for false positive c_{01} and false negative c_{10}; 3) the number of desired base classifiers. *Outputs:* 1) a set of SVM base classifiers {C_t, t =1~M}; 2) the weights for base classifiers.
> *Step 1:* the instance distribution is initialized by Eq.(8)
> *Step 2:* For t=1~M, repeat the following steps
> - Train the t-th SVM classifier, C_t, using data sampled based on distribution D_t.
> - Calculate the relative cost of misclassification of classifier C_t.
> - Set the weight be $\alpha_t = 0.5 \ln((1-\gamma_t)/\gamma_t)$
> - Update the distribution by Eq(12), and then normalized by $D_{t+1}(i) = D_{t+1}(i)/|D_{t+1}|$.
>
> **Classification.** Given M base classification $c_i \in \{-1,+1\}$ (i=1~M), and their associated weights, the expected cost for classifying x as positive or negative are respectively $\varepsilon_+ = \sum_{c_t=-1} \alpha_t c_{01}$ or $\varepsilon_- = \sum_{c_t=+1} \alpha_t c_{10}$. The final classification is $c_{en}(x) = \arg\min_i(\varepsilon_i)$, i.e. the class is positive if $\varepsilon_+ < \varepsilon_-$, otherwise negative.

Fig. 2. Cost-Sensitive Ensemble Learning

4 Performance Evaluation and Discussion

4.1 Experimental Datasets and Results

The proposed approach has been tested by mammogram samples selected from the DDSM Digital Mammography Database at the University of South Florida[1]. All the mammograms selected have a pixel depth of 12 bits. In this study, the emphasis has been put on evaluating the effectiveness of Boosting ensemble of SVMs for the classification of pixels with MC-present and pixels with MC-absent, and so 624 pixels have been carefully selected from the MC-present regions, and 1248 pixels have been randomly picked from the MC-absent regions.

Fig.3 shows the results of three samples of mammogram (the grey images) and the detected MCs (black-white image). The black-white images were produced by masking the MC-absent pixels as black and the MC-present pixel as white.

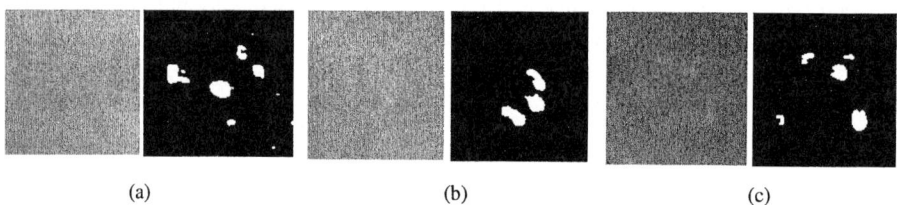

(a)　　　　　　　　　　(b)　　　　　　　　　　(c)

Fig. 3. MC Detection Samples

---
[1] http://marathon.csee.usf.edu/Mammography/Database.html

In the performance evaluation, the data set is divided in a random fashion into two separate subsets; one subset contains 400 (800) MC-present (MC-absent) pixels, and the remaining 224 (448) MC-present (MC-absent) pixels are then used for testing. The ROC cure, a plot of the true-positive rate achieved by a classifier versus the false positive rate, is employed for evaluating the effectiveness of the presented ensemble methods. Provost and Fawcett [20] explained how ROC provides a robust measure for comparing the performance of multiple classifiers in imprecise and changing environments and how ROC graph provides insight into the behavior of a classifier without regard to class distribution and misclassification cost. The area under the ROC curve, abbreviated as AUC [21],[22] provides a single scalar value for comparing the overall performance of classifiers. A classifier having greater AUC has better average performance than one having less AUC.

4.2 Performance of SVM and SVM Ensemble

By using different threshold values, a ROC can be generated by calculating the corresponding *TPR* and *FPR* for a SVM classifier:

$$TPR = \frac{\text{Positives correctly classified}}{\text{Total positives}} \text{ and } FPR = \frac{\text{Negative incorrectly classified}}{\text{Total negatives}}.$$

Fig. 4 (a) shows the ROC curve (solid line) for the ensemble of 7 base SVM classifiers produced by the proposed Boosting algorithms. By comparing it with the ROC curves of single polynomial SVM classifier (dot break line) and SVM ensemble classifier (solid line), it is clearly shown that better performance has been achieved by the SVM ensemble classifier. Fig.4 (b) shows the ROC curves associated with these 7 base classifiers. It has been observed that the ROC of the ensemble is nearly constructed by enveloping the ROC curves of base classifiers, namely each base classifier contributes from different regions to increase the AUC of the ensemble of SVMs. This reveals the reason why the ensemble of SVMs can achieve a better performance than a single SVM classifier.

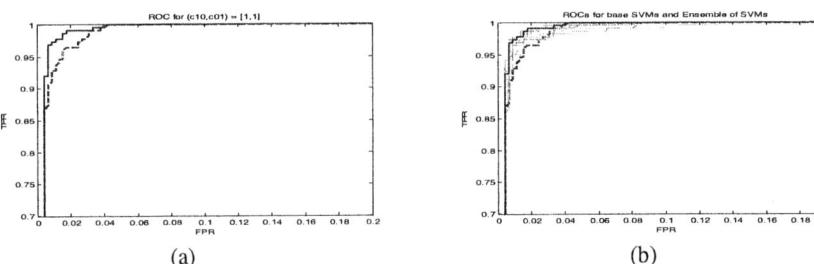

Fig. 4. ROC curves of ensemble SVM classifier.

4.3 Cost-Sensitivity of SVM Ensembles

The cost-sensitivity of the method presented is evaluated by the experiments performed under different cost matrices. Fig. 5(a) and (b) show respectively the results of using two different cost coefficients of false positive and false negative: a) $c_{10}=10$, $c_{01}=1$ and b) $c_{10}=1$, $c_{01}=10$, in which the ROC curves of ensembles and single SVM classifier are shown by the solid and dot break lines respectively. For comparison, the ROC curve of SVM ensemble for $c_{10}=c_{01}=1$ is also shown in Fig.5 in dot line. It is clearly shown, from Fig.6(a), that when increasing the cost coefficient of false positives, the upper part (the region of $FPR=2\%\sim100\%$) of ROC has moved more significantly towards the upper and left-hand corner than the lower part, which results in a significant reduction of false positives. On the other hand, as shown in Fig.6(b), when increasing the cost coefficient of false positives, the lower part of ROC (the region of FPR from 0% to 3%) have moved significantly towards the upper and left-hand corner, which results in a increased true positive rate and a reduction of false negatives rate. These results demonstrate that the method presented is able to increase the AUC and improve the performance for different regions of ROC curve.

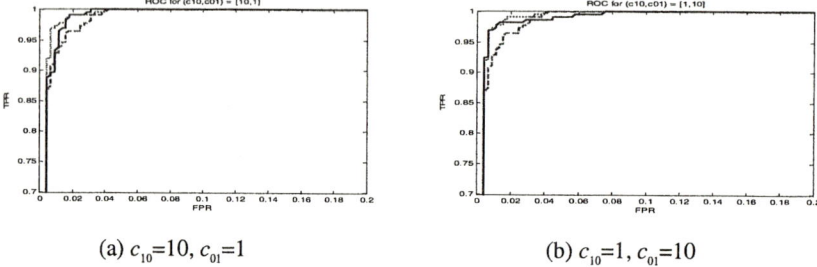

(a) $c_{10}=10$, $c_{01}=1$ (b) $c_{10}=1$, $c_{01}=10$

Fig. 5. ROC curves of SVM ensemble under different cost coefficients

The ROC provides an insight of analyzing the behavior of a classifier. The following experiments are used to evaluate if the method presented can minimize the expected cost of classification. By setting different cost coefficients (Table 2), a set of classification error (false positives and false negatives) and the associated classification cost have been calculated. Experimental results (Fig.6(a)) show (1) increasing the cost of false negative c_{10} results in decreasing FN and increasing FP; (2) increasing the cost coefficient of false positive c_{01} results in reducing FP and increasing FN. These patterns effectively reflect the cost-sensitivity of classification problems and the characteristics of the method presented, i.e. it can adaptively adjust the learning mechanism to focus on learning from instances with the high cost and try to avoid making error for these high cost instances. In Fig.6, the results obtained by the method presented in denoted as CS-enSVM, and the AdaBoost results were obtained by setting $c_{10}=c_{01}=1$.

Table 2. Experimental parameter setup (c_{10}, c_{01})

No.	1	2	3	4	5	6
$c_{10}>=c_{01}$	1,1	2,1	4,1	6,1	8,1	10,1
$c_{10}<=c_{01}$	1,1	1,2	1,4	1,6	1,8	1,10

Fig.6(b) shows the classification costs are calculated based on the false negatives and false positives and the cost coefficients: $c_{01} \times FP + c_{10} \times FN$. Fig.6(b) illustrates that when the cost coefficient changed, the costs of single SVM classifier change significantly; however the cost of SVM ensemble is kept at a minimum level, namely the ensemble of SVM can minimize the expected cost of classification although it may not minimize the classification error rate. Furthermore, the AdaBoost has the similar change pattern as single SVM but with lower cost, which confirms that AdaBoot improves the classification accuracy rather than minimizing the cost of classification.

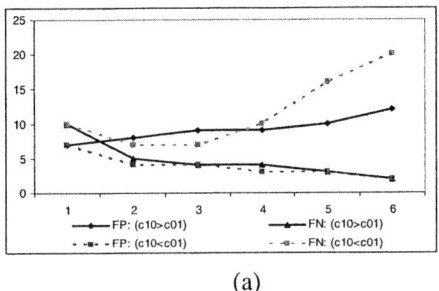
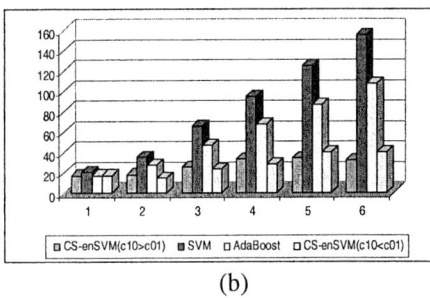

(a) (b)

Fig. 6. Classification costs under different cost coefficients

5 Conclusions and Future Work

This paper presents a cost-sensitive approach for the ensemble of support vector machines so that it can work effectively under different cost conditions. In the proposed method, the sampling probability of training instances is adjusted according to their contributions to the cost of classification. The effectiveness of the proposed method has been demonstrated by a MC detection problem, which is a typical cost-sensitive problem, by means of ROC and the expected cost of classification. Two main results and observations can be summarized as:

1) The ROC of Boosting ensemble of SVM is nearly generated by enveloping the ROC curves of the base SVM classifiers. This indicates the Boosting ensemble of SVMs is able to effectively increase its AUC and thus improve its performance in classification.
2) By analyzing the ROC curves and the expected cost of classification obtained under different cost matrices, it is clearly shown that the method presented can reflect effectively the classification cost and adjust accordingly the learning mechanism (instance sampling) so as to lead to an optimal classifier that minimizes the classification cost.

This research has raised several interesting subjects that are under investigation. One is the use of ROC for guiding the generation of an optimal ensemble of machine learning, i.e. to generate multiple classifiers that have complementary natures so that their combination can maximize the performance using the minimum number of base classifiers. One possible solution is to embed a mode selection mechanism for selecting the sensible base classifiers rather than involving all the base classifiers. Another interesting subject is to develop a cost-insensitive ensemble of SVM that can work robustly under varying cost conditions of classification and class distributions, and it is different from this research that emphasizes on the cost-sensitivity in classification.

References

1. Dietterich, T. G.: Ensemble Methods in Machine Learning, First International Workshop on Multiple Classifier Systems, Springer Verlag 1-15 (2000).
2. Margineantu, D., Dietterich, T. G.: Bootstrap Methods for the Cost-Sensitive Evaluation of Classifiers, *ICML-2000*, 583-590 (2000).
3. Drummond, C., Holte, R. C.: Exploiting the Cost (In)sensitivity of Decision Tree Splitting Criteria, *ICML-2000*, 239-246 (2000).
4. Ting, K. M.: A Comparative Study of Cost-Sensitive Boosting Algorithms, *ICML-2000*, 983-990 (2000).
5. Kim, H. C., Pang, S. et al.: Pattern Classification Using Support Vector Machine Ensemble. *ICPR'02* 20160–20163 (2002).
6. Kim, H. C., Peng S. et al.: Constructing Supporting Vector Machine Ensemble, The Journal of Pattern Recognition 36 (2003) 2757-2767.
7. Buciu, I., Kotropoulos C., Pitas, I.: Combining Support Vector Machines for Accurate Face Detection. Proc. of ICIP'01 1054–1057 (2001).
8. Valentini, G., Dietterich, T. G.: Bias-variance analysis of Support Vector Machines for the Development of SVM-based Ensemble Methods. Journal of Machine Learning Research, 5 (2004) 725–775.
9. Antonie, M. L., Zaiane, O. R., Coman, A.: Application of Data Mining Techniques for Medical Image Classification, MDM/KDD'2001 with ACM SIGKDD (2001).
10. Yu, S., Guan, L.: A CAD System for the Automatic Detection of Clustered Microcalcifications in Digitized Mammogram Films, IEEE Trans. Med. Imag. 19 (2000) 115-126.
11. Sajda, P., Spence, C., Pearson, J.: Learning Contextual Relationships in Mammograms Using a Hierarchical Pyramid Neural Network, IEEE Trans. Med. Imag., 21(3) (2002) 239-250.
12. El-Naqa I., Yang Y., Wernick, M. N. et al: A Support Vector Machine Approach for Detection of Microcalcifications, IEEE Trans. Med. Imag., 21(12) (2002) 1552-1563.
13. Ordonez, C. Santana, C. et al,: Discovering Interesting Association Rules in Medical Data, ACM SIGMOD Workshop on Research Issues in Data Mining and Knowledge Discovery, (2000).
14. Zaiane, O. R., Antonie, M. L., Coman, A.: Mammography Classification by an Association Rule-based Classifier, MDM/KDD 2002 with ACM SIGKDD (2002).
15. Freund, Y., Schapire, R. E.: A Decision-theoretic Generalization of On-line Learning and an Application to Boosting, Journal of Computer and System Sciences, 55(1) (1997) 119-139.
16. Schapire, R. E.: A brief introduction to Boosting. The 16th International Joint Conference on Artificial Intelligence, 1999, 1401–1406.
17. Elkan, C.: The Foundations of Cost-Sensitive Learning, The 17th International Joint Conference on Artificial Intelligence, IJCAI'01 973-978 (2001).

18. Ting, K. M., Zheng, Z.: Boosting Cost-Sensitive Trees, The First International Conference on Discovery Science, 244-255 (1998).
19. Fan, W., Stolfo, S., et al.: Adacost: Misclassification Cost-sensitive Boosting, *ICML-99* 99-105 (1999).
20. Provost F., Fawcett, T.: Robust Classification for Imprecise Environments, Machine Learning Journal, 42(3) (2001) 203-231.
21. Fawcett, T.: ROC Graphs: Notes and Practical Considerations for Researchers, Submitted to Machine Learning Journal, 2004.
22. Bradley, A. P.: The use of the Area under the ROC curve in the Evaluation of Machine Learning Algorithms, The Journal of Pattern Recognition, 30(7), (1997) 1145-1159.

High-Dimensional Shared Nearest Neighbor Clustering Algorithm*

Jian Yin[1], Xianli Fan[1], Yiqun Chen[1,2], and Jiangtao Ren[1]

[1] Zhongshan University, Guangdong, P.R. China, 510275
[2] Guangdong Institute of Education, Guangdong, P.R. China
issjyin@zsu.edu.cn

Abstract. Clustering results often critically depend on density and similarity, and its complexity often changes along with the augment of sample dimensionality. In this paper, we refer to classical shared nearest neighbor clustering algorithm (SNN), and provide a high-dimensional shared nearest neighbor clustering algorithm (DSNN). This DSNN is evaluated using a freeway traffic data set, and experiment results show that DSNN settles many disadvantages in SNN algorithm, such as outliers, statistic, core points, computation complexity etc, also attains better clustering results on multi-dimensional data set than SNN algorithm.

1 Introduction

Cluster analysis mainly originates from many research domains, including data mining, statistics, biology, machine learning and so on. Most data mining researchers focus on the scalability and effectivity on clustering algorithm. Except that, researchers also try to find out methods for cluster data with complex figures and types, high-dimension cluster analysis technology, and methods for hybrid numerical and systematic data.

Clustering is one of the challenging research domains in data mining and is widely applied on all kinds of application. But different application has its special requirement, such as flexibility, domain knowledge minimum to decide input parameter, ability to deal with outliers, not sensitive to input record sequence, high dimensionality, usability and so on.

In this paper, we describe a high-dimensional shared nearest neighbor clustering (DSNN) algorithm, and evaluate it on multi-dimensional spatio-temporal data set. The rest of the paper is organized as follows: In section 2, we will introduce the challenge of high dimensional data and classical shared nearest neighbor clustering algorithm, and analysis its disadvantages. Then by introducing new

---

* This work is supported by the National Natural Science Foundation of China (60205007) , Natural Science Foundation of Guangdong Province (031558,04300462), Research Foundation of National Science and Technology Plan Project (2004BA721A02), Research Foundation of Science and Technology Plan Project in Guangdong Province (2003C50118) and Research Foundation of Science and Technology Plan Project in Guangzhou City(2002Z3-E0017).

relevant similarity and density definitions, we bring out the DSNN algorithm in Section 3. Finally, in Section 4, we evaluated DSNN on American freeway traffic dataset. The experiment results prove that DSNN algorithm has better performance than SNN algorithm on correlative measures, such as in-cluster similarity, between-cluster dissimilarity, spatio-temporal complexity and so on.

2 Related Works

The clustering algorithm is mainly applied on spatio-temporal dataset, especially for traffic dataset.

2.1 Challenge from High Dimensionality

Observe the data samples showed in table 1 and table 2, their Euclidean distance is five. A attribute with a "0" value means it is a useless data(the experiment got no test data) or it is a free traffic situation. Attributes with a value from "1"to "4" stands for a more and more busy traffic situation. That is attribute with a value "4" means the most busy traffic situation. But it is not hard to judge that there is higher similarity between Sample Three and Sample Four than between Sample One and Sample Two, because there are eight shared attributes between the former pair, and none between the latter pair. Generally speaking, if there are redundant "0" attributes in samples, it is sparse; we often define the similarity by ignoring those useless attributes between two samples.

Other measures, such as cosine measure, Jaccard modulus etc, can be used to deal with this problem. If we compute the cosine measures between Sample One and Sample Two, Sample Three and Sample Four respectively, the former is zero and the latter is 0.8286.

Even if cosine and Jaccard measure can provide relevant similarity information about whether there are intact attributes or not, they can't handle high dimensionality very well. By computing cosine measure between high dimensional data, we can find that, if only observing 15-20 percents of attributes of Sample A, those data, which is very similar to Sample A, may belong to other clusters without Sample A, as described in [1]. It doesn't derive from the defective similarity measure, but from the un-trusted direct similarity of high dimensional data, since the similarity between any sample pair is all very low.

Another very important problem derived from similarity measures is *the triangle inequality* doesn't hold. Consider the following example in table 3: Sample

Table 1. data sample 1∼2

Point	Att1	Att2	Att3	Att4	Att5	Att6	Att7	Att8	Att9	Att10	Att11	Att12
1	4	0	0	0	0	0	0	0	0	0	0	0
2	0	0	0	0	0	0	0	0	0	0	0	3

Table 2. data sample 3~4

Point	Att1	Att2	Att3	Att4	Att5	Att6	Att7	Att8	Att9	Att10	Att11	Att12
3	4	2	0	1	4	3	0	2	3	4	1	0
4	0	2	0	1	4	3	0	2	3	4	1	3

Table 3. data sample 5~7

Point	Att1	Att2	Att3	Att4	Att5	Att6	Att7	Att8	Att9	Att10	Att11	Att12
5	3	4	2	4	1	4	3	3	2	2	1	4
6	0	0	2	4	1	4	2	1	4	3	2	3
7	4	2	3	3	0	2	2	1	4	3	3	1

5 is close to sample 6, sample 6 is close to sample 7, but sample 5 and sample 7 have a similarity of 0. The similarities come from different sets of attributes.

2.2 SNN Algorithm

Some researchers provide a *shared k nearest neighbors-based* algorithm (SNN algorithm) [2,3]. Its procedure can be described as: firstly compute the similarity matrix, then sparsify the similarity matrix by keeping only the k most similar neighbors, finally construct a correlative shared nearest neighbor graph. Subsequently, it will find out the SNN density of each point to specify the core points, and form clusters with the core points. Except that, it also discards all noise points and assigns all non-noise, non-core points to clusters.

With the synthesis analysis on this algorithm, it is not difficult to find that there are many disadvantages as follows:

- There are no enough processes about outliers, which results in redundant pointless computations.
  Apparently, there are only a few calculations about outliers in SNN, until finishing constructing SNN graph with all samples. Moreover, you have to begin judging whether it is an outlier until computing link strengths between all points. By analyzing its algorithm, we can find that the complexity of computing similarity matrix and constructing SNN graph is $0(M^2)$, which lead to a high complexity of this algorithm.
- Scientist definitions of thresholds for core points, outliers and filter those link strengths are not clearly provided.
  With the statistics to the sample data, we can find out the thresholds to define core points and outliers. However, there is high spatio-temporal measure in statistics, which undoubtedly add complexity in this total algorithm.
- The procedure of defining core points is not good enough.
  It is not very exact to define core points directly by thresholds, which lead to a vital problem: defined core points may belong to identical cluster. Thus,

if clustering with these core points, it is hard to avoid partitioning born clusters. In other word, in order that the latter clustering can be as accurate as possible, the core points had better to be as disperse as possible.

3 High Dimensional Shared Nearest Neighbor Clustering Algorithm (DSNN)

In order to overcome the disadvantages of the classical SNN algorithm, we bring out the refined DSNN clustering algorithm.

3.1 "Elementary Deletion for Outliers"

Given a distance measure on a feature space, there are many different definitions of distance-based outliers. The simplest type of algorithm based on nested loops in conjunction with randomization and a pruning rule gives state-of-the-art performance [4]. The algorithm not only computes the distance between any two examples using, for example, Euclidean distance for continuous features and Hamming distance for discrete features, but also can be any monotonically decreasing function of the nearest neighbor distances such as the distance to the kth nearest neighbor, or the average distance to the k neighbors.

The main idea in our nested loop algorithm is that for each example in a data set, we keep track of the closest neighbors found so far. When an example's closest neighbors achieve a score lower than the cutoff, we remove the example because it can no longer be an outlier. As we process more examples, the algorithm finds more extreme outliers and the cutoff increases along with pruning efficiency.

What needs to attend is that samples can be put into random order in linear time and constant main memory with a disk-based algorithm. One repeatedly shuffles the data set into random piles and then concatenates them in random order.

As far as the algorithm complexity, because of the nested loops, it could require $0(M^2)$ distance computations and $0(M^2/blocksize)$ data accesses.

3.2 Core Points and Outliers

SNN Algorithm can define the core points and outliers with the help of correlative threshold by users. Those points with a higher link strength than the threshold is defined as core points, vice versa. This method with threshold often have a inferior efficiency, since users are required to have a deep understand on spatio-temporal data set, and also require to take abundant correlative experiments for thresholds setting. There apparently exists the possibility of threshold data error, and it often reflects on spatio-temporal data set, which reduces clustering precision and maneuverability by a long way.

Furthermore, there exists another important problem: All or some of these core points may possibly belong to identical cluster.

DSNN Algorithm has a new idea for defining core points. Firstly, we can confirm a candidate set of core points, which must make certain that all possible core points are included in it. Then, we can choose a core point randomly, and define a candidate point as a new core point by distance measure, whose distance is more than distance threshold. After defining new core points as a seed, we can continuously define other candidate core points in that candidate set until judging all candidate points.

Aimed at outliers, the method in DSNN is based on SNN graph and candidate core point set. If some sample has fewer SNN of other samples, it is defined as an outlier.

3.3 DSNN Algorithm

After discussing the deficiencies of the classical SNN algorithm and providing correlative solutions, we have succeeded in discovering the following DSNN algorithm.

Algorithm 1 DSNN clustering algorithm

Input: spatio-temporal data set
Output: every individual cluster

1. Transfer an algorithm to find distance-based outliers as above, in order to achieve better precision with refined sample set.
2. Based on this sample set, construct the similarity matrix.
3. Sparsify the similarity matrix using k-nn sparsification.
4. Construct the shared nearest neighbor graph from k-nn sparsified similarity matrix.
5. For every point in the graph, calculate the total strength of links coming out of the point.
6. Define noise points by choosing the points that have low total link strength and remove them, and gain a refined dataset again.
7. Based the new dataset, re-implement Step 2 to Step 5 once.
8. Define the candidate set of core points by choosing the points that have high total link strength.
9. For this candidate set, to do some refining as follows: saving those whose relative distance is very high as defined core points, and deleting the remainder.
10. Based on these core points and new SNN graph, to form clusters, where every point in a cluster is either a core point or is connected to a core point.

---

4 Experiments

This traffic dataset is from American freeway network, which is recorded as a matrix, where every row indicates each checkpoint and column to time segment. For every data set $D_{i,j}$, it records traffic state in the jth time segment within the ith checkpoint. Moreover, it has been preprocessed, and records traffic states with 0 to 4, where 0 indicates free, 4 indicates busy.

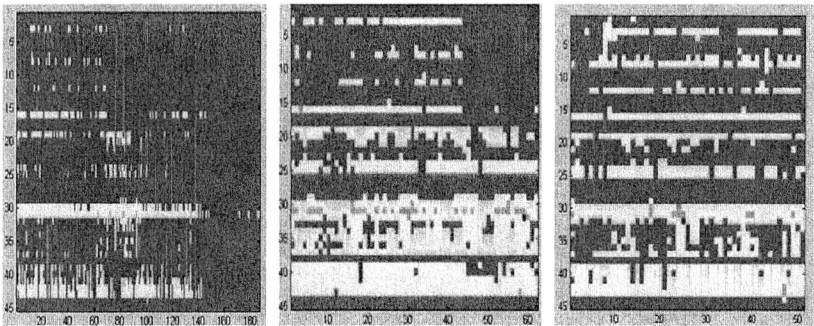

Fig. 1. the same Cluster Results with SNN and DSNN

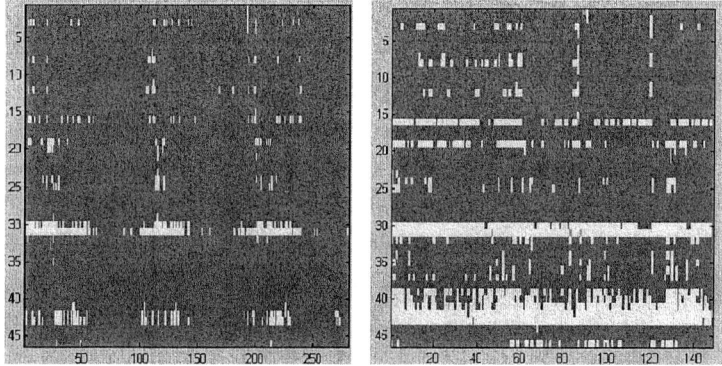

Fig. 2. the Cluster with SNN Algorithm (Left) and DSNN Algorithm (Right)

In this section, we evaluate our DSNN by applying SNN and DSNN on this data set and compare their results on different experiment environment. We will try different attempts with this experiment dataset, such as different amount of data set, and compare their experiment results.

We randomly chose 300 samples from the set, and the clustering result with DSNN method was same as one with SNN method in same time: As the three figure showed in Figure 1, they respectively show a cluster. The number of abscissa stands for the experiment day number, while the ordinate stand for the time period of a day. As we can see from the three figures, the amount and sample data was also same respectively, which shows that they can be used to handle small data set with same precision. We can get the traffic situation from the tint area in the figure: tint area show a free traffic situation, others is a busy traffic situation. For example, the first left figure shows that in the time periods 16, from 29 to 32, from 41 to 44(those tint area), the traffic is in a normal situation, it also shows that the traffic situation has a high similar in these time period.

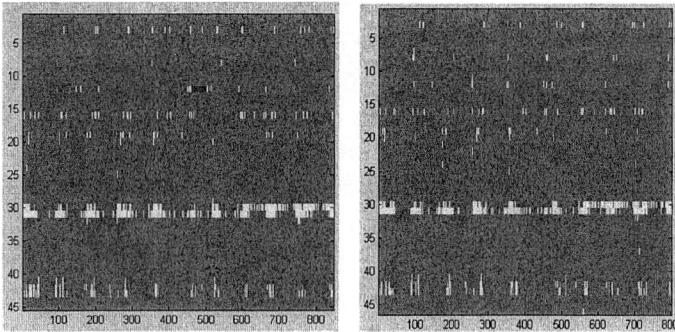

Fig. 3. the Supreme cluster with SNN (left) and DSNN (right) method

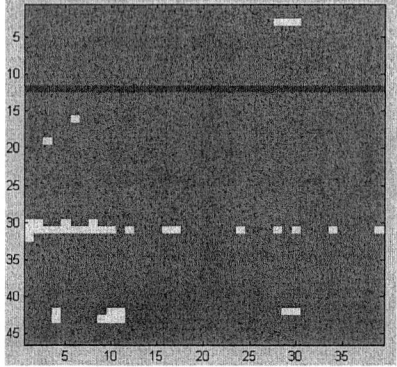

Fig. 4. The partial outliers with DSNN method

What is showed in Figure 2 is the most primary cluster with classical SNN algorithm and DSNN method respectively and about 600 samples. The meaning of abscissa and ordinate is the same with Figure 1. The left figure and the right figure stand for a cluster result respectively. By computing their cluster-in similarities respectively, we can gain the clusters with high cluster-in similarities if adopting the DSNN method. In other word, DSNN clustering algorithm can help with obtaining more accurate clusters on big data sets.

Perhaps there is an incident in the clustering result with six hundred samples, so we decided to cluster with more samples continuously in order to evaluate the performance of the DSNN algorithm.

Figure 3 shows the supreme cluster graphs with the classical SNN method and the DSNN method respectively. The meaning of abscissa and ordinate is the same with Figure 1. The left figure include the data in the right figure, but each of them stand for a cluster result respectively. Obviously, they are same on the whole, but there are more outliers in the former. Moreover, with careful feedback of algorithm implement, we can find that the difference is primarily included in the outliers during the course of deleting outliers, as showed in Figure 4. All the above shows that DSNN has a much better adaptability and performance than SNN.

5 Conclusion

In this paper, we firstly analyzed the classical SNN algorithm in the fields of spatio-temporal data cluster analysis. Then we brought forward the high dimensional nearest neighbor clustering algorithm (DSNN) step by step to overcome SNN's shortcomings. This refined algorithm can reduce the spatio-temporal complexity effectively, and refined many performances, such as outliers, core points, clustering results and so on. With the experiments on American freeway traffic dataset, we prove that DSNN can reduce computation effectively, at the same time, it can accurately judge core points and outliers, and gain better clustering performance than SNN algorithm with better clustering methods.

There are still some limits in DSNN. For example, it can't deal with data flow; it must be used to preprocess sample data. And how to expand its application domains, how to improve its validity and how to delete sensitive data better such as outliers, etc, are our future work.

References

1. Clarke, F., Ekeland, I.: Nonlinear oscillations and boundary-value problems for Hamiltonian systems. Arch. Rat. Mech. Anal. **78** (1982) 315–333
2. S. Guha, R. Rastogi, K. Shim. Cure: An efficient clustering algorithm for large databases. In Proc. 1998 ACM-SIGMOD Int. Conf. Management of Data(SIGMOD'98).(1998)73-84.
3. Levent Ertoz, Michael Steinbach, Vipin Kumar. Finding Clusters of Different Sizes, Shapes, and Densities in Noisy, High Dimensional Data. In Proceedings of Third SIAM International Conference on Data Mining, San Francisco, CA, USA, May 2003.
4. Levent Ertoz, Michael Steinbach, Vipin Kumar. A New Shared Nearest Neighbor Clustering Algorithm and its Applications. Workshop on Clustering High Dimensional Data and its Applications, Second SIAM International Conference on Data Mining, Arlington, VA, USA, 2002.
5. Stephen D. Bay, Mark Schwabacher. Mining Distance-Based Outliers in Near Linear Time with Randomization and a Simple Pruning Rule. In Conference on Knowledge Discovery in Data archive Proceedings of the ninth ACM SIGKDD International Conference (KDD). (2003)29-38.
6. E. Eskin, A. Arnold, M. Prerau, L. Portnov, S. Stolfo. A framework for unsupervised anomaly detection: Detecting intrusions in unlabeled data. Applications of Data Mining in Computer Security, number 10/320,259, filing date: December 16, Kluwer 2002.
7. A. Strehl, J. Ghosh, R. Mooney. Impact of Similarity Measures on Web-page Clustering. In Proceedings of the 17th National Conference on Artificial Intelligence: Workshop of Artificial Intelligence for Web Search, AAAI/MIT Press. (2000)58-64.

8. P.-N. Tan, M. Steinbach, V. Kumar, S. Klooster, C. Potter, and A. Torregrosa. Finding spatio-termporal patterns in earth science data. In KDD Temporal Data Mining Workshop, San Francisco, California, USA, August 2001.
9. M. Steinbach, P.-N. Tan, V. Kumar, S. Klooster, and C. Potter. Temporal data mining for the discovery and analysis of ocean climate indices. In Proceedings of the KDD Temporal Data Mining Workshop, Edmonton, Alberta, Canada, August 2002.
10. S. Shekhar, C. T. Lu, S. Chawla, P. Zhang. Data Mining and Visualization of Twin-Cities Traffic Data. University of Minnesota Academic report, 2001.
11. Vipin Kumar, Michael Steinbach, Pang-Ning Tan. Mining Scientific Data: Discovery of Patterns in the Global Climate System. PAKDD, May 7, 2002.
12. Kitamoto Asanobu. Data mining for Typhoon Image Collection. Journal of Intelligent Information Systems,(2002)Vol.19, No.1.25-41.
13. P.-N. Tan, M. Steinbach, V. Kumar, S. Klooster, C. Potter, and A. Torregrosa. Finding spatio-termporal patterns in earth science data. In KDD Temporal Data Mining Workshop, San Francisco, California, USA, August 2001.

A New Method for Fuzzy Group Decision Making Based on α-Level Cut and Similarity

Jibin Lan[1,2], Liping He[2], and Zhongxing Wang[2]

[1] School of Economics and Management, Southwest Jiaotong University,
610031 Chengdu, Sichuan, P.R. China
lanjibin@gxu.edu.cn
[2] School of Mathematics and Information Science, Guangxi University,
530004 Nanning, Guangxi, P.R. China
billy-heliping@hotmail.com

Abstract. Let opinions of experts among group decision making be represented as L-R fuzzy numbers. The difference of two experts' opinions is reflected by two distances, which are called the left-hand side distance and the right-hand side one. A method to calculate two types of distances based on the same α-level is presented. Then the distances are employed to construct a new similarity function to measure the similarity degrees of both sides which represent the pessimistic and optimistic similarity degrees between the experts, respectively. The degree of importance of each expert among group decision making is obtained by employing Saaty's analytic hierarchy process (AHP). The method of aggregating individual fuzzy opinions into a group consensus opinion by combining similarity degrees and the degree of importance of each expert is proposed. Finally some properties of the proposed similarity measure are proved and some numeric examples are shown to illustrate our method.

1 Introduction

Up to now, several methods have been proposed for drawing consensus from opinions of experts. Methods in [5,9,10,15,17] considered the fuzzy preference relation of each expert. Ishikawa et al.[8] and Xu et al.[19] proposed the individual opinions represented by interval-values and got the consensus from cumulative frequency distribution. Hsu [7] proposed a method called similarity aggregation method (SAM) to aggregate individual opinions of experts based on the similarity degree function. But it is not effective when experts' opinions which are represented as fuzzy numbers have no common intersection. Even when they intersect, the difference between them still exists. For example, let the opinions of expert E_1 and expert E_2 be represented as trapezoidal fuzzy numbers $(1, 2, 4, 5)$ and $(4, 5, 6, 7)$. Obviously, they intersect at the point $P(4.5, 0.5)$. Since the left- and right-hand membership functions are considered to reflect the pessimistic and optimistic opinions of experts, respectively, for expert E_1, the opinion at the point of intersection is optimistic while for expert E_2 it is pessimistic. Thus, they aren't thought to be identical. At the same time, there are a number of approaches to determine the distance between fuzzy sets in the literature [2,11,12,13,14]. Most of them are based on the Hausdorff distance between the α-level cuts. However these approaches can not keep producing output sets with the same nature as the input sets [2,6]. Lee [3] proposed an iterative procedure for approximating the optimal consensus of experts' opinions which are represented by trapezoidal fuzzy numbers. Lee's method based on the distance between $\widetilde{A} = (a_1, a_2, a_3, a_4)$ and $\widetilde{B} = (b_1, b_2, b_3, b_4)$ is defined by:

$$d_p(\widetilde{A}, \widetilde{B}) = (\sum_{i=1}^{4}(|a_i - b_i|)^p)^{1/p} . \qquad (1)$$

This distance satisfies the conditions of non-negativity, symmetry and triangle inequality. However, Lee's method may not be effective in some cases. For example, let $\widetilde{A} = (4, 5, 7, 8)$, $\widetilde{B} = (2, 5, 7, 8)$ and $\widetilde{C} = (5, 6, 8, 9)$ be three trapezoidal fuzzy numbers and assume $p = 2$. According to Lee's method, we obtain $d_2(\widetilde{A}, \widetilde{B}) = 2$, $d_2(\widetilde{A}, \widetilde{C}) = 2$. But in fact the two distances aren't identical.

The main characteristics of the methods mentioned above are synthesized as follows:

(1) most of them are limited to the use of trapezoidal fuzzy numbers;

(2) the shape of the membership function is not taken into consideration or only a part of it is used, which leads to losing much information;

(3) the pessimistic opinion and the optimistic opinion which are two entirely distinct conceptions aren't considered, respectively.

In this paper, an aggregation based on the left- and right-hand side distances between fuzzy numbers for the degrees of similarity among experts whose opinions are represented by L-R fuzzy numbers is presented.

2 L-R Fuzzy Numbers

A fuzzy number is a fuzzy set in the real line and is completely defined by its membership function as follows:

$$\mu(x) : X \to [0, 1] \qquad (2)$$

where X is the subset of real line, $\widetilde{R} = \{(x, \mu(x)) | x \in X\}$ is called a fuzzy set.

The membership degree of x to the fuzzy set $\widetilde{R}$ is expressed by the membership function $\mu(x)$. For computational purpose, this definition is generally restricted to those fuzzy numbers which are both normal and convex:

(1) Normality: $\sup_{x \in X} \mu(x) = 1$, where $x \in X$

(2) Convexity: $\mu\{\lambda x_1 + (1 - \lambda)x_2\} \leq \min\{\mu(x_1), \mu(x_2)\}$, where $x \in \widetilde{R}$ and $\lambda \in [0, 1]$.

The definition and the two requirements imply that the α-level cut $\widetilde{R}_\alpha = \{x : \mu_{\widetilde{R}}(x) \geq \alpha\}$ of such a fuzzy subset $\widetilde{R}$ is a closed interval $[a_l^\alpha, a_r^\alpha]$ for any $\alpha \in (0, 1]$, and the highest membership values are clustered around a given interval (or point). And the 1-level cut $\widetilde{R}_1 = \{x : \mu_{\widetilde{R}}(x) = 1\}$ is not an empty set.

A general type of normal and convex fuzzy number is L-R fuzzy number and is defined by:

$$\mu(x) = \begin{cases} L(\frac{m-x}{\delta}), & x < m, \delta > 0 \\ 1, & m \leq x \leq n \\ R(\frac{x-n}{\gamma}), & x > n, \gamma > 0 \end{cases} \quad (3)$$

where δ, γ are the left-hand and the right-hand spread, respectively; $L(x)$ is a monotone increasing function and satisfies the condition $L(0) = 1$; $R(x)$ which is a monotone decreasing function and also satisfies the condition $R(0) = 1$ is not necessarily symmetric to $L(x)$. If $L(x)$ and $R(x)$ are linear functions, we can call the fuzzy number trapezoidal fuzzy number. A trapezoidal fuzzy number is determined by four parameters a, m, n, b; where $\widetilde{R}_0 = [a, b]$. We denote a trapezoidal fuzzy number as (a, m, n, b), and obtain a triangular fuzzy number while $m = n$.

In this paper, we assume that the pessimistic opinion of an expert's is reflected by the left-hand membership function while the optimistic opinion is reflected by the right-hand one, and the most possible opinion is reflected by the support set which is the closed interval $[m, n]$. Obviously, the closed interval $[m, n]$ is obtained when the pessimistic and optimistic opinions are decided. Since the pessimistic opinion and the optimistic opinion are two entirely distinct conceptions, the pessimistic opinion should be compared with the pessimistic one. So should the optimistic one. Therefore, the similarity degrees between pessimistic opinions and the similarity degrees between optimistic ones are merely considered in this paper, when we consider the similarity degrees of experts.

3 Distance and Similarity Between L-R Fuzzy Numbers

In order to obtain a rational distance, in this paper the left- and right-hand membership functions of L-R fuzzy number are assumed to be strictly monotonic functions. We have the theorem as follows:

Theorem 1. *The inverse function of $g(x)$ is also strictly monotonic if and only if $g(x)$ is a strictly monotonic function.*

Distance is an important concept in science and engineering. In the following, two types of distances' definitions between two L-R fuzzy numbers based on the same α-level are given.

Definition 1. *Let $\widetilde{R}_i$ and $\widetilde{R}_j$, representing the subjective estimate of the ration to an alternative under a given criterion of expert E_i and E_j, be two L-R fuzzy numbers with membership functions $\mu_{\widetilde{R}_i}(x)$ and $\mu_{\widetilde{R}_j}(x)$. The left-hand side distance which reflects the difference of their pessimistic opinions at the same α-level cut, is defined by:*

$$\rho^L_{\widetilde{R}_i,\widetilde{R}_j}(\alpha) = | \mu_{\widetilde{R}_i}^{L^{-1}}(\alpha) - \mu_{\widetilde{R}_j}^{L^{-1}}(\alpha) | \quad (4)$$

The right-hand side distance which reflects the difference of their optimistic opinions at the same α-level cut, is defined by:

$$\rho^R_{\widetilde{R}_i,\widetilde{R}_j}(\alpha) = |\mu^R_{\widetilde{R}_i}{}^{-1}(\alpha) - \mu^R_{\widetilde{R}_j}{}^{-1}(\alpha)| \tag{5}$$

where $\alpha \in [0,1]$; $\mu^L_{\widetilde{R}_i}$ and $\mu^R_{\widetilde{R}_i}$ are the left- and right-hand membership functions of $\widetilde{R}_i$ while $\mu^L_{\widetilde{R}_j}$ and $\mu^R_{\widetilde{R}_j}$ are the left- and right-hand ones of $\widetilde{R}_j$, respectively.

According to the theorem 1, the inverse function has only one value for each α, and both side distances satisfy the conditions of non-negativity, symmetry and triangle inequality. It has to be noted that either of the side distances is equal to zero if and only if $\mu^L_{\widetilde{R}_i}{}^{-1}(\alpha) = \mu^L_{\widetilde{R}_j}{}^{-1}(\alpha)$, or $\mu^R_{\widetilde{R}_i}{}^{-1}(\alpha) = \mu^R_{\widetilde{R}_j}{}^{-1}(\alpha)$. It means that the two experts have the same degrees of membership at the same point with the same opinion that is pessimistic or optimistic.

Now the side similarities between E_i and E_j at the same α-level are considered. It is well known that an exponential operation is highly useful in dealing with a similarity relation, Shannon entropy and in cluster analysis. Therefore, we choose

$$f(x) = e^{-x}. \tag{6}$$

The left- and right-hand side similarities between the two experts at the same α-level are defined by:

$$f^L_{\widetilde{R}_i,\widetilde{R}_j}(\alpha) = e^{-\rho^L_{\widetilde{R}_i,\widetilde{R}_j}(\alpha)}, \quad f^R_{\widetilde{R}_i,\widetilde{R}_j}(\alpha) = e^{-\rho^R_{\widetilde{R}_i,\widetilde{R}_j}(\alpha)} \tag{7}$$

where $f^L_{\widetilde{R}_i,\widetilde{R}_j}(\alpha)$ is named the left-hand side similarity while $f^R_{\widetilde{R}_i,\widetilde{R}_j}(\alpha)$ is named the right-hand one.

Obviously, if either of side distances between them is zero then the corresponding similarity is equal to one. Since the side similarities depend on the side distances and the α-level, the curvilinear integral is taken into account. That is

$$\begin{aligned} S^L(\widetilde{R}_i,\widetilde{R}_j) &= \int_{\rho^L_{\widetilde{R}_i,\widetilde{R}_j}(\alpha)} e^{-\rho^L_{\widetilde{R}_i,\widetilde{R}_j}(\alpha)} ds \Big/ \int_{\rho^L_{\widetilde{R}_i,\widetilde{R}_j}(\alpha)} ds \\ S^R(\widetilde{R}_i,\widetilde{R}_j) &= \int_{\rho^R_{\widetilde{R}_i,\widetilde{R}_j}(\alpha)} e^{-\rho^R_{\widetilde{R}_i,\widetilde{R}_j}(\alpha)} ds \Big/ \int_{\rho^R_{\widetilde{R}_i,\widetilde{R}_j}(\alpha)} ds \end{aligned} \tag{8}$$

where $\alpha \in [0,1]$. $S^L(\widetilde{R}_i,\widetilde{R}_j)$ is named left-hand side similarity measure function while $S^R(\widetilde{R}_i,\widetilde{R}_j)$ is named right-hand side similarity measure function.

We get $S^L(\widetilde{R}_i,\widetilde{R}_j) = 1$ and $S^R(\widetilde{R}_i,\widetilde{R}_j) = 1$ when expert E_i and expert E_j have the same opinions, that is $\widetilde{R}_i = \widetilde{R}_j$. In other words, the opinions of expert E_i and expert E_j are identical, and then the agreement degree between them is equal to one.

Since the left- and right-hand side similarity measure functions have been defined, now the side average agreement degrees of expert $E_i (i = 1, 2, \cdots, n)$ are given by

$$A(E_i^L) = \frac{1}{n-1} \sum_{j=1, j \neq i}^{n} S^L(\widetilde{R}_i, \widetilde{R}_j), A(E_i^R) = \frac{1}{n-1} \sum_{j=1, j \neq i}^{n} S^R(\widetilde{R}_i, \widetilde{R}_j) \quad (9)$$

where $A(E_i^L)$ is named the left-hand side average agreement degree while $A(E_i^R)$ is named the right-hand one.

Now we compute the relative left- and right-hand side agreement degrees of expert $E_i (i = 1, 2, \cdots, n)$ as follows:

$$SA_i^L = A(E_i^L) / \sum_{j=1}^{n} A(E_j^L), \quad SA_i^R = A(E_i^R) / \sum_{j=1}^{n} A(E_j^R) \quad (10)$$

where SA_i^L is named the relative left-hand side agreement degree while SA_i^R is named the right-hand one.

According to the above definitions, an expert's opinion is determined by his pessimistic and optimistic opinions. The relative agreement degree of expert's opinion depends on the side relative agreement degrees. We take the average value of SA_i^L and SA_i^R as the relative agreement degree of expert's opinion by:

$$SA_i = (SA_i^L + SA_i^R)/2 . \quad (11)$$

In practice, the group decision making is heavily influenced by the degrees of importance of participants. Sometimes there are important experts in the decision group, such as the executive manager of a company, or some experts who are more experienced than others. The final decision is influenced by the different importance of each expert. Therefore, a good method to aggregate multi-experts' opinions must consider the degree of importance of each expert in the aggregating procedure. We employ Saaty's [16] analytic hierarchy process (AHP) to deal with the weight of each expert. The construction of the square reciprocal matrix $(a_{ij})_{n \times n}$ is performed by comparing expert E_i with expert E_j, with respect to the degree of importance. The other values are assigned as follows: $a_{ij} = 1/a_{ji}$; $a_{ii} = 1$. To solve the reciprocal matrix, the maximum eigenvalue is cardinal ratio scale for the experts compared. The eigenvector is then normalized and the weight $\gamma_i (i = 1, 2, \cdots, n)$ of each expert is obtained.

As is stated above, we get the relative agreement degree and the degree of importance of each expert. Now the consensus degree coefficients of expert $E_i (i = 1, 2, \cdots, n)$ can be defined by:

$$\omega_i = \beta SA_i + (1 - \beta) \gamma_i \quad (12)$$

where $\beta \in [0, 1]$.

If $\beta = 1$, the degree of importance of expert is not considered in the aggregation process. If $\beta = 0$, only the degree of importance of expert is reflected in the consensus. The membership function of aggregating the consensus opinion $\widetilde{R}$ can be defined by:

$$\mu_{\widetilde{R}}(z) = \sup_{z=\sum_{i=1}^{n} w_i x_i} \min \mu_{w_i x_i}(x_i), (i = 1, 2, \cdots, n) . \tag{13}$$

We summarize the criterion which is discussed above and propose an algorithm to combine all experts' opinion into the consensus opinion of group decision making.

Algorithm 3.1

step 1: For the criterion and an alternative under group decision making environment, each expert E_i ($i = 1, 2, \cdots, n$) proposes his opinion as a L-R fuzzy number denoted by $\widetilde{R}_i$. Suppose the left- and right hand-side membership functions of $\widetilde{R}_i$ are strictly monotonic.

step 2: Calculate the left- and right hand-side distances between $\widetilde{R}_i$ and $\widetilde{R}_j$.

step 3: Calculate the side similarity degrees $S^k(\widetilde{R}_i, \widetilde{R}_j), (k = L, R)$ of the opinions between each pair of experts.

step 4: Calculate the side average agreement degrees $A(E_i^L)$ and $A(E_i^R)$.

step 5: Calculate the side relative agreement degrees SA_i^L and SA_i^R of expert $E_i(i = 1, 2, \cdots, n)$.

step 6: Calculate the relative agreement degree SA_i of expert $E_i(i = 1, 2, \cdots,)$.

step 7: Define the degree of importance of expert $E_i(i = 1, 2, \cdots, n)$ by employing Saaty's AHP.

step 8: Calculate the membership function of the group consensus opinion by equation(13).

The aggregation method preserves some important properties. These properties are as follows:

Corollary 1. *If $\widetilde{R}_i = \widetilde{R}_j$ for all i and j, then $\widetilde{R} = \widetilde{R}_i$. In other words, if all estimates are identical, the combined result should be the common estimate.*

Proof. If all $\widetilde{R}_i$ is equal, then $\widetilde{R} = \sum_{i=1}^{n} w_i \odot \widetilde{R}_i = \widetilde{R}_i \sum_{i=1}^{n} [\beta SA_i + (1-\beta)\gamma_i] = \widetilde{R}_i [\beta \sum_{i=1}^{n} SA_i + (1-\beta) \sum_{i=1}^{n} \gamma_i] = \widetilde{R}_i [\beta + (1-\beta)] = \widetilde{R}_i$.

Agreement preservation is a consistency requirement. □

Corollary 2. *The result of the aggregation method would not depend on the order with which individual opinions are combined. That is, if $\{(1), (2), \cdots, (n)\}$ is a permutation of $\{1, 2, \cdots, n\}$, then $\widetilde{R} = f(\widetilde{R}_1, \widetilde{R}_2, \cdots, \widetilde{R}_n) = f(\widetilde{R}_{(1)}, \widetilde{R}_{(2)}, \cdots, \widetilde{R}_{(n)})$. The result is also a consistency requirement.*

Corollary 3. *Let the uncertainty measure $H(\widetilde{R}_i)$ of individual estimate $\widetilde{R}_i$ be defined as the area under its membership function $\mu_{\widetilde{R}_i}(x)$,*

$$H(\widetilde{R}_i) = \int_{-\infty}^{\infty} \mu_{\widetilde{R}_i}(x) \mathrm{d}x \tag{14}$$

The uncertainty measure H is defined to fulfil the following equation.

$$H(\widetilde{R}) = \sum_{i=1}^{n} \omega_i \times H(\widetilde{R}_i) . \tag{15}$$

Corollary 4. *If an expert's estimate is far from the consensus, then his estimate is less important.*

Corollary 5. *The common intersection of supports of all experts' estimates is the aggregation result, namely $\bigcap_{i=1}^{n} \widetilde{R}_i \subseteq \widetilde{R}$.*

Proof. Let $\alpha - $ cut of $\widetilde{R}_i$ be $\widetilde{R}_i^\alpha = [a_i^\alpha, b_i^\alpha]$; let $\bigcap_{i=1}^{n} \widetilde{R}_i$ be $[a^\alpha, b^\alpha]$, and then
$\widetilde{R}^\alpha = \sum_{i=1}^{n} \omega_i \odot \widetilde{R}_i^\alpha = [\sum_{i=1}^{n} \omega_i a_i^\alpha, \sum_{i=1}^{n} \omega_i b_i^\alpha]$.

Since $\sum_{i=1}^{n} \omega_i a_i^\alpha \leq \max_i\{a_i^\alpha\} \leq a^\alpha$ and $\sum_{i=1}^{n} \omega_i b_i^\alpha \geq \min_i\{b_i^\alpha\} \geq b^\alpha$, we have $\bigcap_{i=1}^{n} \widetilde{R}_i \subseteq \widetilde{R}$. □

Corollary 6. *If $\bigcap_{i=1}^{n} \widetilde{R}_i = \phi$, a consensus also can be derived.*

4 Numerical Example

Example 1. Consider a group decision problem evaluated by three experts. The experts' opinions are represented as trapezoidal fuzzy numbers as follows:

$$\widetilde{R}_1 = (1, 2, 3, 4), \widetilde{R}_2 = (1.5, 2.5, 3.5, 5), \widetilde{R}_3 = (2, 2.5, 4, 6).$$

We employ our method to deal with this problem and consider two cases:

1. Do not consider the degree of importance of expert; i.e. $\beta = 0$
2. Consider the degree of importance of each expert; i.e. $0 < \beta < 1$.

The result of $\widetilde{R}$ is calculated in full details as follows:

Case 1: Do not consider the importance degree of each expert.
step 2: Calculate the left- and right hand-side distances between $\widetilde{R}_i$ and $\widetilde{R}_j$.

The left hand-side distances are as follows:
$\rho^L_{\widetilde{R}_1, \widetilde{R}_1}(\alpha) = \rho^L_{\widetilde{R}_2, \widetilde{R}_2}(\alpha) = \rho^L_{\widetilde{R}_3, \widetilde{R}_3}(\alpha) = 0;$
$\rho^L_{\widetilde{R}_1, \widetilde{R}_2}(\alpha) = \rho^L_{\widetilde{R}_2, \widetilde{R}_1}(\alpha) = | \mu_{\widetilde{R}_1}^{L^{-1}}(\alpha) - \mu_{\widetilde{R}_2}^{L^{-1}}(\alpha) | = \frac{1}{2};$
$\rho^L_{\widetilde{R}_1, \widetilde{R}_3}(\alpha) = \rho^L_{\widetilde{R}_3, \widetilde{R}_1}(\alpha) = | \mu_{\widetilde{R}_1}^{L^{-1}}(\alpha) - \mu_{\widetilde{R}_3}^{L^{-1}}(\alpha) | = 1 - \frac{\alpha}{2};$
$\rho^L_{\widetilde{R}_2, \widetilde{R}_3}(\alpha) = \rho^L_{\widetilde{R}_3, \widetilde{R}_2}(\alpha) = | \mu_{\widetilde{R}_2}^{L^{-1}}(\alpha) - \mu_{\widetilde{R}_3}^{L^{-1}}(\alpha) | = \frac{1}{2} - \frac{\alpha}{2}.$

The right hand-side distances are as follows:

$\rho^R_{\tilde{R}_1,\tilde{R}_1}(\alpha) = \rho^R_{\tilde{R}_2,\tilde{R}_2}(\alpha) = \rho^R_{\tilde{R}_3,\tilde{R}_3}(\alpha) = 0;$

$\rho^R_{\tilde{R}_1,\tilde{R}_2}(\alpha) = \rho^R_{\tilde{R}_2,\tilde{R}_1}(\alpha) = |\mu^R_{\tilde{R}_1}{}^{-1}(\alpha) - \mu^R_{\tilde{R}_2}{}^{-1}(\alpha)| = 1 - \frac{\alpha}{2};$

$\rho^R_{\tilde{R}_1,\tilde{R}_3}(\alpha) = \rho^R_{\tilde{R}_3,\tilde{R}_1}(\alpha) = |\mu^R_{\tilde{R}_1}{}^{-1}(\alpha) - \mu^R_{\tilde{R}_3}{}^{-1}(\alpha)| = 2 - \alpha;$

$\rho^R_{\tilde{R}_2,\tilde{R}_3}(\alpha) = \rho^R_{\tilde{R}_3,\tilde{R}_2}(\alpha) = |\mu^R_{\tilde{R}_2}{}^{-1}(\alpha) - \mu^R_{\tilde{R}_3}{}^{-1}(\alpha)| = 1 - \frac{\alpha}{2}.$

step 3: Calculate the side similarity degrees $S^k(\tilde{R}_i, \tilde{R}_j), (k = L, R)$ of the opinions between each pair of experts.

$S^L(\tilde{R}_1, \tilde{R}_1) = S^L(\tilde{R}_2, \tilde{R}_2) = S^L(\tilde{R}_3, \tilde{R}_3) = 1;$
$S^R(\tilde{R}_1, \tilde{R}_1) = S^R(\tilde{R}_2, \tilde{R}_2) = S^R(\tilde{R}_3, \tilde{R}_3) = 1;$
$S^L(\tilde{R}_1, \tilde{R}_2) = S^L(\tilde{R}_2, \tilde{R}_1) = 0.6065, S^R(\tilde{R}_1, \tilde{R}_2) = S^R(\tilde{R}_2, \tilde{R}_1) = 0.4773;$
$S^L(\tilde{R}_1, \tilde{R}_3) = S^L(\tilde{R}_3, \tilde{R}_1) = 0.4773, S^R(\tilde{R}_1, \tilde{R}_3) = S^R(\tilde{R}_3, \tilde{R}_1) = 0.2325;$
$S^L(\tilde{R}_2, \tilde{R}_3) = S^L(\tilde{R}_3, \tilde{R}_2) = 0.7869, S^R(\tilde{R}_2, \tilde{R}_3) = S^R(\tilde{R}_3, \tilde{R}_2) = 0.4773.$

step 4: Calculate the side average agreement degrees $A(E^L_i)$ and $A(E^R_i)$.

$A(E^L_1) = 0.5419, A(E^L_2) = 0.6967, A(E^L_3) = 0.6321;$
$A(E^R_1) = 0.3549, A(E^R_2) = 0.4773, A(E^R_3) = 0.3549.$

step 5: Calculate the side relative agreement degrees SA^L_i and SA^R_i of expert $E_i(i = 1, 2, \cdots, n)$.

$SA^L_1 = 0.2897, SA^L_2 = 0.3724, SA^L_3 = 0.3379;$
$SA^R_1 = 0.2990, SA^R_2 = 0.4020, SA^R_3 = 0.2990.$

step 6: Calculate the relative agreement degree SA_i of expert $E_i(i = 1, 2, \cdots, n)$.

$SA_1 = 0.2943, SA_2 = 0.3873, SA_3 = 0.3184.$

Because we do not consider the degree of importance of each expert in this case ($\beta = 0$), the consensus degree coefficients of the experts E_1, E_2 and E_3 are

$\omega_1 = SA_1 = 0.2943, \omega_2 = SA_2 = 0.3873, \omega_3 = SA_3 = 0.3184.$

The "overall" fuzzy number of combing experts' opinions is

$\tilde{R} = \omega_1 \odot \tilde{R}_1 + \omega_2 \odot \tilde{R}_2 + \omega_3 \odot \tilde{R}_3 = (1.5121, 2.3528, 3.5121, 5.0241).$

Case 2: Consider the degree of importance of experts.

Suppose that the degrees of importance of each expert are $\gamma_1 = 0.42, \gamma_2 = 0.25,$ and $\gamma_3 = 0.33$ by employing Saaty's AHP. We take $\beta = 0.6$; the aggregation coefficients of the experts E_1, E_2 and E_3 can be computed as

$\omega_1 = 0.6SA_1 + 0.4\gamma_1 = 0.3446, \omega_2 = 0.6SA_2 + 0.4\gamma_2 = 0.3324,$
$\omega_3 = 0.6SA_3 + 0.4\gamma_3 = 0.3230.$

The "overall" fuzzy number of combining experts' opinions is

$\tilde{R} = \omega_1 \odot \tilde{R}_1 + \omega_2 \odot \tilde{R}_2 + \omega_3 \odot \tilde{R}_3 = (1.4892, 2.3277, 3.4892, 4.9785).$

In this example, the width of the result $\tilde{R}$ is smaller than the one using Hus' method, and the uncertainty of the aggregation result for Hus' method in case

1 ($H(\widetilde{R}) = 2.341$) and in case 2 ($H(\widetilde{R}) = 2.33$) is larger than the one using our method in case 1 ($H(\widetilde{R}) = 2.3356$) and in case 2 ($H(\widetilde{R}) = 2.3253$).

Example 2. Consider a group decision making problem with three experts. The datum of the experts' opinions are given as follows:

$\mu_{\widetilde{R}_1}(x) = 1 - 4(5 - 4x)^2, \frac{9}{8} \leq x \leq \frac{11}{8}$; $\mu_{\widetilde{R}_2}(x) = 1 - 4(\frac{5}{2} - 4x)^2, \frac{4}{8} \leq x \leq \frac{6}{8}$;
$\mu_{\widetilde{R}_3}(x) = 1 - (3 - 4x)^2, \frac{1}{2} \leq x \leq 1$.

We employ our method to deal with this problem and assume $\beta = 0$

step 2: Calculate the left- and right hand-side distances between $\widetilde{R}_i$ and $\widetilde{R}_j$.

The left hand-side distances are as follows:

$\rho^L_{\widetilde{R}_1,\widetilde{R}_1}(\alpha) = \rho^L_{\widetilde{R}_2,\widetilde{R}_2}(\alpha) = \rho^L_{\widetilde{R}_3,\widetilde{R}_3}(\alpha) = 0$;

$\rho^L_{\widetilde{R}_1,\widetilde{R}_2}(\alpha) = \rho^L_{\widetilde{R}_2,\widetilde{R}_1}(\alpha) = |\mu^{L\ -1}_{\widetilde{R}_1}(\alpha) - \mu^{L\ -1}_{\widetilde{R}_2}(\alpha)| = \frac{5}{8}$;

$\rho^L_{\widetilde{R}_1,\widetilde{R}_3}(\alpha) = \rho^L_{\widetilde{R}_3,\widetilde{R}_1}(\alpha) = |\mu^{L\ -1}_{\widetilde{R}_1}(\alpha) - \mu^{L\ -1}_{\widetilde{R}_3}(\alpha)| = \frac{1}{2} + \frac{\sqrt{1-\alpha}}{8}$;

$\rho^L_{\widetilde{R}_2,\widetilde{R}_3}(\alpha) = \rho^L_{\widetilde{R}_3,\widetilde{R}_2}(\alpha) = |\mu^{L\ -1}_{\widetilde{R}_2}(\alpha) - \mu^{L\ -1}_{\widetilde{R}_3}(\alpha)| = \frac{1}{8} - \frac{\sqrt{1-\alpha}}{8}$.

The right hand-side distances are as follows:

$\rho^R_{\widetilde{R}_1,\widetilde{R}_1}(\alpha) = \rho^R_{\widetilde{R}_2,\widetilde{R}_2}(\alpha) = \rho^R_{\widetilde{R}_3,\widetilde{R}_3}(\alpha) = 0$;

$\rho^R_{\widetilde{R}_1,\widetilde{R}_2}(\alpha) = \rho^R_{\widetilde{R}_2,\widetilde{R}_1}(\alpha) = |\mu^{R\ -1}_{\widetilde{R}_1}(\alpha) - \mu^{R\ -1}_{\widetilde{R}_2}(\alpha)| = \frac{5}{8}$;

$\rho^R_{\widetilde{R}_1,\widetilde{R}_3}(\alpha) = \rho^R_{\widetilde{R}_3,\widetilde{R}_1}(\alpha) = |\mu^{R\ -1}_{\widetilde{R}_1}(\alpha) - \mu^{R\ -1}_{\widetilde{R}_3}(\alpha)| = \frac{1}{2} - \frac{\sqrt{1-\alpha}}{8}$;

$\rho^R_{\widetilde{R}_2,\widetilde{R}_3}(\alpha) = \rho^R_{\widetilde{R}_3,\widetilde{R}_2}(\alpha) = |\mu^{R\ -1}_{\widetilde{R}_2}(\alpha) - \mu^{R\ -1}_{\widetilde{R}_3}(\alpha)| = \frac{1}{8} + \frac{\sqrt{1-\alpha}}{8}$.

step 3: Calculate the side similarity degrees $S^k(\widetilde{R}_i, \widetilde{R}_j)$, (k = L, R) of the opinions between each pair of experts.

$S^L(\widetilde{R}_1, \widetilde{R}_1) = S^L(\widetilde{R}_2, \widetilde{R}_2) = S^L(\widetilde{R}_3, \widetilde{R}_3) = 1$;
$S^R(\widetilde{R}_1, \widetilde{R}_1) = S^R(\widetilde{R}_2, \widetilde{R}_2) = S^R(\widetilde{R}_3, \widetilde{R}_3) = 1$;
$S^L(\widetilde{R}_1, \widetilde{R}_2) = S^L(\widetilde{R}_2, \widetilde{R}_1) = 0.5353$, $S^R(\widetilde{R}_1, \widetilde{R}_2) = S^R(\widetilde{R}_2, \widetilde{R}_1) = 0.5587$;
$S^L(\widetilde{R}_1, \widetilde{R}_3) = S^L(\widetilde{R}_3, \widetilde{R}_1) = 0.9588$, $S^R(\widetilde{R}_1, \widetilde{R}_3) = S^R(\widetilde{R}_3, \widetilde{R}_1) = 0.5353$;
$S^L(\widetilde{R}_2, \widetilde{R}_3) = S^L(\widetilde{R}_3, \widetilde{R}_2) = 0.6590$, $S^R(\widetilde{R}_2, \widetilde{R}_3) = S^R(\widetilde{R}_3, \widetilde{R}_2) = 0.8130$.

step 4: Calculate the side average agreement degrees $A(E_i^L)$ and $A(E_i^R)$.

$A(E_1^L) = 0.5470, A(E_2^L) = 0.7471, A(E_3^L) = 0.7588$;
$A(E_1^R) = 0.5971, A(E_2^R) = 0.6741, A(E_3^R) = 0.7360$.

step 5: Calculate the side relative agreement degrees SA_i^L and SA_i^R of expert $E_i (i = 1, 2, \cdots, n)$.

$SA_1^L = 0.2665, SA_2^L = 0.3639, SA_3^L = 0.3696$;
$SA_1^R = 0.2975, SA_2^R = 0.3358, SA_3^R = 0.3667$.

step 6: Calculate the relative agreement degree SA_i of expert $E_i (i = 1, 2, \cdots, n)$.

$$SA_1 = 0.2820, SA_2 = 0.3499, SA_3 = 0.3681.$$

Because we do not consider the degree of importance of each expert in this case ($\beta = 0$), the consensus degree coefficients of the experts E_1, E_2 and E_3 are

$$\omega_1 = SA_1 = 0.2820, \omega_2 = SA_2 = 0.3499, \omega_3 = SA_3 = 0.3681.$$

The membership function of the group consensus opinion is

$$\mu_{\widetilde{R}}(x) = 1 - 34.1911(x - 0.8473)^2$$

In this example, the opinions of experts are not expressed by trapezoidal fuzzy numbers, so Lee's method can't deal with it. Meanwhile, Hsu's method can't be also effective because $\widetilde{R_1}$ has obviously no common intersection with $\widetilde{R_2}$ and $\widetilde{R_3}$.

5 Conclusion

In this paper, the problem aggregating individual opinions into group consensus under group decision environment is addressed. A simple similarity measure to deal with the L-R fuzzy numbers has been employed. The distance and similarity function is proposed. The degree of importance of each expert is taken into account further. Meanwhile, the membership function and the pessimistic and optimistic opinions of experts' are considered in the method. This aggregation method preserves some important properties which other aggregation methods processed.

References

1. A.Bardossy, L.Duckstein and Bogardi: Combination of fuzzy numbers representing expert opinions. Fuzzy Sets and Systems. **57**(1993)173-181
2. P.Diamond, P.Kloeden: Metric Spaces of Fuzzy Sets: Theory and Application. World Scientific, Singapore. 1994
3. H.S.Lee: Optimal consensus of fuzzy opinions under group decision making environment. Fuzzy Sets and Systems. **132**(2002)303-315
4. Jiulun Fan and Weixin Xie: Distance measure and induced fuzzy entropy. Fuzzy Sets and Systems. **104**(1999)305-314
5. M.Fedrizzi and J.Kacprzyk: On measuring consensus in the setting of fuzzy preference relations, in: J.Kacprayk and M.Roubens, Eds. Non-conventional preference Relations in Decision Making (Springer, Berlin, 1988). 129-141
6. R.Goetschel, W.Voxman: Topological Properties of Fuzzy Sets. Fuzzy Sets and Systems. **10**(1983) 87-99
7. H.M.Hsu, C.T.Chen: Aggregation of fuzzy opinions under group decision making. Fuzzy Sets and Systems. **79**(1996)279-285
8. A.Ishikawa, M.Ambiguous, T.Shiga, G,Tomizawa, R. Tactic and H.Mileage: The max-min Delpi method and fuzzy Delphi method via fuzzy intergration. Fuzzy sets and Systems. **55** (1993)241-253
9. J.Kacprzyk and M.Federation: A soft measure of consensus in the setting of partial(fuzzy) preferences. Eur.J.OR. **34**(1988)315-325

10. J.Kacprzyk, M.Federation and H.Norm: Group decision making and consensus under fuzzy preferences and fuzzy majority. Fuzzy Sets and Systems. **49**(1992)21-31
11. O.Kaleva, S.Siekkala: On fuzzy metric spaces. Fuzzy Sets and Systems. **12(3)**(1987)301-317
12. G.J.Klir, B.Yuan:Fuzzy Sets and Fuzzy Logic: Theory and Applications, Prentice Hall. Englewood Cliffs, NJ. 1995
13. L.T.Koczy, K.Hirota: Ordering and closeness of fuzzy sets. Fuzzy Sets and Systems. **90**(1997)103-111
14. V.B.Kuz'min: A parametric approach to the description of linguistic variables and hedges. Fuzzy Sets and Systems. **6**(1981)27-41
15. H.Nurmi: Approaches to collective decision making with fuzzy preference relations. Fuzzy Sets and Systems. **6**(1981)249-259
16. T.L. Saaty: Modeling unstructured decision problems-the theory of analytical hierarchies. Math. Comput. Simulation. **20**(1978)147-158
17. T.Tanino: On group decision making under fuzzy preferences, in: J.Kacprzyk, M.Fedrizzi Eds, Multiperson Decision Making Using Fuzzy Sets and Prossibility Theory. Kilowatt Academic Publishers, Dordrecht. (1990)172-185
18. Jon Williams and Nigel Steele: Difference, distance and similarity as a basis for fuzzy decision support based on prototypical decision classes. Fuzzy Sets and Systems. **131**(2002)35-46
19. R.N.Xu and X.Y.Zhai:Extensions of the analytic hierarchy process in fuzzy environment. Fuzzy Sets and Systems. **52**(1992)251-257

Modeling Nonlinear Systems: An Approach of Boosted Linguistic Models

Keun-Chang Kwak[1], Witold Pedrycz[1], and Myung-Geun Chun[2,*]

[1] Dept. of Electrical and Computer Engineering, University of Alberta,
Edmonton, AB, Canada, T6G 2V4
kwak@ece.ualberta.ca
[2] School of Electrical and Computer Engineering, Chungbuk National University,
Cheongju, Korea, 361-763
mgchun@chungbuk.ac.kr

Abstract. We present a method of designing the generic linguistic model based on boosting mechanism to enhance the development process. The enhanced model is concerned with linguistic models being originally proposed by Pedrycz. Based on original linguistic model, we augment it by a bias term. Furthermore we consider the linguistic model as a weak learner and discuss the underlying mechanisms of boosting to deal with the continuous case. Finally, we demonstrate that the results obtained by the boosted linguistic model show a better performance than different design schemes for nonlinear system modeling of a pH neutralization process in a continuous stirred-tank reactor (CSTR).

1 Introduction

We have witnessed a dynamic growth of the area of fuzzy modeling based on the concept of fuzzy models. Several studies have been made on well-established methodologies, design principles, and detailed algorithms [1]. Among various methodologies, the design environments and ensuing architectures of hybrid neuro-fuzzy system have emerged as a useful and comprehensive development paradigm [2]. Furthermore the construction of the interpretable rule-based models is also high on the overall agenda of fuzzy modeling [3][4]. In spite of this profoundly visible diversity of the architectural considerations and ensuing algorithms, a predominant majority of fuzzy models is surprisingly similar in the sense that the final model realizes as a nonlinear numeric mapping transforming multivariable numeric inputs into the corresponding elements of the real line **R**. In essence, fuzzy sets do show up as an integral design element yet the result (fuzzy model) manifests at the numeric level. The principle of linguistic modeling is very much different [5]. We do not look at the minute details of the model but rather start with forming information granules that are reflective of the experimental data at hand and then form a collection of links between them. As the way of building such granules is intuitively appealing, the ensuing links are evident.

---

[*] Corresponding author.

The linguistic model is inherently granular. Even for a numeric input, the output of the linguistic model is inherently granular and comes in the form of some fuzzy set. Information granules are formed through the use of a specialized type of so-called context-based fuzzy clustering [6].

The main goal of this study is to establish a comprehensive design environment of linguistic models with emphasis on their learning enhancements via boosting. The boosting technique has been successful in the development of highly efficient classifiers emerging on a basis of a collection of weak classifiers whose performance is slightly better than random guessing [7][8][9]. As the typical boosting schemes apply to discrete classification schemes, we revisit them and provide with necessary modifications and enhancements so that one could use them to continuous problems (as those addressed by linguistic models). We demonstrate that the results obtained by the boosted linguistic model outperform different design schemes for nonlinear system modeling of a pH process in a continuous stirred-tank reactor (CSTR) [10][11].

2 Boosting-Based Linguistic Models

2.1 The Fundamental of Linguistic Models

We briefly describe the underlying concept and architectural fundamentals of linguistic models as originally introduced by Pedrycz [5]. In contrast to the currently existing plethora of neurofuzzy models, which are in essence nonlinear numeric models, linguistic modeling revolves around information granules – fuzzy sets constructed in input and output spaces. The emphasis is on the formation of these granules while the linkages between them are intuitively straightforward as being the result of the construction of the information granules themselves. The conditional (context-based) fuzzy clustering forms a backbone of the linguistic model. Before moving with the algorithmic details, let us concentrate on the generic architecture and relate it to the development of the linguistic models. Throughout this study, we are dealing with a finite collection of pairs of experimental data of the form $\{(\mathbf{x}_1, \text{target}_1), (\mathbf{x}_2, \text{target}_2),...,(\mathbf{x}_N, \text{target}_N)\}$ where $\mathbf{x}_k \in \mathbf{R}^n$, target $\in \mathbf{R}$. The input space $\mathbf{X}$ is then regarded as a subset of $\mathbf{R}^n$.

The point of departure of all our modeling pursuits is a finite collection of "p" fuzzy sets- linguistic contexts being defined in a certain output space Y. Those are some linguistic landmarks that help organize our view at the multidimensional input data $\mathbf{x}_k$. For the given context W_i, we search for the structure in $\mathbf{X}$ that is implied (or induced) by this context. Such a structure can be easily revealed by a specialized fuzzy clustering called a context-based fuzzy clustering. It is also referred to as a conditional clustering as the structure revealed in $\mathbf{X}$ is conditioned by the given context (W_i). Let us consider that for each context we complete clustering into "c" clusters. This in total, we arrive at c*p clusters. Fig. 1 presents an overall organization of the contexts and induced clusters. The first layer is formed by c*p clusters fully characterized by their prototypes. Each group of these clusters corresponds to the given context and when "activated" by the given input $\mathbf{x}$, the levels of activation are summed up in the successive layer of the model (this results in $z_1, z_2, ..., $ and z_p). Afterwards they are aggregated with the contexts (fuzzy sets) $W_1, W_2, ..., W_p$ at the output layer of the model.

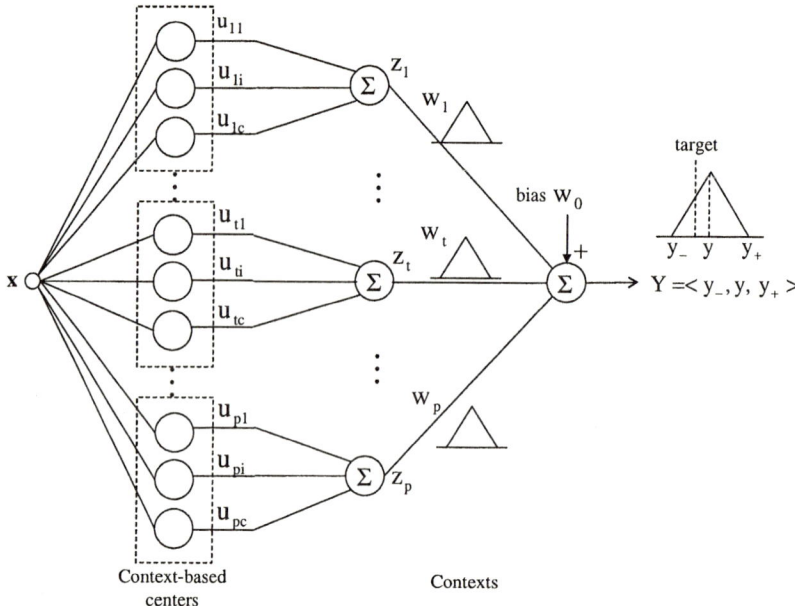

Fig. 1. The general architecture of the linguistic model with bias term regarded as a web of connections between linguistic landmarks

In what follows, we briefly recall the essence of conditional clustering and elaborate on the algorithmic facet of the optimization process. This clustering, which is a variant of the FCM, is realized for individual contexts, $W_1, W_2, \ldots, W_p$. Consider a certain fixed context W_j described by some membership function (the choice of its membership will be discussed later on). Any data point in the output space is then associated with the corresponding membership value, $W_j(target_k)$. Let us introduce a family of the partition matrices induced by the l-th context and denote it by $U(W_l)$

$$U(W_l) = \left\{ u_{ik} \in [0,1] \mid \sum_{i=1}^{c} u_{ik} = w_{lk} \ \forall k \text{ and } 0 < \sum_{k=1}^{N} u_{ik} < N \ \forall i \right\} \quad (1)$$

where w_{lk} denotes a membership value of the k-th datum to the l-th context. The optimization completed by the conditional FCM is realized iteratively by updating the partition matrix and the prototypes. The update of the partition matrix is completed as follows

$$u_{ik} = \frac{w_{lk}}{\sum_{j=1}^{c} \left(\frac{\|x_k - v_i\|}{\|x_k - v_j\|} \right)^{\frac{2}{m-1}}}, \quad i=1,2,\ldots,c, \ k=1,2,\ldots,N \quad (2)$$

where $\|\cdot\|$ denotes a certain distance function. Note that u_{ik} pertains here to the partition matrix induced by the l-th context. The prototypes are calculated in the form

$$v_i = \frac{\sum_{k=1}^{N} u_{ik}^m x_k}{\sum_{k=1}^{N} u_{ik}^m} \tag{3}$$

where i =1,2,…,c. The fuzzification factor (coefficient) is denoted by 'm"; its typical value is taken as 2.0.

For the design of the linguistics model, we consider the contexts to be described by triangular membership functions being equally distributed in Y with the 1/2 overlap between two successive fuzzy sets. Alluding to the overall architecture, Figure 1, we denote those fuzzy sets by $W_1, W_2, …, W_p$. Let us recall that each context generates a number of induced clusters whose activation levels are afterwards summed up as shown in Fig. 1. Denoting those by $z_1, z_2, …, z_p$, the output of the model (network) is granular and more specifically a triangular fuzzy number Y that reads as

$$Y = W_1 \otimes z_1 \oplus W_2 \otimes z_2 \oplus \cdots \oplus W_n \otimes z_n \tag{4}$$

We denote the algebraic operations by $\otimes$ and $\oplus$ to emphasize that the underlying computing operates on a collection of fuzzy numbers. As such, Y is fully characterized by three parameters that are a modal value and the lower and upper bounds. For the k-th datum, x_k, we use the explicit notation $Y(x_k) = <y_{k-}, y_k, y_{k+}>$ which helps emphasize the input-output relationship.

So far the web of the connections between the contexts and their induced clusters was very much reflective of how the clustering has been completed. The emergence of the network structure suggests that we should be able to eliminate possible systematic error and this could be easily accomplished by augmenting the summation node at the output layer by a numeric bias term w_0 as shown in Figure 1. The bias is computed in a straightforward manner

$$w_0 = \frac{1}{N} \sum_{k=1}^{N} (target_k - y_k) \tag{5}$$

where y_k denotes a modal value of Y produced for given input x_k. In essence, the bias term is a numeric singleton which could be written down as $W_0 = (w_0, w_0, w_0)$.

The resulting granular output Y reads in the form

- modal value $\quad \sum_{t=1}^{p} z_t w_t + w_0$

2.2 Boosted Linguistic Models

Boosting is regarded as a commonly used method that helps enhance the performance of weak classifiers [7][8][9]. Originally, boosting was developed and used in the domain of classification problems involving a number of discrete classes. When dealing with continuous problems, it requires further refinement. In this study, we consider a typical variant of the boosting mechanism referred to as AdaBoost introduced by Freund [7] and revisit it to make it suitable for the linguistic models.

We consider a collection of N input-output examples (training data), $\{(\mathbf{x}_1, target_1), \ldots, (\mathbf{x}_k, target_k), \ldots, (\mathbf{x}_N, target_N)\}$ with the continuous output variable. The linguistic model is treated here as a weak learner. Following the essence of the boosting mechanism, the algorithm repeatedly calls the weak learner going through a number of iterations, $t = 1, 2, \ldots, T$. At each iteration, it endows the data set with some discrete probability function, $D_t(k)$ $k=1,2,..,N$. Its role is to selectively focus on some data points which are "difficult" to capture and produce error when handled by the weak learner. Initially, we take $D_1(k)$ as a uniform probability function that is

$$D_1(k) = \frac{1}{N} \quad \text{for all } k \tag{6}$$

Obviously we have $\sum_{k=1}^{N} D_t(k) = 1$. This probability function becomes updated based upon the error of hypothesis produced for each data point. The main objective of the weak learner is to form a hypothesis which minimizes the training error. In our case, the error of hypothesis is computed in the form

$$\varepsilon_t = \sum_{k=1}^{N} D_t(k)(1 - Y(target_k)) \tag{7}$$

Note that the expression in the above sum indicates how much the output fuzzy set $Y(target_k)$ departs from the corresponding numeric datum $target_k$. In essence, we can regard $y(\mathbf{x}_k)$ as the possibility measure of $target_k$ computed with respect to Y being obtained for the input $\mathbf{x} = \mathbf{x}_k$.

In the sequel, we compute the following factor $\beta_t \in [0,1)$

$$\beta_t = \frac{\varepsilon_t}{1 - \varepsilon_t} = \frac{\sum D_t(k)(1 - Y(target_k))}{1 - \sum D_t(k)(1 - Y(target_k))} \tag{8}$$

Based on its values, we adjust the probability function to assume the values

$$D_{t+1}(k) = \frac{D_t(k)}{Z_t} \times \begin{cases} \beta_t & \text{if } (1 - Y(target_k)) < \theta \\ 1 & \text{otherwise} \end{cases} \tag{9}$$

where Z_t provides the required normalization effect. The parameter θ ($0 < \theta \leq 1$) plays a role of the threshold reflecting a way in which the distribution function becomes affected. Once boosting has been completed, the final hypothesis issued for the training data $\mathbf{x}_k$ is taken as a linear combination of the granular outputs of the hypotheses produced in consecutive iterations

$$\tilde{Y}(x_k) = \sum_{t=1}^{T} \overline{\alpha}_t \tilde{Y}_t(x_k) \qquad (10)$$

where $\overline{\alpha}_t = \alpha_t \Big/ \sum_{t=1}^{T} \alpha_t$; here $\alpha_t = \log(1/\beta_t)$. Fig. 2 visualizes the aggregation; here L_t and $\tilde{Y}_t$, t = 1,2,...,T, denote the weak classifier (linguistic model) and their output, respectively. The node labeled as **N** stands for the normalization effect.

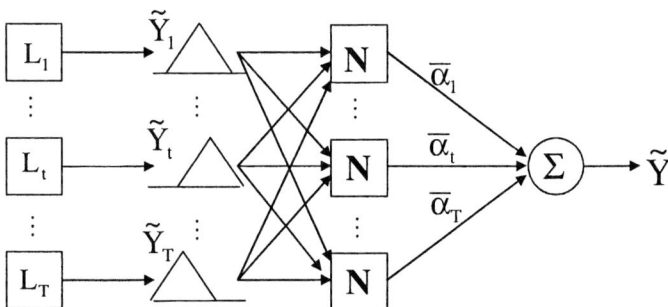

Fig. 2. The final hypothesis realized through a linear combination of the linguistic models

Our objective is to gain a better insight into the performance of the linguistic models, quantify the efficiency of the design process and analyze an impact of the selected design parameters. The performance of the developed models is quantified using the standard RMSE (Root Mean Squared Error) defined in the usual format

$$\text{RMSE} = \sqrt{\frac{1}{N} \sum_{k=1}^{N} (\text{target}_k - \tilde{y}(x_k))^2} \qquad (11)$$

where $\tilde{y}(x_k)$ is the modal value of the fuzzy number produced by the final hypothesis (boosted linguistic model).

3 Experimental Results

We use the well-known benchmark problem with nonlinear dynamics for pH (the concentration index of hydrogen ions) neutralization process in a continuous stirred-tank reactor (CSTR). The input variables are pH(k) and $F_{NaOH}(k)$ in the steady-state process. The output variable to be predicted is pH(k+1). The experimental data set was produced by randomly generating $F_{NaOH}(k)$ in the range of 513-525 l/min. For further details of this dynamic model for a pH in a CSTR, see [10][11]. The dataset includes 2500 input-output pairs. The random split into the training and testing part is the one of 60%-40%. The experiment was repeated 10 times (10 fold cross-validation). The context-based FCM was set up in a standard way: the fuzzification

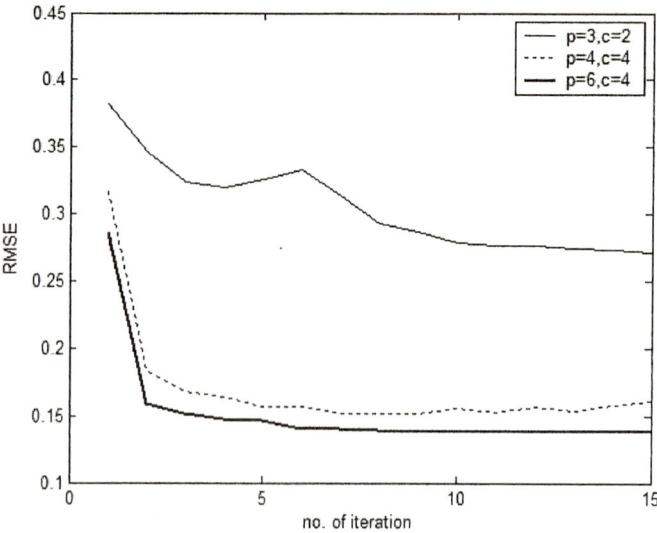

Fig. 3. Changes in the RMSE values in successive iterations of the boosting for training data

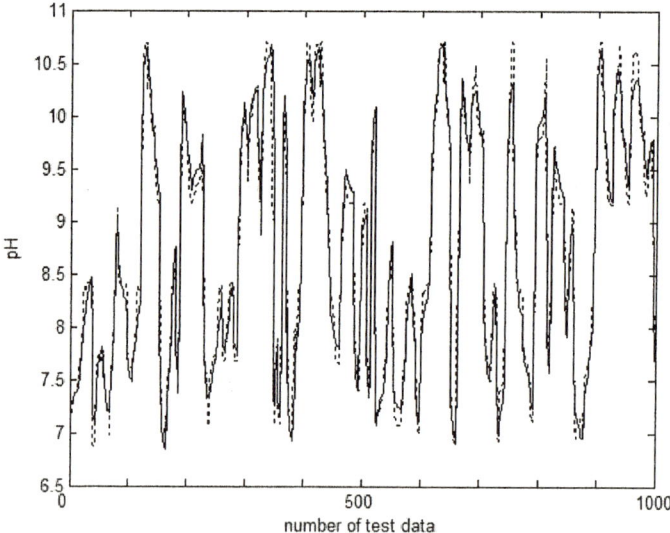

Fig. 4. Comparison of actual output and model output (boosted linguistic models) for test data (p=6,c=4)

factor was equal to 2 while the distance was the Euclidean. Each of the input variables are normalized to within the unit interval [0,1]. The values of the essential parameters of the boosting procedure were set up by trial-and-error; the number of iterations was equal to 15. If θ is close to 0, the resulting distribution function affects a limited number of data. On the other hand, if we admit values of θ close to 1, almost all data

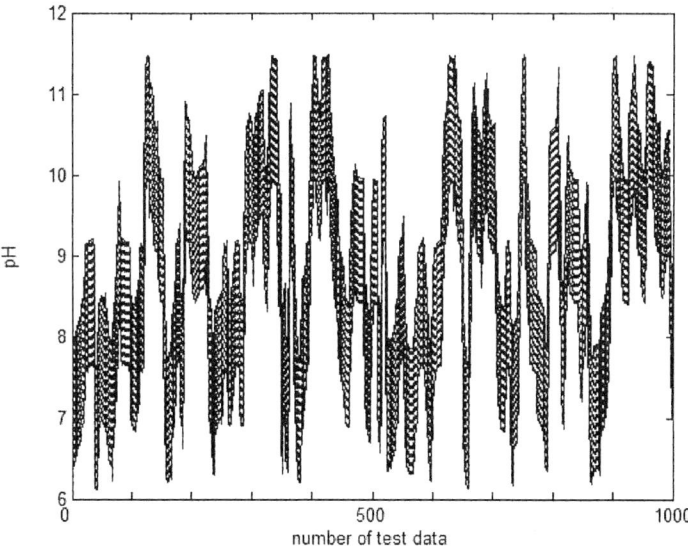

Fig. 5. Uncertain output represented by upper and lower bound (p=6,c=4)

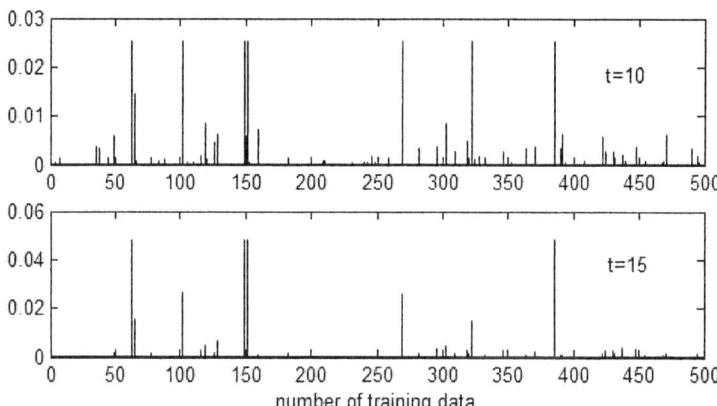

Fig. 6. The changes of the probability function in selected iterations (p=6,c=4) for some training data

become affected. In essence, we can regard this threshold to be a parameter of the boosting mechanism and as such it could be subject to some optimization. The completed experiments reveal that an optimal value of the threshold is in the vicinity of 0.5. The RMSE on the training set gradually reduces in successive iterations as shown in Fig. 3. We obtained the best performance in case of p=6 and c=4 as the number of "p" and "c" increase from 2 to 6. Fig. 4 shows the comparison of actual output and model output obtained by the proposed method for test data (one among 10 runs). The uncertain output represented by the upper and lower bound is visualized in Fig. 5. In essence, the boosting effect translates into a way different data points are treated in the learning process.

Table 1. Comparison of RMSE

	p, c	RMSE (Training data)	RMSE (Test data)
linear ARX model	.	0.547	0.532
Linguistic model	p=6,c=4	0.195 ± 0.005	0.311 ± 0.008
Linguistic model with bias term	p=6,c=4	0.194 ± 0.005	0.302 ± 0.007
Proposed model (with boosting)	p=6,c=4	0.151 ± 0.003	0.157 ± 0.004

As illustrated in Figure 6, over consecutive iterations (t) this distribution changes quite significantly; finally we clearly witness several patterns that deserve more attention (those are the elements with higher values of $D_t(k)$).

The comparative analysis covered in Table 1, shows that the boosted linguistic model yields better performance in compared with linear ARX model and conventional linguistic model.

4 Conclusions

We have developed an augmented design methodology of linguistic models based on the boosting mechanisms. The linguistic models treated as weak learner showed a consistently better performance over the original design strategy. The architectural augmentation of the model by its bias term has also improved its performance. Experimental results clearly demonstrated the improved performance of the linguistic model constructed through boosting over some other models such as linear ARX model and conventional linguistic model.

Acknowledgments

This work was supported by Tgrant No. R01-2002-000-00315-0 from the Basic Research Program of the Korea Science and Engineering Foundation. Support from the Canada Research Chair (CRC) Program (W. Pedrycz), Natural Sciences and Engineering Research Council (NSERC) is gratefully acknowledged.

References

1. Pedrycz, W., Computational Intelligence: An Introduction, Boca Raton, FL: CRC (1997)
2. Jang, S. R., Sun, C. T., Mizutani, E., Neuro-Fuzzy and Soft Computing: A Computational Approach to Learning and Machine Intelligence, Prentice Hall (1997)

3. Abonyi, J., Babuska, R., Szeifert, F., Modified Gath-Geva fuzzy clustering for identification of Takagi-Sugeno fuzzy models, IEEE Trans. on Systems, Man, and Cybernetics-Part B, vol. 32, no. 5 (2002) 612-621
4. Kwak, K. C., Chun, M. G., Ryu, J. W., Han, T. H., FCM-based adaptive fuzzy inferencesystem for coagulant dosing process in a water purification plant, Journal of Knowledge-Based Intelligent Engineering Systems, vol. 4, no. 4 (2000) 230-236
5. Pedrycz, W., Vasilakos, A. V., Linguistic models and linguistic modeling, IEEE Trans. on Systems, Man, and Cybernetics-Part C, vol. 29, no. 6 (1999) 745-757
6. Pedrycz, W., Conditional fuzzy C-Means, Pattern Recognition Letters, vol. 17 (1996) 625-632
7. Freund, Y., Schapire, R. E., A short introduction to boosting, Journal of Japanese Society for Artificial Intelligence, vol. 14, no. 5 (1999) 771-780
8. Hoffmann, F., Combining boosting and evolutionary algorithms for learning of fuzzy classification rules, Fuzzy Sets and Systems, vol. 141 (2004) 47-58
9. Dettling, M., Bühlmann, P., Boosting for tumor classification with gene expression data, Bioinformatics, vol. 19 (2003) 1061-1069
10. Abonyi, J., Babuska, R., Szeifert, F., Fuzzy modeling with multivariate membership functions: gray-box identification and control design, IEEE Trans. on Systems, Man, and Cybernectics, vol. 31, no, 5 (2001)
11. Bhat, N., McAvoy, T., Determining model structure for neural models by network stripping, Computers and Chemical Engineering, vol.16 (1992) 271-281

Multi-criterion Fuzzy Optimization Approach to Imaging from Incomplete Projections

Xin Gao and Shuqian Luo

College of Biomedical Engineering, Capital University of Medical Sciences,
Beijing, 100054, China
singgau@yahoo.com.cn, sqluo@ieee.org

Abstract. To enhance resolution and reduce artifacts in imaging from incomplete projections, a novel imaging model and algorithm to imaging from incomplete projections—multi-criterion fuzzy optimization approach is presented. This model combines fuzzy theory and multi-criterion optimization approach. The membership function is used to substitute objective function and the minimum operator is taken as fuzzy operator. And a novel resolution method was proposed. The result reconstructed from computer-generated noisy projection data is shown. Comparison of the reconstructed images indicates that this algorithm gives better results both in resolution and smoothness over analytic imaging algorithm and conventional iterative imaging algorithm.

1 Introduction

Imaging methods mainly have two categories: analytic algorithm, e.g. Convolution Back Projections (CBP) algorithm etc. and iterative algorithm, e.g. Algebraic Reconstruction Techniques (ART) algorithm.

To imaging from sufficient and exact projections, analytic method is an efficient reconstruction algorithm. To incomplete or/and uneven distributed projections within π (2π) that are necessary conditions for reconstructing certainly precise image by analytic algorithm, however, iterative algorithm is preferred. Single objective optimization imaging is one kind of iterative algorithms. Owing to certain limitation in the single objective optimization, there is a growing trend towards introducing multi-criterion optimization to imaging from incomplete projections.

The starting point of multi-criterion optimization to imaging is that one thinks about many function characters and get their optimum simultaneously as possible so as to ensures various character of reconstructed image to be 'best' [2]. Because multiple objectives are often incommensurable and conflict with each other, the conventional optimality concept of single criterion optimization is replaced by Pareto optimality or efficiency. Therefore, the key of multi-criterion optimization to imaging is how to derive a compromise or satisfactory solution of a decision maker (DM), from a Pareto optimal or an efficient solution set, as final solution.

Along with fuzzy set theory is widely used in various filed, multi-criterion fuzzy optimization (MCFO) has been one wondrously active research region all over the world. We reconstruct image by the aid of the theory and get anticipatory result. The idea of algorithm is that one depicts objective functions with membership functions, then evaluates whole satisfactory degree of every objections corresponding to their optimality respectively by fuzzy operator, and thereby constructs mathematic model, at last presents the solving approach. Results and conclusion are shown separately in the end.

2 Multi-criterion Optimization to Imaging

Let $\mathbf{x} = (x_1, x_2, \cdots, x_n)^T$ denote n dimensions reconstructed image vector, $\mathbf{y} = (y_1, y_2, \cdots, y_m)^T$ is projections vector and $\mathbf{A} = (a_{ij})_{m \times n}$ be projections matrix. We introduce noise vector $\mathbf{e} = (e_1, e_2, \cdots, e_m)^T$ for projection model. Assume that e_i is represented independently of $(0, \sigma_i^2)$ Gaussian distribution.

Aiming at imaging problem from incomplete projections, reconstructed image is expected to be satisfied with, (1) Least error between real projections and re-projections through reconstructed image; (2) Higher whole smoothness in image; (3) Higher local smoothness in image. Whereas there is measurement error during real projection, it works as the constraint to be introduced to the reconstruction model and determines the set of feasible images vector $\mathbf{x}$.

Therefore, the model of vector mathematic programming to imaging is drawn,

$$\min \quad \mathbf{f}(\mathbf{x}) = \left(\| \mathbf{A}\mathbf{x} - \mathbf{y} \|^2, \sum_{j=1}^{n} x_j \ln x_j, \frac{1}{2} \mathbf{x}^T \cdot \mathbf{S} \cdot \mathbf{x} \right)$$
$$\text{s.t.} \quad \sum_{i=1}^{m} \frac{(\mathbf{A}_i \mathbf{x} - y_i)^2}{\sigma_i^2} = m \tag{1}$$

The common resolution to the problem is utility function method [3].

3 Multi-criterion Fuzzy Optimization to Imaging

Due to (1) plenty of stochastic and unsure fuzzy information exist during imaging. (2) It is difficult to equipoise various objective, which sometimes are mutually conflictive and non-commensurable. (3) For handling and tackling such kinds of vagueness in imaging, it is not hard to imagine that the conventional multi-criterion optimization approaches can not be applied. Further more, (4) multi-criterion fuzzy optimization simplifies solving process than conventional trade-off method. This paper presents multi-criterion optimization algorithm under fuzzy rule for imaging from incomplete projections.

There are four essential problems of using multi-criterion fuzzy optimization algorithm to imaging: First, proper membership function is selected to depict fuzzy goal. Second, one or some arithmetic operators are adopted to integrate various objectives, and define a measure of whole estimate in satisfactory degree. Third, mathematical model of MCFO should be established. Fourth, specific algorithm will be deduced.

In this paper, fuzzy goal is depicted by linear membership function.

$$\mu_i(f_i(\mathbf{x})) = \frac{f_i^{\max}(\mathbf{x}) - f_i(\mathbf{x})}{f_i^{\max}(\mathbf{x}) - f_i^{\min}(\mathbf{x})} \qquad (2)$$

Where $f_i^{\max}(\mathbf{x}) = \max_{x \in X} f_i(\mathbf{x})$, $f_i^{\min}(\mathbf{x}) = \min_{x \in X} f_i(\mathbf{x})$ under the given constraints. And minimum operator [4] is selected as fuzzy operator.

$$\mu_D(\mathbf{x}) = \min_{x \in X}(\mu_1(f_1(\mathbf{x})), \mu_2(f_2(\mathbf{x})), \mu_3(f_3(\mathbf{x}))) \qquad (3)$$

Zimmermann algorithm [4] is selected as mathematical model. The final aim of this algorithm is to solve the max satisfactory degree λ in objective set and efficient solution $\mathbf{x}^*$ of original problem. Its mathematical model is expressed as,

$$\begin{cases} \max \quad \lambda \\ \text{s.t.} \quad \lambda \leq 1 - \frac{1}{m}\sum_{i=1}^{m}(\mathbf{A}_i\mathbf{x} - y_i)^2 \quad \lambda \leq \left(\sum_{j=1}^{n} x_j \ln x_j\right)/\ln n \\ \lambda \leq 1 - \frac{\mathbf{x}^T \cdot \mathbf{S} \cdot \mathbf{x}}{n^2} \quad \sum_{i=1}^{m}\frac{(\mathbf{A}_i\mathbf{x} - y_i)^2}{\sigma_i^2} = m \quad \lambda \in [0,1] \end{cases} \qquad (4)$$

We proposed a new method—iterative min-max algorithm to resolve above problem. A summary of the MCFO method to imaging from incomplete projections is as follows.

(1) Choose an initial image vector $\mathbf{x}^0$ that can be initialized by convolution back projection (CBP) for fan-scan projections. And choose a termination error scalar $0 < \varepsilon \ll 1$, a max grade of membership $\lambda = 0, \lambda' = 0$. Let iteration number $k = 0$; (2) $\mu_1(f_1(\mathbf{x})), \mu_2(f_2(\mathbf{x})), \mu_3(f_3(\mathbf{x}))$ corresponding to $\mathbf{x}^k$ are three various grade of membership. One minimum from them is selected as λ; (3) Membership function relating to λ and constraint condition are combined as a new optimization problem under constraint condition. It can be solved by iterating based on Huhn-Tucker condition, and the solution $\mathbf{x}^{k+1}$; (4) $\mu_1(f_1(\mathbf{x})), \mu_2(f_2(\mathbf{x})), \mu_3(f_3(\mathbf{x}))$ corresponding to $\mathbf{x}^{k+1}$ are three various grade of membership. Let $\lambda' = \min\{\mu_1(f_1(\mathbf{x})), \mu_2(f_2(\mathbf{x})), \mu_3(f_3(\mathbf{x}))\}$; (5) If $\lambda' < \lambda$, let $\mathbf{x}^* = \mathbf{x}^k$, output

$\mathbf{x}^*$ and stop. Else go to (6); (6) If $\lambda^{'} - \lambda < \varepsilon$, let $\mathbf{x}^* = \mathbf{x}^{k+1}$, output $\mathbf{x}^*$ and stop. Else go to (7); (7) Let $\lambda = \lambda^{'}$, $k = k+1$, go to (3).

4 Results and Conclusion

The application of the MCFO to imaging from incomplete projections was tested in following imaging situations: computer-generated noisy projections of self-defined Sheep-Logan head model

For comparison, the image from the same projections reconstructed by Convolution Back Projection (CBP), General Algebraic Reconstruction Techniques (ART), Multi-criterion Optimization (MCO) [15] and Multi-criterion Fuzzy Optimization (MCFO) are shown in Fig. 1. (a)~(d) respectively.

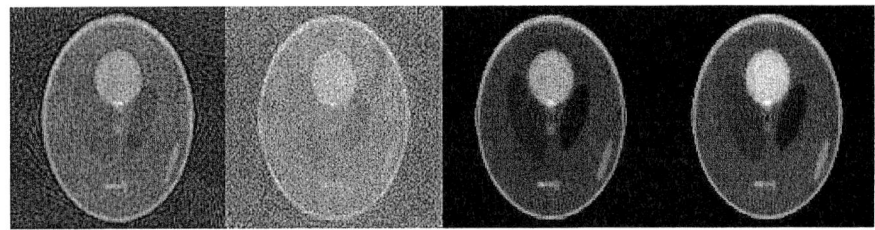

Fig. 1. Reconstructed Image (a) CBP method (b) ART method (c) MCO method (d) MCFO method

Above results showed that the advance of MCFO algorithm compared to other image reconstruction method in error, smoothness and gray resolution, given the same noise and incomplete projections.

References

1. Gao, Xin, Xia, Shunren, Wang, Yuanmei, Luo, Ronglei, "Fast iterative algorithm for image reconstruction from incomplete projections", Journal of Zhejiang University SCIENCE, Vol. 38 No. 9 (2004) 1108-1111.
2. Wang, Yuanmei, Cheng, JP, Heng, PA, "Vector entropy imaging theory with application to computerized tomography", Phys. Med. Biol., 47 (2002) 2301-2310.
3. Fletcher, R., Practical Methods of Optimization, Volume 2 Constrained Optimization, John Wiley & Sons (1981).
4. Bellman, R.E., Zadeh, L.A., "Decision making in a fuzzy environment", Management Science, Vol.17 (1970) 141-164.

Transductive Knowledge Based Fuzzy Inference System for Personalized Modeling

Qun Song, Tianmin Ma, and Nikola Kasabov

Knowledge Engineering & Discovery Research Institute,
Auckland University of Technology,
Private Bag 92006, Auckland 1020, New Zealand
{qsong, mmaa, nkasabov}@aut.ac.nz

Abstract. This paper introduces a novel transductive knowledge based fuzzy inference system (TKBFIS) and its application for creating personalized models. In transductive systems a local model is developed for every new input vector, based on some closest data to this vector from the training data set. A higher-order TSK type fuzzy inference engine is applied in TKBFIS. Some existing formulas or equations, which are used to represent the knowledge and usually have a non-linear form, are taken as consequent parts of the fuzzy rules. The TKBFIS uses a gradient descent algorithm for its training. In this paper, the TKBFIS is illustrated with a case study of personalized modeling for renal function estimation of patients and the result is compared with other transductive or inductive methods.

1 Introduction

1.1 Transductive Versus Inductive Modeling

Most learning models and systems in artificial intelligence developed and implemented so far are based on inductive methods, where a model (a function) is derived from data representing the problem space and this model is further applied on new data [5, 10]. The model is usually created without taking into account any information about a particular new data vector (test data). An error is measured to estimate how well the new data fits into the model. The inductive learning and inference approach is useful when a global model ("the big picture") of the problem is needed even in its very approximate form. In contrast to inductive learning and inference methods, transductive inference methods estimate the value of a potential model (function) only in a single point of the space (the new data vector) utilizing additional information related to this point [9]. This approach seems to be more appropriate for clinical and medical applications of learning systems, where the focus is not on the model, but on the individual patient. Each individual data vector (e.g.: a patient in the medical area; a future time moment for predicting a time series; or a target day for predicting a stock index) may need an individual, local model that best fits the new data, rather than a global model, in which the new data is matched without taking into account any specific information about this data.

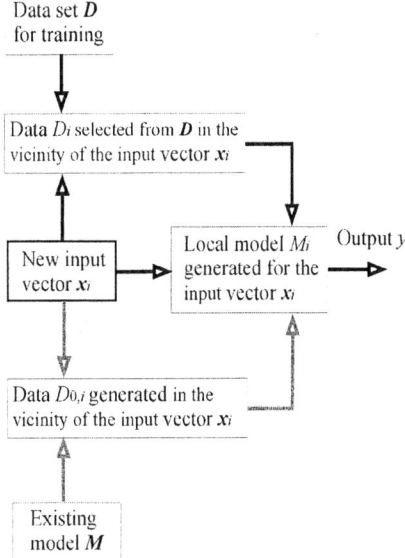

Fig. 1

Fig. 1. A block diagram of a transductive reasoning system An individual model M_i is trained for every new input vector x_i with data use of samples D_i selected from a data set D, and data samples $D_{0,i}$ generated from an existing model (formula) M (if such a model is existing). Data samples in both D_i and $D_{0,i}$ are similar to the new vector x_i according to defined similarity criteria.

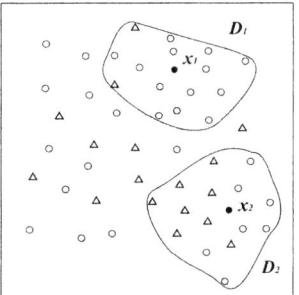

● – a new data vector;   ○ – a sample from D;   △ – a sample from M

Fig. 2

Fig. 2. In the centre of a transductive reasoning system is the new data vector (here illustrated with two of them – x_1 and x_2), surrounded by a fixed number of nearest data samples selected from the training data D and generated from an existing model M

Transductive inference is concerned with the estimation of a function in single point of the space only [9]. For every new input vector x_i that needs to be processed for a prognostic task, the N_i nearest neighbours, which form a sub-data set D_i, are derived from an existing data set D. If necessary, some similar vectors to vector x_i and their outputs can also be generated from an existing model M. A new model M_i is dynamically created from these samples to approximate the function in the point x_i - Fig 1 and Fig 2. The system is then used to calculate the output value y_i for this input vector x_i.

1.2 Knowledge Based Higher-Order TSK Fuzzy Inference Systems

The TSK fuzzy inference system was proposed by Takagi and Sugeno in 1985 [8]. Since then, a lot of concerned research and applications have been developed. The TSK fuzzy model is now established as one of the most powerful fuzzy models and has been widely and successfully used in different research areas such as: adaptive control; classification; prediction and system identification.

A typical TSK fuzzy model consists of If-Then rules that have the following form:

R_l: if x_1 is F_{l1} and x_2 is F_{l2} and ... and x_P is F_{lP},

$$\text{then } y \text{ is } g_l(x_1, x_2, \ldots, x_P) \tag{1}$$

where "x_j is F_{lj}", $l = 1, 2, \ldots m$; $j = 1, 2, \ldots P$, are $m \times P$ fuzzy propositions that form m antecedents for m fuzzy rules respectively; x_j, $j = 1, 2, \ldots, P$, are antecedent variables defined over universes of discourse X_j, $j = 1, 2, \ldots, P$, and F_{lj}, $l = 1, 2, \ldots m$; $j = 1, 2, \ldots, P$ are fuzzy sets defined by their fuzzy membership functions $\mu_{Flj}: X_j \rightarrow [0, 1]$, $l = 1, 2, \ldots m$; $j = 1, 2, \ldots, P$. In the consequent parts of the fuzzy rules, y is the consequent variable, and functions g_l, $l = 1, 2, \ldots m$, are employed.

If $g_l(x_1, x_2, \ldots, x_q) = C_l$, $l = 1, 2, \ldots m$, and C_l are constants, we call this inference a zero-order TSK fuzzy inference system. The system is called a first-order TSK fuzzy inference system if $g_l(x_1, x_2, \ldots, x_q)$, $l = 1, 2, \ldots m$, are linear functions. If these functions are non-linear functions, it is called a higher-order TSK inference system [6].

For an input vector $x^i = [x_1^i\ x_2^i\ \ldots\ x_P^i]$, the result of the inference, y^i, or the output of the system, is the weighted average of each rule's output value indicated as follows:

$$y^i = \frac{\sum_{l=1}^{m} w_l g_l(x_1^i, x_2^i, \ldots, x_P^i)}{\sum_{l=1}^{m} w_l} \tag{2}$$

where, $w_l = \prod_{j=1}^{P} F_{lj}(x_j^i)$; $l = 1, 2, \ldots m$; $j = 1, 2, \ldots, P$.

Because a fuzzy inference system can be considered as an effective aggregator of interconnected subsystems, described with simple models, the first-order TSK models are mostly used so far. In some cases, however, certain kinds of non-linear functions would be more appropriate and more effective to use in a TSK system. In our research, we attempt to find an existing function as the consequent part of a fuzzy rule. Such a function usually has a non-linear form and it represents the knowledge in the certain area.

The paper is organized as follows: Section 2 presents the structure and the algorithm of TKBFIS models. Section 3 illustrates the approach with a case study example. Conclusions are drawn in Section 4.

2 The Learning Procedure for TKBFIS

The TKBFIS is a dynamic neural-fuzzy inference system with a local generalization. Here, the local generalization means that in a sub-space (local area) of the whole problem space a model is crated and it performs generalization in this local area. In TKBFIS models, Gaussian fuzzy membership functions are applied in each fuzzy rule as the antecedent parts. A gradient descent (BP) learning algorithm is used for optimizing the parameters of the fuzzy rules [4, 7].

Suppose there are existing formulas or equations G_h, $h = 1, 2, ..., Q$, that globally represent the knowledge in a certain area. For each new data vector x_q, an individual model is created with the application of the following steps:

1. Search in the training data set in the input space to find N_q training samples that are closest to x_q. The value of N_q can be pre-defined based on experience, or - optimized through the application of an optimization procedure. Here we assume the former approach.
2. All of Q formulas are modified with the gradient descent method on selected N_q training samples (Eq.5 and Eq.6) and the best one (with the minimum *RMSE* – root mean square error) is selected as the consequence for each fuzzy rule.
3. Calculate the distances d_i, $i = 1, 2, ..., N_q$, between each of these data samples and x_q. Calculate the vector weights $v_i = [\max(d) - (d_i - \min(d))]/\max(d)$, $i = 1, 2, ..., N_q$, $\max(d)$ and $\min(d)$ are the minimum value and maximum value respectively in the distance vector $d = [d_1, d_2, ..., d_{Nq}]$.
4. Use a clustering algorithm – *ECM* [2, 6] to cluster and partition the input sub-space that consists of N_q selected training samples.
5. Create fuzzy rules to form a local model M_q and set their initial parameter values according to the *ECM* clustering procedure result and the selected modified function. For each cluster, the cluster centre is taken as the centre of the fuzzy membership function (Gaussian function) and the cluster radius is taken as the width.
6. Apply the gradient descent method (back-propagation) to optimize the parameters of the fuzzy rules in the local model M_q following Eq. (6 – 13).
7. Calculate the output value f_q for the input vector x_q with the local model M_q.
8. End of the procedure.

The parameter optimization procedure is described below:

Consider the system having P inputs, one output and M fuzzy rules defined initially through the *ECM* clustering procedure, the l-th rule has the form of:

R_l : If x_1 is F_{l1} and x_2 is F_{l2} and ... x_P is F_{lP}, then y is G_l . (3)

Here, F_{lj} are fuzzy sets defined by the following Gaussian type membership function:

$$\text{GaussianMF} = \alpha \exp\left[-\frac{(x-m)^2}{2\sigma^2}\right] \tag{4}$$

and G_l are crisp functions selected from Q existing formulas and each of them has parameters b_{pf}.

The TKBFIS is given the selected training input-output data pairs $[x_i, t_i]$, $i = 1, 2, \ldots, N_q$, the function modifying procedure minimizes the following objective function for each existing formulas on the selected training data:

$$E_q = \frac{1}{2}\sum_{i=1}^{N_q}[G_h(x_i) - t_i]^2 \tag{5}$$

The gradient descent algorithm is used to obtain the recursions for updating the parameters b such that E_q of Eq.5 is minimized:

$$b_{pf}(k+1) = b_{pf}(k) - \eta_{bg}[G_h(x_i) - t_i]\frac{\partial G_h}{\partial b_{pf}} \tag{6}$$

Using a *Modified Centre Average Defuzzification* procedure the output value of the system can be calculated for an input vector $x_i = [x_1, x_2, \ldots, x_P]$ as follows:

$$f(x_i) = \frac{\sum_{l=1}^{M} G_l \prod_{j=1}^{P} \alpha_{lj} \exp\left[-\frac{(x_{ij} - m_{lj})^2}{2\sigma_{lj}^2}\right]}{\sum_{l=1}^{M} \prod_{j=1}^{P} \alpha_{lj} \exp\left[-\frac{(x_{ij} - m_{lj})^2}{2\sigma_{lj}^2}\right]} \tag{7}$$

For a training input-output data pair $[x_i, t_i]$, the system minimizes the following objective function (a weighted error function):

$$E = \frac{1}{2}v_i[f(x_i) - t_i]^2 \tag{8}$$

(v_i are defined in Step 3)

The gradient descent algorithm (BP) is used then to obtain the recursions for the optimization of the parameters b, α, m and σ such that the value of E from Eq. (8) is minimized:

$$b_{pf}(k+1) = b_{pf}(k) - \eta_b v_i \Phi(x_i)[f(x_i) - t_i]\frac{\partial G_l}{\partial b_{pf}} \tag{9}$$

$$\alpha_{lj}(k+1) = \alpha_{lj}(k) - \frac{\eta_\alpha v_i \Phi(x_i)}{\alpha_{lj}(k)}[f^{(k)}(x_i) - t_i][G_l(k) - f^{(k)}(x_i)] \tag{10}$$

$$m_{lj}(k+1) = m_{lj}(k) - \frac{\eta_m(k) v_i \Phi(x_i)}{\sigma_{lj}^2(k)}[f^{(k)}(x_i) - t_i][G_l(k) - f^{(k)}(x_i)][x_{ij} - m_{lj}(k)] \tag{11}$$

$$\sigma_{lj}(k+1) = \sigma_{lj}(k) - \frac{\eta_\sigma(k) v_i \Phi(x_i)}{\sigma_{lj}^3(k)}[f^{(k)}(x_i) - t_i][G_l(k) - f^{(k)}(x_i)][x_{ij} - m_{lj}(k)]^2 \tag{12}$$

here,

$$\Phi(x_i) = \frac{\prod_{j=1}^{P} \alpha_{lj} \exp\left\{-\frac{[x_{ij}-m_{lj}(k)]^2}{2\sigma_{lj}^2(k)}\right\}}{\sum_{l=1}^{M}\prod_{j=1}^{P} \alpha_{lj} \exp\left\{-\frac{[x_{ij}-m_{lj}(k)]^2}{2\sigma_{lj}^2(k)}\right\}} \quad (13)$$

where: η_{bg}, η_b, η_α, η_m, and η_σ are learning rates for updating the parameters b, α, m and σ respectively.

In the TKBFIS training–simulating algorithm, the following indexes are used:

- Training data : $i = 1, 2, ..., N$;
- Selected training data set: $i = 1, 2, ..., N_q$
- Input variables: $j = 1, 2, ..., P$;
- Fuzzy rules: $l = 1, 2, ..., M$;
- Number of existing formulas $h = 1, 2, ..., Q$;
- Number of parameters in G_l $pf = 1, 2, ..., L_{pf}$;
- Learning iterations: $k = 1, 2, ...$

3 TKBFIS for Personalized Modeling: A Real World Case Study on Renal Function Evaluation

A real data set from a medical institution is used here for experimental analysis. The data set has 441 samples, collected at hospitals in New Zealand and Australia. Each of the records includes eight variables (inputs): age, gender, serum creatinine, serum albumin, race, blood urea nitrogen concentrations, weight and height, and one output - the glomerular filtration rate value (GFR). One formula is selected as the consequent part for each fuzzy rule in a personalized model from nine existing formulas – Jelliffe(1971), Mawer, Jelliffe(1973), Cockcroft-Gault, Hull, Bjorasson, Gates, Walser and MDRD [3]. These nine formulas have been developed in renal research area during the last thirty years.

Using the proposed model, we have obtained a more accurate result than with the use of the existing formulas, or the use of some other connectionist models. For comparison, the results produced by using the formulas: Gates, Jelliffe73, MDRD and Walser equations; standard NN models, such as MLP and RBF neural networks [3, 5]; the adaptive neural fuzzy inference system (ANFIS) [1]; and the dynamic evolving neural fuzzy inference system (DENFIS) [2], along with the results produced by the proposed TKBFIS model, are listed in Table 1. The results include the number of fuzzy rules (for TKBFIS, ANFIS and DENFIS), or neurons in the hidden layer (for RBF and MLP), the testing RMSE (root mean square error), and the testing MAE (mean absolute error). All experimental results of learning systems reported here are based on leave-one-out cross validation experiments.

Table 1. Experimental results on GFR data

Model	Neurons or rules	RMSE	MAE
Gates	–	7.48	5.63
Jelliffe73	–	7.83	5.87
MDRD	–	7.74	5.88
Walser	–	7.38	5.6
MLP	12	8.44	5.75
ANFIS	36	7.49	5.48
DENFIS	27	7.29	5.29
RBF	32	7.22	5.41
Proposed TKBFIS	6.4 (average)	7.02	5.08

4 Conclusions

This paper presents a transductive knowledge based fuzzy inference system – TKBFIS. The TKBFIS performs a better local generalisation over new data as it develops an individual model for each data vector that takes into account the new input vector location in the space, and it is an adaptive model, in the sense that input-output pairs of data can be added to the data set continuously and immediately made available for transductive inference of local models. This type of modeling is promising for medical decision support systems. As the TKBFIS creates a unique sub-model for each data sample, it usually needs more performing time than an inductive model, especially in the case of training and simulating on large data sets.

Further directions for research include: (1) TKBFIS system parameter optimization such as optimal number of nearest neighbours; (2) applications of the TKBFIS method for other decision support systems, such as: cardio-vascular risk prognosis; biological processes modeling and prediction based on gene expression micro-array data.

Acknowledgement

The research is funded by the New Zealand Foundation for Research, Science and Technology under grant NERF/AUTX02-01. The authors acknowledge the assistance of Dr Mark Marshal from the Middlemore hospital in Auckland for providing data and expertise for the analysis and the validation of the results. The authors would like to thank Janssen-Cilag Pty for their cooperation, and the investigators from the original EPO AUS-14 study for their collaboration and generosity.

References

1. Jang, R.: ANFIS: adaptive network based fuzzy inference system. IEEE Trans. on Syst., Man, and Cybernetics, Vol. 23:3 (1993) 665 – 685.
2. Kasabov, N. and Song, Q.: DENFIS: Dynamic, evolving neural-fuzzy inference systems and its application for time-series prediction. IEEE Trans. on Fuzzy Systems, Vol. 10 (2002) 144 – 154.
3. Levey, A. S., Bosch, J. P., Lewis, J. B., Greene, T., Rogers, N., Roth, D.: for the Modification of Diet in Renal Disease Study Group – A More Accurate Method To Estimate Glomerular Filtration Rate from Serum Creatinine: A New Prediction Equation. Annals of Internal Medicine, Vol. 130 (1999) 461 – 470.
4. Lin, C.T., and Lee, C.S.G.: Neuro Fuzzy Systems. Prentice Hall (1996).
5. Neural Network Toolbox User's Guide. The Math Works Inc, ver. 4 (2001).
6. Song, Q. and Kasabov, N.: ECM - A Novel On-line, Evolving Clustering Method and Its Applications. Proceedings of the Fifth Biannual Conference on Artificial Neural Networks and Expert Systems (ANNES2001), Dunedin, New Zealand, November (2001) 87 – 92.
7. Wang, L.X..: Adaptive Fuzzy System And Control: Design and Stability Analysis. Englewood Cliffs, NJ: Prentice Hall (1994).
8. Takagi, T. and Sugeno, M.: Fuzzy Identification of systems and its applications to modeling and control. IEEE Trans. on Systems, Man, and Cybernetics, Vol. 15 (1985) 116 – 132.
9. Vapnik V.: Statistical Learning Theory, John Wiley & Sons, Inc (1998).
10. Bishop, C., Neural networks for pattern recognition. Oxford University Press (1995).
11. Chakraborty, S., Pal, K.,and Pal, N.R.: A neuro-fuzzy framework for inferencing. Neural Networks, Vol. 15 (2002) 247 – 261.

A Sampling-Based Method for Mining Frequent Patterns from Databases

Yen-Liang Chen and Chin-Yuan Ho

Dept. of Information Management, National Central Univ, Chung-Li, Taiwan 320
{ylchen, chuckho}@mgt.ncu.edu.tw

Abstract. Mining frequent item sets (frequent patterns) in transaction databases is a well known problem in data mining research. This work proposes a sampling-based method to find frequent patterns. The proposed method contains three phases. In the first phase, we draw a small sample of data to estimate the set of frequent patterns, denoted as F^S. The second phase computes the actual supports of the patterns in F^S as well as identifies a subset of patterns in F^S that need to be further examined in the next phase. Finally, the third phase explores this set and finds all missing frequent patterns. The empirical results show that our algorithm is efficient, about two or three times faster than the well-known FP-growth algorithm.

1 Introduction

There have been many algorithms developed for fast mining of frequent patterns, which can be classified into two categories. The first category, candidate generation-and-test approach, such as Apriori [1] as well as many subsequent studies, is directly based on an anti-monotone property: if a pattern with k items is not frequent, any of its super-patterns with $(k+1)$ or more items can never be frequent. The most famous algorithm in this category is the Apriori algorithm, which generates a set of candidate patterns of length $(k+1)$ from the set of frequent patterns of length k and then checks their corresponding occurrence frequencies in the database. Since the algorithm needs multiple passes to generate frequent patterns and each pass requires one full scan of database, its efficiency is not satisfactory when the database contains long patterns or when the number of candidate patterns is huge. Therefore, a number of researches have been proposed to improve its performance by reducing the number of candidate patterns [2], reducing the number of transactions to be scanned [1, 2] or the number of database scans [3, 4].

Recently, another category, compress-and-projection approach, such as the FP-growth algorithm [5], has been proposed. The idea of this approach is, firstly, to build up a compressed data structure to hold the entire database in memory. Then, the database is recursively partitioned into multiple sub-databases according to the frequent patterns found so far and it would search for local frequent patterns to assemble longer global ones. The most famous algorithm in this category is the FP-growth algorithm, which use the FP-tree data structure to store the compressed database.

Other algorithms in this category include the Pattern Repository algorithm [6], the Opportunistic Projection algorithm [7], the H-Mine algorithm [8], the DepthProject algorithm [9] and the MAFIA algorithm [10].

Interestingly but not surprisingly, the weakness of the one approach is the strength of the other approach. The major weakness of the candidate generation-and-test approach is its low efficiency. Two reasons result in this problem: (1) It usually generates a huge set of candidate patterns; (2) it may scan the database many times, especially when the database contains long patterns or dense patterns. In this regard, the compress-and-projection approach performs better because by compressing the entire database into a compressed structure in memory it eliminates the needs to access the database multiple times and by using recursive partition and projection to generate frequent patterns it eliminates the needs to generate the sets of candidate patterns. Therefore, the algorithms in the second category are usually much faster than those in the first category. However, the second approach has its own problems. The difficulty is that they may not fit in the memory when the database is huge. Besides, during the process of partition and projection, multiple copies of the database may be generated and kept in the main memory. This makes the algorithms not scalable in large databases. In this regard, the algorithms in the first category are exempted, because they don't store the compressed database in the main memory.

The comparisons above show a requirement to design an algorithm that has the advantages of both approaches but without their disadvantages. Thus, if the new algorithm is designed based on the first approach, it should meet the following requirements.

1. It needs only few database scans.
2. The set of candidate patterns should be small.
3. The performance should not be inferior to those in the second category.
4. Te entire database should not be kept in the main memory.

The goal of this paper is to propose an algorithm that meets all the above requirements. Basically, our algorithm adopts the framework proposed by Toivonen [11]. This framework consists of three phases. First, it mines a sample S of the database with a lower support threshold than the minimum support to find the frequent patterns local to S (denoted L^s). Then, the second phase scans the whole database once to compute the actual supports of each pattern in L^s. Here, we design a method to determine whether all frequent patterns appear in L^s. If they are, then one scan of database is sufficed. Otherwise, the third phase uses a second scan to find the frequent patterns that were missing in the second phase. In the original paper of Toivonen, he only gave a rough framework without specifying the implementation details.

Although our algorithm has the same framework as Toivonen's algorithm, we use a new advanced data structure to implement the framework. The data structure used in the algorithm is the all-subset tree, where each node corresponds to an itemset. This structure is similar to the lexicographical tree used in the TreeProjection algorithm [12] or the set-enumeration tree used in Max-Miner [13]. By combining sampling technique with the all-subset tree, a novel algorithm that satisfies all the above-mentioned requirements is developed. The major characteristics of this algorithm include: (1) Apart from drawing a sample, the algorithm needs at most two scans of the database; (2) The number of nodes in the all-subset tree is the same as the size of L^s, and this size is much smaller than the size of the set of candidate patterns in the traditional Apriori-like algorithms; (3) Empirical results show that our algorithm can run about two or three times faster than the FP-growth algorithm, and; lastly, (4) The algorithm does not need to keep the database in the main memory.

The rest of the paper is organized as follows. In Section 2, we describe the problem definitions and propose the algorithm. Section 3 runs several simulations to evaluate its performance. Finally, Section 4 is the conclusion.

2 Problem Definitions and the Algorithm

Let $I=\{i_1, i_2, \ldots, i_m\}$ denote all items in database D. Each transaction $T=<e_1,e_2,\ldots,e_n>$ is a set of items, where $e_i \in I$ for all i and e_i is distinct from e_j for $i \neq j$. Let X be a pattern. Then T contains X ($T \supset X$) if every item p in X also appears in T. In database D, the percentage of transactions in database containing X is called the support of X, denoted by *support*(X). A pattern X is frequent if it satisfies *support*(X)$\geq$*minsup*, where *minsup* is specified by the user. Otherwise it is infrequent. We call the number of items in a pattern its size, and call a pattern of size k as a k-pattern. Items within a pattern are kept in lexicographic order. Besides, we use L_k to denote the set of all frequent k-patterns and C_k to denote the set of candidate k-patterns.

Since our algorithm is based on the all-subset tree, we will introduce the tree first. Basically, the tree stores a set of patterns, where each node corresponds to an item set. For example, if the actual frequent patterns for the database include {A}, {B}, {C}, {A, B}, {A, C}, {B, C}, {A, B, C}, {D}, {E}, {A, D}, {A, E}, {D, E}, {A, D, E}, then Fig. 1 shows the corresponding all-subset tree. The tree is named as the all-subset tree because all subsets of a longer pattern must exist in the tree. Note that a count field is associated with each node to record the support of the corresponding pattern. Besides, a hash structure similar to the one used in the hash tree of the Apriori algorithm [1] is attached with each node to accelerate the speed of searching.

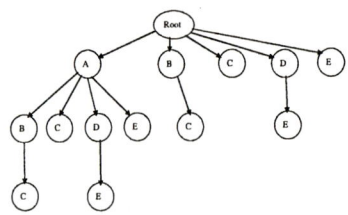

Fig. 1. The all-subset tree constructed from the actual large itemsets

The algorithm contains three phases. In the first phase, we randomly draw a small sample S of transactions from database, from which an approximation set L^s of frequent patterns is obtained. Then, we construct an all-subset tree $T1$ from L^s. Here, two problems arise immediately about $T1$. First, the actual supports of the patterns in $T1$ are still unknown. Second, some frequent patterns in L may not appear in tree $T1$.

For ease of reference, let us call the children of the nodes in $T1$ but not in $T1$ as terminals. To solve the first problem, we need a full scan of the database. After the scan, the supports of all nodes in $T1$ are determined. To solve the second problem, we must check if there exist any terminals of the nodes in $T1$ that are frequent. If all the terminals are infrequent, then all frequent patterns must have already been in $T1$. On the contrary, if some terminals are frequent, then we must further explore these frequent terminals to find if there are any other frequent descendants spawning from them. Due to the consideration above, the second phase of our algorithm extends the tree with terminal nodes. That is, each node will have two kinds of children: those belonging to $T1$ and those not, called terminal nodes. Let us name this extended tree as $T2$. Then the second phase will scan the entire database one time. In processing each transaction, tree $T2$ is traversed and the support counts of the visited nodes, including terminal nodes and the nodes originally in $T1$, are computed. When the traversal is over, the supports of all the nodes are known, and we can find which terminal nodes are frequent. If all terminal nodes are not frequent, we are done. Otherwise, we must scan the database one more time to determine if there are any frequent patterns in the descendants of terminal nodes. That is what the third phase of our algorithm does. The following introduces these three phases in order.

2.1 The First Phase: Building $T1$ by Sampling

We draw x transactions from database in this phase, where x is a parameter specified by users. Furthermore, to ensure that the frequent patterns contained in F^S can be as complete as possible, we lower the minimum support threshold by dividing *minsup* with a parameter α, where $\alpha \geq 1$. After drawing the sample, we use the FP-growth algorithm to find F^S, because this algorithm is famous for its efficiency.

Example 1. Let us consider the database shown in Fig. 2. Suppose we have $x=5$, *minsup*=50% and $\alpha=1.5$. Assume that we draw the first five transactions in the database. By running the FP-growth algorithm with the minimum support threshold 1.67, we find the set L^S as {{A}, {B}, {C}, {D}, {A, B}, {A, C}, {A, D}, {B, C}, {B, D}, {C, D}, {A, B, C}, {A, B, D}, {A, C, D}, {B, C, D}, {A, B, C, D}}. From L^S, we construct the all-subset tree $T1$ as shown in Fig. 3.

TID	Items	TID	Items
001	A, B, C, D	006	A, D, E, F
002	A, B, C	007	A, B, C, D, E
003	A, D, E, G	008	A, D, E, F, G
004	A, B, C, D	009	A, B, C, F
005	C, F	010	A, D, E, G

Fig. 2. The database

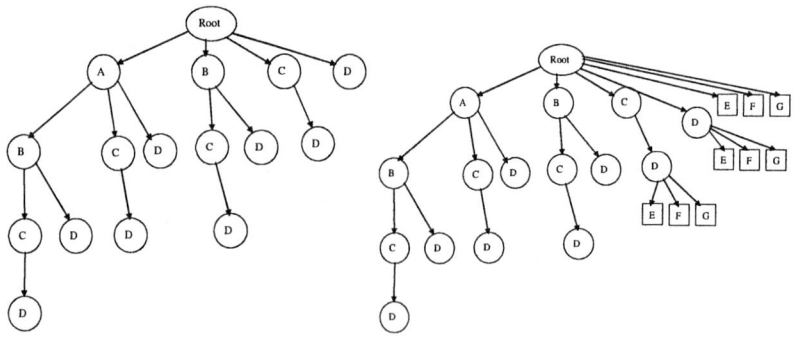

Fig. 3. *T*1 constructed from L^S **Fig. 4.** Extend tree *T*1 with terminal nodes

2.2 The Second Phase: Building *T*2 and the First Scan of Database

As mentioned before, the most difficult problem is that some frequent patterns may be missing in *T*1. To remedy, we need to check if any children of the nodes in *T*1 are frequent. If there are some frequent children, then these children need to be further explored. On the other hand, if all children are infrequent, then all frequent patterns have already been included in *T*1.

Due to the reason above, we need to know not only the supports of the nodes in *T*1, but also the supports of terminal nodes, where the terminal nodes are the children of the nodes in *T*1 but are not in *T*1. Therefore, we extend the original tree *T*1 so that each node in *T*1 is expanded with a set of terminal nodes, which can be used to store the supports of these children. Let us use *T*1 in Fig.3 as an example. Since there are 7 items in the database, i.e., A, B, C, D, E, F and G, the extended tree will look like the one shown in Fig. 4, where we only show the terminal nodes for the root node, node {D} and node {C, D}.

Here, we use a method to speed up the performance. Let us observe the root node in Fig. 4, where we have three pointers pointing to the three terminal nodes and each terminal node needs a count field to store its support value. A smart reader may immediately find that these terminal nodes are not necessary, because we can use the space for storing pointers in the root to store their support values. Thus, we add a Boolean variable preceding each field to tell if this is a pointer pointing to an actual node in *T*1 or the count field of a virtual terminal node. By virtualizing tree *T*1 this way, we obtain a new tree *T*2.

The virtualization benefits the performance, because it makes the tree size very small. Suppose $|L^S|$ denote the number of frequent patterns in L^S. Then the number of the nodes in *T*2 will be also $|L^S|$, because we did not actually generate any terminal nodes. Later, when we need to traverse the tree *T*2 for every transaction, the small tree will make the traversal very efficient.

Having constructed tree *T*2, we need a full scan of the database to determine the supports of all the nodes as well as those of all virtual terminal nodes in *T*2. Once it is finished, we can output all the frequent patterns in *T*2 by depth first search. During the traversal, if we find some virtual terminal nodes that are frequent, then we must store them into set *LT* for further processing in the third phase. Finally, if *LT* is empty, then the algorithm stops. Otherwise, we go to the next phase.

Example 2. Suppose that we use the database in Fig. 2 to traverse the tree shown in Fig. 4. Then, after scanned, the support counts of all nodes will be obtained. By using the depth first search to traverse the tree, we output the frequent patterns including {{A}, {B}, {C}, {A, B}, {A, C}, {B, C}, {A, B, C}, {D}, {E}, {A, D}, {A, E}, {D, E}, {A, D, E}}. Meanwhile, since there are some frequent terminal nodes, we keep them in *LT*={{E}, {D, E}, {A, E}, {A, D, E}} and go to the third phase.

2.3 The Third Phase: Building *T*3 and the Second Scan of Database

We perform this phase only when *LT* is not empty. In this situation, we need to further explore these frequent terminal nodes to see if they have any frequent descendants. To do so, we first construct an all-subset tree from the patterns in *LT*. After that, we insert every transaction into the tree. Note that, if a transaction does not contain any patterns in *LT*, then this transaction will insert nothing into the tree. But if there are some patterns in *LT* that are contained in the transaction, then the nodes corresponding to these patterns will grow descendant nodes.

For ease of presentation, all the nodes are classified into three different classes: non-leaf nodes, terminal nodes and newborn nodes. The terminal nodes correspond to the patterns in *LT*, the non-leaf nodes are the nodes in the paths from the root to the terminal nodes, and newborn nodes are the descendant nodes spawning from terminal nodes. The following steps show how to insert a transaction into the tree.

Subroutine *Traverse*(*u*, *T*, *i*)
Parameters: *u*: the node in the tree where we are currently located
         T: the transaction that we are processing
         i: the last position in the transaction that has been matched
   if *u* is a newborn node then add 1 into the counter of node *u*;
   if *u* is a terminal node or a newborn node then
      for($j = i+1; j \leq length(T); j++$)
         if there is a child *v* of *u* satisfying *item*(*v*) = *item*(*T*(*j*))
         then *Traverse* (*v*, *T*, *j*)
         else create a newborn child *v* of *u* with item label *item*(*T*(*j*))
            Traverse (*v*, *T*, *j*)
   else
      for($j = i+1; j \leq length(T); j++$)
         if there is a child *v* of *u* satisfying *item*(*v*) = *item*(*T*(*j*))
         then *Traverse* (*v*, *T*, *j*);
   return

Example 3. Since Example 2 has *LT*={{E}, {D, E}, {A, E}, {A, D, E}}, we perform the third phase. The first step uses the patterns in *LT* to construct the all-subset tree in Fig. 5. After that, we need to insert every transaction in Fig. 2 into the tree. During the insertion process, only transactions 003, 006 and 008 can insert the tree successfully. The other transactions will fail because they cannot match the patterns in *LT*. Fig. 6 shows the resulting tree, where round, rectangle and round-cornered rectangle nodes denote non-leaf, terminal and newborn nodes, respectively. In addition, the number beside each newborn node is its support count.

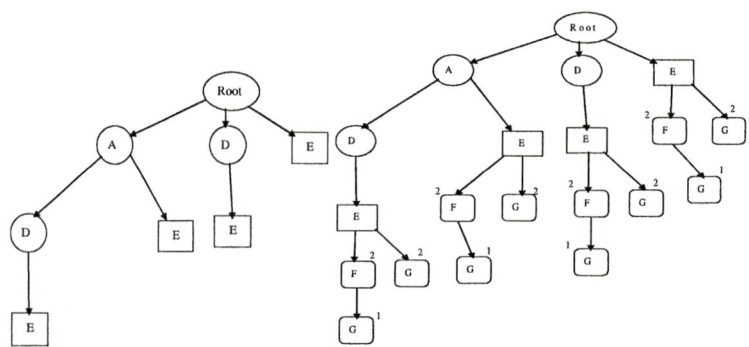

Fig. 5. Building tree from *LT* **Fig. 6.** After insertion

The constructed tree like Fig. 6 may produce a lot of descendents from a single terminal node. The following properties help us to prune the tree.

Property 1. Let u denote the item attached to the current node, and let w denote the item attached to an ancestor node. Then, the pattern corresponding to the current node can be frequent only if (1) $\{w, u\}$ is in L_2 and (2) u is in L_1.

Proof. The pattern corresponding to the current node is the set of items attached to the nodes along the path from the root to the current node. By the anti-monotone property, any subset of a frequent pattern must be also frequent. Therefore, both $\{w, u\}$ and u must be frequent.

Property 2. Let u denote the item attached to the current node and L' the set of frequent patterns obtained after phase 2. Then every frequent pattern in L_1 must be in L'.

Proof. Since *T2* extends every node in *T1* with terminal nodes, every candidate pattern in C_1 must appear in *T2*: either as an actual node or a virtual terminal node in the first level of the tree. Therefore, after computing the supports every frequent pattern in L_1 must exist in L'.

Property 3. Let L' denote the frequent patterns obtained after phase 2. If there are some frequent 2-patterns in L_2, say $\{w, u\}$, that are not in L', then w must be the item attached to a virtual node in the first level of *T2*.

Proof. If y is an actual node with item w in the first level of *T2*, then either y has a virtual child node corresponding to $\{w, u\}$ or has an actual child node corresponding to $\{w, u\}$. In both cases, $\{w, u\}$ must exist in L' after phase 2. This contradiction proves that y must be a virtual node with item w in the first level of *T2*.

Based on the properties above, before we insert a newborn node with item u, we will execute the following test, where w denotes the item attached to an ancestor node of the current node.

1 If u is not in L', then stop the insertion.
2 For every ancestor node with item w
    If (w, u) is not in L' and w is not the item attached to a virtual terminal node in the first level of $T2$, then stop the insertion.
3 Create this newborn node.

3 Experiment Results

To study the performance, our algorithm, named as the All-subset-tree algorithm, and the FP-growth algorithm are implemented by Visual C++ language and tested on a PC with Pentium-III 933 processor and 1.024G main memory under the Window 2000 operating system. We generate the synthetic datasets by applying the well-known synthetic data generation algorithm in [1]. Throughout the simulation, unless stated otherwise, we set the parameters as follows: $|L|$=2000, $|I|$=4, T=10, $|D|$=200000, *minsup* =1% and N=1000. When we need to draw a sample, we draw the data sequentially from the beginning until the number of the data records reaches the target.

At first, we determine the best combinations of the sample size x and the threshold divisor α. So, we execute the All-subset-tree algorithm for 40 different possible combinations. After an extensive study, we find that the better combinations are (4000, 1.6), (6000, 1.5) and (8000, 1.4). Therefore, the following use these three combinations as test cases.

Next, we explore how the performance varies for different minimum support thresholds. Therefore, we generate 30 data sets mentioned before, and vary the minimum support thresholds from 0.5% to 2%. Fig. 7 shows the execution times of the FP-growth algorithm as well as three different versions of the All-subset-tree algorithm. Meanwhile, Table 1 shows the average numbers of database scans required for these three versions of the All-subset-tree algorithm. The results show that all these three are faster than FP-growth, and the average number of scans is close to 1.

Table 1. Average numbers of database scans vs. minimum supports

	(4000, 1.6)	(6000, 1.5)	(8000, 1.4)
0.5%	2.00	1.73	1.77
0.75%	1.27	1.20	1.23
1%	1.03	1.10	1.13
1.5%	1.00	1.00	1.00
2%	1.00	1.00	1.00

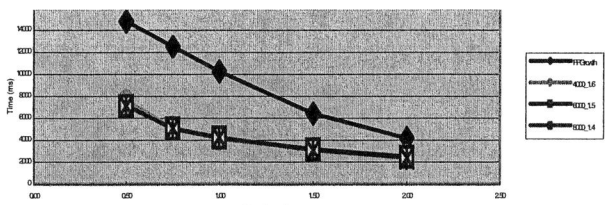

Fig. 7. Execution times vs. minimum supports

Finally, we study how the performance varies when the database size changes. Therefore, we run the comparison 30 times with different numbers of transactions, ranging from 200K to 1000K. Fig. 8 shows the execution times of the FP-growth algorithm as well as three different versions of the All-subset-tree algorithm. Meanwhile, Table 2 shows the average numbers of database scans required for these three versions. The results show that all three versions are faster than FP-growth algorithm. Moreover, it is clear that our algorithm has a better scalability than the FP-growth algorithm, because as the database size gets larger, the ratio of the needed run time of the FP-growth algorithm to that of our method gets larger as well.

Table 2. Average numbers of database scans vs. numbers of transactions

	(4000, 1.6)	(6000, 1.5)	(8000, 1.4)
1000K	1.17	1.07	1.13
800K	1.07	1.03	1.07
600K	1.10	1.03	1.07
400K	1.00	1.03	1.07
200K	1.03	1.10	1.13

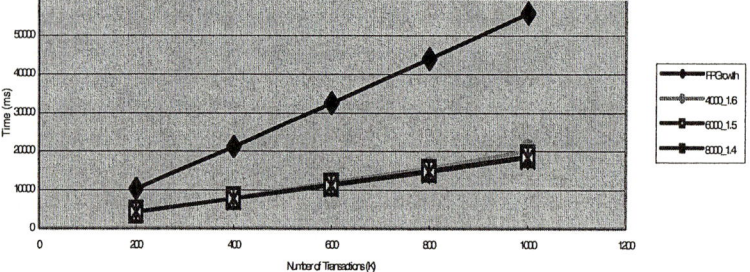

Fig. 8. Execution times vs. numbers of transactions

4 Conclusions

The goal of this paper is to develop a new efficient algorithm for mining frequent patterns. To this end, we use the all-subset tree data structure as the basis to develop a three-phase algorithm. The following are the major characteristics of the proposed algorithm.

1. The algorithm scans the database no more than twice. The experiments show that most of the time one database scan is enough.
2. Since we virtualized the all-subset tree structure, the number of the nodes in the all-subset tree equals to the size of L^S. This is much smaller than the size of the set of candidate patterns produced in the traditional algorithms such as the Apriori algorithm.
3. The algorithm is very efficient. The evaluation shows that it is about two or three times faster than the FP-growth algorithm.

4. The algorithm does not use the main memory to hold the entire database. Due to this reason, it has a better scalability in dealing with large databases.

Acknowledgment

The research was supported in part by the MOE Program for Promoting Academic Excellence of Universities under the Grant Number 91-H-FA07-1-4.

References

1. Agrawal, R., Srikant, R.: Fast algorithms for mining association rules. In: Proceedings of the 20th VLDB Conference. (1994) 478-499
2. Park, J.S., Chen, M.S., Yu, P.S.: Using a hash-based method with transaction trimming for mining association rules. IEEE Transactions on Knowledge and Data Engineering **9** (1997) 813-825
3. Brin, S., Motwani, R., Ullman, J., Tsur, S.: Dynamic itemset counting and implication rules for market basket data. In: Proceedings of the 1997 ACM-SIGMOD Conf. on Management of Data. (1997) 255-264
4. Savasere, A., Omiecinski, E., Navathe, S.: An efficient algorithm for mining association rules in large databases. In: Proceedings of Int'l Conf. Very Large Data Bases. (1995) 432-444
5. Han, J., Pei, J., Yin, Y.: Mining frequent patterns without candidate generation. In: Proceedings of SIGMOD. (2000) 1-12
6. Relue, R., Wu, X., Huang, H.: Efficient runtime generation of association rules. In: Proceedings of the Tenth International Conference on Information and Knowledge Management. (2001) 466-473
7. Liu, J., Pan, Y., Wang, K., Han, J.: Mining frequent item sets by opportunistic projection. In: Proceedings of 2002 Int. Conf. on Knowledge Discovery in Databases. (2002) 229-238
8. Pei, J., Han, J., Lu, H., Nishio, S., Tang, S., Yang, D.: H-mine: hyper-structure mining of frequent patterns in large databases. In: Proceedings of IEEE International Conference on Data Mining. (2001) 441-448
9. Agrawal, R.C., Aggarwal, C.C., Prasad, V.V.V.: Depth first generation of long patterns. In: Proceedings of SIGKDD Conference. (2000) 108-118
10. Burdick, D., Calimlim, M., Gehrke, J.: MAFIA: a maximal frequent itemset algorithm for transactional databases. In: Proceedings of 17th Int. Conf. Data Engineering. (2001) 443-452
11. Toivonen, H.: Sampling large databases for association rules. In: Proceedings of the 22th International Conference on Very Large Databases. (1996) 134-145
12. Agarwal, R., Aggarwal, C., Prasad, V. V. V.: A tree projection algorithm for generation of frequent itemsets. Journal of Parallel and Distributed Computing **61** (2001) 350-371
13. Bayardo Jr., R. J.: Efficiently mining long patterns from databases. In: Proceedings of the ACM-SIGMOD Int'l Conf. on Management of Data. (1998) 85-93

Lagrange Problem in Fuzzy Reversed Posynomial Geometric Programming

Bing-yuan Cao

School of Math. and Inf. Sci., Guangzhou University, 510405 Guangzhou China
Dept. of Math., Shantou University, 515063 Shantou, P.R. China
bycao@stu.edu.cn

Abstract. In this paper, first, the model of a fuzzy reversed posynomial geometric programming is built after introduction of a prime fuzzy posynomial geometric programming. Besides, its fuzzy Lagrange problem is studied. On this basis, a validity of direct algorithm is designed to solve the former programming. Finally, the model and algorithm are testified by a numerical example.

1 Introduction

The author advanced the fuzzy reversed posynomial geometric programming (PGP) based on the prime fuzzy PGP [1][2] by Zadeh fuzzy set theory [5], gives its Lagrange problem and a direct algorithm, which will be wide applied in optimization and classification.

2 Fuzzy Reversed PGP Model

We try to expand the reversed PGP [4] into a fuzzy reversed PGP model.

Definition 1. *We call*

$$(\tilde{P}) \quad \widetilde{\min}\ \tilde{g}_0(x)$$
$$s.t.\ \tilde{g}_i(x) \lesssim 1 (1 \leqslant i \leqslant p'), \tilde{g}_i(x) \gtrsim 1(p'+1 \leqslant i \leqslant p), x > 0$$

a fuzzy reversed PGP, where $x = (x_1, x_2, \cdots, x_m)^T$ is an $m-$dimensional variable vector, 'T' represents a transpose symbol, and all $\tilde{g}_i(x) = \sum_{k=1}^{J_i} \tilde{v}_{ik}(x)(0 \leqslant i \leqslant p)$ are fuzzy posynomials of x, here $\tilde{v}_{ik}(x) = \{\tilde{c}_{ik} \prod_{l=1}^{m} x_l^{\tilde{\gamma}_{ikl}}, (0 \leqslant i \leqslant p'; 1 \leqslant k \leqslant J_i); \tilde{c}_{ik} \prod_{l=1}^{m} x_l^{-\tilde{\gamma}_{ikl}}, (p'+1 \leqslant i \leqslant p; 1 \leqslant k \leqslant J_i)\}$. And for each item $\tilde{v}_{ik}(x)$ $(p'+1 \leqslant i \leqslant p; 1 \leqslant k \leqslant J_i)$ in the reversed inequality $\tilde{g}_i(x) \gtrsim 1$, $-\tilde{\gamma}_{ikl}$ serves as exponents in it. ' $\lesssim$ (or $\gtrsim$)' represents 'approximate $\leqslant$ (or $\geqslant$)'. The membership functions of objective and constraints are defined by

$$\tilde{A}_i(x) = \tilde{B}_i(\bar{g}_i(x)) = \begin{cases} 1, & if\ \bar{g}_i(x) \leqslant b_i, b_i = \{z_0, i = 0; 1, 1 \leqslant i \leqslant p'\}, \\ 1 - \frac{t_i}{d_i}, & if\ \bar{g}_i(x) = b_i + t_i, 0 \leqslant t_i \leqslant d_i, \\ 0, & if\ \bar{g}_i(x) \geqslant b_i + d_i, 0 \leqslant i \leqslant p', \end{cases} \quad (1)$$

$$\tilde{A}'_i(x) = \tilde{B}'_i(\bar{g}_i(x)) = \begin{cases} 1, & if\ \bar{g}_i(x) \geqslant 1, \\ 1 - \frac{t_i}{d_i}, & if\ \bar{g}_i(x) = 1 - t_i, 0 \leqslant t_i \leqslant d_i, \\ 0, & if\ \bar{g}_i(x) \leqslant 1 - d_i, p'+1 \leqslant i \leqslant p, \end{cases} \quad (2)$$

respectively, where z_0 is an aspiration level of the objective function $\tilde{g}_0(x)$, $d_i \geqslant 0 (0 \leqslant i \leqslant p)$ is a flexible index of $i-$th fuzzy function $\tilde{g}_i(x)$. $\tilde{a}$ may be freely fixed in the closed interval $[a^-, a^+]$, and its degree of accomplishment [2] are determined by

$$\tilde{a}(a) = \begin{cases} 0, & \text{if } a < a^-, \\ ((a - a^-)/(a^+ - a^-))^r, & \text{if } a^- \leqslant a \leqslant a^+, \\ 1, & \text{if } a > a^+, \end{cases} \qquad (3)$$

here $\tilde{a}$ represents fuzzy coefficients $\tilde{c}_{ik}$ as well as fuzzy exponents $\tilde{\gamma}_{ikl}$, and a, a^-, a^+, r are all real numbers. $\tilde{A}_i(x) = \tilde{B}_i(\cdot)$ is a fuzzy compound function, $\tilde{g}_i(x) = \sum_{k=1}^{J_i} \tilde{c}_{ik}^{-1}(\beta) \prod_{l=1}^{m} x_l^{\tilde{\gamma}_{ikl}^{-1}(\beta)}, \beta \in [0,1], (0 \leqslant i \leqslant p)$. Symbol $\widetilde{\min}$ is an extended $\min$ operation, and is defined as $\widetilde{\min}\ \tilde{g}_i(x) \leftarrow \tilde{g}_i(x) \lesssim z_0$.

The objective function in $(\tilde{P})$ might be written as a minimizing goal in order to consider z_0 as an upper bound, then $(\tilde{P})$ can be rewritten as

$$\begin{cases} \tilde{g}_0(x) \lesssim z_0, \\ \tilde{g}_i(x) \lesssim 1, (1 \leqslant i \leqslant p'), \tilde{g}_i(x) \gtrsim 1, (p'+1 \leqslant i \leqslant p), x > 0. \end{cases} \qquad (4)$$

Definition 2. $\tilde{A}_i = \{x \in R^m | \tilde{g}_i(x) \lesssim 1, x > 0\}(1 \leqslant i \leqslant p')$ and $\tilde{A}'_i = \{x \in R^m | \tilde{g}_i(x) \gtrsim 1, x > 0\}(p'+1 \leqslant i \leqslant p)$ are fuzzy feasible solution sets corresponding to $\tilde{g}_i(x) \lesssim 1$ and $\tilde{g}_i(x) \gtrsim 1$, respectively. $\tilde{Y} = \tilde{A}_0 \cap (\cap_{1 \leqslant i \leqslant p'} \tilde{A}_i) \cap (\cap_{p'+1 \leqslant i \leqslant p} \tilde{A}'_i)$ is called the fuzzy decision for (4), and so is for $(\tilde{P})$, satisfying

$$\tilde{Y}(x) = \tilde{A}_0(x) \bigwedge \min_{1 \leqslant i \leqslant p'} \tilde{A}_i(x) \bigwedge \min_{p'+1 \leqslant i \leqslant p} \tilde{A}'_i(x), x > 0, \qquad (5)$$

while x^* is called a fuzzy optimal solution to (4), and so is to $(\tilde{P})$, satisfying

$$\tilde{Y}(x^*) = \max_{x>0}\{\tilde{Y}(x) = \min\{\tilde{A}_0(x), \min_{1 \leqslant i \leqslant p'} \tilde{A}_i(x), \min_{p'+1 \leqslant i \leqslant p} \tilde{A}'_i(x)\}\}. \qquad (6)$$

If there exists a fuzzy optimal point set $\tilde{A}_0$ of $\tilde{g}_0(x)$, (5) holds, calling (6) a fuzzy reversed PGP for $\tilde{g}_0(x)$ with respect to $\tilde{Y}$.

Substituting (1)(2)(3) into (5), after some rearrangements [6], then
$$\tilde{Y}(x) = \left(1 - \frac{\tilde{g}_0(x)-z_0}{d_0}\right) \bigwedge \min_{1 \leqslant i \leqslant p'} \left(1 - \frac{\tilde{g}_i(x)-1}{d_i}\right) \bigwedge \min_{p'+1 \leqslant i \leqslant p} \left(1 + \frac{\tilde{g}_i(x)-1}{d_i}\right).$$

Introducing a new variable $\alpha, \alpha \in [0,1]$, the maximization decision of $(\tilde{P})$ can be turned into solving $x(>0)$, that is, maximizing x^* in $\tilde{Y}(x)$.

Let $\alpha = \tilde{Y}(x)$. By using functions (1) and (2), when $\tilde{A}_0(x) = z_0 - d_0$, and $\tilde{A}_i(x) = 1 + d_i$, we have the following theorem.

Theorem 1. The maximizing of $\tilde{Y}(x)$ is equivalent to $\tilde{Y}(x) \geqslant \alpha$, so we arrive at Formula (7)

$$\begin{aligned} &\max\ \alpha \\ &s.t.\ \tilde{g}_0(x) \leqslant z_0 + (1-\alpha)d_0,\ \tilde{g}_i(x) \leqslant 1 + (1-\alpha)d_i, (1 \leqslant i \leqslant p'), \\ &\tilde{g}_i(x) \geqslant 1 + (\alpha - 1)d_i, (p'+1 \leqslant i \leqslant p), x > 0, \alpha, \beta \in [0,1]. \end{aligned} \qquad (7)$$

3 Fuzzy Lagrange Problem and Algorithm

Definition 3. *Assume that $\tilde{g}_i(x)$ is m-dimensional fuzzy differentiable function, and its gradient $\nabla_x \tilde{g}_i(x) = (\frac{\partial}{\partial x_1}\tilde{g}_i(x), \frac{\partial}{\partial x_2}\tilde{g}_i(x), \ldots, \frac{\partial}{\partial x_m}\tilde{g}_i(x))^T$ is equivalent to $\nabla_x \bar{g}_i(x) = (\frac{\partial}{\partial x_1}\bar{g}_i(x), \frac{\partial}{\partial x_2}\bar{g}_i(x), \ldots, \frac{\partial}{\partial x_m}\bar{g}_i(x))^T$.*

Definition 4. *Find fuzzy feasible solution x^* to $(\tilde{P})$ and $\mu^* = (\mu_1^*, \mu_2^*, \ldots, \mu_p^*)^T \geqslant 0$, satisfying $\mu_i^*(\tilde{g}_i(x^*)-1) = 0 (1 \leqslant i \leqslant p)$, such that a fuzzy Lagrange function*

$$\tilde{L}(x,\mu) = \tilde{g}_0(x) + \sum_{i=1}^{p'}(\tilde{g}_i(x) - 1) + \sum_{i=p'+1}^{p} \mu_i(1 - \tilde{g}_i(x))$$

fits $\nabla_x \tilde{L}(x^*, \mu^*) = 0$, called a Lagrange problem in $(\tilde{P})$. Here $\tilde{g}_i(x^*) = 1$ is a fuzzy equality, its membership degree is $\tilde{B}_i(\bar{g}_i(x) - 1) = 1$, and $\tilde{A}_i(x) = \tilde{B}_i(\cdot)$ is a continuous and strictly monotonous (CSM) fuzzy function.

Theorem 2. *Let x^* be a fuzzy feasible solution to PGP $(\tilde{P})$, writing $\mathcal{I} = \{i|\tilde{g}_i(x^*) = 1, 1 \leqslant i \leqslant p\}$ as a subscript set of fuzzy effective constraint at x^* and $\tilde{B}_i(x)(0 \leqslant i \leqslant p)$ are fuzzy functions to CSM. Then μ^* enables (x^*, μ^*) to be a fuzzy solution in Lagrange problem iff all variable vectors $x(> 0)$ satisfy*

$$\sum_{l=1}^{m} \tilde{\Gamma}_{il}(\ln x_l - \ln x_l^*) \lesssim 0 \ (i \in \mathcal{I}), \tag{8}$$

and then
$$\tilde{g}_0(x^*) \lesssim \tilde{g}_0(x), \tag{9}$$

here, $\tilde{\Gamma}_{il} = \sum_{k=1}^{J_i} \tilde{\gamma}_{ikl} \tilde{v}_{ik}(x^*), (i \in \mathcal{I}, 1 \leqslant k \leqslant J_i; 1 \leqslant l \leqslant m)$.

Proof. Let $\tilde{B}_i(x)(0 \leqslant i \leqslant p)$ be a fuzzy function to CSM, then

$$(\tilde{P}) \iff \min \tilde{A}_0(\tilde{g}_0(x))$$
$$\text{s.t.} \ \tilde{B}_i(\tilde{g}_i(x) - 1) \geqslant \alpha, (1 \leqslant i \leqslant p'), \tag{10}$$
$$\tilde{B}_i(1 - \tilde{g}_i(x)) \geqslant \alpha, (p'+1 \leqslant i \leqslant p), x > 0, \alpha \in [0, 1],$$

while $\mathcal{I} \iff \mathcal{I}' = \{i|\tilde{B}_i(\tilde{g}_i(x^*) - 1) = 0, (1 \leqslant i \leqslant p)\}$, with

$$(8) \iff \tilde{B}_i[\tilde{\Gamma}_{il}(\ln x_i - \ln x_i^*)] \geqslant \alpha (i \in \mathcal{I}'), \tag{11}$$

and then
$$(9) \iff \tilde{B}_0[\tilde{g}_0(x^*) - \tilde{g}_0(x)] \geqslant \alpha. \tag{12}$$

From the condition, it is known that x^* is a fuzzy feasible solution to $(\tilde{P})$, which is equivalent that (x^*, α) is a parameter feasible solution to (10)[2]. Therefore, as for subscript set $\mathcal{I}'$ and any $\alpha(\in [0, 1])$, there exists μ^*, enabling (x^*, μ^*, α) to be a Lagrange problem solution with parameter α iff all variable vectors $x > 0$ satisfy (11). And (12) holds from the knowledge of Theorem 4.4.1 in [4]. Hence the theorem holds from arbitrariness of α, β in $[0, 1]$. □

Proposition 1. *Let $\tilde{A}_i(x)(0 \leqslant i \leqslant p)$ be a fuzzy function of CSM. On the assumption of constraint complete lattice, a local optimum solution to $(\tilde{P})$ must be a part of fuzzy solution to the Lagrange problem.*

Proposition 2. Let $\tilde{A}_i(x)(0 \leqslant i \leqslant p)$ be the fuzzy function of CSM and x^* be a part of fuzzy solution to the Lagrange problem. x^* is a global fuzzy optimum solution to $(\tilde{P})$ if $(\tilde{P})$ is fuzzy convex or $p' = p$; x^* is not necessarily a global fuzzy optimum one to $(\tilde{P})$ if $p' \neq p$, not even a local fuzzy optimum one.

4 A Direct Algorithm

If η is a continuous function on $[0,1]$, there exists a unique fixed point $\bar{\alpha} = \eta(\bar{\alpha})$. Let $\tilde{g}_i(x)$ be differentiable. Then steps of a direct algorithm are as follows.

1^0 Let $k = 1$, and determine α_1 as well as h by means of $1 - hd = \alpha_1$.
2^0 Calculate $\eta^{(k)} = \sup_{x \in A_{\alpha_k}} |\tilde{A}_0(\tilde{g}_0(x))|$ and $\tilde{M}^{(k)}(x) = \frac{1}{\eta^{(k)}} \tilde{g}_0(x) \in [0,1]$.
3^0 Calculate $\varepsilon_k = \alpha_k - \tilde{M}^{(k)}(x)$. If $|\varepsilon_k| > \varepsilon$, then go to 2^0, otherwise to 5^0.
4^0 Select $r_k \in [0,1]$ properly. If $\alpha_{k+1} = \alpha_k - r_k \varepsilon_k$, let k be $k+1$, go to 2^0.
5^0 Calculate $\tilde{M}^{(k)}(x^*)$ when $\bar{\alpha} = \alpha_k$, then x^* is an optimal solution to $\tilde{P}$.

Note 1. It is proper to take $\alpha_1 \in [0.9, 1]$ when $\tilde{g}_0(x)$ increases strictly monotonous, otherwise to take $\alpha_1 \in [0.75, 0.9]$. If $b(> 0)$ is very larger, larger, smaller, very small, it is proper to take h to be 0.02, 0.2, 2 and 20, respectively. As for r_k selection, when $\varepsilon_1 \gg \varepsilon_2$, $r_k = 0.5$ may be chosen. If $\varepsilon_1 \frown \varepsilon_2$ changes a little and if $\varepsilon_1 \ll \varepsilon_2$, $r_k \in [0.618, 1]$ and $r_k \in [0.382, 0.4]$ can be properly taken, respectively. Otherwise, a contradictory may appear.

Example 1. Find $\widetilde{\min} \{2x_1 + 3x_2\}$, s.t. $x_1^2 + x_2^2 \gtrsim 1, x_1, x_2 > 0$.

Since its crisp Lagrange solution is $x_1^* = 2/\sqrt{13}, x_2^* = 3/\sqrt{13}$, take $\eta^{(1)} = \sqrt{13}$. Suppose a constraint membership function is $\chi_1(d_1) = \{1 - 0.2h, 0 \leqslant d_1 < 0.25; 0, d_1 \geqslant 0\}$; we can find $x_1^* = 0.915683, x_2^* = 0.555, \alpha_2 = 0.93535, \tilde{M}^{(2)} = 0.969717$ and an objective function represents $S \approx 3.496$ by two steps. A fuzzy optimal solution to the problem is $x^{(0)*} = (1.07075, 0)$; its constraint infimum is $M_{\bar{P}} = 2.1415$; $x^{(2)} = (x_1^*, x_2^*) = (0.915683, 0.555)$ is not global fuzzy optimal solution, nor is a local one. So Proposition 2 is confirmed. But x^* is still a fuzzy optimal solution to all of x satisfying (8). Since $\Gamma_{11} = -2(x_1^*)^2 \approx -1.677, \Gamma_{12} = -2(x_2^*)^2 \approx -0.616$, all of x_1 and x_2 satisfy the problem, such that $-1.677(\ln x_1 - \ln x_1^*) - 0.616(\ln x_2 - \ln x_2^*) \leqslant 0 \Longrightarrow x_1^{-1.677} x_2^{-0.616} \leqslant 0.44572$, then $2x_1 + 3x_2 \geqslant 3.4963$, that is, (x_1^*, x_2^*) is a fuzzy optimal solution to the problem in certain range. So Theorem 2 is confirmed. The property is called tangentially optima of fuzziness.

5 Conclusion

The author gives a direct algorithm to the Lagrange problem of fuzzy reversed PGP. As for its dual programming, it will be built by fuzzy dual theory[3]. Therefore, the fuzzy reversed PGP has great value both in theory and in application.

Acknowledgments

Thanks to the support by National Natural Science Foundation of China (No. 70271047) and " 211" Project Foundation of Shantou University.

References

1. Cao, B.Y.: Solution and Theory of Questions for a Kind of Fuzzy Positive Geometric Program. Proc. of 2nd IFSA Congress, Tokyo, Vol.1(1987) 205-208
2. Cao, B.Y.: Fuzzy Geometric Programming (I). FSS **53**(1993) 135-154
3. Cao, B.Y.: Fuzzy Geometric Programming. Kluwer Academic Pub. Dordrecht (2002)
4. Wu, F., Yuan, Y.Y.: Geometric Programming. Math. Prac. & Theory **1-2**(1982)
5. Zadeh, L.A.: Fuzzy Sets. Inform. and Control. **8**(1965) 338-353
6. Zimmermann, H.-J.: Description and Optimization of Fuzzy Systems. Internat. J. General Systems **2** (1976) 209-215

Direct Candidates Generation: A Novel Algorithm for Discovering Complete Share-Frequent Itemsets

Yu-Chiang Li[1], Jieh-Shan Yeh[2], and Chin-Chen Chang[1, 3]

[1] Department of Computer Science and Information Engineering,
National Chung Cheng University, Chiayi 621 Taiwan
{lyc, ccc}@cs.ccu.edu.tw
[2] Department of Computer Science and Information Management,
Providence University, Taichung 433, Taiwan
jsyeh@pu.edu.tw
[3] Department of Information Engineering and Computer Science,
Feng Chia University, Taichung 407, Taiwan

Abstract. The value of the itemset share is one way of evaluating the magnitude of an itemset. From business perspective, itemset share values reflect more the significance of itemsets for mining association rules in a database. The Share-counted FSM (ShFSM) algorithm is one of the best algorithms which can discover all share-frequent itemsets efficiently. However, ShFSM wastes the computation time on the join and the prune steps of candidate generation in each pass, and generates too many useless candidates. Therefore, this study proposes the Direct Candidates Generation (DCG) algorithm to directly generate candidates without the prune and the join steps in each pass. Moreover, the number of candidates generated by DCG is less than that by ShFSM. Experimental results reveal that the proposed method performs significantly better than ShFSM.

1 Introduction

Data mining techniques have been developed to find new and potentially useful knowledge from data. [11]. Traditional methods for mining association rules are based on the support-confidence framework to discover all relationships among items (each market product is called an *item*) from historical transaction databases.

The support value is applied to measure the importance of itemsets (a group of products bought together in a transaction) in a transaction database. It only reflects the percentage of transactions in which the itemset sold, but neither reveals the profit, the cost nor the real quantity sold of each itemset. For example, in Table 1, the column "Count" indicates the sale volume of each item in each transaction. According to the support value, {A} appears in four transactions, but its real sale volume is 12.

Users are usually more interested in knowing which itemsets are bought in sufficient numbers to gain a certain net profit or attain a given cost. To reveal such knowledge, several issues have been proposed, such as share-confidence framework, profit mining and utility itemsets [7, 8, 16]. In 1997, Carter *et al.* first introduced a share-confidence framework [7]. Instead of discovering frequent itemsets, the method with share-confidence framework finds *share-frequent* (SH-frequent) itemsets. The share measure can provide valuable information of itemsets' net profit or cost, which the support measure cannot [6].

Table 1. Example of a transaction database with counting

TID	Transaction	Count	Total count
T01	{A, B, C, D, G, H}	{1, 1, 1, 1, 1, 1}	6
T02	{A, C, E, F}	{4, 3, 1, 2}	10
T03	{A, C, E}	{4, 3, 3}	10
T04	{B, C, D, F}	{4, 1, 2, 2}	9
T05	{A, B, D}	{3, 1, 2}	6
T06	{B, C, D}	{3, 2, 1}	6

An SH-frequent itemset usually includes some infrequent subsets even no frequent subset. Obviously, an exhaustive search method can be used to generate all SH-frequents, such as the ZP and the ZSP algorithms [4, 6]. However, the exhaustive search method is time-consuming and does not work efficiently in a large dataset. Some algorithms have been proposed to facilitate the extraction of SH-frequents with infrequent subsets, such as SIP, CAC and IAB [4, 5, 6], but they may not discover all SH-frequent itemsets. Recently, Li *et al.* first developed algorithms to swiftly discover all SH-frequent itemsets [12, 13]. Among these algorithms ShFSM is the best. In contrast to the number of SH-frequent itemsets, the performance bottleneck of ShFSM is generating too many candidates in each pass. Accordingly, this work proposes the Direct Candidate Generation (DCG) method to further improve the performance of ShFSM. Our scheme directly generates candidates in each pass without the join and the prune steps. Furthermore, DCG can efficiently lower the number of useless candidates and further accelerate the mining process. For simplicity and without loss of generality, this study supposes that the measure value of each item in each transaction is a non-negative integer.

The rest of this paper is organized as follows. Section 2 reviews support-confidence and share-confidence frameworks. Section 3 explains the proposed Direct Candidate Generation (DCG) algorithm for discovering all SH-frequent itemsets. Section 4 provides experimental results and evaluates the performance of the proposed algorithm. Finally, we conclude in Section 5 with a summary of our work.

2 Review of Support and Share Measures

2.1 Support-Confidence Framework

Agrawal *et al.* first defined the problem of mining association rules [2, 3]. The formal definition is as follows. Let $I = \{i_1, i_2, ..., i_n\}$ be a set of literals with binary attributes, called items. Let X be an itemset, where $X \subseteq I$. Let the transaction database $DB = \{T_1, T_2, ..., T_z\}$ be the set of transactions, where each transaction $T_q \in DB$, $T_q \subseteq I$, $1 \leq q \leq z$. The notation $X \Rightarrow Y$ presents an association rule, where $X \subseteq I$, $Y \subseteq I$ and $X \cap Y = \phi$. The rule $X \Rightarrow Y$ has *support* $s\%$, denoted as $Sup(X \cup Y)$, in the transaction database DB

if the itemset $X \cup Y$ appears in $s\%$ of transactions in DB. The *confidence* of the rule $X \Rightarrow Y$, denoted as $Conf(X \Rightarrow Y)$, is $c\%$ if the set of transactions containing X in DB has $c\%$ transactions also containing Y. Thus, $Conf(X \Rightarrow Y) = Sup(X \cup Y)/Sup(X)$. Given the minimum support (*minSup*) and minimum confidence (*minConf*) thresholds, the process of mining association rules is to generate all rules that satisfy the two certain constraints, respectively.

Apriori is a level-wise (including multiple passes) algorithm. In each pass, Apriori employs the downward closure property to efficiently identify all frequent k-itemsets (itemsets with length k and their supports are greater than or equal to *minSup*). In fact, an arbitrary k-subset of a frequent $(k+1)$-itemset is also frequent; otherwise, the $(k+1)$-itemset is infrequent. This characteristic is called the downward closure property. That is, the itemsets violating the downward closure property will be deleted from the set of candidate $(k+1)$-itemsets. Up to now, there are many algorithms have been proposed to rapidly discover the frequent itemsets, including Apriori and subsequent Apriori-like algorithms [3, 15], and pattern-growth methods [1, 9, 14].

2.2 Share-Confidence Framework

The support measure does not concern the quantity purchased in a transaction. In real circumstances, products may be bought in plural in a transaction. Therefore, the support value of an item usually underestimates the actual frequency of product purchasing. Information derived from the support value may also be misleading [6]. To address this issue, Carter et al. first introduced the share-confidence framework [7]. Instead of a binary attribute, each item i_p involves a numerical attribute in each transaction T_q. The notations and definitions of share measure are described as follows [6, 12, 13].

The *measure value* $mv(i_p, T_q)$ represents the attribute value of i_p in transaction T_q. For example, in Table 1, $mv(C, T02) = 3$ and $mv(D, T04) = 2$. The *transaction measure value* is the total measure value of a transaction T_p, denoted as $tmv(T_p)$, where $tmv(T_p) = \sum_{i_p \in T_q} mv(i_p, T_q)$. For example, in Table 1, $tmv(T02) = 10$ and $mv(T04) = 9$.

The *total measure value* $Tmv(DB)$ represents the total measure value in DB, where $Tmv(DB) = \sum_{T_q \in DB} \sum_{i_p \in T_q} mv(i_p, T_q)$. For example, in Table 1, $Tmv(DB) = 47$.

Let db_X be a set of transactions that contain itemset X in DB. That is each k-itemset $X \subseteq I$ has an associated set of transactions $db_X \subseteq DB$, where $X \subseteq T_q$ and $T_q \in db_X$. For example, in Table 1, $db_{\{AC\}} = \{T01, T02, T03\}$.

Let $X \subseteq T_q$, the *itemset measure value* of an itemset X in a transaction T_q, denoted as $imv(X, T_q)$, is the total measure value of all items of X in T_q. That is, $imv(X, T_q) = \sum_{X \subseteq T_q, i_p \in X} mv(i_p, T_q)$. For example, in Table 1, $imv(\{AC\}, T02) = 7$.

The *local measure value* of an itemset X in DB, denoted as $lmv(X)$, is the sum of the itemset measure values of X in db_X. That is, $lmv(X) = \sum_{T_q \in db_X} imv(X, T_q)$. For example, in Table 1, $lmv(\{AC\}) = imv\{\{AC\}, T01\} + imv\{\{AC\}, T02\} + imv\{\{AC\}, T03\} = 2 + 7 + 7 = 16$.

The *itemset share value* of an itemset X, denoted as $SH(X)$, is the ratio of the local measure value of X to the total measure value in DB. That is, $SH(X) = \frac{lmv(X)}{Tmv(DB)}$. Given a minimum share (*minShare*) threshold $s\%$, A k-itemset X is *share-frequent* (SH-frequent) if $SH(X) \geq s\%$; otherwise, X is infrequent.

Example 2.1 Consider the sample transaction database as shown in Table 1 and $minShare = 30\%$. Table 2 lists the local measure value and the share value of each 1-itemset, where $Tmv(DB) = 47$. Let $X = \{B, D\}$, the local measure value of $\{B, D\}$ is $lmv(X) = imv(X, T01) + imv(X, T04) + imv(X, T05) + imv(X, T06) = 2 + 6 + 3 + 4 = 15$. $SH(X) = lmv(X)/Tmv(DB) = 15/47 = 0.319 > 30\%$. Therefore, $\{B, D\}$ is SH-frequent. Table 3 illustrates all SH-frequent itemsets.

Table 2. Local measure value and itemset share value of each 1-itemset

Item	{A}	{B}	{C}	{D}	{E}	{F}	{G}	{H}	Total
$lmv(i_p)$	12	9	10	6	4	4	1	1	47
$SH(i_p)$	25.5%	19.1%	21.3%	12.8%	8.5%	8.5%	2.1%	2.1%	100%

Table 3. All SH-frequent itemsets of the sample database

SH-frequent itemset	{A, C}	{B, D}	{A, C, E}	{B, C, D}
$lmv(X)$	16	15	18	16
$SH(X)$	34.0%	31.9%	38.3%	34.0%

2.3 Previous Methods of Discovering Share-Frequent Itemsets

The characteristic of downward closure cannot be applied for discovering SH-frequent itemsets, because the subsets of an SH-frequent itemset may be infrequent. For example, in Table 3, $\{A, C, E\}$ is SH-frequent, but its subsets $\{C, E\}$ and $\{A, E\}$ are infrequent. The ZP and ZSP algorithms can be known as the variants of the exhaustive search method. They generate all itemsets to be candidate set except the local measure values of itemsets are exactly zero [6]. Some algorithms have been proposed to extract SH-frequent itemsets with infrequent subsets, such as SIP, CAC, and IAB [4, 5, 6]. However, they do not generate complete SH-frequent itemsets. Recently, Li et al. first proposed the non-exhaustive search method, Fast Share Measure (FSM), to discover all SH-frequent itemsets efficiently [12]. FSM employs the level closure property to decrease the number of candidate itemsets.

Given a transaction database DB and *minShare*, the characteristic of level closure is described as follows. For a candidate k-itemsets X, and an integer *Level*, if $lmv(X) + (lmv(X)/k) \times MV \times Level < min\_lmv$, no superset of X with length $\leq k + Level$ is SH-frequent, where MV is the maximum measure value among all measure values and $min\_lmv = minShare \times Tmv(DB)$. In the case, $Level = ML - k$, where ML is the maximum length among all transactions, the level closure property guarantees that no superset of SH-infrequent if the inequality holds.

Li et al. also developed some more efficient algorithms than FSM to discover all SH-frequent itemset, including EFSM (Enhanced FSM), SuFSM (Support-counted FSM) and ShFSM (Share-counted FSM) [13]. EFSM reduces the time complexity of generating candidate k-itemsets from $O(m^{2k-2})$ to $O(m^k)$, where m is the number of distinct items. SuFSM and ShFSM are based on EFSM. They add the support constraint and the share constraint to the critical function, respectively. The performance of ShFSM is the best among ZSP, FSM, EFSM, SuFSM and ShFSM on some synthetic datasets [13].

3 Direct Candidate Generation (DCG) Method

The key point of previous methods, including FSM, EFSM, SuFSM and ShFSM for discovering all SH-frequent itemsets is to eliminate itemset X from candidate set if X satisfies the inequality $CF(X) < min\_lmv$ [12, 13]. Those methods waste computation time in the join and the prune steps of candidate generation, and generate too many candidates of SH-frequent itemsets. In comparison with previous methods, this study proposes a novel algorithm called Direct Candidates Generation (DCG) method to directly generate a smaller candidate set without the join and the prune steps in each pass. Without lose of generality, we let the literal set I be a totally order set. That is, for any $i, j \in I$, either $i \leq j$ or $j \leq i$. We also denote $i < j$ if $i \leq j$ and $i \neq j$.

Definition 3.1 Let the candidate k-itemset X be $\{i_1, i_2, \ldots, i_k\}$ in the order of literals. Let $i_q \in I$ be an item. If $i_k < i_q$ then the (k+1)-superset of X $\{i_1, i_2, \ldots, i_k, i_q\}$ is defined as the *monotone (k+1)-superset* of X and is denoted as $X_{k+1}^{i_q}$. For example, Let $X = \{A, C, D\}$. $X_{k+1}^E = \{A, C, D, E\}$.

Definition 3.2 Let $X \subseteq I$, the associated set of transactions of X, $db_X = \{T_q \in DB \mid X \subseteq T_q\}$, is the set of transactions that contain X. The *total measure value* of db_X is defined as $Tmv(db_X) = \sum_{T_q \in db_X} \sum_{i_p \in T_q} mv(i_p, T_q)$.

Let $X_{k+1}^{i_q} \subseteq I$ is an arbitrary monotone (k+1)-superset of X. $X_{k+1}^{i_q}$ has an associated set of transactions $db_{X_{k+1}^{i_q}} = \{T_q \in DB \mid X_{k+1}^{i_q} \subseteq T_q\}$. Clearly, $db_{X_{k+1}^{i_q}} \subseteq db_X$. For example, in Table 1, let $X = \{A, C\}$, then $db_X = \{T01, T02, T03\}$ and $db_{X_{k+1}^E} = \{T02, T03\}$.

Theorem 3.1 Given a DB and the minShare, let $min\_lmv = minShare \times Tmv(DB)$. For k-itemset X, if $Tmv(db_X) < min\_lmv$, all supersets of X (including X) are infrequent.

Proof. Let X' be an arbitrary superset of X with length (k+i), where $i \geq 0$. Clearly, $lmv(X') \leq Tmv(db_{X'}) \leq Tmv(db_X)$. So, if the inequality $Tmv(db_X) < min\_lmv$ holds, $lmv(X') < min\_lmv = minShare \times Tmv(DB)$. That is, $SH(X') = lmv(X') / Tmv(DB) < minShare$. X' is infrequent. Q.E.D

By Theorem 3.1, if $Tmv(db_{X_{k+1}^{i_q}}) < min\_lmv$, $X_{k+1}^{i_q}$ and all supersets of $X_{k+1}^{i_q}$ are infrequent; otherwise, $X_{k+1}^{i_q}$ is a candidate (k+1)-itemset.

DCG is also a multiple-pass method for finding all SH-frequent itemsets. In the k-th pass, DCG scans the whole database to count the local measure value of each candidate k-itemset X and counts the potential maximum share value of each monotone (k+1)-superset of X. Next, DCG determinates the SH-frequent k-itemset, where their local measure values are greater than *min_lmv*. Then, DCG selects $X_{k+1}^{i_q}$ to be a candidate $(k+1)$-itemset if the total measure value of $db_{X_{k+1}^{i_q}}$ satisfies the inequality $Tmv(db_{X_{k+1}^{i_q}}) \geq min\_lmv$. The pseudo-code of DCG algorithm is as follows.

Algorithm DCG()
Input: (1) *DB*: a transaction database with counting, (2) *minShare*
Output: All SH-frequent itemsets
Procedure:
```
1. C₁:=I;  // Cₖ: the set of candidate k-itemsets
2. for k:=1 to h
3.    Fₖ:=∅; Cₖ₊₁:=∅; Tmv(db_{X_{k+1}^{i_q}}):=0 for all X in Cₖ;
4.    foreach T∈ DB  // scan DB
5.        if k==1 { count and store tmv(T); }
6.        foreach X∈ Cₖ
7.            count lmv(X);
8.            foreach i_q > iₖ && i_q ∈ T
9.                Tmv(db_{X_{k+1}^{i_q}}) += tmv(T);
10.       foreach X∈ Cₖ
11.           if lmv(X) ≥ min_lmv { Fₖ:= Fₖ+X; }
12.           foreach i_q > iₖ && i_q ∈ I
13.               if Tmv(db_{X_{k+1}^{i_q}}) ≥ min_lmv { Cₖ₊₁:=Cₖ₊₁+ X_{k+1}^{i_q}; }
14.       if Cₖ₊₁==∅ exitfor;
15. return F₁∪F₂∪...∪Fₖ;
```

In line 5, DCG calculates the transaction measure value of each transaction and store it in an array when scanning *DB* first time. The transaction measure value of each transaction will be employed in each pass. In lines 8 to 9, DCG accumulates the total measure value of $db_{X_{k+1}^{i_q}}$. From lines 10 to 13, DCG determines which candidate k-itemsets are SH-frequent and directly generates $(k+1)$-candidates.

Example 3.1 Consider the sample transaction database as listed in Table 1 and *minShare* = 30%. Both $Tmv(DB) = 47$ and $min\_lmv = 15$ can be calculated easily. In Fig. 1, each circle represents candidate itemset X and each number inside the circle is the local measure value of X, $lmv(X)$, such as $lmv(\{A\})=12$ and $lmv(\{BD\}) = 15$. To speed-up counting the total measure value of each monotone $(k+1)$-superset of each candidate X, $Tmv(db_{X_{k+1}^{i_q}})$, we require an array table to store these values for each X. The transaction measure value table is listed in the column "Total count" of Table 1. In the first pass, all items are candidate 1-itemsets. After first scanning *DB*, we can obtain all local measure values of candidates and all $Tmv(db_{X_{2}^{i_q}})$ values, such as $Tmv(db_{\{A\}_2^C}) = 26$ and $Tmv(db_{\{C\}_2^G}) = 6$. No 1-itemset is SH-frequent and there are seven monotone 2-supersets, with values greater than min_lmv, in arrays. These 2-itemsets could be SH-frequent as shown in the dark cells of Fig. 1. Therefore, DCG directly generates the seven 2-itemset candidates {{A, C}, {A, E}, {B, C}, {B, D},

{C, D}, {C, E}, {C, F}}. In the second pass, DCG discovers two SH-frequent 2-itemsets, {{A, C}, {B, D}}, and generates two candidate 3-itemsets, {{A, C, E}, {B, C, D}}, because $lmv(\{A, C\}) = 16$, $lmv(\{B, D\}) = 15$, $Tmv(db_{\{AC\}_3^E}) = 20$ and $Tmv(db_{\{BC\}_3^D}) = 21$ are all greater than min_lmv. No candidate of C_4 can be generated. Therefore, the process terminates after third scanning DB.

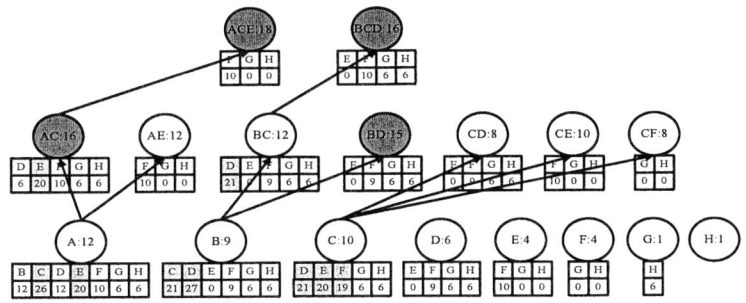

Fig. 1. An example of DCG algorithm

4 Experimental Results

To access the performance of DCG, experiments are conducted to compare its performance with that of FSM, EFSM, SuFSM and ShFSM on artificial and real datasets. All experiments were performed on a 1.5 GHz Pentium IV PC with 1.5 GB of main memory, running Windows XP Professional operating system. All algorithms were coded in Visual C++ 6.0. Each algorithm employed the hash tree structure to count the local measure value of each candidate. All SH-frequent itemsets were output to main memory to eliminate the effect of disk writing.

The artificial datasets were generated by IBM synthetic data generator [18]. To discover SH-frequent itemsets, each item must include an integer attribute. This study modifies these datasets with additional parameter m. The notation $Tx.Iy.Dz.Nn.Sm$ denotes a dataset with five given parameters x, y, z, n and m. The definition of the first four parameters is the same as those in [3]. The parameter m denotes the measure value which was randomly generated between 1 and m, and 50% of measure values in the dataset are set to be 1.

Figures 2 and 3 plot the performance curves of running time associated with these algorithms over various *minShare*, applied to T6.I4.D100k.N200.S10 and T10.I6.D100k.N1000.S10, respectively. Figure 2 uses a logarithmic scale for the y-axis and the range of *minShare* is from 0.1% to 1.2%. Figure 2 demonstrates that DCG performed better than FSM by more than one order of magnitude. DCG had the best performance, followed by ShFSM. For example, the running time of DCG was only 62%, 21%, 9.8 and 0.4% of those of ShFSM, SuFSM, EFSM and FSM, respectively with *minShare* = 0.4%.

In Fig. 3, the range of *minShare* is from 0.01% to 0.12%. FSM and EFSM were not illustrated in Fig. 3 because they generated too many candidates to store in main memory. In *minShare* ≤ 0.4% scenarios, SuFSM also generated too many candidates to run. The running time of DCG was only 59.5% and 16.6% of those of ShFSM and SuFSM with *minShare* = 0.06%, respectively.

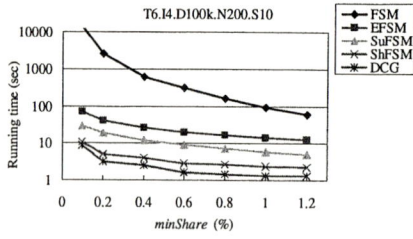
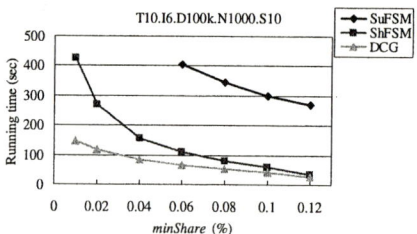

Fig. 2. Comparison of running times **Fig. 3.** Comparison of running times

To compare the difference of candidate numbers among these five algorithms in each pass, Table 4 presents the numbers of C_k and F_k in each pass using the dataset T6.I4.D100k.N200.S10 with *minShare* = 0.1%. DCG generated the fewest candidates among the five algorithms. DCG terminated the process at pass 10 and performed best. ShFSM also terminated the process at pass 10. The others executed the processes to pass 13. All five algorithms can discover all SH-frequent itemsets, even to those SH-frequent k-itemsets had no SH-frequent (k-1)-subset. For example, in the pass 5, all five SH-frequent 5-itemsets have no SH-frequent 4-subset.

Table 4. Comparison of the numbers of candidate sets in each pass

Method Pass (k)	C_k					F_k
	FSM	EFSM	SuFSM	ShFSM	DCG	
k=1	200	200	200	200	200	159
k=2	19900	19900	19701	19306	7200	1844
k=3	829547	829547	564324	190607	9805	101
k=4	3290296	3290296	793042	20913	1425	0
k=5	393833	393833	25003	1050	967	5
k=6	26137	26137	11582	518	510	8
k=7	11141	11141	5940	204	203	7
k=8	4426	4426	2797	58	58	1
$k \geq 9$	2036	2036	1567	12	12	0
Time(sec)	13610.4	71.55	29.67	10.95	8.83	

Figures 4 and 5 compare the scalability of SuFSM, ShFSM and DCG. Figure 4 illustrates the scalability of three algorithms on the transaction numbers of *DB* using T6.I4.Dz.N200.S10 with *minShare* = 0.3%. The range of *DB* size is between 100k and 1000k. The running times of SuFSM, ShFSM and DCG linearly increase with the growth of the *DB* size. Figure 5 presents the scalability of three algorithms on the maximum measure values of *DB* using T6.I4.D100k.N200.S*m* with *minShare* = 0.3%. The *x*-axis represents the maximum measure values between 10 and 60. The running time curves of SuFSM, ShFSM and DCG were found to be flat. The impact of the distinct maximum measure value on these three approaches was insignificant.

BMS-WebView-2 is a real dataset of several months' click stream data from an e-commerce web site [17]. This study modifies these datasets with an additional parameter m. The parameter m denotes the measure value of each item was randomly generated between 1 and m, and 50% of measure values in the dataset are set to be 1. Figure 6 plots the running-time curves associated with ShFSM and DCG. The range of *minShare* is between 0.2% and 1%. In these distinct *minShare* scenarios, FSM, EFSM and SuFSM generated too many candidates to store in main memory. In Fig. 6, when $minShare \leq 0.6\%$, the running times of DCG are only between 59.5% (*minShare* = 0.6%) and 16.6% (*minShare* = 0.2%) of that of ShFSM.

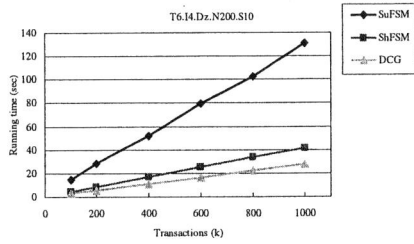
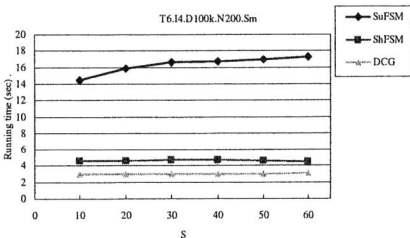

Fig. 4. Scalability of algorithms   **Fig. 5.** Scalability of algorithms

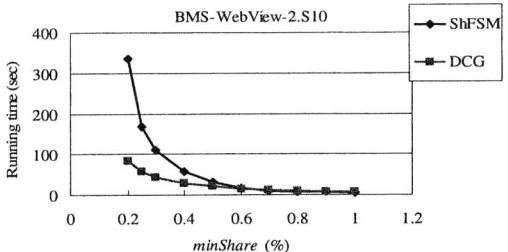

Fig. 6. Comparison of running times on BMS-WebView-2.S10

5 Conclusions

The share measure has been proposed to overcome the drawbacks of the support measure. Therefore, developing an efficient approach for discovering complete SH-frequent itemsets is very valuable. This study proposes the Direct Candidates Generation (DCG) algorithm to avoid the join and the prune steps in each pass. Furthermore, DCG significantly reduces the number of candidates and improves the performance. Experimental results indicate that DCG outperforms all other algorithms in several artificial datasets. Now, we are investigating the development of superior algorithms to efficiently accelerate the process of identifying long SH-frequent itemsets.

Acknowledgements

We would like to thank Blue Martini Software, Inc. for providing the BMS datasets.

References

1. R. C. Agarwal, C. C. Aggarwal, V. V. V. Prasad: A tree projection algorithm for generation of frequent itemsets. Journal of Parallel and Distributed Computing, **61** (2001) 350-361
2. R. Agrawal, T. Imielinski, A. Swami: Mining association rules between sets of items in large databases. In: Proc. 1993 ACM SIGMOD Intl. Conf. on Management of Data, Washington, D.C. (1993) 207-216
3. R. Agrawal, R. Srikant: Fast algorithms for mining association rules. In: Proc. 20th Intl. Conf. on Very Large Data Bases, Santiago, Chile (1994) 487-499
4. B. Barber, H. J. Hamilton: Algorithms for mining share frequent itemsets containing infrequent subsets. In: D. A. Zighed, H. J. Komorowski, J. M. Zytkow (eds.): 4th European Conf. on Principles of Data Mining and Knowledge Discovery. Lecture Notes in Computer Sciences, Vol. 1910. Springer-Verlag, Berlin Heidelberg New York (2000) 316-324
5. B. Barber, H. J. Hamilton: Parametric algorithm for mining share frequent itemsets. Journal of Intelligent Information Systems, 16 (2001) 277-293
6. B. Barber, H. J. Hamilton: Extracting share frequent itemsets with infrequent subsets. Data Mining and Knowledge Discovery, **7** (2003) 153-185.
7. C. L. Carter, H. J. Hamilton, N. Cercone: Share based measures for itemsets. In: H. J. Komorowski, J. M. Zytkow (eds.): 1st European Conf. on the Principles of Data Mining and Knowledge Discovery. Lecture Notes in Computer Science, Vol. 1263, Springer-Verlag, Berlin Heidelberg New York (1997) 14-24
8. R. Chan, Q. Yang, Y. D. Shen: Mining high utility itemsets. In: Proc. 3rd IEEE Intl. Conf. on Data Mining, Melbourne, FL (2003) 19-26
9. J. Han, J. Pei, Y. Yin, R. Mao: Mining frequent pattern without candidate generation: A frequent pattern tree approach. Data Mining and Knowledge Discovery, **8** (2004) 53-87
10. R. J. Hilderman, C. L. Carter, H. J. Hamilton, N. Cercone: Mining association rules from market basket data using share measures and characterized itemsets. Intl. Journal of Artificial Intelligence Tools, **7** (1998) 189-220
11. M. Kantardzic: Data mining: Concepts, models, methods, and algorithms. John Wiley & Sons, Inc., New York (2002)
12. Y. C. Li, J. S. Yeh, C. C. Chang: A fast algorithm for mining share-frequent itemsets. In: Y. Zhang, K. Tanaka, J. X. Yu, etc. (eds.): 7th Asia Pacific Web Conf. Lecture Notes in Computer Science, Vol. 3399, Springer-Verlag, Berlin Heidelberg New York (2005) 417-428
13. Y. C. Li, J. S. Yeh, C. C. Chang: Efficient algorithms for mining share-frequent itemsets. To appear in Proc. 11th World Congress of Intl. Fuzzy Systems Association (2005)
14. J. Liu, Y. Pan, K. Wang, J. Han: Mining frequent item sets by opportunistic projection. In: Proc. 8th ACM-SIGKDD Intl. Conf. on Knowledge Discovery and Data Mining, Alberta, Canada (2002) 229-238
15. J. S. Park, M. S. Chen, P. S. Yu: An effective hash-based algorithm for mining association rules. In: Proc. 1995 ACM-SIGMOD Intl. Conf. on Management of Data, San Jose, CA (1995) 175-186
16. K. Wang, S. Zhou, J. Han: Profit mining: From pattern to actions. In: C. S. Jensen, K. G. Jeffery, J. Pokorný, etc. (eds.): 8th Int. Conf. on Extending Database Technology. Lecture Notes in Computer Science, Vol. 2287, Springer-Verlag, Berlin Heidelberg New York (2002) 70-88
17. Z. Zheng, R. Kohavi, L. Mason: Real world performance of association rule algorithm. In: Proc.7th ACM-SIGKDD Intl. Conf. on Knowledge Discovery and Data Mining, San Francisco, CA (2001) 401-406
18. http://alme1.almaden.ibm.com/software/quest/Resources/datasets/syndata.html

A Three-Step Preprocessing Algorithm for Minimizing E-Mail Document's Atypical Characteristics

Ok-Ran Jeong and Dong-Sub Cho

Department of Computer Science and Engineering,
Ewha Womans University, 11-1 Daehyun-dong,
Seodaemun-ku, Seoul 120-750, Korea
orchung@ewhain.net, dscho@ewha.ac.kr

Abstract. Documents that are widely in use today included many atypical characteristics. In particular, non-standardization appears more frequently in e-mail documents than other documents due to the extensive use of informal expressions such as slang and abbreviation. Automatic document classification may differ significantly according to the characteristics of documents that are subject to classification, as well as classifier's performance. We suggest a three-step preprocessing algorithm by stages for accurate automatic classification for each e-mail category. This research identifies e-mail document's characteristics to apply a three-step preprocessing algorithm that can minimize e-mail document's atypical characteristics.

1 Introduction

Documents that are widely in use today included many atypical characteristics. Accordingly, classification is conducted against refined and standardized document for general automatic document classification. In particular, non-standardization appears more frequently in e-mail documents than other documents due to the extensive use of informal expressions such as slang and abbreviation. Accuracy level of classification decreases accordingly. This research identifies e-mail document's characteristics to recommend a three-step preprocessing algorithm for increased overall classification performance by minimizing e-mail document's atypical characteristics. Automatic document classification process is largely divided into preprocessing stage, feature extraction stage and document classification stage. However, this research classifies preprocessing stage and feature extraction as one into preprocessing stage. This research suggests a three-step preprocessing algorithm by stages for accurate automatic classification for each e-mail category.

In the first stage, learning document for document classification is selected in the feature extraction stage, and applies uncertainty based sampling algorithm based on the use of MAD that re-configures on an intelligence level instead of using arbitrary learning document set as is. In the second stage, preponderance of e-mail document is factored in to endow weighted value by attribute to conduct feature extraction. E-mail document is comprised of title and main text, and expanded Naive Bayesian Classifier is applied to endow greater weight on the title, compared to main text. In the third stage, presumptive algorithm that conducts rule generation is cited as a decision factor for accuracy level of document classification. The role of this algorithm is to form ultimate rule by using learning document set, selected by configuration method for learning document set, and applies Naive Bayesian algorithm that uses dynamic threshold [1].

2 Preprocessing Process of E-Mail Document

The First of all, preprocessing process for the feature extraction removes special symbol, which is an unnecessary element when expressing document information. Then, classification into title and main text takes place, and token is generated and then aligned. During this stage, 1-byte symbol and tag are removed. As for the subsequent stage, the preprocessing stage loads user dictionary on the representative word, an expression that standardizes expressions that belong to abbreviation, slang and specific se, and maintains it at an aligned state.

In the subsequent stage, e-mail document that underwent vectoring, resulting from the removal of unnecessary symbol and tag information, and loaded user dictionary are mapped to convert non-standardized terms that appear on e-mail document into standardized terms. Extraction of characteristics is executed through subsequent Fig.1 in next page. This research extracts token of the title separately as index word, and assigns weighted value, using the method recommended in this research.

In this stage, non-standardized terms that appear on e-mail document are converted into standardized terms by mapping e-mail document that underwent vectoring, resulting from the removal of unnecessary symbol and tag information, and loaded user dictionary. In the feature extraction stage, learning document set, developed as mentioned above, is expressed into vector whereby the learning document set is quantified.

This is obtained by the *tf-idf* value that is calculated based on the total number of documents where frequency and index of the extracted index word that appear on document in turn appear. *tf-idf* is the module that expresses keyword set into quantified vector, and it is the method that is used on most of information search systems or document classification systems.

3 A Three-Step Preprocessing Algorithm

3.1 Uncertainty Based Sampling Algorithm

Active learning algorithm is added onto the learning document set that is necessary for rule generation by selecting document with large amount of information from the overall document set. Effectively developed rule plays the most important role in increasing accuracy level during classification. Accordingly,

Algorithm 1. *Feature_Extraction*
var
  sentence_mark: a set {'.', ',', '!', '(', ')', ...};
  white_chars: a set {' ', '\t', '\n', '\r'};
  text: subject and message of a mail;
  token_list: a token set;
  feature: a pair <word, count>;
  feature_set: a set of feature;
begin
  foreach ch **in** text **do**
  begin
    if ch is in sentence_mark **then** convert it to a space;
    if ch is a uppercase **then** convert it to its lowercase;
  end foreach
  get token_list by breaking text into tokens with the separators white_chars;
  foreach token **in** token_list **do**
  begin
    if token is a word **then**
    begin
      if token exists in feature_set **then**
        increase its count by 1;
      else
        append a new pair <token, 1> to feature_set;
    end if
  end foreach
end.

Fig. 1. Feature Extraction Algorithm

learning document set is considered critical for rule generation. Uncertainty based sampling algorithm is applied at this time. In other terms, there are many diverse standards for judging document with large amount of information among active learning algorithms. Among them, use of the concept of uncertainty is representative[2].

Therefore, this research defined learning document by using MAD. In this measured value, uncertainty is measured by using the distance between above defined values of $P(c|x)$ and these values' average μ. This is defined as follows.

$$U_{MAD}(x) = \frac{1}{|c|} \sum_{i=1}^{|c|} (P(c_i|x) - \mu), \qquad \mu = \frac{1}{|c|} \sum_{i=1}^{|c|} P(c_i|x) \qquad (1)$$

$U_{MAD}(x)$ refers to the average distance that showcases the deviation of these values from their average when it comes to the probability, subject to each category, or member values of document x. Thus, uncertainty is greater as $U_{MAD}(x)$ is lower, and vice versa. Here, average standard deviation of documents that are considered candidate for learning document is calculated, and the learning document is selected with priority set of documents with lower value. The results of this experiment are shown in their entirety in Chapter 4.

3.2 Weighted Value Assigning Method

Weighted value is assigned by factoring in the e-mail document's characteristics during the subsequent stage after learning document is selected by using the algorithm mentioned above. In other terms, weighted value assigning method by attribute is used to decide whether to assign weighted value on the title that stands multifold compared to the main text during the feature extraction. As for the method of assigning weighted value, assumption is made based on Naive Bayesian Classifier to assert that all attribute values are independent, and that they will exert comparable level of influence in the classification for the application. Existing Naive Bayesian Classifier is expanded as follows in order to assign separate weighted value by each attribute that is in the title.

$$V_{NB} = argmax_{c_j \in C}\{P(c_j) \prod_i P((a_k, v_i) \mid C_j)\} \tag{2}$$

sa_k signifies attribute values where applicable v_i belong, and it is materialized as follows to assign weighted value to each attribute.

$$P(v_i \mid c_j, a_k)P(a_k) \leftarrow \frac{n_(a_k, l) + 1}{n_{a_k} + |Doc_{sa_k}|} \times w_{sa_k} \tag{3}$$

n_{a_k} refers to the total number of keyword related to the all attribute values that is applicable to the attribute a_k when it comes to the mail document that is included in the category c_j whining the entire mail document of the learning data. Doc_{sa_k} refers to the number of all keyword attribute values that are selected within the mail document in order to represent classification when it comes to attribute a_k. $n_(a_k, l)$ shows the frequency of appearance of the attribute values keyword v_i that belongs to the attribute a_k within the category c_j whereas w_{sa_k} manifests the individual weighted value for attribute a_k.

3.3 Dynamic Threshold

In this research, e-mail document classification process based Bayesian learning method, a representative supervised learning algorithm, for document classification [3]. Here, supervised learning refers to the learning method that produces a more effective result for subsequent training by setting the objective that enables assessing how well each factor is executed in learning stage, and whether it acted correctly. Measurement of difference between output and target value is used often for the supervised learning training method. This is usually referred to as error rate, and effort is made to approach closer to the objective by measuring the increasebackslashdecrease of this error rate in the subsequent training. This algorithm is appropriate for the classification model, and applies to the feedforward paradigm.

In most systems, document, targeted for classification is classified into the category that assumes the highest probability value. However, this research converts fixed threshold T that was used in the existing Bayesian algorithm into dynamic threshold $T^{'}$ [1].

We already demonstrated improved performance in the research [1] that preceded this research. In other terms, exiting fixed threshold that is used is Naive Bayesian algorithm is improved into a more dynamic threshold to increase filtering's accuracy level. Here, dynamic threshold is applied during the classification stage after applying the two preprocessing algorithm of the previous section.

4 Experiment and Results Analysis

The Performance evaluation of this research is to be applied onto e-mail document [4]. Thus, it is necessary to check precision rate and recall rate first, and then conduct experiment to check whether classification into the applicable category is appropriate when it comes to the contents of mail. Here, F_1 measure value is used to evaluate appropriateness of each category on an individual basis. Macro-averaging method is used to evaluate average performance of all categories. In this formula, recall, precision, and measure value are calculated for each category, and their average is calculated to evaluate the performance of the entire document classification system.

Table 1. Comparison of F_1 Measure Value during Preprocessing Algorithm Application

Num	Category Item	Total	F1 (BA)	F1(SA)	F1(WV)	F1(DT)
1	Sale	38434	0.860	0.870	0.880	0.910
2	Autos	23712	0.920	0.910	0.900	0.920
3	Sports	17124	0.890	0.870	0.880	0.870
4	Electronics	12354	0.940	0.950	0.950	0.990
5	Politics	7971	0.840	0.840	0.830	0.850
6	Computer	573	0.930	0.950	0.940	0.970
7	Graphics	23579	0.910	0.910	0.920	0.860
8	Hardware	1578	0.830	0.830	0.840	0.850
9	Space	11694	0.850	0.870	0.860	0.940
10	Talk	62915	0.890	0.870	0.870	0.850
11	Language	14180	0.880	0.910	0.910	0.910
12	Spam	7260	0.890	0.900	0.920	0.920
	Macro-averaging		0.886	0.888	0.892	0.903

Table 1 shows the value that is calculated by using resume rate and accuracy rate for each category, and shows the improvement rate. As Table 1 shows, F1(BA) is the value that results from using the existing Bayesian algorithm while F1(SA) is the sampling algorithm that results when learning document is selected. F1(WV) results when weighted value is assigned by attribute, and F1(DT) is the value that results after applying active threshold. For each stage, performance improved by 0.2%, 0.6%, and 1.1%, each, and it is possible to witness that the results improved by 1.9% when all three stages of preprocessing algorithm applied.

5 Conclusion

This research subjects mail document to the application of text classification to increase e-mail accuracy level of document classification, and suggests preprocessing algorithm into three stages to prevent incorrect classification. In the learning document stage, documents with extensive amount of information are used to conduct sampling algorithm, using MAD. In the subsequent feature extraction stage, weighted value assigning algorithm by each attribute is conducted to classify weighted value of the title. Lastly, Bayesian algorithm based on the use of dynamic threshold is used during actual document classification. It is possible to manage the mail system with increased convenience if mails are classified into more accurate categories by applying general document classification system onto mail documents. Thus, the e-mail application system based in the use of preprocessing algorithm, proposed by this research, will be used very resourcefully.

References

1. Ok-Ran Jeong, Dong-Sub Cho: A Personalized Recommendation Agent System for E-Mail Document Classification, LNCS 3045, Vol 3, (2004). 558-565
2. David D. Lewis and Jason Catlett.: Heterogeneous Uncertainty Sampling for Supervised Learning. In Proceedings of the 11th international Conference on Machine Learning, (1994) 148-156
3. Tom M., Mitchell : Machine Learning , Kluwer Academic Publishers, (1997)
4. M. Trensh, N. Palmer, and A. Luniewski : Type Classification of Semi-structured Documents. In Proceedings of the 21st ACM SIGMOD International Conference on Management of Data, (1995) 263-274

Failure Detection Method Based on Fuzzy Comprehensive Evaluation for Integrated Navigation System

Guoliang Liu[1], Yingchun Zhang[2], Wenyi Qiang[2], and Zengqi Sun[1]

[1] State Key Laboratory of Intelligent Technology and Systems,
Graduate School at Shenzhen, Tsinghua University,
Shenzhen 518055, China
liugl@sz.tsinghua.edu.cn
http://www.sz.tsinghua.edu.cn
[2] Department of Control Science and Engineering,
Harbin Institute of Technology, Harbin 15001,China
zhang@hit.edu.cn
http://www.hit.edu.cn

Abstract. A failure detection method based on fuzzy comprehensive evaluation for integrated navigation systems was presented in this paper. By using fuzzy comprehensive evaluation, this method judged the measured data of every subsystem of a Kalman filter comprehensively in order to determine their states. This method overcame some shortcomings of conventional failure detection methods. At last, the contrastive experiment among this method, chi-square test and the method of data change rate has indicated that chi-square test could recognize the larger outliers, but couldn't detect the lesser, that the method of data change rate couldn't deal with the longtime continuous failures, and that the method based on fuzzy comprehensive evaluation could recognize all of the failures. So it was a quite practical method.

1 Introduction

How to enhance the performance of integrated navigation systems effectively has become a significant research subject at present [1~3]. Failures may arise during acquiring data due to some uncertain factors in integrated navigation systems, such as imprecise sensors, environment noise and contrived disturbance. If they are applied to integrated navigation systems, those wrong data will degrade the accuracy of them. Therefore, the subsystems with failures must be isolated in time, so that the integrated navigation systems work well. There are several methods to detect failures for integrated navigation systems, such as chi-square test, the method of data change rate and so on[4]. This paper presents a novel failure detection method based on fuzzy comprehensive evaluation, and then compares it with conventional methods. An experiment on an integrated GPS/INS navigation system is performed in the end.

2 Two Conventional Failure Detection Methods

2.1 Chi-Square Test

Chi-square test is one of conventional failure detection methods for integrated navigation systems[5]. This method doesn't concern with the reason of failures, and it only judges validity of measured data. Therefore, it is suitable for system-class failure detection and isolation. Chi-square test includes state chi-square test and residual error chi-square test. In this paper, the latter is used.

In an integrated navigation system, every local filter is an independent Kalman filter. When local filters work properly, the residual error is zero-mean white Gauss noise,

$$\tilde{Z}(k+1) = Z(k+1) - H(k+1)\hat{X}(k+1/k) . \tag{1}$$

Its variance is

$$P_{\tilde{Z}}(k+1) = H(k+1)P(k+1/k)H^T(k+1) + R(k+1) . \tag{2}$$

When the system goes wrong, the mean value of $\tilde{Z}(k+1)$ won't be zero any longer. So it is possible to judge whether the system has failures or not by the mean value of $\tilde{Z}(k+1)$.

2.2 The Method of Data Change Rate

The main idea of the method of data change rate is to use the change rate of the acquired data as detection criteria[6]. The algorithm is showed below:

(1) First, according to the set of the given data, calculate the data change rate by using the equation

$$Rate(k) = (Data(k) - Data(k-1))/T . \tag{3}$$

where T is the sampling period.
(2) Then sort the data change rate by value, weed out the largest and the smallest, and average the rest as the initial value of the data change rate.

$$RateAve(k) = \sum_{k=1}^{SectN-2} Rate(k)/(SectN-2) . \tag{4}$$

(3) If $k > SectN$, calculate the average of data change rate according to the equation (5) while calculating the data change rate at k.

$$RateAve(k) = RateAve(k-1) + (Rate(k) - Rate(k-SectN-1))/SectN . \tag{5}$$

(4) Definite two data zones, $Thr(A)$, $Thr(B)$:

$$Thr(B) = \max(|RateAve(k-1)|, \cdots, |RateAve(k-10)|) . \tag{6}$$

$$Thr(A) = 3*Thr(B) . \tag{7}$$

The data will be considered to be invalid if the current data change rate exceeds the maximal value of the average change rate of the 10 anterior data, that is, the two self-adaptive data zones are used as the criteria to judge failures.

3 The Method Based on Fuzzy Comprehensive Evaluation

The method based on fuzzy comprehensive evaluation is a detection method for multi-proposition. Firstly, select the fuzzy variable set that consists of measured variables of a local filter, such as longitude, latitude, easterly speed and northerly speed of GPS. Define the fuzzy variable set as $U=\{u_1,u_2,\ldots,u_k,\ldots,u_n\}$, in which u_k is the No. k fuzzy variable.

The weight set of the above fuzzy variable set is $\tilde{A}=(a_1,a_2,\ldots,a_k,\ldots,a_n)$, where a_k is the weight of the No. k fuzzy variable u_k, specify

$$\sum_{k=1}^{n} a_k = 1 \cdot$$

How to select a_k is according to the importance of the No. k variable for the judgment result. Therefore, according to the character of modern sensors, we usually select $a_1 \gg a_2 \gg a_3 \cdots a_n$, and weights of the last several variables are all lesser. a_k should be changeable in order to represent the importance of each fuzzy variable and the influence of the actual environment on sensors. Furthermore, wrong judgments and missed judgments should be avoided.

Secondly, we fuzzificate the variables of the fuzzy variable set. $\mu_1, \mu_2, \cdots, \mu_n$ are used to represent the supporting degree of the measured data being right respectively, here $0 < \mu_i < 1$. We select the triangular membership function shown in Figure1.

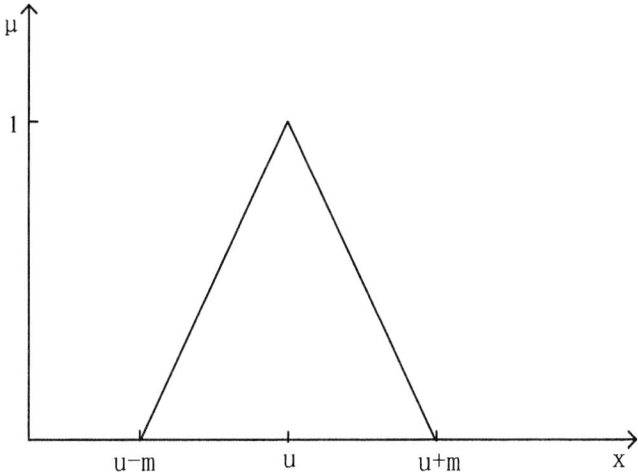

Fig. 1. Triangular Membership Function

The formula is

$$\mu(x) = \begin{cases} 1 - \dfrac{|x-u|}{m} & |x-u| < m \\ 0 & |x-u| > m \end{cases} \qquad (8)$$

Here u that can be determined by prior knowledge is the basis of calculating the degree of membership. m is the boundary of a fuzzy subset, and it is adjusted adaptively according to specific applications.

When the detection method based on fuzzy comprehensive evaluation is applied, an evaluation set and a single variable judgment matrix must be defined besides the fuzzy variables set $U=\{u_1,u_2,\ldots,u_k,\ldots,u_n\}$.

If a judgment result is divided into m levels, the set that consists of these m results is called as the evaluation set. so we have

$$V = \{v_1, v_2, \cdots, v_m\} . \qquad (9)$$

Here $v_l\,(l=1,2,\cdots,m)$ is the judgment result of the No. l level. In fact, the judgment result of any datum is a subset of $V = \{v_1, v_2, \cdots, v_m\}$.

The fuzzy relational matrix defined on the direct product of sets $U \times V$ from U to V is $R = (r_{kl})_{n \times m}$. Where, r_{kl} is the possibility of being the No. l judgment result when the No. k variable is considered, and R is called as the evaluation set. We have

$$R_k = (r_{k1}, r_{k2}, \cdots, r_{km}) \qquad k = 1,2,\cdots,n .$$

where R_k, the judgment result of a single variable for the fuzzy variable u_k, is called as single variable judgment matrix. The comprehensive judgment process is a synthesis operation that combines the weight set A with the single variable judgment matrix R. Given $A \in \mathcal{E}(U)$, $R \in A \in \mathcal{E}(U \times V)$, the fuzzy set B of judgment levels on V is obtained from their synthesis operation,

$$B = A \circ R = (b_1, b_2, \cdots, b_m) . \qquad (10)$$

Here, "∘"denotes the synthesis operation. $b_l, l=1,2,\cdots,m$, denotes the degree of membership for the No. l level v_l.

If $\exists l \in \{1,2,\cdots,m\}$, we have

$$b_l = \max\{b_1, b_2, \cdots, b_m\} . \qquad (11)$$

Then the judgment result is the No. l level according to the principle of maximal degree of membership.

There are a variety of models for the synthesis operation of equation (11). In this paper, the evaluation set V only has both two judgment results: right and wrong. The fuzzy variables set U consists of latitude, longitude, easterly speed and northerly speed, all of which can be considered to be uncorrelated with each other, so all variables can get the same weight. For the specific case in this paper, the weighted average model is selected. The weighted average model is carried out according to the algorithm of general matrixes in linear algebra, i.e.

$$b_l = \sum_{k=1}^{n} a_k r_{bl} \ . \tag{12}$$

4 The Comparative Experiment of Three Detection Methods

An integrated GPS/INS navigation system was used in this experiment, GPS and INS of which were integrated as a loose coupling mode. The INS acted as the primary subsystem, and the GPS as the auxiliary subsystem. 2000 groups of initial data including longitudes, latitudes, easterly speed and northerly speed of GPS and INS were collected, and the sampling period was one second. Longitude data and latitude data were showed in this paper, but easterly speed data and northerly speed data were not. Figure 2 was the initial longitude of the INS, and figure 3 was its initial latitude. Figure 4 was the initial longitude of GPS, and figure 5 was its initial latitude. Figure 6 was the longitude fused by Kalman filter, and figure 7 was the latitude fused by Kalman filter.

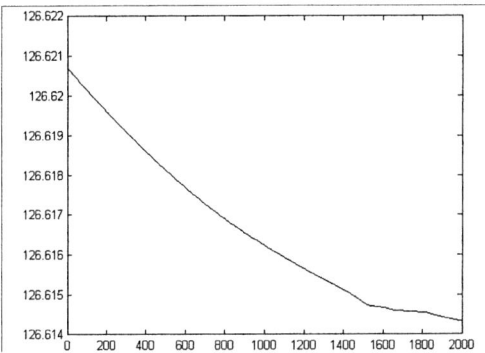

Fig. 2. Initial Longitude of The INS

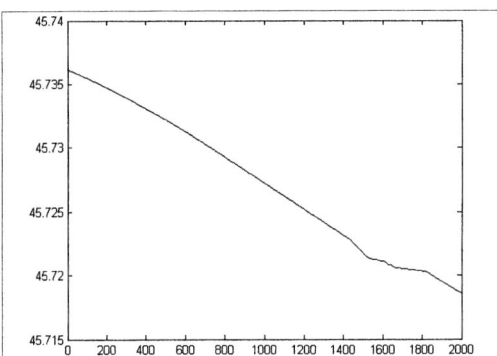

Fig. 3. Initial Latitude of The INS

The effect of failure detection methods was demonstrated by modifying collected data of the GPS. Set the 500th longitude, latitude, easterly speed and northerly speed of the GPS to zero in order to simulate independent large outliers. Set the 1000th longitude of the GPS to 126.6310 degree, and keep rest data unchanged in

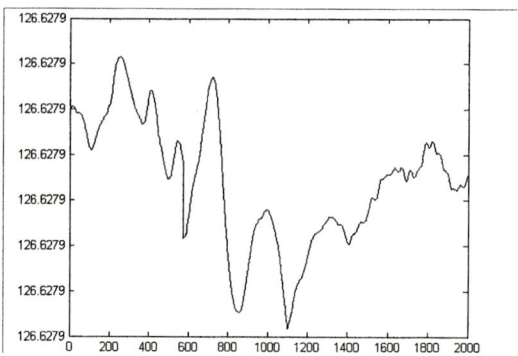

Fig. 4. Initial Longitude of The GPS

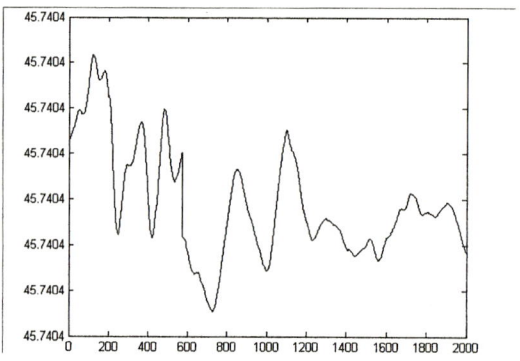

Fig. 5. Initial Latitude of The GPS

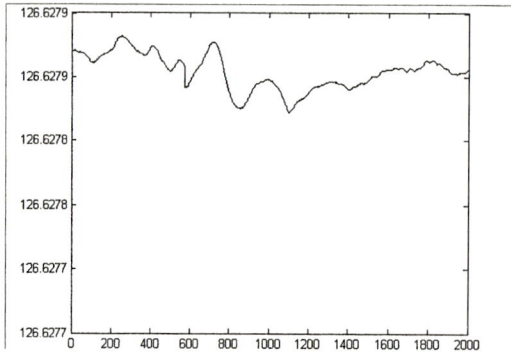

Fig. 6. Longitude Fused by Kalman Filter

order to simulate a small outlier caused by soft-faults. Set all the measured data from the 1400th to the 1600th to zero in order to simulate the longtime continuous failures. Figure 8 was the modified longitude of GPS, and figure 9 was its modified latitude.

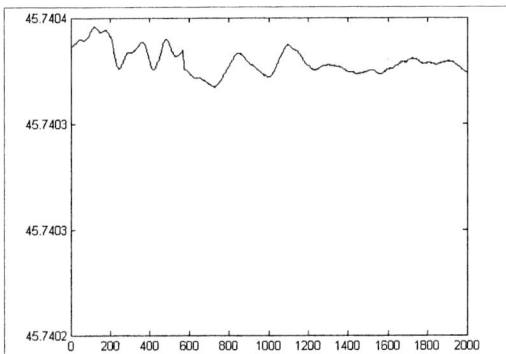

Fig. 7. Latitude Fused by Kalman Filter

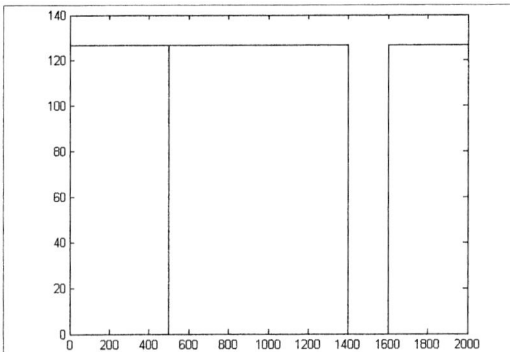

Fig. 8. Modified Longitude of the GPS

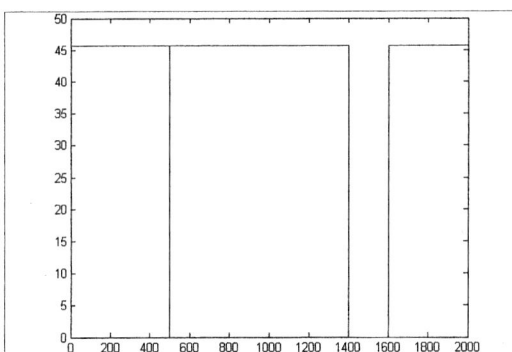

Fig. 9. Modified Latitude of the GPS

Figure10 to figure 15 were the longitude and latitude fused by Kalman filter after failures were detected by three methods respectively. Figure 10 and figure11 were the results after residual error chi-square test was applied, in which it was obvious that the larger outliers could be recognized while the lesser could not. Figure 12 and figure 13

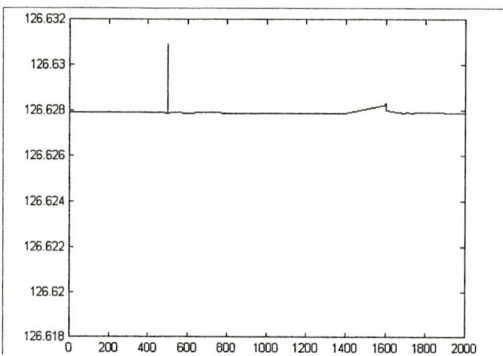

Fig. 10. Fused Longitude after Chi-Square Test was applied

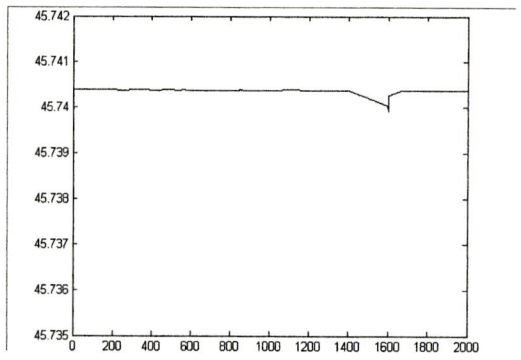

Fig. 11. Fused Latitude after Chi-Square Test was applied

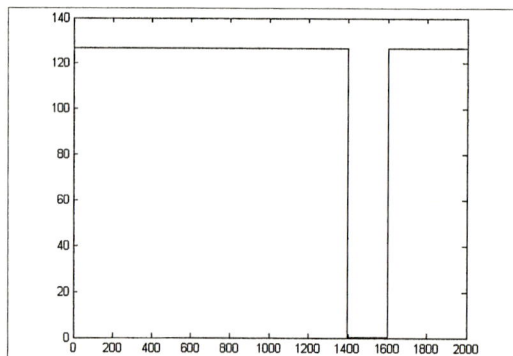

Fig. 12. Fused Longitude after the Method of Data Change Rate was applied

were the results after the method of data change rate was applied, which showed that the failure couldn't be recognized from the 1400th to the 1600th second because the method couldn't deal with the longtime continuous failures. Figure14 and figure15 were the results after the method based on fuzzy comprehensive evaluation was applied, which overcame the shortcomings of the above two methods.

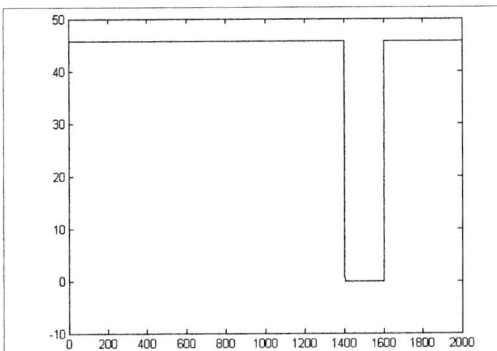

Fig. 13. Fused Latitude after the Method of Data Change Rate was applied

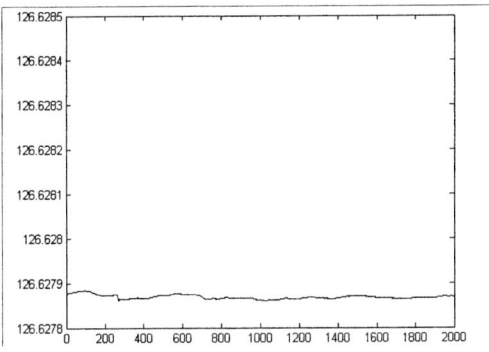

Fig. 14. Fused Longitude after Method Based on Fuzzy Comprehensive Evaluation was applied

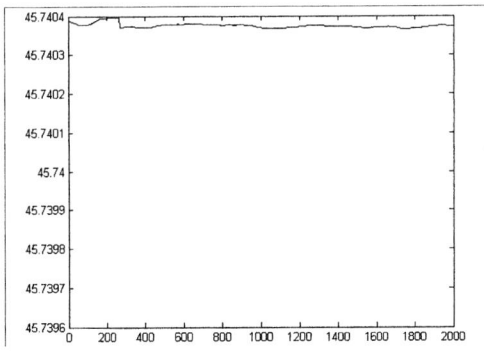

Fig. 15. Fused Latitude after Method Based on Fuzzy Comprehensive Evaluation was applied

5 Conclusion

This paper presented a novel failure detection method based on fuzzy comprehensive evaluation for integrated navigation systems, and compared it with residual error chi-square test and the method of data change rate on an integrated GPS/INS navigation system. The experiment result indicated that residual error chi-square test could recognize the larger outliers, but couldn't detect the lesser. The method of data change rate couldn't deal with the longtime continuous failures. The method based on fuzzy comprehensive evaluation overcame these shortcomings and had the simple algorithm, so it is quite practical for integrated navigation systems.

References

1. He Xiufeng, Chen Yongqi: Application of Interval Kalman Filter to an Integrated GPS/INS System. Transactions of Nanjing University of Aeronautics and Astronautics, Nanjing, Vol.16(1). (1999) 39-45
2. N. A. Carlson: Federated Kalman Filter Simulation Results. Journal of the Institute of Navigation, Vol.41(3). (1994) 297-321.
3. N.A.Carlson: Federated Filter For Failure-tolerant Integrated Navigation System. Processing of IEEE Position Location and Navigation Symposium, Orlando, Florida, (1988) 110-119
4. J.C.Mcmillan: Techniques for Soft-Failure Detection in a Multi-Sensor Integrated System. Journal of Navigation, Vol.40(3). (1993)227-248
5. Ren. D. Failur: Detection of Dynamic Systems with the State Chi-Square Test. Journal of Guidance, Control and Dynamics, Vol.17(2). (1994)271-277
6. Zhu Rongsheng, Shi Xiaocheng: The Algorithm of Weeding Out Failure Data in GPS Data Processing. Journal of Chinese Inertial Technology, Vol.8(2). (2000)27-30

Product Quality Improvement Analysis Using Data Mining: A Case Study in Ultra-Precision Manufacturing Industry

Hailiang Huang[1] and Dianliang Wu[2]

[1] School of Information Management, Shanghai University of Finance & Economics,
Guoding Road 777, 200433 Shanghai, China
hlhuang@mail.shufe.edu.cn
[2] School of Mechanical Engineering, Shanghai Jiaotong University,
200030 Shanghai, P.R. China

Abstract. This paper presents an analysis of product quality improvement in ultra-precision manufacturing industry using data mining for developing quality improvement strategies. Based on 11320 ultra-precision optical products that were produced from the study factory during the period of June 1 and August 31, 2004, important factors impacting the product quality were identified via the decision tree method for data mining. Findings showed that the important factors for the percentage of defectives were type of processing chain, precision requirement, product classes, and raw material. The optimum range of target group in production quality indicators was identified from the gains chart.

1 Introduction

An increasing concern with improving the product quality has led to the adoption of quality improvement approaches in manufacturing industry. These include "total quality management" (TQM)[1] and "continuous quality improvement" (CQI)[2].

In the ultra-precision manufacturing (UPM) industry in China, quality assurance (QA) activity has been launched since 1990s in most state-owned factories. Many UPM companies have developed their own quality improvement (QI) standard and established special QI department for QA activities. But because of inadequate utilization of QI evaluation results and feedback, heavy workloads, and lack of motivation of this endeavor, CQI has not been successfully implemented in most companies. Therefore, there is a need for a decision support system (DSS) that provides quality assessment/feedback information to support the CQI process.

Data mining (DM) is defined as a sophisticated data search capability that uses statistical algorithms to discover patterns and correlations in data [3]. In this case, DM method was employed to identify patterns or rules about various quality problems from the production-oriented data warehouse.

2 Data Mining Methods

The subjects were 11320 ultra-precision optical products that were produced from the study factory during the period of June 1 and 31 August, 2004. Of several quality indicators used in the factory, we focused on the percentage of defectives (POD) of product surface for the decision tree analysis of the influencing factors for quality. Product characteristics such as raw material, product classes, type of processing chain, precision requirements were used in the analysis.

The decision tree was used in the analysis of the factors influencing POD. In our example, the decision tree categorizes the entire subjects according to whether or not they are likely to raise defectives. Chi-squared automatic interaction detection (CHAID) and C5.0 are two popular decision tree inducers, based on the ID3 classification algorithm [4].

3 Results

Among the 11320 products, 6490 (57.3%) were glass-based and 4830 (42.7%) were aluminum-based. Products with precision requirement Ra0.02μm were almost 2 times (7457) more than those of Ra0.01μm (3863). The complete descriptive statistics are shown in Table 1.

Table 1. Characteristics of study subjects

Characteristics	Value	Frequency	%
Raw material	Glass-based	6490	57.3
	Aluminum-based	4830	42.7
Precision requirement	Ra0.02μm	7457	65.9
	Ra0.01μm	3863	34.1
Type of process chain	Grinding only	4754	42.0
	Grinding + Lapping	3357	29.7
	Grinding + Lapping + Ultrasonic processing	3209	28.3
Product class	Optical substrate	4221	37.3
	Particular-use lens	2576	22.8
	Common lens	1285	11.4
	Hard-disk components	1104	9.8
	Laser-disk components	990	8.7
	Digital camera components	734	6.5
	Miscellaneous	410	3.5
	Total	11320	100

The decision tree for POD had 21 statistically significant nodes at 5% level, which are depicted in Fig. 1. Among 11320 products, 208 (1.84%) had surface defectives. The most significant factor explaining the POD was process chain type. Precision requirement was the next significant factor, followed by the product class.

Each node depicted in the decision tree can be expressed in terms of an 'if-then' rule, as follows:

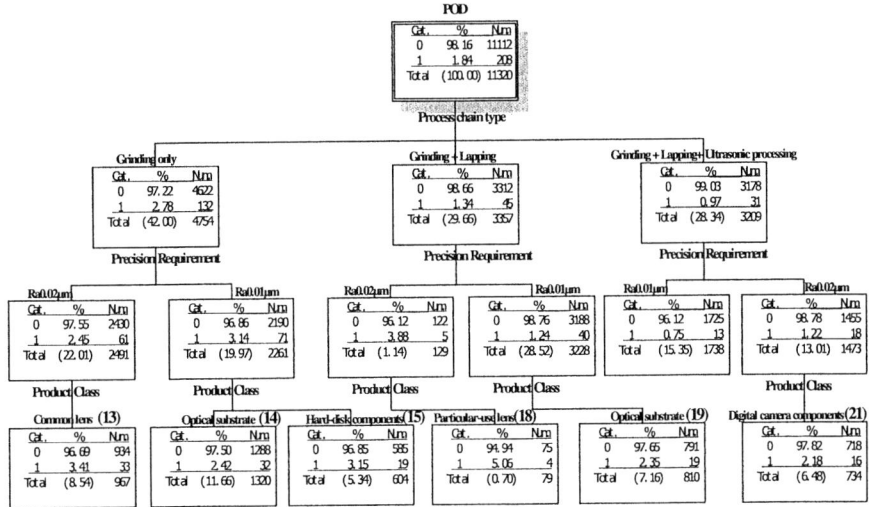

Fig. 1. Decision tree for POD

```
/*Node 18*/

If (process chain is Grinding + lapping) and (precision
requirement is Ra 0.02µm) and (product class is particu-
lar-use lens), then the POD = 5.06%
```

The gains chart produced by the decision tree can be used for developing quality improvement strategies. As shown in Table 2, there are two parts of the gains chart: node-by-node statistics and cumulative statistics. The gains chart shows the nodes sorted by the percentage of cases in the target category for each node (gain percentage). The first node in the table, node 13, contains 33 defective cases out of 967 subjects, or gain percentage is 3.41%. The index percentage shows how the proportion of POD for this particular node compares to the overall POD. For node 13, the index score was 185.5%, meaning that the proportion of respondents for this node is about 2 times the POD for the overall sample.

The cumulative statistics can show us how well we do in finding defective product cases by taking the best segments of the sample. If we take only the best node (node 13), we reach 15.87% (respondent percentage) of all defective cases by targeting only 8.54% (node percentage) of the sample. Similarly, if we include the second best node as well (node 15), then we get 25.0% of the defective cases from only 13.9% of the sample.

The gains chart can also provide valuable information about which segments to target and which to neglect. We might base the decision on the number of prospects we want, the desired POD for the target sample, or the desired proportion of overall POD we want to reach. In this example, suppose we want to investigate the cases with an estimated POD of at least 3%. To achieve this, we would focus on the first three nodes with a gain percentage greater than 3%, namely, node 13, 15, and 18.

Table 2. Gains chart for inpatient mortality

Rule	Node	Node-by-node						Cumulative		
		Node Num.	Node (%)	Resp. Num.	Resp. (%)	Gain (%)	Index (%)	Node (%)	Resp (%)	Gain (%)
1	13	967	8.54	33	15.87	3.41	185.47	8.54	15.87	3.41
2	15	604	5.34	19	9.14	3.15	170.96	13.88	25.00	3.31
3	14	1320	11.66	32	15.38	2.42	131.75	25.54	40.38	2.91
4	19	810	7.16	19	9.13	2.35	127.48	32.69	49.52	2.78
5	21	734	6.48	16	7.69	2.18	118.47	39.18	57.21	2.68
6	18	79	0.70	4	1.92	5.06	275.18	39.88	59.13	2.72
7	16	3510	31.01	47	22.60	1.34	72.77	70.88	81.73	2.12
8	4	2334	20.62	27	12.98	1.16	62.87	91.50	94.71	1.90
9	20	962	8.50	11	5.288	1.14	62.14	100.00	100.00	1.84

4 Conclusions

This study presents an analysis of product quality indicators using data mining for quality improvement. Important factors influencing the percentage of defective were identified using decision tree method based on 11320 ultra-precision optical products that were produced from the study factory during 3 months period. Findings showed that the important factors for the percentage of defectives were type of processing chain, precision requirement, product classes, and raw material. The optimum range of target group in production quality indicators was identified from the gains chart produced by decision tree. The cumulative statistics in the gains chart show us how well we do at finding defective products by taking the best segments of the sample.

References

1. Deming, W. E. Out of the crisis. Cambridge, MIT Press (1986)
2. Juran, J. M. (ed.): Juran's quality control handbook 4th edn. New York, McGraw-Hill (1988)
3. M. S. Chen, J. Han, and P. S. Yu. Data mining: An overview from a database perspective. IEEE Trans. On Knowledge and Data Engineering 8 (1996) 866-883
4. Quinlan, J. R.. C4.5: Programs for machine learning. San Mateo, Morgan Kaufmann (1993)

Two-Tier Based Intrusion Detection System*

Byung-Joo Kim[1] and Il Kon Kim[2]

[1] Youngsan University School of Network and Information Engineering, Korea
bjkim@ysu.ac.kr
[2] Kyungpook National University Department of Computer Science, Korea
ikkim@knu.ac.kr

Abstract. Intrusion detection is a critical component of secure information system. Recently applying artificial intelligence, machine learning and data mining techniques to intrusion detection system are increasing. But most of researches are focused on improving the classification performance of classifier. Selecting important features from input data lead to a simplification of the problem, faster and more accurate detection rates. Thus selecting important features is an important issue in intrusion detection. Another issue in intrusion detection is that most of the intrusion detection systems are performed by off-line and it is not proper method for realtime intrusion detection system. In this paper, we develop the realtime intrusion detection system which combining on-line feature extraction method with Least Squares Support Vector Machine classifier. Applying proposed system to KDD CUP 99 data, experimental results show that it have remarkable feature feature extraction and classification performance compared to existing off-line intrusion detection system.

1 Introduction

Computer security has become a critical issue with the rapid development of business and other transaction systems over the internet. Intrusion detection is to detect intrusive activities while they are acting on computer network systems. Most intrusion detection systems(IDSs) are based on hand-crafted signatures that are developed by manual coding of expert knowledge. These systems match activity on the system being monitored to known signatures of attack. The major problem with this approach is that these IDSs fail to generalize to detect new attacks or attacks without known signatures. Recently, there has been an increased interest in data mining based approaches to building detection models for IDSs. These models generalize from both known attacks and normal behavior in order to detect unknown attacks. They can also be generated in a quicker and more automated method than manually encoded models that require difficult analysis of audit data by domain experts. Several effective data mining

---

* This study was supported by a grant of the Korea Health 21 R&D Project, Ministry of Health & Welfare, Republic of Korea (02-PJ1-PG6-HI03-0004).

techniques for detecting intrusions have been developed[1][2][3], many of which perform close to or better than systems engineered by domain experts.

However, successful data mining techniques are themselves not enough to create effective IDSs. Despite the promise of better detection performance and generalization ability of data mining based IDSs, there are some difficulties in the implementation of the system. We can group these difficulties into three general categories: accuracy(i.e., detection performance), efficiency, and usability. In this paper, we discuss accuracy problem in developing a real-time IDS. Another issue in IDS is that it should operate in real-time. In typical applications of data mining to intrusion detection, detection models are produced off-line because the learning algorithms must process tremendous amounts of archived audit data. These models can naturally be used for off-line intrusion detection. Effective IDS should work in real-time, as intrusions take place, to minimize security compromises. Elimination of the insignificant and/or useless inputs leads to a simplification of the problem, faster and more accurate detection result. Feature selection therefore, is an important issue in intrusion detection.

Principal Component Analysis(PCA)[4] is a powerful technique for extracting features from data sets. For reviews of the existing literature is described in [5][6][7]. Traditional PCA, however, has several problems. First PCA requires a batch computation step and it causes a serious problem when the data set is large i.e., the PCA computation becomes very expensive. Second problem is that, in order to update the subspace of eigenvectors with another data, we have to recompute the whole eigenspace. Finial problem is that PCA only defines a linear projection of the data, the scope of its application is necessarily somewhat limited. It has been shown that most of the data in the real world are inherently non-symmetric and therefore contain higher-order correlation information that could be useful[8]. PCA is incapable of representing such data. For such cases, nonlinear transforms is necessary. Recently kernel trick has been applied to PCA and is based on a formulation of PCA in terms of the dot product matrix instead of the covariance matrix[9]. Kernel PCA(KPCA), however, requires storing and finding the eigenvectors of a $N \times N$ kernel matrix where N is a number of patterns. It is infeasible method when N is large. This fact has motivated the development of on-line way of KPCA method which does not store the kernel matrix. It is hoped that the distribution of the extracted features in the feature space has a simple distribution so that a classifier could do a proper task. But it is point out that extracted features by KPCA are global features for all input data and thus may not be optimal for discriminating one class from others[9]. In order to solve this problem, we developed the realtime intrusion detection system. Proposed real time IDS is composed of two parts. First part is used for on-line feature extraction. To extract on-line nonlinear features, we propose a new feature extraction method which overcomes the problem of memory requirement of KPCA by on-line eigenspace update method incorporating with an adaptation of kernel function. Second part is used for classification. Extracted features are used as input for classification. We take Least Squares Support Vector Machines(LS-SVM)[10] as a classifier. LS-SVM is reformulations to the standard Support Vector Machines(SVM)[11]. Paper is composed of as follows. In Section 2 we will briefly explain the on-line feature extraction method. In Section 3 KPCA is introduced and to make KPCA on-line, empirical kernel map method is is explained. Proposed classifier combining LS-SVM with proposed feature extraction method is described in Section 4. Experimental results to evaluate the performance of proposed system is shown in Section 5. Discussion of proposed IDS and future work is described in Section 6.

2   On-line Feature Extraction

In this section, we will give a brief introduction to the method of on-line PCA alorithm which overcomes the computational complexity and memory requirement of standard PCA. Before continuing, a note on notation is in order. Vectors are columns, and the size of a vector, or matrix, where it is important, is denoted with subscripts. Particular column vectors within a matrix are denoted with a superscript, while a superscript on a vector denotes a particular observation from a set of observations, so we treat observations as column vectors of a matrix. As an example, A_{mn}^i is the ith column vector in an $m \times n$ matrix. We denote a column extension to a matrix using square brackets. Thus $[A_{mn}b]$ is an $(m \times (n+1))$ matrix, with vector b appended to A_{mn} as a last column.

To explain the on-line PCA, we assume that we have already built a set of eigenvectors $U = [u_j], j = 1, \cdots, k$ after having trained the input data $\mathbf{x}_i, i = 1, \cdots, N$. The corresponding eigenvalues are Λ and $\bar{\mathbf{x}}$ is the mean of input vector. On-line building of eigenspace requires to update these eigenspace to take into account of a new input data. Here we give a brief summarization of the method which is described in [12]. First, we update the mean:

$$\bar{x}' = \frac{1}{N+1}(N\bar{x} + x_{N+1}) \qquad (1)$$

We then update the set of eigenvectors to reflect the new input vector and to apply a rotational transformation to U. For doing this, it is necessary to compute the orthogonal residual vector $\hat{h} = (Ua_{N+1} + \bar{x}) - x_{N+1}$ and normalize it to obtain $h_{N+1} = \frac{h_{N+1}}{\|h_{N+1}\|_2}$ for $\|h_{N+1}\|_2 > 0$ and $h_{N+1} = 0$ otherwise. We obtain the new matrix of Eigenvectors U' by appending h_{N+1} to the eigenvectors U and rotating them :

$$U' = [U, h_{N+1}]R \qquad (2)$$

where $R \in \mathbf{R}_{(k+1)\times(k+1)}$ is a rotation matrix. R is the solution of the eigenproblem of the following form:

$$DR = R\Lambda' \qquad (3)$$

where Λ' is a diagonal matrix of new Eigenvalues. We compose $D \in \mathbf{R}_{(k+1)\times(k+1)}$ as:

$$D = \frac{N}{N+1}\begin{bmatrix} \Lambda & 0 \\ 0^T & 0 \end{bmatrix} + \frac{N}{(N+1)^2}\begin{bmatrix} aa^T & \gamma a \\ \gamma a^T & \gamma^2 \end{bmatrix} \qquad (4)$$

where $\gamma = h_{N+1}^T(x_{N+1} - \bar{x})$ and $a = U^T(x_{N+1} - \bar{x})$. Though there are other ways to construct matrix D[13][14], the only method ,however, described in [12] allows for the updating of mean.

2.1 Eigenspace Updating Criterion

The on-line PCA represents the input data with principal components $a_{i(N)}$ and it can be approximated as follows:

$$\widehat{x}_{i(N)} = U a_{i(N)} + \bar{x} \qquad (5)$$

To update the principal components $a_{i(N)}$ for a new input x_{N+1}, computing an auxiliary vector η is necessary. η is calculated as follows:

$$\eta = \left[U \widehat{h}_{N+1} \right]^T (\bar{x} - \bar{x}') \qquad (6)$$

then the computation of all principal components is

$$a_{i(N+1)} = (R')^T \begin{bmatrix} a_{i(N)} \\ 0 \end{bmatrix} + \eta, \qquad i = 1, \cdots, N+1 \qquad (7)$$

The above transformation produces a representation with $k+1$ dimensions. Due to the increase of the dimensionality by one, however, more storage is required to represent the data. If we try to keep a k-dimensional eigenspace, we lose a certain amount of information. It is needed for us to set the criterion on retaining the number of eigenvectors. There is no explicit guideline for retaining a number of eigenvectors. Here we introduce some general criteria to deal with the model's dimensionality:

- Adding a new vector whenever the size of the residual vector exceeds an absolute threshold;
- Adding a new vector when the percentage of energy carried by the last eigenvalue in the total energy of the system exceeds an absolute threshold, or equivalently, defining a percentage of the total energy of the system that will be kept in each update;
- Discarding Eigenvectors whose Eigenvalues are smaller than a percentage of the first Eigenvalue;
- Keeping the dimensionality constant.

In this paper we take a rule described in second. We set our criterion on adding an Eigenvector as $\lambda'_{k+1} > 0.7 \bar{\lambda}$ where $\bar{\lambda}$ is a mean of the λ. Based on this rule, we decide whether adding u'_{k+1} or not.

3 On-line KPCA

A prerequisite of the on-line eigenspace update method is that it has to be applied on the data set. Furthermore on-line PCA builds the subspace of eigenvectors online, it is restricted to apply the linear data. But in the case of KPCA this data set $\Phi(x^N)$ is high dimensional and can most of the time not even be calculated explicitly. For the case of nonlinear data set, applying feature mapping function method to on-line PCA may be one of the solutions. This is performed by so-called *kernel-trick*, which means an implicit embedding to an infinite dimensional Hilbert space[11](i.e. feature space) F.

$$K(x, y) = \Phi(x) \cdot \Phi(y) \tag{8}$$

Where K is a given kernel function in an input space. When K is semi positive definite, the existence of Φ is proven[11]. Most of the case, however, the mapping Φ is high-dimensional and cannot be obtained explicitly. The vector in the feature space is not observable and only the inner product between vectors can be observed via a kernel function. However, for a given data set, it is possible to approximate Φ by empirical kernel map proposed by Scholkopf[15] and Tsuda[16] which is defined as $\Psi_N : \mathbf{R}^d \to \mathbf{R}^N$

$$\begin{aligned}\Psi_N(x) &= [\Phi(x_1) \cdot \Phi(x), \cdots, \Phi(x_N) \cdot \Phi(x)]^T \\ &= [K(x_1, x), \cdots, K(x_N, x)]^T\end{aligned} \tag{9}$$

A performance evaluation of empirical kernel map was shown by Tsuda. He shows that support vector machine with an empirical kernel map is identical with the conventional kernel map[17]. The empirical kernel map $\Psi_N(x_N)$, however, do not form an orthonormal basis in $\mathbf{R}^N$, the dot product in this space is not the ordinary dot product. In the case of KPCA, however, we can be ignored as the following argument. The idea is that we have to perform linear PCA on the $\Psi_N(x_N)$ from the empirical kernel map and thus diagonalize its covariance matrix. Let the $N \times N$ matrix $\Psi = [\Psi_N(x_1), \Psi_N(x_2), \ldots, \Psi_N(x_N)]$, then from equation (9) and definition of the kernel matrix we can construct $\Psi = NK$. The covariance matrix of the empirically mapped data is:

$$C_\Psi = \frac{1}{N}\Psi\Psi^T = NKK^T = NK^2 \tag{10}$$

In case of empirical kernel map, we diagonalize NK^2 instead of K as in KPCA. Mika shows that the two matrices have the same eigenvectors $\{u_k\}$[17]. The eigenvalues $\{\lambda_k\}$ of K are related to the eigenvalues $\{k_k\}$ of NK^2 by

$$\lambda_k = \sqrt{\frac{k_k}{N}} \tag{11}$$

and as before we can normalize the eigenvectors $\{v_k\}$ for the covariance matrix C of the data by dividing each $\{u_k\}$ by $\sqrt{\lambda_k N}$. Instead of actually diagonalize the covariance matrix C_Ψ, the IKPCA is applied directly on the mapped data $\Psi = NK$. This makes it easy for us to adapt the on-line eigenspace update method to KPCA such that it is also correctly takes into account the centering of the mapped data in an on-line way. By this result, we only need to apply the empirical map to one data point at a time and do not need to store the $N \times N$ kernel matrix.

4 Proposed System

In earlier Section 3 we proposed an on-line KPCA method for nonlinear feature extraction. Feature extraction by on-line KPCA effectively acts a nonlinear mapping from the input space to an implicit high dimensional feature space. It is hoped that the distribution of the mapped data in the feature space has a simple distribution so that a classifier can classify them properly. But it is point out that extracted features by KPCA are global features for all input data and thus may not be optimal for discriminating one class from others. For classification purpose, after global features are extracted using they must be used as input data for classification. There are many famous classifier in machine learning field. Among them neural network is popular method for classification and prediction purpose.

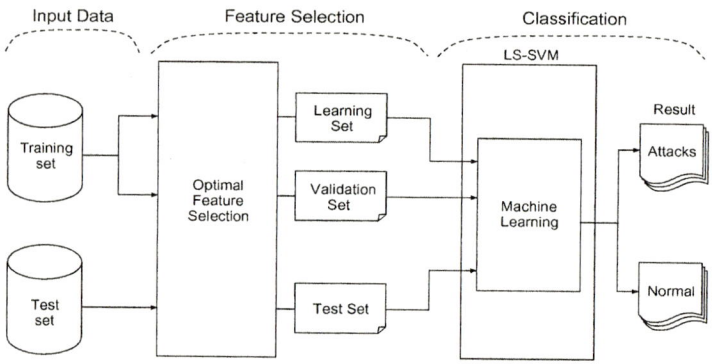

Fig. 1. Overall structure of realtime based realtime IDS

Traditional neural network approaches, however have suffered difficulties with generalization, producing models that can overfit the data. To overcome the problem of classical neural network technique, support vector machines(SVM) have been introduced. The foundations of SVM have been developed by Vapnik and it is a powerful methodology for solving problems in nonlinear classification. Originally, it has been introduced within the context of statistical learning theory and structural risk minimization. In the methods one solves convex optimization problems, typically by quadratic programming(QP). Solving QP problem requires complicated computational effort and need more memory requirement. LS-SVM overcomes this problem by solving a set of linear equations in the problem formulation. LS-SVM method is computationally attractive and easier to extend than SVM. The overall structure and main components of proposed system is depicted in Figure 1. Proposed real time IDS is composed of two parts. First part is used for on-line feature extraction. To extract on-line nonlinear features, we propose a new feature extraction method which overcomes the problem of memory requirement of KPCA by on-line eigenspace update method incorporating with an adaptation of kernel function. Second part is used for classification. Extracted features are used as input for classification. We take Least Squares Support Vector Machines(LS-SVM)[19] as a classifier.

5 Experiment

To evaluate the classification performance of proposed realtime IDS system, we use KDD CUP 99 data[18]. The following sections present the results of experiments.

5.1 Description of Dataset

The raw training data(kddcup.data.gz) was about four gigabytes of compressed binary TCP dump data from seven weeks of network traffic. This was processed into about five million connection records. Similarly, the two weeks of test data yielded around two million connection records. A connection is a sequence of TCP packets starting and ending at some well defined times, between which data flows to and from a source IP address to a target IP address under some well defined protocol. Each connection is labeled as either normal, or as an attack, with exactly one specific attack type. Each connection record consists of about 100 bytes. Attacks fall into four main categories:

- DOS: denial-of-service, e.g. syn flood;
- R2L: unauthorized access from a remote machine, e.g. guessing password;
- U2R: unauthorized access to local superuser (root) privileges, e.g., various "buffer overflow" attacks;
- Probing: surveillance and other probing, e.g., port scanning.

It is important to note that the test data(corrected.gz) is not from the same probability distribution as the training data, and it includes specific attack types not in the training data. This makes the task more realistic. The datasets contain a total of 24 training attack types, with an additional 14 types in the test data only.

5.2 Experimental Condition

To evaluate the classification performance of proposed system, we randomly split the the training data as 80% and remaining as validation data. To evaluate the classification accuracy of proposed system we compare the proposed system to SVM. Because standard LS-SVM and SVM are only capable of binary classification, we take multiclass LS-SVM and SVM. A RBF kernel has been taken and optimal hyperparameter of multiclass SVM and LS-SVM[20] was obtained by 10-fold cross-validation procedure. In [19] it is shown that the use of 10-fold cross-validation for hyperparameter selection of SVM and LS-SVMs consistently leads to very good results.

In experiment we will evaluate the generalization ability of proposed IDS on test data set since there are 14 additional attack types in the test data which are not included int the training set. To do this, extracted features by on-line KPCA will be used as input for multiclass LS-SVM. Our results are summarized in the following sections.

5.3 Evaluate Feature Extraction Performance

Table 1 gives the result of extracted features for each class by on-line KPCA method.

Table 1. Extracted features on each class by on-line KPCA

Class	Extracted features
Normal	1,2,3,5,6,7,8,9,10,11,12,13,14,16,17,18,20,21,22,23,25,27,29,30,31,32,34,37,38,39,41
Probe	3,5,6,23,24,32,33,38
DOS	1,3,6,8,19,23,28,32,33,35,36,38,39,41
U2R	5,6,15,16,18,25,32,33,38,39
R2L	3,5,6,24,32,33,34,35,38

Table 2. Performance of proposed system using all features

Class	Accuracy(%)	Training Time(Sec)	Testing Time(Sec)
Normal	98.55	5.83	1.45
Probe	98.59	28.0	1.96
DOS	98.10	16.62	1.74
U2R	98.64	2.7	1.34
R2L	98.69	7.8	1.27

Table 3. Performance of proposed system using extracted features

Class	Accuracy(%)	Training Time(Sec)	Testing Time(Sec)
Normal	98.43	5.25	1.42
Probe	98.63	25.52	1.55
DOS	98.14	15.92	1.48
U2R	98.64	2.17	1.32
R2L	98.70	7.2	1.08

Table 2 shows the results of the classification performance and computing time for training and testing data by proposed system using all features. Table 3 shows the results of the classification performance and computing time for training and testing data by proposed system using extracted features. We can see that using important features for classification gives similar accuracies compared to using all features and reduces the train, testing time. Comparing

Table 2 with Table 3, we obtain following results. The performance of using the extracted features do not show the significant differences to that of using all features. This means that proposed on-line feature extraction method has good performance in extracting features. Proposed method has another merit in memory requirement. The advantage of proposed feature extraction method is more efficient in terms of memory requirement than a batch KPCA because proposed feature extraction method do not require the whole N × N kernel matrix where N is the number of the training data. Second one is that proposed on-line feature extraction method has similar performance is comparable in performance to a batch KPCA.

5.4 Suitable for Realtime IDS

Table 2 and Table 3 show that using extracted features decreases the training and testing time compared to using all features. Furthermore classification accuracy of proposed system is similar to using all features. This makes proposed IDS suitable for realtime IDS.

5.5 Comparision with SVM

Recently SVM is a powerful methodology for solving problems in nonlinear classification problem. To evaluate the classification accuracy of proposed system it is desirable to compare with SVM.

Table 4. Performance comparision of proposed method and SVM using all features

	Normal	Probe	DOS	U2R	R2L
Proposed method	98.55	98.59	98.10	98.64	98.69
SVM	98.55	98.70	98.25	98.87	98.78

Table 5. Performance comparision of proposed method and SVM using extracted features

	Normal	Probe	DOS	U2R	R2L
Proposed method	98.43	98.63	98.14	98.64	98.70
SVM	98.59	98.38	98.22	98.87	98.78

The disadvantage of incremental method is their accuracy compared to batch method even though it has the advantage of memory efficiency. According to Table 4 and Table 5 we can see that proposed method has similar classification performance compared to batch SVM. By this result we can show that proposed realtime IDS has remarkable classification accuracy though it is worked by incremental way.

6 Conclusion and Remarks

In this paper, we present the realtime intrusion detection system. Applying artificial intelligence, machine learning and data mining techniques to intrusion detection system are increasing. But most of researches are focused on improving the performance of classifier. These classifiers are performed by batch way and it is not proper method for realtime IDS. Applying proposed system to KDD CUP 99 data, experimental result shows that it has remarkable performance in detection rate, computation time and memory requirement compared to off-line IDS. Our ongoing experiment is that applying proposed system to more realistic world data to evaluate the realtime detection performance.

References

1. Eskin, E. :Anomaly detection over noisy data using learned probability distribution. In Proceedings of the Seventeenth International Conference on Machine Learning (2000) 443-482
2. Ghosh, A. and Schwartzbard, A. :A Study in using neural networks for anomaly and misuse detection. In Proceedings of the Eighth USENIX Security Symposium, (1999) 443-482
3. Lee, W. Stolfo, S.J. and Mok, K.:A Data mining in workflow environments. :Experience in intrusion detection. In Proceedings of the 1999 Conference on Knowledge Discovery and Data Mining, (1999)
4. Tipping, M.E. and Bishop, C.M. :Mixtures of probabilistic principal component analysers. Neural Computation 11(2) (1998) 443-482
5. Kramer, M.A.:Nonlinear principal component analysis using autoassociative neural networks. AICHE Journal 37(2) (1991) 233-243
6. Diamantaras, K.I. and Kung, S.Y.:Principal Component Neural Networks: Theory and Applications. New York John Wiley & Sons, Inc (1996)
7. Kim, Byung Joo. Shim, Joo Yong. Hwang, Chang Ha. Kim, Il Kon.: On-line Feature Extraction Based on Emperical Feature Map. Foundations of Intelligent Systems, volume 2871 of Lecture Notes in Artificial Intelligence (2003) 440-444
8. Softky, W.S and Kammen, D.M.: Correlation in high dimensional or asymmetric data set: Hebbian neuronal processing. Neural Networks vol. 4, Nov. (1991) 337-348
9. Gupta, H., Agrawal, A.K., Pruthi, T., Shekhar, C., and Chellappa., R.:An Experimental Evaluation of Linear and Kernel-Based Methods for Face Recognition," accessible at http://citeseer.nj.nec.com.
10. Suykens, J.A.K. and Vandewalle, J.:Least squares support vector machine classifiers. Neural Processing Letters, vol.9, (1999) 293-300
11. Vapnik, V. N.:Statistical learning theory. John Wiley & Sons, New York (1998)
12. Hall, P. Marshall, D. and Martin, R.: On-line eigenalysis for classification. In British Machine Vision Conference, volume 1, September (1998) 286-295
13. Winkeler, J. Manjunath, B.S. and Chandrasekaran, S.:Subset selection for active object recognition. In CVPR, volume 2, IEEE Computer Society Press, June (1999) 511-516
14. Murakami, H. Kumar.,B.V.K.V.:Efficient calculation of primary images from a set of images. IEEE PAMI, 4(5) (1982) 511-515
15. Scholkopf, B. Smola, A. and Muller, K.R.:Nonlinear component analysis as a kernel eigenvalue problem. Neural Computation 10(5), (1998) 1299-1319

16. Tsuda, K.:Support vector classifier based on asymmetric kernel function. Proc. ESANN (1999)
17. Mika, S.:Kernel algorithms for nonlinear signal processing in feature spaces. Master's thesis, Technical University of Berlin, November (1998)
18. Accessable at http://kdd.ics.uci.edu/databases/kddcup99
19. Gestel, V. Suykens, T. J.A.K. Lanckriet, G. Lambrechts, De Moor, A. B. and Vandewalle, J.:A Bayesian Framework for Least Squares Support Vector Machine Classifiers. Internal Report 00-65, ESAT-SISTA, K.U. Leuven.
20. Suykens, J.A.K. and Vandewalle, J.:Multiclass Least Squares Support Vector Machines. In Proc. International Joint Conference on Neural Networks (IJCNN'99), Washington DC (1999)

SuffixMiner: Efficiently Mining Frequent Itemsets in Data Streams by Suffix-Forest[1]

Lifeng Jia, Chunguang Zhou, Zhe Wang, and Xiujuan Xu

College of Computer Science, Jilin University,
Key Laboratory of Symbol Computation and
Knowledge Engineering of the Ministry of Education, Changchun 130012, China
jia_lifeng@hotmail,.com cgzhou@jlu.edu.cn

Abstract. We proposed a new algorithm SuffixMiner which eliminates the requirement of multiple passes through the data when finding out all frequent itemsets in data streams, takes full advantage of the special property of suffix-tree to avoid generating candidate itemsets and traversing each suffix-tree during the itemset growth, and utilizes a new itemset growth method to mine all frequent itemsets in data streams. Experiment results show that the Suffix-Miner algorithm not only has an excellent scalability to mine frequent itemsets over data streams, but also outperforms Apriori and Fp-Growth algorithms.

1 Introduction

A data stream is a continuous, huge, fast changing, rapid, infinite sequence of data elements. The nature of streaming data makes the algorithm which only requires scanning the whole dataset once be devised to support aggregation queries on demand. In addition, this kind of algorithms usually owns a data structure far smaller than the size of whole dataset. As mentioned above, the single scan requirement of streaming data model conflicts with the demand of accurate result. Consequently, an estimation mechanism is proposed in the Lossy Counting algorithm [1] which is based on the well-known Apriori property [2] to harmonize such a conflict. Frequent itemsets in data streams are found when a maximum allowable error ε as well as a minimum support θ is given. The information about the previous mining result is maintained in the lattice which contains a set of entries of the form (e, f, Δ) where e is an itemset, f is the count of itemset e, and Δ is the maximum possible error count of itemsets e. For each entry (e, f, Δ) of an itemset e in lattice, if the itemset e is one of the itemsets identified by new transactions in the buffer, its previous count f is incremented by its count in new transactions. Subsequently, if the estimated count, $f+\Delta$, is less than $\varepsilon \cdot N$, such that N is the number of transactions, its entry is pruned from the lattice. On the other hand, when there is no entry in lattice for this new itemset e identified by the new transactions, a new entry (e, f, Δ) is inserted into the lattice.

---

[1] This work was supported by the Natural Science Foundation of China (Grant No. 60433020) and the Key Science-Technology Project of the National Education Ministry of China (Grant No. 02090).

Its maximum possible error ε is set to $\varepsilon \cdot N'$ where N' denotes the number of transactions that were processed up to the latest batch operation. Han et al. [3] developed a FP-tree-based algorithm, called FP-stream, to mine frequent itemsets at multiple time granularities by a novel titled-time windows technique.

2 SuffixMiner Algorithm

SuffixMiner is single-scan algorithm which utilizes the suffix-forest and is also batch-processed. As mentioned above, SuffixMiner chooses the suffix-forest to store the summary information of data streams. Now, we begin to explain our newly-devised algorithm step by step.

SuffixMiner Algorithm
Input: Minimal support threshold θ, Maximal estimated possible error ε.
Output: frequent itemsets in lattice.
   For each block of transactions in the data stream
   1. Construct the suffix-forest to store the summary information of streaming data;
   2. Generate frequent itemsets in the current batch of transactions from the suffix-forest by counter sequence and depth first itemset growth method.
   3. According to the estimation mechanism, insert the new frequent itemsets into the lattice or update the frequency of old frequent itemsets already in the lattice.
   4. Pruning infrequent itemsets from the lattice based on the Apriori property.

Before explaining how SuffixMiner to avoid traversing suffix-trees during the phase of itemset growth, let introduce an important property of suffix-tree first.

Theorem 1: In the suffix-tree(I_i), its sub-tree whose root is I_k (k>i) must be a sub-tree of suffix-tree(I_k).

Proof: In the suffix-tree(I_i), when constructing a branch of sub-tree whose root is I_k, SuffixMiner must deal with a certain suffix-set$\{I_i,...,I_k,...,I_n\}$. Meanwhile, the suffix-set $\{I_k,...,I_n\}$ must appear together with this suffix-set$\{I_i,...,I_k,...,I_n\}$. Thus, SuffixMiner will construct a new branch of suffix-tree(I_k) to store the suffix-set $\{I_k,...,I_n\}$, or insert it into an already existed branch in suffix-tree(I_k) by updating the frequency of nodes in that branch. Consequently, the theorem 1 is proofed.

   Based on theorem 1, we draw a conclusion: if a certain node(I_j) of suffix-tree(I_m) has a corresponding node(I_j) in the suffix-tree(I_k), this node(I_j) must have an ancestor node(I_k) in the suffix-tree(I_m). In order to recognize the corresponding identical nodes in different suffix-trees, we need to code all nodes of suffix-tree. SuffixMiner algorithm endows every node of suffix-tree the same serial number as that of it which exists in the corresponding complete suffix-tree. The nodes of complete suffix-tree are coded by the depth first. According to the method for coding the nodes of suffix-forest, in the suffix-tree(I_j), the serial number difference of node(I_n) and its ancestor node(I_j) is equal to the serial number of its corresponding node(I_n) in the suffix-tree(I_m). Consequently, we can use this relationship to identify whether two nodes in different suffix-trees are corresponding nodes.

Theorem 2: In an assumed suffix-tree(I_1) with i kinds of child nodes$\{I_2,...,I_{i+1}\}$, let $f(I_{i+1})$ and $c(I_{i+1})$ denote the frequency and the serial number of node(I_{i+1}) respectively. For each node(I_{i+1}) of suffix-tree(I_1), if it has a corresponding node(I_{i+1}) in suffix-tree(I_k), the serial number of this node(I_{i+1}) is stored in the serial number set $C_k(1<k<i+1)$. For all the nodes(I_{i+1}) of suffix-tree(I_1), let the serial number set $C = C_2 \cap ... \cap C_i$. if $C \neq \emptyset$, the frequency of (i+1)-itemset$\{I_1,...,I_{i+1}\} = \sum f(I_{i+1})$, such that $c(I_{i+1}) \in C$; Otherwise, the frequency of (i+1)-itemset$\{I_1,...,I_{i+1}\} = 0$.

Proof: In order to compute the frequency of (i+1)-itemset$\{I_1,...,I_{i+1}\}$, we need to know all the nodes(I_{i+1}) are the common child nodes of node(I_1),..., node(I_i) in the suffix-tree(I_1). According to the conclusion of theorem 1 and the statement of theorem 2, the set C_k stores serial numbers of nodes(I_{i+1}) which have nodes(I_k) as ancestor nodes. Given the set C, $C = C_2 \cap ... \cap C_i$. So, if there is the occurrence of nodes(I_{i+1}) which are the common child nodes of node(I_1),..., node(I_i), the serial numbers of these nodes(I_{i+1}) must be stored in the set C($C \neq \emptyset$). Subsequently, the frequency of (i+1)-itemset$\{I_1,...,I_{i+1}\}$ is equal to the sum of frequencies of nodes(I_{i+1}), i.e., $\sum f(I_{i+1})$, such that $c(I_{i+1}) \in C$. Otherwise, if the set C is empty($C = \emptyset$), the frequency of (i+1)-itemset$\{I_1,...,I_{i+1}\}$ must be equal to zero. Consequently, the theorem 2 is proofed.

Before detailing the counter sequence and depth first itemset growth method, we firstly describe two core operations of it.

Insert Itemset Growth(IIG): When SuffixMiner has already computed the frequency of i-itemset$\{I_1,...,I_m,...,I_k\}$(k>m>i), through the operation of insert itemset growth, SuffixMiner will compute the frequency of (i+1)-itemset$\{I_1,...,I_p,I_m,...,I_k\}$, such that I_p is newly inserted item (i<p<m<k).

Replace Itemset Growth(RIG): When SuffixMiner has already computed the frequency of i-itemset$\{I_1,...,I_m,...,I_k\}$(k>m>i), through the operation of replace itemset growth, SuffixMiner will compute the frequency of i-itemset$\{I_1,...,I_p,...,I_k\}$, such that I_p is newly inserted item to replace the item I_m (i<p<m<k).

We stipulate that the priority of IIG operation is higher than RIG operation. We also stipulate that the sequence of newly-inserted item by both IIG and RIG operations is according to the counter item sequence.

3 Experimental Evaluation

We use synthetic data streams, T5.I4.D1000K and T10.I4D1000K, to simulate the stream environment. The default value of minimum support threshold θ is 0.1%. The maximal estimated possible error threshold ε is a tenth of θ. Results indicate the excellent scalability of SuffixMiner algorithm. We also compare the performance of Apriori and FP-Growth algorithms with that of SuffixMiner algorithm. Since the dataset includes 100 items, so it simulates the environment of dense data stream, and the execution time of SuffixMiner algorithm does not vary a great extent when θ varies. However, SuffixMiner still outperforms both Apriori and Fp-Growth algorithms which can be downloaded in the web: *http://www.cs.umb.edu /~laur/ARtool/*.

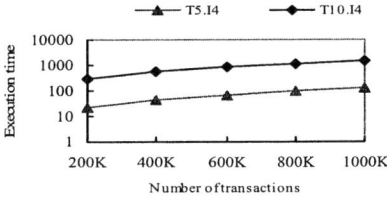

Fig. 1. The performance of SuffixMiner algorithm

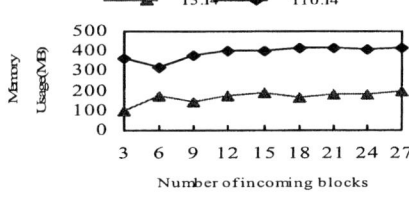

Fig. 2. The performance of SuffixMiner algorithm

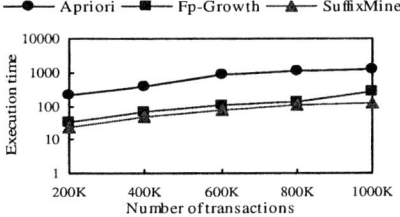

Fig. 3. The comparison of algorithms

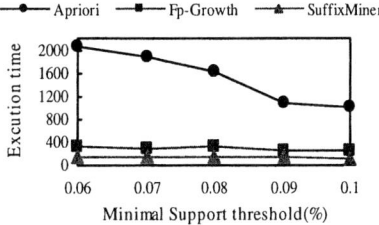

Fig. 4. The comparison of algorithms

4 Conclusion

We present a single-scan algorithm which utilizes the special property of suffix-trees to mine frequent itemsets while it is unnecessary to traverse the suffix-trees. Our algorithm also adopts a novel itemset growth method to avoid generating any candidate itemset. Experiments confirm the excellent abilities of SuffixMiner algorithm.

References

1. Manku, G. S. and Motwani, R.: Approximate Frequency Counts Over Data Streams. In Proceeding of the International Conference on Very Large Data Bases, Hong Kong, China (2002) 346-357
2. Agrawal, R. and Srikant, R.: Fast Algorithms for mining Association Rules. In Proceeding of the International Conference on Very Large Data Bases, Santiago de Chile, Chile (1994) 487-499
3. Giannella, C., Han, J., Pei, J., Yan, X. and Yu, P. S.: Mining Frequent Patterns in Data Streams at Multiple Time Granularities. Next Generation Data Mining, Chapter 3 (2002) 191-211

Improvement of Lee-Kim-Yoo's Remote User Authentication Scheme Using Smart Cards

Da-Zhi Sun and Zhen-Fu Cao

Department of Computer Science and Technology,
Shanghai Jiao Tong University, Shanghai 200030, P.R. China
{sundazhi, zfcao}@cs.sjtu.edu.cn
http://www.cs.sjtu.edu.cn

Abstract. Very recently, Lee, Kim, and Yoo proposed a modified remote user authentication scheme using smart cards [Computer Standards and Interfaces 27 (2) (2005) 181–183] to prevent the parallel session attack. In this paper, we demonstrate that Lee-Kim-Yoo's scheme is vulnerable to a reflection attack. Therefore, an improved scheme is proposed to resolve previous problems. We conclude that our improved scheme not only withstands the reflection attack and the parallel session attack, but also is more efficient compared with Lee-Kim-Yoo's scheme.

1 Introduction

In 2000, Sun [1] gave a remote user authentication scheme using smart cards. This scheme is very efficient because it only requires few hashing operations. In 2002, Chien, Jan, and Tseng [2] pointed out that Sun's scheme cannot achieve mutual authentication and cannot let users freely choose their passwords. An improved scheme, Chien-Jan-Tseng's scheme was therefore proposed. In addition, they provided that the following attributes are important for the remote user authentication scheme using smart cards: (1) *No verification table.* (2) *Human-memorable password.* (3) *Mutual authentication.* (4) *Low cost.* However, Hsu [3] showed that Chien-Jan-Tseng's scheme succumbs to the parallel session attack. Very recently, Lee, Kim, and Yoo [4] proposed a modified scheme to cope with this security flaw. They claimed that their scheme not only prevents a variety of attacks, but also maintains Chien-Jan-Tseng's desirable attributes.

Unfortunately, we find that Lee-Kim-Yoo's scheme is vulnerable to a reflection attack. That is, Lee-Kim-Yoo's scheme violates Chien-Jan-Tseng's desirable attribute of mutual authentication. Therefore, an improved scheme is proposed to overcome this severe weakness.

2 Review of Lee-Kim-Yoo's Scheme

Lee-Kim-Yoo's scheme consists of three phases: the registration phase, the login phase, and the verification phase.

2.1 Registration Phase

Let x be the secret key maintained by the system, and $h()$ be a secure cryptographic hash function. A remote user U_i and the system perform the following steps:

Step R1. U_i submits his identity ID_i and password PW_i to the system for registration in a secure channel.

Step R2. After verifying the qualification of U_i, the system computes $R_i=h(ID_i \oplus x) \oplus PW_i$, where $\oplus$ denotes the bit-wise exclusive-OR operator.

Step R3. The system stores $h()$ and R_i into the memory of the smart card, and issues the smart card to U_i.

2.2 Login Phase

When U_i wants to login to the system, he inserts his smart card into the terminal, and enters his ID_i and PW_i. The smart card then performs the following steps:

Step L1. Compute $C_1=R_i \oplus PW_i$.
Step L2. Compute $C_2=h(C_1 \oplus T_1)$, where T_1 is the current timestamp.
Step L3. Send the message (ID_i, T_1, C_2) to the system.

2.3 Verification Phase

After the authentication request message (ID_i, T_1, C_2) is received, the system and U_i execute the following steps:

Step V1. The system checks the validity of ID_i, and verifies $T_2-T_1 \leq \triangle T$, where T_2 is the current timestamp when receiving the request message and $\triangle T$ is the expected valid time interval for transmission delay.

Step V2. The system computes $C_1'=h(ID_i \oplus x)$, and verifies whether $C_2 ?=h(C_1' \oplus T_1)$. If not equal, then the system rejects U_i's request; otherwise, the system accepts U_i's request.

Step V3. The system computes $C_3=h(h(C_1' \oplus T_3))$, where T_3 is the current timestamp. The system sends the message (T_3, C_3) to U_i.

Step V4. Upon receiving the message (T_3, C_3), U_i verifies $T_4-T_3 \leq \triangle T$, where T_4 is the current timestamp when receiving the message. Then, U_i verifies whether $C_3 ?=h(h(C_1 \oplus T_3))$. If equal, U_i believes that the system is authenticated.

3 Reflection Attack on Lee-Kim-Yoo's Scheme

If an adversary has intercepted and blocked the message transmitting in the Step L3, i.e. (ID_i, T_1, C_2), he can impersonate the system to send the fabrication message $(T_3^a=T_1, C_3^a=h(C_2))$ to U_i in the Step V3. Note that the adversary can easily skip the Step V1 and Step V2. Also note that U_i will accept any timestamp from the system that is within a window around the time on U_i's local clock as long as the system has not used this particular time value before. As a result, U_i will be fooled into believing that the adversary is the system due to the fact that $C_3^a=h(C_2)=h(h(C_1 \oplus T_1))=h(h(C_1 \oplus T_3^a))$.

Since U_i cannot actually authenticate the system, Lee-Kim-Yoo's scheme fails to provide mutual authentication as its authors claimed. Such a weakness may result in serious problems in some electronic commerce applications, e.g. when a user makes an electronic payment via a smart card.

4 Our Improvement

4.1 Our Improved Scheme

We modify the verification phase of Lee-Kim-Yoo's scheme as follows:

After the authentication request message (ID_i, T_1, C_2) is received, the system and U_i execute the following steps:

Step V1. The system checks the validity of ID_i, and verifies $T_2-T_1 \leq \triangle T$, where T_2 is the current timestamp when receiving the request message and $\triangle T$ is the expected valid time interval for transmission delay.

Step V2. The system computes $C_1'=h(ID_i \oplus x)$, and verifies whether $C_2?=h(C_1' \oplus T_1)$. If not equal, then the system rejects U_i's request; otherwise, the system accepts U_i's request.

Step V3. The system computes $C_3=h(C_1' \odot T_3)$, where T_3 is the current timestamp and $\odot$ denotes the bit-wise exclusive-NOR operator. The system sends the message (T_3, C_3) to U_i.

Step V4. Upon receiving the message (T_3, C_3), U_i verifies $T_4-T_3 \leq \triangle T$, where T_4 is the current timestamp when receiving the message. Then, U_i verifies whether $C_3?=h(C_1 \odot T_3)$. If equal, U_i believes that the system is authenticated.

Other phases of our improved scheme are the same as those of Lee-Kim-Yoo's scheme.

4.2 Security Analysis

Now, we analyze the security of our improved scheme as follows:

To prevent the parallel session attack as described in [3], our improved scheme employs asymmetric structure to generate and verify C_2 and C_3. An adversary can obtain the valid message (T_3, C_3) during the Step V3, and impersonate U_i to send the fabrication login message $(ID_i, T_1^a=T_3, C_2^a=C_3)$ to the system in a new session of the Step L3. However, the system will terminate this new session in the Step V2 because $C_2^a=C_3=h(C_1' \odot T_3) \neq h(C_1' \oplus T_3)=h(C_1' \oplus T_1^a)$.

Unlike Lee-Kim-Yoo's scheme, our improved scheme can withstand the reflection attack. Even if an adversary has intercepted and blocked the valid message transmitting in the Step L3, i.e. (ID_i, T_1, C_2), he cannot impersonate the system to send the fabrication message $(T_3^a=T_1, C_3^a=C_2)$ to U_i in the Step V3 due to the fact that $C_3^a=C_2=h(C_1 \oplus T_1) \neq h(C_1 \odot T_1) = h(C_1 \odot T_3^a)$. Furthermore, the adversary has no way to derive another valid message to cheat U_i because of the one-way property of the secure cryptographic hash function.

Our improved scheme is a modification of Lee-Kim-Yoo's scheme. For this reason, the security analysis results in [4], except the point about the parallel session attack, also can be applied to our improved scheme directly.

4.3 Comparison

To be fair, we compare the implementation costs and security concerns of our improved scheme with those of two previous versions. Table 1 only lists the differences for simplicity. We can easily find that our improved scheme is as efficient as Chien-Jan-Tseng's scheme. But, our improved scheme can overcome the reflection attack and the parallel session attack.

Table 1. Comparsion among the similar remote user authentication schemes

Schemes	Computational cost (Smart card)	Computational cost (System)	Reflection attack	Parallel session attack
Our improved scheme	2 hash operations	3 hash operations	Secure	Secure
Lee-Kim-Yoo's scheme	3 hash operations	4 hash operations	Insecure	Secure
Chien-Jan-Tseng's scheme	2 hash operations	3 hash operations	Insecure	Insecure

5 Conclusions

Herein, we have showed that Lee-Kim-Yoo's scheme is vulnerable to a reflection attack. And then, we have proposed an improved scheme to remove this security flaw. According to the security and efficiency analysis, it is obvious that the improved scheme maintains Chien-Jan-Tseng's desirable attributes. Therefore, as a mutual authentication mechanism, our improved scheme is more suitable for real-life authentication applications than previous versions.

References

1. Sun, H.M.: An Efficient Remote User Authentication Scheme Using Smart Cards. IEEE Transactions on Consumer Electronics 46 (4) (2000) 958–961
2. Chien, H.Y., Jan, J.K., Tseng, Y.M.: An Efficient and Practical Solution to Remote Authentication: Smart Card. Computers and Security 21 (4) (2002) 372–375
3. Hsu, C.L.: Security of Chien et al.'s Remote User Authentication Scheme Using Smart Cards. Computer Standards and Interfaces 26 (3) (2004) 167–169
4. Lee, S.W., Kim, H.S., Yoo, K.Y.: Improvement of Chien et al.'s Remote User Authentication Scheme Using Smart Cards. Computer Standards and Interfaces 27 (2) (2005) 181–183

Grapheme-to-Phoneme Conversion Based on a Fast TBL Algorithm in Mandarin TTS Systems

Min Zheng[1], Qin Shi[2], Wei Zhang[2], and Lianhong Cai[1]

[1] Computer Science Department in Tsinghua University
100084, Beijing, China
`kristy99@mails.tsinghua.edu.cn, clh-dcs@tsinghua.edu.cn`
[2] IBM China Research Lab, 100085, Beijing, China
`{shiqin, zhangzw}@cn.ibm.com`

Abstract. Grapheme-to-phoneme (G2P) conversion is an important subcomponent in many speech processing systems. The difficulty in Chinese G2P conversion is to pick out one correct pronunciation from several candidates according to the context information such as part-of-speech, lexical words, length of the word, or position of the polyphone in a word or a sentence. By evaluating the distribution of polyphones in a large text corpus with correct pinyin transcriptions, this paper points out that correct G2P conversion for 78 key polyphones greatly decrease the overall error rate. This paper proposed a fast Transformation-based error-driven learning (TBL) algorithm to solve G2P conversion. The correct rates of polyphones, which originally have high accuracy or low accuracy, are both improved. After compared with Decision Tree algorithm, TBL algorithm shows better performance to solve the polyphone problem.

1 Introduction

The ability to predict the pronunciation of a written word accurately is an important sub-component within many speech processing systems. This task is typically accomplished through explicit pronunciation dictionaries or Grapheme-to-Phoneme (G2P) rule sets. In most of the alphabetic languages such as English, the main problem G2P module is to generate correct pronunciations for words that are out of vocabulary (OOV). Many approaches for letter-to- phoneme conversion have been proposed [1][2]. However, unlike the OOV problem, the difficulty in Chinese G2P conversion is to pick out one correct pronunciation from several candidates according to the context information such as Part-Of-Speech, lexical words or position of the polyphone in a word or sentence. Traditionally, the commonly used method in most Mandarin TTS systems is to list as many as possible the words with polyphonic characters and their pronunciations into a dictionary. But such dictionary can not solve all the problems about polyphones, summarizing pronunciation rules according to the context is needed to handle more complicated problem. Recently, various data-driven methods including neutral network[3], decision trees[4][5], pronunciation-by-analogy models[6] and extended stochastic complexity methods[7] have been tried to solve this G2P problem.

In this paper, the TBL algorithm is proposed to solve the G2P problem in mandarin TTS system. As an automatic rule learning methods, it is proved to be efficient for short distance prediction. TBL is widely used in numerous tasks, including learning rules for part-of-speech tagging (Brill, 1995)[8], prepositional phrase attachment (Yeh & Vilain, 1998)[9] and grammatical relation extraction (Ferro, Vilain, & Yeh, 1999)[10] etc; Now we leverage it to solve the polyphone problem and receive great improvements. Besides, a comparative experiment based-on decision tree is done using the same features and corpus that are used in TBL experiment. The decision tree is successfully dealt with some similar problems like parsing (David M. 1995)[11], prosody labeling (Xijun Ma, 2003)[12] and phrase break prediction (Byeong chang Kim, 2000)[13] etc. Compared the two results, the TBL algorithm is shown better performance to solve this polyphone problem.

The paper is organized as follows: The introduction of the TBL algorithm is explained in Section 2. In this section, a fast TBL algorithm for rules learning is also described. By using the algorithm, the training time of a transformation-based learner is speeded up without sacrificing performance. Section 3 shows the analysis and evaluation process of polyphones, including selection of polyphone candidates and corpus preparation. Section 4 describes the experiment, including features selection, template design and algorithm realization. Section 5 gives a comparative experiment based-on decision tree. Final conclusions are given in Section 6.

2 Transformation-Based Learning Algorithm

2.1 Introduction of TBL algorithm

Transformation-based learning (TBL) (Brill,1995) is one of the most successful rule-based machine learning algorithms. The central idea of transformation-based learning (TBL) is to learn an ordered list of rules which progressively improve upon the current state of the training set. The algorithm is illustrated in Figure 1.

1) *Initial State:* The training corpus is first standardized into initial state. This initial state is annotated with tagging which has been derived from statistical analysis of the training corpus or limited domain knowledge;
2) *Template:* It is composed of several features. The rules search space is limited by the templates. Templates describe valid rules and must describe properties that will reliably indicate when a rule is applicable.
3) *Rules Generating:* The error driven model uses each erroneous tag in the training corpus to propose a set of rules by template instantiation.
4) *Learning Process:* The TBL algorithm attempts to duplicate the training corpus from the initial state by iteratively learning rules which patch errors in the current state. In each iteration, shown highlighted in Figure 1, the highest scoring rule is appended to the learnt sequence and applied to the current state. Iterations continue until no more improvement can be made.

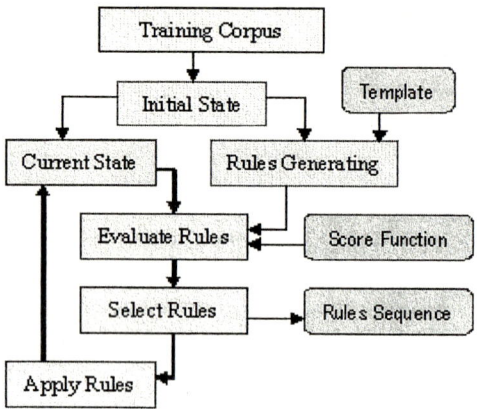

Fig. 1. The Transformation algorithm process

The following definitions are used in the whole paper:

To each transformation "[A->B]" in the corpus: $C(s) = A$, $T(s) = B$; $(C(s)$、$T(s)$: origin and target pronunciation of the polyphone;)

In learning process, it is defined an objective function f for iteration.

$$good(r) = |\{s \mid C(s) = C(r(s)) \,\&\, T(s) = T(r(s))\}| \quad (*)$$
$$bad(r) = |\{s \mid C(s) = C(r(s)) \,\&\, T(s) \,!= T(r(s))\}| \quad (**)$$
$$f(r) = good(r) - bad(r) \quad (***)$$

In each rule r, it will satisfy $C(r(s)) \,!= T(r(s))$.

2.2 Fast TBL Algorithm

One of the most time-consuming steps in TBL algorithm is the updating step. The rules learning process is a greedy search. The iterative nature of the algorithm requires that each newly selected rule should be applied to the whole corpus, then the current state of the corpus and f(r) for remaining rules are updated before the next rule is learned. For example, it almost needs 15min to select one rule for polyphone "为" which means it will take 900min to learn all the rules.

The central idea of this optimizing algorithm is not updating f(r) of all the rules each time. The following analysis will be used in this section:

- Rule r can be generated by the template:Set: $G(r,s) = \{r \mid P(r,s) = true\}$
- Given a newly learned rule b that is to be applied to the corpus S and identify every rule r of which the f(r) has changed: Obviously, if the polyphone sample s hasn't effected when applying rule b, the f(r) of the rules in the set G(r,s') will not change. So I just need to calculate f(r) for the rules in the set G(r,s) that s has effected by applying rule b;
- As polyphone sample s has been effected by the rule b, so $C(s) = C(b(s))$ (1)
  Since to every rule r in the G(r,s), $\quad C(r(s)) \,!= T(r(s))$, (2)
  $\Rightarrow \quad C(s) = C(b(s)) \,!= T(b(s))$; (3)
  After applying rule b, $\quad C'(s) = T(b(s)), T'(s) = T(s)$; (4)

Using equation (*), (**) and (1), (2), (3), (4) to the following four cases:

- **Case I**: $T(s) = C(s)$ & $C(r(s)) = C(s)$:
  Before applying rule b:
    $C(r(s)) = C(s) \Rightarrow C(r(s)) = C(s)$
    $T(r(s)) \mathrel{!=} C(r(s)) = C(s) = T(s) \Rightarrow T(r(s)) \mathrel{!=} T(s)$
  After applying rule b:
    $C(r(s)) = C(s) = C(b(s)) \mathrel{!=} T(b(s)) = C'(s) \Rightarrow C(r(s)) \mathrel{!=} C'(s)$
    $T(r(s)) \mathrel{!=} C(r(s)) = C(s) = T(s) = T'(s) \Rightarrow T(r(s)) \mathrel{!=} T'(s)$
  According to the equation (**) => bad(r) -1;

- **Case II**: $T(s) = C(s)$ & $C(r(s)) \mathrel{!=} C(s)$:
  Before applying rule b:
    $C(r(s)) \mathrel{!=} C(s) \Rightarrow C(r(s)) \mathrel{!=} C(s)$
    $T(r(s)) \mathrel{!=} C(r(s)) \mathrel{!=} C(s) = T(s) \Rightarrow T(r(s)) = T(s)$
  After applying rule b:
    $C(r(s)) \mathrel{!=} C(s) = C(b(s)) \mathrel{!=} T(b(s)) = C'(s) \Rightarrow C(r(s)) = C'(s)$
    $T(r(s)) \mathrel{!=} C(r(s)) \mathrel{!=} C(s) = T(s) = T'(s) \Rightarrow T(r(s)) = T'(s)$
  According to the equation (*) => good(r) +1;

- **Case III**: $T(s) \mathrel{!=} C(s)$ & $C(r(s)) = C(s)$:
  Before applying rule b:
    $C(r(s)) = C(s) \Rightarrow C(r(s)) = C'(s)$
    $T(r(s)) \mathrel{!=} C(r(s)) = C(s) \mathrel{!=} T(s) \Rightarrow T(r(s)) = T'(s)$
  After applying rule b:
    $C(r(s)) = C(s) = C(b(s)) \mathrel{!=} T(b(s)) = C'(s) \Rightarrow C(r(s)) \mathrel{!=} C'(s)$
    $T(r(s)) \mathrel{!=} C(r(s)) = C(s) = T(s) = T'(s) \Rightarrow T(r(s)) \mathrel{!=} T'(s)$
  According to the equation (*) => good(r) -1;

- **Case IV**: $T(s) \mathrel{!=} C(s)$ & $C(r(s)) \mathrel{!=} C(s)$:
  Before applying rule b:
    $C(r(s)) \mathrel{!=} C(s) \Rightarrow C(r(s)) \mathrel{!=} C'(s)$
    $T(r(s)) \mathrel{!=} C(r(s)) \mathrel{!=} C(s) \mathrel{!=} T(s) \Rightarrow T(r(s)) \mathrel{!=} T'(s)$
  After applying rule b:
    $C(r(s)) \mathrel{!=} C(s) = C(b(s)) \mathrel{!=} T(b(s)) = C'(s) \Rightarrow C(r(s)) = C'(s)$
    $T(r(s)) \mathrel{!=} C(r(s)) \mathrel{!=} C(s) \mathrel{!=} T(s) = T'(s) \Rightarrow T(r(s)) \mathrel{!=} T'(s)$
  According to the equation (*) => bad(r) +1;

3 Analysis of Polyphones Effectiveness

There are 12903 sentences in the corpus for polyphones analysis, including 271720 Chinese characters. It is randomly selected from newspapers, novels and oral talk. The corpus is manually checked with correct information of word segmentation, pos tagging (91 kinds) and phonetic notation by listening to the record speech and reading the text transcriptions.

3.1 Key Polyphones Selection

There are 682 polyphones in the Mandarin characters. But many of them have dominating pronunciations or rarely appear in usual articles, which can be solved by dictionary. It's unnecessary to generate rules for all 682 polyphones. The widely used polyphones, which also have high error rate, should be analyzed first. Three factors for selecting key polyphones are considered:

1) Discrepancy among the occurrence frequency of polyphones

There are 682 polyphones defined in the homograph dictionary, but the occurrence of polyphones in the corpus is quite different:

Table 1. Occurrence number of some polyphones

polyphone	Occurrence numbers in the corpus
一	2333
为	775
地	582
冠	38
铺	31

The homograph dictionary is sorted according to the usage frequency of polyphones in descending order. The results of the coverage ratio of the whole polyphones are shown in the following table 2:

Table 2. Coverage ratio of Polyphones

Number of the top polyphones	Coverage ratio of the whole polyphones
50	0.59628
100	0.78404
150	0.88235
200	0.93939
220	**0.95573**

From the table2, the cumulative frequency of the top-220 frequently used polyphones takes up more than 95% of all polyphones appearance. Obviously, it is important to generating pronunciation rules for these 220 polyphones first.

2) Accuracy rate

The pronunciations of some polyphones have already reached a very high accuracy in the current system, such as the right ratio of the pronunciation for the polyphone "会" has already reached 100%. Obviously, it has no use to generate pronunciation rules for these polyphones.

3) Dominating pronunciation rate

Many polyphones have dominating pronunciation. For example, the pronunciation "de0" for the polyphone "的" (de0, di2, di4) takes up 99% ratio, the pronunciation "le0" for the polyphone "了" (le0, liao3) also takes up 98% ratio and special cases could be handled by the homograph dictionary. However, other polyphones like "为" (wei2, wei4) and "长" (chang2, zhang3), who have no significant dominating pronunciations, are the key polyphones that should be processed carefully.

According to the above analysis, a list of key polyphones is generated from the top-220 ones according to the following two criteria:

1) The usage frequency of its dominating pronunciation is lower than 98%;
2) The right ratio of the original pronunciation is lower than 98%.

Only 78 polyphones are left in the list. The detailed statistic analysis results are shown in the following table3:

Table 3. Selected polyphones under different situations

Ratio of the dominating pronunciation	Right Ratio of the original pronunciation	Number of Selected polyphones
0.93	0.94	40
0.95	0.96	53
0.96	0.96	60
0.98	**0.98**	**78**
0.98	0.99	98

3.2 Polyphone Corpus Design

Although the corpus with corrected pinyin scripts is ready, whether it is suitable and enough for learning rules should be studied. Since TBL algorithm is based on error information, the effectiveness of the rules is related both with the size and the error ratio of the corpus. This is illustrated by two experiments on the polyphone "为" who occurs in 5000 sentences.

1) *First experiment*: Increasing the training set gradually from 1000 to 3000 sentences and testing it on 500 sentences disjointed from the training set.
2) *Second experiment*: Increasing the error ratio of the training set gradually from 5% to 20% and testing it on 500 sentences disjointed from the training set.

Table 4. Correct ratio with different corpus capacity

Number of sentences in training set	Error Ratio of the training set	Correct ratio of the testing set
1000	5%	68%
1000	10%	73%
2000	10%	73%
2000	20%	81%
2500	10%	75%
2500	**20%**	**86%**
3000	10%	78%
3000	20%	88%

From the result shown in table4, larger training corpus and higher error ratio will result in higher correct rate. Besides this, the error ratio is discovered has more effect than the size of the training corpus. According to the study, more sentences are collected. The following is the description of TBL training corpus

1) Corpus for rules generating: Averagely, 500 sentences for each polyphone, which have error pronunciations for polyphones from the rules learning corpus,
2) Corpus for rules learning: Averagely, 2500 sentences for each polyphone;
3) Corpus for rules testing: Averagely, 1500 sentences for each polyphone;

4 Experiments

4.1 Experiments Using TBL Algorithm

1) Feature Selection
- Lexical feature: *LC*(character),*LW*(lexical word)
- Syntactic feature: *POS;*
- Other features: *POSITION*(position of the polyphone in a word), *LEN*(length of lexical word), *BEGIN*(whether at the start of the sentence);
- Two special features for "一" and "不": *TONE*(tone of the nearing character);

2) Template Design
The template items are composed of several features such as "POS(Y,-1) & POS(Y,1) & LEN(n,0): A->B", in which "Y" indicates the feature value, the number as "-1" indicates the offset from the polyphone and the letter "A" and "B" indicate the original and standard pronunciation of the polyphone respectively. In the template, the offset is in the range of {-2, 2}. There are two kinds of template for the 78 polyphones:

- One template set is especially designed for "一" and "不": There are 13 templates including 5 features: *"TONE", "POSITION", "LEN", "POS"* and *"LC"*;
- Another template set is designed for other polyphones: 19 templates are designed including 6 features: *"POSITION", "LEN", "POS", "LC", "LW"* and *"BEGIN"*;

3) Rules Generating
Rules are generated by traversing all the incorrect samples according to the rules template. Such as the sentence "对 外(ng) 招牌(ng) 为 [wei4->Wei2](vi)：(w1) 北京(npr) 中医药(ng)" can produce a rule *"POS(ng,-1) & POS(w1,1) & LEN(1,0): wei2->wei4"* according to the template item: *"POS(Y,-1) & POS(Y,1) & LEN(n,0): A->B"*. Generally, it can produce average 9500-10000 rules for each polyphone.

4) Rules Learning
Design an objective function *f* for learning rules. Unlike in many other learning algorithms, the objective function for TBL will directly optimize the evaluation function. Now function f is the difference in performance resulting from applying the rule. Using the equation (*), (**) and (***), the iterative process is as follows:

- If (f(r)<0 ‖ good(r)<threshold) delete the rule;
- Select the rule b with largest f(r);
- Apply rule b to the whole corpus and update f(r) to all the other rules;
- Return 1) until the max f(r) beyond threshold.

5) Experimental Results

The experiment result of TBL for polyphones is shown in the Table5. From the result, the correct rates of polyphones G2P conversion which originally have high accuracy or low accuracy are all improved.

Table 5. Accuracy of some polyphones using TBL

Polyphone	Original accuracy	Accuracy with TBL	Original Runtime	Fast TBL Runtime
为	0.779476	0.901616	42934s	4885s
倒	0.731405	0.828512	20442s	1333s
教	0.836336	0.962462	7551s	557s
冠	0.389744	0.851282	5772s	222s
不	0.867704	0.954270	4434s	132s
朝	0.812963	0.914815	4280s	127s
着	0.944862	0.960526	21667s	1576s
应	0.970354	0.977585	9934s	646s
供	0.926209	0.936387	3900s	126s
还	0.987952	0.988956	3120s	98s

4.2 Comparative Experiment Using Decision Tree

Decision tree (DT) is a decision-making mechanism which assigns a probability to each possible choice based on the context: $P(f|h)$, where f is an element of the feature vocabulary (the set of choices) and h is a history (the context of the decision). This probability $P(f|h)$ is determined by asking a sequence of questions Q1 Q2 ... Qn about the context, where the ith question asked is uniquely determined by the answers to the previous i-1 questions.

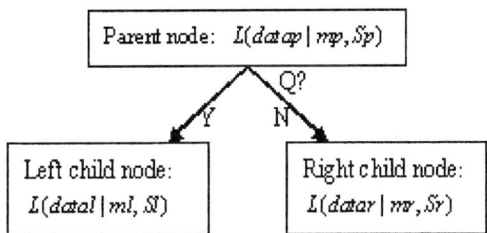

Fig. 2. Splitting node by evaluating question process

Each question asked by the decision tree is represented by a tree node and the possible answers to this question are associated with branches emanating from the node. The best question at a node is the question which maximizes the likelihood of the training data at that node after applying the question (shown in Fig.2).

Each node defines a probability distribution on the space of possible decisions. A node at which the decision tree stops asking questions is a leaf node. The leaf nodes represent the unique states in the decision-making problem, i.e. all contexts which lead to the same leaf node have the same probability distribution for the decision.

The experiment base on decision tree (DT) is used the same corpus and features, comparing with TBL with good initial states, the result is shown in the following table.

Table 6. Comparative results between TBL & DT

Polyphone	Train set	Test set	Rules	Origin accuracy	TBL	Decision Tree
长	2021	1002	109	0.973054	0.978346	0.892495
为	1816	916	102	0.779476	0.901616	0.842623
倒	972	484	117	0.731405	0.828512	0.807453
朝	1099	540	62	0.812963	0.914815	0.795796
重	298	148	39	0.831081	0.908072	0.746575

5 Conclusions

According to the results, the transformation-based error-driven algorithm is very effective for generating rules for polyphones. It improves the performances both for the polyphones which have low original accuracy (such as increases the correct rate to 97.2% from 77.9% for polyphone "为") and for the polyphones which already have high original accuracy (such as increases the correct rate to 97.0% from 97.8% for polyphone "应"). From the comparative experiment with Decision Tree, we analyze the differences and advantages of TBL algorithm:

- The TBL algorithm creates a relatively small number of rules that are linguistically motivated and understandable to both humans and machines;
- Each time the depth of the decision tree is increased, the average amount of training material available per node at that new depth is halved (for a binary tree). In TBL, the entire training corpus is used for finding all transformations, and therefore this method is more resistant to sparse data problems;
- Rules are ordered, with later rules being dependent upon the outcome of applying earlier rules. This allows for intermediate results in classifying one object to be available in classifying other objects; Even if decision trees are applied to a corpus in a left-to-right fashion, they only are allowed one pass in which to properly classify;
- Score function is more powerful than decision trees. In TBL: the objective function used in training is the same as that used for evaluation. In decision-tree: using system accuracy as an objective function for training typically results in poor performance and some measure of node purity (such as entropy reduction). The direct correlation between rules and performance improvement in TBL can make the learned rules more readily interpretable than decision tree rules.

Acknowledgements

Our work is supported by the National Natural Science Foundation of China under Grant No. 60275014 and the National High-Tech Research and Development Program of China under Grant No. 2002AA117010. Special Thanks to Liqin Shen offering me this great opportunity to do some research work in IBM CRL, also thanks to Xijun Ma, Yong Qin, Haiping Li, Yi Liu etc for giving me many useful advices and help.

References

1. Michel Divay, A. J. Vitale.: Algorithms for grapheme phoneme translation for English and French: Applications for database searches and speech synthesis. Computational Linguistics (1997), Vol. 23. 495-523
2. Davel M, Barnard E.: A default-and- refinement approach to pronunciation prediction. In Proceedings of the 15th Annual Symposium of the Pattern Recognition Association of South Africa 119-123
3. T.J. Sejnowski, C.R. Rosenberg.: Parallel networks that learn to pronounce english text. Complex systems (1987), vol. 1. 145–168
4. O. Andersen, R. Kuhn, A. Lazarides. (ed.): Comparison of Two Tree-Structured Approaches for Grapheme-to-Phoneme Conversion. Proceedings of the International Conference on Spoken Language Processing, Philadelphia, USA (1996) 1808-1811
5. Wern-Jun Wang, Shaw-Hwa Hwang, Sin-Horng Chen.: The broad study of homograph disambiguity for mandarin speech synthesis. Proceedings of the International Conference on Spoken Language Processing, Philadelphia, USA (1996), vol.3. 1389-1392
6. F.Yvon.: Grapheme-to-phoneme conversion using multiple unbounded overlapping chunks. In Proceedings of Conference on New Methods in Natural Language Processing (NeMLaP), Ankara, Turkey (1996) 218–228
7. Zi-Rong Zhang, Min Chu.: An efficient way to learn rules for grapheme-to-phoneme conversion in Chinese, Processing of International Symposium on Chinese Spoken Language(ISCSLP 2002) 59.
8. E.Brill.: Transformation-based error-driven learning and natural language processing: A case study in part of speech tagging, Computational Linguistics (1995), vol. 21(4) 543-565
9. Alexander S. Yeh, Marc B. Vilain.: Some Properties of Preposition and Subordinate Conjunction Attachments, 36th Annual Meeting of the Association for Computational Linguistics and 17th International Conference on Computational Linguistics, Montreal, Canada (1998) 1436-1442
10. Ferro, L., Vilain, M., Yeh, A.: Learning transformation rules to find grammatical relations, Workshop on Computational Natural Language Learning (GNLL-1999) 43-52
11. David M. Magerman.: Statistical decision-tree models for parsing, Proceedings of the 33rd Annual Meeting of the ACL (1995) 276 – 283
12. X.J. Ma, W. Zhang.: Automatic Prosody Labeling Using both Text and Acoustic Information, Proc. ICASSP, HongKong (2003)
13. Byeongchang Kim, Gary Geunbae Lee.: Decision-Tree based Error Correction for Statistical Phrase Break Prediction in Korean, Proceedings of the 17th conference on Computational linguistic (2000) 1051-1055

Clarity Ranking for Digital Images

Shutao Li and Guangsheng Chen

College of Electrical and Information Engineering, Hunan University,
Lushan Road, Changsha, 410082, P.R. China
Shutao_li@yahoo.com.cn

Abstract. In this paper, we use three focus measures for ranking digital images, namely, L_2 norm of image gradient, absolute central moment, and spatial frequency. The ranking scores from the focus measures are combined by the sum and product rules to result in the final decision. Experimental results on a photo album demonstrate the proposed method is very useful to users.

1 Introduction

Digital images have significantly increased popularity in recent years due the spread of various powerful but inexpensive digital imaging and storage devices such as digital cameras, reflective scanners, etc [1]. The need for efficient managing the images is emergent [2]. There are many researches on image quality, such as root of mean square error, peak signal to noise ratio, normalized cross-correlation, averaged difference, etc [3-4]. But most of them are based on computable distortion measures, where reference image or original image exists. To the author's knowledge, there is no publication on the image ranking system yet. In fact, this is a very important tool for efficient management of digital album. For example, using the ranking system, a user can delete digital images that receive low-quality ranking values to free up storage space and simplify image organization. Secondly, a user can also use the image ranking values to guide which images are likely to produce good quality prints.

In this paper, we use three focus measures for ranking digital images, namely, L_2 norm of image gradient, absolute central moment, and spatial frequency. The ranking scores from the focus measures are combined to give a final result using the sum and product rule [5].

2 Clarity Ranking Method

For an $M \times N$ image F, with the gray value at pixel position (m,n) denoted by $F(m,n)$, the following three focus measures are calculated.

1) L_2 norm of image gradient(L_2G)

$$L_2G = \sqrt{\sum_{m=1}^{M} \sum_{n=1}^{N} [(L_m * F(m,n))^2 + (L_n * F(m,n))^2]} \qquad (1)$$

where the masks

$$L_m = \begin{bmatrix} -1 & -2 & -1 \\ 0 & 0 & 0 \\ 1 & 2 & 1 \end{bmatrix}, L_n = \begin{bmatrix} -1 & 0 & 1 \\ -2 & 0 & 2 \\ -1 & 0 & 1 \end{bmatrix}.$$

2) Absolute central moment (ACM) [6]

$$ACM = \sum_{i=0}^{I-1} |i - u| p(i) \qquad (2)$$

where μ is the mean intensity value of the image, and i is the gray level.

3) Spatial frequency (SF)[7]

The spatial frequency is defined as

$$SF = \sqrt{RF^2 + CF^2} \qquad (3)$$

where RF and CF are the row frequency $RF = \sqrt{\dfrac{1}{MN} \sum_{m=1}^{M} \sum_{n=2}^{N} (F(m,n) - F(m,n-1))^2}$, and column frequency $CF = \sqrt{\dfrac{1}{MN} \sum_{n=1}^{N} \sum_{m=2}^{M} (F(m,n) - F(m-1,n))^2}$, respectively.

To a large set of digital images, we first change the color images to gray level images according the following transformation.

$$I = [R + G + B]/3 \qquad (4)$$

where R, G, and B are the red, green, and blue band of the color image, and I is the generated gray image.

Then we can compute the above three measures for each image, denoted by $s_{i,j}$, where $i=1,2,3$ is the measure number, $j=1,\ldots,N$, N is the number of the images. To put scores on same scale, all the measured values of each measure are normalized to the range of [0, 1]

$$s'_{i,j} = \frac{s_{i,j} - s_{i,\min}}{s_{i,\max} - s_{i,\min}}, i = 1,2,3, j = 1,\cdots,N \qquad (5)$$

where $s_{i,\min}$ $s_{i,\max}$ are the minimum and maximum values of the ith measure, respectively.

Then each image's three normalized scores are fused into the final decision. Two rules can be used, sum rule and product rule.

The sum rule is

$$SUM_j = \sum_{i=1}^{3} s_{i,j} \qquad j = 1,\cdots,N \qquad (6)$$

The product rule is

$$PROD_j = \prod_{i=1}^{3} s_{i,j} \qquad j=1,\cdots,N \qquad (7)$$

Finally we sort the SUM_j or $PROD_j$ to generate the clarity rank. The lowest value corresponds to the worst focus case, and the largest value corresponds to the best focus case.

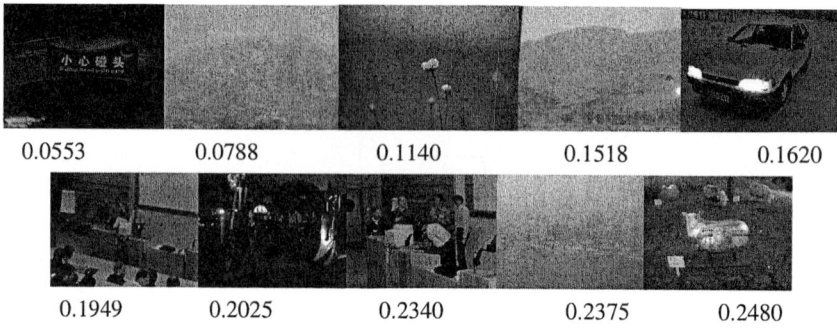

Fig. 1. The top 5% worst clarity images using SUM rule

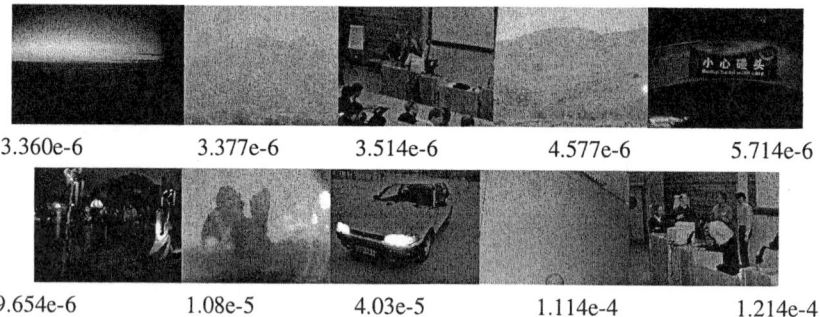

Fig. 2. The top 5% worst clarity images using PROD rule

3 Experimental Results

To test the ranking performance of the system, we use a digital album with 200 images. They are captured under various resolution, different illumination, etc. They contain different subjects, such as outdoor scene, indoor scene, single man, and crowd, etc. Using SUM and PROD fusion rule, the top 5% of the images with lowest ranking values in the album are shown in Fig.1 and Fig.2, respectively. The final scores of the fused focus measures are listed too.

All the images in Fig.1 and Fig.2 are out-of-focus in certain degrees. So the two rules are useful to the users who want to discard unwanted digital images.

4 Conclusions

In this paper, we use three focus measures for ranking digital images, namely, L_2 norm of image gradient, absolute central moment, and spatial frequency. The ranking scores from the focus measures are combined to give a final result. Experimental results on a photo album demonstrate the PROD rule has better performance than the SUM rule. The proposed method is very useful to aid the users. It can be used for selecting digital images for various purposes, such as managing, organizing and recommendations. More works are required for the development of more measures, such as color information, structure information to improve the ranking system.

Acknowledgement

This work is supported by the National Natural Science Foundation of China (No.60402024).

References

1. Rodden, K.: How Do People Organise Their Photographs? In: Proceedings of the Conference on Human Factors in Computing Systems. Florida (2003) 409–416
2. Platt, J.C., Czerwinski, M., Field, B.A.: PhotoTOC: Automatic Clustering for Browsing Personal Photographs. In: Proceedings of Fourth IEEE Pacific Rim Conference on Multimedia. Singapore (2003) 6–10
3. Al-Otum, H.M.: Qualitative and Quantitative Image Quality Assessment of Vector Quantization, JPEG, and JPEG2000 Compressed Images. J. Elec. Imaging 12 (2003) 511–521
4. Avcıbaş, İ., Sankur, B., Sayood, K.: Statistical Evaluation of Image Quality Measures. J. Elec. Imaging 11 (2002) 206–223
5. Kittler, J., Hatef, M., Duin, R.P.W., Matas J.: On Combining Classifiers. IEEE Trans. Patt. Anal. Machine Intell. 20 (1998) 226–239
6. Shirvaikar, M. V.: An Optimal Measure for Camera Focus and Exposure. In: Proceedings of 36th IEEE Southeastern Symposium on System Theory. Atlanta (2004) 472–475
7. Eskicioglu, A.M., Fisher, P.S.: Image Quantity Measures and Their Performance. IEEE Trans. Comm. 43 (1995) 2959–2965

Attribute Uncertainty in GIS Data

Shuliang Wang[1,2], Wenzhong Shi[3], Hanning Yuan[4], and Guoqing Chen[2]

[1] International School of Software, Wuhan University, Wuhan 430072, China
[2] School of Economics and Management, Tsinghua University, Beijing 100084, China
[3] Department of Land Surveying and Geo-Informatics,
The Hong Kong Polytechnic University, Hong Kong
[4] School of Remote Sensing Engineering, Wuhan University, Wuhan 430072, China
slwang2005@whu.edu.cn, wangshl@em.tsinghua.edu.cn

Abstract. Attribute uncertainty may support decision-making and measure the reliability of GIS analysis. This paper presents the attribute uncertainty in GIS data, with a complete perspective of concepts, sources, nature and applicable tools. And some valuable research contents are further given.

1 Introduction

The uncertainty is unavoidable in GIS data sets as the real spatial world is diverse and complex. The indicators of the uncertainty could be modeled for six aspects, i.e., lineage, position uncertainty, attribute uncertainty, logical consistencies, completeness, and temporal uncertainty. Thus, the attribute uncertainty is one of the essential contents of the uncertainty in GIS data. And it has implications for the resulting quality of organizational decision-making, executive behavior, managerial attitudes, and computerized GIS [1].

The uncertainty in GIS data has been studied in accordance with GIS data quality. In the context of GIS data quality, considerable researches and developments have been devoted to describing and assessing the elements, measurement, model, propagation, and cartographic portrayal of the spatial uncertainty. At the beginning, the mathematical solutions are largely hard computing, e.g., the probability theory [2], adjustment of observed data [3], spatial statistics [4], the evidence theory [5], error band, epsilon band, "S" band, and probability vector, all of which deal with random uncertainties. Then, soft computing theories were applied, e.g., fuzzy sets to handle fuzzy uncertainties [6], rough sets on incomplete uncertainties [7], neuron network [8], genetic algorithms [9] etc. When there exist more than one uncertainty at the same time, new techniques were further proposed, e.g. the cloud model integrating randomness and fuzziness [12]. Because most techniques on uncertainties describe some specific situations that may be far beyond the comprehension of the common users, those people without any enough background knowledge may have a difficulty in making sense of the exact nature of uncertainty that an expert specifies. Then, the uncertainties were visualized according to Bertin geographical parameters and their extensions, and virtual reality was also helpful [11]. Moreover, people have also tried to make their studying results available in software, e.g., GIS on the basis of either discrete object model, or continuous field model [12].

Although the attribute uncertainty in GIS data is recognized as an important feature of GIS, little research and discussion can be found on its formal definition, causing sources, external aspects, internal nature, and applicable methods. Without such issues, it will remain difficult to know with a high degree of the confidence, the quality of GIS attribute data and the quality of decisions made on the basis of the attribute data [8]. Furthermore, the attribute uncertainty in GIS data may cause a spatial decision-making to end up with no result or wrong result if they are not paid enough attention to and handled.

The attribute uncertainty in GIS data needs to be further studied. In contrast to the position uncertainty, attribute uncertainty has been paid relatively less attention to for a long time. When the uncertainty of GIS data was studied, data quality mainly focuses on the errors of position data while attribute data uncertainty was studied much later [1] [13]. This leads to that the uncertainty in GIS data mainly emphasizes position uncertainty on the basis of numeric statistics, e.g., root mean squared error (RMSE), variance, entropy, error band, Epsilon error band, and probability density function [14]. Also attribute uncertainty was isolated from position uncertainty, and was only used to describe some characteristics of a spatial entity. In fact, many applicable fields emphasizes particularly on attribute analysis, such as land assessment, soil chemistry, environment protection, agriculture and forestry. In these applicable fields, the quality of attribute data is more important than that of position data, and the impacts of attribute uncertainty are also much bigger than these of position attribute. At the same time, the uncertainty has been an important topic for several centuries. Numerous theories or methods were proposed to model the uncertainties, and even some claimed to be the only proper tools. However, generally speaking, they do not even define sufficiently or only in a very specific and limited sense what is meant by the uncertainty [15]. Most crisp traditional tools are problematic when they are used to handle the attribute uncertainty [8]. The existing tools on the uncertainty often model a specific type of uncertainty under the specific type of circumstances, e.g., the theory of fuzzy sets can only model the fuzzy uncertainty [6]. In this paper, we focus on a complete spectrum of the attribute uncertainty in GIS data.

2 Concepts and Characteristics

GIS represents how the spatial entities are in the real world, via binary digits in the form of zeros and ones to approach them. A spatial entity may be interpreted to the cases, states, processes, and phenomena in the real world, or the natural and artificial objects with geometric features of points, lines, areas, and volume. The attribute is drawn from the facts that can be known about a location. The valued attribute may be quantitative or qualitative [12], and attribute data are the qualitative or quantitative description of points, lines, areas, polygons, and cube with the forms of vector, raster and so on.

2.1 Attribute Uncertainty

An attribute uncertainty in GIS data is the attribute difference between a computerized entity in GIS and what it is in the real world under the objective intrinsic reality. It shows the unknown degree of the apparent features of the entity as recorded in spatial databases and also describes how close the entities described in GIS are compared with what is in the real world. The nature of attribute uncertainty in GIS is complex. Also many factors, e.g. scale, granularity and sampling, may affect the types of uncertainties, the sources of uncertainties, the degrees of uncertainties [3]. The objective properties of the attribute uncertainty are related to the world and data, and the subjective properties are linked to the human opinion on the true value of attributes, as derived from the available data [6] [11].

The attribute uncertainty further refers to the spatiotemporal differences between known attributes and attributes to know in terms of the spatial entity under the subjective informational description. The spatial differences indicate the spatial distribution of the attribute uncertainty, and the temporal differences are the temporal limits, within which the attributes of the entity are valid. Amongst the attribute uncertainty, geometric uncertainty measures how uncertain that the geometric description (point, line, area) depicts the real object is, topological uncertainty shows how doubtful that an object is compatible with other objects is, currency gives how dated that the dataset is under the required up-to-date demands is, and thematic uncertainty is defined by the classification uncertainty, attribute uncertainty, and temporal uncertainty (time point, time interval, time hierarchy). Sometimes, the attribute uncertainty is also taken as the thematic uncertainty in the image classification of remote sensing.

The attribute uncertainty may be discrete or continuous. The discrete attribute data of an object model are discrete variants, the uncertainty of which can be assessed via assessing a group of classification results. And the continuous attribute data of a field model are continuous stochastic variants, the uncertainty of which may be decided by measurement's errors [3]. When the entities in the real world are studied under the umbrella of GIS, the conceptual uncertainty focuses on the deviation of translating the entities into physical and cognitive models, the position uncertainty refers to the difference between their true locations and measured locations in the selected coordinate system, the attribute uncertainty highlights the differences between the true values of their attributes and their measured ones, and the metadata description refers to how the other types of uncertainty are known to users[3][15].

2.2 Attribute Uncertainty vs. Other Related Concepts

- Attribute uncertainty vs. position uncertainty. The position uncertainty highlights the position error associated with a data set, while the attribute uncertainty concerns the attribute error associated with a specific data set. The attribute uncertainty and position uncertainty are both tightly associated with each other. The fundamental atom of the spatial information is the tuple of the spatiotemporal location and a set of attributes, under the umbrella of which the uncertainties in GIS data may be divided into position uncertainty, attribute uncertainty, and their combinations [14]. It is further difficult to position the boundary between neighbor classifications when two objects belong to the same classification but with different spectrums, or two objects belong to different classifications but with the same spectrum. Both of the position uncertainty and attribute uncertainty should be considered, and it is the best approach to studying them together.

- Attribute uncertainty vs. data quality. GIS data quality internally measures the adherence of the specifications of data acquisition, externally indicates how well these specifications fit the current and potential needs of GIS users. An attribute uncertainty is one desirably essential indicator of GIS data quality (Morrison, 1995).
- Attribute uncertainty vs. data error. The attribute data error is the difference between an apparent value recorded in a database and its true value existed in the real world [3]. An attribute uncertainty may be taken as an extension of the attribute data error. But the attribute uncertainty is more complex and generic than attribute error, and it is truth plus distortion disturbed by various kinds of uncertainties. Besides the errors, attribute uncertainties further include randomness, fuzziness, incompleteness, inexactitude, noise, chaos, blunders, omittance, misinterpretation, misclassification, abnomalities and other possibilities [15]. Alternatively, an error can be viewed as a form of inherent uncertainty in some abstracted characteristics of the real world.

3 Sources and Causes

Seen from the abovementioned concepts, the attribute uncertainty pervades the GIS data with the lifecycle of capture, storage, update, transmission, access, archive, restore, deletion, and purge. It mainly arises from the complexity of the real world, the limitation of human recognition, the weakness of computerized machine, or the shortcomings of techniques and methods. In details, they may include instruments, environments, observers, projection's algorithms, slicing and dicing, coordinate system, image resolutions, spectral properties, temporal changes, etc. At the same time, their current limitations might further propagate even enlarge the uncertainty during the process of GIS analysis.

- Objective reality. The world is an infinitely complex system, which is large, changeable, nonlinear and multi-parameter. There are more inexact entities than the exact ones. And the inexact entities are with indeterminacy or inhomogeneity. The information of different entities may be overlapped, mixed, or deformed. Two entities of the same classification may eradiate different spectral information, while two entities that eradiate the same spectral information may belong to different classifications. Traditionally, it was presumed that the spatial world stored in a spatial database was crisply defined, precisely described and accurately measured in computerized GIS [1]. In many cases, there do not exist the pure points, lines, and polygons with geometric definitions [15]. Some true attribute values are even inexact or inaccessible.
- Subjective cognition. As the entity is complex and changeable in the real world, people have to select the most important spatial aspects to approximately approach the realistic entity. All attribute data are acquired with the aids of some theories, techniques and methods that specify implicitly or explicitly the required level of abstraction and generalization. Therefore, the depicted data are less than the total data for the attributes of the spatial entity, and only an essential part of the real variation is described. In consequence, the computerized entities may lose some aspects of the real entities, which make some uncertainties go along with spatial

databases. Furthermore, the unreliability is human opinion on the data sources, and the irrelevance concerns the opinion on data.
- Approximate techniques. The observer cannot perceive the attribute of uncertain spatial entities directly, but only after they have been filtered by the uncertainty theories. Based on the human cognition, the spatial entities in the real world are mapped to the crisp spatial objects in the computerized GIS via the given techniques for the formal modeling, reasoning and computing, e.g. errors adjustment, object model, probability theory and statistics, field model, spatial statistics, error band, epsilon band, "S" band, fuzzy set, rough set, decision theory, cloud theory etc [11]. Because most of them are deterministic while the entity is indeterminate, the techniques and theories are only an approximation when they are used to handle the attribute uncertainty.
- Computerized machine. GIS data in the computerized machine uncertainly reflects about the real world via binary digits in the form of zeros and ones. Some of the uncertainties may arise from the computerized machine, e.g., physical modeling, logical modeling, data encoding, data manipulation, data analysis, algorithms optimization, computerized machine precision, output. And it is a discrepancy between the encoded and actual value of a particular attribute. There is no real measurement with infinite precision, instead of a value with a degree of uncertainty. During the process of machine-based computing analysis, these uncertainties are accumulated and propagated. Moreover the computerized machine may further produce new uncertainties.
- Amalgamating heterogeneity. The attribute uncertainty becomes even more complex when merging different kinds of attribute data, often from different sources and of different reliabilities. When these various local databases are integrated together in the global context, the conflicts among various spatial databases may also cause unpredicted uncertainties, e.g., inconsistency across multiple databases.

4 Nature and Aspects

The aspects of the attribute uncertainty are wide diversity, and the presences of the interrelations between them are complex. Multiple attribute uncertainties often appear simultaneously in the real world. The external aspect shows a mixture of more than one type of uncertainty.

- Inaccuracy and imprecision. Inaccurate attribute data refer to the rate of incorrect values in the data, or the degree of unfitness between the model and the data. Imprecise attribute data are due to a finite representation of spatial entities and depends on the granularity. They both often appear at the same time. Inaccuracy is the difference between the estimate of the measurements over different data sets and the actually dimensional value of the parameters. Imprecision is the degree of spread or deviation between each measurement of the same part or feature. Precision does not mean accuracy. Estimates can be accurate but not precise, or precise but not accurate. A precise but inaccurate estimate of the parameter is usually biased, with the bias equal to the average distance from the real value.

- Inconsistency and conflict. Attribute data are inconsistent or incompatible if there is a conflict between correct and incorrect attribute data of a spatial entity present when several versions of the same object exist because of either different time snapshots, or datasets of different sources, or different abstraction levels. The conflict is either in the context or out of the context of GIS system and its applications. The conflict in the context is the main inconsistency of GIS attribute data, and it is also the objective to examine and distinguish the attributes with the same or related semantic information from the spatial database.
- Incompleteness and absence. Attribute incompleteness in GIS data comes from partial definition, wanting type, lacking value, and deficient model of attributes. The incompleteness can be measured in space, time or theme. It is referred to as how many possible attributes are included in the data set, and whether each decisive rule takes into account various alternatives and factors. The strategies for the removal of incompleteness may include, general issues, using default values, enhancing a data set by information from other data sets or prediction functions, accepting the evaluated incompleteness, relaxing the data definition to increase the degree of completeness, removing missing value or predicting it, and deciding not to perform the application.
- Repetitiveness and redundance. Repetitive attribute data are that there are more than one repeated data on the same spatial entity in a database or heterogenous databases, or two records with the same attribute values, or incomplete matching records are probably repeated because of errors and representations, spelling mistakes, different abbreviations. The repetitive GIS data often lead to the redundant records. The accumulation of repetitiveness will result in more redundance in attribute data when a decisive rule is generated. When testing the repetitiveness and redundance of attributes, it is not a simple arithmetic to predicate whether two values are equal or not, and a group of equivalent rules and uncertain matching techniques should be defined.
- Randomness, fuzziness and possibility. Randomness is the uncertainty included in a case with a clear definition, but not always happens every time, or the instabilities of the membership that an element belongs to a qualitative concept. It is a statistical uncertainty that indicates the likelihood of an event. Fuzziness is the uncertainty included in the case that has happened in the opposed and incomplete world, but cannot be defined exactly. Fuzziness is measured by the fuzzy membership value in the context of fuzzy sets that deal with the similarity of an element to a class. Possibility concerns the ability to occur (happen-ability) or the ability of a propensity to be true. It is the epistemic properties that reflect human opinions on the truth of a statement.
- Chaos and complexity. Chaos is the uncertainty in a complex large system composed of many cell-systems. A single cell may be a simple certain, but the system state shows uncertainties when lots of cells are coupled in a complex system. Chaos refers to the issue of whether or not it is possible to make accurate long-term predictions about the behavior of the system.

5 Applicable Modelling Tools

The attribute uncertainty is less likely to be eliminated even from the best research, but it can be described, sometimes quantified and estimated, and even manipulated in some principles. There are two main directions to develop techniques for controlling and reducing uncertainties in an acceptable degree. One is data acquisition that highlights the information acquired from the process of data collection and data amalgamation, whereas the other is data cognition that emphasizes the knowledge discovered from data extraction process and information generalization.

- Probability theory and statistics. The probability theory is the mathematical study of probabilities, which are taken as numbers in the interval from 0 to 1 assigned to events whose occurrence or failure to occur is random. They study the frequent possibility of many statistical experiments via random sampling. The probability theory and statistics may be used to analyze uncertainties caused by the spatial randomness, especially in applications relating to parameter estimation, hypothesis test, and system identification. Furthermore, the probability theory is extended or developed in the context of different necessities and backgrounds, e.g., Bayesian probability, evidence theory, probability vector, error matrix, error band, epsilon band, and "S" band. However, all kinds of attribute uncertainties are interpreted as randomness in the probability theory and statistic, and their algorithms are most hard computing for representing and manipulating the attribute uncertainty. These result in the limitations or shortcomings. It is worth studying further on the reliability of computer-described spatial objects, uncertainty accumulation; and sampling method of statistical test on uncertainties.
- Fuzzy sets and possibility theory. Fuzzy sets extend the crisp membership set {0,1} with only two values to a fuzzy membership interval [0, 1] with a series of values. \An element is assigned a series of membership values in relation to the respective subsets of the discourse universe since the concept of multiple and partial class membership is fundamental to fuzzy sets. The entity with mixed classification, indeterminate boundary, or gradual change may be described with more than one fuzzy membership values. The possibility theory is a formalism to represent and reason about ignorance or incompleteness. A possibility value of 0 indicates an impossible event, whereas 1 means the completely possible event. Any intermediate value denotes partial possibility. The possibility theory and fuzzy sets provide the principal formal systems explicitly devoted to the representation and manipulation of attribute uncertainty manifested as vagueness. The possibility theory is related to but independent of fuzzy sets, or probability theory.
- Rough sets and geo-rough space. Rough sets [7] specify the uncertainty from incompleteness by giving a pair of exact lower approximation and upper approximation. In the given universe of discourse, rough sets are incompleteness-based reasoning in the form of a decision-making table. Based on whether the statistical information is used or not, the existed rough set models may be grouped into such two major classes as algebraic and probabilistic models. Geo-rough space is a special case of rough sets in geo-spatial science. Spatial entities with indeterminate boundaries[1] may be taken as an embryonic form of rough set

application in geo-informatics. The true spatial entity is the lower approximation, and the spatial entity with a vague boundary is the upper approximation. However, rough sets only represent the uncertainties with the boundary set, while it cannot tell people how the uncertainties are in the boundary set [1].

- GIS data models. A GIS data model mainly includes an object model and a field model. When they are used to describe a spatial entity with spatial data, the object model is for a discrete entity with vector data, and the field model is for a continuous entity with raster data. The object model assumes that the spatial entities may be precisely described via points with the exactly known coordinates, lines linking a series of crisply known points, and areas bounded by sharply defined lines. The field model depicts spatial entities via giving each unit field an attribute value, instead of extracting objects or describing their topological relationships, and it is more suitable to fuzzy, ambiguous spatial entity. A field model is contrast to an object model. When it is used to study the attribute uncertainty, the field model has different characteristics from the object model. Many geographical phenomena, e.g., air pollution, population distribution, are studied as fields. The remote sensing images record the spectral information of spatial entity fields. When the images are used to investigate environments and resources, or produce thematic maps, in GIS, a field model is more suitable to discuss the attribute uncertainty than an object model.

- Handling multiple uncertainties. When multiple uncertainties appears, the classical mathematics gets rid of both randomness and fuzziness, probability and statistics consider only randomness without fuzziness, fuzzy sets consider fuzziness without randomness, and rough sets put randomness and fuzziness into the boundary without considering them in details. Obviously, all of them are inappropriate tools under the conditions with multiple uncertainties. Because they appear simultaneously, the existing uncertainties should be studied together in a mathematical model. The cloud model [12] integrates data with randomness and fuzziness in a unified way. A cloud model has three numerical characteristics, Expected value (Ex), Entropy (En) and Hyper-Entropy (He). In the discourse universe, Ex is the position corresponding to the center of the cloud gravity, whose elements are fully compatible with the spatial linguistic concept; En is a measure of the conceptual coverage, i.e. a measure of the spatial fuzziness, which indicates how many elements could be accepted to the spatial linguistic concept; and He is a measure of the dispersion on the cloud drops, which can also be considered as the entropy of En. Soft computing-based cloud rule is consistent to real data distribution and human thinking, and hard computing is the special case. So the cloud model may compensate the hard-computing deficiency of the probability theory and mathematical statistics, the inherent shortage of the membership function in fuzzy sets.

- Modeling propagation. The attribute uncertainty can be propagated during the process of GIS analysis [10]. The analysis of the error propagation is based on prior knowledge of errors in the sources, in the procedure or in the manipulation of the data. Sometimes little is known. When a GIS operation on the input attributes is a linear function, the error propagation problem is relatively easy. In that case the mean and variance of the output of a GIS operation can be directly and analytically derived. There are three alternative methods to analyze the propagation of the attribute uncertainty, i.e., Taylor series method, Monte Carlo method and sensitivity analysis. Taylor series method is to approximate the function by a linear function that is locally a good approximation of the function. Monte Carlo method uses an entirely different approach to analyze the propagation of error through the GIS operation. Sensitivity analysis can be of great value in acquiring meta-information when results of uncertainty analysis are difficult to be obtained.

6 Valuable Research Contents

The attribute uncertainty has quite a few issues worthy of further study. Hence, we think that the contents may include, for example,

- Characterize the complex of the real world. How does users validly establish the ambiguous definition in entity model, concept model, and physical model under the uncertainty situations? The boundary between neighboring spatial entities is indistinct, and it may be a changeable region, even two entities are mixed together with each other. The topological relationships are with ambiguous definition or semanteme.
- Mathematical model. How is an associated model set up to consider both the position uncertainty and the attribute uncertainty? How do people define the indices to describe the attribute uncertainty?
- Uncertainty in spatial analysis, soft computing, spatial query. How are the properties, accumulation and propagation of attribute uncertainty measured in GIS analysis? How is the spatial distribution of the attribute uncertainty described?
- Sampling test of attribute uncertainty. Reliability is often assessed under some unchangeable conditions, while is seldom under changeable conditions, and
- The influence of the attribute uncertainty on decision-making. What kinds of the attribute uncertainties would influence decisions mostly? How do people assess risks in decisions based on GIS data with attribute uncertainties? How do people measure, convey and describe the uncertainty in terms of the relevance to decision-making, further to determine, document and convey appropriate information about uncertainty, e.g., the spatiotemporal distribution of uncertainty? How are the concepts, metrics, and visualizations of the attribute uncertainty integrated with decision-making processes?

7 Conclusions

This paper presented the attribute uncertainty in GIS data, together with concepts, sources, nature and applicable tools. The study might support decision-making and measure the reliability of spatial analysis. The uncertainty is unavoidable in GIS data sets, and it can never be eliminated completely, even as a theoretical idea. The attribute uncertainty in GIS data is the spatiotemporal differences between known attributes and attributes to know in terms of the spatial entity. It arose from four mainstreams, the complexity of the real world, the limitation of human recognition, the weakness of computerized machine, the shortcomings of techniques and methods, and their amalgamation. The aspects of the attribute uncertainty were in a wide diversity. Different tools are appropriate to different applicable conditions. Given some conditions, certainty and uncertainty can transform each other.

Acknowledgements. This study is supported by National Natural Science Foundation of China (70231010), China Postdoctoral Science Foundation (2004035360), Wuhan University (216-276081), and the CRC scheme, Research Grant Council of The Hong Kong SAR (Project No.: 3_ZB40).

References

1. Burrough, P.A., Frank, A.U. (eds.): Geographic Objects with Indeterminate Boundaries Taylor and Francis, Basingstoke (1996)
2. Arthurs A. M.: Probability Theory. Dover Publications, London (1965)
3. Mikhail, E.M., Ackermann, F.: Observations and Least Squares. IEP-A Dun-Donnelley Publisher, New York (1976)
4. Cressie,N., Statistics for Spatial Data. John Wiley and Sons, New York (1991)
5. Shafer, G.: A Mathematical Theory of Evidence. Princeton University Press, Princeton (1976)
6. Zadeh L.A.: Fuzzy sets. Information and Control. . 8 (1965) 338-353
7. Pawlak, Z.: Rough Sets: Theoretical Aspects of Reasoning About Data. Kluwer Academic Publishers, London (1991)
8. 8 Shi, W.Z., Wang, S.L.: State of the art of research on the attribute uncertainty in GIS data. Journal of Image and Graphics. 6[A](2001) 918-924
9. Zhang, J.X., Goodchild, M.F. Uncertainty in Geographical Information. Taylor & Francis , London (2002)
10. Bonin,O.: Attribute uncertainty propagation in vector geographic information systems: sensitivity analysis. In Proceedings of the Tenth International Conference on Scientific and Statistical Database Management, edited by Kristine KELLY, IEEE Computer Society, Capri. (1998) 254-259
11. Kim, W. et al.: A Taxonomy of Dirty Data. Data Mining and Knowledge Discovery. 7, (2003) 81–99
12. Wang, S.L., et al.: A try for handling uncertainties in spatial data mining. Lecture Notes in Artificial Intelligence, Vol. 3215. Springer, Berlin (2004) 513-520
13. Lodwick, W.A. et al., Attribute error and sensitivity analysis of map operations in GIS: suitability analysis. International Journal of Geographical Information Systems, 4 (1990) 413-428
14. Shi, W.Z., Fisher P.F., Goodchild M. F. (eds.): Spatial Data Quality. Taylor & Francis London (2002)
15. Wang, S.L., et al.: Rough spatial interpretation. Lecture Notes in Artificial Intelligence, Vol. 3066. Springer, Berlin (2004) 435-444

Association Classification Based on Sample Weighting

Jin Zhang[1], Xiaoyun Chen[1,2], Yi Chen[1], and Yunfa Hu[1]

[1] Department of Computer & Information Technology, Fudan University,
200433 Shanghai, China
[2] Institute of Mathematics and Computer Science, Fuzhou University
350002 Fuzhou, China
c_xiaoyun@21cn.com

Abstract. In the territory of text categorization, the distribution and quality of sample set is highly influential to categorization result. Associated rule categorization ARC-BC is effective under common circumstances. The accuracy of categorization obviously falls as distribution of feature words of training samples is uneven. In this paper, a Chinese text classification approach was proposed based on sample weighting associated rules (SW-ARC). The approach improved substantial classification efficiency by performing self-adapting sample weights adjustment. Experiment result shows SW-ARC can solve the quality fall caused by uneven distribution of feature words. Macro-average recall of open test increases from 50% of ARC-BC to 70% of SW-ARC, Macro-average precision increases from 28% of ARC-BC to 70% of SW-ARC.

1 Introduction

Liu, Hsu and Ma firstly present the association classifying method CBA[2] in 1998. CBA integrates classification rule mining procedure and association rule mining procedure. Its classification result is better than decision tree method C4.5 based on rules. Thereafter various improvements for CBA have been developed, such as CMAR[1] and ARC[3]. The basic idea of these methods is to generate frequent feature words or frequent feature item sets of the training set using existing associated rule mining algorithm [5]. These frequent item sets are used to construct classification rules to classify new samples. The more the frequent feature set of a certain class contained in a test sample and the higher the confidence, the greater the probability that the test sample belongs to this class. Otherwise, the smaller the possibility of test samples' belongs to this particular class.

Existing text classification methods based on association rules have an obvious deficiency: the classification accuracy is influenced by the distribution of training samples. When distribution of number and frequency of feature words in training sample sets of various classes is uneven, so is the distribution of generated classification rules, classifier will prefer the class with more rules. In practice, distribution of frequency of features is often uneven among different classes. If the minimum support is set too high, it may be difficult to find sufficient rules for the infrequent class with low frequency of features. If the minimum support is set too low, too many useless or over-fit rules will be generated for the frequent classes with high frequency of features. In order to solve this problem, this paper proposes a text association categorization method (SW-ARC) based on association rules whose sample weight can be adjusted; accuracy of the method is higher than ARC-BC when it applied to classify uneven-distributed Chinese documents set.

2 Existent Problems and Experiment Analysis for ARC-BC

Stability of association classification algorithms is always a focus of research. In relative papers of association classification, they are effective on many famous test set. Especially ARC-BC[3] acquires encouraging result on famous text data set Reuters-21578. Its classification capability and efficiency are both better than usual other text classification methods. However, data sets in these papers are all even-distributed and English corpus. In order to test the classification capability and stability of the algorithm to Chinese text, 600 *Xinhua news* are downloaded from www.xinhuanet.com. These news documents are divided into 6 classes that is *oversea information* (y_1), *news for study abroad* (y_2), *entrance exam for college* (y_3), and *entrance exam for graduate* (y_4), *web-education* (y_5), *academic information* (y_6). In data preparing phase, feature items are extracted adopting N-gram technique and 60 best feature words with the highest χ^2 statistical values are chosen[8]. Then training samples are represented as feature vectors. Lab results show that ARC-BC algorithm isn't effective on the data set. Because all samples of the data set are come from sub-classes of education class, classes are very similar each others. Accuracies are also lower when other classification methods are used. Table 1 gives classification results of the data set adopting kNN [9], NB [9] and ARC-BC [3] respectively.

Table 1. The results of various classification methods

	kNN	Naïve Bayes(NB)	ARC-BC
Micro-average Recall (%)	65	41	25.4
Micro-average Precision (%)	65	42	48.3

The result of research shows that feature words and their frequencies are both uneven-distributed in the data set so that the distribution of classification rules is uneven too, that result to the accuracy of ARC-BC algorithms in the dataset is low. Table 2 shows the number of classification rules in various classes and classification results of closed tests. The number of samples that classified incorrectly from class y_i into class y_j is shown at ith row and jth column (the last row excepted) .Obviously, values at the diagonal are numbers of samples which are classified correctly. The 50 samples of class y_4 are all classified correctly and many samples of other classes are classified into class y_4 incorrectly, so classification ability of rule pointing to class y_4 is strongest. Contrarily, only 13 out of 43 samples are classified correctly in class y_1 and 2 other samples are classified into y_1 incorrectly. Only 18 out of 50 samples are classified correctly in class y_5, 3 other samples are classified into y_5 incorrectly. Only 20 out of 50 samples are classified correctly in class y_6, 3 other sample is classified into y_6 incorrectly. Therefore, the classification ability of rule set for y_1, y_5 and y_6 is weak. Although many other samples are classified into class y_2 and y_3 incorrectly, several samples of this both classes are classified into other incorrectly. So classification ability of rule set for y_2 and y_3 is weaker than y_4. Table 2 shows that number of rules in class y_4 is much more than other classes and the classification ability for samples in this class is also strongest, and y_1, y_5 and y_6 have fewer rules and classification ability for samples in these classes is weak. When rules of different classes are uneven-distributed severely, ARC-BC prefers the class with more rules obviously.

Table 2. The classification results of ARC-BC algorithms

	y_1	y_2	y_3	y_4	y_5	y_6
y_1	**13**	20	1	8	0	1
y_2	1	**44**	0	3	2	0
y_3	0	0	**52**	7	1	0
y_4	0	0	0	**50**	0	0
y_5	1	6	19	6	**18**	0
y_6	0	8	1	21	0	**20**
Number of rules	3	67	39	**107**	14	7

Association classification method ARC-BC uses all classification rules to build classifier, so classification accuracy is affected markedly by distribution of rules. Accuracy is high when rules are even-distributed among different classes. However, when numbers of feature words and word frequencies in various classes is uneven, the accuracy is low. If minimum support threshold is too high, it is difficult to find enough rules in the text subset whose feature words frequencies are commonly low. If minimum support threshold is too low, many useless or over-fit rules will be generated in the text subset whose feature word frequencies are high. So, appropriate rule set is difficult to be generated however minimum support threshold is set.

Numbers and confidences of rules are high in classes whose feature word frequencies are commonly high; numbers and confidences of rules are low in classes whose feature word frequencies are commonly low. Therefore, ARC-BC algorithm obviously prefers classes whose feature word frequencies are high, in other words, classes whose have many rules. These classes are called strong rule classes. Contrarily, classes whose feature word frequencies are low are called weak rule classes. They have fewer rules. Samples in weak rule classes are easy to be classified incorrectly into strong rule classes.

Boosting technique [4]is used here to solve this problem. The weights of training samples which are classified correctly are decreased to reduce numbers and confidences of rules in strong rule classes. The weights of training samples which are classified incorrectly are enhanced to increase numbers and confidences of rules in weak rule classes. Classification accuracy can be improved by performing operations above.

3 Weight Adjustment of Training Samples

3.1 Main Notations and Definitions

In order to describe conveniently, text classification hypothesis is described as follows:

$$h_t : X \times Y \to [0,1].$$

If sample x_i is classified into class y_i correctly, then $h_t(x_i, y_i) = 1$ or else $h_t(x_i) = y_i$; If sample x_i is classified incorrectly, then $h_t(x_i, y_i) = 0$ or else $h_t(x_i) \neq y_i$, the classification hypothesis is generated in the t th iteration.

Definition 1. *Sample Weight Vector* is defined as:

$$W_t = [w_t(1), w_t(2), \ldots, w_t(n)], \; w_t(i) \geq 0, \; \sum_{i=1}^{n} w_t(i) = 1,$$

where W_t is the training sample weight vector after the t-th iteration, $w_t(i)$ is a element of W_t, which is weight of x_i after the t-th iteration.

Initial value of Sample Weight Vector W_1 is $w_1(i) = 1/n$ (i=1,...,n), namely, all training samples have the same weights. New weights of samples are calculated according to result of training samples classification of last iteration. If the training sample is classified correctly by last classification rule set, then weight of the sample is multiplied by a factor less than 1 to decrease its weight. Contrarily, if the sample is classified incorrectly, then its weight is multiplied by a factor larger than 1 to increase its weight. Training sample weights are used to indicate the influence degree of this sample in corresponding classification rule set. There are t classification rule sets generated after t iterations.

3.2 Weight Adjustment of Samples

Weighted training error and *weight adjustment factor* are defined before introducing weight adjustment method.

Definition 2. For the classification hypothesis h_t generated by the tth classification rule set, *weighted training error* is defined as sum of weights of all training samples which are classified incorrectly.

$$\varepsilon_t(h_t) = \sum_{i: h_t(x_i) \neq y_i} w_t(i) \quad (1)$$

Definition 3. *Weight adjustment factor* is defined as:

$$\alpha_t = \frac{1}{2} \ln \frac{1 - \varepsilon_t}{\varepsilon_t} \quad (2)$$

Obviously, if weighted training error $\varepsilon_t < \frac{1}{2}$, $\alpha_t > 0$.

Weight adjustment method for training samples is as follows: weight of each sample which is classified correctly is multiplied by coefficient $e^{-\alpha_t}$, while weights of samples classified incorrectly are multiplied by coefficient e^{α_t}.

If weight adjustment factor α_t is positive, namely $\varepsilon_t < \frac{1}{2}$, $e^{-\alpha_t} < 1$. Weights of samples classified correctly are decreased in the process of adjustment so that its influence is reduced. Contrarily, weights of samples classified incorrectly are increased so that its influence is enhanced ($e^{\alpha_t} > 1$). When $\varepsilon_t \geq \frac{1}{2}$, namely weight adjustment factor $\alpha_t \leq 0$, the iteration process ends.

When the $(t+1)$th iteration starts, the sample weight coefficient $e^{-\alpha_t}$ and e^{α_t} is calculated according to weight adjustment factor α_t generated in last iteration, the detailed formula is as follows:

$$\delta_t(i) = \begin{cases} e^{-\alpha_t} & \text{if } h_t(x_i) = y_i \\ e^{\alpha_t} & \text{if } h_t(x_i) \neq y_i \end{cases} \tag{3}$$

After simplified:

$$\delta_t(i) = \begin{cases} \left(\dfrac{\varepsilon_t}{1-\varepsilon_t}\right)^{1/2} & \text{if } h_t(x_i) = y_i \\ \left(\dfrac{1-\varepsilon_t}{\varepsilon_t}\right)^{1/2} & \text{if } h_t(x_i) \neq y_i \end{cases} \tag{4}$$

After the $(t+1)$ th iteration, training sample weights are adjusted as:

$$W_{t+1}(i) = W_t(i) \times \delta_t(i) \tag{5}$$

In order to ensure that new weight W_{t+1} satisfies $\sum_{i=1}^{n} w_i^{t+1} = 1$, formula (8) is normalized to be:

$$W_{t+1}(i) = \frac{W_t(i)}{Z_t} \times \delta_t(i) \qquad \text{Where } Z^t = \sum_{i=1}^{n} w_i^t \times \delta_t(i) \tag{6}$$

3.3 Training Process of SW-ARC

Training step of associated classification based on sample weight adjustment is an iterative process (As Figure 1) . The detail of the process is as follows: (1) Firstly, associated rule mining algorithm is used to generate classification rules (element classifier). The training samples are classified using the classification rules, and weighted training error and weight adjustment factor are calculated according to classification result. (2) If sample is classified correctly, its weight is multiplied by a coefficient less than 1 to be decreased in the next iteration. Otherwise, its weight is multiplied by a coefficient larger than 1 to be increased. (3) Training samples whose weight are adjusted are used as training set and they are trained again to generate new classification rule set(element classifier). Sample weights are adjusted according to new classification result. Repeat these adjustment steps until one of these conditions is met: (a) the error rate is lower than the given thresholds by user; (b)the iterative time is more than given maximum iterative time; (c) Weighted training error $\varepsilon_t \geq 1/2$. By iterative adjustment, algorithm can focus on those samples which are difficult to be classified.

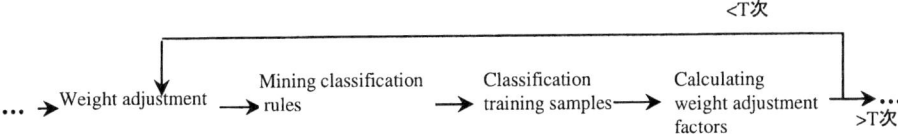

Fig. 1. The training process for SW-ARC algorithm

3.4 Final Classification Hypothesis

Classification hypotheses $h_1, h_2, ..., h_T$ are generated after T iterations. Therefore T classification results $v_1, v_2, ..., v_T$ will be generated after samples are classified using these T classification hypotheses. Final classification hypothesis of samples can be calculated using the maximum of each weighted classification hypothesis:

$$h_{fin}(x) = \arg\max_{y \in Y} \sum_{t=1}^{T} \alpha_t h_t(x, y) \qquad (7)$$

Formula (10) means that the class with the highest score in classification results is chosen as final classification result, Where weight of each classification result is *weight adjustment factor* α_t generated in the training phase. If classification accuracy of the classification hypotheses h_t is high, namely, weighted training error ε_t is small, then α_t is high and weight of h_t is high in final classification hypothesis h_{fin}. Otherwise, if weighted training error ε_t is high, then α_t is small and weight of h_t is low in final classification hypothesis h_{fin}.

4 Generating Classification Rule Based on Weighted Samples

Because samples of existing association classification methods [1, 2, 3] are not weighted, the association rule mining algorithm such as Apriori[5] or FP-tree[6] can be used directly to generate association rules. But in sample weighted association rule classification method SW-ARC, weights of training samples were different since the second iteration. So the original association rule mining algorithm must be modified to make it capable to handle the situation in which samples are weighted. The main changes are the definition of support and confidence measures, they are called weighted support and weighted confidence in SW-ARC.

Definition 4. *The weighted support* of classification rule $T \Rightarrow y_i$ is defined as the ratio of weight sum of training samples which contain item set T in class y_i and weight sum of samples in class y_i.

$$\varphi_w(T \Rightarrow y_i) = \frac{\sum_{\{i|x_i \in y_i \wedge T \in x_i\}} w(i)}{\sum_{\{i|x_i \in y_i\}} w(i)} \tag{8}$$

Definition 5. *The weighted confidence* of classification rule $T \Rightarrow y_i$ is defined as the ratio of weight sum of training samples which contain item set T in class y_i and weight sum of samples containing item T in entire training set.

$$\sigma_w(T \Rightarrow y_i) = \frac{\sum_{\{i|x_i \in y_i \wedge T \in x_i\}} w(i)}{\sum_{\{i|T \in x_i\}} w(i)} \tag{9}$$

Theorem 1. Weighted support $\varphi_w(T \Rightarrow y_i)$ is still closured downward, namely, given item set $T \subset T'$, if $\varphi_w(T \Rightarrow y_i) \leq \min\_\text{sup}$, then $\varphi_w(T' \Rightarrow y_i) \leq \min\_\text{sup}$.

Proof: $\because T' \supset T$, if $T' \in x_i$, $T \in x_i$,

$$\therefore \sum_{\{i|x_i \in y_i \wedge T' \in x_i\}} w(i) \leq \sum_{\{i|x_i \in y_i \wedge T \in x_i\}} w(i)$$

According to definition 7

$$\varphi_w(T' \Rightarrow y_i) = \frac{\sum_{\{i|x_i \in y_i \wedge T' \in x_i\}} w(i)}{\sum_{\{i|x_i \in y_i\}} w(i)} \leq \frac{\sum_{\{i|x_i \in y_i \wedge T \in x_i\}} w(i)}{\sum_{\{i|x_i \in y_i\}} w(i)} = \varphi_w(T \Rightarrow y_i) \leq \min\_\text{sup}$$

So, if item set T is not frequent in class y_i, neither is its superset T'. Thus, sample weighted association rule mining algorithm is similar to common association rule mining algorithm. There is only small difference in the calculated method for support and confidence.

5 Experiment and Result Analyses

Because Reuters-21578 is even-distributed English corpse, association classification ARC-BC acquires encouraging result on this text data set. However, the accuracy of ARC-BC over uneven-distributed Chinese corpse-*Xinhua News* is lower in our experiment (as section 2.4). In order to test the classification capability and stability of our algorithm to uneven-distributed corpse, we still adopt uneven-distributed data set from *Xinhua News* in section 2.4, minimum support threshold is still 10%. Table 3 shows the accuracy of training sample categorization in different turns. Each row represents the accuracy of various classes after each turn except the first row. The last column shows the accuracy of entire training set for each classification hypotheses. When t=1, values in columns $y_1 \sim y_6$ are also accuracies of association rule classification ARC-BC. Here, accuracy of class y_1 is only 0.3 while accuracy of class y_4 is 1. The difference is obvious. When t=1, final accuracy as low as 0.63. After the first weight adjustment (t=2), classification accuracy of training samples improves obviously and reach 0.83. Furthermore, accuracies of y_1, y_5 and y_6 originally low are improved obviously too.

Table 3. The Accuracy of Training Sample Categorization in Different Turns

	y_1	y_2	y_3	y_4	y_5	y_6	accuracy
t=1	0.3	0.88	0.87	1	0.36	0.4	**0.63**
t=2	0.6	0.9	0.9	0.9	0.8	0.9	**0.83**
t=3	0.9	0	0.5	0.4	1	0.9	**0.60**
t=4	0.7	1	1	1	0.2	0.7	**0.78**
t=5	0.8	0	0	0	1	0.9	**0.47**
t=6	0.8	0	0.2	0.2	1	1	**0.53**
t=7	0.6	0.9	0.8	1	0.7	0.6	**0.77**
t=8	0.9	0.9	0.8	0.2	0.8	0.9	**0.75**
t=9	0.6	0	0.8	1	0.7	0.5	**0.60**
t=10	1	1	0.5	1	0.8	1	**0.88**

Where accuracy $= \dfrac{|\{i \mid h(x_i) = y_i\}|}{|D_i|}$

Table 4. The Change of Parameters when t=1~10

	ε	α	$\delta+$	$\delta-$
t=1	0.349	0.312	0.732	1.366
t=2	0.225	0.618	0.539	1.856
t=3	0.323	0.371	0.69	1.4486
t=4	0.289	0.4485	0.639	1.566
t=5	0.42595	0.1492	0.86139	1.1609
t=6	0.40416	0.1941	0.8236	1.21419
t=7	0.33079	0.3523	0.707	1.414
t=8	0.3011	0.421	0.6564	1.5235
t=9	0.37654	0.252	0.7771	1.28676
t=10	0.04669	1.5082	0.2213	4.5186

Table 4 shows the change of variables in each turn of SW-ARC, ε is weighted training error, and α is weighted adjustment factor. During weight adjustment, the weights of training samples which are classified correctly are multiplied by $\delta+$, the weights of training samples which are classified incorrectly are multiplied by $\delta-$.

After sample weight adjustment, macro-average recall and macro-average precision of *training samples classification* are shown in Figure 2 and Figure 3. When iteration time T=1~10, macro-average recall and macro-average precision of *testing samples classification* are shown in figure 3 and figure 4. When T=1, SW-ARC has the same classification precision as ARC-BC.

Figure 2 to Figure 5 indicate that, results of both open test and closed test with weight adjustment are better than the results without weight adjustment. Especially recall and precision of closure tests reach 100% with iteration time increases(see fig.2 and fig.3). On the open test (see fig.4 and fig.5), the recall of SW-ARC is higher 20% than ARC-BC, and the precision of SW-ARC is higher 40% than ARC-BC.

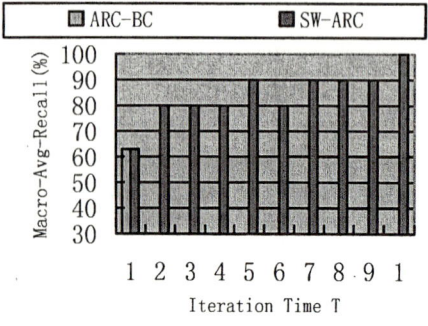

Fig. 2. Macro-Average-Recall on the Close Test When T=1~10

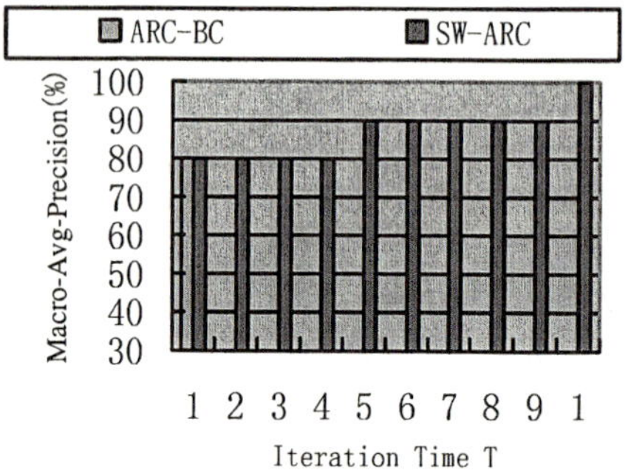

Fig. 3. Macro-Average-Precision on the Close Test When T=1~10

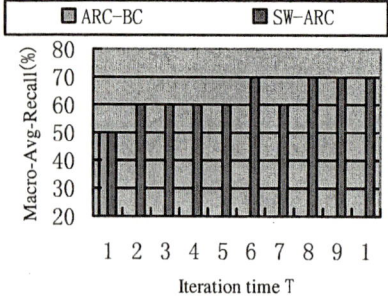
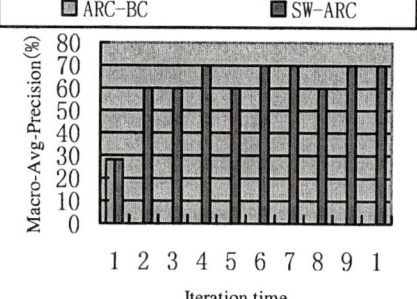

Fig. 4. Macro-Average-Recall on the Open Test When T=1~10

Fig. 5. Macro-Average-Precision on the Open Test When T=1~10

6 Conclusion

In practice application, feature words often are uneven-distributed in different classes. If minimum support threshold is set too high, it is difficult to find enough rules in the text set whose feature word frequencies are commonly low. If minimum support threshold is set too low, many useless or over-fit rules will be generated in the text set whose words frequencies are commonly high. In order to solve this problem, Boosting technique is used to improve the uneven distribution of sample feature words through self-adapt sample weight adjustment. Experiments over Chinese documents data set are carried out, which validate the effectiveness and efficiency of the proposed approach.

Acknowledgments

This work has been supported by the National Nature Science Foundation of China under Grant No.(60473070), Science Foundation of Fujian Province Education Bureau under Grant No.JB02069 and the Science and Technology Development Foundation of Fuzhou University in China under Grant No.2004-XQ-17.

References

1. Li, W., Han, J., Pei, J.: CMAR: Accurate and efficient classification based on multiple classification rules. In: Proceedings of the 2001 IEEE International Conference on Data Mining (ICDM). San Jose, California, US (2001)
2. Liu, B., Hsu, W., Ma, Y.: Integrating classification and association rule mining. In: Proceedings of the 4th ACM International Conference on Knowledge Discovery and Data Mining (SIGKDD). New York City, NY (1998) 80-86
3. Zaïane, O.R., Antonie, M.L.: Classifying text documents by associating terms with text categories. In: Proceedings of the 13th Australasian Database Conference (ADC). Melbourne, Australia (2002) 215-222
4. Freund, Y., Schapire, R.E.: Experiments with a New Boosting Algorithm. In: Proceedings of the 13th International Conference on Machine Learning. Bari, Italy (1996) 148-157
5. Agrawal, R., Srikant, R.: Fast algorithms for mining association rules. In: Proceedings of the 20th International Conference on Very Large Data Bases. Santiago, Chile (1994) 487-499
6. Han, J., Pei, J., Yin, Y.: Mining frequent patterns without candidate generation. In: Proceedings of the 2000 ACM SIGMOD International Conference on Management of Data. Dallas, TX (2000)
7. Yang, Y., Liu, X.: A re-examination of text categorization methods. In: Proceedings of the 22nd ACM Conference on Research and Development in Information Retrieval. Berkeley, CA (1999) 42-49
8. Yang, Y., Pedersen, J.O.: A comparative study on feature selection in text categorization. In: Proceedings of the 14th International Conference on Machine Learning (ICML). Nashville, US (1997)
9. Michell, T.M.: Machine Learning. China Machine Press (2003)
10. Yang, Y., Chute, C.G.: An example-based mapping method for text categorization and retrieval. In: ACM Transaction on Information Systems (TOIS), Vol. 12, No. 3. (1994) 252-277.
11. Wiener, E.: A neural network approach to topic spotting. In: Proceedings of the 4th Annual Symposium on Document Analysis and Information Retrieval (SDAIR). Las Vegas, US (1995)
12. Joachims, T.: Text categorization with support vector machines: Learning with many relevant features. In: Proceedings of the 10th European Conference on Machine Learning (ECML). Springer Verlag, Heidelberg, DE (1998)

Using Fuzzy Logic for Automatic Analysis of Astronomical Pipelines

Lior Shamir and Robert J. Nemiroff

Michigan Technological University, Houghton MI 49931, USA
lshamir@mtu.edu

Abstract. Fundamental astronomical questions on the composition of the universe, the abundance of Earth-like planets, and the cause of the brightest explosions in the universe are being attacked by robotic telescopes costing billions of dollars and returning vast pipelines of data. The success of these programs depends on the accuracy of automated real time processing of the astronomical images. In this paper the needs of modern astronomical pipelines are discussed in the light of fuzzy-logic based decision-making. Several specific fuzzy-logic algorithms have been develop for the first time for astronomical purposes, and tested with excellent results on data from the existing Night Sky Live sky survey.

1 Introduction

In the past few years, pipelines providing astronomical data have been becoming increasingly important. The wide use of robotic telescopes has provided significant discoveries, and sky survey projects are now considered among the premier projects in the field astronomy. In this paper we will concentrate on the ground based missions, although future space based missions like Kepler, SNAP and JWST will also create significant pipelines of astronomical data.

Pan-STARRS [2], a 60 million dollar venture, is being built today and completion is expected by 2006. Pan-STARRS will be composed of 4 large telescopes pointing simultaneously at the same region of the sky. With coverage of 6000 degrees^2 per night, Pan-STARRS will look for transients that include supernovas, planetary eclipses, and asteroids that might pose a future threat to Earth. Similarly but on a larger scale, ground-based LSST [6] is planned to use a powerful 8.4 meter robotic telescope that will cover the entire sky every 10 days. LSST will cost $200M, be completed by 2012, and produce 13 terabytes per night. In addition, many smaller scale robotic telescopes are being deployed and their number is growing rapidly.

However, in the modern age of increasing bandwidth, human identifications are many times impracticably slow. Therefore, efforts toward the automation of the analysis of astronomical pipelines have been gradually increasing. In this paper we present fuzzy logic based algorithms for two basic problems in astronomical pipeline processing which are rejecting cosmic ray hits and converting celestial coordinates to image coordinates.

2 Fuzzy Logic Based Coordinate Transformations

Useful automatic pipeline processing of astronomical images depends on accurate algorithmic decision making. For previously identified objects, one of the first steps in computer-based analysis of astronomical pictures is an association of each object with a known catalog entry. This necessary step enables such science as automatically detected transients and automated photometry of stars. Since computing the topocentric coordinates of a given known star at a given time is a simple task, transforming the celestial topocentric coordinates to image (x, y) coordinates might provide the expected location of any star in the frame. However, in an imperfect world of non-linear wide-angle optics, imperfect optics, inaccurately pointed telescopes, and defect-ridden cameras, accurate transformation of celestial coordinates to image coordinates is not always a trivial first step.

On a CCD image, pixel locations can be specified in either Cartesian or polar coordinates. Let $xzen$ be the x coordinate (in pixels) of the zenith in the image, and $yzen$ be the y coordinate of the zenith. In order to use polar coordinates, it is necessary to transform the topocentric celestial coordinates (Azimuth,Altitude) to a polar distance and angle from $(xzen, yzen)$.

2.1 The Fuzzy Logic Model

The fuzzy logic model is built based on manually identified reference stars. Each identified star contributes an azimuth and altitude (by basic astronomy) and also an angle and radial distance (by measurement from the image). These provide the raw data for constructing a mapping between the two using the fuzzy logic model. In order to transform celestial coordinates into image coordinates, we need to transform the azimuth and altitude of a given location to polar angle in the image from $(xzen, yzen)$ and the radial distance (in pixels) from $(xzen, yzen)$.

In order to obtain the coordinates transformation, we build a fuzzy logic model based on the reference stars. The model has two antecedent variables which are *altitude* and *azimuth*. The *azimuth* is fuzzified using pre-defined four fuzzy sets *North,East, South* and *West*, and each fuzzy set is associated with a Gaussian membership function. The *altitude* is fuzzified using fuzzy sets added to the model by the reference stars such that each fuzzy set is associated with a triangular membership function that reaches its maximum of unity at the reference value and intersects with the x-axis at the points of maximum of its neighboring reference stars.

The fuzzy rules are defined such that the antecedent part of each rule has two fuzzy sets (one for altitude and one for azimuth) and the consequent part has one crisp value which is the distance (in pixels) from $(xzen, yzen)$. The reasoning procedure is based on *product* inferencing and *weighted average* defuzzification, which is an efficient defuzzification method when the fuzzy logic model is built according to a set of singleton values [5].

2.2 Application to the *Night Sky Live* Sky Survey

The algorithm has been implemented for the *Night Sky Live!* [3] project, which deploys 10 all-sky CCD cameras called *CONCAM* at some of the world's premier observatories covering almost the entire night sky. The pictures are 1024 × 1024 FITS images, which is a standard format in astronomical imaging.

The algorithm allows practically 100 percent chance of accurate identification for NSL stars down to a magnitude of 5.6. This accurate identification allows systematic and continuous monitoring of bright star, and photometry measurements are constantly being recorded and stored in a database. The automatic association of PSFs to stars is also used for optical transient detection.

3 Cosmic Ray Hit Rejection Using Fuzzy Logic

The presence of cosmic ray hits in astronomical CCD frames is frequently considered as a disturbing effect. Except from their annoying presence, cosmic ray hits might be mistakenly detected as true astronomical sources. Algorithms that analyze astronomical frames must ignore the peaks caused by cosmic ray hits, yet without rejecting the peaks of the true astronomical sources.

3.1 A Human Perception-Based Fuzzy Logic Model

Cosmic ray hits in astronomical exposures are usually noticeably different then point spread functions of true astronomical sources. Cosmic ray hits are usually smaller than true PSFs, and their edges are usually sharper. An observer trying to manually detect cosmic ray hits would probably examine the edges and the surface size each peak. Since some of the cosmic ray hits have only one or two sharp edges, it is also necessary to examine the sharpest edge of the PSF. For instance, if the surface size of the peak is very small, it has sharp edges and the sharpest edge is extremely sharp, it would be classified as a cosmic ray hit. If the surface size of the peak is large and its edges are not sharp, it would be probably classified as a PSF of an astronomical source.

In order to model the intuition described above, we defined 3 antecedent fuzzy variables: the surface size of the PSF, the sharpness of the sharpest edge and the average sharpness of the edges. The consequent variable is the classification of the peak, and its domain is $\{1,0\}$. Since astronomical images typically contain 1 to 16 million pixels, the triangular membership functions are used for their low computational cost.

The fuzzy rules are defined using the membership functions of the antecedent variables and the domain of the consequent variable ($\{0,1\}$), and are based on the natural language rules of intuition. For instance, the rules of intuition described earlier in this section would be compiled into the fuzzy rules:

small $\wedge$ *sharp* $\wedge$ *extreme* $\longmapsto$ 1
large $\wedge$ *low* $\wedge$ *low* $\longmapsto$ 0

The computation process is based on *product* inferencing and *weighted average* defuzzification [5], and the value of the consequent variable is handled such that value greater than 0.5 is classified as a cosmic ray hits. Otherwise, the value is classified as a non-cosmic ray hits.

3.2 Using the Fuzzy Logic Model

The fuzzy logic model is used in order to classify peaks in the frame as cosmic ray hits or non-cosmic ray hits. In the presented method, searching for peaks in a FITS frame is performed by comparing the value of each pixel with the values of its 8 neighboring pixels. If the pixel is equal or brighter than *all* its neighboring pixels, it is considered as a center of a peak. After finding the peaks in the frame, the fuzzy logic model is applied to the peaks in order to classify them as cosmic ray hits or non-cosmic ray hits.

Measurements of the performance of the algorithm were taken using 24 *Night Sky Live* exposures. Each NSL frame contains an average of 6 noticeable cosmic ray hits brighter than 20σ, and around 1400 astronomical sources. Out of 158 cosmic ray hits that were tested, the algorithm did not reject 4, and mistakenly rejected 6 true astronomical sources out of a total of 31,251 PSFs. These numbers are favorably comparable to previously reported cosmic ray hit rejection algorithms, and the presented algorithm also has a clear advantage in terms of computational complexity.

4 Conclusion

The emerging field of robotic telescopes and autonomous sky surveys introduces a wide range of problems that require complex decision making. We presented solutions to two basic problems, which are star recognition and cosmic ray hit rejection. We showed that fuzzy logic modeling provides the infrastructure for complex decision making required for automatic analysis of astronomical frames, yet complies with the practical algorithmic complexity constraints introduced by the huge amounts of data generated by the astronomical pipelines.

References

1. Borucki, W. J.: Kepler Mission: a mission to find Earth-size planets in the habitable zone. Proc. of the Conf. on Towards Other Earths (2003) 69–81
2. Kaiser, N., Pan-STARRS: a wide-field optical survey telescope array. Proc. of the SPIE **54** (2004) 11–22
3. Nemiroff, R. J., Rafert, J. B.: Toward a continuous record of the sky. PASP **111** (1999) 886–897
4. Salzberg, S., Chandar, R., Ford, H., Murphy, S., K., & White, R., Decision trees for automated identification of cosmic-ray hits in Hubble Space Telescope images. PASP **107** (1995) 279
5. Takagi, T., Sugeno, M.: Fuzzy identification of systems and its applications to modeling and control. IEEE Trans. Syst. Man & Cybern. **20** (1985) 116–132
6. Tyson, J. A., Survey and other telescope technologies and discoveries. Proc. of the SPIE, **48** (2002) 10–20

On the On-line Learning Algorithms for EEG Signal Classification in Brain Computer Interfaces

Shiliang Sun[1], Changshui Zhang[1], and Naijiang Lu[2]

[1] State Key Laboratory of Intelligent Technology and Systems,
Department of Automation, Tsinghua University, Beijing, China, 100084
sunsl02@mails.tsinghua.edu.cn, zcs@mail.tsinghua.edu.cn
[2] Shanghai Cogent Biometrics Identification Technology Co. Ltd., China
lunj@cbitech.com

Abstract. The on-line update of classifiers is an important concern for categorizing the time-varying neurophysiological signals used in brain computer interfaces, e.g. classification of electroencephalographic (EEG) signals. However, up to the present there is not much work dealing with this issue. In this paper, we propose to use the idea of gradient decorrelation to develop the existent basic Least Mean Square (LMS) algorithm for the on-line learning of Bayesian classifiers employed in brain computer interfaces. Under the framework of Gaussian mixture model, we give the detailed representation of Decorrelated Least Mean Square (DLMS) algorithm for updating Bayesian classifiers. Experimental results of off-line analysis for classification of real EEG signals show the superiority of the on-line Bayesian classifier using DLMS algorithm to that using LMS algorithm.

1 Introduction

In the past few years, the research of brain computer interfaces (BCIs), which could enhance our perceptual, motor and cognitive capabilities by revolutionizing the way we use computers and interact with ambient environments, has made significant developments. BCIs can help people with disabilities to improve their quality of life, such as simple communication, operation of artificial limb, and environmental control. Besides, BCIs are also promising to replace humans to control robots that function in dangerous or inhospitable situations (e.g., underwater or in extreme heat or cold) [1][2][3][4]. In this paper, we focus on the research of on-line learning algorithms for BCI applications.

Among the information carriers for BCI utilities, such as electroencephalography (EEG), magnetoencephalography (MEG), functional magnetic resonance imaging (fMRI), optical imaging, and positron emission tomography (PET), EEG is a relatively inexpensive and convenient means to monitor brain's activities. The BCIs using EEG as carriers are often called EEG-based BCIs, which are the object of our current research. Although EEG signals have the above merits, there also exist some disadvantages, e.g. low signal-noise-ratio (SNR) and the time-varying characteristic. As a representation, EEG signals often change due to subject fatigue and attention, due to ambient noises, and with the process of user training [3]. At the present time most BCIs usually require a boring calibration measurement for training classifiers before BCI applications. To make it worst, the time-varying characteristic of electrophysiological signals (e.g. EEG signals) inherently necessitates retraining for BCI utility after a long break, which becomes a big hindrance to the progress of BCI.

To solve the tedious problem caused by retraining, there are usually two kinds of approaches. The first one is model switching, which demands a large amount of information to guide the model selection procedure. The other approach is model tracking, a more practical approach for use when no sufficient information is available [3]. The on-line learning of classifiers studied in this paper belongs to the second category. About the matter of on-line classifier update, Millán et al. have proposed to use Least Mean Square (LMS) algorithm to deal with it recently [5][6][7]. They make an approximation about the gradient of mean value in Gaussian probability density function and a diagonal assumption about the form of covariance matrix. Our current work is an evolution of their LMS algorithm. Based on the Bayesian classifier of Gaussian mixture model, we propose to use Decorrelated Least Mean Square (DLMS) algorithm to develop the LMS algorithm [8]. Besides, we don't make any simplicity about the gradient of mean value and the form of covariance matrix in Gaussian probability density functions. In this case, our approach is much closer to the natural scenario of data distribution. Experimental results with real-world data indicate that the DLMS algorithm is superior to the existed LMS algorithm for on-line EEG signal classification.

The remainder of this article is organized as follows. After introducing the Bayesian classifier of Gaussian mixture model in section 2, we give the formulation of DLMS algorithm in section 3. Section 4 reports experimental results for the problem of on-line EEG signal classification on the real data of BCI applications. Finally, we conclude the paper in section 5.

2   Bayesian Classifier with Gaussian Mixture Model

Although there are some methods existed in the literature, e.g. support vector machine (SVM), Fisher discriminant analysis (FDA), artificial neural network (ANN), which have alleviated the problem of EEG signal classification to some extent, they can't be used for on-line EEG signal classification simply [9][10][11][12]. Throughout this paper, Bayesian classifier is adopted as the prototype of on-line learning for the multiclass categorization issue, as Millán et al. suggested [5][6][7]. In this section, we will first describe the Bayesian classifier with Gaussian mixture model in a systematic way, and then give the optimization object function for on-line learning.

Assume there is a training set comprising N instances which come from K categories, and each class denoted by C_k has prior $P(C_k)$, $(k = 1, ..., K)$, s.t.,

$\sum_{k=1}^{K} P(C_k) = 1$. Under the framework of finite Gaussian mixture model, the conditional probability density function of each category can be approximated through a weighted combination of N_k Gaussian distributions [13][14], i.e.,

$$p(x|C_k) \cong \sum_{i=1}^{N_k} a_k^i G(x|\mu_k^i, \Sigma_k^i), \text{ s.t., } \sum_{i=1}^{N_k} a_k^i = 1 \quad (1)$$

where $G(x|\mu_k^i, \Sigma_k^i)$ is a Gaussian probability density function with mean μ_k^i and covariance Σ_k^i.

According to Bayesian theorem [14], the posterior probability of instance x belonging to class C_k can be given as

$$P(C_k|x) \\ = \frac{P(C_k)p(x|C_k)}{p(x)} \\ = \frac{P(C_k)\sum_{i=1}^{N_k} a_k^i G(x|\mu_k^i, \Sigma_k^i)}{\sum_{j=1}^{K} P(C_j) \sum_{i=1}^{N_j} a_j^i G(x|\mu_j^i, \Sigma_j^i)}. \quad (2)$$

Now we represent the N instances as $\{x_n, y_n\}(n = 1, ..., N)$, where x_n is the feature vector of the nth instance, and y_n is the corresponding label. If $x_n \in C_k$, then y_n has the form of e_K^k, that is, $y_n \doteq e_K^k = \left[0, \ldots, 1_{(k)}, \ldots, 0\right]_{(K)}^T$. Denote $\hat{y}_n$ as the outcome of our Bayesian classifier, i.e.,

$$\hat{y}_n \doteq \left[P(C_1|x_n), P(C_2|x_n), \ldots, P(C_K|x_n)\right]^T.$$

Consequently, under the criterion of least mean square (LMS), the optimization object function for classifying instance x_n becomes

$$\min J(\Theta) \doteq \min E\{\|e_n\|^2\} = \min E\{\|y_n - \hat{y}_n\|^2\} \quad (3)$$

where variable Θ represents any of the parameters $N_k, a_k^i, \mu_k^i, \Sigma_k^i$ in (1). To make our later analysis feasible, we only presume here that parameters N_k, a_k^i are given or obtained from previous training data, while parameters μ_k^i, Σ_k^i would have the most general form (μ_k^i is a general column vector, and Σ_k^i is a symmetric and positive definite matrix) and would be updated through on-line learning.

3 Decorrelated LMS (DLMS) Algorithm for Bayesian Classifier

The Decorrelated LMS (DLMS) algorithm is an improvement of the basic LMS algorithm [8]. Therefore, we start this section with the LMS algorithm. And finally we will present the flow chart of DLMS algorithm for the on-line learning of Baysian classifier.

When using LMS algorithm to solve the problem of (3) for on-line learning, one need first derive the formulation of instantaneous gradient (stochastic gradient) $\nabla_\Theta \|y_n - \hat{y}_n\|^2$. In this paper, Θ just refers to μ_k^i. That is, each time we update μ_k^i via gradient algorithms, but update Σ_k^i using the training data and the update result of μ_k^i.

Because $\|y_n - \hat{y}_n\|^2$ can be rewritten as follows

$$\begin{aligned}
&\|y_n - \hat{y}_n\|^2 \\
&= (y_n - \hat{y}_n)^T(y_n - \hat{y}_n) \\
&= y_n^T y_n - 2y_n^T \hat{y}_n + \hat{y}_n^T \hat{y}_n \\
&= y_n^T y_n - 2\sum_{i=1}^{K} y_n^i P(C_i|x_n) + \sum_{j=1}^{K}(P(C_j|x_n))^2 \\
&= y_n^T y_n + \sum_{j=1}^{K}[(P(C_j|x_n))^2 - 2y_n^j P(C_j|x_n)],
\end{aligned} \qquad (4)$$

where $y_n^T y_n$ is a constant independent of variable μ_k^i, we have

$$\nabla_{\mu_k^i} \|y_n - \hat{y}_n\|^2 = 2\sum_{j=1}^{K}[(P(C_j|x_n) - y_n^j)\nabla_{\mu_k^i} P(C_j|x_n)] . \qquad (5)$$

Moreover, $\nabla_{\mu_k^i} P(C_j|x_n)$ can be derived as

$$\begin{aligned}
&\nabla_{\mu_k^i} P(C_j|x_n) \\
&= \nabla_{\mu_k^i} \frac{P(C_j)p(x_n|C_j)}{p(x_n)} \\
&= \nabla_{\mu_k^i} \frac{P(C_j)\sum_{l=1}^{N_j} a_j^l G(x_n|\mu_j^l, \Sigma_j^l)}{p(x_n)} \\
&= \frac{P(C_k)a_k^i}{p(x_n)}[\delta_{kj} - P(C_j|x_n)]\nabla_{\mu_k^i} G(x_n|\mu_k^i, \Sigma_k^i)
\end{aligned} \qquad (6)$$

where δ_{kj} is a Kronecker delta function which equals 0 for $k \neq j$ and 1 for $k = j$ respectively, and

$$\nabla_{\mu_k^i} G(x_n|\mu_k^i, \Sigma_k^i) = G(x_n|\mu_k^i, \Sigma_k^i)(\Sigma_k^i)^{-1}(x_n - \mu_k^i) . \qquad (7)$$

Define $\Phi \doteq \frac{P(C_k)a_k^i}{p(x_n)} G(x_n|\mu_k^i, \Sigma_k^i)(\Sigma_k^i)^{-1}(x_n - \mu_k^i)$, then according to (5), (6) and (7), we have

$$\begin{aligned}
&\nabla_{\mu_k^i} \|y_n - \hat{y}_n\|^2 \\
&= 2\sum_{j=1}^{K}[(P(C_j|x_n) - y_n^j)(\delta_{kj} - P(C_j|x_n))\Phi] \\
&= 2\Phi \sum_{j=1}^{K}[(P(C_j|x_n) - y_n^j)(\delta_{kj} - P(C_j|x_n))] .
\end{aligned} \qquad (8)$$

Though Millán et al. have given a similar final representation of the instantaneous gradient $\nabla_{\mu_k^i} \|y_n - \hat{y}_n\|^2$, they didn't provide the above systematic derivation process [6]. And during their experiments, they applied an approximated form of instantaneous gradient. Instead, in our paper, we adopt the exact gradient as (8) presents. Now the update equation of LMS algorithm can be formulated as

$$(\mu_k^i)_n = (\mu_k^i)_{n-1} - \gamma_n \cdot \nabla_{(\mu_k^i)_{n-1}} \|y_n - \hat{y}_n\|^2 \tag{9}$$

where γ_n is the learning rate [15]. However, using LMS algorithm directly would take a risk of low convergence rate and poor tracking performance, since stochastic gradient $\nabla_{(\mu_k^i)_{n-1}} \|y_n - \hat{y}_n\|^2$ is only the instantaneous approximation of the true gradient which should be derived from $\nabla_{(\mu_k^i)_{n-1}} E\{\|y_n - \hat{y}_n\|^2\}$. If two consecutive instantaneous gradients correlate with each other, then the mean square error (MSE) might be accumulated and couldn't be corrected in time. Therefore, to get rid of these shortcomings, here we adopt the decorrelated gradient instead of the instantaneous gradient, since it has already earned theoretical support and successful applications [8][15]. Using decorrelated gradient can effectively avoid the case of error accumulation which might arise in instantaneous gradient descent algorithms, and hence, can accelerate the convergence of the adaptive gradient methods.

The decorrelated gradient of $(\mu_k^i)_n$ can be defined as

$$\tilde{\nabla}_{(\mu_k^i)_n} \|y_n - \hat{y}_n\|^2 \doteq \nabla_{(\mu_k^i)_n} \|y_n - \hat{y}_n\|^2 - a_n \cdot \nabla_{(\mu_k^i)_{n-1}} \|y_n - \hat{y}_n\|^2 \tag{10}$$

where a_n is the decorrelation coefficient between instantaneous gradients $\nabla_{(\mu_k^i)_n} \|y_n - \hat{y}_n\|^2$ and $\nabla_{(\mu_k^i)_{n-1}} \|y_n - \hat{y}_n\|^2$. For two vectors w_n and w_{n-1}, the decorrelation coefficient a_n can be defined as

$$a_n = \frac{(w_n - \bar{w}_n)^T (w_{n-1} - \bar{w}_{n-1})}{(w_{n-1} - \bar{w}_{n-1})^T (w_{n-1} - \bar{w}_{n-1})} \tag{11}$$

where $\bar{w}_n$ represents the mean value of w_n [15]. To this end, we can formulate the update equation of DLMS algorithm as

$$(\mu_k^i)_n = (\mu_k^i)_{n-1} - \gamma_n \cdot \tilde{\nabla}_{(\mu_k^i)_{n-1}} \|y_n - \hat{y}_n\|^2 \tag{12}$$

where $\tilde{\nabla}_{(\mu_k^i)_{n-1}} \|y_n - \hat{y}_n\|^2$ is the decorrelated gradient and γ_n is the learning rate [15]. Table 1 gives the flow chart of DLMS algorithm for learning the online Bayesian classifier.

4 Experiments

4.1 Data Description

Here we describe the data set used in our experiments. The data set contains EEG recordings from three normal subjects (denoted by 'A', 'B', 'C' respectively)

Table 1. The flow chart of the decorrelated LMS (DLMS) algorithm for learning on-line Bayesian classifier

> The variable μ_k^i in the following procedure denotes mean value in Gaussian probability density function $G(x|\mu_k^i, \Sigma_k^i)$ with $\{k = 1, \ldots, K; i = 1, \ldots, N_k\}$.
> **Step 1:**
> Initialize μ_k^i with $(\mu_k^i)_0$.
> **Step 2:**
> For $n = 1, 2, \ldots$, calculate the decorrelated gradient $\tilde{\nabla}_{(\mu_k^i)_{n-1}} \|y_n - \hat{y}_n\|^2$ from (10) and (5), and update μ_k^i with $(\mu_k^i)_n = (\mu_k^i)_{n-1} - \gamma_n \cdot \tilde{\nabla}_{(\mu_k^i)_{n-1}} \|y_n - \hat{y}_n\|^2$.

during mental imagery tasks. The subjects sat in a normal chair, relaxed arms resting on their legs. The three tasks are: imagination of repetitive self-paced left hand movements (denoted as class C_1), imagination of repetitive self-paced right hand movements (denoted as class C_2) and generation of different words beginning with the same random letter (denoted as class C_3).

For a given subject, there are three recording sessions acquired on the same day, each lasting about four minutes with breaks of 5-10 minutes in between. The subject performed a given task for about 15 seconds and then switched randomly to the next task at the operator's request. The raw EEG potentials were first spatially filtered by means of a surface Laplacian [16][17]. The superiority of surface Laplacian transformation over raw potentials for the operation of BCI has already been demonstrated [18]. Then, every 62.5 ms, the power spectral density in the band 8-30Hz was estimated over the last second of data with a frequency resolution of 2 Hz for eight centro-parietal channels (EEG signals recorded over this region reflects the activities of brain's sensorimotor cortices). The power spectra in the frequency band 8-30 Hz were then normalized according to the total energy in that band. As a result, an EEG sample is a 96-dimensional vector (8 channels times 12 frequency components). The total number of samples for subject 'A', 'B', and 'C' during three sessions are respectively 3488/3472/3568, 3472/3456/3472, and 3424/3424/3440. For a more detailed description of the data and the brain computer interface protocol, please refer to [19] and the related web page of BCI competition III. In this article, we concentrate on utilizing the 96 dimensional pre-computed features to address the problem of on-line classification.

4.2 Experimental Results

EEG signal classification is conducted for each subject. First of all, to reduce the parameters to be estimated and avoid the over-fitting problem, principal component analysis (PCA) is adopted to reduce the feature dimensions by reserving 90% energy. The threshold 90% is a good tradeoff between dimension reduction and energy preservation for our problem. To initialize the parameters μ_k^i and Σ_k^i

of the DLMS algorithm, we first apply the k-Means clustering algorithm with multiple runs [14], and the result with the least cost value is selected for initialization utility. On the selection of parameters $P(C_k)$, N_k and a_k^i in the Bayesian classifier of Gaussian mixture model, we take the same configuration as [19], for in his research, Millán had shown its effectiveness through cross-validations. Thus, $P(C_k) = \frac{1}{3}$, $N_k = 4$ and $a_k^i = \frac{1}{4}$ ($k = 1, 2, 3; i = 1, 2, 3, 4$).

In this article, the data of session 1 from each mental task of every subject is employed to implement parameter initialization. For class C_k, we first use k-Means clustering algorithm to initialize μ_k^i which comes from one of the N_k cluster centers. Then Σ_k^i can be obtained using the data belonging to the same cluster C_k^i. Subsequently, we update the parameters adaptively on the first one minute data of the next session (the samples are processed sequentially and only once, to completely stimulate the on-line situation). With the final updated parameters, we test the performance of the classifier on the data of the last three minutes from the next session. The same procedure is performed on session 2 and session 3, i.e., we initialize the parameters μ_k^i and Σ_k^i through k-Means clustering on session 2, then update them using the first one minute data of session 3 and test the final classifier on the last three minute data of session 3.

To evaluate the performance of the DLMS algorithm of learning on-line Bayesian classifier, under the same configurations we also carry out on-line classification using the basic LMS algorithm, which adopts instantaneous gradient instead of decorrelated gradient to update parameters μ_k^i. The learning rate γ_n of both LMS algorithm and DLMS algorithm is taken as $1e - 6$, which is found through a small number of grid search to achieve best performance for LMS algorithm. The final classification accuracies for different subjects and sessions using these two algorithms with parameters updated by the corresponding one minute data are given in Table 2. From this comparison, we can see that DLMS algorithm (with higher average value and smaller standard deviation) is better than LMS algorithms, though slightly. Furthermore, for on-line evaluation of learning algorithms, the stability of time course is also an essential factor. Therefore, we show the time courses of classification accuracies of these two algorithms during the on-line update for classifying the last three minutes of session 3 of three subjects in Fig. 1. That is, after every update, we obtain the classification accuracy on the last three minutes of session 3. Furthermore, to give a quantitative comparison the standard deviation (STD) of the classification courses for LMS algorithm and DLMS algorithm are respectively given in Table 3, from which we can see that DLMS algorithm is also superior to the LMS algorithm in the sense of low variance and high stability.

5 Conclusions

In this paper, we address the problem of on-line classification of EEG signals arising in brain-computer interface research. The time-varying characteristic of EEG recordings between experimental sessions makes it a difficult issue to classify different EEG signals, and necessitates learning on-line classifiers. For Bayesian

Table 2. A comparison of classification accuracies for on-line learning Bayesian classifier with LMS algorithm and DLMS algorithm

Subjects	Sessions	LMS	DLMS	Better?
A	2	68.55%	68.31%	×
	3	71.20%	71.20%	=
B	2	48.16%	47.52%	×
	3	50.48%	50.68%	√
C	2	48.82%	48.90%	√
	3	40.12%	41.41%	√
Average		54.56±12.42	54.67±12.13	√

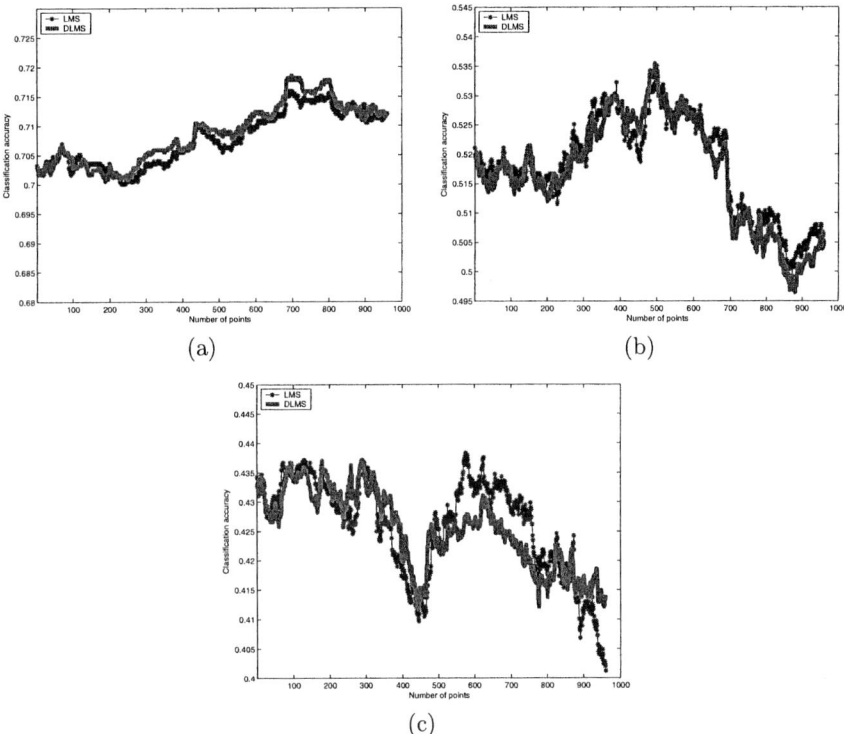

Fig. 1. (a): The time course of classification accuracies on session 3, subject 'A'. (b): The time course of classification accuracies on session 3, subject 'B'. (c): The time course of classification accuracies on session 3, subject 'C'

Table 3. The standard deviations (STDs) (multiplied by $1e+3$ for normalization) of the time courses of on-line classification by LMS algorithm and DLMS algorithm

Subjects	Sessions	LMS STD	DLMS STD	STD Reduced	Better?
A	2	8.0	6.1	23.75%	√
	3	4.5	5.1	−13.33%	×
B	2	18.1	14.6	19.34%	√
	3	8.3	9.4	−13.25%	×
C	2	5.7	4.9	14.04%	√
	3	8.5	6.7	21.18%	√
Average				8.62%	√

classifier with Gaussian mixture model, we systematically derive the formulation of gradient representation with respect to μ_k^i, and propose to use DLMS algorithm to replace LMS algorithm for the on-line learning of Bayesian classifier. Experimental results on real EEG signals also indicate that DLMS algorithm is superior to the existed LMS algorithm for learning Bayesian classifiers.

The computational complexity of DLMS algorithm is quite similar to that of LMS algorithm, for it only adds a step for gradient decorrelation. And thus DLMS algorithm would not influence the real time performance much. In the future, study on the possibility of on-line updating other parameters (e.g. covariance matrix) using gradients and the active selection of training instances would be interesting directions.

Acknowledgements

This work was supported by the National Natural Science Foundation of China under Project 60475001. The authors would like to thank IDIAP Research Institute (Switzerland) for providing the analyzed data.

References

1. Nicolelis, M.A.L.: Actions from Thoughts. Nature, Vol. 409 (2001) 403-407
2. Wolpaw, J.R., Birbaumer, N., McFarland, D.J., Pfurtscheller, G., Vaughan, T.M.: Brain-Computer Interfaces for Communication and Control. Clinical Neurophysiology, Vol. 113 (2002) 767-791
3. Vaughan, T.M.: Guest Editorial Brain-Computer Interface Technology: A Review of the Second International Meeting. IEEE Transactions on Neural Systems and Rehabilitation Engineering, Vol. 11 (2003) 94-109
4. Ebrahimi, T., Vesin, J.M., Garcia, G.: Brain-Computer Interfaces in Multimedia Communication. IEEE Signal Processing Magazine, Vol. 20 (2003) 14-24

5. Millán, J.R., Renkens, F., Mouriño, J., Gerstner, W.: Non-Invasive Brain-Actuated Control of a Mobile Robot. Proceedings of the Eighteenth International Joint Conference on Artificial Intelligence, (2003) 1121-1126
6. Millán, J.R., Renkens, F., Mouriño, J., Gerstner, W.: Brain-Actuated Interaction. Artificial Intelligence, Vol. 159 (2004) 241-259
7. Millán, J.R., Renkens, F., Mouriño, J., Gerstner, W.: Noninvasive Brain-Actuated Control of a Mobile Robot by Human EEG. IEEE Transactions on Biomedical Engineering, Vol. 51 (2004) 1026-1033
8. Doherty, J., Porayath, R.: A Robust Echo Canceler for Acoustic Environments. IEEE Transactions on Circuits and Systems, II, Vol, 44 (1997) 389-398
9. Blankertz, B., Curio, G., Müller, K.R.: Classifying Single Trial EEG: Towards Brain Computer Interfacing. In: Dietterich, T.G., Becker, S., Ghaharamani, Z. (eds.): Advances in Neural Information Processing Systems. MIT Press, Cambridge, MA (2002) 157-164
10. Wang, Y., Zhang, Z., Li, Y., Gao, X., Gao, S., Yang, F.: BCI Competition 2003-Data Set IV: An Algorithm Based on CSSD and FDA for Classifying Single-Trial EEG. IEEE Transactions on Biomedical Engineering, Vol. 51 (2004) 1081-1086
11. Kaper, M., Meinicke, P., Grossekathoefer, U., Lingener, T., Ritter, H.: BCI Competition 2003-Data Set IIb: Support Vector Machines for the P300 Speller Paradigm. IEEE Transactions on Biomedical Engineering, Vol. 51 (2004) 1073-1076
12. Lu, B., Shin, J., Ichikawa, M.: Massively Parallel Classifiation of Single-Trial EEG Signals Using a Min-Max Modular Neural Network. IEEE Transactions on Biomedical Engineering, Vol. 51 (2004) 551-558
13. Mclachlan, G., Peel, D.: Finite Mixture Models. Wiley, New York (2000)
14. Duda, R.O., Hart, P.E., Stork, D.G.: Pattern Classification. 2th edn. John Wiley & Sons, New York (2000)
15. Glentis, G.O., Berberidis, K., Theodoridis, S.: Efficient Least Square Adaptive Algorithms for FIR Transversal Filtering. IEEE Signal Processing Magzine, Vol. 16 (1999), 13-41
16. Perrin, R., Pernier, J., Bertrand, O., Echallier, J.: Spherical Spline for Potential and Current Density Mapping. Electroencephalography and Clinical Neurophysiology, Vol. 72 (1989), 184-187
17. Perrin, R., Pernier, J., Bertrand, O., Echallier, J.: Corrigendum EEG 02274. Electroencephalography and Clinical Neurophysiology, Vol. 76 (1990), 565
18. McFarland, D.J., McCane, L.M., David, S.V., Wolpaw, J.R.: Spatial Filter Selection for EEG-Based Communication. Electroencephalography and Clinical Neurophysiology, Vol. 103 (1997) 386-394
19. Millán, J.R.: On the Need for On-Line Learning in Brain-Computer Interfaces. Proceedings of 2004 International Joint Conference on Neural Networks. Budapest, Hungary (2004)

Automatic Keyphrase Extraction from Chinese News Documents*

Houfeng Wang, Sujian Li, and Shiwen Yu

Department of Computer Science and Technology,
School of Electronic Engineering and Computer Science,
Peking University, Beijing, 100871, China
`wanghf@pku.edu.cn`

Abstract. This paper presents a framework for automatically supplying keyphrases for a Chinese news document. It works as follows: extracts Chinese character strings from a source article as an initial set of keyphrase candidates based on frequency and length of the strings, then, filters out unimportant candidates from the initial set by using elimination-rules and transforms vague ones into their canonical forms according to controlled synonymous terms list and abbreviation list, and finally, selects the best items from the set of the remaining candidates by score measure. The approach is tested on *People Daily* corpus and the experiment results are satisfactory.

1 Introductions

With the rapid growth of available electronic documents, keyphrases, which serve as highly condensed summary, play more and more important role in helping the readers to quickly capture central content of an article and determine whether the article is valuable for them. Keyphrases are also useful to information retrieval, text clustering, text categorization and so on. However, keyphrases are usually not provided by author(s) in most articles, especially in news documents, to which most people always pay a large amount of attentions. Manually assigning keyphrases to articles will be labor-intensive and time-consuming. Therefore, how to automatically supply keyphrase for a document has become an important task.

There exist two general approaches to automatically supplying keyphrases for an article: keyphrase extraction and keyphrase assignment. The fact that most keyphrases will occur somewhere in a source article [4] shows that keyphrases can be directly extracted from an article without a predefined term list.

Supervised machine learning algorithms, such as decision tree, genetic algorithm [4] and Naïve Bayse technique [2][5] have been used in keyphrases extraction successfully. In these methods, a text is treated as a set of words or multi-words, each of which is viewed as either keyphrase or non-keyphrase. Keyphrases extraction therefore becomes a binary-value classification and machine learning techniques are used to classify them. These systems achieved satisfying results. However, they need a

---

* Supported by National Natural Science Foundation of China under Grant No.60473138.

large amount of training documents with known keyphrases to train classifiers. For Chinese Language, keyphrase extraction means that words and phrases need to be recognized before the above mentioned machine learning methods are applied. This is still considered as a difficult question to Chinese Processing. Some researchers tried to avoid Chinese word segmentation and proposed statistics method such as string-frequency. PAT tree is used to represent Chinese character string [1][3] and mutual information is used to evaluate the importance of a string. Unfortunately, the method is not able to extract keyphrases that do not occur sufficiently frequently and resultant string as keyphrase even cannot be ensured as a meaning unit.

For some keyphrases extracted from a source article, there might exist semantic ambiguities. One way of solving this problem is to select terms from a controlled thesaurus that is pre-constructed. This method is called keyphrases assignment. Although it is time-consuming and labor-intensive to build a thesaurus, some agencies, such as *People Daily* in China, have been doing it as a task of local standardization. That makes it possible to automatically assign keyphrases for *People Daily*.

In cooperation with *People Daily* Agency, we combine the two approaches. Our approach directly extracts some important keyphrases from an article no matter whether they occur in predefined closed list of term or not. At the same time, Some keyphrases may be gained by transforming from words or multi-words occurring in the source article to canonical terms according to controlled term list.

In this paper, we discuss some key factors that have an effect on supplying keyphrases and give testing results.

2 Approach

Our approach consists of four main parts: Initialization, filter, transformation and score computation. It can be outlined in Fig.1.

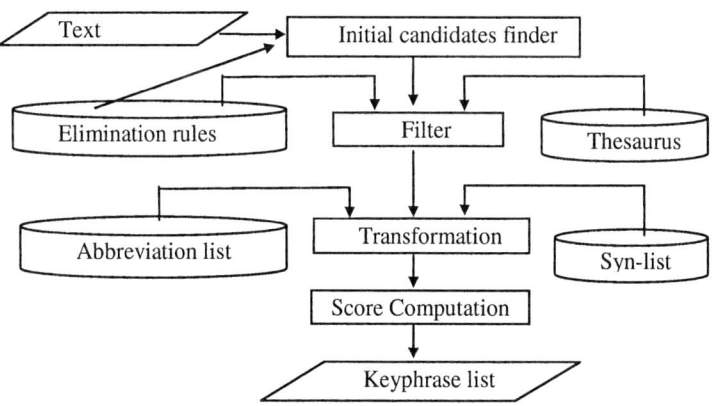

Fig. 1. System structure

A keyphrase is just a Chinese character string in surface form, and to extract keyphrases from an article is to find important strings. By a Chinese character string here we mean a string consisting of only Chinese characters. The total of Chinese character string in an article is usually huge. In order to reduce the complexity of successive processing, obviously unimportant Chinese character string should be filtered out as early as possible.

The approach works as follows: extracts all Chinese character strings from an article as the set of initial keyphrase candidates on condition that they have more than one Chinese character and occurs more than one time in the article, eliminates meaningless strings by deciding whether the string can be segmented into a sequence of words successfully and the POS of the head word is noun, transforms some vague words or multi-words into their canonical forms according to controlled synonymous terms list and abbreviation list, and applies some indicators to score candidates. The strings with the highest scores are selected as resultant keyphrases.

2.1 Initialization and Elimination

Unlike English, there is no boundary between two words except punctuation symbols within Chinese text. Inasmuch as it is difficult to directly initialize keyphrases candidates as list of words or multi-words, we will simply extract Chinese character strings as initial keyphrase candidates. Considering that the size of keyphrase candidate set will seriously affect processing efficiency, those unimportant strings should be eliminated as early as possible. Five elimination rules are thus applied:

E-Rule1: String frequency filter

$$\forall\ S = c_1 c_2 ... c_n \text{ is a Chinese} - \text{character} - \text{string}$$
$$\text{if } freq(S) = 1, \text{ then } S \notin \text{set of keyphrases cadidates}$$

Generally speaking, a word or multi-word will occur many times in an article in order to emphasize the content it expresses if the content is important. This implies that those strings with the occurrence frequency as just one usually do not express the topic and therefore can be filtered out.

E-Rule2: One-character word filter

$$\forall\ S = c_1 c_2 ... c_n \text{ is a Chinese} - \text{character} - \text{string}$$
$$\text{if } S \text{ contains only one Chinese character,}$$
$$\text{then } S \notin \text{set of keyphrases cadidates.}$$

Many Chinese characters can independently act as a word. Most functional words are one-character words and they usually occur frequently in a document, but they seldom express the topic of the document by themselves. The kind of strings also should be eliminated.

A lot of Chinese character strings are filtered out by using *E-Rule1* and *E-Rule2* while the others remain as the initial set of keyphrase candidates.

E-Rule3: Meaningless unit filter

$\forall\ S = c_1 c_2 ... c_n \in$ *the set of keyphrase candidate,*

if $\neg \exists w_1 w_2 ... w_m (S = w_1 w_2 ... w_m)$

then S should be deleted from keyphrases cadidates

where, w_i *is a Chinese word,* $i = 1...m$ *and* $1 \le m \le n$;

c_j *is a Chinese character,* $j = 1...n$

A word is the minimal meaning unit. If a Chinese character string cannot successfully be segmented into any sequence of words, it will be thought as a meaningless string and be eliminated. We use a Chinese word segmentation tool based on a dictionary to do it.

E-Rule4: POS filter

$\forall\ S = w_1 w_2 ... w_m \in$ *the set of keyphrase candidate,*

if $POS(head(w_1 w_2 ... w_m)) \ne Noun$ *and* $m > 1$

then S should be deleted from keyphrases cadidates

Noun words or phrases are commonly believed to be the content bearing units. Therefore, those candidates whose heads are not noun will not thought as keyphrases in our approach. However, the rule do not delete the individual word candidates simply because POS of a single word usually cannot be tagged accurately. They will be treated by rule *E-Rule5*.

In Chinese, the head of a phrase is almost the most right word. After the POS tagging is done, we can easily determine whether a candidate should be filtered out.

E-Rule5: Non-subject word filter

$\forall\ a\ word\ S,$

if $(S \notin subject\ thesaurus)$

then S should be deleted from keyphrases cadidates

There are two obvious disadvantages of single common words: its semantic ambiguity and its too general meaning to convey text content. Therefore, the individual words that do not belong to subject thesaurus will be filtered out. However, multi-word are sometimes better indicators of text content even if they are not the members of the thesaurus. *E-Rule5* has no effect on any candidate consisting of more than one word. It serve as a complement to *E-Rule4*. From this perspective, the subject thesaurus in our approach is only a semi-controlled subject thesaurus.

2.2 Keyphrase Weight

Not all of Keyphrase candidates equally reflect text content. We present some empirical indicators to score the candidates based on our observation. These factors play a decisive role in picking out keyphrases from the set of candidates. Candidates are assigned a score for each indicator; those candidates with the highest total score are proposed as resultant keyphrases. In the following we outline these indicators and assign the empirical scores.

Title and section heading
Title (and section heading) is a condensed description of an article (and section). The words and multi-words appearing in title or headings, in particular, the noun words and multi-words have much closer concept relation with the topic of the article. Their importance should be emphasized and thus are assigned the highest score $w_{t-h} = 7$.

The first paragraph
Many authors like to present topic at the beginning of an article in news documents. Words and multi-words in the first paragraph are much likely the keyphrase. They are also given a high score $w_{f-p} = 3$.

The last paragraph
The conclusion and summary of an article will mainly be in the last paragraph. This is a good indicator of significance. The words or multi-words occur in this paragraph are given a score $w_{l-p} = 2$.

Special punctuations
Some special punctuation marks can hint that some sentences or phrases within an article are important. For example, the symbol dash '——' and pair of close marks such as '( )', ' " " ' can function as emphasis on some topics. The phrases that are bracketed or leaded by special punctuations will be added a score:

$$punctuation(term) = \begin{cases} 1, & \text{if the term with special punctuation} \\ 0, & \text{others} \end{cases} \quad (1)$$

Length of a multi-word
We found that a Chinese keyphrases usually has two characters to eleven ones and the average length of keyphrases is 7 by analyzing the manual keyphrases. In the approach, we use a formula to compute the weight relating to length of a candidate:

$$w_{len} = 1 + \lg \frac{11}{|(length(phrase) - 7)| + 1} \quad (2)$$

Named entity and common phrase
Person name, place name and organization name are very important in news documents and readers are usually much interested in them. We treat them differently from usual candidate as follows:

$$w_{Ne-cp} = \begin{cases} 1.5 & \text{if the phrase a is named entity;} \\ 1 & \text{others} \end{cases} \quad (3)$$

2.3 Transformation

Although we try to avoid parsing and analyzing sentences within the whole text in order to improve efficiency, some deeper processing techniques will still be applied in concerned strings.

2.3.1 Synonymous Phrase Substitution

Writers sometimes use different words or multi-words to express the same or nearly the same meaning in an article. This is an obstruction to string frequency based method. In our approach, a readable dictionary of synonymous terms is built. When different terms express the same meaning, they will be replaced into canonical terms.

For instance, two synonymous words "会见" (meeting) and "会谈"(conferring) appearing in the following example often co-occur in a news document.

在<u>会见</u>澳大利亚总理霍华德时，江泽民说，你与胡锦涛主席和温家宝总理等领导人的<u>会见</u>和<u>会谈</u>，将有助于增加相互了解，扩大双边合作。

(Jiang Zemin said during **meeting** with visiting Australian Prime Minister John Howard that his **meeting** and **conferring** with Chinese President Hu Jintao, Chinese Premier Wen Jiabao and other Chinese leaders will help increase mutual understanding and cooperation between two countries)

Synonymous terms are not contained in subject thesaurus. But each of them corresponds to an entry in subject thesaurus. The structure of the dictionary is:

Norm-term: { syn-item$_1$, syn-item$_2$, …}

Norm-term belongs to the subject thesaurus. Each syn-item$_i$ is a synonymous multi-word of the canonical term. Once a **syn-item** occurs in a source text, it will be transformed into the corresponding canonical term.

2.3.2 Abbreviation Substitution

Abbreviations are everywhere and abbreviations of proper noun are more popular in news document. Meanwhile, named entities usually become focus in news event. Therefore, they will be paid special attention and be preferentially processed. We built an abbreviation dictionary to serve it. Its structure is:

Ex-form: {abbr-form$_1$, abbr-form$_2$, …}

The relationship between some abbreviations and their expansions is many to many. For example, the abbreviation "中印" could be "中国-印度"(Sino-India) or "中国-印度尼西亚" (Sino- Indonesia). They will be disambiguated before they are replaced by the expansion.

2.3.3 Special Phrase Reconstruction

Some words or phrases must collocate others in order to clearly reflect the topic of text even if they are not adjoined. If they are extracted as keyphrases independently, some ambiguities will be caused. We consider the following example:

"钱伟长在上海喜度九十<u>华诞</u>　受江泽民李瑞环李岚清委托，黄菊亲切看望钱伟长"(On Qian Weichang happily celebrating his 90th **birthday** in Shanghai, Huang Ju, on behalf of Jiang Zemin, Li Ruihuan and Li Lanqing, kindly congratulated him).

If the set of keyphrases is {华诞(birthday), 钱伟长(Qian Weichang), 江泽民(Jiang Zemin),…}, it will be difficult to tell whose birthday it is from the set.

Some special words such as "华诞" (birthday), "追悼会"(condolence conference) etc. need to be recombined with other words(or phrases) to form a new multi-word in order to make the meanings clear. In this example, "钱伟长华诞" will be formed and become a keyphrase instead of independent "华诞" and "钱伟长". We have presented some recombination patterns for special words or phrases.

2.4 The Algorithm

The algorithm to supply keyphrases is described informally as follows:

Step1. Find all Chinese character strings with more than one occurrence and more than one character, and generate a set of candidates.

Step2. Recognize named entities occurring in title or heading and the first paragraph, and add them into the set of candidate if they are not in the set.

Step3. Use elimination rules E-Rule3 ~ E-Rule5 to filter some candidates out.

Step4. Transform some words or multi-words into canonical terms and recombine special phrases.

Step5. Assign a score to each occurrence of each candidate and aggregate the total score for each candidate.

Step6. Select those candidates with the highest total score.

The score of each occurrence and total score in step5 are assigned and computed as follows:

$$score(position(term)) = \begin{cases} w_{t-h} = 7 & position\ is\ Title\ or\ heading \\ w_{f-p} = 3 & position\ is\ the\ first\ paragraph \\ w_{l-p} = 2 & position\ is\ the\ last\ paragraph \\ 1 & others \end{cases} \quad (4)$$

If a term occurs in special punctuation, a score will be added:

$$score(occurrence(term)) = score(position(term)) + punctuation(term) \quad (5)$$

A term could occur in different places of an article, so it could be assigned different scores. For each candidate term, its scores in different places are added up and the result is denoted as Sum(term). By combining factors W_{len} and W_{Ne-cp}, the final score of a candidate term is calculated as follows:

$$Total - score(term) = w_{len} * w_{Ne-cp} * Sum(term) \quad (6)$$

Table 1 gives the testing results on an article from *People Daily*. In this table, keyphrases are sorted by total score. The manually assigned keyphrases are underlined. Occurrence frequency of each term in the article is also given.

Although Rank 3, 4, 6 and 7 are not manually assigned in this table, professional(human) indexer admit that Rank 3(中国帆船队), a expansion of the Rank 1(中国队), is acceptable as keyphrases and Rank 4, Rank 6 and Rank 7 （女子欧洲级） are semantically related to the content of the article.

Table 1. Seven keyphrases with the highest scores testing on article "帆船帆板比赛　中国队夺得六金"(Chinese team wins six golds in Sailing Race) from *People Daily*

Rank	Frequency	Score	Keyphrases
1	3	22.14	中国队 (Chinese Team)
2	2	17.40	帆船帆板比赛 (Sailing Race)
3	4	11.73	中国帆船队 (Chinese Sailing Team)
4	5	10.06	奥运会 (Olympic Games)
5	2	8.05	亚运会 (Asian Games)
6	4	6.32	夺金 (wining gold)
7	2	2.53	女子欧洲级 (Women's Europe Class)
…	…	…	…

3 Evaluation

We select 80 articles from *People Daily* with manually assigned keyphrases to test our approach. We classify these articles according to the number of the manually assigned keyphrases. The classes corresponding to the numbers are shown in Table 2. The number of articles in each class is also given in the table. We do not select those articles that have two manually designed keyphrases or more than seven ones due to their small proportion.

Table 2. The Class (Class-i, i =1..5) corresponding to the number of manually assigned keyphrases and the number of articles in each class

	Class-1	Class-2	Class-3	Class-4	Class-5
Number of keyphrases	3	4	5	6	7
Number of articles in each class	12	24	23	15	6
Total number of articles	80				

Table 3 gives the number of correct keyphrases that are automatically indexed by the system for each rank in each class. An indexed keyphrase is defined as correct on Rank i (i= 1.. 7) for an article if it belongs to the set of manually assigned keyphrases. The order of manual keyphrases is not considered in our evaluation. Only the first three ranks in Class-1 are evaluated because these articles in this class have just three manual keyphrases and the rest classes may be deduced by analogy.

In Table 3, The total (the last but three row) shows the total number of keyphrases in each class that are correctly indexed by the system, and the sum (the last but one column) shows the total number of keyphrases on each rank that are correctly indexed by the system. Correspondingly, ratio-C (the last but one row) is the correctness ratio for each class and ratio-R (the last column) for each rank. The ratio-W is the whole correctness ratio of all selected articles. Correctness-ratio is calculated as follows:

$$Correctness-ratio = \frac{the\ number\ of\ keyphrases\ that\ are\ correctly\ indexed}{the\ number\ of\ manual\ keyphrases} \quad (7)$$

Ratio-W indicates that the whole result is satisfactory. Also that the correct ratio Ratio-C for each class is nearly identical with Ratio-W shows that our approach is appropriate for both long news articles and short ones.

Table 3. Testing results: the number of correct keyphrases and the correctness ratio

Rank	Class-1	Class-2	Class-3	Class-4	Class-5	Sum	Ratio-R
1	10	19	19	12	5	65	81.3 %
2	9	15	13	10	4	51	63.8 %
3	3	9	16	7	5	40	50.0 %
4		14	14	8	2	38	55.9 %
5			11	10	2	23	52.3 %
6				7	4	11	52.4 %
7					5	5	83.3 %
Total	22	57	73	54	27		
Manual Total	12*3=36	24*4=96	23*5=115	15*6=90	6*7=42		
Ratio-C	61.1 %	59.4 %	63.5 %	60.0 %	64.3 %		
Ratio-W	61.5 %						

Ratio-R shows that the correct ratios on the first two ranks (Rank1—81.3 % and Rank 2—63.8 %) are obviously higher than those on the following four ranks. An important reason is that some candidates occurring special location like title (heading) are assigned a high score in our approach and they are just manual keyphrases in most cases.

After analyzing the test results, we feel that it is not completely reasonable to evaluate the results only by comparing automatically indexing keyphrases with manual ones. Even different human indexer may assign different keyphrases for the same article and the same people could give different results in different times. The set of keyphrases should not be only one. Just as table 1 shows, Rank 3 can absolutely replace Rank 1. Therefore, we ask people to mark which indexed terms are acceptable, and the result shows acceptable-rate is 75.5%, where,

$$acceptable-rate = \frac{the\ number\ of\ acceptable\ keyphrases}{the\ number\ of\ manual\ keyphrases} \quad (8)$$

4 Conclusion

This paper presented a framework for automatic keyphrase extraction from Chinese news documents. No training is needed and the testing results are satisfactory.

In the future work, we will use a larger Wordnet-like Chinese concept dictionary to process synonymous word. We have finished collecting noun words(phrases) in this dictionary. Also, we will adjust the computation of the score according to empirical observation. For instance, multi-words in thesaurus should be treated differently with those that are not in the thesaurus. Finally, we plan to process more language patterns such as A+ X + B in which A+B belongs to subject thesaurus and is usually separated by other words (X).

References

1. Chien, L. F.: PAT-Tree-based keyword Extraction for Chinese Information Retrieval, Proceedings of the ACM SIGIR International Conference on Information Retrieval (1997) 50-59
2. Frank, E., Paynter, G.W., Witten, I.H., Gutwin, C., and Nevill-Manning, C.G.: Domain-specific keyphrase extraction. Proceedings of the Sixteenth International Joint Conference on Artificial Intelligence (IJCAI-99). California: Morgan Kaufmann. (1999) 668-673
3. Ong T. and Chen H. : Updateable PAT-Tree Approach to Chinese Key Phrase Extraction using Mutual Information: A Linguistic Foundation for Knowledge Management. Proceedings of 2nd Asian Digital Labrary Conference. Taipei, Taiwan, Nov.8-9 (1999) 63-84
4. Turney, P.D.: Learning algorithms for keyphrase extraction. Information Retrieval, 2, (2000) 303-336
5. Witten, I.H., Paynter, G.W., Frank, E., Gutwin, C., and Nevill-Manning, C.G.: KEA: Practical automatic keyphrase extraction. Proceedings of Digital Libraries 99 (DL'99), ACM press (1999) 254-256

A New Model of Document Structure Analysis

Zhiqi Wang, Yongcheng Wang, and Kai Gao

Department of Computer Science and Technology, Shanghai Jiao Tong University,
200030, Shanghai, P.R. China
{shrimpwang, ycwang, gaokai}@sjtu.edu.cn

Abstract. The purpose of document structure analysis is to get the document structure of the source text. Document structure is defined as 3 layers in the paper. A new model of document structure analysis — DLM is proposed. The model is composed of three layers: physical structure layer, logical structure layer and semantic structure layer, which are corresponding to the definition of the document structure. The input, output and operation of each layer are illustrated in details in the paper. The model has the feature of flexible, systematic and extendible. DLM is implemented on the Automatic Summarization System. It shows that the model is feasible and good result can be achieved.

1 Introduction

Document structure analysis is the key method to many research fields. In the process of automatic summarization, document structure analysis is needed because different parts of the document should have different importance so that some important parts can be extracted; in retrieval system, document structure analysis is needed because the system should operate on some certain parts of the document according to the user's requisition; in the process of constructing the information database, document structure analysis is needed because the document contained text noise and distortion need be described with a unified description language. In short, document structure analysis is the base of automatic summarization, automatic classification, automatic indexing and automatic retrieval. Now, a lot of NLP systems are on base of manual-indexing, such as the information databases of China Infobank, some automatic summarization system and some semantic analyzer. Automatic document structure analysis can take place of manual-indexing process and help those systems become practicability. This is the most important significance of document structure analysis.

This paper analyzed the tasks that document structure analysis should fulfill, defined document structure and proposed a new model of document structure analysis — Document Layer Model (DLM). The main features of DLM are flexible, systematic and extendable. The remaining sections of the paper are organized as follows: In section 2, the related work about this field is cursorily described. Section 3 defines document structure and Document Layer Model. Section 4 shows an example of implementing DLM in automatic summarization system. Finally, section 5 concludes with a summary.

2 Related Works

There are some researches focused on the implementation of document structure analysis. Kristen described an implemented approach to discovering full logical hierarchy in generic text documents primarily based on layout information [1]. Salton applying the vector space model to document structure analysis [2], [3]. In his method, vector space model is used to calculate the similarity between the two random paragraphs. The relevance of the paragraphs can be got from the text relationship map constructed according to the similarity, and the document can be segmented into several topic clusters. However, the topic clusters got from this method are usually composed of discontinuous paragraphs. The original sequences of the paragraphs are ignored, which makes the result hard to explain. Hearst also uses vector space model to segment the document into semantic sections [4]. In his method, the document is segmented into several sections containing the same number of sentences. Vector space model is used to calculate the similarity between the two adjacent sections. According to the valley of the similarity curve constructed with the similarities, the document is segmented into several semantic sections. However, the layout information is not used in this method, such as boundary of the paragraphs that can express continuing or turning of the topic. In short, the research mentioned above is focused on some certain aspects of document structure analysis (Kristen focused on document layout analysis. Salton and Hearst focused on document semantic analysis). Our work in this paper is focused on constructing an all-sided document structure analysis model from three layers.

3 Document Layer Model (DLM)

3.1 Definition of Document Layer Model

The problems that should be solved in document structure analysis include: extracting the useful textual information from document; clearly expressing paragraphs and sentences information of document; identifying the feature information including author, abstract, keyword, references etc.; segmenting the semantic sections (usually composed of several paragraphs).

According to these, document structure should be defined as three layers — physical structure, logical structure and semantic structure.

Physical structure reveals the useful textual information of the document. Useful textual information refers to the textual information contained in the document, which is useful to the application. Take an html page of a news report for an instance. The useful textual information refers to the part of title and straight matter of the news report. Related links, advertisement and images don't belong to the useful textual information.

Logical structure defines the logical units of the document. Logical units include title, keywords, author information, reference etc..

Semantic structure explains the organization of document content. There are two kinds of semantic structure. The semantic structure which is expressed with language symbols (e.g. first, second…; 1., 2., …) is defined as apparent semantic structure. On the contrary, the semantic structure which is expressed without any language symbols is defined as latent semantic structure. In latent semantic structure, semantic structure is expressed by the relationship of paragraphs.

Corresponding to the definition of document structure, Document Layer Model is defined as three layers: Physical Structure Layer, Logical Structure Layer and Semantic Structure Layer. The structure of DLM is shown as figure 1.

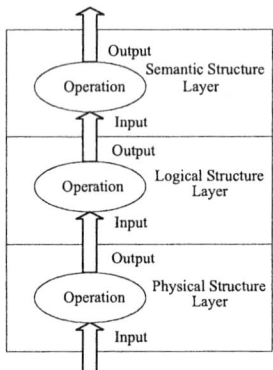

Fig. 1. Document Layer Model

3.2 Physical Structure Layer

The input of physical structure layer is electronic documents with various formats (such as xml, html, doc, wps, pdf etc.). The operation of physical structure layer is extracting useful textual information from document. The output of physical structure layer is the physical structure of the document which is composed of the textual information and its corresponding format information.

The corresponding format information of the textual information can be divided into two kinds: character format and paragraph format. Character format is used to describe each single character including font absolute size, font relative size (the relative font size of character C compared with the main text), font style, font color etc., while paragraph format is used to describe paragraphs including alignment, width, type(representing the paragraph is text, table, image or something other), indent etc..

To get the useful textual information from the document, the structure of the document file should be parsed. Knowledge database of different types of files can be constructed. Machine learning method can be employed to extract the useful textual information of the documents and filter some other information.

3.3 Logical Structure Layer

The main task of logical structure layer is to analyze logical structure from physical structure. The procedure is shown in figure 2.

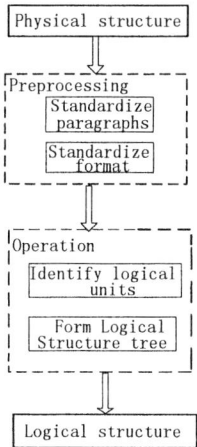

Fig. 2. The procedure of logical structure layer

The input of logical structure layer is the output of the physical structure layer. The granularity of the physical structure may not accord with the requisition of logical structure layer. The physical structure contains two kinds of units: character and paragraph. However, the basic processing unit of logical structure layer is paragraph. Therefore, the physical structure should be transformed into the format that meets the needs of logical structure layer's operations. This becomes the first reason for the preprocessing step. The other reason is the fault use of the carriage return, although the probability of such case is very low. For these two reasons, in the preprocessing, two things need to be done — standardizing paragraphs and standardizing format. The purpose of standardizing paragraphs is to correct the fault use of carriage returns. The method is using the statistics of the document layout information to analyze why there is a carriage return, and then get rid of the excrescent carriage returns. The purpose of standardizing format is to unify the physical structure with the granularity of paragraph. The method is to judge the layout information (character and paragraph) in the physical structure one by one, and then copy attribute, add new attribute and redefine attribute to re-describe this information in a standard form.

The operation of the logical structure layer includes identifying logical units and forming logical structure tree. In the process of identifying logical units, the logical unit such as title, author, keywords, reference etc. is identified by means of the salient feature string or the layout information. Some information identification (e.g. abstract) depends mainly on the salient feature string, while other information identification (e.g. title) depends mainly on layout information. Finally, the logical structure is described with a logical structure tree with the tree nodes representing logical units. A typical tree describing the document logical structure is shown as figure 3.

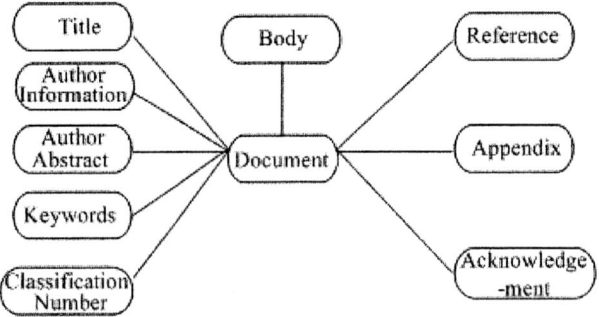

Fig. 3. A typical tree of Document Logical Structure

3.4 Semantic Structure Layer

The main task of semantic structure layer is to analyze the content organization of the document. Usually, a document is composed of several semantic sections. As mentioned above, the semantic structure can be classified into apparent semantic structure and latent semantic structure. Accordingly, the segmenting semantic sections can be classified into apparent semantic section segmenting and latent semantic section segmenting.

(1) Apparent Semantic Section Segmenting

Apparent semantic section segmenting mainly depends on the identification of the apparent semantic structure symbols (headings). In common sense, a heading becomes a paragraph by itself, or it has a special symbol. Therefore, knowledge database can be constructed according to the formal feature of the headings, and then the headings can be identified according to the knowledge in the knowledge database. The hierarchies of the headings need to be identified. This is because that the headings usually have hierarchies, and the headings belong to the same hierarchy express the indication of the semantic sections segmenting. The following example shows the multi-hierarchies of headings.

e.g.:

```
1. XXXXXXXXXXXXXXXXXX
    1.1 XXXXXXXXXXXXXXXXXX
        1.1.1 XXXXXXXXXXXXXXXXXX
        1.1.2 XXXXXXXXXXXXXXXXXX
    1.2 XXXXXXXXXXXXXXXXXX
2. XXXXXXXXXXXXXXXXXX
```

After identifying the headings, the correction should be made to the headings. Through statistics of the proceedings of 16th National Conference of the Computer Information Management, there are a quite number of authors (about 10%) using wrong heading symbols. Heading correction is to correct the clerical error or the edition error of the headings. Finally, the semantic sections are segmented using the headings.

(2) Latent Semantic Section Segmenting.

The latent semantic section segmenting mainly depends on the evaluation of the semantic relationship among the paragraphs. A simple method of semantic section segmenting is described as follow.

Suppose the paragraphs of the document is $P_0, P_1, \ldots, P_{n-1}$. The relevance degree

$$R(i) = Sim(P_i, P_{i+1}) \quad (0 \leq i \leq n-2) \tag{1}$$

The mean of R(i) is

$$\overline{R} = \sum_{i=0}^{n-2} R(i) \bigg/ n-1 \tag{2}$$

If existing R(i) (0<i<n-2) which satisfies the following conditions:

(1) $R(i-1) - R(i) > \xi$ and $R(i+1) - R(i) > \xi$ (ξ is a positive minimum constant),

(2) $R(i) < \overline{R}$

Then R(i) is the division, P_i belongs to a semantic section, and P_{i+1} belongs to the next semantic section. The curve of R(i) is shown in figure 4.

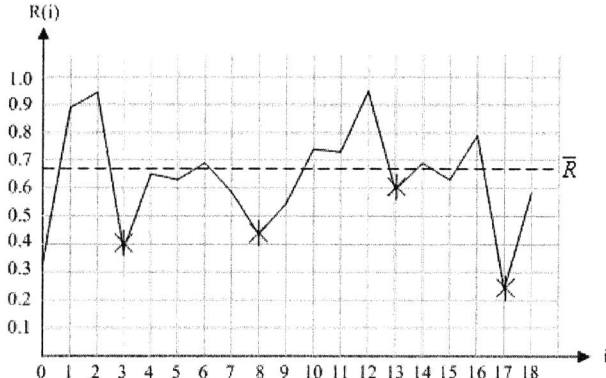

Fig. 4. The R(i) curve. The dots with cross are division dots. They segment all the paragraphs into five sections. P0 to P3 are the first section; P4 to P8 are the second section; P9 to P13 are the third section; P14 to P17 are the forth section; P18 to P19 are the fifth section.

There are many methods to calculate relevance degree of Sim(Pi, Pi+1). One simple method is to evaluate it by the frequency of the co-occurrence words.

3.5 Application Frame of DLM

DLM can be applied in many applications. The architecture of the system with DLM as its core is shown in figure 5.

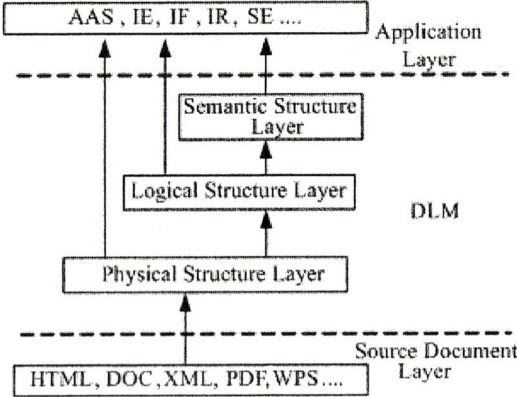

Fig. 5. The system framework based on Document Layer Model

The whole system can be described as three parts: source document, DLM and applications. The source document should be electronic document that can be processed by the system. DLM contains physical structure layer, logical structure layer and semantic structure layer. The applications may be AAS (Automatic Abstract System), IE (Information Extraction), IF (Information Filter), IR (Information Retrieval), SE (Subject Extraction) etc.. Each layer of DLM can offer service respectively to meet the requirement of different application. For example, if the application just needs the count of the words, DLM may offer physical structure layer service; if the application needs title and author name of the document, DLM may offer logical structure service; if the application needs some paragraphs on a certain topic, DLM may offer semantic structure layer service.

4 Implementing DLM in Automatic Summarization System and Experimental Result

4.1 Architecture of Automatic Summarization System

In order to validate that DLM is feasible and practicable, DLM is implemented in the automatic summarization system. The architecture of an automatic summarization system is shown as figure 6.

Document structure analysis is one of the critical modules and the base of other modules. The output of the summary has great attach with the document structure analysis. The main requisite of automatic summarization system can be described as follows:

(1) Only the textual information of the document could be useful to summarizing. Therefore, the useful textual information should be extracted from documents of various formats and some other information such as images should filtered.

(2) Units such as title, keywords, reference etc. play an important role in conveying the main idea of the document. Such units should be identified and located.

(3) Different paragraphs or sentences have different importance. The semantic analysis of the document is needed to help evaluating the importance of different parts.

According to the analysis mentioned above, it can be concluded that automatic summarization system need analysis in physical structure, logical structure and semantic structure. DLM can satisfy the above requirements.

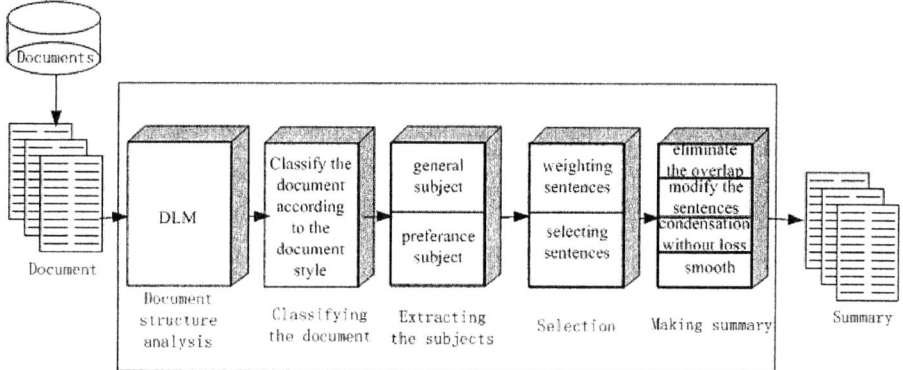

Fig. 6. Architecture of an automatic summarization system

4.2 Experiment

As the trend of natural language processing technology is orienting real language material and practicability, we collected 6000 Chinese document as samples from Sina(http://www.sina.com), China Daily(http://www.chinadaily.com.cn) and CJFD(http://cnki.lib.sjtu.edu.cn/). There are respectively 1500 document of word, txt, pdf, html among them. In the physical structure layer of DLM in automatic summarization system, we only deal with these 3 popular types of files. Some other types of files will be processed in the future work. The experiment is to validate the contribution of DLM to the summarization system's ability of processing various format documents and to improving quality of the summary.

4.3 Experimental Result and Analysis

In the experiment, the automatic summarization system can processed 5994 documents correctly and 6 documents error. The preciseness proportion is 99.9%. The result shows that DLM contributes much for the processing documents with multi-format. The reason for the errors is that the useful textual information of the document can not be extracted correctly (some other part such as advertisement is taken for the useful textual information and be processed by the system).

300 of the 6000 summaries got from the experiment are assessed by human professionals. It is found that the system can generated summary making full use of title, headings and semantic structure of the document. The quality of 91% summaries can be accepted.

Some features of DLM can be concluded from the experiment above:

(1) Systematic. The model systematically and roundly described the task and procedure of document structure analysis.

(2) Flexible. DLM is defined as three layers. Each layer can offer service to the applications respectively.

(3) Extendable. The definition of the three layers of DLM is explicit. Each layer can be respectively extended by means of adding new processing module to improve precision.

It is also shown from the experimental result that the model can be improved in the future work. The analysis of the document semantic structure in current model application is to segment semantic sections. In fact if more semantic information of the document can be mined, the consistency of the sentences and the covering degree can be raised.

5 Conclusions

The purpose of document structure analysis is to get document structure of the original text. Document structure analysis is the base of automatic summarization, automatic classification, automatic indexing and automatic retrieval. A novel document analysis model — DLM is presented. DLM has 3 layers — physical structure layer, logical structure layer and semantic structure layer. The model has the feature of flexible, systematic and extendable. DLM is implemented on automatic summarization system. It makes the system can process documents with multi-format and generate summaries with good quality.

References

1. Kristen Maria Summers: Automatic Discovery of Logical Document Structure, Doctor Dissertation of Cornell University (1998)
2. Gerard Salton, James Allan, Amit Singhal: Automatic Text Decomposition and Structure, Information Processing & Management, Vol.32, No.2 (1996), 127-138
3. Gerard Salton, Amit Singhal, Mandar Mitra, Chris Buckley: Automatic Text Structure and Summarization, Information Processing & Management, Vol.33, No.2 (1997), 193-207
4. Marti A.Hearst: TextTiling: A Quantitative Approach to Discourse Segmentation. http://www.sims.berkeley.edu/~hearst/papers/tiling-tr93.ps

Prediction for Silicon Content in Molten Iron Using a Combined Fuzzy-Associative-Rules Bank*

Shi-Hua Luo, Xiang-Guan Liu, and Min Zhao

Institute of System Optimum Technique, Zhejiang University,
Hangzhou, 310027, China
luoshihua@nbu.edu.cn

Abstract. A general method is developed to generate fuzzy rules from numerical data that collected online from No.1 BF at Laiwu Iron and Steel Group Co.. Using such rules and linguistic rules of human experts, a new algorithm is established to predicting silicon content in molten iron. This new algorithm consists of six steps: step 1 selects some key variables which affecting silicon content in molten iron as input variables, and time lag of each of them is gotten; step 2 divides the input and output spaces of the given numerical data into fuzzy regions; step 3 generates fuzzy rules from the given data; step 4 assigns a degree to each of the generated rules for the purpose of resolving conflicts among the generated rules; step 5 creates a combined Fuzzy-Associative-Rules Bank; step 6 determines a fuzzy system model from input space to output space based on such bank. The rate of hit shot of silicon content is more than 86% in [Si] $\pm 0.1\%$ range using such new algorithm.

1 Introduction

Blast Furnace (BF) ironmaking process is highly complicated; whose operating mechanism is characteristic of nonlinearity, time lag, high dimension, big noise and distribution parameter etc [1]. It has not come true to realize automation of BF ironmaking process in metallurgical technology from the eighteenth of the twentieth century after trying methods of classical cybernetics and modern cybernetics, because of its complexity and no appropriate mathematical models of BF ironmaking process. The quality and the quantity of the different input material as well and many environment factors all influence the quality of the molten iron. Not only is silicon content in molten iron an important quality variable, it also reflects the internal state of the high-temperature lower region of the blast furnace [2], uniformity in silicon content and its accurate and advance prediction can greatly help to stabilize blast furnace operations. In past years, efforts have been made to build up effective model to predict silicon content in molten iron [3-9]. But designing a predictive controller, which can forecast accurately silicon content in molten iron ([Si]) is still a puzzle.

---

* Supported by the National Ministry of Science and Technology (99040422A.) and the Major State Basic Research Development Program of China (973 Program) (No.2002CB312200).

In such real-world prediction control, all of the information can be classified two kinds: numerical information obtained from sensor measurements and linguistic information obtained from human experts. The experience of the human controller is usually expressed as some linguistic "If-Then" rules, which state in what situations which action should be taken. The sampled input-output pairs are some numerical data which give the specific values of the inputs and the corresponding outputs. But each of the two kinds of information alone is usually incomplete. Some information will be lost when the human controller expresses in his/her experience by linguistic rules. On the other hand, the information from sampled input-output data pair is usually also not enough, because the past operations usually cannot cover all the situations the system will face. If such two kinds of information can be gotten, the most interesting case is when the combination of these two kinds of information is sufficient for a successful design [10]. The key idea of the new algorithm is to generate fuzzy rules from numerical data pairs, collect these fuzzy rules and the linguistic fuzzy rules into a combined Fuzzy-Associative-Rules Bank, then, design a predictive controller based on such bank [11].

In the production of molten iron processes, the value of different variables can be collected from different sensor measurements, but not all input variables of the blast furnace are useful for predicting the silicon content in molten iron. Incorporating variables that have little relevance to the particular output variable would cause excessive noise in the model. In [12], Liu selects five variables (see Table1) and proves that such variables are key variables affecting hot metal silicon content.

Table 1. Input variables

VC (t/h)	PI (m^3/min.kPa)	PC (t/h)	BQ (m^3/min)	$[Si]_{n-1}$ (%)
Charging mixture velocity	Permeability index	Pulverized coal injection	Blast quantity	Last [Si]

2 Principle of Algorithm and Model of Predictive Controller

2.1 The Principle of Algorithm

Suppose we are given a set of desired input-output data pairs:

$$(Input;\ Output) = (x_1(t-l_1), \cdots, x_i(t-l_i), \cdots, x_n(t-l_n);\ y(t)) \quad (1)$$

where x_i is input variable and y is output variable, l_i is time lag between each input variable and output variable respectively (i= 1,2..., n). The task here is generate a set of fuzzy rules from the desired input-output pairs of (1), and combines with the linguistic rules coming from human experts into a Fuzzy-Associative-Rules Bank. Then, using such bank to determine a mapping (a fuzzy system) $f: (x_1, x_2, \ldots, x_n) \to y$.

2.2 The Model of Predictive Controller

Being the principle of algorithm, established the model of predictive controller consists of the following six steps:

Step 1: Select some key variables which affecting silicon content in molten iron as input variables, and compute the time lag between each input variable and output variable.

The five variables listed in table.1 are key variables affecting hot metal silicon content, so they are used to be input variables undoubtedly, silicon content in molten iron [Si] is the output. Because the production of hot metal is a complex and length process (the mean interval between hot metal tapped is about 2 hours [12]), it is inevitable that there are time lags between inputs and output.

General Correlation Function R_g, presented by Ding [13] based on Mutual Information $I(X/Y)$, reflects nonlinear correlation of two stochastic variables, and can be used as correlation measure, not only is the correlation coefficient of two stochastic variables calculated by R_g, but time lag of variable can be gotten with R_g approaching its extreme value.

$$R_g = \frac{H(X) - H(X|Y)}{\sqrt{H(X)H(Y)}}, \tag{2}$$

$$H(X|Y) = -\sum_{i=1}^{n} P(y_i) P(x_i | y_i) \lg P(x_i | y_i), \tag{3}$$

where $P(x_i, y_i)$ is the joint distribution of x_i and y_i.

Equal Probability Method is used to calculated R_g, the number of web group is ascertain by following experiential formula:

$$M = 1.87 \times (n-1)^{2/5}, \tag{4}$$

where n is the sample size.

More sufficient sample size is, more adjacently R_g approaches its extreme value [13], shown as fig. 1.

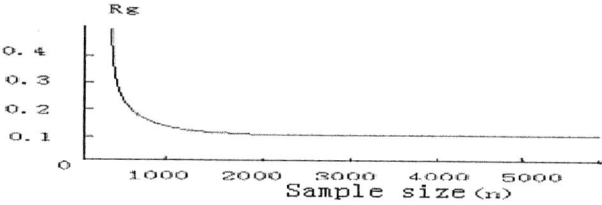

Fig. 1. The average general correlation coefficient of first order varying with sample size

Table 2 gives the time lag of each variable, calculated based on general correlation coefficients.

Table 2. Time lag of each variable

Input variable	Sample size	Web	Time lag (h)	R_g
VC	3000	46×46	2.5	0.107
PI	3000	46×46	1.5	0.103
PC	3000	46×46	2.5	0.106
BQ	3000	46×46	1.5	0.108

After calculating the optimal time lag for each of the input variables, the desired input-output data pairs is adjusted as (5) in order to implement these lags:

$$(Input(n); Output(n+1)) = (VC(n^*), PI(n), PC(n^*), BQ(n), [Si](n); [Si](n+1)) \quad (5)$$

where $VC(n^*) = 0.25VC(n-1) + 0.75VC(n)$ and $PC(n^*) = 0.25PC(n-1) + 0.75PC(n)$.

Step 2: Divide the input and output spaces into fuzzy regions.

Confirm that the domain intervals of every variable respectively, where "domain interval" of a variable means that most probably this variable will be in this internal (the values of a variable are allowed to lie outside its domain interval). Divide each domain interval into $2N+1$ regions (N can be different for different variables, and the lengths of these regions can be equal or unequal), and assign each region a fuzzy membership function μ. Table 3 shows a fuzzy cluster of every variable. The shape of each membership function is triangular, one vertex lies at the center of the region and has membership value unity; the other two vertices lie at centers of the two neighboring regions respectively.

Table 3. Fuzzy cluster of each variable

Fuzzy regions	VC	PI	PC	BQ	[Si]
S2	---	[13, 14.5]	[4, 7]	---	[0.20, 0.30]
S1	[52,70]	[14.5, 15.6]	[7, 10.2]	[1200, 1650]	[0.30, 0.38]
CE	[70,105]	[15.6, 17.8]	[10.2, 13.4]	[1650, 1780]	[0.38, 0.48]
B1	[105,120]	[17.8, 19.5]	[12, 15]	[1780, 1860]	[0.48, 0.60]
B2	---	[19.5, 22]	[15, 17.5]	---	[0.60, 0.90]

Step 3: Generate fuzzy rules from given data pairs.

For every given data pair such as (5), calculate $\mu_i(VC(n^*))$, $\mu_i(PI(n))$, $\mu_i(PC(n^*))$, $\mu_i(BQ(n))$, $\mu_i([Si](n))$ and $\mu_i([Si](n+1))$ (i= S2, S1, CE, B1, B2) respectively, selecting out the maximal value of each of them to create a fuzzy rule with "and". For example, from Fig. 2, two rules can be gained from two different data pairs such as:

(83.35,16.18,13.72,1785.68,0.42; 0.46) ⇒ Rule1: If VC is CE and PI is CE and PC is B1 and BQ is B1 and [Si](n) is CE, then [Si](n+1) is CE.
(100.85,16.5,9.91,1789.52,0.56; 0.78) ⇒ Rule2: If VC is B1and PI is CE and PC is S1 and BQ is B1 and [Si](n) isB1, then [Si](n+1) is B2.
Step 4: Assign a degree to each rule.

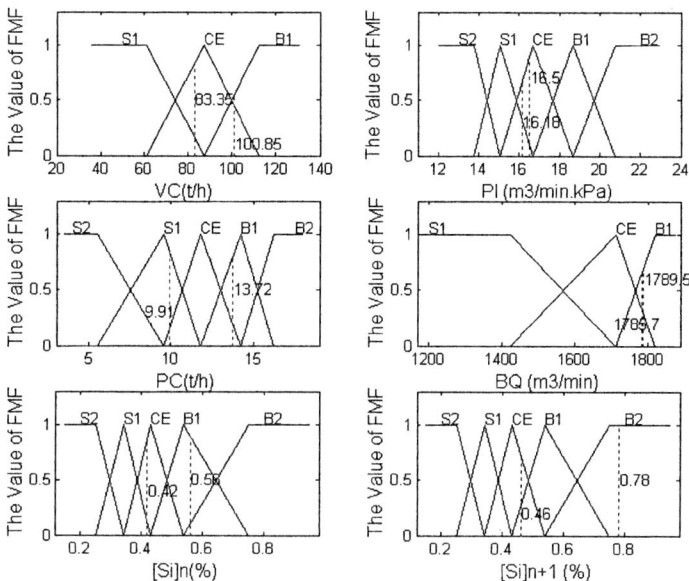

Fig. 2. Divide the input and output spaces into fuzzy regions and assign each region a fuzzy membership function (FMF)

Since there are usually lots of data pairs, and each data pair generates one rule, it is highly probably that there will be some conflicting rules, i.e., rules which have the same IF part but a different THEN part. One way to resolve this conflict is to assign a degree to each rule generated from data pairs, and accepts only the rule from a conflict group that has maximum degree. In this way not only is the conflict problem resolved, but also the number of rules is greatly reduced. The following product strategy is used to assign a degree (denoted by D (Rule)) to each rule, for the rule*: "IF x_1 is B1 and x_2 is CE, THEN y is S1", the degree of this rule is defined as:

$$D(Rule^*) = \mu_{B1}(x_1)\mu_{B2}(x_2)\mu_{S1}(y). \tag{6}$$

As examples: D (Rule1) = $\mu_{CE}(83.35)\mu_{CE}(16.18)\mu_{B1}(13.72)\mu_{B1}(1785.7)$ $\mu_{CE}(0.42)\mu_{CE}(0.46)$ = 0.84×0.68×0.8×0.67×0.89×0.73 = 0.199.
D (Rule2) = $\mu_{B1}(100.85)\mu_{CE}(16.5)\mu_{S1}(9.91)\mu_{B1}(1789.5)\mu_{B1}(0.56)\mu_{B2}(0.78)$ = 0.53×0.88×0.86×0.71×0.91×1 = 0.259.

In practice, some a prior information about the data pairs is existed, if let an expert check given data pairs, the expert may suggest that some are very useful and crucial, but others are very unlikely and may be caused errors. So assign a degree to each data pair which represents the expert's belief of its usefulness is very necessary. Suppose rule* has such degree μ^*_e, then the degree of rule* must be redefined as D (Rule*) $=\mu_{BI}(x_1)\mu_{CE}(x_2)\mu_{SI}(y)\mu^*_e$.

This is important in practical applications. For good data can be assigned higher degree ($\mu^*_e >1$), and for bad data can be assigned lower degree ($\mu^*_e <1$). Rule1 gained in step 3 is a good rule which reflects the traits of BF ironmaking process, so its degree must be recalculate as D(Rule1) = 0.199×1.5 = 0.2985 (let $\mu^*_e = 1.5$). And Rule2 is a bad rule [10], for the relevant data pair may be obtained when BF ironmaking process is very deviant, then we can recalculate its degree as D(Rule1) = 0.259×0.5 = 0.13 (let $\mu^*_e = 0.5$). Some data pairs which even the expert cannot judge would be set the degree μ^*_e to unity.

Step 5: Establish a combined Fuzzy-Associative-Rules Bank.

The form of the Fuzzy-Associative-Rules Bank is a five-dimensional chart, which is composed by some little boxes. The boxes of the bank can be filled with fuzzy rules according to the following strategy: such bank is assigned rules from either those generated from numerical data or linguistic rules (a linguistic rule also has a degree which is assigned by the human expert and reflects the expert's belief of the importance); if there is more than one rule in one box of the bank, use the rule that has maximum degree. In this way, both numerical and linguistic information are codified into a common framework, the combined Fuzzy-Associative-Rules Bank can be established.

In this case, the input space is five-dimensional, but the data pairs and expert rules are limited, as a result, many boxes of such bank may be empty. However, it is possible to fill up these empty boxes based on the limited given rules, and if the number of the data pairs and expert rules is big enough, some empty boxes cannot affect the precision of the model markedly.

Step 6: Determine a fuzzy system based on the combined Fuzzy-Associative-Rules Bank

The fuzzy logic system with combined Fuzzy-Associative- Rules Bank, centroid defuzzifies, product-inference and singleton fuzzifier is adopted such as equation (7) [14], [15]:

$$y = (\sum_{l=1}^{M} \bar{y}^l (\prod_{i=1}^{5} \mu_{A_i^l}(x_i))) / (\sum_{l=1}^{M} (\prod_{i=1}^{5} \mu_{A_i^l}(x_i))), \qquad (7)$$

where $\bar{y}^l$ is the center value of a fuzzy region which subject function is $\prod_{i=1}^{5} \mu_{A_i^l}(x_i)$, and M is the number of fuzzy rules in the combined Fuzzy-Associative- Rules Bank.

3 The Predictive Algorithm Simulation

Using above model to predict the value of sample data which gain from the Intelligent Control Expert System on 750m^3 BF Laiwu Iron & Steel Co. in China.

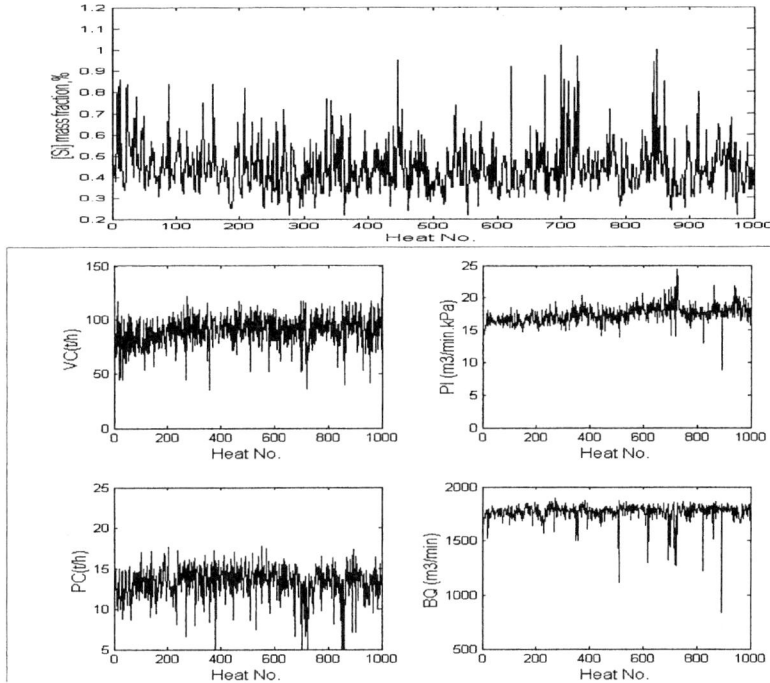

Fig. 3. Time series graph of the every variable in No.1 BF at Laiwu Iron and Steel Group Co

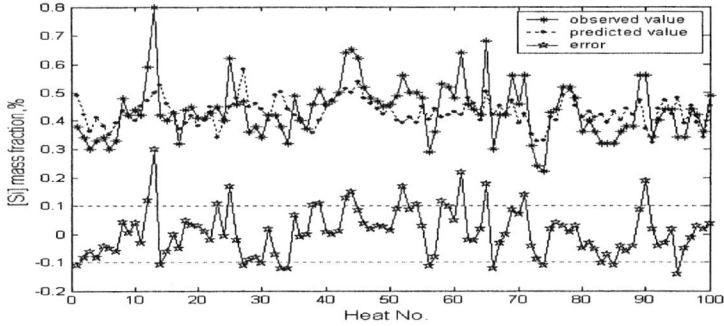

Fig. 4. The predictive result of case(1)

Here we select a sample denoting Heat No. from 30546 to31545. Fig.3 shows the 1000 data pairs that we use to test the new algorithm, every variable's time series is showed detailedly in such figure.

The first 900 points (6-dimension) of the data pairs were used as training data, and the final 100 points were used as test data. Three cases were simulated: (1) 300 training data (from 601-2 to 900) were used to construct the combined Fuzzy-Associative-Rules Bank; (2) 700 training data (from 201-2 to 900) were used; and, (3) 898training data (from 1 to 900) were used. Figs. 4, 5 and 6 show the result of each case. Comparing Figs.4 and 5, we can find that we obtain an evidently improved prediction, but comparing Figs.5 and 6, we obtain a slightly improved prediction. It indicates that if rules are not enough, they cannot reflect the characteristics of BF iron-making process completely, but if rules are enough to do that, more data pairs cannot generate more efficient rules.

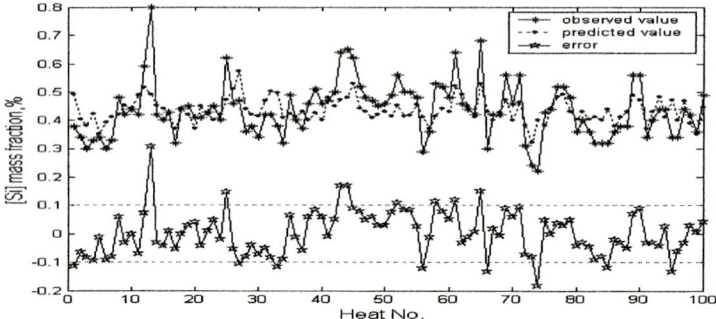

Fig. 5. The predictive result of case (2)

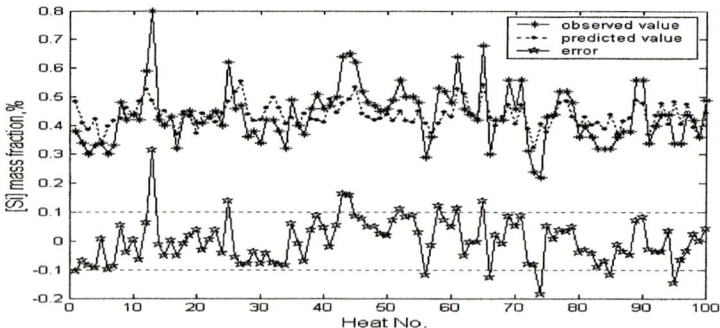

Fig. 6. The predictive result of case (3)

To evaluate the performance of the prediction model, some important criteria used in practice are considered as follows:

$$\text{Perr} = \sum_{j=1}^{N_p} (x'_j - x_j)^2 / \sum_{j=1}^{N_p} x_j^2 \ , \tag{8}$$

where x_j' is the predicted value, x_j the observed value and N_p the total predicted tap numbers.

$$J = \frac{1}{N_p}(\sum_{j=1}^{N_p} H_j) \times 100\%, \qquad (9)$$

where

$$H_j = \begin{cases} 1 & \text{where } |x_j' - x_j| \leq 0.1 \\ 0 & \text{else.} \end{cases} \qquad (10)$$

Generally, if H_j equals to one we say that the prediction hits the target, and J denotes the percentage of prediction hits the target. Analyzing the predicted values and observed values, we can get the result of prediction, shown as Table 4:

Table 4. The result of prediction

Group No.	The number of training	Standard deviation	Perr	J
Case (1)	300	0.119	0.0362	74%
Case (2)	700	0.112	0.0313	84%
Case (3)	898	0.111	0.0301	86%

With these criteria, the percentage of prediction hitting the target is 86% and Perr is in the magnitude of 10^{-2} in case (3), which is helpful for operator to make right decision to operate blast furnace.

One advantage of such new algorithm is that it is very easy to add a new rule to the Fuzzy-Associative-Rules Bank when a new data which creates a new rule gained. If the technics of BF ironmaking process is charged, such work is important for meliorating the predictor.

4 Conclusions

The main conclusions in this paper are: (1) Five key variables affecting silicon content were selected out, and time lag of each of them has been calculated with general correlation coefficient. (2) A general method is developed to generate fuzzy rules from numerical data, using such rules and linguistic rules of human experts, a Fuzzy-Associative-Rules Bank is established. (3) The new algorithm was applied to predict silicon content in molten iron, and good performance is shown due to the high percentage of prediction hitting the target.

References

1. Biswas, A.K.: Principles of Blast Furnace Ironmaking. SBA Publication, Calcutta (1984)
2. Chen, J.: A Predictive System for Blast Furnaces by Integrating a Neural Network with Qualitative Analysis. Engineering application of artificial intelligence, 14 (2001) 77-85
3. Gao, C.H., Zhou, Z.M.: Modified Chaotic Adding Weight One-rank Local-region Forecasting for Silicon Content in Molten Iron of Blast Furnace. Acta Phys. Sin., 53 (2004) 4092-4096
4. Gao, C.H., Zhou, Z.M., Shao, Z.J.: Chaotic Local-region Linear Prediction of Silicon Content in Hot Metal of Blast Furnace. Acta Metall. Sin. 41 (2005) 433-436
5. Singh, H, Sridhar, N.V., Deo, B.: Artificial Neural Nets for Prediction of Silicon Content of Blast Furnace Hot Metal. Steel Research, 67 (1996) 521-527
6. Yao, B., Yang, T. J., Ning, X.J.: An Improved Artificial Neural Network Model for Predicting Silicon Content of Blast Furnace Hot Metal. Journal of University of Science and Technology Beijing, 7(2000) 269-272
7. Miyano, T., Kimoto, S., Shibuta, H. et. al.: Time Series Analysis and Prediction on Complex Dynamical Behavior Observed in a Blast Furnace. Physica D, 135 (2000) 305-330
8. Jose, A.C., Hiroshi, N., Jun-ichiro, Y.: Three-dimensional Multiphase Mathematical Modeling of the Blast Furnace Based on the Multifluid Model. ISIJ International, 42 (2002) 44-52
9. Juan, J., Javier, M., Jesus, S.A., et. al.: Blast Furnace Hot Metal Temperature Prediction Through Neural Networks-based Models. ISIJ International, 44 (2004) 573-580
10. Wang, L.X., Mendel, J.M.: Generating Fuzzy Rules by Learning from Examples. IEEE Trans. on Systems, Man and Cybern., 22(1992) 1414-1427
11. Wang, L.X.: Fuzzy Systems are Universal Approximators. Proc. IEEE Int. Conf. Fuzzy Systems, San Diego, CA, USA, (1992) 1163-1170
12. Liu, X.G., Liu, F.: Optimization and Intelligent Control System of Blast Furnace Ironmaking Process. Metallurgical industry Press, Beijing (2003)
13. Ding, J., Wang, W.S., Zhao, Y.L.: General Correlation Coefficient Between Variables Based on Mutual Information. Journal of Sichuan University (Engineering Science Edition), 34(2002) 1-5
14. Lee, C.C.: Fuzzy Logic in Control Systems, Fuzzy Logic Controller, Part I. IEEE Trans. on SYST., Man, and Cybern., SMC-20 (2) (1990) 404-418
15. Wang, L.X.: A Supervisory Controller for Fuzzy Control Systems that Guarantees Stability. IEEE Trans. On Automatic Control, 39 (1994) 1845-1848

An Investigation into the Use of Delay Coordinate Embedding Technique with MIMO ANFIS for Nonlinear Prediction of Chaotic Signals

Jun Zhang[1], Weiwei Dai[1], Muhui Fan[3], Henry Chung[2], Zhi Wei[3], and D.Bi[4,*]

[1] SUN Yat-sen University, P.R. China,
junzhang@ieee.org
[2] City University of Hong Kong,
[3] School of Mechanical Engineering,
Hebei University of Technology, Tianjin 300130, P.R. China
[4] Box 140, Tianjin University of Science and Technology,
Tianjin, 300222, P.R. China

Abstract. This paper presents an investigation into the use of the delay coordinate embedding technique with multi-input multi-output (MIMO) adaptive-network-based-fuzzy-inference system (ANFIS) to learn and predict the continuation of chaotic signals ahead in time. Based on the average mutual information and global false nearest neighbors techniques, the optimal values of the embedding dimension and the time delay are selected to construct the trajectory on the phase space. The MANFIS technique is trained by gradient descent algorithm. First, the parameter set of the membership functions is generated with the embedded phase space vectors using the back-propagation algorithm. Second, fine-tuned membership functions that make the prediction error as small as possible are built. The model is tested with both periodic and the Mackey-Glass chaotic time series. Moving root-mean-square error is used to monitor the error along the prediction horizon.

1 Introduction

A time series is a sequence of regularly sampled quantities out of an observed system. It provides a useful basis for discovering some of its underlying characteristics, such as periodicity, stochastic distribution, etc. Many time- and frequency-domain prediction methods have been proposed 1. However, in practice, nonlinear chaotic time series are regularly observed in various natural phenomena 13. Since 1970's, chaotic time series prediction has been a popular subject 4 for understanding and controlling the chaotic behaviors to advantage, such as the stock market forecasting 5. Basically, prediction of chaotic time series requires a representative model. In recent years many new prediction approaches, such as the fuzzy predictor 6-13 and time-delay embedding technique 14, have emerged. They provide new insight into this type of systems not available with the traditional methods, such as linear regression technique and auto-regressive integrated moving average models, etc.

---

[*] Corresponding author.

As discussed in 9, the adaptive-network-based fuzzy inference system (ANFIS) with multi-input and single-output in 11 that uses Gaussian membership functions and employs hybrid back-propagation learning of a chaotic time series gives the smallest root-mean-square errors (RMSE) among various fuzzy predictors. However, similar to the other single scale chaotic time series prediction methods, the prediction horizon is usually limited by the fast-varying components, as the methods are based on the finite neighborhood relationships in the prediction. Nevertheless, it forms the best basis for further enhancement. On the other hand, time delay coordinate embedding methods 14-17 use the relationships between the delay coordinates of a point and the points that appear at some time later in the phase space. Its trajectories behave with quasi-periodicity. The nearest trajectories can contribute to the neighboring set with more than one point, resulting in an increased weighting of the contribution coming from the nearest trajectories.

This paper investigates the use of delay coordinate embedding technique with multi-input multi-output (MIMO) ANFIS to learn and predict the continuation of chaotic signals ahead in time. The methodology hybridizes the advantages of ANFIS and the time-delay coordinate embedding technique. The resulting model has better predictive performance with fewer membership functions. Based on the average mutual information and false nearest neighbors technique, the minimum time-delay embedding dimension is selected and phase space is constructed. The ANFIS is trained by gradient descent algorithm. First, the parameter set of the membership functions is generated with the embedded phase space vectors using the back-propagation algorithm. Second, fine-tuned membership functions that make the prediction error as small as possible are built. Section II describes the construction of the chaos-embedded phase space. Section III describes the operations of the MIMO-ANFIS technique. Section IV shows the procedures of the prediction algorithm. Section V gives the simulation results using the proposed method and the one in 11 having different number of membership functions to predict the Mackey-Glass chaotic time series. Moving root-mean-square error is used to monitor the error along the prediction horizon. Section VI contains some concluding remarks.

2 Construction of the Chaos Embedded Phase Space

Section V gives the simulation results using the proposed method and the one in 11 having different number of membership functions to predict the Mackey-Glass chaotic time series. Moving root-mean-square error is used to monitor the error along

A chaotic time series shows stochastic behavior in the time- and frequency-domain and deterministic behaviors in phase space structure 1. The trajectory of the attractors will repeat on the phase space with same initial conditions. Consider a time series $x(1), x(2), x(3), \ldots x(N)$, an embedded vector $y(i)$ is defined as

$$y(i) = [x[i+(D-1)\tau] \quad \ldots \quad x(i+\tau) \quad x(i)] \tag{1}$$

where $1 \leq i \leq N-(D-1)\tau$. $y(i)$ is a D-dimensional vector consisting of past signal samples, in which D is referred to the embedding dimension. τ is the time delay of the samples. $y(i)$ represents one point on a D-dimensional phase space R^D. A trajectory Y

$$Y = [y(1) \quad y(2) \quad ... \quad y(N-(D-1)\tau)]^T \qquad (2)$$

will be formed on R^D. Theoretically, an embedding of the original trajectory can be obtained for sufficiently large value of D with any value of τ. In practice, if D is too large, noise in the data may reduce the density of points defining the attractor and increase the level of contamination of the data unnecessarily [18]. However, if D is too small, the attractor will be folded. Moreover, the choice of τ is another crucial parameter to establish data correlation in the embedded vector. If τ is too small, the elements in each embedded vector will not be independent enough. On the contrary, if τ is too large, the relationships among elements in each embedded vector behave as a set of random data statistically [1]. Thus, optimal values of D and τ have to be determined in order to efficiently extract the original behaviors of the chaotic system. In this paper, optimal values of τ and D are selected by using the average mutual information[20] and the global false nearest neighbors [1], respectively.

A. Determination of τ

Average mutual information $I(\tau)$ is used to determine the nonlinear autocorrelation between $x(i)$ and $x(i+\tau)$ in order to form coordinates in a time delay vector. That is,

$$I(\tau) = \sum_{i=1}^{N-\tau} \sum_{j=\tau+1}^{N} P(x(i), x(j)) \log_2 \frac{P(x(i), x(j))}{P(x(i)) P(x(j))}. \qquad (3)$$

where $P(x(i))$ and $P(x(j))$ are the individual probability densities in values of $x(i)$ and $x(j)$, respectively and $P(x(i), x(j))$ is the joint probability density in values of $x(i)$ and $x(j)$.

The procedures are started with a value of one for τ and the corresponding value of $I(\tau)$ is calculated by (3). τ is then incremented until there is a sign change in the variation of $I(\tau)$ with respect to τ. That is, τ is chosen when

$$\text{sgn}[I(\tau) - I(\tau-1)] \neq \text{sgn}[I(\tau+1) - I(\tau)]. \qquad (4)$$

B. Determination of D

For a given value of τ, a statistical technique is used to determine the minimum value of D that can unfold the reconstructed attractor. The procedures are as follows,

1) A dimension of 2 is firstly chosen as the initial guess.
2) All embedded vectors are formulated by (1) with the assumed phase space dimension.
3) The Euclidean distance between one vector and the others is calculated and a matrix containing all the distances (say $\vec{M}$) for the given dimension is formulated. For example, the distance between the vectors $y(i)$ and $y(j)$ is equal to

$$m_{i,j} = m_{j,i} = \sqrt{\sum_{k=1}^{D} [x(i-(k-1)\tau) - x(j-(k-1)\tau)]^2} \qquad (5)$$

where $1 \leq i \leq N-(D-1)\tau$ and $1 \leq j \leq N-(D-1)\tau$. $m_{i,j}$ is a matrix element at ith row and jth column of $\vec{M}$.

4) By considering the smallest row element in each column of $\vec{M}$, a pair of embedded vectors is chosen for the corresponding column. For example, for a generic column g, an embedded vector pair of $[y(g), y(h)]$ is chosen if the element in the hth row is the smallest along the column. This procedure is performed for all columns in $\vec{M}$.

5) Two modified embedded vectors $y'(g)$ and $y'(h)$ are formulated with a higher dimension of $(D + 1)$. That is,

$$y'(g) = [x(g-D\tau) \quad x[g-(D-1)\tau] \quad \ldots \quad x(g+\tau) \quad x(g)] \quad (6a)$$
$$y'(h) = [x(h-D\tau) \quad x[h-(D-1)\tau] \quad \ldots \quad x(h+\tau) \quad x(h)] \quad (6b)$$

6) A normalized value $\rho(D)$ showing the change in the distance of $[y(g), y(h)]$ and the distance of $[y'(g), y'(h)]$ is used to test the existence of any fold in the trajectory between $y(g)$ and $y(h)$ with D dimensions. That is,

$$\rho(D) = \frac{|x(g-D\tau) - x(h-D\tau)|}{m_{g,h}} \quad (7)$$

If $\rho(D)$ is larger than a threshold value Φ_{thres}, such as 10, it implies that a fold exists between $y(g)$ and $y(h)$. The embedded pair $[y(g), y(h)]$ is a false neighbor. An index λ is used to count the number of false neighbors in $\vec{M}$ for the dimension D. Thus,

$$\text{if } \rho(D) > \Phi_{thres}, \quad \lambda(D) = \lambda(D) + 1. \quad (8)$$

7) The above procedures will be repeated from step 2) with a new dimension of $(D + 1)$. Theoretically, the maximum value of D is up to an integer of ($\frac{N-2}{\tau}$). In this method, D is chosen when λ is in the first minimum. That is, it is in the condition of

$$\text{sgn}[\lambda(D) - \lambda(D-1)] \neq \text{sgn}[\lambda(D+1) - \lambda(D)]. \quad (9)$$

3 Operations of the MIMO-ANFIS

Prediction of chaotic time series using fuzzy neural system has been investigated in 21. In this paper, we showed that better performance can be achieved if prediction is done in embedding phase space instead of time domain. A Multi-Input Multi-Output Adaptive Neural-Fuzzy Inference System (MANFIS) is developed for predicting the chaotic time series in embedding phase space. Based on the ANFIS model 11, the

MANFIS is extended to generate multi-dimensional vector. A fuzzy rule set is applied to model the system, which maps precisely the input vectors (embedding phase space) to the output vectors. The fuzzy rule set consists of a series of IF-THEN rules operating on some fuzzy variables. These fuzzy variables are described by the corresponding membership functions, which is tuned by a gradient descent algorithm using collections of input-output vector pairs. Knowledge acquisition is achieved by multiplying the fuzzy quantities.

The topology of MANFIS is showed in Fig. 1. It can be partitioned, according to functionality, into the following sections.

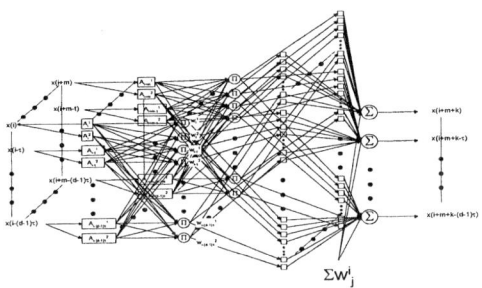

Fig. 1. Proposed Structure of the MANFIS

A. The Input Section

Input to the MANFIS is a matrix with the embedding phase vector as columns and the chaotic time series as rows. An embedding phase space vector, $\overline{y(i)}$, is reconstructed for every single point on the time series $\{x(1), x(2), \cdots x(N)\}$. These vectors are reconstructed by embedding time delay method described in Sec. II. If d is the dimension of the embedding phase space, input to the MANFIS is a $(m+1) \times d$ matrix. For instance, if the embedding time delay is T, the phase space vector is given by:

$$[x(i), x(i-T), \cdots, x(i-(d-1) \cdot T)], \quad \forall i \in [1+(d-1) \cdot T, N] \quad (10)$$

and, the input matrix is:

$$\begin{bmatrix} x(i) & x(i+1) & \cdots & x(i+m) \\ x(i-T) & x(i+1-T) & \cdots & x(i+m-T) \\ \vdots & \vdots & \vdots & \vdots \\ x(i-(d-1) \cdot T) & x(i+1-(d-1) \cdot T) & \cdots & x(i+m-(d-1) \cdot T) \end{bmatrix} \quad (11)$$

where k is the number of prediction steps, and $i \in [1+(d-1) \cdot T, N-m-k]$.

B. The Fuzzifier Section

To be able to utilize fuzzy reasoning on the prepared input data, knowledge representation should be applied to each of the elements of the matrix. This is done by feeding the matrix elements to the fuzzifier that consists of two fuzzy membership functions. Each of these membership functions represents a linguistic label, which will be used in the fuzzy rule to generate the corresponding knowledge. Fuzzifying the matrix elements will quantify how important such a value is inside the particular linguistic label. The fuzzy values will then be used in the knowledge acquisition and reasoning sections. Gaussian distribution fuzzifier is used as the membership function MANFIS. The equation of the membership function is as follows,

$$A_o^p = \mu_p(x_o; c_p, \sigma_p) = \exp\left(-\frac{1}{2}\left(\frac{x_o - c_p}{\sigma_p}\right)^2\right) \tag{12}$$

where, o is the number of element of the input matrix,
p is the number of membership functions,
c and σ are parameters that determine the center and width of the membership function.

The knowledge, which is represented by linguistic labels, is vital to the fuzzy reasoning part, hence, the output of the whole system. Therefore, the choice of membership function and its distribution will have a direct impact on the overall system behavior and performance. Thus, c and σ must be tuned carefully with gradient decent algorithm.

C. The Knowledge Acquisition Section

Since it is very difficult to acquire enough knowledge from the chaotic time series to construct the rule-base, no IF-THEN rule set is used in MANFIS. Instead of relating the linguistic qualifiers with IF-THEN rules, weighted sum of products of the fuzzy values is used. The following equation is used to quantify the knowledge contributed by each element of the embedding phase space vector.

$$w_o^p = \prod_{i+g-(d-1)T \le o \le i+g} A_o^p \tag{13}$$

where g is the number of columns in the input matrix, $(i+g-(d-1)T) \le o \le (i+g)$, $0 \le g \le m$, $1 \le p \le 2$. In fact, w_o^p represents the firing strength of a rule.

D. The Knowledge Reasoning Section

$$f_u^g = \sum_{v=i+g-(d-1)T}^{i+g} a_u^v x(v) \tag{14}$$

where $i+g-(d-1)T \le u \le i+g$, $0 \le g \le m$,

f_u^g is calculated using all elements in the gth vector,

a_u^v is the consequent parameter set,

$x(v)$ is the vth element in g^{th} vector of input matrix.

$$OK = F_{g,u}^{p,h} = w_h^p \cdot f_u^g \tag{15}$$

where $1 \le p \le 2$, $(i+g-(d-1)T) \le u \le (i+g)$, $(i+g-(d-1)T) \le h \le (i+g)$, $0 \le g \le m$.

E. The Defuzzifier Section

The function of this section is to map from the fuzzy set to the real-value points by calculating the centroid of each fuzzy variable. In the MANFIS, it is the output vector elements, which are calculated in (16):

$$OD_j = \frac{\sum\limits_{\substack{0 \leq g \leq m \\ p=1\sim2 \\ i+m-(d-1)T \leq u \leq i+m \\ j+g-(d-1)T \leq h \leq j+g}} F_{g,u}^{p,h}}{\sum\limits_{\substack{o=(i+g-(d-1)T)\sim(i+g) \\ p=1\sim2 \\ g=0\sim m}} w_o^p} \quad (16)$$

where $j=(i+m+k)\sim(i+m-(d-1)T+k)$, F is calculated from (15) and the output vector matrix is

$$\begin{bmatrix} x(i+m+k) \\ x(i+m+k-T) \\ \vdots \\ x(i+m+k-(d-1)\cdot T) \end{bmatrix}, i \in [(d-1)\cdot T, N-m-k] \quad (17)$$

F. The learning algorithm

The MANFIS is trained with data obtained in both time domain and phase space. To achieve accurate prediction gradient descent or back-propagation algorithm is used to tune the membership functions and the consequent parameters. σ and c are the parameters of the membership functions, adjustment to these parameters are determined according to the gradient between the actual and expected output. That is,

$$\sigma_i^p(t+1) = \sigma_i^p(t) - \eta \frac{\partial E_i}{\partial \sigma_i(t)} \quad (18)$$

$$c_i^p(t+1) = c_i^p(t) - \eta \frac{\partial E_i}{\partial c_i(t)} \quad (19)$$

where p is the number of membership functions,
          i is the number of node,
          η is a constant determining the learning rate,
          E is the error measure for the train data.

G. Simulation results and discussion

In order to show that the performance of MANFIS can be improved by applying embedding phase space transformation to the input data, two types of simulations have been carried out. The first one is a periodic time series and the second one is a chaotic time series (Mackey-Glass chaotic time series 20). The input data will be represented in both time domain and embedding phase space domain. Different numbers of training sets (100, 300, and 500) are used. The corresponding prediction errors are compared.

1. *Periodic time series*

The periodic time series equation is under investigation in

$$x(t) = \frac{1}{w}\sum_{i=a}^{b}\sin(k \cdot i \cdot t) \qquad (20)$$

where $w=5$, $k=0.01$, $a=1$ and $b=5$. The time series is shown in Fig. 2.

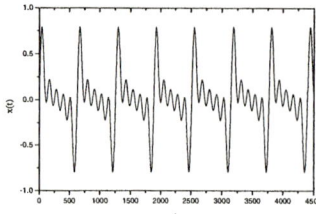
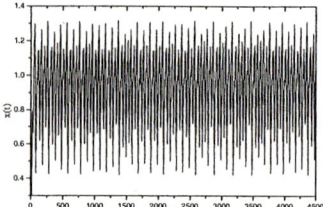

Fig. 2. Periodic Time Series of (20)   **Fig. 3.** Mackey-Glass chaotic time series

2. *Chaotic time series: Mackey-Glass equation*

The following Mackey-Glass equation (21) has been shown to be chaotic in 19 and 20 is investigated in this paper. The time series is shown in Fig. 3.

$$\frac{dx(t)}{dt} = \frac{0.2 \cdot x(t-\tau)}{1+x^{10}(t-\tau)} - 0.1 \cdot x(t) \qquad (21)$$

where $\tau = 17$, $x(0) = 1.2$.

It is a time-delay ordinary differential equation, which displays well-understood chaotic behavior with dimensionality dependent upon the chosen value of the delay parameter. The time series generated by the Mackey-Glass equation has been used as a test bed for a number of new adaptive computing techniques.

3. *Error Estimation:*

To compare the accuracy we compute the normalized mean squared error (NMSE):

$$\text{NMSE}(N) = \frac{\sum_{k\in\Lambda}(actual_k - prediction_k)^2}{\sum_{k\in\Lambda}(actual_k - mean_\Lambda)^2} \approx \frac{1}{\hat{\sigma}_\Lambda^2}\frac{1}{N}\sum_{k\in\Lambda}(x_k - \hat{x}_k)^2 \qquad (22)$$

where x_k is the kth point of the series of length N. $\hat{x}_k$ is the predicted value, and $mean_\Lambda$ and $\hat{\sigma}_\Lambda^2$ denote the sample average and sample variance of the actual values (targets) in Λ.

4. *Comparisons of the prediction errors*

Several simulations on predicting the periodic function in (20) and the chaotic signals in (21) have been performed. All MANFIS coefficients and the prediction results are shown in Fig. 4 – 15 and in Table I & Table II, in which *mf* represents the number of membership functions used, *ts* represents the number of set used, and *te* represents the number of training epoch spent periodic. The following observations can be noted.

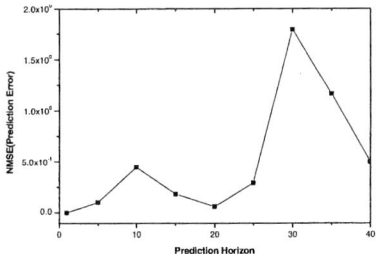
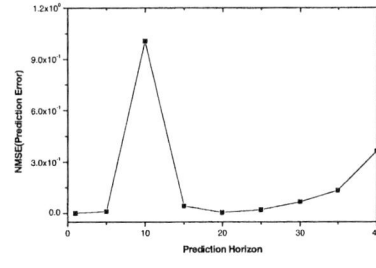

Fig. 4. Periodic time Series Prediction in Time Domain. (mf=2,ts=100,te=500).

Fig. 5. Periodic time Series Prediction in Time Domain. (mf=2,ts=300,te=500).

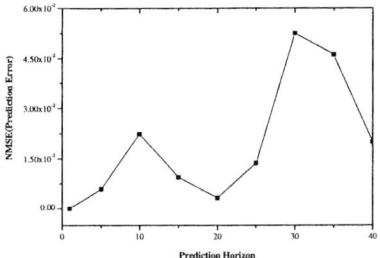
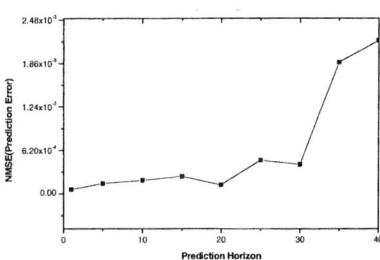

Fig. 6. Periodic time Series Prediction in Time Domain. (*mf=2,ts=500,te=500*).

Fig. 7. Periodic time Series Prediction in Phase Space Domain. (*mf=2,ts=100,te=500*).

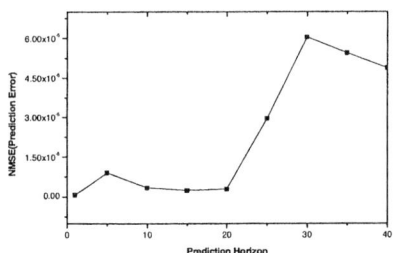
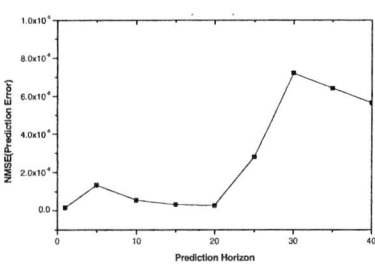

Fig. 8. Periodic time Series Prediction in Phase Space Domain.(mf=2,ts=300,te=500).

Fig. 9. Periodic time Series Prediction in Phase Space Domain. (mf=2,ts=500,te=500).

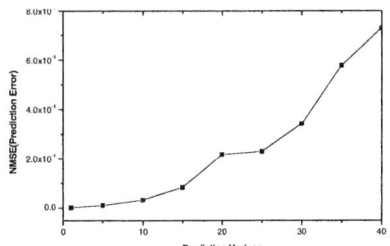
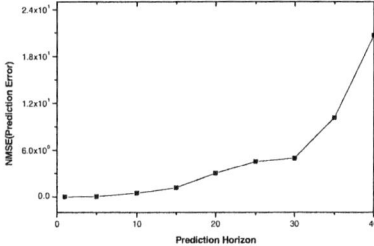

Fig. 10. Chaotic time Series Prediction in Time Domain. (mf=2,ts=100,te=500).

Fig. 11. Chaotic time Series Prediction in Time Domain. (mf=2,ts=300,te=500)/

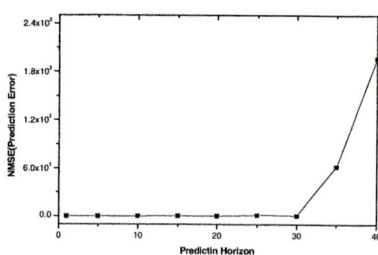
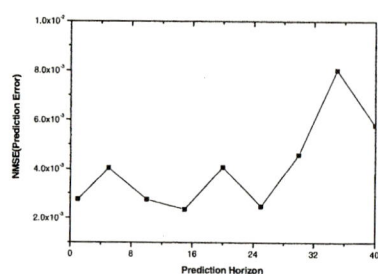

Fig. 12. Chaotic time Series Prediction in Time Domain. (mf=2,ts=500,te=500).

Fig. 13. Chaotic time Series Prediction in Phase Space Domain. (mf=2,ts=100,te=500).

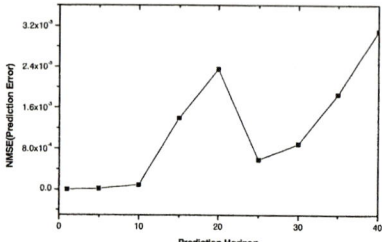
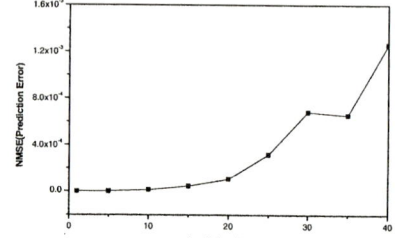

Fig. 14. Chaotic time Series Prediction in Phase Space Domain. (mf=2,ts=300,te=500).

Fig. 15. Chaotic time Series Prediction in Phase Space Domain. (mf=2,ts=500,te=500).

a. Periodic time series prediction

Figs. 4 - 6 show the prediction errors of the periodic time series using the method in 10 with different values of *mf*, *ts*, and *te*. Figs. 7 - 9 show the ones using the proposed method with the same simulation parameters. Compared Figs. 4 – 6 with Figs. 7 – 9, respectively, the prediction errors with embedding phase space preprocessing are lower many times than the ones without embedding phase space preprocessing using the same simulation parameters.

In other words, less number of training sets is required in MANFIS with embedding phase space to achieve the same error in MANFIS without embedding phase space. For example, comparing Fig. 6 and Fig. 7, the former one has 500 training sets and the latter one has 100 training sets. However, the former one gives a maximum error of 0.0525 over the prediction horizon, whilst the latter one gives a maximum error of 0.00218 only.

b. Chaotic time series prediction

Figs.10 - 12 show the prediction errors of the chaotic time series using the method in 10. Figs.13 - 15 show the ones using the proposed method with the same simulation parameters. The prediction errors without embedding phase space preprocessing are high. Even if the training set is large, the chaotic properties of the Mackey-Glass series cannot be predicted. The maximum error in the prediction horizon is 61.973 in Fig. 12. However, with the use embedding phase space, the MANFIS is able to keep the maximum prediction errors below 0.0080 in Fig. 13, 0.0031 in Fig. 14, and 0.00013 in Fig. 15, respectively.

4 Conclusions

It is very difficult to perform accurate prediction on nonlinear or chaotic series such as Mackey-Glass time series. It has been proved that only adaptive fuzzy system cannot give satisfactory prediction results. The use of delay coordinate embedding technique with simple adaptive fuzzy system, as proposed in this paper, can enhance the prediction. The structure of a multi-input multi-output ANFIS (MANFIS) with two membership functions has been investigated. The system was trained with backpropagation learning algorithm. Simulation results show that prediction accuracy of a nonlinear system can be significantly improved by preprocessing the time series data with delay coordinate embedding technique. Moreover, the required training set can also be reduced.

References

1. Brockwell P. and Davis R.: Time Series: Theory and Methods. Springer-Verlag, New York (1987)
2. Abarbanel H.D.I.: Analysis of Observed Chaotic Data. Springer-Verlag: New York (1996).
3. Robert R. Trippi: Chaos & Nonlinear Dynamics in The Financial Markets: Theory, Evidence and Applications. IRWIN Professional Publishing (1995)
4. Weigend A.S. and Gershenfeld N.A.: Time Series Prediction: Forecasting the Future and Understanding the Past. Addison-Wesley, MA (1994).
5. Ye Z. and Gu L.: A fuzzy system for trading the shanghai stock market. In Trading on the Edge, Neural, Genetic, and Fuzzy Systems for Chaotic Financial Markets. G.J. Deboeck, Ed. New York: Wiley (1994) 207-214
6. La Pense A. C. and Mort N.: A Neuro Fuzzy time series prediction application in telephone banking. 1999 Third International Conference On Knowledge-Based Intelligent Information Engineering Systems, Australia (1999) 407-410
7. Kozma R., Kasabov N.K., Swope J.A. and Williams M.J.A.: Combining Neuro-Fuzzy and chaos methods for intelligent time series analysis—case study of heart rate variability. 1997 IEEE Systems, Man, and Cybernetics International Conference on Computational Cybernetics and Simulation 4 (1997) 3025 -3029
8. Wang L.X. and Mendel J.M.: Generating fuzzy rules by learning from examples. IEEE Trans. Syst., Man, Cybern. 22 (1992) 1414-1427
9. Kim D. and Kim C.: Forecasting time series with genetic fuzzy predictor ensemble. IEEE Trans. Fuzzy Systems 5 (1997) 523-535
10. Jang J.R. and Sun C.: Predicting chaotic time series with fuzzy IF-THEN rules. In 2nd IEEE Int. Conf. Fuzzy Syst, San Francisco, CA Mar 2 (1993) 1079-1084
11. Jang J.R.: ANFIS: Adaptive-network-based fuzzy inference system. IEEE Trans. Syst., Man, Cybern. 23 (1993) 665-685
12. Lotfi A. Zadeh: Soft Compting and Fuzzy Logic. IEEE Software (1994) 48-56
13. Lofti A. Zadeh: Fuzzy Logic. IEEE Computer (1988) 83-93
14. Alparslan A.K., Sayar M., and Atilgan A.R.: State-space prediction model for chaotic time series. Physical Review E 58 (1998) 2640-2643
15. Packard N.H., Crutchfield J.P., Farmer J.D., and Shaw R.S.: Geometry from a time series. Phys. Rev. Lett. 45 (1980) 712-716

16. Takens F.: in Dynamical Systems and Turbulence, Warwick. D.A. Rand and L.S. Young eds. Springer, Berlin (1980)366.
17. Eckmann J.P. and Ruelle D.: Ergodic theory of chaos and strange attractor. Rev. Mod. Phys.57 (1985) 617-656
18. Wolf A., Swift J.B., Swinny H.L., and Vastano J.A.: Determining Lyapunov exponents from a time series. Physica 16D (1985) 285-317
19. Edward Ott, Tim Sauer, and James A. York: Coping with Chaos: Analysis of Chaotic Data and The Exploitation of Chaotic Systems. John Wiley & Sons, Inc. (1994) 1-13
20. Fraser A.M. and Swinney H.L.: Independent coordinates for strange attractors from mutual information. Phys. Rev. A 33(1986) 1134-1140
21. Liang Chen and Guanrong Chen: Fuzzy Modeling, prediction, and Control of Uncertain Chaotic Systems Based on Time Series. IEEE Trans. On Circuits and Systems I 47 (2000)
22. Mackey M. C. and Glass L.: Oscillation and chaos in physiological control systems. Science 197 (1977) 287-289

Replay Scene Based Sports Video Abstraction

Jian-quan Ouyang[1,2], Jin-tao Li[1], and Yong-dong Zhang[1]

[1] Institute of Computing Technology, Chinese Academy of Sciences, Beijing, China, 100080
oyjq@cie.xtu.edu.cn, {oyjq, jtli, zhyd}@ict.ac.cn
[2] College of Information Engineering, Xiangtan University, Xiangtan, China, 411105

Abstract. Video abstraction can be useful in multimedia database indexing and querying and can illustrate the important content of a longer video to quick browsing. Further, in sports video, replay scene often demonstrates the highlight of the video. The detection of replay scene in the sports video is a key clue to sports video summarizing. In this paper, we present a framework of replay scene based video abstraction in MPEG sports video. Moreover, we detect identical events using color and camera information after detecting replay scene using MPEG feature. At last, we propose a three-layer replay scene based sports video abstraction. It can achieve real time performance in the MPEG compressed domain, which is validated by experimental results.

1 Introduction

Multimedia analysis and retrieval is one of the hottest issues of the information research. With the development of the artificial intelligent, communication and multimedia technology, the amount of multimedia work including digital video is vast in various fields. However, Traditional text-based information retrieval technology cannot analysis the structure of the multimedia effectively and efficiently. While multimedia analysis and retrieval can provide efficacious framework of retrieving the multimedia by extracting the lower feature and obtain the semantic content and become useful both for research and application.

Among this huge amount of visual information, the need for effective searching, browsing and indexing the videos is obvious in the computer industry and multimedia manufacturer. Fortunately, it can benefit from video abstraction. Video abstraction can be defined as a brief representation of the original video stream. The goal of video abstract is to choose the representative segment from the original long video. Yet the extraction of semantic video information is still a challenge problem. Luckily, in the live sports videos, scenes of important events or highlights repeatedly played using digital video effect or adding "logo". Usually these highlights summarize the essence and exciting actions of the video. So replay scene based sports abstraction can represent the content of the sports video.

Here we present a framework of replay scene based video abstraction in MPEG sports video. The features are directly extracted from the compressed videos. Thereby our method avoids the expensive inverse DCT computation required converting values from the compressed domain to the image domain. Furthermore, the analysis of this allow the macroblock (MB) type and motion vector (MV) information is simple and easy to be programmed, this allows the algorithm to be performed faster than real time video playing. The scheme can also be applicable to multiple types of sports.

The rest of the paper is organized as follows. In section 2 we review previous works related to video abstraction. In section 3, we introduce a framework of replay scene based video abstraction in MPEG sports video. In section 4, we address a new technique of detecting replay scene using MPEG feature. In section 5, we use color and camera information to detect identical events and generate replay based sports video abstract. In section 6, the experimental results with various sports video evaluate the performance of our proposed method. At last, we give conclusions of the paper and future research directions in section 7.

2 Relative Works

Some works on video abstraction have been reported. For instance, Lienhart [1] firstly presented a method of taking visual prosperity to construct a skim video that depicts a synopsis of the video sequence. But it selected the semantic contents relying on the significant visual feature such as faces, motions and verbal information.

Sports video abstraction is also an interesting topic. Li [2] propose a general framework for event detection and summary generation in broadcast sports video. Under this framework, important events in a class of sports are modeled by "plays", defined according to the semantics of the particular sport and the conventional broadcasting patterns. The detected plays are concatenated to generate a compact, time compressed summary of the original video. Obviously, it is about specific tasks including American football, baseball, and sumo wrestling. Lately, Babaguch [3] propose a method of generating a personalized abstraction of broadcasted American football video. Nevertheless, it did not verify the effectiveness of other type of sports.

Moreover, detecting replay events, often representing interesting events, can be used in video summary and content analysis. Kobla [4] used the macroblock, motion and bit-rate information to detect the slow-motion replay sequences. But it can not detect the slow motion generated by the high-speed camera. Babaguchi [5] detect replays by recognize digital video effects (DVE). The model is based on the color and motion of the gradually changing boundary between two overlapped shots. But the features are not robust and need additional computational complexity. Pan [6] detected slow-motion to determine the logo template, then located all the similar frames in the video using the logo template. Finally the algorithm identified segments by grouping the detected logo frames and slow-motion segments. However, the algorithm cannot accurately detect the slow-motion replays generated by a high-speed camera, or slow-motion replays in content whose fields are sub-sampled during encoding. Y. Yasugi [7] proposed a method for detection of identical events by analyzing and matching of the live and replay scenes for broadcasted video of American football, but it ignored zoom motion. Farn [8] proposed two kinds of slow-motion replays detection method. One comes from a standard camera and consists of some repeating or inserted frames. The other is from a high-speed camera with larger variation between two consecutive frames. And yet its experiments mainly are validated on soccer game videos.

The main drawback of the methods above is short of generality and hard to be applicable to other types of sports. Our solution is to make use of replay scene and camera information to generate video abstraction.

3 Framework of Replay Scene Based Video Abstraction in MPEG Sports Video

Here we propose a framework of replay scene based video abstraction in MPEG sports video, which is illustrated in Fig.1. At first, we identify the replay boundary using MPEG feature including macroblock (MB) and motion vector (MV) that is easy extracted from MPEG video, then modify the result of replay boundary detection and recognize replay scenes. Moreover, we use color and camera information to detect identical events. Finally, we introduce a scheme of three-layer replay scene based sports video abstraction.

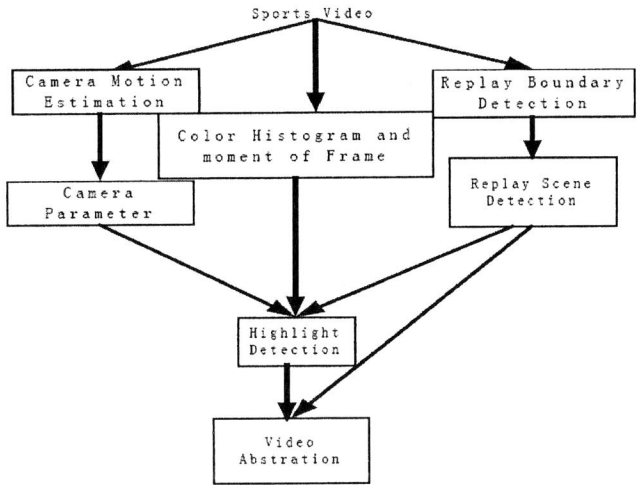

Fig. 1. The flowchart of the proposed framework

4 Replay Scene Detection

In paper [9], we present a model of replay boundary detection, and address a technique of identifying the replay boundary directly from MPEG compressed domain. It uses MPEG feature including macroblock and motion vector that is easy extracted from MPEG video, and then applies the rule of macroblock and motion vector to the detection of replay boundary.

For the reason that it is difficult to distinguish the common gradual change and logo transition, there still remain several false and missed results in replay boundary detection. On the other hand, the scheme of replay boundary detection cannot make use of the temporal information of diving video effectively. Thereby, the followed discussion is based two hypotheses.

I. The logos is symmetry in the broadcast diving video, namely there are logo both in the start and the end of replay scene.
II. The duration of the replay scene is shorter than the interval between the two replay scenes.

So we modify the result of replay boundary detection to detect replay scene based on the hypotheses. Experimental results validate the efficiency of this method.

5 Replay Scene Based Video Abstraction

5.1 Using Color and Camera Information to Detect Identical Events

Color feature is commonly used in video analysis and retrieval. In the MPEG compressed video pixel values are not available directly. The DC terms of I frames can be obtained directly from the MPEG sequence, and the DC terms of P and B frames can be reconstructed in [10].

Moreover, camera motion can reveal the semantic information in the video. In the replay scene, the same event which is often a highlight in the sports video is repeated several times, and often the scene of highlight which is captured by camera in a different perspective view is semantic identical to the replay scene.

We use epipolar line distance based outliers' detection method to estimate the camera motion as a motion feature for detecting identical events. Firstly we choose key frames of the video, and compute the Euclidean distance of the key frames of the replay scene and the shots before the replay scene. If the distance is below a predefined value, then the shot is a candidate shot.

Then we recover the true motion vector by estimating the camera motion, and compute the similarity of the replay scene and the candidate shots.

Similar to [7], we calculate the average true motion vectors in both the candidate and replay shots, respectively. By comparing the average true motion vectors acquired from the candidate and replay shots, if the Euclidean distance between the replay shot and the candidate shot is lower than the threshold, the candidate shot can be recognized as the live identical shot.

The identical events detection algorithm is stated as follows.

Step 1. Estimate camera motion parameter in the MPEG compressed domain.
Step 2. Recover true motion vector,

$$x_i'' = x_i' - \frac{p_1 x_i + p_2 y_i + p_3}{p_5 x_i + p_6 y_i + 1}, y_i'' = y_i' - \frac{-p_2 x_i + p_1 y_i + p_4}{p_5 x_i + p_6 y_i + 1}$$, where (x'', y'') is the image coordinates of recovered motion vector in two neighboring frames, (x, y) and (x', y') are the image coordinates of corresponding points in two neighboring frames, $i = 0, 1, \cdots, m-1$, m is the number of feature points.

Step 3. Calculate average value of the recovered motion vectors in the frame and shot. If both the average value and direction of the shot near to these of the replay scene, then the shot is a candidate shot.

Step 4. Select key frames based on shortest path based algorithm. If it is I frame, then directly extract the DC coefficient, else estimate the DC coefficient using the method in Yeo [11].

Step 5. Compute the Euclidean distance between the replay shot and the candidate shot. If the distance is lower than the predefined value, then the candidate shot can be determined as the identical shot.

5.2 Replay Scene Based Video Abstraction

After detecting the replay scene and linking up live and replay scenes, video abstraction can build on connecting the highlights correspond to replay scene and live scene. We introduce three types of summaries:1) The key frames of the replay and live shot, 2) all replay scenes in a sports game, 3) all live scenes in the same game. The first type of summary is a still-image abstract, the last two type of summaries are moving-picture abstract. Because the users may want to a quickly preview of the video owing to the limited bandwidth, such as in a wireless network, the key frames is the preference. The users can also select the live or replay scene for the rich details of the game.

Furthermore, we propose three-layer replay scene based sports video abstraction. The top layer is the representative scene, which can illustrate the lifespan of the key

Fig. 2. Key frame interface

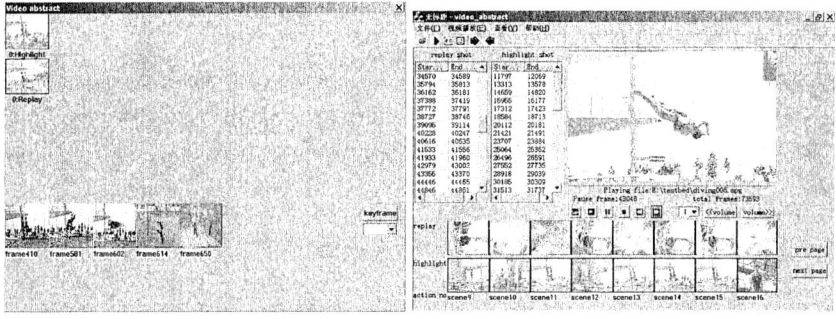

Fig. 3. Representive shot interface   **Fig. 4.** Representative scene interface

actions in the sports video; the middle layer, namely the representative shot, organizes the scene summary; the bottom layer is constituted of key (or representative) frames. Moreover, as shown in Fig.3., the corresponding key frames of replay shot or live highlight is arranged to the bottom of the screen to show the detailed information. When user clicks the replay shot or live highlight, the relevant key frames would be shown. The hierarchical replay based sports video summary at the key frame layer is as shown in Fig.2, representative shot and representative scene interface is also as shown in Fig.3 and Fig.4, respectively.

To present and visualize the sports content for summarization, the representative scene, representative shot and representative frame can express the video abstraction in various granularities. Among the three-hierarchy summarization, representative scene can convey the comprehensive semantic meaning of the sports video, representative shot can demonstrate the exciting action in the video, while key frame address the detailed information. So these three hierarchical summaries can express the various video content in increasing granularity.

6 Experimental Results and Analysis

6.1 Replay Scene Detection

The test data is a set of sports video clip from the live broadcasted TV. They are "The 9th FINA Swimming Championships Fukuoka 2001" including the "3m Synchronized Diving Man (A1)" , "10m Synchronized Diving Women (A2)", "3m springboard Diving Men (A3)" "3m springboard Diving Women (A4)" ,"10m

Table 1. The experimental results of replay scene detection

	A1	A2	A3	A4	A5	A6	A7	B1	B2
Video length	46:54	44:13	1:25:9	1:9:24	1:26:43	49:3	1:9:10	1:1:35	50:37
Total replays	40	40	72	60	72	40	55	36	32
Detect	40	38	66	52	65	40	52	26	25
False alarm	0	0	0	0	0	0	0	0	0
Miss	0	2	6	8	7	0	3	10	7
Recall	100%	95%	91.7%	86.7%	90.3%	100%	94.5%	72.2%	78.1%
Precise	100%	100%	100%	100%	100%	100%	100%	100%	100%

Table 2. Summary of results of table1

	average	The best	The worst
Recall	91.2%	100%	72.2%
Precise	100%	100%	100%

platform Diving Men(A5)", "3m Synchronized Diving Women (A6)","10m platform Diving Women(A7)"; "The 17th FIFA world cup 2002" including the "Brazil vs. Turkey(B1,B2)". Total length of the test MPEG-1 video clip is 9:22:48, each frame is 352x288 pixels in size, and the frame rate is 25 frame/s.

Table 1 lists the experimental results of replay scene detection, and table 2 summarizes the results.

The precise of replay scene detection is desirable, but the recall depends on the result of the method of detecting replay boundary [9], the method of replay boundary detection should be improved further.

For the highlights in the sports video often can be replayed in slow motion, Kobla [4] used the macroblock, motion and bit-rate information to detect the slow-motion replay sequences. But it is only effective for detecting still frames of the replay sequences. Our method can detect the slow motion generated by the high-speed camera.

6.2 Detection Identical Events

We only list the experimental results of identical events detection of A1, A2 and A6. The accuracy and precise of identical events detection is shown in table 3.

Table 3. Eexperimental results of identical events detection

	A1	A2	A6
Video length	46:54	44:13	49:3
Total live scenes	40	40	40
Detected	36	29	36
False alarm	4	6	4
Miss	4	11	4
Recall	90%	72.5%	90%
Precise	90 %	82.8%	90%

As shown in table1, table2 and table3, the accuracy and precise of replay scene detection is fairly good, and the recall of identical events detection is also 100%. But there are still remain some error identical events detection. The false alarm and miss identical events detection mainly result from the approximate error of the camera motion estimation. Compared to the method of Tausig [7], our method can detect highlight by camera motion information, while Tausig ignore the zoom operation.

6.3 Evaluation of Replay Scene Based Video Abstraction

For directly working in the compress domain, the hierarchical summaries can be generated in real-time.

Also, as for the quality of video abstract, He [11] proposed four C's rules to measure the video abstract: conciseness, coverage, context, and context. Conciseness means the selected segment for the video summary should contain only necessary information. Coverage focuses on covering all the "key" points of the video. Context indicates that the summary should be such that prior segments establish appropriate context. Coherence contains the criterion of natural and fluid.

For conciseness and coverage, because replay scenes often drop a hint of the interesting or key the events in the sports video, our abstraction has sufficient in a compact form. Furthermore, the highlight and replay events express the context of sports video. At last, the highlight and replay events in rreplay scene based video abstraction are arranged in original temporal order.

7 Conclusions and Future Research

We have addressed a scheme of replay based video abstraction in MPEG compressed sports video. In sports video, replay scene often implies the emergence of highlight or interesting event of the video. So we apply microblock and motion vector information to detect the replay scene effectively. Moreover, we link up highlight and replay scene using color and camera information. Finally, we propose a three-layer replay-based sports video abstract. For working on features directly from the MPEG compressed domain, it can perform in real time. In the mean time, experiments verify the highlight extraction approach is more robust than current method.

The future work is to integrate our method into semantic sports video abstraction scheme, and apply to the Digital Olympic Project in China.

Acknowledgements

This research has been supported by National Science Foundation of China (60302028, 60473002), Beijing Science Foundation of China (4051004), Scientific Research Project of Beijing (Z0004024040231) and Scientific Research Fund of Hunan Provincial Education Department (03C484).

References

1. R. Lienhart, S. Pfeiffer, and W. Effelsberg.: "Video abstracting," Communications of The ACM, (1997)55–62.
2. B. Li, and MI.Sezan.: "Event Detection and Summarization in Sports Video", Proc. IEEE Workshop on Content-based Access of Image and Video Libraries (CBAIVL'01), (2001)132-138.
3. N. Babaguchi,Y. Kawai, T. Ogura and T. Kitahashi.: ``Personalized Abstraction of Broadcasted American Football Video by Highlight Selection," IEEE Trans. Multimedia,VOL.6,NO.4,AUGUST 2004,(2004)575-586.

4. V Kobla, D DeMenthon, and D. Doermann.: Detection of slow-motion replay sequences for identifying sports videos, Proc. IEEE Workshop on Multimedia Signal Processing,(1999) 135-140.
5. N Babaguchi, Y Kawai, Y Yasugi and T.Kitahashi.: Linking Live and Replay Scenes in Broadcasted Sports Video, ACM International Workshop on Multimedia Information Retrieval, (2000)205-208.
6. H Pan, B Li, and M I Sezan.: Automatic detection of replay segments in broadcast sports programs by detection of logos in scene transitions, IEEE International Conference on Acoustics Speech and Signal Processing, IV, (2002)3385-3388.
7. Y. Yasugi, N. Babaguchi and T.Kitahashi.: Detection of Identical Events from Broadcasted Sports Video by Comparing Camera Works, Proceedings of ACM Multimedia 2001 Workshop on Multimedia Information Retrieval (MIR2001), Ottawa, (2001)66-69.
8. EJ. Farn, LH Chen, and JH Liou.: A New Slow-motion Replay Extractor for Soccer Game Videos, International Journal of Pattern Recognition and Artificial Intelligence, Vol. 17, No. 8,(2003)1467-1482.
9. Jianquan Ouyang, Li Jintao, Zhang Yongdong.: Replay Boundary Detection in MPEG Compressed Video, IEEE The Second International Conference on Machine Learning and Cybernetics, Xi'an China, (2003)2800-2803.
10. B.-L. Yeo and B. Liu.: Rapid scene analysis on compressed videos, IEEE Trans. On Circuits and Systems for Video Technology, Vol. 5, No. 6,(1995)533-544.
11. L. He, E. Sanocki, A. Gupta, and J. Grudin.: Auto-summarization of audio-video presentations, in proceeding of ACM Multimedia, (1999)489–498.

Mapping Web Usage Patterns to MDP Model and Mining with Reinforcement Learning

Yang Gao[1], Zongwei Luo[2], and Ning Li[1]

[1] State Key Laboratory for Novel Software Technology,
Nanjing University, Nanjing, 210093, China
[2] E-business Technology Institute, The University of Hongkong, China

Abstract. For many web usage mining applications, it is crucial to compare navigation paths of different users. This paper presents a reinforcement learning based method for mining the sequential usage patterns of user behaviors. In detail, the temporal data set about every user is constructed from the web log file, and then the navigation paths of the users are modelled using the extended Markov decision process. The proposed method could learn the dynamical sequential usage patterns on-line.

1 Introduction

In general, web visiting pattern is a kind of sequential pattern stored in web server log files. As most users demonstrate these patterns frequently, learning this knowledge will help analyze users' needs and thereby design adaptive web sites. Based on the mined patterns, it will become feasible for us to predict and classify user's actions. Currently, there are a number of different approaches for this purpose, such as web usage based recommendation models [3][4] and collaborative filtering models [2]. In addition, some techniques from association rule analysis [1][8] and clustering [6] have also been successfully used in web log mining. Considering web usage mining is a specific form of sequential pattern mining in sequence data source, many sequential pattern mining methods could be applied to this problem, such as GSP [9] and WAP-tree [7]. Unfortunately, these previous algorithms can hardly adapt to scenarios where dynamical on-line mining is needed. So, some new mining methods are desired.

Markov decision process is a mathematical model for sequence tasks in dynamical surroundings. When the model is explicit, the most optimized action sequence can be obtained through dynamic programming method directly; while when the MDP model is unknown, reinforcement learning technique could be applied to approximate the most optimized policy . It is noteworthy that when the behavior model of user navigation is modelled by Markov decision process, learning the most optimized action sequence could be viewed as being equivalent to the mining of the web sequential usage patterns. Based on this recognition, a new web usage mining method is presented in this paper, where the temporal data set about every user is constructed from the web log file, and then the navigation paths of the users are modelled using the extended Markov decision process.

The rest of this paper is organized as follows. In Section 2, the Markov decision process model and the classical reinforcement learning techniques are introduced. In Section 3, the user behaviors are modelled by extending the Markov decision process. Finally in Section 4, conclusions are drawn.

2 MDP Model and Reinforcement Learning

2.1 MDP Model

Definition 1. MDP Markov decision process is often abbreviated as MDP, which is briefly summarized in Fig. 1. An MDP can be described as a tuple $\langle S, A, T, R \rangle$, where S is a finite set of states of the world, A is a finite set of agent's actions, $T : S \times A \to \Pi(S)$ is the state-transition function, and $R : S \times A \to \Re$ is the reward function. Usually, $T(s, a, s')$ is the probability of ending in state s' when action a is taken in state s, $R(s, a)$ is the reward for taking action a in state s [5].

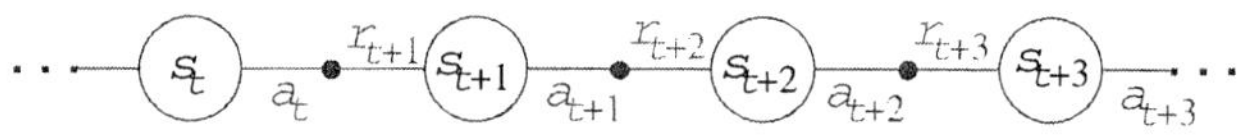

Fig. 1. Markov decision process

Actually, a state-action value function is defined in order to find the most optimal action sequence policy. A policy is $\pi : S \to A$. $Q^\pi(s, a)$ is then defined as the value of taking action a in state s under a policy π, which is the expected return starting from s as Eq. 1, where γ is a discounter factor. The most optimal policy, i.e. Q^*, is defined in Eq. 2 for all $s \in S$ and $a \in A$. If the model is known, i.e. T, R are known in advance, dynamical programming could be applied iteratively to get this optimal policy.

$$Q^\pi(s, a) = E_\pi \{ \Sigma_{i=0}^{\infty} \gamma^i r_{i+1} \} \quad (1)$$

$$Q^*(s, a) = \max_\pi Q^\pi(s, a) \quad (2)$$

2.2 Reinforcement Learning

When the MDP model is unknown, learner(or learning agent) can only get its experiences by trial-and-error and approximate the optimal policy with reinforcement learning. The property of on-line learning distinguishes reinforcement learning from dynamical programming. This is because that reinforcement learning agent faces temporary credit assignment problem in the sequence task learning, where the iteration update method is used to adjust the estimated state-action value function of current state and next state.

Q-learning is one of the most popular reinforcement learning method. In Eq. 3, the Q value is updated by current reward and the maximal Q value of the next state when agent gets a piece of experiences $(s, a, r(s, a), s')$. Since

Q-learning is exploration insensitive, Q-learning method must converge with probability 1 to the optimal Q^* [10].

$$Q(s,a) = Q(s,a) + \alpha[r(s,a) + \gamma \max_{a' \in A} Q(s',a') - Q(s,a)] \qquad (3)$$

The logical structure of reinforcement learning is described in Fig. 2. Learning agent receives input s, and takes action a as output on each step of interaction with the environment. As a result of action a, the environment is transited to new state s'. Meanwhile, learning agent receives reward r. By iterated learning, agent constructs a sequence of actions policy to maximize its reward from the outside environment.

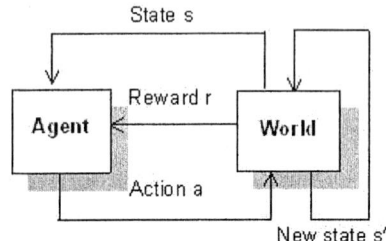

Fig. 2. Logical structure of reinforcement learning

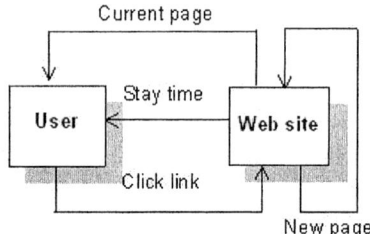

Fig. 3. User navigation model

3 Mapping Web Usage Patterns to Markov Decision Model

Web server's logs store the entire records of clients' visiting of files and most of these logs keep to general log format. Client's IP, user's ID, visiting time, request methods, the URL of visiting pages, the protocol of data transferring, error codes and the bytes transferred etc. are included in general log format. Towards the aims of our mining task, some items such as requesting methods, the protocol of data transferring and so on are not relative to sequential patterns, so these kinds of items should be deleted from the original web server's logs. After the data cleaning, web logs will be transferred to some data format suitable for mining sequential patterns. In this format, only essential attributes related to tasks are included, such as client's IP, URL of request pages, and visiting time.

Definition 2. Web log model Web log model can be described as a tuple $LogSet = \langle Ip, PageURL, Time \rangle$. $LogSet$ is the set of user's action records. In each record, Ip demonstrates the IP address of the client who visits web servers, $PageURL$ represents the current page's URL visited by client, and $Time$ is the moment of this web page visited by the client.

The process of mapping $LogSet = \langle Ip, PageURL, Time \rangle$ to MDP model $\langle S, A, T, R \rangle$ is as follows. Firstly, $PageURL$ corresponds to the state S in MDP model and the transferring from one $PageURL$ to another $PageURL$ depends on the click on hyperlinks and click the back button in the browser's menu. These actions are also demonstrated as $PageURL$ in web log model. Web log model is different from MDP in that the parameter T is ignored because next state in web log model is closely related with the hyperlinks on current pages. The possibility of transferring from current web page to some next state is based on the number of hyperlinks on current pages. Therefore, $PageURL$ not only relates to the state S in MDP model, it also relates to the action A in MDP model. R couldn't be equivalent to $Time$ directly. In our mapping process, we let web user's reward be equivalent to his staying time in current page. The staying time could be calculated by the difference between the $Time$ of current record and the $Time$ of next record. If the current record is the last page, we set the staying time of this last page a statistical value, about 8.4 seconds.

4 Conclusion

This paper discusses how to convert the web log model into MDP model. Based on web log model, reinforcement learning technique is applied to the mining of web usage patterns, where a dynamic on-line method of web sequential pattern mining is brought forward. Because of page limits, experimental results will be shown in another paper. An interesting issue for future research is to apply partial observable Markov decision process to the mining of web pages containing frames, which will be a significant extension of the presented method.

Acknowledgement

This work was supported by the National Natural Science Foundation of China under the Grant No. 60475026, the Natural Science Foundation of Jiangsu Province, China under the Grant No. BK2003409 and the National Grand Fundamental Research 973 Program of China under the Grant No. 2002CB312002.

References

1. R. Agrawal and R. Srikant. Fast algorithms for mining association rules. In *Proceedings of the 20th International Conference on Very Large Data Bases*, Santiago, Chile, pp.487-499, 1994.
2. D. Goldberg, D. Nichols, B. Oki, and D. Terry. Using collaborative filtering to weave an information tapestry. *Communications of the ACM*, 35(12): 61-70, 1992.
3. S. Gündüz and M. T. Özsu. Recommendation models for user accesses to web pages. In *Proceedings of the 13th International Conference on Artificial Neural Networks*, Istanbul, Turkey, pp.1003-1010, 2003.
4. S. Gündüz and M. T. Özsu. A user interest model for web page navigation. In *Proceedings of the International Workshop on Data Mining for Actionable Knowledge*, Seoul, Korea, pp.46-57, 2003.

5. L. P. Kaelbling, M. L. Littman, and A. W. Moore. Reinforcement learning: a survey. *Journal of Artificial Intelligence Research*, 4: 237-285, 1996.
6. B. Mobasher, R. Cooley, and J. Srivastava. Creating adaptive web sites through usage-based clustering of urls. In *IEEE Knowledge and Data Engineering Workshop*, Chicago, IL, pp.19-26, 1999.
7. J. Pei, J. Han, B. Mortazavi-Asl, and H. Zhu. Mining access pattern efficiently from web logs. In *Proceedings of the 4th Pacific-Asia Conference on Knowledge Discovery and Data Mining*, Kyoto, Japan, pp.396-407, 2000.
8. R. Srikant and R. Agrawal. Mining generalized association rules. In *Proceedings of the 21st International Conference on Very Large Databases*, Zurich, Switzerland, pp.407-419, 1995.
9. R. Srikant and R. Agrawal. Mining quantitative association rules in large relational tables. In *Proceedings of the ACM-SIGMOD International Conference on Management of Data*, Mantreal, Canada, pp. 1-12, 1996.
10. R. S. Sutton. Dyna, an integrated architecture for learning, planning, and reacting. In *Working Notes of the 1991 AAAI Spring Symposium*, Palo Alto, CA, pp.151-155, 1991.

Study on Wavelet-Based Fuzzy Multiscale Edge Detection Method

Wen Zhu[1], Beiping Hou[1,2], Zhegen Zhang[1], and Kening Zhou[1]

[1] Department of Automation,
Zhejiang University of Science and Technology, 310012, China
joywenzhu@126.com
[2] Institute of Industrial Process Control,
Zhejiang University, Hang Zhou, 310027, China
houbeiping@126.com

Abstract. A wavelet-based fuzzy multiscale edge detection scheme (WFMED) is presented in this paper. The dyadic wavelet transform is employed to produce the multiscale representation of the image, fuzzy logic is applied in wavelet domain and it can synthesize the information of image across scales effectively, an optimal result of edge detection can be acquired. WFMED method is used to extract the edge of pulp fibre image; the paper compares the performance of WFMED to the Canny edge detector and to Mallat's algorithm. The results show the superiority of WFMED to these other methods.

1 Introduction

Edge detection theory based on multi-resolution plays a very important role in image process field [1]. Canny described the edge detectors by three criteria: good detection, good localization and low spurious response [3], and he combined the edge information in different scales from coarse to fine scale. As suggested by Marr and Hildreth [2], multiscale should be employed to describe the variety of the edge structures. Ziou and Tabbone [4] presented the edge detection method based on Laplacian algorithm. Witkin *et al.* [6] combined the scale information by bayes algorithm. Although edge detection algorithm based on multiscale can partly solve the problem of edge detection and edge localization, it is sensitive to noise for the reason of high pass characteristics.

According to Canny's edge detection criteria, how to represent image in multiscale way, how to combine edge information in different scales and suppress noise effectively become important topics. Dyadic wavelet transform is a proper tool for multiresolution representation of signal. Many researchers, who work in this area, combine the information in different scales. For example, Mallat [7] used evolution across scales modulus maxima. Xu *et al.* [8] used the spatial correlation of the adjacent scales. L.Zhang and P.Bao [9] used the global maximum of the WT scales to locate the important edges in the signal. Generally, all the researchers have used multiresolution representation of the image and tried to find a way to combine the information in different scales of the signal in order to realize more accurate results.

Theory and experiment prove that ambiguity and uncertainty in edge detection and localization confined to each scale at any level. The main ambiguity source is the fact that in finer scales the signal to noise ratio is normally poor whereas in the coarse scales the localization ability is the main reason for uncertainty. In this paper, we aim to solve this problem. For this purpose, we use fuzzy theory and operators to combine the scale information in a fuzzy manner to develop a novel wavelet-based fuzzy multiscale edge detection algorithm (WFMED). The paper is organized as follows. Section 2 introduces dyadic wavelet transform. Section 3 develops the WFMED theory and procedure. Section 4 is the analysis of simulation results.

2 Dyadic Wavelet Transform

A function $\psi(x) \in L^2(R)$ is called a wavelet function if its average is equal to 0, it is dilated by dyadic scale 2^j, then $\psi_{2^j}(x) = \frac{1}{2^j}\psi(\frac{x}{2^j})$. Define $\theta(x)$ is a differentiable smooth function whose integral is 1 and converges to 0 at infinity. Let $\psi(x) = d\theta(x)/dx$, then

$$W_j f(x) = f * \psi_{2^j}(x) = 2^j \frac{d}{dx}(f * \theta_j)(x) \qquad (1)$$

Where * denotes convolution operation, $W_j f(x)$ is proportional to the derivative of $f(x)$ smoothed by $\theta(x)$. The wavelet function used in this paper is the Mallat wavelet [1]. In two-dimension case, two wavelets should be utilized:

$$\psi_{2^j}^1(x,y) = \frac{\partial \theta(x,y)}{\partial x}, \psi_{2^j}^2(x,y) = \frac{\partial \theta(x,y)}{\partial y} \qquad (2)$$

The dyadic wavelet transforms of $f(x, y)$ at scale 2^j along x and y directions are:

$$W_j^1 f(x) = f * \psi_{2^j}^1(x,y), W_j^2 f(x) = f * \psi_{2^j}^2(x,y) \qquad (3)$$

3 Wavelet-Based Fuzzy Multiscale Edge Detection Algorithm

According to the characteristics of dyadic wavelet transform, the local maxima of wavelet coefficients in different scales represent the edge point in signal domain. For this purpose, fuzzy operator is used to combine the edge information in different scales and suppress noise. We propose to manipulate the ambiguous scale information in fuzzy manner to end up with better decisions for maxima corresponding through scales that yields a good trade-off between the localization and detection performances.

3.1 One-Dimension Analysis of WFMED Algorithm

The wavelet-based fuzzy multiscale edge detection algorithm converts edge information among different scales to fuzzy subsets. We can assume each wavelet scale to be a fuzzy subset of the signal denoting the grade of "edginess" for each point in the signal space. The following steps lead us to the definition of the WFMED method.

Step 1: Compute the discrete dyadic wavelet transform of the original signal $f(n)$ in all available scales $2^j (j=1,2,\cdots J)$, then the information matrix $W_j f(n)$ is acquired.

Step 2: Pre-processing scale information

Regard the points in $W_j f(n)$ as noise if they satisfy the following criterions: modulus decreases as the scale increased; the direction of modulus between neighbor scales is oppositive, then eliminate them and the new information matrix $S_j^1(n)$ can be acquired.

Step 3: Segment each signal scale 2^j into its positive and negative components and place them in $f_p(j,n)$ and $f_n(j,n)$, they represent positive edge and negative edge information respectively. The $F_1[\bullet]$ operator is defined as follows:

$$S_j^2(n) = F_1[S_j^1(n)] = \begin{cases} f_p(j,n) & S_j^1(n) \geq 0; \quad n=1,2,\cdots,N \\ -f_p(j,n) & S_j^1(n) < 0; \quad j=1,2,\cdots,J. \end{cases} \quad (4)$$

Step 4: As the range of the membership of a fuzzy subset is the real interval [0,1], normalize each $S_j^2(n)$ to produce $S_j^3(n)$ so that

$$S_j^3(n) = F_2[S_j^2(n)], \quad j=1,2,\cdots J \quad (5)$$

The normalization operator $F_2[\bullet]$ is designed such that: $hgt[S_j^2(n)] = 1$, Where represents the maximum value of the input function [11].

It is observed that Step 5 produces a fuzzy membership function for all the scales. The membership function for each scale is defined as shown in (6):

$$A_j = \{(n, \mu_{A_j}(n)), n=1,2,\cdots N\}, \quad j=1,2,\cdots,J \quad (6)$$

Where $\mu_{A_j}(n) = |S_j^3(n)|$ is the grade of membership for any given point n in the fuzzy subset A_j with magnitude $|S_j^3(n)|$.

Step 5: Since any fuzzy sets A_j denotes the grade of possessing the same property of "edginess" with different scales information for any member n, Step 5 in the creation of the WFMED is to combine the information contained in the different fuzzy subsets of signal.

$$A_D = A_1 \cap A_1 \cap \cdots A_J \quad (7)$$

Where A_D represents the edge fuzzy set of the signal with the following membership value for each point n.

$$\mu_{A_D}(n) = \prod_{j=1}^{J} \mu_{A_j}(n) = \mu_{A_1}(n) \cdot \mu_{A_2}(n) \cdots \mu_{A_J}(n), \quad n = 1, 2, \cdots N \tag{8}$$

The larger the membership value $\mu_{A_D}(n)$ for a point n, the more probable the point belongs to edges.

Step 6: The real edge point can be acquired according to the threshold method [9].

It is obvious that WFMED method not only can improve the localization accuracy and detection efficiency, but also can suppress noise. In Fig. 1, the noisy signal which contains four-step characteristics. WFMED method is used to analyze the results.

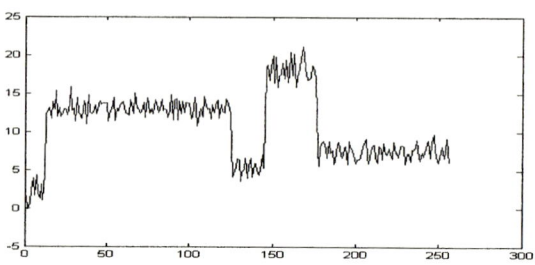

Fig. 1. Sample signal

From Fig.1, the sample signal is the typical signal to verify the efficiency of the WFMED method, so the wavelet transform and fuzzy subsets can be calculated next.

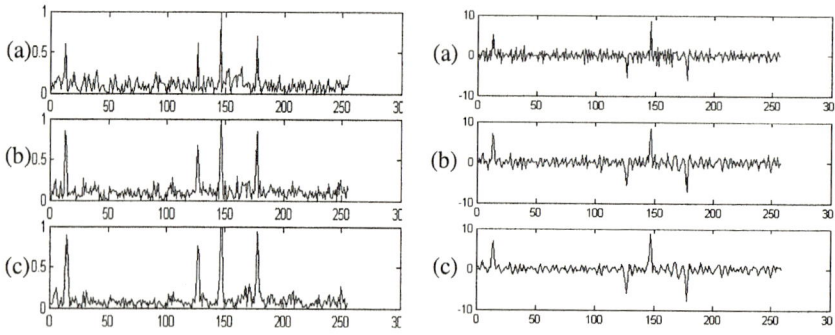

Fig. 2. Wavelet transform of sample signal **Fig. 3.** Fuzzy subsets of sample signal

Fig .2 means the Dyadic Wavelet Transform of sample signal that showed in Fig. 1, Fig.1 (a), Fig.1 (b) and Fig.1(c) mean the first three-order wavelet transform. Fig.3 shows the corresponding fuzzy subsets of the wavelet detail coefficients in Fig.2. According to our WFMED algorithm, the fuzzy decision subsets should be computed to detect the singularities of sample signal.

Fig.4 (a) is the fuzzy decision subsets; it is obvious that the four local maxima mean singularities of sample signal. Fig.4 (b) shows the effective detection of singularities, apparently the singularity points of original signal can be detected correctly, it proves that WFMED method is a useful tool to extract singularity points in one-dimension condition. The two-dimension WFMED algorithm is presented on the basis of one-dimension condition.

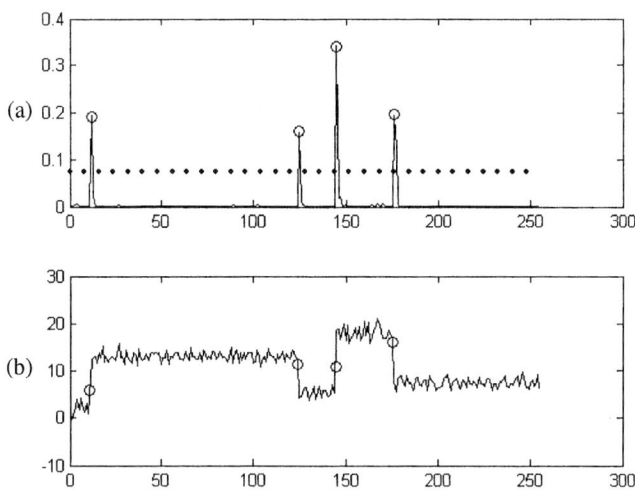

Fig. 4. Singularity detection of sample signal

3.2 Two-Dimension WFMED Algorithm

In two dimensions, two correlation function should be defined in x and y directions. The information matrices $W_j^1 f(n_1, n_2), W_j^2 f(n_1, n_2)$ represent the dyadic wavelet transform of image signal $f(n_1, n_2)$ along horizontal and vertical directions. Similar to 1-D case, after the operations of $F_1 [\,\cdot\,]$ and $F_2 [\,\cdot\,]$, the fuzzy subsets, fuzzy edge sets and membership functions can be acquired as follows:

$$\begin{cases} A_j^1 = \{(n_1, n_2, \mu_{A_j}^1 (n_1, n_2))\}, \\ A_j^2 = \{(n_1, n_2, \mu_{A_j}^2 (n_1, n_2))\}, \end{cases} \quad n_1 = n_2 = 1, 2, \cdots N, j = 1, 2, \cdots, J \quad (9)$$

$$\begin{cases} A_D^1 = A_1^1 \cap A_2^1 \cap \cdots A_J^1, \\ A_D^2 = A_1^2 \cap A_2^2 \cap \cdots A_J^2, \end{cases} \quad j = 1, 2, \cdots, J \quad (10)$$

$$\begin{cases} \mu_{A_D}^1 (n_1, n_2) = \prod_{j=1}^{J} \mu_{A_j}^1 (n_1, n_2), \\ \mu_{A_D}^2 (n_1, n_2) = \prod_{j=1}^{J} \mu_{A_j}^2 (n_1, n_2), \end{cases} \quad n_1 = n_2 = 1, 2, \cdots, N \quad (11)$$

Similar to 1-D case, the singularities can be found along x and y directions, the corresponding image pixel points are edge points. Edge detection of pulp fibre image is implemented to prove the effectiveness of the two-dimension WFMED algorithm.

4 Simulation Research

Fig.5(a) is a 256×256 isolated pulp fibre image. We find edges first by Canny edge algorithm and Mallat wavelet algorithm and then by our scheme.

Canny edge algorithm and Mallat wavelet algorithm are typical edge detection method; the comparisons among three methods are significant. As shown by Fig.5, although Canny edge algorithm and Mallat wavelet algorithm can find edge point, the localization accuracy is poor. From comparisons, our algorithm improves the localization accuracy significantly while keeping high detection efficiency.

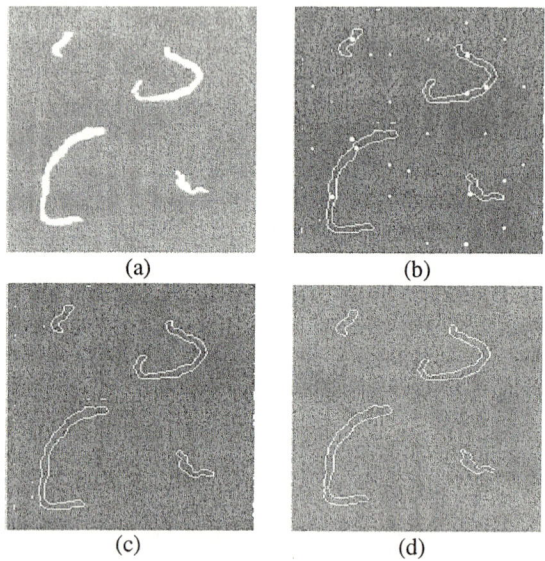

Fig. 5. Edge detection of fibre image

Table 1. Quality values F of edge detection

	Canny algorithm	Mallat algorithm	WFMED
F value	0.8426	0.8503	0.8715

The merit value of Pratt [12] is used to evaluate the performance.

$$F = \frac{1}{\max(N_a, N_d)} \sum_{k=1}^{N_d} \frac{1}{1 + \alpha d^2(k)} \qquad (12)$$

Where N_a is the number of the actual edges and N_d is the number of the detected edges. d_k denotes the distance from the kth actual edge to the corresponding detected edge. α is a scaling constant. F is the largest, which means the best performance.

From Table.1, it is obvious that WFMED method is superior to these other methods.

5 Conclusion

In this paper, a new effective edge detection algorithm is presented. To solved the ambiguity problem of edge detection, we made a novel use of fuzzy logic based on dyadic wavelet transform and associated fuzzy operator to combine scales information in fuzzy manner and developed a Wavelet-based Fuzzy Multiscale Edge Detection (WFMED) scheme. We compared the WFMED to the conventional Canny edge operator and also to the Mallat algorithm by applying them to the test image with noise. The simulation results showed the superiority of WFMED to these other methods. Our algorithm has proved to be more accurate and robust estimate of the edge locations.

References

1. S. Mallat and S. Zhong: Characterization of signals from multiscale edges. IEEE Trans. PAMI, 1992,14(7) 710–732
2. Marr D, Hildreth E: Theory of edge detection. Proc. Royal Soc.London, 1980 187–217
3. J. Canny: A computational approach to edge detection. IEEE Trans. PAMI,1986,8(6) 679–698
4. D.Ziou and S.Tabbone: A multi-scale edge detector. Pattern Recognition, 1993,26 1305–1314
5. Andrew P.Witkin: Scale –space filtering: a new approach to multiscale description. IEEE ICASSP '84, 1984,9 150–153
6. S.Konishi: statistical approach to multi-scale edge detection. Image and vision computing, 2003,21(1) 1–10
7. S. Mallat: Singularity detection and Processing with wavelets. IEEE Trans. Information Theory, 1992,38(2)
8. Y.Xu et al.: Wavelet transform domain filters: a spatially selective noise filtration technique. IEEE Trans. Image Processing, 1994,3 747–58
9. Lei Zhang and Paul Bao: A wavelet-based edge detection method by scale multiplication. Pattern Recognition, 2002,3 501–504
10. Stephane G. Mallat: A theory for multiresolution signal decomposition: the wavelet representation [J]. IEEE Trans on PAMI.,1989,11(7) 674–93
11. D.Dubios and H.prade: Fuzzy sets and systems, Theory and applications. Academic and Harcourt Brace Jovanovich, New York (1980)
12. William K.Pratt: Digital image processing. 3rd edn. John Wiley and Sons (1991)

Sense Rank AALesk: A Semantic Solution for Word Sense Disambiguation*

Yiqun Chen[1,2] and Jian Yin[1]

[1] Zhongshan University, Guangdong, P.R. China, 510275
[2] Guangdong Institute of Education, Guangdong, P.R. China
issjyin@zsu.edu.cn

Abstract. This paper presents an "one fit all" solution for any field's text Word Sense Disambiguation(WSD), with a Sense Rank AALest algorithm derived from the Adapted of Lesk's dictionary-based WSD algorithm. AALesk brings a score for different relationship during gloss comparing, which makes WSD not only based on statistical calculate by process in a semantic way. Rather than simply disambiguate one word's sense one time, our solution considers the whole sentence environment and uses a Sense Rank algorithm to speed up the whole procedure. Sense Rank weights different sense combination according to their importance score. All these contribute to the accuracy and effective of the solution. We evaluated our solution by using the English lexical sample data from the SENSEVAL-2 word sense disambiguation exercise and attains a good result. Additionally, the independence of system components also make our solution adaptive for different field's requirement and can be easily improved it's accuracy by changing its core algorithm AALesk's parameter setting.

1 Introduction

Most words in natural languages are polysemous and synonymy, that is one word has multiple possible meanings or senses and different words may have the same meanings or senses. Humans can understand and using language rarely need to stop and consider which sense of a word is intended. However, computer programs do not have the benefit of a human's vast experience of the world and language, so automatically determining the correct sense of a polysemous word is a difficult problem, which is called **Word Sense Disambiguation(WSD)**. It has long been recognized as a significant component in language processing applications such as information retrieval, machine translation, speech recognition, text mining. Text Mining is about looking for patterns in natural language text, and

---

* This work is supported by the National Natural Science Foundation of China (60205007) , Natural Science Foundation of Guangdong Province (031558,04300462), Research Foundation of National Science and Technology Plan Project (2004BA721A02), Research Foundation of Science and Technology Plan Project in Guangdong Province (2003C50118) and Research Foundation of Science and Technology Plan Project in Guangzhou City(2002Z3-E0017).

may be defined as the process of analyzing text to extract information from it for particular purposes. Text mining is still based on algorithm derived from statistical theory, without the understanding of natural language text, while making a semantic analysis will be a tend in the development of text mining. A good Word Sense Disambiguation solution will help to realize Text mining in semantic way.

In recent years corpus-based approaches to word sense disambiguation have become quite popular. This paper contributes on word sense disambiguation for a adaptable and high effective "one fit all" solution. Firstly, it relies on a tagger tools to tell the words' part of speech, secondly it uses a Sense Rank AALesk algorithm, in which WordNet[7] is a helpful reference corpus for word sense disambiguation. Tagger tool and WordNet is only assistant tools in our system, which can be changed to more high accuracy ones or corpus for particular field text. All the main components in our system are independent to each other, thus the reusable and adaptability is achieved. With a parellel algorithm speed up high corrective word sense disambiguation algorithm, changeable tagger and changeable corpus, our solution contributes a adaptable and high effective solution for all fields word sense disambiguation.

2 Related Works

2.1 Word Sense Disambiguation

There are amount of methods for word sense disambiguation, they can be classified as the following:

- *corpus-based approach* makes use of the information provided by Machine Readable Dictionaries[2].
- *training approach* uses information gathered from training on a corpus that has already been semantically disambiguated (supervised training methods) [3]or from raw corpora (unsupervised training methods)[4].
- *machine learning approach* use machine learning algorithms to learn from user's choice and master telling Part of Speech of a word[5].

Corpus-based approaches to word sense disambiguation have become quite popular. As the others relied on the availability of sense-tagged text, which is expansively manually created and only applicable to text written about similar subjects and for comparable audiences. While the Corpus-based approaches can changes their corpus adapt to particular field, which make it possible to fit for all fields. What's more, corpus for particular field application is easy to find and more corrective. In this paper, our algorithm derived from the Adapted Lesk algorithm, still using WordNet as the assistant dictionary, which will be describe on section 2.3

2.2 The Adapted Lesk Algorithm

The Adapted Lesk algorithm[1] is an adaptation of Lesk's dictionary-based word sense disambiguation algorithm using the lexical database WordNet, which provides a rich hierarchy of semantic relations among words. The algorithm relies upon finding overlaps between the glosses of target words and context words in sentence, also their semantic related words' gloss respectively. To get related words, it use several relationship defined in WordNet according to different Part Of Speech, but not all. what's more, a higher score is given to a n word sequence than what is given to the combined score of those n words, if they were to occur in shorter sequences. This algorithm carry out word sense disambiguation for Noun, Verb and Adjective.

In our system's word sense disambiguation function, we make use of this Adapted Lesk algorithm with some improvements: during calculate the combination score for the target word sense, Adapted Lesk also provides sense tags for the other words in the window of context, which is viewed as a side effect of the algorithm. This contribute to our Sense Rank algorithm, which parallel carry out AALesk for the possible sense combination, and ranking their score to kill candidate process to speed up the system. Adapted Lesk didn't make difference for different relationship and only take a few relationship into account not all. AAlesk adapted it by giving **all** relationship a base score for the related words from WordNet relationship according to their importance. Thus make different influence of different relationship and make our word sense disambiguation not only based on statistic theory but also in a semantic way.

2.3 About WordNet

WordNet is an online lexical reference system. Word forms in WordNet are represented in their familiar orthography; word meanings are represented by synonym sets(**synsets**). Two kinds of relations are recognized: lexical and semantic. Lexical relations hold between word forms; semantic relations hold between word meanings. By above it creates an electronic lexical database of nouns, verbs, adjectives, and adverbs.

Each synset has an associated definition or gloss. This consists of a short entry explaining the meaning of the concept represented by the synset. Each synset can also be refered to by a unique identifier, commonly known as a sense-tag.

Synsets are connected to each other through a variety of semantic relations but do not cross part of speech boundaries. In our experiment, all relationships are taken into consideration and given different weight according to their importance.

As WordNet is a particularly rich source of information about nouns, especially when considering the hypernym and hyponym relations, the ignore of part of speech of a word will lead to a reduce of accuracy for adjectives and verbs, but a little effect on nouns. So we make use of a tool to work out part of speech before the core process Sense Rank AALesk.

3 Solution Description

To bring out an "one design fits all" solution for common word sense disambiguation, there are several problem we need to consider. Firstly, the solution should fit on every field's text. Secondly, when we reform the correctness, efficiency is also important. For the first goal, our solution bring out an component independent system, it can adapt to different field by using different kind of electronic corpus for particular field. For the second goal, we bring out a Sense Rank algorithm to parellel process word sense disambiguation algorithm AALesk.

Given a sentence we will tagged out every words Part Of Speech(POS) and throw away those non-content words, as every word has one or more possible sense in a POS, we then build the sense combination for these words, and finally carry out the parallel Sense Rank algorithm for every combination's AALesk process, until every words' sense is disambiguated.

The biggest contribution of Sense Rank algorithm is make use of the conventional word sense disambiguation's side effect and speed up the process greatly: while calculating the combination score for the target word sense, the program also provides sense tags for the other words in the window of context. While the AALesk algorithm bring a semantic way for word sense disambiguation, besides its conventional statistic way.

3.1 Definitions

Given a sentence $t_1,t_2,...,t_n$,there may be some **non-content words**(words that has no real means such as is pronouns, prepositions, articles and conjunctions, or words not be represented in WordNet), after throw away those non-content words, we got words sequence $w_1,w_2,...,w_k$.

Every words have one or more senses, each of which is represented by a unique synset having a unique sense-tag in WordNet. Let the number of sense-tags of the word w_i be represented by $|w_i|$.Then we got $\Pi_{i=1}^{n}|w_i|$ **sense combination** for a context with n words: every content word in the sentence is given a candidate sense. A **combination score** will be computed for each sense combination. The target word is assigned the sense-tag of the sense combination that attains the maximum score. During the procedure, we need to make the following notions clear.

synset: a lists of synonymous word forms that are interchangeable in some context. every sense of a word is put in a synset, which is the unit in WordNet

candidate synset: is a content word given some sense-tag in the sense combination. It may tell the final sense for the word in the instance.

relationships net: words in WordNet is divided into different synset according to its different sense. There is some relationships between synsets in different POS. For an example, there are relations hyponymy and hypernymy for noun. Those relationship link build a net that you can travel from one word(in some synset) to another word(in some synset) following the link in the net.

assistant synset: is a synset around the candidate synset in the relathionship net, which will be refered to when calculating combination score.

overlap: the longest sequence of one or more consecutive words that occurs in both glosses of a pair of comparing words' relationship

| **overlap** |: the number of words in overlap.

window of context: our choice of small context windows is motivated by Choueka and Lusignan[6], who found that human beings make disambiguation decisions based on very short windows of context that surround a target word, usually no more than two words to the left and two words to the right. So our word sense disambiguation will make reference of the words in the window of 2 in context, let the target word is W_i, then system will consider $W_{i-2}, W_{i-1}, W_{i+1}, W_{i+2}$. If the number of content words in the instance is less than 5, all of the content words will serve as the context.

3.2 Processing Algorithm

The AALesk (see algorithm 1) algorithm was derived form the Adapted Lesk algorithm.

Algorithm 1 AALesk: Adapted Adapted Lesk algorithm

---

for a target word W_i in sentance
    given the sense combination as $W_{i-2}[b]$, $W_{i-1}[c]$, $W_i[a]$, $W_{i+1}[d]$, $W_{i+2}[e]$:
    calculate score for each pair of synset(s_1, s_2)stand for($W_x[m]$, $W_i[a]$)
      if s_1 has r_1 kinds of relationship, s_2 has r_2,
      then there would be s_1*s_2 pair of compare relationship
        for every pair of compare relationship
          find out every overlap between the gloss of the two assistant synset
          (restrict not be made up entirely of non-content words)
          (given Sr1,Sr2 as the base weight of this two relationship)
          the score of this relationship pair=Sr1*Sr2*$\sum |overlap|^2$
    score of this synset pair=$\sum$score of each relationship pair
    score of this sense combination=$\sum$score of each synset pair

---

As we set the context window=2, system will compare the nearest 4 content words beside the target word in the instance. The comparing will be held between two words once. For a example, W_i is the target word, we use $W_i[a]$ to present word W_i given sense-tag a. Then in a sense combination such as $W_{i-2}[b]$, $W_{i-1}[c]$, $W_i[a]$, $W_{i+1}[d]$, $W_{i+2}[e]$. System will calculate the combination score for it by summary the score for each pair's(which is $W_i[a]$ vs $W_{i-2}[b]$, $W_i[a]$ vs $W_{i-1}[c]$, $W_i[a]$ vs $W_{i+1}[d]$, $W_i[a]$ vs $W_{i+2}[e]$) overlap. In AALesk, we firstly give every relationship a based score. Then AALesk will travel around the relationship net to join up the gloss for every assistant synset around the two candidate synset in the comparing pair respectively(we restrict the travel level less than 2, which means only the nearest 2 synset in the net can be reached). Then it will compare the gloss of every comparing relationship pair. The overlap of the gloss contribute to the score of relationship pair, which will be summed to contribute to the score for the candidate synset, the sense combination with the highest score will be choose as the result. The AALesk algorithm only take charge of calculating the score of the sense combination, and the rank and choose function will be held by Sense Rank algorithm(see algorithm 2), during which the sense-tag for the target word in the sense combination with the highest score will be chosen as the sense of the target word.

During the **Sense Rank algorithm**, we firstly carry out parellel **AALesk** process for every sense combination(that means there would be $\prod_{i=1}^{n} |W_i|$ processes), taking the first word as the target word and ranking their sense score, the one with the highest score will be set as the words' sense. When the sense of the first word is fixed, those process for the combination which didn't set this sense for the first word will be stop(this would reduce a lot of work). Then follow the same means, fix for the second word with the condition that the first words sense is fixed...then for the k word with the condition that the words before it has been fixed on a sense..., until finally the sense of every word is found.

Algorithm 2 Parallel Sense Rank algorithm

given content words=t1,...tn
IteratorHeap=Φ
outputHeap=Φ
For each sense combination made out a iterator with size=n
    create a single procedure carry out AALesk algorithm1
    and put it in IteratorHeap,
        the iterators in IteratorHeap are ordered on the score
        of the first words' sense score
While IteratorHeap is not empty and more results required
    Iterator IT=pop the first iterator in IteratorHeap
    u=the first node in IT
    if IT has more nodes beside u,
        remove u, and push IT into IteratorHeap again.
        (iterator in IteratorHeap are still in order of the distance
        between the first and second node in a iterator)
        for every Iterator tmpIT in IteratorHeap
            if u!¡ tmpIT remove tmpIT from IteratorHeap
    Insert u into iterator result,
    if result.size=n
        push it into OutputHeap
        result=blank, result.size=0

---

4 Empirical Evaluation and Discussion

In our experiment, we evaluated our solution using the test data from the English lexical sample task used in the SENSEVAL-2 comparative evaluation of word sense disambiguation systems. As Adapted Lesk using WordNet version 1.7, here we use the newest version 2.0 and also version 1.7 for a comparing. For the base score of relationships between synset in WordNet. We try several settings for test and got different result.

4.1 Different Version in Experiment

Firstly, we set all the relationship a same base score as 1,and got a higher accuracy from WordNet version 2.0 than 1.7 (in table 1). Comparing to the result of

Table 1. Accuracy of different version

version	Nouns	Verbs	Adjectives
Adapted Lesk	32.2%	24.9%	46.9%
our system with WordNet1.7	32.6%	25.1%	47.2%
our system with WordNet2.0	33.3%	26.2%	47.5%

Adapted Lesk, we can see there is also an little improve. Although the Adapted Lesk and AALesk following the same reason and AALesk only enlarge the comparing field in relationship nets. While we achieved a good improved for noun and verb although we use the same setting but with the help from a newest version of WordNet. This maybe because the WordNet 2.0 includes more than 42,000 new links between nouns and verbs that are morphologically related, a topical organization for many areas that classifies synsets by category, region, or usage, gloss and synset fixes, and new terminology, mostly in the terrorism domain. This also telling that a good reference corpus is very important in word sense disambiguation.

4.2 Different Setting in Experiment

Then we use WordNet 2.0 with different base score setting for relationships in WordNet to empirical evaluation(see table 2). The relationship mentioned in the table is what we think is more important and is given a score according to that. Those relationship didn't be mentioned is given a score as 1. as we can see from the table, we got a good result in the experiment. But it is believed that this system can even get a better accuracy than what we have got now, which maybe need more research on the relationship's importance and more experiment to search for the best setting.

Table 2. Result of relationship setting

POS	relationship	base score	accuracy
Nouns	hyponymy, hypernymy	1.7	34.6%
	holonymy, meronymy	1.7	
Verbs	hypernymy, troponymy	1.5	27.3%
	verb in group	1.7	
Adjectives	attribute	1.7	48.1%

5 Conclusion

This paper presents an "one fit all" solution for commonly field's text word sense disambiguation, derived from Adapted Lesk algorithm. While the original algorithm relies upon limited relationship between words and only work out one word's sense one time, our solution take all the relationship into consideration and even make the importance of a relationship inflect the result, which make it more semantic related.

We also bring out a Sense Rank algorithm to make it possible for work out all words in a instance in one procedure, which greatly speed up the system.

Additionally, our solution is built component independently, which make it possible for easily adapted to different particular field's text. Thus settle down the big problem exist in the application of word sense disambiguation: one solution with high accuracy commonly only fit on one field's text but and with a low accuracy in another field's text.

All of above make it possible for text mining to go on the semantic way.

References

1. Satanjeev Banerjee, Ted Pedersen.: An Adapted Lesk Algorithm for Word Sense Disambiguation Using WordNet. In proceedings of the Third International Conference on Intelligent Text Processing and Computational Linguistics.(2002)136-145.
2. Leacock, C.; Chodorow, M. and Miller, G.A.: Using Corpus Statistics and WordNet Relations for Sense Identification, Computational Linguistics. (1998)vol.24 no.1.147-165.
3. Ng, H.T. and Lee, H.B.: Integrating multiple knowledge sources to disambiguate word sense: An examplar-based approach. In proceedings of the 34th Annual Meeting of the Association for Computational Linguistics.(1996)40-47.
4. Resnik, P.: Selectional preference and sense disambiguation. In proceedings of ACL Siglex Workshop on Tagging Text with Lexical Semantics, Why, What and How?.(1997).
5. Yarowsky, D.: Unsupervised word sense disambiguation rivaling supervised methods. In proceedings of the 33rd Annual Meeting of the Association of Computational Linguistics (ACL-95).(1995)189-196.
6. Y.Choueka and S. Lusignan.: Disambiguation by short contexts. Computers and the Humanities. (1985)vol.19.147-157.
7. C.Fellbaum, editor. WordNet: An electronic lexical database. MIT Press.(1998)

Automatic Video Knowledge Mining for Summary Generation Based on Un-supervised Statistical Learning

Jian Ling[1], Yiqun Lian[2], and Yueting Zhuang[1]

[1] Institute of Artificial Intelligence, Zhejiang University,
Hangzhou, 310027, P.R. China
lingjian@zjut.edu.cn
[2] Dept. of Electronic Information, Zhejiang Institute of Media and Communication,
Hangzhou, 3100027, P.R. China
lianyiqun@hzcnc.com

Abstract. The summary of video content provides an effective way to speed up video browsing and comprehension. In this paper, we propose a novel automatic video summarization approach. Video structure is first analyzed by combining spatial-temporal analysis and statistical learning. Video scenes are then detected based on unsupervised statistical learning. The video summary is created by selecting the most informative shots from the video scenes that are modeled as a directed graph. Experiments show that the proposed approach can generate the most concise and informative video summary.

1 Introduction

In recent years, video summarization techniques have gained a lot of attention from researchers. Basically there are two kinds of video summaries: *static video summary* and *dynamic video skimming*. The *static video summary* is composed of a set of key frames extracted from the source video [1][2]. Long video with dynamic visual contents is inadequately represented by the *static video summary* that loses dynamic property and audio track of the source video. In contrast, *dynamic video skimming* generates a new and much shorter video sequence from the source video. Recently, a lot of work has been conducted on *dynamic video skimming*. The Informedia system [3] shows the utilization of integrating language and image-understanding techniques to create the video skimming. As pointed by the authors, the proposed approach is limited due to the fact that it is difficult to clearly understand video frames. A generic framework of video summarization is proposed based on modeling of the viewer's attention [4]. Without fully understanding of video contents semantically, the framework takes advantage of the user attention model and eliminates the needs of complex heuristic rules to construct video skimming.

The video summary is a concise and informative representation of the source video contents. Although many video summary generation techniques have been proposed, few of them have focused on analyzing the spatial-temporal feature of video contents. In this paper, we propose a novel automatic video summarization approach. Video structure is analyzed by combining spatial-temporal analysis and statistical learning.

Video scene is then detected by implementing unsupervised learning. The most informative video shots are then selected from each video scene, which is modeled as a directed graph, to form the video summary.

2 Automatic Video Spatial-Temporal Knowledge Mining

A video is traditionally decomposed into a hierarchical tree structure [5] that consists of four top-down levels such as video scene, group, shot and key-frame.

2.1 Video Shot Detection and Key Frames Extraction

A lot of work has been done in shot boundary detection and many of the approaches achieve satisfactory performance. We use an approach similar to the one used in [6]. For key frame extraction, although more sophisticated techniques were designed, they require high computation effort. We select the beginning and ending frames of a shot as the two key frames to achieve fast processing speed. And color histograms of the beginning and ending frames are used as the feature for the shot.

2.2 Video Group and Scene Detection

Video groups contain temporal feature of the source video. Here, statistical learning is implemented to model the time transition between video groups. Formally, a group G_m transits to another group G_n if there exists a shot in G_m that is temporally adjacent to a shot in G_n. It is modeled by a time transition probability $P(G_m | G_n)$, which can be estimated as follows:

$$P(G_m | G_n) = \frac{1}{|G_n|} \sum_{S_i \in G_m} \sum_{S_j \in G_n} \tau(i - j) \qquad (1)$$

Where $|G_n|$ is the number of shots in G_n, S_i the ith shot ranked in time order. Function $\tau(x)$ is defined as:

$$\tau(x) = \begin{cases} 1, x = 1 \\ 0, x \neq 1 \end{cases} \qquad (2)$$

To measure the probability that a video group G belongs to a video scene SC, the conditional probability $P(G | SC)$ is estimated as follows:

$$P(G | SC) \approx P(G | G_1, G_2, ..., G_n) = \sum_{G_k \in SC} P(G | G_k) \qquad (3)$$

Where n is the number of video groups in scene SC. From equation (1)-(3), we can calculate $P(G | SC)$.

In video, visually similar shots are clustered into a group, but even non-similar groups can be grouped into a single scene if they are semantically related. Here unsupervised clustering is devised to construct video scenes from groups. Given a set of video groups $V = \{G_1, G_2, ..., G_N\}$ and a threshold parameter δ, the clustering procedure can be summarized as follows:

1. Initialization: put group G1 into scene SC1, sceneNum=1;
2. Get the next group Gi. If there is no group left, the procedure exits;
3. Calculate the probability $P(G_i | SC_j)$ for each existing scene SC_j based on equation (1)-(3);
4. Determine which scene is the most appropriate one for group Gi by finding the maximum probability $P_k = \arg\max_j P(G_i | SC_j)$. If $P_k > \delta$, put the group G_i into scene SC_k. Otherwise, it means Gi is not close enough to the existing scenes, put Gi into a new scene and increase sceneNum by 1. Goto 2.

When the clustering procedure stops, each cluster represents one video scene. The next step is to create the video summary.

3 Video Summary Creation

Each video scene consists of a set of groups $SC = \{G_1, G_2, ..., G_n\}$. The more complex a video scene is, the more important it contributes to the video summary. To measure the complexity of a video scene, a complexity parameter (CP) is defined. A directed graph model is easily set up for each video scene to calculate CP based on the equation (1). The directed graph model is defined as $G = (V, E)$, where V is a set of vertices, and each vertex represents a video group in the scene; E is a set of directed edges, and each edge is pointed from group G_m to G_n (or vice versa) with the edge weight $P(G_m | G_n)$ (or $P(G_n | G_m)$). After the graph model is built for each scene, we can find the largest path of the graph and sum up the weight of edges on the path to represent the complexity parameter. We can learn the effectiveness of the CP from two aspects. (1) A complexity scene always consists of more groups than a simple one. It means that the graph model of the complexity scene has a longer largest path than the simple one. (2) A complexity scene always has more time transitions between groups than a simple one. It means that the edge weight of the graph model of the complexity scene is larger than the simple one.

Both aspects make the CP of a complexity video scene to be larger than a simple one. With the complexity parameter and skim ratio R, the video summary can be generated based on the following steps:

1. Build the directed graph model for each scene.
2. Find the largest path of each graph and calculate CP for each scene.
3. Exclude the scenes whose CPs are less than a pre-defined threshold φ. Select the scene SCmax with the maximum CP.
4. Select the group Gmax with the maximum number of shots from SCmax. Remove the vertex corresponding to Gmax and its adjacent edges from the graph.

If the graph becomes unconnected, set up an edge with zero weight between sub-graphs.
5. Select the shots with maximum time duration from each selected groups Gmax. If the skim ratio is less than R, goto step 2. Otherwise rank the selected shots in time order and concatenate them to generate the video summary.

Since video scenes consist of semantically related groups and video groups consist of shots with similar visual contents, the above steps not only select the most informative shots but also greatly reduce the visual content redundancy.

4 Experiments and Discussions

To test the performance of the proposed video summarization approach, we conduct experiments on six new and documentary videos with different time durations.

We first conduct experiment on video structure analysis. Shot detection method [6] is adopted to segment video into shots for its robustness of detecting both cut and transition shots. Some shots whose frame number is less than 25 are discarded because they tend to be transition shots. Key-frame extraction method [2] is implemented to extract one key frame from each shot based on color coherence vector. Then support vector clustering is performed on color coherent vectors of key frames to cluster shots into groups. The resulted numbers of shot and group for each video are shown in Table 1.

Table 1. The numbers of shot and group for each video

No.	Genre	Time	#Shot	#Group	#Shot/#Group
1	News	05:05	44	16	2.8
2	Documentary	04:21	29	12	2.4
3	News	10:41	72	30	2.4
4	Documentary	12:49	78	37	2.1
5	News	29:46	283	71	4.0
6	Documentary	29:47	229	63	3.6
Average	--	15:25	122.2	38.2	3.2

Table 2. The number of generated video scenes with different value of δ

No.	Time	#Scene (δ =0.20)	#Scene (δ =0.30)	#Scene (δ =0.40)
1	05:05	6	6	4
2	04:21	4	4	3
3	10:41	15	13	10
4	12:49	21	17	11
5	29:46	39	33	26
6	29:47	36	29	20

5 Future Work

In this paper, we propose a novel approach to automatically generate the video summary. The video structure is first analyzed by combining spatial-temporal analysis and statistical learning. Video scene is then detected by implementing unsupervised learning. The video summary is generated based on the complexity of the detected video scene that is modeled as a directed graph model.

Future works can be focused on the following topics: (1) Exploit high-level semantics from video contents such as text transcribed from speech, video caption located on the video frames and etc. (2) Integrate multi-model features of the video content into one framework to generate the more effective video summary.

References

1. DeMenthon, D., Kobla, D., Doermann, D.: Video Summarization by Curve Simplification. Proceedings of the ACM International Conference of Multimedia, Bristol, England (1998) 211-218.
2. Zhuang, Y.T., Rui, Y., Huang T.S., Mehrotra S.: Key Frame Extraction Using Unsupervised Clustering. Proceedings of the IEEE International Conference on Image Processing, Chicago (1998) 866-870.
3. Smith, M.A., Kanade, T.: Video Skimming and Characterization Through the Combination of Image and Language Understanding Techniques. Proceedings of the Conference on Computer Vision and Pattern Recognition, Puerto Rico. (1997) 775-781.
4. Ma, Y.F., Lu, L., Zhang, H.J., Li, M.J: A User Attention Model for Video Summarization. Proceedings of the ACM International Conference on Multimedia, Juan les Pins, France (2002) 533-542.
5. Rui, Y., Huang, T.S., Mehrotra, S.: Constructing Table-of-Content for Videos. ACM Multimedia Systems Journal, Special Issue Multimedia Systems on Video Libraries, Vol.7, No.5 (1999) 359-368.
6. Ye, Z.Y., Wu, F., Zhuang, Y.T.: A Robust Fusion Algorithm for Shot Boundary Detection. Journal of Computer Aided Design and Computer Graphics (In Chinese with English Abstract), Vol.15, No.11 (2003) 950-955.

A Model for Classification of Topological Relationships Between Two Spatial Objects

Wu Yang[1,2], Ya Luo[1], Ping Guo[2], HuangFu Tao[2], and Bo He[1]

[1] Department of Computer Science and Engineering,
Chongqing Institute of Technology, 400050 Chongqing, China
`yw@cqit.edu.cn`
[2] Department of Computer Science, Chongqing University,
400050 Chongqing, China

Abstract. From the aspect of basic characteristics of human cognition, a model for classification of topological relationships between two spatial objects, hierarchy model, is proposed. Then, the complete and mutual exclusion properties of the hierarchy model are proved. Finally, the capability of classification for hierarchy model is compared with that of the calculus-based method.

1 Recent Researches of Topological Relationships Between Two Spatial Objects

In the field of classification research of topological relationships between two spatial objects, there are some models, 4IM model, 9IM model, dimension extended method (DEM) and Calculus-based Method (CBM).

This paper presents hierarchy model for topological relationships between two-dimensional spatial objects, and applies hierarchy model to the complete and formal description of spatial topological relationships.

2 Hierarchy Model

The process of human cognition has hierarchy characteristic that is from rough to precise and from simple to complex. Applying those characteristics to the classification of topological relationships, it forms a classification process of thinning step-by-step and extending by hierarchy. From this process, a tree-liked structure is obtained. Five basic topological relationships on leaf nodes are disjoint relationship, touch relationship, overlap relationship, in relationship and equal relationship. Those five relationships and their classification properties will be discussed in this section.

2.1 Hierarchy Characteristic of Cognition

During the analysis of topological relationships between two objects, human cognition is a thinning process step by step. Based on this characteristic, a tree-liked structure is applied to simulate this process, and the root node is a universal set of topological relationships between 2-dimensional objects. This tree-liked structure is represented as figure 1.

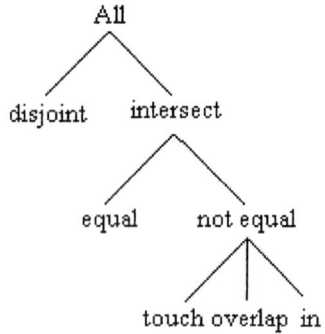

Fig. 1. Tree-liked structure of hierarchy classification

2.2 Formal Definition of Hierarchy Model

Definition 1: In the notation $<\lambda_1, r, \lambda_2>$, r is the relationship between λ_1 and λ_2. In IR^2, five topological relationships are defined as follows:

(1) The disjoint relationship: $<\lambda_1, disjoint, \lambda_2> \Leftrightarrow (\lambda_1 \cap \lambda_2 = \emptyset)$

(2) The touch relationship: $<\lambda_1, touch, \lambda_2> \Leftrightarrow (\lambda_1^\circ \cap \lambda_2^\circ = \emptyset) \wedge (\lambda_1 \cap \lambda_2 \neq \emptyset)$

(3) The overlap relationship: $<\lambda_1, overlap, \lambda_2> \Leftrightarrow$
$$(\lambda_1^\circ \cap \lambda_2^\circ \neq \emptyset) \wedge (\lambda_1^\circ \cap \overline{\lambda_2} \neq \emptyset) \wedge (\overline{\lambda_1} \cap \lambda_2^\circ \neq \emptyset)$$

(4) The in relationship: $<\lambda_1, in, \lambda_2> \Leftrightarrow (\lambda_1^\circ \cap \lambda_2^\circ \neq \emptyset) \wedge (\lambda_1 \cap \lambda_2 = \lambda_1) \wedge (\lambda_1 \cap \lambda_2 \neq \lambda_2)$

(5) The equal relationship: $<\lambda_1, equal, \lambda_2> \Leftrightarrow (\lambda_1 = \lambda_2)$

We call those five basic topological relationships as spatial basic topological relationships based on hierarchy model.

2.3 Classification Capability of Hierarchy Model

In this section, based on hierarchy model, the mutual exclusion quality and complete quality of minimum set for topological relationships between two-dimensional spatial objects will be analyzed, and the classification capability of hierarchy model will be compared with CBM.

Theorem 1: A and B are two two-dimensional spatial objects. It exists and only exists one relationship between A and B in hierarchy model.

Proof: The relationship between spatial objects A and B could be determined by a decision tree of the topological relationships (see Figure 2). In figure 2, each internal node represents a Boolean calculation. If the value of Boolean calculus is true, then the left branch is followed, otherwise the right branch is followed. Repeating this process until a leaf node is reached. Because leaf node indicates to one of 5 basic topological relationships in hierarchy model, one relationship is existed between A and B.

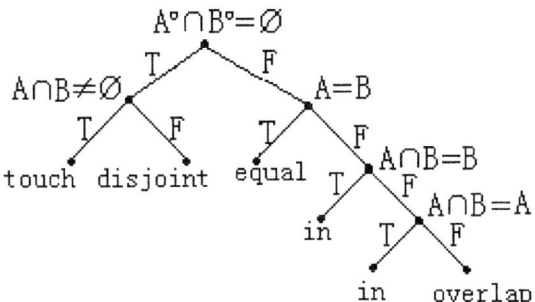

Fig. 2. A topological relations decision tree for 2-dimensional spatial objects

It can be found in this topological relationships decision tree that two different relationships cannot hold between two given two-dimensional spatial objects, because there are only two branches for each non-leaf, and the calculation consequence can and only can be satisfied by only one branch. Each leaf-node can and only can represents one topological relationship of 5 basic topological relationships in hierarchy model.

Theorem 1 explains that the 5 topological relationships in hierarchy model cover any possible topological relationships between spatial objects. From theorem 1, we also get:

Deduction 1: The five topological relationships are mutual exclusive in hierarchy model.

Proof: (The proof is omitted.)

Theorem 2 mentioned below explains that these 5 topological relationships are minimum set of topological relationships based on the human cognition.

Theorem 2: Based on human cognition, topological relationships have five types at least.

Proof: (The proof is omitted.)

From deduction 1 and theorem 2, we have:

Deduction 2: The topological relationships in hierarchy model are complete based on human recognition.

Proof: (This proof is omitted.)

Theorem 3 under mentioned compares the classification capability of hierarchy model with that of CBM.

Theorem 3: Hierarchy model can represent all topological relationships between two-dimensional spatial objects that can be represented by CBM.

Proof: (The proof is omitted.)

From theorem 3, we know hierarchy model is not weaker than CBM in terms of the classification capability.

3 Conclusion

This paper presents hierarchy model, which classifies topological relationships between two-dimensional spatial objects into five basic relationships. The classification capability of CBM is equal to that of 9IM+DEM, so the classification capability of hierarchy model is not weaker than that of 9IM+DEM too.

References

1. Cohn AG, Hazarik AS. Qualitative spatial representation and reasoning: an overview. Fundamenta Informaticae, 2001,46(1-2): 1~29.
2. Schneider M. Uncertainty management for spatial data in databases: fuzzy spatial data types. In: Goos G, Hartmanis J, Leeuwen, JV, eds. Advances in Spatial Databases, the 6th International Symposium, SSD'99. LNCS 1651, Berlin: Springer-Verlag, 1999. 330~351.
3. David V. Pullar and Max J. Egenhofer. Toward formal definitions of topological relations among spatial objects. In Proceedings of the 3rd International Symposium on Spatial Data Handling, Sydney, Australia, pages 225-241, Columbus, OH, August 1988. International Geographical Union IGU.
4. Max J.Egenhofer MJ, Herring JR. Categorizing binary topological relationships between regions, lines and points in geographic database. Technical Report, 91-7, Orono: University of Maine, 1991.
5. Randell DA, Cui Z, Cohn AG. A spatial logic based on regions and connection. In: Nebel B, Rich C, Swartout WR, eds. Proceedings of the 3rd International Conference on Principles of Knowledge Representation and Reasoning. San Francisco: Morgan Kaufmann Publishers, 1992. 165~176.

A New Feature of Uniformity of Image Texture Directions Coinciding with the Human Eyes Perception*

Xing-Jian He[1], Yue Zhang[1], Tat-Ming Lok[2], and Michael R. Lyu[3]

[1] Intelligent Computing Lab, Institute of Intelligent Machines,
Chinese Academy of Sciences, P.O.Box 1130, Hefei, Anhui 230031, China
{xjhe, yzhang}@iim.ac.cn
[2] Information Engineering Dept., The Chinese University of Hong Kong, Shatin, Hong Kong
tmlok@ie.cuhk.edu.hk
[3] Computer Science & Engineering Dept., The Chinese University of Hong Kong,
Shatin, Hong Kong
lyu@cse.cuhk.edu.hk

Abstract. In this paper we present a new feature of texture images which can scale the uniformity degree of image texture directions. The feature value is obtained by examining the statistic characteristic of the gradient information of the image pixels. Simulation results illustrate that this feature can exactly coincide with the uniformity degree of image texture directions according to the perception of human eyes.

1 Introduction

The research of image texture features has been a hot topic for long. Many methods to portray the image texture characteristics have been proposed in a large number of literatures [1]. In resent years, some intelligent methods such as neural network based techniques have been presented [2-6].

An important characteristic of texture image is the distributing trait of texture directions. For images with strong texture structures (e.g., bark, cloth, rock), the statistic properties of texture directions are generally very useful in most practical applications. But peculiar features to reflect the distributing property of texture directions have been seldom studied. The widely used Tamura feature of directionality, which is constructed in accordance with psychological studies on the human perception of texture, is one of these peculiar features. This feature is obtained by examining the sharpness degree of a histogram which is constructed from the gradient vectors of all the image pixels.

But the Tamura feature of directionality behaves not so well in reflecting an important property, i.e., texture direction uniformity, of images. Undoubtedly, a feature reflecting the uniformity degree of image texture directions according to the perception of human eyes is very useful in many fields.

---

* This work was supported by the National Natural Science Foundation of China (Nos.60472111 and 60405002).

2 The Tamura Feature of Directionality

To compute the Tamura feature of directionality, firstly images need to be convoluted with Prewitt masks to obtain the horizontal and vertical differences, ΔH and ΔV, of the image. Then, gradient vector at each pixel can be computed as follows:

$$|\Delta G| = (|\Delta H| + |\Delta V|)/2 \ . \tag{1}$$

$$\theta = \tan^{-1}(\Delta V + \Delta H) + \pi/2 \ . \tag{2}$$

where $|\Delta G|$ is the magnitude and θ is the angle.

Then, by quantizing θ and counting the pixels with the corresponding magnitude $|\Delta G|$ greater than a threshold, a histogram of θ, denoted as H_D, can be constructed. And the Tamura feature of directionality is obtained as follows:

$$F_{dir} = \sum_{p}^{n_p} \sum_{\phi \in w_p} (\phi - \phi_p)^2 H_D(\phi) \ . \tag{3}$$

where p is ranged over n_p peaks; and for each peak p, w_p is the set of bins distributed over the peak; while ϕ_p is the bin that takes the peak value. We can see that this equation reflects the sharpness of the peaks, i.e., a smaller value of F_{dir} will corresponding to the image which gets a sharper peaks histogram.

However, we can also see from eqn. (3) that the angle distances between every two peaks can not be reflected out. So the uniformity of texture directions can not be reflected. In allusion to this shortage, we present a new feature in the next section.

3 The New Feature of Direction Uniformity

Just like the computing of the Tamura feature of directionality, with a selected threshold of $|\Delta G|$ denoted with letter b, we can also construct a histogram of θ. The area $[0, \pi]$ is equally divided into m parts called as m bins indexed with integer 1, 2, 3..., m. Suppose that the initial value of every bin is zero. Examining the gradient vector of each pixel, if the magnitude $|\Delta G|$ is larger than b, then the value of the bin in which the angle θ is contained in will be added by one.

Then the probability histogram is obtained by dividing each value by the sum of the m values of the m bins. Pick out k bins with the greatest k probability values. The integer k is determined by a threshold d with the follow conditions:

$$\sum_{i=1}^{k} p_i \geq d \quad \text{and} \quad \sum_{i=1}^{k-1} p_i < d \ . \tag{4}$$

Generally, the value of d is selected around 0.5. The k bin values are normalized by the following formula:

$$prb_i = p_i / \sum_{i=1}^{k} p_i \qquad i = 1, 2, \ldots, k. \qquad (5)$$

Thus we get k bins with their indexes and normalized probability values prb_i. It should not be forgotten that the index of each bin indicates the direction this bin indicates. In other words, for an image, we then obtained its k texture directions with k largest probability values. For each bin, we consider the acute angle with the first bin, which gets the largest probability value, as the angle distance:

$$a_i = \begin{cases} m - |i_i - i_1| & if\ |i_i - i_1| \geq (m/2) \\ |i_i - i_1| & otherwise \end{cases} \qquad i=1,2,\ldots,k \qquad (6)$$

where m is the fore mentioned number of bins. The presented new feature is computed by the following formula:

$$F_{unf} = k \cdot \sum_{i=1}^{k} [(-prb_i \ln prb_i)] \cdot \sum_{i=1}^{k} a_i^2 prb_i \qquad (7)$$

From the formula we can deduce that the smaller values of F_{unf} correspond to the images which have stronger texture directionality and better direction uniformity.

4 Simulation Results

The texture images in our experiments are all coming from the widely used Brodatz's. The algorithm mentioned above needs three parameters, i.e., the number of bins m, the thresholds b and d. In our experiments we adopted Prewitt masks, assume that the number of bins is 12, the threshold b is 9, and the threshold d is 0.5.

Figure 1 and 2 are the series of the same images sorted by the directionality values and by the new feature values respectively. It can be clearly seen that the new feature has some superiority than the Tamura feature of directionality in reflecting the image texture direction characteristic. Further more, it can also be found that the new feature of texture direction uniformity is to completely coincide with the human eye.

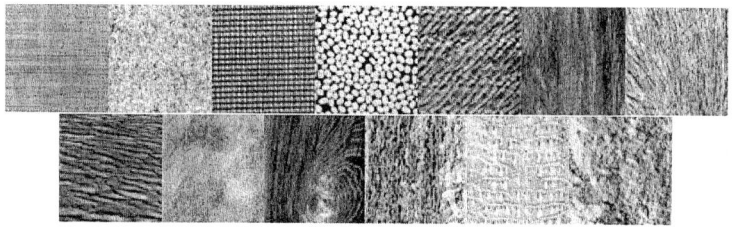

Fig. 1. Texture images sorted by the values of directionality of Tamura feature from smallest to greatest

Fig. 2. The same texture images sorted by the values of the new feature from smallest to greatest

5 Conclusions

This paper has presented a new image texture feature and investigated its implementation ability. The simulation result showed that the new feature is effective in reflecting the degree of the image texture direction uniformity and the strength of image texture directionality. The image sequence sorted by value of this feature of each image also illustrates that the feature can exactly coincide with the uniformity degree of image texture directions in the light of perception of human eyes.

References

1. L.Wang and J. Liu.: Texture Classification using Multiresolution Markov Random Field Models. Pattern Recognition Letters, vol. 20, 1999, pp. 171-182.
2. D.S.Huang.: Intelligent Signal Processing Technique for High Resolution Radars. Publishing House of Machine Industry of China, February 2001.
3. B. Lerner, H. Guterman, M. Aladjem and H. Dinstein.: A Comparative Study of Neural Network based Feature Extraction Paradigms. Pattern Recognition Letters, Vol. 20, 1999, pp. 7-14.
4. D.S.Huang.: Systematic Theory of Neural Networks for Pattern Recognition. Publishing House of Electronic Industry of China, Beijing, 1996.
5. D.S.Huang.: Radial basis probabilistic neural networks: Model and application. International Journal of Pattern Recognition and Artificial Intelligence, 13(7), 1083-1101,1999.
6. D.S.Huang.: The local minima free condition of feedforward neural networks for outer-supervised learning. IEEE Trans on Systems, Man and Cybernetics, Vol.28B, No.3, 1998,477-480.

Sunspot Time Series Prediction Using Parallel-Structure Fuzzy System

Min-Soo Kim[1] and Chan-Soo Chung[2]

[1] Sejong-Lockheed Martin Aerospace Research Center,
Sejong University, 98 Kunja, Kwangjin,
Seoul, Korea 143-747
mskim@sejong.ac.kr
[2] Electrical Engineering Department,
Soongsil University, 1-1 Sangdo, Dongjak,
Seoul, Korea 156-743
chung@ssu.ac.kr

Abstract. Sunspots are dark areas that grow and decay on the lowest level of the Sun that are visible from the Earth. Short-term predictions of solar activity are essential to help plan missions and to design satellites that will survive for their useful lifetimes. This paper presents a parallel-structure fuzzy system (PSFS) for prediction of sunspot number time series. The PSFS consists of a multiple number of component fuzzy systems connected in parallel. Each component fuzzy system in the PSFS predicts future data independently based on its past time series data with different embedding dimension and time delay. An embedding dimension determines the number of inputs of each component fuzzy system and a time delay decides the interval of inputs of the time series. According to the embedding dimension and the time delay, the component fuzzy system takes various input-output pairs. The PSFS determines the final predicted value as an average of all the outputs of the component fuzzy systems in order to reduce error accumulation effect.

1 Introduction

The most visible appearance of solar activity is sunspots on the photosphere. Sunspots are magnetic regions on the Sun with magnetic field strengths thousands of times stronger than the Earth's magnetic field and appear as dark spots on the surface of the Sun and typically last for several days. Sometimes magnetic fields change rapidly releasing huge amounts of energy in solar flares and ejection of material from and through the corona. Solar activity like sunspots tends to vary from a minimum to a maximum and back again in a solar cycle of about 11 years. Planning for satellite orbits and space missions often require knowledge of levels years in advance through prediction of sunspot time series [1]~[4].

There are two basic approaches to prediction: model-based approach and nonparametric method. Model-based approach assumes that sufficient prior information is available with which one can construct an accurate mathematical model for prediction. Nonparametric approach, on the other hand, directly attempts to analyze a sequence of observations produced by a system to predict its future behavior. Though nonparametric approaches often cannot represent full complexity of real systems, many contemporary prediction theories are developed based on the nonparametric approach because of difficulty in constructing accurate mathematical models.

Signals generated from many practical systems show chaotic behaviors due to inherent nonlinear characteristics of the physical system. The fact that chaotic time series are sensitive to initial perturbations makes chaotic time series prediction a difficult task. Therefore, many practical prediction algorithms such as linear prediction, neural networks, and the adaptive algorithms have been used for short-term prediction or trend analysis of chaotic time series data[5]~[8].

This paper presents a parallel-structure fuzzy system (PSFS) for prediction of sunspot data which show characteristic of chaotic time series. This approach corresponds to a nonparametric approach since it generates a prediction result based on past observations of the system output. The PSFS consists of a multiple number of component fuzzy systems connected in parallel. Each component fuzzy system in the PSFS predicts future data independently based on its past time series data with different embedding dimension and time delay.

The embedding dimension determines the number of inputs of each component fuzzy system. According to the time delay, the component fuzzy system takes inputs at different time intervals. Each component fuzzy system produces separate prediction results for a future data at a specific time index. The PSFS determines the final predicted value as an average of all the outputs of the component fuzzy systems.

Each component fuzzy system contains a small number of multiple-input single-output (MISO) Sugeno-type fuzzy rules[9], which are generated by clustering input-output training data. The subtractive clustering algorithm[14] produces fuzzy rules for prediction of sunspot time series. Inputs to the fuzzy system in the parallel-structure fuzzy system are determined according to time delay and embedding dimension. Extensive simulations show that the PSFS with 3 component fuzzy systems successfully predicts chaotic time series data comparing with the prediction results by a single fuzzy system.

2 Prediction of Time Series

Nonparametric approach of time series prediction is based on the assumptions that future behavior of a time series can be represented by a functional relationship of its previous observations. If k previous input data are given at the k th time step, for a time series data $x(k)$, the τ-step-ahead value $\hat{x}(k+\tau)$ can be expressed as

$$\hat{x}(k+\tau) = P[x(k), x(k-1), x(k-2), \cdots, x(2), x(1)] \tag{1}$$

where $P[\cdot]$ denotes a function that represents input-output relationship of nonparametric time series prediction process and positive integer τ is called time delay.

Time series prediction methods can be classified into either one-step-ahead prediction or short-term prediction depending on the fact that predicted values are again used as input values. In one-step-ahead prediction, the predicted value of future data

$\hat{x}(k+\tau)$ is expressed by its previous m inputs with time delay τ of the data sequence as in Eq. (2).

$$\hat{x}(k+\tau) = P[x(k), x(k-\tau), x(k-2\tau), \cdots, x(k-(m-1)\tau)] \quad (2)$$

where positive integer m is time delay. Future value of a time series can be predicted by an output of a linear or nonlinear function $P[\cdot]$ of $m\tau$ previous input data.

In short-term or long-term prediction of time series data, predicted values of the data are again used as inputs for prediction of future data. Eq. (3) shows short-term or long-term prediction of time series. The predicted value of future data $\hat{x}(k+\tau)$ is expressed according to the data previously predicted $\hat{x}(k), \hat{x}(k-\tau), \cdots, \hat{x}(k-(m-1)\tau)$. Therefore, long-term prediction of chaotic time series based on the data previously predicted is a difficult task since small initial error causes enormous error accumulation effects in future values.

$$\hat{x}(k+\tau) = P[\hat{x}(k), \hat{x}(k-\tau), \hat{x}(k-2\tau), \cdots, \hat{x}(k-(m-1)\tau)] \quad (3)$$

The optimal embedding dimension m is determined for a specific time delay τ. For given τ, error performance measures are calculated for the training data and for the validation data. Validation data is a data set not used in training phase for checking if training result is acceptable. Error performance measures are defined as mean-square error and maximum absolute error calculated from the difference between the one-step-ahead prediction results and real data. For a given time series data, MSE and MAE are computed as one increases possible embedding dimension values for a fixed time delay. This process is repeated for training data and for validation data. The optimal embedding dimension corresponds to the embedding dimension whose error measure is minimized both for training and validation data.

3 Parallel-Structure Fuzzy System

3.1 Configuration of a Parallel-Structure Fuzzy System

A parallel-structure fuzzy system (PSFS) predicts future data according to several prediction mechanisms based on different number and different samples of previous data. The PSFS consists of a multiple number of component fuzzy systems connected in parallel for predicting time series. Fig. 1 shows the structure of the parallel-structure fuzzy system. The PSFS contains N component fuzzy systems, *Fuzzy System*$_1$, *Fuzzy System*$_2$, ..., *Fuzzy System*$_N$ connected in parallel. Each component fuzzy system produces independently predicted values of same future data $\hat{x}(k+1)$ at a time index $k+1$ based on previous data. The PSFS produces the final predicted value according to the N prediction results $\hat{x}_1(k+1), \hat{x}_2(k+1), \cdots, \hat{x}_N(k+1)$ of the N component fuzzy systems.

Time series prediction with the PSFS is characterized by the two parameters τ and m. The embedding dimension m defines the number of inputs to each component fuzzy system, and the time delay τ defines the time interval of input data to component fuzzy systems.

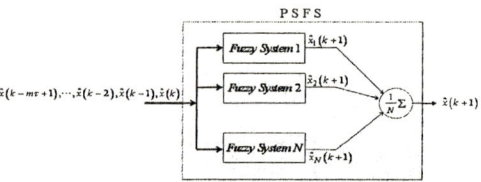

Fig. 1. Structure of the parallel-structure fuzzy system (PSFS)

For the PSFS with N component fuzzy systems in general, each fuzzy system produces prediction results $\hat{x}_1(k+1), \hat{x}_2(k+1), \cdots, \hat{x}_N(k+1)$ based on previous data $\hat{x}(k), \hat{x}(k-1), \hat{x}(k-2), \cdots$, and the final predicted data $\hat{x}(k+1)$ as in Eq. (4) becomes an average of all the prediction results by component fuzzy systems.

$$\hat{x}(k+1) = \frac{1}{N}\left[\sum_{i=1}^{N} \hat{x}_i(k+1)\right] \tag{4}$$

Fig. 2 shows an example of input data used to predict the future data $\hat{x}(k+1)$ using the PSFS. The PSFS consists of 3 component fuzzy systems *Fuzzy System*$_1$, *Fuzzy System*$_2$, and *Fuzzy System*$_3$. The embedding dimensions for each component fuzzy system are assumed 3, 4, and 3 for the time delay 1, 2, and 3, respectively. So the PSFS is characterized by the 3 parameter pairs (τ_i, m_i) of (1,3), (2,4), and (3,3). This means *Fuzzy System*$_1$ predicts $\hat{x}_1(k+1)$ using the input data $\hat{x}(k), \hat{x}(k-1), \hat{x}(k-2)$, *Fuzzy System*$_2$ generates $\hat{x}_2(k+1)$ using $\hat{x}(k-1)$, $\hat{x}(k-3)$, $\hat{x}(k-5)$, and $\hat{x}(k-7)$. *Fuzzy System*$_3$ outputs $\hat{x}_3(k+1)$ using $\hat{x}(k-2), \hat{x}(k-5), \hat{x}(k-8)$. In the PSFS, final predicted value $\hat{x}(k+1)$ are determined by averaging the 3 prediction results $\hat{x}_1(k+1)$, $\hat{x}_2(k+1)$, and $\hat{x}_3(k+1)$ from the 3 component fuzzy systems.

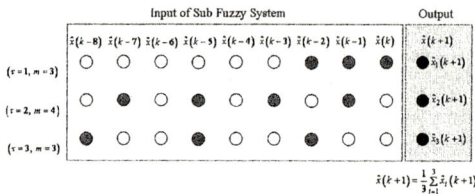

Fig. 2. Inputs of each component fuzzy system in the PSFS for prediction of $\hat{x}(k+1)$

In short-term or long-term prediction, a small amount of prediction error is accumulated to become a big error after several iterations. The PSFS reduces error accumulation effect by averaging the prediction results of all the component fuzzy systems after removing the extreme values of prediction results.

3.2 Component Fuzzy System

Modeling fuzzy systems involves identification of the structure and the parameters with given training data. In the Sugeno fuzzy model[9], unlike the Mamdani method[10], the consequent part is represented by a linear or nonlinear function of input variables. The Sugeno model can represent nonlinear input-output relationships with a small number of fuzzy rules. Each rule in the Sugeno model corresponds to an input-output relationship of a fuzzy partition. Fuzzy rules in the MISO Sugeno model with n inputs $x_1,\cdots,x_n$ and an output variable y_i of the i th fuzzy rule is of the form:

$$\text{IF } x_1 \text{ is } A_{i1} \text{ and } \cdots \text{ and } x_n \text{ is } A_{in}, \text{THEN } y_i = f_i(x_1,\cdots,x_n) \tag{5}$$

where $i=1,\cdots,M$ and A_{ij} is a linguistic label represented by the membership function $A_{ij}(x_j)$, and $f_i(\cdot)$ denotes a function that relates input to output. M denotes the number of rules in the fuzzy system. The Sugeno fuzzy model can easily generates fuzzy rules from numerical input-output data obtained from an actual process. The function $A_{ij}(x_j)$ defines a membership function assigned to the input variable x_j in the i th fuzzy rule. In this paper, Gaussian membership functions and a linear function are used for simplicity.

$$A_{ij}(x_j) = \exp\left(-\frac{1}{2}\left(\frac{x_j - c_{ij}}{w_{ij}}\right)^2\right) \tag{6}$$

$$f_i(x_1,\cdots,x_n) = a_{0i} + a_{1i}x_1 + \cdots + a_{ni}x_n \tag{7}$$

where the parameters c_{ij} and w_{ij} define center and width of the Gaussian membership function $A_{ij}(x_j)$. The coefficients $a_{01}, a_{1i},\cdots,a_{ni}$ are to be determined from input-output training data. In the simplified reasoning method, the output y of the fuzzy system with M rules is represented as

$$y = \frac{\sum_{i=1}^{M} \mu_i f_i(x_1,\cdots,x_n)}{\sum_{i=1}^{M} \mu_i} \tag{8}$$

where μ_i is a degree of relevance. In the product implication method, the degree of relevance is defined as

$$\mu_i = \prod_{j=1}^{n} A_{ij}(x_j) \tag{9}$$

$$= \exp\left(-\frac{1}{2}\sum_{j=1}^{n}\left(\frac{x_j - c_{ij}}{w_{ij}}\right)^2\right) \tag{10}$$

In the Sugeno fuzzy model, the time-consuming rule extraction process from experience of human experts or engineering common sense reduces to a simple parameter optimization process of coefficients $a_{01}, a_{1i},\cdots,a_{ni}$ for a given input-output data set since the consequent part is represented by a linear function of input variables. Using linear function in the consequent part, a group of simple fuzzy rules can successfully approximate nonlinear characteristics of practical complex systems.

Construction of the parallel-structure fuzzy system is basically off-line. The parameters of the PSFS must be determined by the training data before time series prediction operation.

3.3 Modeling of Component Fuzzy System Based on Clustering

The parameters of the component fuzzy systems are characterized using clustering algorithms. The subtractive clustering algorithm[14] finds cluster centers $x_i^* = (x_{1i}^*, \cdots, x_{ni}^*)$ of data in input-output product space by computing the potential values at each data point. The potential value is inversely proportional to distance between data points, which means densely populated data produces large potential values and therefore more cluster centers.

In the Subtractive clustering algorithm, the first cluster center corresponds to the data with the largest potential value. After removing the effect of the cluster center just found, the next cluster center becomes the data with the largest potential value, and so on. This procedure is repeated until the potential value becomes smaller than a predetermined threshold. There are n-dimensional input vectors $x_1, x_2, \cdots, x_m$ and 1-dimensional outputs $y_1, y_2, \cdots, y_m$ forming $(n+1)$-dimensional space of input-output data. For data $X_1, X_2, \cdots, X_N$ in $(n+1)$-dimensional input-output space, the subtractive clustering algorithm produces cluster centers as in the following procedure:

Step 1: Normalize given data into the interval [0,1].

Step 2: Compute the potential values at each data point. The potential value P_i of the data X_i is computed as

$$P_i = \sum_{j=1}^{N} \exp(-\alpha \|X_i - X_j\|^2), \quad i = 1, 2, \cdots, N \tag{11}$$

where a positive constant $\alpha = 4/r_a^2$ determines the data interval which affects the potential values. Data outside the circle with radius over positive constant $r_a < 1$ do not substantially affect potential values.

Step 3: Determine the data with the largest potential value P_1^* as the first cluster center X_1^*.

Step 4: Compute the potential value P_i' after eliminating the influence of the first cluster center.

$$P_i' = P_i - P_1^* \sum_{j=1}^{N} \exp(-\beta \|X_i - X_1^*\|^2) \tag{12}$$

where positive constant $\beta = 4/r_b^2$ prevents the second cluster center from locating close to the first cluster center. If the effect of potential of the first cluster center is not eliminated, second cluster center tends to appear close to the first cluster center, since there are many data concentrated in the first cluster center. Taking $r_b > r_a$ makes the next cluster center not appear near the present cluster center.

Step 5: Determine the data point of the largest potential value P_2^* as the second cluster center X_2^*. In general, compute potential values P_i' after removing the effect of the k th cluster center X_k^*, and choose the data of the largest potential value as the cluster center X_{k+1}^*.

$$P_i' = P_i - P_k^* \exp(-\beta \|X_i - X_k^*\|^2) \tag{13}$$

Step 6: Check if we accept the computed cluster center. If $P_k^*/P_1^* \geq \bar{\varepsilon}$, or $P_k^*/P_1^* > \underline{\varepsilon}$ and $\frac{d_{\min}}{r_a} + \frac{P_k^*}{P_1^*} \geq 1$, then accept the cluster center and repeat step 5. Here $d_{\min}$ denotes the shortest distance to the cluster centers $X_1^*, X_2^*, \cdots, X_k^*$ determined so far. If $P_k^*/P_1^* > \underline{\varepsilon}$ and $\frac{d_{\min}}{r_a} + \frac{P_k^*}{P_1^*} < 1$, then set the X_k^* to 0 and select the data of the next largest potential. If $\frac{d_{\min}}{r_a} + \frac{P_k^*}{P_1^*} \geq 1$ for the data, choose this data as the new cluster center and repeat step 5. If $P_k^*/P_1^* \leq \underline{\varepsilon}$, terminate the iteration.

When determining cluster centers, upper limit $\bar{\varepsilon}$ and lower limit $\underline{\varepsilon}$ allows the data of lower potential and of larger distance $d_{\min}$ between cluster centers to be cluster centers. Step 6 is the determining process of the calculated cluster center according to $d_{\min}$, the smallest distance to the cluster centers $X_1^*, X_2^*, \cdots$ calculated so far. When determining the cluster centers, data with low potential value can be chosen as a cluster center if $d_{\min}$ is big enough due to upper limit $\bar{\varepsilon}$ and lower limit $\underline{\varepsilon}$.

Fuzzy system modeling process using the cluster centers $X_1^*, X_2^*, \cdots, X_M^*$ in input-output space is as follows. The input part of the cluster centers corresponds to antecedent fuzzy sets. In $(n+1)$-dimensional cluster center X_i^*, the first n values are n-dimensional input space $x_i^* = (x_{i1}^*, \cdots, x_{in}^*)$. Each component determines the center of membership functions for each antecedent fuzzy sets. The cluster centers become the center of the membership functions $c_{ij} = x_{ij}^*$. The width of the membership function w_{ji} is decided as

$$w_{ij} = r_a \|\max_i(x_i^*) - \min_i(x_i^*)\| / \sqrt{M} \tag{14}$$

where M denotes the number of cluster centers, $\|\max_i(x_i^*) - \min_i(x_i^*)\|$ denotes the difference between the maximum and the minimum distances between cluster centers. The number of cluster centers corresponds to the number of fuzzy rules. The next process is to compute optimal consequent parameters $a_{0i}, a_{1i}, \cdots, a_{ni}$ in order to produce output y_j of the y_j th rule in the Sugeno fuzzy model. The number of centers equals the number of fuzzy rules. The output of the fuzzy system is defined as a linear function of input variables.

$$y_i = a_{0i} + a_{1i}x_1 + a_{2i}x_2 + \cdots + a_{ni}x_n \tag{15}$$

$$= [a_{1i}, a_{2i}, \cdots, a_{ni}]x + a_{0i} \tag{16}$$

$$= a_i^T x + a_{0i} \tag{17}$$

Compute parameters g_i through linear least-squares estimation, the final output y of the Sugeno fuzzy model is given as

$$y = \frac{\sum_{i=1}^{M}\mu_i(a_i^T x + a_{0i})}{\sum_{i=1}^{M}\mu_i} \qquad (18)$$

This is the final output of the fuzzy system.

4 Simulations

4.1 Sunspot Time Series

Time series data used in this paper is the sunspot number data that is monthly averaged of the number of individual spots through solar observation and consists of monthly sample collected from 1749/1 to 2005/2 like Table 1.

The sunspot number is computed as

$$R = k(10g + s) \qquad (19)$$

where g is the number of sunspot regions, s is the total number of individual spots in all the regions, and k is a scaling factor (usually <1).

Table 1. The sunspot time series

Index	Year / Month	Sunspot Number(R)
1	1749 / 1	58.0
2	1749 / 2	62.6
3	1749 / 3	70.0
...	...	...
3072	2004 / 12	17.9
3073	2005 / 1	31.3
3074	2005 / 2	29.1

Fig. 3 shows the sunspot time series data (an approximate 11-year cycle) used in the prediction with the PSFS. Each data set contains 3,074 samples.

The subtractive clustering algorithm with the parameters $r_a = 0.3, r_b = 0.75, \bar{\varepsilon} = 0.3$, and $\underline{\varepsilon} = 0.1$ generates the cluster centers. In this case, the PSFS with 3 component fuzzy systems ($N = 3$) is applied to time series prediction with 3,074 data. For the modeling of PSFS we use the first 2,924 data samples, except the next 150 test data samples, which divide two parts: one is training data (2,340 samples= $2,924 \times 0.8$) and the other is validation data (584 samples = $2,924 \times 0.2$).

In order to configure the PSFS for time series prediction, several embedding dimensions m must be determined for a specific time delay $\tau = 12$. For given τ, error performance measures are calculated from the difference between the one-step ahead prediction results trained with training data and validation data. The optimal value of m at a fixed τ corresponds to an integer for which the performance measures MSE validation data. Training data are used in constructing the fuzzy system based on the

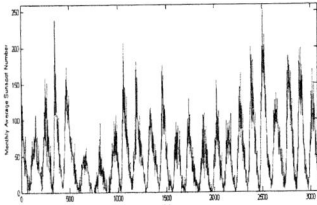
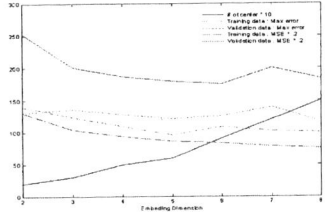

Fig. 3. The sunspot time series data   **Fig. 4.** Determination of embedding dimension

clustering algorithm. Validation data, which is not used to construct the fuzzy system, determines the optimal m according to τ when applied with the one-step-ahead prediction method.

Fig. 4 shows how to select the embedding dimension m for given time delay $\tau = 12$. The proper m values selected as 4, 5, and 6. The sunspot time series data is characterized by the three (τ, m) pairs of (12,4), (12,5), and (12,6) because the PSFS is $N = 3$.

4.2 Prediction with a Single Fuzzy System

Fig. 5 shows the prediction results by a single fuzzy system using each sub fuzzy system of PSFS having the parameter pair (m, τ). It shows the prediction result of a single fuzzy system with $m = 4$, $m = 5$, and $m = 6$ for $\tau = 12$, respectively.

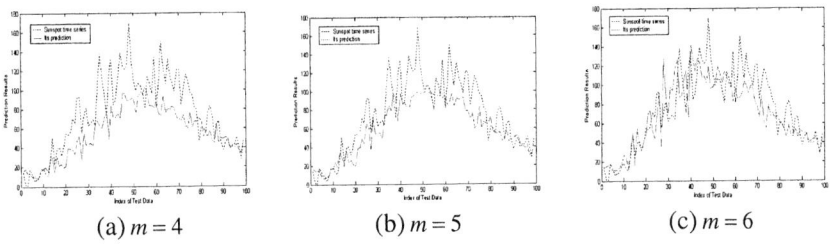

(a) $m = 4$ (b) $m = 5$ (c) $m = 6$

Fig. 5. Prediction result of a single fuzzy system ($\tau = 12$)

Table 2. Performance comparison of component fyzzy systems and PSFS.

	K=10	K=20	K=30	K=40	K=50
$m = 4$	4.74	9.24	15.62	20.78	23.93
$m = 5$	5.14	9.31	12.58	16.17	18.78
$m = 6$	6.35	6.15	10.34	15.74	17.48
$m = 4,5,6$ (PSFS)	4.85	6.89	9.19	12.54	15.01

Table 2 shows the absolute sum of prediction error that defined as the difference between actual values and prediction values. As a result, the single fuzzy system did not produce satisfactory prediction results comparing PSFS. The absolute sum of prediction error is computed as $\frac{1}{K}\sum_{i=1}^{K}|x(i)-\hat{x}(i)|$ where K is the index of sunspot time series.

4.3 Prediction with Parallel-Structure Fuzzy Systems

The PSFS contains three component fuzzy systems ($N=3$) where τ is fixed ($\tau=12$). Each component fuzzy system is characterized by several embedding dimensions for fixed time delay. Three prediction results produced by the component fuzzy systems are averaged at each step.

A 3,074 sunspot time series data are dealt with verification of the prediction performance of the PSFS. The PSFS predicts the 150 test data of the sunspot time series using the values of (τ,m) pairs of (12,4), (12,5), and (12,6).

Fig. 6-(a) shows the prediction result excluding the initial data ($72=12\times6$) by the PSFS. Fig. 6-(b) shows prediction error with the PSFS for the time series data.

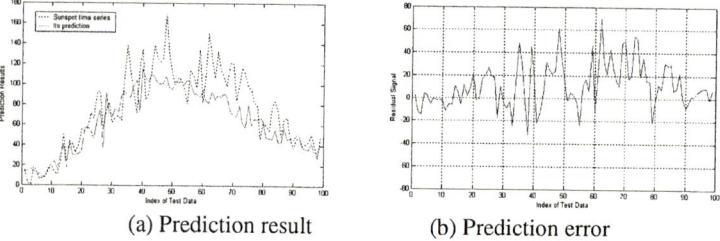

(a) Prediction result   (b) Prediction error

Fig. 6. Prediction result of the PSFS

Next the PSFS is applied to the pure future data represented by the period between 2005/3 and 2013/8. Fig. 7 shows the prediction result of the PSFS.

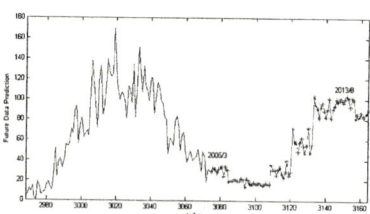

Fig. 7. Prediction result to the future samples of the PSFS

5 Conclusions

This paper presents a parallel-structure fuzzy system (PSFS) for predicting sunspot time series which consist of monthly data. The PSFS consists of a multiple number of component fuzzy systems connected in parallel. Each component fuzzy system is characterized by multiple-input single-output Sugeno-type fuzzy rules, which are useful for extracting information from numerical input-output training data. The component fuzzy systems for time series prediction are modeled by clustering input-output data. Each component fuzzy system predicts the future data at the same time-index with different values of embedding dimension and time delay. The PSFS determines the final prediction value by averaging the results of each fuzzy system in order to reduce error accumulation effect.

Computer simulations showed that the PSFS trained with training and validation data successfully predicted the sunspot time series data. The PSFS produced precise prediction results.

References

1. Kim M. S., Lee H. S., You C. H., Chung C. S.: Chaotic Time Series Prediction using PSFS2, 41st Annual Conference on SICE (2002)
2. Thompson R. J.: A Technique for Predicting the Amplitude of the Solar Cycle, Solar Physics 148 (1993)
3. Hathaway D. H., Wilson, R. M., Reichmann, E. J.: A Synthesis of Solar Cycle Prediction Techniques, Journal Geophys. Res., 104, No. A10 (1999)
4. Li K. J., Yun H.S., Liang H. F., Gu X. M.: Solar activity in Extended Cycles, Journal Geophys. Res., 107, No. A7 (2002)
5. Weigend A. S., Gershenfeld N. A.: Time Series Prediction: Forecasting the Future and Understanding the Past (eds.), Addison-Wesley Pub (1994) 175-193
6. Casdagal M.: Nonlinear Prediction of Chaotic Time Series, Physica D (1989) 335-356
7. Lowe D., Webb A. R.: Time Series Prediction by Adaptive Networks: A Dynamical Systems Perspective, Artificial Neural Networks, Forecasting Time Series, V. R. Vemuri, and R. D. Rogers (eds.), IEEE Computer Society Press (1994) 12-19
8. Broomhead D. S., Lowe D.: Multi-variable Functional Interpolation and Adaptive Networks, Complex Systems (1988) 262-303
9. Sugeno M.: Industrial Applications of Fuzzy Control, Elsevier Science Pub. (1985)
10. Mamdani E. H., Assilian S.: An Experiment in Linguistic Synthesis with a Fuzzy Logic Conroller, Int. J. of Man Machine Studies, Vol. 7, No. 1 (1975) 1-13
11. Kim M. S., Kong S. G.: Time Series Prediction using the Parallel-Structure Fuzzy System, 1999 IEEE Int. Fuzzy Systems Conference Proceedings, Vol. 2 (1999) 934-938
12. Kim M. S., Kong S. G.: Parallel Structure Fuzzy Systems for Time Series Prediction, Int. Jurnal of Fuzzy Systems, Vol. 3, No. 1 (2001)
13. Jang J. S. R., Sun C. T.: Neuro-Fuzzy Modeling and Control, Proceedings of the IEEE (1995)
14. Chiu S.: Fuzzy Model Identification Based on Cluster Estimation, Journal of Intelligent & Fuzzy Systems, Vol. 2, No. 3 (1994)
15. Yager R. R., Filev D. P.: Essentials of Fuzzy Modeling and Control, John Wiley & Sons (1994) 246-264

A Similarity Computing Algorithm for Volumetric Data Sets

Tao Zhang, Wei Chen*, Min Hu, and Qunsheng Peng

State Key Laboratory of CAD&CG, Zhejiang University, Hangzhou, China
{zhangtao, chenwei, humin, peng}@cad.zju.edu.cn

Abstract. Recently, there are remarkable progress in similarity computing for 3D geometric models. Few focus is put on the research of the similarity between volumetric models. This paper proposes a novel approach for performing similarity computation between two volumetric data sets. For each data set, it is performed by four stages. First, the volume data set is resampled into a unified resolution. Second, the data set is band-pass filtered and quantized to reveal its physical attributes. The resulting voxels are then normalized into a canonical coordinate system concerning the center of mass and scale. Subsequently, a series of uniformly spaced concentric shells around the center of mass are constructed, based on which spherical harmonics analysis (SHA) is applied. The coefficients of SHA constitute a rotation invariant spectrum descriptor which are used to measure the similarity between two data sets. The algorithm has been performed on a set of clinical CT and MRI data sets and the preliminary results are fairly inspiring.

1 Motivation

Many attentions have been paid to similarity assessment of 2D medical images[1]. Typically, each 2D image in the database is either an X-ray image or a slice from a 3D image. In the latter case, all slices from one 3D image are correlated spatially and semantically, composing a comprehensive description of the concerned target. Obviously, investigating the 3D images globally is superior to check their individual slice separately. For example, some small differences between two slices of two 3D images might be prominent if their consecutive neighboring slices are taken into consideration.

Although the topic of similarity computing is not new, there is little work which show the importance of similarity computation of volumetric data sets[2]. In our view, this task can be well motivated by considering following properties of volumetric data sets:

– Volumetric data sets are strongly related to medical imaging, scientific visualization, computational biology, mechanical engineering and astrophysics *etc*. Making them available on-line or public will provide a rich resource of references and research materials to scientific community.

---

* Corresponding author.

- A volumetric data set does not correspond to a unique object and hence can not be treated as sets of iso-surfaces. For a 256-level gray CT data set, voxels with density from 60 to 80 usually represent fat tissue; those with values from 80 to 110 constitute the soft tissue while the innermost bone layer has density value ranging from 90 to 255 [3]. A clinical example is shown in Figure 1.

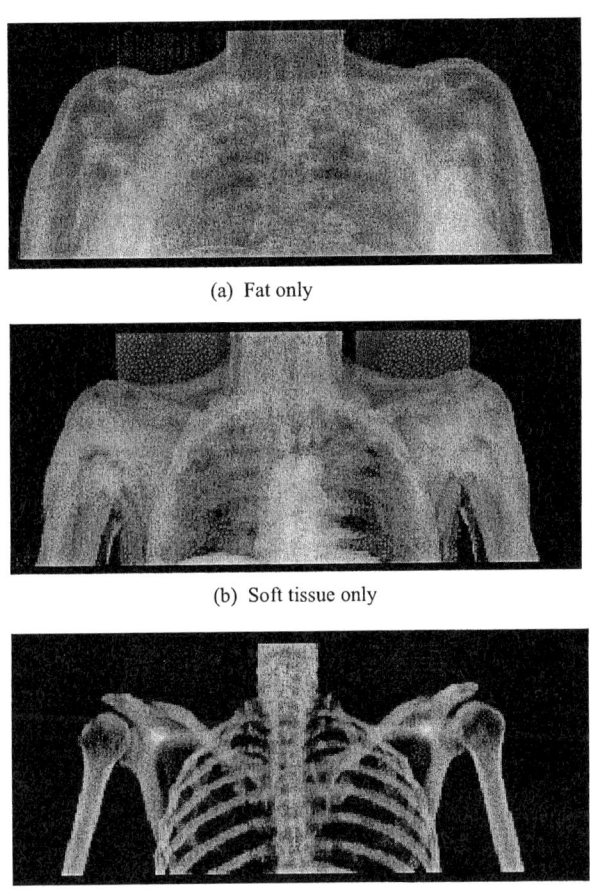

(a) Fat only

(b) Soft tissue only

(c) Bone only

Fig. 1. Volume illustration of different parts within a CT data set

- Volumetric data set can be recognized as a regular sampling of a subjacent, unknown, continuous 3D field. Despite whatever complex scene it depicts, 3D volume has the simplest parameterizations, making it appropriate to a wide variety of analysis tools such as Fourier transformation.

As far as the similarity assessment of volumetric data sets is concerned, we have to answer two questions:

- What is the proper content description for a volumetric data set?
- How can such description be used to compute similarities between two data sets?

The rest of this paper describes our initial solutions. It is organized as follows. In section 2 we summarize previous work on 3D model shape analysis. Our approach towards comparing volumetric data sets is given out in section 3. Section 4 presents experimental results followed by a brief conclusion in section 5.

2   Related Work

To our best knowledge, there are few work on general-purpose similarity computing of volumetric data sets. One pioneer work was introduced by Jain et al.[2], which uses a density histogram of the volume data as its content description. It is simple, robust and east to implement. However, it provides low discrimination power as we can easily find a counterexample that two volumetric data sets have the same density histograms but dramatically different "contents".

On the other hand, there are lots of literatures on shape analysis of 3D geometric models. Different feature descriptors have been proposed to accomplish similarity assessment. They can be roughly categorized as statistical, topological or transformational signatures.

Descriptors based on geometric statistics are probably the most popular methods of shape analysis. Examples include rotation invariant shape descriptor [4], shape histograms[5], discrete moments[6], shape distribution[7], directional histogram model[8], parameterized statistics[9], ray-based descriptor[10][11], and 3D shape spectrum descriptor[12]. They are easy-to-use and put few requirements on model regularity when the discrimination powers need to be enhanced.

Topological matching is particularly useful for the cognitive perception of a visual object. For instance, MRG[13] is constructed on the basis of geodesic distance where a coarse-to-fine comparison strategy establishes the correspondence between objects. In addition, a common technique for classifying objects with genus zero is based on their sphere parametrization forms, such as local curvature distributions method[14]. However, topological analysis is time-consuming and hence infeasible to large scale database.

Common transformation tools include 3D fourier transform[15], wavelet transform[16][17], spherical harmonic transform[18], Hough transform[19] and 3D Zernike moments[20]. Some have well-defined properties such as invariance under translation, rotation and scale. Note that most of them are defined over the binary voxelized volume of geometric models.

3   The Key Idea

We distinguish volumetric data sets from those obtained by **binary** voxelization of polygon models. The former can be regarded as a scalar field while the latter represents the surface of an object. The computation of content descriptor can be divided into four stages:

- The volume data set is preprocessed by resampling it to unified resolution and bucket sorting the resulting volume according to density values.
- The volume data set is filtered based on the user specific grey-level of interest.
- The filtered data is normalized into a canonical coordinate system.
- Sphere harmonics analysis (SHA) is applied to a series of concentric shells of normalized volume.

When the content descriptor are calculated for all data sets, their similarity can be easily estimated by introducing appropriate distance functions. Following subsections elaborate the details of our approach.

3.1 Resampling

Note that, the original resolution of each data set might be different. Typically, the resolution along z axis is less than those along x and y axes. Therefore, each data set is resampled into a unified resolution $N \times N \times N$. Up-sampling and down-sampling are selected accordingly by means of one of nearest neighbor, trilinear and even higher order interpolation methods. Tradeoff is made considering the accuracy, storage and efficiency issues. A typical choice of N is 128.

3.2 Bucket Sorting

For the resampled data, we further bucket sort it according to the density values, resulting in a 256-entry index table. Each entry points to a list of iso-valued voxels. Specifically, every item in the list has the following data structure

$$\text{item} = \langle i_x, i_y, i_z \rangle \qquad (1)$$

where i_x, i_y, i_z are indices of the corresponding voxel along the x, y, z axes.

The conversion of the location index to the position of a voxel in 3D space is straightforward. If the volumetric data set is embedded in a unit bounding box ranging from -0.5 to 0.5, following relationship applies:

$$\text{coordinate}(i_x, i_y, i_z) =$$
$$\langle \frac{2i_x + 1}{2N} - 0.5, \frac{2i_y + 1}{2N} - 0.5, \frac{2i_z + 1}{2N} - 0.5 \rangle \qquad (2)$$

Two stages described in section 3.1 and 3.2 can be accomplished off-line, yielding a preprocessed database.

3.3 Volume Filtering

Different volumetric data sets may be of arbitrary scale, orientation and position in 3D space. In order to achieve correct similarity measurement, it is necessary to locate all data sets in a canonical coordinate system. This step is called normalization. In view that volumetric data set does not correspond to the surface of an object, it is not appropriate to apply principle component analysis for three principal axes, as is usually utilized in 3D model shape analysis.

The normalization of volume data sets should account for the physical properties implied by the data. For example, when examining volume data sets, users can determine gray-level of interest and classify the data sets by band-pass filtering. The users either explicitly appoints the range of gray-level or simply let the system choose it automatically. For example, if the user checks the box denoting the soft tissue, the system will automatically turn this vague verbal description into gray-level on the range of [80, 110].

The normalization for translation and scale is thus query dependent. Let G denote the gray-level of interest, we first check the index table to get all the voxels that have the value $g \in G$. These voxels are classified as valid and constitute a set $V = \{v_i\} = \{i_x, i_y, i_z\}, i = 1, \cdots, n$, where n is the size of V. Other voxels are assigned to value zero.

3.4 Normalization

To put the data set into a canonical coordinates system, we calculate the center of mass of the filtered volume which is defined as:

$$C = \frac{1}{N} \sum_{i=0}^{n} \text{coordinate}(v_i) \qquad (3)$$

V is then translated so that its center of mass coincides with the origin. We denote the resulting voxel set by $V' = \{v_i'\} = \{v_i - C\}$.

The next step is to scale the volume into a unit sphere. Both of translation and scale operations require trilinear interpolation for high quality.

3.5 Spherical Harmonics Analysis

Let $\mathcal{G}$ denote a transformation set. For an arbitrary volumetric data set $\mathcal{M}$ with description $\mathcal{D}$ and every element $g \in \mathcal{G}$, if we have $\mathcal{D}(g\mathcal{M}) = \mathcal{D}(\mathcal{M})$, we state that $\mathcal{D}$ is invariant under the transformation set $\mathcal{G}$. In the context of volumetric data sets, the similarity descriptors should be invariant under the transformation sets of translation, scale and rotation.

One approach to achieve the required transformation invariance is decompose the volume space Ω into subspace Ω_i. Ideally, this partition should meet the following requirements:

- Mutual Orthogonality: $\Omega_i \perp \Omega_j, \quad \forall i \neq j$
- Completeness: $\sum_i \Omega_i = S$
- Invariance: $\mathcal{G}(\Omega_i) = \Omega_i, \quad \forall i$

If we assume $\mathcal{M}$ to be as a function defined on the space Ω, $\mathcal{M}$ can be expressed as a linear combination of the components that are defined on the subspace Ω_i respectively. In other words, we state that $\mathcal{D}$ can be projected onto $l \, \Omega_i$:

$$\mathcal{D}(\Omega) = \langle \mathcal{D}(\Omega_1), \ldots, \mathcal{D}(\Omega_l) \rangle$$
$$= \langle \mathcal{D}_1, \ldots, \mathcal{D}_l \rangle \qquad (4)$$

If $\{\psi_{ij}\}$ forms the orthogonal basis of the subspace Ω_i, $\mathcal{D}_i$ itself is a vector whose jth component is formulated as

$$\mathcal{D}_{ij} = \int_{\Omega_i} \mathcal{D} \cdot \overline{\psi_{ij}} \qquad (5)$$

In the previous subsection, we have already normalized the volumetric data set with translation and scale transformations. Thus we need only concern the transformation of rotation. Fortunately, there are already some available tools. One is spherical harmonics which form a Fourier basis on the sphere similar to the operations of sines and cosines functions on a line. The basis ψ_{ij}, or more specifically Y_l^m, is given by:

$$Y_l^m(\theta, \phi) = N_l^m P_l^m(\cos\theta) e^{im\phi}, m = -l, \cdots, l \qquad (6)$$

where N_l^m is the normalization factor and P_l^m is the associated Legendre functions.

Spherical harmonics have the property that for a given l, $\{Y_l^m, m \in [-l, l]\}$ span an irreducible subspace which is invariant under the operations of the full rotation group. A direct conclusion is that l-th frequency component $h_l = \|\langle h_l^{-l}, \cdots, h_l^l \rangle\|$ defines a rotation invariant where $h_l^m = \int_{\Omega_i} \mathcal{D} \cdot \overline{Y_l^m}$.

The key to applying spherical harmonics to a 3D function is to turn the 3D space into a series of uniformly spaced concentric shells. Therefore, the normalized volume is first decomposed into a sequence of concentric shells based on which spherical harmonics are applied. The conversion between Cartesian coordinates $\langle x, y, z \rangle$ and its spherical coordinates $\langle r, \theta, \phi \rangle$ can be formulated as:

$$\mathcal{M}(x, y, z) = \mathcal{M}(r\sin\theta\sin\phi, r\sin\theta\cos\phi, \cos\theta)$$
$$= \mathcal{M}'(r, \theta, \phi)$$

To sum up, we get the following rotation invariant descriptor:

$$\mathcal{D} = \{h_l^r\}, \quad l \in [0, B), r \in [0, R) \qquad (7)$$

where each h_l^r is the l-th frequency component from the spherical harmonics analysis of the r-th concentric shells. B denotes the bandwidth of the signal represented by volumetric data set. R is the resolution of concentric decomposition.

4 Experimental Results

We tested our algorithm on 10 clinical volumetric data sets including four CT head data sets, three MRI liver data sets and three Ultrasound fetus data sets. Their resolutions are listed in Table 1. The slice thickness for each data set is

Table 1. Volume resolutions of test data sets

Data Sets	Resolutions
CT Head	$208 \times 256 \times 225$, $256 \times 256 \times 203$ $256 \times 256 \times 223$, $256 \times 256 \times 256$
MRI Liver	$256 \times 256 \times 58$, $256 \times 256 \times 58$ $256 \times 256 \times 58$
US Fetus	$214 \times 146 \times 132$, $226 \times 144 \times 128$ $212 \times 148 \times 132$

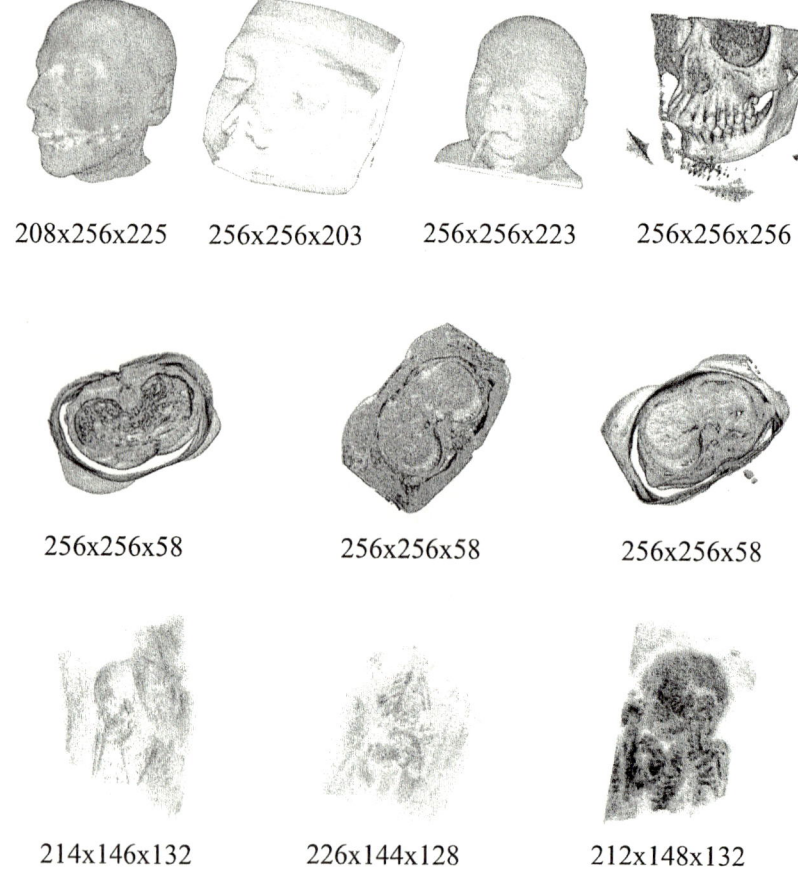

Fig. 2. Illustrations of 10 volume data sets. Top row: four CT head data sets; Middle row: three MRI Liver data sets; Bottom row: three Ultrasound fetus data sets.

$(1.0, 1.0, 1.0)$. We set $R = 64$ and $B = 64$ respectively (refer to Eq(7)) and choose a unified resolution for all these data sets as $128 \times 128 \times 128$.

Two distance measures are implemented for two similarity descriptors f and g. One is the Euclidean distance L_2, the other is the χ^2 distance defined as:

$$D(f,g) = \int \frac{(f-g)^2}{f+g} \tag{8}$$

We list the similarity statistics results in Table 2 for Euclidean distance L_2 and χ^2 distance respectively. The FT (First Tier) column lists the percentage of top k-1 matches (excluding the query) from the query's class, where k is the size of the class. The ST (Second Tier) column lists the same type of result, but for top $2(k$-1) matches. The NN (Nearest neighbor) column lists the percentage of test in which the top match was from the query's class. This validation scheme is typically used in 3D model retrieval[18].

Table 2. Similarity computation results using our algorithm under L_2 and χ^2 distances

Data Sets	CT Head	MRI Liver	US Fetus
FT with L_2	100%	66.7%	50%
FT with χ^2	91.7%	83.3%	33.3%
ST with L_2	100%	100%	100%
ST with χ^2	100%	100%	100%
NN with L_2	100%	100%	50%
NN with χ^2	100%	100%	50%

From Table 2, we conclude that the algorithm provide a favorable similarity measurement for volumetric data sets. In our experiments, the results of χ^2 norm is almost the same as those of L^2 norm.

5 Conclusion and Future Work

Our results confirm that the proposed algorithm is suitable for the similarity computation of volumetric data sets. It is simple, easy to implement, and the result is rotation, translation and scale invariant.

Recent advances in data acquisition devices make the details of the target models more subtle, the mathematical topology of the objects more complex and the data sets larger and time-varied. As a result, structures and pathologies are more and more computationally hard. We would like to investigate more powerful tools to distinguish subtle differences such as a normal and an abnormal anatomy in medical applications. Another important issue for further research lies in similarity computing between a part of and a whole object. We also want to apply the algorithm to protein molecular data sets for structure and functionality forecast.

Acknowledgements

This work is partially supported by 973 program of China (No.2002CB312100), NSF of China for Innovative Research Groups (Grant No.60021201).

References

1. Liu, Y., Dellaert, F.: Classification driven medical image retrieval. (In: Proceedings of the Image Understanding Workshop)
2. Elvins, T., Jain, R.: Web based volumetric data retrieval. In: Symposium on Virtual Reality Modeling Language, (San Diego, CA, USA) 7–12
3. Hubbell, J., Seltzer, S.: Tables of x-ray mass attenuation coefficients and mass energyabsorption coefficients (version 1.03). In: http://physics.nist.gov/xaamdi. (1995)
4. Suzuki, M., Kato, T., Otsu, N.: A similarity retrieval of 3d polygonal models using rotation invariant shape descriptors. In: Proceedings of IEEE International Conference On System, Man, and Cybernetics, (Nashville, USA) 2946–2952
5. Ankerst, M., Kastenmueller, G., Kriegel, H., Seidl, T.: 3d shape histograms for similarity search and classification in spatial databases. In: Proceedings of 6th International Symposium on Large Spatial Databases, (Hongkong, China) 207–228
6. Elad, M., Tal, A., Ar, S.: Content based retrieval of vrml objects - an iterative and interactive approach. In: Proceedings of the Sixth Eurographics Workshop in Multimedia, (Madison, Wisconsin, USA) 97–108
7. Osada, R., Funkhouser, T., Dobkin, D.: Shape distributions. ACM Transactions on Graphics **21** (2002) 93–101
8. Liu, X., Sun, R., Kang, S.B., Shum, H.Y.: Directional histogram model for three dimensional shape similarity. In: Proceedings of IEEE Computer Vision and Pattern Recogintion 2003, (Madison, Wisconsin, USA) 813–820
9. Ohbuchi, R., Otagiri, T., Ibato, M., Takei, T.: Shape similarity search of three-dimensional models using parameterized statistics. In: Proceedings of Pacific Graphics 2002, (Beijing, China) 265–274
10. Vranic, D., Saupe, D.: 3d model retrieval. In: Proceedings of the Spring Conference on Computer Graphics and its Applications, (Budmerice, Slovakia) 89–93
11. Yu, M., Atmosukarto, I., Leow, W.K., Huang, Z., Xu, R.: 3d model retrieval with morphing based geometric and topological feature maps. In: Proceedings of IEEE Computer Vision and Pattern Recognition (CVPR), (Madison, Wisconsin, USA) 656–661
12. Zaharia, T., Preteux, F.: Three-dimensional shape based retrieval within the mpeg-7 framework. In: Proceedings of SPIE Conference 4304 on Nonlinear Image Processing and Pattern Analysis XII, (San Jose, CA, USA) 133–145
13. Hilaga, M., Shinagawa, Y., Kohmura, T., Kunii, T.: Topology matching for fully automatic similarity estimation of 3d shapes. In: Proceedings of ACM SIGGRAPH 2001, (Los Angeles, CA, USA) 203–212
14. Shum, H., Hebert, M., Ikeuchi, K.: On 3d shape similarity. In: Proceedings of IEEE Computer Vision and Pattern Recognition (CVPR), (San Francisco, CA, USA) 526–531
15. Vranic, D.V., Saupe, D.: 3d shape descriptor based on 3d fourier transform. In: Proceedings of the EURASIP Conference on Digital Signal Processing for Multimedia Communications and Services, (Budapest, Hungary) 271–274
16. Gain, J., Scott, J.: Fast polygon mesh query by example. In: Proceedings of SIGGRAPH Technical Sketches. (1999) 241

17. Paquet, E., Rioux, M.: A content-based search engine for vrml databases. In: Proceedings of IEEE Computer Vision and Pattern Recognition (CVPR), (S. Barbara, CA, USA) 541–546
18. Funkhouser, T., Min, P., Kazhdan, M., Chen, J., Halderman, A., Dobkin, D., Jacobs, D.: A search engine for 3d models. ACM Transactions on Graphics **22** (2003) 83–105
19. Zaharia, T., Preteux, F.: Hough transform-based 3d mesh retrieval. In: Proceedings of SPIE Conference 4476 on Vision Geometry X, (San Diego, CA, USA) 175–185
20. Novotni, M., Klein, R.: 3d zernike descriptors for content based shape retrieval. In: Proceedings of the 8th ACM Symposium on Solid Modeling and Applications, (Seattle, WA, USA) 216–225

Extraction of Representative Keywords Considering Co-occurrence in Positive Documents

Byeong-Man Kim[1], Qing Li[1,2], KwangHo Lee[3], and Bo-Yeong Kang[2]

[1] Kumoh National Institute of Technology, Korea
[2] Information and Communications University, Korea
[3] Mokpo National University, Korea

Abstract. In linear text classification, user feedback is usually used to tune up the representative keywords (RK) for a certain class. Despite some algorithms (e.g. Rocchio) deal well with user positive and negative feedback to adjust the RKs, few researches have investigated how to adjust RKs only based on a small positive responses which is a popular case in the real-world application (e.g. users tend to click their interested URL). In this work, we describe a method of extracting representative keywords for a user from a small set of his positive feedback documents. Experiments on the Reuters-21578 collection illustrate that our approach is better than other two famous methods (Rocchio and Widrow-Hoff) with 24.8% and 14.5% improvement, respectively.

1 Introduction

Extracting representative keywords for documents that users interested has a wide application in several areas such as constructing user model, query terms expansion and reweight in information retrieval.

User models can be constructed by hand, or learned automatically with the explicit or implicit user feedback. Some systems require users to explicitly specify their profiles, often as a set of keywords or categories. Studies have shown that such explicit feedback from the user is clearly useful [4],[15]. However, it is difficult for a user to exactly and correctly specify their information needs. Moreover, many users are unwilling to provide relevance judgments on documents in practice [11],[13]. An alternative is to use implicit feedback based user's behavior to automatic construct user models [5],[6],[10]. In this case, system should construct user model, often by extracting automatically representative keywords based on a small set of feedback documents.

In feedback information retrieval environment, a user judges the relevance of one or more of the retrieved documents and these judgements are fed back to the system to improve the initial search result [12]. Buckey et al. [2] experimentally verified that the recall-precision effectiveness is roughly proportional to the log of the number of known relevant documents. In other words, the greater the amount of feedback from the user to the system, the better the search effectiveness of the system. However, this expectation is often not met in a highly interactive situation like Internet surfing.

In this paper, we focus on the extraction of representative keywords of a few documents that might interest a user. For example, around 10 documents. This is a usual case in Web information retrieval because most of users are reluctant to provide hundreds of relevance judgments.

2 Our Approach

We do not consider how to provide relevance judgment on documents and we assume that positive relevant documents are already at hand. What we focus on is how to extract representative keywords from a small set of user positive feedback documents. For instance, in feedback information retrieval, one user marked positive relevant documents from a small set of retrieved documents according to his query. The system will form a new query by expanding and reweighting his old query based on those positive relevant documents. In this case, how to form this new query based on these small positive relevant documents is what we focus in this paper.

For this end, we first extract candidate terms and then choose a number of terms called initial representative keywords (IRKs) from those positive relevant documents. Then, the final representative keywords are extracted by expanding IRKs and reweighting them using term co-occurrence similarity.

2.1 Selection of Initial Representative Keywords

To some extent, initial representative keywords are the primary keywords reflecting preferences of the user who selects these small set of documents as his interested ones. Therefore, it is very important to select IRKs efficiently that finial representative keywords count on. Generally, it is very difficult to select terms representing user information need or a set of documents faithfully. Therefore, in this paper, we decide to take a simple approach to select IRKs because we think that a simple one is enough to show the merit of our approach. The further improvement must be achieved with a better selection method.

At the beginning, we selected a IRK if its weight is higher than a given threshold. However, from the early experiments, we observed that the performance of the approach heavily depended on the threshold value and the performance was also not good especially in the case that some documents do not contain any IRK. Therefore, we take another simple approach where terms are selected as IRKs based on their weights with the constraint that each example document should contain at least one or more IRKs. We select the word with highest weight in each document if it is not listed in the IRKs before. For example, let us assume a small set of documents in which a certain user interested consists of 6 documents: d1,d2,d3,d4,d5, and d6. Each document contains the following terms: d1= {a,b,f}, d2={a,c,d}, d3={d,e,f}, d4={d,f},d5={b,c,e}, d6={e,f}. And the weight of each word is as follows: (a, 0.9), (b, 0.8), (c, 0.7), (d, 0.6), (e, 0.5), (f,

0.4). Since term *a* has the highest weight in d1, it will be put in the IRK list. As for d2, term *a* is also the term with highest weight. However, it is already in the IRK list, we do not need add it to the IRK list anymore. By doing such selection, the final IRK list consists of {a, b,d,e} in this example.

2.2 Selection of Final Representative Keywords

The final representative keywords (FRKs) come from IRKs by expanding IRKs and reweighting them. If 5 terms are required to represent a user's preference and the number of IRKs is 3. Then, 2 terms with highest weights except IRKs are selected additionally. Once obtain the FRKs, we will reweight them by a relevance feedback technique. One of the most popular method is Rocchio's method [9] due to its simple and effective. However, it has some drawbacks.

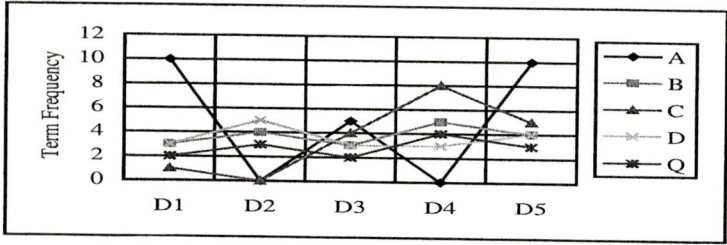

Fig. 1. The Sample of IRKs

Let us consider such a case that the IRK consists of one term Q and $D1$, $D2$, $D3$, $D4$, and $D5$ are positive feedback documents. To be simple, we assume that the candidate terms A, B, C and D have same inverse document frequency (IDF) with value 1. Their term frequencies (TF) are given, as shown in Figure 1. The maximum TF of each document is 10. In Rocchio's method [9], we can obtain the following adjusted weights of the terms. (Since only the positive feedback documents are given, we set $\alpha= 0$, $\beta=1$, and $\gamma=0$ in our application.)

$A = 1.0 + 0 + 0.5 + 0 + 1.0 = 2.5$, $B = 0.3 + 0.4 + 0.3 + 0.5 + 0.4 = 1.9$,
$C = 0.1 + 0 + 0.4 + 0.8 + 0.5 = 1.8$, $D = 0.3 + 0.5 + 0.3 + 0.3 + 0.4 = 1.8$

As shown above, the Rocchio's method gives a higher weight to the candidate term A because it occurs frequently in the relevant documents. The candidate term B is less weighted than A because its sum of term frequencies is lower than A's. However, it would be reasonable that term B has the privilege because the occurrence pattern of term B is more similar to the initial term Q than A.

In this paper, we propose a new variation that considers the co-occurrence similarity between a candidate term and terms in the IRKs. For calculation of the final weights of FRKs, first, their relevance degrees in every positive feedback documents are calculated by

$$RD_{ik} = 1 - \log_p \sqrt{\frac{\sum_{j=1}^{n}(kf_{jk} - tf_{ik})^2}{n} + 1} \qquad (1)$$

where, RD_{ik} is the relevance degree between IRKs and candidate term t_i in document d_k; kf_{jk} is the frequency of IRK j in document d_k; tf_{ik} is the frequency of candidate term t_i in document d_k; n is the number of IRKs, p is a control parameter. In our experiments, p is set to 10. The RD_{ik} is treated as 0 if it has negative value. In this equation, it gives the privilege to those terms which are collocated with IRKs.

For example, let K be a set of IRKs consisting of term $k1$, $k2$ and $k3$ and their frequencies in document $d1$ are 4, 3, and 1, respectively. Also, set the frequency of candidate term $t1$ to be 2. Then, its relevance degree is calculated as follows:

$$RD_{11} = 1 - \log_{10} \sqrt{\frac{2^2 + 1^2 + (-1)^2}{3} + 1} = 1 - 0.238 = 0.762$$

As shown in the above equation, RD_{ik} is inversely proportional to the sum of term frequency difference between initial representative term and candidate term. Therefore, the higher is the value of RD, the more similar the co-occurrence is, that is, the equation reflects the co-occurrence similarity between the initial representative terms and a candidate term appropriately. After calculating the relevance degree of a candidate term, the weight of the term in the set of positive feedback documents is determined by

$$w_{ri} = \sum_{k=1}^{n}(w_{ik} \times RD_{ik}) \qquad (2)$$

where, w_{ri} is the weight of term t_i in the document set; w_{ik} is the weight of term t_i; n is the number of positive feedback documents. Equation 2 is derived from the Rocchio's method by considering the term relevance degree between initial representative terms and a candidate term additionally.

Finally, the weight of FRK is calculated by

$$w_i = w_{ki} + w_{ri} \qquad (3)$$

where, w_{ki}, is the initial weight of term t_i, which is recalculated by

$$w_{ki} = \left(0.5 + \frac{0.5 \times freq_i}{\max_j freq_j}\right) \times \log\left(\frac{N}{n_i}\right) \qquad (4)$$

where, $freq_i$ is the frequency of initial representative keyword t_i; n_i is the frequency of documents in which t_i appear; N is the total number of documents.

2.3 Calculation of Representative Keyword Weight

In Section 2.1, the IRK is selected based on the representative Keyword weight. In Equation 2 in Section 2.2, w_{ik} is the weight of candidate representative keyword t_i. Since those weights should reflect the importance or representative ability of those keywords in the positive feedback documents, several factors should be considered such as distributed term frequency(DTF), document frequency(DF) within positive feedback documents and inverse document frequency(IDF). Because these factors essentially have inexact and uncertain characteristics, we combine them by fuzzy inference instead of a simple equation to obtain the representative keyword weight.

The positive feedback documents are transformed into a set of candidate terms through eliminating stopwords and stemming by Porter's algorithm [1]. The DTF, DF, and IDF of each term are calculated based on this set and used as inputs of fuzzy inference. The DTF (Distributed Term Frequency) reflects the frequency and distributed status of a term in a set of positive feedback documents, which is the ration of total occurrences of the term in a set of documents to the number of documents in the set containing the term. It needs to be normalized for fuzzy inference by

$$NDTF_i = \frac{\frac{TF_i}{DF_i}}{\max_j \left[\frac{TF_j}{DF_j}\right]} \quad (5)$$

where, TF_i is the frequency of term t_i in the example documents; DF_i is the number of documents having term t_i in the documents.

The DF (Document Frequency) represents the frequency of documents having a specific term within the positive feedback documents. Like DTF, DF also provides one measure of how well that term describes the contents of document set. The normalized document frequency, NDF, is defined in equation 6, where DF_i is the number of documents having term t_i.

$$NDF_i = \frac{DF_i}{\max_j DF_j} \quad (6)$$

The IDF (Inverse Document Frequency) represents the inverse document frequency of a specific term over an entire document collection not positive feedback documents. The motivation for usage of IDF factor is that terms which appear in many documents are not very useful. The normalized inverse document frequency, NIDF, is defined as

$$NIDF_i = \frac{IDF_i}{\max_j IDF_j}, IDF_i = \log \frac{N}{n_i} \quad (7)$$

where, N is the total number of documents and n_i is the number of documents in which term t_i appears.

Figure 2 shows the membership functions of the input/output variables - 3 inputs (NDTF, NDF, NIDF) and 1 output (TW:Term Weight) - used in our method. As you can see in Figure 2(a), NDTF variable has S(Small), L(Large) and NDF and NIDF variables have S(Small), M(Middle), L(Large) as linguistic labels (or terms). The fuzzy output variable, TW which represents the importance of a term, has six linguistic labels as shown in Figure 2(b).

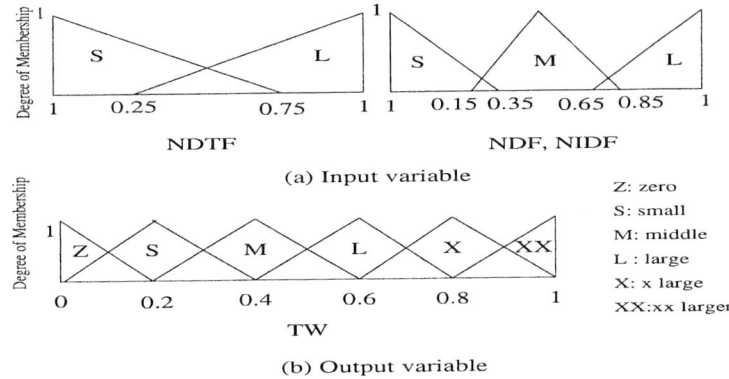

Fig. 2. Fuzzy variables

NDF \ NIDF	S	M	L
S	Z	Z	S
M	Z	M	L
L	S	L	X

NDTF = S

NDF \ NIDF	S	M	L
S	Z	S	M
M	Z	L	X
L	S	X	XX

NDTF = L

Fig. 3. Fuzzy inference rules

The 18 fuzzy rules are involved to infer the term weight. The rules are constructed based on the intuition that the important or representative terms may occur across many positive example documents but not in general documents, i.e., their NDF and NIDF are very high. As shown in Figure 3, the TW of a term is Z in most cases regardless of its NDF and NDTF if its NIDF is S. Because such term may occur frequently in any document and thus its NDF and NDTF can be high. When NDF of a term is high and its NIDF is also high, the term is considered as a representative keyword and then the output value is between X and XX. The other rules are set similarly.

We can get the TW through the following procedure. However, the output is in the form of fuzzy set and thus has to be converted to the crisp value. In this paper, the center of gravity method(COG) is used to defuzzify the output [8].

1. Apply the NDTF, NDF, and NIDF fuzzy values to the antecedent portions of 18 fuzzy rules.
2. Find the minimum value among the membership degrees of three input fuzzy values.
3. Classify every 18-membership degree into 6 groups according to the fuzzy output variable TW.
4. Calculate the maximum output value for each group and then generate 6 output values.

For instance, let us assume, there is one term, whose NIDF is 0.35, NDF is 0.2 and NDTF is 0.3. The degree of membership is determined by plugging the selected input parameter(NIDF, NDF or NDTF) into the horizontal axis and projecting vertically to the upper boundary of the membership function(s) in Figure 2. Therefore the result is as follows.

$$NIDF=0.35 : S=0.00, M=0.57;$$
$$NDF=0.20 : S=0.43, M=0.14;$$
$$NDTF=0.30 : S=0.53, L=0.07.$$

Now referring back to the rules, only 8 rules out of 18 rules need to be selected, which are marked in Figure 3. The effective rules are listed as follows.

1. If(NIDF=S,NDF=S,NDTF=S) then TW=Z
   min{S=0.00,S=0.43,S=0.53}=0.00
2. If(NIDF=S,NDF=M,NDTF=S) then TW=Z
   min{S=0.00,M=0.14,S=0.53}=0.00
3. If(NIDF=M,NDF=S,NDTF=S) then TW=Z
   min{M=0.57,S=0.43,S=0.53}=0.43
4. If(NIDF=M,NDF=M,NDTF=S) then TW=M
   min{M=0.57,M=0.14,S=0.53}=0.14
5. If(NIDF=S,NDF=S,NDTF=L) then TW=Z
   min{S=0.00,S=0.43,L=0.07}=0.00
6. If(NIDF=S,NDF=M,NDTF=L) then TW=Z
   min{S=0.00,M=0.14,L=0.07}=0.00
7. If(NIDF=M,NDF=S,NDTF=L) then TW=S
   min{M=0.57,S=0.43,L=0.07}=0.07
8. If(NIDF=M,NDF=M,NDTF=L) then TW=L
   min{M=0.57,M=0.14,L=0.07}=0.07

Then we calculate the maximum output value for each group and then generate 6 output values, as shown in Figure 4, which consists a fuzzy set of TW as follows.

$$TW=\{Z=0.43,S=0.07,M=0.14,L=0.07,X=0,XX=0\}$$

At last, the COG is used to defuzzify the output into one value.

$$TW = \frac{0.43 \times 0.1 + 0.07 \times 0.2 + 0.14 \times 0.4 + 0.07 \times 0.7}{0.1 + 0.2 + 0.4 + 0.7} = 0.116$$

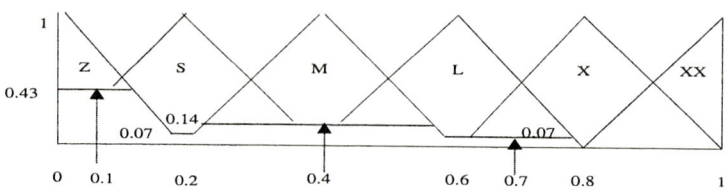

Fig. 4. Defuzzify the Outputs

3 Experiments

We used Reuters-21578 data as our experimental document set. This collection has five different sets of contents related categories: EXCHANGES, ORGS, PEOPLE, PLACES and TOPICS. Some of the category sets have up to 265 categories, but some of them have just 39 categories. We chose the TOPICS category set which has 135 categories. We divided the documents according to the "ModeApte" split. There are 9603 training documents and 3299 test documents. Among the 135 categories, we chose 90 ones that have at least one training example and one testing example. Then, we finally selected 21 categories that

Table 1. Performance of 21 categories in the REUTERS corpus and comparison with two existing algorithms

$No. of FRKs$	Our	$Rocchio$	$W.H$
5	0.596	0.511	0.566
10	0.619	0.496	0.540
15	0.581	0.490	0.529
20	0.577	0.489	0.522
25	0.564	0.491	0.493
30	0.563	0.495	0.500
All	0.504	0.467	0.483

Table 2. The detail result when 10 terms are used for user preferences

	Our	$Rocchio$	$W.H$
lumber	0.756	0.4444	0.6667
dmk	0.444	0.4444	0.4
sunseed	0.62	0.3333	0.3333
lei	1	0.8	1
soy-meal	0.6667	0.5143	0.5185
fuel	0.495	0.4615	0.4615
heat	0.75	0.75	0.75
soy-oil	0.404	0.2692	0.32
lead	0.5775	0.5	0.5
strategic-metal	0.166	0.1053	0.1408
hog	0.8	0.6	0.8
orange	0.9091	0.9091	0.8571
housing	0.5814	0.6667	0.5714
tin	0.98	0.7857	0.9231
rapeseed	0.645	0.5714	0.6154
wpi	0.5833	0.5882	0.5882
pet-chem	0.383	0.2727	0.2759
silver	0.4	0.4	0.5
zinc	0.923	0.6667	0.6842
retail	0.21	0.0548	0.0548
sorghum	0.70	0.2727	0.3871
Average	0.6188	0.4957(+24.8%)	0.5404(+14.5%)

have from 10 to 30 training documents. The 3019 documents of those categories are used as testing documents. The document frequency information from 7770 training documents in 90 categories is used to calculate IDF of terms.

Documents are ranked by the cosine similarity. F-measure which is a weighted combination of recall and precision [1] is used as metric. A higher F-measure means a greater accuracy.

Our method was compared to the Rocchio and Widrow-Hoff algorithms [9]. As we know, different number of FRKs has different representative power. In order to find out the optimal number of FRKs, we carried out a series of experiments by varying the number of FRKs from 5 to 30 with the step 5 and whole terms. Table 1 shows the average result of the proposed method compared to the two existing algorithms for 21 categories. The result shows that our method is better than the others in all cases, especially when 10 FRKs are used. Please note when 5 terms are used to represent user preferences, 19 categories among 21 categories are used because "strategic-metal" and "pet-chem" categories do not satisfy the constraint in our method,that is, 5 terms are too few to cover all training documents.

As shown in Table 2 which shows the detail result in the case that 10 FRKs are applied, our proposed method shows a better performance than the other two traditional algorithm -Rocchio and Widrow-Hoff with an average improvement 24.8 % and 14.5%, respectively.

4 Conclusion

In this paper, we described a method applying a fuzzy inference technique and a term reweighting scheme based on the term co-occurrence similarity to extract important keywords from a small set of positive feedback documents. Though this paper only describes how to extract user preferences from a small document set, the technique is applicable to several areas such as query modification in IR, user profile modification in information filtering, text summarization and so forth directly or with some modifications.

A series of experiments based on the Reuters-21578 collection shows that our method outperforms two well-known feedback algorithms for linear text classifiers -Rocchio and Widrow-Hoff with an average improvement 24.8 % and 14.5%, respectively. However, our method is designed for a small set of documents. Therefore, we could not expect to achieve the same performance improvement as described in this paper when our method is applied to a large set of positive feedback documents. However, such a problem will be alleviated if clustering techniques are used together as in [3],[6],[7].

References

1. Baeza-Yates,R.,B. Ribeiro-Neto; Modern Information Retrieval,ACM Press,(1999).
2. Buckley, C., G. Salton; Optimization of relevance feedback weights,. In Proc. of SIGIR,(1995).

3. Diaz, A., Mana, M., Buenaga, M., Gomez, J.M., Gervas, P.; Using linear classifiers in the integration of user modeling and text content analysis in the personalization of a Web-based Spanish News Service. In Proc. of the Workshop on Machine Learning, Information Retrieval and User Modeling, (2001).
4. Goldberg, D.; Nichols, D.; Oki, B. M.; Terry, Douglas; Using Collaborative Filtering to Weave an Information Tapestry. Commun. ACM 35,(1992).
5. Kim, J., Oard, D.W., and Romanik, K.; User modeling for information filtering based on implicit feedback, In Proc. of ISKO-France, (2000).
6. Konstan, J. A., B. N. Miller, D. Maltz, J. L. Herlocker, L.R. Gordon, and J. Riedl; GroupLens: Applying collaborative filtering to Usenet News, Communication of the ACM, 40(3), p77-87,(1997).
7. Lam, K. and C. Ho; Using a generalized instance set for automatic text categorization, In Proc. of SIGIR,(1998).
8. Lee,C.C.; Fuzzy logic in control systems: fuzzy logic controller-part I, IEEE Trans. On Systems, Man, and Cybernetics, 20 (2), p408-418, (1990).
9. Lewis,D., R. Schapire, J. Callan, and R. Papka; Training Algorithms for Linear Text Classifiers, In Proc. of SIGIR, pp.298-306, (1996).
10. Nichols,D. M.; Implicit ratings and filtering, In Proc. of the 5th DELOS Workshop on Filtering and Collaborative Filtering, p10-12, (1997).
11. Pazzani, M., Billsus, D.; Learning and revising user profiles: the identification of interesting Web sites, Machine Learning, (1997).
12. Schapire,R., Y. Singer, and A. Singal; Boosting and Rocchio Applied to Text Filtering, In Proc. of SIGIR, (1998).
13. Seo, Y., Zhang, B.; Personalized Web Document Filtering Using Reinforcement Learning, Applied Artificial Intelligence, (2001).
14. Soltysiak, S.J. and Crabtree, I.B.; Automatic Learning of User Profiles-Towards the Personalisation of Agent Services, BT Technology Journal, 16(3),(2000).
15. Yan, T. W. and H. Garcia-Molina.; SIFT- A tool for wide-area information dissemination, In Proc. of the 1995 USENIX Technical Conference, (1995).
16. Yang, Y., Pedersen, J.; A comparative study on feature selection in text categorization, In Proc. of ML, (1997).

On the Effective Similarity Measures for the Similarity-Based Pattern Retrieval in Multidimensional Sequence Databases*

Seok-Lyong Lee[1], Ju-Hong Lee[2], and Seok-Ju Chun[3]

[1] School of Industrial and Information Eng., Hankuk University of Foreign Studies, Korea
sllee@hufs.ac.kr
[2] Dept. of Computer Science and Engineering, Inha University, Korea
juhong@inha.ac.kr
[3] Dept. of Computer Education, Seoul National University of Education, Korea
chunsj@snue.ac.kr

Abstract. In this paper, we propose the effective similarity measures on which the similarity-based pattern retrieval is based. Both data sequences and query sequences are partitioned into segments, and the query processing is based upon the comparison of the features between data and query segments, instead of scanning all data elements of entire sequences. We conduct experiments on multidimensional data sequences that are generated by extracting features from video streams, and show the effectiveness of the proposed measures.

1 Introduction

An effective similarity measures is presented in this paper for the similarity-based pattern retrieval in multidimensional sequence databases. First, we define a safe distance measure that guarantees '*no false dismissal,*' to prune irrelevant segments from a database, and then we design a semantic measure that considers the directional and geometric characteristics of a segment such as the moving direction of points and the volume or edge of the segment. The shape of a segment is hyper-rectangular since the current dominant indexing mechanisms such as the R-tree [3] and its variants [8, 2, 1] are based on the minimum-bounding rectangle (MBR) as their node shape, and we can exploit them without (or with slight) modification. Both data sequences and query sequences are partitioned into segments, and the query processing is based upon these segments, instead of scanning data elements of entire sequences.

Various similarity search measures on sequential data have been proposed. To name a few, Rafiei et al. [7] proposed a set of safe linear transformations of a given sequence that can be used as the basis for similarity queries on time-series data. They formulated distance measures considering moving average, reversing, and time warping. These transformations are extended to the multiple transformations in [6], where an index is searched only once and a collection of transformations is simultaneously

---

* This work was supported by the Korea Research Foundation Grant funded by Korean Government (MOEHRD) (R05-2004-000-10972-0).

applied to the index. Yi et al. [9] proposed the distance measure using the arbitrary L_p norms, and Keogh et al. [4] proposed two distance measures in the indexed space that exploit the high fidelity of their search method (APCA) for fast searching: a lower bounding and a non-lower bounding Euclidean distance approximation. However, these methods address the similarity search for one-dimensional time-series data, and thus do not handle multidimensional data sequences (MDS's). In addition, their similarity measures are mostly based on the Euclidean distance and do not consider the semantic aspects such as geometric and directional characteristics of sequences.

Before we start the discussion on the similarity measures, we first describe the segmentation technique [5] briefly, to make the paper self-contained. A hyper-rectangular segment *SEG* with k points, P_j for $j = 1, 2, \ldots, k$ in the n-dimensional space, is represented by two endpoints, L(low point) and H(high point), of its major diagonal, and the number of points in the rectangle as follows: $SEG = \langle L, H, k \rangle$, where $L = \{(L^1, L^2, \ldots, L^n) \mid L^i = min_{1 \leq j \leq k}(P_j^i)\}$, and $H = \{(H^1, H^2, \ldots, H^n) \mid H^i = max_{1 \leq j \leq k}(P_j^i)\}$ for $i = 1, 2, \ldots, n$. Then, the volume $Vol(SEG)$ and the edge $Edge(SEG)$ of a segment *SEG* are computed as:

$$Vol(SEG) = \prod_{1 \leq i \leq n}(SEG.H^i - SEG.L^i). \quad (1)$$

$$Edge(SEG) = 2^{n-1} \cdot \sum_{1 \leq i \leq n}(SEG.H^i - SEG.L^i). \quad (2)$$

As quantitative measures to evaluate the segment characteristics, we use the volume per point (*VPP*) and the edge per point (*EPP*). Suppose MDS *S* is partitioned into p segments, $SEG_1, \ldots, SEG_p$. Then, *VPP* and *EPP* of *S* are defined as follows:

$$VPP = \frac{\sum_{1 \leq j \leq p} Vol(SEG_j)}{\sum_{1 \leq j \leq p} SEG_j.k}, \quad EPP = \frac{\sum_{1 \leq j \leq p} Edge(SEG_j)}{\sum_{1 \leq j \leq p} SEG_j.k} \quad (3)$$

2 Similarity Measures

2.1 Distance-Based Similarity Measure

The distance $D_S(S_1, S_2)$ between two sequences S_1 and S_2 of equal lengths, each of which has k points, is defined as the mean distance of the two, where the mean distance is defined as follows:

$$D_S(S_1, S_2) = D_{mean}(S_1, S_2) = \frac{1}{k} \cdot \sum_{i=0}^{k-1} d(S_1[i], S_2[i]) \quad (4)$$

where $d()$ is an Euclidean distance between two points and $S[i]$ is an ith point of sequence S. Next, let us consider the distance $D_S(S_1, S_2)$ between two sequences S_1 and S_2 of different lengths, each of which has p and q points, respectively. Without loss of generality, we assume $p \leq q$. Then, $D_S(S_1, S_2)$ is defined as the minimum mean distance of every pair, where the minimum mean distance is defined as follows:

$$D_S(S_1, S_2) = \min_{1 \leq j \leq q-p+1} D_{mean}(S_1[1:p], S_2[j:j+p-1]) \quad (5)$$

where $S[a:b]$ is a sub-sequence of S from point $S[a]$ to $S[b]$. To measure the distance between two segments, we define the distance D_{seg} as the minimum Euclidean distance between two hyper-rectangles that bound all points in each segment, respectively. Then the distance D_{seg} is shorter than the distance between any pair of points, one in a segment SEG_1 and the other in a segment SEG_2. That is:

$$D_{seg}(SEG_1, SEG_2) \leq \min_{P_1 \in SEG_1, P_2 \in SEG_2} d(P_1, P_2) \qquad (6)$$

where P_1 and P_2 are the points that are contained in SEG_1 and SEG_2, respectively. Then we are able to derive the following property showing the lower bounding relationships with respect to D_{seg} and D_S, where D_S is the distance between two sequences, S_1 and S_2, that are contained in segments SEG_1 and SEG_2, respectively. The property is: The distance $D_{seg}(SEG_q, SEG_t)$ between a query segment SEG_q and a data segment SEG_t in a database is the lower bound of the distance $D_S(S_q, S_t)$ between two sequences contained in those two segments, respectively. That is: $D_{seg}(SEG_q, SEG_t) \leq D_S(S_q, S_t)$. We omit the proof of this property because of the space limitation. Using this property, we can use the distance $D_{seg}(SEG_q, SEG_t)$ to prune irrelevant segments from a database without 'false dismissals,' since it provides the lower bound with respect to the distance $D_S(S_q, S_t)$ between two sequences of those segments.

2.2 Semantic-Based Similarity Measure

One problem with D_{seg} that guarantees 'no false dismissal' is that it may introduce too many 'false hits', influencing the retrieval efficiency. As the false hits increase, the efficiency degrades since those false segments should be evaluated in the post expensive process. Another problem with the distance measure is that it does not capture enough semantic aspects of segments. To evaluate the similarity between two segments, we extract two features from a segment: *directional* and *geometric*.

Directional feature: Given a segment SEG, the directional vector $SEG.DV$ is defined by a vector from a start point P_{start} to an end point P_{end} of SEG. Let S be a vector from the origin to P_{start}, and E be a vector from the origin to P_{end}, respectively. Then the directional similarity sim_D of two segments, SEG_q and SEG_t, is defined in term of a cosine of two directional vectors of them, known as a *cosine similarity*. Let θ be the angle between two vectors, $SEG_q.DV$ and $SEG_t.DV$. Then the sim_D is represented using the inner product of two vectors as follows:

$$sim_D(SEG_q, SEG_t) = \frac{1}{2}\left(1 + \frac{SEG_q.DV \bullet SEG_t.DV}{|SEG_q.DV| \, \|SEG_t.DV|}\right) \qquad (7)$$

Geometric feature: Consider two objects, o_A and o_B, whose characteristics are represented by two quantitative values, v_A and v_B respectively, with respect to a variable v. Let the domain of v be $dom(v)$ and its upper and lower limits be $max(dom(v))$ and $min(dom(v))$, respectively. Then the similarity between o_A and o_B, $sim_v(o_A, o_B)$, is represented to have the range from 0 to 1, as follows:

$$sim_v(o_A, o_B) = 1 - \frac{|v_A - v_B|}{max(dom(v)) - min(dom(v))} \qquad (8)$$

We represent the geometric similarity sim_G of two segments, SEG_q and SEG_t, in term of VPP and EPP. Let the upper and lower limit of VPP and EPP be: $max(dom(VPP))$, $min(dom(VPP))$, $max(dom(EPP))$, and $min(dom(EPP))$. Assuming that the values follow the Gaussian distribution, we are able to find the range easily in which approximately 99 percent of values fall. This range is found to be $[\mu_v-2.58\sigma_v \leq v \leq \mu_v+2.58\sigma_v]$ since $P[v \leq \mu_v+2.58\sigma_v] = 0.9951$. Namely, we do not take the maximum and minimum value of v as $max(dom(v))$ and $min(dom(v))$. Instead, we take $\mu_v+2.58\sigma_v$ and $\mu_v-2.58\sigma_v$ since they can eliminate extraordinarily large or small values of v from the consideration. We compute the mean μ_{VPP} and the standard deviation σ_{VPP} for VPP values, and μ_{EPP} and σ_{EPP} for EPP values. Using these values, $max(dom(VPP))$ and $max(dom(EPP))$ are computed as follows: $max(dom(VPP))= \mu_{VPP}+2.58\cdot\sigma_{VPP}$, $min(dom(VPP))=\mu_{VPP}-2.58\cdot\sigma_{VPP}$, $max(dom(EPP))=\mu_{EPP}+2.58\cdot\sigma_{VPP}$, and $min(dom(EPP))=\mu_{EPP}-2.58\cdot\sigma_{EPP}$. Thus, the similarity sim_{VPP} and sim_{EPP} of segments, SEG_q and SEG_t are computed as follows:

$$sim_{VPP}(SEG_q,SEG_t) = 1 - \frac{|VPP_q - VPP_t|}{5.16\sigma_{VPP}}, sim_{EPP}(SEG_q,SEG_t) = 1 - \frac{|EPP_q - EPP_t|}{5.16\sigma_{EPP}} \quad (9)$$

By combining these two similarity measures, we define the geometric similarity sim_G of two segments, SEG_q and SEG_t using the weights.

Integrated similarity measure: Using a single attribute for the similarity search may lack sufficient discriminatory information. We therefore integrate two categories of similarity measures discussed above to promote the effectiveness of the retrieval. The integrated similarity measure sim_I between two segments, SEG_q and SEG_t, is defined on the basis of combining directional and geometric similarity measures.

$$sim_I(SEG_q,SEG_t) = \frac{w_D \cdot sim_D + w_G \cdot sim_G}{w_D + w_G} \quad (10)$$

where w_D and w_G are the weights assigned to sim_D and sim_G, respectively. The default values for these weights are 1's and users are allowed to choose the weights based on their application domain.

Similarity retrieval mechanism: In usual similarity-based pattern retrieval for sequences, a sequence is given as a query input to find similar (sub-)sequences in a database. But, we restrict our retrieval mechanism to find similar segments with respect to a query segment, since we focus on verifying the similarity measures in this paper. By utilizing the similarity measure we propose, our mechanism can be extended to support the similarity retrieval for sequences. Before a query is processed, we do some pre-processing that partitions an MDS into segments, extracts the features from each segment for the similarity comparison, and stores them into a database. As an input parameter to the algorithm, a query segment and a similarity threshold (ζ) is given. The similarity threshold is translated to an appropriate distance threshold (ε). Since our space is normalized to an n-dimensional cube, $[0,1]^n$, it is not difficult to transform the ζ-value to the ε-value.

Next, the filtering is done using D_{seg}. It first prunes irrelevant segments from a database with no false dismissal as shown in Section 2.1. Even though D_{seg} guarantees the correctness, too many false hits clearly degrade the retrieval efficiency since irrelevant segments should be evaluated in the subsequent expensive process. The refinement using sim_l, of course, does not guarantee the correctness, however it provides a fairly good precision rate while maintaining the reasonable recall, as we will show in the experiment in Section 3.

3 Experimental Evaluation

We conduct experiments on video streams that are captured from a collection of TV news, dramas, and documentary films. Multidimensional sequences are generated from video clips, by representing each frame of the clips by a point in the multidimensional space. We use a 3-dimensional space, each dimension of which is R, G, and B, respectively, by averaging color values of pixels of a frame. The numbers of database segments and query segments are 8,972 and 50, respectively, and the similarity thresholds are from 0.5 through 0.9. The system is implemented in Microsoft VC++ under Windows Server environment. To observe the effectiveness considering both the precision and recall together, we also include the evaluation by the product of the precision and recall ($P*R$).

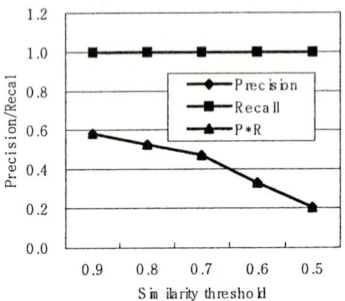
Fig. 1. Precision/Recall for D_{seg}

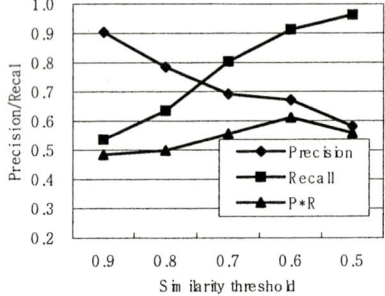
Fig. 2. Precision/Recall for D_{seg} and sim_l

Fig. 1 illustrates the precision, recall, and $P*R$ with respect to various similarity thresholds in case that we use the distance-based measure (D_{seg}) only. Meanwhile, Fig. 2 shows the precision, recall, and $P*R$ for the distance-based (D_{seg}) and semantic (sim_l) measures both. As we can observe in Fig. 1, the recall is 1, since the measure D_{seg} guarantees 'no false dismissal'. However, the precision is relatively low, showing 0.20-0.58, and it decreases as the similarity threshold increases. The value, $P*R$, is of course the same as the precision since the recall is always 1. When we use both measures as shown in Fig. 2, the recall is not 1, since the measure sim_l allows 'false dismissal'. The recall is 0.53-0.96. However, the precision is much better than that of the case using the distance-based measure only. It is 0.58-0.89. By sacrificing the recall a little bit, using sim_l with D_{seg} achieves the better precision, which will improve the retrieval efficiency. The comparison by using the value, $P*R$, also shows much improvement, showing 0.48-0.61. Consequently, we can conclude that using both distance-based and semantic measures improves the overall performance compared to the case using the distance-based measure only, even though the former shows lower recall rate than the latter.

4 Conclusion

In this paper, we propose two effective similarity measures, the distance-based measure that is based on the distance between two segments in the multidimensional space and the semantic measure that captures the semantic aspects of segments such as geometric and directional properties. We have conducted experiments using the data that are extracted from real-world videos to evaluate our proposed measures with respect to the precision and recall. The recall is 0.53-0.96 and the precision is 0.58-0.89, which is quite usable in real-world business environments.

References

1. S. Berchtold, D. Keim, and H. Kriegel: The X-tree: An Index Structure for High-Dimensional Data. Proc. of Int'l Conference on VLDB (1996) 28-39
2. N. Beckmann, H. Kriegel, R. Schneider, and B. Seeger: The R^*-tree: An Efficient and Robust Access Method for Points and Rectangles. Proc. of ACM SIGMOD (1990) 322-331
3. A. Guttman: R-trees: A Dynamic Index Structure for Spatial Searching. Proc. of ACM SIGMOD (1984) 47-57
4. E. J. Keogh, K. Chakrabarti, S. Mehrotra, and M. J. Pazzani: Locally Adaptive Dimensionality Reduction for Indexing Large Time Series Databases. Proc. of ACM SIGMOD (2001) 151-162
5. S. L. Lee and C. W. Chung: Hyper-Rectangle Based Segmentation and Clustering of Large Video Data Sets. Information Science, Vol. 141, No. 1-2 (2002) 139-168
6. D. Rafiei: On Similarity Queries for Time Series Data. Proc. of Int'l Conference on Data Engineering (1999) 410-417
7. D. Rafiei and A. Mendelzon: Similarity-Based Queries for Time Series Data. Proc. of ACM SIGMOD (1997) 13-25
8. T. Sellis, N. Roussopoulos, and C. Faloutsos: The R+ Tree: A Dynamic Index for Multi-Dimensional Objects. Proc. of Int'l Conference on VLDB (1987) 507-518
9. B. K. Yi and C. Faloutsos: Fast Time Sequence Indexing for Arbitrary Lp Norms. Proc. of Int'l Conference on VLDB (2000) 385-394

Crossing the Language Barrier Using Fuzzy Logic

Rowena Chau and Chung-Hsing Yeh

School of Business Systems,
Faculty of Information Technology,
Monash University, Clayton, Victoria 3800, Australia
Rowena.Chau@infotech.monash.edu.au
ChungHsing.Yeh@infotech.monash.edu.au

Abstract. Cross-lingual text retrieval (CLTR) concerns the retrieval of documents across languages. To allow multilingual term matching, a multilingual thesaurus is needed. However, a multilingual thesaurus encoding exact translation equivalent is insufficient for effective CLTR since relevant documents are often indexed by cross-lingual related terms. In this paper, a novel approach for automatically constructing a multilingual thesaurus based on fuzzy set theory is proposed. By introducing a degree of relatedness between multilingual terms using the concept of membership degree, partial match of cross-lingual related terms is facilitated. Development of a fuzzy multilingual news retrieval system using the proposed approach is presented.

1 Introduction

Cross-lingual text retrieval (CLTR) refers to the selection of text in one language based on query in another [3]. Basically, it is a problem of vocabulary mismatch. To solve this problem, a multilingual thesaurus is generally used to suggest corresponding translation equivalents for expanding a query in order to accommodate the vocabulary difference between languages. However, a multilingual thesaurus encoding exact translation equivalents only is insufficient for CLTR. Language is culture bound. Translation equivalents do not always available in a foreign language while co-existing cross-lingual counterparts often varies slightly in meaning. Such intrinsic vagueness of meanings in natural languages implies that reliance on exact match of semantically equivalent terms across languages for CLTR is impractical. Potentially relevant documents are often indexed by semantically similar terms which are partially equivalent. To be effective, a multilingual thesaurus for CLTR should facilitate partial match against cross-lingual related terms. Otherwise, relevant documents indexed by semantically similar terms will be missed out.

In this paper, a novel approach for automatically constructing a multilingual thesaurus based on fuzzy set theory is proposed. By introducing a degree of semantic relatedness between multilingual terms using the concept of membership degree, partial match of terms across languages is made possible. With the support of this fuzzy multilingual thesaurus, recall of CLTR can be reasonably improved as more relevant documents are retrieved. In what follows, Section 2 presents the theoretical background as well as the mathematical model for the construction of a fuzzy multilingual thesaurus. In Section 3, application of the fuzzy multilingual thesaurus in CLTR is discussed. In Section 4, development of a prototype fuzzy multilingual news retrieval system using the proposed approach is presented. Finally, a conclusive remark is included in Section 5.

2 Modeling a Fuzzy Multilingual Thesaurus

A multilingual thesaurus can be considered as a semantic knowledge base consisting of sets of terms in multiple languages and a specification of their cross-lingual semantic relations. Application of fuzzy logic in constructing fuzzy thesaurus for monolingual information retrieval [1,2,4] has been widely discussed over the past three decades as a more realistic approach for semantic knowledge representation. By extending its application to a multilingual environment, it is believed that cross-lingual semantic knowledge may also be effectively represented in a similar way. In this paper, a fuzzy multilingual thesaurus is constructed using a parallel corpus. By analyzing the corpus statistics of term occurrences, concepts relevant to a term's meaning, together with their corresponding degrees of relevance, are extracted. Considering each term's meaning as an integration of its constituent concepts, the lexical meaning of each term is then represented as a fuzzy set of concepts with relevance degrees of all its constituent concepts as membership values. Based on the similarity of meanings, a degree of cross-lingual semantic relatedness is computed. To get a fuzzy thesaurus that will allow partial matching, a fuzzy relation representing the semantic relation of cross-lingual-related-terms is established. Thereby, a fuzzy multilingual thesaurus relating terms across languages with their degrees of semantic relatedness, ranging from *0* to *1*, is constructed. The mathematical model of the proposed fuzzy multilingual thesaurus is presented below:

Given a parallel corpus D in two languages, L_A and L_B, we have:

$$D = \{d_k\} \quad (1)$$

where $d_{k \in \{1,2,...z\}}$ is a parallel document containing identical text in both L_A and L_B versions.

Two sets of terms, A and B, are extracted from the parallel corpus D.

$$A = \{a_i\} \quad \text{where } a_{i \in \{1,2,...x\}} \text{ is a term of } L_A \quad (2)$$

$$B = \{b_j\} \quad \text{where } b_{j \in \{1,2,...y\}} \text{ is a term of } L_B \quad (3)$$

For the establishment of semantic relations, meaning of terms has to be determined. In our approach, each document of the parallel corpus is viewed as a specific concept and each term contained in the document is considered constituting to the totality of the concept represented by the document as a whole. Accordingly, degree of relevance between a term and a concept is revealed by the term's relative frequency within a document. Based on the statistics of relative frequencies, lexical meaning of every term is then represented as a fuzzy set of its constituent concepts with the degrees of relevance between term and concepts as membership values.

For $a_i \in A$, its lexical meaning is represented by:

$$a_i = \sum_{d_k \in D} \mu_{a_i}(d_k)/d_k \qquad (4)$$

where

$$\mu_{a_i}(d_k) = \frac{Frequency\ of\ a_i\ in\ d_k\ written\ in\ L_A}{Length\ of\ d_k\ written\ in\ L_A} \qquad (5)$$

For $b_j \in B$, its lexical meaning is represented by:

$$b_j = \sum_{d_k \in D} \mu_{b_j}(d_k)/d_k \qquad (6)$$

where

$$\mu_{b_j}(d_k) = \frac{Frequency\ of\ b_j\ in\ d_k\ written\ in\ L_B}{Length\ of\ d_k\ written\ in\ L_B} \qquad (7)$$

A fuzzy multilingual thesaurus FT_{AB} involving two languages, L_A and L_B, modeling the semantic relation of cross-lingual-related-terms is expressed as a fuzzy relation $FT(A,B)$ as follows:

$$FT(A,B) = \begin{bmatrix} \mu_{FT}(a_1,b_1) & \mu_{FT}(a_1,b_2) & \mu_{FT}(a_1,b_3) & \cdots & \mu_{FT}(a_1,b_y) \\ \mu_{FT}(a_2,b_1) & \mu_{FT}(a_2,b_2) & \mu_{FT}(a_2,b_3) & \cdots & \mu_{FT}(a_1,b_y) \\ \mu_{FT}(a_3,b_1) & \mu_{FT}(a_3,b_2) & \mu_{FT}(a_3,b_3) & \cdots & \mu_{FT}(a_3,b_y) \\ \cdot & \cdot & \cdot & \cdots & \cdot \\ \cdot & \cdot & \cdot & \cdots & \cdot \\ \mu_{FT}(a_x,b_1) & \mu_{FT}(a_x,b_2) & \mu_{FT}(a_x,b_3) & \cdots & \mu_{FT}(a_x,b_y) \end{bmatrix} \qquad (8)$$

where

$$\mu_{FT}(a_i,b_j) = \frac{|\mu(a_i) \cap \mu(b_j)|}{|\mu(a_i) \cup \mu(b_j)|} = \frac{\sum_{d_k \in D} min(\mu_{a_i}(d_k), \mu_{b_j}(d_k))}{\sum_{d_k \in D} max(\mu_{a_i}(d_k), \mu_{b_j}(d_k))} \qquad (9)$$

is defined as the degree of cross-lingual semantic relatedness between two terms, a_i and b_j, based on the similarity of their meanings. If $\mu_{FT}(a_i,b_j)=1$, then a_i and b_j are translation equivalents of each other.

3 Retrieving Documents Across Languages

By its nature, CLTR is an inference from knowledge whose meaning is not sharply defined. Conventional approach such as classical logic which requires high standards of precision for exact reasoning is thus ineffective because of its inability to grip with the fuzziness involved. On the other hand, fuzzy logic, through the use of approximate reasoning, allows the standards of precision to be adjusted to fit the imprecision of the information involved. Fuzzy reasoning is an inference procedure that uses fuzzy logic to deduce conclusion from a set of fuzzy IF-THEN rules by combining evidence through the compositional rule of inference [7,8]. Based on the mechanism of fuzzy reasoning, a CLTR system is represented by defining fuzzy IF-THEN rules and converting them into corresponding fuzzy relations as follows:

	Propositions		Fuzzy relations
IF	q contains $t \in L_Q$	$\Rightarrow$	$NEED(q, L_Q)$
AND	$t \in L_Q$ is related to $t' \in L_D$	$\Rightarrow$	$FT(L_Q, L_D)$
AND	$t' \in L_D$ is an index term of $d \in D$	$\Rightarrow$	$IND(L_D, D)$
THEN	$d \in D$ is relevant to q	$\Rightarrow$	$REL(q, D)$

Here, $NEED(q, L_Q)$ is a fuzzy query representation function relating a query q and its query terms with weights representing the degree of importance of each query term with respect to that particular query written in L_Q.

$FT(L_Q, L_D)$ is a fuzzy multilingual thesaurus as defined by equation (10). It relates pairs of potential index term and query term to a degree of cross-lingual relatedness based on their semantic "closeness".

$IND(L_D, D)$ is an indexing function relating a document $d \in D$ and its index term to a degree of "aboutness". In other words, it can be considered as a fuzzy indexing function by which documents are described by index terms with term weights.

Finally, $REL(q, D)$ is fuzzy matching function which assigns to each document $d \in D$ a degree of relevance, within the range of 0 and 1, with respect to a particular query q.

Based on the compositional rule of inference, $REL(q, D)$ is inferred by applying the composition operator to the fuzzy relations representing their premises as follows:

$$REL = NEED \circ FT \circ IND \qquad (10)$$

where

$$\mu_{REL}(q,d) = \mu_{NEED \circ FT \circ IND}(q,d)$$
$$= \max_{(t,t') \in FT} [\mu_{NEED}(q,t) \wedge \mu_{FT}(t,t') \wedge \mu_{IND}(t',d)] \qquad (11)$$

is the membership function which gives the degree of membership of (q,d) in REL indicating the extent to which document d is relevant to query q.

4 Developing a Fuzzy Multilingual News Search Engine

To illustrate this approach for CLTR, an application to develop a multilingual news search engine is presented. In a prototype system, the fuzzy logic approach to CLTR is applied to develop a search engine for retrieving online news available in both English and Chinese. An overview of the system is depicted in Figure 1.

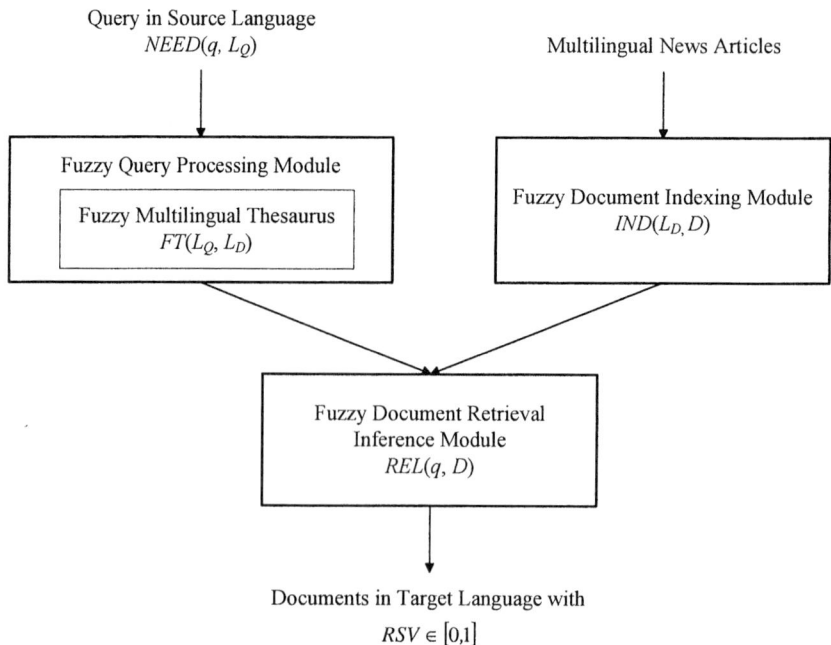

Fig. 1. Architecture of a Fuzzy Multilingual News Search Engine

The multilingual news search engine is composed of three main modules, namely, the *fuzzy query processing module*, the *fuzzy document indexing module*, and the *fuzzy document retrieval inference module*. The fuzzy query processing module will perform the major task of fuzzy cross-lingual query expansion in the course of cross-lingual news retrieval using the fuzzy multilingual thesaurus as the linguistic knowledge base. By accepting a user query in a source language, the fuzzy query processing module will translate the query to another language by expanding it with all its cross-lingual related terms according to the fuzzy multilingual thesaurus. To gather a collection of training documents for the generation of the fuzzy multilingual thesaurus, a set of past parallel online news in both English and Chinese are collected from the Web. To generate the fuzzy multilingual thesaurus $FT(L_Q, L_D)$, a set of multilingual (both English and Chinese) terms are extracted from the training parallel documents using a bilingual wordlist and then the fuzzy multilingual thesaurus construction algorithm is applied.

The fuzzy document indexing module incorporating the fuzzy document indexing function $IND(L_D, D)$ will generate a set of weighted index terms for every document in the multilingual news database. To determine the term weights, standard *TF.IDF* term weighting scheme [5] commonly used in information retrieval is employed. In our approach, query translation is already in place for the cross-lingual text retrieval. Therefore, document indexing only need to be done in a monolingual manner.

Finally, the fuzzy document retrieval inference module with the built-in fuzzy matching function $REL(q, D)$ performs the fuzzy matching between document and query taking the expanded query from the fuzzy query processing module and the fuzzy document indexes from the fuzzy document indexing module as input. A retrieval status value (*RSV*) between *0* and *1* will be computed for every document and a ranked list of relevant documents in a target language will be returned to the user as an output.

5 Conclusion

A multilingual thesaurus specifying lexical relation between pairs of multilingual terms is an important source of semantic evidence for CLTR. However, to make a multilingual thesaurus work effectively for CLTR, the intrinsic vagueness of meaning in natural languages must be well addressed. Otherwise, closely related documents will be missed out. The fuzzy multilingual thesaurus proposed in this paper thus has applied the fuzzy set theory to introduce a degree of cross-lingual semantic relatedness into the lexical relations among multilingual terms. As a result, partial match between multilingual terms is made possible. Closely related documents containing no translation equivalents of the query terms but only semantically similar cross-lingual related terms will then be retrieved. Therefore, with the support of this fuzzy multilingual thesaurus, recall of CLTR will improve. Development of a fuzzy multilingual news search engine using this approach demonstrates its applicability in work.

References

1. De Cock, M. Guadarrama, M. S., Nikravesh, M.: Fuzzy Thesauri for and from the WWW. Soft Computing for Information Processing and Analysis, Studies in Fuzziness and Soft Computing 164, Springer-Verlag, (2005) 275- 284
2. Larsen, H.L. and Yager, R.R.: The use of fuzzy relational thesauri for classificatory problem solving in information retrieval and expert systems. IEEE Trans. Systems, Man and Cybernetics 23 (1993) 31-41
3. Oard, D. W. and Dorr, B. J., A survey of multilingual text retrieval. Technical Report. UMIACS-TR-96-19, University of Maryland, Institute for Advanced Computer Studies. (1996)
4. Radecki, T.: Mathematical model of information retrieval system based on the concept of fuzzy thesaurus. Information Processing and Management. 12 (1976) 313-318
5. Salton, G. and Buckley, C. Term-weighting approaches in automatic text retrieval. Information Processing and Management, 24 (1988) 513-523,.
6. Zadeh, L.A.: Similarity relations and fuzzy orderings. Information Sciences 3 (1971) 177-206
7. Zadeh, L. A., Outline of a new approach to the analysis of complex systems and decision process. *IEEE Trans. Systems, Man and Cybernetics.* **3** (1973) 28-44
8. Zadeh, L. A., The role of fuzzy logic in the management of uncertainty in expert systems. Fuzzy Sets and Systems. 11 (1983) 199-227

New Algorithm Mining Intrusion Patterns*

Wu Liu, Jian-Ping Wu, Hai-Xin Duan, and Xing Li

Network Research Center of Tsinghua University, 100084 Beijing, P.R. China
liuwu@ccert.edu.cn

Abstract. In this paper, we apply data mining techniques to construct intrusion detection patterns. We mine both system audit data and network traffic data for consistent and useful patterns of program and user behavior, and use an iterative low-frequency-finder mining algorithm to find the low frequency but important patterns.

1 Introduction

As the rapid development of computer networks especially the Internet, computer systems have become the target of attackers. So, we need to find best ways to protect our computer systems. Intrusion prevention techniques, such as user authentication, authorization, and access control etc. are not sufficient [4]. Intrusion Detection System (IDS) is therefore needed to protect computer systems.

Currently many intrusion detection systems are constructed by manual and ad-hoc means. In [1] rule templates specifying the allowable attribute values are used to post-process the discovered rules. In [2] boolean expressions over the attribute values are used as item constraints during rule discovery. In [3], a "belief-driven" framework is used to discover the unexpected (hence interesting) patterns. A drawback of all these approaches is that one has to know what rules/patterns are interesting or are already in the belief system. We cannot assume such strong prior knowledge on all audit data.

We aim to develop a systematic framework to semi-automate the process of building intrusion detection systems. A basic premise is that when audit mechanisms are enabled to record system events, distinct evidence of legitimate and intrusive (user and program) activities will be manifested in the audit data. For example, from network traffic audit data, connection failures are normally infrequent. However, certain types of intrusions will result in a large number of consecutive failures that may be easily detected. We there-fore take a data-centric point of view and consider intrusion detection as a data analysis task. Anomaly detection is about establishing the normal usage patterns from the audit data, whereas misuse detection is about encoding and matching intrusion patterns using the audit data.

---

* This work is supported by grants from 973, 863 and the National Natural Science Foundation of China (Grant No. #90104002 & #2003CB314800 & #2003AA142080 & #60203044) and NISAC 2004-R-3-917-A-01.

2 Low-Frequency-Finder Mining

We attempt to utilize the schema level information about audit records to direct the pattern mining process. That is, although we cannot know in advance what patterns, which involve actual attribute values, are interesting, we often know what attributes are more important or useful given a data analysis task.

It is often necessary to include the low frequency patterns. In daily network traffic, some services (for example, gopher), account for very low occurrences. Yet we still need to include their patterns into the network traffic profile (so that we have representative patterns for each supported service). If we use a very low support value for the data mining algorithms, we will then get unnecessarily a very large number of patterns related to the high frequency services, for example, ftp.

We use the following low-frequency-finder mining algorithm for finding frequent sequential patterns from audit data.

Here, we call the essential attribute the Es-attribute when they are used as a form of item constraints in the association rules algorithm. During candidate generation, an item set must contain value(s) of the Es-attribute. We consider the correlations among non- essential attribute as not interesting

The Low-Frequency-Finder Mining Algorithm

Input:
  A_s; /* the initial threshold */
  A_t; /* the terminating threshold */
  X; /* the Es-attribute */;
Output:
  R; /* frequent episode rules */
Begin
$R_0 \leftarrow \{\}$;
Scan database to form $B \leftarrow \{$ 1- item-sets that meet A_t $\}$;
$A \leftarrow A_s$;
while $(A \geq A_t)$ **do**
    Calculate frequent episodes from B: each episode must contain at least one
    X that is not in R_0;
    $R_0 \leftarrow R_0 \cup \{X\}$;
    $R \leftarrow R \cup \{$ episode rules $\}$;
    $A \leftarrow A/2$; /* a smaller support value for the next iteration */
end while
end

Here the idea is to first find the episodes related to high frequency Es-attribute values, for example

(service = ftp; src bytes = 1000);
(service = ftp; src bytes = 1000) → (service = ftp; dst bytes = 1500)

We then iteratively lower the support threshold to find the episodes related to the low frequency Es-attribute values by restricting the participation of the "old" Es-attribute values that already have output episodes. More specifically, when an episode is generated, it must contain at least one "new" (low frequency) Es-attribute value.

For example, in the second iteration, where ftp now is an old Es-attribute value, we get an episoopde rule

(service = ftp; src bytes = 1000);
(service = http; src bytes = 1000) → *(service = ftp; src bytes = 1500)*

The algorithm terminates when a very low support value is reached. In practice, this can be the lowest frequency of all Es-attribute values.

Note that for a high frequency Es-attribute value, we in effect omit its very low frequency episodes (generated in the runs with low support values) because they are not as interesting (i.e., representative). In other words, at each iteration, we have

$$I_A(p) = \begin{cases} 1 & \text{if } p \text{ contains at least one "new" Es-attribute value} \\ 0 & \text{otherwise} \end{cases}$$

We still include all the old Es-attribute values to form episodes along with the new Es-attribute values because it is important to capture the sequential context of the new Es-attribute values. For example, although used infrequently, auth normally co-occurs with other services such as ftp and login. It is therefore imperative to include these high frequency services into the episode rules about auth.

Our approach here is different from the algorithms in [1] since we do not have and can not assume multiple concept levels, rather, we deal with multiple frequency levels of a single concept, e.g., the network service.

3 Experiments

Here we test our hypothesis that the merged rule set can indicate whether the audit data has covered sufficient variations of behavior.

We obtained one month of TCP/IP network traffic data from CCERT_IDS. We segmented the data by day. And for data of each day, we again segmented the data into four partitions: morning, afternoon, evening and night. This partitioning scheme allows us to cross evaluate anomaly detection models of different time segments that

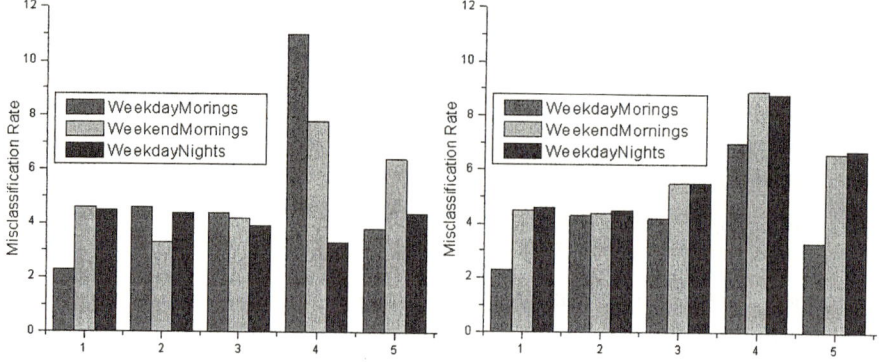

Fig. 1. Misclassification rates of classifier trained on first 8 (left) and 10 (right) weekdays

have different traffic patterns. It is often the case that very little (sometimes no) intrusion data is available when building an anomaly detector. A common practice is to use audit data of legitimate activities that is known to have different behavior patterns for testing and evaluation.

Figure 1 show the performance of these classifiers in detecting anomalies (different behavior) respectively. In each figure, we show the misclassification rate (percentage of misclassifications) on the test data.

4 Conclusion

In this paper we describe data mining techniques for building intrusion detection models. We demonstrated that association rules and frequent episodes from the audit data can be used to guide audit data gathering and feature selection, the critical steps in building effective classification models. We incorporated domain knowledge into the low-frequency-finder mining algorithm.

References

1. M. Klemettinen, H. Mannila, P. Ronkainen, H. Toivonen, and A. I. Verkamo. Finding interesting rules from large sets of discovered association rules. In Proceedings of the 6rd International Conference on Information and Knowledge Management, Gainthersburg, MD, 2002.
2. R. Srikant, Q. Vu, and R. Agrawal. Mining association rules with item constraints. In Proceedings of the 8rd International Conference on Knowledge Discovery and Data Mining, Newport Beach, California, August 2001.
3. B. Padmanabhan and A. Tuzhilin. A belief-driven method for discovering unexpected patterns. In Proceedings of the 4th International Conference on Knowledge Discovery and Data Mining, New York, NY, August 1998. AAAI Press.
4. Wu Liu, Study on Intrusion Detection Technology with Traceback and Isolation of Attacking Sources, PhD Thesis 2004.
5. J. Han and Y. Fu. Discovery of multiple-level association rules from large databases. In Proceedings of the 21th VLDB Conference, Zurich, Switzerland, 1995.
6. K. Ilgun, R. A. Kemmerer, and P. A. Porras. State transition analysis: A rule-based intrusion detection approach. IEEE Transactions on Software Engineering, 21(3):181-199, March 1995.

Dual Filtering Strategy for Chinese Term Extraction

Xiaoming Chen[1,2], Xuening Li[1], Yi Hu[1], and Ruzhan Lu[1]

[1] Dept. of computer science and engineering, Shanghai Jiao Tong Univ., Shanghai 200030
[2] School of computer science, Guizhou Univ, Guiyang 550025, P.R. China
{chen-xm, xuening_li, huyi516, rzlu}@sjtu.edu.cn

Abstract. Automatic term extraction (ATR) is an important problem in natural language processing. But most of extraction methods focus on the extraction of multiword units. Inevitably, many common words (or phrases) as terms are extracted at the same time. In this paper, we propose a hybrid method for automatic extraction of term from domain-specific un-annotated Chinese documents by means of linguistics knowledge and statistical techniques, taking dual filtering strategy and introducing a weight formula to filter term candidates. The results of the research indicate that our system is more efficient and precise than previous methods.

1 Introduction

Terms are known to be linguistic designation of defined concepts in a certain academic or technical domain. In Chinese, the majority of terms are compound words, which are complex in structure and not sufficient in the morpho-syntactic features.

Zhang(2001) analyzed the difference of terms and common words. In brief, we think term has two distinct characteristics: domain relevance and domain universality.

Domain relevance, i.e. domain speciality: Terms of certain domain cannot be used freely in other domains. That is to say, these terms seldom or never appear in other domains. However, two cases need to be taken into consideration.

These terms are professional jargons. However, they are progressively accepted because they are widely used in different media, and enter into the common domain to become common words, such as "股票(*stock*)", "克隆(*clone*)" etc.

These terms are professional jargons. However, they are professional in other domains. So they appear in cross-domains and assume different meanings. They are polysemic terms. A good example is "前锋(*front*)", which appears in game domain and military domain.

Domain universality: Terms of certain domain are current in their own domains. In other words, these terms appear in the major documents of domain instead of appearing in few documents.

Automatic term extraction (or Automatic term recognition, ATR) is an important problem in natural language processing. The goal is to extract domain specific terms from a corpus of a certain academic or technical domain. Applications of automatic term extraction include machine translation, automatic indexing, building lexical knowledge bases or domain ontology, document clustering or classification, and information retrieval.

Building the relation network of domain concepts is our ongoing research. This paper reports a part of our work. In this paper, we propose a novel, domain-independent method for automatic extraction of term from domain-specific unannotated Chinese documents. The focus of the paper will be on dual filtering strategy for our system. Our methodology relies both on linguistic and on statistical knowledge. We defined a weight formula for filtering term candidates with respect to the two characteristics of terms described above. The results of the research indicate that our system is more efficient and precise than previous methods.

The rest of this paper is organized as follows. In Section 2, we review previous studies on term extraction. In Section 3, we give an overview of our term extraction system DSTES. Dual filtering strategies are discussed in Section 4. And Section 5 presents the experimental results on financial corpus, followed by a conclusion and further work in Section 6.

2 Previous Work

In the past years, many papers on term extraction have been published. But most of them deal with foreign languages and only very few with Chinese.

In general, three main approaches have been proposed for term extraction from texts: linguistics-oriented, statistics-oriented and hybrid approaches [1]. On the one hand, linguistic techniques rely on the assumption that terms present specific morpho-syntactic structures or patterns (Bourigault, 1996). The basic strategy of these techniques is to detect and extract the strings whose structure matches some given patterns. Since these patterns are in most cases language-dependent, linguistic techniques demand specific language knowledge processing.

On the other hand, statistical approaches take into account that terms have different statistical features from common words to identify them (for example, the high association grade of multiword constituents). Exactly, in order to estimate the term candidates, we can use statistical models which analyze observed counts of linguistic information related to the candidates. Most of the statistical approaches focus on the extraction of multiword terms, mainly by means of calculating association measures (Chuck & Hanks, 1989; Smadja, 1993; Dias, 1999).

Moreover, some authors adopt hybrid approaches, combining linguistic and statistical techniques. Some of them apply syntactic filters after statistical processing, in order to extract the statistically significant word combinations that match some given morpho-syntactic patterns (Samdja, 1993). In other cases, statistical measures are calculated for a list of term candidates previously selected through linguistic techniques (Justeson, 1993).

In Chinese, only a few papers deal with the extraction of terms, especially domain specific terms. Zheng(2003) first used the statistic method to acquire the rules to combine the segmented characters which should be one word and the rules of semantic distribution. And then she adopted the strategy of co-occurrence, pattern matching, central matching to build the special lexicon on agricultural plant diseases and insect pests. Chen(2003) introduced an automatic learning algorithm based on bootstrapping to acquire field words. But the precision of field words is only 42.8% on financial corpus. Liu(2003) used two improved traditional parameters: mutual information and log-likelihood ratio to extract Chinese terms. Though the precision of the method is 75.4%, there no filtering measures being taken, as a result, many non-terms as terms (such as "记者(*pressman*)", "今天(*today*)", "有一天(*someday*)") are extracted at the same time.

Emphatically, statistics-oriented methods are in the mainstream. But most of these methods focus on the extraction of multiword units. Although some of them are the reputed methods of terms extraction, there are no filtering measures being taken. Inevitably, many common words (or phrases) as terms are extracted at the same time. These words or phrases are also called "term" by error.

Pure linguistics-oriented methods are difficult for Chinese because of some unique characteristics of Chinese terms, such as lack of syntactic information, heuristic information and no rules of word building. However, non-terms are of some rules of word building. So we have adopted a hybrid method by means of linguistics knowledge and statistical techniques, taking dual filtering strategy and introducing a weight formula to filter term candidates.

3 Overview of DSTES

The architecture of our system - DSTES can be illustrated with Fig. 1, and each component is to be explained as follows.

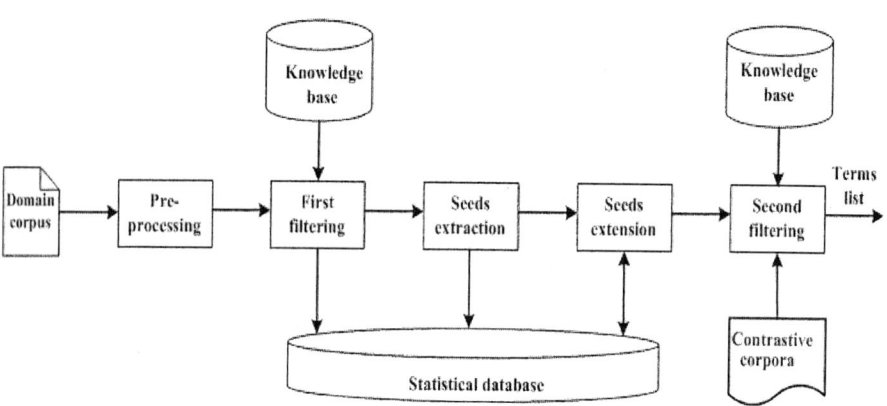

Fig. 1. DSTES architecture

3.1 Preprocessing

The primary function of this module is merging alone raw texts into one text containing document information, that is to say, distinguishing different paragraphs coming from different documents. This kind of documents is conveniently to be dealt with subsequent modules.

3.2 Bi-character Seed Extraction

As illustrated with Fig.1, the preprocessed domain corpus is to be filtered (see Sect. 4), and then to extract the bi-character seeds.

The primary function of this module focuses on counting the frequency of words with single or double Chinese characters and then constructs the statistic database initially. Simultaneously, the corpus going through the first filtering is utilized to extract bi-character seeds conforming to statistic standard.

Two statistic parameters are used in the calculating process: mutual-information (*mi*) and log-likelihood (*logL*). The abilities of extracting words (bi-character) that employ nine statistic models including *mi* and *logL* are compared by Luo(2003). They draw a conclusion that *mi* is the most powerful in extracting words and there is no well inter-complement among each statistic variables.

Though *mi* scales the association degree of each component in a word, it has an obvious disadvantage that does not take the word frequency into account. Thus, even the low occurring frequency of *xy* in corpus, mi (x,y) is high yet if *x* and *y* have the low frequency also. So, it cannot deal with the low frequency and noise just depending on mutual information.

Aiming at this problem, Patrick(2001) integrate another statistical variable having better ability in the case of low frequency, Log-likelihood(*logL*), and propose a new statistic model S(x,y).

To words x and y,

$$S(x, y) = \begin{cases} \log L(x, y) & \text{if } mi(x, y) \geq \text{minMutInfo} \\ 0 & \text{otherwise} \end{cases} \quad (3\text{-}1)$$

This new method is employed in our system.

In terms of *xy* in statistical database (DB), if the following conditions are satisfied, then the *xy* will be stored in the bi-character seed list (SeedList).

$$\begin{aligned} &C(x,y) > minCount \\ &S(x,y) > minLogL \end{aligned} \quad (3\text{-}2)$$

Where, *minCount*、*minLogL* are defined thresholds and $C(x, y)$ is the frequency of *xy* in corpus.

3.3 Bi-character Seed Extension

By the operation of bi-character seed extraction, we acquired the bi-character seeds list (SeedList) from the processed corpus. The function of this module is expanding the statistical database DB further to provide gist for subsequent decision. To every seed *xy* in the seed list, it is to be expanded recursively until K+2-character words or meeting filtering symbols and to acquire multi-character term candidates.

By the processing of bi-character expanded, we acquire the multi-character term candidates list TermCanList. But there exit many non-terms in this list and they will greatly affect the whole performance of the system if they cannot be removed.

4 Dual Filtering Strategy

In order to improve the efficiency of system and the accuracy of term extraction, we integrate linguistic knowledge and statistical techniques. We respectively add filtering module before and after statistical calculation. Compared with previous method, our system has greatly improved performance.

4.1 First Filtering

Our knowledge base used in the first filtering is composed of two parts: filtering words list and pattern base.

The filtering words list can also be divided into two parts: symbol and vocabulary. Symbols in our work include punctuation and all kinds of special symbols. The construction of words in filtering words list is complex. All words in Chinese contain content words and function words. Content words are open and function words are closed. Functional words in Chinese are grammatical, some of which express the logical concepts and assume spurious meanings. Most of them cannot constitute the terms so they are listed into the filtering vocabulary. When listing words, we are prudent to pay attention to exceptions too. At the same time, we list some content words which cannot constitute terms, such as pronouns "我们(we)", "这些(these)" and temporal words "今天(today)", "现在(now)" etc.

Besides the filtering word list, considering the unique characteristics of Chinese terms we also design some patterns for filtering, such as temporal pattern and quantifier pattern and so on. For example, temporal pattern,

$$Word^* Num^+ TimeWord^*$$

Here, $Num = \{$"0","三(three)", "半(half)", etc$\}$, which indicates the quantity.

$Time = \{$"年(year)", "月(month)", "日(day)"$\}$, which indicates time.

$Word$ is any word with single character.

The primary function of this module is replacing the character strings which match the filtering words list or pattern base in preprocessed corpus with the filtering symbols self-defined by system and the output is the document tagged with these filtering symbols.

After processing of the first filtering, the original document are divided into character string parts with filtering symbols and the filtered symbol will not be calculated in counting, extracting and expanding. They are only regarded as the halt conditions, which greatly improved the efficiency of system. The results show that the performing time is shortened to half.

4.2 Second Filtering

As mentioned above, the multi-character term candidates list-TermCanList contains a lot of non-terms that appear frequently also, such as "记者(journalist)","国务院(State

Department)", etc. So it needs the second filtering to improve the accuracy of extracting terms. The concrete function of this module can be illustrated as Fig 2.

In this module, we also conform to the rule of matching pattern firstly in order to remove those non-terms with obvious characteristics.

In Section 1, we mention that terms have two obvious characteristics compared to common words, i.e. domain relevance and domain universality. How to scale these two characteristics is the key in term filtering.

We consider the domain relevance firstly. Through experiments, we find that there also exit many non-terms appearing frequently in extracted candidates list, such as "记者(*journalist*)" and "科技(*science and technology*)". So it cannot scale the relevance between a term and certain domain just depending on word frequency alone. As proposed in [10], the relevance between a term candidate and certain domain can be analyzed by comparing different domain corpus.

Input: term candidate list -TermCanList, DB, Corpus_f, contrastive corpus ParaCorpus and pattern base PDB
Output: terms list TermList
Step 1: remove all candidates satisfying conditions from TermCanList according to matching pattern.
Step 2: calculate the weight of every candidate by the following equation.

$$w_i = \alpha Tr_i + \beta Tc_i$$

If $w_i < countThresh$, then remove it from TermCanList
Step 3: copy the left terms in TermCanList to TermList, which are to be output as final result

---

Fig. 2. Illustration of the second filtering module

We have the following equations:

$$Tr_{i,k} = \frac{P(t \mid D_k)}{\sum_{j=1}^{N} P(t \mid D_j)} \quad (4\text{-}1)$$

$$E(P(t \mid D_k)) = \frac{f_{t,k}}{\sum_{t \in D_k} f_{t',k}} \quad (4\text{-}2)$$

$Tr_{i,k}$ represents the relevance between *i*-th candidate and the certain domain D_k, $P(t \mid D_k)$ the probability of term candidate *t* appearing in the domain, and $P(t \mid D_j)$ the probability of *t* appearing in all kinds of corpus (contrastive and domain corpus). N is the total number of documents and $f_{t,k}$ is the appearing frequency of *t* in domain D_k.

Then we see the domain universality. This characteristic of terms indicates that terms should have its "currency" in its domain. When reflected in domain, it requires that the terms should distribute evenly in domain rather than located in few documents. *m*-degree frequency [12] adequately considers the effect of distribution and integrates the appearing frequency of words and their distribution. So we use it to scale the universality of term candidate in certain domain.

$$Tc_i = \sum_{j=1}^{n} \sqrt[m]{f_{i,j}} \Big/ n^{m-1} \qquad (m \geq 1) \qquad (4\text{-}3)$$

Tc_i represents the universality of the *i–th* term candidate in certain domain, $f_{i,j}$ the appearing frequency of the *i-th* term candidate in document *j*, and *n* is the number of domain corpus.

Experiments show that the higher the degree is, the more the distribution affects Tc_i. In other words, when the distribution is changed, the bigger *m* is, the faster Tc_i shrinks. The value of *m* can be set by need. If the corpus is large, then the distribution affect more and we can use the high degree frequency. If there is no special need, the value of *m* is usually set by 2.

Through the reduplicate experiments, we use the linear combination of (4-1) and (4-3) to decide whether the term candidate *i-th* should be removed.

$$w_i = \alpha Tr_i + \beta Tc_i \qquad \alpha, \beta \in (0,1) \qquad (4\text{-}4)$$

α and β are two thresholds obtained by experiments. The experiment shows that accuracy of terms extracting by calculating every candidate according to (4-4) has been greatly improved when compared with the previous system.

5 Experiment

The test corpus for this system comes from financial, football and transportation domains. The result of the experiment was based on the corpus of financial domain. The financial corpus comes mainly from the news of financial webs, such as Sohu financial, Sina financial, and Chinese financial. From the corpus, we selected randomly 1756 financial articles (including 2, 154, 497 Chinese characters) for a test.

As mentioned before, many words are MWU in the final test including "有一天 (*someday*)"、"上海(*Shanghai*)" and so on. Although they are content phrases, they are not terms. In the previous systems, the distinction has not made between terms and MWU when counting precision and recall, since any content word will be regarded as the right result although the word extracted is in fact a MWU instead of a term.

Therefore, we will give two types of precision in the introduction of the results of our system. One is PM for MWU for the contrast with the performance of other systems. The other is PT for terms.

$$PM = C(M)/C(*) \qquad (5\text{-}1)$$

$$PT = C(T)/C(T_c) \qquad (5\text{-}2)$$

$C(M), C(T_C), C(T), C(*)$ are the number of extracted MWU, candidate terms, terms and total words extracted.

In the process of the test, the system has extracted 6753 term candidates including 5785 MWUs and 971 nonwords. In MWU, 878 words or phrases are the proper names of companies, persons and places. Therefore, PM=85.64%, a great progress compared to the average percentage 74.5% of the literature [7].

After filtering, we output 2305 financial terms from 6753 term candidates. Among 2305, 1381 words or phrases are real financial terms. That is to say, PT=59.91%. It is a much progress compared to 42.8% of the literature [3].

6 Conclusion and Further Work

We adopted dual filtering strategy for Chinese term extraction by means of linguistic knowledge and statistical techniques. Furthermore, we designed some patterns of non-terms for filtering considering the unique characters of Chinese terms. Compared with previous method, our system has greatly improved performance. The results show that the performing time is shortened to half and the precision of term extraction takes a much progress.

However, as mentioned before, many non-terms are extracted by the statistic approach because only the frequency of co-occurrence of word clusters is taken into consideration. Although we have adopted the dual filtering strategy in our system to improve the performance, it cannot separate the terms and non-terms completely.

Therefore, our latest research aims at the inquiry into the intensional and conceptual motivations for compound structure such as terms as the necessary condition for the identification. A wide application will lie in the conceptual alignment of multi-lingual (to be published).

Acknowledgement. This work is supported by NSFC major research program 60496326: Basic Theory and Core Techniques of Non-Canonical Knowledge and by National 863 Project (No. 2001AA114210-11).

References

1. Alegria I., Arregi O., Balza I. (2004). Linguistic and Statistical Approaches to Basque Term Extraction. http://ixa.is.ehu.es
2. Bourigault, D.(1996). Lexter, a Natural Language Processing Tool for Terminology Extraction. In Proceedings of 7th EURALEX International Congress.
3. Chen wenliang, Zhu jingbo, Yao tianshun.(2003).Automatic Learning Field Words by Bootstrapping. Language Computing and Content-based Text Processing. P.67-72
4. Church, K.W. & Hanks, P.P.(1989). Word association norms, mutual information and lexicography. In Proceedings of the 27th Annual Meeting of the ACL. p.:76-83
5. Dias, G.& Guillore, S.& Lopes, J.G.P.(1999). Mutual Expectation: A Measure for Multi-word Lexical Unit Extraction. In Proceedings of VEXTAL Venezia per il Trattamento Automatico delle Lingue

6. Justeson, J.S. & Katz, S.M.(1993). Technical Terminology: Some Linguistic Properties and an Algorithm for Identification in Text. In Natural Language Engineering 1(1): 9-27
7. Liu Jianzhou, He tingting, Ji donghong.(2003). Extracting Chinese Term Based on Open Corpus. Advances in Computation of Oriental Languages. P.43-49
8. Luo shengfen, Sun maosong.(2003). Chinese Word Extraction Based on the Internal Associative Strength of Character Strings. Journal of Chinese Information Processing. 2003(3):P.9-14
9. Patrick Pantel & Dekang Lin.(2001). A Statistical Corpus-Based Term Extractor. Canadian Conference on AI 2001. p. 36- 46.
10. R. Navigli & P. Velardi.(2002). Semantic Interpretation of Terminological Strings. In Proceedings of 4th Conference. Terminology and Knowledge Engineering(TKE 2002). p.325-353
11. Smadja, F.(1993). Retrieving Collocations from Text: XTRACT. Computational Linguistics, 19(1): 143 -177
12. Yin binyong, Fang shizeng.(1994). Word Frequency Counting: A new concept and a new approach. Applied Linguistics 1994(2):P.69-75
13. Zhang pu.(2001). The Application of Circulation to Recognizing Terms in the Field of IT. Proceedings of Conference of the 20th Anniversary of CIPSC. P111-120
14. Zheng Jiaheng, Du yongping, Song lepeng(2003). The Research on Lexical Acquisition of Agricultural Plant Diseases and Insect Pests. Language Computing and Content-based Text Processing. P.61-66

White Blood Cell Segmentation and Classification in Microscopic Bone Marrow Images

Nipon Theera-Umpon

Department of Electrical Engineering, Faculty of Engineering,
Chiang Mai University, Chiang Mai 50200 Thailand
nipon@ieee.org

Abstract. An automatic segmentation technique for microscopic bone marrow white blood cell images is proposed in this paper. The segmentation technique segments each cell image into three regions, i.e., nucleus, cytoplasm, and background. We evaluate the segmentation performance of the proposed technique by comparing its results with the cell images manually segmented by an expert. The probability of error in image segmentation is utilized as an evaluation measure in the comparison. From the experiments, we achieve good segmentation performances in the entire cell and nucleus segmentation. The six-class cell classification problem is also investigated by using the automatic segmented images. We extract four features from the segmented images including the cell area, the peak location of pattern spectrum, the first and second granulometric moments of nucleus. Even though the boundaries between cell classes are not well-defined and there are classification variations among experts, we achieve a promising classification performance using neural networks with five-fold cross validation.

1 Introduction

The differential counts, the counts of different types of white blood cells, provide invaluable information to doctors in diagnosis of several diseases. The traditional method for an expert to achieve the differential counting is very tedious and time-consuming. Therefore, an automatic differential counting system is preferred. White blood cells are classified according to their maturation stages. Even though, the maturation is a continuous variable, white blood cells are classified into discrete classes. Because the boundaries between classes are not well-defined, there are variations of counts among different experts or within an expert himself. In the myelocytic series (or granulocytic series), they can be classified into six classes, i.e., myeloblast, promyelocyte, myelocyte, metamyelocyte, band, and polymorphonuclear (PMN) ordered from the youngest to the oldest cells [1–2]. Samples of all six classes of white blood cells in the myelocytic series are shown in Figure 1. As we can see from the figure, many characteristics of cells change during their maturation.

Most of the previous proposed methods followed the traditional manual procedures performed by an expert, i.e., locating a cell, extracting its features, classifying the cell, and then updating the count [3–8]. It should be noted that most of them were applied to peripheral blood only. The differential counting problem in bone marrow

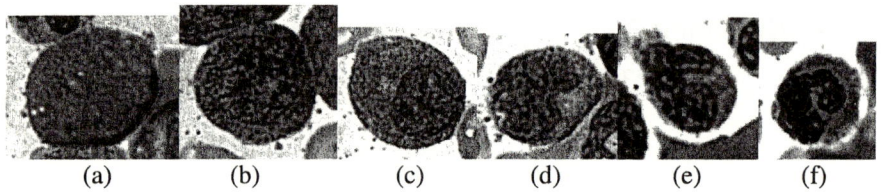

Fig. 1. Examples of cells in the myelocytic or granulocytic series: (a) Myeloblast, (b) Promyelocyte, (c) Myelocyte, (d) Metamyelocyte, (e) Band, and (f) PMN.

is much more difficult due to the high density of cells. Moreover, the immature white blood cells are normally seen only in the bone marrow [2]. There are many types of bone marrow white blood cells that may not be found in the blood. Therefore, the differential counts in peripheral blood may not be enough for doctors to diagnose some certain diseases. Our previous works were all applied to the problem in bone marrow, but were based on an assumption that the manually-segmented images were available [9–13].

To be more specific, we developed the mixing theories of the mathematical morphology and applied them to the bone marrow white blood cell differential counting problem [9–10]. We also developed a new training algorithm for neural networks in order to count numbers of different cell classes, without classification [11,12]. There are several other researches on cell segmentation in literature. Some examples of common techniques used in cell segmentation are thresholding [14,15], cell modeling [15–17], filtering and mathematical morphology [18], watershed clustering [6,17], fuzzy sets [19], etc. It should be noted that only the segmentation techniques performed in [5], [6], and [19] are applied to bone marrow. The other mentioned segmentation techniques are applied to peripheral blood. It should also be noted that most of the researches are emphasized on either segmentation or classification only. There are just a few of them that perform on both segmentation and classification.

In this paper, we propose a technique to segment nucleus and cytoplasm of bone marrow white blood cells. We generate patches in cell images by applying the fuzzy C-means (FCM) algorithm to overly segment cells. The patches in each oversegmented image are then combined to form three segments, i.e., nucleus, cytoplasm, and background. The segmentation errors are evaluated by comparing the automatic segmented images to the corresponding images segmented by an expert using the probability of error in image segmentation. We also apply the outputs of the automatic segmentation technique to the cell classification problem using neural networks with the five-fold cross validation. Four features are extracted from each automatic segmented image based on the area of cell, and shape and size of its nucleus.

This paper is organized as follows. The fuzzy C-means clustering, morphological operations, and morphological granulometries are briefly introduced in the next section. The bone marrow white blood cell data set is described in section 3. Section 4 shows the explanation of the proposed segmentation technique and feature extraction. The experimental results are shown and discussed in section 5. The conclusion is drawn in the final section.

2 Methodology

In this research, we apply the fuzzy C-means (FCM) algorithm and the mathematical morphology to segment white blood cells. The FCM algorithm is applied to overly segment each cell image to form patches. Cell and nucleus smoothing and small patch removal are done by using the binary morphological operations. Morphological granulometies are also applied to extract shape and size of an object.

2.1 Fuzzy C-Means Algorithm

Fuzzy C-means clustering method is a well-known fuzzy clustering technique. It is widely available in literature [e.g., 20,21]. We will briefly introduce it here. Consider a set of data $\mathbf{X} = \{\mathbf{x}_1, \mathbf{x}_2, ..., \mathbf{x}_n\}$, where $\mathbf{x}_k$ is a vector. The goal is to partition the data into c clusters. Assuming that we have a fuzzy pseudopartition $P = \{A_1, A_2, ..., A_c\}$, where A_i contains membership grades of all $\mathbf{x}_k$ to cluster i. The centers of the c clusters can be calculated by

$$\mathbf{v}_i = \frac{\sum_{k=1}^{n}[A_i(\mathbf{x}_k)]^m \mathbf{x}_k}{\sum_{k=1}^{n}[A_i(\mathbf{x}_k)]^m}, \quad i = 1, 2, ..., c, \tag{1}$$

where $m > 1$ is a real number that controls the effect of membership grade. The performance index of a fuzzy pseudopartition P is defined by

$$J_m(P) = \sum_{k=1}^{n}\sum_{i=1}^{c}[A_i(\mathbf{x}_k)]^m \|\mathbf{x}_k - \mathbf{v}_i\|^2, \tag{2}$$

where $\|\bullet\|$ is some inner product-induced norm. The clustering goal is to find a fuzzy pseudopartition P that minimizes the performance index $J_m(P)$. The solution to this optimization problem was given by Bezdek in [21] and is now widely available in several textbooks.

2.2 Mathematical Morphology

Mathematical morphology was first introduced by Matheron in the context of random sets [22,23]. Morphological methods are used in many ways in image processing, for example, enhancement, segmentation, restoration, edge detection, texture analysis, shape analysis, etc. [24,25]. Morphological operations are nonlinear, translation invariant transformations. Because we consider only binary images in this research, we only describe binary morphological operations. The basic morphological operations involving an image S and a structuring element E are

$$\text{erosion: } S \ominus E = \cap \{S - e : e \in E\}, \tag{3}$$

$$\text{dilation: } S \oplus E = \cup \{E + s : s \in S\}, \tag{4}$$

where ∩ and ∪ denote the set intersection and union, respectively. $A + x$ denotes the translation of a set A by a point x. The closing and opening operations, derived from the erosion and dilation, are defined by

$$closing: S \bullet E = (S \oplus (-E)) \ominus (-E), \qquad (5)$$

$$opening: S \bigcirc E = (S \ominus E) \oplus E, \qquad (6)$$

where $-E = \{-e: e \in E\}$ denotes the 180° rotation of E about the origin.

We successively apply the opening operation to an image and increase the size of structuring element in order to diminish the image. Let $\Omega(t)$ be area of $S \bigcirc tE$ where t is a real number and $\Omega(0)$ is area of S. $\Omega(t)$ is called a size distribution. The normalized size distribution $\Phi(t) = 1 - \Omega(t)/\Omega(0)$, and $d\Phi(t)/dt$ are called granulometric size distribution or pattern spectrum of S. The moments of the pattern spectrum are called granulometric moments.

3 White Blood Cell Data Set

In the experiments we use grayscale bone marrow images collected at the University of Missouri Ellis-Fischel Cancer Center. Each white blood cell image was cropped manually to form a single-cell image. Then, each single-cell image was segmented manually into nucleus, cytoplasm, and background regions. The images were classified by Dr. C. William Caldwell, Professor of Pathology and Director of the Pathology Labs at the Ellis-Fischel Cancer Center. The data set consists of six classes of white blood cells – myeloblast, promyelocyte, myelocyte, metamyelocyte, band, and PMN – from the myelocytic series. After eliminating the images that do not contain the entire cells, we end up with 20, 9, 116, 31, 38, and 162 manually-segmented images for all six cell classes, respectively. Each manually-segmented image is composed of three regions – nucleus, cytoplasm, and background – with gray levels of 0, 176, and 255, respectively. The manually-segmented images corresponding to the cells shown in Figure 1 are shown in Figure 2.

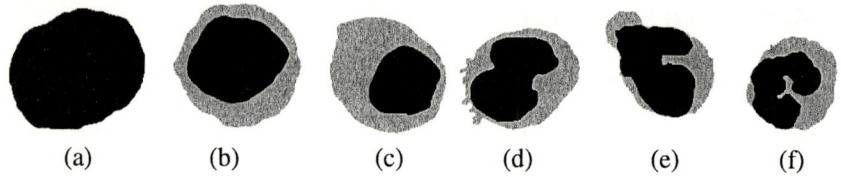

Fig. 2. Corresponding manually-segmented images of cells shown in Figure 1: (a) Myeloblast, (b) Promyelocyte, (c) Myelocyte, (d) Metamyelocyte, (e) Band, and (f) PMN.

4 Proposed Techniques

We propose a white blood cell segmentation technique that segments an image into three regions, i.e., nucleus, cytoplasm, and background. In this research, everything in an image except the cell of interest is considered background. We also introduce the features extracted from each segmented cell image.

4.1 White Blood Cell Segmentation Technique

In this research we apply a 15×15 median filter to each cell image to ease the problem of intensity inconsistency in each region of a cell. This is a big problem, particularly in this data set, because the images are grayscale. The filtered images are then overly segmented using the FCM clustering. We heuristically set the parameter m to 2 and the number of clusters c to 10. Each patch is formed by connected pixels that belong to the same cluster. After overly segmentation, the patches in the oversegmented images are combined to form images with three segments – nucleus, cytoplasm, and background. The patch combining is achieved by considering the FCM centers. If the center of the patch is less than 60% of the mean of all centers (dark), then the patch is labeled as nucleus. If the center of the patch is less than 150% of the mean of FCM centers but greater than 60% of that (somewhat dark), then the patch is labeled as cytoplasm. Otherwise (bright), it is labeled as background. It should be noted that the list of the FCM centers is dynamic. If a patch is considered nucleus or cytoplasm but it touches the image border, then it will be labeled as background (this patch belongs to another cell) and the corresponding FCM center will be discarded from the list. This helps in the segmentation in which the cell of interest is brighter than the surrounding cells. The morphological operations, i.e., opening following by closing, both with a disk structuring element with the diameter of five pixels, are applied in the final step to remove the small patches and smooth the edges. The algorithm of the proposed technique is summarized as follows:

Apply median filter to input image
Apply FCM algorithm to the filtered image
Sort FCM centers in ascending order
For each patch corresponding to sorted FCM centers (from dark to bright)
    If (FCM center of patch) < (60% of mean of centers), then label patch as nucleus
    If (60% of mean of centers) < (FCM center of patch) < (150% of mean of centers), then label patch as cytoplasm
    If patch is labeled as nucleus or cytoplasm but it touches image border, then label patch as background and discard the FCM center from the list
End (For each patch)
Apply opening following by closing to nucleus region
Apply opening following by closing to cytoplasm region
Combine nucleus and cytoplasm regions

4.2 Features of Segmented Cell Images

After segmenting each cell image, four features are extracted from each segmented image to form a feature vector for a classifier. As we know that the cell size becomes smaller and the size and shape of its nucleus changes when it becomes more mature, we extract the features accordingly. One feature is extracted from the entire cell segmentation, i.e., entire cell area. Three remaining features are extracted from the pattern spectrum of the nucleus of each cell, i.e., pattern spectrum peak location, first granulometric moment, and second granulometric moment. These last three features possess the size and shape information of the cell's nucleus. In the experiments, we use a small disk with the diameter of four pixels as the structuring element.

5 Experimental Results

Figure 3 shows examples of the outputs at each stage of our proposed cell segmentation technique. The original grayscale image, the corresponding oversegmented, and final automatic segmented images are depicted in Figure 3(a)-(c), respectively. We also show examples of automatic segmentation results along with original grayscale and expert's manually-segmented images of all cell classes in Figure 4. By visualization, we achieve good overall segmentation results. In some cases, our results differ from the expert's manually-segmented images but they are acceptable. For example, the output of the promyelocyte shown in Figure 4, it is hard to define the real boundary of the nucleus. To numerically evaluate the segmentation technique, we use the probability of error (PE) in image segmentation defined as

$$PE = P(O)P(B|O) + P(B)P(O|B), \qquad (7)$$

where $P(O)$ and $P(B)$ are *a priori* probabilities of objects and background in images, $P(B|O)$ is the probability of error in classifying objects as background, and $P(O|B)$ is the probability of error in classifying background as objects [26],[27]. This is basically the degree of disagreement between the algorithm and an expert. In the experiment, we compute the PE in segmentation of each segmented image compared to the corresponding expert's manually-segmented image. We consider two objects of interest, i.e., nucleus and entire cell (nucleus+cytoplasm.) The overall segmentation error is calculated by averaging those of all 376 cell images. From the experiment, we achieve the overall PE in segmentation of 9.62% and 8.82% for nucleus and cell. segmentation, respectively. To evaluate the segmentation performance in each cell class, we calculate the class-wise PE in segmentation by averaging the errors in each class. The segmentation error for nucleus and entire cell segmentation in each class are shown in Tables 1 and 2, respectively.

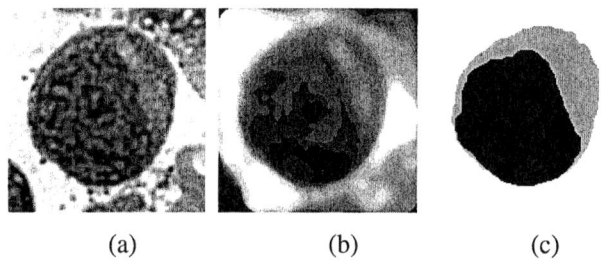

(a)   (b)   (c)

Fig. 3. Examples of (a) grayscale image of a myelocyte, (b) corresponding oversegmented images, and (c) corresponding algorithm's segmented images

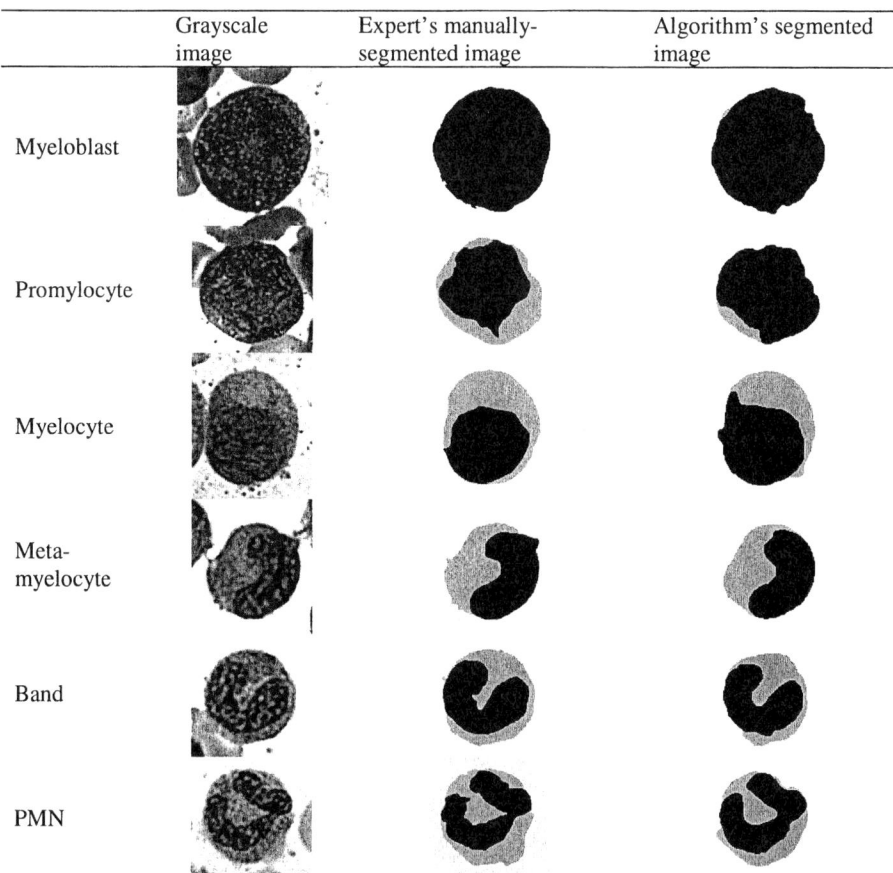

Fig. 4. Examples of grayscale, corresponding manually-segmented, and automatic segmented images of six classes of bone marrow white blood cells

Table 1. Class-wise probability of error in nucleus segmentation (%)

Cell class	Myeloblast	Promyelocyte	Myelocyte	Metamyelocyte	Band	PMN
PE	10.01	16.75	13.89	9.26	7.60	6.69

Table 2. Class-wise probability of error in entire cell segmentation (%)

Cell class	Myeloblast	Promyelocyte	Myelocyte	Metamyelocyte	Band	PMN
PE	6.88	8.77	8.38	8.90	9.15	9.29

From Table 1, the PE in nucleus segmentation is smaller for a more mature class. This is because the nucleus boundary of a more mature cell is better defined than that of a younger cell. The intensity contrast between nucleus and cytoplasm is higher when a cell becomes more mature. The PE in the entire cell segmentation are similar among all six classes. However, the overall PE in the entire cell segmentation is smaller than that in the nucleus segmentation. This is because we try to discriminate the entire cells from the background. The similarity between a cell region and background is less than that between nucleus and cytoplasm. It should be noted that, in this case, background means everything except the cell of interest. Hence, parts of other cells and red cells can also cause the nucleus and entire cell segmentation errors.

The good segmentation performance is not yet our final goal. We further apply the automatic segmentation results to the automatic cell classification. To justify the use of the derived automatic segmented images, we classify the cells using one of the most popular classifiers, i.e., neural networks. The neural networks used in the experiments consist of one hidden layers with ten hidden neurons. Because the cell data set is not divided into the training and test set, we perform the five-fold cross validation. We calculate the pattern spectrum of the nucleus of each segmented image. Four features, i.e., cell area, pattern spectrum's peak location, first and second granulometric moments, as described in section 4.2, are extracted. The classification rates achieved by using the automatic segmented images are 70.74% and 65.69% on the training and test sets, respectively. While the classification rates achieved by using the manually-segmented images are 71.81% and 69.68% on the training and test sets, respectively. We can see that the classification rates achieved by using the automatic segmented images are close to that achieved by using the images segmented manually by the expert. These results show the promising classification performance based on the results of the automatic cell segmentation.

6 Conclusion

We develop an automatic segmentation technique for microscopic bone marrow white blood cell images which is an important step in an automatic white blood cell differential counting. Each cell image is segmented into three regions, i.e., nucleus, cytoplasm, and background. The proposed segmentation technique is evaluated by comparing the results with the manually segmented images performed by an expert using the probability of error (PE) in image segmentation. We consider the entire cell and its nucleus as the objects of interest. From the experiments, we achieve good segmentation performances of less than 10% PE in the entire cell and nucleus segmentation. We further investigate the application of the automatic segmented images to the classification problem. Neural networks are chosen to be our classifier with four features extracted from the segmented images including the cell area, the peak location of pattern spectrum, the first and second granulometric moments of nucleus. The promising performance is achieved for this six-class classification problem with highly overlapping of cell from the adjacent classes because the boundaries are weak defined. One possible improvement is the acquisition of color microscopic images which will ease the segmentation problem very much, and, therefore, ease the classification problem.

Acknowledgments

This work is supported by the Ministry of University Affairs and the Thailand Research Fund under Contract MRG4680150. The author would like to thank Dr. C. William Caldwell of Ellis-Fishel Cancer Center, University of Missouri, for providing the data and the ground truth. We acknowledge the contribution of Dr. James Keller and Dr. Paul Gader through many technical discussions on this research.

References

1. Diggs L.W., Sturm D., and Bell A.: The Morphology of Human Blood Cells, Abbott Laboratories, Abbott Park (1985)
2. Minnich V.: Immature Cells in the Granulocytic, Monocytic, and Lymphocytic Series, American Society of Clinical Pathologists Press, Chicago (1982)
3. Beksaç M., Beksaç M.S., Tipi V.B., Duru H.A., Karakas M.U., Çakar A.N.: An Artificial Intelligent Diagnostic System on Differential Recognition of Hematopoietic Cells From Microscopic Images. In: Cytometry, Vol. 30. (1997) 145–150
4. Harms H., Aus H., Haucke M., Gunzer U.: Segmentation of Stained Blood Cell Images Measured at High Scanning Density With High Magnification and High Numerical Aperture Optics. In: Cytometry. Vol. 7. (1986) 522–531
5. Park J., Keller J.: Fuzzy Patch Label Relaxation in Bone Marrow Cell Segmentation. In: IEEE Intl Conf on Syst, Man,Cybern. (1997) 1133–1138
6. Park J., Keller J.: Snakes on the Watershed. In: IEEE Trans Pattern Anal Mach Intell. Vol. 23. No. 10. (2001) 1201–1205
7. Poon S.S.S., Ward R.K., Palcic B.: Automated Image Detection and Segmentation in Blood Smears. In: Cytometry. Vol. 13 (1992) 766–774
8. Sohn S.: Bone Marrow White Blood Cell Classification, Master's Project, University of Missouri-Columbia, (1999)
9. Theera-Umpon N., Gader P.D.: Counting White Blood Cells Using Morphological Granulometries. In: Journal of Electronic Imaging. Vol. 9. No. 2. (2000) 170–177
10. Theera-Umpon N., Dougherty E.R., Gader P.D.: Non-Homothetic Granulometric Mixing Theory with Application to Blood Cell Counting. In: Pattern Recognition. Vol. 34. No. 12. (2001) 2547–2560
11. Theera-Umpon N., Gader P.D.: Training Neural Networks to Count White Blood Cells via a Minimum Counting Error Objective Function. In: Proc 15th Intl Conf on Pattern Recog, (2000) 299–302
12. Theera-Umpon N., Gader P.D.: System Level Training of Neural Networks for Counting White Blood Cells. In: IEEE Trans Systems, Man, and Cybern Part C: App and Reviews. Vol. 32. No. 1. (2002) 48–53
13. Theera-Umpon N.: Automatic White Blood Cell Classification using Biased-Output Neural Networks with Morphological Features. In: Thammasat Intl Journal of Sci and Tech. Vol. 8. No. 1. (2003) 64–71
14. Cseke I.: A Fast Segmentation Scheme for White Blood Cell Images. In: Proc 11th IAPR Intl Conf on Image, Speech and Signal Analysis. (1992) 530–533
15. Liao Q., Deng Y.: An Accurate Segmentation Method for White Blood Cell Images. In: IEEE Intl Sym on Biomedical Imaging. (2002) 245–248

16. Nilsson B., Heyden A.: Model-Based Segmentation of Leukocytes Clusters. In: Proc 16th Intl Conf on Pattern Recognition (2002) 727–730
17. Jiang K., Liao Q., Dai S.: A Novel White Blood Cell Segmentation Scheme Using Scale-Space Filtering and Watershed Clustering. In: Proc 2nd Intl Conf on Machine Learning and Cybern. (2003) 2820–2825
18. Anoraganingrum D.: Cell Segmentation with Median Filter and Mathematical Morphology Operation. In: Proc Intl Conf on Image Anal and Proc. (1999) 1043–1046
19. Sobrevilla P., Montseny E., Keller J.: White Blood Cell Detection in Bone Marrow Images. In: Proc 18th Intl Conf of the North American Fuzzy Info Proc Soc (NAFIPS). (1999) 403–407
20. Klir G.J., Yuan B.: Fuzzy Sets and Fuzzy Logic: Theory and Applications. Prentice Hall, New Jersey (1995)
21. Bezdek J.C.: Pattern Recognition with Fuzzy Objective Function Algorithms. Plenum Press, New York (1981)
22. Matheron G.: Random Sets and Integral Geometry. Wiley, New York (1975)
23. Serra J.: Image Analysis and Mathematical Morphology. Academic Press, New York (1983)
24. Dougherty E.R.: An Introduction to Morphological Image Processing. SPIE Press, Bellingham, Washington (1992)
25. Dougherty E.R.: Random Processes for Image and Signal Processing. SPIE Press, Bellingham, Washington, and IEEE Press, New York (1999)
26. Lee S.U., Chung S.Y., Park R.H.: A Comparative Performance Study of Several Global Thresholding Techniques for Segmentation. In: Computer Vision, Graphics, and Image Processing. Vol. 52. No. 2. (1990) 171–190
27. Zhang X.-W., Song J.-Q., Lyu M.R., Cai S.-J.: Extraction of Karyocytes and Their Components from Microscopic Bone Marrow Images Based on Regional Color Features. In: Pattern Recognition. Vol. 37. No. 2. (2004) 351–361

KNN Based Evolutionary Techniques for Updating Query Cost Models

Zhining Liao[1], Hui Wang[1], David Glass[1], and Gongde Guo[2]

[1] School of Computing and Mathematics, University of Ulster,
BT37 0QB, Northern Ireland, UK
{Z.Liao, H.Wang, Dh.Glass}@ulster.ac.uk
[2] Department of Computer Science, Fujian Normal University, 350007, China
G.Guo@bradford.ac.uk

Abstract. Data integration system usually runs on unpredictable and volatile environments. Query cost model should be update with the changes of the environment. In this paper, we tackle this problem by evolving the cost model so that it can adapt to the environment change and keep up-to-date. Firstly, the factors causing the system environment to change are analyzed and different methods are proposed to deal with these changes. Then an architecture for evolving a cost model in dynamic environment is proposed. Our experimental results show the architecture of evolving a cost model in dynamic environment can well capture changes of environment and keep cost models up-to-date.

1 Introduction

The key challenges arise in the query optimization in a data integration system due to the dynamics and unpredictability of the workloads of both the network and the autonomy of remote data sources. Therefore, some methods of deriving cost models for autonomous data sources at a global level are extremely important in order to process queries accurately [6]. The methods discussed in [1, 3, 8] assume that the system environment does not change significantly over time. In [11], the effects of the workload of a server on the cost of a query are investigated and a method to decide the contention states of a server is developed. In [4, 10], the importance of coping with the dynamic network environment is addressed but not to consider the complexity of queries. In [7], we combined two factors (network congestion situation, server contention states) together as system contention states and construct a set of cost formulae by using a multiple regression model [2]. The rest of paper is organized as follows. Section 2 analyses the factors that can affect the environment. Three approaches to evolving query cost model are presented in section 3. In section 4, the architecture of evolving a cost model in a dynamic data integration system environment is shown. The experimental results are presented in section 5. Finally some conclusions are drawn in section 6.

2 Factors Affecting the Environment

There are two sets of parameters in our cost models. The first one is parameters $[X_1, X_2 ... X_p]$, which are p explanatory variables. The other set is regression coefficients

$[B_0, B_1, B_2... B_p]$. In our cost models, the explanatory variables are about the data states in the database and query result. We call them *data parameters* in this paper. The regression coefficients are calculated from explanatory variables and query cost Y. Different system situation results in different regression coefficients. So these coefficients will be called *system parameters* in the rest of this paper. Factors change that affect the accurately estimating of the cost of queries will be revealed from relevant parameters in the cost model. So, the factors can be classified into the following two types based on their affect to the parameters in the cost formulae.

1. *Factors affecting the data parameters:* This type of factor contains data volume of data sources, such as the number of tuples in a table, the number of tuples changing in the result of queries.
2. *Factors affecting the system parameters:* This type of factor includes server workload and network speed, such as configuration parameters in local database management system (DBMS), physical distribution of data on a hard disk. Obviously, factors affecting the *system parameters* can be classified into three sub-classes based on their changing frequencies.
    - (I) *Rapidly changing factors.* This kind of factors includes server workload (such as, CPU load, number of I/Os per second, etc. They are put together as server workload) and network speed. The major feature of these factors is that they can change significantly within a short period of time.
    - (II) *Slowly changing factors.* Factors, such as configuration parameters in local database management system (DBMS), usually change little by little, and significant change may be accumulated after a certain period of time.
    - (III) *Steady factors.* Some factors, such as local DBMS type, local database location, may stay unchanged for a long time. If this kind of factor is changed, the existing cost model will be discarded and a new one should be built up.

3 Approaches to Evolve Query Cost Model

In section 2, the environment variants have been discussed. Our cost models are built based on the specific environment in which the sample queries were executed. If the environment changes, the cost model may become out of date. This section aims to discuss the approaches to deal with these system changes by updating the cost model in a dynamic environment. In our cost model, there are several approaches to deal with these changes of the environment, which are caused by different kinds of factors, such as sample query method, cost model rebuilding and *Cwk*NN algorithm (*cover (counting) weighted k nearest neighbours, details are presented in [5,9]*).

(1) Sample query method: Database parameters contain the number of tuples in an operand table, the number of tuples in the result table, etc. The sample query method is employed to detect the variation of the data parameter, and then the query cost model can be adjusted according to the detected value. This approach can be used to solve the case of data parameters changing.

(2) Model rebuilding: There are two ways to rebuild the cost formulae. The first one is to use new data to totally rebuild the model. When all collected statistics cannot

be used, what we can do is to recollect the necessary statistics data to rebuild new cost models for the system. This cost model upgrading by this way is suitable in the situation that the steady factors changed. The other way is to Input new data to update cost model. In this way, we remove the part of oldest data and input some new data to build the new formula for the cost model to capture the changing of environment. By using this way, the changes of environment that are cased by the slowly changing factors can be captured.

(3) *CwkNN*-based of clustering method: When new data are obtained, we wonder if the system environment changed from one contention state to other contention state. To deal with this problem, we propose a novel *kNN-CwkNN*-by-tree algorithm to reclassify the new data to determine the system contention states at these time points in a dynamic environment. By employed the method, these changes of environment that caused by the rapidly changing factors can be captured.

4 The Architecture of Evolving a Cost Model

The details of the process of evolving a cost model are described as follows:

1) Input the observed cost of query.
2) To classify the cost of the sample query.
3) Compare the observed cost of query to the estimated cost of query. If the error rate of the estimated cost is lower than a predefined threshold, go to step 6. If the data have been dealt with the sample query method, go to 5
4) If the error rate is not lower, employ the sample query method to detect the parameters for the cost formula. Then go to 3.
5) Input the new data and remove part of the oldest data to rebuild the cost formula. Then go to 3.
6) Output the cost model, end of process

5 Experimental Results

To effectively simulate changing dynamic environment, we artificially generate different numbers of concurrent processes with various work/sleep ratios to change the system contention level. Note that, unlike scientific computation in engineering,

Table 1. The percentages of good and very good cost estimates for test queries

Contention state	Static: Good%	Static: Very good%	Evolutionary: Good%	Evolutionary: Very good%
1	21%	7%	85%	76%
2	17%	12%	72%	64%
3	34%	22%	81%	77%
4	56%	23%	88%	71%
Average	32%	16%	81.5%	72%

the accuracy of cost estimation in query optimization is not required to be very high. Table 1 shows the percentages of good and very good cost estimates for test queries for contention states at four contention states In the table, cost estimated from initial cost model and the evolutionary cost estimated by the architecture are listed.

6 Summary

In this paper, we analyze the factors to affect the system and classify the factors into two classes based on our cost model. Three methods are suggested to deal with the different cases of system environment changes: the sample query method, $CwkNN$-by-tree method and rebuilding cost model method. An architecture for evolving a cost model in dynamic environment is proposed. The experimental results demonstrate that the proposed techniques are quite promising in maintaining accurate cost models efficiently for dynamically changing data integration systems.

References

1. Adali, S., Candan, K.S., Papakonstantinou, Y., and Subrahmanian, V.S: Query Caching and Optimization in Distributed Mediator Systems. In Proc. of ACM SIGMOD (1996), 137–148
2. Chatterjee, S. and Price, B. Regression Analysis by Example (2nd ed.) John Wiley & Sons. Inc. (1991)
3. Du, W., Krishnamurthy, R., and Shan, M.C. Query Optimization in Heterogeneous DBMS. In Proc. of VLDB (1992) 277–291
4. Gruser, J.R., Raschid, L, Zadorozhny, V., and Zhan, T.: Learning Response Time for Web-Sources Using Query Feedback and Application in Query Optimization. VLDB Journal, 9(1), (2000) 18-37
5. Liao, Z., Wang, H., Glass, D. and Guo, G.: KNN-based Approach to Cost Model Updating. 2005. to appear
6. Ling, Y. and Sun, W.: A Supplement to Sampling-based Methods for Query Size Estimation in a Database System. SIGMOD Record, 21(4), (1992) 12-15
7. Liu, W., Liao, Z., Hong, J. and Liao, Z.F.: Query Cost Estimation through Remote Server Analysis Over the Internet. In Proc. Of WI (2003) 345-355
8. Roth, M.T., Ozcan, F. and Haas, L.M.: Cost Models DO Matter: Providing Cost Information for Diverse Data Sources in a Federated System. In Proc. of VLDB (1999) 599–610
9. Wang, H.: K-nearest Neighbours by Counting: to appear
10. Zadorozhny, V., Raschid, L., Zhan, T. and Bright, L. Validating an Access Cost Model for Wide Area Applications. Cooperative Information Systems, Vol 9, (2001) 371-385
11. Zhu, Q., Motheramgari, S. and Sun, Y.: Developing Cost Models with Qualitative Variables for dynamic Multidatabase Environments. In Proc. of ICDE (2000)

A SVM Method for Web Page Categorization Based on Weight Adjustment and Boosting Mechanism

Mingyu Lu[1,2], Chonghui Guo[2], Jiantao Sun[2], and Yuchang Lu[2]

[1] Institute of Computer Science and Technology,
Dalian Maritime University, Dalian, China 116026
[2] Department of Computer Science and Technology,
Tsinghua University, Beijing, China 100084
my-lu@tsinghua.edu.cn

Abstract. Web page classification is an important research direction of web mining. In the paper, a SVM method of web page classification is presented. It include four steps: (1) using analysis module to extract the core text and structural tags from a web page; (2) adopting the improved VSM model to generate the initial feature vectors based on the core text of web page; (3) adjusting weights of the selected features based on structural tags in web page to generate the base SVM classifier; (4) combining the base classifiers produced by iteration based on Boosting mechanism to obtain the target SVM classifier. The experiment of web page classification shows that the approach presented is efficient.

1 Introduction

Most web pages are still in HTML format. The characteristics of current web pages is of free style, including rich media information such as images, sounds, banners and flashes in addition to text and hyperlinks, and lack uniform pattern. Comparing to text classification, web page classification faces more difficulties and challenges.

Web page classification methods usually take text classification method as foundation in conjunction with structure analysis and link analysis technologies. Past researches indicate that direct application of text classification method to web page classification is not satisfactory due to the existence of noises [1].

This paper proposes a web page classification approach based on weight adjustment and boosting mechanism. The approach takes full advantage of text and structure information of web page. The main idea of the approach includes:

a) Prepares web page training set and web page test set for every information category.
b) Extracts core text and special tags useful for classification from every page;
c) Applies an improved VSM model [2,3] to core text of web pages and produces initial feature vectors for every information category.
d) Adjusts weights of initial feature vectors using the structure tag information of web page and produces a base SVM classifier;
e) Generates multiple base SVM classifiers using boosting principle iteratively and combines them to produce the final SVM classifier.

During the classification, we need to perform operations such as Chinese word segmentation, POS (part-of-speech) tagging and feature extraction, and recognize type of web page, delete index pages because we only classify content web pages.

2 Web Page Recognition and Segmentation– Web Page Analysis Module [4]

Generally, core text is the most important part to web page classification, then the structure tag information and hyperlink information, and the last multimedia information. Ad, flash and navigation links enticing user to visit other contents in web page are noises that would degrade accuracy of web page classification and must be deleted. Classification methods taking whole page as text suffer from accuracy loss mostly due to the presentation of those noises.

Core text of content web page is the essential part of text the page trying to express. Usually it resides on central part of the web page taking the form of a paragraph of text in most situations, including tables probably, or an image and its explanatory note, or a word, pdf, ps document, or an image(s) without any text or multimedia stream file in some extreme situations.

Structure tag information can be used in web page classification include <Head>,<Title>,<description> in <meta> tag, <Keyword> and font position, size and bold. Of course categories of links to other pages (i.e. outbound links) and links to this page from other pages (i.e. inbound links) should be considered. However, a constraint exists in the method discussed in this paper: Web pages to be classified are collected by web spider (web pages to be classified in most web site collected this way). For a web page collected this way and stored in local, original inbound or outbound links may be incomplete, in consequence hyperlink information in the web page lost its value. We thus do not consider hyperlink information in this discussion.

Recognition and utilization of multimedia information need techniques from image retrieval or speech recognition. And because there are no good research results at present, multimedia information is seldom considered.

In order to extract core text (incl. tables) and special tags useful for classification from web pages to be classified (target web pages), a web page analysis module is designed. The module analyzes source code of web page, outputs information needed by its upper modules (i.e., information retrieval, classification and information extraction, etc). The module supports HTML4.0 standard, and is compiled as a dynamic linked library named HTMLAnalysis.dll using VC6.

The library provides following functions for web page analysis.

1) BOOL SetSource(Cstring Url, CString HtmlSourceCode)
   //Specifies html source code and its URL to be analyzed.
2) BOOL GetTitle(Cstring*)
   //Returns the title of the html file specified by SetSource.
3) BOOL GetText (Cstring*)
   //Returns the text displaying in a browser of the HTML file specified by SetSource.

4) BOOL GetTextWithExtra (Cstring*)
   //Returns the text displaying in a browser, alternative text for image and corresponding text for option list, of the HTML file specified by SetSource
5) BOOL GetAllLinkUrlArray(CstringArray*)
   //Returns all urls in the web page, including both links pointing towards intra-website and inter-websites. Related links are changed to absolute links. All links are put into a CstringArray. The address of the CstringArray is returned. Intra-page links are ignored.
   //Note: the following links include both links pointing towards intra-website and inter-websites.
6) Int GetLinkTotal()
   //Returns total number of links, denoted as N.
7) BOOL GetLinkUrl(int LinkIndex, Cstring *LinkUrl)
   //Returns the LinkIndex-th url, where LinkIndex<=N.
8) BOOL GetLinkAnchor(int LinkIndex,Cstring *LinkAnchor)
   // Returns the LinkIndex-th anchor, where 1=<LinkIndex<=N.
9) BOOL GetLinkNeighborText(int LinkIndex, Cstring *LinkNighborText)
   //Returns the text information around the LinkIndex-th link, where 1=<LinkIndex<=N. The text information around a link includes paragraph text arount the link and title of the paragraph.
10) BOOL GetAllHiText(Cstring *AllHiText)
    //Returns all text of H1,H2,...
11) BOOL GetMetaKeyword(Cstring *MetaKeyword)
    //Returns corresponding text of keyword in meta tag.
12) BOOL GetMetaDesc(Cstring *MetaDesc)
    //Returns corresponding text of description in meta tag.
13) BOOL GetKeyText(Cstring * KeyText)
    //Returns core text.

The function of the module can be changed to meet the requirement of its upper modules. When an upper module (such as a module of web page classification) needs to use the module, HTMLAnalysis.dll, HTMLImport.h and HTMLAnalysis.lib should be copied to local directory, and HTMLImport.h and HTMLAnalysis.lib should be added to the Project of the upper module.

3   Computation of Initial Weight for Feature Vectors Using an Improved VSM

Assume p as the target page to be classified, T as core text extracted form p. Using a improved VSM model to extract feature vector from T, we get feature vector $F=\{(T_i, w_i) \mid i=1,2,\ldots,n\}$, where f_i is the i-th feature after word segmentation, POS tagging, stemming and word frequency count; $w(f_i)$ is the weight of f_i, and $tf(f_i)$ is the frequency of f_i in core text T.

Formula for $w(f_i)$ in classical VSM is

$$w_i(f_i) = tf(f_i) \times idf(f_i)$$
$$= tf(f_i) \times \log(\frac{N}{N_i} + 0.5). \quad (1)$$

where N is the number of documents in training set, N_k is the number of documents containing item f_i, and $idf(f_i) = \log(N/N_i + 0.5)$ is inversed document frequency of f_i.

After in-depth analysis of above VSM model, we found that inversed document frequency $idf(f_i)$ is the main cause of degraded accuracy of the model. We proposed an improved VSM model using feature evaluation function (i.e., information gain, mutual information, odd ratios, etc) to replace inversed document frequency function[3]. For example, if we use odd ration to replace inversed document frequency function, we have

$$w_i(f_i) = tf(f_i) \times oddsRatio(f_i)$$
$$= tf(f_i) \times \log \frac{odds(f_i|pos)}{odds(f_i|neg)} \quad (2)$$
$$= tf(f_i) \times \log \frac{P(f_i|pos)(1 - P(f_i|neg))}{P(f_i|neg)(1 - P(f_i|pos))}$$

The above proposed improved VSM model is used in this paper to compute initial weight $w(f_i)$ of f_i. Next we would adjust $w(f_i)$ according to structure tag information in the web page.

4 Feature Vector Adjustment Using Structure Tag Information

In addition to the core text of the web page, useful information for web page classification include structure information such as <Head>, <Hm>, <Meta> tag, and font emphasis information as <B>/<STRONG>, <I>/<EM>, <U>, <STRIKE>, <CENTER> tag. These parts can be extracted from the web page using functions in above section.

All above four kinds of tag are useful for web page classification but differ in power. We assign different coefficient for the four kinds of tag for weight adjustment.

Rules used in this paper for weight adjustment are as follows.

(1) Weight adjustment for font emphasis
For the j-th occurrence of feature item f_i in text T ($1 \leq j \leq w_i$), if the tag is one of <B> (or <Strong>), <I> (or <EM>) , <U>, <STRIKE>, <CENTER>, the weight here increases from 1 to $(1+\alpha)$ ($0<\alpha<1$) . α should be different for different tag. For the sake of convenience, we set α to 0.5 in our experiment. We denote adjusted weight for font emphasis for feature item f_i as $w_i^{'}$;

(2) Weight adjustment for hierarchy caption
If frequency of feature item f_i at hierarchy caption <H_m> in web page p is w_m, the weight here increase from w_m to $\beta * w_m$. In the experiment, β is set as follows.

$$\beta = \begin{cases} 1 & \text{if } H_m = H4 \\ 2 & \text{if } H_m = H3 \\ 3 & \text{if } H_m = H2 \\ 4 & \text{if } H_m = H1 \\ 1 & \text{if } H_m = H5 \text{ or } H6 \end{cases}$$

After above adjustment, weight of feature item f_i for hierarchy caption is denoted as w_i'':

$$w_i'' = w_i' + \sum \beta * w_m$$

(3) Weight adjustment for other important positions

If w_t, w_k, w_d are corresponding weight of feature item f_i at <TiTle>, <Meta> and the summary of web page p, the total weight of f_i after above weight adjustment for these position is

$$w_i''' = w_i'' + \gamma_1 * w_t + \gamma_2 * w_k + \gamma_3 * w_d$$

In the experiments, we set γ_i as:

$\gamma_1 = 5$; $\gamma_2 = \gamma_3 = 4$

After three kinds of adjustment as above, the feature vector of web page p is:

$F' = \{(f_i, w_i''') \mid i=1,2,\ldots,l\}$ $(l=n+n_l)$.

Where n is number of feature item extracted from core text, n_l is number of feature item extracted from special tags. The feature vector is used in web page classification.

5 Assembling of SVM Classifiers Using Boosting Principle

In 1995, Vapnik introduced SVM (Support Vector Machine) for two-class pattern recognition problem, and Joachims first applied SVM to text classification [2,5,6,7].

SVM turns text classification into a series of two-class classification problem. It is one of most accurate classifiers for web page classification. We use it as base classifier for web page classification.

SVM first maps the original sample space to a high-dimensional linear space via a nonlinear mapping defined by a dot product function, and then finds a separating hyperplane, which is optimal and is decided by support vectors, in the linear space.

Finding the hyperplane can be casted into a quadratic optimization problem. It is can be proved that any SVM can be simplified to a nearest neighbor classifier with a positive example and a negative example as representation points. The two points are determined by support vector of positive example and negative example respectively [8].

In the field of web page classification, the problem space is usually represented by high dimensional vectors, and usually a great number of examples are needed to establish the training set prepared for learning process. In many real world problems, the number of Support Vectors is much smaller than that of training samples. So it would be a good idea to only use support vectors for simplifying the training process of SVM.

Boosting is an efficient way to improve prediction ability of learning system, and a representative method of assembly learning. Combining multiple base SVM classifiers with boosting is an interesting research area that may result high performance text classifier [7,9].

Boosting is an iterative learning procedure that successively classifies a weighted version of the sample by its base classifiers generated one for each time of iteration, and then re-weights the sample dependent on how successful the classification was. In the process correctly classified samples receive smaller weight, thus the generated new base classifier could focus on those bigger weighted, hard-to-classify samples. Through combining multiple base classifiers, boosting can find high accuracy classification rules while base classifier can be a weak learner with a low accuracy [9].

There are two approaches to boost SVM, i.e., BSVM and Sboost [9]. BSVM directly boost weighted SVMs, while Sboost integrates the constituent classifiers of boosting using SVM. However, The computational complexity of SVM is substantial, and so is the boosting iteration. BSVM and Sboost are effective but not efficient.

We proposed an efficient approach IBSVM (Interactively boosted SVM) to boost SVM. Its learning time is much shorter than BSVM and Sboost while it has comparable classification accuracy with them.

We use fixed size subset of training set (usually much smaller than training set) to train base SVM classifier for boosting. The subset is dynamically adjusted through each iteration of boosting. Outside pages important to current classification are added into the subset while inside pages unimportant to the current classification are removed from the subset to maintain the size of the subset. Once the subset is modified, new base SVM classifier can be generated. At last we combine those base classifiers use boosting principle. The subset is kept small to ensure quick generation of base SVM classifier. Adjustment to the subset is to generate multiple base SVM classifiers.

6 The IDSVM Algorithm

Boosting algorithm here is based on AdaBoost.M1.
 Given original training sample set:

$$\{(x_1, y_i), (x_2, y_2), \cdots, (x_N, y_N)\},$$

where x_i is any training sample, $y \in \{-1,+1\}$ denotes the classes x_i belongs to N is the number of training samples. $i = 1, 2, \cdots, N$

This algorithm maintains a set of weights as a distribution W over samples, i.e., for each $x_i \in x$, and at each boosting round s, there is an associated real value $w_s[i]$.

Given a fixed number of training samples used for finding support vectors. We choose $n = \frac{1}{100} N$ in the experiment.

Given a fixed number of boosting rounds T.

Choose a kernel function $K(x, x')$ to compute the dot product of two vectors.
Algorithm Description:

Step 1:
Initialize $w_1[i] = 1/N$ for all $i = 1, 2, \cdots, N$, set $s = 0$.

Step 2:
Select randomly n samples from original sample set according to current weight distribution W (if we regard the normalized weight of each sample as its apriori probability). These selected samples establish a training set of size n for finding support vectors. We call this small training set as *STS*.

Step 3:
Use each sample $(x, y) \in STS$ to find the support vectors.
We first find the coefficients $\alpha^* = \{\alpha_1^*, \ldots, \alpha_n^*\}$ that maximize

$$W(\alpha) = \sum_{i=1}^{n} \alpha_i - \frac{1}{2} \sum_{i,j=1}^{n} \alpha_i \alpha_j y_i y_j K(x_i, x_j)$$

if $\alpha_i^* \neq 0$, then x_i is the support vector we are looking for.

Set $s = s + 1$ to compute the number of base classifiers already generated. If $s \geq T$ then go to step 8.

Step 4:
Combine these support vectors to two representative points: $x^+ = \frac{1}{C} \sum_{(x_i, y_i) \in SV^+} \alpha_i^* x_i$

for positive samples, and $x^- = \frac{1}{C} \sum_{(x_i, y_i) \in SV^-} \alpha_i^* x_i$ $C = \sum_{y_i = -1} \alpha_i^* = \sum_{y_i = +1} \alpha_i^*$.

Where SV^+ and SV^- denote all the positive and negative support vectors respectively.

Based on these two points, we establish a nearest neighbor classifier whose decision rule is, if an unknown point is nearer to x^+ (or x^-) than to x^- (or x^+), it will be classified as the positive (or negative) class.

Step 5:
Classify all the samples of original sample set.
Compute the weighted error:

$$\varepsilon = \sum_{i=1}^{N} w[i] \cdot [\![(x_i, y_i) \text{ is } misclassified]\!].$$

$[\![\bullet]\!]$ is a function that maps its content to 1 if it is true, otherwise, maps to 0. Increase the samples weights if they are misclassified. Otherwise, decrease:

$$w[i] = w[i] \div \begin{cases} 2\varepsilon & \text{if } (x_i, y_i) \text{ is } misclassified \\ 2(1-\varepsilon) & \text{otherwise} \end{cases}$$

Normalize these weights because we hope they also can be regarded as probability distribution.

Step 6:
If, between any two support vectors of negative positive pair, there are samples not in the *STS*, then we choose the sample nearest to the separating hyperplane to join in

the *STS*. Before that, the sample in the *STS* with smallest weight value will be deleted from *STS*. Go to step 3. if there are no sample in the margin, go to step 7.

Step 7:

Select randomly from the original sample set according to current weight distribution W. If the sample is not in *STS* and is misclassified by current base classifier, then the sample with smallest weight value in the *STS* will be deleted, and the selected sample will be added to *STS*. Go to Step 3; otherwise, go to step 7.

Step 8:

Combine the *T* base classifiers by weighted votes:

$$H_B(x) = \text{sgn}\left(\sum_{s=1}^{T} \ln(\varepsilon_s/(1-\varepsilon_s)) \cdot H_s(x)\right).$$

It is important to note here that the distance between any two points is decided by the kernel function, say, all the dot products in the original space are replaced by kernel functions to compute them. i.e.

$$\|x - x'\| = K(x,x) - 2K(x,x') + K(x',x')$$

denotes the distance between x and x' in the new space

7 Web Page Classification Experiments

The training web page set and test web page set are extracted from Yibao Chinese news website manually. In order to test our proposed approach, three experiments are performed using unprepared raw web pages for comparison. Results are shown in table 1.

Table 1. Web page classification experiment results

Category	Accuracy1 (Web page)	Accuracy2 (Core text)	Accuracy3 (Core text and tags)
General	0.51	0.69	0.75
Economics	0.58	0.75	0.79
Education	0.64	0.80	0.84
Sports	0.63	0.82	0.86
International	0.61	0.78	0.83
Politics	0.59	0.76	0.81

Three experiments:

Experiment 1: Web pages are treated as texts and classified using our text classification system CZW [10].

Experiment 2: Core texts are extracted from web pages using web page analysis module, and then CZW system is used to perform classification.

Experiment 3: Core texts are extracted from web pages using web page segmentation module, and then CZW system is used to perform classification using core text and web page tag information.

Notes on experiment data:

Experiment data: as shown in table 1
Data source: Yibao Chinese news
Training set: 10000 web pages
Test set: 1045 web pages
Feature items kept after feature selection: 30%
Language: Chinese
Category: International, economics, sports, Education, Politics, General
Classification method: IBSVM
Classification measure: Feature adjustment and Boosting

From table 1, we can see that treating web pages as texts yields the worst result, indicating huge influence of lots of useless information in web pages. Accuracy of classification based on core text is not good either because some problems existing in our web page segmentation module prevent it from accurately recognition of all core texts of all web pages. As a result some noisy tags were kept and the accuracy was affected.

Although classification based on core text and useful tag was affected by noisy tags too, accuracy of the classification stays high because noisy tags are included in core text which receive relative lower weights. The result shows that useful tags in web page helps a lot to classification accuracy as well as efficiency of our proposed approach.

8 Conclusions

With the rapid growth of World Wide Web, We are facing increasing huge web information. Web page classification is a fundamental approach to huge web page information processing as well as an important research topic in text mining and web mining. It has wide application in traditional information retrieval, website index structure establishment and web information retrieval and filtering.

In March 2003, our research group attended the first national symposium of search engine and web mining, and took part in a contest of web page classification organized by the symposium. Our web page classification won runner-up.

The web page classification approach proposed in this paper was also used in our research project "Automatic acquisition of specific domain information based on user customization" and "Recommendation of Web Pages Based on Concept Association" [10]

Acknowledgments. This research is supported by Natural Science Foundation of China (No. 60473115) and. Postdoctoral Science Foundation of China (No. 2003034147).

References

1. Yiming Yang. An evaluation of statistical approach to text categorization. In Technical Report CMU-CS-97-127, Computer Science Department, Carnegie Mellon University (1997).
2. Joachims T. A Probabilistic Analysis of the Rocchio Algorithm with TF-IDF for Text Categorization, Proc of ICML97, Morgan Kaufmann Publishers (1997) 143~151.

3. Lu Yu-chang, Lu Ming-yu, Li fan, Zhou Li-zhu. Analysis and construction of word weighing function in VSM, Computer Research and Development, Vol. 39(10), (2002) 1205-1210.
4. Xia Xianjun, Sun jiantao, Lu yuchang. The research and implementation of a new result-faced methods for webpage information extraction, Computer Engineering and Application, Vol. 38: (2002) 87-91.
5. Kecman, V. Learning and Soft Computing, Support Verctor Machines, Neural Networks and Fuzzy Logic Models, The MIT Press, Cambridge, MA (2001).
6. Wang, L.P.(Ed.): Support Vector Machines: Theory and Application. Springer, Berlin Heidelberg New York (2005).
7. Sun Jiantao, Shen Dou, Lu Yuchang. Web document classification techniques, Journal of Tsinghua University, Vol. 44(1), (2004) 65-68.
8. Diao, Lili, Lu, Mingyu; Hu, Keyun; Lu, Yuchang; Shi, Chunyi, New boosting algorithms for text categorization, Proceedings of the World Congress on Intelligent Control and Automation (WCICA), vol. 3 (2002) 2326-2329.
9. Diao Lili, Lu Yuchang, Shi Chunyi. A Method to Boost Support Vector Machines, Proc. of PAKDD 2002, Springer-verlag, Lecture Notes in Artificial Intelligence (LNAI) (2002) 463-468.
10. Lu Mingyu, Zhou Qiang, Li Fan, et al. Recommendation of Web Pages Based on Concept Association, Proc. of 4th IEEE International Workshop on Advanced Issues of E-Commerce and Web-Based Information Systems, IEEE Computer Society Press (2002) 221-227.

Feature Selection for Specific Antibody Deficiency Syndrome by Neural Network with Weighted Fuzzy Membership Functions

Joon S. Lim[1], Tae W. Ryu[2], Ho J. Kim[3], and Sudhir Gupta[4]

[1] College of Software, Kyungwon University, Sungnam 461-701, Korea
jslim@kyungwon.ac.kr
[2] Department of Computer Science, California State Univ., Fullerton, CA 92834, USA
tryu@fullerton.edu
[3] School of CSEE, Handong University, Pohang, 791-940, Korea
hjkim@handong.edu
[4] Department of Medicine, University of California, Irvine, CA 92868, USA
sgupta@uci.edu

Abstract. Fuzzy neural networks have been successfully applied to analyze/generate predictive rules for medical or diagnostic data. This paper presents selected membership functions extracted by a fuzzy neural network named NEWFM. The selected membership functions can capture the concentrated and essential information without sacrificing the classification capability. To verify the performance of the NEWFM, the well-known data set of Wisconsin breast cancer is performed. We applied NEWFM model to extract fuzzy membership functions for the UCI antibody deficiency syndrome diagnosis. Then selected features obtained by non-overlapped area measurement method are presented.

1 Introduction

Data collected from clinical settings may have variations due to similar (or the same) clinical tasks being performed in varying demographic settings or clinical procedures differing from region to region. Therefore clinical data analysis and related research is in great need of an adaptive decision support tool that can handle ambiguous and noisy input data and generate predictive rules for diagnostic information.

Neural network and fuzzy set theory can be effectively used for this type of application as a major pattern classification and predictive rule generation tool. The goal of pattern classification is to partition the feature space into decision regions. Artificial neural networks have been successfully used in many pattern classification problems [3] [5]. Fuzzy set theory was introduced by Zadeh [22] as a means of representing and processing data by allowing partial set membership rather than crisp set membership or non-membership. As a combined approach of neural network and fuzzy set theory, fuzzy neural networks (FNN) have been proposed as an adaptive decision support tool [2] [7] [10] [11] [12] [13] [14] [17] [19]. Various architectures of FNN have been introduced along with algorithms for learning, adaptation and rule extraction [6] [9]

[15] [16]. In medical diagnosis, high accuracy of acceptable regular patterns with simple types of knowledge representation in linguistic terms is desirable. In addition, the way of obtaining the optimal structure for fast adaptation without sacrificing the accuracy and performance is needed. The capability of fuzzy *if-then* rule extraction is one of the advantages of FNN for medical decision-making. For this, various approaches has been proposed. Among those approaches, fuzzy neural networks with self-organizing systems were developed by [8] [18] [20] to extract knowledge from a given set of training data. Setnes [16] presented a compact and accurate fuzzy rule-based model using a genetic algorithm. However, most approaches proposed so far have not considered the weights for the membership functions.

In this paper, we present a neural network model with weighted fuzzy membership functions (NEWFM) to predict diagnoses of antibody deficiency syndrome. The extracted weighted fuzzy membership functions can preserve the disjunctive fuzzy information and characteristic input patterns in the pattern space, which can result in reducing the number of membership functions. The effectiveness of NEWFM is first validated using the popularly used data set, Wisconsin breast cancer [21], for the benchmarking of pattern classifications. The NEWFM is then applied to the antibody deficiency syndrome data set obtained from the University of California, Irvine (UCI) immunology laboratory. Then selected features obtained by non-overlapped area measurement method are presented.

2 Neural Network with Weighted Fuzzy Membership Functions (NEWFM)

2.1 The Structure of NEWFM

The structure of NEWFM is illustrated in Fig. 1. The NEWFM comprises three layers, namely input, hyperbox, and class layer. The input layer contains n input nodes for n featured input patterns. The hyperbox layer consists of m hyperbox nodes. Each hyperbox node B_l to be connected to a class node contains n fuzzy sets for n input nodes. The output layer is composed of p class nodes. Each class node is connected to one or more hyperbox nodes. An hth medical pattern can be recorded as $I_h = \{A_h = (a_1,$

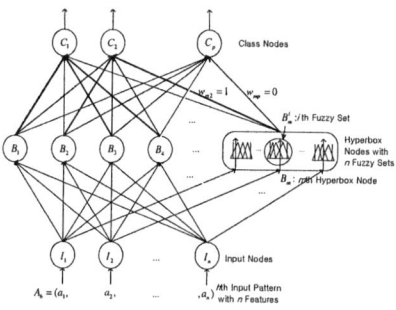

Fig. 1. Structure of NEWFM

$a_2, \ldots, a_n$), *diagnosis*}, where *diagnosis* is the result of diagnosis and A_h is the pattern on n different features. 'Unknown' features are represented as NULL. The ith fuzzy set of B_l, denoted by B_l^i, has three weighted membership functions as shown in Fig. 2. These three weighted fuzzy membership functions are adjusted by the learning algorithm, which will be discussed in Section 2.3. After learning, the weighted membership functions for classification are located in the hyperbox layer. The details of the processes are described in the Section 2.4.

2.2 Definitions and Operations

1) w_{li}: The connection weight between a hyperbox node B_l and a class node C_i is represented by w_{li}, which is initially set to 0. If there is a connection from a hyperbox node B_l to a class node C_i, the w_{li} will be set to 1 from 0. C_i can have more than one connection from hyperbox nodes, whereas B_l is restricted to have one connection to a class node.

2) v_i: The v_1, v_2, and v_3 represent the *center vertices* of the small, medium, and large membership functions respectively in Fig. 2. The *center vertices* can be adjusted during learning, while the other vertices v_0 and v_4 are fixed. It is assumed that the input feature value a_i ranges from v_{min} to v_{max} as shown in Fig. 2.

3) μ_j and W_j: μ_js are ith set of membership functions of a hyperbox node B_l for $j=1,2,3$, representing small, medium, and large respectively. The shape of each membership function μ_j is triangular, which is characterized by three vertices (v_{j-1}, v_j, v_{j+1}) with its *membership function weight* W_j ($0 \leq W_j \leq 1$, random weights in the range of $0.45 \leq W_j \leq 0.55$ are initially set) that represents the strength of the membership function determined through learning. The shaded triangles in Fig. 2, called *weighted fuzzy membership functions*, can be formed by (v_{j-1}, W_j, v_{j+1}).

4) $Len(\mu_j)$: To measure the size of a μ_j, a length function, Len is defined as follow:

$$Len(\mu_j) = ((v_{j+1} - v_{j-1})/2)W_j \quad (2.1)$$

This function is similar to the length function, $Len(\mu_j)$, defined by Lee et al. [13] except that the $Len(\mu_j)$ is multiplied by the weights W_js.

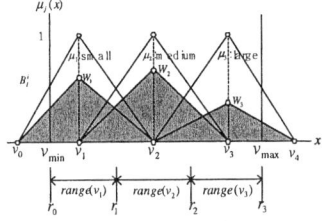

Fig. 2. An ith Set of Weighted Membership Functions of B_l

5) $Adjust(B_l)$: This operation adjusts membership functions and their weights for n fuzzy sets in the hyperbox node B_l according to input $A_h=(a_1, a_2, \ldots, a_n)$. For each ith set of membership functions with the input a_i, the v_js and W_js of the membership functions are adjusted by the value of $\mu_j(a_i)$, where $j=1, 2, 3$. As a result of $Adjust(B_l)$ operation, the new vertices $new(v_j)$s and new weights $new(W_j)$s are set by the following expression:

$$new(v_j) = v_j \pm \alpha E_j \mu_j(a_i) W_j \qquad (2.2)$$
$$new(W_j) = W_j + \beta(\mu_j(a_i) - W_j) \qquad (2.3)$$

In these expressions, the α and β are the learning rate in the range from 0 to 1, and the variable E_j is the difference between v_j and input a_i. If E_j is bigger than the adjacent $E_{j\pm1}$, the smaller one is selected. The detail process of the $Adjust(B_l)$ operation is described in the learning algorithm in the next section. Fig. 3 shows the result of the $Adjust(B_l)$ operations for the input a_i and the ith set of fuzzy weighted membership functions in B_l.

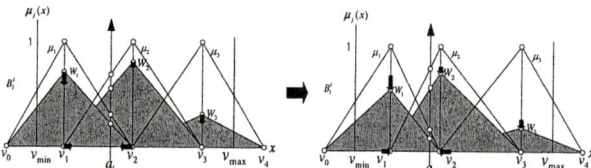

Fig. 3. Example of Before And After $Adjust(B_l)$ Operation

As shown in Fig. 3, the weights and the center of each membership functions are adjusted by the Adjust(B_l) operation, e.g., W_1, W_2, and W_3 are moved down, v_1 and v_2 are moved toward a_i, and v_3 remains in the same location.

6) *Random(B_l)*: This operation makes the hyperbox node B_l with n (number of inputs) sets of randomly distributed membership functions with their random weights ranging from 0.45 to 0.55. The random *center vertices* v_is should be in their ranges such that
$$r_{i-1} \le v_i < r_i, \qquad \text{where } i = 1, 2, 3$$
while v_0 and v_4 are fixed as in Fig. 2. Also, the connection weights w_{li}s are set to 0.

7) *Output(B_l)*: For the hth input $A_h=(a_1, a_2, \ldots, a_n)$ with n features to the hyperbox B_l, output of the B_l is calculated by

$$Output(B_l) = \frac{1}{n} \sum_{i=1}^{n} \sum_{j=1}^{3} B_l^i(\mu_j(a_i))W_j. \qquad (2.4)$$

2.3 Learning Algorithm for NEWFM

This section describes the learning algorithm for NEWFM. The algorithm uses the Learning(B_l,C_i) procedure to adjust the locations of vertices and weights, and to connect the hyperbox nodes to class nodes.

Algorithm *NEWFM*;
1   **While** (result is satisfied)
1.1   **for** $l = 1$ **to** m // m is number of hyperboxes, usually start from number input
              Random(B_l);
1.1.1   **for** $j = 1$ **to** p // p is number of class nodes

1.1.2.1	$w_{lj} = 0$; // initial connection weight between B_l and C_j
1.2	**for** $k = 1$ **to** h // h is number of input patterns
1.2.1	find B_l that has the maximum value of **EnhOutput**(B_l)
	//see the formula 2.6 among m hyperbox nodes from the input A_k;
	// input vector: $I_k = \{A_k = (a_1, a_2, \ldots, a_n), diagnosis\}$
1.2.2	**Learning**(B_l, C_i); // C_i is a *diagnosis* in I_k

Procedure *Learning*(B_l, C_i);
// m is number of hyperboxes
1   Case 1: $\forall m, w_{mi} = 0$, where $m \neq l$;
1.1 $w_{li} = 1$;
1.2   **Adjust**(B_l);

2   Case 2: $\exists m$ satisfying $w_{mi} = 1$, where $m \neq l$;
2.1 $w_{li} = 1$;
2.2   **Adjust**(B_l);

Procedure Adjust(B_l);
1   **for** $i=1$ **to** n // for each ith set of membership function in B_l
1.1   **for** $j=1$ **to** 3 // for each membership function
1.1.1   **if** $v_{j-1} \leq a_i < v_j$ // for left side of μ_j
1.1.1.1 $E_j = \min(|v_j - a_i|, |v_{j-1} - a_i|)$;
1.1.1.2 $new(v_j) = v_j - \alpha E_j \mu_j(a_i) W_j$;
1.1.2   **else** $v_j \leq a_i \leq v_{j+1}$ // for right side of μ_j
1.1.2.1 $E_j = \min(|v_j - a_i|, |v_{j+1} - a_i|)$;
1.1.2.2 $new(v_j) = v_j + \alpha E_j \mu_j(a_i) W_j$;
1.1.3 $new(W_j) = W_j + \beta(\mu_j(a_i) - W_j)$;

2.4 Fuzzy Rule Extraction

The learned NEWFM can be used for fuzzy rule extraction in *if-then* form to classify input patterns. After learning, each of n fuzzy sets in hyperbox node B_l contains three *weighted fuzzy membership functions* (WFMs, grey membership functions in Fig. 4), where n is the number of input nodes.

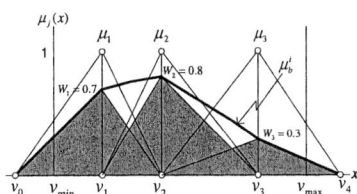

Fig. 4. Example of Bounded Sum of the 3 Weighted Fuzzy Membership Functions (BSWFM, Bold Line)

The rules can be extracted directly from the WFMs. We suggest a rule extraction strategy as described below.

1) The *bounded sum*(one of operations on fuzzy set) of WFMs (BSWFM) in the *i*th fuzzy set of $B_l^i(x)$, denoted as $\mu_b^i(x)$ (bold line in Fig. 4), is defined by

$$\mu_b^i(x) = \sum_{j=1}^{3} B_l^i(\mu_j(x)). \tag{2.5}$$

The BSWFM combines the fuzzy characteristics of three WFMs in Fig. 4.

2) We use an inference mechanism that performs the reasoning process to derive an *Output*(B_l) (2.4). In addition, we suggest *enhanced BSWFM* that uses a heuristic of *Len*(μ_j) defined in 2.2 to enhance the reasoning capability such that:

$$EnhOutput(B_l) = \frac{1}{n}\sum_{i=1}^{n}\sum_{j=1}^{3} B_l^i(\mu_j(a_i))W_j / Len(\mu_j) \tag{2.6}$$

The heuristic of *Len*(μ_j) is based on the concept that smaller *Len*(μ_j) has more concentrated information that is essential for classification of the given pattern.

3) The rules for a class C_i is the fuzzy membership functions represented by enhanced BSWFMs in B_l such that:

if (enhanced BSWFMs in B_l *and* w_{li}=1)*or*…*or*(enhanced BSWFMs of B_m *and* w_{mi}=1)
 then C_i

3 Experimental Results

In this section, we present the experimental results for Wisconsin breast cancer data set, which can be obtained from the UCI data set archive [1], to evaluate the accuracy and rule extraction capability of the NEWFM. We apply the NEWFM to our target clinical data set obtained from the UCI Immunology Laboratory to generate fuzzy rules for antibody deficiency syndrome (ADS). Then selected features obtained by non-overlapped area measurement method are presented.

3.1 Wisconsin Breast Cancer Classifications

We use the Wisconsin breast cancer (WBC) dataset [21] that was obtained from the University of Wisconsin hospital to classify two possible classes, *benign* or *malignant*. Although the NEWFM can handle missing values, we use 683 cases excluding the 16 cases that have missing values for fair comparison with other approaches. We use 683 cases for training and testing. The enhanced BSWFMs learned by NEWFM are illustrated in Fig. 5. In this experiment, two hyperboxes are created for classification. While a hyperbox contains a set of 9 bold lines (enhanced BSWFMs) in Fig. 5 is a rule for benign class, the other hyperbox contains a set of 9 light lines (enhanced BSWFMs) is another rule for malignant class. Each graph in Fig. 5 shows the difference between benign and malignant classes for each input feature graphically. Table 1 shows the comparison results of NEWFM and other classifiers in [15] and [4] for the WBC data set. As shown in the result, NEWFM outperforms the methods we compared with in terms of the classification rate and the number of rules used.

Table 1. Performance Results on Wisconsin Breast Cancer

Algorithm	Recognition Rate	No. of Rules
Nauck [15]	96.5%	4
Gomez [4]	99.12%	11
NEWFM	**99.56%**	**2**

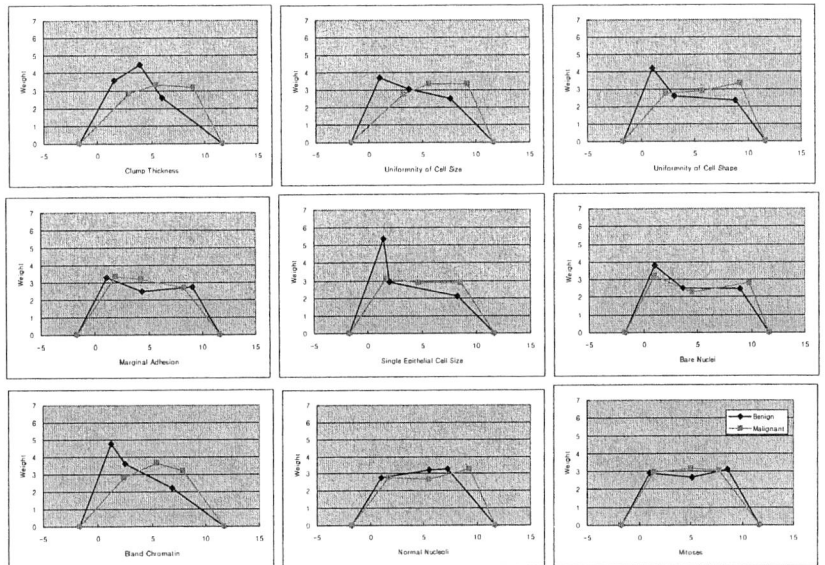

Fig. 5. Enhanced BSWFMs for Wisconsin Breast Cancer Data Set

3.2 Finding Fuzzy Rules for UCIIL Data Set

The NEWFM is applied to a clinical data set of antibody deficiency syndrome. The data set used in this study consists of 153 patient records, each comprising 23 features from experimental data collected in the UCI (University of California, Irvine) Immunology Laboratory (UCIIL). The samples were taken from patients in the UCI Hospital between 1996 ~ 2003. The UCIIL dataset contains 90 control cases and 63 cases of antibody deficiency syndrome. Table 2 provides the definition of features.

The followings are the descriptions of diagnoses in the UCIIL data.

1) Control: Healthy volunteer donors between ages of 18 and 35 years. They included students, residents/fellows, and staff members at UCI.
2) Antibody Deficiency Syndrome: Patients who have poor antibody responses to pneumococcal polysaccharide antigens and/or to tetanus toxoid; however their serum immunoglobulin isotypes and subclasses were normal or near normal.

Table 2. Definitions of Features Used in Clinical Diagnoses and Their Normal Ranges and Feature Ranks

Group	Feature	Normal Range	Rank
Lymphocyte Subsets	*CD3+ (Total T Cells)	65.0-80.0%	10
	CD3+/CD4+ (Helper/Inducer T Cells)	31.0-56.0%	21
	*CD3+/CD8+ (Suppressor/Cytotoxic T Cells)	17.0-34.0%	1
	CD4+/CD8+ Ratio (Helper/ Suppressor)	0.78-2.2	22
	CD19+(Total B Cells)	4.0-17.0%	19
	CD3-/CD16+/CD56+ (Natural Killer Cells)	4.0-16.0%	17
Lymphocyte Transformation - Mitogens	*Phytohemaglutinin (PHA)	123,609-193,794(counts per minute)	2
	*Concanavalin A (Con A)	110,829-184,949(counts per minute)	3
	Pokeweed Mitogen (PWM)	27,965-71,961(counts per minute)	20
Lymphocyte Transformation - Antigens	Mumps Virus	21,153-69,093(counts per minute)	15
	*Tetanus Toxoid	6,983-66,297(counts per minute)	12
	PPD	423-3,411(counts per minute)	23
	*C. albicans	21,494-78,166(counts per minute)	7
Activation Marks	CD4+/CD25+	6.0-19.0%	18
	*CD8+/CD25+	1.0-4.0%	8
	CD4+/CD26+	15.0-49.0%	13
	*CD8+/CD26+:	4.0-19.0%	6
	*CD4+/CD38+:	12.0-49.0%	11
	CD8+/CD38+	8.0-30%	16
	CD4+/CD69+	1.0-8.0%	14
	*CD8+/CD69+	1.0-5.0%	4
	*CD4+/CD97+	1.0-5.0%	9
	*CD8+/CD97+	1.0-5.0%	5

* The result is obtained from 12 selected features out of 23 features.

NEWFM is run with the UCIIL data to extract fuzzy rules from the enhanced BSWFMs after learning. As a result, 153 input patterns in the UCIIL data are classified by two rules. From the extracted rules, accuracy of 92.2% (12 misclassifications) was achieved. The features that have overlapped shapes of enhanced BSWFMs for both control and antibody deficiency syndrome could be less important.

3.3 Non-overlapped Area Measurement Method for Feature Selection

The ranks of feature evaluation can be obtained from the BSWFMs. Non-overlapped area heuristic is applied to measure importance of a feature. The larger area, the more feature characteristic is implied. Fig. 6 is an example to measure non-overlapped area of CD3+/CD8 feature.

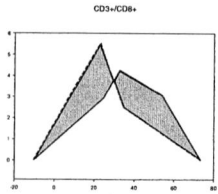

Fig. 6. Non-Overlapped Area of CD3+/CD8

Table 2 shows the importance of feature ranked by the non-overlapped area measurements. The 12 features selected from 23 features by the measurements, accuracy of 99.35% (1 misclassification from 153 input patterns) was achieved (Table 3).

Table 3. Performance Results on Antibody Deficiency Syndrome with Salient Features

Algorithm	Recognition Rate	No. of Rules
NEWFM	92.41%	2
NEWFM with Non-Overlapping Area Measurement Method	99.35%*	2

* The result is obtained from 12 selected features out of 23 features.

4 Concluding Remarks

In NEWFM model, a fuzzy set in a hyperbox node implies information of characteristics for all input features. The *enhanced bounded sums of weighted fuzzy membership functions* (*enhanced BSWFMs*) in hyperbox nodes amplify the characteristics of input patterns with the following advantages.

1) The enhanced BSWFMs in a hyperbox can maintain complementary information which handles local maxima problems, ambiguous subsets of feature space, and disjunctive fuzzy information in the pattern space.
2) Rules can be extracted easily since a set of enhanced BSWFMs in a hyperbox is direct interpretation of a fuzzy rule.
3) The number of hyperbox nodes can be reduced, since the enhanced BSWFMs represent the characteristics of input patterns by preserving the fuzziness of small, medium, and large.
4) The enhanced BSWFMs show important feature values visually. The heuristic of non-overlapped area is applied to measure importance of features.

By the above advantageous properties of NEWFM, we extract human comprehensive rules for UCIIL data set to predict antibody deficiency syndrome diagnosis with 99.35% accuracy using only 12 features.

References

1. Blake, C. L. and Merz, C. J.: UCI Repository of machine learning databases [http://www.ics.uci.edu/~mlearn/MLRepository.html], Irvine, CA: University of California, Department of Information and Computer Science (1998)
2. G. A. Carpenter, S. Grossberg, and J. H. Reynolds, "ARTMAP: Supervised real-time learning and classification of nonstationary data by a self-organizing neural network," Neural Networks, Vol. 4, pp. 565-588 (1991)
3. R. O. Duda and P.E. Hart, "Pattern Classification and Scene Analysis," New York: Wiley (1973)

4. A. F. Gomez-Skarmeta, M. V. F. Jimenez, J. G. Marin-Blazques, "Approximative Fuzzy Rules Approaches for Classification with Hybrid-GA Techniques," Information Sciences, Vol. 136, pp.193-214 (2001)
5. S. Haykin, "Neural Networks, a comprehensive foundation," Prentice Hall, New Jersey (1999)
6. H. Ishibuchi and T. Nakashima, "Voting in Fuzzy Rule-Based Systems for Pattern Classification Problems," Fuzzy Sets and Systems, Vol. 103, pp. 223-238 (1999)
7. R. Jang, "ANFIS: Adaptive network-based fuzzy inference system," IEEE Trans. Syst., Man, Cybern., Vol. 23, pp. 665-685 (1993)
8. C. F. Juang and C. T. Lin, "An On-Line Self-Constructing Neural Fuzzy Inference Network and Its Applications," IEEE Trans. Fuzzy Systems, Vol. 6(1), pp. 12-32 (1998)
9. N. Kasabov, Foundation of Neural Networks, Fuzzy Systems and Knowledge Engineering, The MIT Press, Cambridge, MA (1996)
10. T. Kasuba, "Simplified Fuzzy ARTMAP," IEEE AI Expert, pp. 19-25 (1993)
11. Ho J. Kim, Tae W. Ryu, Thai T. Nguyen, Joon S. Lim, and Sudhir Gupta, "A Modified Fuzzy Min-Max Neural Network for Pattern Classification," Computational Science and Its Applications – ICCSA 2004 (LNCS 3046), pp. 792-798, Springer-Verlag (2004)
12. B. Kosko, Neural networks and Fuzzy Systems: A Dynamical Systems Approach to Machine Intelligence, Englewood Cliffs, NJ:Prentice-Hall (1992)
13. H.-M. Lee, K.-H. Chen, and I-F. Jiang, "A Neural Networks with Disjunctive Fuzzy Information," Neural Networks, Vol. 11, pp. 1113-1125 (1998)
14. C. T. Lin and C. S. George Lee, "Neural-network-based fuzzy logic control and decision system," IEEE Trans. Computers, Vol. 40, No. 12 (1991)
15. D. Nauck and R. Kruse, "A Neuro-Fuzzy Method to Learn Fuzzy Classification Rules from Data," Fuzzy Sets and Systems, Vol. 89, pp. 277-288 (1997)
16. M. Setnes and H. Roubos, "GA-Fuzzy Modeling and Classification: Complexity and Performance," IEEE Trans. Fuzzy Systems, Vol. 8(5), pp. 509-522 (2000)
17. P. Simpson, "Fuzzy min-max neural networks-Part 1: Classification," IEEE Trans. Neural Networks, Vol. 3, pp. 776-786 (1992)
18. K. Tanaka, M. Sano, and H. Watanabe, "Modeling and Control of Carbon Monoxide Concentration Using a Neuro-Fuzzy technique," IEEE Trans. Fuzzy Systems, Vol. 3, pp. 271-279 (1995)
19. C. Z. Ye, J. Yang, D. Y. Geng, Y. Zhou, N. Y. Chen, Fuzzy Rules to Predict Degree of Malignancy in Brain Glioma," Medical and Biological Engineering and Computing, Vol.40, pp. 145-152 (2002)
20. J. S. Wang and C. S. G. Lee, "Self-Adaptive Neuro-Fuzzy Inference System for Classification Applications," IEEE Trans. Fuzzy Systems, Vol. 10(6), pp. 790-802 (2002)
21. W. Wolberg, O. Mangasarian, "Multisurface Method of Pattern for Medical Diagnosis Applied to Breast Cytology," Proc. National Academy of Sciences, Vol. 87, pp. 9193-9166 (1990)
22. L. Zadeh, "Fuzzy sets," Information and Control, Vol. 8, pp. 338-353, 1965.

Evaluation and Fuzzy Classification of Gene Finding Programs on Human Genome Sequences

Atulya Nagar, Sujita Purushothaman, and Hissam Tawfik

Intelligent and Distributed Systems (IDS) Research Group,
Liverpool Hope University, Hope Park, Liverpool L16 9JD
{nagara, tawfikh}@hope.ac.uk

Abstract. This paper presents an evaluation of the four of the more common gene finding programs. The evaluation was conducted on a new data set consisting of only human genome sequences extracted from GenBank. Newest sequences were used to avoid overlap with the training sets of the gene-finding programs. The results of this evaluation are then used to classify the gene finding programs using fuzzy logic. The programs are classified into three fuzzy sets of high, mediocre and low accuracy. The results are then presented in the form of words so as to be easily understood by humans.

1 Background

With many sequencing projects underway, a number of them in different stages of completion, the current challenge is to make sense of the enormous amount of information coming out of these projects. One of the main areas of research is in the identification of actual genes in the sequences. In April 2003, The Human Genome Project announced the completion of the DNA reference sequence of humans. One major task ahead is to find and understand the working of the human genes, since only about 15,000 of an estimated 30,000 to 40,000 of which have been found. Many programs have been developed from more than a decade ago, and many more are in development today to computationally find the genes in the DNA sequences. Previous evaluations of computational gene finding programs have either used genome sequences of more than one organism [11], [3]) or used semi-artificial sequences [7]. This evaluation uses only human genome sequences extracted from GenBank and uses the traditional measures of accuracy. Fuzzy set theory is then used to classify these programs as having high, mediocre or low accuracy.

2 Gene Finding

This work deals only with protein coding genes. Computer based gene finding is essentially based on two methods: homology searches and ab initio prediction. Homology searching programs compare genomic sequence data to gene, protein sequences, etc. of the same or other organisms that have already been identified and are present in databases. Programs that use ab initio methods are often divided into two categories, signal sensors and content sensors. Signal sensors attempt to identify sequence signals such as start and stop codons, promoter sequences, etc. They mainly comprise of pattern recognition methods such as weight matrices, decision trees and neural networks. Content sensors work by trying to find genes by detecting the content of the sequence. Haussler [9] states that the most important and most studied content sensor is the sensor that predicts coding regions. Current gene finding programs range from simple open reading frame (ORF) finders to complex programs that construct gene models by using results generated from a diverse set of resources such as the Combiner [1]. Li [10] maintains a good list of current gene finding programs.

Computational gene-identification will however be useful only if it is proven to be able to achieve useful levels of prediction accuracy. Also, there currently are a range of gene finding programs with different strengths and weaknesses and different capabilities and limitations. Hence independent evaluations to determine the accuracy of gene finding programs are necessary. Burset and Guigo's work [3] evaluated the programs that predicted protein coding genes in genomic DNA sequences. That work has now become the de facto standard for evaluation of the accuracy of gene-identification programs. Rogic [11] also performed an evaluation of computational gene finding programs, based on Burset and Guigo's [3] standards. Guigo et al. [7] have also done an evaluation of gene-prediction accuracy on large semi-artificial DNA sequences while Zhang and Zhang [17] have conducted an evaluation of gene finding algorithms by a content-balancing accuracy index.

Current computational methods for gene finding have several drawbacks. All predictions are of protein coding regions: non-coding regions are not detected. Also, since only protein genes are detected, non-protein coding RNA genes are not found. Predictions are generally for 'typical' genes, and must have a beginning and an end. Partial and multiple genes are often missed, and methods are sensitive to G+C content [2]. It has been found that important structural properties of genes are strongly correlated with the amount of the bases Guanine and Cytosine (G+C content) in the genomic sequences [6]. Genes from G + C-poor regions code for proteins that are on average longer then those from G + C-rich regions. With just this list of drawbacks, it can be seen that there still is a long way to go to achieve highly reliable computational gene finding. It is hoped that this evaluation using random real sequences, while not taking into account most of these constraints, will give us an idea of the current state of computational gene finding and thus enable us to form an idea of the progress that needs to be made to be able to achieve the ultimate goal of computational gene finding with results that very closely match biological laboratory methods.

2.1 Motivation

Although there have been a few independent evaluations of computational gene finding programs, there hasn't been one done solely to evaluate the accuracy of these programs on only human genomic sequences. Guigo et al. [7] used a semi-artificial test set. Since their work, the Human Genome Project has in April 2003 announced the completion of sequencing of the human genome. Much more actual data has therefore become available. It is now possible to use the gene finding programs on substantial real genomic sequences. This became the motivation to conduct this evaluation of gene finding programs with test sequences specifically from the human genome.

2.2 The Data Set

The data set used for this evaluation consists of exactly one hundred human genomic gene sequences extracted from Genbank using Entrez Gene. Only sequences of protein coding genes that were created between the 1st of June 2003 and the 1st of January 2005 with validated RefSeqs were chosen. No further criterion was considered, to allow the evaluation to be conducted on random real world data.

3 Test Method

Only predictions for the Watson or plus strand were taken into account for this evaluation. The predictions of the gene finding programs were saved into local files. The CDS section of the GENBANK entry for each gene contains the actual coding sequences and this information was also annotated and stored locally. Rogic's [11] script for calculating the accuracy measures was modified and rewritten into C. The accuracy measures were calculated by reading data from the files containing the predicted data and actual data. The results are shown in Table 2.1.

4 Accuracy Measures

The accuracy of a gene finding program can measured at nucleotide level and exon level. This evaluation uses the accuracy measures that have now become the *de facto* standard for measuring the accuracy of gene finding programs.

The nucleotide level accuracy gives a measure of the search by content ability of the gene finding program, while the exon level accuracy gives a measure of the search by signal ability of the program.

4.1 Nucleotide Level Accuracy

At the nucleotide level, accuracy is measured by comparing the predicted status (coding or non-coding) and actual status of each nucleotide along the test sequence. Here both prediction and reality are binary variables whose values (coding or non-coding) have been observed along the nucleotides in the sequence. The following values are then defined:

1. TP - the number of coding nucleotides predicted correctly as coding
2. TN - the number of coding nucleotides predicted correctly as non-coding
3. FP - the number of non-coding nucleotides predicted incorrectly as coding
4. FN - the number of non-coding nucleotides predicted incorrectly as non-coding

Then we can find the proportion of coding nucleotides that are correctly predicted as coding, sensitivity (Sn), with the formula:

$$Sn = \frac{TP}{(TP+FN)} \quad (1)$$

and the proportion of nucleotides that are predicted as coding and are actually coding, specificity (Sp), with the formula:

$$Sp = \frac{TP}{(TP+FP)} \quad (2)$$

Sensitivity and specificity are in fact conditional probabilities. Sn is the probability of a nucleotide that is actually a coding nucleotide being predicted as coding, and Sp is the probability of a nucleotide that has been predicted as a coding nucleotide to be actually a coding nucleotide.

Neither Sensitivity nor Specificity alone constitute good measures of global accuracy, therefore a single scalar measure that captures both sensitivity and specificity is used. This is called correlation coefficient and is defined as :

$$Cc = \frac{(TP*TN) - (FN*FP)}{\sqrt{(TP+FN)*(TN+FP)*(TP+FP)*(TN+FN)}} \quad (3)$$

The Correlation Coefficient has been widely used to evaluate gene finding programs [8], [13],[4],[15],[14]. However it is not defined in cases such as those where the input sequence has no coding nucleotides or the prediction contains no coding nucleotides. A similar measure that always defined is the approximate correlation, AC:

$$AC = (ACP - 0.5) * 2 \quad (4)$$

Where ACP is the average conditional probability defined as:

$$ACP = \frac{1}{4}\left(\frac{TP}{TP+FN} + \frac{TP}{TP+FP} + \frac{TN}{TN+FP} + \frac{TN}{TN+FN}\right) \quad (5)$$

4.2 Exon Level Accuracy

Exon level accuracy of the gene finding programs is evaluated by comparing predicted and true exons along the test sequence. This is also a widely used approach. However, the prediction of exons is not a crisp value as with nucleotide level accuracy. Exons can be exactly predicted, with both boundaries correctly predicted, partially correct with either one boundary predicted correctly, partially correct with neither boundary correctly predicted but having a region of overlap, and wrong exons. The next section of this work discusses the accuracy in terms of fuzzy logic to take into account this fuzzy concept of correctly predicted exons. Here we continue with the traditional method of evaluating accuracy by determining the exon level sensitivity (ESn) and (ESp) :

$$ESn = \frac{NT}{NA} \quad (6)$$

and

$$ESp = \frac{NT}{NP} \quad (7)$$

Table 1. Results from the current evaluation

Programs	Nucleotide Level Accuracy		Exon Level Accuracy				
	Sn	Sp	ESn	ESp	ME	WE	OL
HMMGene	0.81	0.78	0.53	0.55	0.22	0.22	0.04
GeneMark	0.75	0.76	0.42	0.42	0.22	0.25	0.10
FGENES	0.75	0.73	0.54	0.49	0.19	0.26	0.08
FGENESH_GC	0.51	0.46	0.37	0.33	0.47	0.5	0.08

Table 2. Results from Rogic's (2000) HMR 195 dataset

Programs	Nucleotide Level Accuracy		Exon Level Accuracy				
	Sn	Sp	Sn	Sp	ME	WE	OL
HMMGene	0.93	0.93	0.76	0.77	0.12	0.07	0.02
GeneMark	0.87	0.89	0.53	0.54	0.13	0.11	0.09
FGENES	0.86	0.88	0.67	0.67	0.12	0.09	0.02

Where

1. NT = the number of exactly predicted exons, also called true exons.
2. NA = the number of annotated exons
3. NP = the number of predicted exons

As with nucleotide level accuracy, these measures are not wholly correct when used alone, and hence usually their average measure is used as a reliable measure of the program's exon level accuracy. The tables also show values for:

1. OL - proportion of predicted exons that overlap actual exons
2. ME - proportion of missed exons
3. WE - proportion of wrong exons.

(Burset and Guigo, 1996 and Rogic, 2000)

5 Discussion

The results of this evaluation are shown in table 2.1. The results of the evaluation conducted by Rogic [11] using the HMR195 data set are shown in table 2.2 to allow for easy comparison. It can be seen that the accuracy in each case is lower than the accuracy achieved with the previous evaluations of the corresponding programs conducted by Rogic [11]. This can be attributed to the difference in data sets. Rogic's [11] dataset was a carefully considered set, where many criteria were used to select the genomic sequences such as excluding sequences containing alternatively spliced genes, sequences whose coding region contains in-frame stop codons, etc., whereas the data set used in this evaluation can be said to be partially 'blind', i.e. any protein coding genic sequence with a suitable length for the gene finding programs were accepted. Also, Rogic's data set consisted of human, mouse and rat genomic sequences while the data set for the current evaluation consists of only human genomic sequences. Thus it can be seen that while the current range of gene finding programs are able to find genes in genomic sequences, there is still a long way to go to achieve reliable prediction on random real genomic sequences.

6 Classification Using Fuzzy Logic

6.1 Introduction

The previous section presented the results of evaluation of several gene finding programs. The accuracy was then computed using the conventional accuracy measures of nucleotide and exon level sensitivities and specificities, and the proportions of missed, wrong and overlapping exons. This information would be useful to both biologists looking to use the gene finding programs in a biological context, as well as to computer scientists looking to understand and work on the gene finding programs in a computational context. However, to neither of these groups of people would the accuracy measures be immediately useful: they would have to study the meaning of the specific measures in order to form an idea of the accuracy. This is because it would not be immediately apparent if, for example, 0.71 is considered to be a high value for sensitivity or the opposite, a low value for sensitivity. One would need to look at the values for sensitivity for a number of programs with a range of sensitivities before being able to guess if one particular value is high or low. Hence it can be said that accuracy measures presented in the previous section are just data - they will not give us knowledge without deeper study. Therefore we propose the use of fuzzy logic to classify these programs, with the aim of being able to convey sufficient information to both biologists and computer scientists regarding the accuracy of the gene finding programs.

Zadeh in 1965 [16] introduced the concept of fuzzy sets. He defined the term fuzzy logic as a set of mathematical principles for knowledge representation based on degrees of membership rather than on crisp membership of classical binary logic. According to this definition, a temperature of zero degrees would have the lowest degree of membership (i.e. not a member at all) in the fuzzy set of hot temperatures, and would have the highest degree of membership in the fuzzy set of cold temperatures. Fuzzy logic thus, unlike conventional Boolean logic, is multi-valued and deals with degrees of membership and degrees of truth. This makes it better suited for modelling real world knowledge and problems.The problem of classifying the gene finding programs in terms of levels of accuracy is similar to the classification of hot and cold temperatures. There is no sharp distinction between any program that defines it to be accurate or not. In fact, accuracy is a 'fuzzy' value. It is therefore apt to model the accuracy of the gene finding programs using fuzzy sets. If we go one step further and label the fuzzy sets using human-understandable words, instead of using crisp numbers, we are then able to easily convey sufficient information to the persons interested in simply knowing which program is more accurate, and in what ways.

Fuzzy logic has been used in classification in many other applications. Different and more complex fuzzy methods such as rule-based and decision tree fuzzy classification methods have been used. Examples range from Roubos *et al.* [12] who have developed an automatic rule-based classification system for classifying the Wine data, to Driese [5] who uses a fuzzy classification for developing and measuring the accuracy of a vegetation map.

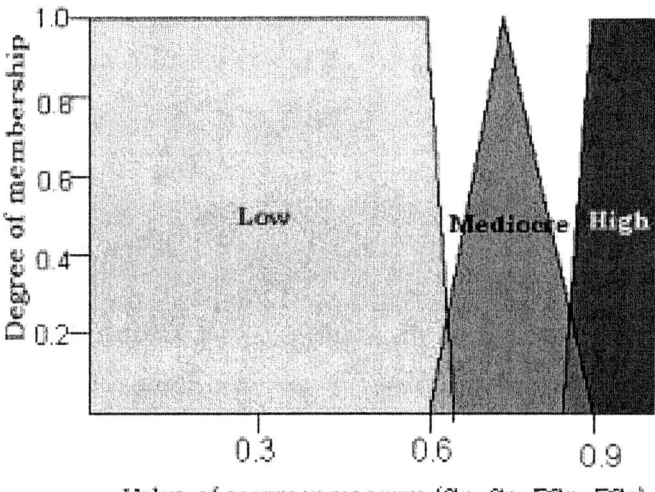

Fig. 1. Fuzzy sets Low, Mediocre, High

6.2 Accuracy Measures Classified Using Fuzzy Sets

In order to classify the accuracy measures, we define three fuzzy sets for each accuracy measure: high, mediocre and low as shown in Fig. 1. We then determine the membership of each member of the data set used in the evaluation using the membership function shown below.

Fuzzy set Low of the data set X is defined by the function $\mu_m(x)$ such that

$$\mu_m(x) : X \rightarrow [0, 1] \qquad (8)$$

Where

$$\mu_m(x) = 1 \quad \text{if } x \text{ is totally in Low} \qquad (9)$$
$$\mu_m(x) = 0 \quad \text{if } x \text{ is not in Low} \qquad (10)$$
$$0 < \mu_m(x) < 1 \quad \text{if } x \text{ is partly in Low} \qquad (11)$$

Identical membership functions are used for the fuzzy sets Mediocre and High.

For each of the accuracy measures for each program, the membership of each member of the data set is then determined; then the set with the highest number of members for each accuracy measure determines the characteristic of that accuracy measure for that program. For example, for the program FGenes, for the first sequence in the data set, the value Sn, nucleotide level sensitivity is determined. If this value is higher than or equal to 0.9, then the first sequence belongs to the fuzzy set High with a membership degree of 1. Otherwise, if this value is less than 0.9 and greater than 0.85, then it partly belongs in the fuzzy set High and partly belongs to the fuzzy set Mediocre. The membership degree is arbitrarily assigned to the value 0.2. These membership values are then totalled and the set with the highest value determines the

Table 3. Results of the current evaluation for GeneMark

Programs	Nucleotide Level Accuracy		Exon Level Accuracy				
	Sn	Sp	Sn	Sp	ME	WE	OL
GeneMark	0.75	0.76	0.42	0.42	0.22	0.25	0.10

characteristic of that particular accuracy measure. The specific values which determine the set are in this instance, determined by a detailed comparison of the raw data. So, continuing with the example for FGenes, if the data set contains 5 members each with the following memberships and degrees of memberships for the accuracy measure nucleotide level sensitivity:

1. Member no. 1 belongs to fuzzy set High with a membership degree of 1.0
2. Member no. 2 belongs to fuzzy set High with a membership degree of 1.0
3. Member no. 3 belongs to fuzzy set High with a membership degree of 0.5 and belongs to the fuzzy set Mediocre with a membership degree of 0.5
4. Member no. 4 belongs to fuzzy set Mediocre with a membership degree of 1.0
5. Member no. 5 belongs to fuzzy set Low with a membership degree of 1.0

Thus the total for the fuzzy set High is 2.5, for the fuzzy set Mediocre is 1.5 and for the fuzzy set Low is 1.0. Hence the program FGenes can be said to have 'high' nucleotide level sensitivity. In this way, the characteristic of each of the accuracy measures is determined. This result is then displayed in words, for example the output from the program GeneMark is :

``This program has high sensitivity and high specificity at the nucleotide level, and has low sensitivity and low specificity at the exon level. ''

Now, if compared with the results in the form above, and the 'conventional' results, in the table form as shown in table 3.1, we can easily see the usefulness of the fuzzy classification. With the results from the fuzzy classification, one would be able to get an idea of the accuracies of each program at a glance.

6.3 Results from Fuzzy Classification

The fuzzy classification method presented in this work gave the following results:

1. HMMGene:
   "This program has high sensitivity and high specificity at the nucleotide level, and has low sensitivity and low specificity at the exon level. "
2. FGENES:
   "This program has high sensitivity and high specificity at the nucleotide level, and has low sensitivity and low specificity at the exon level. "
3. FGENESH_GC:
   "This program has low sensitivity and low specificity at the nucleotide level, and has low sensitivity and low specificity at the exon level."

4. GeneMark:
"This program has high sensitivity and high specificity at the nucleotide level, and has low sensitivity and low specificity at the exon level. "

References

1. Allen, J. E., Pertea, M. and Salzberg, S.L.: Computational gene prediction using multiple sources of evidence. Genome Research, **14(1)** (2004)
2. Baxevanis, A.: Predictive Methods using DNA and Protein Sequences Current Topics in Genome Analysis, National Institute of Health, US. (2000) [Available online at http://www.genome.gov/Pages/Hyperion/COURSE2000/Pdf/baxevanis_lec1.pdf last accessed Feb 2005]
3. Burset,M.and Guigo,R.: Evaluation of gene structure prediction programs. Genomics **34** (1996) 353 -367.
4. Dong, S., and Searls, D. B.: Gene structure prediction by lin- guistic methods. Genomics **23** 540-551 (1994)
5. Driese,K.L. : A Vegetation Map for the Catskill Park, NY, Derived from Multi-temporal Landsat Imagery and GIS Data Northeastern Naturalist (2004)
6. Duret, L., Mouchiroud, D., and Gautier, C.: Statistical analysis of vertebrate sequences reveals that long genes are scarce in GC-rich isochores. J. Mol. Evol. **40** (1995) 308-317
7. Guigo et al: An Assessment of Gene Prediction Accuracy in Large DNA Sequences Genome Research **10(10)** (2000)
8. Guigo et al. (1992) *Prediction of gene structure*. J. Mol. Biol. 226: 141-157.
9. Haussler, D.: Computational Genefinding. Trends in Biochemical Sciences, Supplementary Guide to Bioinformatics (1998) 12-15
10. Li, W.: Computational Gene Recognition Programs (2004) [Available at: http://www.nslij-genetics.org/gene/programs.html last accessed Nov 2004]
11. Rogic, S.: Evaluating and Improving the Accuracy of Computational Gene Finding on Mammalian DNA Sequences, MSc. thesis, University of British Columbia, Canada (2000) [Available online at : http://www.cs.ubc.ca/~rogic/MSCthesisWeb.pdf last accessed Nov 2004]
12. Roubos, J.A., Setnes, M. and Abonyi, J.: Learning Fuzzy Classification Rules From Data Developments in Soft Computing, Springer - Verlag Berlin/Heidelberg, 108-115, 2001
13. Snyder, E. E., and Stormo, G. D. (1993). Identification of coding regions in genomic DNA sequences: An application of dynamic programming and neural networks. Nucleic Acids Res. 21: 607-613
14. Solovyev, V. V., Salamov, A. A., and Lawrence, C. B.: Predicting internal exons by oligonucleotide composition and discriminant analysis of spliceable open reading frames. Nucleic Acids Res. **22** 5156-5163. (1994)
15. Xu, Y., Mural, R. J., and Uberbacher, E. C. (1994b). Constructing gene models from accurately predicted exons: An application of dynamic programming. Comput. Appl. Biosci. **10** 613-623.
16. Zadeh, L.: Fuzzy Sets. Information and Control **8(3)** 338-353. (1965)
17. Zhang, C. and Zhang, R.: Evaluation of Gene-Finding algorithms by a Content-Balancing Accuracy Index J. Biomol. Struct. Dyn. **19** (2002) 1045-52

Application of a Genetic Algorithm — Support Vector Machine Hybrid for Prediction of Clinical Phenotypes Based on Genome-Wide SNP Profiles of Sib Pairs

Binsheng Gong[1], Zheng Guo[1,2], Jing Li[1], Guohua Zhu[3], Sali Lv[1], Shaoqi Rao[1,4,*], and Xia Li[1,2,*]

[1] Department of Bioinformatics, Harbin Medical University, Harbin 150086, P.R. China
gongbinsheng@gmail.com, markgz@0451.com,
{lijing, zhugh, lvsl}@ems.hrbmu.edu.cn
[2] Department of Computer Science, Harbin Institute of Technology,
Harbin 150080, P.R. China
[3] Department of Epidemiology and Biostatistics, Case Western Reserve University,
Cleveland, OH 44106, USA
[4] Departments of Cardiovascular Medicine and Molecular Cardiology,
Cleveland Clinic Foundation, Cleveland, Ohio 44195, USA
raos@ccf.org

Abstract. Large-scale genome-wide genetic profiling using markers of single nucleotide polymorphisms (SNPs) has offered the opportunities to investigate the possibility of using those biomarkers for predicting genetic risks. Because of the special data structure characterized with a high dimension, signal-to-noise ratio and correlations between genes, but with a relative small sample size, the data analysis needs special strategies. We propose a robust data reduction technique based on a hybrid between genetic algorithm and support vector machine. The major goal of this hybridization is to fully exploit their respective merits (e.g., robustness to the size of solution space and capability of handling a very large dimension of features) for identification of key SNP features for risk prediction. We have applied the approach to the Genetic Analysis Workshop 14 COGA data to predict affection status of a sib pair based on genome-wide SNP identical-by-decent (IBD) informatics. This application has demonstrated its potential to extract useful information from the massive SNP data.

1 Introduction

The linkage analysis based on sib-pair's identical-by-decent statistics is one of the most popular methods for mapping genes for complex human diseases. Nevertheless, application of the sib-pair informatics to predict the disease status of a sib pair has not seen in the literature. The vast amount of data being acquired in the large-scale genome-wide genetic profiling using makers of single nucleotide polymorphisms

---

[*] Corresponding authors.

(SNPs) such as for the Genetic Analysis Workshop (GAW) 14 COGA data has provided ideal opportunities to address this issue. Its medical implications include a new molecular way to early prediction and reliable diagnosis.

The high-throughput SNP profile data have the characteristics of the high dimension (over ten thousands in this GAW data), noise, and relative small sample sizes. Therefore, data reduction (or called feature selection) techniques for analysis of the high dimensional data have been one of the focuses in the methodological development in recent years. Their utilizations are not simply to reduce the dimension of SNP features and to avoid the curse of dimension. More importantly, their applications can improve the prediction performance of the resulted classifiers, and exclude the interferences of a large number of irrelevant features with hunting for the molecular signatures for clinical implications.

A procedure for feature selection can be divided into two distinct steps: the search step and the evaluation step. Although an exhaustive search over the entire feature space, and branch-and-bound algorithm [3] can lead to an optimal solution, the two approaches have rarely been used in analysis of the high-dimension biological data because of computational costs. Thus, heuristic search methods such as Greedy Climbing Hill [4], the Best First Method and Genetic Algorithm(GA) [5,6] have received increased attentions for dimension reduction. The first two methods search for an optimum by changing the local search space though The Best First Method allows backtracking along the search path. They thus fail to capture many important feature subsets. In either approaches, once a feature is taken in (or removed out), it will never be considered again. Genetic algorithm is an adaptive search engine that emulates the natural selection process in genetics [5]. It employs a population of competing solutions—evolved over time by crossover and mutation; and selection—to converge to an optimal solution. The solution space is efficiently searched in parallel and a set of solutions instead of a solution are computed to avoid becoming stuck in a local optimum that can occur with other search techniques. In addition, its robustness to size of search space and the underlying multivariate distribution assumptions has made it a promising method for feature selection over a high-dimension space. Nevertheless, genetic algorithm itself is merely a searching algorithm. To apply this algorithm to biological applications, several issues need to be resolved. First, a suitable fitness function has to be defined to map the biological reality. Second, the number of features contained in the optimal feature subsets can be large at early generations. The coupled classifier evaluating fitness of a candidate feature subset must have the capacity for handling the data of a very high-dimension feature space but of a limited sample space. There are many potential choices for a classifier, but majority can only deal with a limited dimension of features. Support Vector Machine (SVM) [2], resulted from recent advances in statistical learning theory and machine learning, is an exception. Its unique advantages for treating this particular data structure, for avoiding overfitting and dimensional curse, and for nonlinear modeling, have made it a popular tool in pattern recognitions. In this study, we evaluate a hybrid between genetic algorithm and support vector machine (termed GA-SVM) that can fully utilize the unique merits of the two data mining tools. Genetic algorithm is used as the search engine, while support vector machine is used as the classifier (the evaluator). We apply the proposed approach to predict the affection status of a sib pair based on genome-wide IBD profiling for the COGA study. This application has demonstrated its potential to extract useful information from the massive SNP data.

2 Methods

2.1 Defining the Phenotype(s) of a Sib Pair

Consider a binary disease trait. Each sib can take any one of two possible value, say, c (c = 1, 2). We define the phenotype of a sib pair (a specific combination of two values of a sib pair) as: Gi (i = 1, 2, 3):

G2 = concordant affected, both sibs in a sib pair are affected,
G1 = discordant, only one in a sib pair is affected,
G0 = concordant unaffected, no sibs in a sib pair are affected.

In this study, we exclude the data for discordant sib pairs so that we have a population consisting of two mutually exclusive groups.

2.2 Defining the IBD Feature Vector

Next, for each sib pair we define a feature vector (X) which can include the estimated proportions of alleles (x) shared IBD by the sib pair at L markers along a chromosome (segment). The feature vector can also include clinical features to construct a clinical-genomic model. To investigate the genetic effects of the underlying molecular signatures, we are searching for a cluster of SNP markers whose IBD distributions among the disease affection groups lead to the best-fit partition (grouping) to the observed one. Consider a general setting of K disease affection groups for an ordinal trait (here K = 2) and M feature variables (L SNPs plus M-L clinical covariates). Let $N_1, N_2, \ldots, N_K$ be sample sizes for Gi (i = 1, 2, ..., K). Then, the feature vector data for N (N = ΣNi) sib pairs can be expressed as:

$$\begin{array}{cccc} x_1^{(1)}, & x_2^{(1)}, & \cdots, & x_{N_1}^{(1)} \\ x_1^{(2)}, & x_2^{(2)}, & \cdots, & x_{N_2}^{(2)} \\ \cdots, & \cdots, & \cdots, & \cdots \\ x_1^{(K)}, & x_2^{(K)}, & \cdots, & x_{N_K}^{(K)} \end{array}$$

2.3 The GA-SVM Algorithm

First, we generate randomly the fixed-length binary strings for n individuals of feature IBDs to build up the initial population. Each string represents a (IBD) feature subset (coding of the feature subset) and the values at each position in the string are coded either presence or absence of a particular IBD feature. Then, we calculate the fitness (i.e., how well a feature subset survives over the specified evaluation criteria) for each feature subset. Here, we adopt classification accuracy as the fitness index (eval), evaluated using a linear SVM. Better feature subsets have a greater chance of being selected to form a new subset through crossover or mutation. Mutation changes some of the values (thus adding or deleting features) in a subset randomly. Crossover combines different features from a pair of subsets into a new subset. The algorithm is an iterative process where each successive generation is produced by applying genetic operators to the members of the current generation. In this manner, good subsets are "evolved" over time until the stopping criteria are met. The detailed computational procedures and the Matlab source codes can be found at www.biocc.net/ga-svm. We classify the phenotypes of sib pairs with the IBD features contained in each individual using a linear SVM. The accuracy of classification is estimated:

$$acc = (\sum_{t=1}^{T} I(y_t, \hat{y}_t))/T \qquad (1)$$

where T is the number of test samples and,

$$I(y_t, \hat{y}_t) = \begin{cases} 1 & \text{if } y_t = \hat{y}_t \\ 0 & \text{otherwise} \end{cases} \qquad (2)$$

In this study, a five-fold cross-validation (CV) resampling approach is used to construct the learning and test sets. First, the two-class samples are randomly divided into 5 non-overlapping subsets of roughly equal size, respectively. A random combination of the subsets for the two classes constitutes a test set and rest of subsets are totally used as the learning set. The 5-fold CV resampling produces 25 pairs of learning and test sets. Each individual is evaluated by the averaged value over the 25 pairs, i.e.,

$$eval_j = (\sum_{k=1}^{25} acc_k)/25 \qquad (3)$$

where k is the replicate number and acc_k is the classification accuracy for the kth replicate. The data reduction process stops when the drop (or difference) in classification accuracy in the best individuals at successive generation reaches to 0.001.

3 Application To GAW 14

We perform analysis of all the Affymetrix SNP markers from chromosomes 1-22 in the GAW 14 COGA data, simultaneously. First, using a t-test ($\alpha=0.01$), we have filtered out 117 markers of differential IBD using 790 sib pairs of two affection groups (concordant affected or concordant unaffected for ALDX1, an alcoholic dependence measure). The resulted IBD matrix is of 790×117 dimensions. We conduct the SNP feature selection on the data by using the GA-SVM algorithm.

We start with all the 117 markers and then step down to reduce the dimension of the markers successively for 8 generations. For each generation, we first randomly halve the number of markers of the previous one in each individual. As shown in Figure 1, the accuracy of GA-SVM for classification of a sib-pair affection status drops slightly from 80.48% (with all 117 SNPs as predictors) to 76.00% (with the converged subset of 18 SNPs (Table 1) as the prediction set) and then is stabilized. For comparison, a counterpart hybrid between genetic algorithm and K-nearest neighbors (KNN) is also performed, which has generally significant reduced performance than our proposed hybrid. This application implicate that there is a space for further improvement in predicting the alcoholism affection status of a sib pair based on genome-wide SNP IBD profiles, perhaps with the contributed other clinical risk factors, to some extent which have reflected the complex nature of the behavior disorder.

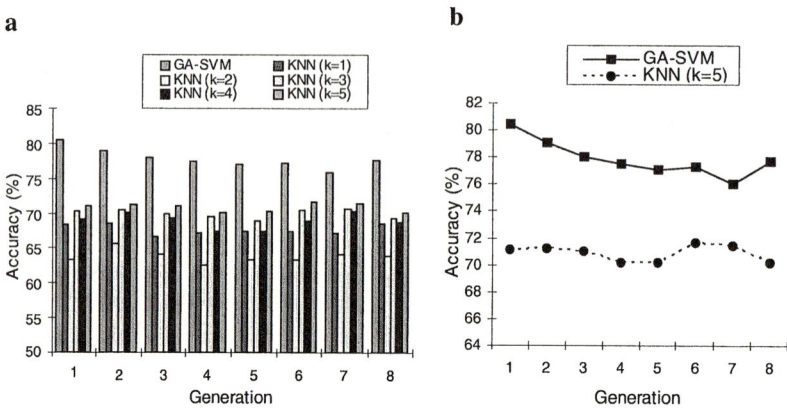

Fig. 1. Comparison of GA-SVM and GA-KNN for classification of alcoholism status of a sib pair based on SNP IBD profiles. Shown are the averaged accuracies of the best individual trained at each generation and evaluated with 25 replicates produced from a 5-fold cross-validation procedure.

Table 1. The list of SNPs in the best converged individual trained with GA-SVM for 5 generations

CHR01	CHR02	CHR03	CHR04	CHR05	CHR06	CHR07
tsc0239498**	tsc0515224**	tsc0049652**	tsc0272738**	tsc0624855	tsc0525458	tsc0049271
tsc1482739	tsc1531602*		tsc1210823**	tsc0797361	tsc0593649	tsc0051005*
tsc0302182**						
tsc0611403						

CHR08	CHR10	CHR12	CHR13	CHR15	CHR16	CHR22
tsc0671100*	tsc0499894	tsc0060814*	tsc0249697	tsc0149546**	tsc0446682	tsc1721389
tsc0697651*	tsc0811622					

Note: Six shaded markers are deleted by GA-SVM at the last generations and the final individual contains 18 markers.
* and ** $P < 0.05$ and $P < 0.01$ with Haseman-Elston regression test, as implemented in SIBPAL2 of the S.A.G.E. package.

4 Discussions

A promising direction of research for analysis of the large-scale SNP data is to perform a feature selection procedure for shrinking the feature space, and achieving a high generalization performance for biological application, such as genetic risk prediction in this study. Current methods for the feature selection can be divided into three groups: marginal filtering, sequential selection, and strictly multivariate modeling. Marginal filtering [1] has computational advantages of speed and simplicity, but it can not utilize the information hidden in gene interactions. Theoretically, distribution-based multivariate modeling is the most powerful and can take into account the multivariate correlation structures of the high-dimension SNPs simultaneously. Nevertheless, its computational complexities confine its application to a very limited dimension. Sequential selection approaches lie in between the above two methods and enjoy extensive application for feature selection, particular in machine learning and pattern recognitions.

Acknowledgements

This work was supported in part by the National Natural Science Foundation of China (30170515, and 30370798), the Chinese 863 Program (2003AA2Z2051), the 211 Project, the Tenth 'Five-year' Plan, Harbin Medical University and Heilongjiang Province Science and Technology Key Project (GB03C602-4 and 1055HG009).

References

1. Blum, A.L., Langley, P.: Selection of relevant features and examples in machine learning. Artificial Intelligence. 97 (1997) 245-271
2. Brown, M.P., Grundy, W.N., Lin, D., Cristianini, N., Sugnet, C.W., Furey, T.S., Ares, M., Jr., Haussler, D.: Knowledge-based analysis of microarray gene expression data by using support vector machines. Proc Natl Acad Sci U S A. 97 (2000) 262-267
3. Brusco, M.J.: An enhanced branch-and-bound algorithm for a partitioning problem. Br J Math Stat Psychol. 56 (2003) 83-92
4. Holte, R.C. Combinatorial auctions, Knapsack problems, and Hill-Climbing search. In Proceedings of the 14th Biennial Conference of the Canadian Society on Computational Studies of Intelligence: Advances in Artificial Intelligence, June 07-09. Ottawa, Canada: Springer; 2001. 57-66 p.
5. Houck, C.R., Joines, J.A., Kay, M.G., Wilson, J.R.: Empirical investigation of the benefits of partial Lamarckianism. Evol Comput. 5 (1997) 31-60
6. Stefanini, F.M., Camussi, A.: The reduction of large molecular profiles to informative components using a genetic algorithm. Bioinformatics. 16 (2000) 923-931

A New Method for Gene Functional Prediction Based on Homologous Expression Profile

Sali Lv[1], Qianghu Wang[1], Guangmei Zhang[3], Fengxia Wen[4], Zhenzhen Wang[1], and Xia Li[1,2]

[1] Department of Bioinformatics, Harbin Medical University, Harbin, China 150086
lvsali@163.com, strongtiger2000@sina.com,
{wangzz, lixia}@ems.hrbmu.edu.cn
http://www.biocc.net
[2] Department of Computer Science, Harbin Institute of Technology, Harbin, China 150001
[3] Heilongjiang University of Chinese Medicine, Harbin, China 15040
yzgm66@yahoo.com.cn
[4] Department of Mathematics, Baotou Medical College, NeiMeng, China 014010
fengxiawen@126.com

Abstract. It is a project with significant challenge to predict functions of genes in the post-genomics era. Most function annotation systems is available to predict functions of part genes, but the rate of annotation is very low and a large number of genes can not be annotated, what's more, the measure of credibility isn't quite sure. Aiming at the problem, we address the new concept of functional expression profile using knowledge system existed, and consider the method of gene cluster mapping associated with hub genes to predict functions of genes which are not still annotated based on the association between gene expression and gene function. At last we applied the method to colon data set. The results implied the prediction efficiency and credibility have been improved significantly by the method.

1 Introduction

Genome researchers have shifted their focus from structural genomics to functional genomics[1].The availability of a growing number of completely sequenced genomes opens new opportunities for understanding of complex biological systems. The sheer amount of data obtained by microarray experiments and complexity of relevant biological knowledge present a number of challenges in functional prediction in the post-genomics era. Its major goal is to understand the functions of organism system by analyzing genome data, so that it is one of the most important task to predict functions of large numbers of genes. It's a deeply significative research project that how to mine the functional information by the methods of bioinformatics facing the vast data. At present, some familiar functional annotation system and related database could predict the function of genes, such as GO,KEGG[2] and Genbank[3], and so on. However, the genes whose function is not known are still in the majority after all. Only annotation information of minority genes can be inferred using annotation system or database query system existed. As a result, the rate of annotation is too small For those genes which can't be annotated, we can realize functional prediction by sequence align and homologous comparison[4].And yet the efficiency of annotation is low by the way and many databases couldn't provide batch process service but submitting record by record. Clearly, it has became a bottleneck for the functional prediction of high-throughout gene. Aiming at the problem, we begin with the association between gene expression and gene function to construct the homologous expression profile using annotated information of genes. We proceed clustering analysis by introducing the measure of functional similarity of genes. As mentioned above the hub genes are selected and then the functional prediction is carried out. The results show that the prediction efficiency and credibility can be improved significantly.

The paper is organized as follows. In Section 2, we describe how to construct functional expression profile, how to construct the matrix of related genes via the cluster results from functional expression profile, how to extract the hub genes and how to predict functions of genes which are not annotated with the cluster involving hub genes. In Section 3, we describe the data experiments and discussion. In this section, data set and experiments are introduced. The experiments provide the results of functional prediction by annotation information from Pathway[5]and GO respectively, comparing the differences between two results and explaining the cause produced. At the same time, we give the credibility of prediction a simple validation. We raise several issues for future work in the last section.

2 Method

2.1 Construct the Homologous Expression Profile Under Different Types of Annotation Systems

Usually, we use matrixes to store the information about the gene expression profiles. Each gene is represented by a row and each example or a type of experimental condition is represented by a column. The element at the crossing indicates the expression value of gene which is corresponding to the row under the sample or the condition represented by the corresponding column. Similar to the way that using matrix to describe gene expression profiles, we can store the annotation information coming from different annotation systems in a matrix. But there are some preliminary works to do, that is, to alternate the annotation results into numerical forms. Because the storage form of this kind of information is similar to the gene expression profile, we call it homologous expression profile. We'll illustrate the steps of developing homologous expression profile by taking the annotation results of genes in GO as an example.

We denote the set of genes annotated and the node set obtained by annotating from GO as $G = \{g_1, g_2, \cdots g_m\}$ and $V = \{v_1, v_2, \cdots v_k\}$ respectively. Then we can represent the annotation result of every gene by using $S_i = \{v_j | 1 \le j \le k\}$ $(1 \le i \le m)$, and all the node sets of annotation results for all of genes can be expressed as

$S = S_1 \cup S_2 \cup \cdots S_i \cdots \cup S_m$. Regarding genes in G as row of matrix and nodes in S as column, we denote the element in a matrix by using the symbol x'_{ij}, where

$$x'_{ij} = \begin{cases} 1, \text{gene } i \text{ is annotated on node } j, i \neq j \\ 0, \text{otherwise} \end{cases} \quad (1)$$

In the matrix of homologous expression profile, each row vector representing every gene suggests the result of gene annotated using GO terms, which shows the nodes that gene is annotated from GO. Each row vector implies the function of a gene. So that, a homologous expression profile reveals the function of every gene in the sense and we also call the homologous expression profile functional profile.

Equivalently, we can gain the homologous expression profile of annotation results corresponding to the genes annotated from Pathway, and both of the two profiles are matrixes with 0 and 1 as the element.

2.2 Functional Prediction

2.2.1 Clustering Based on Homologous Expression Profile

We annotate the genes to be studied in GO and Pathway, and transform the annotation results in homologous expression profile form. Thus, we can store the annotation information of genes with matrix. We can analyze the homologous expression profile data using the methods which have been implemented in analysis for expression profile data.

There are more several methods for clustering, and here the hierarchical-clustering method is applied[6]. As we know Hamming-distance is suitable for processing binary-value, mainly 0 and 1. So we select it as metric for calculating distance in clustering. Hamming-distance is defined by $d_{rs} = (\#(x'_{rj} \neq x'_{sj})/k), (j = 1,2,\cdots k)$, where x'_{rj} and x'_{sj} represent the homologous expression value (0 or 1) of gene r and gene s under the sample j or condition j respectively. $\#$ implies the number of gene pairs which are satisfied some condition and k implies the number of nodes annotated. Hamming distance reveals the difference between annotation results of two genes. The smaller d_{rs} is, the more similar between functions of two genes. In this way, we can get clustering results for annotation information from GO and Pathway.

In addition, clustering is applied to the information of gene expression, which is more easier with Euclidean distance as cluster metric. According to be mentioned above, the clustering results with three ways can be obtained. The clustering results for annotation information from GO and Pathway respectively, and the clustering results for gene expression information, altogether three clustering results are given.

2.2.2 Associated Intensity Matrix of Genes

According to the analysis of cluster based on above, we can gain m cluster results at most by clustering with m genes. If let C be the number of clusters for every cluster result, then we can conclude $C=1,2,\cdots m$. And for every cluster result, an associated intensity matrix of genes can be obtained as $T_C = (t_{ij})_{n \times n}, (C = 1,2\cdots m)$, where

$$t_{ij} = \begin{cases} 1, \text{gene } i \text{ and gene } j \text{ are grouped in the same cluster, } i \neq j \\ 0, \text{otherwise} \end{cases} \quad (2)$$

Under different number of cluster when clustering is applied, the more the associated intensity between two genes is strong, the more the possibility that these two genes are expected to be grouped into the same cluster is high, vice versa. So that we can add all the associated intensity matrix of genes corresponding to every number of cluster, that is $T = \sum_{C=1}^{m} T_C$, which implies the associated intensity between every two genes. But the number of genes annotated is so variational across different annotation system that T obtained from different annotation system are incomparable. It's necessary to make T normalized, the result is given by formula $T' = \frac{\sum_{C=1}^{m} T_C}{m}$, let t'_{ij} be the new form of t_{ij} after T is normalized, and $t'_{i,j} \in [0,1], (i,j = 1,2,\cdots m)$.

For normalized associated intensity matrix of genes T', in which every numeric element reveals the intensity of association between genes standing corresponding row and column. But the elements which can imply significant associated intensity are only part of the whole. Statistical significance test is needed to associated intensity in the associated intensity matrix of genes. Let α be the one-tailed significance level, then all the elements in the matrix are ranked in order of ascending their value, get the element at percentile of $1-\alpha$ as the threshold at last. Thus we can set all of the elements which are smaller than the threshold as zeros. Like this, all the nonzero elements in the matrix can be seen completely significant associated intensity. Let T'' be the associated intensity matrix of genes after intensity value is filtered.

2.2.3 Correlated Gene Matrix

In the associated intensity matrix of genes, nonzero element t''_{ij} represents the significant associated intensity between gene i and gene j. For every gene i ($i = 1,2,\cdots m$) in the associated intensity matrix of genes, let nonzero element t''_{ij} in the row vector corresponding to gene i be replaced by some identifier of gene j ($1,2,\cdots m, j \neq i$), such as gene ID, gene accession, and so on. We call the matrix correlated gene matrix, represented by T_{ID}.

2.2.4 Construct Functional Prediction Profile of Correlated Genes

We can construct related gene network considering the associated intensity matrix of genes as adjacency matrix[7,8]. In the associated intensity matrix of genes T'', every row vector reveals the connection between the gene represented by the row vector and the other genes. In the correlated gene matrix, the genes involved in every row are connected with the gene represented by the row vector. And the genes may have same or similar function. A related gene network can be obtained from an associated intensity matrix of genes. Furthermore, we can also seek the hub genes (connecting with other genes in high frequency) utilizing the correlated gene network.

Equivalently hub genes are also identified by correlated gene matrix, based on the corresponding relation between known correlated gene matrix and related network. According with the concept 'degree' in gene network, degree also exists in correlated gene matrix. Degree of every gene indicates the number of nonzero elements in the row vector representing the gene. Here there is not any difference in in-degree and out-degree, namely no direction. Being similar to the approach that confirms significant associated intensity in the associated intensity matrix, the genes with significantly high degree can be identified. These genes are defined hub genes.

In the correlated gene matrix T_{ID}, extract every row vector corresponding to every hub gene and let the row vectors constitute a new correlated gene matrix T_{hub}. For genes in every row and the corresponding hub gene represented by each row vector in matrix T_{hub}, the probability that they are always clustered together is more big, which reveals they have much more similar functional annotation results. And it is shown that genes in every row have homologous function in which the function of hub gene takes up the principal part.

Recounted above all, no matter which annotation system we chose, can we get a associated intensity matrix of genes, next get the correlated gene matrix. Note that all genes researched can be involved in the correlated gene matrix computed by expression data, but only just part of all genes researched are involved in the correlated gene matrix computed by homologous expression profile data from annotation results. Part or most part of the whole genes are not annotated yet. Next we'll predict the function of the genes not annotated considering the association between gene expression and gene function.

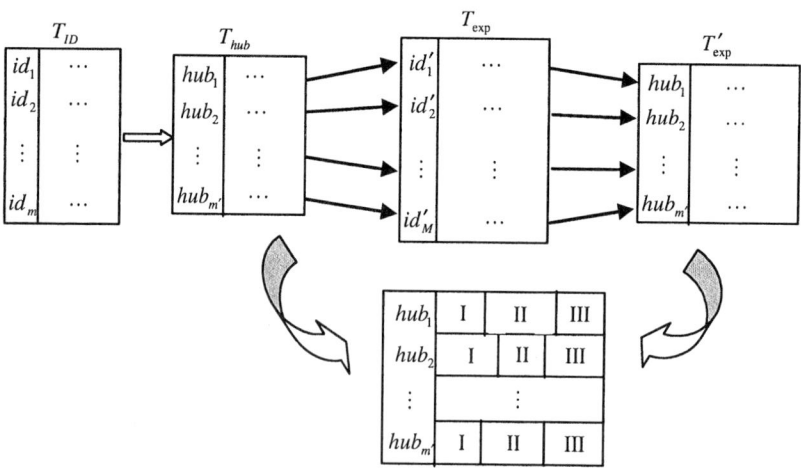

Fig. 1. Flow chart of Constructing correlated gene functional prediction matrix

Generally, we explored a hypothesis that genes with same or similar expression profiles are likely to have similar function. Expression data of all genes researched can be obtained through microarray technology[9]. Then we can get a correlated gene matrix T_{exp} according to the method mentioned. Map the hub genes in the correlated gene matrix T_{hub} to the correlated gene matrix T_{exp} one by one, regarding the hub genes ID as mapping index. Extract row vectors corresponding to the hub genes mapped from T_{exp} and these row vectors extracted combine to form a new matrix T'_{exp}. Finally, merge T_{hub} and T'_{exp} becoming a new matrix T_{pre} called gene functional prediction matrix. Every row vector in T_{pre} consists of three parts: The first part implies the public genes shared by the corresponding row in T_{hub} and T'_{exp}. The second part implies the genes which are involved in corresponding row in T_{hub} but not in T'_{exp}. The third part implies the genes which are involved in corresponding row in T'_{exp} but not in T_{hub}. Therefore we can get the functional prediction results of genes not annotated under the support that the more similar gene expression is, the more gene function is correlated. The function of genes in part three is predicted the same with hub genes and the cluster which the hub genes involved in. The complex prediction process is summarized in Fig.1.

3 Experimental Results and Analysis of Results

3.1 Data Set

The data set used in the study is related to Colon cancer gene expression containing 2000 genes. There are 40 samples included tumor tissue and 22 samples included normal colon tissue. The data set can be downloaded from Affymetrix company web site. The raw array contains about 65,000 Oligonucleotide probes from Affymetrix Oligonucleotide arrays. Only gene expression data including 2000 genes selected by U.ALON is analyzed[10]. All arrays in this dataset have been normalized.

3.2 Experiments and Discussion

Utilizing SOURCE database online, the functional annotation information of 2000 genes by GO is queried. As a result, functional annotation information of 1421 genes among 2000 genes can be obtained by GO. In addition, the annotation information of 2000 genes in Pathway is queried via KEGG database, and only 339 genes are annotated. The number 339 is obviously decreased compared with the former 1421. It's mainly because that the annotation results in KEGG are inferred through circumspect experiments. Consequently, low rate of annotation is concluded.

Construct the corresponding homologous functional expression profile with the annotation results in GO and Pathway according to the approach mentioned above. Then the two homologous expression profiles obtained from the genes annotated and the expression profile obtained from total genes are processed by cluster analysis. Here hamming distance is chosen as a criterion of cluster in homologous expression profiles data and Euclidean distance is chosen in expression profile data. After that, construct associated intensity matrix of genes based on the analysis results of clustering, and correlated gene matrix is constructed then. Next, seek the hub genes and predict the function of genes not annotated with correlated gene matrix, the results in detailed is described in Table 1.

Table 1. Comparison of annotation results obtained from Pathway and GO

Annotation System	Number of Gene Annotated	Matrix of homologous expression profile	Clustering Criteria	Number of Hub Gene	Number of Gene Predicted
Pathway	339	339×101	Hamming	40	1105
GO	1421	1421×1395	Euclidean	224	475

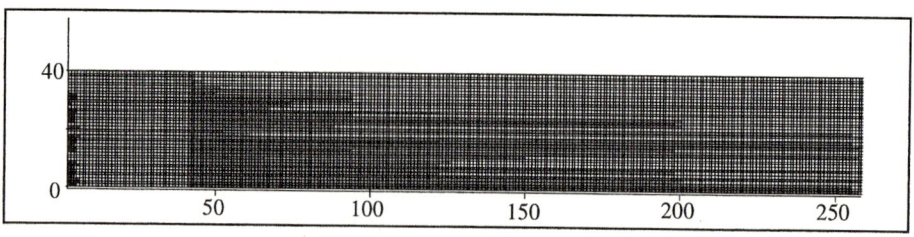

Fig. 2. Functional prediction profile of correlated genes from Pathway annotation system

In figure2 we show a functional prediction profile of correlated genes from Pathway annotation system. In the picture ,the part colored blue reveals the public genes which are shared by the gene cluster corresponding to the hub genes annotated by Pathway and genes cluster mapped to the whole 2000 genes by hub genes annotated in Pathway. The part colored green reveals the genes which belong to the former but not to the latter. The part colored red reveals the genes which are involved in the latter but not in the former. And the genes colored red are just what we are predicting. We think the function of the predicted genes is similar to the function of hub genes. Yellow is as background color, no genes involves in it. The functional prediction profile can be inferred by GO annotation system at the same theory. It can be observed that the number of genes predicted using GO annotation information is much fewer than the number of genes predicted using Pathway annotation information from Table 1. The primary cause is number of genes annotated by Pathway is much fewer than the genes annotated by GO. Based on such reasoning the number of genes predicted across Pathway annotation information will be increased relatively, as depicted in Figure 3.On the other hand, although the number of genes predicted using Pathway annotation information is large, its credibility is a bit low compared with the credibility from GO.

Additional analysis shows that the number of genes predicted is correlated with the threshold of associated intensity between genes. Figure 3 shows the difference in the numbers of the genes predicted under different significant level.

In order to quantify the prediction quality, we predict function by randomly selection of pseudo hub genes out of true hub genes. Among the whole genes predicted, regard those genes annotated in Pathway or GO as test samples, and make 'recall' [11] as the measure of predicted credibility. Figure 4 shows the comparison of the predicted results between true prediction and random prediction by 50 permutations[12].We can see from the figure that predicted credibility by true hub genes is evidently high compared with the results under random permutation.

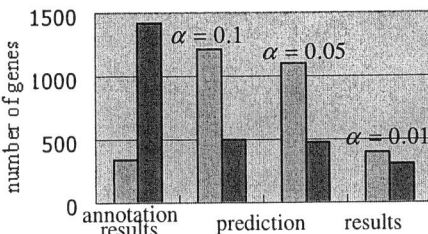
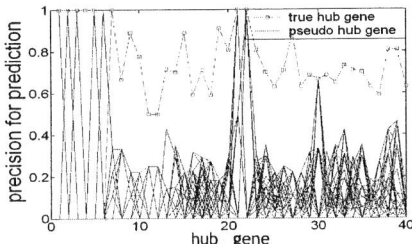

Fig. 3. Comparison of predicted results under different level. Histogram colored blue represents the results of annotation or prediction in Pathway and histogram colored purple represents the same in Go.

Fig. 4. Comparison of the predicted results between prediction by true hub genes and the prediction by pseudo hub genes sampled in random

4 Future Work

Based on those experiments and the analysis we believe our approach is a very effective tool for predicting functions of genes as more diverse experiments are performed. The most important work in the future is to construct gene network based on the hub genes which are regarded as the center of network, and to explore the association between clusters with hub gene as its center. In particular, it's attempt to find disease-related gene clusters according to the differences between gene clusters in normal sample and gene clusters in abnormal sample because of correlation existed among genes.

Acknowledgements

This work was supported in part by the National Natural Science Foundation of China (Grant Nos. 30170515 and 30370798), the National High Tech Development Project of China, the 863 Program (Grant Nos. 2003AA2Z2051 and 2002AA2Z2052), the 211 Project, the Tenth 'Five-year' Plan, Harbin Medical University and Heilongjiang Province Science and Technology Key Project (Grant No. GB03C602-4 and 1055HG009).

References

1. Hieter, P., Boguski, M.: Functional genomics: It's all how you read it. Science. 278 (1997) 601–602
2. Kanehisa, M., Goto, S.: KEGG: kyoto encyclopedia of genes and genomes. Nucleic Acids Res. 28 (2000) 27-30
3. Benson, D.A., Karsch-Mizrachi, I., Lipman, D.J., Ostell, J., Wheeler, D.L.: GenBank. Nucleic Acids Res. 31 (2003) 23-27

4. Haiyuan, Y., Nicholas, M., Luscombe: Annotation Transfer Between Genomes: Protein-Protein Interologs and Protein-DNA Regulogs. Genome Research. 14 1107-1118.
5. Karp, P.D.: Pathway databases: a case study in computational symbolic theories. Science. 293 (2001) 2040-2044
6. Eisen, M.B., Spellman, P.T., Brown, P.O., Botstein, D.: Cluster analysis and display of genome-wide expression patterns. Proc Natl Acad Sci U S A. 95 (1998) 14863-14868
7. Watts, D.: Networks, dynamics and the small world phenomenon. American Journal of Sociology. 105 (1999) 493-527
8. Bornholdt, S., Schuster, H.: Handbook of Graphs and Networks: from Biological Nets to the Internet and WWW, Oxford University Press (2003)
9. Rouse. R., Hardiman, G.: Microarray technology - an intellectual property etrospective. Pharmacogenomics. 4 (2003) 623-632
10. Alon, U., Barkai, N., Notterman, D.A., Gish, K., Ybarra, S., Mack, D., Levine, A.J.: Broad patterns of gene expression revealed by clustering analysis of tumor and normal colon tissues probed by oligonucleotide arrays. Proc Natl Acad Sci U S A. 96 (1999) 6745-6750
11. Kuramochi, M., Karypis, G.: Gene Classification using Expression Profiles: A Feasibility Study. 2nd IEEE International Symposium on ioinformatics and Bioengineering (BIBE'01) (2001) 04-06
12. Landgrebe, J., Wurst, W., Welzl, G.: Permutation-validated principal components analysis of microarray data. Genome Biol. 3 (2002) RESEARCH0019

Analysis of Sib-Pair IBD Profiles and Genomic Context for Identification of the Relevant Molecular Signatures for Alcoholism

Chuanxing Li[1], Lei Du[1], Xia Li[1,2,*], Binsheng Gong[1], Jie Zhang[1], and Shaoqi Rao[1,3,*]

[1] Department of Bioinformatics, Harbin Medical University, Harbin 150086, P.R. China
{licx, dulei, lixia, gongbs, zhangjie}@ems.hrbmu.edu.cn
[2] Department of Computer Science, Harbin Institute of Technology,
Harbin 150080, P. R. China
[3] Departments of Cardiovascular Medicine and Molecular Cardiology, Cleveland Clinic Foundation, Cleveland, Ohio 44195, USA
raos@ccf.org

Abstract. Recent advances in SNPs that allow genome-wide profiling of complex biological phenotypes have offered the golden opportunities to unravel the high-order mechanisms and have also motivated development of the corresponding analysis strategies. Here, we design four novel comprehensive association criteria concerning both informatics of IBD statistic and genomic context. Application of these criteria along with sliding window and permutation test to 100 simulated replicates for two American populations to extract the relevant SNPs for alcoholism from sib-pair IBD profiles of pedigrees demonstrates that the proposed new approaches have successfully identified most of the simulated true loci, thus implicating that IBD statistic and genomic context could be used as the informatics for mining the underlying genes for complex human diseases. Compared with the classical Haseman-Elston method, our strategy is more efficient and simpler.

1 Introduction

Single-nucleotide polymorphism (SNP) is the most widespread form of DNA polymorphism in human genome. SNPs are generally considered the ideal genetic markers, as they are common, stable and increasingly amenable to automated large-scale methods. Searching for disease relevant SNPs is an important task of human genomics. Many methods have been suggested and are under development for detection and analysis of relevant SNPs, but no optimal method(s) for association analysis of SNPs has been found so far [1].

Many complex human diseases such as behaviors of alcoholism investigated by the Genetic Analysis Workshop 14 (GAW 14, http://www.gaworkshop.org/) are not simple Mendelian disorders. Instead, they may have mixed contributions of genes,

---

* Corresponding authors.

environments and their interactions. A sophisticated mathematical model(s) is thus desirable to map the epidemiological complexities.

Recent advances in identity-by-descent (IBD) linkage analysis, chromosome structure analysis (the Z curve method for computing the G+C content) [2], disease gene mining [3-7], adjacent and co-expressed genes along chromosome discovery (sliding window method) [8], and permutation test [9] give us insights for association study. An ideal measurement of correlations between molecular signatures and phenotypes should reflect both the signature's effects on phenotypes and its interactions with others. Here, we design novel comprehensive criteria concerning both informatics of IBD statistic and genomic context. In this investigation, we use these criteria along with sliding window and permutation test to extract the relevant SNPs for alcoholism using sib-pair IBD profiles of pedigrees generated for GAW14.

2 Methods

2.1 Defining the Phenotypic Attribute of a Sib Pair

First, we define the phenotypic attribute of a sib pair, the affection status of a sib pair for alcoholism. For the binary trait, there are three possible attributes, of which two attributes are chosen to be the phenotypes for learning: concordant affected, both sibs in a sib pair are affected; and concordant unaffected, no sibs in a sib pair are affected.

2.2 Defining the Features to Be Mined

The genetic features are defined to be the estimated proportions of alleles shared IBD by the sib pair at the SNP positions, provided by the GENIBD of the SAGE package [10]. Because our main interest is to explore the utility of the proposed method for extracting useful genetic information from the large-scale SNP data, we do not model the second-moment quantities of clinical covariates for the sib pairs.

2.3 Four Statistics of Association Between Molecular Signatures and Phenotypes

The IBD values can reflect the proportion of alleles identical by descent at the putative locus, for sibling pairs. The higher the IBD differences between concordant affected and concordant unaffected sib pairs, the stronger the SNP's association with disease. Here, we will define four criteria to measure the association of molecular signatures with phenotypes.

IBD difference. The IBD difference (DF) of a single marker (i) measures the discrepancy in its two means of IBD values in all concordant affected and concordant unaffected sib pairs and is defined as:

$$DF_i = mean(IBD_i^{disease}) - mean(IBD_i^{normal}) \tag{1}$$

Average IBD differences for window. The measurement of average IBD differences for window (ADF) is a composite index of the IBD information and genomic context, and is derived based on the thought that adjacent, co-expressed and functional associated genes are inclined to cluster along the chromosome. It reveals both association of the signatures with a disease and the interaction effects between adjacent SNPs. ADF of the i th marker (ADF_i) is the mean IBD differences of signatures within a window, which contains w markers and is centered by the i th signature. ADF is calculated for signatures in a sliding window across the genome, and this process is tried for windows of different sizes. Then a suitable window size is selected for subsequent analyses. It is calculated using the equation:

$$ADF_i = \frac{\sum_{window(i)} DF_j}{w}, \quad j \in window(i), \quad i=1,2,\ldots N. \tag{2}$$

Z curve. This approach is analogous to the Z curve method originally designed for analysis of the G+C content in the human genome. Consider a SNP profile (or called sequence) with N molecular signatures. From the first (putative) SNP signature, we inspect the sequence by one signature at a time. Let assume that n ($n = 1, 2, \ldots N$) SNPs (or called steps) have been inspected. In the n th step, we calculate the cumulative IBD difference of all the previous signatures (denoted by Z_n). We further denote the genomic location of the n th signature by X_n, and define $X_1=0$. Then, the Z curve consists of a series of nodes P_n, where $n = 1, 2, \ldots N$, whose coordinates are determined by X_n and Z_n, which are calculated:

$$\begin{cases} X_n = Location(n) \\ Z_n = \sum_{i=1}^{n} DF_i \end{cases} \quad n=0,1,2\ldots N. \tag{3}$$

As implied, Z_n profiles the distribution of DF along a sequence (here chromosome). Usually, for a DF-rich genome, Z_n is approximately a monotonously increasing linear function of X_n. It is convenient to fit the curve of $z_n \sim x_n$ by a straight line using the least square technique,

$$z = kX_n, \tag{4}$$

where (z, X_n) is the coordinate of a point on the straight line fitted and k is its slope. In this study, however, we use the $z'_n \sim x_n$ curve, where

$$z'_n = Z_n - z = Z_n - kX_n. \tag{5}$$

Let $\overline{DF}$ denote the average DF within a region ΔX_n in a sequence; we find from Eqs. (3)–(5) that

$$\overline{DF} = k + \frac{\Delta Z'_n}{\Delta X_n} = k + k', \tag{6}$$

where $k' = \Delta z'_n / \Delta x_n$ is the average slope of the $z'_n \sim x_n$ curve within the region Δx_n. As seen from Eq.(6), an up jump in the $z'_n \sim x_n$ curve, i.e., $k' > 0$, indicates an increase of the average *DF* between concordant affected and concordant unaffected sib pairs within a region, and vice versa.

Average slope for window (*AS*). it is calculated using the equation:

$$AS_i = \frac{\sum_{window(i)} k'_j}{w}, \qquad j \in window(i), \quad i = 1, 2, \ldots N. \tag{7}$$

2.4 Permutation Test

To evaluate the correlations between markers and phenotypes, permutation tests are applied. Permutation (or randomization) tests have the advantage that a particular data distribution is not assumed. They rely solely on the observed data examples and can be applied with a variety of test statistics. To evaluate a test statistic, its empirical distribution should be constructed. The statistical significance can be determined by comparing the test statistic of the original data with the distribution derived from permuted data [9].

In detail, the null hypothesis assumes the independence between SNPs and a disease phenotype. The empirical distribution for a test statistic is produced by using 100 randomly permuted replicates. By comparing the observed statistic with its empirical distribution, we can evaluate the relevance of a SNP with alcoholism. To assess its significance ($p-value$), we define $p = m/s$, where m is the number of the test statistic for the permutated samples larger than the estimated value for the SNP using the original data and s is the total number of test statistic in the empirical null distribution. We set the level of significance for a SNP association $p = 0.01$.

3 Results

In this study, we use the 100 GAW14 simulated replicates to demonstrate the behaviors and properties of the proposed methods for mining alcoholism relevant SNPs. The dataset that we analyze contains a total of 917 SNPs located on ten chromosomes (with an average spacing of 3 cM) and 100 simulated replicates for two hypothetical American populations (Aipotu, and Karnagar, each with 100 pedigrees). There are nine simulated answers distributed on seven chromosomes (chromosomes 1, 2, 3, 5, 8, 9 and 10). We perform the same analysis procedures, separately for each chromosome. The four statistics for each SNP are calculated. After exploring different window sizes ($w = 3, 5, 7$), we select $w = 3$ in the computations of *ADF* and *AS*. Finally, we use the means of each statistic derived from 100 simulated replicates as the overall index to assess significance of the association ($p-value$). We have successfully identified 70% and 78% (i.e. realized statistical powers) of simulated true answers for Aipotu and Karnagar populations, respectively. It is not surprising that the proposed method has also identified the clusters of SNPs that are nearby the true trait loci as the proposed method is designed for extracting "redundant" or interacting features (here, the reason for redundancy can be close linkage or association of linkage disequilibrium between the markers within the cluster).

Efficiency of each method. Efficiency evaluation is equivalent to the measuring of the strength of the resulted classifier. Here, we choose four measures: accuracy (the proportion of the total number of predictions that are correct), precision (the proportion of the predicted positive cases that are correct), recall (the proportion of the total number of positive cases that are correctly identified) and F value (a secondary measure derived from precision and recall, and suitable for the unbalanced cases between positives and negatives). For all the four measures, a higher value of the measures relates to a better performance that the resulted classifier can achieve. The results for the novel approaches and the Haseman-Elston (HE) method are shown in Fig. 1. Comparing with HE, we find that all the four novel approaches have significant improvement in terms of recall rate (with an increase of 46% and 60%, respectively for two human populations), but have no significance between the investigated methods in terms of other three evaluation criteria. This is particularly because of the serious imbalance between positives and negatives, 9 true loci versus a large number of irrelevant regions (a total of 908 SNPs). Among the proposed four methods, IBD difference and Z curve are more powerful than the two window sliding approaches. Logically, the sliding window methods may have advantages when dense markers are used. To enhance performance of the novel methods, we increase the level of the cut-off threshold by selecting only five top significant SNPs (all $p-value < 0.01$). This strategy yields substantial improvements (increases of 5-20%) in terms of rates of accuracy, precision and F value for both populations (Fig. 2.), and no drops in the recall values. As a result, all the proposed methods achieve higher classification (prediction) efficiency than the traditional HE linkage analysis and are therefore deemed to be powerful alternatives for locating subtle genetic determinants that are thought to underlie most complex diseases.

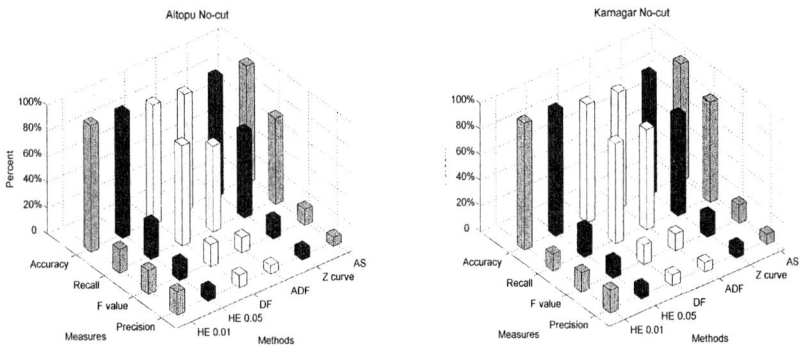

Fig. 1. Comparison of four novel methods and the traditional Haseman-Elston sib-pair regression (with two feature selection criteria, 0.05 and 0.01) for identification of alcoholism-relevant SNPs.

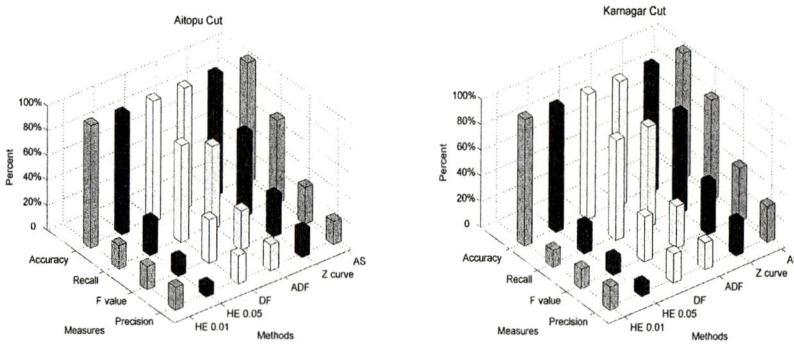

Fig. 2. Comparison of four novel methods and the traditional Haseman-Elston sib-pair regression (with two feature selection criteria, 0.05 and 0.01) for identification of alcoholism-relevant SNPs. A more strict criterion is used for SNP feature selection, in which only five top significant SNPs are include in the predictive models.

4 Discussion

In this paper we have demonstrated the potential of four novel methods for hunting for relevant SNPs using the high-dimension biomarker data. The numerical applications to 100 simulated replicates for two American populations to extract the relevant SNPs for alcoholism from sib-pair IBD profiles of pedigrees prove that IBD statistic and genomic context could be used as the informatics for mining the underlying genes for complex human diseases. The higher cut-off threshold(s) for feature SNP selection is recommended to achieve higher discriminative power of the resulted classifier(s), although there is no universal agreed threshold because of its data-dependence. Relevance is an established concept in the fields of data mining and machine learning, but its applications to the large-scale genomics (e.g. genome-wide SNP profiles) remain to be fully explored. In the context of this study, many causes can contribute the relevance of a SNP to alcoholism, for example, the close linkage between the SNP and a nearby putative trait locus or between SNPs, linkage disequilibrium between loci and gene-gene interactions, which require further genetic analysis to be clarified. This application also suggests that the sib-pair IBD statistics can be good genetic features to be mined. However, the IBD feature vectors within a pedigree tend to be correlated, effects of which on the proposed data mining approaches remain unknown.

Acknowledgements

This work is supported in part by the National Natural Science Foundation of China (Grant Nos. 30170515 and 30370798), the National High Tech Development Project, the Chinese 863 Program (Grant No. 2003AA2Z2051), the 211 Project, the Tenth 'Five-year' Plan, Harbin Medical University, and Heilongjiang Province Science and Technology grants (Grant Nos. GB03C602-4 and 1055HG009).

References

1. Zabarovsky, E.R.: Novel strategies to clone identical and distinct DNA sequences for several complex genomes. Molecular Biology. 34 (2000) 612-625
2. Zhang, C.T., Zhang, R.: An isochore map of the human genome based on the Z curve method. Gene. 317 (2003) 127-135
3. Li, X., Rao, S., Zhang, T., Guo, Z., Zhang, Q., Moser, K.L., Topol, E.J.: An ensemble method for gene discovery based on DNA microarray data. Sci China C Life Sci. 47 (2004) 396-405
4. Li, X., Rao, S., Wang, Y., Gong, B.: Gene mining: a novel and powerful ensemble decision approach to hunting for disease genes using microarray expression profiling. Nucleic Acids Res. 32 (2004) 2685-2694
5. Li, X., Rao, S., Moser, K.L., Elston, R.C., Olson, J.M., Guo, Z., Zhang, T.: Genetic mapping of complex discrete human diseases by discriminant analysis. Progress In Natural Science 12 (2002) 431-437
6. Bell, D.A., Wang, H.: A formalism for relevance and its application in feature subset selection. Machine Learning. 41 (2000) 175-195
7. Li, X., Rao, S., Elston, R.C., Olson, J.M., Moser, K.L., Zhang, T., Guo, Z.: Locating the genes underlying a simulated complex disease by discriminant analysis. Genet Epidemiol. 21 Suppl 1 (2001) S516-521
8. Spellman, P.T., Rubin, G.M.: Evidence for large domains of similarly expressed genes in the Drosophila genome. J Biol. 1 (2002) 5
9. Futschik, M., Crompton, T.: Model selection and efficiency testing for normalization of cDNA microarray data. Genome Biol. 5 (2004) R60
10. S.A.G.E.: Statistical Analysis for Genetic Epidemiology, 4.4. A computer program package available from Statistical Solutions. Cork, Ireland (2003)

A Novel Ensemble Decision Tree Approach for Mining Genes Coding Ion Channels for Cardiopathy Subtype

Jie Zhang[1], Xia Li[1,2,*], Wei Jiang[1], Yanqiu Wang[1], Chuanxing Li[1], Qiuju Wang[3], and Shaoqi Rao[1]

[1] Department of Bioinformatics, Harbin Medical University, Harbin 150086, P.R. China
zhangjie_qd@hotmail.com
{lixia, jiangw, yqwang, licx}@ems.hrbmu.edu.cn; raos@ccf.org
[2] Department of Computer Science,
Harbin Institute of Technology, Harbin 150080, P.R. China
[3] Institute of Otolaryngology, Chinese PLA General Hospital, Beijing 100853, P.R. China
wqcr@301ent.org

Abstract. Ion channels are critical for normal physiological function of humans and their functional abnormality may cause many disorders named channelopathy. Meanwhile, they are one of the few proteins that can be efficiently regulated by small molecule drugs, so they are ideal candidates for drug targets. Upon these viewpoints, it is known that research on ion channels will bring great scientific and practical value. Here, we applied a novel ensemble decision tree approach based on mining genes encoding the ion channels. Using this ensemble method, we analyzed an oligo array data set concerning the human cardiopathy which investigated by Medical College of Harvard University. By analyzing 57 samples and 1172 genes related to ion channels and other transmembrane proteins, we demonstrated that the ensemble approach can efficiently mine out disease related CACNA genes.

1 Introduction

Ion channels are special proteins embedded in the plasma membrane and participate in the cellular electrophysiological activities. They are critical for the normal physiological function of an organism. A previous study [14] revealed that when the genes encoding subunits of ion channels come into being mutations, their functions will become abnormality which may cause many disorders[2], named channelopathy. Up to date, 30 kinds of channelopathies have been discovered, which are involved in the cardiovascular, neural, endocrine and urinary systems and are deemed to be responsible for the functional abnormality of the sodium channels, potassium channels, calcium channels, chlorine channels and some receptors related to ligand gated channels. Chanelopathies have seriously threatened the life and health of humans. Meanwhile, ion channels are one of the few proteins that can be efficiently regulated by small molecule drugs[2,4], so they are ideal candidates for drug targets.

---

[*] Corresponding author.

Upon these viewpoints, it is known that research on ion channels will bring great scientific and practical value[4,12].

Many studies[12,14] indicated that the functional abnormality of ion channels can cause the cardiopathy. But these studies mainly paid attention to the index of histology and pathology or focused on a single gene(s). Although some previous researchers focused on ion channels, the technologies they used were patch clamp or voltage clamp. Up to date, few studies have used pattern recognition for analysis of ion channel gene expression profiles. In this study, we applied a novel ensemble decision tree approach[3,16] to mining disease relevant genes encoding ion channels related to cardiopathy subtype. Cross-validation[1] and permutation test[13] are used to evaluate the built trees [6,7]. This application showed that the novel ensemble approach yielded high performance for analysis of the ion channel gene expression profiles.

2 Methods

2.1 Decision Tree

Ion channel and other transmembrane protein gene expression data with p probes and n DNA samples can be described as a $n \times p$ matrix $X = (x_{ij})$, $X = (x_{ij})$, where x_{ij} represents the expression level for the j th gene (I_j) on the i th sample (X_i). The procedure of tree building is as follows. First, we split the data set X into two subsets, training set and test set. Then, a binary tree is grown on training set by a recursive partition algorithm. The search for feature genes starts at the root of the tree and proceeds to its leaves. At each internal node, a decision is made with regard to the choice of a feature gene and a threshold value (cutoff) such that the class impurity is reduced to a minimum when a branch is made by an induction rule. After the optimal bifurcation is made, the samples are divided into two non-overlapping subsets (two child nodes). For each subset the same process is conducted successively until a leaf is reached or stopping criteria for tree growth are satisfied. Here the impurity is assessed by Gini inequality index[15]. We proposed several evaluation criteria[15] for validating the built trees using test data sets, including accuracy-acc, error rate-e, precision-p, recall-r. To test whether the extracted subset of genes (G_d) is able to significantly distinguish the phenotypes of cardiopathy subtype, we also calculated the χ^2 statistic [15].

2.2 Algorithm Flow of the Ensemble Decision Tree Method (Fig. 1)

This study deals with multiple (say K) biological types. First, we collapse K-type data into 2-type data, without the loss of generality, by labeling 1 for one type and 2 for all the remaining types. By working on every type in turn, we produce K collapsed data sets. For each newly generated 2-type data set, we randomly divide the samples of two types into n non-overlapping subsets of roughly equal size, respectively. A random combination of type 1 samples and type 2 samples constitutes a test set and the rest of the subsets are used as the training set. This n-fold cross-validation resampling produces $n \times n$ pairs of training and test sets. We repeat the cross-validation resampling M times and obtain $n \times n \times M$ pairs of training and test sets. For each pair of data sets, we grow and evaluate a binary recursive partition tree. In this way, we build a gene forest of a total of $k \times n \times n \times M$ trees.

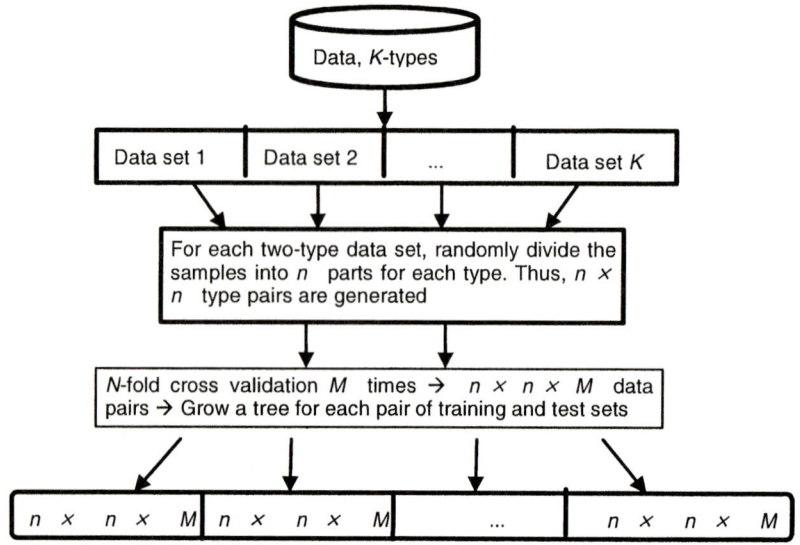

Fig. 1. Algorithm flow of the iterative ensemble decision tree approach for multiple-type data

The ion channel feature genes are extracted and evaluated using an ensemble voting approach and a permutation test as described previously [15].

3 Experiment

3.1 Data Set and Pro-process

The data set analyzed here is GSE1145, which was submitted to NCBI by Martina Schinke from the Medical College of Harvard University (available at http://cardiogenomics.med.harvard.edu/public-data).

The original data set consists of absolute measurements from an oligo arrays with 132 samples of 54675 human probe sets. Because of our aims, we select three phenotypic categories, with the largest sample sizes (11 normal samples, 15 idiopathic dilated cardiomyopathy samples, 31 ischemic cardiomyopathy samples, respectively). The selected data set including measures for all the genes is standardized using the BRB-ArrayTools Version 3.0.1(Dr. Richard Simon, Biometrics Research Branch, National Cancer Institute, 2003). We then filtered out 1172 probe sets, encoding ion channels or other transmembrane proteins, for ensemble decision tree analysis.

3.2 Results and Discussion

We use the newly proposed ensemble decision tree approach coupled with performing 5-fold cross-validation 20 times, 20 permutation tests. Here, The number of iterative and permutation test has been tested several times, and we find that it can get a steady result. But for cross-validation, we will continue to validate the choice of the number in future work.

First, translate the data set into two-classes. Then we can get 3 new data sets, named Sample 1, Sample 2 and Sample3. Process each data set according to the algorithm flow as Fig 1, represent the results in table 1. Each data set can produce 500 trees, sort these trees at a given set by a degressive χ^2 value, then select out the trees with largest χ^2 value, named them fine trees. The same process is carried out in the three datasets separately. Put the results into Table 1. For an obvious observation to the significance of the feature ion channel genes in the fine trees, we draw Fig 2. This figure is based on the frequency of the genes encoding ion channels and other transmembrane protein appeared in fine trees from different data sets. And it can show that the genes with Gene ID 788, 680, 1050, 384, 443 relevantly not only have high frequency, but also apparent in two sets, Sample 2 and Sample 3(the details of genes see table 2). It can prove that these genes have a special meaning and function. This phenomenon may relate to the use of the local space. CACNA has been proved to has a close relationship with cardiopathy in some literatures[8,9]. Further, reports about CACNA1C (788) also have pointed out that it correlated to the idiopathic dilated[10,11] and ischemic[16] cardiopathy.

Table 1. Fine trees selected by largest χ^2 value on three data sets

Data Set	Tree ID	Gene ID	χ^2
Sample1	T1	340 1094 544 270	5.8800
	T2	733 1094 544 270	5.8800
Sample2	T3	70 972 384* 788* 1050* 680* 441	5.4857
	T4	1008 732 788* 384* 443* 680* 585	5.4857
Sample3	T5	788* 855 680* 1050* 29 1150 52 515 62 65 981 107 123 1021	6.5360
	T6	788* 680* 855 1050* 29 1150 52 515 62 65 83 69 107 123 1021	6.5360
	T7	443* 680* 541 788* 29 1150 52 515 62 65 981 83 384* 69 107 123 1021	6.5360
	T8	788* 680* 443* 541 29 1150 52 515 62 65 107 111 64 123 1021	6.5360
	T9	443* 788* 541 680* 29 1150 52 515 64 257 65 107 111 123 1021	6.5360

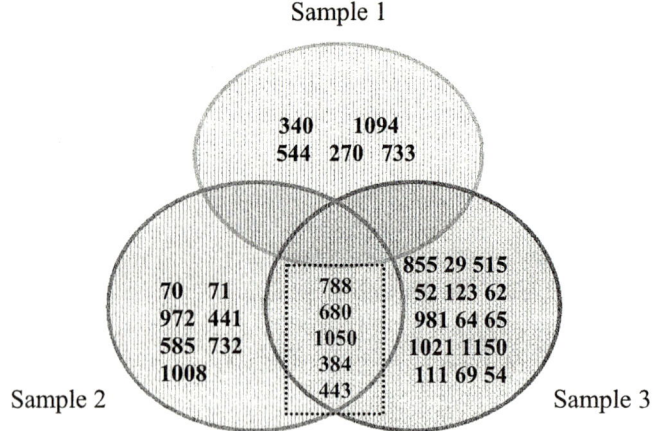

Fig. 2. Relationships between each two data set selected from the fine trees

Make a use of permutation test, we process the three data sets and make the result as *result*2. Each data set can gain 5×5×20 trees. Choose a frequency threshold $f*(i)$(i=1, 2, 3, three data sets) as a cut off at the significant level of 0.05 from the three data sets separately. Look for the *result* 1, select out all genes whose frequency value is not smaller than $f*(i)$ respectively. The selection of the χ^2 value can prevent irrelevant genes selected as feature genes and bring a weaken result. At the next step, to get an annotation about feature ion channel and other transmembrane protein genes, map them onto the Source website which is developed by Stanford University (http://source.stanford.edu). The source results see table 2.

Table 2. The feature gene sets based on frequency from the permutation test ($P<0.05$)

Data Set	Gene ID	Frequency	Gene Symbol
Sample1	1094	0.5920	KCNJ15
	733	0.6380	TMC5
Sample2	972	0.3160	CACNA1D
	384	0.3400	CNGB1
	788	0.4880	CACNA1C
	70	0.5020	KCNH7
Sample3	541	0.3540	EPB42
	1150	0.3840	TRPM8
	29	0.4740	SLC17A8
	1050	0.5300	TMEM2
	443	0.5780	CACNA1E
	680	0.6980	SLC7A8
	788	0.9520	CACNA1C

3.3 Validation by Other Classification Methods

Based on each nine fine trees, we select out nine feature genes sets. Pick the data from the primal data set X according to these gene sets, then we will get nine new data sets $S^*(i)(i = 1, 2,...,9)$. Each set has 57 samples and the number of genes is equal to the number of feature genes contained in the relevant fine trees. To validate the results get from the novel ensemble method, we propose the other six classification methods provided by BRB-ArrayTools Version 3.0.1, Compound Covariate Predictor, Diagonal Liner Discriminant Analysis, 1-Nearest Neighbor, 3-Nearest Neighbors, Nearest Centroid and Support Vector Machines. Make a use of the six classification methods on the nine data sets at the significance level of 0.05, and put the validate results into table 3. The results show that there is a high classification accuracy of the feature gene sets gained from the nine fine trees when used the six classification methods. And the classification just based on the ion channel genes can get a credible and fine result. Based on the validation results of table 3, we draw the figure 3 based on the validation results of table 3. From the histograms, we can see that the classification results of the six validate methods seem to have the same trend in Sample 1 and Sample 3 separately, see Fig.3 (A) and (B). From the left to right, we sort them as the ascending χ^2 values. One can find that in Sample 1, the order is 3-NN, SVM, NC, CPP, 1-NN and DLDA in turn, and NC, CPP, DLDA, SVM, 1-NN and 3-NN in Sample 3. This phenomenon can say feature genes from the nine fine trees we have mined out have a steady performance. But there has no such orderliness in Sample 2, to some extent it can reflect two problems. One is that it is related to the stability of the fine trees selected by the ensemble decision tree, the other one is that it may rely on the classification methods we chose for validation.

Table 3. Accuracy values get from the fine trees by six classification methods

Tree ID	CCP (%)	DLDA (%)	1-NN (%)	3-NN (%)	NC (%)	SVM (%)	Average Acc	Order
T1	86	89	88	12	82	81	73.0	28
T2	88	91	88	9	84	81	73.5	30
T3	79	81	77	79	81	60	76.2	21
T4	72	79	51	81	67	54	67.3	17
T5	68	74	82	86	54	74	73.0	21
T6	68	74	82	86	54	74	74.7	21
T7	60	68	81	84	51	74	6937	15
T8	60	70	82	86	51	74	70.5	18
T9	58	68	82	86	51	74	69.8	16

In table 3, 1st column is the Tree ID same as that in table 1; 2nd to 7th columns contain the *acc* value(%) get from Compound Covariate Predictor, Diagonal Liner Discriminant Analysis, 1-Nearest Neighbor, 3-Nearest Neighbors, Nearest Centroid and Support Vector Machines respectively; 8th column is a average value from the six *acc* values of each tree; Order column is a rank sum for evaluate each tree. give an integral number to each column according the *acc*. For example, in column CPP, number T9 as 1, T7 and T8 as 2, and so on. A same process performed in the other column, sum the rank in each row and a sum in column Order. Based on Average Acc and Order, Trees like T2, T3, T1, T5 and T6 are the better trees. T1 and T2 belong to Sample 1, T3 belongs to Sample 2, T5 and T6 belong to Sample 3. It is seen that the distributing of the better trees are balanceable (The detail refers to table 1). The only difference between T1 and T2 is that the genes with the Gene ID 340 and 733 belong to each tree separately, and the gene with ID 340 is KCNJ2 which is critical for a new genotype (LQT7) of LQTS[5]. Also, T1 with a highest χ^2 value has appeared twice in *result*1. So it is more stable. T3 belongs to Sample 2 and owns 4 important feature genes appeared in Fig 2 whose Gene ID are 788, 1050, 680, 384. T6 and T5 belong to Sample 3 own more common feature genes which can make the trees more stably.

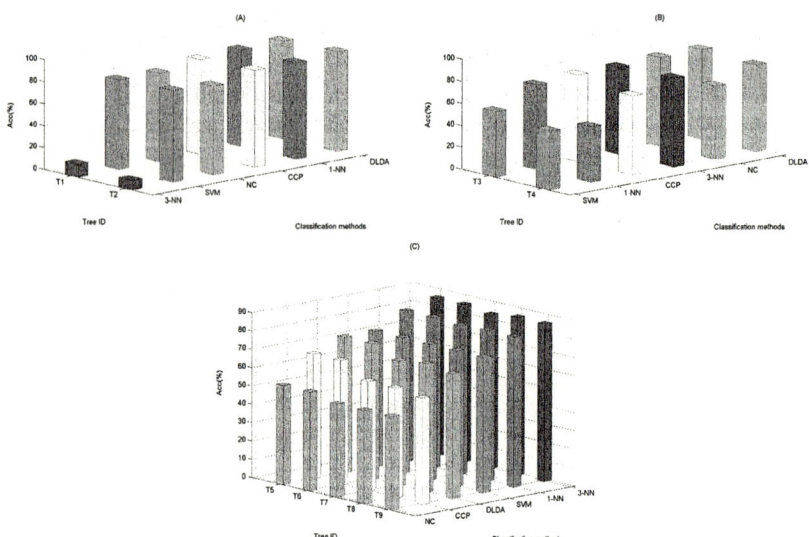

Fig. 3. Comparison between six validation classification methods

4 Conclusion

This study suggests that the proposed ensemble decision tree approach holds the promise of deciphering some of the secrets of the cardiopathy from microarray data set of ion channels. It is a powerful tool not only for classification but also for disease related gene discovery. And the reliability and stability have been proved by other six classification methods. More amazingly, we extract the critical disease-relevant genes, CACNA1C, SLC7A8, CACNA1E, TMEM2, etc. Some studies have pointed out that the abnormal function of CACNA has a close relationship with idiopathic dilated[10,11] and ischemic[16] cardiomypathy. However, there is no report to expatiate the more true details about the relationship between the two diseases and the subunits of L-type voltage-gated calcium channel. But in our study, three subunits CACNA1C, CACNA1D and CACNA1E have been extracted, as for whether the abnormal function of the three subunits can cause the idiopathic or ischemic cardiomyopathy, this need a farther biological experiment to validate the true mechanism.

Acknowledgements

This work was supported in part by the National Natural Science Foundation of China (Grant Nos. 30170515 ; and 30370798), the National High Tech Development Project, the Chinese 863 Program (Grant No. 2003AA2Z2051), the 211 Project, the Tenth 'Five-year' Plan, Harbin Medical University and Heilongjiang Province Science and Technology Key Project (Grant Nos.GB03C602-4 and 1055HG009).

References

1. Breiman, L.: Bagging predictors. Machine Learning. 24 (1996) 123-140
2. Felix, R.: Channelopathies: ion channel defects linked to heritable clinical disorders. J Med Genet. 37 (2000) 729-740
3. Fengjing Shao, Z.Y.: Principle and Algorithm of Data Mining. (2003) 126-169
4. Katoh, M.: Bioinformatics for Oncogenomic Target Identification. Genome informatics. 14 (2003) 348-349
5. Koumi, S., Backer, C.L., Arentzen, C.E.: Characterization of inwardly rectifying K+ channel in human cardiac myocytes. Alterations in channel behavior in myocytes isolated from patients with idiopathic dilated cardiomyopathy. Circulation. 92 (1995) 164-174
6. Kuo, W.J., Chang, R.F., Chen, D.R., Lee, C.C.: Data mining with decision trees for diagnosis of breast tumor in medical ultrasonic images. Breast Cancer Res Treat. 66 (2001) 51-57
7. Li T, Z.C., Ogihara M: a comparative study of feature selection and multiclass classificaition methods for tissue classification based on gene expression. (2004)
8. Li, X., Huang, C.X., Jiang, H., Cao, F., Wang, T.: The beta-adrenergic blocker carvedilol restores L-type calcium current in a myocardial infarction model of rabbit. Chin Med J (Engl). 118 (2005) 377-382
9. Lin, L., Conti, J.B., Curtis, A.B.: Termination and suppression of idiopathic left ventricular tachycardia by diltiazem. Clin Cardiol. 20 (1997) 890-893
10. Richter, A., Loscher, W.: Antidystonic effects of L-type Ca2+ channel antagonists in a hamster model of idiopathic dystonia. Eur J Pharmacol. 300 (1996) 197-202
11. Sander, T., Peters, C., Janz, D., Bianchi, A., Bauer, G., Wienker, T.F., Hildmann, T., Epplen, J.T., Riess, O.: The gene encoding the alpha1A-voltage-dependent calcium channel (CACN1A4) is not a candidate for causing common subtypes of idiopathic generalized epilepsy. Epilepsy Res. 29 (1998) 115-122
12. Swindells, R.F.a.M.: Bioinformatics,target discovery and the pharmaceutical/ biotechnology industry. Current Opinion in Molecular Therapeutics. 2 (2000) 655-661

13. Tan, A.C., Gilbert, D.: Ensemble machine learning on gene expression data for cancer classification. Appl Bioinformatics. 2 (2003) S75-83
14. Willumsen, N.J., Bech, M., Olesen, S.P., Jensen, B.S., Korsgaard, M.P., Christophersen, P.: High throughput electrophysiology: new perspectives for ion channel drug discovery. Receptors Channels. 9 (2003) 3-12
15. Xia Li, S.R., Yadong Wang and Binsheng Gong: gene mining: a novel and powerful ensemble decision approach to hunting for disease genes using microarray expression profiling. Nucleic Acids Research. 32 (2004) 2685-2694
16. Zhao, P., Ise, H., Hongo, M., Ota, M., Konishi, I., Nikaido, T.: Human amniotic mesenchymal cells have some characteristics of cardiomyocytes. Transplantation. 79 (2005) 528-535

A Permutation-Based Genetic Algorithm for Predicting RNA Secondary Structure—A Practicable Approach

Yongqiang Zhan and Maozu Guo

School of Computer Science and Technology,
Harbin Institute of Technology,
Harbin 150001, PRC
{yqzhan, maozuguo}@hit.edu.cn

Abstract. The paper presents a permutation-based algorithm for predicting RNA secondary structure. It is practicable, and can be used to predict real RNA molecules. The conception of permutation is introduced, which is the start point of our algorithm. Individual is represented as a permutation of stem list. Crossover operator, mutation operator, and selection strategy are designed to be compatible with such an individual representation. At the end of the paper, a comparison between our result and that from RNAstructure is outlined. It is proved that our algorithm has achieved comparable or better result than RNAstructure.

1 Introduction

RNAs act as either the blueprint or other fundamental roles for building proteins. So many scientists who studied proteins before begin to study RNA now, hoping to find a deeper relation between RNAs and proteins. In the field of RNA biology, RNA structure prediction is an active direction. Because secondary structure is the foundation of tertiary structure and a tertiary structure is difficult to predict directly, most of algorithms about RNA structure prediction focus on secondary structure. There are a lot of approaches on this topic, such as phylogeny approach based on multiple sequence alignment, dynamic algorithm based on minimal free energy [1], and all kinds of heuristic algorithms including genetic algorithm (GA) [2], [3], simulate annealing algorithm and artificial neural net algorithm. In this paper, a genetic algorithm based on permutation is presented to predict a real RNA secondary structure, not like in [3], which is just a study in theory.

In Section 2, we give an introduction to RNA structures including primary structure and higher structures, introduce the conception of permutation-based genetic algorithm and describe how to translate the problem of RNA structure prediction to a TSP. In Section 3, we elucidate the GA proposed by us for precise prediction of the RNA secondary structure. Section 4 contains an application. A conclusion is given in Section 5.

2 RNA Secondary Structure

A RNA primary structure is a sequence which is composed of four kinds of bases, adenine (A), cytosine (C), guanine (G) and uracil (U). A RNA secondary structure is resulted from the formation of hydrogen bonds between bases. A RNA tertiary structure is the 3-dimension spatial structure of the secondary structure. In this paper, a RNA secondary structure is viewed as a set of stems, which is formed by some consecutive stacked loops. Wiese and Glen [3] introduce the conception of permutation to the individual representation of RNA secondary structure in GA. If a permutation is considered to be an individual in GA, RNA secondary structure prediction is essentially a TSP in mathematical language.

3 Method

3.1 Fitness Function

We define the negative value of the free energy of a RNA secondary structure as the fitness function. The structure with the lowest free energy has the highest fitness. We compute the free energy of RNA secondary structure according to the energy rule described in [4] and the energy data coming from "http://rna.chem.rochester.edu", which are also the default resources used in RNAstructure 4.11. Thus, we can directly compare the results from the two programs (plus ours).

3.2 Crossover and Mutation

In this paper, we use the CX (cycle crossover) [3] as the crossover operator. This operator preserves the absolute position of elements in the parent sequence. A parent sequence and a cycle starting point are randomly selected. The element at the cycle starting point of the selected parent is inherited by the child. The element which is in the same position in the other parent can not then be placed in the position so its position is found in the selected parent and is inherited from that position by the child. This continues until the cycle is completed by encountering the initial item in the unselected parent. Any elements which are not yet present in the offspring are inherited from the unselected parent. The mutation operation is implemented by randomly switching positions of two stems.

3.3 Selection Strategy

In the step of selection, we use both the Standard Roulette Wheel Selection (STDS) and a strategy called KBR [5]. With STDS, all individuals are assigned a pie-shaped slice on a roulette wheel proportional in size to their fitness as compared to the sum of fitnesses for all individuals in the population. Individuals are chosen for recombination by spinning the roulette wheel. An individual is kept if it is located in the slice (bin) where the wheel stops and culled otherwise. This strategy gives a higher chance for the individuals of high fitness. In KBR, parents are chosen in the same way as in STDS. Then, crossover and mutation are performed. The set of parents and offspring therefore undergo an additional selection step to achieve the goal that the fittest offspring generated from the fittest parent survive to the next generation.

3.4 Algorithm

```
Input: a RNA string, population size n, crossover prob-
ability p_c, and mutation probability p_m.
Output: stem list, and its energy.
Program Begin
Create a stem list with all possible stems;
Create a population with size of n.
Get the biggest fitness value in population;
While (the fitness<a certain value) {
      Do STDS on the current population;
      Do crossover on the current population;
      Do mutation on the current population;
      Do KBR on the current population;
      Get the biggest fitness value in population;}
Output the stem list of the fittest individual and its
energy.
End
```

4 Result

We use JAVA to realize our algorithm, and RNA RD0260 (a sample RNA sequence in RNAstructure 4.11) to test our algorithm. We compare our result with that from RNAstructure 4.11. The lowest energy predicted by RNAstructure is -28.3. Under the parameters setting in Fig. 1a, we get an energy value of -28.2. Under the parameters setting in Fig. 1b, our program give an energy estimate of –28.4, which demonstrates that our algorithm has achieved comparable or better result than RNAstructure 4.11.

```
209
lenght of individual:6
stem id:8,first pair of stem:(1,73),length of stem:8
stem id:209,first pair of stem:(40,64),length of stem:4
stem id:116,first pair of stem:(18,29),length of stem:4
stem id:84,first pair of stem:(13,33),length of stem:2
stem id:225,first pair of stem:(46,57),length of stem:4
stem id:55,first pair of stem:(9,37),length of stem:3
energy=-28.199997
            a
```

```
10
lenght of individual:6
stem id:58,first pair of stem:(22,51),length of stem:3
stem id:36,first pair of stem:(12,65),length of stem:3
stem id:45,first pair of stem:(16,59),length of stem:4
stem id:73,first pair of stem:(28,46),length of stem:3
stem id:2,first pair of stem:(1,73),length of stem:8
stem id:79,first pair of stem:(32,42),length of stem:4
energy=-28.400002
            b
```

Fig. 1. RD0260 prediction result from our algorithm: (a) Population size is 1000. The shortest stem length is 2. (b) Population size is 1000. The shortest stem length is 3.

5 Conclusions

Application of our newly developed algorithm to RD0260 supports that the permutation-based GA is a good competitor to the popular software tool RNAstructure 4.11 for precisely predicting RNA secondary structure. It has a high caliber to be used in RNA biology.

Acknowledgements

The work was supported by the Natural Science Foundation of Heilongjiang Province (Grant No. F2004-16) and the National 863 Hi-Tech Project of China (Grant No. 2003AA118030).

References

1. Zuker, M., Stiegler, P.: Optimal computer folding of large RNA sequences using thermodynamics and auxiliary information. Nucleic Acids Res (1981) 9, 133-148
2. Gultyaev, A.P., Van Batenburg, F.H.D., Pleij, C.W.A.: The computer simulation of RNA folding pathways using a genetic algorithm. J. Mol. Biol. (1995) 250, 37–51
3. Wiese, K.C., Glen, E.: A Permutation Based Genetic Algorithm for the RNA Folding Problem: A Critical Look at Selection Strategies, Crossover Operators and Representation Issues. BioSystems-Special Issue on Computational Intelligence in Bioinformatics (2003)
4. Mathews, D.H., Sabina, J., Zuker, M., Turner, D.H.: Expanded sequence dependence of thermodynamic parameters improves prediction of RNA secondary structure. J. Mol. Biol. (1999) 288, 911-940
5. Wiese, K.C., Goodwin, S.D.: Keep-Best Reproduction: A Local Family Competition Selection Strategy and the Environment it Flourishes in. Constraints (2001) 6, 399–422

G Protein Binding Sites Analysis

Fan Zhang[1], Zhicheng Liu[1], Xia Li[1,2,*], and Shaoqi Rao[2]

[1] Department of Bioinformatics, Capital University of Medical Sciences,
Beijing 100054, China
`fanzhang@tsinghua.edu.cn zcliu@cpums.edu.cn`
[2] Department of Bioinformatics, Harbin Medical University,
Harbin 150086, China
`lixia@ems.hrbmu.edu.cn, raos@ccf.org`
`http://www.biocc.net`

Abstract. Protein active sites control nearly all protein functions and determine the interactions upon which biological pathways and cellular networks are built. Characterization of the active sites in a protein would therefore lead to new methods of controlling proteins and ultimately controlling cells. This paper uses evolutionary trace method to analyze the binding sites of G Protein and finally give an example of 1A80. Results show that the method can be helpful in understanding how proteins carry out certain biological functions.

1 Introduction

In order to understand how proteins carry out a highly integrated biological function(s), for example, how they recognize ligands and form protein-protein or protein-ligand complexes, it is essential to identify the protein functional residues and the interaction interface(s) (e.g. the active sites or binding sites [1-3]). The functional interfaces can serve as targets for structural based drug design or to guide the site-directed mutagenesis in studying the protein structure-function relationship. X-ray crystallography is a powerful tool for studying protein function and structure. However, the number of protein sequences is discovered with much higher speed than the number of protein structures are [4]. Even when the structural information is available, it is still difficult to infer the functional roles of some specific residues in protein function directly from the structure data. Exhaustive mutational analysis is one of common instruments for binding site characterization. With the increasing size of sequence databases, a wealth of mutational data is already available in the databases, where sequences homologous to the protein of interest record mutation "experiments" that have passed the test of natural selection. In this paper, we apply the newly developed *evolutionary trace method* [5-7] to extract the mutational data embedded in the large number of DNA sequences and to infer which residues are likely to be important to protein function.

---

* Corresponding author.

2 Evolutionary Trace Method

In general, functional residues undergo fewer mutations during evolution. On the other hand, certain functional residues are forced to mutate to achieve the selectivity of subgroups within the protein family. The *evolutionary trace method* makes a direct connection between residue conservation in aligned sequences and their functional importance [8-10]. It also identifies functional specificity by partitioning the protein sequence alignment into subgroups according to the sequence similarity. Furthermore, with the 3D structure of the protein of interest, exposed functional residues which are more likely to be responsible for binding or enzymatic activity can be distinguished from the buried residues which are more important for maintaining the structural integrity. This method can be applied to proteins with known experimental structures as well as theoretical models.

A family of homologous sequences can be aligned using a multiple sequence alignment program. Based on the alignment, the sequences are then clustered using a hierarchical clustering method according to the average percentage sequence identity. The distance between two nodes A and B in the sequence cluster is calculated as follows:

$$d_{AB} = \sum_{ij} \frac{d_{ij}}{m \cdot n}, \qquad (1)$$

where d_{ij} is the distance between sequence i in node A and sequence j in node B. There are m and n sequences in node A and B, respectively.

The sequence cluster, or dendrogram, is a representation of the evolutionary or functional relationship of the sequences in the sequence family. Based on the dendrogram, sequences in a family can be divided into subfamilies at a given sequence identity cutoff. At different cutoffs, the subfamily represents different levels of functional resolution. At high sequence identity cutoffs, the sub-family consists of smaller groups of sequences and shows more functional specificity, while at low sequence identity cutoffs, the subfamilies are larger with less specificity.

From the multiple sequence alignment, residues shared by the family of protein sequences are identified as conserved residues and are assumed to be essential for maintaining protein functions. Based on the dendrogram, sequences can be partitioned into subgroups at selected Partition Identity Cutoffs (PICs). PICs are defined as boundaries that partition the dendrogram into clusters at certain values of percentage sequence identity. Clusters of sequences which share their most distant common node at levels of sequence identity less than a given PIC value are considered as individual sets of sequences in the Evolutionary Trace (ET) analysis. If a sequence is not similar to any other sequence and itself is a subgroup isolated at selected PIC, this sequence will be ignored in the analysis. It will be included in the analysis until the PIC is low enough to make it a member of a subgroup. For each such partition, an 'evolutionary trace' is constructed. First, a consensus sequence is constructed for each subgroup containing two or more sequences. Each position in the consensus sequence is either variable within the subgroup, in which case it is considered 'neutral' (indicated by an underscore: '_'), or it is invariant in the subgroup, and takes on the single-letter code for the side-chain at that position.

The resulting consensus sequences are then aligned, and the ET sequence' for the partition is obtained. Each position in the ET sequence can be 'neutral', 'conserved' or 'class-specific'. A position is 'neutral' in the ET sequence if it is neutral in any of the consensus sequences, and 'conserved' (indicated by the side-chain letter) if it is conserved as the same side-chain in all consensus sequences. A position is labelled 'class-specific' (indicated by 'X') if it is conserved in each consensus sequence, but the side-chain varies between consensus sequences.

3 An Example: Protein 1A80 and Conclusions

Protein 1A80 has 73 homologues sequences. Based on the above method, we analyze the binding sites of 1A80 shown in figure 2.

The 3-D views of three molecular modes for 1A80 are shown in figures 3, 4, and 5, respectively.

The evolutionary trace method makes a direct connection between residue conservation in the aligned sequences and their functional importance. It can also identify

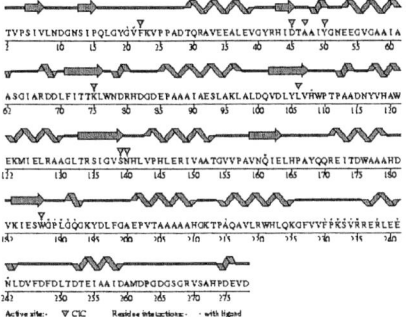

Fig. 2. 1A80 Sequence

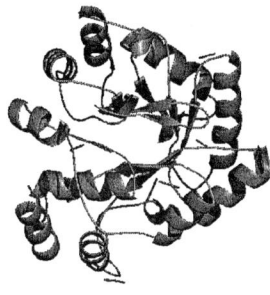

Fig. 3. Binding Site Cartoon Mode. PIC=90% (Red=Conserved; Magenta=Class Specific).

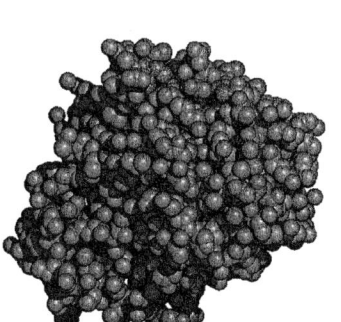

Fig. 4. Binding Site Sphere Mode. PIC=90% (Red=Conserved; Magenta=Class Specific).

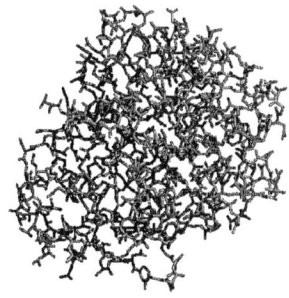

Fig. 5. Binding. Site Sticks Mode. PIC=90% (Red=Conserved; Magenta=Class Specific).

the functional specificity by partitioning the protein sequences into sub groups according to the sequence similarity. Furthermore, with the 3D structure of the protein of interest, exposed functional residues which are more likely to be responsible for binding or enzymatic activity can be distinguished from the buried residues which are more important for maintaining the structural integrity. This method can be applied to proteins with known experimental structures as well as theoretical models.

Acknowledgement

This work was supported in part by the National Natural Science Foundation of China (Grant Nos. 30170515 and 30370798), the National High Tech Development Project of China, the 863 Program (Grant Nos. 2003AA2Z2051 and 2002AA2Z2052) and the 211 Project, the Tenth 'Five-year' Plan, Harbin Medical University, and Device And Lab Platform Department of Capital University of Medicine Sciences.

References

1. Ravi Iyengar and John D.Hildebrandt, Methods in Enzymology, Volume 343: G Protein Pathways, Part A: Receptors, Academic Press 2001.
2. Ravi Iyengar and John D.Hildebrandt, Methods in Enzymology, Volume 344: G Protein Pathways, Part B: G Proteins and Their Regulators, Academic Press 2001.
3. Ravi Iyengar and John D.Hildebrandt, Methods in Enzymology, Volume 345: G Protein Pathways, Part C: Effector Mechanisms, Academic Press 2001.
4. 4.R.Durbin, S.Eddy, A.Krogh, and G.Mitchison, Biological sequence analysis: probabilistic models of proteins and nucleic acids, Cambridge University Press 2002.
5. Lichtarge O, Bourne HR, and Cohen FE, "Evolutionarily conserved G Alpha Beta Gama binding surfaces support a model of the G protein-receptor complex," Proc Natl Acad Sci USA, Vol. 93, No. 15, 1996, pp. 7507-7511.
6. Lichtarge O, Bourne HR, and Cohen FE, "An evolutionary trace method defines binding surfaces common to protein families," J Mol Biol, Vol. 257, No. 2, 1996, pp. 342-358.
7. Lichtarge O, K.R.Yamamoto, and F.E.Cohen, "Identification of functional surfaces of the zinc binding domains of intracellular receptors," J Mol Biol, Vol. 274, No. 3, 1997, pp. 325-337.
8. Hisako Ichihara and Hiromi Daiyasu, "Analysis of Asymmetry of Membrane proteins by Evolutionary Trace," Genome Informatics, Vol. 13, 2002, pp. 422-423.
9. Innis C.A., Shi J., and lundell T.L., "Evolutionary trace analysis of TGF-Beta and related growth factors:implications for site-directed mutagenesis," Protein Eng., Vol. 13, No. 12, 2000, pp. 839-847.
10. Leighton Pritchard and Mark J.Dufton, "Evolutionary Trace Analysis of the Kunitz/BPTI Family of Proteins: Functional Divergence May Have Been Based on conformational Adjustment," J. Mol. Biol., Vol. 285, 1999, pp. 1589-1607.

A Novel Feature Ensemble Technology to Improve Prediction Performance of Multiple Heterogeneous Phenotypes Based on Microarray Data

Haiyun Wang[1,2], Qingpu Zhang[3], Yadong Wang[3], Xia Li[1,2,3,*],
Shaoqi Rao[4], and Zuquan Ding[1,*]

[1] Institute of Life Science and Medical Engineering, TongJi University, ShangHai 200092, China
fly_why@tom.com, dingzuquan@mail.tongji.edu.cn
[2] Department of Bioinformatics, Harbin Medical University, Harbin 150086, China
lixia@ems.hrbmu.edu.cn
[3] Harbin Institute of Technology, Harbin 150001, China
zzqp2002@126.com, ydwang@hit.edu.cn
[4] Departments of Cardiovascular Medicine and Molecular Cardiology, Cleveland Clinic Foundation, Cleveland, Ohio 44195, USA
raos@ccf.org

Abstract. Gene expression microarray technology provides the global information on transcriptional activities of essentially all genes simultaneously, and it thus promotes the new application of traditional feature selection methods in the fields of molecular biology and life sciences. The basic strategy for the traditional feature selection methods is to seek for a single gene subset that leads to the best prediction of biological types, for example tumor versus normal tissues. Because of complexities and genetic heterogeneities of biological phenotypes (e.g. complex diseases), robust computational approaches are desirable to achieve high generalization performance with multiple classifiers and perturbations of the data structures. The purpose of this study is to develop an ensemble decision approach to analysis of multiple heterogeneous phenotypes. The results from an application to a lymphoma data of five subtypes indicate that the proposed analysis strategy is feasible and powerful to perform biological subtype.

1 Introduction

Gene expression microarray technology has inspired much new hope for the research of complex diseases, for it can provide the global information on transcription activities of essentially all genes simultaneously. Therefore, genetic dissection of complex diseases should be carried out in a new global view, and the patterns of up-regulation or down-regulation of gene activities can serve as secondary endpoints of biomarkers[1]. The method of finding such novel biomarkers is actually the application of traditional feature selection methods in the field of molecular biology and life sciences. Feature

---

[*] Corresponding authors.

selection aims to pick out d features, the subset of D features (D>d), discriminate heterogeneous samples best [2]. We call these selected features as feature genes for their fine discrimination in different biological samples.

In the current literature three basic approaches to feature selection predominate [3,4,5,6]: filter approaches, wrapper approaches and embedded approaches. Traditional feature selection methods devote to finding a best feature subset, however based on it we always can't get good classifying performance because such feature selection methods focus on finding a best feature subset which contains few genes and is also sensitive to many factors such as learning algorithm, training samples and so on. Using ensembles of base classifiers to improve classifying performance has been the hot topic of current machine learning [7,8,9]. In this paper, we apply Feature Ensemble Technology, which is similar to ensembles of base classifiers to integrate a series of fine local feature subsets selected by embedded approach decision tree. Under the supervision of classifying performance when the composition of training samples is changed, ultimately we get some stable feature subsets which don't depend on the composition of training samples.

Presently, in the process of embedded approaches or wrapper approaches researchers always pay more attention to the learning algorithm, thus in a sense the result of feature selection is the byproduct of such learning algorithm. In this paper, we focus on local feature selection and let classifying performance to guide the result of feature selection, namely we pick out a series of fine feature subsets under the supervision of classifying performance when changing the composition of the training samples, then we use Feature Ensemble Technology to select stable feature subsets from fine feature subsets, and goodness of these fine feature subsets and stable feature subsets is assessed by classifying performance again. Thus being a scale classifying performance run through the whole process of feature selection. The purpose of this study is to extend our newly developed ensemble decision approach [10] to analysis of multiple heterogeneous phenotypes (for example, the numerous subtypes of lymphoma).

2 Methods

2.1 Local Feature Selection and Feature Ensemble Technology (LFSE)

We put forward the method named Local Feature Selection and Feature Ensemble Technology (LFSE) to pick out fine stable local feature subsets. LFSE is shown in figure1.

The lymphoma dataset we use is a multidimensional data (data with multi-class), and is feature-space heterogeneous so that individual features have unequal importance in different sub areas of the whole feature space. To make the feature selection locally, we first apply a technology for transforming samples with multi-class into two-class, that is to say, all samples belonging to the target class are positive samples while the others are negative samples. Secondly, when disturbing the composition of training samples with Sample Disturbing Technology, we obtain original feature subsets set that each feature subset G_j^c ($c = 1, 2 \cdots C$, $j = 1, 2 \cdots k_c$), where C is the class number,

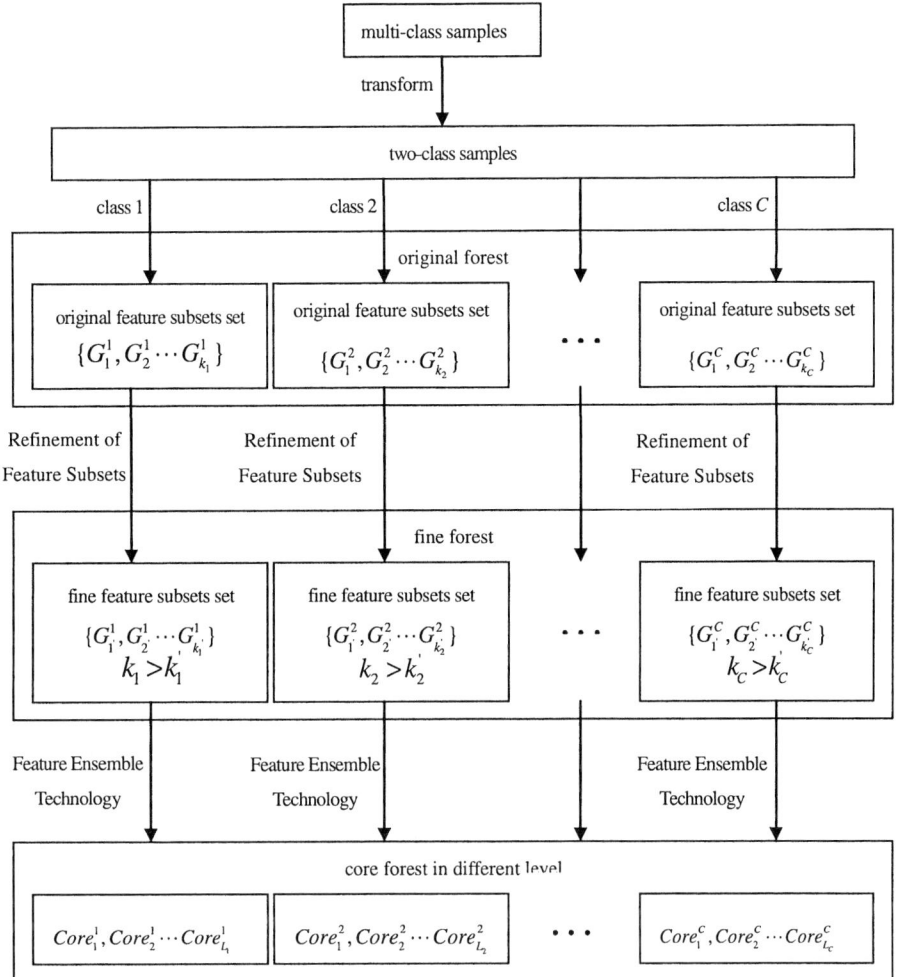

Fig. 1. LFSE

and *j* is the serial number of feature subset. Each feature subset is corresponding to all split nodes and their split rule in one decision tree and can discriminate samples between positive and negative classes (in this paper, we don't differentiate between feature subset and decision tree clearly because one decision tree is corresponding to one feature subset). C original feature subsets sets constitute the original forest.

Then under the supervision of classifying performance(ε see the following text) of each feature subset we pick out fine feature subsets set from original feature subsets set that element in fine feature subsets set, each feature subset, is noted as $G_j^{c'}$ ($c = 1, 2 \cdots C$, $j = 1, 2 \cdots k_c'$). Fine feature subsets set is a subset of original feature subsets set, so $k_c > k_c'$. C fine feature subsets sets constitute fine forest.

Thirdly, we apply Feature Ensemble Technology to see if there are some genes frequently come in groups (indicating strong gene-dependence) or some individual genes frequently come in these fine feature subsets set. We name genes frequently come in groups gene cores, and genes frequently come stable individual genes. In figure1, *Core* denote gene cores set selected by algorithm Gene Cores Finding, $Core_j^c$ ($c = 1, 2 \cdots C$, $j = 1, 2 \cdots L_c$) is gene cores set in the *j*th level belonging to class *c*. *C* fine feature subsets sets including gene cores constitute core forest.

Finally Two-Level Integrating Evaluation Machine (see the following text) and other four classifiers are used to test classifying performance of genes selected by LFSE. Following is the sub-algorithm in LFSE and classifying evaluation methods of LFSE:

2.2 Sample Disturbing Technology

Sample Disturbing Technology, including sample random grouping, cross validation and boosting, can change the composition of training samples.

Sample random grouping and cross validation: samples are randomly split into *n* groups. Samples in *n-1* groups are used to establish feature subsets, and those in the other one group are used to assess the goodness of feature subsets and pick out fine feature subsets.

Boosting [11]: this method repeatedly builds different feature subset using decision tree by adjusting the weights of training samples according to their classifying performance. Misclassified samples are given higher weight while those classified correctly are given lower weight.

2.3 Refinement of Feature Subsets

Some indexes such as accuracy and error are always used to assess the classifying performance of certain feature subset [12]. However, when using these two indexes we are sometimes inclined to pick out feature subsets that don't have good classifying ability for high false positive rate or false negative rate of classifying results especially when samples have a highly unbalanced class distribution. In this paper we transform multi-class samples into two-class samples, which ineluctably result in a highly unbalanced class distribution. So we adopt another index ε:

$$\varepsilon = 2p \times r / (p + r) \quad (1)$$

Where $p = TP/(FP + TP)$, $r = TP/(FN + TP)$, *TP* is true positive, *FP* is false positive, *TN* is true negative, and *FN* is false negative. This index will get a high value when both false positive rate and false negative rate are low. Now we let 0.6 as the threshold, and those feature subsets with $\varepsilon > 0.6$ are picked out as fine feature subsets.

2.4 Feature Ensemble Technology

We design two kinds of feature ensemble technology:

One kind is based on individual genes, this defines Stable Individual Genes Selection: some individual genes frequently come are selected in light of the frequency of each gene in fine feature subsets.

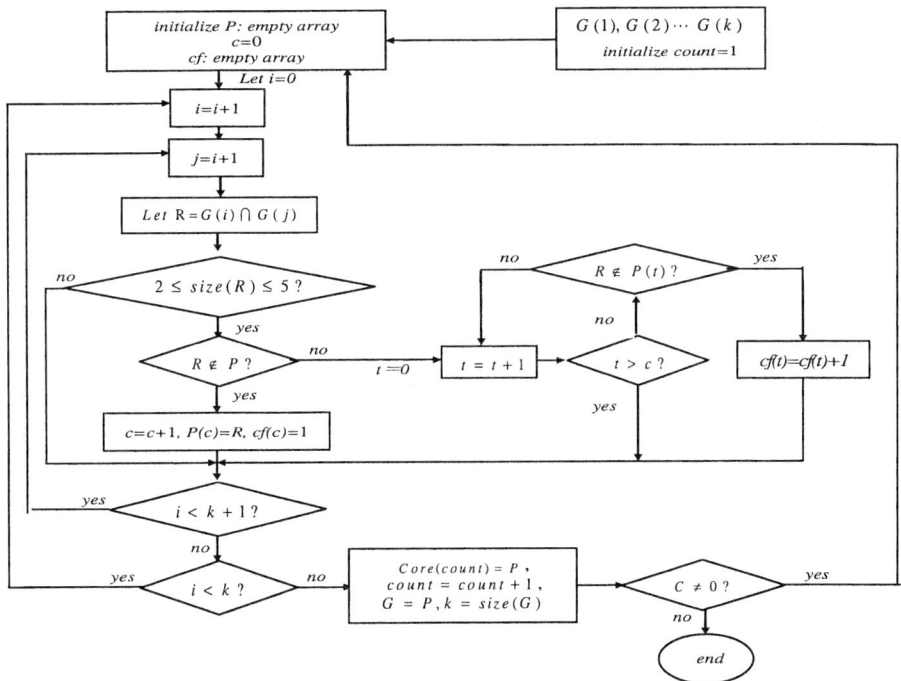

Fig. 2. Flow chart of Gene Cores Finding algorithm

In addition, another kind is Gene Cores Finding, an algorithm we develop on the basis of algorithm Mining Core [13], see if there are some genes frequently come in groups.

Input is the set of fine feature subsets belonging to class c $\{G_1^c, G_2^c \cdots G_{k_c}^c\}$. The algorithm can deal with fine feature subsets belonging to different classes in parallel, so in figure2 $\{G(1), G(2) \cdots G(k)\}$ replace $\{G_1^c, G_2^c \cdots G_{k_c}^c\}$ as input, where $k = k_c'$. In flow chart, P is an array and it contains the set of gene cores in certain level; Core is a cell array and it contains all sets of gene cores in different levels; count memorizes the level of gene cores set, cf memorizes the frequency of each gene core. We limit the size of gene core between 2 to 5 ($2 \leq size(R) \leq 5$, namely if there exists same genes in two feature subsets and number of same genes is between 2 to 5, these same genes constitute one gene core.) We limit the size of gene core Num≤5 for feature subset probably overfit the training samples when size of gene core is too large. The outputs of algorithm are hierarchical gene cores sets belonging to class c: $Core(1), Core(2) \cdots Core(size(Core))$, which are corresponding to $Core_1^c, Core_2^c \cdots Core_{L_c}^c$ in figure2. Where the level number of gene cores set is equal to size(Core), and {all genes in $Core(1)$} ⊃ {all genes in $Core(2)$} ⊃ ... ⊃ {all genes in $Core(size(Core))$}.

2.5 Classifying Performance Evaluation of LFSE

We apply Two-Level Evaluation Machine and other four classifiers to assess the classifying performance of feature genes selected by LFSE.

Two-Level Integrating Evaluation Machine: Evaluation Machine is shown in figure3.

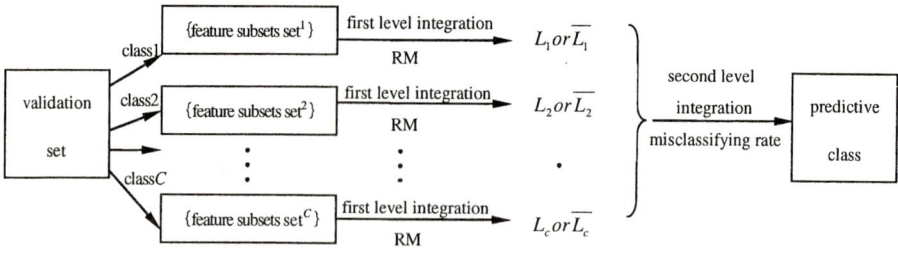

Fig. 3. Two Level integrating Evaluation Machine

when we use Evaluation Machine to assess classifying performance of original forest, { feature subsets set^1 } is corresponding to original feature subsets set{ $G_1^1, G_2^1 \cdots G_{k_1}^1$ } ,{ feature subsets set^2 } is corresponding to original feature subsets set{ $G_1^2, G_2^2 \cdots G_{k_2}^2$ } and so on. In the same way, when fine forest and core forest are assessed, { feature subsets set^c } ($c = 1, 2 \cdots C$) is corresponding to fine feature subsets set{ $G_1^{c'}, G_2^{c'} \cdots G_{k_c}^{c'}$ } and fine feature subsets set including $Core_j^c$. It is necessary to uniform level of gene cores of different class, for example, when we assess the classifying performance of core forest in N level, we must pick out fine feature subsets including $Core_N^1, Core_N^2, Core_N^3 \ldots Core_N^C$. If the deepest level of gene cores belonging to certain class $L_c < N$, $Core_{L_c}^c$ replaces $Core_N^c$.

Evaluation Machine is a two-level integrating classifier: at the first level, the classifier integrates classifying results of each feature subset in feature subsets set belonging to certain class {feature subsets set^c}, where one feature subset is one base classifier, Whose weight is decided by ε, and these feature subsets constitute a decision committee. At the second level, Evaluation Machine integrates classifying results of feature subsets set belonging to different class, and we use weighted nearest neighbor (WNN) integration [7], where one {feature subsets set^c} is one base classifier, whose weight depend on their classifying performance to testing sample's nearest K neighbors which will check in misclassifying rate table(see next text). After the first level integration, each sample will be given C class labels that each class label decides whether it belongs to target class ($L_c or \overline{L_c}$). Then after the second level integration, just one class label is for one sample, which decides which class it is predicted to belong to.

Other four classifiers: Other four classifiers such as Fisher linear discriminate, Logit nonlinear discriminate, Mahal distance and K-nearest neighbor classifier are used to assess classifying performance of gene cores in deepest level. These four classifiers are described in another study [12].

3 Results

3.1 Lymphoma Dataset

The lymphoma expression profile dataset we have used [14] includes nine kinds of biological samples. In addition, one kind of samples diffuse large B-cell lymphoma(DLBCL) has been found including two subtypes: One type expressed genes characteristic of germinal centre B cells (germinal centre B-like DLBCL, GCB-like-DLBCL); the second type expressed genes normally induced during *in vitro* activation of peripheral blood B cells (activated B-like DLBCL, AB-like-DLBCL). We combine the samples in different normal classes for their small samples and delete some samples without typical expression profile. Finally, eighty-six samples of five classes and 4026 genes are used to analysis: 21 GCB-like-DLBCL samples belong to class 1, 21 AB-like-DLBCL samples belong to class 2; 11 CLL samples belong to class 3; 9 FL samples belong to class 4 and 24 normal samples belong to class 5.

3.2 Experiment

Acquiring misclassifying rate table: As figure 3 shows, the second integration of Evaluation Machine is WNN integration based on misclassifying rate table. But how can we get this misclassifying rate table? First we transform samples with five classes into those with two classes. Then boosting builds a series of feature subsets. Terminating condition of boosting is set to misclassifying rate of base classifier equaling to zero or being greater than 0.35, and when convergence of boosting isn't able to reach, we set the number of base classifiers on ten. Finally, predictive class of testing samples will be acquired by ensemble of feature subsets, whose weight is ε. We apply samples random grouping 20 times and three-fold cross validation. Comparing predictive class and true class, we will acquire misclassifying rate tables whose row is sample, column is class, and element is misclassifying rate of certain sample classified by feature subsets belonging to certain class.

LFSE and its classifying performance: This work can be summarized as follows:

Step1. Randomly split 86 samples into five parts. One part is left for validation set, which never takes part in feature selection and is used to assess classifying performance of LFSE. The other four parts are used to establish original feature subsets, select fine feature subsets, find gene cores and stable individual genes.

Step2. Randomly split samples in the other four parts into three parts, and two parts as training set aim to find original feature subsets, the rest as testing set aim to select fine feature subsets. Three-fold cross validation technology is used.

Step3. Transform sample with multi-class into two-class. Boosting builds a series of feature subsets belonging to each class while training set is fixed. Parameters of boosting are the same as those in acquiring misclassifying rate table.

Step4. Pick out fine feature subsets set belonging to each class under the direction of testing set's classifying performance ($\varepsilon > 0.6$).

Step5. Repeat from step2 to step4 twenty times to achieve a series of fine feature subsets sets belonging to each class { $G_1^{c'}, G_2^{c'} \cdots G_{k_c}^{c'}$ } ($c = 1, 2 \cdots 5$).

Step6. Repeat from step1 to step5 ten times.

Step7. Achieve ten series of { $G_1^{c'}, G_2^{c'} \cdots G_{k_c}^{c'}$ } after above steps from step1 to step6. Calculate frequency of each gene in each series of { $G_1^{c'}, G_2^{c'} \cdots G_{k_c}^{c'}$ } and pick out genes with frequency $f > 1$ to enter feature genes set $\{F_c\}$ (we randomly pick out four genes from total genes to make up of one simulated feature subset for one real feature subset containing four genes averagely. Then we repeat above work k_c times and calculate each gene's accumulative frequency appearing in these simulated feature subsets. Results show there are few and far between genes with frequency $f > 1$). Then we calculate accumulative frequency of genes appeared in ten series of $\{F_c\}$ and pick out genes with frequency higher than nine times to enter final stable individual genes set $\{fF_c\}$. In addition, we use Gene Cores Finding algorithm to acquire 10 series of hierarchical gene cores set of different class $Core_j^c$ ($c = 1, 2 \cdots C$, $j = 1, 2 \cdots L_c$) and the frequency of each gene core. High frequency gene cores of each class in ten series of $Core_j^c$ are selected.

Step8. Classifying performance evaluation of LFSE using validation set.

All algorithms in this paper are achieved using MATLAB 6.5和JAVA 1.4. we download algorithm of decision tree written in Matlab designed by statistics department of Carnegie-Mellon University

Result

Finding stable genes. For every class we acquire final stable individual genes sets and gene cores.

Gene cores of each class are shown in table 1. Result shows: there exit gene cores belonging to each class. Collaborating with other genes, these gene cores can successfully discriminate samples of target class from other samples. We also find stable individual genes belonging to different lymphoma subtype, and we try to find the biological meanings of these stable genes and gene cores, which are mentioned in another paper.

Table 1. Gene cores of different class with high frequency. Number on the white background is geneID, and different geneIDs in the same column appear in the same gene core. Number of a grey background is frequency of gene core on a white background in the same column.

Class 1	3	2	25	1	2	1032	161	2	5	5	1	
	1835	1835	1835	2	5	1835	1835	75	1835	6	8	
	22	19	19	14	11	8	8	8	8	8	8	
Class 2	75	2	1	1	18	1	45	2	5			
	3791	3791	3791	75	3791	7	3791	4	3791			
	24	22	22	12	10	13	9	9	8			
Class 3	1	425	1	2	17							
	678	678	425	61	639							
			678	697								
	23	17	7	7	5							
Class 4	5	1	1	19	1	1						
	812	19	19	801	1008	760						
			1070									
	9	7	7	5	5	5						
Class 5	2	16	1	7	1	8	3	6	2	4	1	3	18	17
	236	236	236	236	2	236	236	235	16	236	235	235	235	236
	56	51	36	28	28	26	24	22	20	18	13	13	12	10

Multi-class classifying performance evaluation using Evaluation Machine. When ten series of validation sets are fixed, we apply Two-Level Integrating Evaluation Machine to assess the classifying performance (accuracy) of original forest, fine forest and hierarchical core forest. Results show in figure 4: average accuracy of fine forest reach 86.50%, which is remarkably higher than that of original forest 54.32%. Such classifying performance is also high compared with that of other classifying algorithm for multi-class samples [7]. Moreover, accuracy of deeper-level core forest doesn't reduce obviously along with the decrease of amount of genes in core forest.

Two-class classifying performance evaluation using other classifiers. When other four classifiers are available, we apply three fold cross validation to assess two-class discriminating ability of local feature genes included in gene cores at the deepest level(we call them core genes). At the same time, same number of genes are sampled

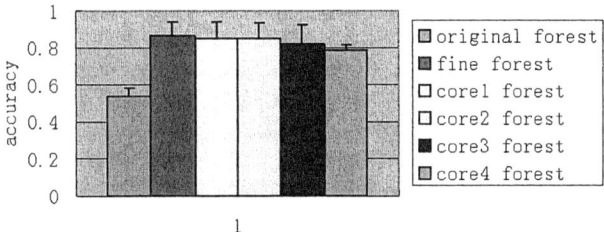

Fig. 4. Classifying performance evaluation of LFSE. CoreN forest (N=1,2...4) is the core forest in N level, which means that feature subsets for any class used to classify are fine feature subsets including gene cores in the Nth level.

randomly from all genes to do the same work, and in order to assure the randomicity, such random sampling repeats 10 times. Results show as table 2: two-class discriminating ability of core genes is higher than that of randomly sampling genes markedly, especially when Fisher linear discriminate, Logit nonlinear discrimination and Mahal distance classifier being used, the accuracy of core genes can reach around 90%. While we use K-nearest neighbor classifier to assess core genes of class 3, accuracy of them is a little lower than that of randomly sampling genes. We suggest it due to the flexibility of K-nearest neighbor classifier or sampling bias.

Table 2. Two-class classifying performance evaluation of LFSE. std is abbreviation of standard deviation

class		Fisher linear discriminate Mean(std)	Logit nonlinear discriminate Mean(std)	K-nearest neighbor Mean(std)	Mahal distance Mean(std)
1	Core genes	0.8536 (0.0996)	0.8248 (0.1104)	0.9114 (0.0348)	0.6746 (0.1174)
	Random genes	0.7449 (0.1020)	0.7368 (0.0921)	0.7658 (0.0740)	0.5096 (0.1844)
2	Core genes	0.9116 (0.0631)	0.9233 (0.0625)	0.8678 (0.0839)	0.7531 (0.0755)
	Random genes	0.7574 (0.1050)	0.6975 (0.1018)	0.7863 (0.0823)	0.4573 (0.2164)
3	Core genes	1 (0)	0.9976 (0.0118)	0.9672 (0.0482)	0.8156 (0.1070)
	Random genes	0.8997 (0.0821)	0.8929 (0.0636)	0.9117 (0.0679)	0.3384 (0.1647)
4	Core genes	0.9275 (0.0514)	0.9299 (0.0379)	0.8836 (0.0705)	0.5531 (0.0792)
	Random genes	0.8451 (0.0932)	0.8883 (0.0662)	0.9182 (0.0618)	0.3493 (0.2209)
5	Core genes	0.8869 (0.0925)	0.8663 (0.0921)	0.8389 (0.0601)	0.8834 (0.0727)
	Random genes	0.7453 (0.1071)	0.8413 (0.0965)	0.799 (0.0742)	0.4121 (0.1239)

4 Conclusion

In this paper, we focus on LFSE. Results show: Local Feature Selection optimize feature subsets, for these fine feature subsets achieve higher classifying performance to multi-class samples. Gene cores found by Feature Ensemble Technology also have good classifying performance.

Acknowledgements

We thank two anonymous reviewers for their comments on an early version of the manuscript. This work was supported in part by the National High Tech Development Project of China (Grant Nos. 2003AA2Z2051 and 2002AA2Z2052), the National Natural Science Foundation of China (Grant Nos. 30170515 and 30370798) and the Cardiovascular Genetics Funds from Cleveland Clinic Foundation of USA.

References

1. Gu CC, Rao DC, Stormo G, Hicks C, and Province MA.: Role of gene expression microarray analysis in finding complex disease genes. *Genet Epidemiol.* 23 (2002) 37-56.
2. Bian Z and Zhang X.: Pattern Recognition. TsingHua Press,Beijing. (2000) pp. 198;87-90;113-116;120-121.
3. Kohavi R. and John G.: Wrappers for feature subset selection. Artificial Intelligence. 97 (1997) 273-324.
4. Furlanello C, Serafini M, Merler S, and Jurman G.: Entropy-based gene ranking without selection bias for the predictive classification of microarray data. BMC Bioinformatics. 4 (2003) 54
5. John GH, Kohavi R, and Pfleger K.: Irrelevant features and the subset selection problem. Machine Learning: Proceedings of the 11th International Conference. (1994) 121-129.
6. Blum AL and Langley P.: Selection of relevant features and examples in machine learning. Artificial Intelligence. 97 (1997) 245-271.
7. Puuronen S and Tsymbal A.: Local feature selection with dynamic integration of classifiers. Fundamenta Informaticae. 47 (2001) 91-117.
8. Hansen JV : Combining predictors: comparison of five meta machine learning methods. Information Science. 119 (1999) 91-105.
9. Opitz DW and Maclin RF : An empirical evaluation of bagging and boosting for artificial neural networks. International conference on neural networks. 3 (1997) 1401-1405.
10. Li X, Rao S, Wang Y, and Gong B.: Gene mining: a novel and powerful ensemble decision approach to hunting for disease genes using microarray expression profiling. Nucl Acids Res. 32 (2004) 2685-2694.
11. Zheng Z, Webb G, and Ting K.: Integrating boosting and stochastic attribute selection committees for further improving the performance of decision tree learning. 10th International Conference on Tools With Artificial Intelligence TAI '98, edited by Society IC, Los Alamitos, USA. (1998) 216-223.
12. Wang HY, Li X, and Guo Z.: Research on pattern classification methods using gene expression data. Biomedical Engineering Journal. (2005) in press.
13. Kurra, G., Niu, W. & Bhatnagar, R.: Mining microarray expression data for classifier gene-cores. Proceedings of the Workshop on Data Mining in Bioinformatics. (2001) 8-14.
14. Alizadeh AA, Eisen MB, Davis RE, Ma C, Lossos IS, Rosenwald A, Boldrick JC, Sabet H, Tran T, Yu X, Powell JI, Yang L, Marti GE, Moore T, Hudson J, Jr., Lu L, Lewis DB, Tibshirani R, Sherlock G, Chan WC, Greiner TC, Weisenburger DD, Armitage JO, Warnke R, Levy R, Wilson W, Grever MR, Byrd JC, Botstein D, Brown PO, and Staudt LM.: Distinct types of diffuse large B-cell lymphoma identified by gene expression profiling. Nature. 403 (2000) 503-511.

Fuzzy Routing in QoS Networks

Runtong Zhang[1,2] and Xiaomin Zhu[1]

[1] Institute of Information Systems, School of Economics and Management,
Beijing Jiaotong University, Beijing, 100044, P.R. China
rtzhang@center.njtu.edu.cn
[2] Lab of Compter & Network Architectures, Swedish Institute of Computer Science,
P.O. Box 1263, Kista 16429, Stcockholm, Sweden
runtong@sics.se

Abstract. QoS (Quality of Service) routing is a key network function for the transmission and distribution of digitized audio/video across next-generation high-speed networks. It has two objectives: finding routes that satisfy the QoS constraints and making efficient use of network resources. The complexity involved in the networks may require the consideration of multiple constraints to make the routing decision. In this paper, we propose a novel approach using fuzzy logic technique to QoS routing that allows multiple constraints to be considered in a simple and intuitive way. Simulation shows that this fuzzy routing algorithm is efficient and promising.

1 Introduction

In the current Internet, data packets of a session may follow different paths to the destinations, and the network resources (e.g., router buffer and link bandwidth) are fairly shared by packets from different sessions. However, this architecture does not meet the QoS (quality of service) requirements of future integrated services networks that will carry heterogeneous data traffic.

QoS routing consists of two basic tasks. The first task is to collect the state information and keep it up to date. The second task is to find a feasible path for a new connection based on the collected information. A routing algorithm generally focuses on the second task, i.e., it assumes that a global state is well detected and the present work falls into this category.

One of the biggest difficulties in QoS routing area is that multiple constraints often make the routing problem intractable. This normally includes things such as node buffer capacities, residual link capacities, and the number of hops on the path (i.e., the number of nodes a packet must pass through on the route). Many common routing algorithms require that these factors be expressed together in a closed, analytical form for evaluation. Fuzzy control is a control technique based on the principles of fuzzy set theory [11,22]. Fuzzy control systems are designed to mimic human control better than classical control systems by incorporating expert knowledge and experience in the control process.

Normally, a good Internet service requires several criteria simultaneously, and depends on the network situations (e.g., the structure or load), which are generally not available or dynamically changed. Fuzzy control is an intermediate approach between complicated analysis and simple intuition. Some successful examples of implying fuzzy approach to the network optimization can be found in [17-21]. It could also allow a means of expressing complex relationships and dependencies predicted to be evident in future QoS enable communication networks that support various applications. This could have a great impact on the performance of the routing algorithm and consequently, the network performance. A survey of recent advances in fuzzy logic in telecommunications networks can be found in [12].

The organization of this paper is as follows. Sections 2 and 3 provide some tutorial information on QoS routing and fuzzy control. Section 4 describes the fuzzy controller to the QoS routing problem. Section 5 gives a numerical example to illustrate the implementation of the fuzzy approach. System simulations and comparisons are also outlined in section 5. The final section concludes this work.

2 QoS Routing

The notion of QoS has been proposed to capture the qualitatively or quantitatively defined performance contract between the service provider and the user applications. The QoS requirement of a connection is given as a set of constraints. A link constraint specifies a restriction on the use of links. A bandwidth constraint of a unicast connection requires. A path constraint specifies the end-to-end QoS requirement on a single path. A feasible path is one that has sufficient residual (unused) resources to satisfy the QoS constraints of a connection.

The basic function of QoS routing is to find such a feasible path. In addition, most QoS routing algorithms consider the optimization of resource utilization measured by an abstract metric in cost [6]. The cost of a link can be defined in dollars or as a function of the buffer or bandwidth utilization. The cost of a path is the total cost all links on the path. The optimization problem is to find the lowest-cost path among all feasible paths.

The problem of QoS routing is difficult for a number of reasons. First, distributed applications such as Internet phone and distributed games have very diverse QoS constraints on delay, delay jitter, loss ratio, bandwidth, and so on. Multiple constraints often make the routing problem intractable. Second, any future integrated services network is likely to carry both QoS and best-effort traffic, which makes the issue of performance optimization complicated. Third, the network state changes dynamically due to transient load fluctuation, connections in and out, and links up and down.

The routing problems can also be divided into two major classes: unicast routing and multicast routing. A unicast QoS routing problem is defined as follows: given a source node a, a destination node b, a set of QoS constraints C, and possibly an optimization goal, find the best feasible path from a to b which satisfies C. The multicast routing problem is defined as follows: given a source node a, a set R of destination nodes, a set of constraints C and possible an optimization goal, find the best feasible tree covering a and all nodes in R which satisfies C. Multicast routing can be viewed as a generalization of unicast routing in many cases.

The task of admission control [8, 17 and 19] is to determine whether a connection request should be accepted or rejected. Once a request is accepted, the required resources must be guaranteed. The admission control is often considered a by-product of QoS routing and resource reservation. In addition to the rejection of a connection request, a negotiation with the application for degrading the QoS requirements may be conducted. This motivates the concept of differentiated services [14]. QoS routing can assist the negotiation by finding the best available path and returning the QoS bounds supported. If the negotiation is successful according to the provided bounds, the best available path can be used immediately.

There are three routing strategies: source routing, distributed routing and hierarchical routing. They are classified according to how the state information is maintained and how the search of feasible paths is carried out. Many routing algorithms in this area are proposed in the literature. Generally, most source unicast routing algorithms transform the routing problem to a shortest path problem and then solve it by Djikstra's algorithm [10]. The Djikstra's algorithm is also known as the shortest path routing algorithm, and we will use it as one of the reference frameworks to test our work in section 5. The Ma-steenkiste algorithm [16] provides a routing solution to rate-based networks; The Guerin-Orda algorithm work with imprecision information, and hence is suitable to be used in hierarchical routing; The performance of the Chen-Nahrstedt algorithm [5] is tunable by trading overhead for success probability; The Awerbuch et. al. algorithm [1] takes the connection duration into account, which allows more precise cost-profit comparison. All the above algorithms are executed at the connection arrival time on a per-connection basis. Path precomputation and caching were studied to make a trade-off between processing overhead and routing performance.

3 Fuzzy Logic Control

A fuzzy control system [11,21] is a rule-based system in which a set of so-called fuzzy rules represents a control decision mechanism to adjust the effects of certain causes coming from the system. The aim of a fuzzy control system is normally to substitute for or replace a skilled human operator with a fuzzy rule-based system. Specifically, based on the current state of a system, an inference engine equipped with a fuzzy rule base determines an on-line decision to adjust the system behavior in order to guarantee that it is optimal in some certain senses.

The design process of a fuzzy control system consists of a series of steps. The first step in fuzzy control is to define the input variables and the control variables. The input variables may be crisp or fuzzy.

Once these membership functions have been defined for each quantification of the input and control variables, a fuzzy rule base must be design. This rule base determines what control actions take place under what input conditions. The rules are written in an if-then format.

Once the rule base is established, an approximate reasoning method must be used to determine the fuzzy control action. The approximate reasoning method provides a means of activating the fuzzy rule base. An implication formula is used to evaluate the individual if-then rules in the rule base. A composition rule is used to aggregate the rule results to yield a fuzzy output set. The implication formula provides a membership function that measures the degree of truth of the implication relation (i.e., the if-then rule) between the input and output variables. One frequently used implication formula is that of Mamdani. Let a fuzzy rule be stated as follows: if x is A, then y is N. The implication formulas of Mamdani is as follows:

$$\mu_{A \to N}(x,y) = \mu_A(x) \wedge \mu_N(y) \tag{1}$$

where $\mu_A(x)$ is the membership of x in A, $\mu_N(y)$ is the membership of y in N, $\mu_{A \to N}(x,y)$ is the membership of the implication relation between x and y, and $\wedge$ is the minimum operator.

A defuzzification method is then applied to the fuzzy control action to produce a crisp control action. One simple and frequently used defuzzification method is the Height method. Let $c^{(k)}$ and f_k be the peak value and height, respectively, of the kth fuzzy set of the fuzzy output. Then by the Height method, the defuzzified crisp output u^* is given

$$u^* = \frac{\sum_{k=1}^{n} c^{(k)} f_k}{\sum_{k=1}^{n} f_k} \tag{2}$$

where n is the total number of the fuzzy sets of the fuzzy output.

There are generally two kinds of fuzzy logic controllers. One is feedback controller, which is not suitable for the high performance communication networks. Another one, which is used in this paper, is shown in Figure 1.

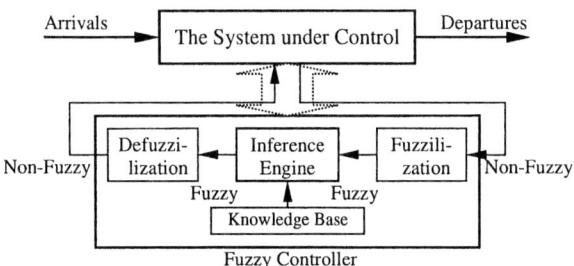

Fig. 1. The fuzzy controller

In this paper, most of the membership functions for the fuzzy sets are chosen to be triangular. We make this choice because the parametric, functional descriptions of triangular membership functions are the most economic ones. In addition, it has been proven that such membership functions can approximate any other membership function. To describe the fuzzy rules, we use ZO, PS, PM, PB to indicate "zero", "positive small", "positive medium" and "positive big".

We simulate and control queueing systems in C++ language. Mamdani implication is used to represent the meaning of "if-then" rules. The height method of defuzzification is used to transform the fuzzy output into a usable crisp one. For more information on the implement of fuzzy control, refer to [11, 21].

4 The Fuzzy Routing Algorithm

The network model used for testing the fuzzy QoS routing algorithm is adapted from Balakrishnan et. al. [2], and it is shown in Figure 2. For the sake of easy illustration, it is assumed that the links between the nodes are with same transmission bandwidth, and their length are all the same. These assumptions are logical because the propagation delay of a traffic flow in the high performance communication is normally very small compared with its queueing delay at the switching nodes. Each node has incoming packet buffer with a maximum capacity of B. Nodes one, five, six seven, eight nine, and ten act as both traffic generating nodes and switching nodes. Nodes two three and four are pure switching nodes. According to the QoS requirements, a traffic route should be determined before a traffic flow is going to be sent off at its generating node, and the chosen route will not be changed afterward. The problem is to determine the optimal QoS routing policy for each traffic flow at its generating node based on the state of the system. The optimal criteria are multiple, which are the minimal percentage of connections rejected at the generating nodes, the minimal percentage of connections lost along the routes, and the minimal mean packet delay in the network.

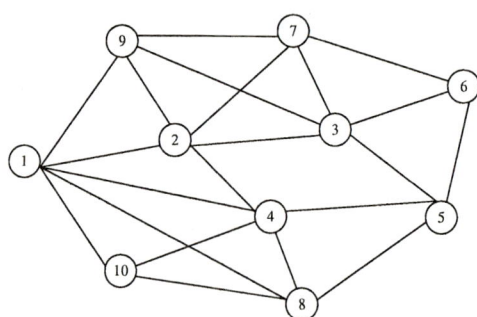

Fig. 2. Experimental communication network topology

For a given traffic flow at its generating node, the state of each eligible path is described by (s, n_i), where $s \subset \{1,2,3,4\}$ is the number of hops on the path, and $n_i = 0,1,2,\ldots,B$, $i=1,2,\ldots,s$, is the number of packets currently in buffer i on the given path. The state of the system changes whenever an arrival or departure at any nodes along the given path occurs. Without loss of generality, the decision epochs are the time instances that a new traffic flow is being generated and sent to the network.

We use fuzzy control technique to solve this QoS routing problem and the algorithm is referred to fuzzy routing algorithm. The algorithm first determines the crisp path ratings for all eligible paths between the source and destination nodes from the view point of fuzzy inference. The path with the highest rating is then chosen to route the traffic flow. The path rate in this paper represents the degree of the path usability in the sense of the multiple criteria required. The connection is only rejected if all of the buffers on the chosen are currently full. Otherwise, the connection traffic is routed over the chosen path for the duration of the connection. Whenever traffic flow is routed to a chosen path, a packet is dropped when it arrives at a full buffer.

We choose as fuzzy inputs: the number of hops on the path s and the path utilization $\bar{\rho}$. The fuzzy output is the path rating r. The fuzzy rule base is shown in Table 1.

Table 1. Fuzzy rule base

	r	$\bar{\rho}$			
		ZO	PS	PM	PB
s	ZO	PB	PM	PS	ZO
	PS	PM	PS	ZO	ZO
	PM	PS	ZO	ZO	ZO
	PB	ZO	ZO	ZO	ZO

The path utilization $\bar{\rho}$ is calculated by the following series of steps. First, the utilization of each buffer ρ_i on the path is calculated as in (3). The sum of these utilization measures is taken and used to generate a weighting measure λ_i for each buffer I as in (4) and (5). Finally, the estimated path utilization $\bar{\rho}$ is calculated by multiplying the umber of packets in each buffer by its corresponding weight factor, which is shown in (6).

$$\rho_i = \frac{n_i}{B}, i=1 \text{ to } s \quad (3)$$

$$P = \sum_{t=1}^{t} \rho_i \quad (4)$$

$$\lambda_i = \frac{\rho_i}{P}, i=1 \text{ to } s \quad (5)$$

$$\bar{\rho} = \sum_{i=1}^{t} \lambda_i \cdot n_i \quad (6)$$

The membership functions for the fuzzy variables s, $\bar{\rho}$ and r are shown in Figures 3 (a), (b) and (a), respectively. The universes of discourse for the fuzzy variables are all chosen [0, 6]. The sojourn time of a packet in the system increases with the total number of packets in the system as the sequence 1,3,6,..., which is given by $t_j=t_{j-1}+j$, $t_0=0$, $j=1,2,...$, thus we choose the fuzzy membership functions for $\bar{\rho}$ as shown in Figure 3(b).

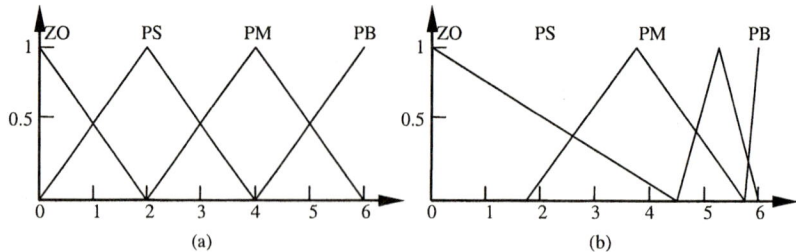

Fig. 3. Membership functions

To sum up, the fuzzy QoS routing algorithm is outlined as follows.

(a) All eligible paths between the source and destination nodes and corresponding state information are collected. This work is needed for all source routing related algorithms.

(b) Calculating the values of s and $\bar{\rho}$ for each eligible path by (3-6).

(c) Using the calculated values pair of s and $\bar{\rho}$ as crisp inputs, we determine the crisp path ratings r for each eligible path via fuzzification (based on the membership functions shown in Figure 3), fuzzy inference (based on the rule base shown in Table 1 and the Mamdani implication) and de-fuzzification (based on the High method of de-fuzzification).

(d) The path with highest rating is chosen to route the traffic flow.

5 Simulation Results

We examine a QoS network model shown in Figure 1. It is assumed that the links between the nodes are all 2 km in length, and the bandwidth of the links are all 100 Mbps. Each node has incoming packet buffer with a maximum capacity of 50 packets. The interarrival rate of connection attempts is assumed to be exponential. The mean of this exponential variable varies from 0.5 to 1.0 in increments of 0.05. We wish to determine the optimal QoS routing policy for each traffic flow at its generating node based on the state of the system. The optimal criteria are multiple, which are the minimal percentage of connections rejected at the generating nodes, the minimal percentage of connections lost along the routes, and the minimal mean packet delay in the network. The fuzzy routing scheme is tested against three other routing algorithms: a fixed directory routing algorithm [2], a shortest path routing algorithm [10], and a "crisp" or non-fuzzy version of the fuzzy routing scheme.

The fixed directory routing algorithm [2] is a simplified version of the shortest path problem, and is also based on the number of hops on the path. All of the one two, three, and four hop paths for a given source/destination pair are listed in a directory. The directory gives preference to the minimum hop paths. When a connection is requested, it is made on the first path in the directory that can accommodate the connection. The only reason that a path cannot accommodate a connection is if all buffers on the path are full.

The shortest path routing algorithm calculates the shortest delay path. Once again, only the one, two, three, and four hop paths for each source/destination node pair are considered. The path with the shortest estimated delay is chosen to route the connection.

The "crisp" non-fuzzy version of the fuzzy routing algorithm (henceforth referred to as the crisp routing algorithm) utilizes the path utilization calculation presented above in the breakdown of the fuzzy control routing algorithm.

The simulations are executed on an IBM T23 ThinkPad. Each interarrival rate is simulated ten times. Each simulation run simulates the network for 300 seconds.

The graphs for the percentage of connections rejected, the percentage of packets lost, and the mean packet delay (in second) in the network are shown in Figures 4, 5 and 6, respectively. These statistic parameters are plotted in the value axis of the three figures, respectively, while the category axis are all mean call interarrival time.

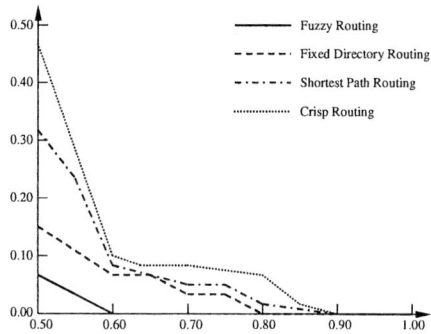

Fig. 4. QoS network percentage of connections rejected

Figure 4 illustrates that the fuzzy routing algorithm rejects a smaller percentage of connections than the other three routing algorithms. Recall that the only reason for which connections are rejected is if all buffers on the route chosen are full; therefore, the fuzzy routing algorithm appears to outperform the others at dispersing traffic in the network (to avoid extreme congestion on individual paths).

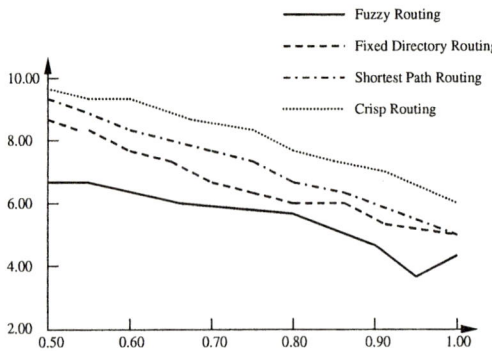

Fig. 5. QoS network of packets lost percentage

Figure 5 reveals that the fuzzy routing algorithm also loses a smaller percentage of packets than the other routing algorithms. This is another illustration of the fuzzy routing algorithm's ability to outperform the other routing algorithms at dispersing traffic in the network. The fact that a fewer percentage of packets are lost under the fuzzy scheme means that not as many packets are approaching full buffers under this scheme.

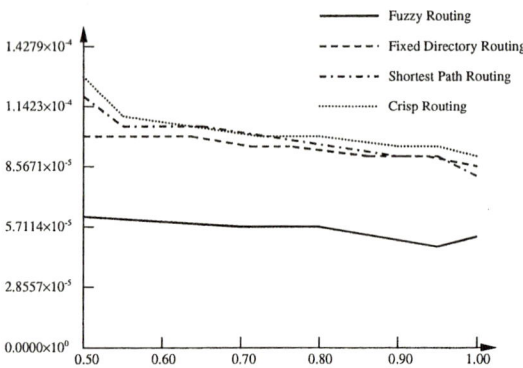

Fig. 6. QoS network mean packet delay in the network

Figure 6 reveals that the fuzzy routing algorithm results in a smaller mean packet delay in the network than the other routing algorithms. The fuzzy algorithm also does not experience as sharp an increase in mean packet delay as the other algorithms when the call arrival rate increases. This illustrates the ability of the fuzzy routing algorithm to handle an increased traffic load better than the other three algorithms.

Overall, the fuzzy routing algorithm outperforms the other three routing algorithms with regard to all of the measures collected. The results shown in the graphs indicate that the fuzzy routing algorithm does a better job at dispersing traffic in a more uniform manner. It also handles an increased traffic load more efficiently.

6 Conclusions

We propose a routing algorithm based on fuzzy control for QoS communication networks. The benefits of such an algorithm include increased flexibility in the constraints that can be considered in the routing decision and the ease in considering multiple constraints. The computational burden of a fuzzy control routing system is not severe enough to rule it out as a viable option. This is heavily due to the simple if-then structure of the rule base.

The design of a simple fuzzy control routing algorithm is presented and tested on an experimental QoS network. The results of this experiment prove favorable for the fuzzy control routing algorithm. The fuzzy algorithm displayed better performance than its "crisp" counterpart, the fixed directory routing algorithm and the classic shortest path routing algorithm. The results of this research indicate a promising future for fuzzy control in the world of communication network routing.

References

1. Awerbuch, B.: Throughput-competitive on-line routing, Proc. 34th Annual Symp. Foundations of Comp. Sci. (1993)
2. Balakrishnan, K., Tipper D. Medhi, D.: Routing strategies for fault recovery in wide area packet networks. Proc. MILICOM'95, (1995) 1139-1143
3. Bell, P. R., Jabbour, K.: Review of point-to-point network routing algorithms. IEEE Communications Magazine, 24(1), (1986) 34-38
4. Braden, R., Zhang, L., Berson, S., Herzog, S., Jamin, S.: Resource Reservation Protocol (RSVP) Version 1 Functional Specification, RFC 2205, (1997)
5. Chen, S., Nahrstedt, K.: On finding multi-constrainted paths. Proc. IEEE/ICC'98 (1998)
6. Chen, S. Nahrstedt, K.: An overview of quality of service routing for next-generation high-speed networks: problems and solutions, IEEE Network, 12(6), (1998) 64-79
7. Cheng, R., Chang, C.: Design of a fuzzy traffic controller for ATM networks, IEEE/ACM Trans. Networking, 4(3), (1996) 460-469
8. Courcoubetis, C., Kesidis, G., Ridder, A., Walrand, J., Weber, R.: Admission control and routing in ATM networks using inferences from measured buffer occupancy, IEEE Trans. Communications, 43(4), (1995) 1778-1784
9. Douligeris, C., Develekos, G.: A fuzzy logic approach to congestion control in ATM networks, Proc. 1995 IEEE International Conference on Communications, (1995) 1969-1973
10. Dijkstra, E.: A note on two problems in connection with graphs, Numerische Mathematik, (1959) 169-271
11. D. Driankov, H. Hellendoorn and M. Reinfrank, An introduction to fuzzy control, Springer-Verlag, Berlin, New York, 1993.
12. Ghosh, S., Razouqi, Q., J. Schumacher H., Celmins, A.: A survey of recent advances in fuzzy logic in telecommunications networks and new challenges, IEEE Trans. Fuzzy Systems, 6(3) (1998)
13. Guerin, R., Orda, A: QoS based routing in networks with inaccurate information: theory and algorithms, Proc. IEEE/ INFOCOM'97, Japan, (1997)
14. Kilkki, K.: Differentiated services for the Internet, Macmillan Technical Publishing, Indianapolis, USA, (1999)

15. Schwartz M., Stern, T. E.: Routing techniques used in computer communication networks, IEEE Trans. Communications, COM-28(4), (1980) 265-277
16. Ma Q., Steenkiste, P.: Quality of service routing with performance guarantees, Proc. 4th IFIP Workshop QoS, (1997)
17. Zhang R., Phillis, Y.: A fuzzy approach to the flow control problems, J. Intelligent and Fuzzy Systems, (6), (1998) 447-458
18. Zhang R., Phillis, Y.: Fuzzy control of queueing systems with heterogeneous servers", IEEE Trans. Fuzzy Systems, 7 (1), (1999). 17-26
19. Zhang R., Phillis, Y.: Fuzzy control of arrivals to tandem queues with two stations, IEEE Trans. Fuzzy Systems, 7 (3), (1999) 161-167
20. Zhang R., Phillis, Y.: Fuzzy control of two-station queueing networks with two types of customers, J. Intelligent and Fuzzy Systems, 8, (2000) 27-42
21. Zhang R., Phillis, Y., Ma, J.: A fuzzy approach to the balance of drops and delay priorities in differentiated services networks, IEEE Trans. Fuzzy Systems, 11(6), (2003) 840 – 846
22. Zimmermann, H. J.: Fuzzy Set Theory – and Its Applications (second edition), Kluwer Academic Publishers, Boston, (1991).

Component Content Soft-Sensor Based on Adaptive Fuzzy System in Rare-Earth Countercurrent Extraction Process*

Hui Yang[1], Chonghui Song[2], Chunyan Yang[3], and Tianyou Chai[4]

[1] School of Electrical and Electronics Engineering,
East China Jiaotong University, Nanchang 330013, China
yhshuo@263.net
[2] Department of Information Science and Engineering,
Northeastern University, 110004
[3] Mechatronics Research Center,
Jiangxi Academy of Science, Nanchang 330029, China
[4] Research center of Automation, Northeastern University, Shenyang, 110004, China

Abstract. In this paper, fusion of the mechanism modeling and the fuzzy modeling, a component content soft-sensor, which is composed of the equilibrium calculation model for multi-component rare earth extraction and the error compensation model of fuzzy system, is proposed to solve the problem that the component content in countercurrent rare-earth extraction process is hardly measured on line. An industry experiment in the extraction Y process by HAB using this hybrid soft-sensor proves its effectiveness.

1 Introduction

China has the most abundant rare-earth resource in the world. But the process automation mostly is still in the stage that component content is measured off-line, the process is controlled by the experience and the process parameters are regulated by hands. This situation leads to low efficient production rate, high resource consumption and unstable production quality [1]. To implement automation in the rare-earth extraction process, the component content on-line measuring must be achieved at first. The present chief methods for the component content on-line measurement in rare earth extraction process include UV-VIS, FIA, LaF3 ISE, Isotopic XRF etc [2, 3]. Because of high cost of the equipments, low reliability and stability, their usage in industry are generally limited. The soft sensor method provides a new way to measure component content in countercurrent rare earth extraction production on line. We further our research of paper [4] by fusion of the mechanism modeling and the intelligent modeling and propose a hybrid soft-sensor of the rare earth component content which contributes to better prediction accuracy and wider applicability.

---

* The work is supported by the National Natural Science Foundation of China (50474020),the National Tenth Five-Year-Plan of Key Technology (2002BA315A).

2 Component Content Soft-Sensor in Rare Earth Countercurrent Extraction Process

2.1 Description of Rare Earth Countercurrent Extraction Process

The two component A and B extraction process is shown in figure 1, where A is easy extracted component and B is the hard extracted component. In figure 1, u_1 is the flow of rare earth feed, u_2 is the flow of extraction solvent, u_3 is the flow of scrub solvent, u_4 and u_5 are the distribution of A and B in the feed respectively, where $u_4 + u_5 = 1$. ρ_A is organic phase product purity of A at the exit and ρ_B is aqueous phase product purity of B at the exit. $\rho_{A,k}$ is organic phase component content at the specified sampling point in scrub section and $\rho_{B,k}$ is aqueous phase component content at the specified sampling point in extraction section.

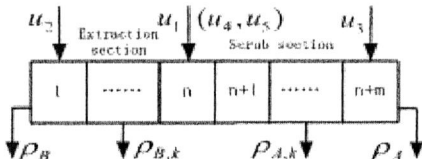

Fig. 1. Rare earth extraction process

Since the whole process is composed from few decades to one hundred stages, the flow regulation of extraction solvent, scrub solvent and the feed could influence the product purity at the exit after a long-time (often few decade hours) step by step delivery. For the above reason, the sampling point is set near the exit and the exit product purity (ρ_A, ρ_B) is guaranteed by measuring and control of the component content (ρ_{AK}, ρ_{BK}) at the sampling point. How to measure the parameters (ρ_{AK}, ρ_{BK}) has became the key point of rare earth extraction process control.

2.2 Profile of Component Content Soft-Sensor in Rare Earth Extraction Process

Via the mechanism analysis of countercurrent extraction process, parameters $\rho_{A,k}$, $\rho_{A,k}$, u_1, u_2, u_3, u_4 (or u_5) have following relationships

$$\begin{cases} \rho_{A,k} = f_{A,k}\{u_1, u_2, u_3, u_4, \omega\} \\ \rho_{B,k} = f_{B,k}\{u_1, u_2, u_3, u_5, \omega\} \end{cases} \quad (1)$$

where $f_{A,k}\{\cdot\}$ and $f_{B,k}\{\cdot\}$ are some form nonlinear functions, ω is the influence of the outside factors such as extraction solvent concentration, feed concentration *etc.* in component content. Since the soft-sensor principle is same for $\rho_{A,k}$ and $\rho_{B,k}$, we only describe one component content soft-sensor and use ρ instead of $\rho_{A,k}$ or $\rho_{B,k}$.

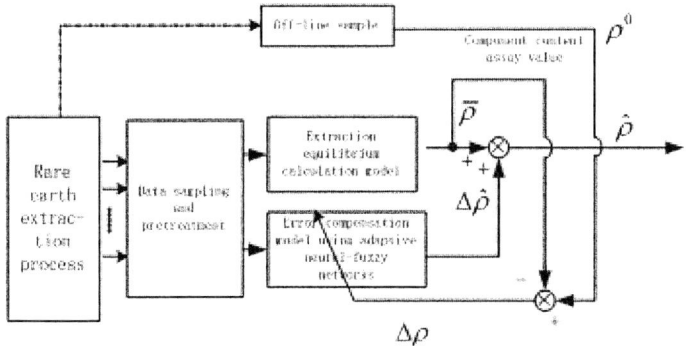

Fig. 2. Framework of rare earth extraction component content soft-sensor

The framework of rare-earth extraction component content soft-sensor system is described in figure 2.

Parameter ρ^0 is the component content of off-line assay value, $\bar{\rho}$ is the output of countercurrent extraction equilibrium calculation model, parameter $\Delta\tilde{\rho} = \rho^0 - \bar{\rho}$ is the error in modeling, $\Delta\hat{\rho}$ is output of error compensation model. The $\Delta\rho_d = \Delta\tilde{\rho} - \Delta\hat{\rho}$ is used to correct the error compensation model. Then the output of the component content error compensation model based on adaptive fuzzy neural networks. Then the component content of soft sensor measurement in detecting point will be:

$$\hat{\rho} = \bar{\rho} + \Delta\hat{\rho} \qquad (2)$$

The deduction of equilibrium calculation model for countercurrent extraction sees also in [5].

2.3 Component Content Error Compensation Model Based on Adaptive Fuzzy System

The error $\Delta\rho$ between countercurrent equilibrium calculation model output $\bar{\rho}$ and the real sampled output ρ^0 has the following relationship

$$\Delta\rho = \rho^0 - \bar{\rho} = g\{u_1, u_2, u_3, u_4\} \qquad (3)$$

where $g\{\cdot\}$ is some form nonlinear function.

Using the adaptive fuzzy system [6] implements the component content error compensation model. The training input-output pairs are constructed by $[u_1, u_2, u_3, u_4, \Delta\rho]^T$. The output of the error compensation model is $\Delta\hat{\rho}$. The whole error compensation mode can be described by following rules

$$R^i : \text{if } (u_1 \text{ is } F_1^i) \text{ and } (u_2 \text{ is } F_2^i) \text{ and } (u_3 \text{ is } F_3^i) \text{ and } (u_4 \text{ is } F_4^i)$$
$$\text{then } \Delta\hat{\rho}_i = g_i(u) = p_0^i + p_1^i \cdot u_1 + p_2^i \cdot u_2 + p_3^i \cdot u_3 + p_4^i \cdot u_4, \ i = 1, 2, \cdots, M \qquad (4)$$

where R^i denote the ith fuzzy rules F_j^i denote the ith fuzzy set of u_j; the membership function $\mu_{ij}(u_j) = exp[-\frac{(u_j - m_{ij})^2}{2\sigma_{ij}^2}]$, m_{ij} and σ_{ij} are the center and the width of the membership function and are called precondition parameters, p_j^i is called conclusion parameter, $j = 1,2,3,4$.

The output of the error compensation model can be written into a compact form

$$\Delta\hat{\rho} = (\sum_{i=1}^{M} g_i(u) \prod_{j=1}^{4} \mu_{ij}(u_j)) / (\sum_{i=1}^{M} \prod_{j=1}^{4} \mu_{ij}(u_j))$$
$$= (\sum_{i=1}^{M} g_i(u) exp[-(\sum_{j=1}^{4} \frac{(u_j - m_{ij})^2}{\sigma_{ij}^2})]) / (\sum_{i=1}^{M} exp[-(\sum_{j=1}^{4} \frac{(u_j - m_{ij})^2}{\sigma_{ij}^2})]) \quad (5)$$

In order to set up the extraction component content error compensation model, we need to decide the structure and parameters of this model, i.e. the rule number M, precondition parameters m_{ij} and σ_{ij} and the conclusion parameters p_j^i.

According to equation (3), we construct the training input-output pairs which is sampled in the product line and denote these data by $\{X_1, X_2, \cdots, X_N\}$, where

$$X_l = [u_1(l), u_2(l), u_3(l), u_4(l), \Delta\rho(l)] = [U_l, \Delta\rho(l)], \quad l = 1, 2, \cdots, N.$$

By building density function and calculating sample data density index, we use the subtraction clustering method [7] to adaptively confirm the initial model structure. After the subtraction clustering finished, we can get the clustering center $(U_c^i, \Delta\rho_c^i)$ and get the initial network structure

R^i: IF u is close to U_c^i then $\Delta\hat{\rho}$ is close to $\Delta\rho_c^i$, $i = 1, 2, \cdots, M$

2.4 Parameters Optimization of Error Compensation Model

We use the gradient descent algorithm and the least squares estimate algorithm to optimize the prediction parameters and the conclusion parameters.

At first, fix the prediction parameters m_{ij}, σ_{ij}, $i = 1, 2, \cdots, M$, $j = 1,2,3,4$ and use the least squares estimate algorithm to identify those parameters. Transform equation (5) to an equivalent form

$$\Delta\hat{\rho} = g(u) = \Psi^T(u) \cdot P \quad (6)$$

where $\varphi_k^i(u) = (u_k \prod_{j=1}^{4} \mu_{ij}(u_j)) / (\sum_{i=1}^{M} \prod_{j=1}^{4} \mu_{ij}(u_j))$, $\Psi(u) = [\varphi_k^i(u)]$, $P = [p_k^i(u)]$, $k = 0, 1, \cdots, 4$. Let

$$\Phi = \Psi^T(u) \quad (7)$$

Define error index $J(P) = \frac{1}{2}\|\Delta\hat{\rho} - \Phi \cdot P\|^2$. Then according to the least squares estimate, the parameter P which the minimized $J(P)$ is

$$P = [\Phi^T \cdot \Phi]^{-1} \cdot \Phi^T \cdot \Delta\rho \tag{8}$$

Then, fix the prediction parameters p_k^i and use the gradient descent algorithm to obtain the prediction parameters. Consider error cost index $E = \frac{1}{2}\sum_{i=1}^{N}(\Delta\rho(i) - \Delta\hat{\rho}(i))^2$, we can get the parameter regulation algorithm

$$m_{ij}(k+1) = m_{ij}(k) - \alpha_m(\Delta\hat{\rho} - \Delta\rho)(g_i(u) - \Delta\hat{\rho})\frac{(u_j - m_{ij}(k))}{\sigma_{ij}^2(k)}\phi_i(u) \tag{9}$$

$$\sigma_{ij}(k+1) = \sigma_{ij}(k) - \alpha_\sigma(\Delta\hat{\rho} - \Delta\rho)(g_i(u) - \Delta\hat{\rho})\frac{(u_j - m_{ij}(k))^2}{\sigma_{ij}^3(k)}\phi_i(u) \tag{10}$$

where $\phi_i(u) = (\prod_{j=1}^{4}\mu_{ij}(u_j)) / (\sum_{i=1}^{M}\prod_{j=1}^{4}\mu_{ij}(u_j))$, $g_i(u)$ the is consequent value of the *ith* rule, $0 < \alpha_m < 1$ and $0 < \alpha_\sigma < 1$ is the study rate.

Repeat steps above until the criteria satisfied. After the prediction parameters and conclusion parameters optimized, the component content compensation value $\Delta\hat{\rho}$ can be calculated by equation (5). Use equation (2) to get the soft sensor output $\hat{\rho}$.

3 Industry Experiment of Soft-Sensor

A company extracted high purity yttrium from ionic rare earth, in which the content of Y_2O_3 is more than 40%, adopting new extraction technique of HAB dual solution.

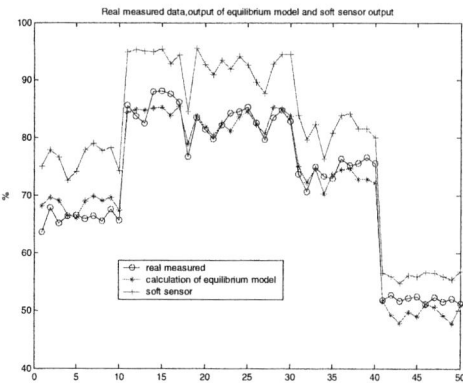

Fig. 3. Y component content curves of the experiment research

The extraction process consists of sixty stages of mixed extractors. The feed is in the 22th stage. To guarantee the product purity requirement at each exit, the sampling point is set at 15th stage according to extraction product process automation requirement. The experiment results are shown in figure 3.

From figure 3, it shows that the varied trends of the equilibrium model output and the soft-sensor output are identical with the real sampling data. At the points 13, 19 and 30 in figure 3, the errors between the output of the equilibrium calculation model and the real sampling data are 12.58, 11.99 and 11.63, but the errors between the soft-sensor output and the real sampling data are 2.365, -0.320 and 0.955. *RMSE* and *MAXE* for the equilibrium calculation model are *RMSE*=2.918 and *MAXE*=12.58. *RMSE* and *MAXE* for the soft-sensor are *RMSE*=2.315 and *MAXE*=4.509. It satisfies the process control requirement and has higher estimate precision.

4 Conclusions

The component content soft-sensor model of the rare earth extraction process proposed in this paper is a hybrid model composed of the concurrent extraction equilibrium calculation model and the error compensation model using adaptive fuzzy system. The proposed hybrid model can be used in the case that dynamic disturbance exists. When dynamic disturbance exists, the original equilibrium calculation model has larger errors. The successful application of the soft-sensor model in the extraction Y product line by HAB shows that the proposed soft-sensor method are effective to solve the component content online measurement problem.

References

1. Xu G.X.: Rare Earths, Metallurgical Industry Press, Beijing (1995) 612-727
2. Yan C.H., Jia J.T.: Automatic Control System of Countercurrent Rare Earth Extraction Process. Rare Earths, 18 (1997) 37-42
3. Chai T.Y., Yang H.: Situation and Developing Trend of Rare-earth Countercurrent Extraction Processes Control. Journal of Rare Earths, 22 (2004) 590-596
4. Yang H., Chai T.Y.: Neural Networks Based Component Content Soft-sensor in Countercurrent Rare-earth Extraction. Journal of Rare Earth, 21 (2003) 691-696
5. Yang H.: Component Soft Sensor for Rare Earth Countercurrent Extraction Process and Its Applications. Northeastern University, Doctor Dissertation, 2004.
6. Roger Jang J.S.: ANFIS: Adaptive-Network-based Fuzzy Inference System. IEEE Trans. on System, Man, and Cybernetics, 23 (1993) 665-685
7. Yager R.R., Filev D.P.: Approximate Clustering via the Mountain Method. IEEE Trans. on Systems, Man and Cybernetics, 24 (1994) 1274-1284

The Fuzzy-Logic-Based Reasoning Mechanism for Product Development Process

Ying-Kui Gu[1], Hong-Zhong Huang[1,2], Wei-Dong Wu [3], and Chun-Sheng Liu [3]

[1] Key Lab. for Precision and Non-traditional Machining Technol. of Ministry of Education, Dalian University of Technology, Dalian, Liaoning, 116023, China
guyingkui@163.com
[2] School of Mechatronics Engn, University of Electronic Science and Technology of China, Chengdu, Sichuan, 610054, China
hzhuang@uestc.edu.cn
[3] Department of Mechanical Engineering, Heilongjiang Institute of Science and Technology, Harbin, Heilongjiang, 150027, China

Abstract. Product development process can be viewed as a set of sub-processes with stronger interrelated dependency relationships. In this paper, the quantitative and qualitative dependency measures of serial and parallel product development processes are analyzed firstly. Based on the analysis results, the process net is developed where the processes are viewed as nodes and the logic constraints are viewed as verges of the net. The fuzzy-logic-based reasoning mechanism is developed to reason the dependency relations between development processes in the case that there is no sufficient quantitative information or the information is fuzzy and imprecise. The results show that the proposed method can improve the reasoning efficiency, reduce the cost and complexity degree of process improvement, and make a fast response to the dynamic development environment.

1 Introduction

Process is the basic unit of activity that is carried out during a product's life cycle (Yu, 2002). In this sense, the whole development process can be viewed as a set of sub-processes that their physical meanings are varied continuously along with time. The process in the net is not isolated. There exist complicated relationships among processes, where the logic relationship is one of the important relationships. The purposes to analyze and program the logic relationship are as follows:

1. It is to improve the concurrency degree of processes by arranging the serial and parallel modes of development process reasonably.
2. It is to decrease the cost and complexity of process improvement.
3. It is to program the whole development process effectively and reduce float processes to a great degree by identifying the important and unimportant processes.
4. It is to increase the amount of information provided to the designers for making decisions.

Product development is a dynamic process. In order to make the development process optimal, it needs essential process improvement or process re-organizing activities. Therefore, it is very necessary to analyze the relationships among processes in the case that design information is incomplete and imprecise.

Allen (1984) first defined the temporal logic relations of tasks. According to him, there is a set of primitive and mutually exclusive relations that could be applied over time intervals. The temporal logic of Allen is defined in a context where it is essential to have properties such as the definition of a minimal set of basic relations, the mutual exclusion among these relations and the possibility to make inferences over them. For this reason, Raposo, Magalhaes, Ricarte and Fuks (2001) made some adaptations to Allen's basic relations, adding a couple of new relations and creating some variations of those originally proposed. Cruz, Alberto and Leo (2002) used seven basic relations of Allen to develop a logic relation graph of tasks and proposed two properties of relation graph. Li, Liu and Guo (2002) defined the concept of process templet and classified the logic relations of processes into five categories, i.e., before, meet, start, equal and finish. Gu, Huang and Wu (2003) analyzed the preconditions for executing the logic relations strictly.

However, it is very difficult to program and execute the logic relations between development processes because of the complexity, fuzziness and dynamic uncertainty of product development process. Graph theory and fuzzy logic provide stronger tools for process modeling and analyzing. Alocilja (1990) presented process network theory (PNT). According to him, the generic properties of process networks can provide a practical analytical framework for the systematic analysis, design and management of physical production system, including material flows, technical costs, etc. Kusiak (1995) used graph theory as a tool to develop a dependency network for design variables and analyze the dependencies between design variables and goals. Cruz, Alberto and Leo (2002) presented a methodology to express both analytically and graphically the interdependencies among tasks realized in a collaborative environment. Fuzzy logic coupled with rule-based system is enabling the modeling of the approximate and imprecise reasoning processes common in human problem solving. Zakarian (2001) presented an analysis approach for process models based on fuzzy logic and approximate rule-based reasoning. He used possibility distributions to represent uncertain and incomplete information of process variables, and developed an approximate rule based reasoning approach for quantitative analysis of process models. Kusiak (1995) developed fuzzy-logic-based approach to model imprecise dependencies between variables in the case when no sufficient quantitative information is available. Sun, Kalenchuk, Xue and Gu (2000) presented a approach for design candidate identification by using neural network-based fuzzy reasoning.

In this paper, we present an approach to analyze the dependency relations between processes based on graph theory and fuzzy logic. First, a process net is developed by using graph theory. Second, we develop a fuzzy-logic-based reasoning mechanism to analyze the dependency relations between processes under the fuzzy and imprecise design environment, which can be used to increase the amount of information provided to the designers for making decisions.

2 Developing Network of Development Processes

The set of logic relations adopted in this work is based on the temporal logic proposed by Allen (1984). That is before, meets, stars, finishs, equals, overlaps and during. The relationships above could be expressed by $\langle b \rangle$, $\langle m \rangle$, $\langle s \rangle$, $\langle e \rangle$, $\langle f \rangle$, $\langle o \rangle$, $\langle d \rangle$ respectively.

The whole development process consists of many interrelated sub-processes. These sub-processes form a net, and the net is called as process net. The process net can be expressed by graphical representation. There exist matching relationships between elements and nodes of the graph, and the process and edge exist corresponding relationships. $P_{net} = \langle P, R \rangle$. $P = \langle P_1, P_2, \cdots, P_n \rangle$ is the node set, and represent the non-empty set of processes and their information. $R = \langle R_1, R_2, \cdots, R_m \rangle$ is the edge, and represent the constraint relationships among process nodes.

The steps of developing a process net are as follows:

(1) Developing graph of binary logic relations for all processes.
(2) Developing relationship matrix of processes based on the logic relationships among processes. In the matrix, "0" represents that there hasn't direct logic relationships between the two processes, and "1" represents that there has direct logic relationships between the two processes.
(3) On the basis of the relationship matrix, the process net can be developed through connecting all the edges in turn. Where the processes variables are viewed as nodes, and the logic constraints among variables are viewed as verges.

3 Reasoning Mechanism for Process Dependency

There exist collaborative or conflict relationships between development processes. Because of the interrelations of process net, there have direct or indirect effect relationships between processes, and the strength degree of these relationships has very important influence on the performance and improvement of processes. Therefore, to develop the reasoning mechanism of process dependency not only is avail to arrange the process modes reasonably and improve the concurrency degree of processes, but also is avail to decrease the complexity and the cost of process improvement. The physical meanings of symbols are list in Table 1.

Table 1. The physical meanings of symbols

ψ_{ij}	Physical meanings
+	The change of process p_i has good influence on the improvement of process p_j
-	The change of process p_i has bad influence on the improvement of process p_j
0	The change of process p_i has not influence on the improvement of process p_j
?	Otherwise

3.1 The Reasoning Rules for Serial Process

As shown in Fig. 1(a), the quantitative and qualitative dependency measures for serial processes can be denoted as follows:

$$\psi_{i,j \to k} = \psi_{ij} \otimes \psi_{jk}, \quad \delta_{i,j \to k} = \delta_{ij} \times \delta_{jk}. \quad (1)$$

The reasoning rules are developed and listed as follows:

(1) If ψ_{ij} ="+" and ψ_{jk} ="+", then $\psi_{ij \to k}$ ="+";
(2) If ψ_{ij} ="+" and ψ_{jk} ="−", then $\psi_{ij \to k}$ ="−";
(3) If ψ_{ij} ="−" and ψ_{jk} ="−", then $\psi_{ij \to k}$ ="+";
(4) If ψ_{ij} ="−" and ψ_{jk} ="+", then $\psi_{ij \to k}$ ="−";
(5) If ψ_{ij} ="+"("−") and ψ_{jk} ="0", then $\psi_{ij \to k}$ ="0";
(6) If ψ_{ij} ="0" and ψ_{jk} ="+"("−"), then $\psi_{ij \to k}$ ="+"("−").

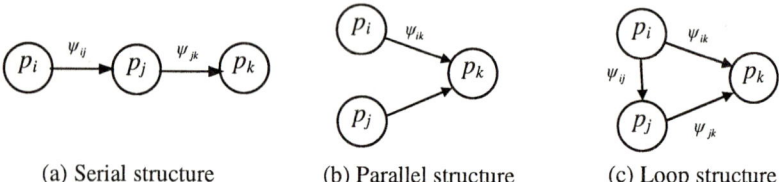

(a) Serial structure (b) Parallel structure (c) Loop structure

Fig. 1. Structure of process

3.2 The Reasoning Rules for Parallel Process

As shown in Fig. 1(b), the quantitative and qualitative dependency measures for parallel processes can be denoted as follows:

$$\psi_{i,j \to k} = \psi_{ik} \oplus \psi_{jk}, \quad \delta_{i,j \to k} = \delta_{ik} + \delta_{jk}. \quad (2)$$

The reasoning rules are developed and listed as follows:

(1) If ψ_{ik} ="+" and ψ_{jk} ="+", then $\psi_{ij \to k}$ ="+";
(2) If ψ_{ik} ="−" and ψ_{jk} ="−", then $\psi_{ij \to k}$ ="−";
(3) If $\psi_{ik}(\psi_{jk})$ ="0" and $\psi_{jk}(\psi_{ik})$ ="+"("−"), then $\psi_{ij \to k}$ ="+"("−");
(4) If $\psi_{ik}(\psi_{jk})$ ="+"("−") and $\psi_{jk}(\psi_{ik})$ ="−"("+"), then $\psi_{ij \to k}$ = ?.

There exist loop structure as shown in Fig. 1(c). If ψ_{ik} ="0" (i.e., there doesn't exist direct dependency relations between processes p_i and p_k), the loop structure transforms into serial structure. If ψ_{ij} ="0" (i.e., there doesn't exist direct dependency relations between processes p_i and p_j), the loop structure transforms into parallel structure. If there exist direct dependency relations among the three processes, the quantitative and qualitative dependency measures for loop structure can be denoted as follows:

$$\psi_{i,j \to k} = (\psi_{ij} \otimes \psi_{jk}) \oplus \psi_{ik}, \quad \delta_{i,j \to k} = (\delta_{ij} \times \delta_{jk}) + \delta_{ik}. \tag{3}$$

3.3 Fuzzy-Logic-Based Reasoning for Process Dependency Relations

3.3.1 Fuzzy Logic (Zakarian, 2001; Dutt, 1993; Zadeh, 1983)

A fuzzy logic consists of IF-THEN fuzzy rules, where IF portion of the fuzzy rule includes the premise part and THEN portion, the consequence part. The premise and consequence of fuzzy rules contain linguistic variables. An inference process of fuzzy logic takes the fuzzy sets representing the rules and the facts and produces a resultant fuzzy set, over the domain of discourse of the consequent.

Fuzzy linguistic are used to represent dependencies between processes in an uncertain and imprecise design environment and can be described as

$$L = (U, T, E, N)$$

where U is a universe of discourse. T is the set of names of linguistic terms. E is a syntactic rule for generating the terms in the term set T. N is a fuzzy relation from E to U, and its membership function can be denoted by:

$$N: \text{Supp}(T) \times U \to [0,1]. \tag{4}$$

It is a binary function. That is to say, to $x \in \text{Supp}(T)$ and $y \in U$, the degree of membership $N(x, y) \in [0,1]$

The operations of fuzzy set include fuzzy intersection, fuzzy union and fuzzy complement.

(1) Fuzzy intersection. The intersection of fuzzy sets A and B is a function of the form:

$$\mu_{A \cap B}(x): [0,1] \times [0,1] \to [0,1]. \tag{5}$$

and can be obtained from:

$$\mu_{A \cap B}(x) = \min\{\mu_A(x), \mu_B(x)\} = \cap\{\mu_A(x), \mu_B(x)\}. \tag{6}$$

by taking the minimum of the degrees of membership of the elements in fuzzy sets A and B.

(2) Fuzzy union. The union of fuzzy sets A and B is a function of the form:

$$\mu_{A \cup B}(x): [0,1] \times [0,1] \to [0,1]. \tag{7}$$

and can be obtained from:

$$\mu_{A \cup B}(x) = \max\{\mu_A(x), \mu_B(x)\} = \cup\{\mu_A(x), \mu_B(x)\}. \tag{8}$$

by taking the maximum of the degrees of membership of the elements in fuzzy sets A and B.

(3) Fuzzy complement. The union of fuzzy sets A is a function of the form

$$\mu_{-A}(x):[0,1]\to[0,1]. \qquad (9)$$

and can be obtained from:

$$\mu_{-A}(x)=1-\mu_A(x). \qquad (10)$$

3.3.2 Fuzzy Reasoning Approach for Process Dependency

The dependency between processes can be described as a linguistic variable characterized by a quintuple $(V,T(V),U,G,M)$. Where V is the linguistic variable "dependency"; $T(V)$ is the set of names of linguistic terms of V; U is the universe of discourse; G is the syntactic rule for generating terms in the term set $T(V)$; M is the semantic rule that assigns a meaning, i.e., a fuzzy set, to the terms.

Let V is the rate of change of process p_k that caused by the change of processes p_i and p_j.

$T(V)$ = {PL, PM, PS, NL, NM, NS}
    = {Positive Large, Positive Same, Positive Small, Negative Large, Negative Same, Negative Small}

The membership functions of the linguistic terms can be represented as shown in Fig. 2

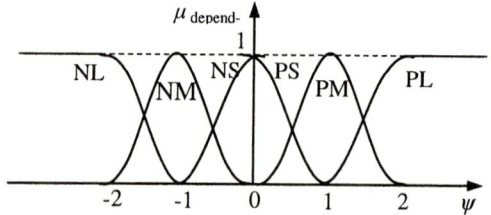

Fig. 2. The membership functions of the linguistic terms

The membership functions of the linguistic terms can be defined as follows:

$$\mu_{PS}(\psi)=\frac{1}{1+100\psi^4}, \quad \psi\geq 0$$

$$\mu_{PM}(\psi)=\frac{1}{1+100(\psi-1)^4}, \quad \psi\geq 0$$

$$\mu_{PL}(\psi)=\begin{cases}\dfrac{1}{1-100(\psi-2)^4}, & 0\leq\psi\leq 2 \\ 1, & \psi\geq 2\end{cases}$$

$$\mu_{NS}(\psi)=\frac{1}{1+100\psi^4}, \quad \psi<0$$

$$\mu_{NM}(\psi) = \frac{1}{1+100(\psi+1)^4}, \quad \psi < 0$$

$$\mu_{NL}(\psi) = \begin{cases} \frac{1}{1-100(\psi+2)^4}, & -2 \le \psi \le 0 \\ 1, & \psi \le -2 \end{cases}$$

The fuzzy rule is represented as follows:

IF $\psi_{ik} = V_m$ and $\psi_{jk} = V_n$, THEN $\psi_{ik} \oplus \psi_{jk} = V_k$

where V_m, V_n and V_k are fuzzy linguistic terms.

Thirty-six fuzzy rules can be developed to represent the dependencies among process p_i, p_j and p_k, as shown in Table 2.

Table 2. Fuzzy rules

Then $\psi_{ij \to k}$		IF ψ_{ik}					
		PS	PM	PL	NS	NM	NL
IF ψ_{jk}	PS	PS	PM	PL	PS or NS	NS	NL
	PM	PM	PL	PL	PM	NS or PS	NL or NM
	PL	PL	PL	PL	PL	PL	?
	NS	PS or NS	PM	PL	NS	NM	NL
	NM	NM	PS or NS	PL	NM	NL	NL
	NL	NL	NL or NM	?	NL	NL	Very NL

In fuzzy reasoning, logical "and" and "or" operations produce the minimum and maximum membership function values. The output value can be obtained by calculating the center of gravity of the output membership function. As shown in Fig. 8, for each rule, we can obtain the membeship function measures for the two input variables ψ_{i0} and ψ_{j0} first. The smaller value can be selected as the measure to evaluate the matching of the rule. The result membership function of fuzzy reasoning considering only one rule is the minimum of the membership function at the THEN part of the rule and rule matching measure. The result membership function of fuzzy reasoning $\mu_{i,j \to k}(\psi)$ considering all the relevent rules can be achived by obtaining the maximum value of these result membership functions for these rules.

The output value of variable is the gravity center of the output membership function $\mu_{i,j \to k}(\psi)$, calculated by (Zhong, 2000)

$$\psi_w = \frac{\int_{\psi_{min}}^{\psi_{max}} \mu_{i,j \to k}(\psi) \psi d\psi}{\int_{\psi_{min}}^{\psi_{max}} \mu_{i,j \to k}(\psi) d\psi}. \tag{11}$$

Assume the following two fuzzy rules and the values of input variables

Rule 1: IF ψ_{ik} ="PS", and ψ_{jk} ="NS", then $\psi_{i,j\to k}$ ="PS" or "NS"

Rule 2: IF ψ_{ik} ="PS", and ψ_{jk} ="NM", then $\psi_{i,j\to k}$ ="NM"

Input: $\psi_{ik} = \psi_{i0}$, $\psi_{jk} = \psi_{j0}$

The fuzzy reasoning process can be illustrated by Fig. 3.

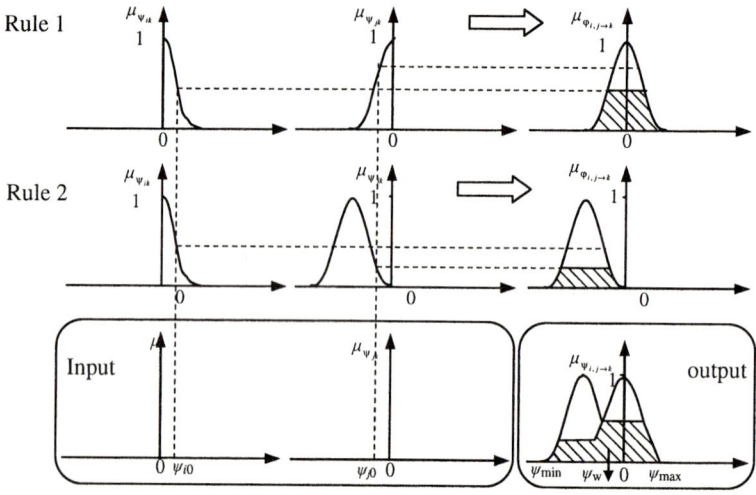

Fig. 3. Fuzzy-logic-based fuzzy reasoning process

4 Case Study

Take the design process of worm drive as an example to analyze the dependency relationships using the proposed neural-network-driven fuzzy reasoning mechanism. Requirement analysis (p_1), cost analysis (p_2) and scenario design (p_3) are the three sub-processes of the whole development process. The process structure of the three processes is parallel. First, we pick up the key process variables of each process respectively,

v_1 = transmission efficiency,

v_2 = cost,

v_3 = satisfaction degree of scenario.

The change of key process variables is the dominant factors to result in the change of the process and its relative processes. The key of improving the process is to improve its key variables. Take the change rate of key process variables as fuzzy variable (i.e., ψ_1, ψ_2 and ψ_3). The membership function of variable can be developed and shown in Fig. 4. Thirty-six fuzzy rules can be developed as shown in Table 3.

Now, the process variable v_1 (transmission efficiency) will increase 20% because of the change of working conditions of worm, which will make v_2 (cost) increase

30% at the same time. The fuzzy reasoning mechanism illustrated in Fig. 2 can be used to analyze the change of v_3 (satisfaction degree of scenario) which results from the change of transmission efficiency and cost. If we input $\psi_1 = 0.20$ and $\psi_6 = 0.30$, we can calculate the gravity center of the output membership function $\mu_3(\psi_3)$, $\psi_3^w = -0.125$.

Table 3. Fuzzy rules

Then ψ_3		IF ψ_1					
		PS	PM	PL	NS	NM	NL
IF ψ_2	PS	PS	NS	NM	PS	PM	PL
	PM	PM	PS	NM	PM	PL	PL
	PL	PL	PS	NM	PL	PL	very PL
	NS	NS	NM	NL	PS	PM	PL
	NM	NM	NL	NL	NS	PS	PM
	NL	NL	NL	very NL	NL	NL	NM or NL

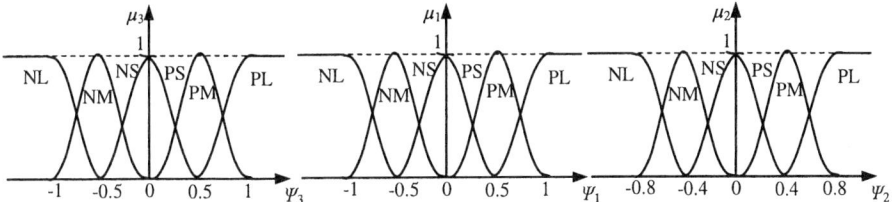

Fig. 4. Membership function of ψ_1, ψ_2 and ψ_3

Given the membership function of $\mu_3(\psi_3)$

$$\mu_{NS}(\psi_3) = \frac{1}{1+100\psi^4}, \quad \psi < 0$$

$$\mu_{NM}(\psi_3) = \frac{1}{1+100(\psi+0.5)^4}, \quad \psi < 0$$

We can get

$$\mu_{NS}(\psi_3) = 0.9762, \quad \mu_{NM}(\psi_3) = 0.3360$$

From these we can see, the satisfaction degree of scenario decrease a little when the transmission efficiency increases 20%. That is to say, the process p_3 changes along the negative direction in some sort when p_1 and p_2 change according to the design requirements.

5 Conclusions

The characteristics of product development process, such as evolvement, dependency and irreversibility decide that there exist stronger logic relations between processes. In order to arrive at the product development goal which is to develop the product fast and optimal, the graph theory was used as a tool to develop a process net, and the fuzzy logic was used to develop the reasoning mechanism to analyze the dependency relations in the case that the design information is fuzzy and incomplete. The proposed method can provide theory foundation for process programming and improving, and also provide a method for realizing automatization of product development at the same time.

Acknowledgements

This research was partially supported by the National Natural Science Foundation of China under contract number 50175010, and the National Excellent Doctoral Dissertation Special Foundation under contract number 200232 (China).

References

1. Allen, J. F.: Towards A General Theory of Action and Time. Artificial Intelligence. 23 (1984) 123-154
2. Alocilja, E. C.: Process Network Theory. Proceedings of IEEE International Conference on Systems, Man and Cybernetics. (1990) 66-71
3. Cruz, A. J. A., Raposo, A. B., Magalhaes, L. P:. Coordination in Collaborative Environments-A Global Approach. Proceedings of the 7th International Conference on Computer Supported Cooperative Work in Design. (2002) 25-30
4. Dutt, S.: Fuzzy Logic Applications: Technological and Strategic Issues. IEEE Transactions on Engineering Management. 40 (1993) 237-254
5. Gu, Y.K, Huang, H.Z., Wu, W.D.: Product Development Microcosmic Process Model Based on Constraint Nets. Journal of Tsinghua University. 43 (2003) 1448-1451
6. Kusiak, A.: Dependency Analysis in Constraint Negotiation. IEEE Transaction on System, Man, and Cybernetics. 25 (1995) 1301-1313
7. Li, S. P., Liu, N. R., Guo, M.:. Data Standard of Product and PDM. Press of Tsinghua University, Beijing (2002)
8. Raposo, A. B., Magalhaes, L. P., Ricarte, I. L. M., Fuks, H.: Coordination of Collaborative Activities: A Framework for the Definition of Tasks Interdependencies. Proceedings of Seventh International Workshop on Groupware. (2001) 170-179
9. Sun, J., Kalenchuk, D. K., Xue, D., Gu, P.: Design Candidate Identification Using Neural Network-Based Fuzzy Reasoning. Robotics and Computer Integrated Manufacturing. 16 (2000) 383-396
10. Yu, J.: China Mechanical Design Canon: Vol. 1, Modern Method of Mechanical Design. Nanchang Press of Science and Technology, Nanchang (2002)
11. Zadeh, L. A.: The Role of Fuzzy Logic in The Management of Uncertainty in Expert Systems. Fuzzy sets and System. 11 (1983) 199-227
12. Zakarian, A.: Analysis of Process Models: A Fuzzy Logic Approach. The International Journal of Advanced Manufacturing Technology. 17 (2001) 444-452
13. Zhong, S. S.: Fuzzy Theory and Technology in Engineering Concept Design. Harbin Institute of Technology, Harbin (2000)

Single Machine Scheduling Problem with Fuzzy Precedence Delays and Fuzzy Processing Times

Yuan Xie[1], Jianying Xie[1], and Jun Liu[2]

[1] Department of Automation, Shanghai Jiao Tong University, Shanghai, 200030, China
hei_xieyuan@sjtu.edu.cn
[2] First Research Institute of Corps of Engineers,
General Armaments Department, PLA, Wuxi, 214035, China

Abstract. A single machine scheduling problem with fuzzy precedence delays and fuzzy processing times is considered. A kind of precedence delays scheduling problem with special structure is investigated because general single machine scheduling problem with precedence delays is NP-hard. And the objective is to minimize the maximum complete time of all jobs.

1 Introduction

There are many scheduling problems [1]-[3] of interest in which there are precedence delays between jobs. Precedence delay means a job can start its execution only after any of its predecessors has been completed and the delay between the two jobs has elapsed. Most of single machine scheduling problems with precedence delays are computationally intractable in the sense that is strongly NP-hard [1], [3], even if in the case of unit execution times and integer lengths of delays [2]. In real world, input data may be uncertain or imprecise. Recently, Muthusamy et al. [4] considered fuzzy version of Wikum's problem [3] with upper bounds of precedence delays are infinite and proposed an $O(n^8)$ algorithm. In our research, we take both precedence delays and processing times as fuzzy numbers because processing times may be uncertain or imprecise in practical environment. The objective is to minimize the makespan.

2 Fuzzy Numbers and Ranking Fuzzy Numbers

A fuzzy number [5] is defined by convex normalized fuzzy set $\tilde{A}$ of the real line with a membership function $\mu_{\tilde{A}}(x): R \to [0,1]$. *L-R* type fuzzy number $\tilde{A}$ [5], has following form (Please refer [5] for more details about *L-R* type fuzzy number):

$$\tilde{A} = (\underline{m}, \overline{m}, \alpha, \beta)_{LR} \tag{1}$$

The addition and subtraction operations of *L-R* type fuzzy numbers can be calculated as:

$$(\underline{m}, \overline{m}, \alpha, \beta)_{LR} \mp (\underline{n}, \overline{n}, \gamma, \delta)_{LR} = (\underline{m} + \underline{n}, \overline{m} + \overline{n}, \alpha + \gamma, \beta + \delta)_{L'R'} \tag{2}$$

$$(\underline{m},\overline{m},\alpha,\beta)_{LR} \dotplus (\underline{n},\overline{n},\gamma,\delta)_{RL} = (\underline{m}-\overline{n},\overline{m}-\underline{n},\alpha+\delta,\beta+\gamma)_{L'R'} \quad (3)$$

λ-cut ($\lambda \in (0,1]$) of a fuzzy number $\tilde{A}$ of the L-R type is mathematically stated:

$$\tilde{A}^{\lambda} = [\underline{m}(\lambda),\overline{m}(\lambda)] = [\underline{m} - L^{-1}(\lambda)\alpha, \overline{m} + R^{-1}(\lambda)\beta] \quad (4)$$

L^{-1} and R^{-1} denotes the reverse function of L and R. The expected value[6] of fuzzy number $\tilde{A}$ is a real number $E(\tilde{A})$ determined by the following formula:

$$E(\tilde{A}) = \frac{1}{2}\int_0^1 (\underline{m}(\lambda) + \overline{m}(\lambda))d\lambda \quad (5)$$

Let $\tilde{A}$ and $\tilde{B}$ be any fuzzy numbers. The following conditions holds:

$$E(\tilde{A} \dotplus \tilde{B}) = E(\tilde{A}) + E(\tilde{B}) \quad (6)$$

$$E(\tilde{A} \dotplus \tilde{B}) \geq E(\tilde{A}) \quad (7)$$

3 Problem Formulation

Suppose that there are $n+1$ independent jobs $J_1, J_2,\ldots, J_n, J_{n+1}$ to be processed on a single machine nonpreemptively. The processing time of job J_i is $\tilde{p}_i$ ($\tilde{p}_i$ are L-R fuzzy numbers). Jobs form the set $\mathbf{J}=\{J_1, J_2,\ldots, J_n\}$ can be processed in arbitrary order with respect to ordinary precedence. But job J_{n+1} cannot be started until all jobs from $\mathbf{J}$ are completed. There are time delay between the completion time of J_i from $\mathbf{J}$ and the starting time of J_{n+1}. This delay is a prescribed fuzzy quantity $\tilde{d}_i$, which can be formulated as L-R fuzzy number $(l_i, u_i, \alpha, \beta)_{LR}$ defined on R^+. In fuzzy precedence delay, l_i and u_i can be considered as lower and upper bounds of delays in crisp situation showed by Wilum et al. [3]. The crisp single scheduling problem with precedence delays is NP-hard when the condition of either $l_i=0$ or $u_i=\infty$ doesn't hold. Sometime, precedence delays are not such accurate and scheduling decision-makers just know they will be in certain interval before scheduling.

The problem considered here can be formulated as follow. A feasible schedule is represented by a pair $(\pi, \tilde{C}_{n+1})$ of a permutation π of $\{1,2,\ldots,n\}$ and the completion time $\tilde{C}_{n+1}$ of job J_{n+1}. Each scheduling $(\pi, \tilde{C}_{n+1})$ determines the position of job $J_1, J_2,\ldots, J_n, J_{n+1}$. for jobs from $\mathbf{J}=\{J_1, J_2,\ldots, J_n\}$, Job's completion time is the sum of pre-executed jobs' processing times. Then in schedule $(\pi, \tilde{C}_{n+1})$, $\tilde{C}_{\pi(j)}$, which denotes the completion time of j-th ($j=1, \ldots,n$) job, is defined by:

$$\tilde{C}_{\pi(j)} = \sum_{k=1}^{j} \tilde{p}_{\pi(k)} \quad (8)$$

$\tilde{C}_{\pi(j)}$ also becomes L-R type fuzzy number. For the last job J_{n+1}, its completion time, i.e. the makespan of a schedule, is determined by completion time and fuzzy precedence delay of every job J_i in $\mathbf{J}$. That is, it equals processing time of J_{n+1} plus maximum sum of completion time and precedence delay of J_i:

$$\tilde{C}_{n+1} = \max\{\tilde{C}_i \dotplus \tilde{d}_i \dotplus \tilde{p}_{n+1} \mid i=1,\ldots,n\} = \max\{\tilde{C}_i \dotplus \tilde{d}_i\} \dotplus \tilde{p}_{n+1} \quad (9)$$

A general problem of minimizing a maximum completion cost, denoted by **P**, consists in determining a feasible schedule S of jobs with respect to the precedence delays such that: **P**: $\tilde{C}_{n+1}(S) \to \min$, We call the schedule construct for problem $1 \mid prec(\tilde{d}_i) \mid \tilde{C}_{\max}$. Next, we will give the procedure of solving the problem above.

4 Solution Procedure

With the presence of ordinary precedence constraints (no delay) for jobs $\mathbf{J}=\{J_1, J_2,\ldots, J_n\}$, the following algorithm (Algorithm 1) is proposed to minimize the makespan of our single machine scheduling problem with precedence delays. To simplify the notation, we denoted $\tilde{C}_i \tilde{+} \tilde{d}_i$ by $\widetilde{Cd}_i$.

1. $\mathbf{J} \leftarrow \{J_1,\cdots,J_n\}$ /* All jobs preceding job J_{n+1} */
2. $S \leftarrow \varnothing$ /* An optimal schedule */
3. while $\mathbf{J} \neq \varnothing$, do
4. $\mathbf{J}' \leftarrow \mathbf{J} \setminus \{J_i \in \mathbf{J} \mid \exists (J_i, J_k \in prec\}$ /* Jobs processed at the end of schedule with respect to ordinary precedence*/
5. Find $J_i \in \mathbf{J}'$ such that $E(\tilde{d}_i)$ is minimal
6. $S \leftarrow J_i \cup S$
7. $\mathbf{J} \leftarrow \mathbf{J} \setminus J_i$
8. end while
9. $S \leftarrow S \cup J_{n+1}$
10. $\tilde{C}_{n+1} = \max\{\widetilde{Cd}_i \mid i=1,\ldots,n\} \tilde{+} \tilde{p}_{n+1}$

Because cost function, the expected value of fuzzy number, satisfies inequation (7), step 3 to 8 is a variant of the algorithm designed by Lawler [7]. After that, step 9 to 10 minimizes completion of J_{n+1}. Late on we will give the proof of optimization of Algorithm 1.

Theorem 1. Algorithm 1 constructs an optimal sequence for problem **P**.

Proof. Because job J_{n+1} must be processed last, we only need to prove that Algorithm 1 constructs an optimal sequence for first n jobs. Enumerate the jobs in such a way that $1,2,\cdots,n$ is the sequence constructed by the Algorithm 1. Let $\sigma : \sigma(1),\ldots,\sigma(n)$ be an optimal sequence with $\sigma(i) = i$ for $i = n, n-1, \ldots, r$ and $\sigma(r-1) \neq r-1$ where r is minimal. Suppose that $\sigma(k) = r-1$, where $1 \leq k < r-1$. This means there is no an optimal sequence with less value of r.

$\sigma : [\sigma(1),\ldots,\sigma(k-1), \sigma(k) = r-1, \sigma(k+1),\ldots,\sigma(r-1), \sigma(r)\ldots,n]$
$\mu : [\sigma(1),\ldots,\sigma(k-1), \sigma(k+1),\ldots \sigma(r-1), \sigma(k), \sigma(r),\ldots,n]$

It is possible to schedule $J_{\sigma(k)}$ (i.e. J_{r-1}) immediately before $J_{\sigma(r)}$ because $J_{\sigma(k)}(J_{r-1})$ and $J_{\sigma(r)}(J_r)$ have no successor in the set $\{J_{\sigma(1)},\ldots,J_{\sigma(r-1)}\}$ and $1,2,\cdots,n$ is constructed by the Algorithm 1. We can create the feasible sequence μ

by moving $J_{\sigma(k)}$ immediately before $J_{\sigma(r)}$. We obtain the sequence μ in which $\mu(i) = \sigma(i)$ ($i=1,\ldots,k-1$), $\mu(i) = \sigma(i+1)$ ($i=k,\ldots,r-2$), $\mu(r-1) = \sigma(k)$, $\mu(i) = \sigma(i)$ ($i = r,\ldots,n$). It is obvious that for $i=1,\ldots,k-1$ and $i = r,\ldots,n$

$$\widetilde{Cd}_{\mu(i)} = \widetilde{Cd}_{\sigma(i)} \tag{10}$$

Because $\mu(r-1) = \sigma(k) = r-1$ and $1,2,\cdots,n$ is constructed by the algorithm 1. So

$$\widetilde{Cd}_{\mu(r-1)} \leq \widetilde{Cd}_{\sigma(r-1)} \tag{11}$$

For each $i = k,\ldots,r-2$, according to inequation (8) we obtain

$$\widetilde{Cd}_{\mu(i)} = \tilde{C}_{\mu(i)} \mp \tilde{d}_{\mu(i)} = \tilde{C}_{\sigma(i+1)} \mp \tilde{p}_{\sigma(k)} \mp \tilde{d}_{\sigma(i+1)} \leq \tilde{C}_{\sigma(i+1)} \mp \tilde{d}_{\sigma(i+1)} \leq \widetilde{Cd}_{\sigma(i+1)} \tag{12}$$

From (10)-(12), we can conclude that the sequence μ is not worse than σ, thus it is also optimal. This contradicts the minimality of r.

Algorithm 1 proposes a useful scheme of the Lawler's method modified for the problem **P**, and its complexity is as simple as Lawler's. In Algorithm 1, complexity from step 1 to step 8 is $O(n^2)$, complexity of step 9 is $O(n)$. Therefore, the total complexity of Algorithm 1 is $O(n^2)$.

5 Conclusion

This paper has investigated a single machine scheduling problem with fuzzy precedence delays and fuzzy processing times. The upper and lower bounds showed by Wikum et al. are regarded as parameters of *L-R* type fuzzy numbers. Then original NP-hard problem can be solved by modified Lawler's algorithm in $O(n^2)$.

References

1. Brucker, P., Hilbig, T., Hurink, J.: A Branch and Bound Algorithm for a Single-machine Scheduling Problem with Positive and Negative Time-lags. Discrete Applied Mathematics, 94 (1999) 77–99
2. Finta, L., Liu, Z.: Single Machine Scheduling Subject to Precedence Delays. Discrete Aplied Mathematics, 70 (1996) 247-266
3. Wilum, E. D., Llewellyn, D. C., Nemhauser, G. L.: One-machine Generalized Precedence Constrained Scheduling Problems. Operations Research Letters, 16 (1994) 87-99
4. Muthusamy, K., Sung, S. C., Vlach, M., Ishii, H.: Scheduling with Fuzzy Delays and Fuzzy Precedences. Fuzzy Sets and Systems, 134 (2003) 387-395
5. Dubois, D., Prade, H.: Fuzzy Sets and Systems. Academic Press, New York (1980)
6. Dubois, D., Prade, H.: The Mean Value of a Fuzzy Number. Fuzzy Sets and Systems, 24 (1987) 279-300
7. Lawler, E. L.: Optimal Sequencing of a Single Machine Subject to Precedence Constraints. Management Science, 19 (1973) 544-546

Fuzzy-Based Dynamic Bandwidth Allocation System

Fang-Yie Leu, Shi-Jie Yan, and Wen-Kui Chang

[1] Department of Computer Science and Information Engineering,
Tunghai University, Taiwan
leufy@mail.thu.edu.tw
[2] Department of Computer Science and Information Engineering,
Tunghai University, Taiwan
youngman@dorm.thu.edu.tw
[3] Department of Computer Science and Information Engineering,
Tunghai University, Taiwan
wkc@mail.thu.edu.tw

Abstract. This article proposes a network Quality of Service (QoS) system named Fuzzy-based Dynamic Bandwidth Allocation (FDBA) system which partitions network traffic into several channels and then recommends a suitable queuing strategy to network administrator so that user traffic and heterogeneous data packets of different classes can be properly multiplexed. Class Based Weight Fair Queuing (CBWFQ) is deployed as the congestion resolution mechanism. Markov Modulate Poisson Process (MMPP) is encompassed to model network traffic. Expectation Maximization is used to estimate MMPP parameters. Fuzzy is also invoked to calculate the bandwidth reserved for a channel. Reserving bandwidth guarantees the least network traffic, while declaring threshold of queue limit defines the priority of a class. A packet with higher priority will be delivered with higher probability. Moreover, FDBA can automatically tune its QoS parameters to adapt various network traffic, thus decreasing managerial load and providing user a higher network service quality.

1 Introduction

As computers are widely used in various domains, network becomes a key issue in today's scientific, engineering and business environments. People have built a huge Internet and relayed it for many ways, such as communication, e-commerce and entertainment. To meet users' abundant requirements, network facilities have evolved to much more complex than usual, but its basic technology, e.g., TCP/IP, does not change.

Also, multi-media and voice applications have been extensively developed, distributed and used. People can conveniently access them worldwide through Internet. However, many Internet providers do not restrict users' information contents and amount of data transferred. A mass of useless and dirty data, such as Spam, worms and illegal p2p sharing file, are transmitted within Internet which has limited bandwidth. That is why network quality of service (QoS) has significantly attracted researcher's eyes in recent years. As users access network wantonly, network congestion causes packet loss and damages network reliability. Hence, network managers have two major challenges. One is how to defend against injurious traffic in order to keep network operate safely. However, this is a network security problem and is beyond the scope of this article. The other is how to maximize resource utilization and dynamically share resources in order to guarantee promised QoS for users.

Standardized by the Internet Engineering Task Force (IETF) in mid-1997, Differentiated Services (DiffServ), combining edge policing, provisioning, and traffic prioritization to achieve service differentiation [1], deploys per hop behavior (PHB) to guarantee resources for end users and to manage congestion, especially when network congests frequently. DiffServ does not request all network nodes of concerned system to implement PHB, therefore it is an adequate QoS policy for large scale networks. DiffServ involves several queuing strategies, such as First In First Out (FIFO) and Weighted Fair Queuing (WFQ), each has its own way to manage packets. However, most routers only support routing platform. When to use which queuing and how to setup and tune parameters are definitely depending on manager's experience.

This article proposes a QoS system named Fuzzy-based Dynamic Bandwidth Allocation (FDBA) system which divides network traffic into channels, each consists of several classes. A class serves users of a small group in an enterprise or an institute, like a department or a bureau, i.e., packets coming from a user group is assigned to a specific class. Class Based Weight Fair Queuing (CBWFQ) is deployed as the congestion resolution mechanism to reserve bandwidth for channels.

Furthermore, FDBA can automatically tune QoS parameters to adapt various network traffic, thus decreasing managerial load and providing users a higher network service quality.

2 Survey

Cisco supports several DiffServ mechanisms to fulfill QoS requirements [2, 3]. WFQ is a special one offering dynamic fair queuing that, depending on weights, partitions bandwidth across channel queues to ensure all traffic to be treated fairly.

There are three common network service levels, expedited forwarding (EF), assured forwarding (AF), best effort forwarding (BE), where EF has highest priority and BE the lowest. The main issue of CBWFQ is how to allocate bandwidth to traffic of different service levels. [4] proposed a dynamic DiffServ bandwidth allocation model (DDBAM), that dynamically changes the bandwidth individually allocated to EF, AF and BE. Exponential average (EA) is used to predict traffic of next moment. However, EA shapes traffic, especially bursty traffic. This may influence its predictive accuracy. DDBAM first satisfies EF's required bandwidth, and assigns remaining bandwidth to AF. If AF does not use up the remainder, BE can be then severed. Due to absolute precedence, this model may cause starvation.

[5] proposed a fuzzy-based Dynamic Resource allocation approach that allocates bandwidth to ATM network. They deploy fuzzy logic controller (FLC) to assign variable priorities to classes depending on cell loss ratio, cell transfer delay, cell delay variation and total number of cells generated in each class to solve starvation problem. FLC is embedded in router kernel so that changing bandwidth can quickly respond. But, embedding FLC into router is a crucial work due to performance reduc-

tion. Situation becomes worse, especially in a high speed network. Remote control may be a feasible and easy way.

[6] proposed a traffic shaping algorithm, which restricts end user side network traffic from reaching its maximum reserved bandwidth, and monitors current transfer rates with neural network to change shaping parameters. It has no feedback mechanism, i.e., no historical consideration. Also, neural network can not explain network behavior.

[7,8] surveyed and examined several traffic models, such as Renewal Traffic Models, Markov-Modulated Traffic Models (MMTM), Fluid Traffic Models and Autoregressive Traffic Models, and explained which one is suitable for which kind of network. MMTM is an application of Hidden Markov Model (HMM) [9-12]. [5] employed Markov Modulated Poison Process (MMPP), which is the most commonly used MMTM having represented a variety of traffic models, to simulate traffic as experimental input. [13,14] estimated parameters for MMTM.

3 FDBA Architecture

FDBA consists of four subsystems, Traffic Collector (TC), Traffic Predictor (TP), Fuzzy Controller (FC) and Bandwidth Allocator (BA), as shown in Fig. 1. FDBA uses MMPP to model network traffic. TC is responsible for collecting traffic volume transmitted via router as current observation for MMPP. TP predicts traffic of next moment by training and tuning MMPP. Each class within a channel conducts an independent traffic. By calculating the maximum probability for MMPP, we can more accurately predict traffic for each class. FC foretells and distributes finite bandwidth to channels. Based on this distribution, BA dynamically changes CBWFQ configuration to increase CBWFQ's runtime flexibility.

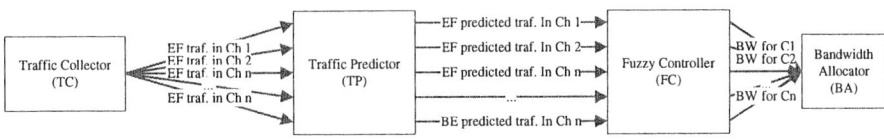

Fig. 1. FDBA architecture

3.1 CBWFQ

CBWFQ assigns a channel Ch a weight proportional to Ch's bandwidth in order to ensure the corresponding queue being served fairly. In a router, each channel has its own FIFO queue.

CBWFQ reserves the minimum bandwidth for each channel. The sum of all bandwidth allocated to an interface, e.g., Ethernet port or a serial port, should not exceed total available bandwidth. [2] recommended that total allocated bandwidth should not exceed 75 percent of total bandwidth. The remainder is for other overhead, including Layer 2 overhead, routing traffic, and best-effort traffic.

3.2 Traffic Collector

Network traffic can be collected at least in two ways: Simple Network Management Protocol (SNMP) and NetFlow. The former is a well-known network management tool, providing network manager a protocol to retrieve information from Management Information Base (MIB) in a router, switch or server. The latter is a powerful tool proposed by Cisco for increasing router's performance by ways of caching packet information into buffer. Netflow has a bonus function to export header information of a packet flowing through underlying router [15].

We retrieve traffic of an interface from standard MIB, and traffic of each channel from Cisco proprietary MIB so as to realize the traffic volume belonging to each class. However, SNMP is unable to read class traffic. NetFlow is then invoked. It collects traffic for each user, then storing the collection into database and finally summarizing the traffic for each class. Nevertheless NetFlow is not a real-time system, predicting traffic from history becomes important.

3.3 Traffic Prediction

We employ MMPP to model network traffic as stated above, and use Expectation Maximization (EM) algorithm [9] to estimate MMPP parameters, such as state transition probability and probability distribution of traffic volume for each state, i.e., each state has its own traffic probability density function (PDF).

Assume $O = \{o_1 o_2 ... o_T\}$ is the observed traffic sequence from time t_1 to t_T and $q = \{q_1 q_2 ... q_T\}$ the hidden state sequence for each observation. MMPP consists of state transition probabilities $\Pi = \{\pi_{ij}; i, j = 1, 2, ... N\}$ from state S_i to state S_j at time t where $\pi_{ij} = P(q_{t+1} = S_j | q_t = S_i)$ and N is the number of states. Traffic PDF $\lambda_i(x) = \frac{1}{\sqrt{2\pi\sigma_i^2}} e^{\frac{-(x-\mu_i)^2}{2\sigma_i^2}}$ is a Gaussian distribution with mean μ_i and variance σ_i^2 for state S_i, $i = 1, 2, ... N$ and π_i is the probability of the initial state S_i (at time t_1), i.e., $\pi_i = P(q_1 = i | O)$, $i = 1, 2, ... N$. MMPP is defined by the parameter set Ω:

$$\Omega = \{\pi_i, \pi_{ij}, \lambda_i(\mu_i, \sigma_i^2); i, j = 1, 2, ... N\}$$

Our goal is to estimate Ω for predicting traffic volume of next moment given O. The problems raised are four-fold: 1. How to efficiently compute the probability that the predicted sequence generated by MMPP is same as the observed sequence? 2. How to choose a corresponding state sequence q that is optimal (maximum probability) for the observation? 3. How to adjust Ω to maximize $P(O|\Omega)$? 4. How to predict next observation o_{T+1} with MMPP?

The first problem can be solved by computing the forward variable $\alpha_t(i) = P(o_1 o_2 ... o_t, q_t = i | \Omega)$ and backward variable $\beta_t(i) = P(o_{t+1} o_{t+2} ... o_T, q_t = i | \Omega)$ with forward procedure and backward procedures [9] respectively, where $\alpha_t(i)$ ($\beta_t(i)$) is the probability of the partially observed sequence $o_1 o_2 ... o_t$ ($o_{t+1} o_{t+2} ... o_T$) from the very beginning to time t (from time t to t_T), given the model Ω and time t in state S_i.

Since $P(O|\Omega) = P(o_1 o_2 ... o_t o_{t+1} ... o_T | \Omega) = \sum_{i=1}^{N} P(o_1 o_2 ... o_t, q_t = i | \Omega) P(o_{t+1} o_{t+2} ... o_T, q_t = i | \Omega)$,

by definition $P(O|\Omega) = \sum_{i=1}^{N} \alpha_t(i) \beta_t(i)$.

The second problem can be solved by Viterbi algorithm [12] which finds out the best state sequence with dynamic programming method. The quantity is defined as follows:

$$\delta_t(i) = \max_{q_1 q_2 ... q_{t-1}} P(q_1 q_2 ... q_{t-1}, q_t = i, o_1 o_2 ... o_t | \Omega)$$

which is the best score (the highest probability) along sequence $q_1 q_2 ... q_t$ at time t. The complete Viterbi algorithm is as follows.

(1). Initiate $\delta_t(i)$ at time t_1, and reset the array argument ψ_1, i.e.,

$$\delta_1(i) = \pi_i \lambda_i(o_1) \quad, \quad \psi_1(i) = 0 \quad, \quad 1 \le i \le N$$

(2). Compute $\delta_t(i)$ from time t_2 to time t_T, and store the state of each t into ψ_t.

$$\delta_t(j) = \max_{1 \le i \le N}[\delta_{t-1}(i) \pi_{ij}] \lambda_j(o_t) \quad, \quad \psi_t(j) = \arg\max_{1 \le i \le N}[\delta_{t-1}(i) \pi_{ij}] \quad t_2 \le t \le t_T \quad, \quad 1 \le j \le N$$

where arg is the operation storing the state which has maximum probability causing $\max_{1 \le i \le N}[\delta_{t-1}(i) \pi_{ij}]$.

(3). Select the maximum probability of $\delta_T(i)$ as the final probability P^*, and retrieve final state q_T^*.

$$P^* = \max_{1 \le i \le N}[\delta_T(i)] \qquad q_T^* = \arg\max_{1 \le i \le N}[\delta_T(i)]$$

(4). Backtrack to get the complete state sequence.

$$q_t^* = \psi_{t+1}(q_{t+1}^*) \qquad t = t_{T-1}, t_{T-2}, ..., t_2, t_1$$

The third problem is a key issue in tuning MMPP to meet real traffic. Reestimating parameters is performed by EM algorithm [9]. First, we define $\gamma_t(i) = P(q_t = i | O, \Omega)$ as the probability of being in state S_i at time t, given O and Ω. By induction, we get

$$\gamma_t(i) = P(q_t = i | O, \Omega) = \frac{P(O, q_t = i | \Omega)}{P(O | \Omega)} = \frac{P(O, q_t = i | \Omega)}{\sum_{i=1}^{N} P(O, q_t = i | \Omega)}$$

since $P(O, q_t = i | \Omega) = P(o_1 o_2 ... o_t, q_t = i | \Omega) \times P(o_{t+1} o_{t+2} ... o_T, q_t = i | \Omega) = \alpha_t(i) \beta_t(i)$, we can rewrite $\gamma_t(i)$,

$$\gamma_t(i) = \frac{\alpha_t(i) \beta_t(i)}{\sum_{i=1}^{N} \alpha_t(i) \beta_t(i)}$$

In order to reestimate (iteratively update and improve) Ω, we define $\xi_t(i, j)$

$$\xi_t(i, j) = P(q_t = i, q_{t+1} = j | O, \Omega)$$

which is the probability of being in state S_i at time t and state S_j at time t+1, given Ω and O.

$$\xi_t(i,j) = \frac{P(q_t=i, q_{t+1}=j, O\mid\Omega)}{P(O\mid\Omega)} = \frac{\alpha_t(i)\pi_{ij}\lambda_j(o_{t+1})\beta_{t+1}(j)}{P(O\mid\Omega)} = \frac{\alpha_t(i)\pi_{ij}\lambda_j(o_{t+1})\beta_{t+1}(j)}{\sum_{i=1}^{N}\sum_{j=1}^{N}\alpha_t(i)\pi_{ij}\lambda_j(o_{t+1})\beta_{t+1}(j)}$$

By definition, we got $\gamma_t(i) = \sum_{j=1}^{N}\xi_t(i,j)$.

Summing up $\gamma_t(i)$ over time t can get the expected number of transitions outgoing from S_i, and summing up $\xi_t(i,j)$ can derive expected value of transitions from S_i to S_j. The reestimation formulas for Ω are

$\overline{\pi_i}$ = expected frequency (number of times) in state S_i at time $t_1 = \gamma_1(i)$

$$\overline{\pi_{ij}} = \frac{\text{expected number of transitions from state } S_i \text{ to state } S_j}{\text{expected number of transitions from state } S_i} = \frac{\sum_{t=t_1}^{T-1}\xi_t(i,j)}{\sum_{t=t_1}^{T-1}\gamma_t(i)}$$

$$\overline{\mu_i} = \text{mean of the observation at states } S_i = \frac{\sum_{t=t_1}^{T}P(q_t=i\mid O,\Omega)o_t}{\sum_{t=t_1}^{T}P(q_t=i\mid O,\Omega)} \quad [14]$$

$$\overline{\sigma_i^2} = \text{variance of the observation at states } S_i = \frac{\sum_{t=t_1}^{T}P(q_t=i\mid O,\Omega)(o_t-\overline{\mu_i})^2}{\sum_{t=t_1}^{T}P(q_t=i\mid O,\Omega)} \quad [14]$$

The reestimation formulas can be directly derived by maximizing Baum's auxiliary function [11].

$$Q(\Omega,\overline{\Omega}) = \sum_q P(O,q\mid\Omega)\log[P(O,q\mid\overline{\Omega})]$$

where $\overline{\Omega}$ denotes a new estimate for Ω. Baum and his colleagues has proven that maximizing $Q(\Omega,\overline{\Omega})$ leads to increased likelihood, i.e.,

$$\max_{\overline{\Omega}}[Q(\Omega,\overline{\Omega})] \Rightarrow P(O\mid\overline{\Omega}) \geq P(O\mid\Omega)$$

We can solve the final problem by computing the maximized probability for next observation with the tuned model. The algorithm is as follows:

(1) Compute the probabilitty for the transition from q_T to q_{T+1}, $q_{T+1}=1,2,...,N$, given $q_T=i$ at time t_T and choose the maximum one. The corresponding q_{T+1}, say K, will be the next state, i.e.,

$$q_{T+1} = \arg\max_{1\leq j\leq N}\overline{\pi_{ij}} = K$$

(2) Retrieve the traffic volume having the maximum probability as the predicted traffic at t_{T+1}, i.e.,

$$o_{T+1} = \arg\max_{\text{all traffic volume } x}\overline{\lambda_j(x)}$$

3.4 Fuzzy Controller

FDBA follows the following procedure to dynamically assign bandwidth to CBWFQ: determining scores (weights) for each channel and allocating bandwidth based on the scores. Three trapezoidal membership functions (TMFs) namely high, medium and low are defined for each of EF, AF and BE. There are a total of 27 output scores whose corresponding fuzzy rules are listed in Table 1. Bandwidth allocated to channel i, say BW_i, is calculated as follows.

$$BW_i = \frac{Score_i}{\sum_{j=1}^{n} Score_j} \times BW_{total}$$

where n is the total number of channels managed by FDBA, $Score_i$ score of channel i and BW_{total} the total bandwidth of underlying interface.

Table 1. Fuzzy rules for bandwidth sharing

SN.	EF	AF	BE	score
1	low	low	low	1
2	low	low	medium	2
3	low	low	high	3
4	low	medium	low	4
5	low	medium	medium	5
6	low	medium	high	6
7	low	high	low	7
8	low	high	medium	8
9	low	high	high	9
10	medium	low	low	10
11	medium	low	medium	11
12	medium	low	high	12
13	medium	medium	low	13
14	medium	medium	medium	14
15	medium	medium	high	15
16	medium	high	low	16
17	medium	high	medium	17
18	medium	high	high	18
19	high	low	low	19
20	high	low	medium	20
21	high	low	high	21
22	high	medium	low	22
23	high	medium	medium	23
24	high	medium	high	24
25	high	high	low	25
26	high	high	medium	26
27	high	high	high	27

4 Experiments

Four experiments are performed. In the first, every 60 observations are treated as a unit. A total of 115 units are prepared. We use the first 60 to initiate MMPP and tune it by Baum's algorithm. 55 predictions are generated. The remaining 55 observations are the reference list used to compare with the 55 predictions. The accuracy is calculated by the following formula:

$$accuracy = \left[1 - \left(\frac{\sum_{t=61}^{115}|o_t - p_t|}{\sum_{t=61}^{115} o_t}\right)\right] \times 100\%$$

where o_t is the tth observation, p_t the tth prediction.

Table 2. The accuracy for FDBA, Qiu and OP

Samples \ model	FDBA	Qiu	OP
60 observations as a unit	88.59%	87.60%	85.72%
30 observations as a unit	83.04%	80.14%	77.72%
10 observations as a unit	81.29%	71.27%	67.70%
bursty traffic	60.55%	-37.00%	-68.74%

The second and the third experiments are the same as the first except "60 observations as a unit" is respectively replaced by "30 observations as a unit" and "10 observations as a unit". The fourth experiment deploys bursty traffic as the observed data. Table 2 shows the accuracies of three prediction approaches, FDBA, Qiu and OP (observation as prediction).

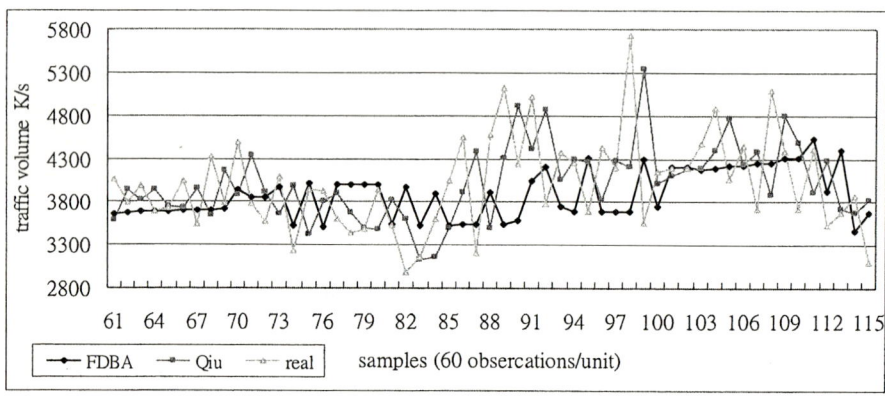

Fig. 2. Comparison on sampling size=60 observations/unit

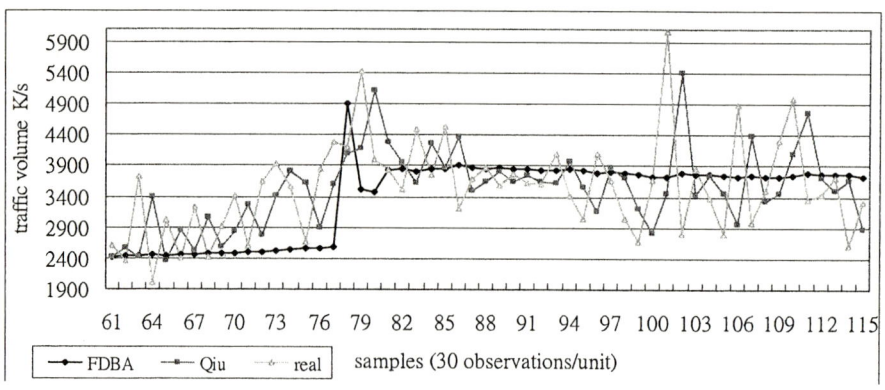

Fig. 3. Comparison on sampling size=30 observations/unit

FDBA is better than the other two especially in a short term sampling. When bursty is given, its accuracy is 60.55%, while the other two become negative, i.e., $\sum_{t=61}^{115}|o_t - p_t| > \sum_{t=61}^{115} o_t$, due to Qiu's shaping traffic and OP's tracking traffic. That means the latter two bring forth serious errors. Fig. 2 to Fig. 5 show four predictions for FDBA, Qiu and real traffic. OP's results are not illustrated to simplify the scopes. Its values can be obtained by right shifting "real" values to their next adjacent observations.

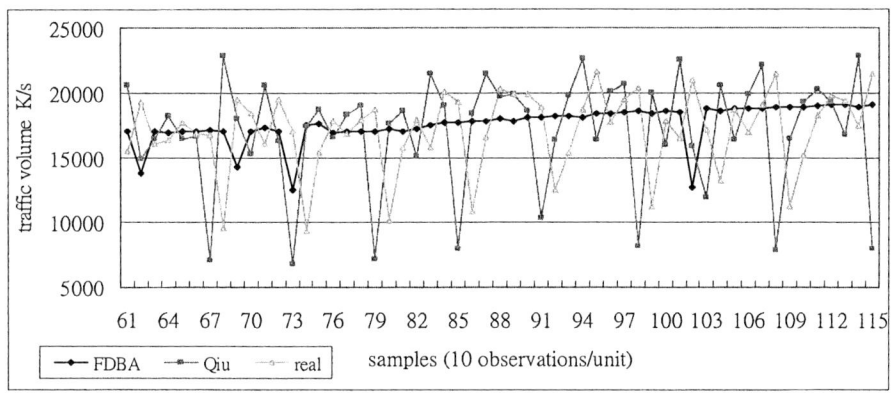

Fig. 4. Comparison on sampling size=10 observations/unit

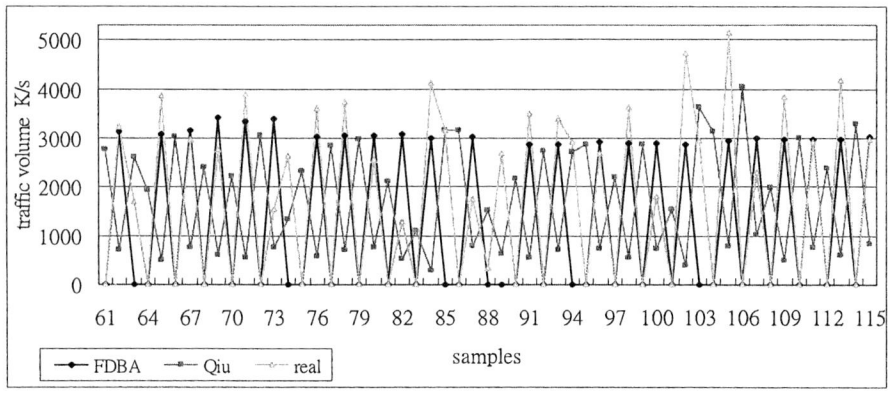

Fig. 5. Comparison on bursty traffic

5 Conclusion

DiffServ is simple and suitable for large scale networks. It supplies different queuing strategies to routers. This is useful especially when congestion frequently occurs. However, once set, DiffServ's parameter values are fixed so that it is inadequate for dynamic network traffic. Queuing scheduling and buffering, such as fair queuing and shaping, are ways to increase its flexibility. However, these methods are suitable for short term variance, like bursty traffic. Long term variance, such as different characteristics between midnight and midday or weekend and weekdays, is difficult to predict.

Dynamically changing QoS parameters can solve long term variance problem. This article proposes a Fuzzy-based dynamic bandwidth allocation mechanism to improve DiffServ performance and reduce managerial burden. In order to implement this system on existing platform, we use external approach to remotely control router settings. However, remote control is not sensitive enough to achieve real-time regulation due to router's slow response. Rewriting router's kernel and constructing real-time controller can improve its behavior.

References

1. Hai, D.Q., Vuong, S.T.: Dynamic-Distributed Differentiated Service for Multimedia Applications. IEEE Dependable Systems and Networks (2000) 586-594
2. Cisco IOS Quality of Service Solutions Configuration Guide http://www.cisco.com/en/US/products/sw/iosswrel/ps1835/products_configuration_guide_book09186a00800c5e31.html
3. Cisco IOS Quality of Service Solutions Command Reference http://www.cisco.com/en/US/proucts/sw/iosswrel/ps5207/products_command_reference_book09186a00801a7ec7.html
4. Qiu, W.B.: Exponential Average-based dynamic DiffServ tuning model. master's thesis of TungHai University (in Chinese) (2002)
5. Chandramathi, S., Shanmugavel, S.: Fuzzy-based Dynamic Resource allocation for heterogeneous sources in ATM networks. Soft Computing (2003) 53-70
6. Kaul, M., Khosla, R., Mitsukura, Y.: Intelligent Packet Shaper to Avoid Network Congestion for Improved Streaming Video Quality at Clients. IEEE 2003 International Symposium on Computational Intelligence for Robotics and Automation (2003) 988-993
7. Frost, V.S., and Melamed, B.: Traffic Modeling For Telecommunications Networks. IEEE Communications Magazine (1994) 70-81
8. Adas, A.: Traffic Models in Broadband Networks. IEEE Communications Magazine (1997) 82-89
9. Rabiner, L.R.: A Tutorial on Hidden Markov Models and Selected Applications in Speech Recognitions. Proceedings of the IEEE, Vol. 77, No. 2 (1989) 257-286
10. Dempster, A.P., Laird, N.M., Rubin, D.B.: Maximum-LikeHood from Incomplete Data via the EM algorithm (with Discussion). Journal of the Royal statistical society, series, 39 (1997) 1-38
11. Baum, L.E., Sell, G.R.: Growth functions for transformations on manifolds. Pac. J. Math., vol. 27, no. 2 (1968) 211-227
12. Forney, G.D., Jr: The Viterbi Algorithm. Proceedings of the IEEE, vol. 61 (1973) 268-278
13. Krishnamurthy, V.: Adaptive estimation of hidden nearly completely decomposable markov chains. IEEE Acoustics, Speech, and Signal Processing (1994), vol.4 IV/337-IV/340
14. Yegenoglu, F., Jabbari, B.: Maximum likelihood estimation of ATM traffic model parameters. IEEE Global Telecommunications Conference (1994) 34-38
15. NetFlow Services Solutions Guide http://www.cisco.com/en/US/products/sw/netmgtsw-/ps1964/ proucts_implementation_design_guide09186a00800d6a11.html

Self-localization of a Mobile Robot by Local Map Matching Using Fuzzy Logic*

Jinxia Yu [1,2], Zixing Cai [1], Xiaobing Zou [1], and Zhuohua Duan [1,3]

[1] College of Information Science & Engineering, Central South University,
410083 Changsha Hunan, China
[2] Department of Computer Science & Technology, Henan Polytechnic University,
454003 Jiaozuo Henan, China
melissa2002@163.com
[3] Department of Computer Science, Shaoguan University,
512003 Shaoguan Guangdong, China

Abstract. Reliable localization is a fundamental issue in robot navigation techniques. This paper describes an apporach for realizing self-localization of mobile robot by matching the local map generated from a 2D laser scanner. Environment map is represented by occupancy grids and it fuses the information of the robot's pose using dead-reckoning method and the range to obstacles by laser scanner using maximum likehood estimation. After a current laser scan, the positon of mobile robot, in relation to a previous scan and pose estimates, is computed by matching the local map using fuzzy logic method. The effectiveness of this method is demonstrated by experiments.

1 Introduction

Reliable localization is a fundamental issue in robot navigation techniques[1]. Usually, The localization methods can be categorized into two groups: relative and absolute position measurements[2]. Because of the lack of a single good method, developers of mobile robots usually combine two methods. As a kind of absolute position methods, map matching is widely used to correct accumulated errors of relative localization.

In our research, a 2D laser scanner LMS291 is utilized as exteroceptive sensor to build the environment map for its advantages. Environment map is represented by occupancy grids and it fuses the information of the robot's pose using dead reckoning and the range to obstacales by laser scanner using maximum likehood estimation. Due to the presence of random noise, dynamic disturbance and self-occlusion, map matching according to the results of maximum likelihood estimation cannot completely be used to find the most likely positon of mobile robot. While fuzzy techniques have been proved to be effective in addressing the self-localization problem. So we apply fuzzy logic into the results of maximum likelihood estimation in order to improve the robot localization performance. By experiments of the robot platform equipped with LMS291, the effectiveness of this method is demonstrated and the localization performance of mobile robot is validated.

---

* This work is supported by the National Natural Science Foundation of China (No. 60234030).

2 Local Map Building

Kinematic model of robot platform can be described by its rigid constraints. Similar to reference 3, we only focus our interest in the *x-y* plane and in the θ direction here. Let $X=(x_r, y_r, \theta_r)$ denotes the robot's pose in a 2D space, then the kinematic equations in the discrete time domain can be got by pose sensors such as odemetry, fiber optic gyros(FOG). A laser scanner LMS291 made by SICK corporation[4] is mounted on the rotating table of mobile robot to detect the environment. Each scan point is represented by the laser beam direction, and the range measurement along that direction.

The operating environment of mobile robot is described by 2D Cartesian grids, and a 2D array is taken to storage the environment map. The konwledge about occupancy condition of a given cell at time *k* is stored as probability of two states, empty or occupied, given all the prior sensor observations. When an obstacle is observed at some position, the corresponding array value is changed from 0 to 1. Since the size of grids has an effect on the resolution of control algorithms directly, each grid cell corresponding to 5×5cm² in real enviroment is adopted in view of the measurement resolution and response time of LMS291. If doing so, the local map of mobile robot can be created in real time and keep the high accuracy.

3 Self-localization Using Fuzzy Logic

3.1 Maximum Likelihood Estimation

The inherent uncertainty of environment grids is mainly derived from the accumulated pose errors. The errors between the estimated and the real values of robot's pose is represented by $\delta X = \{\delta x, \delta y, \delta \theta\}$. By searching for the optimal parameters $\delta x, \delta y, \delta \theta$ under map matching, we will find the most likely postion of mobile robot in the global coordinate system. Therefore the estimation for the parameters $\delta x, \delta y, \delta \theta$ is a maximum likelihood estimation problem.

$$(\delta x^*, \delta y^*, \delta \theta^*) = \arg \max_{\delta x, \delta y, \delta \theta} l(\delta x, \delta y, \delta \theta) \tag{1}$$

For estimating the parameters $\delta x, \delta y, \delta \theta$, the map matching is formulated in terms of maximum likelihood estimation using the distance from the occupied cells in the current local map to their cloest occupied cells in the previous map with respect to some relative postion between the maps as measurements. We denote the distances for these occupied grid cells at some position of mobile robot by $D_1, \cdots, D_n$, thus the likelihood function for the parameters $\delta x, \delta y, \delta \theta$ is formulated as follows. The map matching is limited in a 3×3 grid area centered by the matching grid cell.

$$l(\delta x, \delta y, \delta \theta) = \sum_{i=1}^{n} \ln p(D_i; \delta x, \delta y, \delta \theta) \tag{2}$$

3.2 Self-localization Based on Fuzzy Logic

The initial postion of mobile robot is given by dead reckoning so that we have an initial position to compare agasist. The new robot pose and its asssociated uncertainty from the previous pose, given dead-reckoning information, is estimated. Then the ambient environment is completely scanned by the laser scanner with the rotation of the rotating table. The errors between the estimated and the real values of robot's pose is computed using maximum likelihood estimation during matcing the map between scans. Since the scan range of LMS291 is from −90° to 90° and the rotating table rotates from −150° to 150° in horizontal direction, the perceptual environment around the robot can be completely sensed by 3 scans of LMS291 with the rotation of the rotating table at least which would be devided into corresponding scan sectors in the same position. On account of the analysis in introduction, it is different for the results of maximum likelihood estimation in various sector whether the robot moves or not. Hence, we assign a fuzzy degree of membership to each estimated parameter between scans and define its memebership function is {low, rather low, median, rather high, high}. Next, we make a fuzzy logic inference according to the matching results, and then the crisp number is gained by a defuzzification based on the centroid average method. The algorithms is described in figure 1.

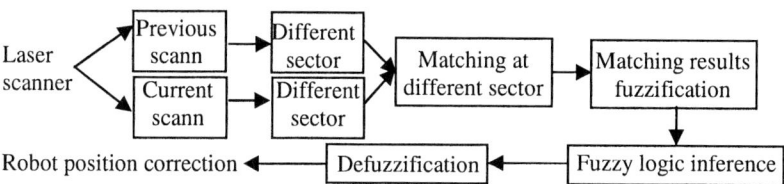

Fig. 1. Algorithms description based on fuzzy logic

4 Experiments

We implemented experiments in our office using mobile robot IMR-01 with two rocker-bogie suspension, four dirve wheels and an omnidirectional wheel as experimental platform. Experimental results are shown in figure2. Figure 2 (a) depicts the robot uses pose sensors and laser scanner to determinate its pose and build an initial local map so as to compare against from which we can see black grids are obstacles, a door is on the left and researcher are on the right as dynamic obstacles; (b) shows the robot moves along the wall to scan the environment in which blue curve is its trajectories and we can know the accumulate errors increases over time; (c) and (d) are the local map before and after correction. Compared the left and right side of office, the results using maximum likelihood estimation exist discrepancy owing to the dynamic disturbances. After the robot wander for some time, we can find the pose of mobile robot changed due to the accumulate errors and caused the local map uncertainty from (a) and (c). By the correction for self-localization by the local map matching using fuzzy logic, the random noise and dynamic disturbances can be reduced in relation to (a) and (d).

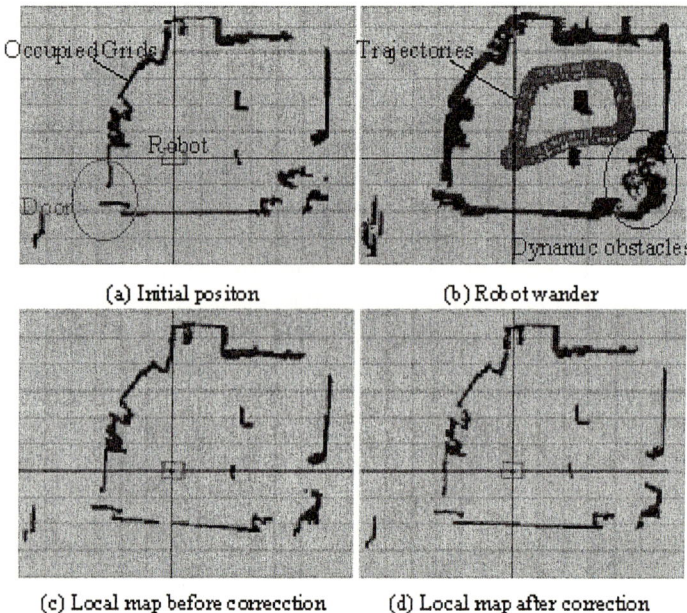

Fig. 2. Local map matching

5 Conclusions

An approach for a mobile robot equipped with a 2D laser scanner to realize its self-localization by matching the local map is described in this paper. Environment map is represented by occupancy grids. Because the uncertainty of environment grids is mainly caused by the noise of pose sensors, fuzzy logic is applied into the map matching so as to limit the robot's pose errors to small enough. Experimental results demonstrate the effectiveness of this method.

References

1. Cai, Z.X., He, H.G., Chen, H.: Some issues for mobile robot navigation under unknown environments (in Chinese). Control and Decision, 17(4) (2002) 385-391
2. Borenstein, J., Everett, H.R., Feng, L., et al.: Mobile Robot Positioning-Sensors and Techniques. Journal of Robotic Systems, 14(4) (1997) 231-249
3. Olson, C.F., Matthies, L.H.: Maximum Likelihood Rover Localization by Matching Range Maps. In: Proceedings of IEEE Int. Conf. on Robotics and Automation, Leuven Belgium, (1998) 272-277
4. SICK AG Corp. Technical Description: LMS200/ LMS211/ LMS220/ LMS221/ LMS291 Laser Measurement Systems. //http: www.sick.com

Navigation of Mobile Robots in Unstructured Environment Using Grid Based Fuzzy Maps

Özhan Karaman[1] and Hakan Temelta[2]

[1] Istanbul Technical University, Faculty of Electrical and Electronics Engineering,
Maslak, 34469, Istanbul Turkey
ozhankaraman@yahoo.com
[2] Istanbul Technical University, Faculty of Electrical and Electronics Engineering,
Maslak, 34469, Istanbul Turkey
temeltas@elk.itu.edu.tr

Abstract. This article proposes a navigation technique based on fuzzy maps of the environment for real-time applications. Particularly choosing the right operator plays important role in obtaining good results rapidly and accurately for different kind of environments. Mainly, A^* algorithm is used for navigating the mobile robot in the fuzzy mapped environment defined by Yager Union operator. Experiments are implemented on Nomad 200 mobile robot successfully.

1 Introduction

Many robotic systems are widely using multi sensory systems rather than using one single smart sensor (such as cameras, laser or GPS systems). Instead of using improved complex configurations, simple sensors require generally less system resources and less cost. They can smoothly be distributed over a larger area which makes the sensing system less prone to errors. If a sensor is broken down, the other sensors can handle broken sensors' operation. This feature makes multi sensory systems more robust. Shortly multi sensory systems are cheaper, robust, distributed and flexible. The main problems in multi sensor systems are spectacular reflections, angular uncertainty and fusion of all sensors measurements. Researchers developed many techniques to handle these problems. Commonly used sensor fusion methods are Kalman filtering[1], Dempster Shafer's evidence theory [2], Neural Networks [3] and Fuzzy Systems[4].

In this paper, the map building results are used in navigation process using fuzzy logic base environment maps is constructed. Mobile robot finds its way using these maps. If the mobile robot cannot reach to the goal point map building and navigation processes are repeated recursively. The relationship between map building and navigation results are shown in Fig. 1.

The overall concept of map building algorithm is when ultrasonic range reading process is applied by robotic system, range reading values coming from sensors are processed by several different fuzzy union operators. These operators process each sensors value by its sensors model and get their values in two fuzzy sets. These sets show local occupancy and emptiness value of the cells under the radiation cones.

After gathering local values of the cells, these local values are aggregated by the same fuzzy union operator to get the global map of the surrounding. For detailed information about the process can be found from [5]. Dombi, Dubois-Prade, Yager, Sugeno and Einstein union operators applied to the map building process and their results are compared with speed and successfully mapping criteria's. Navigation process activates after getting 360 degrees surrounding map data from map building process. A* path finding algorithm is used for navigation process. A* path finding algorithm finds paths from explored environment. If mobile robot reaches to the goal point mobile robot stops and it finishes its current work. But if it cannot reach to the goal point, recursive map building processes are repeated and at the end of processes mobile robot reaches to the goal point.

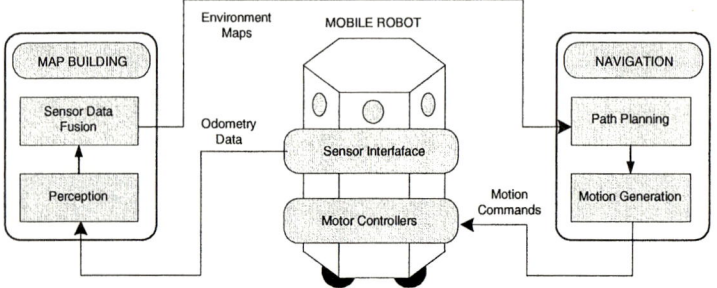

Fig. 1. Functional diagram of the map building and navigation processes

This paper is organized as follows. Section 2 gives the background of sensory model of system. In section 3 we will give different fuzzy operators time based and performance based comparison results for map building process. In section 4 and 5 we will introduce navigation process and its experimental results are given. In section 6 conclusions are given.

2 Sensor Model

We have used fuzzy sensory model for fusion of sensory range readings. Ultrasonic sensors' measures distance values between obstacles and sensor by the time of flight of the ultrasonic waves in the air. This range values has sometimes problems like angular uncertainties and spectacular reflections. Fuzzy logic offers a natural framework in which uncertain information can be handled. Nomad 200 mobile robot has equipped with 16 Poloroid ultrasonic range finders [6]. Poloroid ultrasonic range finders can measure distance value with %1 accuracy by approximately 0.12 - 6.5 meters. For each perception phase turret system rotates 7.5 degrees for two times and takes three consecutive range readings. Totally we get 3x16=48 range values for each perception process. To eliminate false reflections we will not consider the values higher than 1.5 meters. All the points inside 1 meter radius circle falls into minimum of 3 radiation cones.

48 sensory range reading values give 360 degrees surrounding scene of the robot. From these range readings two fuzzy sets empty ($\mathcal{E}$) and occupied (O) are formed. These sets are not complementary and all the cells under the radiation cones have different membership values to these sets. To integrate multiple sensory range readings on the evidence grid, we must use a sensor model to convert the range information into evidence values. Cells located near the cone axis and near the circumference of radius r have more evidence for being occupied. Cells which are far from cone axis and far from circumference of radius r have more evidenced for being empty. To model these circumstances, the sensory model give in Eq. (1).

$$f_\varepsilon(c,r) = \begin{cases} k_\varepsilon & 0 \leq c < r - \Delta r \\ k_\varepsilon \left(\frac{r-c}{\Delta r}\right)^2 & r - \Delta r \leq c < r \\ 0 & c \geq r \end{cases}; f_O(c,r) = \begin{cases} 0 & 0 \leq c < r - \Delta r \\ k_O \left[1 - \left(\frac{r-c}{\Delta r}\right)^2\right] & r - \Delta r \leq c < r + \Delta r \\ 0 & c \geq r + \Delta r \end{cases} \quad (1)$$

Where c identifies the cells distance from sensor, r identifies the range reading, k_ε and k_O identifies maximum values of empty and occupied sets, $2\Delta r$ identifies length of proximal area to the arc of radius r. Detailed sensor model is given in [7].

3 Map Building

Map building process consists of 3 sub-processes. Every process has different tasks to complete. These 3 tasks are repeated recursively and the environment map is formed. Fig. 2 shows the structure of 3 sub-processes. The selection of proper fuzzy union operator is very important for overall system performance. Because of choosing incorrect union operators; some regions in the environment map can be identified incorrect by the map building phase. For example empty regions can be identified occupied or occupied regions can be identified empty. These false data could become map building and navigation processes totally inefficient. Navigation process can find long paths to the goal or the process cannot find paths to reach to the goal.

Perception phase takes sensory range reading values from multi sensory system. Processing phase gets sensory values and processes these values using chosen sensory model and generates ε^k and O^k fuzzy sets., so that, $\varepsilon^k = U_i \varepsilon_i^k; O^k = U_i O_i^k$ and these fuzzy sets gather all the emptiness and occupancy information about k^{th} range reading process. ε^k and O^k sets are generated by Dombi, Dubois-Prade, Yager, Sugeno, Einstein's union operators. Strength values of the operators can be tuned by their own constants while computing. Fusion phase gets local information from ε^k and O^k sets and generates global ε and O fuzzy sets. Fusion phase aggregates local in formation to global information using union operators which are used in processing phase given by $\varepsilon := \varepsilon \cup \varepsilon^k; O := O \cup O^k$. Fuzzy logic map building techniques has superior performance on detection of conflicting or insufficient information and they are highly computational efficient. Using ε and O fuzzy sets surrounding map M is obtained by $A = \varepsilon \cap O$ and fuzzy sets represents the membership of ambiguous cells in set ε

and O. Another fuzzy sets determine intermediate cells given by $I = \bar{\varepsilon} \cap \bar{O}$ and I safe for planning cells given by $S = \varepsilon^2 \cap \bar{O} \cap \bar{A} \cup I$. It clear that $M = S$ is written. Every map building process updates the M map. For every map building operation is done M environment map is generated.

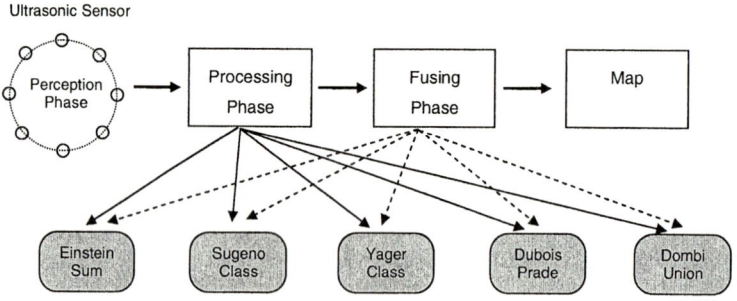

Fig. 2. Map Building Process

4 Navigation

Navigation process is finding a suitable and shortest path to the goal. The chosen path must be safe inside the area so far explored and at same time should provide directions for further exploration to reach goal. Navigation process finds paths to the goal by processing incoming map data from map building process. Mobile robot reaches to the goal point by the repetition of navigation and map building processes. In this work A* path finding algorithm is chosen for Navigation process. A* algorithm is widely used by many robotic researchers [9]. A* algorithm generates sub paths from current location to the goal. A* algorithm uses all the environment knowledge coming from map building phase. Navigation process uses this knowledge to find paths in the explored areas and robot follows these paths to reach the goal point. If the reached point is the real goal point mobile robot stops and navigation process finishes but on the other hand if the reached point is different from the goal point, mobile robot starts a new map building and navigations process till reaching to the goal point.

A* algorithm can find suitable paths in shorter time than other navigation techniques like Dijkstra or Best First Search Techniques. A* is like other graph-searching algorithms in that it can potentially search a huge area of the map. It's like Dijkstra's Algorithm in that it can be used to find a shortest path. It's like BFS in that it can use a heuristic to guide itself. A* navigation techniques behavior can be tuned by changing its heuristic function it can be act like Dijksta, Best First Search or mixture of both two algorithms. Heuristic function is estimated cost function; which calculates the estimated cost from working cell to the goal cell. A* algorithms cost function f(n) is given by f(n) = g(n) + h(n) in which g(n) represents the cost of the path from the starting cell to the working point n, and h(n) represents the heuristic estimated cost from working point n to the goal point. h(n) heuristic functions structure is chosen from mobile robots movement characteristics. These movement characteristics are mobile robots ability of horizontal, vertical and diagonal movement.

5 Experimental Results for Navigation

Navigation results are taken on the Nomad 200 Simulation Program. 8 neighbor cell working and Manhattan Heuristic Function is used for A* path finding algorithm. Formulation about Manhattan Heuristic Function is given in Eq. (2):

$$h(n) = D*(abs(n_x - goal_x) + abs(n_y - goal_y)) \quad (2)$$

For the above heuristic function D is cost function for horizontal, vertical and diagonal movements. $goal_x$ and $goal_y$ are goal cells coordinates.

The navigation results for simulation environment are shown in Fig. 3. For simulation system, Yager Union Operator is used for map building process whose outputs are used by navigation processes. Mobile robot reaches to the goal point without bumping to obstacles in the environment by using the incoming data from map building process. The path followed by mobile robot becomes different by the changing of heuristic function and Δr length of proximal area value. By degreasing the Δr value, narrow spaces can be identified and used by navigation process.

6 Conclusions

In this study A* technique is successfully used by navigation process on fuzzy mapped environment. A* technique can find short and safe paths by using Yager, Sugeno union operators in map building phase. Einstein operator can determine some wrong areas in environment map. Einstein operator can determine empty spaces as occupied or occupied spaces as empty and this behavior result of finding long paths and Dombi Union Operator is good at navigation processes but its computational time is too long so it is not preferable.

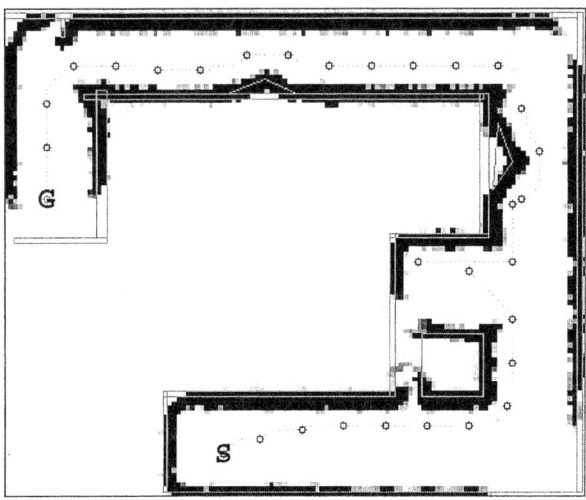

Fig. 3. The navigation results for simulation environment. Goal point is signed with capital G letter and starting point is signed with capital S letter.

References

1. B.S.Y. Rao, H.F. Durrant-Whyte and J.A.Sheen.: A Fully Decentralized Multi-Sensor Systems for Tracking and Surveillance. The International Journal of Robotics Research Vol.12, No.1 (1993) 20-44
2. Daniel Pagec. Eduardo M. Nebot and Hugh Durrant-Whyte.: An Evidential Approach to Map-Building for Autonomous Vehicles, IEEE Transactions on Robotics and Automation, Vol 14, No.4 (1998) 623-629,
3. Kristof Van Laerhoven, Kofi A. Aidoo and Steven Lowette.: Real-time Analysis of Data from Many Sensors with Neural Networks, Proceedings of the fourth International Symposium on Wearable Computers Zurich (2001)
4. Alesandro Saffiotti.: The Uses of Fuzzy Logic in Autonomous Robot Navigation, Technical Report TR/IRIDIA/97-6 (1997)
5. Giuseppe Oriolo, Giovanni Ulivi and Marilena Vendittelli.: Real-Time Map Building and Navigation for Autonomous Robots in Unknown Environments, IEEE Transactions on Systems, Man, Cybernetics Volume 28, Issue 3 (1998) 316-333
6. P. E. Hart, N. J. Nilsson, and B. Raphael.: A formal basis for the heuristic determination of minimum cost paths. IEEE Trans. Syst.,Sci.,Cybern., vol. SSC-4, no. 2, (1968) 100–107
7. Karaman, H. Temeltas.: Comparison of Different Grid Based Techniques for Real-Time Map Building. IEEE International Conference on Industrial Technology, December Hammamet, Tunisia (2004)

A Fuzzy Mixed Projects and Securities Portfolio Selection Model*

Yong Fang[1], K.K. Lai[2,3,**], and Shou-Yang Wang[1]

[1] Institute of Systems Science,
Academy of Mathematics and Systems Science,
Chinese Academy of Sciences, Beijing 100080, China
yfang@amss.ac.cn
swang@iss.ac.cn
[2] College of Business Administration,
Hunan University, Changsha, China
[3] Department of Management Sciences,
City University of Hong Kong, Kowloon, Hong Kong
mskklai@cityu.edu.hk

Abstract. The business environment is full of uncertainties. Investing in various asset classes may lower the risk of overall portfolio and increase the potential for greater returns. In this paper, we propose a bi-objective mixed asset portfolio selection model involving projects as well as securities. Furthermore, based on fuzzy decision theory, a fuzzy mixed projects and securities portfolio selection model is also proposed. A numerical example is given to illustrate the behavior of the proposed fuzzy mixed asset portfolio selection model.

1 Introduction

During the past decade there has been a dramatic increase in the institutional investment. Although most of those investments remain focused on the traditional securities investment, there is an increase in various forms of alternative investment classes, e.g., venture capital, private equity, private debt and real estate, etc. With the extension of investment asset classes, the overall portfolio risk can be lowered while the potential for more benefits can be increased over the long term.

The mean variance methodology for portfolio selection proposed by Markowitz in [5] has been central to research activities in the traditional securities investment field and has served as a basis for the development of modern financial theory over the past five decades. Konno [4] used the absolute deviation risk function to replace the risk function in Markowitz's model to formulate a mean absolute deviation portfolio optimization model. In today's extremely

---

* Supported by the National Natural Science Foundation of China under Grant No. 70221001 and City University of Hong Kong under Grant No. 7100289.
** Corresponding author.

competitive business environment, investors may consider investing their funds in other kinds assets besides securities. Byrne and Lee [2] found that the mixed asset portfolio including listed property trusts, direct property and financial assets always dominated the financial asset portfolio. In recent years, some researchers studied project portfolio selection problems by using mathematical programming methods, e.g., Coffin and Taylor III [3], and Ringuest, Graves and Case [7], etc. Reyck, Degraeve and Gustafsson [6] proposed a mixed asset portfolio selection model involving projects and securities.

In this paper, considering the proportional transaction costs, we will use the expected return and the semi-absolute deviation risk as objective functions and propose a bi-objective programming model for mixed asset portfolio selection problem. Furthermore, we use fuzzy numbers to describe investors' vague aspiration levels for the expected return and the semi-absolute deviation risk and propose a fuzzy mixed asset portfolio selection model.

The paper is organized as follows. In Section 2, we present a mixed asset portfolio selection model involving projects and securities without transaction costs. In Section 3, regarding investors' vague aspiration levels for the expected return and the semi-absolute deviation risk as fuzzy numbers, we propose a fuzzy mixed projects and securities portfolio selection model. In Section 4, a numerical example is given to illustrate the behavior of the proposed mixed asset portfolio selection model. Some concluding remarks are given in Section 5.

2 Bi-objective Programming Model for Mixed Asset Portfolio Selection

In this paper, we assume that an investor allocates his/her wealth among traditional securities and projects. Hence, in the mixed asset portfolio selection problem, available investment assets classes are divided into two types, i.e., traditional securities and projects. There is a capital budget for each project before investment. The capital budgets can be given by investors or some experts. We assume that the cost of carrying out a project will be the corresponding capital budget once the project is started. The investment in these projects cannot be reallocated at any time, while the investment in securities can be reallocated in any amount at any time.

We assume that the securities component of mixed asset is composed by n risky securities $S_i, i = 1, 2, \cdots, n$, offering random rates of returns and a risk-free security S_{n+1} offering a fixed rate of return. The projects component is composed by m projects $P_j, j = 1, 2, \cdots, m$. Assume the investor starts with an existing portfolio which only includes securities, and then decides how to reconstruct a new mixed asset portfolio with securities and projects. We introduce some notations as follows.

$\tilde{r}_i$: the random variable representing the rate of return on the security $S_i, i = 1, 2, \cdots, n$ without transaction costs;

r_{n+1}: the rate of return of the risk-free security S_{n+1};

r_i: the rate of expected return on the security $S_i, i = 1, 2, \cdots, n$ without transaction costs;
$\widetilde{R_j}$: the random variable representing the random net return on the project $P_j, j = 1, 2, \cdots, m$ after removing the cost (the budget);
R_j: the expected net return on the project $P_j, j = 1, 2, \cdots, m$ after removing the cost (the budget);
M: the total amount of assets owned by the investor;
X_i: the amount of the total investment devoted to the risky security $S_i, i = 1, 2, \cdots, n$ and the risk-free security S_{n+1};
x_i: the proportion of the total investment devoted to the risky security $S_i, i = 1, 2, \cdots, n$ and the risk-free security S_{n+1}, i.e., $x_i = \frac{X_i}{M}$;
X_i^0: the amount of the total investment devoted to the risky security $S_i, i = 1, 2, \cdots, n$ and the risk-free security S_{n+1} in the existing portfolio;
x_i^0: the proportion of the total investment devoted to the risky security $S_i, i = 1, 2, \cdots, n$ and the risk-free security S_{n+1} in the existing portfolio;
k_i: the rate of transaction costs for the risky security $S_i, i = 1, 2, \cdots, n$ and the risk-free security S_{n+1};
z_j: the binary variable indicating whether project $P_j, j = 1, 2, \cdots, m$ is started or not,

$$z_j = \begin{cases} 1 & \text{if project } P_j \text{ is selected for funding,} \\ 0 & \text{otherwise.} \end{cases}$$

We assume that the vector of random variables $(\widetilde{r_1}, \widetilde{r_2}, \cdots, \widetilde{r_n}, \widetilde{R_1}, \widetilde{R_2}, \cdots, \widetilde{R_m})$ is distributed over the finite sample space $\{(r_{1t}, r_{2t}, \cdots, r_{nt}, R_{1t}, R_{2t}, \cdots, R_{mt}), t = 1, 2, \cdots, T\}$ and the probabilities

$$p_t = P_r\{(\widetilde{r_1}, \cdots, \widetilde{r_n}, \widetilde{R_1}, \cdots, \widetilde{R_m}) = (r_{1t}, \cdots, r_{nt}, R_{1t}, \cdots, R_{mt})\}, \ t = 1, 2, \cdots, T$$

are known. Then the expected rate of return r_i of the risky security $S_i, i = 1, 2, \cdots, n$ without transaction costs is given by

$$r_i = \sum_{t=1}^{T} p_t r_{it}, \ i = 1, 2, \cdots, n,$$

where r_{it} can be determined by forecast data. The expected net return R_j on the project $P_j, j = 1, 2, \cdots, m$ is given by

$$R_j = \sum_{t=1}^{T} p_t R_{jt}, \ j = 1, 2, \cdots, m,$$

where R_{jt} can be determined by forecast data.

Given a mixed asset portfolio $(x_1, x_2, \cdots, x_n, x_{n+1}, z_1, z_2, \cdots, z_m)$, the expected return of the portfolio without transaction costs can be expressed by

$$\sum_{i=1}^{n+1} r_i X_i + \sum_{j=1}^{m} R_j z_j = \sum_{i=1}^{n+1} \sum_{t=1}^{T} p_t r_{it} X_i + \sum_{j=1}^{m} \sum_{t=1}^{T} p_t R_{jt} z_j,$$

where $X_i = Mx_i$, $i = 1, 2, \cdots, n+1$ and $r_{n+1,t} = r_{n+1}$, $t = 1, 2, \cdots, T$.

We use a V shape function to express the transaction costs. Specifically we let the transaction costs of the security S_i, $i = 1, 2, \cdots, n, n+1$ be given by

$$C_i(X_i) = k_i|X_i - X_i^0|.$$

Hence the total transaction costs of the mixed asset portfolio are expressed as

$$\sum_{i=1}^{n+1} C_i(X_i) = \sum_{i=1}^{n+1} k_i|X_i - X_i^0|.$$

Let $x = (x_1, x_2, \cdots, x_{n+1})$, $z = (z_1, z_2, \cdots, z_m)$ and $X = (X_1, X_2, \cdots, X_{n+1})$. Then the expected net return on the mixed asset portfolio after paying the transaction costs is given by

$$f(X, z) = \sum_{i=1}^{n+1} r_i X_i + \sum_{j=1}^{m} R_j z_j - \sum_{i=1}^{n+1} C_i(X_i)$$

$$= \sum_{i=1}^{n+1}\sum_{t=1}^{T} p_t r_{it} X_i + \sum_{j=1}^{m}\sum_{t=1}^{T} p_t R_{jt} z_j - \sum_{i=1}^{n+1} k_i|X_i - X_i^0|.$$

If we use $x_1, \cdots, x_n, x_{n+1}, x_1^0, \cdots, x_n^0, x_{n+1}^0$ instead of $X_1, \cdots, X_n, X_{n+1}, X_1^0, \cdots, X_n^0, X_{n+1}^0$, respectively, then the expected net return on the mixed asset portfolio after paying the transaction costs is also given by

$$f(x, z) = \sum_{i=1}^{n+1}\sum_{t=1}^{T} p_t r_{it} x_i M + \sum_{j=1}^{m}\sum_{t=1}^{T} p_t R_{jt} z_j - \sum_{i=1}^{n+1} k_i M|x_i - x_i^0|.$$

Maximizing the expected net return $f(x, z)$ on the mixed asset portfolio after paying the transaction costs can be considered an objective of the mixed asset portfolio selection problem.

The semi-absolute deviation of return on the mixed asset portfolio below the expected return at state t, $t = 1, 2, \cdots, T$ can be represented as

$$W_t(X, z) = \Big|\min\{0, \sum_{i=1}^{n}(r_{ti} - r_i)X_i + \sum_{j=1}^{m}(R_{tj} - R_j)z_j\}\Big|.$$

So the expected semi-absolute deviation of the return on the mixed asset portfolio below the expected return can be represented as

$$W(X, z) = \sum_{t=1}^{T} p_t W_t(X, z)$$

$$= \sum_{t=1}^{T} p_t \Big|\min\{0, \sum_{i=1}^{n}(r_{ti} - r_i)X_i + \sum_{j=1}^{m}(R_{tj} - R_j)z_j\}\Big|.$$

Let $w(x, z) = \frac{W(X,z)}{M}$. Then we have

$$w(x, z) = \sum_{t=1}^{T} p_t \Big|\min\{0, \sum_{i=1}^{n}(r_{ti} - r_i)x_i + \sum_{j=1}^{m} z_j \frac{R_{tj} - R_j}{M}\}\Big|.$$

In this paper, we adopt the function $w(x,z)$ to measure the risk of the mixed asset portfolio. Minimizing the risk of the mixed asset portfolio can be considered another objective of the mixed asset portfolio selection problem.

In the mixed asset portfolio selection problem with securities and projects, we consider the following constraints. First, we introduce some notations.

B_j: the capital budget of project $P_j, j = 1, 2, \cdots, m$, i.e., the cost that the investor pays once the project is decided to start;

B: the maximum amount of investment devoted to projects component in the mixed asset portfolio;

Y_j: the amount of the total investment devoted to the project P_j, i.e.,

$$Y_j = B_j z_j, \quad j = 1, 2, \cdots, m;$$

S: the maximum amount of investment devoted to securities component in the mixed asset portfolio;

- Capital budget constraint on projects component:

$$\sum_{j=1}^{m} Y_j = \sum_{j=1}^{m} B_j z_j \leq B.$$

- Capital constraint on securities component:

$$\sum_{i=1}^{n+1} x_i M \leq S.$$

- Total capital constraint:

$$\sum_{j=1}^{m} Y_j + \sum_{i=1}^{n+1} x_i M = \sum_{j=1}^{m} B_j z_j + \sum_{i=1}^{n+1} x_i M \leq M.$$

- No short selling of securities:

$$x_i \geq 0, \quad i = 1, 2, \cdots, n+1.$$

Using the objectives and the constraints introduced in the previous subsection, the mixed asset portfolio selection problem can be formally stated as follows:

(BOP) $\max f(x,z) = \sum_{i=1}^{n+1} \sum_{t=1}^{T} p_t r_{it} x_i M + \sum_{j=1}^{m} \sum_{t=1}^{T} p_t R_{jt} z_j - \sum_{i=1}^{n+1} k_i M |x_i - x_i^0|$

$\min w(x,z) = \sum_{t=1}^{T} p_t |\min\{0, \sum_{i=1}^{n} (r_{ti} - r_i) x_i + \sum_{j=1}^{m} z_j \frac{R_{tj} - R_j}{M}\}|$

s.t. $\sum_{j=1}^{m} B_j z_j \leq B,$

$\sum_{i=1}^{n+1} x_i M \leq S,$

$\sum_{j=1}^{m} B_j z_j + \sum_{i=1}^{n+1} x_i M \leq M,$

$x_i \geq 0, \quad i = 1, 2, \cdots, n+1,$

$z_j = \{0, 1\}, \quad j = 1, 2, \cdots, m.$

The problem (BOP) is a bi-objective mixed-integer nonlinear programming problem. The problem can be reformulated as a bi-objective mixed-integer linear programming problem by using the following technique. Note that

$$\left|\min\{0, \sum_{i=1}^{n}(r_{ti}-r_i)x_i + \sum_{j=1}^{m} z_j \frac{R_{tj}-R_j}{M}\}\right|$$
$$= \left|\sum_{i=1}^{n} \frac{(r_{ti}-r_i)x_i}{2} + \sum_{j=1}^{m} \frac{z_j(R_{tj}-R_j)}{2M}\right| - \sum_{i=1}^{n} \frac{(r_{ti}-r_i)x_i}{2} - \sum_{j=1}^{m} \frac{z_j(R_{tj}-R_j)}{2M}.$$

Then, by introducing auxiliary variables $a_i^+, a_i^-, i = 1, 2, \cdots, n+1$, and ξ_t^+, ξ_t^-, $t = 1, 2, \cdots, T$, such that

$$a_i^+ + a_i^- = |x_i - x_i^0|,$$
$$a_i^+ - a_i^- = x_i - x_i^0,$$
$$a_i^+ \geq 0, \ a_i^- \geq 0, \ i = 1, 2, \cdots, n+1,$$
$$\xi_t^+ + \xi_t^- = \left|\sum_{i=1}^{n} \frac{(r_{ti}-r_i)x_i}{2} + \sum_{j=1}^{m} \frac{z_j(R_{tj}-R_j)}{2M}\right|,$$
$$\xi_t^+ - \xi_t^- = \sum_{i=1}^{n} \frac{(r_{ti}-r_i)x_i}{2} + \sum_{j=1}^{m} \frac{z_j(R_{tj}-R_j)}{2M},$$
$$\xi_t^+ \geq 0, \ \xi_t^- \geq 0, \ t = 1, 2, \cdots, T,$$

we may consider the following bi-objective mixed-integer linear programming problem:

(BILP) $\max f(x,z) = \sum_{i=1}^{n+1}\sum_{t=1}^{T} p_t r_{it} x_i M + \sum_{j=1}^{m}\sum_{t=1}^{T} p_t R_{jt} z_j - \sum_{i=1}^{n+1} k_i M(a_i^+ + a_i^-)$

$\min w(x,z) = \sum_{t=1}^{T} 2p_t \xi_t^-$

s.t. $a_i^+ - a_i^- = x_i - x_i^0, \ i = 1, 2, \cdots, n+1,$
$\xi_t^+ - \xi_t^- = \sum_{i=1}^{n} \frac{(r_{ti}-r_i)x_i}{2} + \sum_{j=1}^{m} \frac{z_j(R_{tj}-R_j)}{2M}, \ t = 1, 2, \cdots, T,$
$\sum_{j=1}^{m} B_j z_j \leq B,$
$\sum_{i=1}^{n+1} x_i M \leq S,$
$\sum_{j=1}^{m} B_j z_j + \sum_{i=1}^{n+1} x_i M \leq M,$
$x_i \geq 0, \ i = 1, 2, \cdots, n+1,$
$a_i^+ \geq 0, \ a_i^- \geq 0, \ i = 1, 2, \cdots, n+1,$
$\xi_t^+ \geq 0, \ \xi_t^- \geq 0, \ t = 1, 2, \cdots, T,$
$z_j = \{0, 1\}, \ j = 1, 2, \cdots, m.$

It is not difficult to see that (BILP) is equivalent to (BOP). Thus the investor may determine his/her investment strategies by computing efficient solutions of (BILP).

3   Fuzzy Mixed Asset Portfolio Selection Model

In an investment, the knowledge and experience of experts are very important in an investor's decision-making. Based on experts' knowledge, the investor may decide his/her levels of aspiration for the expected return and the risk of mixed asset portfolio. In [8], Watada employed a non-linear S shape membership function, to express aspiration levels of expected return and of risk which the investor would expect and proposed a fuzzy active portfolio selection model. The S shape membership function is given by:

$$f(x) = \frac{1}{1 + \exp(-\alpha x)}.$$

In the bi-objective programming model of mixed asset portfolio selection proposed in Section 2, the two objectives, the expected return and the risk, are considered. Since the expected return and the risk are vague and uncertain, we use the non-linear S shape membership functions proposed by Watada to express the aspiration levels of the expected return and the risk.

The membership function of the expected return is given by

$$\mu_f(x,z) = \frac{1}{1 + \exp\left(-\alpha_f \left(f(x,z) - f_M\right)\right)},$$

where f_M is the mid-point where the membership function value is 0.5 and α_f can be given by the investor based on his/her own degree of satisfaction for the expected return.

The membership function of the risk is given by

$$\mu_w(x,z) = \frac{1}{1 + \exp(\alpha_w(w(x,z) - w_M))},$$

where w_M is the mid-point where the membership function value is 0.5 and α_w can be given by the investor based on his/her own degree of satisfaction regarding the level of risk.

According to Bellman and Zadeh's maximization principle [1], we can define

$$\lambda = \min\left\{\mu_f(x,z), \mu_w(x,z)\right\}.$$

The fuzzy mixed asset portfolio selection problem can be formulated as follows:

(FP)   max λ
   s.t. $\mu_f(x,z) \geq \lambda$,
      $\mu_w(x,z) \geq \lambda$,
      and all constraints of (BILP).

Let $\eta = \log\frac{1}{1-\lambda}$, then $\lambda = \frac{1}{1+\exp(-\eta)}$. The logistic function is monotonously increasing, so maximizing λ makes η maximize. Therefore, the above problem can be transformed to an equivalent problem as follows:

(FLP) max η
s.t. $\alpha_f (f(x,z) - f_M) - \eta \geq 0$,
$\alpha_w (w(x,z) - w_M) + \eta \leq 0$,
and all constraints of (BILP),

where α_f and α_w are parameters which can be given by the investor based on his/her own degree of satisfaction regarding the expected return and the risk.

(FLP) is a standard linear programming problem. One can use one of several algorithms of linear programming to solve it efficiently, for example, the simplex method.

4 Numerical Example

Assume that the total amount of assets M owned by an investor is \$300000. Originally, all the assets are invested in the risk-free security S_6. The rate r_6 of return of the risk-free security is 1.5%. Generally, the return of the risk-free security is less than those of risky securities and projects. To obtain more profits, the investor reallocates his/her wealth among five risky securities $S_1, \cdots, S_5$, five projects $P_1, \cdots, P_5$ and the risk-free security S_6, where the rate $k_i, i = 1, 2, \cdots, 6$ of transaction costs is 0.4% for all securities.

Assume that there are eight states of business environment in the future. Possible rates of returns on the five risky securities in these states and the corresponding state probabilities are listed in Table 1. Possible net returns on the five projects in these states and the corresponding state probabilities are listed in Table 2. The capital budgets of projects are listed in Table 3. The investor may give the values of the maximum amounts of the investment, B and S, devoted to projects component and securities component according to his/her investment preference. In the following we examine a case.

Table 1. Possible rates of returns and the expected rates of returns on the securities

State t	p_t	r_{t1}	r_{t2}	r_{t3}	r_{t4}	r_{t5}
1	0.100	-0.089	-0.007	-0.020	-0.011	-0.022
2	0.120	-0.042	0.043	0.036	-0.117	-0.053
3	0.120	0.120	0.047	0.128	-0.054	0.008
4	0.125	-0.062	-0.126	-0.090	0.109	0.057
5	0.125	0.147	0.230	-0.018	0.368	0.124
6	0.130	0.210	0.640	0.271	-0.135	0.277
7	0.130	0.011	-0.053	0.047	0.060	-0.105
8	0.150	0.005	-0.051	-0.017	0.014	-0.060
Expected Return (r_i)		0.041	0.092	0.043	0.030	0.028

Suppose that the maximum amount S of the investment devoted to securities component is \$240000 and the maximum amount B of the investment devoted to projects component is \$150000. We give the values of f_M and w_M, i.e., $f_M =$

Table 2. Possible net returns and the expected net returns on the projects

State t	p_t	R_{t1}	R_{t2}	R_{t3}	R_{t4}	R_{t5}
1	0.100	$20000	-$20000	$2000	-$10000	$20000
2	0.120	$0	$10000	-$10000	$20000	-$20000
3	0.120	$40000	-$15000	$10000	-$20000	$10000
4	0.125	-$10000	$35000	$40000	$40000	$30000
5	0.125	-$5000	$40000	-$15000	$60000	$40000
6	0.130	-$15000	-$18000	$50000	-$15000	-$10000
7	0.130	$15000	$20000	-$8000	$10000	-$40000
8	0.150	$10000	$8000	$16000	-$30000	-$25000
Expected Return (R_j)		$6425	$8235	$10625	$10000	$11875

Table 3. The capital budgets of projects

B_1	B_2	B_3	B_4	B_5
$40000	$50000	$80000	$100000	$120000

Table 4. Membership grade λ, obtained risk and obtained expected return

λ	η	α_f	α_w	obtained risk	obtained expected return
0.999	14.0529	600	800	0.0146	0.0629
1.000	16.0195	500	1000	0.0215	0.0720
1.000	16.2212	400	1200	0.0240	0.0790

Table 5. The allocations of three optimal portfolios and the corresponding returns

Portfolio	x_1	x_2	x_3	x_4	x_5	x_6
Portfolio 1	0.0000	0.0316	0.2880	0.0000	0.3363	0.0440
Portfolio 2	0.0000	0.2343	0.0505	0.0000	0.3105	0.1048
Portfolio 3	0.0000	0.3862	0.0000	0.0000	0.1743	0.1395

Portfolio	z_1	z_2	z_3	z_4	z_5
Portfolio 1	1	1	0	0	0
Portfolio 2	1	1	0	0	0
Portfolio 3	1	1	0	0	0

$0.04, w_M = 0.0375$. In the example, we give three different values of parameters α_f and α_w. Using the above data, we computed satisfactory investment strategies by solving (FLP). All computations were carried out on a WINDOWS PC using the LINDO solver. The membership grade, the obtained risk and the obtained return are listed in Table 4. The detailed allocations of three optimal portfolios are shown in Table 5.

Since it is possible that the non-linear S shape membership function changes its shape according to the values of the parameters, the non-linear membership function can reflect the investor's mind accurately and suitably. From the above results, we can find that we get the different investment strategies by solving (FLP) in which the different values of the parameters α_f and α_w are given. Through choosing the values of the parameters α_r and α_w according to the investor's frame of mind, the investor may get a favorite investment strategy.

5 Conclusion

In today's extremely competitive business environment, investors have already invested in various asset classes to keep their competitive advantage. The mixed asset portfolio increases the investors' benefit opportunities. Regarding the expected return and the risk as two objective functions, we have proposed a bi-objective programming model for the mixed asset portfolio selection problem with transaction costs. Furthermore, investors' vague aspiration levels for the return and the risk are considered as fuzzy numbers. Based on fuzzy decision theory, we have proposed a fuzzy mixed projects and securities portfolio selection model. The computation results of the numerical example show that the proposed model can generate a favorite mixed asset portfolio strategy according to the investor's satisfactory degree.

References

1. Bellman, R., Zadeh, L.A.: Decision Making in a Fuzzy Environment. Management Science 17 (1970) 141–164.
2. Byrne, P., Lee, S.: The place of property in an Australian multi-asset portfolio: a comparison of MPT and MAD optimisation. Australian Land Economics Review 5 (1999) 21–28.
3. Coffin, M.A., Taylor III, B.W.: Multiple criteria R&D project selection and scheduling using fuzzy logic. Computers and Operations Research 23 (1996) 207–220.
4. Konno, H., Yamazaki, H.: Mean Absolute Portfolio Optimization Model and Its Application to Tokyo Stock Market. Management Science 37(5) (1991) 519–531.
5. Markowitz, H.M.: Portfolio Selection. Journal of Finance 7 (1952) 77–91.
6. Reyck, B.D., Degraeve, Z., Gustafsson, J.: Project valuation in mixed asset portfolio selection. Working Paper of Systems Analysis Laboratory, Helsinki University of Technology, June 2003.
7. Ringuesta, J.L., Graves, S.B., Case, R.H.: Mean-Gini analysis in R&D portfolio selection. European Journal of Operational Research 154 (2004) 157–169.
8. Watada, J.: Fuzzy Portfolio Model for Decision Making in Investment. In: Yoshida, Y. (eds.): Dynamical Asspects in Fuzzy Decision Making. Physica-Verlag, Heidelberg (2001) 141–162.

Contract Net Protocol Using Fuzzy Case Based Reasoning

Wunan Wan, Xiaojing Wang, and Yang Liu

Chengdu Institute of Computer Applications, Chinese Academy of Sciences,
610041Chengdu, China
{nan_wwn}@hotmail.com

Abstract. This paper describes a new communication load reduction method ontask negotiation with Contract Net Protocol (CNP) in a multi-agent system. For coordination the agents, the CNP is widely used during the problem solving. But its performance degrades drastically when communicating agents and tasks increase. In order to overcome this problem, Fuzzy Case Based Reasoning (FCBR) is used in CNP, it can reduce the scope of competitors and communication load on task negotiation, make a distinct improvement to the efficiency. At the same time, in FCBR system a new cases maintaining method based on time-serial is proposed in order to ensure a better cooperation between agents and adapt to higher flexibility of system. Experimental results show the performance of the protocol based on FCBR improves significantly.

1 Introduction

Negotiation between intelligent agents is one of the fundamental issues of research in Distributed Artificial Intelligence (DAI). In the last several years, the contract net protocol (CNP) [1] is a widely used coordination mechanism in multi agent systems. Most task negotiation systems with *CNP* have run into message congestion problem, thus this limits its usability in a large scale multi agent system. It is observed that a high proportion of the negotiation communication load comes from the first step (announcement broadcasting of tasks). Methods to solve the problem include focused addressing earliest [1] and audience restrictions [2]. Later, Takuya [3] presented an attempt to extract useful knowledge by Case-based reasoning (CBR).By the method, the tasks can be assigned more directly in order to avoid announcing too much message. The same idea was adopted by Umesh Deshapande [4]. But his method does not use CBR, but uses previous instances, called Instance Based Learning (IBL) mechanism. It is well-known that there are vagueness and uncertainty in CBR. Especially, when dealing with the similarity assessment it is difficult to find the case. And learning makes the case base expand quickly which enables the search time become longer and longer. Thus, tradition *CBR* technique is not completely suitable to CNP. Recently, a fuzzy logic based methodology for knowledge acquisition is developed in *CBR* systems [5]. Thus, this paper proposes *CNP* based on *FCBR*, fuzzy logic method and the concept time are introduced into case definition, case base maintenance algorithms. This will help to make negotiation more efficient between agents, and make communication load lower.

The rest of the paper is organized as follows. Section 2 introduces Contract Net Model based on *FCBR*. Section 3 illustrates Task Negotiator Process with *CNP* based on *FCBR*. The experimental results are presented in Section 4. Finally, Section 5 concludes the paper.

2 Contract Net Model Based on FCBR

CBR is a method of making use of past experience to solve newly encountered problems. When a new problem appears, the CBR system retrieves from the case base to find similar candidate cases to the new problem, and the best of candidate cases will be modified to fit the new problem. Fuzzy theory is studied and applied in the all process of many CBR systems [6]. They may convert numerical feature into fuzzy term and have greater flexibility in the retrieval candidate cases, inference cases. We use CBR and fuzzy logic techniques into case representation, similarity-measuring and cases maintaining.

2.1 Case Definition

Definition of cases is significant which effects on reasoning [5]. In multi agent systems, agents negotiate tasks with CNP by message interaction, thus useful knowledge may be acquired by messages between agents, some of these are often vague and uncertain, thus fuzzy logic is employed in cases definition. And negotiation message has greater relation with time series, the case description with time series may examine the developing process of objects. According to CNP, there are descriptions of four message types [3]. Next, we define a case according to message description. Time concept is introduced into case representation. A case C is represented as a set of nine parts:

Definition 1: $C=(C_{id}, Task, Announce, Bid, Award, Report, v, \mu, C_0)$
 C_{id} the id number of case;
 $Task = (A_1 : V_1, A_2 : V_2, ..., A_m : V_m, TASK\ ID : tasked\ )$
 Where A_i (i =1, 2, ..., m) is i^{th} attribute, and V_i (i =1, 2, ..., m) is the value corresponding to A_i, taskid is an identifier of the task.
 $Announce = (ann_1, ann_2, ..., ann_m)$; $Bid = (bid_1, bid_2, ..., bid_m)$;
 $Award = (award_1, award_2, ..., award_m)$; $Report = (rep_1, rep_2, ..., rep_m)$;
 v forget coefficient; μ time coefficient;
 v and μ is two parameters, they are defined as the relation between the current task and the cases.

Definition 2: $C_0=\{ ann_1, ann_2, ...ann_m\}$
C_0 is *is a special case, other attributes id NULL except a attribute of Announce.* Attribute-values of Announce are some managers which take part *in* system recently and whose capabilities change as go on.

2.2 Cases Maintaining

In a multi-agent system, learning makes the case base expand quickly, which enable the search time becoming longer and longer. At this time, good practical maintenance strategies are crucial to sustain and improve the efficiency and solution quality of *CBR* systems as their case-bases grow and their tasks or environments change over long-term use. In general, there are some methods about maintenance [6]. We discuss emphatically new deletion method during maintaining cases base. The below is the full description of the proposed method. Case deletion algorithms:

Step1. Let threshold of Case base is the length K
Step2. Let threshold of Utility α;
Step3. Let $C = C_1, C_2, ..., C_n$ be the set of n past cases; Compute Utility according to the function is defined by Minton[7]as following:

$$Utility = (ApplicationFreq * AverageSavings) - Match - Cost; \quad (1)$$

where ApplicationFreq is $1/\nu$, AverageSavings is $1/\mu$;
Step4. Order the value of the function Utility (descending sequence)in the list, which constraint the length K;
Step5. Return the result.

3 Task Negotiation Process with CNP Based on FCBR

When UA interfaces with the user and receives the new tasks for the multi-agent system, it submits the new tasks to a manager. Fig.1 shows the detail task negotiator process.

Negotiator: Negotiator is responsible for establishing the negotiation with agents and controlling task's execution or Contractors who perform tasks.

Reasoner: Reasoner selects similar case (according to *KNN* algorithm),and Reasoner calculates"*suitabilities* " of the agents from the similar cases[3]. Then the Reasoner chooses some best contracts according to *suitabilities*, and returns a list of contractors to the Negotiator.

Knowledge Manager: Knowledge Manager in Reasoner makes case according the task negotiation processes. Knowledge Manager maintains case base according to maintains case base according to Case deletion algorithm (section 2.2), and can supervise work of the Reasoner and the Negotiator.

4 Experimental Results

To evaluate performance of *CNP* based on *FCBR*, we tested these behaviors with 2 manager and 8 bidders when moving the tasks from 4 to 30(cf. Figure 2). The experiments show that the time performance of the protocol proposed based on *FCBR* is superior to other methods. It is proved that *FCBR* can provide more suitable agents in order to spent less time and reduce invalid negotiation processes.

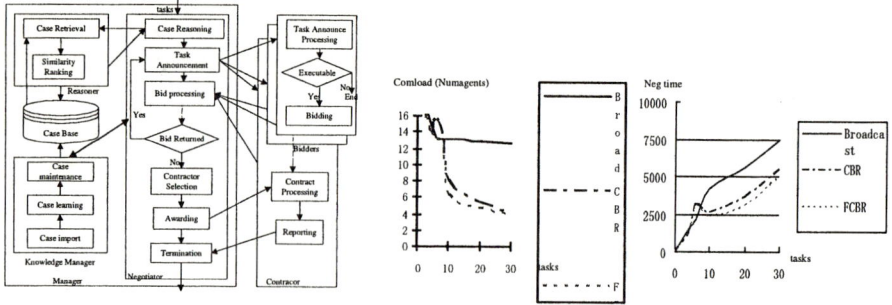

Fig. 1. Task negotiation process

Fig. 2. Communication load and negotiation time

5 Conclusion and Future Work

In this article, we have shown the use of *CNP* solving agent coordination problems by task allocation. Due to the broadcast of task announcement, communication load is intensive. A task negotiation model with the low communication load for multi-agent system, i.e. *CNP* based on *FCBR* is presented. The performance of a system using *CNP* based on can be improved.

References

1. Reid G,Smith.: The Contract Net Protocol :High-Level Communication and Control in a Distributed Problem Solver. IEEE Tran.on Computers.29 (1980) 1104–1113
2. Sandholm, T.,Lesser,V.:Issues in automated negotiation and electronic commerce: Extending the contract net framework. AAAI Press, USA.(1995)328-335
3. T. Ohko, K. Hiraki.:LEMMING:A Learning system for Multi Robot Environments. IEEE/RSJ Conference. Japan July. (1993) 26-30
4. Umesh Deshpande.: Performance Improvement of the Contract Net Protocol Using Instance Based Learning. IWDC 2003, LNCS 2918(2003)290-299
5. Kraslaw skia , Pedrycz W. :Fuzzy neural network as instance generator for case-based reasoning system. Neural Computing Applications.8 (1999) 106-113
6. Azuaje F et al.: Discovering relevance know ledge in data: a growing cell structure approach. IEEE Tran. on Systems, Man and Cybernetics,June.30 (2000) 448-460
7. Minton S.:Qualitative results concerning the utility of explanation based learning. Artificial Intelligence.42 (1990)363-391

A Fuzzy Approach for Equilibrium Programming with Simulated Annealing Algorithm*

Jie Su[1], Junpeng Yuan[2], Qiang Han[1], and Jin Huang[2]

[1] School of Mathematics and System Science, Shandong University, Jinan, P.R. China
[2] School of Management & Economics,
Beijing Institute of Technology, Beijing, P.R. China
sujie@sdu.edu.cn

Abstract. Equilibrium programming (EP) is an active topic in mathematical programming. For the EP problem with the fuzzy optimal membership function in objectives and the penalty function in constraints, we propose a new fuzzy approach with simulated annealing algorithm to get the fuzzy optimal solution. The approach provides the decision-makers with a group of fuzzy optimal solutions on similarly fuzzy optimal grade, so that the equilibrium decision can be obtained easily and more reasonably. The numerical simulation of example demonstrates the feasibility and effectiveness of the approach.

1 Introduction

Recently, the research for equilibrium point of Equilibrium programming (EP) has become a favorite topic [1, 2, 3], as general economic equilibrium theory was proposed [4]. Due to NP-hardness of EP, we study the fuzzy optimality of EP based on fuzzy theory [5]. Simulated annealing (SA) algorithm, as a kind of stochastic global optimization algorithm, has broad applications in optimization problems [6, 7].

The fuzzy optimal solution of EP is proposed and a homogeneous SA algorithm is put forward in Sec. 2. And computational results show the feasibility and efficiency of the SA algorithm in Sec. 3.

2 The Fuzzy Optimal Solution of EP

An equilibrium system is described as follows [1]: there are M factors in the system, the $i^{th}(i = 1, \ldots, M)$ factor's decision variable is $x^i \in \Re^{n_i}$, and the decision variable of system is $x = (x^i, x^{-i}) \in \Re^n$, where $x^{-i} = (x^1, \ldots, x^{i-1}, x^{i+1}, \ldots, x^M)$; for a given x^{-i}, the ith factor seeks x^i to optimizing EP as follows:

$$\max_{x^i} f^i(x^i, x^{-i})$$
$$\text{s.t. } g^i(x^i, x^{-i}) \geq 0, \quad (1)$$
$$h^i(x^i, x^{-i}) = 0,$$

---

* This work is supported by YuMiao Foundation of Beijing Institute of Technology.

where functions $f^i : \Re^n \to \Re^1$, $g^i : \Re^n \to \Re^{k_i}$ and $h^i : \Re^n \to \Re^{m_i}$.

Suppose the ith factor optimizes his own objective function satisfying all the behavior constrains of all M factors in the system, but does not consider the behalf of the other factors. Then his decision model is:

$$\max_x f^i(x^i, x^{-i})$$
$$\text{s.t. } g^i(x^i, x^{-i}) \geq 0, \ i = 1, \ldots, M; \qquad (2)$$
$$h^i(x^i, x^{-i}) = 0, \ i = 1, \ldots, M.$$

Denote the optimal solution of problem (2) by (x_0^i, x_0^{-i}).

In EP, the ith factor is not directly affected by the other factors' constrains, but affected indirectly via the value x^{-i}. So the optimal value for problem (2) is an upper bound of the equilibrium value for the ith factor. On the other hand, $(x_0^i, x_0^{-i})(i = 1, \ldots, M)$ can not be equal generally, for there are some difference and conflict among the factors in equilibrium system [8]. Then it will be worse if the ith factor only obeys the other factors' decisions, but not makes his own decision. So the minimum value of $f^j(x_0^j, x_0^{-j})$ for all $j \neq i$ is a lower bound of the equilibrium value for the ith factor, i.e.,

$$f_U^i = f^i(x_0^i, x_0^{-i}), \ f_L^i = \min_{j \neq i} f^j(x_0^j, x_0^{-j}); \ i = 1, \ldots, M.$$

Thus the ith factor's fuzzy membership function for solution (x^i, x^{-i}) can be defined as:

$$\mu^i(f^i(x^i, x^{-i})) = \begin{cases} 1, & \text{if } f^i(x^i, x^{-i}) \leq f_L^i; \\ \frac{f_U^i - f^i(x^i, x^{-i})}{f_U^i - f_L^i}, & \text{if } f_L^i < f^i(x^i, x^{-i}) < f_U^i; \\ 0, & \text{if } f^i(x^i, x^{-i}) \geq f_U^i. \end{cases} \qquad (3)$$

Consequently, the fuzzy optimal solution of EP is defined as the solution satisfying all constraints and maximizing the minimum value of membership grade μ^i for all factors. So the model for the fuzzy optimal solution of EP is:

$$\max_{x,\mu} \mu$$
$$\text{s.t. } \mu^i(f^i(x^i, x^{-i})) \geq \mu, \ i = 1, \ldots, M; \qquad (4)$$
$$g^i(x^i, x^{-i}) \geq 0, \ i = 1, \ldots, M;$$
$$h^i(x^i, x^{-i}) = 0, \ i = 1, \ldots, M.$$

Note that there are multi-constraints in problem (4), we introduce the penalty function method to solve it. The penalty function is chosen as

$$p(x, \mu) = P \sum_{i=1}^{M} \{\min\{\varphi^i(x^i, x^{-i}, \mu), 0\} + \min\{g^i(x^i, x^{-i}), 0\} + |h^i(x^i, x^{-i})|\},$$

where P is the penalty parameter and $\varphi^i(x^i, x^{-i}, \mu) = \mu^i(f^i(x^i, x^{-i})) - \mu$. Then the generalized programming of problem (4) with penalty parameter P is

$$\max_{x,\mu} F(x, \mu) = \mu - p(x, \mu). \qquad (5)$$

3 The SA Algorithm for Fuzzy Optimal Solution of EP

We propose a homogeneous SA algorithm to obtain the fuzzy optimal solution of EP, by suitably designing search operators.

3.1 The Search Operators of the SA Algorithm

A new solution $z' = (x', \mu')$ is obtained from the current solution $z = (x, \mu)$ based on the neighborhood function which is created by Gaussian distribution disturbance [9], i.e., $z' = z + \eta \cdot \xi$, where random variable $\xi \sim N(0,1)$ and η is the step length.

The initial temperature is chosen as $t_0 = \frac{F_{min} - F_{max}}{\ln p_0}$, where $p_0 \in (0,1)$ is the initial acceptance probability, F_{max} and F_{min} are the maximum and minimum objective values of initial solutions, respectively.

The temperature descending coefficient is $\alpha (0 \leq \alpha \leq 1)$, and $t_{r+1} := \alpha t_r$.

The circular number in the same temperature is a fixed positive integer s which is decided by the size of problem.

There are two stopping criteria in the SA algorithm. One is Zero Method–that is, the algorithm stops if temperature $t_r \leq \varepsilon$, where ε is a given small positive number. The other is Immovable Rule–the algorithm stops if the current best solution cannot be improved within Q consecutive generations.

3.2 The Procedure of the SA Algorithm

The procedure of the SA algorithm is described as follows:

Step 1 Set up the fuzzy membership function for the solution.

Step 2 Initialize the search parameters. Produce the initial solutions randomly and choose the optimal one to be the current solution z. Initialize the best solution $z^* := z$. Let $k := 0$ and $r := 0$.

Step 3 Produce new solution z' from the current solution z, and let $k := k+1$.

Step 4 If $k = s$, go to Step 6; else, go to Step 5.

Step 5 If $\exp[(F_{z'} - F_z)/t_r] > random(0,1)$ or $F_{z'} \geq F_z$, let $z := z'$; and farther if $F_{z'} > F_{z^*}$, let $z^* := z'$. Else, return to Step 3.

Step 6 If z^* does not change in Q consecutive generations, the algorithm stops; else, let $t_{r+1} := \alpha t_r$ and $r := r + 1$, go to Step 7.

Step 7 If $t_r \leq \varepsilon$, the algorithm stops; else, return to Step 3.

4 Computational Results and Discussion

An example of EP with three factors is given as follows:

$$\max_{x_1} f_1(x_1, x_2, x_3) = 2x_1 + x_2 - x_3$$
$$\text{s.t. } x_1^2 + x_2^2 + x_3^3 \leq 20,$$
$$x_1 \geq 0.$$

Table 1. The results of simulations of the SA algorithm

Simulation	Fuzzy best solution			Membership of fuzzy
order	x_1	x_2	x_3	best solution μ
No.1	2.739295	3.422193	0.779938	0.812537
No.2	2.970856	3.475086	1.466626	0.824372
No.3	0.859941	3.575023	1.444010	0.933182
No.4	1.298875	1.632071	5.038921	0.957633

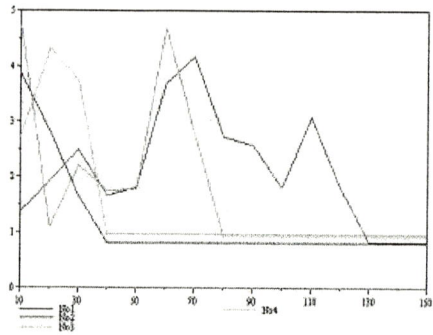

Fig. 1. The iteration results of SA algorithm: abscissa expresses the times of temperature descending, ordinate expresses the fuzzy membership of optimal solution

$$\max_{x_2} f_2(x_1, x_2, x_3) = (x_1 - 2)^2 + (x_2 - 1)^2 + x_3^2$$
$$\text{s.t. } x_1 x_2 x_3 \geq 16,$$
$$x_2 \geq 0.$$

$$\max_{x_3} f_3(x_1, x_2, x_3) = (x_1 - 3)(4 - x_2)(x_3 + 1)$$
$$\text{s.t. } x_1 - 2x_2 + x_3 \leq 4,$$
$$x_3 \geq 0.$$

The search operators are chosen as $t_0 = 100$, $\varepsilon = 2$, $\alpha = 0.9$, $s = 10$, $Q = 100$, $P = 150$, $L = 50$ and $p_0 = 0.1$, then a series of numerical simulations, using the SA algorithm in this paper, are shown in Table 1 and Fig. 1.

From the simulation results, we can get the advantages of the SA algorithm:

1. Optimality. All the values of fuzzy membership grade μ in the results in Table 1 exceed 0.8, which shows that the algorithm can obtain the optimal solutions for all factors in equilibrium system with high satisfactoriness.
2. Multi-solutions. The different simulation results show that the algorithm can provide the factors with a group of approximate fuzzy optimal solutions on similarly fuzzy optimal grade. So it is easier and more reasonable to achieve the equilibrium decision for the factors.

3. Robustness. Although the initial solutions and search operators are stochastic (as shown in Fig. 1), the best solutions are consistent on a similarly fuzzy optimal grade. It illuminates that the SA algorithm has preferable robustness on the initial value.
4. Speediness. It runs in less than 0.6 second by Visual Basic program on Pentium III of the SA algorithm for the examples. Furthermore, in Fig. 1, the best results are achieved within 80 times of temperature descending averagely, so the average rate of convergence is speedy.

So, the SA algorithm in this paper is feasible and effective to find the fuzzy optimal solution of EP.

5 Conclusion and Future Research

In this paper, a new fuzzy approach with SA algorithm to get the fuzzy optimal solution for EP is proposed, by brought forward the fuzzy optimal membership function for objectives and the penalty function for constraints of EP. And the numerical simulation of example demonstrates the effectiveness and feasibility of the algorithm.

For future research, we can try to find the solution of EP by hybrid optimal algorithms, such as GA algorithm incorporated with SA, neural network with hybrid SA algorithm, etc. Furthermore, we can study the property of the fuzzy optimal solution for EP.

References

1. Zangwill, W.I., Garcia, C.B.: Equilibrium programming: the following approach and dynamics. Math. Programming **21** (1981) 262–289
2. Flam, S.D., Antipin, A.S.: Equilibrium programming using proximal-like algorithms. Math. Programming. **78** (1997) 29–41
3. Antipin, A.S.: Methods for bilinear equilibrium and game problems. Internat. Conf. on Optimization and Optimal Control. (2002) 77–84
4. Walras, L.: Elements d'economie politique pure, Trans. by William Jaffe as Elements of pure economics Homewood, III: Richard D. Irwin (1954)
5. Wang, P., Li, H.: Fuzzy system theory and fuzzy computer (in Chinese). Publishing Company of Science, Beijing (1996)
6. Feng, T., Cheng, Y., Ching, C.: Applying the genetic approach to simulated annealing in solving some NP-hard problems. IEEE Trans. on SMC. **23** (1993) 1752–1767
7. Kang, L., Xie, Y., You, S., et al.: Simulated annealing algorithm (in Chinese). Publishing Company of Science, Beijing (1998)
8. Ma, J., Cui, Y., Su, J.: The optimal models of complex decision systems. Internat. Conf. on Modeling and Simulating of Complex system. (2003) 146–151
9. Wang, Z., Zhang, T., Wang, H.: Simulated annealing algorithm based on chaotic variable (in Chinese). Control and Decision. **14** (1999) 381–384

Image Processing Application with a TSK Fuzzy Model

Perfecto Mariño, Vicente Pastoriza,
Miguel Santamaría, and Emilio Martínez

University of Vigo, Spain

Abstract. The authors have been involved in developing an automated inspection system, based on machine vision, to improve the repair coating quality control (RCQ control) in can ends of metal containers for fish food. The RCQ of each end is assesed estimating its average repair coating quality (ARCQ). In this work we present a fuzzy model building to make the acceptance/rejection decision for each can end from the information obtained by the vision system. In addition it is interesting to note that such model could be interpreted and supplemented by process operators. In order to achieve such aims, we use a fuzzy model due to its ability to favour the interpretability for many applications. Firstly, the easy open can end manufacturing process, and the current, conventional method for quality control of easy open can end repair coating, are described. Then, we show the machine vision system operations. After that, the fuzzy modeling, results obtained and their discussion are presented. Finally, concluding remarks are stated.

1 Introduction

In the food canning sector, in the easy open can end manufacturing process, to guarantee the desired product lifespan, a manual, nondestructive testing (NDT) procedure is carried out. Due to the high processing rate, only an small part of each lot is verified. Therefore, it is important to develop an automated inspection system to improve the easy open can end repair coating quality control process (all can ends are checked, and inline). It is for this reason that we had been involved in the design and implementation of an inline, automated machine vision system to evaluate the repair coating of the can ends, that we have named end repair coating inspection system (ERCIS). In this work we explore the use of fuzzy models to make the acceptance/rejection (A/R) decision for each can end, due to its ability to favour the interpretability for many applications [4]. A Takagi-Sugeno-Kang (TSK) fuzzy model is developed using a neuro-fuzzy modeling. The remainder of this paper is organized as follows: in the next section we provide an overview of the easy open can end manufacturing process and its repair coating quality control process. Section 3 shows the machine vision system operations. Then, Section 4 describes the fuzzy modeling. The results obtained and their discussion are presented in Section 5. Finally, we state concluding remarks in Section 6.

2 Background

A can consists of can body and can end, which are made from aluminum or steel. There are cans of different shapes and sizes. Interior and exterior coatings are applied to protect the can from corrosion. Metal cans are used to contain a wide variety of products, including beverages, foods, aerosol products, paints, and many other contents.

Can ends, from henceforth ends, are used for all type of cans and can be standard or easy open. In this paper, we study a specific end format named 1/4 Club, with an easy-open tab in one of its corners (Fig. 1).

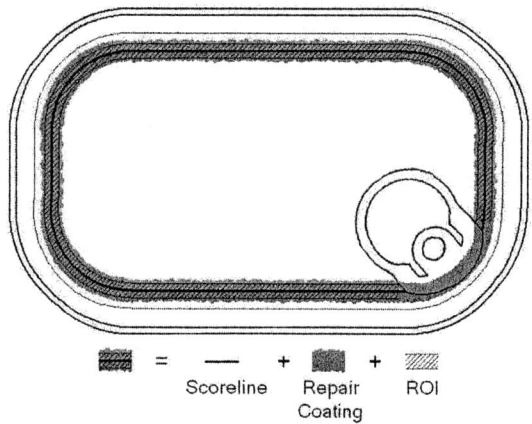

Fig. 1. Repair coating in 1/4 Club easy open can end

2.1 Easy Open End Manufacturing Process

Easy open ends are made from pre-coated metal coils or sheets. Ends are stamped from coil or sheets in a press. After stamping, the ends are scored in a predefined geometric shape (scoreline) intended to ease the end opening. Finally, a tab is attached to form an easy open end. These steps are performed after the end piece has been coated and therefore damage the coating, especially on the scoreline. Repair coating, which has a fluorescent pigment, is applied after these steps on the required area to restore the integrity of the coating.

2.2 Easy Open End Repair Coating Quality Control

Presently a visual inspection of the easy open ends is carried out [1,10], where the inspectors assess the repair coating on the scoreline. This visual inspection is a manual and NDT procedure. To assist in this end inspection the repair coating has a fluorescent pigment that stands out as a bright light blue when excited by an ultraviolet black light, while the background color remains unchanged. This inspection is based on a statistical sampling. It is because of this and the high rate of the repair coating process (100 to 500 ends per minute, depending on the end format) that only a small part of each lot is verified. Therefore, defective can ends can be sent to the canneries.

3 Machine Vision System Operations

The vision algorithm running on ERCIS inspects the repair coating quality (RCQ) on each end. The vision algorithm has two parts: one offline and other inline. The flowchart of ERCIS is shown in Fig. 2.

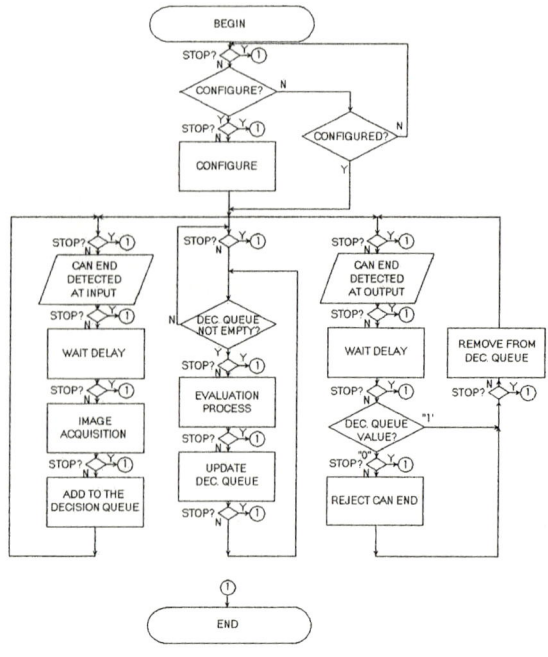

Fig. 2. Flowchart of ERCIS

3.1 Offline

Before the ERCIS begins the continuous or online inspection of ends is necessary to configure or reconfigure a series of parameters that will be used later in the inline processing.

Time Delay Configuration. An offline adjustment can be necessary to set the delay, input delay time (IDT), between the sensor at the ERCIS input detecting the end and the end reaching the camera, and the delay, output delay time (ODT), between the ERCIS output sensor detecting the end and the end arriving next to the rejection system.

Region of Interest Definition. the scoreline is enclosed in the region of interest (ROI) (Fig. 1). Each quadrant of this ROI, and by simmetry, the whole ROI is geometrically modeled by the parameters b and r (Fig. 3). Besides these parameters it is necessary to add an e width to the ROI (Fig. 3).

$$ROI_{\text{HOR}} = f(b, r, e) \tag{1}$$

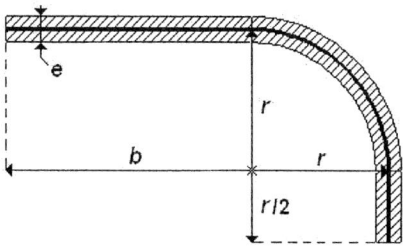

Fig. 3. End ROI quadrant geometrical model

Parameters will be adjusted before the line continuous working, and they depend directly on the distance between and and camera (working distance).

Parameter Configuration to Decide the End Rejection. In order to take the A/R decision for each end during the inline processing is necessary to offline configure the Minimum Average Repair Coating Quality (MARCQ) of the end to not reject it. The object of this parameter can be seen on Subsection 3.2.

3.2 Inline

The end continuous inspection can only start after the offline parameter configuration. This inline process has for each end the following sequence of steps:

Acquisition. This process undertakes the detection of the ends and acquisition of images of them. It is divided into the subprocesses:

- End detection before the ERCIS: A sensor located before the ERCIS warns said system that an end for inspection is approaching.
- End acquisition: A tiem IDT after the end has been detectedthe camera acquires the image and the decision queue size is incremented. The end is totally included in the image if, offline, IDT and working distance have been properly adjusted..

Evaluation. If the decision queue is not empty then an image has already been acquired and can be processed. This process has the following steps:

- End center location and end orientation: Equation 1 defines de geometric shape of the ROI given that the end is centered on the image and its mayor axis is parallel to the image horizontal axis. Nevertheless, during an usual inline working, the end can be not centered and show a light inclination with respect to the camera at the moment of image capture. It is necessary, though, to determine the center C and inclination α of the end in the image to rightly position the ROI (Fig. 4 and (2) and (3)).

$$\left.\begin{array}{l} \tan \alpha_1 = \frac{y_A - y_B}{x_A - x_B} \\ \tan \alpha_2 = \frac{y_E - y_D}{x_E - x_D} \end{array}\right\} \Rightarrow \alpha = \frac{\alpha_1 + \alpha_2}{2} \quad (2)$$

$$\left.\begin{array}{l} F = \{x_A, y_A + \frac{y_D - y_A}{2}\} \\ G = \{x_B, y_B + \frac{y_E - y_B}{2}\} \\ C = \{x_H + \frac{x_I - x_H}{2}, y_H + \frac{y_I - y_H}{2}\} \end{array}\right\} \Rightarrow s \Rightarrow H, I \Rightarrow C \quad (3)$$

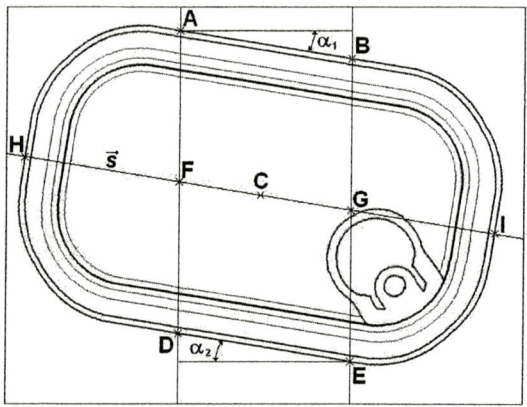

Fig. 4. End center location and end orientation parameters

Thus, we get the following expression:

$$ROI = g\left(ROI_{HOR}, C, \alpha\right) \quad (4)$$

- ROI rectification. The Region of Interest (ROI) is converted into a straight line strip to ease its analysis. This strip is a Look-up-Table (LUT) whose size is $n \times e$ pixels (Fig. 5). The length n, in which is divided the ROI perimeter, depends on the selected resolution. The rectification method employed selects for each one of $n \times e$ pixels the nearest 4-neighbour [3].

$$LUT = rectificate\left(ROI, resolution\right) \quad (5)$$

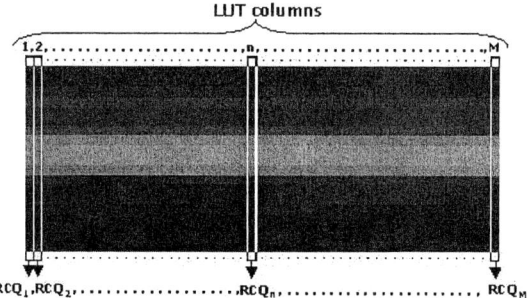

Fig. 5. RCQ_n of each LUT column

- Repair coating quality: The RCQ is assessed analyzing one-by-one all the n positions of the LUT by means of the model obtained in Section 4. The model gives an estimation of the RCQ at each position from several attributes of said position. These attributes are computed from each e-pixel group belonging to each one of the n positions. The three most influential attributes are the maximum, standard deviation, and center of mass (Section 4).
- End A/R decision: The average repair coating quality (ARCQ) is computed from the RCQ values at the n positions in which have been divided the ROI. Then, an end have to be rejected when the condition ARCQ<MARCQ is given (see the meaning of MARCQ in Subsection 3.1).
- Update the decision queue: The decision queue must be updated after making the A/R decision of each end.

Expulsion. This process undertakes the detection of the ends after the ERCIS and looks up the decision queue to expulse the end or not. This process is subdivided in:

- End detection after the ERCIS: A sensor placed after the ERCIS let the system know that an end is passing it by.
- End expulsion: A time lapse ODT after being detected the end by the sensor at the ERCIS output, if it has decided that the end is defective then it activates the expulsion system. ODT has to be well adjusted to be able to reject the end. After, the decision queue is decremented.

Stop. If stop signal is activated then the process goes to the flowchart end with independence of the current state of the process.

4 Modeling

In order to evaluate the RCQ on each one of n positions of the LUT, a set of a few attributes that contain most of the relevant information on each one of the n positions is studied. The 9 attributes computed from each e-pixel group of nth

LUT position are: Maximum pixel intensity (Max), Minimum pixel intensity, Mean pixel intensity is a measure of central tendency (location), Median pixel intensity is a measure of central tendency (location), Pixel intensity standard deviation (Std) is a measure of dispersion, Pixel intensity skewness is a measure of the asymmetry, Pixel intensity center of mass (CoM), Pixel intensity moment of inertia about an axis passing through the CoM, and Pixel intensity bisector.

A fuzzy inference system (FIS) [8,9,13], whose inputs are the selected attributes, will be used to evaluate the RCQ on each of the n positions. As an excessive number of inputs prevents the interpretability of the underlying model and increases the computational burden, we look for a model with a trade off between high accuracy and reduced number of inputs. We got a modeling problem with 9 candidate inputs and we want to find the 3 most influential inputs as the inputs of the model. We so can build 84 fuzzy models, each one with a different combination of 3 inputs. The proposed FIS model is a TSK inference system [11,12]. These models are suited for modeling non-linear systems by interpolating multiple linear models. The TSK model is designed with zero order (singleton values for each consequent), 3 of 9 attributes as inputs and RCQ as output. The TSK model is developed using the adaptive neuro-fuzzy inference system (ANFIS) algorithm [5,7].

We use a quick and straightforward way of neuro-fuzzy modeling input selection using ANFIS to improve the interpretability [6]. This input selection method is based on the hypothesis that the ANFIS model with smallest root mean squared error (RMSE) after one epoch of training has a greater potential of achieving a lower RMSE when given more epochs of training.

Representative input-output data set of the system should be selected to tune a model. We have only worked with a specific end format named 1/4 Club, with an easy-open tab in one of its corners (Fig. 1). We have selected a collection of 11 ends that agglutinate all possible end repair coating defects. The obtained LUT for each end has a length n of 702 positions, with a width e of 19 pixels. After removing instances with outlier values, the data set was reduced to 6669 entries. This data set is divided into training and testing sets of size 3335×19 and 3334×19 respectively. The testing set is used to determine when training should be terminated to prevent overfitting.

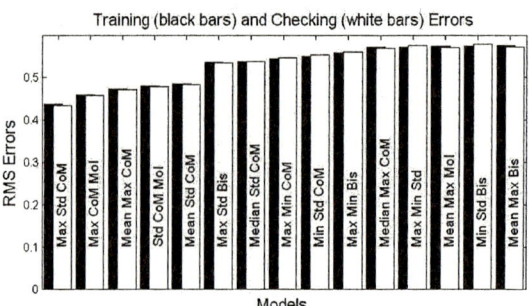

Fig. 6. The best fifteen 3-input fuzzy models for end RCQ prediction

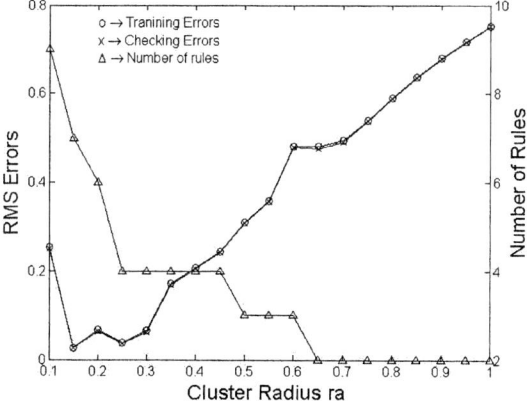

Fig. 7. Model size and prediction error as function of cluster radius

It has been selected grid partitioning as the ANFIS partition method. The best model after one epoch of training selects as input attributes the maximum (Max), the standard deviation (Std), and the center of mass (CoM) (Fig. 6). The problem is that this partitioning leads to a high number of rules, $2^3 = 8$ rules for each model.

In order to reduce the model complexity we use subtractive clustering [2] for the 3 inputs previously selected. The results, after one-epoch training, as a function of the cluster center's range of influence applying substractive clustering, are shown in Fig. 7. From that we have selected the model with range of influence 0.5 that has 3 rules that gives a RMSE of 0.3106 for training and 0.3091 for testing.

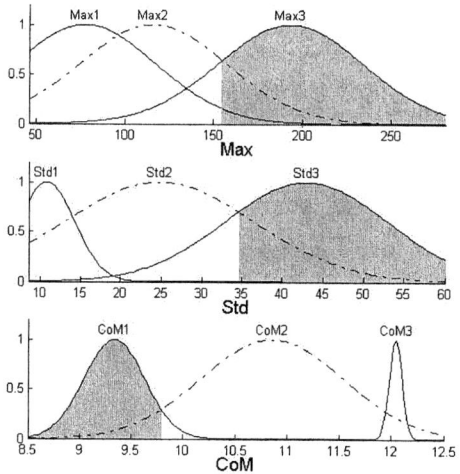

Fig. 8. Fuzzy model membership functions

Table 1. Fuzzy Models Rules

Rules	Repair Coating Quality
If Max is Max1 & Std is Std1 & CoM is CoM2	S1[5]* Defective
If Max is Max2 & Std is Std2 & CoM is CoM3	S1[5] Defective
If Max is Max3 & Std is Std3 & CoM is CoM1	S2[10] Acceptable

*S[m] = singleton (m=mean)

We can further refine said model performance applying extended ANFIS training. The final model obtained uses Max, Std, and CoM as model inputs, RCQ as output, and 3 rules (Table 1) to define relationships among inputs and output. The membership functions for each input feature are shown in Fig. 8 and singleton values for each consequent in Table 1.

This model gives a RMSE of 0.0102 for training and 0.0101 for testing, more similar values which indicate that there is no overfitting. Regarding the interpretability of the model and from its rules, is deduced that the RCQ at n^{th} LUT position is acceptable if and only if at said position, see Fig. 8 and Table 1, the maximum pixel intensity is higher than 150 and the standard deviation is higher than 35 and the center of mass is close to 9.5, which is the e-pixel group center. The interpretation of this is the following:

- The higher the maximum pixel intensity, the higher the lacquer quantity.
- As a defect region has little or no lacquer and is more uniformly distributed than an acceptable region, then the pixel intensity is less scattered (less Std) in defect regions.
- As the ROI is positioned in the way that the scoreline is at its center zone, and as an acceptable end has the highest lacquer level at scoreline, then at each one of n LUT positions the nearer of the e-pixel group center is the CoM, the better the RCQ.

5 Results

The RCQ of each end is assessed estimating its ARCQ. This average quality is computed from the RQ of the n positions in which has been divided the ROI, and where the RCQ at each n position is analyzed by means of the FIS obtained.

Figure 9 shows the ARCQ classification of the 11 ends previously used to tune the fuzzy model. The ends were sorted in descending order of ARCQ.

As each end have to be rejected when ARCQ<MARCQ is given (see the meaning of MARCQ in Subsection 3.1) then what ends are rejected will depend on the MARCQ value selected. For example, if MARCQ is 80 then all 11 ends are rejected. This flexibility to be able to modify the rejection threshold is an important property of the ERCIS.

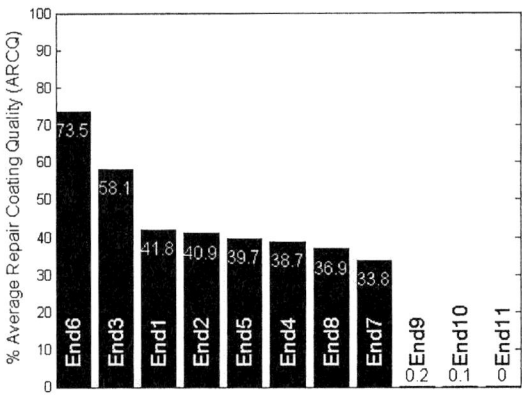

Fig. 9. ARCQ classification of the ends

But the most important result is that, with independence of the MARCQ selected, the ARCQ classification agrees with the one made by an expert human inspector.

6 Concluding Remarks

We have been involved in the implementation of a machine vision system to improve the repair coating quality control of the easy open can end manufacturing process. The system has the following properties:

- End classification in agreement with the one made by an expert human inspector.
- Flexibility to be able to modify the rejection threshold.
- Interpretability supplied to the operators in order to find out the failure causes and reduce mean time to repair (MTTR) during failures.
- Total inspection of 100% end production.

In spite of the fact that the end repair coating process of only one end format (1/4 Club) has been studied, as this process is common to all formats, it is reasonable to think that fuzzy models like the found model can be obtained to make the A/R decision for another end format. All this leads to the conclusion that is possible to design an inline, automated machine vision system, which only extracting the ARCQ from each end, makes a right A/R decision. ANFIS, the neuro-fuzzy modeling technique used to optimize the fuzzy model, provided excellent prediction accuracy. In the future we will study the existence of models that estimate the easy open can end repair coating process failure causes. Furthermore, we will research the application of coevolutionary genetic fuzzy modeling techniques that improve the interpretability without a significant loss of accuracy.

References

1. CFIA. Metal can defects; identification and classification manual. Technical report, Canadian Food Inspection Agency (CFIA) (1998).
2. Chiu, S.: Fuzzy model identification based on cluster estimation. *J. of Intelligent & Fuzzy Systems* **2(3)** (1994) 267–278.
3. González, R., Woods, R.: Digital Image Processing. Prentice Hall, Upper Saddle River, NJ, 2nd edition (2002).
4. Halgamuge, S. K., Wang, L., editors: Computational Intelligence for Modelling and Prediction Series: Studies in Computational Intelligence, Vol. 2. Springer, Berlin Heidelberg New York (2005).
5. Jang, J.: Anfis: Adaptive network based fuzzy inference system. *IEEE, Transactions on Systems, Man, and Cybernetics*, **23(3)** (1993) 665–685.
6. Jang, J.: Input selection for anfis learning. *Proceedings of the IEEE International Conference on Fuzzy Systems* (1996) 1493–1499.
7. Jang, J., Sun, C.: Neuro-fuzzy modeling and control. *The Proceedings of the IEEE* **83(3)** (1995) 378–406.
8. Klir, G., Yuan, B.: Fuzzy Sets and Fuzzy Logic: Theory and Applications. Prentice Hall, Upper Saddle River, NJ (1995).
9. Kosko, B.: Neural Networks and Fuzzy Systems: A Dynamical Systems Approach to Machine Intelligence. Prentice Hall, Englewood Cliffs, NJ (1992).
10. Lin, R., King, P., Johnston, M.: Examination of metal containers for integrity. In Merker, R., editor, *FDA's Bacteriological Analytical Manual Online*. Center for Food Safety and Applied Nutrition (CFSAN), U.S. Food & Drug Administration (FDA) (1998).
11. Sugeno, M., Kang, G.: Structure identification of fuzzy model. *Fuzzy Sets and Systems*, **28(1)** (1988) 15–33.
12. Takagi, T., Sugeno, M.: Fuzzy identification of systems and its applications to modeling and control. *Transactions on Systems, Man, and Cybernetics*, **15(1)** (1985) 116–132.
13. Yager, R., Zadeh, L., editors: Fuzzy Sets, Neural Networks, and Soft Computing. Van Nostrand Reinhold, New York (1994).

A Fuzzy Dead Reckoning Algorithm for Distributed Interactive Applications*

Ling Chen and Gencai Chen

College of Computer Science, Zhejiang University, Hangzhou 310027, P.R. China
lingchen@cs.zju.edu.cn

Abstract. A fuzzy Dead Reckoning (DR) algorithm for distributed interactive applications is proposed in this paper. Since fixed threshold cannot adequately handle the dynamic relationships between moving entities, some multi-level threshold DR algorithms were proposed in the past few years. In these algorithms the level of threshold is adaptively adjusted based on the distance between entities. The proposed fuzzy DR algorithm is based on multi-level threshold DR algorithm and takes all properties of entity (e.g. position, size and view angle etc.) into consideration when adjusting the level of threshold. This algorithm employs fuzzy correlation degree to measure the relationships between entities and determine the level of threshold for DR algorithm. Fuzzy consistent relation is used to distribute weight for each property. Simulation results indicate that fuzzy DR algorithm can achieve a considerable reduction in the number of state update messages while maintaining adequate accuracy in extrapolation.

1 Introduction

Thanks to the development of network, processor and graphics chip techniques, distributed interactive applications have a rapid development in the past few years. Distributed Interactive Simulation (DIS) and Collaborative Virtual Environments (CVEs) all can be regarded as distributed interactive applications. DIS is a technology for linking simulations of various types at multiple locations to create a realistic, complex, "virtual world" for the simulation of highly interactive activities [1]. CVEs is a computer based, distributed, virtual space or set of places. In such places, people can meet and interact with others, or with virtual objects to perform tasks [2].

In order to promote application reuse and interoperability, the High Level Architecture (HLA) [3] was proposed as general purpose architecture. The requirements of application reuse and interoperability are achieved by the concept of the federation in HLA, which is a composable set of interacting simulations. Although the DIS standard [4] has been supplanted by the HLA, the DIS community has developed the real-time platform reference federation object model (RPR FOM) [5] to provide the functionality of the DIS standard within the HLA environment.

---

* This work was supported by Zhejiang Provincial NSF of China (Grant No. Y104199).

Since entities of distributed interactive applications are physically distributed, for a large scale distributed interactive application, state update messages of the entities may generate a large amount of communication and thus saturate network bandwidth. To reduce the amount of communication, some message filtering techniques, such as the dead reckoning (DR) technique which is a fundamental feature of the DIS standard [4] and relevance filtering technique [6], were developed. Using a dead reckoning model, the state of an entity will be extrapolated. Therefore, instead of emitting a state update message after each movement of the entity, state update messages are only transmitted when the difference between the extrapolated and the true state exceeds a predefined threshold [7]. Relevance filtering is concerned with eliminating the transmission of irrelevant state update messages, but it does not employ any prediction techniques.

In general, DR technique uses a fixed threshold, regardless of the relationships between entities, to control errors in extrapolation. In order to maintain an adequate accuracy, a small threshold is usually used and many redundant state update messages are still be transmitted. In order to reduce the number of state update messages, some multi-level threshold DR algorithms were proposed. Dynamic message filtering technique [8] introduces Flexible Threshold Mechanism (FTM) and Update Lifetime (UL) to DR algorithm. UL is the time duration between two sequential state update messages and is employed to determine the level of threshold for DR algorithm. If the UL is short, a large threshold is assigned to maintain filtering performance as a normal level. On the contrary, a small threshold is used to improve the accuracy. Orientation update message filtering algorithm [9] introduces multi-level threshold to orientation DR algorithm. Variable Thresholding for Orientation (VTO), which is based on the average recent angular velocity of entity's rotation, is introduced and the threshold level is adjusted with the average recent angular velocity.

As mentioned in relevance filtering technique, some criterions can be introduced to determine whether a state update message is "important". For example, distance between two entities is the most obvious criterion. Auto-adaptive DR algorithm [10] is the combination of relevance filtering technique and multi-level threshold DR algorithm. It employs the distance between two entities to determine the threshold level for DR algorithm. If the distance is short, a small threshold is assigned to improve the accuracy. On the contrary, a large threshold is used to filter more messages and improve the scalability.

To the characteristics of human perception, distance is just one of the criterions that determine whether a state update message is "important". For example, the size and the view angle of entity all can influence the essentiality of a state update message. Therefore, just taking distance into consideration is not very reasonable when determining the level of threshold. Therefore, a fuzzy DR algorithm is proposed in this paper. This algorithm introduces fuzzy correlation degree to take all properties into consideration when measuring the relationships between entities. The level of threshold for DR algorithm is determined based on fuzzy correlation degree. Fuzzy consistent relation is used to distribute weight for each property.

The rest of the paper is arranged as follows: Section 2 gives a brief introduction on DR technique; fuzzy correlation degree, which takes all properties of entity into consideration when determining the relationships between entities, is introduced in Section 3; Section 4 describes the fuzzy DR algorithm; The simulation and results discussion are presented in Section 5; finally, Section 6 will conclude the paper.

2 DR Technique

One of the important aspects in distributed interactive applications is the ability of each site to represent accurately in real-time the state of all entities (both position and orientation), including both local and remote, participating in the same exercise. To reduce the number of state update messages, the DR technique is used. In addition to the high fidelity model that maintains the accurate state about its own entities, each site also has a dead reckoning model that estimates the state of all entities (both local and remote). The anticipated state of an entity is usually calculated based on the last (or past) accurate state information of the entity using an extrapolation equation. To maintain accuracy, after each update of its own entity, a site needs to compare the true state of the entity obtained from the high fidelity model and its extrapolated state. If the difference between the true and the extrapolated state is greater than a pre-defined threshold, a state update message will need to be sent to other sites. Extrapolation for the entity will then be corrected by all dead reckoning models at all sites, based on the updated state of the entity. So, instead of transmitting state update messages, the estimated state of a remote entity is readily available through a simple, local computation.

Threshold is an important parameter in the DR algorithm. It is used to control the accuracy of extrapolation, and affects the number of transmitted state update messages. A small threshold makes the DR algorithm generate state update messages at a higher frequency, but results in higher accuracy in the estimation of the entity's state. On the other hand, the DR algorithm using a large threshold generates fewer update messages, but its accuracy is also lower.

Table 1. Extrapolation equations

	One-Step	Two-Step
1st order	$x_t = x_{t'} + v_{t'}\tau$	$x_t = x_{t'} + \dfrac{x_{t'} - x_{t''}}{t' - t''}\tau$
2nd order	$x_t = x_{t'} + v_{t'}\tau + 0.5 a_{t'}\tau^2$	$x_t = x_{t'} + v_{t'}\tau + 0.5 \dfrac{v_{t'} - v_{t''}}{t' - t''}\tau^2$

Table 1 shows the position (x dimension) extrapolation equations. $x_{t'}$, $v_{t'}$ and $a_{t'}$ represent respectively the position, velocity and acceleration of the entity as found in the last state update message. Similarly $x_{t''}$, $v_{t''}$ and $a_{t''}$ are the position, velocity and acceleration of the second last state update message. τ is the elapsed time from the last update. The equations are used to extrapolate the position of the entity at time $t = t' + \tau$. Corresponding extrapolation equations are available for y, z dimension position estimation and orientation estimation. Current extrapolation equations can be divided into two groups: one-step equations and multi-step equations [11]. One-step equations only use the last state update message to extrapolate the state of an entity, whereas, multi-step equations use the last two or more state update messages in the extrapolation. A multi-step equation needs less data from the state update message than the one-step equation of the same order. But, in multi-step equations, high-order derivatives, such as velocity (or acceleration), are calculated from positions (or velocities). They are not as accurate as that used in one-step equations. Therefore, to track the state of the same entity, using multi-step equations may introduce more errors and generate more state update messages than using one-step equations of the same order for a given threshold. Conventionally, first order extrapolation is used for orientation, and second order for position [12].

For simplicity, the simulation of the proposed fuzzy DR algorithm only employs 2nd order one-step extrapolation equation to estimate two-dimension position of an entity. However, the proposed algorithm can also be applied to multi-dimension position estimation and orientation estimation.

3 Fuzzy Correlation Degree

Most current multi-level threshold DR algorithms take distance between entities as criterion to determine the level of threshold. However, the accuracy of such algorithms cannot be guaranteed, for there are many distributed interactive applications in which other properties of the entity (e.g. the size and the view angle of entity) are as important as the position property. For example, in the aviation DIS application, not only the distance between the radar station and the plane but also the angle property of the radar, the size and even material properties of the plane should be considered to determine whether a state update message should be sent from the plane to the radar station.

The fuzzy correlation degree [13] was proposed to take all properties into consideration when grouping entities for Data Distribution Management (DDM). The grouping results of the approach which employs fuzzy correlation degree are more accurate than traditional grouping approaches (e.g. region-based approach [14], hybrid approach [15]). Because fuzzy correlation degree can take all properties of an entity into consideration, fuzzy DR algorithm employs it to determine the relationships between entities. If the relationship between two entities is tightly-coupled, a small threshold can be used to guarantee high accuracy. On the other hand, a large threshold can be used to filter more messages. In order to distribute weights for properties, fuzzy consistent relation [16] is employed. Through fuzzy consistent relation, the priority of properties can be transferred to particular weights.

Domain U includes all properties that exist in a distributed interactive application and the length of U is N. Aggregate A is the properties of an entity and A is not a void generally.

$$A = \{(u_1, u_2, ..., u_k) \mid u_1, u_2, ..., u_k \in U, k \leq N\} \quad (1)$$

Definition 1: Property $x, y \in U$, the property correlation degree $d(x, y)$ is determined by a real value function d,

$$d : U \times U \to [0,1] \quad (2)$$

For example, x, y are position property of entity a, b, D is the distance between entity a and b. Through position property x and y, D can be calculated. The property correlation degree can be determined by following function which takes D as variable.

$$d(x,y) = \begin{cases} 1 & 0 \le D \le 10000 \\ \left[1 + \left(\dfrac{D-10000}{2000}\right)^2\right]^{-1} & D \ge 10000 \end{cases} \quad (3)$$

Definition 2: For two entities A and B whose properties are defined by equation 1, the property correlation degrees of them are: $d_1, d_2, ..., d_k$, the weight of each property $u_1, u_2, ..., u_k$ is $w(u_1), w(u_2), ..., w(u_k)$. Fuzzy correlation degree between entity A and B is determined by following function:

$$d(A,B) = \overset{k}{\underset{i=1}{V}} (w(u_i) \cdot d_i) \quad (4)$$

In a particular distributed interactive system it is always easy to qualify the priority of properties, e.g. property A is more important than property B, but it is difficult to precisely quantify it. The fuzzy DR algorithm employs the fuzzy consistent relation to distribute weight for each property. It transfers the priority relation of the properties to the fuzzy precedence degree, and then the weights of properties are calculated from the fuzzy precedence degree.

Domain $U = \{u_i | i \in I\}, I = \{1, 2, \cdots\}$, u_i represents the properties of the entity.

Definition 3: R is a fuzzy relation based on the kronecker products $U \times U$:

$$R: U \times U \to [0,1] \quad (5)$$

if any two elements: $u_i \in U, u_j \in U$ satisfy the following equation:

$$\mu_R(u_i, u_j) = \mu_R(u_i, u_k) - \mu_R(u_j, u_k) + 0.5 \quad \forall k \in I \quad (6)$$

Then R is the fuzzy consistent relation in domain U. For example, a domain $U = \{u_1, u_2, \cdots, u_m\}$, elements of the domain represents the properties of an entity. R is a fuzzy consistent relation that is used to judge two properties which one is more important than the other. Following equations must be met:

(1) If u_i is as important as u_j, $\mu_R(u_i, u_j) = 0.5$.
(2) If u_j is more important than u_i, $0 \le \mu_R(u_i, u_j) \le 0.5$.
(3) If u_i is more important than u_j, $0.5 \le \mu_R(u_i, u_j) \le 1.0$.

The fuzzy DR algorithm uses fuzzy consistent relation to get the fuzzy precedence degree from the priority of properties in a particular distributed interactive application, and then calculate the weights. Following is the steps:

(1) Decide the priority of properties in the domain U, and construct the fuzzy precedence relation. μ_B represents the fuzzy precedence relation.

$$B: \mu_B(u_i, u_j) \quad (7)$$

If u_i is more important than u_j, $\mu_B(u_i, u_j) = 1.0$.

If u_i is as important as u_j, $\mu_B(u_i, u_j) = 0.5$.
If u_j is more important than u_i, $\mu_B(u_i, u_j) = 0.0$.

(2) Convert the fuzzy precedence relation to fuzzy consistent relation. μ_R represents the fuzzy consistent relation.

$$R : \mu_R(u_i, u_j) \tag{8}$$

$$\mu_R(u_i, u_j) = (r_i - r_j)/2m + 0.5 \tag{9}$$

$$r_i = \sum_{k=1}^{m} \mu_B(u_i, u_k) \tag{10}$$

(3) Calculate fuzzy precedence degree c_i ($i = 1,..., m$) for each property.

$$c_i = [\prod_{j=1}^{m} \mu_R(u_i, u_j)]^{1/m} \tag{11}$$

(4) Calculate weight w_i ($i = 1,..., m$) for each property.

$$w_i = \frac{c_i}{\sum_{j=1}^{m} c_j} \tag{12}$$

4 Fuzzy DR Algorithm

Because fuzzy correlation degree takes all properties and their priority into consideration, the measurement of relationships between entities through fuzzy correlation degree is more accurate than the distance based measurement and other one property based measurement. Therefore, Fuzzy DR algorithm employs fuzzy correlation degree to measure the relationships between entities and the level of threshold is adjusted based on it. In order to distribute weight for each property, fuzzy consistent relation is used to get the fuzzy precedence degree from the priority of properties in a particular distributed interactive application and then calculate the weights.

Following is the steps of fuzzy DR algorithm:

(1) Determine threshold distribution function. Threshold distribution function is used to map fuzzy correlation degree to threshold and it can be designed according to the characteristics of particular distributed interactive application. For example, following 1st order function can be used as threshold distribution function:

$$\lambda(A,B) = \begin{cases} 0.05 & d(A,B) > 0.7 \\ 0.25 & 0.7 \geq d(A,B) > 0.5 \\ 1.25 & 0.5 \geq d(A,B) > 0.35 \\ 6.25 & 0.35 \geq d(A,B) \end{cases} \tag{13}$$

$d(A,B)$ is the fuzzy correlation degree between two entities, $\lambda(A,B)$ is the threshold which is used to determine whether send state update messages.

(2) Estimate the state of the entity and calculate the difference between the extrapolated state and the real one.

(3) Calculate fuzzy correlation degrees between local entity and remote entities, and then the threshold distribution function is used to determine the threshold λ.

(4) Judge whether a state update message should be transmitted based on the difference and the threshold λ.

5 Simulation

Simulation was carried out to evaluate the performance of the proposed fuzzy DR algorithm. It compared the performance of fuzzy DR algorithm with fixed threshold DR algorithm and distance based multi-level threshold DR algorithm. For simplicity, the 2nd order one-step extrapolation equations listed in Table 1 was used to estimate two-dimension position of the entity. The position, size and view angle properties were employed in the simulation. The view angle property included view direction information. The priority of these properties was defined as follows: view angle > size > position. Through fuzzy consistent relation, the weights were distributed as follows: W (view angle) = 0.45; W (size) = 0.34; W (position) = 0.21. Equation 3 was used to calculate property correlation degree and equation 13 was used as threshold distribution function. In the simulation, the distance unit was meter and the state update message included following information: position; orientation; speed and acceleration.

In the simulation, some entities were distributed in a rectangle (500×700 two-dimension space). The movement of entities had following characteristics: initial speed of entity was a random value between 0 m/s and 2 m/s; initial acceleration of entity was a random value between 0 m/s^2 and 1 m/s^2; initial orientation was a random value between -180° and 180°; every 1 s the speed and acceleration of an entity would be reselected in the range mentioned above and the orientation of an entity would add a angle ranging from -7.5° to 7.5°; if an entity collided the border it would continue its movement in the negative direction. The simulation lasted 500 s. The simulation was run 8 times and the number of entities ranged from 8 to 128 (8, 16, 24, 32, 48, 64, 96 and 128). In addition, fixed threshold DR algorithm (threshold were 0.25 and 1.25) and multi-level threshold DR algorithm (AOI and SR were 90, 20; 165, 50; 250, 50; if the distance between two entities was less than SR, threshold was set to 0.05; if the distance between two entities was less than AOI, threshold was set to 0.25; if the distance between two entities was in the range of AOI and AOI + AOI, threshold was set to 1.5; if the distance between two entities was large than AOI + AOI, threshold was set to 6.25) were simulated using the same parameters to carry out performance comparison.

In order to compare performance, three metrics (filtering rate, accuracy rate and accuracy filtering rate) were introduced. Following are the definitions of them.

Filtering rate: if the number of generated state update messages is AM and the number of transmitted state update messages is TM, the Filtering Rate (FR) is determined by following equation.

$$FR = (AM - TM) / AM \qquad FR \in [0, 1] \qquad (14)$$

Accuracy rate: if the number of generated state update messages is AM and the number of filtered state update messages that are necessary for the purpose of accuracy is EM, the Accuracy Rate (AR) is determined by following equation.

$$AR = (AM - EM) / AM \qquad FR \in [0, 1] \qquad (15)$$

At the beginning of the simulation, the EM was set to 0. All generated state update messages were judged by the threshold and the difference between real state and estimated state to determine whether filtering the message. When a message was determined to be filtered, it was judged by some criterions to determine whether it was a necessary state update message for purpose of accuracy. If the message was necessary, EM increased by 1. The criterions are very important in determine AR. In this simulation, the fuzzy correlation degree and threshold distribution function, which employs equation 13, were selected as the criterion to determine the necessity of a state update message, for it takes all properties of entity into consideration and consists with human perception. Therefore, the AR of fuzzy DR algorithm was 100%.

Accuracy filtering rate: If the filtering rate of the simulation is FR and the accuracy rate of the simulation is AR, the Accuracy Filtering Rate (AFR) is determined by following equation.

$$AFR = AR \times FR \qquad AFR \in [0, 1] \qquad (16)$$

AFR is an integrated evaluation of scalability and accuracy. FR is an evaluation of scalability and AR is an evaluation of accuracy. The ideal DR algorithm should achieve a high AFR and that means a good tradeoff between scalability and accuracy.

Table 2. FR under different DR algorithms and entity numbers

Entity number	Fixed threshold Threshold = 1.25	Fixed threshold Threshold = 0.25	Multi-level threshold SR=20 AOI=90	Multi-level threshold SR=50 AOI=250	Multi-level threshold SR=50 AOI=165	Fuzzy DR
8	75.4%	54.8%	85.1%	69.2%	76.3%	80.8%
16	75.6%	54.6%	86.6%	69.6%	78.0%	80.2%
24	75.8%	54.3%	88.4%	70.1%	79.5%	79.8%
32	76.0%	54.5%	88.1%	70.2%	79.4%	81.0%
48	75.5%	54.6%	88.0%	71.0%	79.8%	80.0%
64	76.0%	54.7%	88.2%	71.3%	80.4%	80.6%
96	75.8%	54.6%	88.0%	70.5%	79.7%	79.4%
128	75.7%	54.6%	87.9%	70.3%	79.4%	78.9%

Table 2 – 4 shows FR, AR and AFR of different DR algorithms under different entity numbers. First of all, it can be seen that FR, AR and AFR of different DR algorithms are very stable and the incensement of entity number has no influence on them.

As table 2 shows, multi-level threshold DR algorithm (SR = 20, AOI = 90) achieves highest FR (about 88%) and fixed threshold DR algorithm (threshold = 0.25) gets the worst FR (about 54%). As table 3 shows, Fuzzy DR algorithm and fixed threshold DR algorithm (threshold = 0.25) achieve highest AR (almost 100%) and multi-level threshold DR algorithm (SR = 20, AOI = 90) gets the worst AR (about 69%). As table 4 shows, fuzzy DR algorithm achieves best AFR (about 80%) and fixed threshold DR algorithm (threshold = 0.25) gets the worst AFR (about 54%). Because AFR is an integrated evaluation of scalability and accuracy, fuzzy DR algorithm offers the best performance, fixed threshold DR algorithm is the worst and multi-level threshold DR algorithm is among them.

Table 3. AR under different DR algorithms and entity numbers

Entity number	Fixed threshold Threshold = 1.25	Fixed threshold Threshold = 0.25	Multi-level threshold SR=20 AOI=90	Multi-level threshold SR=50 AOI=250	Multi-level threshold SR=50 AOI=165	Fuzzy DR
8	90.6%	99.7%	72.0%	88.9%	79.2%	100.0%
16	91.0%	99.7%	69.3%	89.3%	78.5%	100.0%
24	90.5%	99.6%	70.0%	91.0%	80.6%	100.0%
32	92.2%	99.8%	71.8%	92.2%	81.4%	100.0%
48	92.1%	99.7%	69.2%	89.4%	78.7%	100.0%
64	92.4%	99.7%	69.7%	90.8%	79.4%	100.0%
96	91.5%	99.7%	67.5%	90.8%	78.8%	100.0%
128	90.8%	99.6%	66.4%	90.4%	78.0%	100.0%

Table 4. AFR under different DR algorithms and entity numbers

Entity number	Fixed threshold Threshold = 1.25	Fixed threshold Threshold = 0.25	Multi-level threshold SR=20 AOI=90	Multi-level threshold SR=50 AOI=250	Multi-level threshold SR=50 AOI=165	Fuzzy DR
8	68.3%	54.6%	61.3%	61.5%	60.5%	80.8%
16	68.8%	54.4%	60.1%	62.1%	61.2%	80.2%
24	68.6%	54.1%	61.9%	63.8%	64.1%	79.8%
32	70.1%	54.4%	63.3%	64.7%	64.6%	81.0%
48	69.5%	54.5%	60.9%	63.5%	62.8%	80.0%
64	70.2%	54.5%	61.5%	64.7%	63.8%	80.6%
96	69.4%	54.4%	59.4%	64.0%	62.8%	79.4%
128	68.7%	54.4%	58.4%	63.6%	62.0%	78.9%

Above simulation results indicate that fuzzy DR algorithm can filter redundant state update messages and keep all messages that are necessary for accuracy. Fuzzy DR algorithm gives a mechanism that takes all properties of entity into consideration when measuring the relationships between entities. When implementing fuzzy DR algorithm in a particular distributed interactive application, designer must determine what properties should be taken into consideration and the priority of them. In addition, the threshold distribution function is very import and it should be carefully adjusted according to the characteristics of the application. After carefully selection and adjusting, fuzzy DR algorithm can filter most state update messages which are useless for human perception.

6 Conclusions

This paper describes a new fuzzy DR algorithm for distributed interactive applications. This algorithm is based on multi-level threshold DR algorithm and takes all properties of entities into consideration. Fuzzy correlation degree is employed to measure the relationships between entities and fuzzy consistent relation is used to distribute weights for each property. Stimulation results indicate that fuzzy DR algorithm keeps all messages that are necessary for system accuracy and filters the redundant messages. It achieves a better tradeoff between scalability and accuracy than fixed threshold DR algorithm and multi-level threshold DR algorithm.

References

1. DIS Steering Committee. The DIS vision, a map to the future of distributed simulation. Technical Report IST-SP-94-01, Institute for Simulation and Training, Orlando FL. 1994.
2. J. Leigh. A review of tele-immersive applications in the CAVE research network. In Proc. of IEEE International Conference on Virtual Reality. 1999: 180-187.
3. Department of Defense. High Level Architecture programmers guide. 1998.
4. IEEE 1278-1993. Standard for information technology - protocols for distributed interactive simulation applications. 1993.
5. G.C. Shanks. The RPR FOM, a reference federation object model to promote simulation interoperability. In Proc. of Spring Simulation Interoperability Workshop, 1997.
6. M. Bassiouni, M.H. Chiu, M. Loper, M. Garnsey. Performance and reliability analysis of relevance filtering for scalable distributed interactive simulation. ACM Transaction on Modeling and Computer Simulation, Vol. 7, No. 3, 1997: 293-331.
7. A.R. Pope. The SIMNET network and protocols. Report No. 7262, BBN Systems and Technologies, Cambridge MA. 1991.
8. S.J. Yu, Y.C. Choy. A dynamic message filtering technique for 3D cyberspaces. Computer Communications, Vol. 24, 2001: 1745-1758.
9. M.J. Zhang, N.D. Georganas. An orientation update message filtering algorithm in collaborative virtual environments. Journal of Computer Science and Technology, Vol. 19, No. 3, 2004: 423-429.
10. W. Cai, F.B.S. Lee, L. Chen. An auto-adaptive dead reckoning algorithm for distributed interactive simulation. In Proc. of Workshop on Parallel and Distributed Simulation, 1999: 82-89.

11. K.C. Lin. Dead reckoning and distributed interactive simulation. In Proc. of DIS Systems in the Aerospace Environment/Critical Reviews, 1995: 16-36.
12. A. Katz. Synchronization of networked simulations. In Proc. of DIS Workshop on Standards for the Interoperability of Distributed Simulation, 1994: 81-87.
13. X.P. Qian, Q.P. Zhao. An approach to data filtering based on fuzzy correlation space. Chinese Journal of Computers, Vol. 25, No.7, 2002: 723-729.
14. T.C. Lu, C.C. Lee, W.Y. Hisa. Supporting large-scale distributed simulation using HLA. ACM Transactions on Modeling and Computer Simulation, Vol. 10, No. 3, 2000: 268-294.
15. G. Tan, Y.S. Zhang, R. Ayani. A hybrid approach to data distribution management, In Proc. of IEEE International Workshop on Distributed Simulation and Real-Time Applications, 2000: 55-61.
16. M. Yao, Y.J Huang. Fuzzy consistent relation and its applications. Jounal of USET of China, Vol. 26, No. 6, 1997: 632-635.

Intelligent Automated Negotiation Mechanism Based on Fuzzy Method

Hong Zhang and Yuhui Qiu

Faculty of Computer & Information Science,
Southwest-China Normal University, Chongqing, 400715, China
{zhangh, yhqiu}@swnu.edu.cn

Abstract. Negotiation is an important function for e-commerce system to be efficient. However, negotiation is complicated, time-consuming and difficulty for participants to reach an agreement. This paper aims to establish an automated negotiation mechanism based on fuzzy method in order to alleviate the difficulty of negotiation. This automated negotiation is performed by autonomous agents that use fuzzy logic and issue-trading strategies in finding mutually-agreed contracts.

1 Introduction

E-commerce refers to completing every section of commercial activities by digital electronic means. It includes publication and search of commercial information, electronic advertisements, subscriptions of electronic contracts, payments in electronic currency, before-sales and after sales services and many other processes. Problems of coordination and cooperation are not unique to automated commerce systems; they exist at multiple levels of activity in a wide range of commerce. People pursue their own goals through communication and cooperation with other people or machine. Since people like to use negotiation as a means to compromise in order to reach mutually beneficial agreements in E-commerce.

The reminder of the paper is organized as follows.Section 2 introduce the workflow of automated negotiation,section 3 illustrate how to get satisfactory new offer which maximize the current utility and comparability.Section 4 summarizes our main contributions and indicates avenues of further research.

2 Negotiation Workflow

In automated negotiation, the primary workflow can described as follow:
  Step 0: Negotiation Start.
  Step 1: Present the Initial Proposal.
  Step 2: Evaluating the Opponent's Proposal.
  Step 3: If $U_s(x^B) > T_s$ or $(U_s(x^S) > T_B)$ Then Accept Proposal, go to Step6.
    Else if $N_{time} > NegotiationNumber$ Then Reject Proposal, go to Step6.
    Else go to Step4 to Search New Proposal.
  Step 4: Search New Proposal based on Multi-Agent Negotiation System.

Step 5: if $U_{t+1} > U_t$, then $t \leftarrow t+1$, go to Step2.
  Else if $N_{proposal} >$ Proposal time, Then Reject Proposal, go to Step6.
  Else go to Step4 Search New Proposal.
Step 6: Negotiation End and Notify Both Parties.

3 Producing New Offer

In this section, we illuminated how to get the better new offer. The aim of new offer is maximize own utility and comparability between the new offer and buyer's last offer.

3.1 Fuzzy Reasoning Rule

Preference of issue is vital factor in automated negotiation system. Generally speaking, value of preference is lager, it also represent this issue is more important, vice versa. Whether or not the new offer is effective depend on adjusting the important issue. So exact affirm the preference of each issue is very important. In our research, we divided preference into three aspects: careless, neutral, important. Their subjection function can be found as follow:

$$\mu_{careless}(w) = \begin{cases} 1, & w \leq CL, \\ 1 - \frac{w-CL}{CH-CL}, & CL \leq w \leq CH. \end{cases} \quad (1)$$

$$\mu_{neutral}(w) = \begin{cases} \frac{w-NL}{NM-NL}, & NL \leq w \leq NM, \\ 1 - \frac{w-NM}{NH-NM}, & NM \leq w \leq NH. \end{cases} \quad (2)$$

$$\mu_{important}(w) = \begin{cases} \frac{w-IL}{IH-IL}, & IL \leq w \leq IH, \\ 1, & w \geq IH.[0.2cm] \end{cases} \quad (3)$$

where, CL denotes Careless lower limit, CH denotes careless upper limit NL denotes Neutral lower limit, NM denotes neutral middle value, NH denotes neutral upper limit, IL denotes important lower limit, LH denotes important upper limit.

3.2 Concession Space

Each negotiation round, seller agent can afford new suggestion which will descend some issue offer, it also called concession, and there are functions to decide the concession space. Firstly, concession rate C_r will be presented as follow:

$$C_r = \frac{10 - w_i^s}{(10 - w_i^S) + (10 - w_i^B)} \quad (4)$$

then, we defined the concession space as $C_r \times x_{i,t}^d$.

Where, $x_{i,t}^d$ denotes distance between seller offer and buyer offer, it can defined as follow:

Continual issue: $x_{i,t}^d = |x_{i,t}^S - x_{i,t}^B|$

Discrete issue: $x_{i,t}^d = \phi_s(x_{i,t}^S) - \phi_s(x_{i,t}^B)$

thereinto $\phi_s(x_{i,t}^B)$ denotes the order of $x_{i,t}^B$ in value fields of issue i.

3.3 Fuzzy Strategy of Issue

Δx_i denotes the current concession value of issue i, so concession rules based on fuzzy methods of the all kinds of issue can be found in follow table 1.

Table 1. Fuzzy Strategy of Sequence issue

NO.	W_i^S	W_i^B	Δx_i	
			gain-driven	cost-driven
1	Important	Important	-rand(0, $c_r, x_{i,t}^d$)	rand(0, $c_r, x_{i,t}^d$)
2	Important	Neutral	-rand(0, $k_1 c_r, X_{i,t}^d$)	rand(0, $k_1 c_r, x_{i,t}^d$)
3	Important	Careless	rand(0, $c_r, x_{i,t}^d$)	-rand(0, $c_r, x_{i,t}^d$)
4	Neutral	Important	-rand(0, $k_2 c_r x_{i,t}^d$)	rand(0, $k_2 c_r x_{i,t}^d$)
5	Neutral	Neutral	rand($-k_1 c_r x_{i,t}^d, k_1 c_r x_{i,t}^d$)	
6	Neutral	Careless	rand(0, $k_1 c_r x_{i,t}^d$)	-rand(0, $k_1 c_r x_{i,t}^d$)
7	Careless	Important	-rand(0, $k_3 c_r x_{i,t}^d$)	rand(0, $k_3 c_r x_{i,t}^d$)
8	Careless	Neutral	-rand(0, $k_2 c_r x_{i,t}^d$)	rand(0, $k_2 c_r x_{i,t}^d$)
9	Careless	Careless	rand($-c_r x_{i,t}^d, c_r x_{i,t}^d$)	
Thereto $k_1 \in [0,1], k_3 > k_2 > 1$.				

3.4 Producing New Offer

Now, we can give method to produce new offer for $issue_i$, it can be presented as follow:

$$\Delta x_i = \frac{\sum_{j=1}^{9} Z_j R_j}{\sum_{j=1}^{9} Z_j}, \text{ and } x_{i,t+1}^s = x_{i,t}^s + \Delta x_i \quad (5)$$

Where, Z_j denotes start-up degree of fuzzy strategy i, it can be decided by the preference of seller and buyer, for example start-up degree of No.1 strategy is as follow:

$$z_1 = \mu_{important}(w_t^S) \cdot \mu_{important}(w_t^B)$$

R_j denotes the result of strategy i from table 1.

3.5 Studying Rules of Preference of Issue

Negotiation Agent can't get exact preference weight directly and timely, it also establish studying mechanism to get real-time preference of negotiation other sides. The rule is

$$r = \frac{|x_{i,t}^B - x_{i,t+1}^B|}{|x_{i,t}^S - x_{i,t+1}^S|} \tag{6}$$

Where, $x_{i,t}^B$ is buyer's offer of issue i at time t. $x_{i,t}^S$ is seller's offer of issue i at time t. Tell it like it is, r¿1 denotes buyer preference when issue i is increase , r=1 denoted keep value, r¡1 denotes value decrease. So the weight of preference can be adjust as follow: $W_{i,t+1}^B = W_{i,t}^B * r$

4 Conclusion

In [1], a model for bilateral multi-attribute negotiation is presented, where attributes are negotiation sequentially. The issue studied is the optimal agenda for such a negotiation under both incomplete information and time constraints. However a central mediator is used and the issues all have continuous values. In earlier research [2] a slightly different model is proposed, but the focus of the research is still on time constraints and the effect of deadlines on the agents' strategies. The argumentation approach to negotiation [3] allows the agents to exchange not only bids but also arguments that influence other agents' beliefs and goals, which allows more flexibility. Another important direction in multi-attribute negotiation is presented by [4], which propose models that overcome the linear independence assumption between attribute evaluations. There contrast with our negotiation mechanism, where efficiency of the outcome and not time is the main issue studied. This is because we found that, due to our assumption; a deal is usually reached in maximum 8-12 steps. on the other hand, our model is more flexible in specifying attribute values and better explainable. To our knowledge, there has not been work which completely addresses these problems. Therefore, the research on resolving them will be of great challenge and significance.

References

1. S.S.Fatima, M.J.Wooldridge and N.R.Jennings, An Agenda-Based Framework for Multi-issue Negotiation. Artificial Intelligence, 152: 1-45,2004.
2. S.S.Fatima, M.J.Wooldrige and N.R.Jennings, Multi-issue Negotiation under time constraints. Proc.1st International Joint Conference on Autonomous Agents and Multi-Agent Systems, Bologna, Italy, 143-150,2002
3. S.Parson,N.R.Jennings, Negotiation through Argumentation-a preliminary report, Proceeding of the International Conference on Multi-Agent Systems, Kyoto, 1996.
4. M.Klein, P.Faratin, H.Sayana and Y.Bar-Yam, Negotiation Complex Contracts, MIT Slon Working Paper No.4196, 01,2001.

Congestion Control in Differentiated Services Networks by Means of Fuzzy Logic

Morteza Mosavi[1] and Mehdi Galily[2]

[1] Azad University of Arak Branch, Arak, Iran
m_mosavi@iau-arak.ac.ir
[2] Young Researchers Club, Azad University
m.galily@gmail.com

Abstract. A fuzzy logic based intelligent controller is design to congestion control and avoidance in differentiated computer services networks. The proposed controller provide a robust active queue management system to secure high utilization, bounded delay and loss, while the network complies with the demands each traffic class sets.

1 Introduction

The aim of this paper is to design a robust active queue management system for a differentiated services computer network [1,2]. Most proposed schemes for queue management are developed using intuition and simple nonlinear control designs. These have been demonstrated to be robust in a variety of scenarios that have been simulated [1]. The interaction of additional nonlinear feedback loops can produce unexpected and erratic behavior [2]. In [1,2] a very useful model is developed to tackle the flow control problem in differentiated services architecture, which divides traffic into three basic types of service (in the same spirit as those adopted for the Internet by the IETF Diff-Serv working group, i.e. Premium, Ordinary, and Best Effort). We will apply Fuzzy Logic Controller (FLC) to such system [3]. The proposed control strategy is shown via simulations to be robust with respect to traffic modeling uncertainties and system non-linearities, yet provide tight control (and as a result offer good service).

2 Dynamic Network Model

A diagram of a sample queue is depicted in Fig.1. Let $x(t)$ be a state variable denoting the ensemble average number in the system in an arbitrary queuing model at time t. Furthermore, let $f_{in}(t)$ and $f_{out}(t)$ be ensemble averages of the flow entering and exiting the system, respectively. $\dot{x}(t) = dx(t)/dt$ can be written as:

$$\dot{x}(t) = f_{in}(t) - f_{out}(t) \tag{1}$$

The above equation has been used in the literature, and is commonly referred to as fluid flow equation [1,2]. To use this equation in a queuing system, C and λ have been defined as the queue server capacity and average arrival rate, respectively. Assuming that the queue capacity is unlimited, $f_{in}(t)$ is just the arrival rate λ. The flow going out of the system, $f_{out}(t)$, can be related to the ensemble average utilization of the queue, $\rho(t)$, by $f_{out}(t)=\rho(t)C$. It is assumed that the utilization of the link, ρ, can be approximated by the function $G(x(t))$, which represents the ensemble average utilization of the link at time t as a function of the state variable. Hence, queue model can be represented by the following nonlinear differential equation:

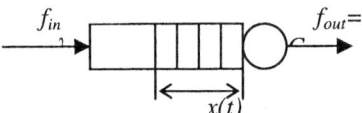

Fig. 1. Diagram of sample queue

$$\dot{x}(t) = -CG(x(t)) + \lambda \qquad (2)$$

Utilization function, $G(x(t))$, depends on the queuing in the under study system. If statistical data is available, this function can be empirically formulated. This, however, is not the general case and $G(x(t))$ is normally determined by matching the results of steady state queuing theory with (2). M/M/1 has been adopted in many communication network traffics. For M/M/1 the state space equation is:

$$\dot{x}(t) = -C\frac{x(t)}{1+x(t)} + \lambda \qquad (3)$$

The validity of this model has been verified by a number of researchers [17,18]. It is noticeable that (3) fits the real model, however there exists some mismatch. In order to include the uncertainties, (3) can be modified as:

$$\dot{x}(t) = -\rho C \mu (\frac{x(t)}{1+x(t)} + \Delta)C(t) + \lambda \qquad (4)$$

where Δ denotes model uncertainties and

$$\|\Delta\|^2 \le \Delta_{max} \qquad (5)$$

Consider a router of K input and L output ports handling three differentiated traffic classes mentioned above. At each output port, a controller is employed to handle different classes of traffic flows entering to that port. An example case of the controller is illustrated in Fig. 2. The incoming traffic to the input node includes different classes of traffic. The input node separates each class according to their class identifier tags and forwards the packets to the proper queue. The output port can transmit packets at maximum rate of C_{server} to destination where

$$C_{server} = C_p + C_r + C_b \qquad (6)$$

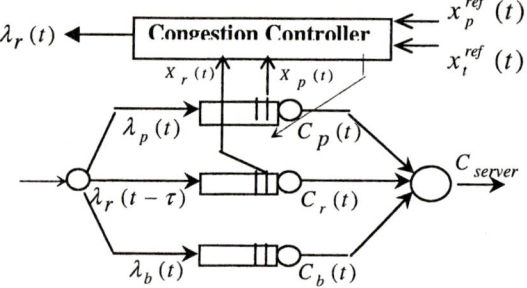

Fig. 2. Traffic flow control scheme

Premium traffic flow needs strict guarantees of delivery. Delay, jitter and packet drops should be kept as small as possible. The queue dynamic model can be as follows:

$$\dot{x}_p(t) = -C_p(t)\frac{x_p(t)}{1+x_p(t)} + \lambda_p(t) \qquad (7)$$

Here, the control goal is to determine $C_p(t)$ at any time and for any arrival rate, $\lambda_p(t)$, in which the queue length, $x_p(t)$, is kept close to a reference value, $x_p^{ref}(t)$, which is determined by the operator or designer. The objective is to allocate minimum possible capacity for the premium traffic to save extra capacity for other classes of traffic as well as providing a good QoS for premium flows. Note that we are confined to control signals as

$$0 < C_p(t) < C_{server} \qquad (8)$$

3 Fuzzy Logic Controller (FLC) Design

FLC is a knowledge-based control that uses fuzzy set theory, fuzzy reasoning and fuzzy logic for knowledge representation and inference [3]. In this paper a fuzzy system consisting of a fuzzifier, a knowledge base (rule base), a fuzzy inference engine and defuzzier will be considered. The controller has two inputs, the error and its derivative and the control input. Five triangular membership functions are defined for error (Fig. 3), namely, Negative Large (NL), Negative Small (NS), Zero, Positive Small (PS), and Positive Large (PL). Similarly three triangular membership functions are defined for derivative of the error and there are as follows, Negative Small (NS), Zero, and Positive Small (PS). Also five triangular membership functions are defined for the control input and there are Zero, Small, Medium, Large and Very Large. The complete fuzzy rules are shown in Table 1. The first rule is outlined below,

Rule 1: *If* (e) *is* **PL** *AND* $(\dot{e})$ *is* **Zero** *THEN* (p) *is* **Large**.

We have made the following assumptions for controller design throughout this paper:

C_{max}=300000 Packets Per Second
λ_{max}=280000 Packets Per Second

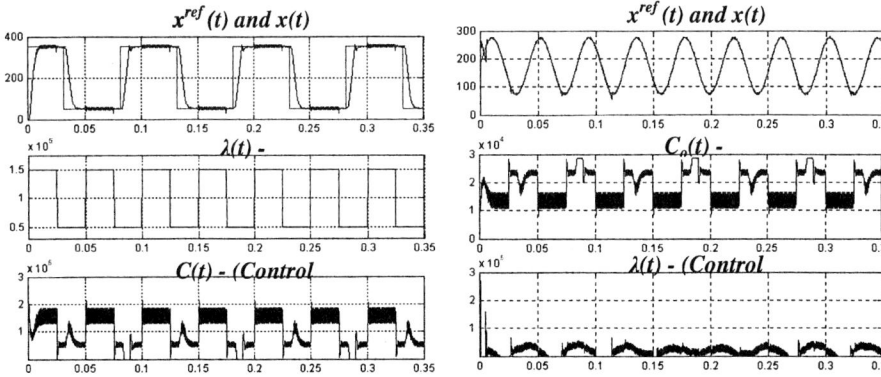

Fig. 3. Behavior of the time evaluation of Premium traffic using proposed robust control strategy

Fig. 4. Behavior of the time evaluation of Ordinary traffic using proposed robust control strategy

The simulation results are depicted in Figs. 3 and 4 for Premium and Ordinary traffics, respectively. As it can be seen, the performance of the controller is satisfactory and the output can follow the reference trajectory.

4 Conclusion

This paper proposes a robust scheme for congestion control based on fuzzy control theory, which uses an integrated dynamic congestion control approach (IDCC). We divide traffic into three basic types of service (in the same spirit as those adopted for the Internet by the IETF Diff-Serv working group, i.e. Premium, Ordinary, and Best Effort). The controller works in an integrated way with different services and has simple implementation and low computational overhead, as well as featuring a very small set of design constants that can be easily set (tuned) from simple understanding of the system behavior.

References

1. Chrysostomou, C., Pitsillides, A., Hadjipollas, G., Sekercioglu, M., Polycarpou, M.: Fuzzy Explicit Marking for Congestion Control in Differentiated Services Networks, in Proc. IEEE Sym. Computers and Communications (2003) 312-319
2. Jalili-Kharaajoo, M., Araabi, B.N.: Application of Predictive Control Algorithm to Congestion Control in Differentiated Service Networks, LNCS, 3124 (2004) 1157-1162
3. Jalili-Kharaajoo, M.: Application of Robust Fuzzy Adaptive Second-Order Sliding-Mode Control to Active Queue Management, LNCS, 2957 (2004) 109-119

Fault Diagnosis System Based on Rough Set Theory and Support Vector Machine[*]

Yitian Xu and Laisheng Wang[**]

College of Science, China Agricultural University, 100083, Beijing, China
{xytshuxue, wanglaish}@126.com

Abstract. The fault diagnosis on diesel engine is a difficult problem due to the complex structure of the engine and the presence of multi-excite sources. A new kind of fault diagnosis system based on Rough Set Theory and Support Vector Machine is proposed in the paper. Integrating the advantages of Rough Set Theory in effectively dealing with the uncertainty information and Support Vector Machine's greater generalization performance. The diagnosis of a diesel demonstrated that the solution can reduce the cost and raise the efficiency of diagnosis, and verified the feasibility of engineering application.

1 Introduction

In order to raise the efficiency and reduce the cost of fault diagnosis, intelligent identification of faults is desired in engineering application. Some theories or methods in computational intelligence are applicable to this task, such as neural networks, fuzzy set theory, genetic algorithm and so on. Considering the vagueness and uncertainty information in the process of fault diagnosis, a kind of hybrid fault diagnosis system based on Support vector machine (SVM) and Rough Set Theory(RS) is proposed in the paper.

Support vector machine is a new and promising machine learning technique proposed by Vapnik and his group at AT Bell Laboratories, It is based on VC dimensional theory and statistical learning theory. Classification is one of the most important applications. It is widely applied to machine learning, data mining, knowledge discovery and so on because of its greater generalization performance. But there are some drawbacks that it doesn't distinguish the importance of sample attributes, computation rate is slow and takes up more data storage space because of a large number of sample attributes. Moreover, It doesn't effectively deal with vagueness and uncertainty information. In order to resolve those problems, A kind of SVM fault diagnosis system based on Rough set pre-processing is proposed in the paper, Making great use of the advantages of Rough Set theory in pre-processing large data, eliminating redundant information and overcoming the disadvantages of slow processing speed causedby SVM approach. A

---

[*] This work was Supported by the National Natural Science foundation of China (No.10371131).
[**] Corresponding author.

hybrid fault diagnosis system based on Rough set and Support Vector Machine is presented in the paper too. It is more suitable to multi-classification. It may decrease fault diagnosis system complexity and improve fault diagnosis efficiency and accuracy.

2 Support Vector Machine[1][2][3]

Consider the problem of separable training vectors belonging to two separate classes,

$$T = \{(x_1, y_1), \cdots, (x_l, y_l)\}, x_i \in R^n, y_i \in \{-1, 1\}, i = 1, \cdots, l \quad (2.1)$$

with a hyperplane

$$(w \cdot x) + b = 0 \quad (2.2)$$

The set of vectors is said to be optimally separated by the hyperplane if it is separated without error and the distance between the closest vectors to the hyperplane is maximal. where the parameters w, b are constrained by

$$min |(w \cdot x) + b| = 1 \quad (2.3)$$

we should find a linear function:

$$f(x) = (w \cdot x) + b \quad (2.4)$$

that is to say ,we should make the margin between the two classes points as possible as big , it is equal to minimize $\frac{1}{2}\|w\|^2$, we should be according to structure risk minimum principle not experiential risk minimization principle, that is to minimize equation (2.5) upper bound with probability $1 - \sigma$,

$$R[f] \leq Remp[f] + \sqrt{\frac{1}{2}(h \ln(\frac{2l}{h} + 1)) + ln(\frac{4}{\sigma})} \quad (2.5)$$

the optimal classification function is transformed into a convex quadratic programming problem:

$$\min \quad \frac{1}{2}\|w\|^2 \quad (2.6)$$
$$\text{s.t.} \quad y_i((w \cdot x_i) + b) \geq 1, i = 1, 2, \cdots, l \quad (2.7)$$

when the training points are non-linearly Separable, (2.6)-(2.7) should be transformed into(2.8)-(2.9).

$$\min \quad \frac{1}{2}\|w\|^2 + c\sum_{i=1}^{l} \xi_i \quad (2.8)$$
$$\text{s.t.} \quad y_i((w \cdot x_i) + b) \geq 1 - \xi_i, i = 1, 2, \cdots, l \quad (2.9)$$

The solution to the above optimization problem of equation (2.8)-(2.9) is transformed into the dual problem(2.10)-(2.12) by the saddle point of the Lagrange functional

$$\min_{\alpha} \frac{1}{2}\sum_{i=1}^{l}\sum_{j=1}^{l} y_i y_j \alpha_i \alpha_j K(x_i, x_j) - \sum_{j=1}^{l} \alpha_j \qquad (2.10)$$

$$\text{s.t.} \quad \sum_{i=1}^{l} y_i \alpha_i = 0 \qquad (2.11)$$

$$0 \leq \alpha_i \leq c, \ i = 1, 2, \cdots, l \qquad (2.12)$$

We can get the decision function:

$$f(x) = \sum_{i=1}^{l} y_i \alpha_i K(x_i, x) + b \qquad (2.13)$$

kernel function $K(x_i, x) = (\Phi(x_i) \cdot \Phi(x))$ is a symmetric function satisfying Mercer's condition, when given the sample sets are not separate in the primal space, we can be used to map the data with mapping Φ into a high dimensional feature space where linear classification is performed.

There are three parameters in svm model that we should choose, they make great impact on model's generalization ability, It is well known that svm generalization performance (estimation accuracy) depends on a good setting of hyperparameters C, the kernel function and kernel parameter. moreover, kernel function and kernel parameter's selection connects with feature selection in svm, so feature selection is very important.

3 Rough Set Theory[4]

Rough sets theory has been introduced by Zdzislaw Pawlak (Pawlak, 1991) to deal with imprecise or vague concepts. It has been developed for knowledge discovery in databases and experimental data set, It is based on the concept of an upper and a lower approximation of a set.

Rough set theory deals with information represented by a table called an information system. This table consists of objects (or cases) and attributes. The entries in the table are the categorical values of the features and possible categories. An information system is composed of a 4-tuple as following:

$$S = \langle U, A, V, f \rangle \qquad (3.1)$$

where U is the universe, a finite set of N objects$\{x_1, x_2, \cdots, x_N\}$(a nonempty set), $A = C \bigcup D$ is condition attribute and decision attribute. V is attribute value. $f : U \times A \longrightarrow V$ is the total decision function called the information function.

For a given information system S, a given subset of attributes $R \subseteq A$ determines the approximation space $RS = (U, ind(A))$ in S, For a given $R \subseteq A$

and $X \subseteq U$ (a concept X), the R-lower approximation $\underline{R}X$ of set X in RS and the R upper approximation $\overline{R}X$ of set X in RS are defined as follows:

$$\underline{R}X = \{x \in U : [x]_R \subseteq X\}, \overline{R}X = \{x \in U : [x]_R \bigcap X \neq \phi\} \quad (3.2)$$

where $[X]_R$ denotes the set of all equivalence classes of $ind(R)$ (called indiscernibility relation). The following ratio defines an accuracy of the approximation of $X (X \neq \phi)$, by means of the attributes from R:

$$\alpha_R = \frac{|\underline{R}X|}{|\overline{R}X|} \quad (3.3)$$

where $|\underline{R}X|$ indicates the cardinality of a (definite) set $\underline{R}X$. Obviously $0 \leq \alpha_R \leq 1$. If $\alpha_R = 1$, then X is an ordinary (exact) set with respect to R; if $\alpha_R < 1$, then X is a rough (vague) set with respect to R.

Attribute reduction is one of the most important concept in RS. the process of finding a smaller set of attributes than original one with same classification capability as original sets is called attribute reduction. A reduction is the essential part of an information system (related to a subset of attributes) which can discern all objects discernible by the original information system. Core is the intersection of all reductions. Given an information system S, condition attributes C and decision attributes D, $A = C \bigcup D$, for a given set of condition attributes $P \subseteq (C)$, we can define a positive region $pos_p(D) = \bigcup_{X \in U/D} \underline{P}X$, The positive region $pos_p(D)$ contains all objects in U, which can be classified without error (ideally) into distinct classes defined by $ind(D)$ based only on information in the $ind(P)$. Another important issue in data analysis is discovering dependencies between attributes. Let D and C be subsets of A. D depends on C in a degree denoted as

$$\gamma_C(D) = |pos_C(D)|/|U| \quad (3.4)$$

It was shown previously that the number $\gamma_C(D)$ expresses the degree of dependency between attributes C and D, It may be now checked how the coefficient $\gamma_C(D)$ changes when some attribute is removed. In other words, what is the difference between $\gamma_C(D)$ and $\gamma_{C-\{\alpha\}}(D)$. Attribute importance $\{\alpha\}$ about decision attribute is defined by

$$\sigma_{CD}\{\alpha\} = \sigma_C(D) - \sigma_{C-\{\alpha\}}(D) \quad (3.5)$$

4 Fault Diagnosis System Based on Rough Set Theory and Support Vector Machine[5][6]

In support vector machine, the solution of the model is transformed into a quadratic programming problem and we will achieve global solution but not local solution. it will produce good generation performance, But it is difficult to resolve a large number of training sample sets and not to deal with the vagueness and uncertainty information. Rough Set Theory is a data analysis tool in pre-processing imprecise or vague concepts. It is only based on the original data and does not need any additional information about data like probability in statistics or grade of membership in the Fuzzy set theory, it can reduce the attributes without decreasing its discriminating capability.

Integrating the advantages of RS and SVM, a kind of support vector machine fault diagnosis systemon the Rough Sets pre-processor is presented in the paper. When given a training sample set, we firstly discretize them if the sample attributes values are continuous and we can get a minimal feature subset that fully describes all concepts by attribute reduction, constructing a support vector machine fault diagnosis system . When given a testing set, we reduce the corresponding attributes and then put into SVM fault diagnosis system, then acquire the testing result. The whole process as fig 1.

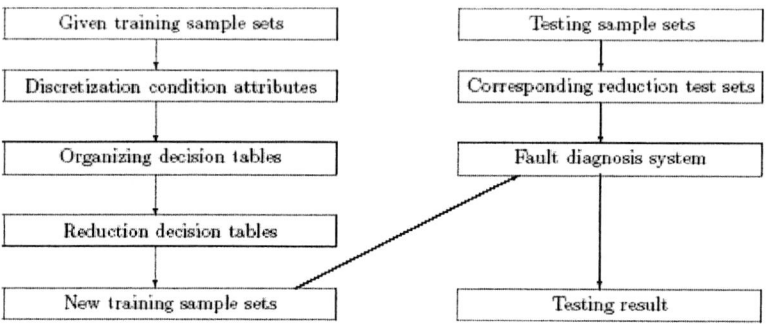

Fig. 1. Fault diagnosis system based on Rough set theory and support vector machine

When the training samples are the two classes separable, we can achieve them by above the fault diagnosis system. At first, we may preprocess training samples sets by Rough set theory, then classify them by support vector machine. Only need one classifier. When the training samples are multi-class (such as k classes), we often resolve them according to blow 3 method.

(1) One versus the rest: one class sample are signed "+1", the rest classes samples are signed "-1", Its need to construct k classification hyperplane to achieve them, that is to need to resolve k quadratic programming problem, but there is drawback that is produce multi classes to some samples, or some samples don't belong to any classes.

(2) One versus one: we can select 2 classes from all classes at random, thus it needs k(k-1)/2 classifier, calculating capacity is very larger. but there is drawback that may produce multi-classes to some samples.

(3) Layer classification method: This method is a improved One versus one method, we may combined K classes into 2 classes at first, and so on, different layers at last, we can classify by support vector machine in each layer.

In order to avoid multi-class to some samples, a kind of hybrid fault diagnosis system based on RS and SVM is presented in the paper. Utilizing the advantages of Rough set theory in extraction of rules, in order to improve classification accuracy, We can classify them exactly by support vector machine.

5 Fault Diagnosis About 4153 Diesel Engine[7][8]

Fault diagnosis on machinery has been researched presently. In this paper, we will take the 4153 diesel engine fault diagnosis for example. The fault diagnosis on diesel engine is a difficult problem due to the complex structure of the engine and the presence of multi-excite sources. The vibration signal of a 4135 diesel under normal and fault states is acquired(i=1,2,3) is the symptom, and represents the waveform complexity in frequency domain, center frequency of spectrum, waveform complexity in time domain, nonperiod complexity, variance of time series, and kurtosis of time series of the signals from measurement point 1, 2 and 3 respectively. They are the first cylinder head, the second cylinder head and another one that is at the center of the piston stroke, on the surface of the cylinder block. D is the fault reason, and the associated integers 1, 2, 3 and 4 represent normal state, intake valve clearance is too small, intake valve clearance is too large, and exhaust valve clearance is too large respectively.

5.1 Continuous Attributes Discretization Based on Fuzzy K-Means

Rough set theory only analyzes the discrete data, but the fault diagnostic data is continuous. so they must be quantized before extraction of rules from the original data. continuous attribute discretization directly affects the analysis result. Considering the vagueness and uncertainty diagnosis data in the process of fault diagnosis, Fuzzy K-means discretization method is proposed in the paper, it is an objective function based on fuzzy clustering algorithm, this algorithm typically converges to the local minimums, and possesses better robustness. In the course of clustering, the number selection of clusters is important. If the clusters are few, incompatible decision system will be resulted, and decision can not be made in the applications. If the clusters are too much, overdiscretization will be resulted, so that match of the condition for every rule will become too complicate. In this paper, corresponding to the states of the engine, 4 clusters are determined for each attribute. we select sample 8,9,10,11,12,13,25,26,27,28,29,30 as testing set, and the rest samples as training set, and getting the discretization decision Table 1.

5.2 Attributes Reduction and Rules Attraction Based Rough Set Theory

By attributes reduction, we can get a reduction of fault diagnosis system decision table as Table 2. certainly it isn't the only reduction table. In Table 2, more redundant values can be reduced, from decision and more concise rules can be

Table 1. Training sample continuous attributes values discretization besed on Fuzzy k-means, the number of clustering is 4

U	a1	b1	c1	d1	e1	f1	a2	b2	c2	d2	e2	f2	a3	b3	c3	d3	e3	f3	D
1	1	3	1	1	1	1	1	1	3	3	3	1	3	1	1	3	1	1	1
2	1	3	1	1	1	1	2	1	3	3	3	1	3	1	1	3	1	1	1
3	1	3	1	1	1	2	3	2	3	1	3	1	1	1	1	4	2	1	1
4	1	3	1	1	1	2	2	2	3	1	3	1	2	1	1	3	2	1	1
5	1	1	3	1	1	2	1	1	3	2	3	1	1	1	1	4	1	1	1
6	1	1	3	1	1	2	1	1	3	2	3	1	1	1	1	3	1	1	1
7	3	3	1	1	1	2	4	3	1	4	1	1	4	1	1	4	2	1	1
14	3	3	1	1	3	4	1	2	3	3	3	1	4	2	1	4	2	1	2
15	3	4	2	1	4	3	1	2	3	3	3	3	4	2	2	4	2	2	2
16	3	3	1	1	4	3	1	2	3	3	3	4	4	2	3	4	2	2	2
17	3	4	2	2	3	1	1	3	1	3	1	2	4	3	4	1	4	4	2
18	3	4	2	1	3	1	1	3	1	3	1	2	4	4	4	1	4	4	2
19	1	3	1	1	1	4	1	4	4	3	4	1	4	1	1	4	3	1	3
20	1	3	1	3	1	4	1	4	4	3	4	1	4	1	1	4	3	1	3
21	1	3	1	1	1	3	1	3	1	1	1	1	1	3	1	1	1	1	3
22	1	3	1	1	1	1	1	3	1	3	1	1	4	1	1	4	3	1	3
23	1	3	1	1	1	3	1	3	1	3	1	1	4	1	1	4	3	1	3
24	1	4	2	1	1	3	1	3	1	3	1	1	4	1	1	4	3	1	3
31	1	1	3	1	1	1	1	3	1	1	1	1	1	3	1	1	1	1	4
32	1	3	1	1	1	1	1	3	1	3	1	1	4	4	4	4	2	1	4
33	1	3	1	1	1	1	1	3	1	3	1	1	4	1	1	4	2	1	4
34	1	1	3	2	2	1	1	3	2	3	2	1	4	2	1	4	2	1	4
35	1	1	3	2	2	4	1	3	1	3	1	1	4	2	1	4	2	1	4
36	3	2	4	4	1	1	1	3	1	3	1	1	4	2	1	4	2	1	4
37	3	1	3	4	1	1	1	3	1	3	1	1	4	2	1	4	2	1	4

Table 2. The fault diagnosis decision system table after attributes reduction

U	e1	f1	e3	D
1	1	1	1	1
3	1	2	2	1
5	1	2	1	1
15	4	3	2	2
17	3	1	4	2
19	1	4	3	3
21	1	3	3	3
31	1	1	2	4
34	2	1	2	4
35	2	4	2	4

generated. we can know that attribute $\{e1, f1, e3\}$ are the most important attributes in the fault diagnosis system. we will get the same fault diagnosis result without losing any information by decision Table 2.

We can achieve some decision rules by the decision Table 2, such as
(1) If $e1 = 1$, $f1 = 1$ and $e3 = 1$ then $D = 1$;
(2) If $e1 = 1$, $f1 = 2$ and $e3 = 2$ then $D = 1$;
(3) If $e1 = 1$, $f1 = 2$ and $e3 = 1$ then $D = 1$;
(4) If $e1 = 1$, $f1 = 1$ and $e3 = 2$ then $D = 4$;
and so on.

As we can learn that the decision rules will be different because of different attribute value about f1 or e3. It will cause much difficult in the course of fault diagnosis. The rules generated by the Rough set theory are often unstable and have low fault diagnosis accuracy. In order to improve the fault diagnosis accuracy, we will further diagnose them by support vector machine because of it's greater generalization performance.

For example, if e1=3 or 4 then we can learn D= 2, if e1=2 then D=4. such can reduce the diagnosis time, but if e1=1, it will cause difficult to our diagnosis because of being to D=1 or D=3 or D=4. It need us to further diagnosis, we can classify them by the second or third attribute values, but we can't classify them exactly sometimes, we can classify them by support vector machine and by using the first multi-class method.

Certainly, we can construct the support vector fault diagnosis system based on the attribute reduction by Rough set theory. There are 18 attributes in the fault diagnosis system before the attribute reduction, but there are only 3 attributes after attribute reduction, it will bring us convenience to fault diagnosis and overcome effectively the drawback of support vector machine.

5.3 Multi-classification Based on Support Vector Machine

We can separately construct the fault diagnosis system on the conditions of before and after attribute reduction of the original data. we select one against the rest method in the multi-classes method. In 25 training samples, we separately train and test. Firstly, sample 1-7 as "+1" and the rest samples as "-1"; second, sample 14-18(original data) as "+1" and the rest samples as "-1"; third, sample 19-24(original data) as "+1" and the rest samples as "-1"; the last, sample 31-37(original data) as "+1" and the rest samples as "-1".

Choosing the parameter C=10, kernel function $K(x_i, x) = e^{-\gamma \|x - x_i\|^2}$ and kernel parameter $\gamma = 0.05$, getting decision function (2.15). then testing the 12 testing sets, fault diagnosis results(average accuracy) as Table 3.

Table 3. The fault diagnosis result comparative table about multi-classification, first-rest is the first class versus the rest classes

methed	first-rest	second-rest	third-rest	fourth-rest
Svm(%)	100	100	91.6	100
Rsvm(%)	100	100	100	100

As we can learn that great diagnosis accuracy has been produced only by support vector machine, and only one sample testing error. but fault diagnosis accuracy based on support vector machine and Rough set pre-processing is 100%. At the same time, reducing fault diagnosis system complexity, reducing training time and data storage space. generally speaking, it contributes us to diagnose the fault on time and reduce the cost for machine fault.

6 Conclusions

On the one hand, on the condition of keeping with same diagnosis ability, making great use of the advantage of Rough set theory in pre-processing, eliminating redundant information and reducing the training sample's dimension, a kind of support vector machine fault diagnosis System based on Rough set pre-processing is proposed in the paper. On the other hand, utilizing the advantage of Rough sets theory in acquiring diagnosis rules and combining with support vector machine greater generalization performance, a kind of hybrid fault diagnosis system based on Rs and SVM is proposed too in the paper. The diagnosis of a diesel demonstrated that the solution can reduce the cost and raise the efficiency of diagnosis.

References

1. Kecman,v. Learning and Soft Computing, support vector machine, Neural Networks and Fuzzy Logic Models.The MIT Press, Cambridge, MA (2001)
2. Deng Naiyang, Tian Yingjie. A new method of Data Mining-Support Vector Machine, Science Press (2004)
3. Wang, L.P.(Ed.): Support Vector Machines: Theory and Application. Springer, Berlin Heidelbrg New YorK (2005)
4. Zhang Wenxiu, Wu Weizhi. Rough Set Theory and Application, Science Press (2001)
5. Li Bo, Li Xinjun. A kind of hybrid classification algorithm based on Rough set and Support Vector Machine, Computer Application. 3 (2004) 65-70
6. Renpu Li, Zheng-ou Wang. Mining classification rules using rough sets and neural networks. European Journal of Operational Research. 157 (2004) 439-448
7. Lixiang Shen, Francis E.H. Tay, Liangsheng Qu, Yudi Shen. Fault diagnosis using Rough Sets Theory. Computers in Industry. 43 (2000) 61-72
8. FENG Zhi-peng, DU Jin-lian, SONG Xi-geng, CHI Zhong-xian, GE Yu-lin, SUN Yu-ming. Fault diagnosis based on integration of roughsets and neural networks.Journal of Dalian University of Technology. 1 (2003) 70-76

A Fuzzy Framework for Flashover Monitoring

Chang-Gun Um, Chang-Gi Jung, Byung-Gil Han, Young-Chul Song,
and Doo-Hyun Choi*

School of Electrical Engineering and Computer Science, Kyungpook National University,
Daegu, 702-701, South Korea
dhc@ee.knu.ac.kr

Abstract. This paper presents a new analysis method of the leakage current on contaminated polymer insulators under salt-fog conditions. The proposed method tries to combine frequency-domain information with time-domain information using the framework of the fuzzy inference engine. Experimental results show that the unified approach of different domain data using fuzzy framework is available for flashover prediction and monitoring the contamination conditions of outdoor insulator.

1 Introduction

Although polymer insulators are increasingly being used in power distribution lines, they have several disadvantages like aging, unknown long-term reliability, and difficulties in detecting defection. Aging is the main cause of registered failure for polymer insulators and leads to a flashover under contaminated conditions even at a normal operating voltage. Leakage current is the most crucial cause and consequence of aging in the contaminated insulator. There are so many researches on nondestructive testing are performed [1-3] and also many researches on the leakage current analysis are carried out to predict flashover of an insulator [4]. However, in most cases, only the low-frequency components, namely, the fundamental, 3rd, and 5th harmonic components are used for a spectral analysis of the leakage current. It is based on the assumption that the low-frequency components contain more important information than the high-frequency components [4, 5].

In this paper, a new framework to combine time-domain information with frequency-domain hints. The proposed framework is basically the fuzzy inference engine. It uses two inputs: one is the high-frequency energy and the other is the standard deviation of leakage current at a predetermined interval.

2 Proposed Fuzzy Flashover Monitoring System

A new framework based on fuzzy inference engine is presented in this paper. It is designed to analyze leakage current on a contaminated insulator. The proposed framework uses different domain signals as inputs and expert knowledge is used to construct fuzzy rule base. Fig. 1 shows all the architecture of the proposed framework.

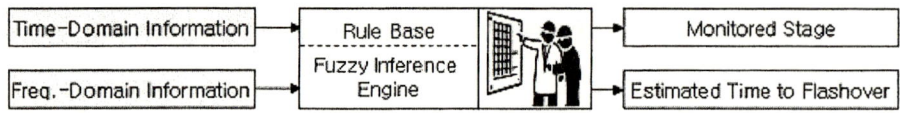

Fig. 1. Proposed Fuzzy Flashover Monitoring Framework

3 Experimental Setup and Typical Stages of Leakage Current

The experiment based on the salt-fog test method was carried out using the equipment shown in Fig. 2. An EPDM-distribution-suspended insulator with a diameter of 100mm was used in the experiment. It was contaminated in coastal areas from 1998 to 2002. The leakage current on the insulator was measured simultaneously with the fog application. The NaCl content in deionized water was adjusted to 25g, 50g, and 75g per liter. Applying 18kV on the insulator in a laboratory fog chamber, several tests were conducted and the leakage current was measured. The measurement was continued until the flashover occurred. During the measurement, a video camera was used to record the flashover behavior on the surface of the insulator. The measured leakage currents were stored in a PC via a 12-bit A/D converter with a sampling period of 0.1 ms [6].

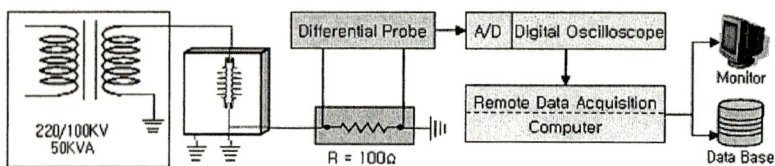

Fig. 2. Experimental Setup

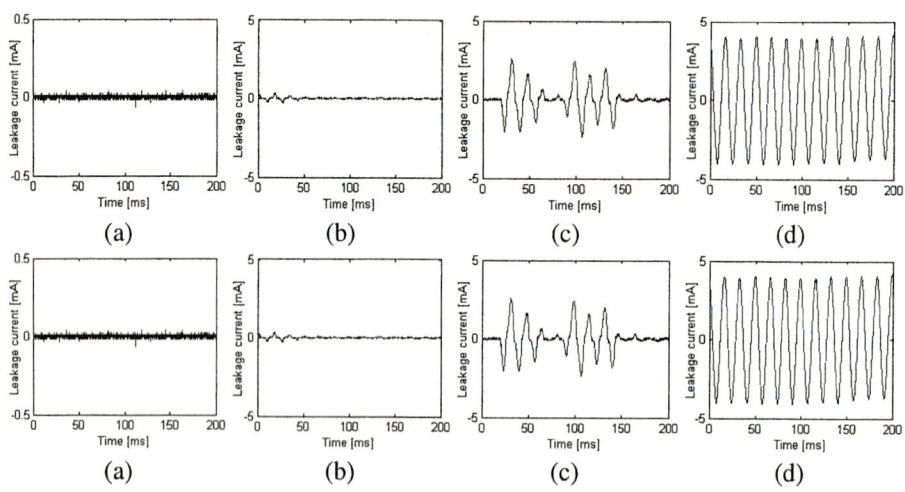

Fig. 3. Leakage current at each stage. (a) Initial stage, (b) middle stage, (c) final stage, (d) flashover stage.

Fig. 3 presents the typical waveforms of the leakage current. Based on the waveforms, the progress of flashover on an insulator can be categorized as the 4 stages: initial, middle, final, and flashover stages. As close to the flashover stage, the waveform of the leakage current is shaped like complete sine wave and the amplitude is increased [7].

4 Experimental Result and Discussion

At first, it is tried to find which bend is useful to monitor the flashover progressing. It is found that the accumulated power of high frequency components (especially between 3950-4910Hz) shape up uniformly stable steps from initial stage to flashover stage. The power is a useful hint to find the status of a contaminated insulator.

Also, it is found that the feature of standard deviation increases gradually. The standard deviation of leakage current doesn't divide clearly each stage of insulator condition. But it tends to increase continuously and doesn't diverge suddenly. It can be used as another hint for flashover monitoring.

To use the two different domain features at a framework, the fuzzy inference engine is used. Experts' knowledge is modeled to construct the fuzzy rule base. Table 1 represents the implemented rule base and fuzzy membership functions are represented in Fig. 4.

Table 1. Fuzzy rule base

Input feature & Stage		Standard Deviation			
		Initial	Middle	Final	Flashover
Accumulated Power	Initial	Initial	Initial	Initial	Middle	Middle
	Middle	Middle	Middle	Middle	Middle	Final
	Final	Final	Final	Final	Final	Final
	Flashover	Flashover	Final	Final	Final	Flashover

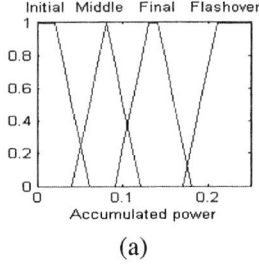
(a)

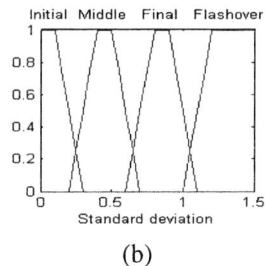
(b)

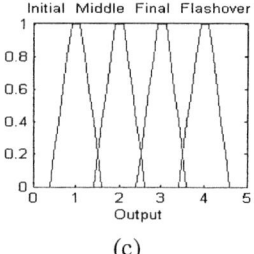
(c)

Fig. 4. Fuzzy membership functions for (a) Accumulated power, (b) Standard deviation, and (c) Output Stage

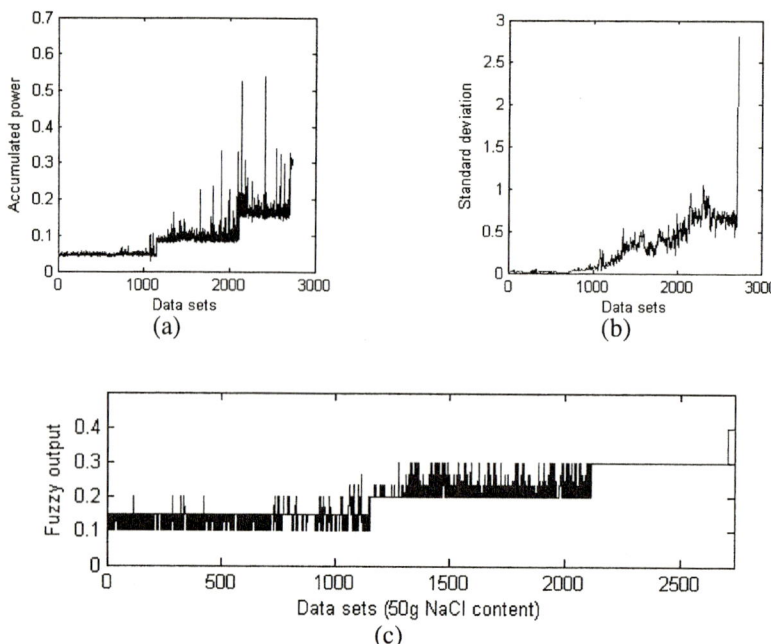

Fig. 5. Input features and fuzzy output for 50g NaCl contents. (a) Accumulated power, (b) Standard deviation, (c) Output stage.

Fig. 5 shows two input features and the determined output stage for 50g NaCl contents. It is easy to find that the fuzzy monitoring system provides stable status information. Moreover, it is very useful that the proposed system provides all the same criterion for status decision. This is a very good side effect of the proposed fuzzy system. Using this fact, it might be possible to make a general tool to estimate the flashover time without curse of too many parameters.

Recently, a paper on the aging of polymer insulators is published [8]. But it uses only high frequency componets of leakage current and does not use any knid of time-domain information.

References

1. Doo-Hyun Choi, Young-Chul Song, Ji-Hong Kim, and Jung-Gu Kim: Hand-Eye Coordination of a Robot for the Automatic Inspection of Steam-Generator Tubes in Nuclear Power Plants. Key Engineering Materials – Advances in Nondestructive Evaluation (Aug. 2004) 2256-2261
2. Jung-Gu Kim, Hong Jeong, Doo-Hyun Choi, and Ji-Hong Kim: Stereo Vision System for Welding Line Tracking Robot. Key Engineering Materials – Advances in Nondestructive Evaluation (Aug. 2004) 2315-2320

3. Young-Chul Song, Doo-Hyun Choi, and Kil-Houm Park: Morphological Blob-Mura Defect Detection Method for TFT-LCD Panel Inspection. Lecture Notes in Artificial Intelligence, Vol. 3215, (Sep. 2004) 862-868
4. Suda, T. : Frequency Characteristics of Leakage Current Waveforms of a String of Suspension Insulators. IEEE Transactions on Power Delivery, Vol. 20, Issue 1, (2005) 481-487
5. A. H. Ei-Hag, S. H. Jayaram, and E. A. Cherney : Fundamental and Low Frequency Harmonic Components of Leakage Current as a Diagnostic Tool to Study Aging of RTV and HTV Silicon Rubber in Salt-Fog. IEEE Transactions on Dielectrics and Electrical Insulation, Vol. 10, No. 1, (2003) 128-136.
6. Young-Chul Song, Jan-Jun Park and Doo-Hyun Choi: A Flashover Prediction Method for Insulators Stochastic Analysis of Leakage Current. Japanese Journal of Applied Physics. Vol. 43, No. 5A, (2004) 2693-2696
7. M. A. R. M. Fernando and S. M. Gubanski : Leakage Current Patterns on Contaminated Polymeric Surfaces. IEEE Transactions on Dielectrics and Electrical Insulation, Vol. 6, Issue 5, (1999) 688-694
8. Young-Chul Song and Doo-Hyun Choi: High-frequency components of leakage current as diagnostic tool to study aging of polymer insulators under solt fog. Electronics Letters, vol. 41, no. 12, (Jun. 2005) 17-18.

Feature Recognition Technique from 2D Ship Drawings Using Fuzzy Inference System

Deok-Eun Kim[1], Sung-Chul Shin[2], and Soo-Young Kim[1]

[1] Pusan National University, Jangjeon-Dong, Geumjeong-Gu, Busan, South Korea
{punuri, sykim}@pusan.ac.kr
[2] Mokpo National Maritime University, Jukkyo-Dong, Mokpo, SouthKorea
scshin@mmu.ac.kr

Abstract. This paper presents the feature recognition technique that recognizes the features from 2D ship drawings using the fuzzy inference system. Generally, ship drawings include a lot of symbols and texts. They were complicatedly combined each other. So, it is very difficult to recognize the feature from 2D ship model. The fuzzy inference system is suitable to solve these problems. Input information for fuzzy inference is connection type of drawing elements and properties of element. Output value is the correspondence between target feature and candidate feature. The recognition rule is the fuzzy rule that has been predefined by designer. In this study, the midship section drawing of general cargo ship was used to verifying suggested methodology. Experimental results showed that this approach is more efficient than existing methods and reflects the human knowledge for recognition of the feature.

1 Introduction

In the initial stage of ship design, the concept of product is embodied and function, arrangement and simple feature of product are defined. And the best-optimized design plan is confirmed through the inspection of scheme. This process has to be accomplished rapidly. But design knowledge is poor in this stage. Therefore, 2D drawings are usually used in this stage. However, ship drawings in the detail stage are represented as 3D model. Because that it makes possible to check the interference of parts, analyze the structural safety and the hydrodynamic performance etc. Currently, the processes that transform 2D model into 3D model are performed by designers. But many problems have been occurred in this process such as the delay of design time, the omission of parts information, and the mistake of input etc.

Therefore, the automatic transform technique that transforms the 2D model into the 3D model is required. The first step of automatic transformation technique is to recognize the part feature from 2D ship model.

Shin applied the feature recognition method. He used the recognition rule such as table 1 to recognize the feature of 2D ship drawing [5]. However, '10% inclined', '50% inclined', '1st UP' and others are crisp value. If the feature properties are fuzzy and have some range, this recognition rule cannot recognize the ship feature. Therefore, we introduced fuzzy inference system to overcome these problems. Fuzzy rule is suitable for these problems [3] [6].

Table 1. The recognition rule of ship design feature that used by Shin.

FEATURE	H-LOCATION	V-LOCATION	LENGTH	DIRECTION
DECK	1st UP	LEFT	LONGEST	10% INCLINED
SSHELL	1st DOWN	LEFT	LONGEST	VERTICAL
BOTTOM	1st DOWN	LEFT	LONGEST	HORIZONTAL
IN-BOTTOM	2nd DOWN	LEFT	LONGEST	HORIZONTAL
GIRDER	DOWN	LEFT		VERTICAL
TTOP	2nd UP	LEFT		50% INCLINED
SLANT	DOWN	LEFT		50% INCLINED

2 Methodology

First step is to read entity from 2D drawings. Secondly, it constructs graph structure based on read entity. Finally, it finds target feature on constructed graph structure using fuzzy inference system and graph matching algorithm.

2.1 Read Entity from 2D Drawings

The In this study, the format of input drawings is DXF(Data eXchange Format) file format. For the feature recognition system has wide use, the input drawing format has to be shared data structure between different CAD systems. Now, most of CAD system support DXF file format. A DXF file is divided into six sections. They are HEADER, CLASSES, BLOCKS, TABLES, OBJECT and ENTITIES. Through the division of sections, user can get easily desired information from DXF file. Among the six sections, only ENTITIES section contains geometric information. So, feature information is extracted from this section.

2.2 Filtering of Input Data

Generally, drawing composed of a number entity. A drawing of midship section includes about 3000 ~ 4000 entity. If all entities are read into memory, recognition process become very complex and calculation performance is deteriorated. In this study, it solved this problem using filtering process. Filtering criterion can be color of entity, layer information, and line type etc.

In the drawing of midship section, a longitudinal part and a transverse part are drawn together. However, these parts are classified by color information. In the drawing that was used this study, longitudinal parts are drawn red color and transverse parts are drawn white color. Before filtering, the number of entity was 828. After filtering, the number of entity was 88. Amount of processing data decreased over 80%.

2.3 Construction of Graph Structure

Feature is an assembly of basic entity. Accordingly, the relation of entities must be considered for feature recognition as well as property of entity [1] [2].

Therefore, entities that were filtered are constructed as an upper level data structure. The data structure has to include the property of entity and the relation of entities. In this study, graph data structure used for upper level data structure [4]. Graph data structure composed of vertices and edges. The properties of entity are saved vertices. The relation of entities is saved edge. The properties of entity are length, angle, point, and color etc. The relations of entities are the type of connection, the angle of connection etc.

2.4 Feature Recognition Using Fuzzy Inference

The feature recognition is achieved by graph matching algorithm and fuzzy inference system. First, select a candidate sub-graph on the whole graph structure. Secondly, evaluate a score of candidate sub-graph through the fuzzy inference system. Input valuables of fuzzy inference module are the properties of entities and the relations of entities. This is extracted from graph data structure. The evaluation function is a fuzzy rule for feature recognition. Fuzzy rule is predefined by user. And, repeat this process continuously. Finally, sub-graph that gets a best score is confirmed as the desired feature.

3 Application Result

In this section, we present a set of examples to illustrate and verify the suggested method. We experimented to recognize bottom plate and longitudinal stiffener from the midship section drawing of cargo tanker using suggested method. The recognition method that Shin suggested can recognize bottom, side but can't longitudinal stiffener. The method of this study can recognize both of bottom and longitudinal stiffener.

3.1 Recognition of Bottom Plate

The parts of parts that would be evaluated are 87. However, representative parts were only represented in table 2, 3 for the rack of paper space. Input valuables of fuzzy inference are a length of a part, an angle of a part, and a horizontal position of a part on whole drawing. The horizontal position of center on drawing is zero. According to table 2 Maximum score is 81.5(part 4). Therefore, part 4 was recognized as the bottom plate.

Table 2. Result of bottom plate recognition

Part	Length(mm)	Angle(degree)	Horizontal Position (mm)	Score
1	500	0	2709	18.3
2	1000	90	3459	14.2
3	5418	90	-1331	15.8
4	5319	0.88	-199	81.15
5	4922	34.8	-379	37

3.2 Recognition of Longitudinal Stiffener

Table 3 shows the result of longitudinal stiffener recognition. Input valuables of fuzzy inference are lengths of two parts, an angle between two parts. Maximum score is 83.9(part 3). Therefore, part 3 was recognized as longitudinal stiffener

Table 3. Result of longitudinal stiffener recognition

Part	Length1(mm)	Angle(degree)	Length2 (mm)	Score
1	500	90	1000	18.3
2	5418	88.1	3116	18.7
3	250	90	90	83.9
4	150	88.2	6503	15.1
5	150	0.88	90	15.3

4 Conclusions

In this study, we have introduced the fuzzy theory in the feature recognition technique. It makes the recognition of complex parts that have been recognized in the existed method from 2D ship drawings. Also, we have introduced the filtering process for decrease of calculation amount. It makes possible to decrease the amount of data processing over 80%.

In conclusion, this study has presented a basic method for transforming 2D model into 3D model and improved the recognition performance of part feature form 2D ship drawings.

References

1. B. Aldefeld: On automatic recognition of 3D structures from 2D representations. Computer Aided Design, Vol.15. Elsevier (1983)
2. G. Little, R. Tuttle, D.E.R. Clark, J. Corney: A Graph-based Approach to Recognition. Proceedings of DETC97, ASME Design Engineering Technical Conferences September (1997)
3. Mamdani E.H.: Applications of fuzzy algorithms for simple dynamic plant. IEE Proceedings, vol. 121 (1974) 1585-1588
4. M. R. Henderson, S. H. Chuang, P. Ganu, P. Gavankar: Graph-Based Feature Extraction, Arizona State University (1990)
5. Y.J., Shin: Data Enhancement for Sharing of Ship Design Models. Computer Aided Design, Vol. 30. Elsevier (1998) 931-941
6. Zadeh L.A.: Fuzzy sets. Information and Control, vol. 8 (1965) 338-353

Transmission Relay Method for Balanced Energy Depletion in Wireless Sensor Networks Using Fuzzy Logic*

Seung-Beom Baeg and Tae-Ho Cho

School of Information and Communication Engineering, Sungkyunkwan University
{envy100, taecho}@ece.skku.ac.kr

Abstract. Wireless sensor networks will become very useful in the near future. The efficient energy consumption in wireless sensor network is a critical issue since the energy in the nodes is constrained resource. In this paper, we present a transmission relay method of communications between BS (Base Station) and CHs (Cluster Heads) for balancing the energy consumption and extending the average lifetime of sensor nodes by the fuzzy logic application. The proposed method is designed based on LEACH protocol. The area deployed by sensor nodes is divided into two groups based on distance from BS to the nodes. RCH (Relay Cluster Head) relays transmissions from CH to BS if the CH is in the area far away from BS in order to reduce the energy consumption. RCH decides whether to relay the transmissions based on the threshold distance value that is obtained as a output of fuzzy logic system. Our simulation result shows that the application of fuzzy logic provides the better balancing of energy depletion and prolonged lifetime of the nodes.

1 Introduction

Recent advances in micro-electro-mechanical systems and low power and highly integrated digital electronics have led to the development of micro-sensors [1]. These sensors measure ambient conditions in the environment surrounding them and then transform these measurements into signals that can be processed to reveal some characteristics about phenomena located in the area around these sensors. A large number of these sensors can be networked in many applications that require unattended operations, hence producing a wireless sensor network (WSN) [2]. These systems enable the reliable monitoring of a variety of environments for applications that include home security, machine failure diagnosis, chemical/biological detection, medical monitoring, habitat, weather monitoring and a variety of military applications [3-5].

To keep the cost and size of these sensors small, they are equipped with small batteries that can store at most 1 Joule [6]. This puts significant constraints on the power

---

* This research was supported by the MIC (Ministry of Information and Communication), Korea, under the ITRC (Information Technology Research Center) support program supervised by the IITA (Institute of Information Technology Assessment).

available for communications, thus limiting both the transmission range and the data rate. The cost of transmitting a bit is higher that a computation [7] and hence it may be advantageous to organize the sensors into clusters. Since the sensors are now communicating data over short distance in the clustered environment, the energy spent in the network will be much lower than energy spent when every sensor communicates directly to the information-processing center. Therefore, it is important to consider the balanced energy dissipation in WSN. In this paper, we propose a method for balancing the energy depletion by improving LEACH protocol with the application of fuzzy logic system.

The remainder of the paper is organized as follows. Section 2 gives a brief description of the LEACH and motivation of this work. Section 3 shows the details of the transmission relay using fuzzy logic. Section 4 reviews the simulation results and comparisons. Conclusions are made in Section5.

2 Related Work and Motivation

In general, routing in WSNs can be classified into three types. These are *flat-based* routing, *clustering-based* routing, and *direct communication-based* routing. In flat-based routing the roles of all nodes are identical, whereas, that of clustering-based routing are different. In direct communication-based routing, a sensor node sends data directly to the Base Station (BS) [8]. The detailed explanation on these routing protocols can be found in [9][10][11][12][13][14].

Clustering routing is an efficient way to lower energy consumption within a cluster, performing data aggregation and fusion in order to decrease the number of transmitted message to the BS [2].

2.1 LEACH

LEACH is a clustering-based protocol that utilizes randomized rotation of the cluster-heads (CHs) to evenly distribute the energy load among the sensor nodes in the network. LEACH is organized into rounds, where each of them begins with a set-up phase, and is followed by a steady-state phase [8]. In the set-up phase, clusters are organized and CHs are selected. In the steady-state phase, the actual data transfer to the BS takes place. The duration of steady-state phase is longer than the duration of the set-up phase in order to minimize overhead. During the set-up phase, a predetermined fraction of nodes, p, elect themselves as CH as follows. A sensor node chooses a random number, r, between 0 and 1. If this random number is less than threshold value, $T(n)$, the node becomes a CH for the current round. The threshold value is calculated based on the equation (1) that incorporates the desired percentage to become a CH, the current round, and the set of nodes that have not been selected as a CH in the last $(1/P)$ rounds, denoted G. It is given by

$$T(n) = \frac{p}{1 - p(r \bmod (1/p))} \text{ if } n \in G. \tag{1}$$

G is the set of nodes that are involved in the CH election. All elected CHs broadcast an advertisement message to the rest of the nodes in the network that they are the new CHs. All the non-CH nodes, after receiving this advertisement, decide on the cluster to which they want to belong. This decision is base on the signal strength of the advertisement. The non-CH nodes inform the appropriate CHs that they will be a member of the cluster. After receiving the entire message from the nodes that would like to be included in the cluster, the CH node creates a TDMA schedule assign each node a time slot when it can transmit. This schedule is broadcast to all the nodes in the cluster [1][2].

During the steady-state phase, the sensor nodes can begin sensing and transmitting data to the CHs. The CH node, after receiving all data, aggregates it before sensing it to BS. After a certain time, which is determined a priori, the network goes back into the set-up phase again and enters another round of selecting new CHs. Each cluster communicates using different CDMA codes to reduce interference from nodes belonging to other clusters [15].

2.2 Motivation

Radio model use in LEACH as followed [15]:
For transmitting

$$E_{Tx}(k,d) = E_{Tx-elec}(k) + E_{Tx-amp}(k,d)$$
$$E_{Tx}(k,d) = E_{elec} * k + \epsilon_{amp} * k * d^2 \quad (2)$$

For receiving

$$E_{Rx}(k) = E_{Rx-elec}(k)$$
$$E_{Rx}(k) = E_{elec} * k \quad (3)$$

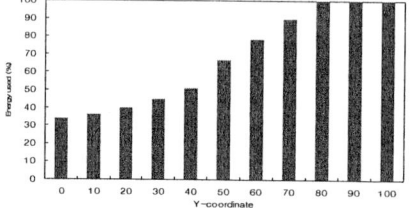

Fig. 1. Energy histogram in LEACH

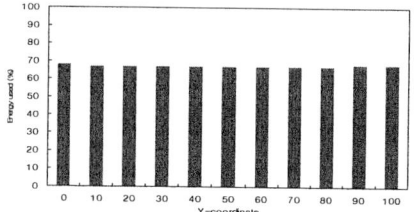

Fig. 2. Ideal energy histogram

Figure 1 present a LEACH energy consumption histogram at a certain point in time. Nodes, which are near by BS, have hardly been used, while others, which are far from BS, have almost completely drained their energy. The unbalance of energy depletion is caused by different distance from BS [16]. If we assume that all the nodes are equally important, no nodes should be more critical than other one. At each moment every node should, therefore, have used about the same amount of energy, which should also be minimized [17]. We improve the balanced energy depletion in LEACH as Figure 2.

3 Transmission Relay Using Fuzzy Logic

The transmission relay is designed by enhancing LEACH protocol with the application of fuzzy logic which takes remaining energy level of RCH, average distance of CHs and number of alive nodes as inputs. Output of the fuzzy logic system is a threshold distance value for deciding whether to relay transmissions or not.

The following subsections explain how the transmission relay is done and how to calculate the threshold distance value.

3.1 Transmission Relay

The sensor network being considered in this paper has the following properties:

- The BS is located far from the sensors.
- All nodes in the network are homogenous and energy-constrained.
- Symmetric propagation channel is employed.
- CH performs data compression.
- All nodes have location information about themselves.

Figure 3 shows a 100-node sensor network in a play field of size 100m x 100m. A typical application in a sensor network is gathering of sensed data at a distant BS [9]. We assume that all nodes have location information about themselves. The location of nodes may be available directly by communicating with a satellite using GPS if nodes are equipped with small low-power GPS receiver [18].

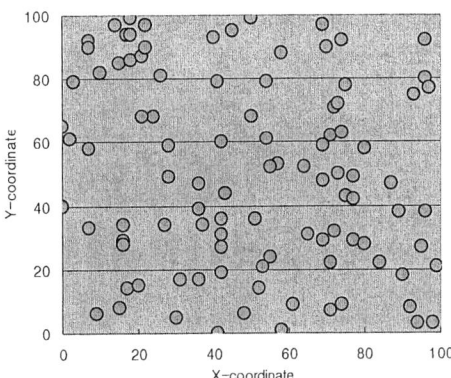

Fig. 3. Random 100-node topology for 100m x 100m network. BS is located at (50, 200), which is at least 100m from the nearest nodes.

Based on equation (2) and (3), CH located at (40, 0) spends 8.1 mJ energy for a transmission, whereas, CH at (40, 100) spends 2.1 mJ energy for a transmission, hence, the cost to transmit a data will be four times that of CH at (40, 100). However, if a node located at (40, 100) relays transmission of CH that is located at (40, 0), the CH spends 2.1 mJ and the node spends 3.25 mJ which was included transmitting and receiving energy. Consequently, the consumption of transmission energy is balanced and the total energy spent is less for the relaying case.

Figure 4 shows how data messages are transmitted from CHs to BS in LEACH protocol and the transmission relay method. In LEACH, two CHs located beyond the threshold value transmit collected data directly to BS as shown in dotted lines. In transmission relay method, the two CHs transmit collected data to RCH as shown in dashed lines and then RCH relays to BS.

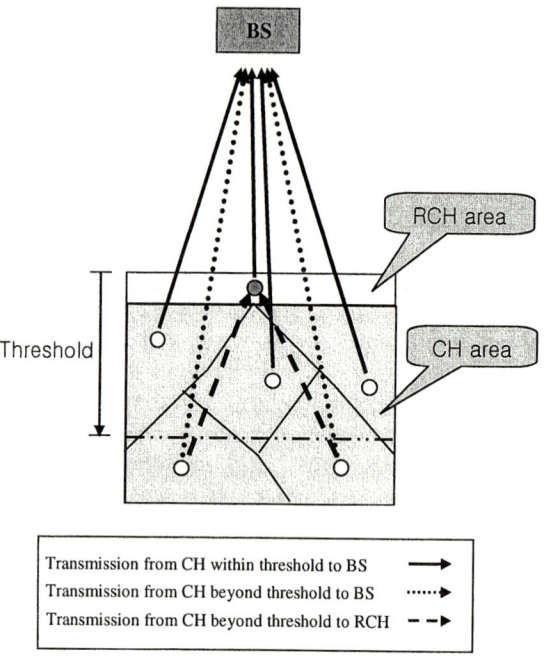

Fig. 4. RCH (shadowed circle) relays transmission of CH located at beyond the threshold distance value

We divide network into two groups based on the distance from BS: RCH area and CH area. RCH area is from the nearest BS to position which was decided network operator. If RCH area is very close to BS, nodes in RCH area are dying quickly. On the other hand, if RCH area is very far from BS, CH is dying quickly. This value depends on Y-coordinate of plot, sensing area and number of nodes.

By equation (1), RCH and CH are elected. The election of RCH and CH are done from the nodes in RCH area and CH area respectively. CH acts as the same manner as in LEACH. RCH relays transmission from CH to BS so that transmission energy of CH, which is far from BS, can be reduced. However, if RCH relays all of CH's transmission, nodes in RCH area are very quickly dying since RCH nodes spend a lot of energy in receiving transmissions. Thus, it doesn't expect balanced energy depletion.

3.2 Threshold Distance Value

RCH decides whether to relay transmissions or not based on the threshold distance value. RCH relays transmission of CH beyond the threshold distance value. The threshold value is the distance from the node closest to BS to a certain position that is calculated by fuzzy application. When all elected CHs broadcast an advertisement message to the rest of the nodes in the network, the location information of each CH is included in the advertisement messages. RCH compares the location information of CH with threshold value so that RCH can decide weather relaying is needed or not. If CHs are located at beyond the threshold value, RCH reports its position to these CHs. Then the CHs transmit the collected data to RCH instead of BS.

Since the CHs are elected from the various positions within CH area the threshold value should be dynamically decided for the balanced energy depletion as well as extending the lifetime of sensor nodes. In our method, RCH dynamically calculates the threshold value based on remaining energy level of RCH, average distance of CHs, number of alive nodes.

Fuzzy input sets are the energy of RCH (represented by ENERGY), average distance of CHs (represented by DISTANCE) and number of alive nodes in RCH area (represented by ALIVE). Figures 5-7 illustrates the mapping of inputs of the fuzzy logic into some appropriate membership functions for the remaining energy of RCH, average distance of CHs and the number of nodes in RCH area, are presented in Figure 5-7, respectively. Where ENERGY = {VLOW, LOW, LMEDIUM, HMEDIUM, HIGH, VHIGH}, DISTANCE = {VSHORT, SHORT, NORMAL, LONG, VLOG}, ALIVE = {LOW, MEDIUM, HIGH}. The output parameter of the fuzzy logic THRESHOLD is defined as the dynamic threshold of our method. The fuzzy linguistic variables for the output are 'VSHORT', 'SHORT', 'SNORMAL', 'LNORMAL', 'LONG', 'VLONG', which is represented by the membership functions as shown in Figure 8. The rules are created using the fuzzy system editor contained in the *Matlab Fuzzy Toolbox*.

There are two different fuzzy logic applications for deciding dynamic threshold values. These are called as ED and EDA. ED considers ENERGY and DISTANCE. If ENERGY is HIGH and DISTANCE is NORMAL the threshold value is set to a value below the NORMAL value of DISTANCE, e.g. VSHORT, SHORT, SNORMAL. On the contrary, if ENERGY is LOW and DISTANCE is NORMAL the threshold value is set to the value above the NORMAL value of DISTANCE, e.g., LNORMAL, LONG, VLONG. Figure 9 illustrate control surface of ED based on fuzzy if-then rules. In the figure, ENERGY is the remaining energy of RCH and DISTANCE is the distance from BS to the average distance of CHs. If the energy level at RCH is low, i.e. below 0.1 J, the threshold distance is decided at its maximum allowed value for the most of CH's average distance.

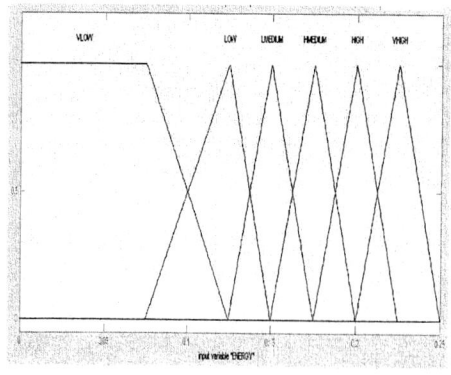

Fig. 5. Membership function for energy of RCH (ENERGY)

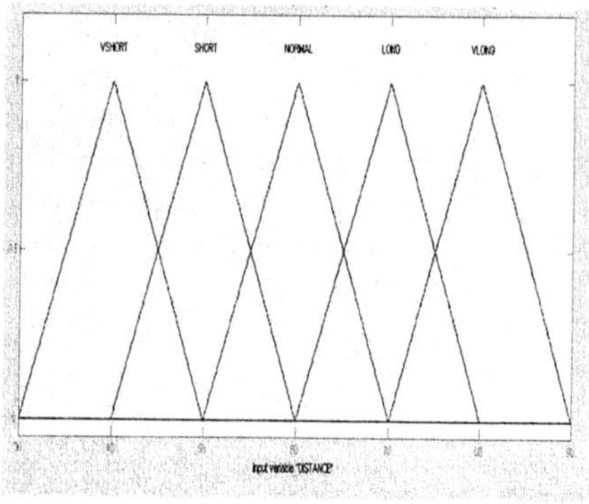

Fig. 6. Membership function for average distance of CHs (DISTANCE)

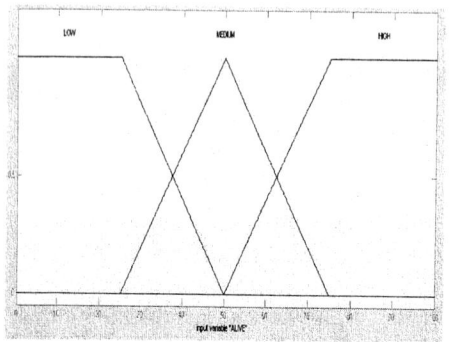
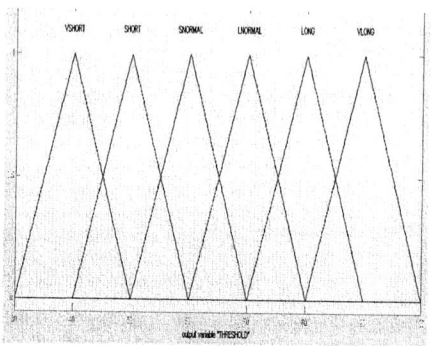

Fig. 7. Membership function for number of alive node (ALIVE)

Fig. 8. Output membership function

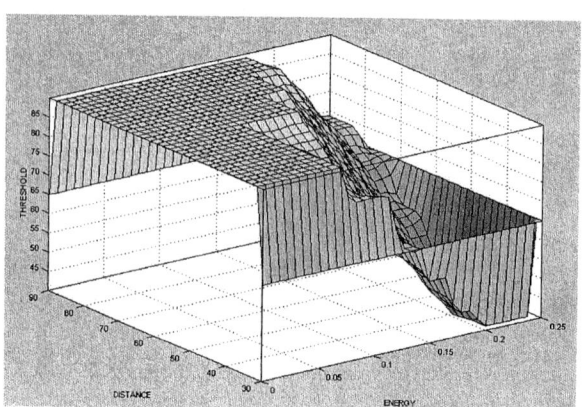

Fig. 9. Control surface for ED (Energy, Distance) fuzzy logic application to decide threshold

EDA considers one more variable than ED, which is ALIVE. If ENERGY is LOW and DISTANCE is NORMAL, the threshold value can take on a value below or above NORMAL value of DISTANCE depending on the value of ALIVE. Some of the example rules are shown below.

```
R22: IF ENERGY is LOW AND DISTANCE is NORMAL
AND ALIVE is LOW THEN THRESHOLD is VLONG

R24: IF ENERGY is LOW AND DISTANCE is NORMAL
AND ALIVE is HIGH THEN THRESHOLD is LNORMAL

R67: IF ENERGY is HIGH AND DISTANCE is NORMAL
AND ALIVE is LOW THEN THRESHOLD is SNORMAL

R69: IF ENERGY is HIGH AND DISTANCE is NORMAL
AND ALIVE is HIGH THEN THRESHOLD is SHORT
```

4 Simulation Results

In our simulation, RCH area's Y-coordinate ends at 15m from the beginning of the deployment area and there are 100 nodes within the area. The area is 100x100 meters. Each node is equipped with an energy source whose total amount of energy accounts for 0.25 J at the beginning of the simulation.

We ran simulation for LEACH, ED and EDA. Figure 10 illustrates simulation results regarding the number of alive nodes in three different protocols. LEACH protocol is compared with our two fuzzy protocols. EDA is approximately two times better than LEACH in alive node counting at 450 rounds and increases the network lifetime by 10% over LEACH. ED increases the number of alive nodes by 60% and the network lifetime by 15% over LEACH at 450 rounds.

Figure 11 shows the number of initial nodes according to Y-coordinate in the network. Figure 12 shows the distribution of alive nodes when 60 nodes are dead. In LEACH protocol, the nodes far from BS are almost dead and about half of the nodes

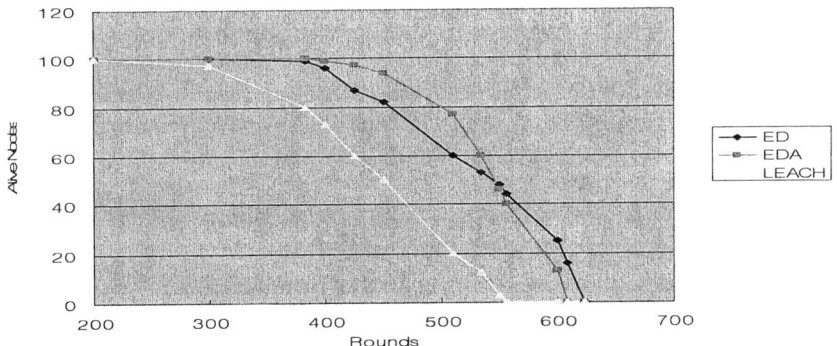

Fig. 10. Simulation result of alive nodes / rounds

near by BS are alive. Whereas, the energy distribution is well balanced for ED and EDA case compared to LEACH.

EDA shows similar results to ED in terms of balanced energy depletion. The speed of energy depletion in EDA, however, is faster than ED since EDA has to broadcast messages in order to find out the alive nodes within RCH area.

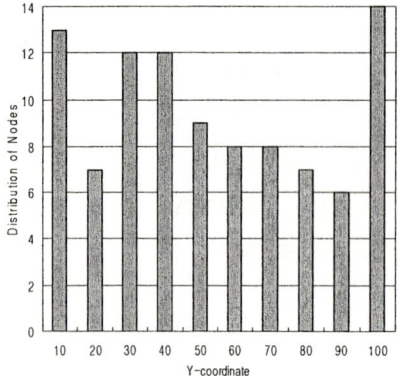 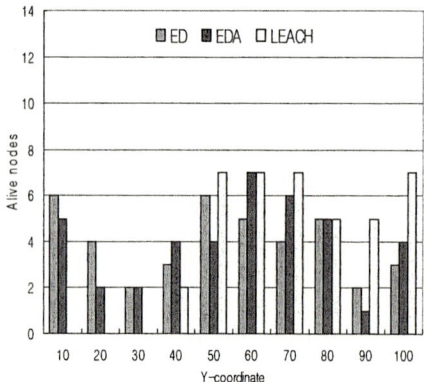

Fig. 11. Initial nodes distribution in the network

Fig. 12. Distribution of alive nodes when 60 nodes are dead

5 Conclusions and Future Work

Since the distances from BS to the sensor nodes are different the energy consumption among the nodes tends to be unbalanced, which decreases availability of the sensor network. We have proposed the transmission relay method between BS and CHs in order to enhance the balancing problem with the application fuzzy logic. Fuzzy logic decides the threshold distance value dynamically based on the energy level in each node, average distance from the BS to CHs, and number of alive nodes within RCH area. RCH decides whether to relay a specific transmission or not according to a threshold value and location information of CH. Simulation result shows that the proposed method enhanced the balancing of the energy consumption and lengthened the average lifetime of the deployed nodes.

The further improvement can be made with the consideration of density dependent clustering of the nodes so that CHs are elected within their clusters for efficient energy use.

References

1. K. Akkaya and M. Younis, "A survey on routing protocols for wireless sensor networks," *Ad hoc networks*, vol. 3, no. 3, pp. 325-349, 2004.
2. J.N. Al-Karaki and A.E. Kamal, "Routing techniques in wireless sensor networks: a survey," *Wireless Communications*, vol. 11, issue 6, Dec. 2004.

3. A. Mainwaring, J. Polastre, R. Szewczyk, D. Culler, and J. Anderson, "Wireless Sensor Networks for Habitat Monitoring," *Wireless sensor networks & applications*, pp. 88-97, 2002.
4. D. Estrin, R. Govindan, J. Heidemann, and S. Kumar, "Scalable Coordination in Sensor Networks," *Proc. Mobicom*, pp. 263 -270, Aug. 1999.
5. J. Kahn, R. Katz, and K. Pister, "Mobile Networking for Smart Dust," *Proc. Mobicom*, pp. 271-278, Aug. 1999.
6. J.M. Kahn, R.H. Katz, and K.S.J. Pister, "Next Century Challenges: Mobile Networking for Smart Dust," *Proc. MobiCom*, pp. 271-278, Aug. 1999.
7. G.J. Pottie and W.J. Kaiser, "Wireless Integrated Network Sensors," *Communications of the ACM*, vol. 43, no. 5, pp 51-58, May 2000.
8. Q. Jiang and D. Manivannan, "Routing protocols for sensor networks," *Proc. CCNC*, pp. 93 - 98, Jan. 2004.
9. D. Braginsky and D. Estrin, "Rumor Routing Algorithm for Sensor Networks," *Proc. 1st Wksp. Sensor Networks and Apps.*, pp. 22 - 31, Oct. 2002
10. R. C. Shah and J. Rabaey, "Energy Aware Routing for Low Energy Ad Hoc Sensor Networks," *IEEE WCNC*, vol. 1, pp. 350 - 355, Mar. 2002.
11. W. Heinzehman, A. Chandrakasan, and H. Balakrishnan, "Energy-Efficient Communication Protocol for Wireless Microsensor Networks," *Proc. Hawaii Conf. on Sys. Sci.*, pp. 3005 - 3014, Jan. 2000
12. S. Lindsey and C. Raghavendra, "PEGASIS: Power-Efficient Gathering in Sensor Information Systems," *IEEE Aerospace Conf. Proc.*, vol. 3, pp. 1125 – 1130, 2002.
13. Y. Xu, J. Heidemann, and D. Estrin, "Geography-informed Energy Conservation for Adhoc Routing," *Proc. 7th Annual ACM/IEEE Int'l. Conf. Mobile Comp. And Net.*, pp. 70 – 84, 2001
14. I. Stojmenovic and X. Lin, "GEDIR: Loop-Free Location Based Routing in Wireless Networks," *Int'l. Conf. Parallel and Distrib. Comp. and Sys.*, pp. 1025 - 1028 Nov. 1999.
15. W. Heinzehman, A. Chandrakasan, and H. Balakrishnan, "Energy-Efficient Communication Protocol for Wireless Microsensor Networks," *Proc. Hawaii Conf. on Sys. Sci.*, pp. 3005 - 3014, Jan. 2000.
16. S. Lee, J. Yoo, and T. Chung, "Distance-based Energy Efficient Clustering from Wireless Sensor Networks," *Proc. IEEE Int'l Conf. on Local Comp. Net.*, pp. 567 - 568, Nov. 2004.
17. C. Schurgers and M.B. Srivastava, "Energy efficient routing in wireless sensor networks," *Proc. MILCOM*, vol. 1, pp. 357 - 361, Oct. 2001.
18. Y. Xu, J. Heidemann, and D. Estrin, "Geography-informed Energy Conservation for Adhoc Routing," *Proc. ACM/IEEE int'l. Conf. on Mobile Comp. and Net.*, pp. 70-84, Jul. 2001.

Validation and Comparison of Microscopic Car-Following Models Using Beijing Traffic Flow Data*

Dewang Chen[1,2], Yueming Yuan[3], Baiheng Li[3], and Jianping Wu[2,3]

[1] School of Electronics and Information Engineering,
Beijing Jiaotong University, 100044, Beijing, China
dwchen76@yahoo.com.cn
[2] UK China Joint ITS Center, Southampton Univ., UK and Beijing Jiaotong Univ.,China
j.wu@soton.ac.uk
[3] School of Traffic and Transportation, Beijing Jiaotong Univ., 100044, Beijing, China
lerrygreen@126.com, libaiheng@tom.com

Abstract. In this oppaper, camera calibration and video tracking technology are used to get the vehicle location information so as to calibrate the Gazis-Herman-Rothery (GHR) model and fuzzy car-following model. The detail analyses about the models' parameters and accuracy show that the fuzzy model is easy to understand and have better performance.

1 Introduction

Traffic flow theory is the foundation of traffic science and engineering. Car-following models, a part of traffic flow theory, are attributed to microscopic approaches, which are based on the assumption that each driver reacts to a stimulus from the cars ahead in a single-lane [1].

There are several classical car-following models such as GHR model, safety distances model, fuzzy logic-based model and so on. The study of car-following model not only can help us understand traffic stream characteristic well, but also can play fundamental role in study of microscopic traffic flow simulation. In this paper, we use vehicle tracking and camera calibration technology to get the vehicle location data [2], so to calibrate fuzzy car-following model and compare it with other models.

2 Microscopic Car-Following Models

2.1 Gazis-Herman-Rothery Model

GHR model is one of the most well known car-following models in late 50s and early 60s in last century [3]. Its assumption is that the acceleration of vehicle is proportional to relative speed and distance with the one in front. At the same time the speed of itself will have impact on it. We can describe it in the following equation:

---

* This paper is supported by National Science Foundation China, Under Grant 50322283.

$$\alpha_{n+1}(t+T) = cv_{n+1}{}^{m}(t)\frac{\Delta v(t)}{\Delta x^{l}(t)} \quad (1)$$

2.2 Fuzzy Logic Based Model

In 2000, J.Wu proposed a microscopic fuzzy logic-based simulation model, which can well describe driver's behavior during car-following [4]. The car-following model has two principal premise variables to the decision making process, which are relative speed (DV) and the distance divergence, DSSD (the ratio of vehicle separation, DS, to the driver's desired following distance, SD). Each premise variable consists of several overlapping fuzzy sets. A triangular membership function has been assumed for all the fuzzy sets in the car-following model.

3 Data Collection

Traffic flow data were collected using digital video camera at the top of a four-lane expressway roadside building. The observed road section covered about 100m and 1 hour of data was gotten for analysis. The output of the video processing software includes the positions of each vehicle at each one second sample time. We total got 322 sets of location data of a pair of leading-following vehicles. We can easily get the speed and acceleration from location data according to (2) and (3). The computed data include the RD (relative distance), RS (relative speed), speed of FV(following vehicles), speed of LV(leading vehicles) and the acceleration of FV.

$$v = \frac{\Delta x}{\Delta t} \quad (2)$$

$$\alpha = \frac{\Delta v}{\Delta t} \quad (3)$$

4 Model Calibration

4.1 Model Calibration for GHR Model

Firstly, we use the data to match the simplest car-following model at m=0 and l=0, which is believed that the acceleration is proportional to the relative speed between two vehicles. Scatterplots and the regression line are depicted in Fig.1. Proportional coefficient c and some performance index of the regression function are reported in Table 1.

Secondly, we use our data to match the m=0, l=1 model which describe the following car's acceleration is proportional to RS and RD between FL and FV, then (1) can be rewritten in (4). The validation results are illustrated in Fig.2, and the parameter c and performance index is reported in Table 2.

$$\alpha_{n+1}(t+T) = c\frac{\Delta v(t)}{\Delta x(t)} \quad (4)$$

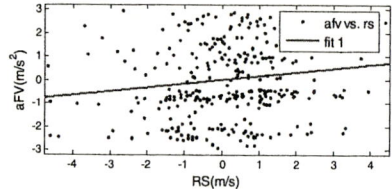

Fig. 1. Validation of GHR model at m=0 and l=0

Table 1. The parameter and performance of GM model at m=0 and l=0

C	SSE	R-square	RMSE
0.1595	782.8	0.01886	1.562

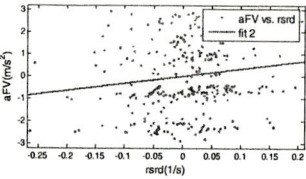

Fig. 2. Validation of GHR model at m=0 and l=1

Table 2. The parameter and performance of GM model at m=0 and l=1

C	SSE	R-square	RMSE
3.22	784.6	0.01661	1.563

4.2 Fuzzy Logic-Based Model Calibration

Wang-Mendel method [5] is used to calibrate this Fuzzy model proposed by Wu. We use triangle MF (membership function) for every fuzzy set. There are total 15 fuzzy set, 25 fuzzy rules and 45 parameters in the fuzzy car following model. We total got 22 fuzzy rules, which have some corresponding relationship with the common sense

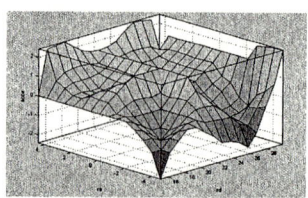

Fig. 3. The surface of fuzzy car-following model

when driving a car. For example, rule 1 means "if relative distance is much too close and relative speed is closing fast, then the acceleration is strong". The fuzzy surface is illustrated in Fig.3. The SSE and RMSE of this fuzzy model is 403.1958 and 1.119 respectively.

5 Conclusions

The model calibration results show that the fuzzy model is better than GHR model. The different kinds of GHR model show similar performance index, so maybe there is no more improvement room for GHR model. However, the fuzzy model is easy to understand and similar with the reasoning process of drivers. If adaptive technologies to adjust the parameters of MF are used, better results could be gotten. Because the number of samples is still few at present, so the statistic results are not very satisfied. In the future, we will use more data to validate more models to find which one is the best.

References

1. Mark Brackstone, Mike McDonald: Car-following: a historical review. Transportation Research Part F 2 (1999) 181-196
2. Hongliang Bai, Jianping Wu and Dewang Chen, "A Flexible Camera Calibration for Freeway Scenes", Technical report, Institute of Automation, Chinese Academy of Sciences, 2005.
3. Gazis, D. C., Herman, R., & Potts, R. B. (1959). Car following theory of steady state traffic flow. Operations Research,7, 499~505.
4. Jianping Wu, Mark Brackstone, Mike McDonald: Fuzzy sets and systems for a motorway microscopic simulation model. Fuzzy Sets and Systems 116 (2000) 65-76.
5. L.X.Wang and J.M.Mendel, "Generating fuzzy rules by learning from examples," IEEETrans.Syst.,Man,Cybern.,vol.22,pp.1414–1427, Dec.1992.

Apply Fuzzy-Logic-Based Functional-Center Hierarchies as Inference Engines for Self-learning Manufacture Process Diagnoses

Yu-Shu Hu[1,2] and Mohammad Modarres[1]

[1] The Center for Risk and Reliability, University of Maryland,
College Park, MD 20742, U.S.A.
[2] Digital Content School, Beijing Technology and Business University,
33 Fucheng Road, Beijing 100037, China

Abstract. In a production process, there are numerous systems that provide information/reports for various purposes. However, most of the knowledge for decision-making is kept in minds of experienced employees rather than exists in IT systems that can be managed systematically. Even experienced managers may make flaw/improper decisions due to the lack of must-known information, not to mention what those who are less experienced or have been urged by the pressure of time will probably do. In this paper, a fuzzy-logic-based functional center hierarchical model named Dynamic Master Logic (DML) is designed as an interview interface for representing engineers' tacit knowledge and a self-learning model for tuning the knowledge base from historical cases. The DML representation itself can also be the inference engine in a manufacture process diagnoses expert system. A semiconductor Wafer Acceptance Test (WAT) root cause diagnostics which usually involves more than 40,000 parameters in a 500-step production process is selected to examine the DML model. In this research, it has been proven to shorten the WAT diagnostics time from 72 hours to 15 minutes with 98.5% accuracy and to save the human resource form 2 senior engineers to one junior engineer.

1 Introduction

The major challenge for manufacture process diagnostics is to handle the complexity of abnormal variance. A complex production (e.g., a semiconductor manufacture) requires hundreds of steps to complete the final product. In each step, hundreds of parameters involved will cause thousands of different types of failure modes. In a modern production factory, there are numerous systems that provide information/ reports for various purposes, for instance, Engineering Data Analyses (EDA) system, Manufacturing Execution System (MES) …etc. However, most of the knowledge for decision-making is kept in minds of experienced employees rather than exists in IT systems that can be managed systematically. Even experienced managers may make flaw/improper decisions due to the lack of must-known information, not to mention what those who are less experienced or have been urged by the pressure of time will probably do.

One good example is the diagnostics of the Wafer Acceptance Test (WAT). At the end of a semiconductor wafer manufacture process, WAT will be performed to check the quality of wafers. If there is any critical failure measured, diagnostics will be performed by experts to identify the root causes. Current Engineering Data Analyses (EDA) system can detect abnormal parameters and hold lot in WAT step automatically, but engineer have to find out the abnormal process machine in working day manually based on his experience. However, as a roughly estimation, there are 100,000 rules involved in the WAT diagnostics for a given wafer discrepancy which could be caused by an incorrect recipe or operation related to 40,000 parameters in 500 processes. The diagnostics know-how exists in various engineers in different positions with different expertise.

To develop the knowledge base of an expert system to support the decision making in the WAT diagnostics, there are various potential solutions. One potential solution is to design a mechanism to collect engineers' tacit knowledge for WAT diagnostics. The knowledge collection model must allow engineers to focus on their own rules separately, and then to organize and to integrate them in a consistent representation. The other potential solution is to design a self-learning engine which could create diagnostics rules from the historical cases. The major challenge of this solution is the huge numbers of cases required to generate 100,000 rules. The verification and validation of the knowledge base (rules) is another major challenge in both solutions.

In this paper, a fuzzy-logic-based functional center hierarchical model named Dynamic Master Logic (DML) [1]-[3] is introduced as an interview interface for representing engineers' tacit knowledge and a self-learning model for tuning the knowledge base from historical cases. The DML representation itself can also be the inference engine in a manufacture process diagnoses expert system. In this research, it has been proven to shorten the WAT diagnostics time from 72 hours to 15 minutes with 98.5% accuracy and to save the human resource form 2 senior engineers to one junior engineer.

2 Dynamic Master Logic (DML) Diagram

DML is a hierarchical knowledge representation with a top-down and outside-in logic structure (containing: elements, operators and relations). In this DML concept, logical and uncertain connectivity relationships are directly represented by time-dependent fuzzy logic [1]. Physical connectivity relationship is represented by fuzzy-logic-based interactions between various levels of a hierarchy. The degree of the fuzzy integration is governed by the physics laws which describe the integration. Accordingly, the logical relationships modeled in a DML are accompanied by corresponding physical relationships by fuzzifying such relationships. Combined the concept of Functional-Center Modeling, DML can represent a complex personal or enterprise knowledge from Goal/Condition, Event, Functional, Structural and Behavioral points of views. In this section, the DML concept is introduced briefly. For more discussions about specific applications of the DML, see papers of Hu and Modarres [1]-[3].

2.1 Notations and Basic Structure

When we talk about the DML representation, we are talking about a family of models (diagrams). That is, for representing a physical behavior, there is no unique DML model; rather, experts can come up with varieties of DML and different numbers of nodes and layers, fuzzy sets, and transition logic. However, these DML should yield approximately similar results.

The basic notations of the DML are summarized in [1]. Four types of logic gates are designed to represent the fuzzy logic rules. Additionally, five different dependency-matrix nodes are used in DML to describe the probability and the degree of truth in relationships. To convert different numbers of modes in a node, the direction of fuzzification and defuzzification is indicated by the location of the name noted. Using these notations and symbols, we may organize a DML to represent different types of connectivity relations, time dependency and uncertainty as shown in Fig. 1. The basic structures in DML can be grouped into eight basic classes: static, uncertain (or priority) output, uncertain (or weighted) input, scheduled, time-lagged, auto-correlated, feedback, and comparison [2],[3]. A DML model (diagram) is assembled from these basic structures.

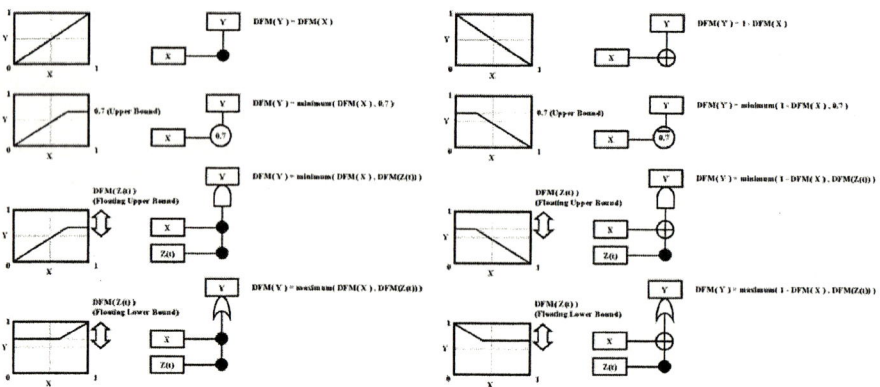

Fig. 1. Examples for representing different types of connectivity relations, time dependency and uncertainty in DML

One of the major advantages of the DML modeling is physical connectivity representation. Physics plays a fundamental role in most important fields of science and engineering. Since most of the known physical models can be represented by mathematical relations. A DML must be able to describe mathematic relations. Two major types of physical models are discussed in this section: solved model and row model.
- Solved models are represented by functions in which the solution of the target variable can be computed straight forward, for example, polynomials. As shown in Fig. 2, for a known solution of solved model, critical points of the relation should be chosen as fuzzy modes. The states between modes will be approached by membership

functions automatically and linearly. To adapt the minor curve shape difference, membership functions can be relocated as Fig. 3. Optimization algorithm, such as Least Squares Method, can be applied to identify the best fuzzy-set family. Theoretically, by adjusting the number of fuzzy sets and the shape of membership function, one might approach a curve to any acceptable accuracy. However, because of the uncertain nature of fuzzy logic modeling, not all objects require such accuracy.

- Row models are represented by unsolved equations in which the target variable (i.e., the supported DML node) cannot be computed straight forward. For example, connectivity includes linear/nonlinear equations, integration, differentiation, or differential equation. The DML and its estimations of a simple harmonic motion example are discussed in [3]. The known equation of row model is represented in a DML. Initial values are given to trigger the temporal behavior of the model.

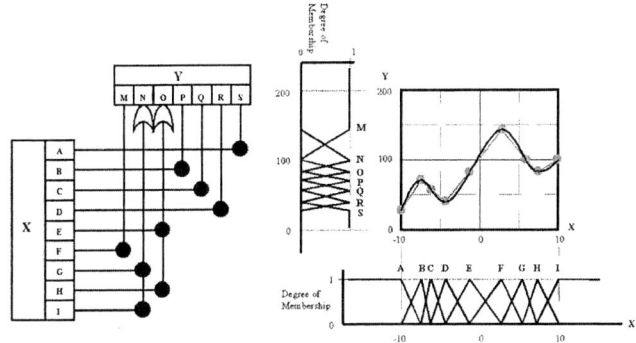

Fig. 2. DML as a representation of physical solved models

2.2 Diagnostics Engine

Because of its hierarchy, the DML of a complex system provides an excellent model to describe causal effects (downward and upward causations) in a complex system. Two important causal relations can be extracted from the DML. The first is to identify the ultimate effect of a disturbance (such as a failure), and the second is to determine the ways that a goal or function can be realized or structural organizations that would be needed.

DML Diagnostics Engine is formed by a DML representation which contains inference logics for failure diagnostics. By tracing the elements, operators and relations in a DML, the root causes of specific symptom (i.e., WHY it happened) and the potential recovery paths (i.e., HOW to do) can be identified as shown in Fig. 3.

Functional-Center Decomposition

To show the complexity of systems from different point of views, a complete DML could be combined by, but not required to, goal, event, function, behavior and structure hierarchies. The details of DML functional-center model is discussed in [5].

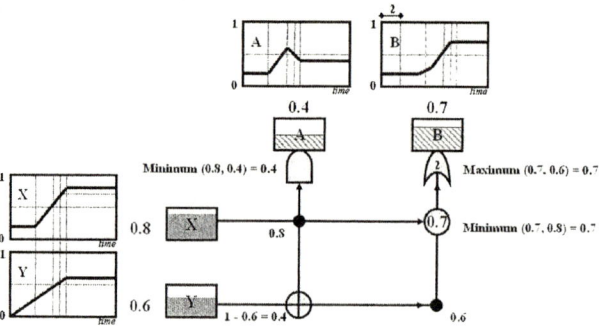

Fig. 3. DML with time-dependent fuzzy logic could show the dynamic behavior of systems

One simple but clear example to apply DML for representing the knowledge related to a door is introduced in this section.

As shown Fig. 4, the goals of a door are for space connection, privacy, security ... or esthetic. Such goals are achieved by the sub-functions of flow controlling, such as flow management, flow prevention ...etc. The functions are performed by the physical structure of the door and the operating behaviors of users. Failure of functions will cause the happening of events. Such logic relations could be organized by the correlation lattice in the middle of the hierarchies. The interactive cause-impact behavior will be shown by failing the function, the physical structure or the operating procedures.

Generally, a DML approach can improve the traditional hierarchical models in various thought. From state determining point of view, since DML applies fuzzy sets that have overlapped and full-scale membership functions, it allows floating threshold with fault tolerance and preventive warning. For connectivity relationship representing, DML can model not only full-scale physical and logical connectivity but also probabilistic, linguistic and resolutional uncertainty. Transition effects of a system (e.g., partial success/failure, auto-correlation, feedback, schedule and time-lagged dynamics) can also be well represented in a DML hierarchy.

On one hand, since the DML estimation is based on logic, the speed of the estimation is much faster than a numerical simulator. On the other hand, DML provides full-scale logical reasoning information that cannot be concluded in the classical logic-based systems. As such, a DML-based expert system, which has capability of full-scale logical reasoning and rapid simulation, can be implemented efficiently and economically.

3 Construction of the Diagnostics Knowledge Base

The construction for a DML-based knowledge base is a multiple-step process. Firstly, the DML is applied as an interview interface to organize engineers' tacit know-how as shown in Fig. 5. Secondly, historical cases are introduced to adapt the fuzzy relations and to cluster new fuzzy sets. Finally, the fuzzy relations are normalized to reduce the statistical impact as shown in Fig. 6. The details of the DML construction are discussed below.

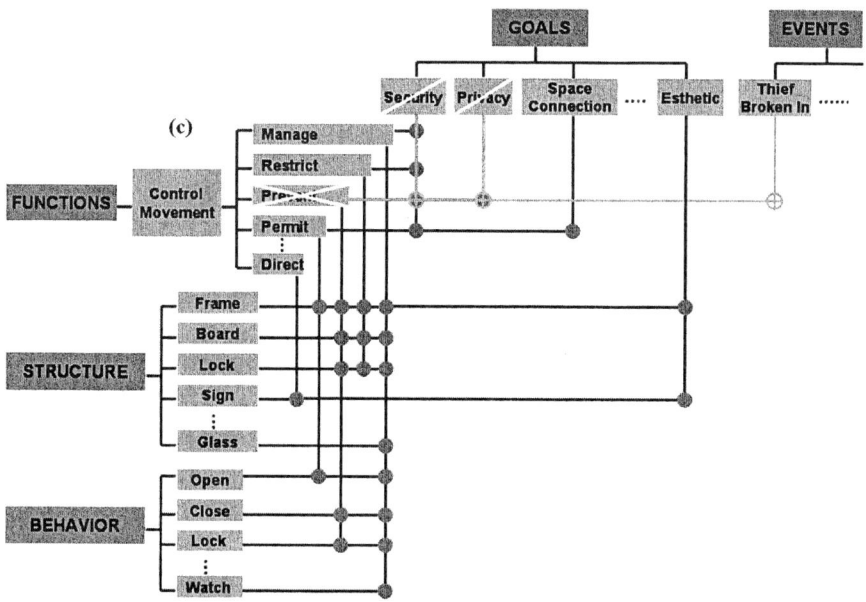

Fig. 4. An example to apply DML for representing the knowledge related to a door

STEP 1: DML Interview Processes

As shown in Fig. 5(b), engineers will have the capability to organize the basic logic relations and decision-making flows by DML after a one-hour training. Since the know-how of the details connectivity (i.e. the dark block in the DML) is owned by different engineers in different positions with different expertise, Interviews are performed to different engineers to focus on different topics before reorganized the complete DML. Fig. 5(a) shows an example of the further decomposition of the connectivity in Fig. 5(b) with fuzzy sets and relations introduced.

Based on the proven projects, the accuracy of the DML from interviews is only around 70%. To improve the usability of the DML as a diagnostics engine, historical cases are required to tune the fuzzy sets and fuzzy relations in a DML.

STEP 2: Adapting Relations and Clustering Fuzzy Sets from Historical Cases

Engineers' interview can give a brief picture of the diagnostics logic. To enhance the accuracy, historical cases are introduced to tune the fuzzy sets and fuzzy relations in a DML. The tuning process is a simple statistical concept as shown in Fig. 6 and Fig.7. However, the major problem of this statistical concept is the degree of the fuzzy relation of rare cases will be small and be ignored. STEP 3 will fix this issue. Various researches have extensively proven to solving real-life optimization and control problems by applying neural networks and neuro-fuzzy systems [6]-[10].

STEP 3: Relation Normalization for Significant Cases

For rare historical cases, the degree of fuzzy relations will be small and be ignored based on the nature of statistical estimation. Thus, degree of fuzzy relations in the horizontal direction will be normalized. In other words, the fuzzy relation will become obvious even only one historical case existing in such normalization, if the scenario group is not a sub set of other scenario groups.

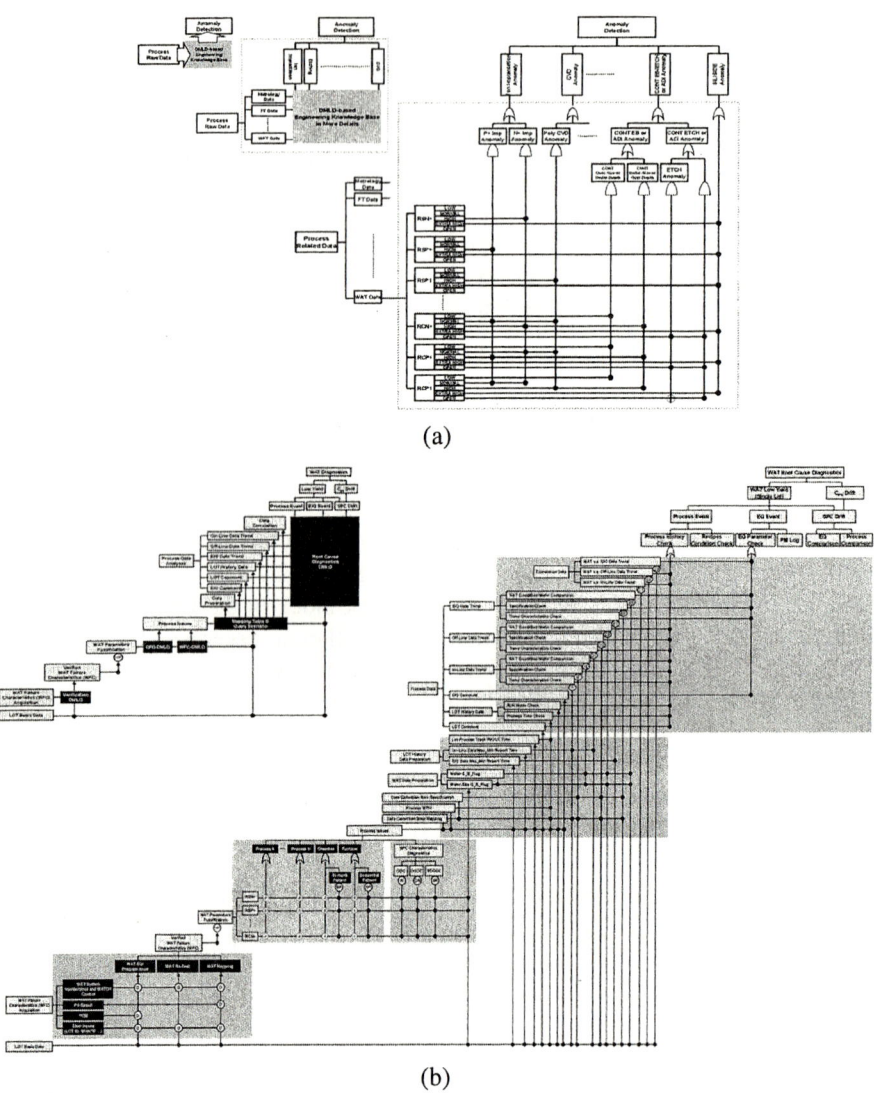

Fig. 5. (a). DML is applied as an interview interface to organize the detail fuzzy relations of engineers' tacit know-how. (b). DML is applied as an interview interface to organize the basic logic relations and decision-making flows of engineers' tacit know-how.

4 Validation and Verification of the Diagnostics Knowledge Base

As shown in Fig. 8, a case generation module is designed to create testing cases based on a target knowledge base. For each rule stored in the target knowledge base, a testing case could be created. The result knowledge base should be identical to the target knowledge base if the self-learning module and the diagnostics module are designed properly. The following characteristics are required and are confirmed in the DML V&V process:

- Correctness and Completeness
- Repeatability
- Consistency
- Converge
- Learning Tolerance
- Multiple Root Causes Learning Tolerance

Fig. 9 shows an example of the V&V results of the Converge and the Learning Tolerance. On the left side of the Fig. 9, the difference spectrum is lighter after more testing cases feed. This result shows the self-learning module is converge. On the right side of the Fig. 9, the correct result will be approached after more testing cases feed even the initial relations are incorrect. This is an example test result for the Learning Tolerance proven.

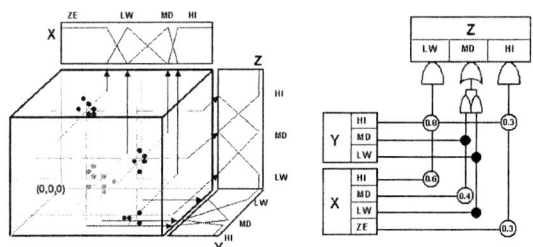

Fig. 6. historical cases are introduced to cluster the fuzzy sets and relations in a DML

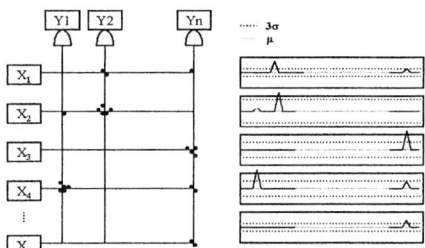

Fig. 7. Historical cases are introduced to cluster the fuzzy sets and relations in a DML

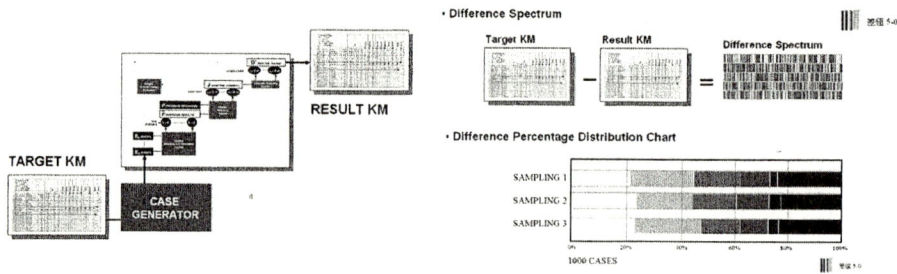

Fig. 8. The DML V&V module is a case generator based on a target knowledge base

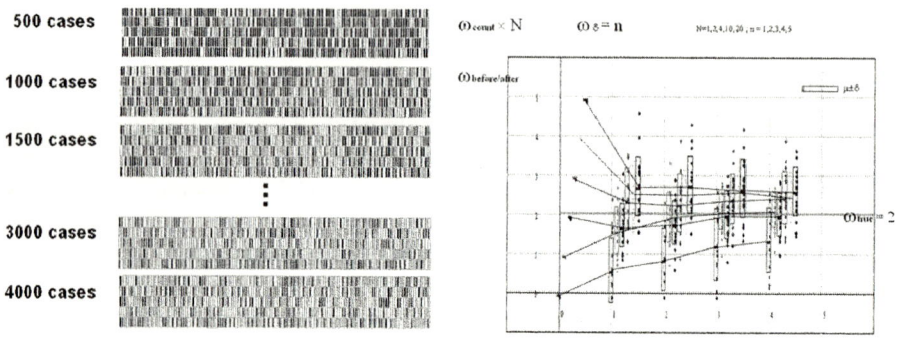

Fig. 9. An example result of DML V&V

5 Conclusion

In this research, a functional-center hierarchical model named DML is introduced to design a process diagnostics expert system. In this DML concept, logical and uncertain connectivity relationships are directly represented by time-dependent fuzzy logic. How to automatic check WAT test result by statistic control limit, find out the abnormal parameter, and correlate with relates process step to find out the suspect abnormal machines is applied to examine the DML algorithm.

The WAT diagnostics expert system is characterized by:

- Capable of diagnosing historical data with built-in intelligence to help users analyze and find out root causes
- Capable of suggesting corrective actions to reduce the potential impact of latent problems and optimize the benefits of business operation flow
- Detect potential problem actively instead of display status / information passively
- Provide not only "What's Happened/Happening" but also "What to Do" information
- Capable of integrating information kept in various systems
- Use the "Dynamic Master Logic Diagram" theory by University of Maryland to build a knowledge inference engine

- Co-work with domain experts to extract their knowledge
- Construct systematic knowledge framework for diagnostics
- Develop system from execution level to decision-making level (bottom-up)
- In proven projects, the DML-based diagnostics is proven to
- Shorten the WAT root cause diagnostics time from 72 hours to 15 minutes with accuracy 98.5% and save the human resource form 2 senior engineers to one junior engineer
- Accelerate the R&D SPICE QA speed to provide corrective actions for existing problems from 2 man months to 30 man minutes
- Detect potential problems and reduce the latent impact to save cost and maximize the benefit in a Super Hot Lot 48-hour Rolling Monitoring System
- Deliver environment and reusable components that can be integrated easily and meet fast-changed business environment/requirements (from 0.18-based to 0.13-based in 3 months)

References

1. Hu, Yu-Shu and Modarres, M.: Time-dependent System Knowledge Representation Based on Dynamic MPLD, Control Engineering Practice J., Vol. 4, No 1, (1996) 89-98.
2. Hu, Yu-Shu and Modarres, M.: Evaluating System Behavior through Dynamic Master Logic Diagram (DML) Modeling, Reliability Engineering and System Safety J., Vol. 64 (1999) 241-269.
3. Hu, Yu-Shu and Modarres, M.: Logic-Based Hierarchies for Modeling Behavior of Complex Dynaimc Systems with Applications, in: Fuzzy Logic Application in Nuclear Power Plant, Chapter 17, Physica-Verlag (2000)
4. Chang, Y.-J., Hu, Y.-S. and Chang, S.-K.: Apply a Fuzzy Hierarchy Model for Semiconductor Fabrication Process Supervising, SEMI Technical Symposium, Zelenograd, Moscow, Russia (1999)
5. Modarres, M.: Functional Modeling of Complex Systems (Editorial), Reliability Engineering and System Safety J., Vol 64 (1999)
6. Jang, J. R.: ANFIS: Adaptive-network-based Fuzzy Inference System. IEEE Trans. Syst., Man, Cybern., 23 (1993) 665¨C685
7. Frayman, Y., Wang, L.P.: Data Mining using Dynamically Constructed Recurrent Fuzzy Neural Networks. Proc. 2nd Pacific-Asia Conference on Knowledge Discovery and Data Mining, LNCS Vol. 1394 (1998) 122-131
8. Wai, R.-J., Chen, P.-C.: Intelligent Tracking Control for Robot Manipulator Including Actuator Dynamics via TSK-type Fuzzy Neural Network. IEEE Trans. Fuzzy Systems 12 (2004) 552-560
9. Kiguchi, K., Tanaka, T., Fukuda, T.: Neuro-fuzzy Control of a Robotic Exoskeleton with EMG signals. IEEE Trans. Fuzzy Systems 12 (2004) 481-490
10. Wang, L.P., Frayman, Y.: A Dynamically-generated Fuzzy Neural Network and its Application to Torsional Vibration Control of Tandem Cold Rolling Mill Spindles. Engineering Applications of Artificial Intelligence 15 (2003) 541-550

Fuzzy Spatial Location Model and Its Application in Spatial Query*

Yongjian Yang[1] and Chunling Cao[2]

[1] College of Computer Science and Technology, Jilin University, ChangChun 130012
yyj@jlu.edu.cn
[2] Mathmatics Department, Jilin University, ChangChun 130012
caocl@jlu.edu.cn

Abstract. To study the spatial relationships with the instability is becoming one of the hot spots and the difficulties in studying the spatial relationships. This paper express and apply the information of relationships among spatial objects in the real world in computer system from the cognitive view, study the fuzzy extension about description of spatial relationships at the base. Guided by the spatial query, we makes the model on the base of regular indefinite spatial inferring, applies fuzzy theory and spatial relationship theory in the spatial query and solves the fuzzy location problems in applying GIS network resource management.

1 Introduction

We can use fuzzy reasoning to solve the fuzzy location problem in applying GIS. The position information of the query point provided by the users is unlikely sufficiently definite when we want to inquiry data from the spatial database. The system will proceed on the description of the position and the evaluation of asserting degree to infer and assert the nicety position.

Here, we solve the problem under the regular indefinite spatial inferring. For specialty in solving problems, the definition of the fact, the rule and the inference machine algorism are different from the traditional model[1].

2 Definition

Objects can be obtained from any facts related to spatial phenomena and processes, including the following six types: numerals, binary, monodromy, multivalue, ambiguity and expression.

We define two facts. One is the description to location (Loc), the other is the aim point of location (Aim).

Loc (Ref_ obj,Ori,Dis)Ref_obj: referent objects; Ori(Ox,Oy): Orientation ;
Aim(pos);pos: position; including coordinate information of location : Pos.X, Pos.Y

---

* This work was supported by ZHSTPP PC200320001 of China

Generally speaking, in system the Following Action of inference machine is to insert new facts. Following Action is various due to the specificity of problems. Four sorts of the Following Actions are defined as follows:

Add fact [c]: Add fact, degree of belief is c
Del fact [c]: Delete fact
Update fact [c1]: update fact, degree of belief is c1
Show Message : Show warning message to users

During the operation of ADD and UPDATE, the degree of belief is not determined by the algorithm of inference machine but the rules. This is also the difference between the traditional inference machine and the current one.

Fuzzy Type:
Format: DISTANCE (aim1, aim2). Aim1 and aim 2 are two aim points; the fuzzy item DISTANCE shows the degree of the distance between them.

Fuzzy Operator:
Membership Function of Degree of Distance Operator near, mid, far:
Suppose the farthest distance is y, then membership functions of near, mid, far distance are:

$$\text{near}(x) = \begin{cases} 1, & x < 0.2y \\ 2 - x/2y, & 0.2y \leq x < 0.4y \\ 0 & 0.4y \leq x \leq y \end{cases} \quad (1)$$

$$\text{mid}(x) = \begin{cases} 0, & x < 0.2y \\ x/2y - 1, & 0.2y \leq x \leq 0.4y \\ 1 & 0.4y \leq x < 0.6y \\ 3 - x/2y & 0.6y \leq x < 0.8y \\ 0 & 0.8y \leq x \leq y \end{cases} \quad (2)$$

$$\text{far}(x) = \begin{cases} 1, & x < 0.2y \\ x/2y - 3, & 0.2y \leq x < 0.4y \\ 0 & 0.4y \leq x \end{cases} \quad (3)$$

Membership Function of Tone Operators: most, highly, very, relatively, some, little:

Suppose H_λ is tone operator, $(H_\lambda A)(x) = [A(x)]^\lambda$. When $\lambda > 1$, H_λ is called centralized operator. When $\lambda < 1$, H_λ is called diffused operator. Generally H_4 is most, H_2 is highly, $H_{1.25}$ is very, $H_{0.75}$ is relatively, $H_{0.5}$ is some, $H_{0.25}$ is little.

Membership Function of Fuzzy Operator app_ (approximately), pro_ (probably), abo_(about):

A common form of fuzzy operator

$$F(\underset{\sim}{A})(x) = (\underset{\sim}{E} \bullet \underset{\sim}{A})(x) = \bigvee_{y \in U} [\underset{\sim}{E}x, y) \wedge \underset{\sim}{A}(y)] \quad (4)$$

$\underset{\sim}{E}$ is the similar relationship in U, when $U = (-\infty, +\infty)$, generally we suppose:

$$\underset{\sim}{E}(x, y) = \begin{cases} e^{-(x-y)^2}, & |x-y| < \delta \\ 0, & |x-y| \geq \delta \quad \delta > 0 \end{cases} \quad (5)$$

We have different δ values due to the variety of fuzzy degree. Generally we get δ of app_ (approximately) 0.9, δ of pro_ (probably) 0.75, δ of abo_(about) 0.5. Fuzzy item can be constituted by aspiration and conjunction of operators above..

Functions below are applied in the rules:

1) Calculate Position (ro,o,d)
Return value: position coordinate Position(x,y) types
Function: return the reference object of distance R_o(x_{ro}, y_{ro}) the orientation of geometric center O(O_x, O_y), aim point coordinate with distance d.
Condition: $d \geq 0$

$$x = x_{ro} + \frac{O_x \times d}{\sqrt{o_x^2 + o_y^2}} \quad (6)$$

$$y = y_{ro} + \frac{O_y \times d}{\sqrt{o_x^2 + o_y^2}} \quad (7)$$

2) Dis(pos1,pos2)
Return value: numerical type
Function : return the geometric distance between pos1 and pos2

$$d = \sqrt{(pos1x - pos2x)^2 + (pos1y - pos2y)^2} \quad (8)$$

3 Rule Sets and Inference Machine

Rules are statements that represent the relationship between priori proposition and posteriori proposition [2][3]. There is a fuzzy or non-fuzzy deterministic factor, which marks the degree of belief of the rules for each rule. The previous action has one or several propositions which connected with AND or OR. The following action has only one proposition.

```
{RULE name of rule
    IF (previous action)    THEN following action }
  CERTAINTY is determined factor
```

Weight value and threshold value also can be added into rules in addition to fuzzy comparison. Rules together with weight value represent the importance degree of proposition. Weight value may be fuzzy value, users can add a language variable in a weight value bracketed with parentheses. Threshold value (system threshold value) can be defined in determining whether rules should be triggered or not. If the determinacy of previous actions is greater than the threshold value, rules will be triggered. If we hope that a certain rule has its own threshold value, we can add it in previous matters marked with {}. Deterministic factors can also be added into rules to represent the determinacy of rules, in addition to weight value and threshold value.

Rules Summary:

(I) Rule no.1 to no.5 is the preprocessing to location description, which avoids obvious contradiction in it.
(II) Rule no.6 to no.7 is to form aim points with location description.
(III) Rule no.8 to no.11 is to adjust belief degree of aim points.

At the beginning, the inference facts are completely facts of location. A normal result after inference is to produce an Aim fact with higher degree of belief, while others are quite low. Otherwise inference is failed. Algorism of inference machine can be described as follows:

(1) Apply rule no.1 to no.5 into all location facts until no suitable rules are available.
(2) Apply rule no.6 and no.7 into all location facts until no suitable rules are available.
(3) Apply rule no.8 to no.11 into all facts until no suitable rules are available.
(4) Sequence degree of belief of all Aim facts in descending and forming an array Aims.

If the degree of belief of Aims [0] is greater than 0.8 and that of Aims [1] is smaller than 0.5, algorism will be succeed and quit then output position information Aims [0]. Otherwise algorism fails and returns, and asks for users to input facts again. The determination of belief degree of rules for previous actions can be sorted into two cases. One is actually a pattern matching with regard to factors such as Location (ro1, o1, d1) [c1] or Aim (pos) [c2]. Another one is that degree of belief is that of membership determined by fuzzy set in case of "DISTANCE(aim1,aim2) is near". Conjunction of previous matters are "and" and "or". Rules do not concern uncertainty, that is, the intensities of all rules are 1. Triggering threshold controls the trigger of rules. If the degree of belief of previous matters is greater than that of the following matters, following matters will be executed. Different with traditional inference machine, the degree of belief of following facts is determined by rules but not the belief degree of previous matters or intensities of rule [4][5].

4 Realization

In order to realize the model above, we need to apply diverse technologies such as expert system, GIS, database management, and so on. Actually, it can also be regarded as a small space decision support system [6].

The kernel of SDSS is constituted by facts base, rule base and inference machine. And it utilizes production rule as the basic knowledge expression mode.[7] Knowledge base stores all knowledge from field experts that can be represented as rules, semantic net, frame or objects. Inference machine collects knowledge from knowledge base to make inference. With regard to production system, inference machine determine which rules can be triggered according to the facts in global database. These rules are sequenced according to priority. We construct SDSS with expert system shell and involve utilizing external communication mechanism to enlarge applied environment. So it is necessary to connect with database, GIS and mathematical calculation programs. SDSS can meet the satisfaction and deal with uncertainty.

Fuzzy spatial position model is an important part of the whole system. It can connect well with other parts. When users input fuzzy query statements, system begins keywords match first. These keywords include location, distance, orientation and fuzzy operator. And then, system makes explanation and inference calculation through spatial decision support system to get users' query demands. While system calls for corresponding fuzzy operator to make calculation and transforms them into precise query demands, puts them into data query module to visit spatial data base, pick up corresponding position data which meet demands. System returns result set finally.

References

1. Chun Liu, Dajie Liu. On the spot Study of GIS and its Application [J], Modern Mapping, Vol. 36 (2003)
2. (American) Shashi Shekhar, SanjayChawla, Kunqing Xie, Xiujun Ma, Dongqing Yang, etc translation.Spatial database[M]. Beijing: Engineering Industry Publishing Company, 1(2004
3. The Open GISTM Abstract Specification ,Version 4[J] ,Open GIS Consortium ,1999
4. KangLin Xie, JinYou Fu Nerve Fuzzy Logical Control System Membership Function and Inferring Rule Assertion[J].Shanghai Traffic University Transaction Vol. 8,31 (1997)
5. Qiao Wang,JiTao Wu Research the Standardization Problem in Spatial Decision Support System [J].Mapping Transaction Vol 2 (1999)
6. Cohn AG, Hazarika SM. Qualitative Spatial Representation and Reasoning: An overview. Fundamental Informatics, 2001, 46
7. Yu QY, Liu DY, Xie Q. A Survey of Analysis Methods of Topological Relations between Spatial Regions. Journal of software, 2003, 14(4)

Segmentation of Multimodality Osteosarcoma MRI with Vectorial Fuzzy-Connectedness Theory*

Jing Ma, Minglu Li, and Yongqiang Zhao

Department of Computer Science and Engineering, Shanghai Jiao Tong University
Shanghai, China
julery@sjtu.edu.cn

Abstract. This paper illustrates an algorithm for osteosarcoma segmentation, using vectorial fuzzy-connectedness segmentation, and coming up with a methodology which can be used to segment some distinct tissues of osteosarcoma such as tumor, necrosis and parosteal sarcoma from 3D vectorial images. However, fuzzy-connectedness segmentation can be successfully used only in connected regions. In this paper, some improvements have been made to segment the interested tissues which are distributed in disconnected regions. And the paper speeds up the process of segmentation by segmenting two osteosarcoma tissues simultaneously. The methology has been applied to a medical image analysis system of osteosarcoma segmentation and 3D reconstruction, which has been put into practical use in some hospitals.

1 Introduction

Osteosarcoma is a type of bone cancer that occurs most often in children, adolescents, and young adults. It is a desperate disease endangering people's life. However, the biologic heterology brings on much diversity among patients both in clinical symptoms and radiological features, so it is difficult to obtain a good information extraction with normal segmentation methods like threshold and edge detection. In the paper, the fuzzy-connectedness segmentation is taken into consideration. Some improvements are proposed to reduce computational time and to evaluate a better volume of disconnected regions. We have realized the segmentation in an osteosarcoma image analysis system, and our experiments have proven that it is an efficient method which can be applied practically.

2 Vectorial Fuzzy-Connectedness Image Segmentation

The fuzzy-connectedness segmentation aims at capturing the fuzzy notion via a fuzzy topological notion which defines how the image elements hang together

---

* This research is supported by Dawning Program of Shanghai, China (grant ♯ 02SG15).

spatially in spite of their gradation of intensities. Jayaram K. Udupa and his cooperators have done a great deal of research to apply the fuzzy digital topology to image processing, and have constructed a self-contained theory[1].

The theory has been utilized in many medical applications and proved to be a precise and efficient method. However, the process of segmentation is sometimes time-consuming in practical applications, and some disconnected regions can not be better extracted. In this paper some work has been done to address these two problems. When segmenting two objects simultaneously, some changes of conventional segmentation can be taken to reduce the computational time.

3 Methodology

The whole process of segmentation includes the following 7 steps:

Step1: Image data acquirement.
The data of 2-D images - MRI data of 10 patients with osteosarcoma were provided by Tongji Hospital Affiliate to Tongji University.

Step2: Format unification.
Adjusting some parameters of T1WI, T2WI, STIR sequences to make the format uniform.

Step3: Selection of interested region.
Constraining the region to be analyzed to limited areas, so the processing time was considerably reduced (shown in Fig.1(c)).

Step4: Selection of seed points.

Step5: Information fusion of multimodality MRI.

$$f(x, y, z) = (f(T1WI), f(T2WI), f(STRI)). \qquad (1)$$

Step6: Segmentation of tumor and parosteal sarcoma tissues.
Before the process of segmentation, there are 3 important concepts need to be specified. All the definitions are according to[2].

(1) Local relationship: The affinity between each pair of voxels.

(2) Global relationship: The strength of every path.

(3) Relative fuzzy connectedness of multiple objects:
The process of the two objects segmentation is a process of competition. In this competition, every pair of voxels in the image will have a strength of connectedness in each object.

The algorithm is essentially the same as the algorithm κVMRFOE presented in [2] except the simultaneous extraction of the two tissues. In the algorithm, O_1 represents tumor while O_2 represents parosteal sarcoma. s_1 is one of the seeds of O_1, and s_2 is one of the seeds of O_2. Each voxel of images is added a flag which tags every voxel to identify which object it belongs to. All the flags are set to 0 at the beginning. If the voxels have the fuzzy affinity of s_1, they will be regarded as that they belong to O_1, the flag of which will be set to 1. At the same time, if the voxel belongs to O_2, the flag of it will be set to 2. The usage of the flag is to decrease the computation of the affinity value. If a voxel belongs to O_1, it is unnecessary to compute the affinity value. The improvements will increase the speed of computation if we use the algorithm to simultaneously segment the two objects.

Step7: Detection of necrosis regions.

Necrosis is distributed as some spots in the region of osteosarcoma, so it is hard to specify all the seeds for them. In this step, a method was come up with to detect all potential necrosis spots would be detected automatically.

Suppose that there are k seed points for necrosis and the set of them are defined as $S = \{s_1, s_2, \cdots, s_k\}$. Define s as the average intensity value of the set. The object-feature-based component of affinity with the only feature considered here is the voxel intensity itself. So the seed points can be specified by computing μ_{Φ_S}(object-feature-based component of affinity) between the manually selected seed point and each voxel. The new specified seeds would be added to S.

4 Result

The processes of segmentation were applied to the system mentioned above, and some experiments were done with several series of MR images. Fig.1 (a), (b), (c) are some slices respectively in T1, T2, and STIR sequences. Fig.2 (a),(b),(c) are the segmented results. As shown in Fig.2 (b), the necrosis was clearly segmented

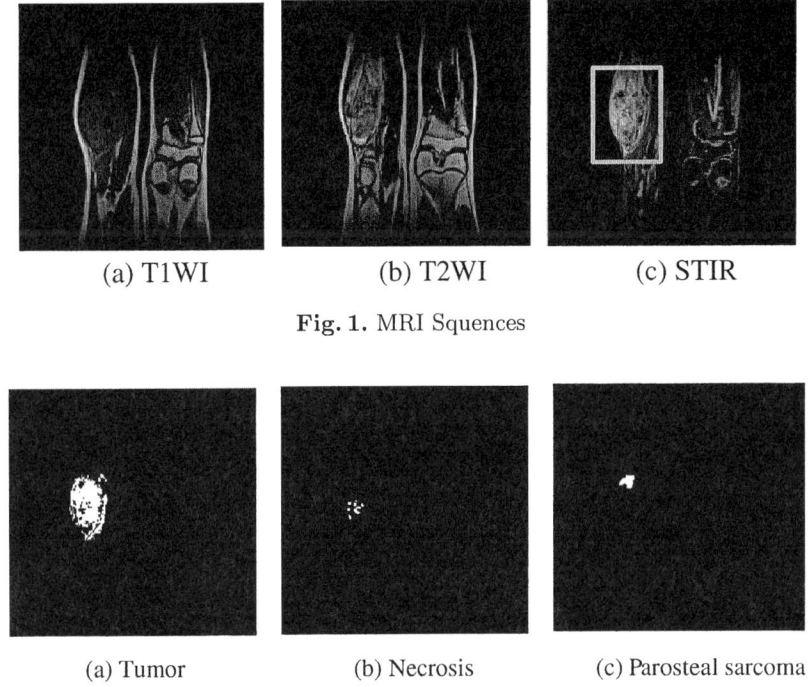

(a) T1WI (b) T2WI (c) STIR

Fig. 1. MRI Squences

(a) Tumor (b) Necrosis (c) Parosteal sarcoma

Fig. 2. The Segmented Tissues

from disconnected regions, proving that the improvement of the algorithm is feasible and effective.

The system was run on a computer of Pentium IV 1.8G HZ CPU and 512 M RAM with Windows 2000. Two experiments are presented to illustrate the efficiency of the algorithm. Table I shows the information of the image data and the computational time of segmentation. In the first experiment, we choose the information of T1WI sequences as the standard to adjust the size and spacing of T2WI sequences. "Sep Time" is an abbreviation for the computational time which adds the time of separately segmenting tumor and parosteal sarcoma tissues. "Sim Time" is an abbreviation for the computational time of simultaneously segmenting the tissues. "Total Time" is an abbreviation for the computational time of generating all the volumes with the improved fashion. As shown in Table I, it is obvious that the method accelerates the speed of segmenting the two tissues and saves the required time by 20%.

Table 1. Image Data Information and Computational Time of Segmentation

Experiment	1	2
Size	$256 \times 256 \times 16$	$256 \times 256 \times 16$
Spacing	$1.289 \times 1.289 \times 5.492$	$1.563 \times 1.563 \times 6.499$
Sep Time(m)	5.82	5.26
Sim Time(m)	4.03	3.49
Total Time(m)	5.64	5.02

5 Conclusion

This paper presents a methodology for osteosacoma segmentation with fuzzy-connectedness theory. The input image data is vectorial and the output data are three sets of binary 3-D images, which denote tumor, necrosis and parosteal sarcoma tissues respectively. We mark every voxels to speed up the segmentation and automatically identify the seeds of the disconnected region. The results show that the improvements can speed up the segmentation and get a better estimation of disconnected regions. The method may be used for routine segmentation in the hospital.

References

1. Udupa, J.K. and Saha, P.K.: Fuzzy connectedness and image segmentation. Proceedings of the IEEE. **91** (2003) 1649–1669
2. Y. Zhuge, J. K. Udupa, and P. K. Saha: Vectorial scale-based fuzzy connected image segmentation. Proceedings of SPIE: Medical Imaging. **4684** (2002) 1476–1487

A Global Optimization Algorithm for Protein Folds Prediction in 3D Space

Xiaoguang Liu, Gang Wang, and Jing Liu

Department of Computer Science, Nankai University, Tianjin, 300071, China
{liuxg74, wgzwp}@hotmail.com, jingliu@nankai.edu.cn

Abstract. Protein folds prediction is one of the most important problems in computational biology. In previous works, local optimization algorithms were used mostly. A new global optimization algorithm is presented in this paper. Compared with previous works, our algorithm obtains much lower energy states in all examples with a lower complexity.

1 Introduction

Predicting the structure of proteins, given their sequence of amino acid, is one of the core problems in computational biology. With the rapid advances in DNA analysis, the number of known amino acid sequences has increased enormously. However, the progress in understanding their 3D structure and their functions has lagged behind owing to the difficulty of solving the folding problem.

Since the problem is too difficult to be approached with fully realistic potentials, many researchers have studied it in various degrees of simplifications. By the simplifications, protein fold prediction is converted to a combinatorial optimization problem. Its main target is to design algorithms which can find the lowest energy states of the amino acid sequences in three-dimensional space. The most popular model used in related works is HP model [1,2] which only consider two types of monomers, H (hydrophobic) and P (polar) ones. Hydrophobic monomers tend to avoid water which can only attract mutually by themselves. All the monomers are connected like a chain. There are repulsive or attractive interactions among neighboring monometers. The energies are defined as $\varepsilon_{HH} = -1$, and $\varepsilon_{HP} = \varepsilon_{PP} = 0$.

Many computational strategies have been used to analyze these problems, such as Monte Carlo simulations[3], chain growth algorithms[4], genetic algorithms[5], PERM and improved PERM[6], etc. Most models mentioned above are discrete. It's possible that some potential solutions are missed by the discrete models in 3D space. In reference 7, Huang devised a continuous model for 3D protein structure prediction. But the results from reference 7 had some errors owing to the defects in algorithm . Following the idea of Huang's model, we present a continuous optimization algorithm in the paper.

2 The Algorithm

In HP model, all amino acid monomers are connected and form a n-monomer chain. It's easy to understand that every monomer can be considered as a rigid ball. In order to present more succinctly, hydrophobic monomers are denoted as H balls and polar monomers are denoted as P balls in the following sections.

If the number of the balls in the chain is n and the radius of every ball is one, then protein folds prediction can be transformed into discovering the fit positions of these balls in 3D Euclidean space. It requires all the neighboring balls connected each other are tangent and all H balls are close as much as possible.

More precisely, the algorithm wants to obtain a n-dimensional position vector $P(P_1, P_2...P_n)$ in 3D Euclidean space satisfying the following three conditions:

$$d_{i,j} \geq 2 \qquad (1)$$

Where $d_{i,j}$ is the distance between position P_i and P_j, ($i \neq j, i, j = 1, 2...n$)

$$d_{i,i+1} = 2 \qquad i = 1, 2, ..., n-1 \qquad (2)$$

Minimized E where $E = \sum_{i=1}^{n} \sum_{j>i}^{n} \varepsilon / d_{i,j}$, $\varepsilon_{HH} = -1$, $\varepsilon_{HP} = \varepsilon_{PP} = 0$ \qquad (3)

E is the gravitational energy of all balls.

We can consider all the balls in the chain are connected by a spring. Thus there are three types of forces in the n-ball chain, the pull forces of spring between the adjacent balls, the repulsion forces between two embedded balls and the gravitational forces between two H balls (since $\varepsilon_{HP} = \varepsilon_{PP} = 0$).

At any time, the external force that each ball received is the sum of forces that all the other balls in the same chain imposing on it. From the initial state, all the balls in the chain will be moved continuously driven by the external force. The n-ball system keeps moving until all the forces reach the equilibrium. During the process, the pull and repulsion forces drive the system to meet the requirements of equation (1) and (2), the gravitational forces among all H balls pull them together as close as possible. In the equilibrium state, the position vector of all the balls $P(P1, P2...Pn)$ represent a best fit approximation to 3D protein structure prediction. The value of P can be determined according to equations (1) ,(2) and (3).

Considering the pull forces that ball i put on ball j,

$$\vec{F_{pij}} = \begin{cases} \dfrac{k_p \times (d_{ij} - 2) \times (\vec{r_i} - \vec{r_j})}{d_{ij}} & \text{if } d_{ij} > 0 \\ 0, & \text{if } d_{ij} = 0 \end{cases} \qquad (4)$$

Where $\vec{r_i}$ ($\vec{r_j}$) is the vector pointing to the position of ball i (j) from grid origin, d_{ij} is the distance between ball i and j, and k_p is the elastic coefficient of the spring in the chain. It's easy to understand that there is only one pull force to the first and the last ball in the n-ball chain. To the others, the pull forces will be produced by the previous and the following balls. Obviously, the pull forces will be changed into push forces if $0 < d_{ij} < 2$ according to equation (4).

To the repulsion forces between ball i and j,

$$\overrightarrow{F_{rij}} = \begin{cases} \dfrac{k_r \times (2 - d_{ij}) \times (\vec{r_i} - \vec{r_j})}{d_{ij}}, & d_{ij} < 2 \\ 0, & d_{ij} \geq 2 \end{cases} \quad (5)$$

Where k_r is the repulsive coefficient of the balls in the case that two balls are embedded each other. To the gravitational forces between two H balls,

$$\overrightarrow{F_{gij}} = \begin{cases} k_g \times (\vec{r_i} - \vec{r_j})/d_{ij}^3, & \text{if } d_{ij} \geq 2 \\ 0, & \text{if } d_{ij} < 2 \end{cases} \quad (6)$$

According to equations (4), (5) and (6), the force $\overrightarrow{F_i}$, which exerted to ball i at any time, is the composition of the forces giving by the other balls in the chain.

$$\overrightarrow{F_i} = \sum_{j=i-1,i+1}^{n} \overrightarrow{F_{pji}} + \sum_{j=1,j\neq i}^{n} \overrightarrow{F_{rji}} + \sum_{\substack{j=1,j\neq i, \\ i,j \in H-balls}}^{n} \overrightarrow{F_{gji}} \quad (7)$$

Our algorithm can be described as following,

Initially, all the balls in the chain are distributed orderly on the surface of a virtual sphere in the 3D Euclidean space as even as possible. Therefore every ball will be coequal in the initial state. In the next period, each ball is moved in a small distance by the composition of external forces. This process repeats continuously until the n-ball system reaches the equilibrium. The positions of all the balls in the equilibrium state should be the solution to 3D protein structure prediction.

The pseudocode of the algorithm

```
Initialization.
for (t=0;t<t_MAX; t++)
   for (i=0;i<n; i++)
   {
   F_i =Compute_Force(i);//Computing the Force to ball i;
   r_i^{t+1} = r_i^t + λ × F_i ;
   };
```

Where t_{max} the upper bound of the periods, and λ is is the movement coefficient in the iterative equation.

3 Experimental Results

Four 3D HP sequences with length N equals to 58,103,124 and 136 respectively were described in reference 6 as models of actual proteins. Using our algorithm, we recalculated the four sequences and found much lower energy states for all these sequences.

In our experiments, we set the maximal number of periods $t_{max} = 3.0 \times 10^9$, set the movement coefficient $\lambda = 2 \times 10^{-7}$.

To the other coefficients used in the algorithm,

$$K_P = \begin{cases} 500 + 0.5 \times t, & n = 58 \\ 800 + 0.5 \times t, & n = 103,124 \\ 1000 + 0.5 \times t, & n = 136 \end{cases} \quad (8)$$

$$K_r = \begin{cases} 500 + 0.6 \times t, & n = 58 \\ 800 + 0.6 \times t, & n = 103,124 \\ 1000 + 0.6 \times t, & n = 136 \end{cases} \quad (9)$$

$$K_g = \begin{cases} 30, & n = 58 \\ 40, & n = 103,124 \\ 50, & n = 136 \end{cases} \quad (10)$$

Table 1. Experimental results

N	($\varepsilon_{HH}, \varepsilon_{HP}, \varepsilon_{PP}$)	Sequence	E_{min}^a	E_{min}^b	CPU time^c	CPU time^d
58	(-1,0,0)	PHPH$_3$PH$_3$P$_2$H$_2$PHPH$_2$PH$_3$P HPHPH$_2$P$_2$H$_3$P$_2$HPHP$_4$HP$_2$H P$_2$H$_2$P$_2$HP$_2$H	-63	-44	3.81	0.19
103	(-1,0,0)	P$_2$H$_2$P$_5$H$_2$P$_2$H$_2$PHP$_2$HP$_7$HP$_3$H $_2$PH$_2$P$_6$HP$_2$HPHP$_2$HP$_5$H$_3$P$_4$H$_2$ PH$_2$P$_5$H$_2$P$_4$H$_4$PHP$_8$H$_5$P$_2$HP$_2$	-88	-54	10.39	3.12
124	(-1,0,0)	P$_3$H$_3$PHP$_4$HP$_5$H$_2$P$_4$H$_2$P$_2$H$_2$P$_4$ HP$_4$HP$_2$HP$_2$H$_2$P$_3$H$_2$PHPH$_3$P$_4$ H$_3$P$_6$H$_2$P$_2$HP$_2$HPHP$_2$HP$_7$HP$_2$ H$_3$P$_4$HP$_3$H$_5$P$_4$H$_2$PHPHPHPH	-109	-71	24.25	12.3
136	(-1,0,0)	HP$_5$HP$_4$HPH$_2$PH$_2$P$_4$HPH$_3$P$_4$H PHPH$_4$P$_{11}$HP$_2$HP$_3$HPH$_2$P$_3$H$_2$P $_2$HP$_2$HPHPHP$_8$HP$_3$H$_6$P$_3$H$_2$P$_2$ H$_3$P$_3$H$_2$PH$_5$P$_9$HP$_4$HPHP$_4$	-117	-80	36.01	110

^{a}Lowest energies found in present work
^{b}Lowest energies found in reference 6
^{c}CPU times (hours) cost on 3.0 GHz Intel P4(results in our experiments)
^{d}CPU times (hours) cost on 667 MHz DEC ALPHA 21264 (results from reference 6)

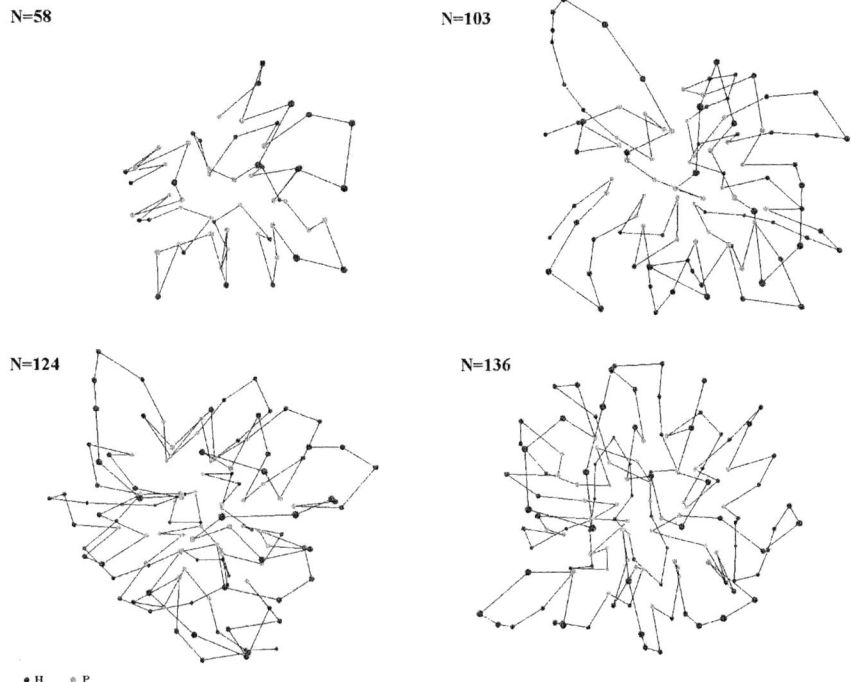

Fig. 1. The experimental results shown in 3D space

Where t is the current period.

Furthermore, the precision is set to 1×10^{-7} in order to determine whether the two H balls are tangent or not. The details of the results are shown in table 1.

Among all previous work, the growth algorithm in reference 6 provided the best results. Compared with the results from reference 6, our algorithm can find much lower energy states for all the four HP sequences. The big gap is mainly caused by the differences in algorithm.

The growth algorithm used in reference 6 is a depth-first *implementation* of the "go-with-the-winners" strategy. Actually, it can be regarded as a special type of greedy algorithm. It always takes the local optimal solution while resolving the problem. As we known, a winner of a battle may not be a winner of the whole war. The solutions found by the growth algorithm may not be the global optimal to the problems.

On the contrary, the algorithm in present paper is a global optimization algorithm. All the monomers in the HP sequences have the same weights initially. The n-monomer chain can move in many directions. After long-time iterations, many possible positions can be reached following our algorithm.

For the shorter chains, our algorithm is more time consuming than growth algorithm used in reference 6, but the situation inverses to the longest chain. Indeed, growth algorithm requires exponential time consuming with the increase of the chain's length. Comparing the results, it is apparently that the times increase much slower in our experiments which demonstrate that our algorithm has better performance when applied to longer HP sequences.

4 Discussion

In this paper we present a new continuous optimization algorithm for 3D protein structure prediction. The main idea of the algorithm is that all monomers share the same initial weight and will move in continuous three-dimensional space following certain physical theories. As a global optimization algorithm, our algorithm can search many potential solutions to find the optimum solution to the problems.

Comparing our results to the best results in previous works, we obtain lower energy states in all 3D cases. Moreover, our algorithm has lower time complexity than previous work. It will show more advantages to the proteins which have longer amino acid sequence.

Following the way of previous work, we used HP model, an abstract model of protein folds prediction, to study the problem. Actually, our algorithm can be used for a much wider range of applications. We anticipate it can be applied to more realistic protein models.

In the future work, we will try to add more information about the proteins into our algorithm, such as molectronics, experiential data, and examine the improvement on performance of the enhanced algorithm.

Acknowledgements

This paper is sponsored by NSF of China (No. 60273031), Education Ministry Doctoral Research Foundation of China (No. 20020055021) and Nankai university ISC. And the proofreading by Dr. Gu Dayong

References

1. K.F.Lau, K.A.Dill, A lattice statistical mechanics model of the cobformation and sequence space of proteins, Macromolecules, 22, 2002
2. Shortle, D., H.S. Chan, and K.A. Dill, Modeling the Effects of Mutations on the Denatured States of Proteins, Protein Science, 1 (1992) : 201-215.
3. J. M. Deutsch, Long range moves for high density polymer simulations, J. Chem. Phys. 1997,106,8849-8856
4. E.M. O'Toole,A.Z. Panagiotopoulos, Effect of sequence and intermolecular interactions on the number and nature of low-energy states for simple model proteins,J. Chem. Phys.1993,98(4),3185-3190.
5. R. König and T. Dandekar Solvent entropy driven searching for protein modeling examined and tested in simplified models. Protein Eng.,2001, **14**, 329-335.
6. Hsiao-Ping Hsu, Vishal Mehra, Walter Nadler and Peter Grassberger, Growth Algorithms for Lattice Heteropolymers at Low Temperatures, J. Chem. Phys.2003(118): 444-448 .
7. Huang Wen-Qi, Huang Qin-bo, Shi He, An Quasiphysical Algorithm for 3D Protein Structure Prediction,J. of Wuhan University,2004,50(5): 586-590

Classification Analysis of SAGE Data Using Maximum Entropy Model

Jin Xin[1,2] and Rongfang Bie[2]

[1] Department of Physics, Beijing Normal University, Beijing 100875, China
xinjin796@126.com
[2] College of Information Science and Technology, Beijing Normal University, China
rfbie@bnu.edu.cn

Abstract. SAGE data can be used to learn classification models to aid cancer classification. In this paper, maximum entropy models are built for SAGE data classification by estimating the conditional distribution of the class variable given the samples. In experiments we compare accuracy and precision to SVMs (one of the most effective classifiers in performing accurate cancer diagnosis from *microarray* gene expression data) and show that maximum entropy is better. The results indicate that maximum entropy is a promising technique for SAGE data classification.

1 Introduction

Cancer types classification is crucial to diagnosis, but traditional methods are based basically on morphological appearance of the tumor that provide limited and unreliable information about the disease. This fact makes very important a correct cancer classification. SAGE (Serial Analysis of Gene Expression) technology provides researchers with a new powerful tool to solve the problem [1]. The SAGE data can be used to learn classification models to aid cancer classification. Here we propose to use maximum entropy techniques for SAGE data classification.

Several classification methods have been done on *microarray* gene expression data. These methods include decision trees, Fisher linear discriminant, Multi-Layer Perceptrons (MLP), Nearest-Neighbor classifiers, linear discriminant analysis, Parzen windows, Bayesian Network and Support Vector Machines (SVMs)[2,3,4]. However, to the best of our knowledge, no supervised classification analysis have been reported on SAGE data owing to a lack of appropriate statistical methods that consider the specific properties of SAGE data.

2 Maximum Entropy Models for SAGE Data

The basic idea behind maximum entropy is that one should prefer the most uniform models that also satisfy any given constraints [5]. For example, consider a three-class SAGE library classification task where we know advance that on average 60% of library with the tag CTCTAAGAAG in them are in the Astrocytoma class. Intuitively, when given a library with CTCTAAGAAG in it, we would say it has a 60% chance of being an Astrocytoma library, and a 20% chance for each of the other two classes. If a library does not have CTCTAAGAAG we may prefer the uniform class distribution of 1/3 each. Calculating the model is easy in this example, but when there are many constraints to satisfy, rigorous techniques are needed to find the optimal solution.

In general, the SAGE data classification problem can be described as follows. Taking into account that one library only belongs to one class (type of cancer), for a given library L with N tags we search for a class c that maximizes the posterior probability p(c | L) which is estimated by:

$$P_\Lambda(c \mid L) = \frac{\exp(\sum_i \lambda_i \times f_i(L,c))}{Z(L)}, \tag{1}$$

where $f_i(L,c)$ is a feature function with a real-value, {λ_i} describes the feature weights vector to be estimated, and $Z(L)$ is a normalizing factor, called the partition function, which is defined as follows:

$$Z(L) = \sum_j \exp(\sum_i \lambda_i \times f_i(L,c_j)), \tag{2}$$

Each λ_i weight is selected so that the expectation of each feature equals its observed frequency in the training data. We estimate the expectation for the feature empirically with the training libraries T.

$$\forall_i \quad \frac{1}{|T|} \sum_{L \in T} f_i(L, c(L)) = \frac{1}{|T|} \sum_{L \in T} \sum_j P(c_j \mid L) f_i(L, c_j), \tag{3}$$

where $c(L)$ is the correct classification of library L specified in the training set, and P is the probability calculated from the statistical model, and |T| is the number of instances in the training set.

To apply maximum entropy to SAGE data classification, we use tag counts as the features to set the constraints. For each tag t and cancer class c combination we instantiate a feature f as the relative frequency of tag t in a library L:

$$f_{t,c'}(L,c) = \begin{cases} 0 & c' \neq c \\ \dfrac{num(t,L)}{|L|} & c' = c, \end{cases} \tag{4}$$

where num(t, L) is the count of the tag t in the library L, and |L| is the number of different tags in the library L. we use real-value feature instead of binary feature that is used in most maximum entropy applications in order to improve classification.

3 Experiments

The experiments are based on 52 Hs SAGE libraries which were publicly available on the NCBI SAGE website. For multicategory classification, we used all of the 52 libraries, which fall in to four categories: Astrocytoma (12 libraries), Ependymoma (8 libraries), Glioblastoma (8 libraries) and Medulloblastoma (24 libraries). For binary classification, we used Astrocytoma and Medulloblastoma. One problem with SAGE data is that each library has a too large volume of features, we simply remove all but the top 3000 tags by selecting tags with highest information gains.

Maximum entropy and SVMs experiments are performed with six trials of randomly selected train-test splits of the set of SAGE libraries. Half the samples in each class are held-out for training. Table 1 and 2 shows the average SAGE data classification performance results. Maximum entropy achieved a prediction accuracy of 98% for binary classification and 89% for multicategory classification. The corresponding results achieved by SVMs are 77% and 86%.

Table 1. Performance for SAGE data *Binary Classification*. (M.E.=Maximum Entropy)

Method	Precision		Accuracy
	Astrocytoma	Medulloblastoma	
M.E.	97%	97%	98%
SVMs	100%	64%	77%

Table 2. Performance for SAGE data classification: *Multicategory Classification*. Astr. =Astrocytoma, Epen.=Ependymoma, Glio.=Glioblastoma, Med.=Medulloblastoma.

Method	Precision				Accuracy
	Astro.	Epen.	Glio.	Med.	
M.E.	85%	100%	58%	97%	89%
SVMs	72%	96%	54%	100%	86%

4 Conclusions

In this paper, we proposed a maximum entropy method for SAGE data classification analysis and made comparisons with SVMs. We demonstrated that the method could classify cancer types accurately based on gene expression profiles produced by SAGE technique.

Acknowlegments

This work was supported by the National Science Foundation of China under the Grant No. 10001006 and No. 60273015. We thank Lin Kui for his helpful comments on a draft of this paper.

References

1. Velculescu V. E., Zhang L., Vogelstein B., Kinzler K.W.: Serial Analysis of Gene Expression. Science, Vol. 270, October 20 484-487 (1995)
2. S. Dudoit, J. Fridlyand, T. Speed: Comparison of Discrimination Methods for the Classification of Tumors Using Gene Expression Data. JASA 97(457) 77–87 (2002)
3. J. Khan et al.: Classification and Diagnostic Prediction of Cancers Using Gene Expression Profiling and Artificial Neural Networks. Nature Medicine, 7(6) 673–679 (2001)
4. Helman, P. et al.: A Bayesian Network Classification Methodology for Gene Expression Data. Journal of Computational Biology, Vol. 11, No. 4, August 581-615 (2004)
5. Kamal Nigam, John Lafferty, Andrew McCallum: Using Maximum Entropy for Text Classification. In IJCAI Workshop on Machine Learning For Information Filtering, 61-67 (1999)

DNA Sequence Identification by Statistics-Based Models

Jitimon Keinduangjun[1], Punpiti Piamsa-nga[1], and Yong Poovorawan[2]

[1] Department of Computer Engineering, Faculty of Engineering, Kasetsart University,
Bangkok, 10900, Thailand
{jitimon.k, punpiti.p}@ku.ac.th
[2] Department of Pediatrics, Faculty of Medicine, Chulalongkorn University,
Bangkok, 10400, Thailand
yong.p@chula.ac.th

Abstract. Basically, one of the most important issues in identifying biological sequences is accuracy; however, since the exponential growth and excessive diversity of biological data, the requirement to compute within a considerably appropriate time span is usually in conflict with accuracy. We propose a novel approach for accurate identification of DNA sequences in shorter time by discovering sequence patterns -- signatures, which are sufficiently distinctive information for the identity of a sequence. The approach is to discover the signatures from the best combination of n-gram patterns and statistics-based models, which are regularly used in the research of Information Retrieval, and then use the signatures to create identifiers. We evaluate the performance of all identifiers on three different types of organisms and three different numbers of identification classes. The experimental results showed that the difference of organisms has no effect on the performance of the proposed model; whereas the different numbers of classes slightly affect the performance. The sole use of *Information Gain* is changed in a small range of n-grams since the use of its pattern absence brings the unbalanced class and pattern score distribution. However, several identifiers provide over 95% and up to 100% of accuracy, when they are constructed by signatures using the appropriate n-grams and statistics-based models. Our proposed model works well in identifying DNA sequences accurately, and it requires less processing time.

1 Introduction

The rapid growth of genomic and sequencing technologies during the past few decades has facilitated the incredibly large size of diverse genome data, such as DNA and protein sequences. However, most biological sequences contain very little known meaning. Therefore, more knowledge is waiting to be discovered. Techniques to acquire knowledge from biological sequences become more important in transforming unknowledgeable, diverse and huge data into useful, concise and compact information. These techniques generally consume long computation time; their accuracies usually depend on the data size; and there is no best-known solution in any particular circumstances. Many research projects on biological sequence processing still share some common stages in their experiments. An important research project of computational biology is sequence identification.

Ideally, the success of research in sequence identification depends on the method employed to find any short informative data (signatures) that can identify the type of the sequences. The recognition of a small, informative signature is useful to reduce the computation time, as data are more compact and more precise [4]. All the current techniques require long computation time as categorized into three groups, namely:

Inductive Learning: this technique takes learning tools, such as Neural Network and Bayesians, to derive rules that are used to identify biological sequences. For instance, gene identification uses a neural network [14] to derive a rule that verifies whether or not a subsequence is a gene. The identification generally uses small contents, such as frequencies of two or three bases, instead of plain sequences as inputs of the learning process. Although the use of the rule derived from the learning process attains low processing time, the procedure used as a pre-process of the systems still needs too high computation time to finalize in the identification process.

Sequence Alignment: the tools that use this technique, such as BLAST [7] and FASTA [10], align uncharacterized sequences with all existing sequences in genome database and then assign the uncharacterized sequences to the same class with one of the sequences in database that gets the best alignment score. The processing time of this alignment technique is much longer than other techniques since the process has to be performed directly on all the sequences in database, whose sizes are usually large.

Consensus Search: this technique takes a multiple alignment [13] as a pre-process of systems for finding a "consensus" sequence, which is used to derive a rule. Then, the rule is used to identify sequences in uncharacterized biological sequences. The consensus sequence is nearly a signature, but it is created out of majority bases in the multiple alignments of a sequence collection. Because of its high computation on the multiple alignment, the consensus sequence is not widely used in biological tasks.

Our approach constructs the identifiers on the same way as the pre-process in the inductive learning and the multiple alignments. However, the proposed approach requires much less pre-processing time than the learning technique and the multiple alignment technique. First, we apply an *n*-gram method to transform DNA sequences into a pattern set. Second, the signatures are discovered from the pattern set using statistics-based models. Third, identifiers are constructed using the signatures. Finally, the performances of the identifiers are estimated by testing on DNA sequences of three different organisms: *Influenza A virus*, *Rotavirus* and *Hepatitis B virus*.

Our identification system and methods are described in Section 2. This section has details on transforming DNA sequences into pattern sets, employing statistics-based models to discover signatures, constructing identifiers from the signatures and evaluating the performances of identifiers. Our experiments on the three organisms are described in Section 3. Finally, we summarize the proposed approach in Section 4.

2 System and Methods

The identification is a challenging research in the field of Computational Biology since the biological sequences are zero-knowledge based data, unlike text data of the human language. Biological data are totally different from the text data which basically use words as their data representation; whereas we do not know any "words" of the biological data. Therefore, a process for generating the data representation is needed. Our sequence identification framework [6] is depicted in Fig. 1:

Step I: *Pattern Generation* is to transform training data into a pattern set as the data representation of DNA sequences. In this paper, we create training data from DNA sequences and then use them to generate all n-gram patterns as the n-gram model.

Step II: *Signature Discovery* is a process of finding the signatures of DNA sequences. This process is divided into two consecutive parts: 1) *Pattern Score Computation* aims to measure the significance of each n-gram pattern, called "Scores", using variously different Statistical Scoring Models; and 2) *Pattern Subset Discrimination* aims to select the best patterns as "Signatures".

Step III: *Identifier Construction* is a process of creating identifiers using the discovered signatures. This process formulates a similarity scoring function. If there is a query sequence, the function is used to detect whether it is the target sequence.

Step IV: *Performance Evaluation* is to estimate the performances of identifiers using test data, sequences that are not members of the training data. The performance estimator compares a goodness of each identifier, generated by the different sets of signatures, to find the best one for accurate identification of unseen data, sequences that are not members of both training data and test data.

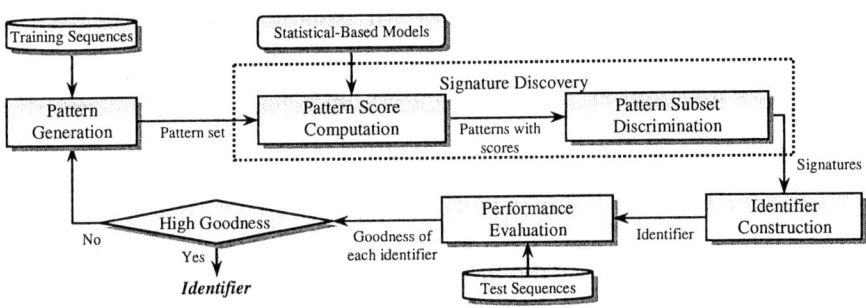

Fig. 1. Common identification framework

2.1 Pattern Generation

Finding a data representation of DNA sequences is performed by a pattern generating process using training data. The data representation, called a pattern, is a substring of the DNA sequences. DNA sequences include four symbols {A, C, G, T}; therefore, the members of a pattern are also restricted to the four symbols. Let Y be a sequence $y_1 y_2 ... y_M$ of length M over the four symbols of DNA. A substring $t_1 t_2 ... t_n$ of length n is called an n-gram pattern [3]. The n-gram methodology generates n-gram patterns representing the DNA sequences with different n-values. There are $M-n+1$ patterns in a sequence of length M for generating the n-gram patterns, but there are only 4^n possible patterns for any n-values.

For example, let Y be a 9-base-long sequence CTCGATCGA. Then, the number of 4-gram patterns generated from the sequence of length 9 is 6 (9-4+1): CTCG, TCGA,

CGAT, GATC, ATCG, TCGA. There are 256 (4^4) possible patterns; however, only 5 patterns occur in this example: CTCG (1), TCGA (2), CGAT (1), GATC (1) and ATCG (1). From the five patterns, one pattern occurs twice.

Our experiments found that the use of too long patterns (high n-values) produces poor identifiers. Notice that 1-gram patterns have 4 (4^1) possible patterns; whereas 24-gram patterns have 2.8×10^{14} (4^{24}) possible patterns. The numbers of possible patterns obviously vary from 4 to 2.8×10^{14} (4^{24}) patterns. Too high n-values may not be necessary since each generated pattern does not occur repeatedly and statistics-based models cannot discover signatures from the set of n-gram patterns. However, in our experiments, the pattern sets are solely generated from 1- to 24-grams since the higher n-values do not yield any improvement of the performance. The discovery of the signatures from the pattern sets is discussed in the following section.

2.2 Signature Discovery

The signature discovery is a process of finding signatures from the n-gram patterns. We propose statistics-based models for evaluating the goodness of each n-gram pattern called a score to discover the signatures. The statistics-based models are described in Section 2.2.1. Section 2.2.2 describes the process of signature discovery.

2.2.1 Statistics-Based Models

The statistics-based models are well-known metrics in the research of Information Retrieval [12]. We evaluate six statistics-based models how statistically efficient they can discover signatures for identifying biological sequences. The six statistics-based models are roughly divided into two groups, namely

Group I: *Based on common patterns*, such as *Term Frequency (TF)* [1] and *Rocchio (TF-IDF)* [5].

Group II: *Based on class distribution*, such as *DIA association factor (Z)* [11], *Mutual Information (MI)* [8], *Cross Entropy (CE)* [8] and *Information Gain (IG)* [11].

The models based on common patterns (Group I) are the simplest techniques in the computation of pattern score. They independently compute scores of patterns from the frequencies of each pattern; whereas the models based on class distribution (Group II) compute the scores of patterns from the frequencies of co-occurrences of patterns and classes. Each identification class comprises sequences which are unified to be of the same type. Most models based on class distribution consider only the presence (p) of pattern in a class, except the *"Information Gain"* that considers both the presence and absence ($\bar{p}$) of a pattern in a class [9]. Like the models based on the common patterns, they independently compute the score of each pattern.

Statistics-based models concentrated on pattern and class value are shown in Table 1. Where *Freq(p)* is the Term Frequency (the number of times pattern p occurred); *IDF* is the Invert Document Frequency; *DF(p)* is the Document Frequency (the number of sequences in which pattern p occurs at least once); $|D|$ is the total number of sequences; $P(p)$ is the probability that pattern p occurred; $\bar{p}$ means that pattern p does not occur; $P(C_i)$ is the probability of the ith class value; $P(C_i|p)$ is the conditional probability of the ith class value given that pattern p occurred; and $P(p|C_i)$ is the conditional probability of pattern occurrence given the ith class value.

Table 1. Mathematical forms of statistics-based models concentrated on pattern and class value

Statistics-Based Models	Common Patterns	Class Distribution	
		Pattern Presence	Pattern Absence
$TF(p) = Freq(p)$.	Yes	-	-
$TF - IDF(p) = TF(p) \cdot \log(\frac{|D|}{DF(p)})$.	Yes	-	-
$Z(p) = P(C_i \mid p)$.	-	Yes	-
$CE(p) = P(p) \sum_i P(C_i \mid p) \log \frac{P(C_i \mid p)}{P(C_i)}$.	-	Yes	-
$MI(p) = \sum_i P(C_i) \log \frac{P(p \mid C_i)}{P(p)}$.	-	Yes	-
$IG(p) = P(p) \sum_i P(C_i \mid p) \log \frac{P(C_i \mid p)}{P(C_i)} + P(\overline{p}) \sum_i P(C_i \mid \overline{p}) \log \frac{P(C_i \mid \overline{p})}{P(C_i)}$.	-	Yes	Yes

2.2.2 Signature Discovery Process

The signature discovery process is a process of finding any short informative data (signatures) that can identify types of DNA sequences efficiently. The process applies the statistics-based models to evaluate the significance of each n-gram pattern. Then, the best-score patterns are selected to be the signatures.

The algorithm of signature discovery is given in Fig. 2. Firstly, it transforms all DNA sequences of each class into the set P of n-gram patterns of each class (step 1-4) using the n-gram model as discussed above. Secondly, the algorithm measures the significance of each n-gram pattern, called a score, using the statistics-based models (step 5-7). Thirdly, all n-gram patterns of each class are sorted in descendent order of the scores (step 8). Finally, the k highest-score patterns of each class are selected as the set S of signatures of each class (step 9-12). The signatures of each class are returned as the output of algorithm (step 13).

Algorithm 1. Signature Discovery: Discovering the signatures in the set of n-gram patterns.
Input: Training sequences which are separated into several classes as the same type of sequences.
Output: Signatures of each class (S).
1) $P \in \theta$
2) for all DNA sequences of each class do
3) P = generate_n-gram_pattern(DNA sequences) U P
4) end for
5) for each pattern $p \in P$ do begin
6)   score(p) = Compute the significance of each pattern p using a statistics-based model
7) end for
8) P = sort(P) as the descendent scores of every pattern p of each class
9) $j = 1$
10) while $j <= k$ do begin (k is a defined value by heuristics)
11) S = insert(p_j)
12) end while
13) Return S

Fig. 2. An algorithm of signature discovery based on statistics-based models

The question is how many the highest-score patterns the model should select as signatures (or what value k is in the step 10). We use heuristics to select the optimal number of signatures. Following our previous experiments [6], the model gets the highest efficiency when the number of signatures is ten.

2.3 Identifier Construction

The identifier construction uses signatures to formulate a similarity scoring function. When there is a query sequence, the scoring function is used to estimate the significance of each identification class (*SimScore*). If the *SimScore* of any class is maximal, the query is identified as a member of the class. On the other hand, if *SimScore* of every class is zero; the query is not assigned to any classes. Assume that we pass a query through an identification system; the system is performed as follows:

Step I: Set the *SimScore* of every identification class to zero.

Step II: Generate patterns of the query sequence according to the *n*-gram method using the same *n*-value as the one used in the signature generation process.

Step III: Match between patterns generated by the query sequence and signatures of every identification class. If a match occurs in any identification class; the *SimScore* of the class is added one point (the matching is accepted with one mutation). For example, a pattern ACGTC matches to a signature ACGAC, with one mutation at position 4 (*T* of the pattern and the second *A* of the signature).

Step IV: Assign the query sequence to a class which has the highest *SimScore*, or not assign it to any classes, if the *SimScore* of every class is zero.

For instance of measuring the *SimScore*, let X be a query sequence with cardinality m. Let $p_{x_1}, p_{x_2} \ldots p_{x_d}$ be a set of d $(m-n+1)$ n-gram patterns generated by the query sequence. Let $s_{y_1}, s_{y_2} \ldots s_{y_e}$ be a set of e signatures in an identification class Y. Then, $sim(p_x, s_y)$ is a similarity score of a pattern p_x and a signature s_y, where $sim(p_x, s_y)$ is 1, if p_x is similar to s_y (with one mutation accepted), and $sim(p_x, s_y)$ is 0, if not similar. The similarity score of an identification class Y, denoted *SimScore(X,Y)*, is a summation of similarity scores of every pattern p_x and every signature s_y as

$$SimScore(X,Y) = \sum_{1 \leq i \leq d, 1 \leq j \leq e} sim(p_{x_i}, s_{y_j}). \tag{1}$$

2.4 Performance Evaluation

Evaluation process is to assess the performances of identifiers over the set of test data. Our studies evaluate the performances of the identifiers by applying them to distinguish the diverse types of DNA sequences in the test data. We use a common strategy, *Accuracy*, for evaluating the identification system [12]. The *Accuracy* is the ratio of the number of sequences correctly identified to the total number of test sequences formulated as follows:

$$Accuracy = \frac{Number\ of\ Sequences\ Correctly\ Identified}{Total\ Number\ of\ Test\ Sequences} \times 100\%. \tag{2}$$

3 Experiments

3.1 Datasets

The datasets of our experiments are selected from the Genbank genome database of National Center for Biotechnology Information (NCBI) [2]. We use DNA sequences of three organisms, *Influenza A virus*, *Rotavirus* and *Hepatitis B virus*, as the datasets for evaluating the performances of the identifiers. There are three datasets of the experiments created by the three different organisms. A characteristic of these organisms are suitable for testing the performances, because each organism is distinctly divided into various segments which have different gene in each segment. The approximate length of each complete gene is between 500-2,400 bases.

Dataset I composes of 960 (120*8) sequences which are randomly picked from 120 sequences of eight inner genes of Influenza A virus. *Dataset II* comprises 240 (60*4) sequences which are randomly picked from 60 sequences of four inner genes of Rotavirus. *Dataset III* composes of 120 (60*2) sequences which are randomly selected from 60 sequences of two inner genes of Hepatitis B virus. That the numbers of sequences of *Dataset II* and *III* are less than the *Dataset I* is due to the lower actual data collection in the Genbank genome database.

Each dataset is separated into two groups: training set and test set. One-third of each dataset is generally used for the test set and the rest for the training set. Our identification system discovers signatures and constructs the identifiers using the training set only and then evaluates the performances of identifiers in the test set.

3.2 Results and Discussion

In our experiments, we compare the performances of the identifiers by distinguishing the inner genes of three organisms. Each identifier is produced by different sets of signatures which each set derives from the use of one n-value of n-gram and one statistics-based model. The n-values used in experiments are ranged from 1 to 24 and there are six statistics-based models. First, we evaluate the performances of identifiers on the *Dataset I* of *Influenza A virus*. The experimental results as illustrated in Fig. 3 show that the identifiers are averagely reliable with 95% accuracy, when n-values of n-gram is between 5 and 14 in almost all models, except the *Information Gain*. For shorter than 5-gram, the accuracy varies and is poor in every model; whereas in the case of longer than 14-gram, the accuracy is 10% less than the one used in the 5- to 14-gram signatures in every model, nevertheless they are quite stable and good. Additionally, the peak accuracy provided is up to 100% at 11-gram in several models.

Following the experimental results, the models solely based on the frequencies of patterns, such as *TF* and *TF-IDF*, achieve good results; whereas the ones based on class distribution depend on the length of n-gram signatures and the criteria used in the models. The models using only the pattern presence produce good identifiers, while the model based on the pattern absence, such as *IG*, produces identifiers which provide lower performance than the other models, when the lengths of signatures are ranged from 1- to 14-grams; those which are longer than the 14-gram, the *IG* model gets the same performance as the others. Comparing the significance of every pattern, generated by the 1- to 14-grams, uses frequencies of both pattern presence (p) and pattern absence ($\bar{p}$) which bring about unbalanced class and pattern score distribution, while those which are longer than 14-gram, the comparison mainly uses frequencies of the pattern presence only since the frequencies of pattern absence of all the patterns are similar that are very high, because of too high pattern distribution. The high pattern distribution in long patterns (>14-gram) occurs from the large number of possible patterns (>4^{14} possible patterns) and the low probability of the same pattern occurrence. Hence, the performance of model using pattern absence in the long signatures is comparable with the performance of the other models based on class distribution owing to depending on the same only pattern presence.

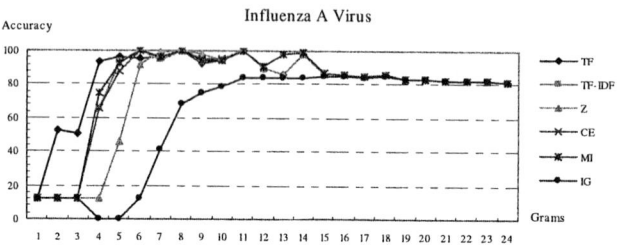

Fig. 3. The performances of identifiers constructed by signatures using diverse n-grams and six statistics-based models on distinguishing the eight inner genes of *Influenza A virus*

The statistical analysis of signature discovery works well to identify the eight inner genes of *Influenza A virus*. The interesting question is whether the performance changes for arbitrarily chosen DNA sequences or the different numbers of classes of identification or not. Next, we test our model on the *Dataset II* of *Rotavirus* which is four-class identification. The experimental results are depicted in Fig. 4. The results show that our model succeeds in the same way as tested on the Influenza sequences. The identifiers, constructed by 5- to 9-gram signatures, are achievable with approximately 95% accuracy in almost all models, except the *Information Gain*. Moreover, the peak accuracy rises up to 100% at 7-gram in several models. For shorter than 5-gram, the accuracy varies and is poor in every model; while for longer than 9-gram, the accuracy is less than the 5- to 9-grams, but it is still stable and good.

Notice that, following the results of *Rotavirus*, the use of *IG* model gets the same performance as the use of the other models, when the signatures are longer than 9-gram, while in the *Influenza virus*, *IG* model is comparable with the other models, when the signatures are longer than 14-gram. As discussed above, that the *IG* gets the same performance as other models is due to the too high pattern distribution. Too high pattern distribution occurs from the low probability of the same pattern occurrence. We found that the less the number of classes is, the lower the same pattern occurrence gets. Hence, the pattern distribution of four-class identification of *Rotavirus* is higher than the pattern distribution of eight-class identification of *Influenza virus* at the same n-gram and too high since 10-gram. That is why the performance of use of *IG* model in the *Rotavirus* is comparable with the use of the other models since the 10-gram.

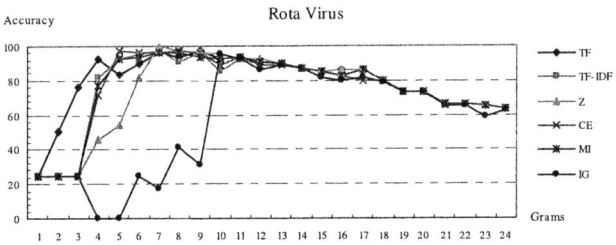

Fig. 4. The performances of identifiers constructed by signatures using diverse n-grams and six statistics-based models on distinguishing the four inner genes of *Rotavirus*

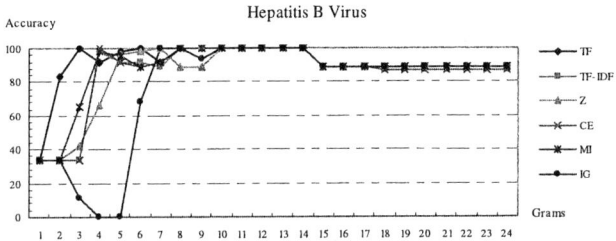

Fig. 5. The performances of identifiers constructed by signatures using diverse n-grams and six statistics-based models on distinguishing the two inner genes of *Hepatitis B virus*

Finally, we test our model on two-class identification of *Hepatitis B virus* in the *Dataset III*. The experiments also succeed in the same way as two previous organisms as illustrated in Fig. 5. The identifiers, constructed from 5- to 14-gram signatures, are achievable with approximately 95% accuracy in most statistics-based models, except the *Information Gain* owing to the previous reason. In addition, the accuracy provides up to 100% between 7- to 14-grams in several models. For shorter than 5-gram, the accuracy varies and is poor in every model; while longer than 14-gram, the accuracy is less than the use of the 5- to 14-grams, nevertheless it is also stable and good.

From all the experiments, the difference of organisms has no effect to the performances of identifiers; while the different numbers of classes slightly affect. Solely the use of *IG* is changed in a small range of n-grams. Our model works well on the diverse DNA sequences and the different numbers of identification classes.

Additionally, we compare our model with BLAST [7], the most well-known tool using an alignment technique. We use 640 training sequences for identifier construction and then distinguish 2,000 test sequences of inner genes of *Influenza virus*. The results showed that an accuracy of BLAST is 2% higher than our model; however its processing time is obviously higher, 87.96 seconds. The total time of our model is only 4.53 seconds, pre-processing time (1.28 seconds) and processing time (3.25 seconds). The speedup of our model is 27 times as fast as the BLAST.

4 Conclusion

The acquisition of an intelligent and low computation method is one of the most significant tasks of identification since the existing research projects apply a high computation method for accomplishing in the target concept of identification, such as projects based on alignments. We hereby propose a new model for accurately identifying DNA sequences in shorter time. The model applies an n-gram method and statistics-based models to discover one of the intelligent data in biological sequences, called "*Signatures*", as inputs of the identification system. The signatures derive from the best combination of the n-grams and the statistics-based models. The experimental results showed that several identifiers provide accuracy over 95% and up to 100%, when they are constructed by signatures using the appropriate n-grams and statistics-based models. Moreover, the identifiers work well on various organisms and the different numbers of identification classes, and also require low computation time. The proposed model is accomplished in the construction of powerful identifiers.

References

1. Aalbersberg, I.: A Document Retrieval Model Based on Term Frequency Ranks. Proceedings of the 7th Annual International ACM SIGIR Conference on Research and Development in Information Retrieval (1994) 163-172
2. Benson, D.A., Karsch-Mizrachi, I., Lipman, D.J., Ostell, J., Rapp, B.A., Wheeler, D.L.: GenBank. Nucleic Acids Research 28(1) (2000) 15-18
3. Brown, P.F., de Souza, P.V., Della Pietra, V.J., Mercer, R.L.: Class-Based N-Gram Models of Natural Language. Computational Linguistics 18(4) (1992) 467-479
4. Chuzhanova, N.A., Jones, A.J., Margetts, S.: Feature Selection for Genetic Sequence Classification. Bioinformatics Journal 14(2) (1998) 139-143
5. Joachims, T.: A Probabilistic Analysis of the Rocchio Algorithm with TFIDF for Text Categorization. Proceedings of the 14th International Conference on Machine Learning (1997) 143-151
6. Keinduangjun, J., Piamsa-nga, P., Poovorawan, Y.: Models for Discovering Signatures in DNA Sequences. Proceedings of the 3rd IASTED International Conference on Biomedical Engineering, Innsbruck, Austria (2005) 548-553
7. Krauthammer, M., Rzhetsky, A., Morozov, P., Friedman, C.: Using BLAST for Identifying Gene and Protein Names in Journal Articles. Gene 259(1-2) (2000) 245-252
8. Mladenic, D., Grobelnik, M.: Feature Selection for Classification Based on Text Hierarchy. In Working Notes of Learning from Text and the Web. Conference on Automated Learning and Discovery, Carnegie Mellon University, Pittsburgh (1998)
9. Mladenic, D., Grobelnik, M.: Feature Selection for Unbalanced Class Distribution and Naïve Bayes. Proceedings of the 16th International Conference on Machine Learning (1999) 258-267
10. Pearson, W.R.: Using the FASTA Program to Search Protein and DNA Sequence Databases. Methods Molecular Biology 25 (1994) 365-389
11. Sebastiani, F.: Machine Learning in Automated Text Categorization. ACM Computing Surveys 34(1) (2002) 1-47
12. Spitters, M.: Comparing Feature Sets for Learning Text Categorization. Proceedings on RIAO (2000)
13. Wang, J.T.L., Rozen, S., Shapiro, B.A., Shasha, D., Wang, Z., Yin, M.: New Techniques for DNA Sequence Classification. Journal of Computational Biology 6(2) (1999) 209-218
14. Xu, Y., Mural, R., Einstein, J., Shah, M., Uberbacher, E.: Grail: A Multiagent Neural Network System for Gene Identification. Proceedings of the IEEE 84(10) (1996) 1544-1552

A New Method to Mine Gene Regulation Relationship Information

De Pan[1,2], Fei Wang[1,2,*], Jiankui Guo[3], and Jianhua Ding[1,2]

[1] Shanghai key laboratory of Intelligent Information Processing
[2] Department of Computer Science and Engineering
[3] Department of Computer Information and Technique,
Fudan University, Shanghai 200433, P.R. China
{pande, wangfei}@fudan.edu.cn

Abstract. It is difficult to build a gene regulatory network directly. So the main interest focuses on the gene-gene regulation relationship mining, which reveals an active or repressive action from one gene to another. The previous methods, such as *Event Method, Edge Detection Method, q-cluster method*, didn't solve the gene regulatory relationship with a great succeed. In this paper, we propose a new method by introducing several more relational techniques. The results demonstrate the complete and detailed information between the genes. The data set and software will be available upon request.

1 Introduction

With the explosively rapid development of DNA microarray technologies in recent years, biologists have been able to attain the expression levels of thousands of genes simultaneously during biological processes, which contain information to facilitate the study of genome. In order to build an integrated and comprehensive gene regulatory network, we need to know the regulatory relationship between genes by a manual experimental method in the earlier times or the widely used DNA microarray method contrasted much more efficient than the former. A genetic regulatory network is a system in which proteins and genes bind to each other to form regulators, and affect the expression levels of the regulated genes by binding to their promoters. The regulations are mainly divided into two types: activation and inhibition. In an activation process, the increase of certain genes' expression level will increase some other down-stream genes' expression level in a certain biological process, while in an inhibition process, an increase of some genes' expression level will result in a decrease in the expression level of some other genes. Intuitively, there exists a time lag in the previous two regulatory processes, but a co-regulatory mechanism also works widely, in which some genes express simultaneously with an increase or decrease, and an obvious time lag may not exist.

---

* Correspondence author.

There have been a number of approaches for extracting regulation information from microarray data.The *Cross-Correlation Function* method [1] makes modification of the traditional *Pearson Correlation Coefficient Method* for time factors, but this method can only give the global similarity between gene variables, not sometime the local similarity. The *Bayesian Network*[2] tries to directly build a gene regulatory network, but its high computational cost make it impractical. The *Edge Detection Method*[3] tries to find the main changing in expression level (*edge*) and gives a score by comparing between two genes, which will lose information when two edges are far apart. Another method is *Dominant Spectral Component Method*[4], which decomposes the time series data into frequency components and attains the correlation by summing all the scaled sub-correlation. The *Event method* [5] discretizes time series expression level into a string of discrete events— R(Rising), F(Falling) & C(No changing), and gives each gene pair a score by applying *Needleman-Wunsch* alignment algorithm, which can lose useful information when discretizing and only give the global similarity score. And the last method is called *q-clusters method*[6], by using a similar discretizing method in*Event Method*, which puts all genes to all acceptable gene clusters, predefined by event pattern, to form the q-clusters. Genes in the same q-cluster will share a local event pattern. But this method will generate too much redundant information for a small fixed-length event pattern.

Our work tries to give a better method to mine gene regulatory relationship information. In order to consider both the global and local regulation possibility and time series factor, we will give a matrix to mine the regulation pattern, which can find the regulation pattern as long as possible, ranging from the minimum length to the global length. And this method is based on the introduction of a new standard to discretize the gene expression level data for preventing the lost of information, from which we define a match matrix for later matching. And in fact, by applying our method on the microarray data, we get a good experiment result and the analysis will be presented in detail. Compared with the *q-clusters method* and the *event method*, our method can produce all the gene regulatory information they generated, and beat them when comparing the results with the known gene regulatory information attained by manual biological experiments.

2 Methods to Mine Gene Regulatory Relationships

In this section we will present our method in detail. We will introduce the data source and some techniques to preprocess the given microarray data in the first subsection. In the following subsection, we will transform the original data by the new standard. After that, we will perform the matching between genes, and at last, the method will generate the required information.

2.1 Data Source

We use the earliest gene expression data which is accessible at *Paul T. Spellman's* website [7]. The data are mainly attained by three independent experiments for synchronized reasons: factor arrest, elutriation and arrest of a cdc15 temperature-sensitive mutant, which consists of all the 6178Yeast ORFs. By using the *Fourier* algorithm and the *co-relation function* we get 800 target genes' expression level data involved in cell-cycle. With additional 9 genes reported by other papers but not found here, we have a DNA microarray data consisting of 809 genes with 73 condition points totally for 4 independent synchronized biological experiments to run our method.

2.2 Data Transformation

The preprocessing of the data is to deal with the NaN value, generated for reasons of sampling, experiments or other noises. We will discard the genes with too much NaN and fill the genes with the number of NaN less than a customized $MaxNaNNumber$. We can fill a NaN with a randomized value in a range, or calculate a *local average value* for it which takes the nearby neighboring values into account, or a *regression method* will works better but costs high. Here we apply the local average value method, which will calculate a local average to fill the NaN and relatively is more rational.

We denote the gene expression level data to be matrix $G_{m \times n}$ with float point values, and each row of G is the expression level data of a gene, which has n condition points. Our aim is to transform $G_{m \times n}$ to be matrix $E_{m \times (n-1)}$. E stands for the changing tendency matrix with symbol values.

The first step of transformation is to discretization. We define the set of changing tendencies, denoted by $1, 0, -1, P, N$; 1 means an increase of gene expression level between two neighboring condition points, 0 means no change, -1 means a decrease.And compared with methods in previous papers, we define two more changing tendency classes:the P means a increase but not strong, and N means a decrease but weak. Such that we can deal with the noisy signal and preserve valuable information. There is another reason to support such an introduction. For example, if the expression of gene i at condition t is as high as 1, and at condition $t+1$ is 1.9, then by equation (1) which is used to calculate the changing tendency by comparing the result with the thresholds. the result (1.9-1)/1=0.9 is smaller than the threshold 1(usually 1 for positive result and -1 for negative result), *no changing* will be assigned, but obviously it is an increase, so here we think a P will be more rational.

$$Tendency = \frac{G_{i,j+1} - G_{i,j}}{|G_{i,j}|} \quad (1)$$

It is the same reason for the introduction of N. The results of our method at the end of this paper also show that it is reasonable to introduce such a fuzzy concept to deal data discretizaion.

The second step is applying the rules in (2) on matrix G and attain the temporary matrix E' and the last step is applying (3) on the matrix E' and finally get the matrix E.(t in (3) is an empirically assigned value, and usually set 1.0.)

$$E'_{i,j} = \begin{cases} 1, & \text{if } G_{i,j} = 0 \ \& \ G_{i,j+1} > 0, \\ -1, & \text{if } G_{i,j} = 0 \ \& \ G_{i,j+1} < 0, \\ 0, & \text{if } G_{i,j} = 0 \ \& \ G_{i,j+1} = 0, \\ \frac{G_{i,j+1} - G_{i,j}}{|G_{i,j}|}, & \text{if } G_{i,j} <> 0. \end{cases} \quad (2)$$

$$E_{i,j} = \begin{cases} 1, & \text{if} E'_{i,j} >= t, \\ -1, & \text{if } E'_{i,j} <= -t, \\ 0, & \text{if } E'_{i,j} == 0 \\ P, & \text{if } 0 < E'_{i,j} < t, \\ N, & \text{if } -t < E'_{i,j} < 0. \end{cases} \quad (3)$$

A simple example explains the operation of the data transformation as follows. The first row of Tab.1 is the conditions of the gene expression level, and here means different time points. The first column of Tab.1 list two genes AAC3 and ABF1 from previously described *Yeast* gene expression level data with the first 11 condition points for editing reasons. And the value in the table represents gene expression level of corresponding gene in corresponding time point. And in Tab.2 we get the transformed matrix.

Table 1. Original Matrix G

	C_1	C_2	C_3	C_4	C_5	C_6	C_7	C_8	C_9	C_{10}	C_{11}
AAC3	-0.01	-0.41	0.08	0.09	0.41	-0.1	0.43	-0.71	0.22	0.27	-0.08
ABF1	-0.99	-0.47	-0.13	0.42	-0.08	0.57	-0.01	0.26	-0.2	0.24	-0.38

Table 2. Transformed Matrix E

	$C_{1,2}$	$C_{2,3}$	$C_{3,4}$	$C_{4,5}$	$C_{5,6}$	$C_{6,7}$	$C_{7,8}$	$C_{8,9}$	$C_{9,10}$	$C_{10,11}$
AAC3	-1	1	P	1	-1	1	-1	1	P	-1
ABF1	P	P	1	-1	1	-1	1	-1	1	-1

2.3 Pattern Matching

In order to find the regulation relationship between genes, we need to compare the genes in the transformed matrix E, so that we can get the local or global matching information and mine the regulation relationship. We call our method to be *matching matrix method*, because we'll mine the regulation information by using a matching matrix. We define the pattern matching matrix in Tab.3 which will be used later.

Here we give a brief explanation, that the first column and the first row of Tab.3 are the five kinds changing tendency. And we define the $T_{i,j}$ in Tab.3 to be the corresponding matching result between the ith changing tendency in the first column and the jth changing tendency in the first row. The matching results form a set $+, -, S, D, 0$, here $+$ means a exact positive match and $-$ means a

Table 3. Matching Table

	1	-1	P	N	0
1	+	-	S	D	0
-1	-	+	D	S	0
P	S	D	S	D	S
N	D	S	D	S	D
0	0	0	S	D	+

Table 4. An example of pattern matching

	-1	1	P	1	-1	1	-1	1	P	-1
P	D	S	**+**	D	S	D	S	D	S	D
P	D	S	**S**	D	S	D	S	D	S	N
1	-	+	S	**+**	-	+	-	+	S	-
-1	+	-	D	-	**+**	-	+	-	D	+
1	-	+	S	+	-	**+**	-	+	S	-
-1	+	-	D	-	+	-	**+**	-	D	+
1	-	+	S	+	-	+	-	**+**	S	-
-1	+	-	D	-	+	-	+	-	D	+
1	-	+	S	+	-	+	-	+	S	-
-1	+	-	D	-	+	-	+	-	D	+

exact negative match, S means a fuzzy positive a match while D means a fuzzy negative match, and the 0 means no match. Our aim is to build a matching matrix for each pair of genes. Tab.4 gives an example of the usage of Tab.3. The first row is the gene AAC3 and the first column is gene ABF1.

2.4 Regulation Relationship Mining

In this section we will mine the regulation relationship between two genes. By using Tab.3, we can get a matching table of each gene pair, as Tab.4. We define two languages: $L_1 = \{+, S\}^*$ and $L_2 = \{-, D\}^*$, and our aim is to find all possible pattern p belonging to L_1 or L_2. The length of p is the number of symbols, noted by $|p|$, which must satisfy a customized minimum. For example, $p_1 = \{+++++SS+++\}$ belongs to L_1, and $p_2 = \{DDD-----DD---\}$ belongs to L_2, here $|p_1| = 10$ and $|p_2| = 13$. If we find that a pattern p belongs to L_1 or L_2, the two genes may have a potential regulation relationship. Usually 6 is the shortest requirement for an acceptable pattern[8], because a pattern with length lessen than 6 has a high random probability between two genes, which may result a high imprecise.

Using a table like Tab.4, we can mine all possible regulation information. Denote Tab.4 to be $R_{m \times m}$, here $m=11$ and the index is based on 0. We check all the cells on all slope lines of the matrix parallelized to the diagonal from $R[0,0]$ to $R[10,10]$ separately. For example we find two $p_1 = "SS++++"$, p_1 belonging to L_1 and $|p_1| = 7$; the cells are bolded in Tab.4. And we can further mine some useful information about regulation such as the start position of the genes and the original pattern. Taking p_1 for example, $R[1,2] =' S'$, we can infer the start position of the two genes are 1 and 2 respectively. Because the pattern length is 7, so the original pattern of the two genes are $"PP1(-1)1(-1)1"$ and $"1P1(-1)1(-1)1"$ respectively. And we can put all the genes sharing a common pattern together to form a gene cluster for further analysis.

Because we use symbols S and D here, which means a potential match or potential opposite match respectively. Consider such a pattern $"SSSS+SS+"$, which contains too much S and causes a doubt about its precise. A brute method to distinguish a "good" or a "bad" pattern is to constrain the number of $'S'$ or $'D'$ in the pattern, if the number exceed a customized threshold, drop it out. Another method is to calculate a *confidence degree* for each found pattern by scoring differently. We assign 1 to each $'+'$ and $'-'$, and 0.5 $'S'$ and $'D'$, such that a pattern's *confidence degree* is calculated as follows:first we score the pattern we found and then divide it by the length of the pattern.For example, the pattern p is $"+++ +SSS+ +++"$, then the score is $1 \times 8 + 0.5 \times 3 = 9.5$, while $|p|$ is 11, such that the confidence is $9.5/11 \approx 0.86$. We can set a confidence threshold to skip some results you don't need.

3 Experiment Results

The method can find potential co-regulatory gene pairs. The co-regulatory gene pairs usually have a series of common changing tendencies, globally or locally, which have the same start condition point and a similar changing tendency follows at each corresponding point. If there are some genes having a co-regulatory gene relationship with the same start condition point and length of common pattern, we can put them into the same gene cluster, such that we can infer these genes involving in the same biological process with the similar biological functions. Fig.1 demonstrates that the three genes—— ADH2, ASN2 and ERG10—— are co-regulated all with start condition point 1 and last 18 points with their original expression level, scaling from -1 to 1. The common pattern between ADH2 and ASN2 is $++S+++SS+++S+++$, while the common pattern between ADH2 and ERG10 is $+++S++SS+S+S++++$, such that a gene cluster with a same expression profile, globally or locally, will be naturally founded.

The method can find potential activation and inhibition regulation types, and also give the time lag between the regulator and the regulated or the inhibitor and the inhibited. The potential activation regulation relationship between gene pairs must have a matching result only containing symbols $'+'$ and $'S'$ and

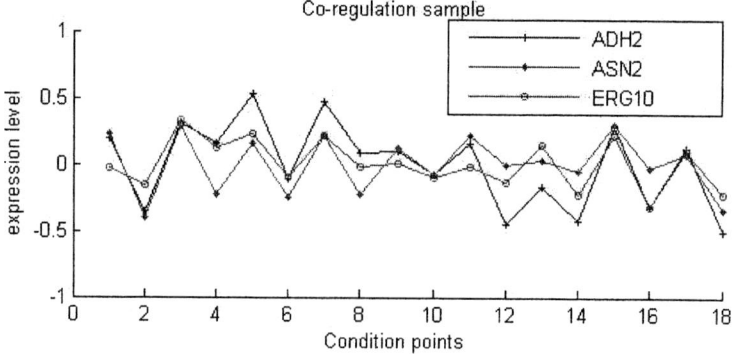

Fig. 1. Gene co-regulation sample

Table 5

GeneName	Start	Length	Matched Pattern
ADH2	1	15	$++++++SS+SSS+++$
HNM1	3		

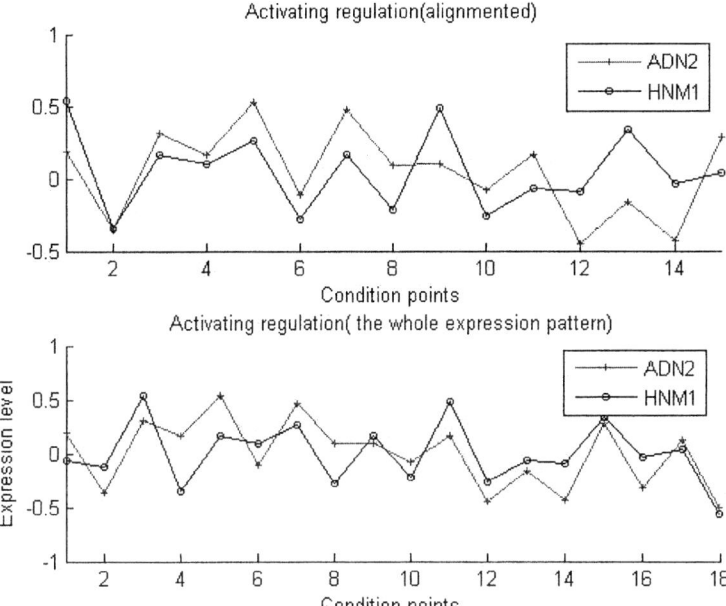

Fig. 2. Gene activating regulation sample

satisfying a minimum matching length, while a potential inhibition regulation relationship means a matching pattern only consisting of $'-'$ and $'D'$. And usually the gene with the smaller start regulatory point should be the activator and the gene with bigger start time corresponds to the activated, and it is the same with the inhibition regulation relationship. Tab.5 records the matching result of gene ADH2 and HNM1. Obviously, this is the activating regulation with the activator being gene ADH2, which has a smaller start condition point compared with the activated, and there is a 2 time unit delay between the genes.

The upper of the Fig.2 is the matching result of the locally matched pattern with the length 15 and the bottom is the whole expression profile of the two genes.

Tab.6 records the matching result between gene HIS5 and gene RIM101. The matched pattern show that it's an inhibitory regulation, with the inhibitor gene being RIM101 and the inhibited HIS5, and the time lag is 1 because the start point of gene RIM101 is 1 time unit earlier than the start point of gene HIS5.

The plot of the two genes' expression profile and regulatory pattern is showed in Fig.3. The upper of Fig.3 is the locally matched pattern of the two genes, which

Table 6

GeneName	Start	Length	Matched Pattern
HIS5	2	16	$DD---D----DD-D--$
RIM101	1		

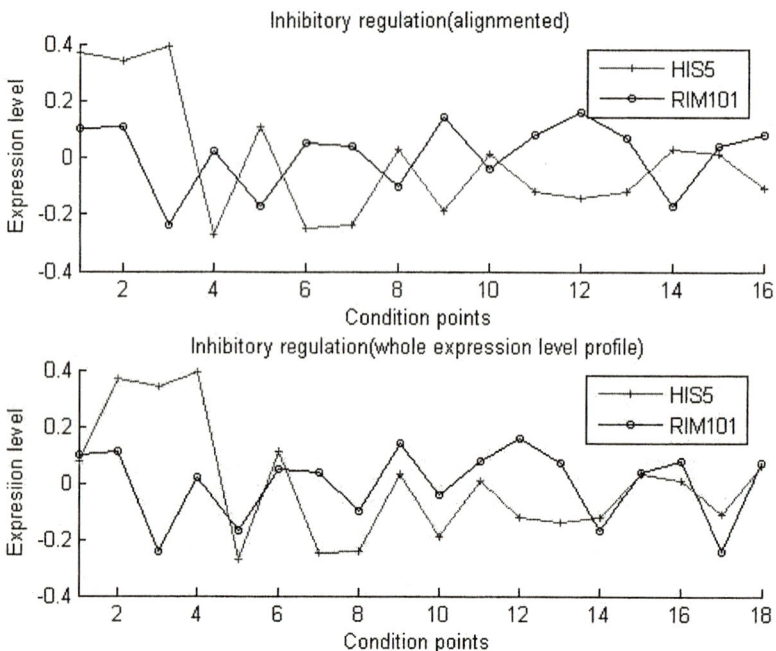

Fig. 3. Gene inhibitory regulation sample

is a inhibitory regulation with regulatory length 16, while at the bottom is the whole expression level profile of the two genes. And the start regulation point of gene RIM101 is condition point 1, while the start regulation point of gene HIS5 is 2.

The method can use the microarray data more efficient by introducing a more detailed discretized standard compared with traditional approaches only with 1, -1, & 0 used in the paper [6] or the three Event kinds R, F, C used in [5], where the three event kinds correspond to the 1, -1 and 0 respectively. And also our method can concurrently gain the global and local regulation information, without a fixed length pattern compared with other methods, for example in [6] the pattern length is fixed to 7 and in [5] the detection is applied to the whole expression level. We specially apply our method to the alpha data set which includes 288 genes and 18 condition points with the minimum pattern length being 6 and no $'S'$ and $'D'$ restriction, and there are totally 439 proved regulation relationships by biologist, including 343 positive regulations and 96 negative regulations[8]. The results shows we can find more than 70% of these

Type	Regulator	Regulated	Start1	Start2	Pattern	Type	Inhibitor	Inhibited	Start1	Start2	Pattern
'+'	'GCN4'	'ADE4'	8	8	PPPP1P1P1P	'-'	'ACC1'	'PHO5'	1	1	NNNNNNNN
'+'	'GCN4'	'ARG1'	1	8	PPP1PP11	'-'	'CBF1'	'CYC1'	8	11	N------
'+'	'GCN4'	'ARG2'	3	8	1111PP	'-'	'CBF1'	'CYT1'	7	8	NN--N---
'+'	'GCN4'	'ARG3'	4	4	11111PP	'-'	'CBF1'	'MET17'	7	10	NNNNNNN-
'+'	'GCN4'	'ARG4'	11	11	PPP1P1P	'-'	'CBF1'	'QCR8'	6	9	NNNN-NNN-
'+'	'GCN4'	'ARO3'	1	8	11111111P	'-'	'FUS3'	'CDC28'	3	6	NNN---NN
'+'	'GCN4'	'ARO4'	1	6	111PPPP1	'-'	'GCN3'	'GCD1'	6	9	N----N--N
'+'	'GCN4'	'ARO7'	3	3	1PPP111	'-'	'MIG1'	'CAT8'	1	5	NN-N--N
'+'	'GCN4'	'ASN2'	1	8	11111111	'-'	'RAS2'	'TRX2'	3	1	N----N
'+'	'GCN4'	'GCV3'	1	12	11111P	'-'	'RGR1'	'IME1'	7	2	N------NN
'+'	'GCN4'	'GLN1'	2	5	P11PPP	'-'	'SIN4'	'IME1'	1	2	N------N
'+'	'GCN4'	'GLT1'	3	7	11P111	'-'	'TUP1'	'CYC1'	2	10	N-----N
'+'	'GCN4'	'HIS3'	10	10	P11PPPP	'-'	'TUP1'	'HXT4'	3	3	-N---N-
'+'	'GCN4'	'HIS4'	1	2	PP111P	'-'	'TUP1'	'ROX1'	1	11	NN-----

Fig. 4. Some mined regulation samples

regulations, and a less than 30% miss. It is similar when use the method to the cdc28 data set, consisting of 352 genes and 17 condition points. But for the *q-cluster method* in [6] and the *Event Method* in [5], this is not the case, and the results show that they may find nothing. The reason is that the found patterns always include S and D, so that it is not suitable for these methods to mine these information by a rigid discretization method and the results also prove our theoretical analysis.

Fig.4 shows some picked regulation between genes which have been proved and found by our methods. Each row of the Fig.4 represents a regulation between two genes except the first row. And in the first row, the symbols $'+'$ and $'-'$ in the $'Type'$ column mean activating regulation and inhibitory regulation respectively. The following two columns stand for the regulator and the regulated or the inhibitor and the inhibited for the negative regulation. Then follow the two start regulation points of the regulator and the regulated respectively. And the last column is the corresponding matched pattern.

4 Discussion

In this paper, we have proposed a new method to mine the gene regulation relationship to serve for the building of a genetic regulation network, which can systematicly explain some biological process and help people comprehend the function of certain genes. Compared with the previous methods, we improve the precise of the discretization of the gene expression level values by introducing a new classifying standard to gain more information between gene pairs, as well as focus our method on the local expression profile without a fixed length pattern but satisfying a minimum threshold, for it is time series data and a regulatory pattern's varies among a range. And we have run experiments on the *Yeast* gene expression level data set and compared the result with the known regulation types attained by biologists' manual experiments. The results show our approach delivers useful results and gives a competitive tool that facilitates the further study of gene network.

Acknowledgement

This work is supported by grants 60303009, 60496325 of Chinese National Natural Science Foundation.

References

1. Mamoru Kato, Tatsuhiko Tsunoda,Toshihisa Takagi. Lag Analysis of Genetic Networks in the Cell Cycle of Budding Yeast. Genome Informatics**12**(2001) 266–277.
2. Barash, Y. & Friedman, N. Contest-specific Bayesian clustering for gene expression data. In Proceedings of the Third Annual International Conference on Research in Computational Molecular Biology(2001)12-21.
3. Chen, T., Filkov, V. & Skiena, S. Identifying gene regulatory networks from experimental data. In Proceedings of the Third Annual International Conference on Research in Computational Molecular Biology (1999)94–103.
4. Yeung,L.K, Szeto, L.K.,Liew,A.W.,& Yan, H. Dominant spectral component analysis for transcriptional regulations using microarray time-series data. Bioinformatics**20**(2004) 742–749.
5. Andrew T. Kwon, Holger H. Hoos abd Raymond Ng, Inference of transcriptional regulation relationships from gene expression data. Bioinformatics **19(8)** (2002) 905–912.
6. Liping Ji, and Kian-Lee Tan , Identifying Time-Lagged Gene Clusters on Gene Expression Data. Bioinformatics **21(4)**(2005) 509–516.
7. Spellman,P., Sherlock,G., Zhang,M., Iyer,V., Anders,K., Eisen,M., Brown,P., Botstein,D. and Futcher,B. Comprehensive identification of cell cycle-regulated genes of the yeast Saccharomyces cerevisiae by microarray hybridization. Mol. Biol. Cell, **9**(1998), 3273C3297.
8. Filkov,V., Skiena,S. and Zhi,J. Analysis techniques for microarray time-series data. J. Comput. Biol., **9**(2002), 317C330.
9. Liping Ji and Kian-Lee Tan.Mining gene expression data for positive and negative co-regulated gene clusters. Bioinformatics, **20(16)**(2004) 2711–2718.
10. Chad Creighton , and Samir Hanash. Mining gene expression databases for association rules. Bioinformatics ,**19(1)** (2003) 79–86.

Shot Transition Detection by Compensating for Global and Local Motions

Seok-Woo Jang[1], Gye-Young Kim[2], and Hyung-Il Choi[2]

[1] Senior Researcher, Korea Institute of Construction Technology
swjang@kict.re.kr
[2] Professor, Soongsil University, Seoul, Republic of Korea
{gykim, hic}@computing.soongsil.ac.kr

Abstract. This paper proposes a method of detecting shot transitions by compensating for motions contained in video sequences. The proposed method detects shot transitions including cuts, fades, and dissolves after compensating for global motions (camera motions) and eliminating local motions (moving objects), so that our approach prevents false positives caused by those motions. Experimental results show that our method works as a promising solution.

1 Introduction

For the automatic content analysis and efficient browsing of video, it is necessary to split the video sequences into more manageable segments. A camera shot may be a good candidate for such segmentation. To detect the shot transitions automatically, various computer vision techniques have recently received increasing attention [1].

Multimedia researchers, working on shot transition detection, proposed different detection techniques [2-4]. The histogram-based method computes gray or color histograms of consecutive frames to detect shot transitions [2]. The edge-based method detects various shot transitions by looking at the relative values of entering and exiting edge percentages [3]. In the method using motion vectors, the motion compensation error provides a distance measure between two successive frames for detecting shot transitions [4]. However, most of the existing techniques work in restricted situations and lack robustness, since they do not consider motions included in video. These motions make chromatic changes between two consecutive frames and they do not mean shot transitions. However, many existing detection methods regard them as shot transitions. Therefore, we may have false positives if we do not compensate for those motions before detecting shot transitions.

To deal with these limitations, we propose a motion-compensated approach for detecting shot transitions in video sequences. The proposed method detects and classifies shot transitions including cuts, fades, and dissolves after compensating for global motions and eliminating local motions. Our algorithm has three main modules: a motion estimation module, a feature extraction module, and a shot transition detection module.

2 Feature Extraction

To estimate global and local motions, we first extract motion vectors with the size-variable block matching [5], and the extracted motion vectors are then applied to the adaptive robust estimation method [6]. After estimating motions from images, we extract image features. We use the HSI color model to represent a color in terms of hue, saturation, and intensity. Our feature set includes three different types of measures on frame differencing and changes in color attributes, and one coherency measure on camera motions. The first feature is the correlation measure of hues and intensities between two consecutive frames as in (1). If two consecutive frames are similar in terms of the distribution of hues and intensities, it is very likely that they belong to a same scene. In (1), the α and β are weights that control the importance of related terms. We assign a higher weight to α, as hues are less sensitive to illumination than intensities are. Block Hue Mean (*BHM*) denotes the average of block hues and Image Hue Mean (*IHM*) denotes the average of image hues, respectively.

$$F_{corr} = \alpha \times H_{corr} + \beta \times I_{corr} \qquad (1)$$

$$H_{corr} = \frac{1}{2}\left(1 + \frac{\sum_{i=1}^{N}(BHM_{t-1}^{i} - IHM_{t-1})(BHM_{t}^{i} - IHM_{t})}{\sqrt{\sum_{i=1}^{N}(BHM_{t-1}^{i} - IHM_{t-1})}\sqrt{\sum_{i=1}^{N}(BHM_{t}^{i} - IHM_{t})}}\right)$$

where $0 \leq F_{corr} \leq 1,\ 0 \prec \beta \leq \alpha \prec 1,\ \alpha + \beta = 1$

(2) represents *BM* that are used to define H_{corr}. In (2), $BH_t^i(x, y)$ denotes the hue value of the starting position of the *i*-th block. *n* denotes the size of a block, *N* denotes the total number of blocks, and w_i denotes the weight of the *i*-th block that was computed in the adaptive robust estimation. The weight of non-outlier blocks has a value of 1, and the weight of outlier blocks has a value of 0. u_i and v_i denote the horizontal and vertical components of the motion vector of the *i*-th block, which are to compensate camera motions. I_{corr} is defined in the similar way.

$$BHM_t^i = \frac{w_i \cdot \left(\sum_{k=1}^{n}\sum_{l=1}^{n} BH_t^i(x+k, y+l)\right)}{n^2} \qquad (2)$$

$$BHM_{t-1}^i = \frac{w_i \cdot \left(\sum_{k=1}^{n}\sum_{l=1}^{n} BH_{t-1}^i(x+u_i+k, y+v_i+l)\right)}{n^2}$$

The second feature is used to evaluate how intensities of successive frames vary in the course of time. During fade, frames have their intensities multiplied by some value of δ. A fade in forces δ to increase from 0 to 1, while a fade out forces δ to decrease from 1 to 0. In other words, overall intensities of frames transit toward a constant. The speed with which δ changes controls the fade rate. To detect such a variation, we define a ratio of overall intensity variations as in (3).

$$F_{ratio} = \frac{I_{diff}}{I_{Adiff}} \qquad (3)$$

$$I_{diff} = \frac{\sum_{i=1}^{N} w_i \cdot \left(\sum_{k=1}^{n} \sum_{l=1}^{n} BI_t^i(x+k, y+l) - BI_{t-1}^i(x+u_i+k, y+v_i+l) \right)}{n^2 \cdot N \cdot I_{max}}$$

$$I_{Adiff} = \frac{\sum_{i=1}^{N} w_i \cdot \left(\sum_{k=1}^{n} \sum_{l=1}^{n} \left| BI_t^i(x+k, y+l) - BI_{t-1}^i(x+u_i+k, y+v_i+l) \right| \right)}{n^2 \cdot N \cdot I_{max}}$$

In (3), I_{max} denotes the maximum value of intensity. The F_{ratio} ranges from -1 to 1, revealing whether frames become bright or dark. It has negative values during a fade out and positive values during a fade in, while the magnitude of the values approaches to 1. On the other hand, the ratio remains unvaried during a normal situation.

The third feature is used to evaluate the differences of hues, intensities and saturations between two consecutive frames. A dissolve occurs when one scene fades out and another scene fades in. During a dissolve, hues, intensities and saturations change very slowly. To detect such a variation, we define a sum of differences of HIS color components as in (4).

$$F_{diff} = \alpha \cdot H_{Adiff} + \beta \cdot I_{Adiff} + \gamma \cdot S_{Adiff} \qquad (4)$$

$$H_{Adiff} = \frac{\sum_{i=1}^{N} w_i \cdot \left(\sum_{k=1}^{n} \sum_{l=1}^{n} \left| BH_t^i(x+k, y+l) - BH_{t-1}^i(x+u_i+k, y+v_i+l) \right| \right)}{n^2 \cdot N \cdot H_{max}}$$

In (4), $BS_t^i(x, y)$ denotes the saturation value of the starting position of the i-th block, and S_{max} denotes the maximum value of saturation. F_{diff} has a value between 0 and 1. If it is close to 1, the probability of a dissolve gets high. On the other hand, if it is close to 0, the probability of a dissolve gets low. I_{Adiff} and S_{Adiff} are defined in the similar to H_{Adiff}.

The fourth feature is used to evaluate the coherency of motion vectors obtained between two consecutive frames. In general, motion vectors from images containing camera motions show regular patterns. On the other hand, if images contain shot transitions, their motion vectors show irregular patterns. To reflect such a situation, we define a coherency measure of camera motions as in (5). For that measure, we use the error of camera motion parameters estimated in the adaptive robust estimation.

$$F_{camera} = \sum_{i=1}^{N} w_i \cdot \left[y_i' - y(x_i, y_i, \mathbf{a}) \right]^2 \qquad (5)$$

In (5), F_{camera} has a value greater than or equal to 0. The probability of a shot transition occurrence gets high when F_{camera}'s value is large.

3 Shot Transition Inference

In this section, we present a technique to infer shot transitions by using rules containing certain factors. The certainty factor was pioneered in the MYCIN system, which attempts to recommend appropriate therapies for patients with bacterial infections [7]. We first set five types of hypotheses as shown in the following:

(1) H1: Shot transitions do not occur   [STAY]
(2) H2: A cut occurs   [CUT]
(3) H3: A fade in occurs   [FADE IN]
(4) H4: A fade out occurs   [FADE OUT]
(5) H5: A dissolve occurs   [DISSOLVE]

We then define the certainty factor of each feature to generate rules on shot transitions. We use a heuristic method to define them. Certainty factors of F_{corr}, $F_{dissolve}$ and F_{camera} are defined with two levels, low and high, and those of F_{fade} are defined with three levels, low, medium, and high.

We generate the rules on each shot transition using certainty factors as in (6)

$$\begin{aligned}
&\text{IF High}(F_{corr}) \text{ AND Low}(F_{camera}) \text{ THEN STAY} && CF = 0.8 \\
&\text{IF High}(F_{corr}) \text{ AND Low}(F_{diff}) \text{ THEN STAY} && CF = 0.5 \\
&\text{IF Low}(F_{corr}) \text{ AND High}(F_{camera}) \text{ THEN CUT} && CF = 0.8 \\
&\text{IF High}(F_{corr}) \text{ AND Low}(F_{ratio}) \text{ THEN FADE IN} && CF = 0.95 \\
&\text{IF High}(F_{corr}) \text{ AND Low}(F_{ratio}) \text{ THEN FADE OUT} && CF = 0.95 \\
&\text{IF High}(F_{corr}) \text{ AND High}(F_{diff}) \text{ THEN DISSOLVE} && CF = 0.8 \\
&\text{IF High}(F_{diff}) \text{ AND High}(F_{camera}) \text{ THEN DISSOLVE} && CF = 0.5
\end{aligned} \quad (6)$$

In (6), we do not consider the coherency feature of camera motions in generating the rules on fades, since the feature shows an irregular pattern when a fade occurs. The certainty factor of each rule is also defined through a heuristic method. We compute certainty factors on all the rules and select the hypothesis of the rule that has the maximum probability of occurrence as the final shot transition.

4 Experimental Results and Conclusions

To verify the performance of the proposed method, a number of experiments had been carried out with various video data including music videos, documentaries, movies, news, cartoons, advertisements, and so on. We selected a number of mpeg files that contained shot transitions with camera motions and moving object motions, then concatenated them into one mpeg file. It includes 65 cuts, 8 fades, and 8 dissolves.

We compared the performance of our suggested approach with the approaches based on the histogram-based method [2], the edge-based method [3], and the motion vector-based method [4]. Table 1 summarizes the comparison in terms of accuracy, where Nc denotes the number of shot transitions that are detected correctly, Nm denotes the number of misses, and Nf denotes the number of false positives. The experiment results showed that the proposed method performed better than others. We attribute excellent image features extracted from motion-compensated image sequences as the main reason for the better performance.

Table 1. Accuracy of shot transition detection

Method	Cut			Fade in(out)			Dissolve		
	N_c	N_m	N_f	N_c	N_m	N_f	N_c	N_m	N_f
Histogram-based	61	4	7	4(4)	4(4)	3(2)	2	6	2
Edge-based	62	3	6	7(7)	1(1)	2(3)	5	3	2
MV-based	62	3	4	5(6)	3(2)	4(3)	4	4	3
Proposed method	64	1	3	8(8)	0(0)	0(0)	7	1	1

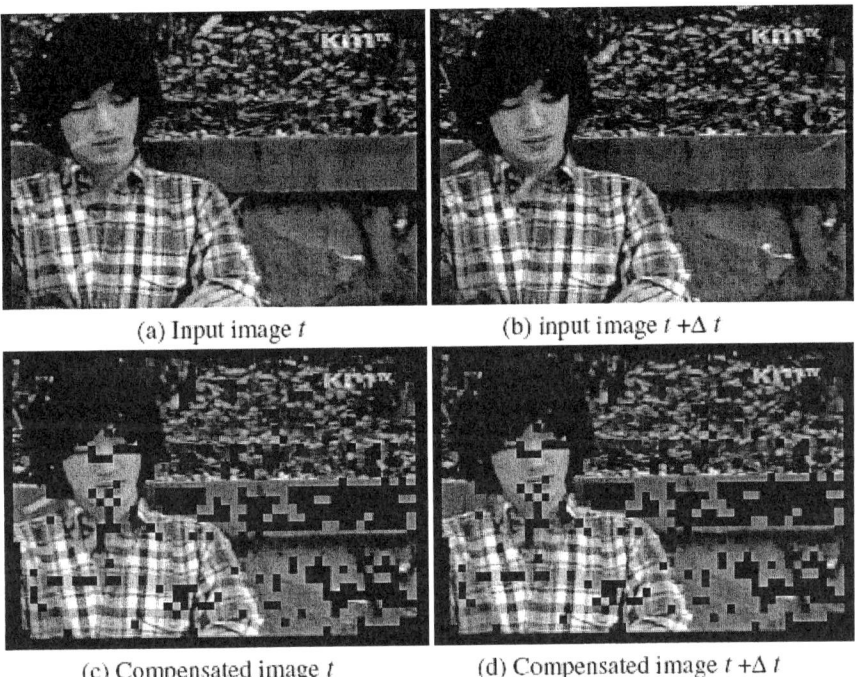

(a) Input image t (b) input image $t + \Delta t$

(c) Compensated image t (d) Compensated image $t + \Delta t$

Fig. 1. Compensating camera motions

Fig. 1 shows one example of compensating camera motions. Fig. 1 (a) and (b) are two consecutive images including camera tilting operation. We can clearly see that the positions of the person's chin in Fig. 1 (a) and (b) are not the same, but they become the same in Fig. 1 (c) and (d), where we compensated the image t by the camera tilting motion obtained with the adaptive robust estimation. With the help of such motion compensation, our detection method prevents false positives.

We proposed a shot transition detection method that compensates motions included in video. We first extracted motion vectors from two consecutive frames by using the size-variable block matching and then applied them to the adaptive robust estimation method to estimate the global motions and eliminate local motions. We extracted image features after compensating input images by using the estimated global motion, and then inferred shot transitions using certainty factors.

To sum up, our proposed approach is a promising solution for detecting shot transitions, although results can vary depending on the involved features. In the future, we plan to generate the certainty factors of features and the rules on each shot transition automatically.

Acknowledgement

This work was supported by the Korea Research Foundation Grant(KRF-2004-005-D00198).

References

1. Dan Lelescu, Statistical Sequential Analysis for Scene Change Detection on Compressed Multimedia Bitstream, IEEE Trans. on Multimedia, Vol. 5, No. 1 (2003) 106-117
2. Mee-Sook Lee, Seong-Whan. Lee, Automatic Video Parsing Using Shot Boundary Detection and Camera Operation Analysis, Pattern Recognition, Vol. 34, No. 3 (2001) 711-725
3. Ramin Zabih, Justin Miller, Kevin Mai, A Feature-Based Algorithms for Detecting and Classifying Production Effects, Multimedia Systems, Vol. 7 (1999) 119-128
4. Chong-Wah Ngo, Ting-Chuen Pong, Hong-Jiang Zhang, R. T. Chin, Motion-Based Video Representation for Scene Change Detection, ICPR, Vol. 1 (2000) 827-830.
5. Seok-Woo Jang, Hyung-Il Choi, A Strategy of Matching Blocks at Multi-Levels, International Journal of Intelligent Systems, Vol. 17, No. 10 (2002) 965-975
6. Seok-Woo Jang, Marc Pomplun, Hyung-Il Choi, Extracting Motion Model Parameters with Robust Estimation, Lecture Notes in Computer Science, Vol. 2667 (2003) 633-642
7. Joseph Giarratano, *Expert Systems*, Third Edition, PWS Publishing Company (1998)

Hybrid Methods for Stock Index Modeling

Yuehui Chen[1], Ajith Abraham[2], Ju Yang[1], and Bo Yang[1]

[1] School of Information Science and Engineering,
Jinan University, Jinan 250022, P.R. China
yhchen@ujn.edu.cn
[2] School of Computer Science and Engineering,
Chung-Ang University, Seoul, Republic of Korea
ajith.abraham@ieee.org

Abstract. In this paper, we investigate how the seemingly chaotic behavior of stock markets could be well represented using neural network, TS fuzzy system and hierarchical TS fuzzy techniques. To demonstrate the different techniques, we considered Nasdaq−100 index of Nasdaq Stock Market^{SM} and the S&P CNX NIFTY stock index. We analyzed 7 year's Nasdaq 100 main index values and 4 year's NIFTY index values. The parameters of the different techniques are optimized by the particle swarm optimization algorithm. This paper briefly explains how the different learning paradigms could be formulated using various methods and then investigates whether they can provide the required level of performance, which are sufficiently good and robust so as to provide a reliable forecast model for stock market indices. Experiment results reveal that all the models considered could represent the stock indices behavior very accurately.

1 Introduction

During the last decade, stocks and futures traders have come to rely upon various types of intelligent systems to make trading decisions [1]. Several intelligent systems have in recent years been developed for modelling expertise, decision support and complicated automation tasks etc[2][3]. In this paper, we analyzed the seemingly chaotic behavior of two well-known stock indices namely Nasdaq−100 index of Nasdaq^{SM} [4] and the S&P CNX NIFTY stock index [5]. The Nasdaq-100 index reflects Nasdaq's largest companies across major industry groups, including computer hardware and software, telecommunications, retail/wholesale trade and biotechnology [4]. The Nasdaq-100 index is a modified capitalization-weighted index, which is designed to limit domination of the Index by a few large stocks while generally retaining the capitalization ranking of companies. Through an investment in Nasdaq-100 index tracking stock, investors can participate in the collective performance of many of the Nasdaq stocks that are often in the news or have become household names. Similarly, S&P CNX NIFTY is a well-diversified 50 stock index accounting for 25 sectors of the economy [5]. It is used for a variety of purposes such as benchmarking fund portfolios, index based derivatives and index funds. The CNX Indices are computed using market

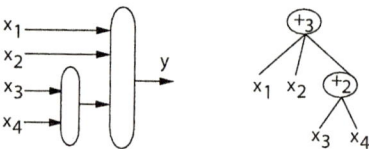

Fig. 1. Left: An example of possible hierarchical TS-FS model with 4 inputs and 3 hierarchical ayers, Right: the tree structural representation of the corresponding hierarchical TS-FS model, where the used instruction set is $I = \{+_2, +_3, x_1, x_2, x_3, x_4\}$

capitalization weighted method, wherein the level of the Index reflects the total market value of all the stocks in the index relative to a particular base period.

Our research is to investigate the performance analysis of neural network, TS Fuzzy system and hierarchical fuzzy system [6] for modelling the Nasdaq–100 and NIFTY stock market indices. We analyzed the Nasdaq–100 index value from 11 January 1995 to 11 January 2002 [4] and the NIFTY index from 01 January 1998 to 03 December 2001 [5]. For both the indices, we divided the entire data into almost two equal parts. No special rules were used to select the training set other than ensuring a reasonable representation of the parameter space of the problem domain [1].

2 Hierarchical TS-FS

A Hierarchical TS Fuzzy Systems (H-TS-FS) not only provide a more complex and flexible architecture for modelling nonlinear systems, but can also reduce the size of rule base to some extend. In this paper, finding a proper hierarchical TS-FS model can be posed as a search problem in the structure and parameter space. Fig.1(left) shows an example of possible hierarchical TS-FS models with 4 input variables and 3 hierarchical layers. A hybrid automatic approach has been proposed to optimize the hierarchical TS-FS with a Ant Programming (AP) and PSO algorithms [6]. In this research, a modified tree-structure based GP-like evolutionary algorithm was employed to find a optimal architecture of the H-TS-FS. A tree-structural based encoding method with specific instruction set is selected for representing a hierarchical TS-FS in this research. The output of each non-leaf node is calculated as a single TS fuzzy sub-model. For this reason the non-leaf node $+_2$ is also called a two-inputs TS fuzzy instruction/operator. Fig.1(right) shows the corresponding tree structural representation of the hierarchical TS-FS model.

It should be noted that in order to calculate the output of each TS-FS sub-model (non-leaf node), parameters in the antecedent parts and consequent parts of the TS-FS submodel should be encoded into the tree. The output of a hierarchical TS-FS tree can be calculated in a recursive way. In this work, the fitness function used for GP-like evolutionary algorithm and PSO is given by Root Mean Square Error.

The proposed method interleaves both optimizations. Starting with random structures and rules' parameters, it first tries to improve the hierarchical structure and then as soon as an improved structure is found, it fine tunes its rules'

Table 1. Empirical comparison of RMSE results for three learning methods

	Train			Test		
	NN-PSO	Fuzzy-TS	H-TS-FS	NN-PSO	Fuzzy-TS	H-TS-FS
Nasdaq	0.02573	0.02634	0.02498	0.01864	0.01924	0.01782
NIFTY	0.01729	0.01895	0.01702	0.01326	0.01468	0.01328

Table 2. Statistical analysis of three learning methods (test data)

	Nasdaq			NIFTY		
	NN-PSO	Fuzzy-TS	H-TS-FS	NN-PSO	Fuzzy-TS	H-TS-FS
CC	0.997704	0.997538	0.997698	0.997079	0.997581	0.0997685
MAP	141.363	156.464	138.736	27.257	30.432	27.087
MAPE	6.528	6.543	6.205	3.092	3.328	3.046

parameters. It then goes back to improving the structure again and, provided it finds a better structure, it again fine tunes the rules' parameters. This loop continues until a satisfactory solution is found or a time limit is reached.

3 Experiments

We considered 7 year's stock data for Nasdaq-100 Index and 4 year's for NIFTY index. Our target is to develop efficient forecast models that could predict the index value of the following trade day based on the opening, closing and maximum values of the same on a given day. We used the same training and test data sets to evaluate the fuzzy TS and hierarchical TS fuzzy models. The assessment of the prediction performance of the different soft computing paradigms were done by quantifying the prediction obtained on an independent data set. The Root Mean Squared Error (RMSE), Maximum Absolute Percentage Error (MAP) and Mean Absolute Percentage Error (MAPE) and Correlation Coefficient (CC) were used to study the performance of the trained forecasting model for the test data. We used instruction sets $I = \{+_2, +_3, +_4, x_0, x_1, x_2\}$ and $I = \{+_2, +_3, \cdots, +_6, x_0, x_1, x_2, x_3, x_4\}$ for modeling the Nasdaq-100 index and the NIFTY index, respectively. We used two NN models with network architecture {3-12-1} and {5-12-1} for modeling the Nasdaq-100 index and the NIFTY index, respectively. Table 1 summarizes the training and test results achieved for the two stock indices using the three different approaches. The performance analysis is shown in Table 2. Figures 2 depicts the test results for the one day ahead prediction of Nasdaq-100 index and NIFTY index respectively.

4 Conclusions

In this paper, we have demonstrated how the chaotic behavior of stock indices could be well represented by different hybrid learning paradigms. Empirical results on the two data sets using three different learning models clearly reveal the

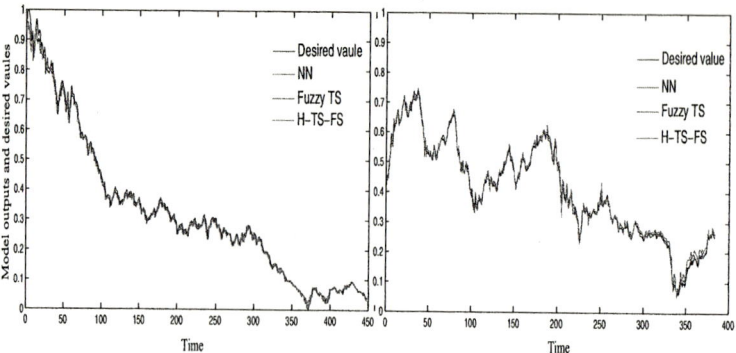

Fig. 2. Test results show the performance of the different methods for modeling Nasdaq-100 index (left) and the NIFTY index (right)

efficiency of the proposed techniques. A low MAP value is a crucial indicator for evaluating the stability of a market under unforeseen fluctuations. In the present example, the predictability assures the fact that the decrease in trade is only a temporary cyclic variation that is perfectly under control. Our research was to predict the share price for the following trade day based on the opening, closing and maximum values of the same on a given day. Our experiment results indicate that the most prominent parameters that affect share prices are their immediate opening and closing values. The fluctuations in the share market are chaotic in the sense that they heavily depend on the values of their immediate forerunning fluctuations. Long-term trends exist, but are slow variations and this information is useful for long-term investment strategies. Our study focus on short term, on floor trades, in which the risk is higher. However, the results of our study show that even in the seemingly random fluctuations, there is an underlying deterministic feature that is directly enciphered in the opening, closing and maximum values of the index of any day making predictability possible.

References

1. Abraham A., Philip N.S., and Saratchandran P.: Modeling Chaotic Behavior of Stock Indices Using Intelligent Paradigms, International Journal of Neural, Parallel and Scientific Computations, USA, Volume 11, Issue (1,2), (2003)143-160.
2. Leigh W., Modani N., Purvis R. and Roberts T.: Stock market trading rule discovery using technical charting heuristics, Expert Systems with Applications 23(2), (2002)155-159.
3. Leigh W., Purvis R. and Ragusa J.M.: Forecasting the NYSE composite index with technical analysis, pattern recognizer, neural network, and genetic algorithm: a case study in romantic decision support, Decision Support Systems 32(4),(2002)361-377.
4. Nasdaq Stock Market^{SM}: http://www.nasdaq.com
5. National Stock Exchange of India Limited: http://www.nse-india.com
6. Chen Y., Yang B. and Dong J.: Automatic Design of Hierarchical TS-FS Models using Ant Programming and PSO algorithm, LNCS, 3192, (2004)285-294.

Designing an Intelligent Web Information System of Government Based on Web Mining

Gye Hang Hong[1] and Jang Hee Lee[2]

[1] Industrial System and Information Engineering,
Korea University, Seoul, Republic of Korea
kistduck@korea.ac.kr
[2] Industrial Management, Korea University of Technology and Education,
Chunan, Republic of Korea
janghlee@kut.ac.kr

Abstract. A purpose of web information service in government agencies and public institutions is to provide various kinds of public information to support good decision-making of people. However, people, the users of web information system of government agencies and public institutions, have different information access environments, ability to understand the served information and information pursuit desire, and so on. We present a desirable web information system of government agency and public institution for providing the class of information weakness, disadvantaged users, with the personalized web information that make a more profits in their economic behaviors.

1 Introduction

The website construction of government agencies and public institutions is actively propelled because of the e-government implementation. It offers civilians various information and service and plays the most important means in the collecting of public opinion. According as electronic public information service are expanded, government agencies and public institutions should provide public information so that all people can easily access electronic administration and public information service and take advantage of necessary information regardless of the economical, physically handicap and region difference. However, in case the universality of access, use and benefit about this public information service is not secured, the information gap is happened. It is mainly considered that the causes of information gap are the differences of information access environment, ability to understand information and information pursuit desire, and so on, which come from the economical, education level, physically handicap and regional difference.

When the contents provided in the website are hard to understand and not utilized beneficently, more serious information gap can be produced by combining with those causes such as the differences of information access environment, ability to understand information and information pursuit desire. In order to remove that possibility, various and useful web contents should be designed and managed so that the disadvantaged users can understand easily and utilize beneficently. This paper discusses the desirable web information service system of government agency and public institution based on web mining tools for providing the disadvantaged users with the requisite web contents that support their successful use.

2 Literature Reviews

A purpose of the access to information services of the categories is to provide various kinds of public information to support decision-making of various users such as researchers, public servants in a web service government agency and other government agency, citizens and so on. Because various users having different education level, information pursuit desire, etc, government agencies should collect various data, transform them into various kinds of information, and deliver the information to them.

Marchionini G. and M. Levi suggested design issues in their BLS (the Bureau of Labor Statistics) project [1]. They emphasized a web design method that can contain maximum information in minimum pages and cover various different types of information to understand for various users.

A personalization can give benefit for reducing searching time of information to users. However, pubic-sector such as government agency and public institute don't apply the technology to their web service system because of legal restrictions regarding the use of personal information and privacy. Increasing the user demands can overcome the legal restrictions and make the technology to apply to their system [2], [3]. Therefore, the public-sector's information system must deliver the beneficial information considering user level, so that user will make the more profits by referring the information to his/her decision-making and trust the public-sector system.

3 Intelligent Web Information System (IWIS)

Identifying key summarization schema and decision variable module (IKSSDVM)
The value of information can be different according to the user's power of understanding. *IKSSDVM* evaluates user's behavior in market after referring to the information to measure the value of information. *IKSSDVM* transforms user's behavior data into the total profit (TP), probability for profit (PFP) and probability for loss (PFL) metrics. *IKSSDVM* segments all users into several clusters having similar features through the use of SOM (Self-Organizing Map), which is one of the clustering methods [4].

Table 1. Summary of cluster characteristics

Cluster	TP	PFP	PFL	Evaluation
1	38,725	0.86	0.04	Advantaged Users
2	7,082	0.63	0.33	Advantaged Users
3	-25,213	0.63	0.39	Disadvantaged Users
4	-33,404	0.33	0.94	Disadvantaged Users

As shown in table 1, we can see that the users belonging to cluster 1 and 2 make more total profits than the users belonging to cluster 3 ands 4. Especially, the users belonging to cluster 1 have very higher probability for profit per transaction than the others, and consequently cluster 1 is the best advantaged cluster. On the other hand, the users belonging to cluster 4 have very lower probability of profit per transaction and total profits. These differences among clusters are caused by user's power of understanding to understand information and kinds of utilizing information for his/her decision-making. So, we must find the web pages that they mainly used and their summarization schema and decision variables.

IKSSDVM identify differences of web usage patterns between advantaged group and disadvantaged group. The difference is the kinds of web page accessed for decision-making and the level of understanding information. The following sub-steps identify the difference of web pages accessed between two groups.

1) It calculates an access count of user to web page by analyzing web-log data. The access count is transformed into one of the five levels: High(*H*), Low(*L*), Average(*A*), Above Average(*AA*), and Below Average(*BA*). Table 2 shows the calculated access count of user to web page.

2) It identifies the important web pages which can distinguish two groups. We use the C4.5 which is one of classification methods [5]. We defined input data of the C4.5 as a level of access frequency of all web pages and output as the number of group.

Table 2. Access count of user to web page

	Web page 01	Web page 02	......	Web page l	Group
User 1	H (5)	L (1)		BA (2)	Cluster1
User 2	H (5)	L (1)		L (1)	Cluster1
...	...	...	...	...	...
User k-1	L (1)	H (5)		AA (3)	Cluster 4
User k	L (1)	H (5)		H (5)	Cluster 4

IKSSDVM analyzes the structure of the web pages which each group mainly accesses. It decomposes the important web pages of each group into decision-variables and summarization schema. The decomposition process is needed to understand decision-variables which advantaged people consider for decision-making and a schema by which the decision-variables are summarized. The schema is defined as a cuboid, a combination of dimensions, and concept levels of dimensions, summary level of a dimension in a cuboid, of the cuboid [6].

Fig. 1 shows the structure of web pages. The web page web-05 is made by summarizing transaction data according to two decision variables of price and quantity and two dimensions of time and market.

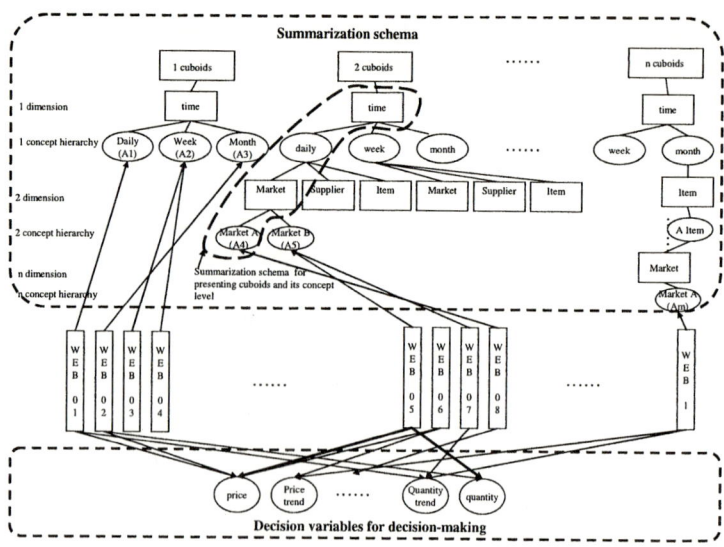

Fig. 1. Schema and decision variables of which web pages consist

Table 3 shows the schema and the decision-variables of which the important web pages of each group consist. The cluster 1 has an ability to understand information summarized with four kinds of dimensions and consider the two decision variables of quantities and its trend. The values of the decision variables are summarized into average values per a week. In the other hand, the cluster 4 has an ability to understand information summarized with three kinds of dimensions. So, if we delivery information summarized according to four kinds of dimensions or more to the cluster 4, the group may not understand means of the information and may be

Table 3. Summary of the schema and decision variables (DV) of each group

	No. of Schema (Degree of reference)	Description	Characteristics
Cluster1	A04 (> AA)	Cubiod : week, market, and item DV : price and quality	Maximum degree of dimension : 4 Key summarization : 1) Summarizing data of quantities and its trend according to week, market, supplier and item 2) Summarizing data of price and quality according to week, market, and item
	A15 (> A)	Cubiod : week, market, supplier, item DV : quantities and its trend	
Cluster4	A01 (> AA)	Cubiod : daily, item, market DV : price and quality	Maximum degree of dimension : 3 Key summarization : Summarizing data of price and quality according to daily, market and item

misled to error in decision-making. Therefore, if the users of cluster 4 refer the information comparing with the other suppliers' quantity and its trend by week importantly, they may improve total profit and probability for profit. We need to redesign the information to a type of information which they can understand.

Redesigning web-pages Module(RWM)
The *RWM* redesigns the web information. *RWM* finds the best web page among all web pages for the best advantaged users' decision-making, which are generated by the feasible value of summarization schema of best advantaged cluster. In order to find the best web page, *RWM* calculates the total values for all web pages generated by the feasible value of summarization schema of best advantaged cluster, which are calculated from the linear weighting model [5] as following:

$$TV_i = W_{i1} * \mu_{i1} + W_{i2} * \mu_{i2} + \ldots + W_{ij} * \mu_{ij}$$

where *TV* means total value, *i* is the label of generated web page, *j* is decision variable, *Wij* is the weight value of decision variable *j* in the web page *i*, and *µij* is the average value of decision variable j in the web page *i*

RWM compares the total values of all web pages and chooses the web page with the highest total value, which is the best web page. By the same way, *RWM* also finds the best web page among all web pages generated by the feasible value of summarization schema of each cluster except the best advantaged cluster. And then *RWM* converts the information of best web page into the information summarized by the summarization schema of best advantaged cluster.

The web page in the Fig. 2 (a) is the best web page to which the users belonging to the best advantaged cluster refer for their decision-makings, which has the information summarized by the two decision variables, quantity and its trend, and the schema having four dimensions, item, market, week and supplier.

The left web page of Fig. 2 (b) is the best web page to which the users of disadvantaged cluster (i.e., cluster 4) refer, which has the information summarized by the two decision variables, price and quantity and the schema having three dimensions, daily, item, and market. The disadvantaged users can not understand the best web page that the best advantaged users used in their decision-makings because it has the schema having four dimensions, item, market, week and supplier. As shown in table 3, the disadvantaged users can understand the information summarized by three kinds of dimensions.

The right web page of Fig. 2 (b) shows the redesigned web page of 2(a) for the disadvantaged users cluster, which has their understandable type of web information summarized by the decision variables and schema of the best advantaged cluster. The redesigned web page has the type of information summarized by three dimensions, item, market, and week and one fixed dimension, supplier.

Monitoring the change of market environment Module (MCMEM)
The environment variables of market are changed over the period in time. However, users may not be aware of the change of market environment and consider the same kinds of information that were used for their decision-makings in the previous time

period. It is why they suffer great losses and have lower probability for profit per transaction. Therefore, the *MCMEM* can monitor market environment and inform the change of environment variables to users. *MCMEM* finds the important variables of market environment from main customers' behaviors in market and compares them with the important variables of previous market environment.

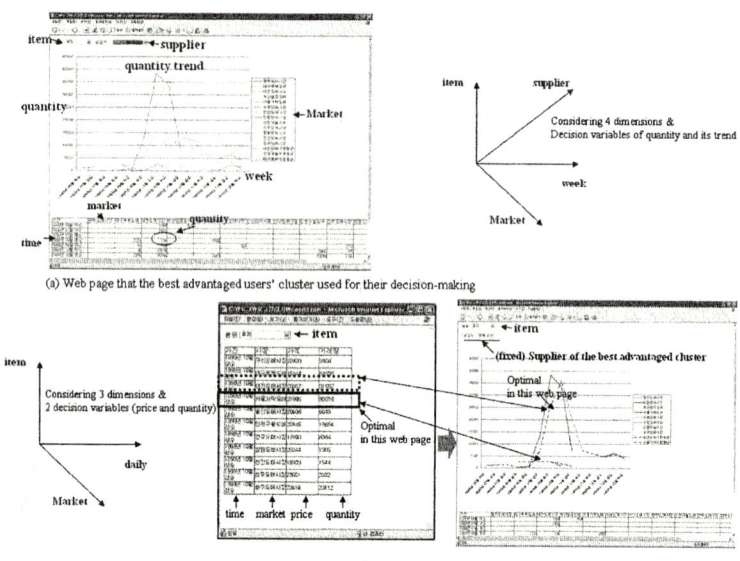

Fig. 2. Redesigning a web-page (Some Korean words were translated into English for readers' understanding)

4 Validation Study

To evaluate the validation of proposed system, we used three evaluation measures, total profit, and probability for profit and for loss. We randomly selected twenty users from the users of the disadvantaged cluster. And then we divide them into two equal groups: 'control' group and 'treatment' group. The proposed system redesigned the web information, web pages, summarized by the schema and decision variables of the best advantaged cluster. The web information has the type of information that the users of treatment group can easily understand. The proposed system served the information to the users of treatment group.

After monitoring the profit performance of all groups during ten weeks, we obtained the following results. As shown in Fig. 3 (a), the users of the best advantaged cluster had the higher profits than the two groups during almost all periods. The users of treatment group didn't have any difference of profit with the control group in the first period but they had the higher profits than the control group

During the rest periods. In the point of total profit, the total profits of treatment group were sharply improved (refer to Fig. 3 (b)). In the point of probability for profit and loss, the probability for profit of treatment group was increased and the probability for loss of treatment group was decreased (refer to Fig. 3 (c)). From these results, we could see that the difference of user's ability to understand information and information used for his/her decision-making have an effect on the success/failure of his/her economic behavior in market.

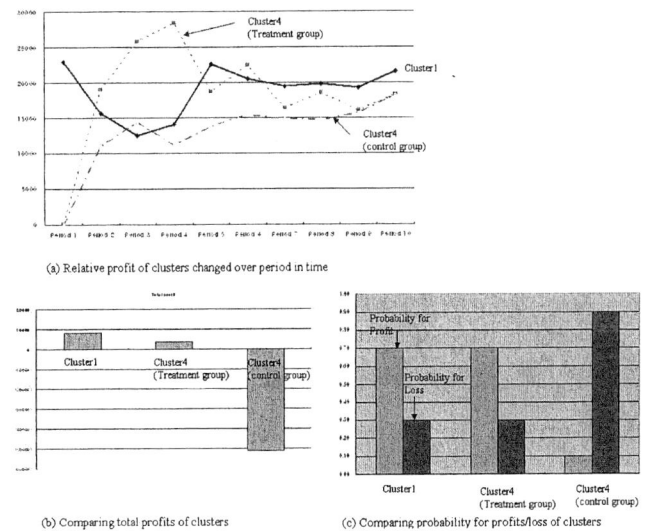

Fig. 3. Results of validation study

5 Conclusion

We proposed an intelligent web information system of government for helping disadvantaged users make more profit in their economic behaviors. The proposed system collects users' transaction data in markets, evaluates users' decision-makings in the profit/loss aspect and then classifies users into two user groups: advantaged and disadvantaged user group. It also identifies the difference of web information used between the two groups, redesigns the web information to help disadvantaged users make good decision-makings for their profit improvement.

From the validation study, we could see that the difference of user's ability to understand information and information used for his/her decision-making have an effect on the success/failure of his/her economic behavior. If a web information system of government can help disadvantaged users understand the information that the best advantaged users make good use of their decision-makings, it may be trusted to all users including disadvantaged users because they can make a more profits.

References

1. Marchionini, G. and Levi, M.: Digital Government Information Services: The Bureau of Labor Statistics Case, Interactions, Vol. 10 (2003) 18-27
2. Hinnant, C. C. and O'Looney, J.A.: Examining pre-adoption interest in on-line innovations: An exploratory study of e-service personalization in public sector, IEEE Transactions on engineering management, vol.50,No.4 (2003) 436-447
3. Medjahed, B., Rezgui, A., Bouquettaya, A. and Ouzzani, M.: Infrastructure for e-government web services, IEEE Internet Computing, Vol.7 no.1 (2003) 58-65
4. Kohonen, T.: Self-oranized formation of topologically correct feature maps, Biolog. Cybern., vol.43, (1982) 59-69
5. Quinlan, J.R.: C4.5 : Programs for machine learning, San Mateo, CA: MaGrw-Hill (1993)
6. Han, J. and Kamber, M.: Data Mining – concepts and techniques, Morgan Kaufmann (2001)
7. de Boer, L., Labro, E., and Morlacchi, P.: A review of methods supporting supplier selection, European Journal of Purchasing and Supply Management, vol.7 (2001) 75-89

Automatic Segmentation and Diagnosis of Breast Lesions Using Morphology Method Based on Ultrasound

In-Sung Jung, Devinder Thapa, and Gi-Nam Wang

Department of Industrial and Information Engineering, Ajou University, South Korea
{gabriel7,debu}@ajou.ac.kr, gnwang@madang.ajou.ac.kr

Abstract. The main objective of this paper is to use the auto segmentation with morphological technique to find out predictable region of interest (ROI), especially the center and margin area of the tumor. The proposed method has employed moving average method for detecting edge of tumor after estimating the corresponding center using the aid of medical domain knowledge. In our re-search, after computing distance between center and edge of tumor we get factual and numerical data of tumor to calculate multi-deviation and circularity test. It is useful to construct tumor profiling by splitting up the lesion into 4 divisions with the mean of multi-standard deviation (benign: 13.7, malignancies: 38.32) and 8 divisions with the mean of multi-standard deviation (benign: 3.36, malignancies: 15.29) with equal segments. We used K-means algorithm to make classification between benign and malignance tumor. This technique has been fully validated by using more than 100 ultrasound images of the patients and found to be accurate with 90% degree of confidence. This study will help the physicians and radiologist to improve the efficiency in accurate detection of the image and appropriate diagnosis of the cancer tumor.

1 Introduction

According to recent surveys, there has been a sudden increase in cases of breast cancer diagnosis. This proliferation has given rise to advanced research in breast cancer treatment. So far, there has been a lot of research in the area of mammography; however,there is a considerable need to conduct research in ultrasound diagnosis as well. Recent statistics show that breast cancer is the leading cause of cancer disease among women. In Korea, the incidence of breast cancer is escalating over the years and accounted as the prime cause of the women cancer [1]. According to the National Health Institute report in 1997, cervical cancer was the highest proportion (22.8%), followed by stomach cancer (15.2%) and breast cancer. However, in 2001, within the span of 4 years National Health Institute declares the breast cancer has been the high-est rated (16.1%) cancer among women followed by stomach cancer (15.3%) and cervical cancer [2]. Mammography has been shown to be effective in screening asymptomatic women to detect occult breast cancers. This positive benefit resulted in US government recommending that all women every year or every second year to be screened using mammography. On the contrary, it was difficult for a radiologist to interpret screening mammography in large numbers. Sometimes the radiologist failed to detect the cancer that was evident retrospectively. These sorts of incidents happened due to the lack of numerical reference data. Diagnosis with X-rays has many associated problems like radiation risks, and cancer risks from breast compression etc. Alternatively, detection and diagnosis using ultrasound is less risk prone than mammography. However, it was quite difficult to get the understandable image by using ultrasound due to sensitivity to noise while transferring to computers. To overcome this insufficiency of data and reduce the risk of mammography, we have approached a novel method of Computer Aided Diagnosis system with ultrasound. The main objective of this method is to use the auto segmentation with morphological technique to find out estimated region of interest (ROI), especially the center and margin area of the tumor. The proposed method has employed moving average method for detecting edge of tumor after estimating the corresponding center using the aid of medical domain knowledge. After computing distance between center and edge of tumor we get factual and numerical data of tumor to calculate multi-deviation and circularity test. It is useful to construct tumor profiling by splitting up the lesion into 4 divisions with the mean of multi-standard deviation (benign: 13.7, malignancies: 38.32) and 8 divisions with the mean of multi-standard deviation (benign: 3.36, malignancies: 15.29) with equal segments. We have applied K-means algorithm to make classification between benign and malignance tumor. Our paper is based on the preliminary work done by Karla Horsch [11] and E. Tohno [18], related to detecting breast cancer with ultrasonic image using image processing technique, and B. Sahiner [17], Sheng - Fang, Ruey-Feng Woo Kyung Moon's research [19] related to breast cancer diagnosis. This paper has been arranged as per order, section 2 gives a detailed description of the material and methods used to design the framework and image processing techniques, section 3 describes about the comparative paradigm and finally concluded with future perspective of the system with conclusion.

2 Proposed Method

Our database consists of 100 consecutive ultrasound cases, being represented by 200 images. These images were obtained during diagnostic breast exams at the several breast cancer surgery Hospitals. Most of the cases were collected retrospectively and had been either biopsy or aspirated. Out of the images of 100 cases, 50 were benign lesions; the other 50 were malignant lesions which have been chosen for the purpose of testing. The size of the tumors were between $1.5 \sim 2.5cm$.

In this section, we describe the various methods of detecting, segmenting and enhancing the quality of sonogram images. In this process we are using the CAD (Computer Aided Diagnosis), it is made up of the following two processes:

1. Detection process is the act of sensing doubtful tumor or detection of breast cancer lesions.
2. Diagnosis process is the act of analysis and diagnosis, using domain knowledge with circularity of tumor and multi standard deviation. In this section we explained about the automatic lesion segmentation algorithm. It involves the following steps:
   (a) Image pixel acquisition from original image
   (b) Preprocessing for ROI (region of interest) and estimate of tumor area using morphology,
   (c) Stepwise segmentation based on one of domain knowledge
   (d) Determine the tumor centre and identify shape of tumor type,
   (e) Analysis of the tumor using circulation rate and standard deviation
   (f) Diagnosis using K-Means using the factors.

Detail descriptions of the steps are as follows:

Step 1: Image pixel acquisition

In each original image (Fig. 3-a), level of pixel value information are reveal by the use of matrix (Fig. 3-b)

Step 2: Pre-Processing

This process consists of 4 steps, which are as follows:

1. First step is to remove some spot in the image with a 10 by 10 median filtering
2. Second step is calculating of image histogram for making threshold value (Fig. 3-c)
3. Third step is to remove most of noise from threshold value
4. Fourth step is negative transformation. (Fig. 3-d)

$$I(x,y): \text{original gray scale image; T: threshold Value} \\ B(x,y): \text{Blank matrix(row*column)} \\ [row, column] = size(I) \\ \text{If } I(x,y) \leq T, B(x,y) = 0, else, B(x,y) = I(x,y), end \\ I(x,y) = -I + 255 : \text{image negative transformation}$$ (1)

Step 3: Estimation of tumor area with morphology

We approached morphological combination of dilation and erosion technique. Dilation can be said to add pixels to an object, or to make it bigger, and erosion will make an image smaller. In the simplest case, binary erosion will remove the outer layer of pixels from an object. An image Dilation is using a structuring element.

Dilation Process

The origin of the structuring element is placed over the first black pixel in the image, and the pixels in the structuring element are copied into their corresponding positions in the result image. Then the structuring element is placed over the next black pixel in the image and the process is repeated. This is done for every black pixel in the image.

$$A \otimes B = \{x : B + x < A\} \tag{2}$$

Erosion Process

The structuring element is translated to the position of a black pixel in the image. In this case all members of the structuring element correspond to black image pixels so the result is a black pixel. Now the structuring element is translated to the next black pixel in the image, and there is one pixel that does not match. The result is a white pixel. The remaining image pixels are white and could not match the origin of the structuring element; they need not be considered.

$$A \oplus B = [A^c - (B^c)]^c \tag{3}$$

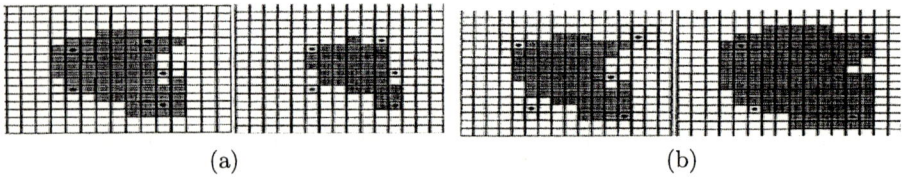

(a)   (b)

Fig. 1. Erosion and Dilation Example

Combination of Dilation and Erosion

Dilation and erosion are often used in combination to implement image processing operations with the same structuring element for removal some small size of noise. However, in this time we have been creating large ball type structuring element 15 by 15 for removal large scale of noise.

Step 4: Estimation of tumor center area

Detecting the ROI center of the tumor area is very important. Due to the irregular shape of the tumor, it was very difficult to find the tumor area completely. Therefore, we tried to find the estimated area of with the combination of dilation and erosion image and then find out approximate center of tumor. We computed the sum of pixels row and column of each matrix, then maximize the matrix row and column. After that we found out the maximum position.

$$[value_x, index_x] = max(sum(Im, row))$$
$$[value_y, index_y] = max(sum(Im, col)) \tag{4}$$

$$C_T(r, c) : center of tumor$$

Step 5: Image segmentation

The proposed segmentation algorithms first approach was to detect the coarse edge of tumor. The moving average is employed to distinguish tumor from background objects with the aid of medical domain knowledge. The algorithm includes two steps; first step is to identify the approximate center and second step is to detect fine edge of the tumor described as follows.

1. Identify center of the tumor $C_T(r,c)$: $center of tumor$ pixel value, S = Start point value
2. Setup the mask size is 3 by 3, M = mask, MCT = Mask center
3. Set cut off boundaries
    (a) Gap between start point value (S) and mask center (MCT)
    (b) difference boundary upper and lower boundary value each mask center
4. Scanning of the tumor area from inside to outside for computing of moving average Compare the MMVt to S and MCT. If the value is inside of boundary, it can do scan-ning continue, otherwise break
n=3
Mean of mask value (MMVt) =

$$(MAt-1+At-2+...At-n)/n. \quad (5)$$

If $(lowerbound \leq MMV \leq upperbound)$
Blank matrix (I ,J) = 255 (the brightest)
Else , Blank matrix (I ,J) = 0 (the darkest)
5. Edge detection with 'Sobel' algorithm for detection coarse tumor shape (Fig. 3-g) The 'Sobel' method finds edges using the Sobel approximation to the derivative. It returns edges at those points where the gradient of I is maximum.

sobel apporximation to derivate = $\begin{pmatrix} -1 & -2 & -1 \\ 0 & 0 & 0 \\ 1 & 2 & 1 \end{pmatrix}$

6. Changing coordinate from rectangular to polar for gathering angle and distance $C_T(r,c)$: center of tumor, $X = I(i,1), Y = I(i,2), I(X,Y)$: rectangular coordinate

$$angle = tan^{-1}(I(c,2) - C_T(1,2)/I(r,1) - C_T(1,1)) \quad (6)$$

distance=$\sqrt{(I(i,1-C_T(1,1))^2 + (I(i,2)-C_T(1,2))^2)}$

7. Scanning of the tumor area from 0 angles to 360 to use moving average

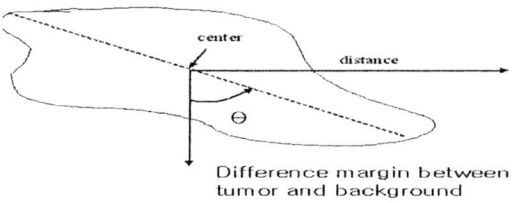

Fig. 2. Scanning example

8. Edge detection with 'Sobel' algorithm for detection of fine tumor shape (Fig. 3-h) Fig. 3 represents detection steps to find out tumor shape from original ultrasound image (Fig. 3-a)

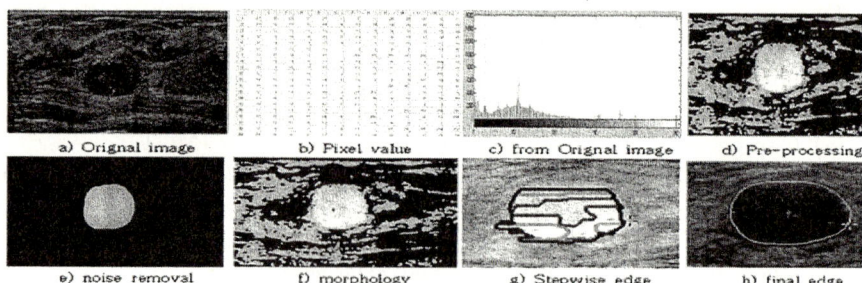

Fig. 3. Detection of the Tumor

Step 6: Analysis of the tumor shape

We want to obtain numerical data of the shape because a shape is one of the important factors based on domain knowledge. Therefore we computed the tumor's distance from center to edge with Euclid method as follows:

$$distance = \sqrt{(((y - y_y)^2 + (x - x_x)^2)} \quad (7)$$

This calculated value can be plotted to the graph (Fig. 4(a-c)). If the curve looks like Fig. 4(b), then it will be a benign, or if the plotted line seems to be like in Fig. 4(c), it will be a cancer tumor. X-axis is angle and Y-axis is deviation

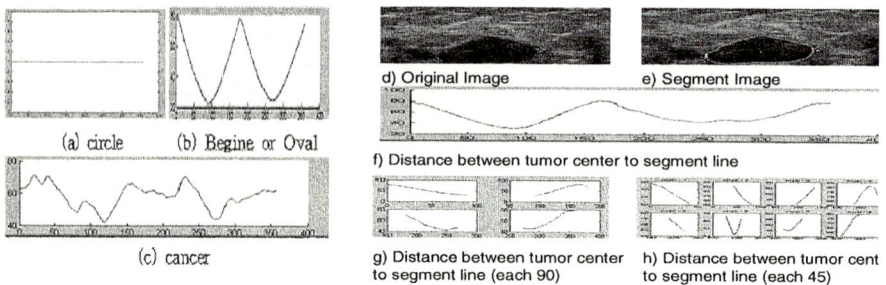

Fig. 4. Analysis of the Typical Tumor

Furthermore, we split the lesion into 4 and 8 divisions for analysis, which will more clearly enumerate and identify the deviation. This approach is based on one of the malignance factor, which is called distortion. Fig. 4(d-h) right side described distance from the tumor center to outline of tumor by original ultrasound image but there is some difference in each of them. Graph Fig. 4(f) shows better identification of the tumor type. In addition Fig. 4(g) and Fig. 4(h) are used to acquire close examination from different angles (Fig. 4(g) used 90 and Fig. 4(h) used 45 degree of angles)

Step 7: Analysis of the tumor circularity.
Circularity is one of the important factors for the diagnosis. If shape of the tumor is similar to a circle, it can be a benign but if the shape is irregular, it can be a malignance. Hence, we computed the tumor circularity for analysis. In our paper we have tried to analyze the tumor circularity(C) (eqn. 8) as well as compute standard deviation (eqn. 9) for more detail study of the image. As per computation, If C is closed to zero that means there is the chances or possibility that the tumor is benign. According to Fig. 5, it shows important boundary between benign and malignance circularity and standard deviation are numerical analysis of image information, if circularity and standard deviation is small, the tumor would seem to be benign otherwise if it is bigger than 20 that means it will be acute tumor or malignance

C=Circularuty; P=Perimeter; A=Area

$$C = P^2/4 * A \qquad (8)$$

$$\sigma = \sqrt{1/(N-1)} * \Sigma_{k=1}^{N}(distance_i - distance_{average}) \qquad (9)$$

	Name	Type	Image	C	Standard Deviation
Benign	Smooth		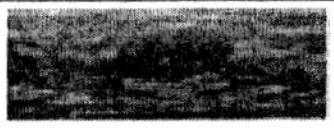	10.80	1.62
	Lobulate	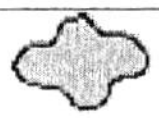		14.32	1.83
	Micro lobulate		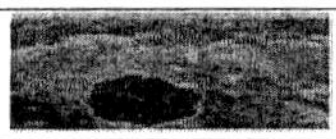	14.78	2.22
	Irrgulate	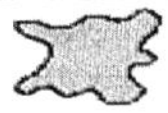	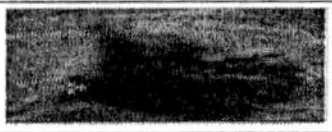	21.99	23.2
Cancer	Spiculate			77.21	24.3

Fig. 5. Analysis of the typical Tumor

3 Discussion

It was quite difficult to get an understandable image by using ultrasound due to sensitivity to noise while transferring to computers. Ultrasound creates a lot of noise for getting ROI(region of interest) image to detect auto segmentation, but it is almost same as anatomical feature image. The image is extremely important for a surgical operation of breast cancer. However, it is really difficult to find out region of interest area due to much noise generated with automatic segmentation. In this paper, we presented auto segmentation with morphological technique to find out the estimated central area of the tumor. Sometimes it is difficult to differentiate between background color and tumor color, so we can

Table 1. Diagnosis Table

	Sum of Multi standard Deviation (Distortion)					
	Benign			Cancer		
division	Sum of Multi standard Deviation			Sum of Multi standard Deviation		
4	13.70			38.32		
8	3.36			15.29		
Analysis with K-Means						
	Discriminate of benign			Discriminate of cancer		
Division	standard	Hit	Error	standard	Hit	Error
4	16.78	82%	20%	45.45	80%	18%
8	9.30	90%	20%	42.06	80%	10%
Table 1. diagnosis result						

detect the edge of tumor with moving average based on domain knowledge acquired from the tumor center. This method will help us to discriminate between tumor and other objects. The algorithm includes domain knowledge of margin, density and shape and they are the helpful factors for diagnosis of the image. Previous work based on domain knowledge, which computed tumor speculation on ultrasound Sheng-Fang Huang, Ruey-Feng Chang ([19]), described one plot so it was kind of rough plotting. Furthermore we analyzed classification using circularity and shape by K-means be-tween benign and malignance (benign: hit 90% error 10%, malignance hit 80% error 20%) as follow table 1. Even though we were using small sample size, if we simulate a lot of cases and analyze more data, it will form reliable information for diagnosis people, such as medical doctor and radiologist. Sheng-Fang Huang with 4 people approached diagnosis to use image enhancement then threshold making binary mode, next to selected region. Fig. 6 and Fig. 7 describe the method and the result after processing. Fig. 6 shows the flowchart of the process. Notably, the lesion must be identified by a physician, and then the 3-D US volume that contains the lesion is analyzed using a computer. Histogram equalization compensates for the non-uniform luminance and the contrast of the original image. Consequently, the lesion and dense adipose tissue can be highlighted. An adaptive threshold-ing method (N. Otsu[20]) is then applied to obtain a binary image to separate the tumor from its background.

Fig. 9 Detection of the tumor image Other related work has been done by Karla Horsch([12])work by using segmentation approach. The algorithm begins

Fig. 6. The flowchart of detection of the location of tumor

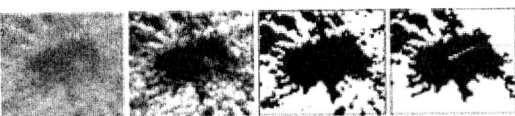

Fig. 7. Detection of the tumor image

with the preprocessing of median filtering and then select ROI area manually. They used 10 by 10 median filter then Gaussian, Laplacian method for image segmentation. This research seems to be an extension of the previous work (K. Horsch [12]), but our methodology is completely different from them in that we can find the estimated ROI area and center of tumor with morphology rather than scanning from the middle of the tumor to out of the threshold value or margin area. In addition, it is possible to analyze about tumor using multi-standard deviation and circulatory analysis. It can also identify distorted area by distance plotting and numerical data obtained from multi-standard deviation and circulatory analysis, which can be good information about malignance to medical doctor. Comparatively, our approach is good for segmentation and supportive to medical doctor for meticulous analysis.

4 Conclusions

Recently, it has been observed that ultrasound could be utilized for breast cancer diagnosis. It can be better conducted by using ultrasound technology with computer aided diagnosis. The main objective of this paper is to use the auto segmentation with morphological technique to find out predictable region of interest (ROI), especially the center and margin area of the tumor. In our approach we have used the moving average method for detecting edge of tumor after estimating the corresponding center using the domain knowledge obtained from tumor center. After computing distance between center and edge of tumor we get factual and numerical data of tumor to calculate multi-deviation and circularity test. It is useful to construct tumor profiling by splitting up the lesion into 4 divisions with the mean of multi-standard deviation (benign: 13.7, malignancies: 38.32) and 8 divisions with the mean of multi standard deviation (benign: 3.36, malignancies: 15.29) with equal segments. We have used the K-means algorithm to make classification between benign and malignance tumor to get the more reliable results. This study will help the physicians and radiologist to improve the efficiency in accurate detection of the image and appropriate diagnosis of the cancer tumor. Our future endeavor is to develop all the technique and tools, which can originate high reliable CAD system. We are looking into malignant factors like tumor shadow-ing, distortion style, elasticity mapping and elasticity image. Elasticity image will be effective information to diagnose difficult cases in the near future.

References

1. kyoung,Moon,woo :Screening of Breast . *Lecture note in Breast cancer image, Seoul National University.* (2004)1-8
2. Ministry of health and welfare :www.ncc.re.kr. *Korean cancer report (2001.1.1-2001.12.31)*,(2003)
3. Karvonen,J.,A.: Baltic Sea Ice SAR Segmentation and Classification Using Modified Pulse-Coupled Neural Networks . *IEEE Trans. on Geoscience and Remote Sensing, vol.42, No.7,* (2004)
4. Bick,U. ,Giger, M., L. , Schmidt,R. ,A. , and Doi,K. : A new single-image method for computer-aided detection of small mammography masses.*Proceedings of CAR 95,* (1995) 357-363.
5. Tsai,D.,Y., et al.: A Computer - Aided System for Discrimination of Dilated Cardiomyopathy Using Echocardiograph Images. *IEICE Trans. Fundamentals,*Vol. E78-A,(1995)1649-1654
6. Tsai,D.,Y. , et al.: Comparative Performance Study of BP-and GA-based Neural Networks for Automated Classification of heart Diseases from Ultrasound Images. *CAR'98 Computer Assisted radiology and Surgery,*(1998) 248-253
7. Haykin,Simon : Neural Netwroks A Comprehensive Foundation. *Prentice Hall International. Inc.* ,(1999)156-479
8. Jung,In-Sung, and Wang,Gi-Nam : Development of an adaptive-intelligent CAD framework . *Proceedings of HCI 2004* , (2003)
9. Jung,In-Sung, and Wang,Gi-Nam : CAD system framework. *Proceedings of HCI 2004* , (2003)
10. Acharya,M. ,De,R.,K., and Kundu,M.,K.: Extraction of Features Using M-Band Wavelet Packet Frame and Their Neuro-Fuzzy Evaluation for Multitexture Segmentation . *IEEE Trans. on Pattern Analysis and Machine Intelligence, vol.25, No.12,* December (2003)
11. Horsch,K.,M., Gifer,L. ,Luz, A. ,V., Vyborny,C., J. : Computerize diagnosis of breast lesions on ultrasound. *Medical Physics, Vol. 29, No. 2,* (2002)157-164
12. Horsch,K.,M., Gifer,L. ,Luz, A. ,V., Vyborny,C., J. : Automatic segmentation of breast lesions on ultrasound . *Medical Physics, Vol. 29, No. 2,* (2002)
13. Stavros,T., Thickman,D., Ra,C. ,L., Dennis,M. ,A. ,Parker, S., H. , and Sisney,G., A. : Solid breast nodule. *Use of sonography to distinguish between benign and malignant lesions. Radiology 196,* (1995) 123-134.
14. Giger,M. ,L. ,Al-Hallaq, H. ,Huo, Z. , Moran,C. , Wolverton,D., E. ,Chan, C., W. , and Zhong,W. : Computerized analysis of lesions in us image of the breast. *Acad. Radiol 6,* (1999) 665-674
15. Garra,S. , Krasner,B., H. , Horii,S., C. , Ascher,S. , Mun,S., K. , and Zeman,P., K. : Improving the distinction between benign and malignant breast lesion. *The value of sonographic texture analysis. Ultrason. Imaging 15,* (1993) 267-285
16. Golub,R., M., et al.: Differentiation of breast tumors by ultrasonic tissue characterization. *J. Ultrasound Med .* 12, (2004) 601-608
17. Sahiner,B., et al.: Computerize characterization of breast masses three-dimensional ultrasound images. *in proceeding of the SPIE (SPIE, Bellingham,WA,1988,Vol.3338)* , (1988)301-312
18. Tohno,E. , Cosgrove,D., O. , and Sloane,J., P.: ultrasound Diagnosis of Breast Disease . *ChurchillLivingstone, Edinburgh, Scotland,*(1994)
19. Huang,S.,F. , Chang,R.,F., Member, IEEE, Chen,D.,R., and Moon,W. K. :Characterization of Spiculation on Ultrasound Lesions .*IEEE TRANSACTIONS ON MEDICAL IMAGING,VOL. 23* , (2004), 1 111-121
20. Otsu,N. : A threshold selection method from gray-level histograms .*IEEE Trans. Syst. Man Cybern., vol. SMC-9,* (1979), 62-66

Composition of Web Services Using Ontology with Monotonic Inheritance

Changyun Li [1,2], Beishui Liao [2], Aimin Yang [1], and Lijun Liao [1]

[1] Department of Computer, Zhuzhou Institute of Technology, Zhuzhou 412008, P.R. China
[2] Institute of Computer Software, Zhejiang University, Hangzhou 310027, P.R. China
lcy469@163.com

Abstract. To realize the automatic and on-demand integration of web services, it is necessary to resolve such issues as the high-level task decomposing and the goal planning at the concept level. We propose a process ontology model by adopting an inheritance mechanism with overriding declaration and monotonic inheritance. The inherited relation among processes implies a layered structure of task decomposing. A framework of automatic composition of web services at knowledge level is proposed.

1 Introduction

Due to their support of flexible connection, and high agility, web services are increasingly becoming a prevailing technology for the integration of distributed and heterogeneous applications [1]. However, the more challenging problem is how to dynamically integrate web services on demand, and especially how to compose them to meet some requirements automatically that are not realized by the existing services. This is the research content of semantic web services [2-6].

The Semantic Web is an extension of the current web in which information is given a well-defined meaning; better enabling computers and people to work in cooperation [7]. Currently, the popular markup language for semantic web is DAML+OIL [8]. Its revision now has been accepted by W3C as an international standard, namely, OWL (Web Ontology Language). The sub-language of OWL, OWL-S [3], is used to describe the semantics of web services, i.e., the semantic web services, which support automated service discovery, invocation and composition.

In OWL-S, the gap between the notions used by human and the data interpreted by machines has not been bridged completely. The reason is that RDF and OIL, based on by OWL, adopt a monotonic inheritance mechanism without the ability of overriding declaration. In addition, every class defined in RDF and OIL can be instantiated to get individuals, while the virtual class similar to an object-oriented method does not exist.

Focusing on the shortcomings mentioned above, we have established a process ontology model by adopting an inheritance mechanism with overriding declaration and monotonic inheritance. This model is the extension of OWL-S. On the basis of this model, we set up a framework for the composition of semantic web services at the knowledge level to further facilitate the automatic composition and invocation of web services.

2 Process Ontology Model with Monotonic Inheritance

Commonly, we define ontology as follows:

Definition 1. An ontology is a quadruple (C, I, P, R), and

C — a set of classes, which consists of Cr and Cv. Cr denotes common classes that can be instantiated, while Cv denotes virtual classes that can't be instantiated. Each class has a globally unique identifier;

I — a set of individuals. Each individual has an individual identifier;

P — a set of properties. The properties are independent of classes. Each property has a property identifier;

R — a set of relations, $R \subseteq (C \cup I \cup P) \times (C \cup I \cup P)$. The main relations include subClassOf, sameClass, hasPropertyOf, sameIndividual, Itype, Ptype, hasValue etc.

Process ontology used a declaration method to described correlative process model of domain, and it provided sharing knowledge about bussiness process for finding, running and composing of Web Service.

Definition 2. Process ontology is a 4-tuple (pC, pI, pP, pR), where

pC — a set of classes of business processes;

PI — a set of process instances;

pP — a set of process properties, including purposes, tasks, categories and performances of a process.

pR — a set of relations, including inheritance relationships between process classes, instances of processes, and belonging relationships between process classes.

Definition 3. Suppose that c1,c2 are process classes, if(c1 subClassOf c2) and ($\exists$ p.((c1 hasProperty p) $\land \neg$ (c2 hasProperty p))), p $\in$ subprocs $\cup$ trans, then we call c1 inherits from c2 in the form of process increment, denoted as c1 pAsubClassOf c2. This definition means that c1 contains not only all the properties of c2, but also the sub-processes or the transitions that do not exist in c2.

It means c1 not only has all the properties of c2, it also contain some sub-process or transformation that c2 does not have. In such situation, derivation class is a proper subset of super class, and it means $c1 \subset c2$.

Precondition of process port, input parameters, output parameters effects and transformation condition also can adopt other domain's ontology.

Definition 4. Suppose c1、c2 $\in$ pC, c3、c4 $\in$ C, If(c1 subClassOf c2)and(($\exists$ p.(c2 hasProperty p)) $\land$ (c1.p Ptype c3) $\land$ (c2.p Ptype c4)), p $\in$ inputs $\cup$ outs $\cup$ pres $\cup$ effects, c3 subClassOf c4, then we call c1 inherits from c2 in the form of parameter extension, denoted as c1 pEsubClassOf c2.

Definition 5. Suppose c1、c2、c3、c4 $\in$ pC, if(c1 subClassOf c2)and(($\exists$ p.(c2 hasProperty p)) $\land$ (c1.p Ptype c3) $\land$ (c2.p Ptype c4)), then we call c1 inherits from c2 in the form of sub-process extension, denoted as c1 sEsubClassOf c2, where p $\in$ subprocs, c3 is a complex process, c4 is a simple process, and c3 is extended from c4.

Definition 4 described that derivation process class changed abstract parameter and variable type, which adopted by super process, to concrete behavior; definition 5 explained that derivation process extended simple process, which is contain in the super process, to the behavior of composition process. But at the same time, these two kinds of inheritance is an overriding declaration of sub process, which is not supported by OWL-S, OWL-S only supports the inheritance form of process increment. Different from common inheritance overriding declaration, the property of derivation class will not contradict with the mark property of super class, between them there is a compatible relation. And this means $c1 \subseteq c2$ in subClassOf c2's semantic still can be satisfied. As a result, this is a kind of monotonic inheritance.

3 Ontology-Based Integration Framework of Web Services

To realize the automatic, on-demand integration of web services, it is necessary to resolve such issues as the high-level task decomposition and the goal planning at the concept level. Our process ontology model is oriented to problem solutions, so it is adaptable to the cross-domain process reuse and the high-level task decomposition. Based on our ontology model, we propose an integration framework for web services as shown in Figure 1.

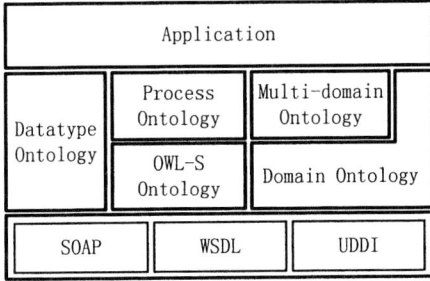

Fig. 1. Integration framework for web services

In this figure, user's applications describe these problems and tasks that the user needs to resolve. The Process ontology describes how the user's task is decomposed into a series of activities, and how a process forms through the transitions among activities. It is oriented to users and problems for high-level goal planning and task decomposing. The solution of the user's task may be cross-domain, so process ontology should be able to express the abstract, application-independent solving process of problems. In order to reduce the complexity of task decomposing, a layered structure is adopted. Meanwhile, the description of a process is independent of web services. The process ontology model mentioned in the section 2 can meet this need.

The Data type ontology contains the basic data types of the concept properties of other ontologies, such as integer, numerical value, and string etc. The domain ontology defines entities of a specific application domain, and the relationships among them. Domain ontology describes the special vocabulary and the classification architecture of a specific application domain. The multi-domain ontology defines a classification for the value types of parameters in an abstract process. It depends on the existence of the process ontology, and is on the basis of various domain ontologies. From the perspective of process, the multi-domain ontology links several domain ontologies together.

OWL-S ontology describes the semantics of web services by using OWL-S, and establishes a classification of web services. During of the process of solving a task, the simple process can be mapped to the port operation of web services. A series of web services perform this task according to the process control flow and the data flow. In the lowest layer, there are industry standards of web services, including SOAP, WSDL, and UDDI.

4 Conclusions

Our method presented in this paper enables the service integration at the knowledge level, with a formal theoretical basis and an explicit hierarchy, and is understandable and highly automatic. Currently, we have developed a prototype system to test the integration framework and the methods proposed in this paper. Our future researches will be done on continually improving the ontology model described in this paper, especially on its reasoning mechanism, the theory of semantic inconsistency checking.

References

1. Tsalgatidou, A., Pilioura, T.: An Overview of Standards and Related Technology in Web services. Distributed and Parallel Databases, Kluwer Academic Publishers, VOL.12, NO.3 (2002) 135-162
2. McIlraith, A., CaoSon, T., Honglei, Z.: Semantic Web Services. IEEE INTELLIGENT SYSTEMS,VOL.16,NO.2 (2001) 46-53
3. Dean, M. (eds): OWL-S: Semantic Markup for Web Services. http://www.daml.org/services/owl-s/1.0/owl-s.pdf (2004)
4. Gomez-Perez , A., Cabero, R.G.: A Framework for Design and Composition of Semantic Web Services, 2004 AAAI Spring Symposium Series. Palo Alto, California (2004)
5. Sirin, E., Hendler, J., Parsia,B.: Semi Automatic Composition of Web Services using Semantic Descriptions. In Proceedings of the ICEIS-2003 Workshop on Web Services:Modeling, Architecture and Infrastructure. Angers, France (2003)
6. Peer, J.: Towards Automatic Web Service Composition using AI Planning Techniques. http://sws.mcm.unisg.ch/docs/wsplanning.pdf (2003)
7. Berners-Lee, T., Hendler, J., Lassila,O.: The Semantic Web. Scientific American, VOL.284,NO.5 (2001) 34-43
8. Horrocks, I., Harmelen, F.: Reference Description of the DAML+OIL Ontology Markup Language. http://www.daml.org/2001/03/reference.html (2001)
9. Dean, M., Schreiber, G.: OWL Web Ontology Language Reference. W3C Candidate Recommendation. http://www.w3c.org/TR/owl-ref/ (2003)

Ontology-DTD Matching Algorithm for Efficient XML Query*

Myung Sook Kim and Yong Hae Kong

Division of Information Technology Engineering, Soonchunhyang University,
Asan-si, Choongnam-do, 336-745, Korea
{krhkms, yhkong}@sch.ac.kr

Abstract. XML queries are often expanded based on ontology for broad and in-depth search. But, queries generated from ontology itself are not specific to target documents. Accordingly, the overall search efficiency will deteriorate with those superfluous queries that are not succinct to the target. We suggest an ontology reduction algorithm where the target DTD is matched to ontology such that queries can be minimally expanded. The matched and reduced ontology is successively reusable for the document of a kind. This target-fitted query expansion method is expected to be more efficient than conventional methods in query processing.

1 Introduction

The structural variety of XML documents makes query to them difficult: [1],[2],[3]. If XML query is expanded based on ontology, the expanded ones may tolerate the structural variance to some extent: [4],[5]. Since W3C and others are making ontology standards on application areas, we can assume that necessary ontology will be soon available: [6]. Even though necessary ontology is at hand and queries are expanded based on it, there occurs an efficiency problem. When queries are independently expanded without considering target document structures, only a few of the expanded ones will fit to the documents while the rest will not. This will cause a low hit-ratio in query process and the overall search efficiency will deteriorate.

One way to remedy this inefficiency is to make query-expansion specific to target documents. Unfortunately, matching ontology and DTD is difficult since the two entities are not homogeneous. That is, a DTD is essentially a structural entity while ontology is a graph having relational entity and inheritance characteristics. We propose an ontology-DTD matching algorithm that results in a reduced ontology, with which queries are expanded specific to target documents.

If a web site provides information which interests one, one tends to keep on searching the site in regular fashion. A DTD of a web site usually remains unchanged for a period of time and so dose the document structure, while the documents may frequently change. Once a matching is done, the reduced ontology can be quickly

---

* University Fundamental Research Program supported by Ministry of Information & Communication in Republic of Korea.

referenced whenever one tries to access the same site without any further matching, because the reduced ontology is successively reusable for the documents of a kind. Consequently, XML search efficiency can be improved by target-succinct queries through a single matching step.

We implement an XML search system consisting of ontology-DTD match, query-expansion and a query engine as shown in fig. 1. Ontology concepts and DTD elements are matched and the missing concepts are eliminated or deactivated. Then, ontology attributes are matched with DTD and unnecessary attributes are removed. After the two matches, ontology shrinks in scale. Finally, queries are minimally expanded with the reduced ontology and the expanded ones are applied to the target documents. We experiment the efficiency and effectiveness of this system with a set of sample XML documents.

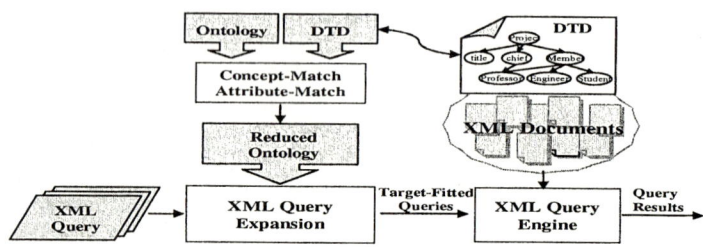

Fig. 1. Target-Dependent XML Queries by Ontology-DTD Match

2 Ontology-DTD Matching Algorithm

Expanding XML query may enable broader and in depth search on XML documents. However, such expanded queries are superfluous because they lack target document dependency: [7]. We suggest an algorithm that matches ontology and a target DTD. Since the matched ontology enables query-expansion succinct to target documents, we can minimize useless search with these queries.

Our matching algorithm consists of two steps. The first step matches ontology concepts and target DTD elements and the subsequent step matches ontology attributes with DTD. When there are ontology concepts that do not correspond to DTD elements, such concepts are removed or deactivated depending on their inheritance characteristics. This match results in a conceptually-reduced ontology. Since attributes of a concept are inherited to its sub-concepts, matching attributes needs a special attention on their inheritance characteristics. The above conceptually-reduced ontology is again reduced by removing those attributes that do not match a target DTD and the matched attributes are associated to the corresponding concepts.

After the two-step matching, the finally reduced ontology can be used in minimal query expansion. The matching algorithm is depicted in fig. 2. Two algorithms are elaborated in successive sections using ontology "College-Research-Center" of fig. 3 and a DTD of fig. 4. Fig. 3(a) is a concept hierarchy, and fig. 3(b) is the attributes of each concept and their relations between attributes. Fig. 4 shows an example DTD.

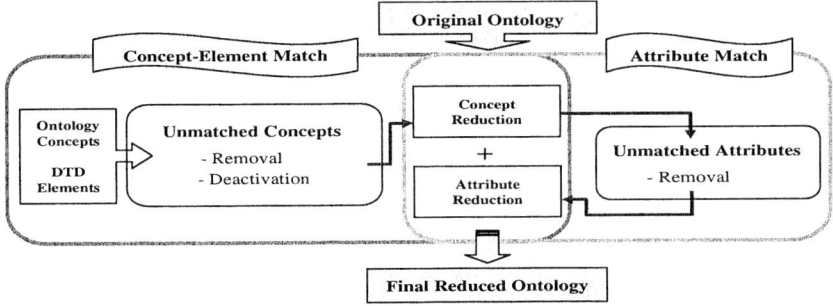

Fig. 2. Ontology-DTD Match Diagram

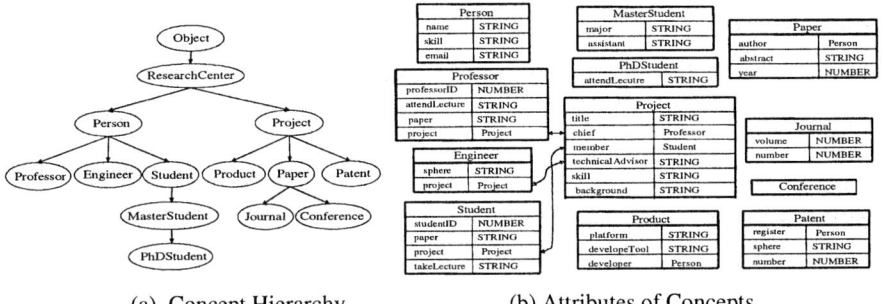

(a) Concept Hierarchy (b) Attributes of Concepts

Fig. 3. 'College-Research-Center' Ontology

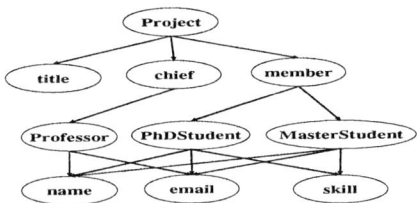

Fig. 4. Example DTD Structure

2.1 Concept Matching Algorithm

Using ontology in fig. 3, query //Project [title="XML"] is expanded to the six queries, //(Project | Product | Paper | Patent | Journal | Conference)[title="XML"]. When the six queries are applied to the XML documents having DTD of fig. 4, only the query for Project is valid and the rest five queries are vain due to the discrepancy of ontology and the DTD. To minimize such unnecessary queries, we want to reduce ontology such that only the minimal concepts are preserved. Selecting and removing concepts in ontology that are not part of the DTD is a simple matter. But, removal of a concept out of ontology not only disconnects the concept hierarchy but also eliminates its attributes that must be inherited to its sub-concepts. Leaf concepts may be simply removed, while intermediate concepts need to be made inactive. Note that inactive concepts can only be used for query-expansion purpose only because they are in fact removed. Concept matching algorithm is summarized in table 1.

Table 1. Concept Matching Algorithm

For all the ontology concepts, select a concept.
Compare the concept with elements of DTD.
If the concept is not part of the DTD,
If the concept is a leaf concept,
Remove the concept and its attributes from ontology.
Otherwise, make the concept inactive.

For ontology of fig. 3 and DTD of fig. 4, mismatched concepts Journal, Conference, Products, Paper, and Patent are successively removed. On the other hand, mismatched concepts Person and Student are made inactive. Fig. 5 is the reduced ontology after concept-element match.

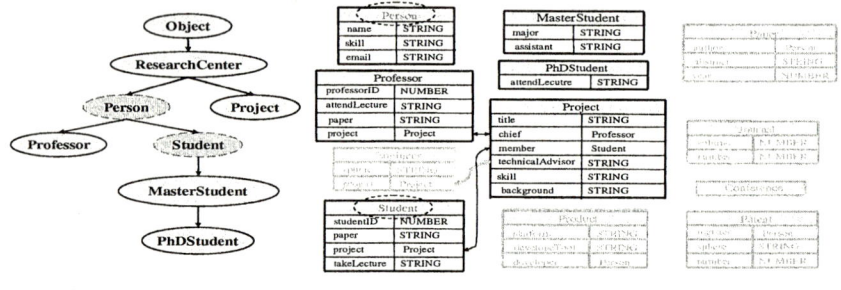

(a) Reduced Ontology Hierarchy   (b) Attributes of Concepts

Fig. 5. Conceptually Reduced "College-Research-Center" Ontology

2.2 Attribute Matching Algorithm

The reduced ontology by section 2.1 is certainly helpful in minimizing unnecessary queries. However, it is not sufficient since discrepancy between ontology attributes and DTD elements/attributes still exists. Since a concept has its own attributes plus its ancestors, matching them to DTD is not trivial. For example, concepts Student, PhDStudent and MasterStudent in fig. 3 are interrelated as sub-concepts, while elements PhDStudent and MasterStudent are siblings in fig. 4. When the attributes of concept MasterStudent are matched to those of element MasterStudent, removal of missing attributes of concept MasterStudent will prevent its attributes from being inherited to all its sub-concepts.

Ontology attribute matching algorithm is the following. Given the reduced ontology by section 2.1 such as fig. 5, suppose ontology concept A and its corresponding DTD element B. If attribute P of concept A is an element or a sub-attribute of element B, set association between attribute P and concept A. Otherwise, consider a case where attribute or sub-element Q of element B is not contained in concept A. If Q is included in the attributes of ancestor concept C of concept A, set association between concept C's attribute Q and concept A. This association is made for every attribute and sub-element of element B. Repeat this process for all the ontology concepts in bottom-up fashion. When this matching process is completed, the attributes that do not have any association are removed from ontology. Since the attributes that have any association indicate the DTD dependency, they can be used for adequate query-expansion.

Ontology attribute matching further reduces the ontology and this fully reduced ontology can optimize query-expansion. Ontology attribute matching algorithm is in table 2 and the matched result for fig. 5 is shown in fig. 6.

Table 2. Attribute Matching Algorithm

For all the ontology concepts in bottom-up order,
    For ontology concept A and its corresponding DTD element B,
        1. If attribute P of concept A is an attribute or a sub-element of element B,
            Set association between attribute P and concept A.
        2. If attribute or sub-element Q of element B is not contained in concept A,
            For each ancestor concept C of concept A,
                If Q is included in concept C's attributes,
                      Set association between C's attribute Q and concept A.

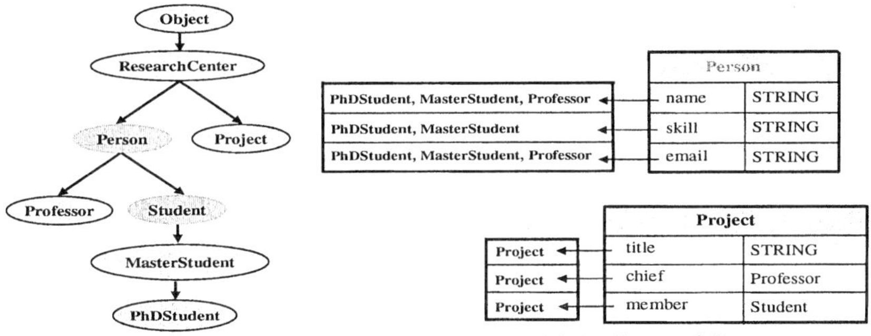

(a) Concept Hierarchy   (b) Attributes of Concepts

Fig. 6. Attribute-Reduced "College-Research-Center" Ontology

3 Minimal Query-Expansion with Reduced Ontology

Raw queries themselves are not enough to retrieve semantic information from XML documents: [8]. Therefore, queries are expanded either by explicit conceptual inheritance or implicit conceptual association or both. Furthermore, for the expanded queries to be specific to target documents, we minimally expand queries using the reduced ontology in chapter 2.

Structural variations of documents can be tolerated by considering explicit inheritances by the algorithm in table 3. By analyzing the reduced concept hierarchy, the algorithm finds out all the corresponding sub-concepts of the given query. For example, query for concept Person in fig. 6 is expanded to all the sub-concepts, Professor, MasterStudent and PhDStudent, excluding Person and Student which are inactive concepts.

Table 3. Conceptual Inheritance Expansion Algorithm

---
1. If a query contains no attribute at all,
    For all the concepts of a query,
       1.1 Select a concept.
       1.2 Search for all its descendant concepts.
       1.3 If the selected concept and its descendant concepts are active,
          Add the descendant concepts to a query set.
2. If a query contains any attribute,
    For all the concepts associated with the selected attribute,
       If a concept is a descendant of the concepts in a query set, add the concept to a query set.

We inferred association between related concepts by special attention to the ontology attributes as in table 4 algorithm. These inferred associations are expressed as rules and they are used as a quick reference in query-expansion.

Table 4. Association Rule Inferring Algorithm

For all the ontology concepts,
   1. Select a concept A.
   2. For all the attributes of concept A, select an attribute B.
   3. Search concept C that matches attribute B.
   4. Set association between concepts A and C.

In fig. 5, through implicit association, two associations are established between concepts Student and Project and concepts Professor and Project. These associations generate the below inference rules. The former rule states that project of Student and member of Project is semantically identical.

FORALL Pers1, Proj1   Proj1 : Project[member ->> Pers1] <-> Pers1:Student[project ->> Proj1]
FORALL Prof1, Proj1   Proj1 : Project[chief ->> Prof1] <-> Prof1 : Professor[project ->> Proj1]

With the inferred rules, a query can be semantically expanded. For example, since query //Student[project] is associated with //Project[member], query //Student[project] is expanded to concept Project and all its descendant concepts. Table 5 is a query expansion algorithm using the inferred association rules.

Table 5. Query Expansion Algorithm using Inferred Association Rules

---
For all the concepts in a query set, select a concept,
   For the selected concept,
      If an association rule exists, add the associated concept to a query set recursively.
Expand a query set using table 3 algorithm.
---

4 Experiments

Query expansion by ontology-DTD match is experimented by the DTD samples in fig. 7(a), 8(a), 9(a) and 10(a). Even though the four DTDs seem structurally different, they convey similar information. Ontology-DTD matching algorithm of chapter 2 successfully reduces ontology of fig. 3 to fig. 7(b), 8(b), 9(b) and 10(b) respectively.

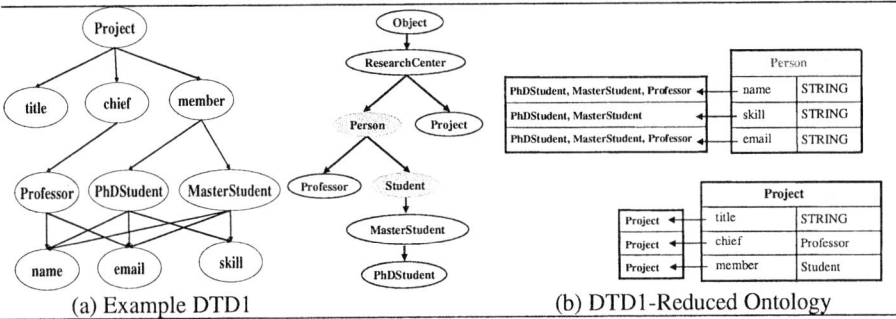

Fig. 7. Ontology-DTD1 Match

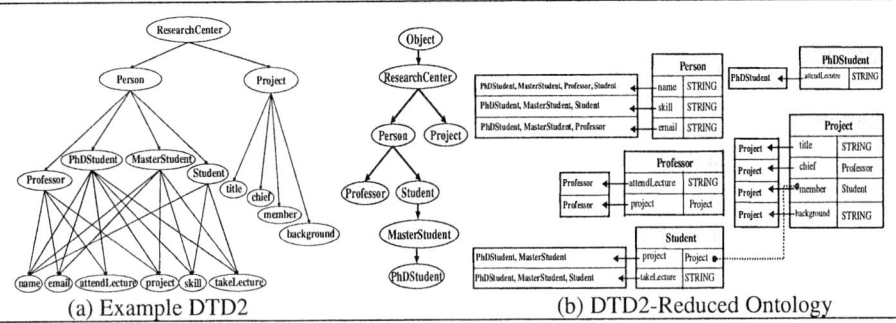

Fig. 8. Ontology-DTD2 Match

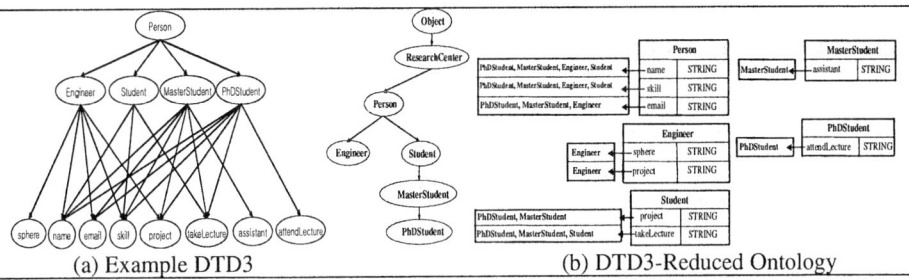

Fig. 9. Ontology-DTD3 Match

Using both original ontology of fig. 3 and the reduced ontology of fig. 7(b), 8(b), 9(b) and 10(b), three sample queries are separately expanded. Original ontology expands query //Project[member] into 36 queries, while the reduced ontology generates only 8 queries as in table 6. Original ontology produces 24 and 28 queries for queries //Person[skill] and //Project[chief], but each of the reduced ontology generates only 11 and 4 queries as in table 7 and table 8 respectively. Note also that the associated rules are used in query expansion of //Project[member] and //Project[chief] as in table 6 and table 8.

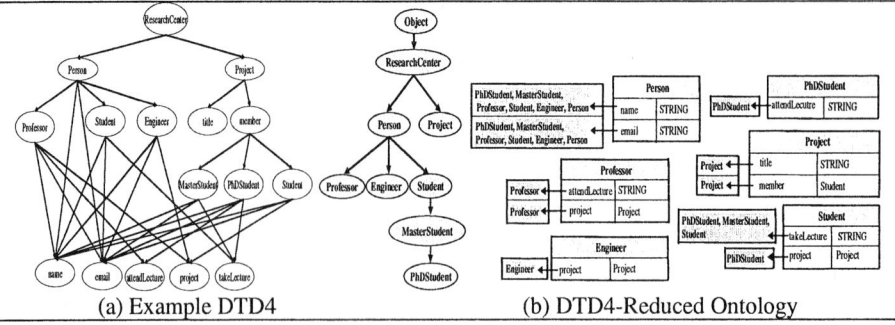

(a) Example DTD4   (b) DTD4-Reduced Ontology

Fig. 10. Ontology-DTD4 Match

Table 6. Query //Project[member] Expansion

DTD \ Query	Associated Rule, //Project[member] <==> //Student[project], is used			
	By Original Ontology			By Reduced Ontology

DTD1	//Project[member]
	 //Patent[member]
	 //Student[project]
		//Product[member]
		 //Journal[member]
		 //MasterStudent[project]
			//Paper[member]
			 //Conference[member]
			 //PhDStudent[project]
				//Project[member]

DTD2	//Project[member]
	 //Patent[member]
	 //Student[project]
		//Product[member]
		 //Journal[member]
		 //MasterStudent[project]
			//Paper[member]
			 //Conference[member]
			 //PhDStudent[project]
				//Project[member]
				 //MasterStudent[project]
				 //PhDStudent[project]

DTD3	//Project[member]
	 //Patent[member]
	 //Student[project]
		//Product[member]
		 //Journal[member]
		 //MasterStudent[project]
			//Paper[member]
			 //Conference[member]
			 //PhDStudent[project]
				//MasterStudent[project]
				 //PhDStudent[project]

DTD4	//Project[member]
	 //Patent[member]
	 //Student[project]
		//Product[member]
		 //Journal[member]
		 //MasterStudent[project]
			//Paper[member]
			 //Conference[member]
			 //PhDStudent[project]
				//Project[member]
				 //Student[project]

Total Queries	36			8

Table 7. Query //Person[skill] Expansion

DTD \ Query	No Association Rule is used		
	By Original Ontology		By Reduced Ontology

DTD1	//Person[skill]
	 //Engineer[skill]
	 //MasterStudent[skill]
		//Professor[skill]
		 //Student[skill]
		 //PhDStudent[skill]
			//MasterStudent[skill]
			 //PhDStudent[skill]

DTD2	//Person[skill]
	 //Engineer[skill]
	 //MasterStudent[skill]
		//Professor[skill]
		 //Student[skill]
		 //PhDStudent[skill]
			//Student[skill] //Professor[skill]
			 //MasterStudent[skill]
			 //PhDStudent[skill]

DTD3	//Person[skill]
	 //Engineer[skill]
	 //MasterStudent[skill]
		//Professor[skill]
		 //Student[skill]
		 //PhDStudent[skill]
			//Student[skill] //Engineer[skill]
			 //MasterStudent[skill]
			 //PhDStudent[skill]

DTD4	//Person[skill]
	 //Engineer[skill]
	 //MasterStudent[skill]
		//Professor[skill]
		 //Student[skill]
		 //PhDStudent[skill]
			No query

Total Queries	24		10

Table 8. Query //Project[chief] Expansion

Query \ DTD	Associated Rule, //Project[chief] <==> //Professor[project], is used		
	By Original Ontology		By Reduced Ontology
DTD1	//Project[chief] //Product[chief] //Paper[chief] //Patent[chief] //Journal[chief] //Conference[chief] //Professor[project]		//Project[chief]
DTD2	//Project[chief] //Product[chief] //Paper[chief] //Patent[chief] //Journal[chief] //Conference[chief] //Professor[project]		//Project[chief] //Professor[project]
DTD3	//Project[chief] //Product[chief] //Paper[chief] //Patent[chief] //Journal[chief] //Conference[chief] //Professor[project]]		No query
DTD4	//Project[chief] //Product[chief] //Paper[chief] //Patent[chief] //Journal[chief] //Conference[chief] //Professor[project]		//Professor[project]
Total Queries	28		4

Table 9. Hit Ratio Comparison Summary between Original and Reduced Ontology

Queries \ DTD	Hit Ratio (hit queries / total queries)					
	//Project[member]		//Person[skill]		//Project[chief]	
	Original Ontology	Reduced Ontology	Original Ontology	Reduced Ontology	Original Ontology	Reduced Ontology
DTD1	1/9	1/1	2/6	2/3	1/7	1/1
DTD2	2/9	2/3	2/6	2/4	2/7	2/2
DTD3	1/9	1/2	4/6	4/4	0/7	0/0
DTD4	1/9	1/2	0/6	0/0	1/7	1/1
Hit Ratio	5/36 (14%)	5/8 (63%)	8/24 (33%)	8/10 (80%)	4/28 (14%)	4/4 (100%)
Ave. Original Ontology	17/88 (19%)					
Ave. Reduced Ontology	17/22 (77%)					

Table 9 summarizes the three query expansion by ontology of fig. 3 and the correspondingly reduced ones. As a result, the hit ratio substantially increases from 19% to 77% through ontology-DTD match

Our experiments confirm that ontology-DTD matching algorithm effectively reduces ontology such that the reduced ontology enables queries to be minimally expanded specific to target documents.

5 Conclusions

Since XML query is difficult due to the variety of their structures and attributes, queries are often expanded based on ontology for broad and in-depth search. But, expansion solely with ontology generates excessive queries since it dose not consider target structures. These superfluous queries will deteriorate the overall search efficiency.

Search efficiency can be substantially improved if queries are expanded succinct and specific to target documents. Matching ontology and target DTD is not trivial since the two entities are essentially different in structure, attribute, and inheritance. We propose an ontology-DTD matching algorithm that considers target document characteristics. Our method consists of two steps. The first step matches ontology concepts and DTD elements and the subsequent step matches ontology attributes with DTD. Each step reduces ontology and the final ontology can be used in minimal query expansion. Moreover, the matched and reduced ontology are successively reusable for the document of a kind.

By implementing ontology-DTD matching algorithm, query-expansion algorithm, and an XML query engine, we experiment the proposed method on various sample XML documents. The results show that our method effectively reduces ontology and generates succinct queries fitted to target documents, which is more efficient than conventional methods in query processing.

References

1. Fhur, N., Großjohann, K.: XIRQL: An XML Query Language Based on Information Retrieval Concepts. ACM Transaction on Information Systems, Vol. 22. No. 2. (2004) 313-356
2. Kamps, J., Marx, M., de Rijke, M., Sigurbjornsson, B.: Best-Match Querying from Document-Centric XML. WebDB (2004)
3. Florescu, D., Kossmann, D., Manolescu, I.: Integrating Keyword Search into XML Query Processing. Computer Networks, Vol. 33. (2000) 119-135
4. Erdmann, M., Studer, R.: How to Structure and Access XML Document with Ontologies. Data & Knowledge Engineering, Vol. 36. No. 3. (2001) 317-335
5. Theobald, A.: An Ontology for Domain-oriented Semantic Similarity Search on XML Data. Datenbanksysteme für Business, Technologie und Web (BTW) (2003) 217-226
6. McGuinness, D. L., van Harmelen, F.: OWL Web Ontology Language Overview. W3C Recommendation (2004)
7. Erdmann, M., Decker, S.: Ontology-award XML queries. WebDB (2000)
8. Cohen, S., Mamou, J., Kanza, Y., Sagiv, Y.: XSEarch: A Semantic Search Engine for XML. VLDB (2003)

An Approach to Web Service Discovery Based on the Semantics*

Jing Fan, Bo Ren, and Li-Rong Xiong

College of Software, Zhejiang University of Technology,
Zhaohui Liuqu, Hangzhou, 310014, P.R. China
fanjing@zjut.edu.cn

Abstract. The research work in this paper focuses on solving the critical problems in Web Service discovery such as how to locate functionality-desired Web Services and how to select the best one from large numbers of functionality-similar Web Services. The semantic description and quality description of Web Service based on ontology proposed in this paper, provides a consistent description of different kinds of Web Service, which is used as the basis of service matching. Then the matching model of Web Service is discussed in detail, which consists of semantic similarity matching according to functionality of Web Service, and semantic filtering according to quality of Web Service. Using the descriptions and matching approach, the performance of Web Service discovery can be improved by increasing the precision and recall of Web Service searching.

1 Introduction

With the rapid growing number of the services that can be obtained, people have to select a proper service from thousands of Web Service groups to construct the application. Therefore the capability of the Web Service discovery becomes more important. There are two key problems to be solved in the research of service discovery. One is how to give out the semantic description of services, and the other is how to match the search conditions with the service descriptions.

After analyzing different kinds of service description language and take the Web Service Ontology OWL-S as reference [1], this paper presents the Semantics of Web Service Description ontology (SWSD) and the Quality of Web Service description Ontology (QoWSO). The semantic description of primary information, functional information, and other non-functional information of Web Services provided by the SWSD, and the quality description of Web Services provided by the QoWSO, present a standard form for the providers and requestors to describe Web Services and support the exact location of Web Services.

The service matching model introduced in this paper is constructed on the description of SWSD and QoWSO. It mainly depends on the function-based semantic simi-

---

* The research work in this paper is supported by the project of National 863 High Technology Planning of China (No. 2003AA413320), and the project of Science and Technology Planning of Zhejiang Province (No. 2004C31099).

larity matching for Web Services, and also takes the filtering result of service quality as reference. By extracting the potential semantics in Web Services to do semantic similarity matching for the input and output of service functionality, it can improve the precision and recall of Web Service searching. Applying the filtering of service quality, best service will be found from a lot of functionally similar services.

2 Semantic Descriptions of Web Services

In the Web Service Semantics Description Ontology (SWSD), the semantics of Web Service is described by relating the input and output of services to the concepts in ontology. And the attribute *certifiedID* of the service description in SWSD is used to relate to the Quality of Web Service Description Ontology (QoWSO), to support the service quality description. SWSD is defined in ontology language OWL.

The information for service discovery provided by SWSD is consisted of the following four types:

- Primary Information and Provider Information: Primary information tells name and brief description of the service, Provider Information contains the contact information of the entity that provides the service.

- Functional Description: It expresses the conversion of the information as the content of the service's functionality. The conversion of the information is represented by input and output.

- Quality Description: It is described by attribute *certifiedID* relating to QoWSO.

- Other Attributes Description: It is described by the elements as the category that the service belonging to and other extensible attributes.

This paper also defines a Quality of Web Service Description Ontology (QoWSO) to construct the consistent description of service quality, which takes both the historical statistical information and the up-to-minute information of Web Service quality into consideration [2]. QoWSO is described by the class *QoWSProfile*, while class *QoWSProfile* is described by five quality parameters. QoWSO is also defined in ontology language OWL.

The definitions of five quality parameters are as follows:

- Stability: difference in response time of calling the same service at different moment.

- ResponseTime: time from the moment that service requestor sends the request to the moment that he receives the response.

- Reliability: possibility that Web Service can work in order when the user requests.

- AcessedTimes: times that Web Services are invoked.

- Grade: evaluation from the users to the Web Services that they invoke.

3 Publication of Web Services

Finishing the description of service, the next work is the publication of service on the web in order to implement sharing. This paper applies the method of mapping WSDL service description to UDDI Registry to implement the registration of semantic description in UDDI Registry [3]. It first extends the types of tModel in UDDI to provide type *swsdSpec* and *qowsSpec*, which are similar as type *wsdlSpec*. And type *swsdSpec* and *qowsSpec* are corresponding to the services defined using SWSD and QoWSO respectively.

Having the definition of type *swsdSpec* and *qowsSpec* tModel, the detail semantic service description of SWSD and QoWSO can be registered in UDDI Registry as tModel.

Take the mapping from DAML-S Profile to UDDI as reference [4], which is used in the registration of WSDL in UDDI Registry, the mapping from SWSD to UDDI includes: the mapping from Provider Information in SWSD to Contacts of Business Entity in UDDI, the mapping from Primary Information as *serviceName* and *textDescription* in SWSD to *name* and *description* in Business Service, the mapping from SWSD service semantic description to tModel, and the mapping from service quality description to tModel.

4 Service Matching on Semantics

Applying the SWSD service semantic description ontology, service matching on semantics uses matching the content of service description in tModel, to improve the service matching ability. In the service finding, the tModels in which SWSD semantic description registered are listed according to the type *swsdSpec*, then service matching are done based on the service functionality, after that the matched Web Services are searched in UDDI. At last the best service is chosen from the found Web Services using the semantic filtering based on QoWSO. This approach may have higher efficiency since the number of Web Services to be processed is relatively smaller. It is because that the search work is first done based on the functions of service, and then the detailed matching is performed on the traversal.

4.1 Function-Based Semantic Similarity Matching of Web Service

Function-based semantic similarity matching of Web Service includes the semantic similarity matching of input and output, which is called IO matching for short. Assume that the input set and output set of service request are R_I and R_o respectively, and the input set and output set of advertisement service are A_I and A_o respectively, therefore if $R_o \subseteq A_o$, which means that advertisement service can provide all the outputs required by the service request, then output matching is successful; and if $A_I \subseteq R_I$, which means that service request can provide all the inputs needed by the advertisement service, then input matching is successful. Obviously, the IO matching can be regarded as the containing problem in Set.

Since the input parameter and output parameter of service, which represent the service functionality, is described with class in domain ontology defined by OWL, "$a=b$"

can be explained as that the class related to a is the same as the class related to b. But in the function-based semantic similarity matching, the definition that a and b belong to the same class is not suitable. The semantic similarity between the class related to a and the class related to b is used to determine the matching of a and b. And the functional semantic similarity can be obtained through the computation of class similarity [5] and set similarity.

4.2 Semantic Filtering on Quality of Web Service

Semantic filtering on Quality of Web Service (QoWS) is proposed to find the best service in a lot of functional similar Web Services. The key in Semantic filtering on QoWS is to compare the measured value of five service quality parameters in the QoWSO between different services.

According the meaning of measurement, the five service quality parameters described in section 2.2 are classified as two kinds: one is that the bigger measured value means the better quality parameter, such as Reliability, AcessedTimes and Grade; the other is the smaller measured value means the better quality parameter, such as ResponseTime and Stability.

The performance of quality parameter for the two services can simply be obtained by comparison with the two measured value of the quality parameter.

References

1. Deborah, L., Harmelen, F.: OWL Web Ontology Language Overview [EB/OL]. http://www.w3.org/ TR/owl-features, 2003-12-15
2. Cardoso, J., Miller, J.: Modeling Quality of Service for Workflows and Web Service Processes [J]. The International Journal on Very Large Data Bases (VLDBJ) (2002)
3. Chinnici, R., Gudgin, M., Moreau, J., Weerawarana, S.: Web Services Description Language (WSDL) Version 1.2, W3C Working Draft 24 January 2003, Available at http://www.w3.org/TR/2003/WD-wsdl12-20030124/ (2003)
4. Paolucci, M., Kawamura, T., Payne, T. R., Sycara, K.: Importing the Semantic Web in UDDI. In Web Services, E-Business and Semantic Web Workshop (2002)
5. Rodriguez, A., Egenhofer, M.: Determining Semantic Similarity Among Entity Classes from Different Ontologies. IEEE Transactions on Knowledge and Data Engineering (2002)

Non-deterministic Event Correlation Based on C-F Model*

Qiuhua Zheng[1,2], Yuntao Qian[1,2], and Min Yao[1]

[1] Computational Intelligence Research Laboratory, College of Computer Science
Zhejiang University, Hangzhou, Zhejiang Province, China
[2] State key Laboratory of Information Security,
Institute of Software of Chinese Academy of Sciences
zheng_qiuhua@163.com, ytqian@zju.edu.cn

Abstract. This paper proposes a non-deterministic event correlation technique for diagnosis problem of end-to-end service in network, which uses an event – fault model based on path domain of events. The technique utilizes a refined heuristic approach to create and update fault hypotheses that can explain these events received by system, and computes these hypotheses' belief by C-F model method. The result of event correlation is the hypothesis with maximum belief. Simulation shows this approach can get a high accuracy even in the case of low observability events ratio.

1 Introduction

Event correlation, a central aspect of network fault diagnosis, is a process of analyzing alarms received to isolate possible root cause responsible for network's symptoms occurrences. Because failures in large network are unavoidable, an effective correlation can make network system more robust, more reliable, and ultimately increasing the level of confidence in the services they provide. So far, a number of event correlation techniques[1-10] have been proposed. These techniques mainly focus on the following aspects: (1) Model of event-fault relationship (EFM): Many EFMs have been proposed, such as causality graph model[9], dependence graph model[1], context-free grammar model[1], etc. These models have been applied into several network fault management applications, and have a better performance. However, they need to be built in advance. It depresses system's flexibility and adaptability, especially in an ever-variable situation. To overcome this difficulty, we extend an EFM based on event path domain[10] through introducing "logic node" and "logic link" in this paper. (2) Event correlation algorithm: Event correlation algorithm can be divided into deterministic and non-deterministic. The deterministic techniques mainly focused on diagnosing problems related to network connection, and on low-level network protocols. Deterministic techniques are sufficient for diagnosis related to the availability of services offered by lower layers, but they are difficult to diagnosis problems related to service performance. To improve the accuracy of diagnosis on troubles related to network performance, several non-deterministic algorithms have been proposed in [1, 4, 6-8, 11]. On the whole, we considered that

---

* This research was supported part by Huawei Technologies.

IHU technique is the best one among these techniques. Simulation shows that the algorithm has a good performance on accuracy and performance. Whereas, this algorithm has the following disadvantages: (a) When network event loss ration is high, the confidence factor of the positive events which are taken into account by this approach will become low, (b) This approach need to obtain system's events loss ratio, but it is difficult in real-life network.

In this paper, we presented a new non-deterministic event correlation technique. This technique improves the EFM based on path domain of events and IHU algorithm, including: (a) Extending the EFM by introducing "logic node" and "logic link". (b) Improving the IHU algorithm through adding a constraint to maximum fault number of the system. (c) Improving the robustness of algorithm by revising the range of positive events. We compute the belief of fault hypotheses with C-F model, and choose the hypothesis with maximum belief as the result of event correlation.

The rest of this paper is organized as follows. In section 2, we describe the extension to EFM based on event's path domain. In section 3, we introduce the basic concept of certainty factor model. In section 4, we propose our improvement on IHU and our event correlation approach using C-F model in detail. In section 5, we evaluate our techniques through simulation study and compare it with IHU algorithm. Finally, in section 6, we conclude our work and present the direction for future research in this area.

2 Refined EPM Model

An EFM based on path domain of events has been proposed in literature[10]. The model assumes that a unique path can be determined according to event's source and destination, and the faulty NE must be included in this path domain, where the term "path domain" represents the set of nodes and links in the faulty path. By this way, system's EFM can be established during run time. An example is shown as follows. In Fig.1, node D emitted an event E_1 that indicated that node D fails to visit node E, and node A reported an event E_1 which implies that node A fails to visit with node C. According to event E_1, we can ascertain the unique path PM_{E1}, which consists of node D, link D-B, node B, link B-E and node E. According to event E_2, we can ascertain path PM_{E2} which is composed of node A, link A-B, node B, link B-C and node C. By synthesizing the path of these events, a real-time and partial EFM can be built. (Fig.2)

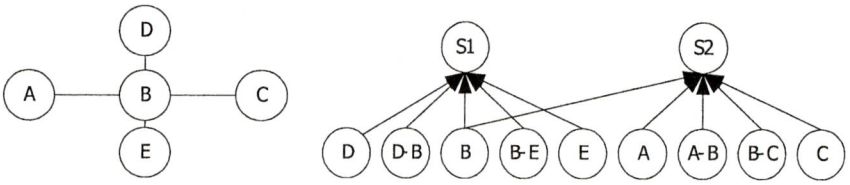

Fig. 1. A simple tree-shaped network **Fig. 2.** An EPM based-on path domain

For tree-shaped and other networks in which there is a unique path between two nodes, the model is simple and high efficient for implement. But the limitation makes its application scope narrow. To overcome this difficulty, we extend this method through introducing "logic node" and "logic link". This extension replaces these "multi-path" parts in the network with logic node, and stores these links between nodes in the "multi-path" with a logic link table related to the logic node. After the multi-path parts replaced by logic node, the EFM is built as follows: First we get the un-extended path domain model; Then look up the corresponding logic link in the logic link table according to the conjoint node, and replace the part related to the logic node in the original path with the corresponding logic link. An example is shown as following figures.

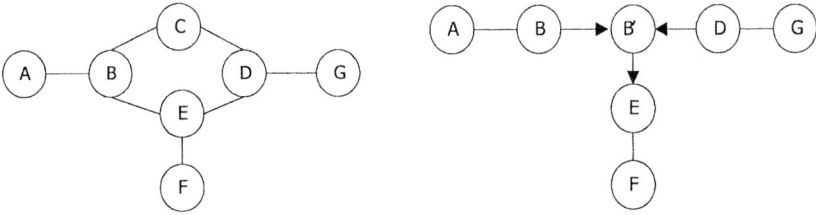

Fig. 3. An example that network exists multiple paths between nodes

Fig. 4. The network after replacing the "multi-path" part by logic node B'

Table 1. Nodes and logic links table of logic node B'

Conjoint nodes	Logic links
B、E	(B-E)'
B、D	(B-D)'
D、E	(D-E)'

In Fig.3, when node A fails to visit node F, the simple path domain model method cannot be taken for solving the problem because the constraint that there is a unique path between network nodes is not fulfilled. After using logic node and link, this requirement has been meet (Fig. 4). The path includes node A, link A-B, node B, link B-B', node B', link B'-E, node E, link E-F, and node F. According to conjoint node B and node E with logic node B', logic link $(B$-$E)'$ is get from the table of logic node B' to replace the part related to logic node B' in the original path. At last the path solved is composed of node A, link A-B, node B, link $(B$-$E)'$, node E, link E-F and node F.

Our EPM can also be used to perform non-deterministic event correlation by assigning the prior fault probability and the conditional probability that faulty NEs lead to the corresponding path fail. The prior fault probability can be computed according to the reliability and runtime of NE. The conditional probability can be gained from the history data of this kind of equipment. Compared with other EPM, the EPM can be established based on the topology configuration and equipment data even without the history fault data of itself.

3 C-F Model Basic

Certainty factor (C-F) theory was proposed by Shortliffle and Buchannen etc. when they developed MYCIN system[12]. Now it has been an important technique for uncertainty reasoning. We introduce C-F model in brief in this section.

(1) Representing Uncertain Rules
In the C-F theory, CF values can be attached to rules to represent the uncertain relationship between the evidence E given in the rule's premise and the hypothesis H given in the rule's conclusion. The basic structure of a rule used in the certainty model is as: IF E THEN H CF(H,E), where: $CF(H,E)$ is the level of belief of H when given E and is called certainty factor. This number provides a range of –1(definitely false) to +1(definitely true). A positive value represents a degree of belief, while a negative value indicates a degree of disbelief.

In this theory, $CF(H,E)$ value is computed from MB and MD values by way of the following equation: $CF(H,E)=MB(H,E)-MD(H,E)$. where: MB called measure of belief, it reflects the measure of increased belief in hypothesis H based on evidence E. MD called measure of disbelief, it reflects the measure of increased disbelief in hypothesis H based on evidence E.

The MB is defined as the following equations.

$$MB(H,E) = \begin{cases} 1 & \text{if } P(H)=1 \\ \dfrac{\max\{P(H|E), P(H)\} - P(H)}{1 - P(H)} & \text{else} \end{cases} \quad (1)$$

The MD is defined as the following equations.

$$MD(H,E) = \begin{cases} 1 & \text{if } P(H)=0 \\ \dfrac{\min\{P(H|E), P(H)\} - P(H)}{-P(H)} & \text{else} \end{cases} \quad (2)$$

CF values can be computed by way of the following equation.

$$CF(H,E) = \begin{cases} MB(H,E) - 0 = \dfrac{P(H|E) - P(H)}{1 - P(H)} & \text{if } P(H|E) > P(H) \\ 0 & \text{if } P(H|E) = P(H) \\ 0 - MD(H,E) = -\dfrac{P(H) - P(H|E)}{P(H)} & f\ P(H|E) < P(H) \end{cases} \quad (3)$$

where: $P(H)$ is prior probability of H, $P(H|E)$ is probability that H is true given evidence E.

(2) Representing Uncertain Evidence
In the C-F theory, a CF value is assigned to the uncertain evidence, which indicates the belief in evidence E. If evidence E is definitely true, the $CF(E)$ value is +1. If evidence E is definitely false, on the contrary, the $CF(E)$ value is –1. And if evidence E is unknown, the $CF(E)$ value is 0. If evidence E is true probably, the $CF(E)$ value is in the range of 0 to +1. Otherwise, the $CF(E)$ value is in the range of -1 to 0.

(3) Certainty for Combining of Multiple Evidences
For conjunctive rules, i.e. the rule like "If $E=E_1$ AND E_2 AND ... E_n", the approach used in the certainty model is as follows:

$$CF(E) = \min\{CF(E_1), CF(E_2), ..., CF(E_n)\} \quad (4)$$

For disjunctive rules, i.e. the rule like *"If $E=E_1$ OR E_2 OR ... E_n"*, the approach taken to determine belief is as follows:

$$CF(E) = \max\{CF(E_1), CF(E_2), ..., CF(E_n)\} \quad (5)$$

(4) Certainty Propagation for Premise Rules

Certainty factor propagation is concerned with establishing the level of belief in a rule's conclusion when the available evidence contained in rule's premise is uncertain. $CF(H)$ value is computed by way of the following equation:

$$CF(H) = CF(H,E) * \max\{0, CF(E)\} \quad (6)$$

(5) Certainty Propagation for Similarly Concluded Rules

For a hypothesis concluded by more than one rule, the various *CF* values are combined using a technique called "incrementally acquired evidence". This technique can be divided into two steps. First step: to compute $CF(H)$ according to each rule. Second step: to compute the combined $CF(H)$ value by way of the following equation:

$$CF_{1,2}(H) = \begin{cases} CF_1(H) + CF_2(H) - CF_1(H)CF_2(H) & \text{both} > 0 \\ \dfrac{CF_1(H) + CF_2(H)}{1 - \min\{|CF_1(H)|, |CF_2(H)|\}} & \text{one} > 0 \\ CF_1(H) + CF_2(H) + CF_1(H)CF_2(H) & \text{both} < 0 \end{cases} \quad (7)$$

4 Event Correlation Using C-F Model

Section 2 has explained how to expand the simple EFM model based on path domain so that it can be applied to the network in which multiple paths exist between nodes. This section introduces a new event correlation algorithm, which uses a refined heuristics approach to create and update possible fault hypothesis, then compares the possibilities of fault hypotheses according to their belief. The algorithm can be briefly stated as follows.

When an event E_i is received, firstly, the algorithm generates the path domain PM_{Ei} of the event according to the event's source and destination, and then adds the NEs of PM_{Ei} into fault NEs set Ψ. Then, the algorithm computes the increased fault belief of NEs to update the fault belief of NEs, and creates a set of fault hypothesis FHS_i in which each hypothesis can explain event $E1,...,Ei$, by updating FHS_{i-1} with a NE in the path domain of event E_i. After dealing with all events, the algorithm chooses the hypothesis h_k with the highest belief value as a result of event correlation. In the following parts, we discuss the event correlation algorithm in detail.

4.1 The Refined Heuristic Approach for Creating the Fault Hypothesis Set

Fault hypothesis set is a set of hypothesis in which each hypothesis is a subset of Ψ that explains all events in E_O. The meaning that hypothesis h_k can explain event $E_i \in E_O$ is hypothesis h_k includes at least one NE which faults can lead to the event E_i occurrence.

At the worst case, there may be $2^{|PM|}$ fault hypothesis can responsible for fault event E_O, so fault hypothesis set don't contain all subsets that can explain event E_O. To limit the number of fault hypothesis, a heuristic approach is presented in literate[8]. But when network become large, number of hypothesis still grows rapidly. So, we improved this heuristic approach by constraining maximum number of faulty NE.

When event E_i is received, this algorithm creates a set of fault hypothesis FHS_i by updating FHS_{i-1} with an explanation of event E_i. The way for appending the explanation of event E_i is that: To analyze each fault hypothesis h_k of FHS_{i-1}, if h_k can explain event E_i, h_k can be appended into fault hypothesis FHS_i, otherwise, h_k need to be extended with a NE in path domain PM_{Ei}. As discussed above, the greedy algorithm will result in fast growth of fault hypothesis set's size then lead to the computational complexity of event correlation algorithm is unacceptable. So literature [8] proposed a heuristics approach which uses a function $u(l)$ to determinate whether NE $l \in PM_{Ei}$ can be added into hypothesis $h_k \in FHS_{i-1}$. NE $l \in PM_{Ei}$ can be appended into $h_k \in FHS_{i-1}$. only if the size of h_k, $|h_k|$, is smaller than $u(l)$, where function $u(l)$ is defined as the minimal size of a hypothesis in FHS_{i-1} that contains NE l and explains event E_i. The usage of this heuristic comes from the following assumption: In most event correlation problems, the probability of multiple simultaneous faults is smaller than the probability of any single fault. Thus, in these hypotheses containing NE l, the fewest size of hypothesis is the one most likely to be the optimal event explanation. The heuristic approach reduces the number of hypothesis during the event correlation to a great extent. However, while the network's size is large, the size of fault hypothesis created by this approach is still very large. Therefore, we refined the heuristic approach with a constraint f_{max}, where f_{max} is defined as the maximum fault NE's number occurred in the network. Since in most of event correlation problems, the number of fault NE is usually small, and the number of fault simultaneous beyond a certain value is also very small. So this limitation is reasonable to make the size of fault hypothesis in an acceptable range even the network's size is large.

4.2 The Influence of Events on Fault Belief of NE

When a fault event E_i is received, the event will produce an increment of fault belief to these NEs in path domain PM_{Ei}. Below, we discuss the impacts of negative event and positive event respectively. In this paper, negative event is defined as the event that reports fault in end-to-end network service; and positive event is defined as the event that indicates end-to-end network service in good condition.

If these prior fault probabilities of NEs in path PM are known, the fault probability of path PM is computed as following equation.

$$P(PM=0) = \sum_{i=0}^{2^n} P(S_1, S_2, ..., S_n) P(PM=0 | S_1, S_2, ..., S_n) \tag{8}$$

It is difficult to get the condition probability in real-life communication network. So the calculation of $P(PM=0)$ is also very difficult through equation (6). However, if there are two or more NEs that in a special path occurs fault simultaneously, the path will fail in most cases. Therefore, we can calculate the approximation of $P(PM=0)$, $P'(PM=0)$, by following equation.

$$P'(PM=0) = 1 - P(\forall S_{NE} = 1) - \sum_{i=0}^{n} P(only\ S_i = 0)P(PM = 1|only\ S_i = 0) \quad (9)$$

The Influence of Negative Events on Fault Belief of NE

When a negative event E_i is received, the fault probability of NE in path PM_{Ei} is computed as following equation.

$$P(S_j = 0 | PM = 0) \approx \frac{P(PM = 0 | only\ S_j = 0)P(only\ S_j = 0)}{P'(PM = 0)} \quad (10)$$

Thus, the increment of fault belief of NE j by event E_i is computed as following equation.

$$CF(S_j = 0, PM_{E_i} = 0) \approx \frac{\frac{P(PM_{E_i} = 0 | only\ S_j = 0)P(only\ S_j = 0)}{P'(PM_{E_i} = 0)} - P(S_j = 0)}{1 - P(S_j = 0)} \quad (11)$$

The Influence of Positive Events on Fault Belief of NE

When a positive event E_i is received, the fault probability of path PM is computed as following equation.

$$P(S_j = 0 | PM_{E_i} = 1) \approx \frac{P(PM_{E_i} = 1 | only\ S_j = 0)P(only\ S_j = 0)}{1 - P'(PM_{E_i} = 0)} \quad (12)$$

Thus, the increment of fault belief of NE j by event E_i is computed as follows.

$$CF(S_j = 0, PM_{E_i} = 1) \approx -\frac{P(S_j = 0) - \frac{P(PM_{E_i} = 1 | only\ S_j = 0)P(only\ S_j = 0)}{1 - P'(PM_{E_i} = 0)}}{P(S_j = 0)} \quad (13)$$

4.3 Fault Belief of NE

The total fault belief of NE j, produced by event $E_1, E_2, \ldots, E_i$, is calculated as following equation.

$$CF_i(S_j) = \begin{cases} CF_{i-1}(S_j) + CF_i(S_j, PM_{E_i}) - CF_{i-1}(S_j)CF_i(S_j, PM_{E_i}) & both > 0 \\ \dfrac{CF_{i-1}(S_j) + CF_i(S_j, PM_{E_i})}{1 - \min\{|CF_{i-1}(S_j)|, |CF_i(S_j, PM_{E_i})|\}} & one > 0 \\ CF_{i-1}(S_j) + CF_i(S_j, PM_{E_i}) + CF_{i-1}(S_j)CF_i(S_j, PM_{E_i}) & both < 0 \end{cases} \quad (14)$$

4.4 Belief of Fault Hypothesis

We compare the probability of hypotheses with their belief, which are defined as the product of all NE's fault belief in these hypotheses. The greater belief of hypothesis is, the more probability hypothesis holds. The belief of hypothesis h_k is computed as following equation.

$$CF_i(h_k) = \begin{cases} \prod_{j=0}^{|h_k|} CF_i(S_j) & \forall\ CF_i(S_j) > 0 \\ -\prod_{j=0}^{|h_k|} |CF_i(S_j)| & \exists\ CF_i(S_j) < 0 \end{cases} \quad (15)$$

The algorithm is defined by the following pseudo-code.

Algorithm 1(C-F Model Algorithm)
```
set f_max, let FHS_0={Φ}, Ψ={Φ}
for every observed events E_i
    compute PM_Ei
    for all NE j ∈PM_Ei do
        compute CF_i(S_j, PM_Ei)
        add j to Ψ
    for all NE k ∈Ψ do
        compute CF_i(S_k)
    let FHS_i={Φ}
    for all l ∈PM_Ei
        let u(l)=f_max
    for all hj ∈FHS_{i-1} do
        for all NE l ∈h_j such that l ∈PM_Ei
            set u(l)=min(u(l),|h_i|)
            add h_i to FHS_i and calculate CF(S_l)
    for all h_i ∈FHS_{i-1}\FHS_i do
        for all j ∈PM_Ei such that u(l)>|h_i| do
            add h_i∪{l} to FHS_i compute CF(h_i∪{l})
choose h_i ∈FHS_{|Eo|} such that CF_{|So|}(h_i) is maximum
```

5 Simulation Study

In this section, we describe the simulation study performed to evaluate the technique presented in this paper. In our simulation, we use OR to represent the negative event observed ratio, i.e., $OR=|E_{io_N}|/|E_{ic_N}|$ and n to represent the number of network nodes. Given parameter of OR and n, we design K_n simulation cases as follows:

First, we create a random tree-shaped n-node network $N_i(1 \leq i \leq K_n)$.

Then, we randomly generate prior fault probability distribution $P(S_i=0)$ of NEs, $P(S_i=0) \rightarrow [0.001\ 0.01]$, and conditional probability distribution $P(PM=0|Si=0)$ of NEs, where $P(PM=0|Si=0)$ defined as the probability of NE's fault leading to corresponding path fail, $P(PM=0|Si=0) \rightarrow [0\ 1]$.

For i-th simulation case($1 \leq i \leq K_n$), we create M_s simulation scenarios as follows.

(1) Using prior fault probability of NEs, we randomly generate the set $F_{iC}^k (1 \leq k \leq M_s)$ of faulty NEs in network N_i.

(2) Using conditional probability distribution of NEs, we generate the set of events E_{iC}^k resulting from faults in F_{iC}^k.

(3) We randomly generate the set of negative event $E_{iO_N}^k$ such that on average $E_{iO_N}^k / E_{iC_N}^k = OR$.

(4) We randomly generate the set of positive events $E_{iO_P}^k$. The size of $E_{iO_P}^k$ is varied 2 to 8 depending on the size of N_j.

(5) We set the set of events $E_{iO}^k = E_{iO_N}^k + E_{iO_P}^k$, E_{iO}^k is the set of events received by fault management.

(6) Using the event correlation algorithm proposed above, we compute F_{iD}^k, the most likely explanation of events in E_{iO}^k. Detection rate (DR_i^k) and false positive rate (FPR_i^k) are computed as the following equations.

$$DR_i^k = \left|F_{iD}^k \cap F_{iC}^k\right|/\left|F_{iC}^k\right| \quad FPR_i^k = \left|F_{iD}^k \setminus F_{iC}^k\right|/\left|F_{iD}^k\right| \qquad (16)$$

For i-th simulation case, we calculate the mean detection rate $DR_i = \sum_{k=1}^{M_s} DR_i^k/M_s$ and mean false positive rate $FPR_i = \sum_{k=1}^{M_s} FPR_i^k/M_s$. Then, we calculate the expected values of detection rate $DR(n)$ and false positive rate $FPR(n)$, respectively. In our simulation, we used $K_n=100$, $M_s=100$. We varied n from 40 to 120. The result of experiment is shown as follows.

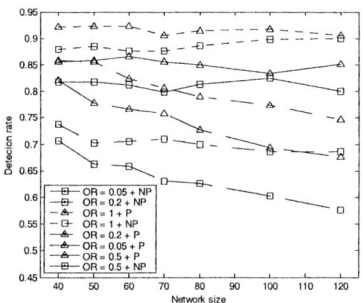 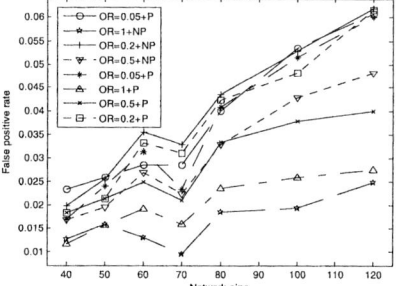

Fig. 5. Detection rates   **Fig. 6.** False positive rates

As shown in Fig.5 and Fig.6, the accuracy of algorithm depends on event observability ratio and network size. The higher *OR* the event correlation process is with, the higher accuracy can get; the larger network size is, the lower accuracy can get. The reason is that: firstly, high *OR* means more information about faults, and, in consequence, higher accuracy; secondly, large network means more multiple faults will occur in system, and these fault can be explain some hypotheses with smaller size. Besides, the inclusion of positive events in the event correlation allows the detection to be substantially improved. The improvement is bigger for low observability ratio; with high observability ratio (e.g., *OR=1*), the number of negative events is typically large enough to allow quite accurate fault localization without considering positive events. However, such high events observability is unlikely in real-life systems. So the inclusion of positive events in the event correlation process is an effective method to improve the accuracy of event correlation. As mentioned in section 1, several approaches have been proposed in non-deterministic fault diagnosis. Because of the difference of approaches for building EPM, it is difficult to select an event correlation algorithm as a baseline. In our simulation, we choose the IHU as the baseline. Compared with IHU, the detection rate of our algorithm (taking positive events into account) is higher than that of IHU (disregarding positive events), while lower than IHU (taking positive events into account). There is no statistically significant difference in the false positive rate. The algorithm IHU (taking positive events into account) assumes that the positive events certainty is in a high level. However, the assumption is not always true in the case of high event loss ratio. In our algorithm, we only consider that positive events which are confirmed, so it is more reliable. The high event loss ratio was resolved by adding the parameter of event loss ratio in IHU, but it is difficult to obtain the parameter in real network. Especially when there is difference between estimation and real value, the inclusion of positive event will not increase the accuracy of event correlation; on the contrary, it will decrease the accuracy. Therefore, we consider that IHU (taking positive events into account) is more suitable as an ideal algorithm. Compared with it, our algorithm is more robust and more practical.

6 Conclusion

This paper proposed a non-determinate event correlation technique to perform fault localization in communication network. The technique builds system's EFM during run-time by extending the EFM model method proposed in [10], and creates the set of fault hypothesis with a refined heuristics approach, and chooses the optimum fault hypothesis by C-F model. In diagnosis of end-to-end service failure in communication network, our method can build EFM during run time, and perform event correlation for multiple simultaneous faults. Simulation study shows our approach can get high accuracy, and is robust for events loss. However, the accuracy of our algorithm in very large network will deteriorate gradually. In future research, we would study event correlation techniques for large network.

References

1. A. T. Bouloutas, S. Calo, and A.Finkel, Alarm Correlation and Fault Identification in Communication Networks, IEEE Trans. on Communications, vol. 42, (1994) 523-533.
2. A. T. Bouloutas, G. W. Hart, and M. Schwartz, Fault Identification Using a Finite State Machine Model with Unreliable Partially Observed Data Sequences, IEEE Trans. on Communications, vol. 41, (1993) 1074-083.
3. Hasan, B. Sugla, and R. Viswanathan, A Conceptual Framework for Network Management Event Correlation and Filtering Systems, in IM1999, Boston, MA USA (1999).
4. Hong and P. Sen, Incorporating Non-deterministic Reasoning in Managing Heterogeneous Network Faults, Integrated Network Mangement II, (1991) 481-492.
5. I. Katzela and M. Schwartz, Schemes for Fault Identification in Communication Networks, IEEE/ACM Trans. on Networking, vol. 3, (1995) 753-764.
6. M. Steinder and A. S. Sethi, Non-deterministic Diagnosis of End-to-end Service Failures in a Multi-layer Communication System, in ICCCN, Scottsdate, AZ (2001).
7. M. Steinder and A. S. Sethi, Increasing Robustness of Fault Localization through Analysis of Lost, Spurious, and Positive Symptoms, in INFOCOM 2002, New York (2002)

8. M. Steinderand A. S. Sethi, Probabilistic Fault Diagnosis in Communication Systems through Incremental Hypothesis Updating, Computer Networks, vol. 45, (2004) 537-562.
9. S. A. Yemini, S. Kliger, E. Mozes, Y. Yemini, and D. Ohsie, High Speed and Robust Event Correlation, IEEE Communications Magazine, vol. 34, (1996) 82-90.
10. S. Yu-bei, Research on Event Correlation and IP Network Fault Simulation, Model, vol. PHD: Wuhan University (2001).
11. M. Steinder and A. S. Sethi, End-to-End Service Failure Diagnosis Using Belief Networks, in NOMS2002, Florence, Italy (2002).
12. E. H. Shortliffe, Computer-based medical consultation: MYCIN. New York: American Elsevier (1976).

Flexible Goal Recognition via Graph Construction and Analysis

Minghao Yin[1,2], Wenxiang Gu[1], and Yinghua Lu[1]

[1] College of Computer Science, Northeast Normal University,
Postcode 130024,Changchun, China
[2] College of Computer Science, Jilin University, Postcode 130012, Changchun, China
{mhyin, gwx, lyh}@nenu.edu.cn

Abstract. Instead of using a plan library, the recognizer introduced in this paper uses a compact structure called flexible to represent goals, actions and states of the world. This method doesn't suffer the problem of acquisition and hand-coding a larger plan library as traditional methods do. The recognizer also extends classical methods in two directions. First, using flexible goals and actions via fuzzy sets, the recognizer can recognize goals even when the agent has not enough domain knowledge. Second, the recognizer offers a method for assessment of various plan hypothesis and eventual selection good ones. Since the recognizer is domain independent the method can be adapted in almost every domain. Empirical and theoretical results also show the method is efficiency and scalability.

1 Introduction

Plan recognition involves inferring an agent's goal from a set of observed actions and organizing the actions into a plan structure for the goal [1]. Wherever a system is expected to produce a kind of cooperative or competitive behavior, plan recognition is a crucial component. For example, using plan recognition mechanism, an intelligent user agent can observe a user's actions, jumping in when his operation is on a sub-optimal way. However, inferring an agent's goal without different semantics is rather a difficult task because the observed actions are always fractional and the same action may appear in some different plans. On the other hand, algorithms using formalized description developed recently, though sound in semantics, often produce combination-exploded problems [2], [3].

In order to solve such problems, researchers have made many attempts. Vilain advanced a grammatical analysis paradigm that is super in searching speed but poor in plan representation [4].

Kautz in his pioneering paper presented a hierarchical event based framework, which has been widely used in most plan recognition based systems for its advantage in plan representation [5], [6]. Jiang etc. used AND/OR graph like structure to represent plan based on this framework [7]. And Lesh etc. in [2] improved speed over this method. But using a plan library often makes the system suffer problems in the acquisition and hand coding of large plan libraries that is unimaginable in a huge plan system [1], [8].

Bauer etc. made attempts to apply machine learning mechanics to automated acquisition and coding of plan libraries, but searching in the plan space is still exponential in the number of actions [9], [10]. What's more, most recognizers using a plan library can only solve problems with fewer than 100 plans and goals [2].

Since using a plan library often makes systems suffer problems in acquisition and hand coding of large plan libraries; further, it often leads to searching the plan space of exponential size. Hong Jun in our project advanced an algorithm [1], [8] based on graph construction and analysis. This kind of algorithm doesn't need a plan library to make goal recognition applicable in practical systems [11], but it is based on an ideal assume that the "actor" should have complete knowledge of the planning environment. This is always not accord with the facts and is always limited in application. Allen etc. in [12] advanced a statistical, corpus based method for goal recognition without plan library, but this method is domain dependent, and can't work until sufficient statistic information is available.

In this paper, we introduce a totally different compact structure called flexible goal graph for plan recognition. Compared to goal graph, first, this method does not need a plan library, and thus it inherits the advantage of goal graph such as avoiding problems of hand coding a plan library; second, the behaviors of the agent observed can be more flexible and adventurous, which is often a feature of agents in a state of partially or totally ignorance about the environment; third, our method provides a qualitative criterion to assess plan quality, thus better plan can be selected while "bad" ones can be excluded.

The rest of the paper is organized as followed. We first describe syntactic and semantics of flexible goal recognition problems; and then we introduce how to use flexible goal graph to implement flexible goal recognition; theoretical and empirical results will also be shown to prove our algorithm efficient and sound, and in the last section we conclude the paper and point out our future work.

2 Flexible Goal Recognition

Specifically, in this paper, we focus on goal recognition, a special case of plan recognition, as goal graph does. And the task of this section is to define a flexible goal recognition problem, the framework that can capture the inherent "softness".

2.1 Why to Introduce Flexible Goal Recognition?

There are at least two reasons for us to focus on flexible goal recognition. First, classic goal recognition based systems, goal graph for example, assume that the actor (the agent observed) should have complete knowledge about the state of the world thus the agent always performs perfectly during execution of the actions. This is not necessarily the case. A good case may make things clear. Taking into account of Andrew's famous UM-Translog example [13], the instantiate of a LOAD operator requires that all of the preconditions to be implemented. In other words, a load action can be executed only when a) the truck and package are collocated; b) the truck is well armored; c) there is a guard on the truck before loading. Obviously, a) is an imperative constraint, yet b) and c) are both flexible constraints that can be relaxed. In the real world, a robot that does not know that the package is valuable may load the package when the truck is only an unarmored truck and (or) even when there is no guard on the truck. Allowing this kind of flexibility is particularly useful in those human-machine collaboration systems. For instance, a user may close a document without noticing that the document has been changed. Since classic goal recognition system unrealistically assumes that the agent is both rational enough and has total knowledge about the environment, the user is assumed to never close an unsaved document. Flexible goal recognition then offers a method to recognize the goals of those agents who are in a state of ignorance about the planning environment.

Second, actions are viewed identically contributing to construct a plan in most plan recognition algorithms. However, in real world domains, each action play different role in the plan and behaviors of the agent may even cause the damage of plan quality. For example, moving a valuable package from location l_1 to l_2 through a dangerous road decreases the successful degree of implementing the goal of keeping the package in l_2. Thus the recognizer needs some qualitative criterion to assess the plans recognized, and then makes rational decision to choose "good" ones and exclude those "bad" ones. This is particularly useful in human machine collaboration systems, because using this mechanism the recognizer can judge whether the user's actions are on the optimal way. What's more, our recognizer can even find which action mainly harms the plan quality. In this sense, advice can be given for the agent to plan better next time.

2.2 Representation of Flexible Goal Recognition

Now we discuss the both the syntactic and semantics of the flexible goal recognition problems.

Problem of flexible goal recognition. Generally speaking, the problem of flexible plan recognition consists of:

1. A set of flexible operator that can be instantiated to actions,
2. A finite, dynamic universe of typed objects,
3. A set of flexible propositions called the initial conditions,
4. A set of flexible goal schemata specifying flexible goals,
5. A set of observed flexible actions,
6. An explicit notion of discrete time step.

States and Flexible Propositions. A state is a complete description of the world at a single point in time, and it is described by a set of propositions in which every proposition appears only once. In flexible goal recognition, a Flexible state S is then composed of a set of flexible propositions, of the form $(p, \Phi_1, \Phi_2, \Phi_3, ..., k_i)$, where Φ_i denotes a plan object and k_i is an element of totally ordered set, K, which denotes the subjective degree of truth of the proposition, p. K is composed of a finite number of membership degrees, $k_\perp, k_1, ..., k$. $(p, k_\perp)$ and (p, k) respectively denote that the proposition p is totally true and totally false. So a flexible proposition can be viewed as a fuzzy relation R, whose membership function is $f_R(.): \Phi_1 \times \Phi_2 \times \Phi_3 \times ... \times \Phi_j \to K$.

Flexible Actions. Specifically, a flexible action can be regarded as representing sets of state transitions, with a state being a particular assignment to the set of state variables.

The action consists of a flexible precondition, characterizing the state where the action is applicable in, and a set of flexible effects. Flexible actions can also be viewed as a fuzzy relationship mapping from the precondition space to a particular set of flexible effects and a totally ordered satisfaction scale L. L is also composed of a finite number of membership degrees, $1_\perp, 1_1..., 1$, where $1_\perp, 1$ respectively denote complete unsatisfied and complete satisfied. In the following we show the BNF grammar for definition of a flexible operator that can be instantiated into flexible actions.

BNF grammar of actions:

```
<action-def> ::= (: action <action symbol>)
<action-def> ::= (: action <action symbol>
                [:parameters (<typed list (variable)>)]
                <action-def body>)
<action symbol>::= <name>
<action-def body> ::= [:precondition <GD>]
                     [:effect <effect>]
<GD> ::= <atomic formula (term)>|(and <GD>*
         |(not <term>)|(not <GD>)|(or <GD>*)
<atomic formula (x)> ::= <predicate> <x>*
                        |<predicate>
<term> ::= <name>|<variable>|<const>
<effect> ::= <flex-effect>|and <effect>*)
<flex-effect> ::= <f-effect><sat-degree>
<f-effect> ::= <atomic formula (term)>
              |(not <atomic formula (term)>)
<predicate> ::= <name>
<sat-degree>    ::= <element of L>.
```

Flexible Goals. Flexible goal schemata is a fuzzy relationship from the descriptions of the goal to the satisfaction scale set L. Descriptions of the goal schemata are defined the same as the preconditions in flexible operators.

3 Flexible Goal Graph

In this section, we mainly discuss how to implement our recognizer. But before that, we introduce some useful concepts.

3.1 Useful Concepts

Definition 1 (Flexible Casual Link). Let a_i and a_j be two flexible actions at time steps i and j respectively, where i <j. There exists a causal link between a_i and a_j, written as $a_i < a_j$, if and only if one of the effects of a_i has the form (p, l), and the execution of the a_j requires p.

Definition 2. (Flexible valid plan) Beliefs concerning how to reach a flexible goal from the initial state Δ are formalized as a triple <A, Б, C>, which formalizes a valid plan, where Б denotes the set of flexible casual link relations in A, the set of flexible actions, and C denotes a set of constraints. A flexible plan is valid if and only if the goal is achieved after the actions are executed from Δ in an order consistent with C over a given satisfaction degree, e.g. l_3.

Definition 3. (Flexible Relevant Action) Given a flexible goal g and a set of observed flexible actions, A, we said an action a, a∈ A, is relevant to the flexible goal if and only if there exists a flexible causal link between a and g or there exists a flexible causal link between a and b, where b is flexible relevant to g.

Definition 4. (Flexible Consistent Goal) A flexible goal is consistent with a set of flexible actions, if and only if the strict majority of actions in the set are flexible relevant to the goal.

Definition 5. (Flexible Valid Plan for Consistent Goal) Let A a set of observed actions, Δ be the initial state, g be a flexible goal, p = <A, Б, C> is a flexible valid plan; we said p is a flexible valid plan for g if and only if g is achieved after A over a given satisfaction degree.

3.2 Algorithm

Generally speaking, our method can be regarded as a counter part of Miguel's flexible planning graph [18], which has been proved to have great success in planning fields. However, although graph structures are used in both approaches, they are composed of different kind of nodes and edges. In a time step, a flexible planning graph represents all the possible propositions either added by actions or brought forward by no-ops in previous time step and represents all the possible actions whose preconditions can be satisfied. Yet in our approach flexible goal graph not only represents the propositions added by plausible actions, but also represents all the plausible actions and the goals partially or fully achieved. So flexible goal graph not only represents relations between flexible propositions and flexible actions, but also represents relations between flexible propositions and flexible goals.

We now describe our goal recognition algorithm that runs in a two-stage cycle at each time step. In the first stage, the "constructor" procedure takes the observed actions to extend a flexible goal graph; in the second procedure the "analyzer" procedure analyzes the constructed flexible goal graph to recognize the goals achieved and then find the flexible valid plan consistent with the goals. We use a 5-tuple <Γ,A,Б,Ω,C> to represent the graph, where Γ denotes the set of current world states which includes initial states, Δ∈ Γ; A is a set of action nodes, Б is a set of edges, Ω is a set of goal nodes, C is the constraint relationships implied in the graph. Flexible proposition node, flexible action node and flexible goal node are respectively represented as prop(p, i), action(a, i), goal(g, i), where p is a flexible proposition, a is a flexible action, g is a recognized flexible goal, and i is time step. And flexible goal graph is a leveled directed graph including flexible proposition levels with flexible proposition nodes, flexible actions levels with flexible action nodes and flexible goal levels with flexible goal nodes.

Constructor Procedure. The constructor procedure shows how to generate a flexible goal graph and how the actions and goals nodes and edges are generated dynamically. Flexible goal graph constructor starts with a graph <Δ, { }, { }, { }, { }> that consists of only flexible proposition level 1 with nodes representing the initial conditions. Let operatori() be a function which takes the ith operator of the problem and the predict instantiation() to instantiate an operator into an action during running time. Similarly, let goali() be a function which takes the ith goal of the problem, and implement() be the predict implementing a goal. Then the result of the progression of Θ from proposition level to action level through the problem is {∪instantiate(operatori(),Θ, AO)| $1 \leq i \leq$ a_leng(P)}. After the action level is constructed, by calling goal expansion procedure, the goals recognized should be {∪(implement (goali(),R)),$1 \leq i \leq$ g_leng (P)}}. Action expansion and goal expansion procedure explains how the graph is constructed level by level more explicitly.

Constructor procedure of the recognizer

```
Constructor(<Γ,A,Β,Ω,C>,P,n,A₀)
  begin
    if n < 1 or n > N
      then return false
    else  Θ = proposition_level (Γ, n-1)
    A<n> = instantiate(A₀,Θ,Operatorᵢ()|1 ≤ i≤ a_leng(P))
    C<n> = constraint_check(A<n>)
    A<n> = action_check(C<n>, A<n>)
    Δ = initial_state(Γ)
    Ω<n>= {∪implement(goalᵢ(),Γ)|1≤ i ≤ g_leng(P)}
    Ω<n> = goal_check(C<n>,Ω<n>)
    for each g∈Ω<n>
      plan = ANALYSER(Δ,g ,<Γ,A,Β,Ω,C>, n)
      if (!plan)
        then  Ω<n> =Ω<n> - {g}
        else return plan
    Constructor(<Γ,A,Β,Ω,C>, n+1)
  end.
ACTION_EXPANSION(<Γ,A,Β,Ω,C>, n)
  begin
    for each a∈A<n>
      for each pp∈prec(a)
        if prop(pp)∈Γ<n> then
          Β<n> =Β<n>∪ {pos_prec_edge(prop(pp,n),
action(a,n))}}
      for each pe∈effec(a)
        Γ<n+1> = prop(pe, n+1)∪Γ<n+1>
        Β<n> =Β<n>∪{effec_edge(a, prop(pe, n+1))}
      for each prop(p,n)∈ Γ<n>
        if prop(p,n+1)∉Γ<n+1> then
          Γ<n+1> =Γ<n+1>∪prop(p,n+1))
          Β<n> =Β<n>∪
            {pers_edge(prop(p,n), prop(p,n+1))}
    return <Γ,A,Β,Ω,C>
  end.
```

```
GOAL_EXPANSION(<Γ,A,Б,Ω,C>, n)
  begin
    for each g∈Ω<n>
      for each pg∈des(g)
        if prop(pg,n)∈P then
          Б<n> =Б<n>∪{des_edge(prop(pg,n))}
    return <Γ,A,Б,Ω,C>
  end.
```

Analyzer Procedure. Given a flexible goal graph constructed, the analyzer then analyses the graph to recognize consistent goals and valid plans. Analyzer procedure include 4 arguments, Δ –the initial states set, g -the goal recognized, <Γ,A,Б,Ω,C>-the graph, n-the current total time steps. Interestingly, as is shown in the pseudo-code of construction procedure, two kinds of information are propagated through the graph. The first kind propagated by causal links between actions tells whether an action is relevant to the goal; in other words, the recognizer knows whether an action contributes a valid plan consistent with the goal recognized. The second kind of information propagated by satisfaction degree of actions tells to what extent an action contributes to the plan. Specifically the satisfaction degree is propagated though the graph as followed: (1) a proposition is labeled with maximum satisfaction degree of those actions that assert it as an effect; (2) an action is labeled as minimum satisfaction degree of those propositions that attached it as a precondition and its own satisfaction degree. In this way, we can compute satisfaction degree of a flexible valid plan consistent with a given goal recognized, which can be viewed as assessment of the plan quality. Since the satisfaction degree of a plan is computed as the conjunctive combination of the satisfaction degrees of actions, it's easy to find which action may harm the plan quality.

Analyzer procedure of the recognizer

```
ANALYZER(Δ, g, <Γ,A,Б,Ω,C>, n)
  begin
    if n<1
      then return false
    S = des(g)
    if (S⊆Δ) then return plan;
    Λ = ∪{a|∃p, p∈effect(a), p∈S,a∈A<n>}}
    S_previous = {∪regress(S,a) | a∈Λ}
    plan = ANALYSER(Δ, S_previous, <Γ,A,Б,Ω,C>, n-1)
    sat-degree = satisfaction(plan)
    if satisfied (sat-degree, plan)
      then return plan
    else return false
  end.
```

Empirical and Theoretical Results

We now introduce several experimental domains and show the results of running our algorithm on them. We test our algorithm on an IBM RS6000 type machine. The purpose of these experiments is to examine the speed of the proposed recognizers.

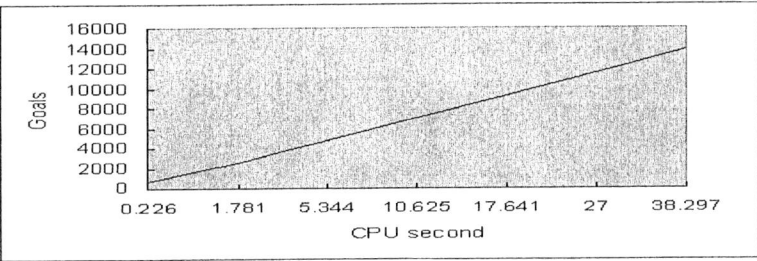

Fig. 1. Empirical results of flexible block domains

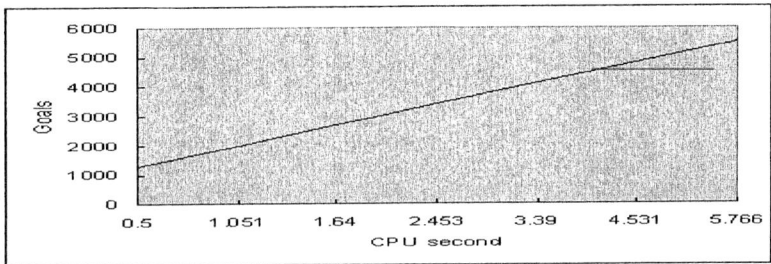

Fig. 2. Empirical results of flexible briefcase domains

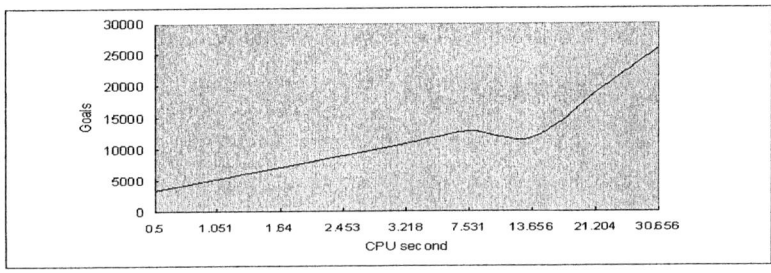

Fig. 3. Empirical results of flexible transportation domains

We first extend the famous blocks domain by introducing a flexible precondition of load operator: flat (?blocktop), which means to what extent the top of the block is flat. We also introduce several flexible goals and operators by allowing the robot to move a block to the other without checking the block's top. It's interesting in this domain because many goals recognized currently have been already recognized previously. Results of this domain are shown in figure 1. Flexible briefcase domain is similar to the flexible transport domain, which is first introduced in [18]. A briefcase with a check should be moved with a guard to protect it. If not, satisfaction degree of the move action should be decreased. Figure 2 and figure 3 shows results of the two domains. As is shown in the results, CPU second in these domains taken to process the problems is approximately linear in the numbers of goals. Indeed, we can prove that our algorithm is polynomial-space and polynomial-time.

Theorem 1 (Polynomial Size). Consider a t-level flexible goal graph, the space of the graph is polynomial size.

Proof. Consider a problem with n objects, p propositions in the initial conditions, m operators and each operator has no more than k parameters and q effects, r goals and each goal has no more than o description. Simply the most number of propositions created by an operators is $O(qn^k)$, so the largest number of an propositions level is $O(tmqn^k)$, and the largest number of an action level is $O(mn^k)$. Since a goal can be regarded as an action having no effects, so the largest number is $O(rn^o)$. Since k and o are both constant numbers, the size of the graph is polynomial size in k, m, n, o, q and t. According to theorem 1 we have theorem 2:

Theorem 2 (Polynomial Time). Consider a t-level flexible, the constructing time and analyzing time are polynomial size.

Proof. It's clear that the time needed to create both the nodes and edges in each level is polynomial in the number of nodes and edges in the level. So the time of constructor procedure is polynomial size. Note that only actions observed branch in the graph. Let l_1 be the goals recognized in goal level t, m-the largest number of a goal's description, l_2- -the numbers of actions observed, k-the largest number of an action's precondition. Then for each goal in goal level t, the maximum number of paths searched for relevant actions to the goal are $O(m+l_2kt)$, so the time needed to recognize all the consistent goals is $O(l_1(m+l_2kt))$. Obviously computing the satisfaction degree of a valid plan is also polynomial in the size of actions in the plan. So time to constructing and analyzing flexible goal graph is polynomial size.

4 Discussion

Although the results shown in this work is encouraging, there are still significant challenges for scaling up the system to sufficiently complex domains.

First, although our recognizer is domain independent, this doesn't guarantee that our goal recognizer will be effective in every domain. But we believe our recognizer can be more effective by using domain knowledge. Second, although our recognizer handles actions and goal abstraction, it does not handle hierarchical plans. In this sense we should enhance the representing ability of our method.Third, Our method makes the assumption that the initial state of the world is completely known and the effects of the actions are deterministic, so next time we can extend this recognizer for probabilistic domains.Last, our recognizer can be viewed as a counterpart of [18], which has been proved to have great success in planning fields. So the advance made about [11] can also be used in our recognizer.

Acknowledgements

This work is fully supported by National Natural Science Foundation of China (Grant No.60273080, 60473003, 60473042); Ministry of Education of China (Grant No.

02090), Natural Science Foundation of Northeast Normal University (Grant No. 20041001).

References

1. Hong, J.: Goal Recognition through Goal Graph Analysis, Journal of Artificial Intelligence Research, 15 (2001) 1-30.
2. Lesh N.: Scalable and adaptive goal recognition, [Ph. D. Thesis], University of Washington (1998).
3. Lesh N., etc.: Scaling up goal recognition, proceedings of the 5th Int. Conf. on Principles of Knowledge Representation and Reasoning (1996), 178-189.
4. Vilain, M: Getting serious about parsing plans: a grammatical analysis of plan recognition, proceedings of the National Conference on Artificial Intelligence, Boston (1990) 190-197, AAAI Press.
5. Kautz, H.: A formal theory of plan recognition [Ph. D. Thesis], Rochester: University of Rochester (1987).
6. Kautz, H. and Allen, J. Generalized plan recognition, Proceedings of National Conference on Artificial Intelligence, AAAI press (1986).
7. Jiang, Y. H., Ma N.: A plan recognition algorithm based on plan knowledge graph, Journal of Software, 13 (2001) 686-692.
8. Hong, J.: Graph construction and analysis as a paradigm for plan recognition, proceedings of National Conference on Artificial Intelligence (2000) 774-779, AAAI Press.
9. Bauer, M., "Acquisition of abstract plan descriptions for plan recognition", proceedings of National Conference on Artificial Intelligence, (1998) 936–941, AAAI press.
10. Albrecht, D. W, etc., Bayesian model for keyhole plan recognition in an adventure game, User Modeling and User-Adapted Interaction, 8(1-2), (1998) 5-47.
11. Yin, M. H.: Incorporating goal recognition into human-machine collaboration, proceedings of ICMLC (2004).
12. Blaylock, N. and Allen, J.: Corpus-based, Statistical Goal Recognition, proceedings of Int. joint conf. of Artificial Intelligence (2003).
13. Andrew, S. etc.: UM translog: A planning domain for the development and benchmarking of planning systems, Tech. Rpt., Univ. of Maryland (1995).
14. Wilensky, R. etc.: The Berkeley UNIX consultant project, Computational Linguistics 14(4), (1988) 35-84.
15. Avirm L.Blum, Merrick L. Furst: Fast planning though planning graph analysis, Artificial Intelligent 90(1997) 281-300.
16. Miguel I., Shen Q.: Solution techniques for constraint satisfaction problems: Advanced approaches, Artificial Intelligence Review 15 (2001) 269-293.
17. Miguel I., Shen Q.: Dynamic flexible constraint satisfaction, Applied Intelligence 13(3) (2000) 231-245.
18. Miguel I., Shen Q.: Fuzzy rrDFCSP and planning, Artificial Intelligence 148 (2003) 11-52.
19. Pednault, E. P. D.: Synthesizing plans that contain actions with context-dependent effects, Computational Intelligence, 14(4), (1988) 483-509.

An Implementation for Mapping SBML to BioSPI*

Zhupeng Dong[1], Xiaoju Dong[1], Xian Xu[1], Yuxi Fu[1],
Zhizhou Zhang[2], and Lin He[2]

[1] BASICS, Department of Computer Science and Engineering,
Shanghai Jiao Tong University, Shanghai 200030, China
{dongzp, xjdong, xuxian, yxfu}@sjtu.edu.cn
[2] BDCC, College of Life Science and Biotechnology,
Shanghai Jiao Tong University, Shanghai 200030, China
{zhangzz, helin}@sjtu.edu.cn

Abstract. The Systems Biology Markup Language(SBML) is an XML-based format for representing models of Systems Biology. BioSPI is a formal model to simulate biological systems, which is evolved from process calculi. Based on the previous research on modeling Systems Biology using process algebra, we propose a method to map SBML to BioSPI. The motivation of the work is to make full use of BioSPI to analyze biological systems described by SBML. In this paper, the mapping rules are presented and an example is given to show the simulation results.

1 Introduction

The number of complete genome sequences, together with data arising from post-genomic investigation, are so huge that the biologists are facing a data explosion problem. How do we integrate all the data to explain the interaction between the components of any biological system and understand the system as a whole? Systems Biology is being developed to solve the problem.

There exist many computational models and tools for studying Systems Biology. However all current models and tools have their particular focus of interest. To understand a biological system, several models or tools with different formats might be used. The idea is to develop some tools to do model conversion automatically.

The Systems Biology Markup Language (SBML) [1], based on eXtensible Markup Language, is a kind of language for representing models of biological systems. It is applicable to metabolic networks, cell-signaling pathways, regulatory networks, and many others. SBML has strong description capability, but weak in analysis.

A recent developed field of research on Systems Biology is to describe and analyze biological systems using formal models studied in computer science, such

---

* The work is supported by The National Distinguished Young Scientist Fund of NNSFC (60225012), BDCC (03DZ14025), The National Nature Science Foundation of China (60473006), MSRA, and The BoShiDian Research Fund (20010248033).

as Process Calculi and Petri Net [2]. BioSPI [3] abstracts and models biological processes and pathways with the pi calculus and the ambient calculus. It has advantages in formally representing the complex networks, simulating and monitoring the behavior and formally verifying their properties and compare networks across organisms. BioSPI is good at simulation and analysis, but its description is hard to understand.

Since the software of BioSPI has been used widely and become more and more mature, we give a method to implement the convertion from SBML to BioSPI. One can use the automata conversion tool to simulate and analyze the data stored in the format of SBML.

2 Mapping SBML to BioSPI

In this section we present the main mapping rules and the implementation. An example shows the result of the conversion.

2.1 The Mapping Rules

We could use the following rules to map a SBML file to a BioSPI program:

- A whole SBML Biology model is defined as a system process.

$$system \Leftarrow \langle sbml \rangle \langle model \rangle \cdots \langle /model \rangle \langle /sbml \rangle \qquad (M1)$$

- Different compartments correspond to different processes.

$$\begin{aligned}
system ::= &\ (new(\,)\,C1)\,|\,(new(\,)\,C2)\,|\,\cdots \\
\Leftarrow &\ \langle listOfCompartments \rangle \\
&\ \langle compartment\ id\ =\ \text{``}C1\text{''}\ \cdots /\rangle \\
&\ \langle compartment\ id\ =\ \text{``}C2\text{''}\ \cdots /\rangle \qquad (M2) \\
&\ \vdots \\
&\ \langle /listOfCompartments \rangle
\end{aligned}$$

- A type of species is expressed by the same name, as shown in $M3$. The $EnvC1$ process here is very useful when reactants are not enough to formally describe a reaction, as shown in $M4$. The name of a reaction is used as a channel name for the BioSPI program in $M4$. Communications just happen on channels.

$$\begin{aligned}
C1 ::= &\ \{EnvC1\ \|\ S1\ \|\ S2\ \|\ \cdots\} \\
\Leftarrow &\ \langle listOfSpecies \rangle \\
&\ \langle specie\ name\ =\ \text{``}S1\text{''}\ compartment\ =\ \text{``}C1\text{''}\ \cdots /\rangle \\
&\ \langle specie\ name\ =\ \text{``}S2\text{''}\ compartment\ =\ \text{``}C1\text{''}\ \cdots /\rangle \qquad (M3) \\
&\ \vdots \\
&\ \langle /listOfSpecies \rangle
\end{aligned}$$

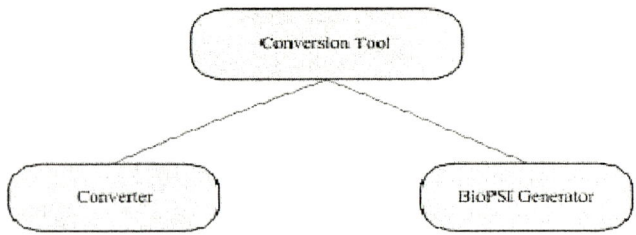

Fig. 1. The Architecture of the Implementation

$$\{EnvC1 \parallel S1\} ::= \overline{R_E}\langle\rangle EnvC1 \mid R_E(\,)S1''$$
$$\{S1 \parallel S2\} ::= R1(\,)S1' \mid \overline{R1}\langle\rangle S2'$$
$$\Leftarrow \langle listOfReactions \rangle$$
$$\langle reation\ name\ =\ ``R1"\cdots/\rangle$$
$$\langle listOfReactants \cdots/\rangle\langle listOfProducts \cdots/\rangle \quad (M4)$$
$$\langle listOfModifiers \cdots/\rangle \cdots$$
$$\langle /reation \rangle$$
$$\vdots$$
$$\langle /listOfReactions \rangle$$

Hence, the whole BioSPI program can be generated from the above rules:
-language(psifcp).
export(System).
global(R1(base rate1), R2(base rate2), $\cdots$, dummy(infinite)).
System ::= C1 | $\cdots$.
C1::=⟨⟨ Create-S1(initial amout1) | Create-S2(initial amount2) | EnvC1 | $\cdots$.
        Create-S1(C) ::= { C <= 0}, true; { C > 0}, { C − −} | S1 | self.
        Create-S2(C) ::= { C <= 0}, true; { C > 0}, { C − −} | S2 | self⟩⟩.
S1 ::= R1?[], S1'; R2?[], S1''.   S1' ::= dummy![],0.   S1'' ::= dummy![],0.
S2 ::= R1![], S2'.   S2' ::= dummy![],0.
EnvC1 ::= R2![], EnvC1.

2.2 The Implementation

The conversion tool is composed of two major modules: the Converter module and the BioSPI Generator, as shown in Figure 1.

The conversion procedure starts from the Converter module. According to the above mapping rules, the Converter module preprocesses the source SBML file. Then the BioSPI module generates different BioSPI code fragments and constitutes the whole BioSPI file.

Figure 2 shows the "MAPK cascade with negative feedback" chart generated from a SBML file. The conversion tool automatically maps it to a BioSPI program. The simulation result of the system is presented.

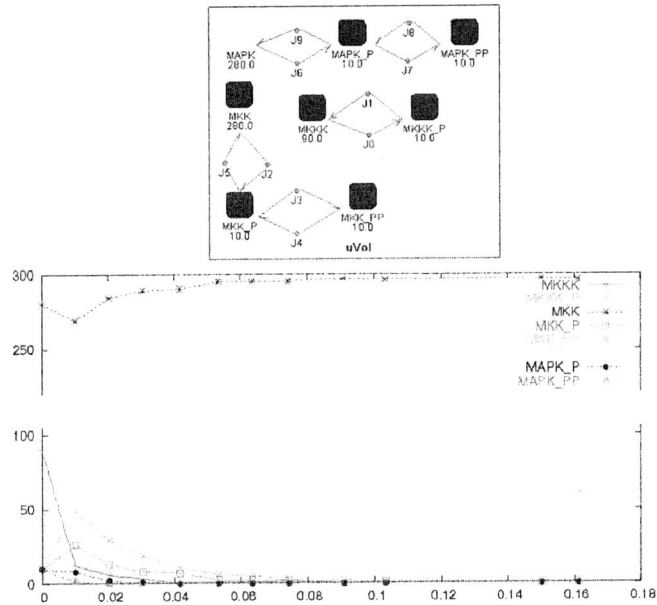

Fig. 2. MAPK cascade with negative feedback and the simulation result

3 Remarks

The two models, SBML and BioSPI, come from different research point of the same biological systems. SBML is targeted for biology, while BioSPI for Computer Science. Our conversion tool makes it possible that BioSPI can simulate and analyze biological systems described by SBML.

Compared with the biological systems modeled by BioSPI, Systems Biology described by SBML is closer to reality. During the conversion, some details of the source SBML file are ignored. Hence, after a SBML file has been converted to a BioSPI program, there is no way to convert it back.

The conversion tool is based on SBML level 1. High level SBML will provide more detailed and accurate models for Systems Biology. Our further work is to make the conversion tool compatible with high level SBML.

References

1. Hucka, M., Finney, A., Sauro, H., etc.: Systems Biology Markup Language (SBML) Level 1: Structures and Facilities for Basic Model Definitions. Available via the World Wide Web at http://www.sbml.org.
2. Heiner, M., Koch, I., Voss, K.: Analysis and Simulation of Steady States in Metabolic Pathways with Petri Nets. 3rd Workshop and Tutorial on Practical Use of Coloured Petri Nets and the CPN Tools (2001) 15–34.
3. Priami, C., Regev, A., Silverman, W., etc.: Application of a stochastic name passing calculus to representation and simulation of molecular processes. Information Processing Letters **80** (2001) 25–31.

Knowledge-Based Faults Diagnosis System for Wastewater Treatment

Jang-Hwan Park[1], Byong-Hee Jun[2,*], and Myung-Geun Chun[3]

[1] School of Electrical and Electronic Engineering,
Chungju National University, Chungju, Korea
[2] Institute of Construction Technology, Chungbuk National University, Cheongju, Korea
[3] Dept. of Electrical and Computer Engineering, Chungbuk National University,
Cheongju, Korea
bhjun@chungbuk.ac.kr

Abstract. This paper proposed a knowledge-based fault diagnosis system using ORP (Oxidation-Reduction Potential) and DO (Dissolved Oxygen) values which usually applied as control parameters in wastewater treatment plants. If the basic control parameters such as ORP and DO can be applied to operation diagnosis, the stability of process will be remarkably improved without additional expenses. This proposed diagnosis method uses only the ORP and DO values obtained from full-scale SBR (Sequencing Batch Reactor). For the classification and diagnosis of these statues, a sequenced process of preprocessing, dimension reducing using PCA and feature extraction with ORP, DO and a synthetic parameter of [ORP DO] were proposed and applied. As results, the synthetic parameter of [ORP DO] shows better fault recognition rate than that of independent application of each parameter. It was considered that this diagnostic system using control parameters could be used to support small-scale wastewater treatment management.

1 Introduction

A biological nitrogen removal in wastewater consists of two major steps : nitrification, where inorganic ammonium ion is oxidized to nitrate ion via nitrite by autotrophic bacteria under aerobic conditions; and denitrification, where the oxidized nitrate ion is reduced to nitrogen gas by hetetrophic bacteria under anoxic condition. Even though almost wastewater systems required at least two reactors with distinguished operation modes, in SBR (Sequencing Batch Reactor), these two distinct biological reactions go though a time sequence of treatment process in the same reactor. This research was focused on the development of intelligent diagnosis based on the on-line data such as ORP and DO which are routinely adapted for process control purpose. If a reliable operation diagnosis using DO and ORP is developed, this technique will contribute to small-scale wastewater treatment plant operation without additional instrument.

---

[*] Corresponding author.

2 SBR (Sequencing Batch Reactor) Process and Faults

2.1 SBR Process

As shown in Fig. 1, a whole-cycle of SBR consisted of 4 sub-cycles with 1hr anoxic and 3hr aerobic period. Because the most operation time is occupied with aeration time, the optimization and control of aeration are important. The full-scale SBR with effective volume of $20m^3$ and digester of $30\ m^3$ for swine wastewater treatment were installed in Kimhae City. The profiles of DO and ORP were obtained from full-scale SBR. In previous study, the process control using ORP or DO was performed with threshold method including set point of dORP/dt or dDO/dt. However, since these set points are affected by reactor and influent conditions, a periodical fine tuning of set point was required for stable control [1]. This study includes the evaluation and diagnosis of set points in threshold type controller.

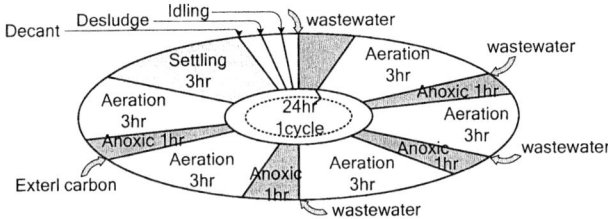

Fig. 1. Schematic diagram of overall SBR operating cycle.

2.2 Fault Selection

For the purpose of this study, fault types in three categories-controller malfunction, influent disturbance and instrument trouble were selected. Table 1 lists the detailed definition of the selected malfunctions.

Table 1. Selected malfunctions

Location	Malfunctions		Fault no.
Set value in controller (dORP/dt)	High		F2
	Low		F3
Influent	loading rate	extremely high	F4
		high	F5
		no feeding	F6
	quality	scraper type	F1(normal)
		slurry type	F7
Instruments	chemical pump trouble		F8

A typical DO and ORP profiles corresponding to each fault case named F1 to F8 were shown in Fig. 2.

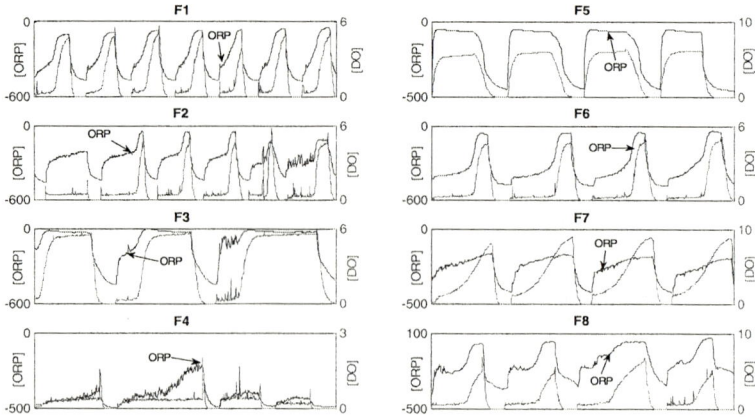

Fig. 2. ORP, DO profiles in each normal or fault case

Fig. 3 showed the diagram of fault diagnosis. ORP and DO values acquired from full-scale SBR were preprocessed by resampling, low-pass filtering and normalization. After the competitive learning clustering, the normalized data were reduced by PCA (Principal component analysis) and the feature vectors were produced [2]. Using the Euclidean distance measure, the test data were compared with the feature vectors and classified to each class from F1 to F8.

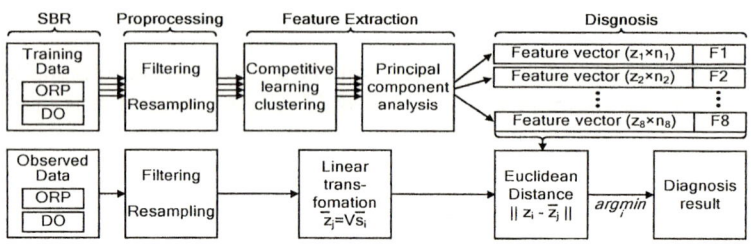

Fig. 3. Diagram of fault diagnosis

3 Knowledge-Based Faults Diagnosis

During the operation, reactor status can be divided to 8 cases of F1-F8 as shown in Table 1. The F1-F8 statuses have 18, 7, 9, 6, 17, 5, 18 and 12 data and the total number of training data were 92. Each status data were preprocessed and normalized to 100×1 vector as described earlier. Fig. 4 shows the preprocessing results of ORP for training data.

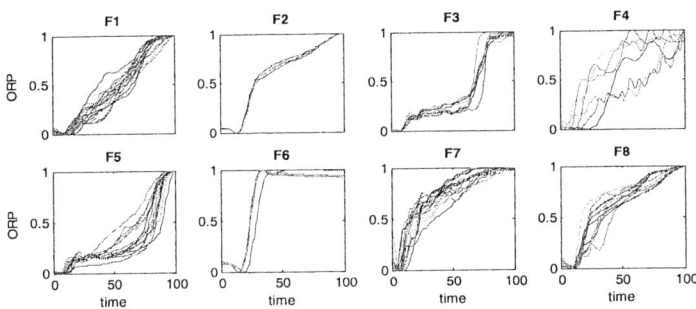

Fig. 4. Preprocessed ORP data of each fault mode

Table 2 summarizes the diagnosis results. The fault recognition rate of ORP and DO showed similar range and the results of [ORP DO] were superior to others. Moreover the fault recognition rates were increased with the number of center.

Consequently, this proposed diagnosis system based on ORP and DO should improve the stability of process management without additional costs, since the ORP and DO values are the most popular parameter in control purpose and also almost of wastewater treatment plants have these sensors. Moreover, the rapid fault detection by using only aeration phase signal made it possible to response against each fault case. The improved fault recognition rate in synthetic parameter of [ORP DO] was thought to creative results in this study. A further study to develop feed-back control and management system based on this diagnosis technique is required.

Table 2. Diagnosis results

centers no. / clusters no.	ORP	DO	[ORP DO]
1 / 8	81.52 %	82.61 %	91.30 %
2 / 8	88.04 %	86.96 %	98.91 %
3 / 8	90.22 %	91.30 %	100 %

Acknowledgements

This work was supported by Korea Research Foundation Grant (KRF-2003-050-D00010) and Ministry of Environment (071-041-069).

References

1. Jun, B. H., Kim, D. H., Chio, E. H., Bae, H., Kim, S. S., Kim, C. W.: Control of SBR operation for piggery wastewater treatment with DO and ORP. J. of Kor. Soc. on Water Quality. **18** (2002) 545-551
2. Park, J.H., Park, S.M., Lee, D.J., Kim, D.H., and Chun, M.G.: Fault Diagnosis of voltage-fed inverters using pattern recognition techniques for induction motor drive. J. of Kor. Inst. of IEIE. **19** (2005) 1-8

Study on Intelligent Information Integration of Knowledge Portals

Yongjin Zhang[1,2], Hongqi Chen[2], and Jiancang Xie[2]

[1] North-West University, Xi'an, ShannXi, China, 710000
yjzhang@xaut.edu.cn
[2] Xi'an University Of Technology, Xi'an, ShannXi, China, 710048
{hqchen, jcxie}@ xaut.edu.cn

Abstract. Web-based information portals provide a point of access onto an integrated and structured body of information about some domain. Knowledge portals are information portals, which make an important contribution to enabling enterprise knowledge management by providing users with a consolidated, personalized user interface that allows efficient access to various types of information. Portlets are mainly ways to present contents in knowledge portals. They are a group of components, which can be involved by a portal container. However, there are lacks no interaction between those portlets. This paper discusses information integration aspects within knowledge portals and presents an approach for communicating the user context (revealing the user's information need) among portlets, utilizing ontologies technologies.

1 Introduction

A major challenge of today's information systems is to provide the user with the right information at the right time. Using Web-based technologies, portals are an emerging approach for providing a single point of access to various types of information. It represents an important area for investigation and development. Essentially, this allows us to treat all web accessible content as objects that can be wrapped inside standard XML interfaces. Through a standard web service framework such as WSRP, portal containers will be able to dynamically discover and bind to desired content portlets. From this aspect, portals become an aggregate of distributed portlets, each of which in turn is a client to one or more web services. All components describe themselves and communicate with XML. However, portal systems just provide a way to integrate portlets loosely. A portlet represents an information resource; it does not know the other portlets, which means that each source has to be searched individually for relevant information. In this paper, authors discuss integration aspects of portals.

We base our approach on integrated metadata, using ontology for concept mapping information integration based on portals platform. The rest of this paper is organized as follows: In section 2 enterprise knowledge portals and knowledge management systems are introduced. The main contribution of this paper is, however, the information integration approach presented in section 3 and Section 4. Finally, section 5 concludes the paper and discusses remaining open issues and possible future work.

2 Knowledge Portals and Knowledge Management System

Knowledge management is the process of creating value from an organization's intangible assets. It deals with how best to leverage knowledge internally in the organization and externally to the customers and stakeholders. As such, knowledge management combines various concepts from numerous disciplines, including organizational behavior, human resources management, artificial intelligence, information technology, and the like. The focus is how best to share knowledge to create value-added benefits to the organization.

Since Knowledge Management becomes more and more important. Today the term knowledge portal is more and more used instead of information portal. Knowledge portals in the Web are intended to help people achieve a particular task taking place in a complex setting, e.g. learning about solutions and pitfalls of Knowledge Management. Hence, knowledge portal providers must act as intermediaries that structure relevant aspects of the info world for presentation on the portal in order to allow people flexible and easy access to all the contents.

At this point artificial intelligence comes in as a key enabler for helping to structure, access and provide information that has been aggregated by collaboration of people. Advanced techniques try to help the user with accessing the right information at the right time. This implies the support of organizational learning and corporate knowledge processes. Therefore knowledge portals are the ideal user interface to a Knowledge Management System[5].

The overall architecture of a KMS (Knowledge Management System) using a knowledge portal as a user interface is shown in figure 1. We divide the architecture of a KMS into three layers. The memory repositories layer includes all the data stores that together build the organization's knowledge base. The knowledge administration layer contains the software components that are used to access and interpret the different data sources. Difference from others, we consider these components are some portlets (i.e. CMS portlets for different news group, IR portlets for different type of documents). Finally, the presentation layer is responsible for transporting the information to the end user. Our proposal is to use a web-based portal for this purpose.

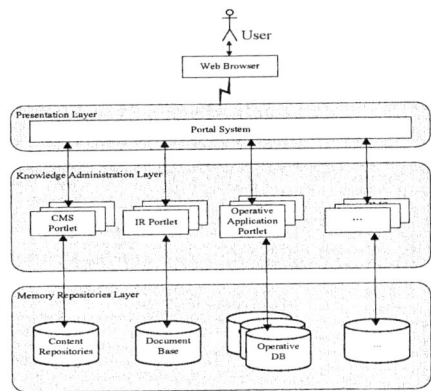

Fig. 1. Knowledge Management System with Knowledge Portal

Today, there are a number of commercial portal platforms available (e.g. IBM, BEA, etc.). The individual portal components (representing different information sources) which are rendered together to a portal webpage are called portlets[1]. Portal systems provide an integration platform on user interface level (i.e. within the presentation layer). The problem of available standard solutions is, however, the lack of interaction (or integration) between the individual portlets both in knowledge administration layer and presentation layer. We will address this issue in the following of paper.

Integration in the memory repositories layer obviously means data integration. In this paper we just introduce it shortly. One of the most prominent examples is the ETL (extract, transform, and load) process which extracts data from different source databases, and feeds it into a data warehouse. Additional integration initiatives on this layer would, among others, involve metadata integration.

3 Integration in Knowledge Administration Layer

As mentioned earlier, also, many portal platforms exist as commercial systems. The problem of these standard solutions is the lack of interaction between the individual portlets. In knowledge administration layer, portlets just descript themselves with deployment descriptor[4]. These descriptors are lack of semantic. Based on these, we can not build the relationship of portlets. In this section, we first descript the standard deployment descriptors of portlets, and then provide an approach to expand these descriptors.

3.1 Standard Deployment Descriptor of Portlets

The deployment descriptor of portlets conveys the elements and configuration information of a portlet application between Application Developers, Application Assemblers, and Deployers. Portlet applications are self-contained applications that are intended to work without further resources. Portlet applications are managed by the portlet container.

In the case of portlet applications, there are two deployment descriptors: one to specify the web application resources (web.xml) and one to specify the portlet resources (portlet.xml).

We can set portlet application name in the web.xml using the <display-name> tag. For example, we can declare a portlet application whose name is MyPortlet as following in web.xml:

<display-name>MyPortlet</display-name>

The portlet deployment descriptor in portlet.xml includes configuration and deployment information:

- Portlet Application Definition
- Portlet Definition

Portlets container distinguishes different portlets from their portlet application name and portlet name (e.g. MyPortlet:JSPPortlet).

Although portal platforms enable users to select the right portlets which they want, typically it is difficult for users to make a decision just on portlet name, especially when there are too many portlets that users can select. Even they are very familiar with the services that portlets provide, it is time consuming to select among many portlets.

3.2 Extending Deployment Descriptor of Portlets

It is an effective way to help users easily select portlets by using intellectual technologies to do automatic reasoning, giving some advice and helpful information when they chose portlets. But the prerequisite of automatic reasoning is to carry on further descriptions of portlets, and this kind of descriptions should be suitable for the automatic reasoning mechanism.

Ontologies based on Description Logics is a concept descript language which has clear semantic and standard, and is easy to share, at present, there are already such standards as RDF, RDFS, DAML+OIL, etc. We can adopt RDF[2] and RDFS[3] as describing portlets characteristic knowledge expression methods, and base on this to do automatic reasoning.

To describe portlets, RDFS should be set up firstly. From the point of application, RDFS can be divided into two parts. One part is to the descriptions of portlet' essential features, which basic model is shown in figure 2, i.e. it is thought that one portlet is a processing system that can accept inputs and produce outputs. To integrate a group of portlets is just to describe the outputs that each of them offers, which is the foundation of integration. The other part is correlated with this field. To describe the content of outputs needs to define the standard vocabularies of this field, in which portlets suitable for a certain field describes its characteristics. Later, on the basis of RDFS, RDF description of portlets should be set up.

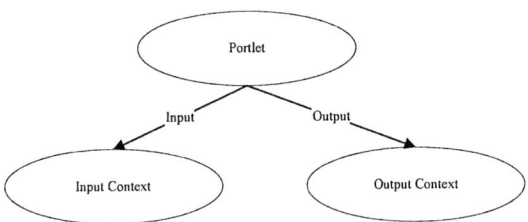

Fig. 2. Description of Portlets

We can set up the connection between portlet and its description by the <init-param> element in portlet deployment descriptor as showing in figure 3. For example, we can add the following description to show that its corresponding ontology name is "test ontology" in JSPPortlet.

```
<init-param>
    <name>ontology_description</name>
    <value>test ontology</value>
</init-param>
```

Fig. 3. Example of Portlet Deployment Descriptor

On the basis of describing portlets, it can improve user's efficiency of choosing portlets from two aspects. On one hand, based on RDFS description of Output Context, it should classify and delaminate portlets for user's selection, according to standard field vocabulary; On the other hand, after users choose a certain portlet, it uses the reasoning mechanism to look for relevant portlets to recommend users in the collection of knowledge.

4 Integration of Portlets in Presentation Layer

In knowledge administration layer, by setting up the collection of knowledge to portlets based on ontologies, it can realize the static integration, namely organizational progress the integration of portal pages. In presentation layer, there is also the lack of interaction between the individual portlets. When a user navigates within a certain portlet, the other portlets remain static. In order to provide an efficient knowledge access it would be desirable that the user's information need, revealed by his navigation within one portlet, could be provided to the other portlets enabling them to automatically find related information. The approach presented in this paper develops a framework for communicating the user context between portlets to provide such integration in a generic way.

As mentioned in section 3, we can realize the integration in knowledge administration layer through describing Output Context of one portlet. Similarly, we can realize the integration in presentation layer through describing Input Context. In fact, in regular portal systems, portlets can get input information by rendering request/action request object. The problem is the heterogeneity of the portlets and the underlying systems that manage the information displayed by them. We propose to describe a portlet based on the semantic of ontology, and that can realize the exchange and transformation of inputting information among different portlets thus can solve this problem. Its structure is shown in figure 4.

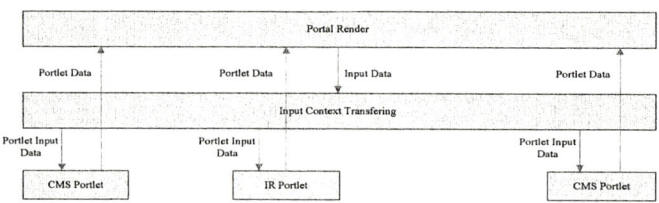

Fig. 4. Architecture of Input Context Integration

From the point of implement, specification portlet applications are generally subclass of GenericPortlet[4] according to JSR168. Therefore, we can encapsulate general treatment methods of input context semantic transferring by adding an inheriting layer. The construction of inheriting relation among classes is shown in figure 5.

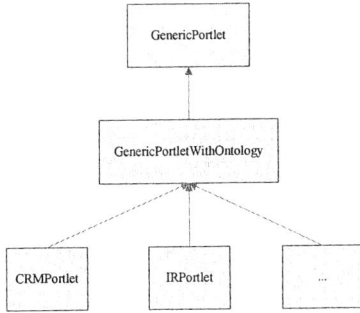

Fig. 5. Heterogeneity of Portlets Class

5 Conclusions

Today, efficient access to information of all kinds is becoming more an more important. Knowledge management systems and enterprise knowledge portals provide a means of addressing this issue. In this paper we discussed integration aspects in enterprise knowledge portals in two layers of KMS. In particular we presented approaches for implementing integrating within portal system. Using this approach an information retrieval system can, for example, automatically provide the user with documents from the organization's document management system that are related to what he is currently viewing in a CRM report.

We think that our work on knowledge portals is only one very early starting point towards the Semantic Web, which will provide machine-readable information for all kinds of web-based applications. In particular, future applications will need to integrate more automatic techniques — for building ontologies, for providing meta data, and for learning from the usage of the Semantic Web.

References

1. Wege, C. Portal Server Technology. IEEE Internet Computing, May/June 2002.
2. Brickley, D. and Guha, R.V. RDF Vocabulary Description Language 1.0: RDF Schema. W3C Working Draft, January 2003.
3. O. Lassila and R. Swick. Resource description framework (RDF). model and syntax specification. Technical report, W3C, 1999. W3C Recommendation.
4. Alejandro Abdelnur and Stefan Hepper. Java Portlet Specification 1.0, Technical report, W3C, 2003. W3C Recommendation.
5. S. Decker, D. Fensel, F. van Harmelen, I. Horrocks, S. Melnik, M. Klein, and J. Broekstra. Knowledge representation on the web. In Proceedings of the 2000 International Workshop on Description Logics (DL2000), Aachen, Germany, 2000.

The Risk Identification and Assessment in E-Business Development

Lin Wang[1] and Yurong Zeng[2]

[1] College of Management, Huazhong University of Science and Technology,
Wuhan, 430074, China
wanglin@mail.hust.edu.cn
[2] Department of Computer and Electronics,
Hubei University of Economics, Wuhan, 430205, China
zengyurong@sohu.com

Abstract. Any development of electronic business entails some level risks. Reasonable risk analysis can enhance the chance of successful electronic business project implementation. In this paper, related risk assessment indexes associated with electronic business development are put forward from the aspect of technology, organization and environment. At the same time, a risk assessment model is proposed by using of fuzzy set and grey theory and its effectiveness and feasibility are tested in a practical example. The model can assist the decision-makers to understand the current risks more intensively and assess the overall risks in a more exact way.

1 Introduction

E-Business (EB) is "a modern business methodology that addresses the needs of organizations, merchants, and consumers to cut costs while improving the quality of goods and services and increasing the speed of service delivery [1]. In fact, EB involves the use of information technology to enhance communications and transactions with a company's stakeholders, and includes activities such as establishing a web page to facilitate those communications [2]. It has been adopted widely in most enterprises and is growing rapidly all over the world.

Although EB can offer various business opportunities, EB development is puzzled by various kinds of risk and effective risk management is necessary to deal with these problems. Indeed, a task that is vital to the proper management of EB development is the assessment of risk. Proper risk management is an essential element of project success because without appropriate risk management it fails to achieve significant return on investment or competitive purpose [3]. One of the important phases in risk management is risk analysis, which involves a process of risk identification and risk assessment. It is a fact that EB development is relatively new to most companies, and only limited information is available on the associated risks. The application of fuzzy and grey theory to risk analysis seems appropriate, as such analysis is highly subjective and related to inexact and grey information. This paper is trying to establish a systematic and comprehensive evaluation model, which will be useful for the risk evaluation associated with an EB project development.

2 Risks Identification for EB Development

An important step in advancing our knowledge requires that we understand and address these prevail risks in detail. In this paper, risks associated with EB development are defined as the risks of direct or indirect loss to the organization in development an EB project, which refers to the development stages as planning, analysis, design and implementation of an EB system [3]. At present, there is no agreed upon universal definition of EB risk but information security is a widely recognized aspect of EB risk. Some papers have discussed this problem from different aspect [4-7]. Among these papers, the risks given by E.W.T Ngai and T.K.T Wat (2005) and Tom Addison (2003) are relative detailed and reasonable. On the cornerstones of their works, the risk experts' ideas are analyzed and necessary information is gathered by using group-discussing and anonymous questionnaire methods. According to the principle of rationality, comparability and maneuverability, the following factors are selected as the risk indexes.

2.1 Technical Risks

It is necessary to create an environment which secures the trust of the user and hence the user's commitment. So, we must try our best to minimize current and future threats to the successful completion of the EB project. Moreover, there is a need to understand the user/customer requirements correctly and adopt the appropriate technology available. As mentioned above, related risks are listed as follows.

1) Client-server Security Risk
P_1: Absence of reliable firewall
P_2: Hacker gaining unauthorized access
P_3: Lack of using reliable cryptography
P_4: Poor "key" management
P_5: Malicious code attacks
2) Physical Security Risk
P_6: Threat of sabotage in internal network
P_7: Loss of audit trail
P_8: Natural disaster-caused equipment failure
P_9: Human factor-caused equipment failure
P_{10}: Poor design, code or maintenance procedure
3) Requirement Risk
P_{11}: Wrong functions and properties development
P_{12}: Wrong user interface development
P_{13}: Underestimated project complexity
P_{14}: Wrong project size estimation
P_{15}: Rapid technological renovation
P_{16}: Continuous change of system requirements

2.2 Organizational Risks

Successful EB project need all kinds of resources and some inevitable modifications of the current business process. All of these works must be planed scientifically. So, there are also many risks from the aspect of organizational to be considered in advance.

1) Resources Risk
P_{17}: Wrong schedule estimation
P_{18}: Project behind schedule
P_{19}: Project over budget or inadequate cash flow
P_{20}: Depression economy or vague industry policy about EB
P_{21}: Necessary personnel shortfalls
P_{22}: Loss of key experienced person
2) Managerial Risk
P_{23}: Lack of top management support
P_{24}: Poor project planning
P_{25}: Unclear project objectives
P_{26}: Indefinite project range
P_{27}: Lack of contingency plans
3) Reengineering Risk
P_{28}: Business process redesign
P_{29}: Organizational restructuring

2.3 Environmental Risks

It is familiar for EB outsourcing. So, there are some problems to be dealt with to control the quality of the EB project construction. At the same time, the risks from legal and culture can't be neglected. The main risks are as follows.

1) Legal Risk
P_{30}: Lack of international legal standards about EB
P_{31}: New laws and regulations constantly change the online legal landscape
P_{32}: Uncertain legal jurisdiction
2) Cultural Risk
P_{33}: Difference users with different in culture customers and business styles
P_{34}: Language obstacle
3) Outsourcing Risk
P_{35}: Incompletion of contract terms
P_{36}: Difficult to change outsourcing decision/vendor
P_{37}: Loss of data control
P_{38}: Loss of control over vendor
P_{39}: Underestimated hidden cost
4) Vendor Quality Risk
P_{40}: Lack of vendor expertise and experience
P_{41}: Vendor offers outdated technology solution
P_{42}: Vendor provides poor quality service

3 Risk Assessment Model

The bases of this model are fuzzy and grey theory. Fuzzy mathematics pays more attention to objects that intension clear while extension vague. It depicts fuzzy phenomenon abstractly with mathematics ways and provide an effective bridge between classic mathematic and realistic fuzzy world [8-9]. At the same time, grey theory pays more attention to objects that intension vague while extension clear [10-11]. It adopts the way to provide a supplement to the small sample data and have practical value while the sample is small and can't meet the requirement of statistic. In the following model, we will use different theory to learn from fuzzy theory's advantage to offset grey theory's weakness. So, the evaluation result will be more reasonable.

3.1 Confirm the Index Set

From what has been discussed above, we can identify risk index set $P=\{P_1,P_2,...,P_n\}$.

3.2 Confirm the Weight of Each Index

The weight of each index can be confirmed by many ways, such as entropy value, DELPHI and AHP method. AHP is an effective multi-criteria decision making tool to find out the relative priorities to be assigned to different criteria and alternatives which characterize a decision [12-13]. We can identify the weight of each index using AHP, i.e. $W=\{W_1, W_2,..., W_n\}$. Note: W_i is the corresponding weight for index P_i ($i=1,2,...,n$).

3.3 Confirm the Sample Matrix

Though a certain index may be evaluated quantitatively, it is also rather difficulty to judge its direct effect on the development of EB. So, we can give a score (1 to 10) to evaluate the risk in EB development according to the practical condition for every index. The score of a certain index will be high if the risk is small. That is to say, this EB development can be completed on-time and on-budget from this aspect. Supposing the number of experts is r, so $E=\{E_1,E_2,...,E_r\}$. Assuming the value of index i that given by expert L is d_{li}, then the sample matrix for all the experts can be expressed as following:

$$D = \begin{bmatrix} d_{11} & d_{12} & \cdots & d_{1n} \\ d_{21} & d_{22} & \cdots & d_{2n} \\ \vdots & \vdots & \vdots & \vdots \\ d_{r1} & d_{r2} & \cdots & d_{rn} \end{bmatrix} \quad (1)$$

3.4 Confirm the Evaluation Degree

According to scientific estimate theory, we confirm the degree of risks as m. So, the comprehensive evaluate standard matrix is: $V=\{V_1,V_2,...,V_m\}$.

3.5 Confirm the Evaluation Grey Number

According to the evaluation degree, grey number can be confirmed by qualitative analysis. There are three functions that are widely used.

1) Upper end level, grey number $\otimes \in [0, \infty)$, the corresponding whitening function (WF) is:

$$f_1(d_{ki}) = \begin{cases} d_{ki}/d_1 & d_{ki} \in [0, d_1] \\ 1 & d_{ki} \in [d_1, \infty] \\ 0 & d_{ki} \in (-\infty, 0) \end{cases} \quad (2)$$

2) Middle level, grey number $\otimes \in [0, d_1, 2d_1]$, the corresponding WF is:

$$f_2(d_{ki}) = \begin{cases} d_{ki}/d_1 & d_{ki} \in [0, d_1] \\ 2 - d_{ki}/d_1 & d_{ki} \in [d_1, 2d_1] \\ 0 & d_{ki} \notin (0, 2d_1] \end{cases} \quad (3)$$

3) Low end level, grey number $\otimes \in (0, d_1, d_2)$, the corresponding WF is:

$$f_3(d_{ki}) = \begin{cases} 1 & d_{ki} \in [0, d_1] \\ \dfrac{d_2 - d_{ki}}{d_2 - d_1} & d_{ki} \in [d_1, d_2] \\ 0 & d_{ki} \notin (0, d_2] \end{cases} \quad (4)$$

The turning point for whitening weight function can be confirmed by two ways. Firstly, it can be confirmed by analogy according to rule and experience. On the other hand, we can regard maximum, minimum and middle value as upper limit, lower limit and middle value.

3.6 Calculate Grey Statistics

We can obtain $f_j(d_{ki})$ which represents the "weight value" of d_{ki} belongs to risk degree j ($j=1,2,\ldots,m$) by grey theory, then n_{ij} and n_i can be calculated by the formulas 5 and 6.

$$n_{ij} = \sum_{k=1}^{r} f_j(d_{ki}) \quad (5)$$

$$n_i = \sum_{j=1}^{m} n_{ij} \quad (6)$$

3.7 Calculate Grey Evaluation Value and Fuzzy Matrix

Then, r_{ij} can be calculated by formula $r_{ij}=n_{ij}/n_i$, thus:

$$R = \begin{bmatrix} r_{11} & r_{12} & \cdots & r_{1m} \\ r_{21} & r_{22} & \cdots & r_{2m} \\ \vdots & \vdots & \vdots & \vdots \\ r_{n1} & r_{n2} & \cdots & r_{nm} \end{bmatrix} \quad (7)$$

3.8 Calculate Fuzzy Comprehensive Matrix

$$B = (b_1, b_2, \ldots, b_m) = (w_1, w_2, \ldots, w_n) \cdot \begin{bmatrix} r_{11} & r_{12} & \cdots & r_{1m} \\ r_{21} & r_{22} & \cdots & r_{2m} \\ \vdots & \vdots & \vdots & \vdots \\ r_{n1} & r_{n2} & \cdots & r_{nm} \end{bmatrix} \quad (8)$$

Note: $b_j = \sum_{i=1}^{n} w_i \cdot r_{ij}$, Let $b_i^1 = b_i / \sum_{j=1}^{m} b_j$, then $\sum_{i=1}^{m} b_i^1 = 1$.

3.9 Calculate the Result of Risk Assessment

Firstly, we should confirm the degree of risk: $C=(V_1,V_2,\ldots,V_m)^T$. Thus, Z can be obtained by formula $Z=(W \cdot R) \cdot C$, that is the result of risk assessment for an EB development project.

4 Example

Relative data for a risk assessment to the development of EB are as follows:
Using AHP, we can identify the $W=\{W_1, W_2,\ldots, W_{42}\}$ (omitted);
Sample matrix: $D=[\ldots]_{15 \times 42}$ (omitted). That is to say, 15 experts provided the scores for 42 indexes according to the actual development facts about an EB project. For example, an expert can give a score of "9.5" according to index P_1 because he thinks a certain EB project have a rather reliable firewall.

According to the experts' advices, we divide risk into five degree: lowest risk, low risk, usual risk, high risk and highest risk. At the same time, we confirm corresponding $V=\{9,7,5,3,1\}$ and identify corresponding grey number and white functions depicted in Figure 1. That is to say, for a score of "6", the grey number that represents the value of d_{ki} belongs to the five risk degree is {6/9,6/7,4/5,0/3,0/2} respectively.

Thus, n_{ij}, n_i and r_{ij} ($i=1,2,\ldots,42$; $j=1,2,\ldots,5$) can be calculated step by step, and R can be obtained. We can obtain B=(0.3352,0.4177,0.2471,0,0) and $Z=B \cdot V^T = 7.1762$. So, the risk of this EB project can be regarded as the "low" class.

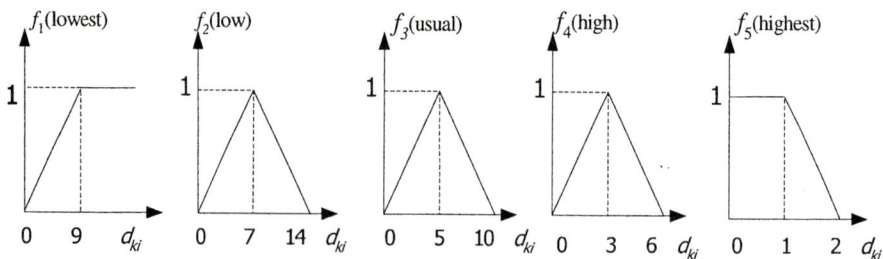

Fig. 1. five weight function for risk assessment

5 Conclusions and Further Enhancements

Proper risk assessment can enhance the chance of successful EB project implementation. A method to identify and assess the risks in EB development is proposed by using fuzzy mathematics and grey theory. It is evident that this method of risks analysis's useful and reliable. It utilizes the classic method to the optimum degree and makes up its disadvantage. It can be widely used in the field of risk evaluation of the EB project development and deal with other evaluations that include fuzzy and grey information.

However, there are many business models of E-Business. For example the B2C (Business to Consumer) and the B2B (Business to Business) situations are different, and there is a need for further research to determine the major risks for different trading modes. Moreover, there is a need to design and development a web-based decision support system incorporate the proposed risks evaluation model to provide the convenient aided decision help.

Acknowledgements

The authors are grateful for the constructive comments of the referees and the editors.

References

1. R. Kalakota, A.B. Winston.: Frontiers of the electronic commerce. Addison-Wesley, Reading, MA (1996)
2. Tom Addison.:E-commerce project development risks: evidence from a Delphi survey. International Journal of Information Management, 23 (2003) 25-40
3. E.W.T Ngai, T.K.T Wat.:Fuzzy decision support system for risk analysis in e-commerce development. Decision Support System, 40(2005) 235-255
4. R.K.J.R.Rainer,C.A.Snyder, H.H. Carr.:Risk analysis for information technology. Journal of Management Information Systems, 8(1994) 129-147
5. T. Stoehr.: Managing e-business projects: 99 key success factors. Springer, Hamburg (2002)

6. Marilyn Greenstein, Todd.M. Feinman.:Electronic commerce: security, risk management and control. Beijing, China Machine Press (2000)
7. C. Mceachern.:Technology risks: don't panic- financial services firms seem to have cyber risk under control. Wall Street Technology (2001)
8. L.A.Zadeh.:Fuzzy sets and their applications. New York, Academic Press (1975)
9. Helmer S.:Evaluating different approaches for indexing fuzzy sets. Fuzzy Sets and Systems, 140(1)(2003) 167-182
10. Deng Julong. :A basic method of grey system. Wuhan, Huazhong University of Science and Technology Press (1987)
11. Hsu Chaug-lng, Wen Yuh-Horng.:Application of grey theory and multiobjective programming towards airline network design. European Journal of Operational Research, 127(1)(2000) 44-68
12. Stam, Antonie, Duarte Silva, A Pedro.: On multiplicative priority rating methods for the AHP. European Journal of Operational Research, 145(1)(2003) 92-108
13. Byun Dae-Ho.:The AHP approach for selecting an automobile purchase model. Information and Management, 38(5)(2001) 289-297

A Novel Wavelet Transform Based on Polar Coordinates for Datamining Applications*

Seonggoo Kang[1], Sangjun Lee[2], and Sukho Lee[1]

[1] School of Electrical Engineering and Computer Science,
Seoul National University, Seoul 151-742, Korea
exodus@db.snu.ac.kr, shlee@cse.snu.ac.kr
[2] School of Computing, Soongsil University, Seoul 156-743, Korea
sjlee@computing.ssu.ac.kr

Abstract. In this paper, we propose a novel wavelet transform based on the polar coordinates for datamining applications. In general, the Harr wavelet transform has been popularly used for data decomposition. However, the Harr wavelet transform shows the poor performance for the locally distributed data which are clustered around certain values, since it uses the averages as representatives for data decomposition. The proposed wavelet transform is based on the the polar coordinates which is not affected by the averages and is more suitable than the Harr wavelet transform for data decomposition of the locally distributed data.

1 Introduction

Similarity search is the essential operation in datamining[1] such as pattern matching, time series analysis. The main issue of similarity search is to improve the search performance. Although the sequential scanning can be used to perform similarity search in large databases, it is obvious that the sequential scanning scales poorly as the size of databases increases. In general, an indexing scheme is used to support fast similarity search in datamining applications. However, a naive indexing of high-dimensional data using a spatial access method such as R-tree[2] suffers from performance deterioration due to the *dimensionality curse*[3] of an index structure.

Various methods have been proposed to process similarity search in large databases. The most popular method is to extract the feature extraction via data decomposition, and then uses a spatial access method to index these features. Among data decomposition methods, the Harr wavelet transform[4] has been popularly used. However, the Harr wavelet transform shows the poor performance for locally distributed data which are clustered around certain values, since it uses the averages as representatives for data decomposition.

In this paper, we propose a novel wavelet transform based on polar coordinates for datamining applications. The proposed wavelet transform is based

---

* This work was supported in part by the Brain Korea 21 Project and in part by the Ministry of Information and Communications, Korea, under the Information Technology Research Center (ITRC) Support Program in 2005.

on the polar coordinates which is not affected by the averages and is more suitable than the Harr wavelet transform for data decomposition of the locally distributed data.

2 Overview of the Harr Wavelet Transform

Given the data $X = (x_1, x_2)$ of length 2, the Harr wavelet transform uses the average of x_1 and x_2 and the difference between x_1 and the average($= \frac{x_1+x_2}{2}$) for data decomposition. Figure 1 shoes the Harr wavelet transform for the data of length 2.

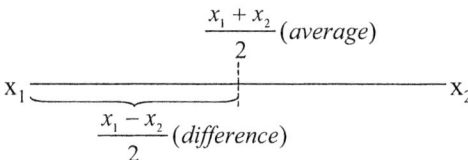

Fig. 1. Harr wavelet transform for the data of length 2

The original data $X = (x_1, x_2)$ can be obtained by the following equations using the average and the difference.

$$x_1 = \frac{x_1 + x_2}{2} + \frac{x_1 - x_2}{2}$$
$$x_2 = \frac{x_1 + x_2}{2} - \frac{x_1 - x_2}{2}$$

The Harr wavelet transform can be regarded as a series of multi-level operations for the average and the difference. Since the Harr wavelet transform uses the averages and the differences for data decomposition, it is not suitable for the locally distributed data which are clustered around certain values. Moreover, the Harr wavelet transform can be defined only for data of length of 2^n.

3 Proposed Wavelet Transform

Compared with the Harr wavelet transform, the proposed wavelet transform uses the radius and the radian value as the unit of angle in the polar coordinates for data decomposition instead of the average and the difference. Given the data $X = (x_1, x_2)$ of length 2, the proposed wavelet transform uses the radius $r = \sqrt{x_1^2 + x_2^2}$ and the radian value $\theta = \cos^{-1}(\frac{x_1}{r}) = \sin^{-1}(\frac{x_2}{r})$ for data decomposition. Figure 2 shows the proposed wavelet transform based on the polar coordinates for the data of length 2.

The original data $X = (x_1, x_2)$ can be obtained by the following equations using the radius and the radian value in the polar coordinates.

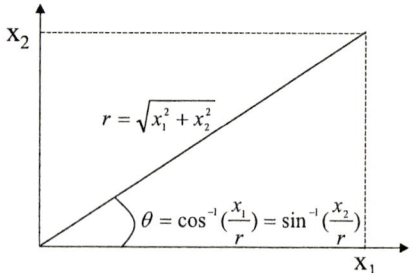

Fig. 2. Proposed wavelet transform for the data of length 2

Resolution	Radius	Radian
4	(3 4 6 8)	
2	(5 10)	(1.03 1.03)
1	$(5\sqrt{5})$	(1.23)

Fig. 3. Decomposition procedure using the proposed wavelet transform

$$x_1 = r \cdot \cos\theta$$
$$x_2 = r \cdot \sin\theta$$

The proposed wavelet transform can be regarded as a series of multi-level operations for the radius and the radian value in the polar coordinates. The decomposition of the data $S = (3, 4, 6, 8)$ using the proposed wavelet transform is shown in Figure 3. The resolution 4 is the full resolution of the data $S = (3, 4, 6, 8)$. In resolution 2, $(5,10)$ are obtained by taking the radiuses of $(3, 4)$ and $(6,8)$ at the resolution 4 respectively. $(1.03, 1.03)$ are the radian values of $(3, 4)$ and $(6, 8)$ respectively. This procedure is continued until the resolution 1 is obtained.

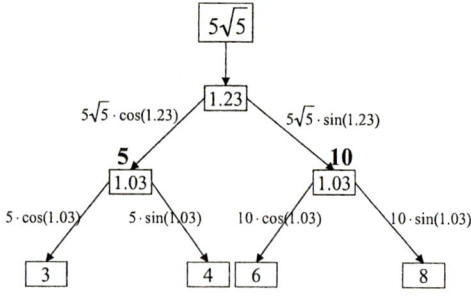

Fig. 4. Obtaining the different resolution in the proposed wavelet transform

The different resolution can be obtained by multiplying $cos(radianvalue)$ to the radius for the left side and by multiplying $sin(radianvalue)$ to the radius for the right side. For example, $(5, 10) = (5\sqrt{5} * \cos(1.23), 5\sqrt{5} * \sin(1.23))$ where $5\sqrt{5}$ and 1.23 are the first and second coefficient respectively. This procedure can be done recursively until the full resolution is obtained. The procedure of obtaining the different resolution is shown in Figure 4.

4 Conclusion

In this paper, we have proposed the novel wavelet transform based on the polar coordinates which it is not affected by the averages. The proposed wavelet transform is more suitable than the Harr wavelet transform for data decomposition of the locally distributed data.

References

1. Rakesh Agrawal, Tomasz Imielinski, and Arun N. Swami,"Database Mining: A Performance Perspective", *IEEE TKDE, Special issue on Learning and Discovery in Knowledge-Based Databases*, 5(6), pp. 914-925, 1993.
2. Antonin Guttman,"R-trees: A Dynamic Index Structure for Spatial Searching", In *Proc. of ACM SIGMOD International Conference on Management of Data*, pp. 47-57, 1984.
3. Rakesh Agrawal, Christos Faloutsos, and Arun N. Swami, "Efficient Similarity Search In Sequence Databases", In *Proc. of International Conference on Foundations of Data Organization and Algorithms*, pp. 69-84, 1993.
4. Kin-pong Chan, Ada Wai-chee Fu, "Efficient Time Series Matching by Wavelets", In *Proc. of International Conference on Data Engineering*, pp. 126-133, 1999.

Impact on the Writing Granularity for Incremental Checkpointing*

Junyoung Heo, Xuefeng Piao, Sangho Yi, Geunyoung Park, Minkyu Park[1], Jiman Hong[2], and Yookun Cho[1]

[1] Seoul National University
[2] Kwangwoon University

Abstract. Incremental checkpointing is an cost-efficient fault tolerant technique for long running programs such as genetic algorithms. In this paper, we derive the equations for the writing granularity of incremental checkpointing and find factors associated with the time overhead and disk space for incremental checkpoint. We also verify the applicability of the derived equation and the acceptability of the factors through experiments.

1 Introduction

Checkpointing mechanism can be employed to provide fault tolerance for long running process like genetic algorithms However, this checkpointing method has overhead to save memory state of a process.

To reduce the overhead, incremental checkpointing[1] was proposed. Incremental checkpointing saves the modified portion of memory state. If only a few words within a page are modified, then would be more efficient to store those words itself rather than the entire page[2]. Generally, the smaller the block size for incremental checkpointing, the more checkpoint file size is reduced, and hence getting better performance[3]. Therefore, word-level granularity is optimal for incremental checkpointing[4]. However, word-level granularity is not always more efficient than page-level granularity, because word-level granularity sometimes increases the overhead for finding the differences between two checkpoints and for writing the address of the modified word to detecting exact changed words of memory.

In this paper, we derive the equations for the writing granularity of incremental checkpointing and find factors associated with the time overhead and disk space for a incremental checkpoint. We also show that the derived equation the factors are fairly reasonable through experiments performed on Linux Kernel-level incremental checkpointing facility.

---

* The present Research has been conducted by the Research Grant of Kwangwoon University in 2004 and supported by the Brain Korea 21 Project in 2005.

2 Analysis of Writing Granularity of Incremental Checkpointing

Some common notations are presented in Table 1.

Table 1. List of notations used in this paper

N	The number of pages in a process.
R	Ratio of modified page of a process since last checkpointing.
r_p	Ratio of modified portion in a page p
$\overline{C}$	Average time to check the difference of same-addressed pages in the successive checkpoints.
$\overline{D}$	Average duplication time of one page
$\overline{F}$	Average processing time of write-protection fault in fault handler.
$\overline{S}$	Average time to write a word into stable storage.
W	The number of word in one page.
$f_b(x)$	1 if $x > 0$. 0 otherwise.

2.1 Analysis of Checkpointing Time Overhead

In this section, we derive the time overhead in case of both the word-level and page-level incremental checkpointing. First, we consider the time overhead for saving a page in word-level incremental checkpointing, $T_{word}(p) = \overline{D} + 2r_p W \overline{S} + \overline{C}$, where $2r_p W \overline{S}$ is the time overhead for saving modified words.

In word-level incremental checkpointing, time overhead for saving is computed in twice because the address of modified words must also be saved. Thus, the time overhead for saving a process in word-level incremental checkpointing is given by

$$T_{word} = \sum_{p=1}^{N} T_{word}(p) = \sum_{p=1}^{N} (\overline{D} + 2r_p W \overline{S} + \overline{C})$$

$$= N(\overline{D} + \overline{C} + 2W\overline{S}\overline{r}_p), \text{ where } \overline{r}_p \text{ is } \frac{\sum_{p=1}^{N} r_p}{N}.$$

Then, the time overhead for saving a page in page-level incremental checkpointing, $T_{page}(p) = (\overline{F} + W\overline{S})f_b(r_p)$. Thus, the time overhead for saving a process in page-level incremental checkpointing is given by

$$T_{page} = \sum_{p=1}^{N} T_{page}(p) = \sum_{p=1}^{N} (\overline{F} + W\overline{S})f_b(r_p)$$

$$= RN(\overline{F} + W\overline{S}) = N(R\overline{F} + RW\overline{S}), \text{ where } RN \text{ is } \sum_{p=1}^{N} f_b(r_p).$$

To compare the time overhead for saving a process, we compute the difference T_{diff}, between the word-level and page-level incremental checkpointing as follows.

$$T_{diff} = T_{word} - T_{page} = N((\overline{D} + \overline{C} - R\overline{F}) + W\overline{S}(2\overline{r}_p - R))$$

If $\overline{D} + \overline{C} > R\overline{F}$ and $2\overline{r}_p > R$ then the time overhead of word-level incremental checkpointing has larger than that of page-level.

2.2 Analysis of Disk Space Overhead

In this section, we derive the space overhead in case of both the word-level and page-level incremental checkpointing. First, we consider the space overhead for saving a page in word-level incremental checkpointing, $S_{word}(p) = 2r_pW$. Thus, the space overhead for saving a process in word-level incremental checkpointing is given by

$$S_{word} = \sum_{p=1}^{N} S_{word}(p) = \sum_{p=1}^{N} 2r_pW = 2\overline{r}_pNW.$$

Then, the space overhead for saving a page in page-level incremental checkpointing, $S_{page}(p) = Wf_b(r_p)$. Thus, the space overhead for saving a process in page-level incremental checkpointing is given by

$$S_{page} = \sum_{p=1}^{N} S_{page}(p) = \sum_{p=1}^{N} Wf_b(r_p) = RNW.$$

To compare the space overhead for saving a process, we compute the difference S_{diff}, between the word-level and page-level incremental checkpointing as follows.

$$S_{diff} = S_{word} - S_{page} = 2\overline{r}_pNW - RNW = NW(2\overline{r}_p - R)$$

If $2\overline{r}_p > R$ then the space overhead of word-level incremental checkpointing has larger than that of page-level.

3 Verification of the Overhead Model

In this section, we present the verification of mathematical analysis. According to the analysis of time and disk space overhead in Section 2, if $2\overline{r}_p > R$ and $\frac{\overline{D}+\overline{C}}{\overline{F}} > R$, then the time and disk space overhead of page-level is less than that of word-level incremental checkpointing.

To verify our analysis of checkpointing overhead based on equation (T_{diff}), we used our *pickpt*, a Linux kernel-level incremental checkpointing facility which provide both the page-level and word-level incremental checkpointing. We used Quick Sort program to compare the overhead.

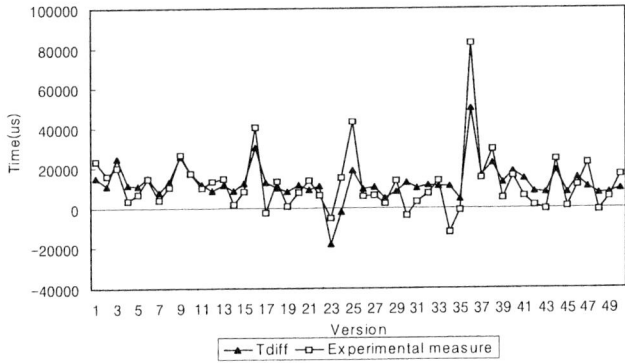

Fig. 1. The overhead difference of Quick Sort

First, we measured $\overline{D}$, $\overline{C}$ and $\overline{F}$ on this system. $\overline{D}$, $\overline{C}$ and $\overline{F}$ were $0.578\mu s$, $5.898\mu s$ and $5.898\mu s$ respectively.

Before taking a checkpoint, we calculated $\overline{r}_p$ and R. Then, we calculated T_{diff} with these values. Figure 1 shows the T_{diff} and the difference obtained from the experimental results. In Fig. 1, the analyzed time overhead in Section 2 is much closer to the experimental results.

4 Conclusions

In this paper, we derived equations for the writing granularity of incremental checkpointing and presented certain factors associated with the time overhead and disk space in incremental checkpoint. We also showed that the factors are fairly reasonable through experiments performed on Linux Kernel-level incremental checkpointing facility.

References

1. Plank, J., Beck, M., Kingsley, G., Li, K.: Libckpt:transparent checkpointing under unix. In: Usenix Winter Technical Conference. (1995) 213–223
2. Plank, J., Chen, Y., K. Li, M.B., Kingsley, G.: Memory exclusion: optimizing the performance of checkpointing systems. Software Practice and Experience **29** (1999) 125–142
3. Agarwal, S., Garg, R., Gupta, M.S., Moreira, J.E.: Adaptive incremental checkpointing for massively parallel systems. In: 18th annual international conference on Supercomputing. (2004) 277–286
4. Plank, J., Xu, J., Netzer, R.: Compressed differences: An algorithm for fast incremental checkpointing. Technical Report CS-95-302, University of Tennessee (1995)

Using Feedback Cycle for Developing an Adjustable Security Design Metric*

Charlie Y. Shim[1], Jung Y. Kim[2], Sung Y. Shin[1], and Jiman Hong[3]

[1] South Dakota State University
{yong_shim, sung_shin}@sdstate.edu
[2] University of Wyoming
kimj@uwyo.edu
[3] Kwangwoon University
gman@daisy.kw.ac.kr

Abstract. In this paper, we develop a security design metric that can be used at system design time to build more secure systems. This metric is based on the system-wide approach and adopt a reliability model and scenario testing technique to produce a feedback cycle.

1 Introduction

New approach based on the idea that security violation is an instance of system failure leads to the use of reliability model for the development of security model[1]. It is known that no reliability model is applicable at system design phase because there is no system testing data available at that time. Even though some design metrics are available at design time, they are not based on system operation and usually known as inaccurate. Because of this reason, it is not easy to build an accurate security design metric out of reliability model. Through out this paper, we provide a general methodology for developing a more accurate security design metric. The proposed methodology adopts a software reliability model and scenario testing.

2 System-Wide Approach for the Security

Previously, system security has been viewed as an attachment for existing systems. "Though some security concerns are addressed during the requirement analysis phase, most security requirements come to light only after functional requirements have been completed. As a result, security policies are added as an afterthought to the standard (functional) requirements[2]." This approach, however, didn't provide efficient security mechanism as we can see from many cases of disastrous security violations. Due to this reason, many different approaches to achieve more system security have been tried. One of the major

---

* The present Research has been conducted by the Research Grant of Kwangwoon University in 2005.

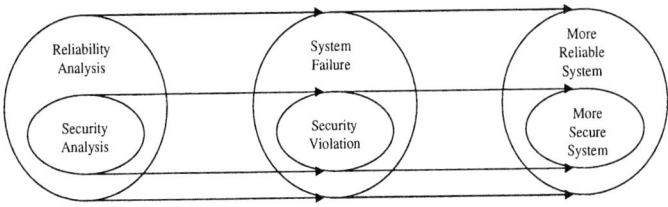

Fig. 1. Relationship between reliability and security in system-wide approach

approaches is system wide approach. In this system-wide approach, we view security as a system requirement and consider it from the design phase. System designers have long recognized the need to incorporate reliability into system design processes[2]. Figure 1 summarizes the system-wide approach for the security. In this approach, we view security violations as one kind of failure, security analysis as related to reliability analysis, and a secure system as a part of a reliable system..

3 A Methodology for Developing a Security Design Metric

A method of developing a security design metric by using a reliability model has been previously proposed[1]. In this method, we used Goel's Nonhomogeneous Poisson Process Reliability model[3] with collected security violation data

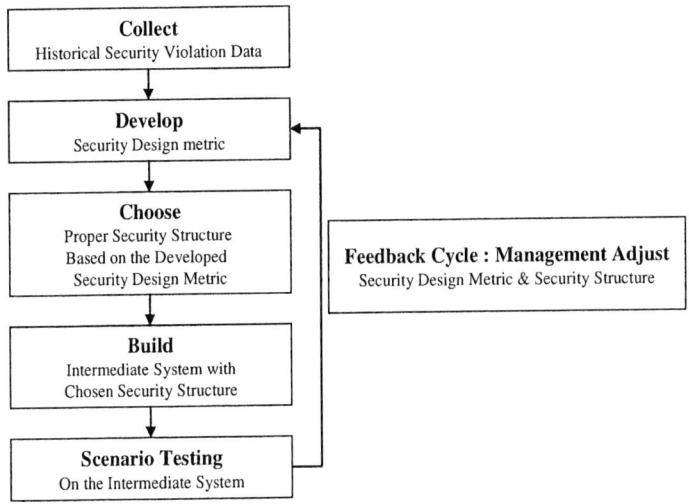

Fig. 2. The procedure for developing an adjustable security design metric

instead of system testing data. "Much of software development today is largely a matter of integrating off-the-shelf components: rarely are new systems built entirely from scratch [2]." This might be also true for system security. In order to achieve system security, many systems use combinations of available security countermeasures. Many of security violation cases for specific security countermeasures are even reported to and available at some security organizations such as CERT (Computer Emergency Response Team). Because of this reason, using security violation data instead of security testing data solves the problem that no security testing data is available at design phase. However, the security design metric made of collected security violation data still has a limit. Those security violation data collected are from the same kind of security countermeasure that is included in different system's security structure. Thus, there could be a variation between security structure in the system to be built and that of reported system. In order to solve the variation problem, the developed security design metric need to reflect the security structure of the system to be built. We suggest performing more testing with the intermediate system and adjust the developed security design metric according to the testing result.

4 Adjusting the Security Design Metric by Adopting Software Testing Technique

As we mentioned in section 3, viewing each security countermeasure as a component of a security structure allows us to interpret the collected security violation data as the result of unit testing. The next level of testing is integrating test which focuses on design and the construction of the system architecture. Scenario testing will be a good candidate for this level of testing. Scenario testing concentrates on what the user does, not what the product does[4]. Performing scenario testing to intermediate system will produce some testing data. Feed back the testing data to the developed security design metric. In other words, adjust the security design metric based on the scenario testing data. Then, the security design metric is based on both unit testing and integrating testing data and reflects the nature of the system's security structure more. Figure 2 explains the procedure for developing an adjustable security design metric. We can continue this feedback cycle until the metric shows certain satisfactory level of security. Since the metric can be updated continuously, this is also good for the continuous maintenance of system security. Due to the changes in security attack trends as time passes, system security should be updated regularly by applying proposed feedback cycle regularly with new data.

5 Conclusions

In this paper, we proposed a methodology for developing a security design metric by using software engineering techniques: reliability model and scenario testing. Since the proposed method is a repetitive procedure, the security design metric will be updated continuously and the metric will become more accurate. As the metric updated, the security structured also considered for updating in manageable way and this repetitive procedure helps continuous security maintenance. The continuous security maintenance eventually makes systems more secure.

References

1. Shim, C., Shin, S. : An Approach from Software Reliability Modeling to Security Modeling. In: Proc. ISCA International Conference on Computer and Their Applications(CATA) (2002) 63–66
2. Devanbu, P., Stubblebine, S. : Software Engineering for Se-curity: a Roadmap. In: Proc. conference on The Future of Soft-ware Engineering (2000)
3. Goel,A. L. : Software reliability models: assumptions, limitations, and applicability. : IEEE Transactions on Software Engineering **11** (1985) 1411-1423
4. Pressman, R. S. : Software reliability models: assumptions, limitations, and applicability. : Software Engineering: A Practitioner's Approach (4th Ed.) McGraw-Hill Companies, Inc.

w-LLC: Weighted Low-Energy Localized Clustering for Embedded Networked Sensors[*]

Joongheon Kim[1], Wonjun Lee[1,**], Eunkyo Kim[2], and Choonhwa Lee[3]

[1] Department of Computer Science and Engineering,
Korea University, Seoul, Republic of Korea
wlee@korea.ac.kr
[2] LG Electronics Institute of Technology,
LG Electronics Co., Seoul, Republic of Korea
[3] College of Information and Communication,
Hanyang Univerity, Seoul, Republic of Korea

Abstract. This paper addresses a weighted dynamic localized clustering unique to a hierarchical sensor network structure, while reducing the energy consumption of cluster heads and as a result prolonging the network lifetime. *Low-Energy Localized Clustering*, our previous work, dynamically regulates the radii of clusters to minimize energy consumption of cluster heads while the network field is being covered. We present *weighted Low-Energy Localized Clustering (w-LLC)*, which consumes less energy than LLC with weight functions.

1 Introduction

Recently, many research efforts on sensor networks have become one of the most active research topics on network technologies [1]. Due to limited power source of sensors, the main components of sensor networks, energy consumption has been concerned as the critical factor when designing sensor network protocols. Facing this challenge, several approaches to prolong lifetime of the networks, including clustering schemes [2] and structured schemes with a two-tiered hierarchy [3], have been investigated. It consists of the upper tier to communicate among cluster heads (CHs) and the lower tier to sense events and transmit them to CHs. However, in traditional clustering scheme and two-tiered hierarchical structuring scheme, CHs cannot adjust their radius. If the cluster range is larger than optimal one, a CH consumes more energy than required. On the other hand, a smaller range than necessary results in the entire sensing field not being covered. Therefore, we proposed a novel localized clustering algorithm [4]. Our proposed scheme, *Low-Energy Localized Clustering (LLC)*, can regulate the cluster radius by communicating with CHs for energy savings.

We extend the LLC to *weighted Low-Energy Localized Clustering (w-LLC)* to cope with the case where events further frequently occur in a certain area.

---

[*] This work was supported by grant No. R01-2005-000-10267-0 from Korea Science and Engineering Foundation in Ministry of Science and Technology.
[**] Corresponding author.

2 w-LLC: Weighted Low-Energy Localized Clustering

w-LLC aims to minimize energy consumption of CHs. It consists of two phases and one policy with the assumption that a sink knows the position of CHs.

2.1 Initial Phase

In this phase, the CHs deployed at random construct a triangle to determine a Cluster Radius Decision Point (CRDP) that is able to minimize energy consumptions of CHs.

The distance between CRDP and each point can be estimated as the radius of each cluster. *Delaunay* triangulation, which guarantees the construction of an approximate equilateral triangle, is used for constructing a triangle consisting of three CHs.

2.2 Modified Cluster Radius Control Phase

The cluster radius of three points including the construction of a triangle can be dynamically controlled by using a CRDP as a pivot. The goal of cluster radius control phase of LLC is to determine the CDRPs which minimize the energy consumption of each CH by finding optimal cluster radii. However, in some cases where subsets of CHs are assigned to specific tasks, they need to play a special role than other CHs. Therefore, we suggest a w-LLC.

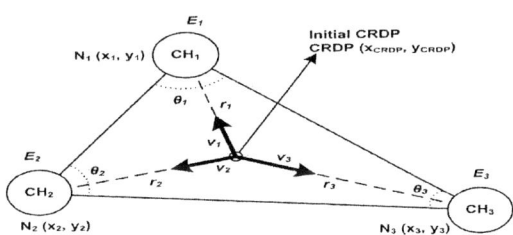

Fig. 1. Notations for w-LLC

NLP-Based Approach for w-LLC. By using NLP-based approach, a CRDP is determined by an objective function as the Eq. (1) with an iteration policy.

minimize: $f(r_1, r_2, r_3, \theta_1, \theta_2, \theta_3, \phi_{1,1}(\vec{x}), \ldots, \phi_{m,3}(\vec{x}), \psi_{1,1}(\vec{x}), \ldots, \psi_{n,3}(\vec{x}))$

$$= \frac{1}{2} \sum_{k=1}^{3} \theta_k r_k^2 \prod_{l=1}^{m} \frac{\phi_{l,k}(\vec{x})}{\frac{1}{3}\sum_{l=1}^{3} \phi_{l,i}(\vec{x})} \prod_{g=1}^{n} \frac{\frac{1}{\psi_{g,k}(\vec{x})}}{\frac{1}{3}\sum_{i=1}^{3} \frac{1}{\psi_{g,i}(\vec{x})}} - S_{triangle} \quad (1)$$

s.t. $r_i^2 = (x_{CRDP} - x_i)^2 + (y_{CRDP} - y_i)^2$

In Eq. (1), θ_k and denotes the angle value of CH_k, r_k means the distance between CRDP and CH_k, and E_k denotes the energy state of CH_k. $S_{triangle}$ is the area of triangle. A weight function is assigned to each CH where $\phi_{i,j}(\vec{x})$ and $\psi_{i,j}(\vec{x})$ represents a *penalty function* and *reward function*, respectively. A smaller penalty function value means a higher priority, while a smaller reward function value indicates a lower priority.

VC-Based Approach for *w*-LLC. In NLP-based approach, it may generate additional computation overheads due to an iterative NLP method to solve the objective function in Eq. (1). We thus consider another method based on vector computation to reduce the computation overheads to find the optimal solution. The factors of vector values for calculating the vector coordination is added. It initially executes an NLP. Weight factors do not need to be considered at first. The equation to obtain the initial position of CDRP is an Eq. (2).

$$\textbf{minimize:}\ f(r_1, r_2, r_3, \theta_1, \theta_2, \theta_3) = \frac{1}{2}\sum_{k=1}^{3}\theta_k r_k^2 - S_{triangle} \qquad (2)$$

$$\text{s.t.}\ r_i^2 = (x_{CRDP} - x_i)^2 + (y_{CRDP} - y_i)^2$$

As time goes, however, the objective function may move a CRDP towards the CH that has consumed the most amount of energy. In the objective function of VC-based approach, we do not consider the weight factors including energy state. In the iterative procedure, therefore, we need to consider the weight factors. Therefore, the next position of CRDP is determined by using Eq. (3) under the iteration policy.

$$CRDP_{i+1} = CRDP_i - \sum_{k=1}^{3}\prod_{l=1}^{m}\frac{\phi_{l,k}(\vec{x})}{\frac{1}{3}\sum_{l=1}^{3}\phi_{l,i}(\vec{x})}\prod_{g=1}^{n}\frac{\frac{1}{\psi_{g,k}(\vec{x})}}{\frac{1}{3}\sum_{i=1}^{3}\frac{1}{\psi_{g,i}(\vec{x})}}\cdot\frac{v_k}{\|v_k\|} \qquad (3)$$

$$\text{s.t.}\ CRDP_k = (x_{CRDP_k}, y_{CRDP_k})$$

The notations of Eq. (3) are same with the notations of Eq. (1).

2.3 Iteration Policy

When events occur frequently in some areas where a certain CH consumes its energy a lot. Then, the CHs in the region will consume more energy and the operation will be more important than the other CHs. Then a sink has to change the radius of CHs to balance energy consumption and to preserve the energy of the CH which has higher priority than the other CHs. Moreover, since the sink has an iteration timer, if no events occur until the timer is expired, the sink starts the *w*-LLC.

3 Simulation Results

We evaluate the performance of w-LLC. We consider clustering scheme with fixed radius (FR) and LLC for the comparison studies with w-LLC. We consider two scenarios, (1) sensing events occur around in certain hot spots, named *focused sensing*; and (2) events occur evenly across the sensor network, named *fair sensing*. As shown in Fig. 2, residual energies in FR are further quickly consumed than LLC and w-LLC. The CHs in the hot spots of *focused sensing* exhaust their own energies rapidly. In *focused sensing*, w-LLC is the most energy efficient, since it uses weight functions that reflect territorial characteristics.

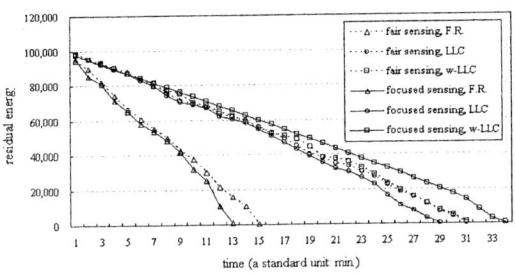

Fig. 2. Comparison of the residual energy

4 Conclusions and Future Work

We extended our previous research [4] to w-LLC to cope with the situation where events frequently occur in a certain area of a sensor network. To improve LLC, we apply the concept of weight functions to the LLC. As the technologies for wireless sensor computing improve, more complicated situations in which sensors have mobility will be considered as our future work.

References

1. I. F. Akyildiz, W. L. Su, Y. Sankarasubramaniam, and E. Cayirci, "A Survey on Sensor Networks," IEEE Communications Magazine, Vol. 40, No. 8, pp. 102-114, 2002.
2. V. Mhatre and C. Rosenberg, "Design Guidelines for Wireless Sensor Networks: Communication, Clustering and Aggregation," Ad Hoc Networks Journal, Elsevier Science, Vol. 2, No. 1, pp. 45-63, 2004.
3. J. Pan, Y. T. Hou, L. Cai, Y. Shi, and S. X. Shen, "Topology Control for Wireless Sensor Networks," in Proc. of ACM MobiCom, 2003.
4. J. Kim, E. Kim, S. Kim, D. Kim, and W. Lee, "Low-Energy Localized Clustering: An Adaptive Cluster Radius Configuration Scheme for Topology Control in Wireless Sensor Networks," in Proc. of IEEE VTC, May 2005.

Energy Efficient Dynamic Cluster Based Clock Synchronization for Wireless Sensor Network*

Md. Mamun-Or-Rashid, Choong Seon Hong**, and Jinsung Cho

Dept. of Computer Engineering, Kyung Hee University,
Seocheon, Giheung, Yongin, Gyeonggi, 449-701 Korea
mamun@networking.khu.ac.kr, cshong@khu.ac.kr,
chojs@khu.ac.kr

Abstract. Core operations (e.g. TDMA scheduler, synchronized sleep period, data aggregation) of many proposed protocols for different layer of sensor network necessitate clock synchronization. Our paper mingles the scheme of dynamic clustering and diffusion based asynchronous averaging algorithm for clock synchronization in sensor network. Our proposed algorithm takes the advantage of dynamic clustering and then applies asynchronous averaging algorithm for synchronization to reduce number of rounds and operations required for converging time which in turn save energy significantly than energy required in diffusion based asynchronous averaging algorithm.

1 Introduction

Wireless sensor network is inherently distributed in nature. Time synchronization is a critical and well studied issue in distributed system, but sensor networks differ substantially in many ways from traditional distributed systems. Typical applications of sensor networks are environmental monitoring which detects several environmental parameters such as fire, oil slicks or animal herds. In order to determine the happening time and decision making in accordance with the happenings, synchronization of different devices has to be determined. Our work is motivated by [2], fully localized diffusion based technique for global clock synchronization. Keen observation about diffusion based method is it can achieve synchronization but it needs the participation of all the nodes in the network which in turn effectively reduce the lifetime of the nodes as the nodes consumes energy for the synchronization process. Also the time convergence requires large number of rounds and incurs huge amount of flooding overhead to exchange the clock collection and updates. Clustering can efficiently reduce flooding overhead and the node participation significantly which, in turn save energy and enlarge the lifetime of the network. In this paper we study how diffusion based technique can be benefited from a dynamic clustering to reduce its flooding overhead during clock exchange and thus make the synchronization process energy efficient. We have used passive clustering [3] which can avoid potential long setup time and reduce re-forwarding significantly.

---

* This work was supported by University ITRC project of MIC.
** Corresponding author.

2 Related Work

A lot of research works [4] [5] [6] [7] [8] available in the field of clock synchronization in distributed systems. However, techniques described for synchronization in distributed system do not take into account the limited resource availability for sensor networks and other dynamic constraints such as mobility. Different works on clock synchronization in sensor network has been introduced in [1], and [8]. Our work is motivated by [2]. Three methods have been proposed for global clock synchronization in [2]. Those are (1) All-Node-Based, (2) Cluster Based and (3) Diffusion Based method. The idea of "All-Node-Based" method is impractical to implement due to its assumption of finding a single cycle which includes all the nodes at least once. To implement the idea of "All-Node-Based" method in a small manageable set they proposed "Cluster Based" method which create and maintain clusters and use "All-Node-Based" method to synchronize among the nodes of clusters. But still "Cluster Based" method is having a great amount of overhead for cluster maintenance which, ascertain its limitation to be implemented in energy constraint sensor network. Finally, the proposed fully localized "Diffusion Based" method for clock synchronization in which nodes can be synchronized at any time with its neighbor.

3 Dynamic Cluster Based Algorithm for Synchronization

Work In [2], Qun Li et al have proposed fully localized diffusion based technique for global clock synchronization in sensor networks with synchronous and asynchronous implementations. The technique is based on exchange and update clock information locally among the neighbor nodes. Synchronous rate based algorithm exchange clock reading values proportional to their clock difference in a set order. On the other hand, asynchronous method can synchronize with its neighbor at any time in any order. For detailed understanding of we refer [2].

Considering the reduced operation and less overhead of cluster head maintenance of passive clustering, we change the asynchronous algorithm in the following way:

Like other traditional synchronization algorithms our algorithm is having two major operations:

1. Collecting clock information and averaging
2. Sending the new clock value to be updated

In our algorithm cluster head and gateway nodes are responsible for initiating clock collection, averaging and updating. All gateway nodes and cluster heads are having equal probability to execute averaging operation, while asynchronous diffusion in based algorithm all nodes are responsible for clock averaging operation. Synchronization process is as follows

for each cluster head node hn_i and gateway node gn_i in the network {
  if the node is a cluster head{
    collect clock information from the member nodes of the cluster
    compute the average of the clocks of the nodes in a cluster
    send new clock to the members of the cluster }

```
if the node is a gateway node{
  collect clock information from cluster head nodes within the gateways
  transmission range
  compute the average clocks of the cluster head nodes
    send new clock information to the cluster head nodes}}
```

4 Simulation

To justify our claim of reduction of operation and thus reduction of energy, we implement two different algorithms and compare the results. Implemented algorithm executed by varying number of nodes and transmission range. In case of asynchronous averaging algorithm for each time slot, each node executes the averaging operation and the order of the node is randomized. Similarly, our algorithm for each time slot cluster head and gateway nodes execute averaging operation once and the operation of those nodes are randomized. Real time deployment of our algorithm needs the time interval of synchronization operation for the sensor nodes which depends on the clock drift of the nodes in the sensor network.

Our first and second set of experiment has the following parameters

    a. Roaming Space 200 X 200
    b. Number of Nodes = 200

The experiment shows the number of rounds required for asynchronous averaging algorithm and passive clustering averaging algorithm. Both the algorithms require comparatively large number of rounds at lower transmission range and fewer numbers of rounds at high transmission range. At low transmission range the network nodes supposed to have fewer neighbors and the effect of diffusion operation effects slowly. Number of round requirement at higher transmission range can be explained with inverse logic.

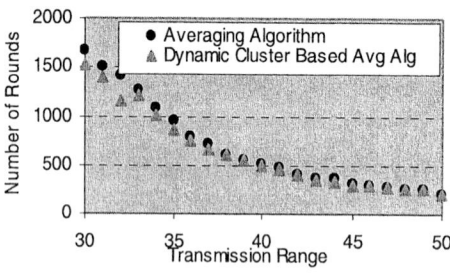
Fig. 1. Number of rounds---Fixed number of nodes---Varying transmission range

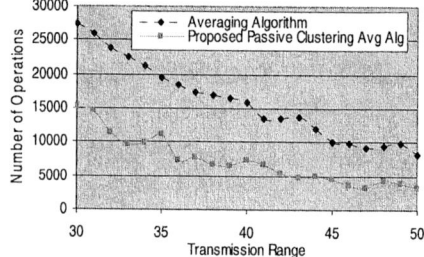
Fig. 2. Number of operations---Fixed number of nodes---Varying transmission range

Next set of experiments compares our proposed algorithm with asynchronous averaging algorithm under increasing number of nodes. Our proposed algorithm performs better both in terms of number of rounds and operations required for synchronization.

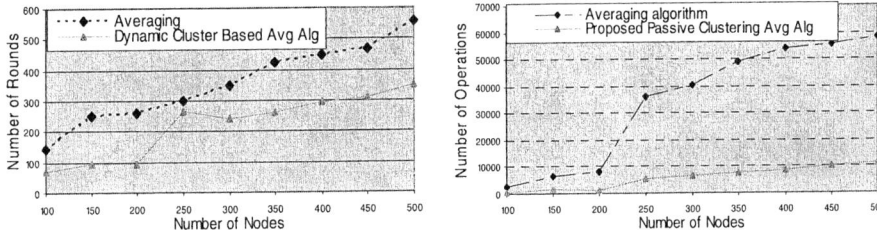

Fig. 3. Number of rounds --- varying number of nodes --- Fixed transmission range

Fig. 4. Number of operations --- varying number of nodes --- Fixed transmission range

5 Conclusion

Diffusion based asynchronous averaging algorithm each node performs its operation locally and diffused to whole network to achieve the global synchronization. Albeit Cluster based method proposed in the same work have a better performance but suffered from the overhead of cluster creation and maintenance. One of the factors that affects the performance of diffusion based asynchronous algorithm is all node have the equal probability to execute averaging operation. Our proposed algorithm takes the advantage of passive clustering which reduces the chance of executing averaging operation in all nodes. Rather our proposed algorithm executes averaging operation only in cluster head and gateway nodes which makes it energy efficient.

References

1. J. Elson and D. Estrin, Time synchronization for wireless sensor networks, in Proc. of the 2001 International Parallel and Distributed Processing Symposium (IPDPS), Workshop on Parallel and Distributed Computing Issues in Wireless Networks and Mobile Computing, San Francisco, CA, (2001)
2. Qun Li, Daniela Rus, Global Clock Synchronization in Sensor Networks, IEEE INFOCOM Vol. 1, Hong Kong, China, (2004) 564-574
3. Kwon, T.J., Gerla, M. Efficient flooding with passive clustering (PC) in ad hoc networks. ACM SIGCOMM Computer Communication Review 32 (2002) 44–56
4. Chalermek Intanagonwiwat, Ramesh Govindan, and Deborah Estrin. Directed diffusion: A scalable and robust communication paradigm for sensor networks. In MobiCOM, Boston, Massachusetts, (2000) 56-67
5. K. Arvind, Probabilistic Clock Synchronization in Distributed Systems, IEEE Transactions on Parallel and Distributed Systems, Volume 5, Issue 5, (1994) 474 – 487
6. M.Lemmon, J. Ganguly, and L. Xia, Model-based clock synchronization in networks with drifting clocks, in Proc Of the 2000 Pacific Rim International Symposium on Dependable Computing, Los Angeles, CA, (2000) 177–185
7. B. Hofmann-Wellenhof, H. Lichtenegger, and J. Collins, Global Positioning System: Theory and Practice, 4th ed. Springer- Verlag, 1997
8. K. Romer, Time synchronization in ad hoc networks, in Proc. of ACM Mobihoc, Long Beach, CA, (2001) 173-182

An Intelligent Power Management Scheme for Wireless Embedded Systems Using Channel State Feedbacks*

Hyukjun Oh[1], Jiman Hong[1], and Heejune Ahn[2]

[1] Kwangwoon University, Seoul, Korea
{hj_oh, gman}@kw.ac.kr
[2] Seoul National University of Technology, Seoul, Korea
heejune@snut.ac.kr

Abstract. In this paper, an intelligent power management scheme for embedded systems with wireless applications is proposed to reduce the power consumption of the overall system. The proposed method is based on the feedback of the extreme channel state indicator that is designed to detect the extremely bad channel condition. The considerable power reduction is achieved by turning off modules within the embedded system related to the information transmissions under such an unreliable channel condition. A simple extreme channel state detector is also proposed.

1 Introduction

Minimizing power and energy dissipation is a key factor in wireless embedded system designs. Therefore, the system should be designed with respect to that particular application to be more power-efficient. This has led to a significant research effort in power-efficient designs in such applications [1]-[4]. It is natural to think of ways to reduce the power consumption considering and utilizing inherent properties of the wireless applications. The use of such properties requires power-efficient design from the inside of the communication module, and it will provide satisfactory power reduction performance inherently. However, most previous works to date have treated it as a given and fixed module and have focused on the power reduction from the top of it [1]-[4].

In this paper, an intelligent power management scheme for embedded systems with wireless communication applications is proposed using such inherent properties of the wireless applications. Any transmission related module is required to be on in the wireless applications only when the transmission is active and reliable. The proposed method is based on the feedback of the extreme channel state indicator from the inside of the baseband transmission module. Under the unreliable channel condition, carrying user information over the air link is completely impossible. The considerable power reduction can be achieved by

---

* The present research has been conducted by the Research Grant of Kwangwoon University in 2004.

turning off several modules within the embedded system related to the information transmission like LCD, image encoder, voice encoder, and power amplifier under this condition. A simple software code can serve as a brain of such tasks given primitive feedback signals from the detector. A simple and efficient signal processing algorithm to detect the extreme channel condition is also proposed.

2 System Model

A usual embedded system comprises a processor core, an instruction cache, a data cache, a main memory, and several custom blocks [1]. Such a custom block has been regarded as a simple one that adds just a constant amount of power dissipation to the overall system. Most previous works have tried to minimize power consumption from the perspective of the rest components other than those ones [1]-[4]. The communication module in the system with the wireless application has been considered as one of this kind. Considerable portion of total power consumption is caused by the transmission related functionality itself in the wireless embedded system. Therefore, it is desirable to include this module in optimizing the power consumption of the overall system. Because our interest is focused on the system with the application of wireless communications, the best target system would be cellular phone. Fig. 1 shows simplified block diagram of a simple cellular phone. It shows only major components and it mainly consists of three parts: application related, basedband modem related, and RF/IF related components. It is assumed that the dedicated application or modem processor is equipped with an operating system and it serves as a brain for performing the proposed power management scheme in this paper. A real time operating system (RTOS) would be appropriate for this purpose, but we do not limit the form of possible operating system in our study. It simply provides a room for central control of the power management scheme.

3 Proposed Power Management Scheme

Applications in wireless embedded systems are inevitably closely related to the propagation channel and channel conditions. In normal channel condition, there is nothing special to consider for the purpose of reducing power dissipation from the point of the overall system. In this case, several existing system models are useful and many previous works on top of them can be applied. In other words, any power saving scheme assuming the transmission related blocks as simple constant power consuming black-boxes can be used. However, it is no longer valid in unsatisfactory channel condition. In the extremely bad channel condition, the real time transmission over the poor air link is not possible. It means that operating any transmission related code, function, module, logic, and component in the system is meaningless. This observation leads to the design of the power saving scheme considering the transmission related blocks beneficial resources for our purpose.

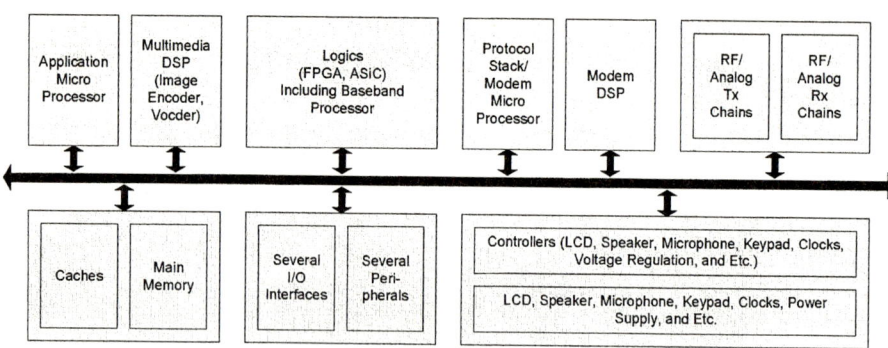

Fig. 1. A system model of a simple cellular phone

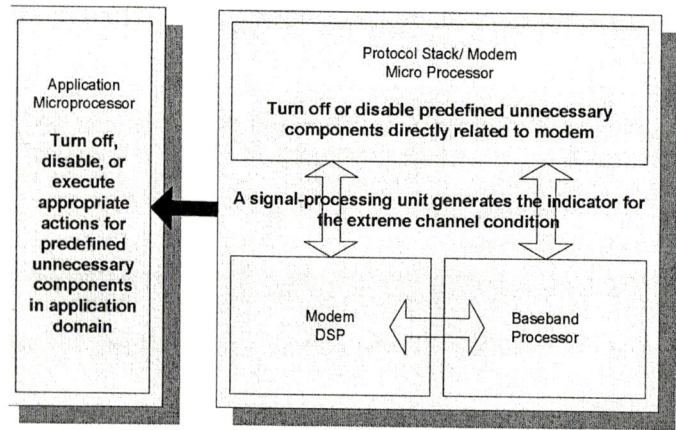

Fig. 2. The proposed power saving scheme using the extreme channel state indicator

Such a channel condition mentioned above can happen often in wireless environment due to the characteristic of mobile channels of wide and fast variations. In the case of channel being unrecoverable or/and a long-lasting bad channel condition, a dedicated higher layer function in the protocol stack kicks in to control radio link in most popular wireless communication system [5]. However, when the propagation channel falls in and out such an extremely bad conditions quite often or repeatedly. the dedicated function does not help. In this paper, we propose to disable or turn off the meaningless components (codes, functions, modules, logics, and blocks) to minimize the power consumption when the extreme channel condition is detected. Moreover, some components do not need to operate in full performance under such a condition for power efficiency. The operating system on the dedicated processor serves as a central control unit of managing and performing the required operations for given extreme channel condition indicator. First of all, we need to identify which components can be disabled or put on much lower performance under this situation. Those components are dependent on the configuration of the target embedded system. In our system model, LCD, image/video encoder, vocoder, and RF chains including power amplifier can be candidates for such components. Furthermore, the data memory can be flushed and/or the number of accessing internal or external memory can be minimized. Note that RF chains should be turned on during the pilot transmission to get recovered for the possible forthcoming normal channel condition. Several variants are also possible from the proposed method as per the architecture of the target embedded system. The proposed power-efficient scheme is summarized in Fig. 2.

4 Detection of Extreme Channel Condition

A computationally efficient signal-processing algorithm is proposed to detect the extremely unstable channel condition. The algorithm should be simple to implement so that it does not add another noticeable power dissipation to the overall system. The proposed algorithm is based on the use of channel state information (CSI) that is available in most wireless communication system for the reliable transmissions.

We consider the CDMA cellular systems that are the most widely accepted third generation standards for the cellular communications in the world [5]. However, the proposed scheme is not limited to the applications with the CDMA systems. It can be generalized easily to other systems like OFDM based ones. In CDMA systems, binary CSI's of the transmitted power control bits are available in plenty of time. A value of the bit is indicating that the current channel quality is good enough to satisfy the required quality of service. The other is representing the channel condition that is not good enough to achieve the reliable transmissions. The proposed detection scheme simply estimates the frequency of occurrences of this unsatisfactory channel state by counting the number of the second state of the power control bits during the given time period. Then, the current channel condition is regarded as the extremely bad state if the majority of the bits are the second state in the given time frame. Several simulation results showed that the proposed algorithm worked very well as expected. The details and the performance of the proposed scheme are not addressed in this paper due to the space limitation.

References

1. Li, Y., Henkel, J.: A framework for estimating and minimizing energy dissipation of embedded HW/SW systems. Proc. Int. Conf. Design Automation (1998) 188–193
2. Tiwari, V., Malik, S., Wolfe, A.: Power analysis of embedded software: A first step towards software power minimization. IEEE Trans. VLSI Systems (1994) 437–445
3. Benini, L., Micheli, G.: System-level power optimization: Techniques and tools. Proc. Int. Symp. Low Power Electronics and Design (1999) 288–293
4. Simunic, T., Micheli, G., Benini, L.: Energy efficient design of battery-powered embedded systems. Proc. Int. Symp. Low Power Electronics and Design (1999) 212–217
5. 3GPP Specifications, Release 5.

Analyze and Guess Type of Piece in the Computer Game Intelligent System

Z.Y. Xia[1], Y.A. Hu[2], J. Wang[1], Y.C. Jiang[1], and X.L. Qin[1]

[1] Department of computer science, Nanjing University of Aeronautics and Astronautics, China
zhengyou_xia@yahoo.com
[2] IBM Corporation, China
yahoo@gbai.info

Abstract. Siguo game is an interesting test-bed for artificial intelligent research. It is a game of imperfect information, where completing players must deal with possible knowledge, risk assessment, and possible deception and leaguing players have to deal with cooperation and information signal transmission. Since Siguo game is imperfect information game that the player doesn't know the type of piece and strategy that opponent moves, to exactly guess type of opponent' piece is a very important parameter to evaluate the capability of Siguo game program. In this paper, we first construct a fuzzy type table by analyzing more than one thousand different embattle lineups (i.e. chess manuals) of Siguo game, and then we present a algorithm that updates type table by using information from opponent during playing game. The updating type of pieces algorithm is designed by considering the two strategies, i.e. optimism and pessimism based on the fuzzy notion. At last we give a method to guess the type of piece by using fuzzy type proximity relation between two neighboring pieces.

1 Introduction

The field of computer strategy games is an intriguing one. Building strong game-playing programs has been an important goal for artificial intelligence researchers for more than half a century [1] [2] [3]. The principal aim is to witness the intelligence of computers. The artificial intelligence community has benefited from the positive publicity generated by chess (e.g. Blue Deep [4] [5][6][18] that have defeated the human world chess champion Garry Kasparov), checker [7], backgammon [8], Shogi [9] and Othello [10] programs that capable of defeating the best human player. Of all the classic games of mental skill, only card game, Chinese chess [11] [12], and Go (WeiQi) [13] [14] [15] have yet to see the appearance of serious computer challengers.

In the Chinese chess and Go, players have complete knowledge of the entire game state, since everything is visible to both participants. Therefore we called these games (e.g. chess, Go, checker, .etc.) two-player perfect-information games. Since these game can get the full information of opponent, to solve these games adopts methods of knowledge database (including endgame database), Evaluation function + search algorithms (including brute-force, heuristic game tree search algorithms, etc.). In contrast, bridge [16] and Siguo involve imperfect information, since the other players' cards or type of piece are not known. We called Bridge and Siguo game imperfect-information games. Traditional methods for perfect-information games have not been sufficient to play Bridge [16] and Siguo games well. Therefore, it is also the reason these games promise greater potential research benefits.

During playing Siguo game, to exactly guess type of opponent's pieces is a very most parameter to evaluate capability of computer Siguo game system. To improve the capability of guessing the type of opponent's piece can validly help computer Siguo game system to forecast and evaluate the strategy that opponent will move. However, few studies have been done on computer Siguo game. In this paper, we construct fuzzy type table for the pieces of opponent in the opening phase by analyzing more one thousand lineups of Siguo game and present an algorithm to update fuzzy type table during playing game by using information from opponent. We also construct type proximity relation matrix between the two neighboring pieces to guess the type of neighboring piece unknown by using type of piece known.

2 Introduction to Siguo Game

Siguo Junqi game (abs. Siguo game) is the most popular game with imperfect information in china. Siguo game is different to western Kriegspiel in the rules and chessboard architecture. Siguo game is developed by Chinese people. The player doesn't know the type of piece and strategy of opponent and confederate during playing game. With development of Internet, there are more four million players to play Siguo game on the line. In china, there are more fifty thousand players to simultaneously play Siguo game on the line in the several game stations like as LianZhong[19], etc. The Siguo game can be classified two kinds. it is 1vs1 shown in the Figure 1 and 2 vs 2 shown in the Figure 2. The most popular kind is 2 vs 2 on the internet. In Siguo game, per player has twenty-five pieces. These pieces are consisted of three Sappers (Chinese people call it Xiao bing), three Second Lieutenants (Pai zhang), two Captains (Lian zhang), two Majors (Ying zhang), two Lieutenant Colonels (Tuan zhang), two Colonels (Lv zhang), two Senior Colonels (Shi zhang), one Lieutenant Generals (Jun zhang), one Marshals (Si ling), three bombs (Zha dan), three mines (Lei), one oriflamme (Jun qi).

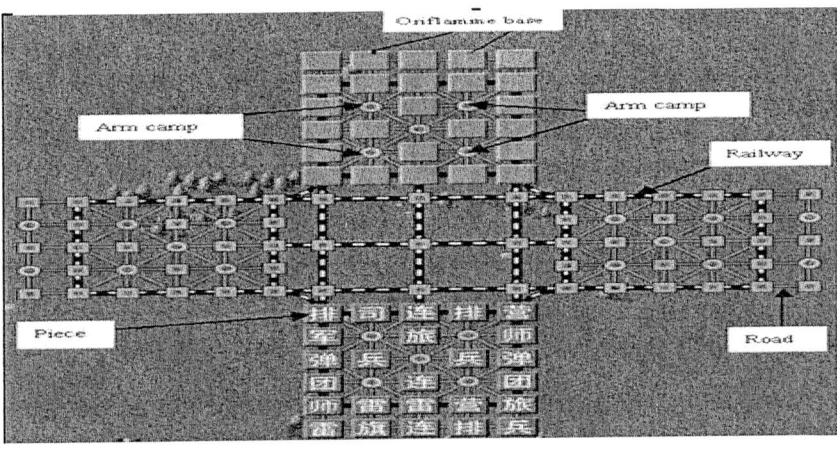

Fig. 1. Siguo game: 1 vs 1

Rules: embattle lineup

Per player has to embattle lineup of pieces before playing Siguo game. Embattling lineup of pieces is to put pieces in the different places of Siguo chessboard according to intention of player. The rules of embattling lineup of pieces are as the follows.

1. The Bomb cannot be placed in the first row and the mine can only be placed in the last two rows.
2. The oriflamme can only placed in the First or Second oriflamme base.
3. The Piece of Siguo game cannot be placed in the arm camp at embattling lineup of pieces.

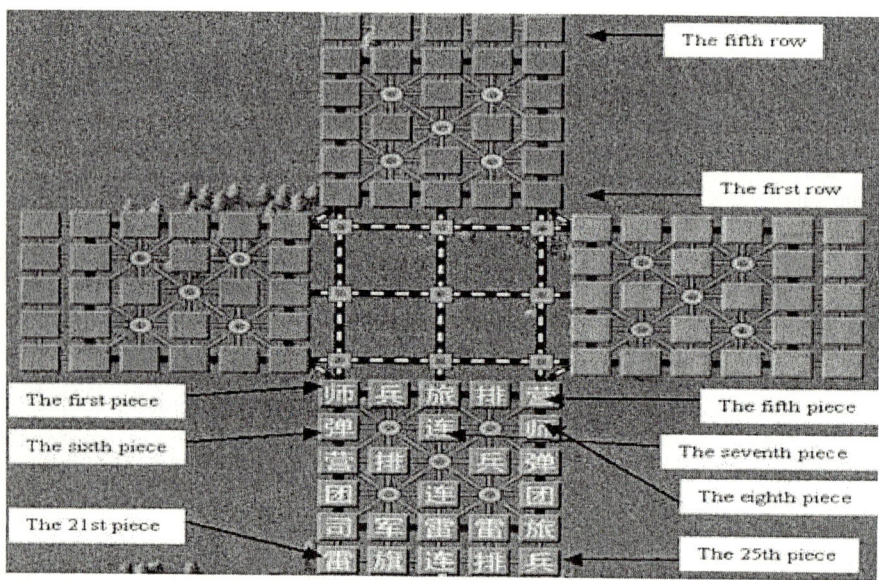

Fig. 2. Siguo game: 2 vs 2

Rules: moving piece

1. The mines and pieces at oriflamme base aren't allowed to move.
2. The pieces can only be moved by one step on the road.
3. At the any time, there is no more than one piece in the same place.
4. The pieces of Siguo game can only be moved along the straight line excepting sapper.
5. The sappers can turn the corner to move on the railway as long as there is no piece to obstruct it to move to the destination address.
6. Steps of moving piece aren't limited on the railway as long as there is no piece to obstruct it to move to the destination address.

Win or Lose Rules: Who first can kill out all pieces of the rival that can move or capture rival oriflamme, and one will win this game.

Rules: Killing piece
The rank from high to low is as the following: sapper < Platoon leader< Company commander< battalion< Colonel< Brigadier< Division commander< Army corps commander< Chief of staff.
1. The high rank piece can kill the low rank piece.
2. When the two same rank pieces meet, the two pieces is killed each other.
3. Any piece except sapper meets mine, the piece will be killed. Only sapper can kill mine.
4. Any piece meet the bomb, the two pieces will be killed by each other.
5. If any piece of player can capture oriflamme of opponent, the player wins this game.
6. Piece in the arm camp cannot be killed by any pieces.

3 Constructing Fuzzy Type Table

The notion of fuzzy set first appeared in the papers written by Zadeh [17]. The notion tries to show that an object more or less corresponds to a particular category. The degree to which an element belongs to a category is an element of the continuous interval [0, 1] rather than the Boolean pair {0, 1}. In the Sgiuo game, since the player doesn't know the type of opponent' piece, to exactly guess about the type of opponent' piece is a very important parameter to evaluate the capability of computer Siguo game program. In the Siguo game, the players have to embattle lineup of pieces before Siguo game is begun. We analyze one thousand three hundred different lineups of pieces from the country Siguo game competition, league matches of ourgame.com[21], and then construct the original fuzzy type table about the opponent' pieces in the opening phase. The type table can be used to analyze and guess the type of opponent 'piece during playing Siguo game.

The types of pieces are classified the twelve different types. Bomb, mine, and oriflamme of Siguo game are the especial types and the rank of other types is from low to high: sapper < Platoon leader< Company commander< battalion< Colonel< Brigadier< Division commander< Army corps commander< Chief of staff. We use the statistic membership function Equation 1 to analyze type of the opponent's pieces.

$$\mu_{x_i}(P_j) = \frac{times \quad of \quad x_i \quad belong \quad to \quad P_j}{the \quad total \quad times \quad be\log \quad to \quad P_j}, (i \in (1,\cdots,12), j \in (1,\cdots,25)) \quad (1)$$

$where \; x_i \; denotes \; type \;, P_j \; denotes \; piece$

The fuzzy set of twenty five pieces in Siguo game is illustrated in Table1.
In the Table 1, for example, the fuzzy set of the first piece of opponent is:

$P1 = Sapper/0 + Second\ Lieutenant/0.123 + Captain/0.102 + Major/0.188 + Lieutenant\ Colonel/0.147$
$+ Colonel/0.07 + Senior\ Colonel/0.164 + Lieutenant\ General/0.048 + Marshal/0.158 + min\ e/0 + bomb/0 + oriflamme/0$

This denotes that the sapper doesn't belong to the first piece of opponent, and the membership value of captain belonging to the first piece of opponent is 0.102.

The original fuzzy type state about the opponent' piece in the opening phase is shown in the Table 1. During playing game, the player can gradually get some information from opponent, which is caused by fighting with opponent and intention that opponent moves piece. Therefore, we can use information from the opponent to update the fuzzy type table, exactly predicate the type of opponent' piece and search the optimum move strategy to win over the opponent. There are two strategies to predicate the type of opponent's piece. One is the optimistic strategy that the player considers the type of biggest membership value to be the type of opponent's piece (i.e. the most possible thing can happen by all means). The other is pessimistic strategy that the most impossible thing may be changed as the most possible one. This membership function represents the degree of player confidently predicate what's type of opponent' piece and we define the optimistic membership function as Equation 2.

$$\mu_{dop}(x_i, p_j) = \frac{\mu_{x_i}(P_j) - \min}{\max - \min}, (i \in (1,\cdots,12), j \in (1,\cdots,25)) \text{ and } \mu_{x_i}(P_j) \rangle 0 \qquad (2)$$
$$where \quad \min = 0, \max = \max(\mu_{x_1}(P_j),\cdots,\mu_{x_{12}}(P_j)) \in (0,1)$$

The membership function Equation 2 is linear increase with $\mu_{x_i}(P_j)$ increase.

Table 1. Type table for twenty-five pieces of Siguro game at open phase

$\mu_i(P_j)$	Sapper	Platoon leader	Company commander	battalion	Colonel	Brigadier	Division commander	Army corps commander	Chief of staff	Mine	Bomb	Oriflamme
1	0	0.123	0.102	0.188	0.147	0.07	0.164	0.048	0.158	0	0	0
2	0.416	0.169	0.177	0.054	0.056	0.032	0.029	0.027	0.04	0	0	0
3	0.137	0.115	0.145	0.088	0.083	0.193	0.038	0.088	0.113	0	0	0
4	0.426	0.166	0.174	0.046	0.035	0.04	0.04	0.027	0.046	0	0	0
5	0	0.115	0.102	0.187	0.149	0.054	0.168	0.051	0.174	0	0	0
6	0.013	0.113	0.115	0.071	0.091	0.064	0.184	0.07	0.075	0	0.204	0
7	0.284	0.07	0.18	0.064	0.067	0.045	0.053	0.056	0.013	0	0.168	0
8	0.016	0.139	0.083	0.091	0.075	0.09	0.162	0.075	0.064	0	0.205	0
9	0.021	0.105	0.105	0.046	0.073	0.101	0.182	0.094	0.091	0	0.182	0
10	0.18	0.131	0.115	0.083	0.078	0.048	0.059	0.043	0.021	0	0.242	0
11	0.276	0.112	0.118	0.046	0.067	0.075	0.051	0.043	0.016	0	0.19	0
12	0.029	0.105	0.129	0.091	0.075	0.124	0.181	0.043	0.054	0	0.169	0
13	0.024	0.142	0.161	0.089	0.158	0.088	0.105	0.064	0.035	0	0.134	0
14	0.287	0.089	0.292	0.08	0.051	0.04	0.016	0.035	0.003	0	0.107	0
15	0.064	0.129	0.206	0.104	0.164	0.048	0.078	0.032	0.038	0	0.137	0
16	0.075	0.051	0.091	0.113	0.094	0.174	0.072	0.054	0.018	0.198	0.06	0
17	0.129	0.035	0.028	0.091	0.054	0.107	0.11	0.038	0.005	0.4	0.003	0
18	0.155	0.04	0.097	0.105	0.118	0.129	0.107	0.043	0.008	0.123	0.075	0
19	0.19	0.046	0.097	0.07	0.067	0.145	0.086	0.019	0	0.271	0.011	0
20	0.083	0.043	0.102	0.097	0.11	0.134	0.059	0.04	0.021	0.236	0.075	0
21	0.054	0.121	0.062	0.091	0.075	0.083	0.038	0.005	0.002	0.45	0.019	0
22	0	0.238	0	0	0	0	0	0	0	0.137	0	0.625
23	0.064	0.078	0.051	0.043	0.067	0.054	0.003	0.003	0	0.635	0.002	0
24	0	0.4	0	0	0	0	0	0	0	0.225	0	0.375
25	0.075	0.12	0.255	0.064	0.059	0.062	0.016	0.003	0.003	0.324	0.019	0

If the player attitude is pessimistic and the type of least membership value is considered to be possible type of opponent' piece, the membership function is present in Equation 3.

$$\mu_{dpp}(x_i, p_j) = \frac{\mu_{x_i}(P_j) - \max}{\min - \max}, (i \in (1,\cdots,12), j \in (1,\cdots,25)) \text{ and } \mu_{x_i}(P_j) \rangle 0 \quad (3)$$

$$where \ \min = 0, \max = \max(\mu_{x_1}(P_j),\cdots,\mu_{x_{12}}(P_j)) \in (0,1)$$

The membership function Equation 3 is linear decrease with $\mu_{x_i}(P_j)$ increase.

We let maximum be $\max(\mu_{x_1}(P_j),\cdots,\mu_{x_{12}}(P_j))$ and the minimum be 0. In our computer Siguo game system, we use the equation (4) to update type table.

$$\mu_{dp}(x_i, p_j) = w_1 \mu_{dop}(x_i, p_j) + w_2 \mu_{dpp}(x_i, p_j), w_1 + w_2 = 1, w_1 \in (0,1), w_2 \in (0,1) \quad (4)$$

The fuzzy type table is updated by the following algorithm using information from opponent during playing game with opponent.

Update type table algorithm

1) Check the result of our piece Vs opponent's piece.
2) If our piece is killed by opponent's piece then update type table according to the following operations:
   a) the type of opponent' piece less than and equal to our piece type is set membership value down 0;
   b) Set the maximum to be the max membership value among type membership value of the opponent' piece.
   c) Compute fuzzy membership value according to equation (4)
3) If our piece kills the opponent's piece then update opponent's type table according to the following operations:
   a) the type of opponent' piece more than and equal to our piece type is set membership value down 0;
   b) Set the maximum to be the max membership value among type membership value of the opponent' piece.
   c) Compute fuzzy membership value according to equation (4)
4) If pieces are all removed from chessboard then update opponent's piece according to the following operations:
   a) the type of opponent' piece more than and less than our piece type except bomb type is set membership value down 0;
   b) Set the maximum to be the max membership value among type membership value of the opponent' piece.
   c) Compute fuzzy membership value according to equation (4)

---

4 Guess Type of Neighboring Piece Based on Type Proximity Relation

In Siguo game, the embattling process is that the players put the pieces in different places of chessboard by using some relation according to players' intention. The relation between two neighboring pieces can provide information that helps us to analyze and guess the type of opponent's piece. Therefore, when we exactly know the type of some piece, we can guess the type of their neighboring pieces according to the relation. Since the information about piece of opponent is uncertain and fuzzy, we use fuzzy notion to analyze the relation. The relation is called fuzzy type proximity relation. According to the excellent human player experience, it is not necessary to guess the exact type of neighboring piece of opponent. It is enough for excellent human player to well play Siguo game by knowing the range about the type of neighboring piece.

Let U= {Sapper, Second Lieutenant, Captain, Major, Lieutenant Colonel, Colonel, Senior Colonel, Lieutenant General, Marshal} be nine elements set. L = {Minimal, Small, Medium, Large, Very large} denotes the five different levels. We survey evaluation about the type levels of piece from the one hundred human players of Siguo game and construct the membership function illustrated Figure 3a. According to membership function in Figure 3a, we can get the following fuzzy set of five levels illustrated Figure 3b. The fuzzy set of twenty five pieces in Siguo game is described in Table 2.

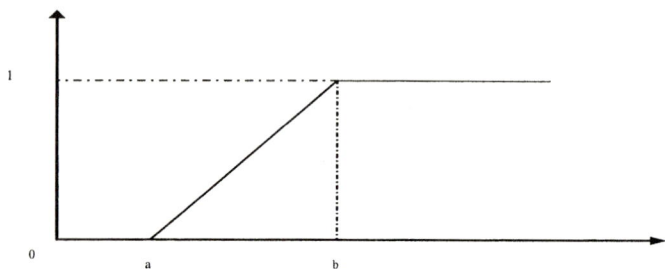

Fig. 3a. Membership function, a = 20, b = 80

Minimal (Sapper/1+ Second Lieutenant/0+ Captain/0+ Major/0+ Lieutenant Colonel/0+ Colonel/0+ Senior Colonel/0+ LieutenantGeneral/0+ Marshal/0).
Small ((Sapper/0+ Second Lieutenant/1 +Captain/1+ Major/0.775+ Lieutenant Colonel/0+ Colonel/0+ Senior Colonel/0+ LieutenantGeneral/0+ Marshal/0).
Medium (Sapper/0+ Second Lieutenant/0+Captain/0 + Major/0.225+ Lieutenant Colonel/1+ Colonel/0+ Senior Colonel/0+ Lieutenant General/0+ Marshal/0).
Large (Sapper/0+ Second Lieutenant/0+Captain/0 + Major/0+colorel/1+ Senior Colonel/1+ Lieutenant General/0+ Marshal/0).
Very Large (Sapper/0+ Second Lieutenant/0+Captain/0 + Major/0+colonel/0+ Senior Colonel/0+Lieutenant General/1+ Marshal/1).

Fig. 3b. Fuzzy set of five levels

During playing Siguo game, we can conjecture type of the neighboring piece by using fuzzy type proximity relation. We use the following membership Equation 5.

$$\mu_s(x, y) = e^{-k^2|x_i - y_j|}, 0 < x_i \text{ and } x_i \in P_k \text{ fuzzy set}, 0 < y_j \text{ and } y_j \in P_l \text{ fuzzy set}, \quad (5)$$

P_k and P_l neighboring pieces. x_i, y_j respectively denotes type of piece

We use the experience value $k^2 = 5.18$ to construct fuzzy type proximity relation between two neighboring pieces. For example, the type proximity relation between first piece and sixth piece of opponent is described in Figure 4 and Equation 6.

Analyze and Guess Type of Piece in the Computer Game Intelligent System   1181

Table 2. Pieces of Siguo game membership $\mu_p(x)$

	minimal	small	medium	large	Very large	Mine	Bomb	oriflamme
1	0	0.371	0.189	0.234	0.206	0	0	0
2	0.416	0.388	0.068	0.061	0.067	0	0	0
3	0.137	0.368	0.103	0.231	0.221	0	0	0
4	0.426	0.376	0.045	0.08	0.073	0	0	0
5	0	0.362	0.191	0.222	0.225	0	0	0
6	0.013	0.283	0.107	0.248	0.145	0	0.204	0
7	0.284	0.3	0.081	0.098	0.069	0	0.168	0
8	0.016	0.293	0.095	0.252	0.139	0	0.205	0
9	0.021	0.246	0.093	0.273	0.185	0	0.182	0
10	0.18	0.31	0.097	0.107	0.064	0	0.242	0
11	0.276	0.272	0.077	0.126	0.059	0	0.19	0
12	0.029	0.315	0.095	0.295	0.097	0	0.169	0
13	0.024	0.372	0.178	0.193	0.099	0	0.134	0
14	0.287	0.443	0.069	0.056	0.038	0	0.107	0
15	0.064	0.416	0.187	0.126	0.07	0	0.137	0
16	0.075	0.23	0.119	0.246	0.072	0.198	0.06	0
17	0.129	0.134	0.074	0.217	0.043	0.4	0.003	0
18	0.155	0.218	0.142	0.236	0.051	0.123	0.075	0
19	0.19	0.196	0.082	0.231	0.019	0.271	0.011	0
20	0.083	0.22	0.132	0.193	0.061	0.236	0.075	0
21	0.054	0.254	0.095	0.121	0.007	0.45	0.019	0
22	0	0.238	0	0	0	0.137	0	0.625
23	0.064	0.162	0.077	0.057	0.003	0.635	0.002	0
24	0	0.4	0	0	0	0	0.225	0.375
25	0.075	0.425	0.073	0.078	0.006	0.324	0.019	0

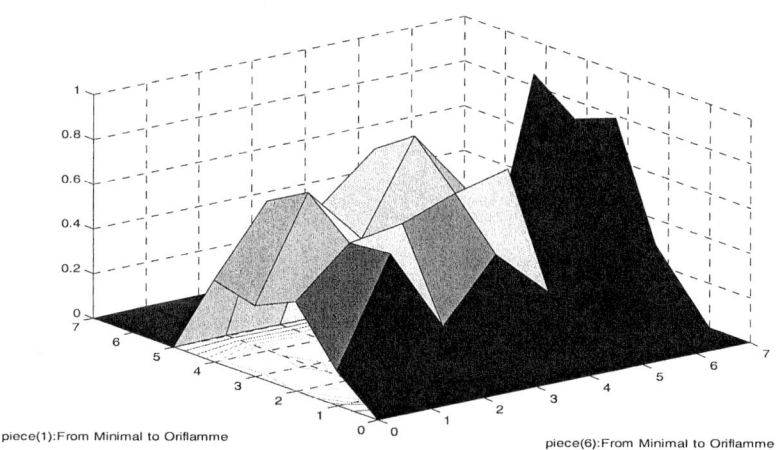

piece(1):From Minimal to Oriflamme   piece(6):From Minimal to Oriflamme

Fig. 4. Fuzzy type proximity relation between Piece(1) and Piece(6)

During opening phase and middle game, the type proximity relation is often used. For example, if we know the opponent's first piece is large type (i.e. colonel or senior colonel), we can use the type proximity relation between the first piece and sixth piece to conjecture type of the sixth piece. According to Figure 4, we can get the sixth piece fuzzy set, which is P6:minimal/0.32+small/0.76+medium/0.52+large/0.93+ verylarge/0.63+bomb/0.86+mine/0+oriflamme/0),). If we use the $\lambda = 0.8$ -cut, we can get the following type proximity relation matrix between the first piece and sixth piece showed in Equation 7. Therefore, when the fist piece is large type, the sixth piece is possible large and bomb type at $\lambda = 0.8$ -cut.

$$R_{type(1,6)} = \begin{cases} P_1\text{:min} & 0 & 0 & 0 & 0 & 0 & 0 & 0 & 0 \\ \text{small} & 0.16 & 0.63 & 0.25 & 0.53 & 0.31 & 0 & 0.42 & 0 \\ \text{medium} & 0.40 & 0.61 & 0.65 & 0.74 & 0.80 & 0 & 0.93 & 0 \\ \text{large} & 0.32 & 0.76 & 0.52 & 0.93 & 0.63 & 0 & 0.86 & 0 \\ \text{very large} & 0.37 & 0.67 & 0.60 & 0.80 & 0.73 & 0 & 0.98 & 0 \\ \text{mine} & 0 & 0 & 0 & 0 & 0 & 0 & 0 & 0 \\ \text{bomb} & 0 & 0 & 0 & 0 & 0 & 0 & 0 & 0 \\ \text{oriflamme} & 0 & 0 & 0 & 0 & 0 & 0 & 0 & 0 \\ P_6: & \text{min} & \text{small} & \text{medium} & \text{large} & \text{very large} & \text{mine} & \text{bomb} & \text{oriflamme} \end{cases} \quad (6)$$

$$R_{\lambda=0.8} = \begin{cases} 0 & 0 & 0 & 0 & 0 & 0 & 0 & 0 \\ 0 & 0 & 0 & 0 & 0 & 0 & 0 & 0 \\ 0 & 0 & 0 & 0 & 1 & 0 & 1 & 0 \\ 0 & 0 & 0 & 1 & 0 & 0 & 1 & 0 \\ 0 & 0 & 0 & 1 & 0 & 0 & 1 & 0 \\ 0 & 0 & 0 & 0 & 0 & 0 & 0 & 0 \\ 0 & 0 & 0 & 0 & 0 & 0 & 0 & 0 \\ 0 & 0 & 0 & 0 & 0 & 0 & 0 & 0 \end{cases} \quad (7)$$

In the actual playing Siguo game, if the first piece of player is large type, the sixth piece is often large or bomb type. This verifies that the conjecture type of piece based on fuzzy type proximity relation between neighboring two pieces is accord with actual result. To use the fuzzy type proximity relation, 82% exact guessing type of neighboring piece has been achieved by our computer Siguo game system in the actual playing game with human players.

5 Conclusion

The good computer game programs with perfect information like as international chess, Chinese chess is showed by strong compute capability. We consider the good Siguo game system with imperfect information should be showed by the compute+ psychology capability. To win over the best human player, we design the fuzzy type table, updating type table algorithm and guessing type of piece based on type proximity relation matrix between the two neighboring pieces in this paper. These methods validly help our computer Siguo game system to evaluate and conjecture the type of piece and strategy that opponent moves. Presently, the rank of our computer Siguo game system is equal to the second rank of human player in the two-person competition (i.e.1V1). However, the rank of our system is equal to novice of human player in four-person competition (i.e.2V2). Therefore, in our future works, we will research the fuzzy cooperation between leagues and construct the better evaluation function to improve our computer Siguo game capacity.

Acknowledgments

ZhengYou Xia's research was supported by grants from the JiangSu High Technology Research Plan Project No.BG2004005 and Aeronautics science fund No.02F52033.

References

1. C.E. Shannon, Programming a computer for playing chess, Philos. Mag. 41, 7, 1950 256-275.
2. H.J.van den Herik,Jack van Rijswijck,etc., Games solved: Now and in the future, artificial Intelligence 134 ,277-311,2002.
3. Aviezri S. Fraenkel: Complexity, appeal and challenges of combinatorial games. Theor. Comput. Sci. 303(3): 393-415, 2004.
4. F.-h. Hsu, Behind Deep Blue, Princeton University Press, Princeton, NJ, 2002.
5. McGrew, T,Collaborative intelligence. The Internet Chess Club on game 2 of Kasparov vs. Deep Blue, Internet Computing,IEEE,May/Jun 1997,38-42.
6. M. Newborn: Kasparov versus DEEP BLUE:Computer Chess Comes of Age.Springer-Verlag ,New York, ISBN 0-387-94820-1.
7. J. Schaeffer, J. Culberson, N. Treloar, B. Knight, P. Lu, D. Szafron, A world championship caliber checkers program, Artificial Intelligence 53 (2–3) (1992) 273–289.
8. Tesauro.G, TD-Gammon, a self-teaching Backgammon program achieves master-level play, Neural,Comput. 6 (2) (1994) 215–219.
9. Iida, H., Sakuta, M., Rollason, J. Computer shogi. Artificial intelligent, vol.134, Nos.1-2, pp121-144.
10. M. Buro, How machines have learned to play Othello, IEEE Intelligent Systems J. 14 (6) (1999) 12–14.
11. H.-r. Fang, T.-s. Hsu, S.-c. Hsu, Construction of Chinese Chess endgame databases by retrograde analysis, LNCS 2063, Springer, New York, 2001, pp. 99–118.
12. R.Wu, D.F. Beal, Computer analysis of some Chinese Chess endgames, Advances in Computer Games, Vol. 9, Universities Maastricht, 2001, pp. 261–273.
13. T. Thomsen, Lambda-search in game trees—with application to Go, ICGA J. 23 (4) (2000) 203–217.
14. J.W.H.M. Uiterwijk, Knowledge and strategies in Go-Moku, in: H.J. van den Herik, L.V. Allis (Eds.),Heuristic Programming in Artificial Intelligence 3: The Third Computer Olympiad, Ellis Horwood,Chichester, 1992, pp. 165–179.
15. E.O. Thorp, W.E. Walden, A computer-assisted study of Go on M × N boards, Information Sciences 4,(1972) 1–33.
16. M. Ginsberg, GIB: Imperfect information in a computationally challenging game, J. Artificial Intelligence Res. 14 (2001) 303–358.
17. L. A. Zadeh, "Fuzzy sets," Inform. Control, vol. 8, pp. 338–353, 1965.
18. Yngvi Bjornsson, Tony Marsland, Jonathan Schaeffer and Andreas Junghanns, Searching with Uncertainty Cut-Offs, Advances in Computer Chess 8, pp: 167-179.
19. www.ourgame.com.

# Large-Scale Ensemble Decision Analysis of Sib-Pair IBD Profiles for Identification of the Relevant Molecular Signatures for Alcoholism

Xia Li[1,2,*], Shaoqi Rao[2,3,*], Wei Zhang[2], Guo Zheng[1,2], Wei Jiang[2], and Lei Du[2]

[1] Department of Computer Science,
Harbin Institute of Technology, Harbin 150080, P.R. China
{lixia, weizhang, guozheng, jiangwei, dulei}@ems.hrbmu.edu.cn,
raos@ccf.org
[2] Department of Bioinformatics, Harbin Medical University, Harbin 150086, P.R. China
[3] Departments of Cardiovascular Medicine and Molecular Cardiology,
Cleveland Clinic Foundation, Cleveland, Ohio 44195, USA

Abstract. The large-scale genome-wide SNP data being acquired from biomedical domains have offered resources to evaluate modern data mining techniques in applications to genetic studies. The purpose of this study is to extend our recently developed gene mining approach to extracting the relevant SNPs for alcoholism using sib-pair IBD profiles of pedigrees. Application to a publicly available large dataset of 100 simulated replicates for three American populations demonstrates that the proposed ensemble decision approach has successfully identified most of the simulated true loci, thus implicating that IBD statistic could be used as one of the informatics for mining the genetic underpins for complex human diseases.

1 Introduction

Many complex human diseases such as behaviors of alcoholism investigated by the Genetic Analysis Workshop 14 (GAW 14, http://www.gaworkshop.org/) are not simple Mendelian disorders. Instead, they may have mixed contributions of genes, environments and their interactions. A sophisticated mathematical model(s) is thus desirable to map the epidemiological complexities. Recent advances in single nucleotide polymorphisms (SNPs) and microarrays that allow for genome-wide profiling of complex biological phenotypes have offered the golden opportunities to unravel the high-order mechanisms and have also motivated development of the corresponding analysis strategies.

Recently, we have developed an ensemble decision approach [1, 2] to mining disease genes using genome-wide gene expression profiles on microarrays. In that work, we were the first group to explicitly formalize the task – mining disease

---

[*] Corresponding authors.

relevant genes using microarrays. Relevance at large has been studied extensively over the last three decades and there is an increasing interest in applications to a wide range of areas, in particular, in the area of machine leaning for feature subset selection, possibly owing to the advent of computational ability to handle massive high-dimension data sets. A nice review and study of the relevance concepts in Bell & Wang's work [3] reveal that either definitions or interpretations of relevance evolve considerably, from a simple and intuitive relevance concept for marginally filtering a feature to the sophisticated mathematical formalism of the concept that is quantitative and normalized. We quantified with a relevance intensity the relevance formalism of Kohavi and John's [4] because their formalism is easily generalized to our specific biological applications and is appealing to capture the reality of biological complexities (e.g., epistasis or gene-gene interactions). In this investigation, we extend our ensemble decision approach to mining alcoholism relevant SNPs using the genome-wide SNP data generated for GAW14.

2 Methods

2.1 Defining the Relevance of a SNP with Alcoholism

A SNP is said to be completely relevant if it is included in all the classifiers for alcoholism and the removal of the SNP alone will result in performance deterioration (in term of misclassification rate) of all the classifiers, obtained by learning on a number of training sets generated by a proper resampling technique. A SNP is partially relevant if it is not completely relevant and there exists a subset of SNP features, G, such that the performance of the induced classifier on G is worse than the performance on the union of G and the SNP. A SNP feature is irrelevant if it is neither completely nor partially relevant.

2.2 Defining the Phenotypic Attribute of a Sib Pair

We extend our ensemble decision approach to sib-pair analysis of pedigrees. First, we define the phenotypic attribute of a sib pair, the affection status of a sib pair for alcoholism. For the binary trait, there are three possible attributes, of which two attributes are chosen to be the phenotypes for learning: concordant affected, both sibs in a sib pair are affected; and concordant unaffected, no sibs in a sib pair are affected.

2.3 Defining the Features to Be Mined

The genetic features are defined to be the estimated proportions of alleles shared IBD by the sib pair at the SNP positions, provided by the GENIBD of the SAGE package [5]. Because our main interest is to explore the utility of the proposed method for extracting useful genetic information from the large-scale SNP data, we did not model the second-moment quantities of clinical covariates for the sib pairs.

2.4 Mining Alcoholism Relevant SNPs

In this study, we use the 100 GAW14 simulated replicates to demonstrate the behaviors and properties of the proposed method for mining alcoholism relevant SNPs. For computational convenience, we perform the same analysis procedures separately for each investigated chromosome.

Construction of training sets and test sets: We employ a five-fold cross-validation procedure to construct training sets and test sets. For each chromosome of each simulated replicate, we randomly divide the data into five equal parts. One part of data is used as the test set and the rest are used as the training set. We repeat the procedure to the investigated chromosomes of each replicate to generate 100 pairs of training data sets and test data sets for the 100 replicates.

Induction algorithm for building a SNP tree for alcoholism: Our proposed ensemble decision method is a supervised-learning approach based on a recursive partition tree. The basic procedures are as follows. First, a resampling technique as the above is employed to build up pairs of training sets and test sets for learning and testing, respectively. Then, a binary tree is being grown on a learning data set by a recursive partition algorithm. At each branching of the tree, the best SNP feature is selected such that it leads to the minimal impurity at the node. This binary recursive partition continues until the tree growth is stopped. For each grown tree, one subset of SNPs is extracted and is tested on the sister test set. This process for feature selection is repeated on each pair of learning and test data sets, which consequently results in an ensemble of SNP feature subsets, $M_1, \cdots, M_d, \cdots, M_m$, $M_d = \{m_1^d, m_2^d, \cdots, m_k^d\}$, or called SNP forests. The details for the tree-building algorithm have been given previously [2].

Evaluation of significance of each grown SNP tree: To test whether an extracted subset of SNPs (Tree M_d) is able to significantly distinguish the affection status of a sib pair using the sister test sample (T_d), we propose a χ^2 statistic:

$$\chi^2 = \frac{[|n_{00}n_{11} - n_{01}n_{10}| - n/2]^2 n}{(n_{00}+n_{01})(n_{10}+n_{11})(n_{00}+n_{10})(n_{01}+n_{11})} \qquad (1)$$

where $n = n_{00} + n_{01} + n_{10} + n_{11}$ and n_{00}, n_{01}, n_{10} and n_{11} are the frequencies for true negative, false positive, false negative and true positive, respectively. This statistic follows an asymptotic Chi-squared distribution with one degree of freedom.

Ensemble decision for selecting the alcoholism relevant SNPs: Based on the constructed SNP forests, we make an ensemble decision for selecting an alcoholism relevant SNP. Whether a SNP feature is relevant to the disease depends on the magnitude of its relevance intensity (or called an ensemble vote), FV.

For a particular SNP feature m_k, define:

$$FV(m_k) = F(M_1, M_2, \cdots M_m) = \frac{\sum_d \varpi_d I(m_k, M_d)}{\sum_d \varpi_d}, \qquad (2)$$

where $I(m_k, M_d)$ is an indicator function:

$$I(m_k, M_d) = \begin{cases} n_d^k & m_k \in M_d \\ 0 & \text{otherwise} \end{cases} \qquad (3)$$

where n_d^k is the frequency of SNP feature m_k appearing in tree M_d.

A weight, ϖ_d, can be a measure for the classification performance of M_d, for example, $\varpi_d = \chi_d^2$ or set $\varpi_d = 1$ for an equal weight for all the SNP feature subsets. We resort to a permutation approach to generate the null distribution of FV, for which we randomly assign a phenotypic attribute to a sib pair and $FV^0(m_k)$ is computed using the permutated data. Then, the empirical null distribution of $FV^0(m_k)$ is constructed. Given the empirical $FV^0(m_k)$ and a specified significance level, β (e.g., 0.1 or 0.05), a critical value FV_β^0 is obtained. A SNP feature is selected if its $FV \geq FV_\beta^0$ (one-tailed).

3 Results

We perform genetic mining of the relevant SNPs for alcoholism (the variable of Kofendrerd Personality Disorder (KPD) as the measure for alcoholic dependence) using the 100 simulated replicates of SNP data for three simulated populations (Aipotu, Karnagar and Danacaa, each with 100 pedigrees). Each of four chromosomes (chromosomes 2, 4, 5 and 9) is selected with the simulation answers and is analyzed separately, using the SNP data, with markers 3 cM apart, on the average. The critical values FV_β^0 ($\beta = 0.1$) for the four chromosomes, obtained from 1000 permutations are 2.83, 2.63, 2.78 and 3.08, respectively. The highly relevant SNP features for KPD trained and evaluated with the SNP forests obtained by analysis of the 100 simulated replicates are shown in Table 1. We have successfully identified all the simulation 'answers' or the SNP(s) that are closest to an answer or another: C02R0097, next to the answer B02T1050 on chromosome 2; C04R0282, next to the answer B04T3068 on chromosome 4; C05R0380, next to B05T4136 on chromosome 5; and C09R0765, one marker spaced between it and the true answer B09T8334 (< 0.5 cM) on chromosome 9. It is not surprising that the proposed method has identified also the clusters of SNPs that are nearby the true trait loci as the proposed method was designed for extracting "redundant" features (here, the reason for redundance can be close linkage or association of linkage disequilibrium between the markers within the cluster).

Table 1. The highly relevant SNPs for alcoholism (KPD), identified by ensemble decision analysis of three simulated populations (Aipotu, AISNP; Karnagar, KASNP; and Danacaa, DASNP). The bold markers denote the ones that are closest to the true trait loci.

Chromosome	population	SNP markers			
Chr 2	AISNP	C02R0092	C02R0093	C02R0094	C02R0101
	FV	4.57	3.94	3.71	3.68
		C02R0104	**C02R0098**	C02R0095	**C02R0097**
		3.54	3.36	3.35	2.94
	KASNP	C02R0092	C02R0101	**C02R0098**	C02R0093
	FV	4.60	3.76	3.74	3.68
		C02R0094	C02R0109	C02R0104	C02R0108
		3.57	3.52	3.51	3.19
	DASNP	C02R0092	C02R0104	C02R0109	C02R0101
	FV	4.79	4.27	4.04	3.99
		C02R0108	**C02R0098**	C02R0093	**C02R0097**
		3.80	3.72	3.70	3.69
Chr 4	AISNP	**C04R0282**	C04R0284	**C04R0283**	C04R0290
	FV	4.46	3.98	3.63	3.33
		C04R0289	C04R0285	C04R0288	C04R0295
		3.32	3.18	3.13	3.06
	KASNP	**C04R0282**	C04R0284	**C04R0283**	C04R0289
	FV	4.61	4.15	3.87	3.65
		C04R0295	C04R0290	C04R0291	C04R0288
		3.51	3.19	3.15	3.10
	DASNP	C04R0289	C04R0284	**C04R0282**	C04R0295
	FV	4.63	4.49	4.23	4.03
		C04R0283	C04R0288	C04R0339	C04R0321
		3.97	3.61	3.58	3.51
Chr 5	AISNP	**C05R0378**	C05R0382	**C05R0379**	C05R0383
	FV	5.12	4.11	4.03	3.74
		C05R0381	**C05R0380**	C05R0391	C05R0384
		3.43	3.42	2.94	2.87
	KASNP	**C05R0378**	C05R0382	C05R0383	**C05R0379**
	FV	5.51	4.28	3.85	3.84
		C05R0381	C05R0390	C05R0389	C05R0391
		3.22	3.20	3.08	3.07
	DASNP	**C05R0378**	C05R0382	C05R0383	C05R0396
	FV	5.75	4.92	4.37	3.99
		C05R0390	**C05R0379**	C05R0401	**C05R0380**
		3.88	3.75	3.72	3.62
Chr 9	AISNP	C09R0763	C09R0767	**C09R0766**	C09R0773
	FV	4.08	3.87	3.86	3.65
		C09R0768	**C09R0764**	C09R0769	**C09R0765**
		3.52	3.48	3.32	3.25
	KASNP	C09R0763	**C09R0766**	C09R0767	**C09R0764**
	FV	4.07	3.83	3.80	3.57
		C09R0768	**C09R0765**	C09R0779	C09R0773
		3.39	3.39	3.36	3.27
	DASNP	**C09R0766**	C09R0767	C09R0763	C09R0769
	FV	4.82	4.80	4.38	4.13
		C09R0773	C09R0779	C09R0768	**C09R0765**
		4.03	3.95	3.84	3.84

4 Discussion

Caution should be taken in interpretation of the SNP feature selection for alcoholism. The SNPs relevant to alcoholism should be considered as important biomarkers which may also be the disease genes (if the genetic marker is located within a gene and its polymorphisms can directly influence protein structure or expression level of the host gene). Many causes can contribute the relevance of a SNP to alcoholism, for example, the close linkage between the SNP and a nearby putative trait locus or between SNPs, linkage disequilibrium between loci and gene-gene interactions, which require further genetic analysis to be clarified. This application suggests that the sib-pair IBD statistics can be good genetic features to be mined. However, the IBD feature vectors within a pedigree tend to be correlated, effects of which on the proposed data mining approach remain unknown. Because of time limitation, we analyzed each chromosome separately. To address gene-gene interactions between different chromosomes and to construct the global-view SNP relevance network(s) for alcoholism, we are undertaking the ensemble decision analysis of the data on a genome-wide scale.

Acknowledgements

This work was supported in part by the National Natural Science Foundation of China (Grant Nos. 30170515 and 30370798), the National High Tech Development Project, the Chinese 863 Program (Grant No. 2003AA2Z2051), the 211 Project, the Tenth 'Five-year' Plan, Harbin Medical University and Heilongjiang Province Science and Technology Key Project (Grant No. GB03C602-4 and 1055HG009).

References

1. Li, X., Rao, S., Zhang, T., Guo, Z., Moser, K.L., Topol, E.J.: [An ensemble method for gene discovery based on DNA microarray data]. Science in China (C), Chinese Edition. 34 (2004) 195-202
2. Li, X., Rao, S., Wang, Y., Gong, B.: Gene mining: a novel and powerful ensemble decision approach to hunting for disease genes using microarray expression profiling. Nucleic Acids Res. 32 (2004) 2685-2694
3. Bell, D.A., Wang, H.: A formalism for relevance and its application in feature subset selection. Machine Learning. 41 (2000) 175-195
4. Kohavi, R., John, G.H.: Wrappers for feature subset selection. Artificial Intelligence. 97 (1997) 273-324
5. S.A.G.E.: Statistical Analysis for Genetic Epidemiology, 4.4. A computer program package available from Statistical Solutions. Cork, Ireland (2003)

A Novel Visualization Classifier and Its Applications

Jie Li[1], Xiang Long Tang[1], and Xia Li[1,2,*]

[1] Department of Computer Science and Technology,
Harbin Institute of Technology, Harbin 150001, China
{Jie Li, Xianglong Tang, jielee}@hit.edu.cn
[2] Department of Bioinformatics, Harbin Medical University, Harbin 150086, China
{Xia Li, Lixia6}@yahoo.com

Abstract. Classifiers, as one of the important tools of analyzing gene expression data in the post-genomic epoch, have been used widely in the classification of different cancer types in the past few years. Although most existing classifiers have high classification accuracy, the process of classification is a black box and they can not give biologists more information and interpretable results of classification. In this paper, we propose a novel visualization cancer classification method. Besides offering high classification accuracy, the method can help us identify complex disease-related genes and assess gene expression variation during the process of classification. The results of classification are natural and interpretable and the process of classification is visible. To evaluate the performance of the method we have applied the proposed method to three public data sets. The experimental results demonstrate that the approach is feasible and useful.

1 Introduction

The classification of different tumor types is of great importance in cancer diagnosis and drug discovery. Most previous cancer classifications are morphological and clinical-based and have limited diagnostic ability. The recent advent of DNA microarray technique has made simultaneous monitoring of thousands of gene expression levels possible [1]. With this abundance of gene expression data, researchers have started to classify cancer patients (including subtypes) using gene expression data. Most of classification methods used can give high classification accuracy, such as hierarchical clustering [2,3,4], Singular Value Decomposition [5,6], Principle Component Analysis(PCA) [7], the decision-tree methods [8], the linear discrimination analysis, the Bayesian network, K-nearest neighbor [9,10], neural network approach [11,12], Support Vector Machine (SVM) [13], etc. due to the fact that biologists still have limited knowledge about the cancer disease, they expect that, besides high classification accuracy, classifiers can give more information during the process of classification and help them understand the development of cancer. The majority of above classification methods can not meet this expectation. Existing several visualization classification methods [3, 14] serve only as assistant tools or graphical displays of major clustering results. They can not offer further help for cancer-related studies. Here, we present a visualization classification (VC) method. Using the method, biologists can attain very high classification accuracy, interpret the results, find complex disease-related genes easily, and observe gene expression variation conveniently.

This paper is organized as follows. The method fulfilling cancer classification is elaborated in the section 2. Section 3 gives experimental results. In section 4, we use our method to assess gene expression variation in different samples and find disease-related genes. Finally the discussions are given.

2 Method

2.1 Feature Gene Selection

Of the thousands of genes measured in a microarray experiment, most of them show little variations across the tissue samples. In order to improve the accuracy of classifier, Selection of feature genes which show large variance among the experiments is very important in cancer classification.

For cancer classification based on gene expression data, there are many feature gene selection methods, such as statistical tests (t-test [2,15,16], F-test [17]), information gain, Markov blanket and Wrapper method [18,19]. In this paper, we use the "signal-to-noise" ratio (SNR) to select feature gene [9], because SNR can describe difference of the gene expression in two classes very well.

Assume that we are given gene expression data with p probes and $n(n=n_1+n_2)$ DNA samples that belong to 2 categories, ω^+, ω^- and let each sample be labeled with either "1" or "-1", where n_1 is the number of samples with label "1", n_2 is the number of samples with label "-1". The gene expression data can be described by a $p \times n$ matrix.

For gene j, we calculate its SNR_j.

$$SNR_j = (\mu_j^+ - \mu_j^-)/(\sigma_j^+ + \sigma_j^-) \tag{1}$$

Where $\mu_j^+, \mu_j^-, \sigma_j^+$ and σ_j^- are given by the following formula.

$$\mu_j^+ = \frac{1}{n_1}\sum_{i=1}^{n_1} y_{ij}, \quad \mu_j^- = \frac{1}{n_2}\sum_{i=1}^{n_2} x_{ij}, \quad p \geq j \geq 1 \tag{2}$$

$$\sigma_j^+ = \frac{1}{n_1}\sum_{i=1}^{n_1}(y_{ij} - \mu_j^+)^2, \quad \sigma_j^- = \frac{1}{n_2}\sum_{i=1}^{n_2}(x_{ij} - \mu_j^-)^2, \quad p \geq j \geq 1 \tag{3}$$

Where y_{ij} represents the expression level of the jth gene in the ith sample with label "1", x_{ij} denotes the expression level of the jth gene in the ith sample with label "-1". The higher the absolute value of the SNR is, the more the expression level of the gene differs on average in the two classes.

Thousands of genes are measured in a microarray experiment. Usually there are parts of genes whose SNR is higher. It is a difficult problem to decide upon the number of the feature genes. In this study we sort genes by *SNR* in ascending and plot their *SNR* curve. According to the curve we decide the number of feature genes. For example, for colon cancer data we can get the *SNR* curve of 2000 genes (Fig.1). Fig.1 shows clearly that about one hundred genes have higher *SNR*. We choose 130 genes as feature genes.

After choosing feature genes, we can describe the gene expression data using a $q \times n$ matrix.

$$\begin{matrix} & s_{11} & s_{12} & \cdots & s_{1n_1} & s_{21} & s_{22} & \cdots & s_{2n_2} \\ g_1 & y_{11} & y_{21} & \cdots & y_{n_1 1} & x_{11} & x_{21} & \cdots & x_{n_2 1} \\ g_2 & y_{12} & y_{22} & \cdots & y_{n_1 2} & x_{12} & x_{22} & \cdots & x_{n_2 2} \\ & \cdots & \cdots & \cdots & \cdots & \cdots & \cdots & \cdots & \cdots \\ g_q & y_{1q} & y_{2q} & \cdots & y_{n_1 q} & x_{1q} & x_{2q} & \cdots & x_{n_2 q} \end{matrix}$$

q is the number of feature genes. Each sample with label "1" ("-1") is represented by a vector $Y_i = (y_{i1}, y_{i2}, ..., y_{iq})$ ($X_i = (x_{i1}, x_{i2}, ..., x_{iq})$).

2.2 Visualization Classifier Algorithm

The visualization classifier algorithm can be stated as follows:

(1) Calculate the *SNR* of all genes and select feature genes according to the *SNR* curve.

(2) Divide n samples into two parts randomly: N training samples and M testing samples, $n = N + M$, N training samples include N_1 samples with label "1" and N_2 samples with label "-1".

(3) Calculate mean vector, μ^+ and μ^-, of two categories of training samples.

$$\mu^+ = (\mu_1^+, \mu_2^+, ..., \mu_q^+), \mu_j^+ = \frac{1}{N_1}\sum_{i=1}^{N_1} y_{ij}, \quad q \geq j \geq 1, \tag{4}$$

$$\mu^- = (\mu_1^-, \mu_2^-, ..., \mu_q^-), \mu_j^- = \frac{1}{N_2}\sum_{i=1}^{N_2} x_{ij}, \quad q \geq j \geq 1, \tag{5}$$

(4) Plot all "*" spots defined by μ_j^+ versus μ_j^- pairs (Fig. 2).

(5) Suppose we select a random sample, $S_k = (g_1, g_2 ..., g_q)$ from the testing set, g_j is the expression level of jth gene.

(6) Plot all "o" spots defined by μ_j^+ versus g_j pairs overlaid on the Fig. 2.

(7) Assign S_k in ω^+, if the "o" spotters scatter both sides of diagonals (S_k and μ^+ are highly correlated, as shown in Fig.3), otherwise assign S in ω^- (S_k and μ^- are highly correlated, as shown in Fig. 4).

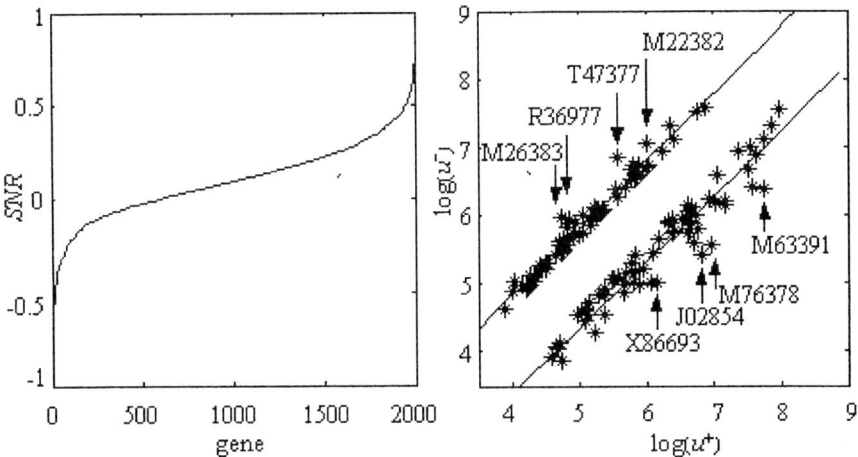

Fig. 1. Sort 2000 genes from colon cancer data

Fig. 2. The correlation between log(μ^+) by SNR in ascending and log(μ^-) from colon cancer data

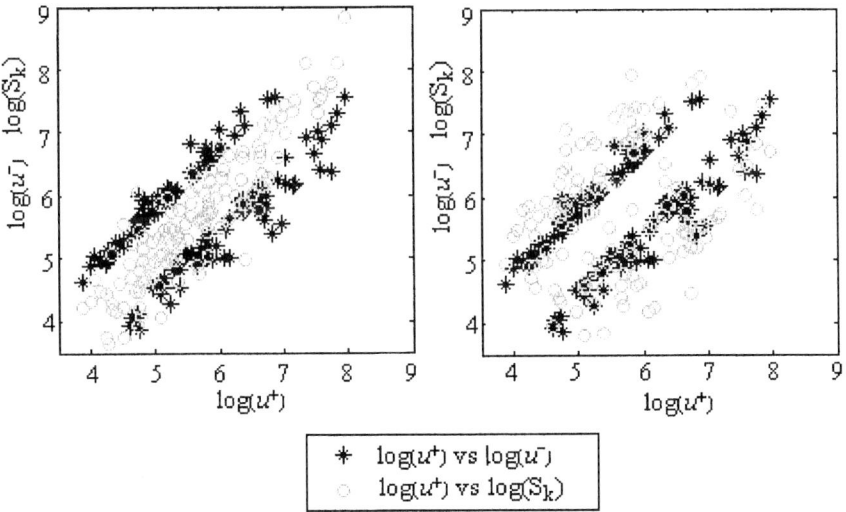

Fig. 3. The correlation between log(μ^+) log(S_k) from colon cancer data, $S_k \in \omega^+$ k=22 (normal 11)

Fig. 4. The correlation between log(μ^+) and, log(S_k) from colon cancer data, $S_k \in \omega^-$, k= 28 (tumor 16).

3 Experimental Results

In the section, we use the proposed method to classify colon cancer data [20]. Leave-one-out cross-validation experiments were made, because we have only a few samples. Colon cancer data consists of 22 normal and 40 tumor colon tissue samples and the number of genes is 2000. The data were downloaded from http://microarray.princeton.edu/oncology/affydata/index.html. First, we calculate the SNRs of all genes and select 130 feature genes according to the Fig.1, 65 genes of them have higher expression level ($\mu_j^+ \geq \mu_j^-$) in normal samples. 65 genes of them have higher expression level ($\mu_j^+ \leq \mu_j^-$) in tumor samples. Then, we plot all "*" spots defined by μ_j^+ versus μ_j^- pairs (Fig.2). Fig. 2 shows quite clearly that feature genes are distributed in two area. The genes which scatter in the top left have higher expression level in tumor samples, and down right genes have lower expression value in tumor samples. For the top left (or down right) genes, it can be seen that variables μ_j^+ and μ_j^- are correlated. We use two regression lines to describe relation of variables μ_j^+ and μ_j^-. One line describes the top left spots, while the other one describes down right spots. At the same time, we find there are several spots which lie outside the regression lines. These spots are called outliers. In the following section, we will analyze these outliers. It is just the predominance of our method to help biologists to find the correlation between the genes and these outliers. Finally, we classify samples selected from testing sample set according to step 5, 6 and 7. The experimental results are shown in Fig. 5.

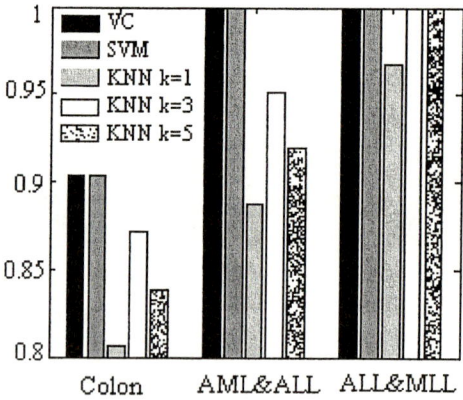

Fig. 5. Comparison of classification accuracy between VC (our method), SVM and KNN using three public gene expression data sets: Colon cancer data, AML&ALL and ALL&MLL.

In order to further evaluate the performance of the classifier, we use our method to classify other two public gene expression data sets: leukemia data [9, 21]. The first leukemia data (AML&ALL) includes 38 acute leukemia patients (27 acute myeloid leukemia (AML) and 11 acute lymphoblastic leukemia(ALL)); the second leukemia data (ALL&MLL) consists of 44 patients (24 ALL samples and 20 mixed-lineage leukemia (MLL) samples). From Fig. 5 we can see that the accuracy of our classifier is high enough to classify these samples. For the colon cancer data, the accuracy is 90.32%. The performance of the classifier is much better for the other two datasets, with classification accuracy 100%, 100%, respectively. In order to further evaluate the generalization ability of our classifier, we have also compared the performances of the method with other classification methods: such as SVM and KNN using the same test sets. For the three data sets, SVM (liner kernel) [22] has the same accuracy as ours. In the case of KNN classifier, we performed a sequence of experiments in which we set the number of neighbors, K, to be 1, 3 and 5 respectively. The precision achieved in this sequence of experiments was 80.65%, 88.71%, 96.77%; 87.1%, 95.16%, 100% and 83.87%, 91.94%, 100% respectively. Thus with 3 neighbors KNN shows the best results, however its precision is lower than ours for the colon and AML&ALL data.

4 Applications

4.1 Mining Disease-Related Genes

It is known that genes interact with each other. The gene whose expression level changes will affect the expression levels of other genes. Compared with genes in normal cell, many genes in cancer cell have changed in expression level. There are always some genes which have very significant changes and are very different from their "neighbors". We can find these spots easily from Fig.2. We called them disease-related genes. In linear regress model, they are called outliers. The outliers were detected using studentized residuals with the given alpha level. For colon data 8 disease-relate genes were identified with the 95% confidence intervals (alpha=0.05) about their residuals (Fig.2). M76378, J02854, M63391and X86693 have lower mean expression levels in tumor samples; T47377, M26383, R36977 and M22382 have higher mean expression levels in tumor samples. We further analyzed the genes identified as most important.

M76378, human cysteine-rich protein (CRP) gene, exons 5 and 6, whose ratio of μ_j^- and μ_j^+ is 0.25. Given so low expression level, it can be assumed that this gene is highly correlated with colon cancer. CRP gene, proved as a primary response gene[23], is widely expressed in a variety of human primary and established cell lines and plays a pivotal role in cell function. The LIM/double zinc finger motif found in cysteine-rich protein is found in an expanding group of proteins with critical functions in gene regulation, cell growth, and somatic differentiation. Within the family of proteins CRP's structure is highly similar to rhombotin, a putative oncogene; The structural similarity between the proteins in this gene family suggests that they play fundamental and interrelated roles in cell function and development. A high correlation between CRP gene and colon cancer also has been proved by the following example.

The expression of both the mouse and human CRP genes is induced as a primary response to serum in quiescent Balb/c 3T3 cells and in human fibroblasts. The profile of this primary response is remarkably parallel to that of c-myc in the Balb/c 3T3 cell line. The coordinate regulation of CRP gene with c-myc suggests that these two genes respond to the same regulatory pathways and may share transcription control features during the G0 to S transition [24]. It is known c-myc is a putative oncogene, overexpression of this gene has been reported in numerous cancers, including breast, cervical, lung, leukemia, lymphoma, prostate and ovarian tumors. Surprisingly CRP gene is not among the top of 500 genes identified by Student's t-test. It was also not included in the best three subsets determined by ensemble decision approach [25] and the two classification trees [8]. However, our method identifies it successfully. The other two most significant genes identified by our method are J02854 and M63391. It was not at all surprising that these two genes were also identified by RankGene [26] to be either the top one using the measures sum minority, majority or as the second using the measures information gain, sum of variances. In addition, M63391 is also found to be significantly down-expressed in other cancer types such as the melanoma cell line [27].

M26383 (IL-8), human monocyte-derived neutrophil-activating protein (MONAP) mRNA, complete cds, whose ratio is 3.46. Given so significant difference, it can be postulated that this gene plays a pivotal role as a central hub for the gene network that maps to the underlying pathological complexity of colon cancer. It was reported that IL-8 was correlated with the development of colon cancer [28], the migration of human colonic epithelial cell lines [29], and metastasis of bladder cancer [30]. Molecular experiments have proved that MONAP is constitutively overexpressed by human tumor lines [31]. Strikingly, MONAP was ranked as the top colon tumor gene using information gain, sum of variances, classification tree [8], ensemble decision trees [25], but surprisingly not among the top of 100 genes identified by Student's t-test. The second gene is the human mitochondrial matrix protein P1. Molecular experiments suggested this gene played an important role in the replication of mammalian DNA [32] and certain autoimmune disease [33], but at present the exact biological role of P1 in mammalian systems is not clear. Therefore, further studies on the cellular function of this highly conserved protein are of great significance.

4.2 Assessing Gene Expression Variations in Samples

Gene expression profiling has become a widely used tool to identify genes that are linked to particular phenotype, such as disease state, drug resistance, or toxicity. Interpretation of results from these expression studies is critically dependent on the choice of samples used in the analysis, and it is important to understand the variations in the sample population itself. Our method is very efficient for displaying the variations of samples. For example there are two samples: tumor 28 and tumor 31 from colon cancer data. Fig. 6 and Fig. 7 show their gene expression variations. For the 28th tumor sample expression levels of most genes are lower than the mean expression levels, and for 31th tumor sample most genes have higher expression levels.

Identifying these variations is very important for calculating drug doses and giving different treatments. In our results, two tumor samples (tumor 30, tumor 33) had the same distribution with normal samples. This is worth further investigation. For general classifiers it is impossible to find these information.

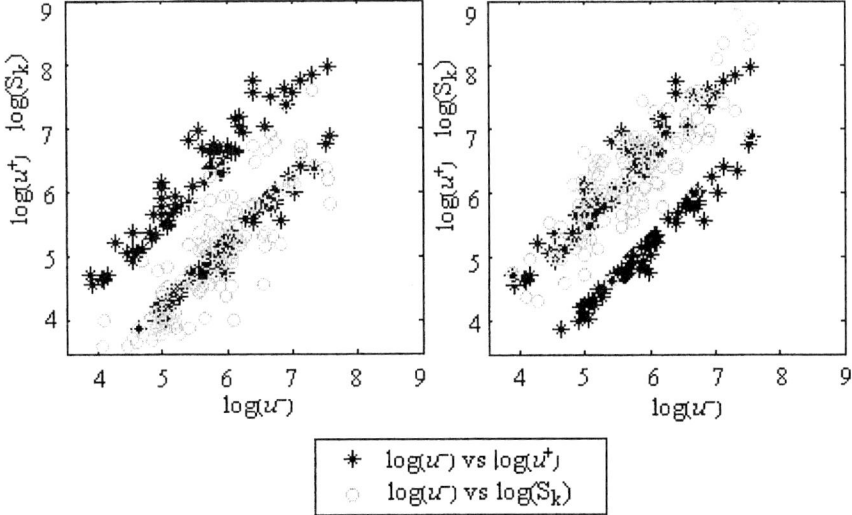

Fig. 6. The correlation between the average vectors from nomal and tumor ($S_k \in \omega^-$), $k=41$ (tumor 28)

Fig.7. The correlation between the average vectors from nomal and tumor ($S_k \in \omega^-$), $k=46$ (tumor 31)

5 Discussion

Applications to several cancer data support our speculation that the new visualization classification method can help biologists find important information about samples and genes. Experimental results have proved that the method also has very high classification accuracy. So our approach is promising for analyzing and visualizing of gene expression data and further development is worthwhile. Difference between the classifier and other classifiers from statistical and machine learning area is that the classifier only classifies one sample every time and sometimes it is difficulty to classify some samples according to the distribution of data. Integrating the classifier and other classifiers from statistical and machine learning area is our next work.

Acknowledgement

This work was supported in part by the National Natural Science Foundation of China (Grant Nos. 30170515 and 30370798), the National High Tech Development Project of China, the 863 Program (Grant Nos. 2003AA2Z2051 and 2002AA2Z2052) and the 211 Project, the Tenth 'Five-year' Plan, Harbin Medical University.

References

1. Brazma A and Vilo J. Minireview: Gene expression data analysis. Federation of European Biochemical Societies, 480 (2000), 17–24.
2. Welsh JB, Zarrinkar PP, et al. Analysis of gene expression profiles in normal and neoplastic ovarian tissue samples identifies candidate molecular markers of epithelial ovarian cancer. Proc.Natl Acad. Sci.USA 98 (2001), 1176–1181.
3. Michael B.Eisen, Paul T.Spellman, et al. Cluster analysis and display of genome-wide expression patterns. Proc.Natl Acad. Sci. USA, 95 (1998), 14863–14868.
4. Ash A. Alizadeh, Michael B.Eisen, et al. Distinct types of diffuse large B-Cell lymphoma identified by gene expression profiling. Nature, 403 (2000), 503–511.
5. Orly Alter, Patrick O.Brown, et al. Singular value decomposition for genome-wide expression data processing and modeling. Proc.Natl Acad. Sci. USA, 97 (2000), 10101–10106.
6. Neal S.Holter, Amos Maritan, et al. Dynamic modeling of gene Expression Data. Proc.Natl Acad. Sci. USA, 98 (2001), 1693–1698.
7. Xiling Wen, Stefanie Fuhrman, et al. Large-scale temporal gene expression mapping of central nervous system development. Proc.Natl Acad. Sci. USA, 95 (1998), 334–339.
8. Heping Zhang, Chang-Yung Yu, et al. Recursive partitioning for tumor classification with gene expression microarray data. Proc.Natl Acad. Sci. USA, 98 (2001), 6730–6735.
9. T.R.Golub, D.K.Slonim, et al. Molecular classification of cancer: class discovery and class prediction by gene expression monitoring. Science, 286 (1999), 531–537.
10. Ka Yee Yeung, David R. Haynor,et al. Validating clustering for gene expression data. Bioinformatics, 17 (2001), 309–318.
11. Neal S.Holter, Madhusmita Mitra, et al. Fundamental patterns underlying gene expression profiles: simplicity from complexity. Proc.Natl Acad. Sci. USA, 97 (2000), 8409–8414.
12. Pablo Tamayo, Donna Slonim, et al. Interpreting patterns of gene expression with self-organizing maps: methods and application to hematopoietic differentiation. Proc.Natl Acad. Sci.USA, 96 (1999), 2907 – 2912.
13. Terrence S. Furey, Nello Cristianini, et al. Support vector machine classification and validation of cancer tissue samples using microarray expression data. Bioinformatics, 16 (2002), 906–914.
14. Rob M. Ewing and J. Michael Cherry. Visualization of expression clusters using Sammon's no-linear mapping. Bioinformatics, 17 (2001), 658–659.
15. A. Ben-Dor, L. Bruhn, et al. Tissue classification with gene expression profiles, in: Proceedings of the Fourth Annual International Conference on Computational Molecular Biology, Tokyo, (2000), Japan.
16. Model F, Adorjan P, et al. Feature selection for DNA methylation based cancer classification. Bioinformatics, 17 (2001), S157–S164.
17. Chris H.Q. Ding. Analysis of gene expression profiles: class discovery and leaf ordering. RECOMB 2002, 127–136.
18. Li Wuju and Xiong Momiao. Tclass: tumor classification system based on gene expression profile. Bioinformatics, 18 (2002), 325–326.
19. Leping Li, Clarice R, et al. Gene selection for sample classification based on gene expression data: study of sensitivity to choice of parameters of the GA/KNN method. Bioinformatics, 17 (2001), 1131–1142.
20. U. Alon, N. Barkai, et al. Broad patterns of gene expression revealed by clustering analysis of tumor and normal colon tissues probed Oligonucleotide arrays. Proc.Natl Acad. Sci. USA, 96 (1999), 6745–6750.

21. Scott A. Armstrong, Jane E Staunton, et al. MLL translocations specify a distinct gene expression profile that distinguishes a unique leukemia. Nat Genet, 30 (2002), 41–47.
22. T. Joachims. Making large-Scale SVM Learning Practical. Advances in Kernel Methods - Support Vector Learning, B. Schölkopf and C. Burges and A. Smola (ed.), MIT-Press, 1999.
23. Wang X, Lee G, et al. A member of the LIM/double-finger family displaying coordinate serum induction with c-myc. J Biol Chem., 267(1992), 9176–9184.
24. Heikkila R, Schwab G, et al. A c-myc antisense oligodeoxynucleotide inhibits entry into S phase but not progress from G0 to G1. Nature, 328 (1987),445-449.
25. Li Xia, Shaoqi Rao, et al. Gene mining: a novel and powerful ensemble decision approach to hunting for disease genes using microarray expression profiling. Nucleic Acids Res., 32 (2004), 2685–2694.
26. Yang Su, T.M. Murali, et al. RankGene: Identification of diagnostic genes based on expression data. Bioinformatics. 19 (2003), 1578–1579.
27. Anne Gutgemann, Michaela Golob, et al. Isolation of invasion-associated cDNAs in melanoma. Arch Dermatol Res, 293 (2001), 283-290.
28. Fox SH, Whalen GF, et al. Angiogenesis in normal tissue adjacent to colon cancer. J. Surg. Oncol., 69 (1998), 230–234.
29. Toshina K, Hirata I, et al. Enprostil, a prostaglandin-E(2) analogue, inhibits interleukin-8 production of human colonic epithelial cell lines. Scand. J. Immunol,., 52 (2000), 570–575..
30. Inoue K, Perrotte P, et al. Gene therapy of human bladder cancer with adenovirus-mediated antisense basic fibroblast growth factor. Clin. Cancer Res., 6 (2000), 4422–4431.
31. J Kowalski and D T Denhardt. Regulation of the mRNA for monocyte-derived neutrophil-activating peptide in differentiating HL60 promyelocytes. Mol Cell Biol, 9 (1989), 1946–1957.
32. P Thommes, R Fett, et al. Properties of the nuclear P1 protein, a mammalian homologue of the yeast Mcm3 replication protein. Nucleic Acids Res. 20 (1992), 1069-1074.
33. Satish Jindal, Anil K. Dudani, et al. Primary structure of a human mitochondrial protein homologous to the bacteriacterial and plant chaperonins and to the 65-Kilodalton mycobacterial antigen. Mol. Cell Biol. 9 (1989), 2279–2283.

# Automatic Creation of Links:
An Approach Based on Decision Tree

Peng Li[1] and Seiji Yamada[2]

[1] CISS, IGSSE, Tokyo Institute of Technology
4259 Nagatuta, Midori-ku, Yokohama, Japan
liorlee@nii.ac.jp
[2] National Institute of Informatics
2-1-2 Hitotsubashi, Chiyoda-ku, Tokyo, Japan
seiji@nii.ac.jp

Abstract. With the dramatic development of web technologies, tremendous amount of information become available to users. The great advantages of the web are the ease with which information can be published and made available to a wide audience, and the ability to organize and connect different resources in a graph-based structure using hyperlinks. However, most of these links are created manually and the page that the link represents must be known to the author of the link. In this paper, we propose a decision-tree-based approach to solve this problem. We set up a system that gathers information about the candidate pages, evaluates them and creates links to them automatically.

1 Introduction

The World Wide Web contains over 2 billion pages and this number is growing at a breakneck speed. Links between these pages are generated and removed very constantly. Over the past years, a number of approaches such as Adaptive Web sites [1], FWEB [2], ARC [3] and some link- or topology-based approaches[4][5][6] have been developed to improve link structure.

In most of the cases, *official sites* (organizations, products, people, services, facilities, events, etc) possess links to the other sites that deal with the same topics. These links are created manually. This can be a disadvantage if information in a particular field is incomplete and expanding rapidly over time and where a page author cannot be expected to know which pages are the most appropriate to link to and when they become available.

Our aim is to create these links automatically. We focus on official sites because these sites are usually so popular that other pages which deal with the same topics are willing to be linked from them. In this research, our targets are the pages that have sent link requests to the specified official sites. We call these pages *link request pages*. When receiving a request, our system will gather information about this page such as its access amount, maintenance frequency, page similarity, etc. We set up a decision tree to handle these information and determine whether the link request page should be linked or not. Links are created, modified and deleted automatically with the changes of link request pages.

2 Automatic Link Creation System Based on Decision Tree

2.1 System Overview

The purpose of our system is to create and modify links on the links page automatically. Broken links are removed and new links are added. Moreover, with the growing popularity and richness of the contents, the position of a link will rise and vice versa.

2.2 Decision Tree

A decision tree is a tree-shaped structure that represents a set of decisions. These decisions generate rules for the classification of a dataset. In a decision tree, each "branch node" (feature) represents a choice between a number of alternatives, and each "leaf node" (class) represents a class or decision. OC1 (Oblique Classifier 1) [7] is a decision tree induction system designed for applications where the instances have numeric (continuous) feature values. OC1 builds decision trees that contain linear combinations of one or more attributes at each internal node; these trees then partition the space of examples with both oblique and axis-parallel hyperplanes. OC1 has been used for classification of data representing diverse problem domains, including astronomy, DNA sequence analysis and others. In this research, we use OC1 to construct decision trees because the values of our features and classes are numeric.

2.3 Evaluation Features

We use 4 types of features in this system, access amount, maintenance frequency, page similarity and contents richness. These features are chosen because they are very important characteristics for the evaluation of a page. On the other hand, these information are extractable with current web technologies. We plan to add a few more features to raise the accuracy of our system in the evaluation experiment.

2.4 System Construction

Fig.1 shows the system construction. The details of several modules are described below.

Training Data. Training data are used to construct a decision tree. They are information of a set of valued pages. Each page has its class value given by the administrator and 4 attributes values gathered by web robots.

Test Data. Classification is performed according to test data. They are information of a set of unvalued pages. Each page only contents 4 attributes values.

Decision tree. A decision tree is constructed by OC1 through training data. When unvalued pages are given, the decision tree classifies these pages to give each page a class value.

Check Engine. When check engine receives requests from link request pages, it sends web robots to gather 4 types of attributes from these pages. Usually, this information is sent to the decision tree as test data. But if the administrator evaluates some pages, they are given as training data to construct a decision tree. The check engine will gather information from link request pages regularly, even if no new request is detected.

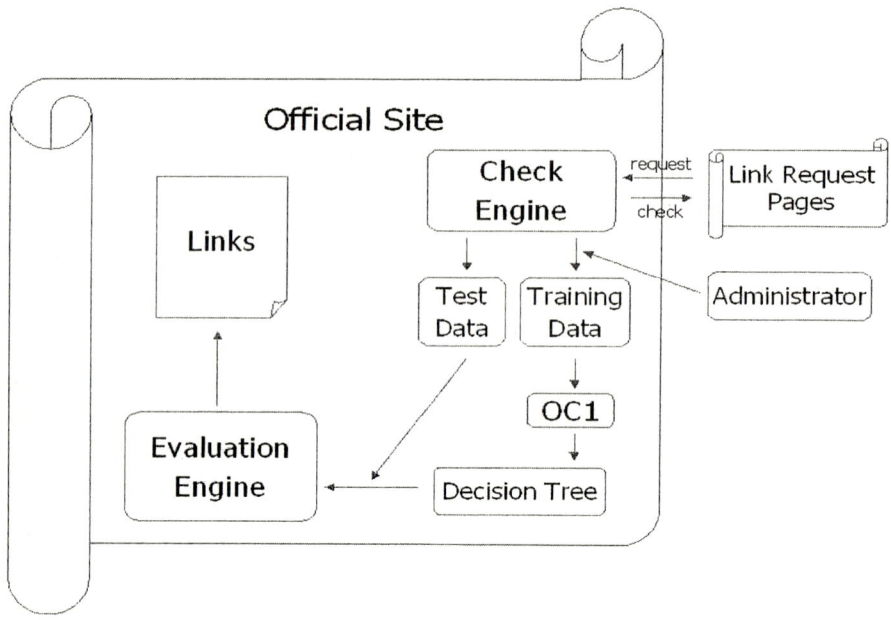

Fig. 1. System Construction

Evaluation Engine. Evaluation engine sorts the link request pages by their class values and creates Links.

2.5 Links Creation Flow

(1) Link request pages send requests to the Official Site.
(2) Check engine sends web robots to collect information about the link request pages.
(3) If the decision tree has been constructed, go to (6).
(4) The administrator evaluates some pages.
(5) Construct a decision by OC1.
(6) Use the decision tree to classify unvalued pages.
(7) Links are created by evaluation engine.

This Flow starts regularly from (2) even if there is no request detected.
It will work from (4) if the administrator gives new evaluations.

3 Conclusion

We proposed an automatic link creation system base on decision tree. We focused on the links page that is a part of nearly all of the official sites. Our system gathers information about the link request pages, evaluates them and creates links to them. We are preparing an evaluation experiment to inspect the effectiveness of our system. More evaluation features will be added at that time. Further discussion about our evaluation metrics need to be performed.

References

1. Perkowitz, M., Etizoni, O.: Toward Adaptive Web Sites: Concept and Case Study. Artificial Intelligence Journal, vol.118 (2000) 245-275
2. Courtenage, S.,Williams, S.: Automatic Hyperlink Creation Using P2P and Publish/Subscribe. Workshop on Peer-to-Peer and Agent Infrastructures for Knowledge Management (PAIKM) (2005)
3. Chakrabarti, S., Dom, B., Raghavan, P., Rajagopalan, S., Gibson, D, Kleinberg, J.: Automatic Resource Compilation by Analyzing Hyperlink Structure and Associated Text. Proceedings of the Seventh International World Wide Web Conference (WWW7) (1998)
4. Page, L., Brin, S., Motwani, R., Winograd, T.: The Pagerank citation ranking: Bringing order to the Web. Technical report, Stanford University (1998)
5. Kleinberg, J.: Authoritative sources in a hyperlinked environment. In Proceedings of the 9th Annual ACM-SIAM Symposium on Discrete Algorithms (1998) 668—677
6. Phelan, D., Kushmerick, N.: A descendant-based link analysis algorithm for web search (2002)
7. Murthy, S., Kasif, S., Salzberg, S., Beigel, R.: Randomized induction of oblique decision trees. In Proceedings of the Eleventh National Conference on Artificial Intelligence (1993) 322-327

Extraction of Structural Information from the Web

Tsuyoshi Murata

Department of Computer Science,
Graduate School of Information Science and Engineering, Tokyo Institute of Technology,
2-12-1 W8-59 Ookayama, Meguro, Tokyo 152-8552 Japan
murata@cs.titech.ac.jp

Abstract. The Web can be regarded as a huge graph when each Web page is regarded as a node and each hyperlink as an edge. There are several attempts for visualizing the structure of the Web, such as touchgraph or KartOO. In order to achieve visualization that assists users' information acquisition from the Web, two constructs (keywords and pages) are required in the visualization. In this paper, a cluster of keywords and Web pages is regarded as "structural information" in the Web. We have developed a visualization system that shows clusters of Web pages and keywords. Based on online Web resources, appropriate relations can be visualized without analyzing the contents of Web pages.

1 Introduction

Several attempts have been made for Web mining, discovery of useful knowledge from huge Web information. There are three main approaches for Web mining as claimed by Kosala [5]. Among the approaches, Web structure mining based on hyperlink graph topology is important not only for engineering such as information retrieval and Web page crawling, but also for sociology such as prediction of Web growth and detection of trends of real human world.

The graph structure of the Web is like a bow-tie from macroscopic view [2], and bipartite core graph from microscopic view [6]. There are groups of related Web pages whose hyperlinks are densely connected with each other, which are called Web communities [3][6][7]. Modeling the structures and detecting relations among Web communities will give insight to the characteristics of real human communities. The Web is huge and is growing. It is expected that Web communities may change over time in the processes of birth, grow and death. And they may merge or split dynamically. Such dynamic changes of Web communities often reflect needs for new information of human communities. Methods for detecting such dynamic changes of Web communities are important as well as static analysis of the structures of the communities.

As an approach for Web structure mining, the author has developed systems for discovering and visualizing Web communities based on hyperlink information [8][9]. The systems utilize a search engine as a resource for Web data acquisition. In general, a search engine plays an important role for users' Web watching; it collects huge Web data by crawling, and users perform search in order to find Web pages related to input keywords.

This paper describes an attempt for visualizing the Web. In order to achieve visualization that assists users' information acquisition from the Web, two constructs (key-words and pages) are required in the visualization. In this paper, a cluster of keywords and Web pages is regarded as "structural information" in the Web. We have developed a visualization system that shows clusters of Web pages and keywords. Based on online Web resources, appropriate relations can be discovered without analyzing the contents of Web pages.

2 Related Work for Visualizing Web Pages

There are two approaches for extracting structure of the Web; decomposing Web network into smaller clusters, and composing a cluster of neighboring pages based on local connectivity. Broder's macroscopic analysis of the Web [2] is an example of the former approach. As the latter approach, TouchGraph (http://www.touchgraph.com/), Google set vista (http://www.langreiter.com/space/google-set-vista), Anacubis (http://www.anacubis.com/googledemo/google/index.asp), and KartOO (http://www.kartoo.com/) are famous examples. Many of these systems employ online resources such as Google Web API in order to acquire relations among Web pages. Google Web API (http://www.google.com/apis/) is a Web service that provides accesses to Google's Web data from users' computer programs. At the current stage, the number of queries per day is limited to 1,000 in order to avoid heavy load to Google.

3 Visualization of Pages and Terms Based on Google Web API

As described above, Google Web API is an interface that allows our computer programs to access Google resources. Google contains more than 8 billion Web pages as of April 2005. Utilizing this huge search engine as a resource is important for developing intelligent systems. In this section, our visualization system based on Google Web API is described. Relations of Web pages and terms are shown in the form of graph in the system. Our system generates graph whose node are either Web page or term, and edge is relation between a page and a term. A term can be regarded as label of connecting Web page, and it clarifies the characteristics of neighboring page cluster.

In our system, the following options are employed in order to acquire needed information for graph drawing:

1. "related" option: search for Web pages related to specific page
2. "info" option: search for terms regarding specific page

The former option is used for search of related Web page. For example, results of the search "related:www.yahoo.com" contain sites such as "www.google.com", "www.msn.com", and "www.altavista.com". The latter option is used for the search of terms that can be used for labels for specific Web page. For example, results of the search "info:www.yahoo.com" contain "Searching", "Internet", "Directories" and "Computers".

At the present stage, a Java program that combines Google Web API and Web community visualization is developed. Fig. 1 shows the snapshot of the system. The system accepts URLs as rectangle nodes. By clicking "info" button, a search is performed by using Google API and labels about the URLs are attached as ellipse nodes. In Fig. 1, www.ft.com, www.cbsnews.com, www.nytimes.com and www.chicagotribune.com are connected with each other via "News" node. On the other hand, slashdot.org is located separately from the group since its labels are not the same as those of the group. This result shows that labels obtained from Google API can be useful for relating or discriminating input URLs.

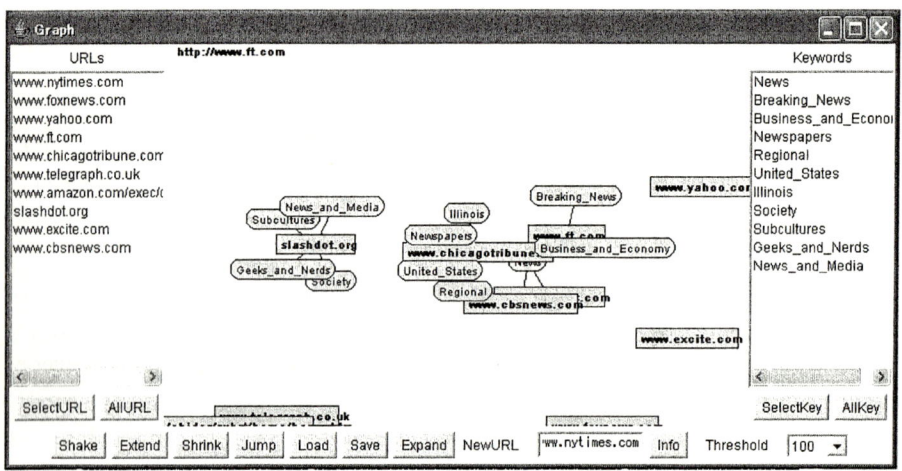

Fig. 1. Visualization of Web pages using Google Web API

Visualization performed by previous systems (such as touchgraph and anacubis) contain Web pages only – terms for explaining pages are not included. Although visualization of KartOO contain terms, relations between pages and terms are not clear. The system described in this section is the first step for visualizing a kind of map in the Web.

4 Concluding Remark

As everybody knows, the Web is huge and its quality is not uniform. For the discovery and visualization of Web communities, the following requirements should be considered:

1. Partiality of input data: Nobody can collect data of the whole Web. Algorithms for the discovery of Web communities need to handle partial Web data. Suitable strategies for collecting Web data have to be considered.
2. Quantities of input data: On the contrary to the above, Web data are still huge even though they are partial. Capabilities for handling large-scale data are required for Web community discovery methods.
3. Qualities of input data: Depending on the network conditions, some of the Web pages may not be accessible. Robustness for missing or noisy data is necessary for Web community discovery.

4. Various structures of Web communities: Although Kumar regards a bipartite graph as a characteristic structure for Web communities, there might be other characteristic graph structures. Search algorithms for specific graph structure, such as clique or bipartite graph, are important. However, they are not enough for discovering real complicated Web communities.
5. Post processing of discovered Web communities: When fixed graph structure is searched from given Web data, many overlapping graphs will be found. Post processing of discovered Web communities such as clustering or labeling is necessary to assist users' understanding.
6. Interactive discovery of Web communities: Discovered Web communities are not always satisfactory to users since there are several criteria for "relatedness" among Web pages. It is preferable if users can control the strategies for searching Web communities by giving examples or negative examples.

Online Web resources are expected to meet some of the above requirements. For example, use of Google Web API may compensate partiality of Web data. At the present stage, only two options (related and info) are employed in our system. Information obtained by search using other options is expected to enrich our visualization further.

References

1. Barabasi, A.-L., "LINKED – The New Science of Networks", Perseus Publishing, 2002.
2. Broder, A., Kumar, R., Maghoul, F., Raghavan, P., Rajagopalan, S., Stata, R., Tomkins, A., Wiener, J.: "Graph Structure in the Web: Experiments and models", Proc. of the 9th WWW Conference, pp.309-320, 2000.
3. G. W. Flake, S. Lawrence, C. L. Giles, F. M. Coetzee: "Self-Organization and Identification of Web Communities", IEEE Computer, Vol.35, No.3, pp.66-71, 2002.
4. J. Kleinberg, R. Kumar, P. Raghavan, S. Rajagopalan, A. Tomkins: "The Web as a Graph: Measurements, Models, and Methods", Proc. of COCOON'99, LNCS 1627, pp.1-17, 1999.
5. R. Kosala, H. Blockeel: "Web Mining Research: A Survey", ACM SIGKDD Explorations, Vol.2, No.1, pp.1-15, 2000.
6. R. Kumar, P. Raghavan, S. Rajagopalan, A. Tomkins: "Trawling the Web for Emerging Cyber-Communities", Proc. of the 8th WWW Conference, 1999.
7. T. Murata: "Discovery of Web Communities Based on the Co-occurrence of References", Proc. of DS2000, LNAI 1967, pp.65-75, Springer, 2000.
8. T. Murata: "Finding Related Web Pages Based on Connectivity Information from a Search Engine", Poster Proc. of 10th WWW conference, pp.18-19, 2001.
9. T. Murata: "Visualizing the Structure of Web Communities Based on Data Acquired from a Search Engine", IEEE Transactions on Industrial Electronics, Vol. 50, No. 5, pp.860-866, 2003.
10. L. Page, S. Brin, R. Motwani, T. Winograd.: "The PageRank Citation Ranking: Bringing Order to the Web", Online manuscript, http://www-db.stanford.edu/~backrub/pagerank sub.ps, 1998.

Blog Search with Keyword Map-Based Relevance Feedback

Yasufumi Takama[1,2], Tomoki Kajinami[1], and Akio Matsumura[2]

[1] Tokyo Metropolitan Institute of Technology
[2] Tokyo Metropolitan University, 6-6 Asahigaoka, Hino, 191-0065 Tokyo, Japan
ytakama@cc.tmit.ac.jp

Abstract. In this paper, keyword map-based relevance feedback is applied to interactive Blog search. There exists vast amount of information in the Web, from which users usually gather information without definite information needs. In particular, when exploring the Blog space, the range of user's interests is expected to be broader than usual Web browsing process. The relevance feedback techniques have been studied in the field of document retrieval, aiming to generate appropriate queries for users' information needs. Although this approach is effective when the assumption that a user has a concrete criteria on the relevance of retrieved documents holds, it could not always hold when searching Blog, which consists of vast number of short articles about various topics. Compared with the previous work on keyword map-based relevance feedback, the algorithm proposed in this paper can consider multiple topics, in which a user is interested on the keyword map.

1 Introduction

Recent growth of the Web has provided us with the variety of information resource. Among them, Blog is getting to attract people as one of the new information resources[2]. As it is assumed that the persons who browse the Blog space do not usually have definite information needs, because the Blog consists of vast number of short articles about various topics. Therefore, interactive facility should be given to the systems that support searching in the Blog space.

The relevance feedback (RF) techniques have been studied in the field of document retrieval, aiming to generate appropriate queries for users' information needs. Although the effectiveness of RF techniques have been shown in studies of various applications, conventional RF techniques seem to have some drawbacks. That is, conventional RF techniques assume that a user can estimate the relevance/irrelevance of a given document, even if he cannot represent his information needs as queries. This assumption is not always valid, especially when users do not have enough background knowledge about the topics of the target documents. Furthermore, the conventional RF employs the Vector Space Model (VSM), which does not seem to be suitable for Blogs consisting of the documents with short length.

A method for performing relevance feedback on keyword space has been proposed [6]. This method provides users with two kinds of retrieved results, including keyword map as well as retrieved documents. The keyword map visualizes the relationship between keywords extracted from retrieved documents, on which users can edit the keyword arrangement according to their interests. User's interests are estimated from his modification of keywords arrangement on the map. Although the method is expected to be effective for interactive Blog search, the previous paper[6] has considered only a simple case that a user represents a single interest on the map. It is assumed that a user is exploring the Blog space while having multiple interests at the same time. Therefore, the algorithm is improved so that it can consider multiple topics, in which a user is interested on the keyword map.

2 Fundamental Technology and Related Work

2.1 Blog as Information Resource on the Web

It is difficult to give the exact definition of Blog (Weblog), but it is usually defined as the homepages that have the following features.

- It consists of entities, each of which denotes the short, personal comment about a certain topic or event.
- A single author, or a few restricted members write the entities.
- It employs a dated log format, in which the most recent entity is added at the top of the page.
- It employs trackback, which is a framework for peer-to-peer communication and notifications between web sites.
- Newly added comments are summarized with RSS format, and distributed over the Web.

Compared with the usual homepages, Blogs are said to contain new topics as well as subjective information (personal comments), which attracts us as a new kind of information resource on the Web[2]. From the viewpoint of accessibility, the tools such as RSS feeders / aggregators, which read and parse RSS, can access uniformly to a number of Blog sites. Furthermore, there also exist Blog search engines, such as Bulkfeeds[1] and Feedster[2]. Both contents and accessibility are important factors for information resources. From this viewpoint, Blog is one of the promising information resources on the Web.

2.2 Keyword Map Visualization System

In the field of Web information retrieval, the information visualization systems based on clustering techniques have been researched in order to help users explore

---
[1] http://bulkfeeds.net/
[2] http://www.feedster.com/

retrieved results[7]. In order for users to understand context information from the visualized results, presenting only document clusters is not enough, but the relationship among clusters is also important.

In such a case, visualizing keyword space is also effective. From this viewpoint, we have developed the **keyword map** [5], on which the keywords extracted from documents are arranged so that the pair of keywords that frequently appears in the same documents can be arranged closely to each other. The developed system called TMIT (Topic Map Idea Tool) employs the spring model to arrange keywords on 2D space. The basic algorithm of TMIT is as follows.

1. Define the distance l_{ij} between keyword i and j based on their similarity $R_{ij} (\in [-1,1]^3)$ by Eq. (1) (m is positive constant).

$$l_{ij} = m(1 - R_{ij}). \tag{1}$$

2. The moving distance of keyword i in each step, $(\delta_{xi}, \delta_{yi})$, is calculated by Eq. (2).

$$(\delta_{xi}, \delta_{yi}) = \left(c\frac{\partial E}{\partial x_i}, c\frac{\partial E}{\partial y_i} \right), \tag{2}$$

$$E = \sum_i \sum_j \frac{1}{2} k_{ij} (d_{ij} - l_{ij})^2, \tag{3}$$

$$d_{ij} = \sqrt{(x_i - x_j)^2 + (y_i - y_j)^2}. \tag{4}$$

3. In each step, the center of gravity of the map is adjusted to the center of 2D space.

As for the interactive functions, a user can drag any keywords and fix them to arbitrary positions, while other keywords are arranged automatically based on spring model.

2.3 Relevance Feedback

Interaction should be bidirectional. That is, interactive interface should not only provide users with information in understandable manner, but also get their intentions and preferences. Relevance feedback is one of major approaches for implicitly obtaining the users' preferences.

A basic algorithm of Conventional RF is as follows:

1. Submits an initial query to a search engine.
2. Judges each document in the retrieved result as relevant / irrelevant.
3. The current query is modified based on the user's judgment in Step 2.
4. Submits new query to the search engine, and goes to Step 2.

---

[3] In the current keyword map system, $R_{ij} \in (0,1]$ when keywords co-occur within documents, and $R_{ij}=-1$ when they do not appear within the same document.

The majority of existing RF algorithms employ VSM for the retrieval, of which the basic modification formula is represented as Eq. (5).

$$q(t) = q(t-1) + \alpha \sum_{d_p \in D_p} d_p - \beta \sum_{d_n \in D_n} d_n, \qquad (5)$$

where $q(t)$ indicates the query vector for t-th retrieval, d indicates the document vector of d, in which TFIDF weighting is usually employed. The α and β are coefficients, which vary according to the algorithms. One of the most famous formulas is called Rocchio's one[1], in which $\alpha = 1/|D_p|$ and $\beta = 1/|D_n|$ are used.

There also exist other RF algorithms with VSM. Onoda[3] employs SVM (Support Vector Machine) for query modification. The FISH View system[4] obtains the feedback from users in different ways. Instead of judging the relevance of each document, a user groups documents hierarchically with information visualization tool, from which results the system extracts the user's viewpoint based on a concept dictionary.

3 Relevance Feedback Based on Keyword Map

3.1 Concept of RF on Keyword Map

Although RF is expected to be useful for improving the Web interaction, there are still a few things to be considered. Nowadays, it is rational that the interaction between humans and the Web involves existing search engines. Therefore, the conventional RF, which is basically based on VSM, should be combined with widely-used search engines such as Google. That is, a query vector as a result of relevance feedback should be converted to a set of keywords, which can be submitted as a query to the search engines. An easy solution for that is to select a couple of keywords that have higher weights in the query vector than others. However, it can not make use of the expressive power of Boolean logic, which is available in most of existing Web search engines.

Furthermore, conventional RF usually assumes that users can judge the relevance of documents even if they cannot represent their information needs as explicit queries. However, this assumption is not always valid, especially when users do not have enough background knowledge about the target areas.

The conventional RF performed on document space also has the problem that the queries which can be generated is strictly restricted by the current document set. As for a typical example, it cannot generate the queries for a topic that is not contained in the current document set. Of course, selecting relevant/irrelevant documents is efficient when a topic of interest and others are mentioned in the distinct set of documents. However, let us consider the case that two words A and B often co-appear in the current document set, as they correspond to the same topic in the document set. When a user wants to search documents about more specific topic, say, "the documents about A but not B", it is difficult to generate such a query by selecting relevant/irrelevant documents from the current document set[6]. In particular, these problems of the conventional RF seems to be serious when searching the Blog, which consists of vast number of short articles about various topics. That is, it is assumed that a users is exploring the Blog space without definite information need, while having multiple interests at the same time.

To handle above-mentioned problems, the concept of keyword map-based RF have been proposed[6]. Compared with conventional RF, of which the retrieved results are only documents, our approach provides users with two kinds of retrieved results, including the keyword map generated from the retrieved documents as well as the documents themselves. Therefore, the keyword map-based RF can infer the user's intention from the keyword space. That is, the relationship between keywords, which the VSM does not consider, could be estimated from their distance on the map, and could be converted to the Boolean expressions.

This paper applies the keyword map-based RF to interactive Blog Search. The processes of the proposed system are as follows.

1. Submits the initial query to Blog search engine.
2. Generates a keyword map by TMIT from the retrieved result and presents to a user.
3. The user modifies the keyword arrangements as he likes.
4. Based on the modification, new query is generated and submitted to the Blog search engine. Goes to step 2.

In step 1 and 4, We use Bulkfeeds because APIs for searching Blogs are available. When a keyword map is presented to a user in step 2, he usually finds the difference between the keyword arrangement on the map and his background knowledge. Therefore, he wants to modify the arrangement, as he likes. If the system can infer the user's intention from the keyword map modified by him, relevance feedback can be available.

In step 4, the previous paper[6] has focused on how to extract important keyword pairs that reflect user's interests. That is, the pair of keywords, of which the distance has been changed from far to near or vice versa by the user's modification, is extracted based on keywords' coordinates.

In this paper, the TMIT is modified so that the information about focused keywords [4] can be available. As the focused keywords reflect user's interests, we focus on only them for generating queries.

3.2 Query Generation from Keyword Map Modification

As the Blog search engine employed in this paper accepts Boolean queries, a user's interests should be translated into such forms in step 4 in Sec. 3.1. The basic idea of the proposed method is to find the keyword clusters generated by a user on the keyword map. Let us consider the following two cases:

---

[4] A focused keyword is the keyword, of which a user fix the position on the map, as noted in Section 2.2.

- A user rearranges the keyword A close to keyword B.
- A user moves apart keyword A and B.

The former case will generate a single keyword cluster around keyword A and B, while the latter case will generate two distinct clusters, one cluster around keyword A and another around keyword B. The former case has already considered in the previous paper[6]. In this case, collecting new document that contain both keywords should satisfy the user's interest. The latter case might be more complicated, and there will be several possibilities. In this paper, we consider the case as that the user wants to divide the topic represented by keyword A and B into two detailed topics.

The algorithm for generating queries from the user's modification of keyword arrangements is as follows.

1. The coordinates of focused keywords are given by TMIT.
2. Clusters the focused keywords based on their distance. The results is represented as the set of keyword groups, $C_q = \{c_1, \ldots, c_n\}$, $c_i = \{word_1^i, \ldots, word_{m_i}^i\}$.
3. Each cluster c_i is translated into AND-connected sub queries, $q_i = word_1^i$ AND ... AND $word_{m_i}^i$.
4. Submits each sub query q_i and obtains the corresponding set of documents (retrieved result) D_i.
5. Obtain a set of documents that are contained in only a single retrieved result.

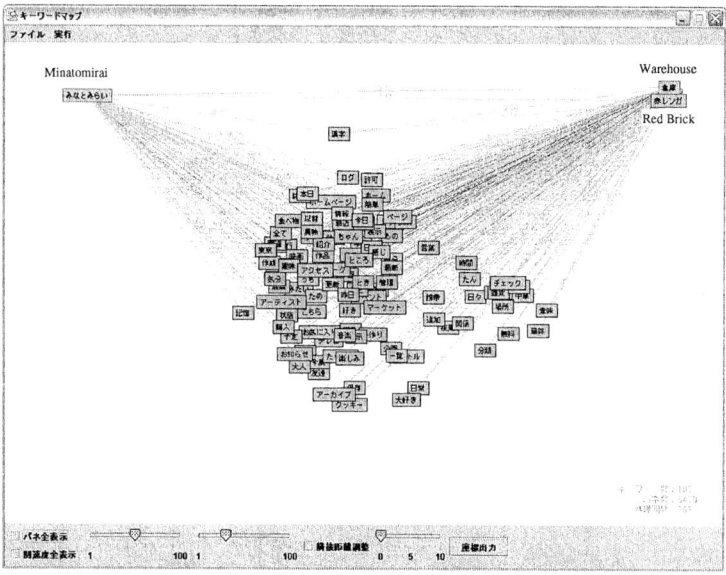

Fig. 1. Keyword Map Edited by User

4 Experiments on Keyword Map-Based Relevance Feedback

This section shows the examples how the proposed algorithm works on a keyword map actually generated from the retrieval results of existing Blog search engines. It should also be noted that the experiments are performed on Japanese Blog sites, and results are translated from Japanese into English hereafter. Regarding the step 2 of the algorithm proposed in Sec. 3.2, it is assumed in the experiments that the keywords within the rectangle, of which the width and height are one tenth of the map's width and height, belong to the same cluster.

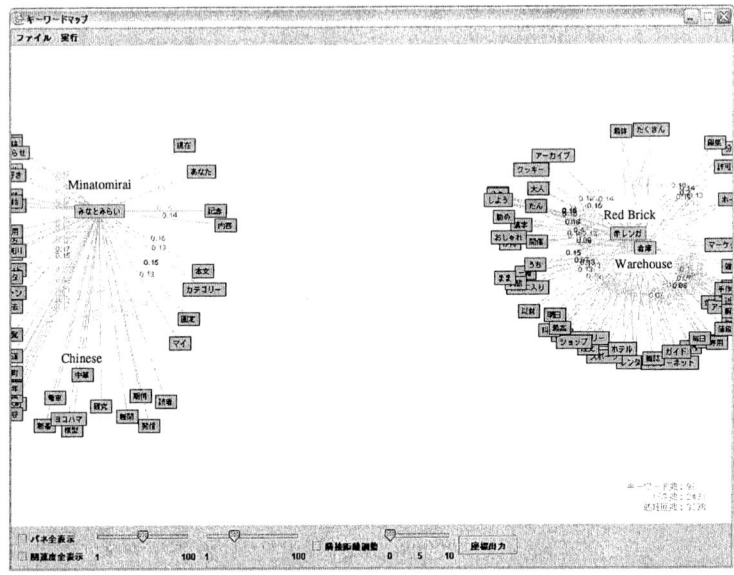

Fig. 2. Keyword Map Generated from 2nd query

In the experiments, we submit the initial query "Red Brick" and "Warehouse" to Bulkfeeds. Fig. 1 shows the map after a user modifies it so that "Red Brick" and "Warehouse" are arranged closely (upper-right corner of the map), whereas "Minatomirai" (left-upper corner) is away from those. From this result, a user is supposed to be interested in two distinct topics, "Minatomirai" (a location name) and "Red Brick Warehouse" (a sight spot). This might correspond to the user's intention that he wants to know more information about "Red Brick Warehouse" as well as other site spots in "Minatomirai" area. Fig. 2 show the keyword map generated from the second retrieval based on the algorithm proposed in the previous section. It can be seen that above-mentioned topics can form distinct keyword clusters on the map. In particular, let us focus on the "Chinese" (Chuka) that has appeared in the cluster of "Minatomirai". Although this keyword also appeared in Fig. 1, it can be clearly understood in Fig. 2 as other site spots than "Red Brick Warehouse". Therefore, by handling multiple topics as distinct clusters, the readability of the keyword map can be improved.

5 Conclusion

The relevance feedback based on interactive keyword map system is applied to interactive Blog search. The proposed algorithm can consider multiple topics, in which a user is interested, and perform the subsequent retrieval based on those topics. The experimental results show how the developed system works. In the experiments, the detection of keyword clusters is performed based on the fixed threshold of distance, which is determined empirically. The experiment with test subjects in order to examine the appropriate threshold is now in progress, of which the results will be reported soon.

References

1. Baeza-Yates, R. and Ribeiro-Neto, B., "5. Query Operations" in Modern Information Retrieval, Addison Wasley, 1999.
2. Fujiki, T., Nanno, T., Suzuki, Y., Okumura, M., "Identification of Bursts in a Document Stream", First International Workshop on Knowledge Discovery in Data Streams (in conjunction with ECML/PKDD 2004), pp.55-64, 2004.
3. Onoda, T., Murata, H. and Yamada, S., "Document Retrieval based on Relevance Feedback with Active Learning," SIG-KBS-A301 (JSAI), pp.13-18, 2003.
4. Takama, Y. and Ishizuka, M., "FISH VIEW System: A Document Ordering Support System Employing Concept-structure-based Viewpoint Extraction," J. of Information Processing Society of Japan, Vol. 41, No. 7, pp.1976-1986, 2000.
5. Takama, Y. and Tetsuya, H., "Application of Immune Network Metaphor to Keyword Map-based Topic Stream Visualization," Proc. 2003 IEEE Int'l Symp. on Computational Intelligence in Robotics and Automation (CIRA2003), pp. 770-775, 2003.
6. Takama, Y., "Consideration of Relevance Feedback on Keyword Space for Interactive Information Retrieval," 2004 IEEE Conf. on Cybernetics and Intelligent Systems (CIS2004), WP7-6, 2004.
7. Zamir, O. and Etzioni, O., "Grouper: A Dynamic Clustering Interface to Web Search Results," Proc. 8th International WWW Conference, 1999.

An One Class Classification Approach to Non-relevance Feedback Document Retrieval

Takashi Onoda[1], Hiroshi Murata[1], and Seiji Yamada[2]

[1] Central Research Institute of Electric Power Industry,
System Engineering Research Laboratory, 2-11-1 Iwado Kita,
Komae-shi, Tokyo 201-8511 Japan
{onoda, murata}@criepi.denken.or.jp
[2] National Institute of Informatics, 2-1-2 Hitotsubashi,
Chiyoda-ku, Tokyo 101-8430 Japan
seiji@nii.ac.jp http://research.nii.ac.jp/~seiji/index-e.html

Abstract. This paper reports a new document retrieval method using non-relevant documents. From a large data set of documents, we need to find documents that relate to human interesting in as few iterations of human testing or checking as possible. In each iteration a comparatively small batch of documents is evaluated for relating to the human interesting. The relevance feedback needs a set of relevant and non-relevant documents to work usefully. However, the initial retrieved documents, which are displayed to a user, sometimes don't include relevant documents. In order to solve this problem, we propose a new feedback method using information of non-relevant documents only. We named this method *non-relevance feedback document retrieval*. The non-relevance feedback document retrieval is based on One-class Support Vector Machine. Our experimental results show that this method can retrieve relevant documents using information of non-relevant documents only.

1 Introduction

As the Internet technology progresses, accessible information by end users is explosively increasing. In this situation, we can now easily access a huge document database through the Web. However it is hard for a user to retrieve relevant documents from which he/she can obtain useful information, and a lot of studies have been done in information retrieval, especially document retrieval [1]. Many works for such document retrieval have been reported in TREC (Text Retrieval Conference) [2] for English documents, IREX (Information Retrieval and Extraction Exercise) [3] and NTCIR (NII-NACSIS Test Collection for Information Retrieval System) [4] for Japanese documents.

In most frameworks for information retrieval, a vector space model in which a document is described with a high-dimensional vector is used [5]. An information retrieval system using a vector space model computes the similarity between a query vector and document vectors by the cosine of the two vectors and indicates a user a list of retrieved documents.

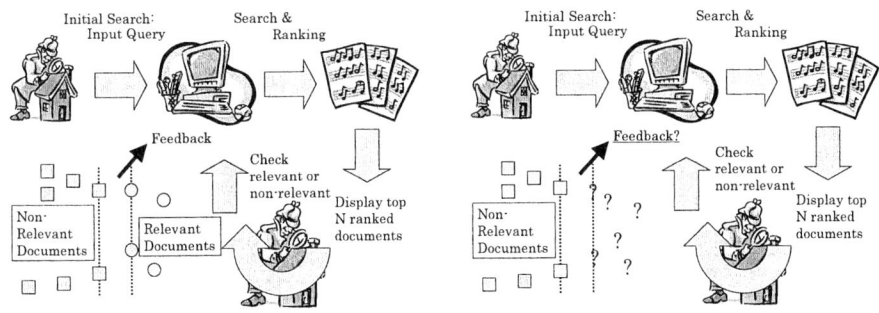

Fig. 1. Outline of the relevance feedback documents retrieval(left side) and Image of a problem in the relevance feedback documents retrieval(right side): The gray arrow parts are made iteratively to retrieve useful documents for the user. This iteration is called feedback iteration in the information retrieval research area. But if the evaluation of the user has only non-relevant documents, ordinary relevance feedback methods can not feed back the information of useful retrieval.

In general, since a user hardly describes a precise query in the first trial, interactive approach to modify the query vector using evaluation of the documents on a list of retrieved documents by a user. This method is called *relevance feedback* [6] and used widely in information retrieval systems. In this method, a user directly evaluates whether a document in a list of retrieved documents is relevant or non-relevant , and a system modifies the query vector using the user evaluation. A traditional way to modify a query vector is a simple learning rule to reduce the difference between the query vector and documents evaluated as relevant by a user (see Figure 1 left side).

In another approach, relevant and irrelevant document vectors are considered as positive and negative examples, and relevance feedback is transposed to a binary class classification problem [7]. For the binary class classification problem, Support Vector Machines (which are called SVMs) have shown the excellent ability. And some studies applied SVM to the text classification problems [8] and the information retrieval problems [9]. Recently, we have proposed a relevance feedback framework with SVM as *active learning* and shown the usefulness of our proposed method experimentally [10].

The initial retrieved documents, which are displayed to a user, sometimes don't include relevant documents. In this case, almost all relevance feedback document retrieval systems do not work well, because the systems need relevant and non-relevant documents to construct a binary class classification problem (see Figure 1 right side).

While a machine learning research field has some methods which can deal with one class classification problem. In the above document retrieval case, we can use non-relevant documents information only. Therefore, we consider this retrieval situation is as same as one class classification problems.

In this paper, we propose a framework of an interactive document retrieval using non-relevant documents information only. We call this interactive docu-

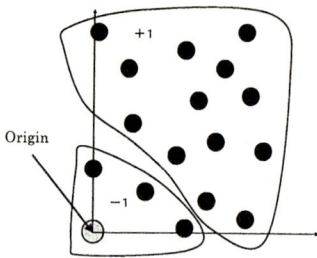

Fig. 2. One-Class SVM Classifier: the origin is the only original member of the second class

ment retrieval as *non-relevance feedback document retrieval*, because we can use non-relevant documents information only. Our proposed non-relevance document retrieval is based on One Class Support Vector Machine(One-Class SVM) [11]. One-Class SVM can generate a discriminant hyperplane that can separate the non-relevant documents which are evaluated by a user. Our proposed method can display documents, which may be relevant documents for the user, using the discriminant hyperplane.

In the remaining parts of this paper, we explain the One-Class SVM algorithm in the next section briefly, and propose our document retrieval method based on One-Class SVM in the third section. In the fourth section, in order to evaluate the effectiveness of our approach, we made experiments using a TREC data set of Los Angels Times and discuss the experimental results. Finally we conclude our work and discuss our future work in the fifth section.

2 One-Class Support Vector Machine

Schölkopf et al. suggested a method of adapting the SVM methodology to one-class classification problem. Essentially, after transforming the feature via a kernel, they treat the origin as the only member of the second class. The using "relaxation parameters" they separate the image of the one class from the origin. Then the standard two-class SVM techniques are employed.

One-class SVM [11] returns a function f that takes the value +1 in a *small* region capturing most of the training data points, and -1 elsewhere.

The algorithm can be summarized as mapping the data into a feature space H using an appropriate kernel function, and then trying to separate the mapped vectors from the origin with maximum margin (see Figure 2).

Let the training data be

$$\mathbf{x}_1, \ldots, \mathbf{x}_\ell \tag{1}$$

belonging to one class X, where X is a compact subset of R^N and ℓ is the number of observations. Let $\Phi : X \to H$ be a kernel map which transforms the training examples to feature space. The dot product in the image of Φ can be computed by evaluating some simple kernel

$$k(\mathbf{x}, \mathbf{y}) = (\Phi(\mathbf{x}) \cdot \Phi(\mathbf{y})) \tag{2}$$

such as the linear kernel, which is used in our experiment,

$$k(\mathbf{x}, \mathbf{y}) = \mathbf{x}^\top \mathbf{y}. \tag{3}$$

The strategy is to map the data into the feature space corresponding to the kernel, and to separate them from the origin with maximum margin. Then, to separate the data set from the origin, one needs to solve the following quadratic program:

$$\min_{\mathbf{w} \in H, \xi \in R^\ell, \rho \in R^N} \frac{1}{2}\|\mathbf{w}\|^2 + \frac{1}{\nu\ell}\sum_i \xi_i - \rho$$
$$\text{subject to } (\mathbf{w} \cdot \Phi(\mathbf{x}_i)) \geq \rho - \xi_i, \quad \xi_i \geq 0. \tag{4}$$

Here, $\nu \in (0,1)$ is an upper bound on the fraction of outliers, and a lower bound on the fraction of Support Vectors (SVs).

Since nonzero slack variables ξ_i are penalized in the objective function, we can expect that if $\mathbf{w}$ and ρ solve this problem, then the decision function

$$f(\mathbf{x}) = \text{sgn}\left((\mathbf{w} \cdot \Phi(\mathbf{x})) - \rho\right) \tag{5}$$

will be positive for most examples $\mathbf{x}_i$ contained in the training set, while the SV type regularization term $\|w\|$ will still be small. The actual trade-off between these two is controlled by ν. For a new point $\mathbf{x}$, the value $f(\mathbf{x})$ is determined by evaluating which side of the hyperplane it falls on, in feature space.

In our research we used the LIBSVM. This is an integrated tool for support vector classification and regression which can handle one-class SVM using the Schölkopf etc algorithms. The LIBSVM is available at
http://www.csie.ntu.edu.tw/~cjlin/libsvm.

3 Non-relevance Feedback Document Retrieval

In this section, we describe our proposed method of document retrieval based on Non-relevant documents using One-class SVM.

In relevance feedback document retrieval, the user has the option of labeling some of the top ranked documents according to whether they are relevant or non-relevant. The labeled documents along with the original request are then given to a supervised learning procedure to produce a new classifier. The new classifier is used to produce a new ranking, which retrievals more relevant documents at higher ranks than the original did (see Figure 1 left side) [10].

The initial retrieved documents, which are displayed to a user, sometimes don't include relevant documents. In this case, almost all relevance feedback document retrieval systems do not contribute to efficient document retrieval, because the systems need relevant and non-relevant documents to construct a binary class classification problem (see Figure 1 right side).

The One-Class SVM can generate discriminant hyperplane for the one class using one class training data. Consequently, we propose to apply One-Class SVM in a *non-relevance feedback document retrieval method*. The retrieval steps of proposed method perform as follows:

Step 1: Preparation of documents for the first feedback
The conventional information retrieval system based on vector space model displays the top N ranked documents along with a request query to the user. In our method, the top N ranked documents are selected by using the cosine distance between the request query vector and each document vectors for the first feedback iteration.

Step 2: Judgment of documents
The user then classifiers these N documents into relevant or non-relevant. If the user labels all N documents non-relevant, the documents are labeled "+1" and go to the next step. If the user classifies the N documents into relevant documents and non-relevant documents, the non-relevant documents are labeled "+1" and relevant documents are labeled "-1" and then our previous proposed relevant feedback method is adopted [10].

Step 3: Determination of non-relevant documents area based on non-relevant documents
The discriminant hyperplane for classifying non-relevant documents area is generated by using One-Class SVM. In order to generate the hyperplane, the One-Class SVM learns labeled non-relevant documents which are evaluated in the previous step (see Figure 3 left side).

Step 4: Classification of all documents and Selection of retrieved documents
The One-class SVM learned by previous step can classifies the whole documents as non-relevant or not non-relevant. The documents which are discriminated in *not non-relevant are* are newly selected. From the selected documents, the top N ranked documents, which are ranked in the order of the distance from the non-relevant documents area, are shown to user as the document retrieval results of the system (see Figure 3 right side). These N documents have high existence probability of initial keywords. Then return to Step 2.

The feature of our One-Class SVM based non-relevant feedback document retrieval is the selection of displayed documents to a user in Step 4. Our proposed method selects the documents which are discriminated as *not non-relevant* and near the discriminant hyperplane between non-relevant documents and not non-relevant documents. Generally if the system got the opposite information from a user, the system should select the information, which is far from the opposite information area, for displaying to the user. However, in our case, the classified non-relevant documents by the user includes a request query vector of the user. Therefore, if we select the documents, which are far from the non-relevant documents area, the documents may not include the request query of the user.

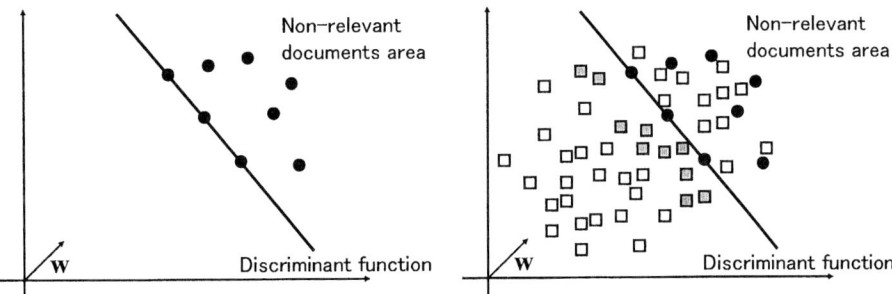

Fig. 3. Generation of a hyperplane to discriminate non-relevant documents area and Mapped non-checked documents into the feature space: Circles denote documents which are checked non-relevant by a user. The solid line denotes the discriminant hyperplane. Boxes denote non-checked documents which are mapped into the feature space. Gray boxes denotes the displayed documents to a user in the next iteration. These documents are in the *not non-relevant document area* and near the discriminant hyperplane.

Our selected documents (see Figure 3 right side) is expected that the probability of the relevant documents for the user is high, because the documents are not non-relevant and may include the query vector of the user.

4 Experiments

4.1 Experimental Setting

We made experiments for evaluating the effectiveness of our interactive document retrieval based on non-relevant documents using One-Class SVM described in section 3. The document data set we used is a set of articles in the Los Angels Times which is widely used in the document retrieval conference TREC [2]. The data set has about 130 thousands articles. The average number of words in a article is 526. This data set includes not only queries but also the relevant documents to each query. Thus we used the queries for the experiments. We used three topics for experiments and show these topics in Table 1. These topics do not have relevant documents in top 20 ranked documents in the order of the cosine distance between the query vector and document vectors. Our experiments set the size of N of displayed documents presented in Step 1 in the section 3 to 10 or 20.

Table 1. Topics, query words and the number of relevant documents in the Los Angels Times used for experiments

topic	query words	# of relevant doc.
306	Africa, civilian, death	34
343	police, death	88
383	mental, ill, drug	55

We used TFIDF [1], which is one of the most popular methods in information retrieval to generate document feature vectors, and the concrete equation [12] of a weight of a term t in a document d w_t^d are in the following.

$$w_t^d = L \times t \times u \tag{6}$$

$$L = \frac{1 + \log(tf(t,d))}{1 + \log(\text{average of } tf(t,d) \text{ in } d)} \text{ (TF)}, \quad t = \log(\frac{n+1}{df(t)}) \text{ (IDF)}$$

$$u = \frac{1}{0.8 + 0.2 \frac{uniq(d)}{\text{average of } uniq(d)}} \text{ (normalization)}$$

The notations in these equation denote as follows:

- w_t^d is a weight of a term t in a document d,
- $tf(t,d)$ is a frequency of a term t in a document d,
- n is the total number of documents in a data set,
- $df(t)$ is the number of documents including a term t,
- $uniq(d)$ is the number of different terms in a document d.

In our experiments, we used the linear kernel for One-class SVM learning, and found a discriminant function for the One-class SVM classifier in the feature space. The vector space model of documents is high dimensional space. Moreover, the number of the documents which are evaluated by a user is small. Therefore, we do not need to use the kernel trick and the parameter ν (see section 2) is set adequately small value ($\nu = 0.01$). The small ν means hard margin in the One-Class SVM and it is important to make hard margin in our problem.

For comparison with our approach, two information retrieval methods were used. The first is an information retrieval method that does not use a feedback, namely documents are retrieved using the rank in vector space model (VSM). The second is an information retrieval method using the conventional Rocchio-based relevance feedback [6] which is widely used in information retrieval research.

The Rocchio-based relevance feedback modifies a query vector Q_i by evaluation of a user using the following equation.

$$Q_{i+1} = Q_i + \alpha \sum_{x \in R_r} x - \beta \sum_{x \in R_n} x, \tag{7}$$

where R_r is a set of documents which were evaluated as relevant documents by a user at the ith feedback, and R_n is a set of documents which were evaluated as non-relevant documents at the i feedback. α and β are weights for relevant and non-relevant documents respectively. In this experiment, we set $\alpha = 1.0, \beta = 0.5$ which are known adequate experimentally.

4.2 Experimental Results

Here, we describe the relationships between the performances of the proposed method and the number of feedback iterations. Table 2 left side gave the number

Table 2. The number of retrieved relevant documents at each iteration: the number of displayed documents is 10 at each iteration

	# of displayed doc. is 10			# of displayed doc. is 20		
topic 306	# of retrieved relev. doc.			# of retrieved relev. doc.		
# of iter.	Proposed	VSM	Rocchio	Proposed	VSM	Rocchio
1	1	0	0	1	1	0
2	–	0	0	–	–	0
3	–	1	0	–	–	0
4	–	–	0	–	–	0
5	–	–	0	–	–	0
topic 343	# of retrieved relev. doc.			# of retrieved relevant doc.		
# of iter.	Proposed	VSM	Rocchio	Proposed	VSM	Rocchio
1	0	0	0	1	0	0
2	1	0	0	–	0	0
3	–	0	0	–	0	0
4	–	0	0	–	1	0
5	–	0	0	–	–	0
topic 383	# of retrieved relev. doc.			# of retrieved relevant doc.		
# of iter.	Proposed	VSM	Rocchio	Proposed	VSM	Rocchio
1	0	0	0	1	0	0
2	1	0	0	–	1	0
3	–	0	0	–	–	0
4	–	1	0	–	–	0
5	–	–	0	–	–	0

of retrieved relevant documents at each feedback iteration. At each feedback iteration, the system displays ten higher ranked *not non-relevant* documents, which are near the discriminant hyperplane, for our proposed method. We also show the retrieved documents of the Rocchio-based method at each feedback iteration for comparing to the proposed method in table 2 left side.

We can see from this table that our non-relevance feedback approach gives the higher performance in terms of the number of iteration for retrieving relevant document. On the other hand, the Rocchio-based feedback method cannot search a relevant document in all cases. The vector space model without feedback is better than the Rocchio-based feedback. After all, we can believe that the proposed method can make an effective document retrieval using only non-relevant documents, and the Rocchio-based feedback method can not work well when the system can receive the only non-relevant documents information.

Table 2 right side gave the number of retrieved relevant documents at each feedback iteration. At each feedback iteration, the system displays twenty higher ranked *not non-relevant* documents, which are near the discriminant hyperplane, for our proposed method. We also show the retrieved documents of the Rocchio-based method at each feedback iteration for comparing to the proposed method in table 2 right side.

We can observe from this table that our non-relevance feedback approach gives the higher performance in terms of the number of iteration for retrieving

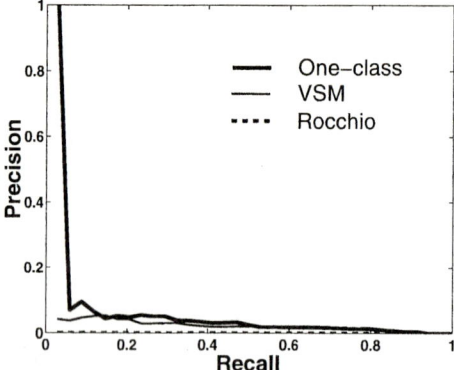

Fig. 4. The precision and recall curve of topic no. 304 at the second iteration

relevant documents, and the same experimental results as table 2 left side about the Rocchio-based method and VSM.

In table 2 left side, a user already have seen twenty documents at the first iteration. Before the fist iteration, the user have to see ten documents, which are retrieved results using the cosine distance between a query vector and document vectors in VSM. In table 2 right side, the user also have seen forty documents at the first iteration. Before the fist iteration, the user also have to see ten documents to evaluate the documents, which are retrieved results using the cosine distance between a query vector and document vectors in VSM. When we compare the experimental results of table 2 left side with the results of table 2 right side, we can observe that the small number of displayed documents makes more effective document retrieval performance than the large number of displayed documents. In table 2 left side, the user had to see thirty documents by finding the first relevant document about topic 343 and 383. In table 2 right side, the user had to see forty documents by finding the first relevant document about topic 343 and 383. Therefore, we believe that the early non-relevance feedback is useful for an interactive document retrieval.

We also show a precision and recall curve in figure 4. This figure is the precision and recall curve of topic no. 306 at the second iteration. From this figure, we can understand that all precision-recall curves are not good. However, our proposed approach is more efficient than the two other approaches.

5 Conclusion

In this paper, we proposed the non-relevance feedback document retrieval based on One-Class SVM using only non-relevant documents for a user. In our non-relevance feedback document retrieval, the system use only non-relevant documents information. One-Class SVM can generate a discriminant hyperplane of observed one class information, so our proposed method adopted One-Class SVM for non-relevance feedback document retrieval.

This paper compared our method with a conventional relevance feedback method and a vector space model without feedback. Experimental results on a set of articles in the Los Angels Times showed that the proposed method gave a consistently better performance than the compared method. Therefore we believe that our proposed One-Class SVM based approach is very useful for the document retrieval with only non-relevant documents information.

This paper proposed that the system should display the documents which are in the *not non-relevant* documents area and near the discriminant hyperplane of One-Class SVM at each feedback iteration. However, we do not discuss how the selection of documents influence both the effective learning and the performance of information retrieval theoretically. This point is our future work.

References

1. Yates, R.B., Neto, B.R.: Modern Information Retrieval. Addison Wesley (1999)
2. TREC Web page: (http://trec.nist.gov/)
3. IREX: (http://cs.nyu.edu/cs/projects/proteus/irex/)
4. NTCIR: (http://www.rd.nacsis.ac.jp/~ntcadm/)
5. Salton, G., McGill, J.: Introduction to modern information retrieval. McGraw-Hill (1983)
6. Salton, G., ed. In: Relevance feedback in information retrieval. Englewood Cliffs, N.J.: Prentice Hall (1971) 313–323
7. Okabe, M., Yamada, S.: Interactive document retrieval with relational learning. In: Proceedings of the 16th ACM Symposium on Applied Computing. (2001) 27–31
8. Tong, S., Koller, D.: Support vector machine active learning with applications to text classification. In: Journal of Machine Learning Research. Volume 2. (2001) 45–66
9. Drucker, H., Shahrary, B., Gibbon, D.C.: Relevance feedback using support vector machines. In: Proceedings of the Eighteenth International Conference on Machine Learning. (2001) 122–129
10. Onoda, T., Murata, H., Yamada, S.: Relevance feedback with active learning for document retrieval. In: Proc. of IJCNN2003. (2003) 1757–1762
11. Schölkopf, B., Platt, J., Shawe-Taylor, J., Smola, A., Williamson, R.: Estimating the support for a high-dimensional distribution. Technical Report MSR-TR-99-87, Microsoft Research, One Microsoft Way Redmon WA 98052 (1999)
12. Schapire, R., Singer, Y., Singhal, A.: Boosting and rocchio applied to text filtering. In: Proceedings of the Twenty-First Annual International ACM SIGIR. (1998) 215–223

Automated Knowledge Extraction from Internet for a Crisis Communication Portal

Ong Sing Goh and Chun Che Fung

School of Information Technology, Murdoch University,
Murdoch, Western Australia, 6150, Australia
`os.goh@murdoch.edu.au, l.fung@murdoch.edu.au`

Abstract. This paper describes the development of an Automated Knowledge Extraction Agent (AKEA) which was designed to acquire online news and document from the internet for the establishment of a knowledge based crisis communication portal. It was recognized that in times of crisis, an effective communication mechanism is essential to maintain peace and calmness in the community by providing timely and appropriate information. It is proposed that the incorporation of software agents into the crisis communication portal will be capable to send alert news to subscribed users via internet and mobile services. The proposed system consists of crawler, wrapper, name-entity tagger, AIML (Artificial Intelligence Markup language) and an animated character is used in the front-end for human computer communication.

1 Introduction

With the acceptance and increasingly reliance of the Internet, the Internet has now become "the" repository of human knowledge and information for the 21st century. On the other hand, advancements in internet and mobile communication technologies have provided effective and cheap means of communication for the modern society. The global implications of such technologies are unparalleled in the history of human civilization. Hence, the Internet now serves two of the most important functions in the modern world – as a giant virtual storehouse of data, information and knowledge, and, as the true information superhighway whereby delivery of all kinds of data and information can be done cheaply and quickly.

The potential of effective use of these two aspects are particularly important in times of crisis. Within the context of this paper, crisis may be referred to events or incidents that have the potential to cause national panic, confusion, unrest and possible catastrophe. These crises may be due to health epidemic, natural disasters and man-made tragedies such as terrorist attacks. Examples of these events that happened in the recent past are Severe Acute Respiratory Syndrome (SARS), bird flu, mad cow disease, September 11, earth quakes and tsunami. In these cases, accurate information delivered within the shortest duration of time at the lowest costs would be essential in informing the affected communities and the relevant authorities. In particular, if decisions are made quickly and appropriately, this will have the benefits of reducing the potential damages and will lead to better manage of the situations.

This paper reports the development an Automated Knowledge Extraction Agent (AKEA) which was designed to establish the knowledge base for a global crisis communication system called CCNet. CCNet was proposed during the height of the SARS epidemic in 2003. It was aimed at providing up-to-date information to its users via a conversational software robot called AINI (Artificial Intelligence Neural-network Identity). The purpose of AINI is to deliver essential information from trusted sources and is able to interact with its users by animated characters. The idea is to rely on a human-like communication approach thereby providing a sense of comfort and familiarity. The functionalities of AINI have been reported in the past and development on AINI is ongoing [1]. It is foreseeable that the combination of AINI and AKEA will produce a more natural means of communication and computing in the near future.

1.1 Objectives of the Research

The objectives of this research are:

a. To develop a global crisis communication portal (CCNet portal) in order to provide the latest information for public awareness on the knowledge about and how to respond to a particular crisis.
b. To establish a new and effective human-computer communication approach by transforming traditional websites with static text and images to "humanized" websites by deploying Artificial Intelligent Neural-network Identity (AINI).
c. To develop an Automated Knowledge Extraction Agent (AKEA) to automatically build and enhance the knowledge base for the AINI conversation software robot.
d. To develop new approaches in internet and communication technologies for the effective distribution of information to the global communities.

2 Architectural Overview

The architectural design of the proposed system is shown in Figure 1. The CCNet Portal can be divided into two main parts plus a middle-tier of multiple knowledge bases. The two main parts are termed the *Front-End*, responsible for interaction with the user, and, the *Back-End*, which is designed to establish the knowledge bases in the background.

The AINI Server and Mobile Gateway are located in the middle. They function as the interconnection linkage between the Front-End (Client) and the Back-End (Server) of the system. They process the communication between the users of the system and the CCNet Portal. The AINI's engine comprises of an intelligent agent framework. All communications with AINI are carried out through a natural human-machine interface that uses natural language processing and speech technologies via a 3D animated character. AINI's engine carries out the sophisticated decision making process based on the information it interprets from the knowledge bases. These decision-making capabilities are based on the knowledge embedded in the XML specifications. The input and output of the modules in the AINI knowledge bases such as Expression Emotion, Customers and AlertNews are stored in XML-encoded data structure. These modules are representations of the knowledge conceptualized in the format of XML data structure. From the perspective of the users, the CCNet system accepts questions and requests, and it is also capable to process the queries based on the information contained in AINI's knowledge base.

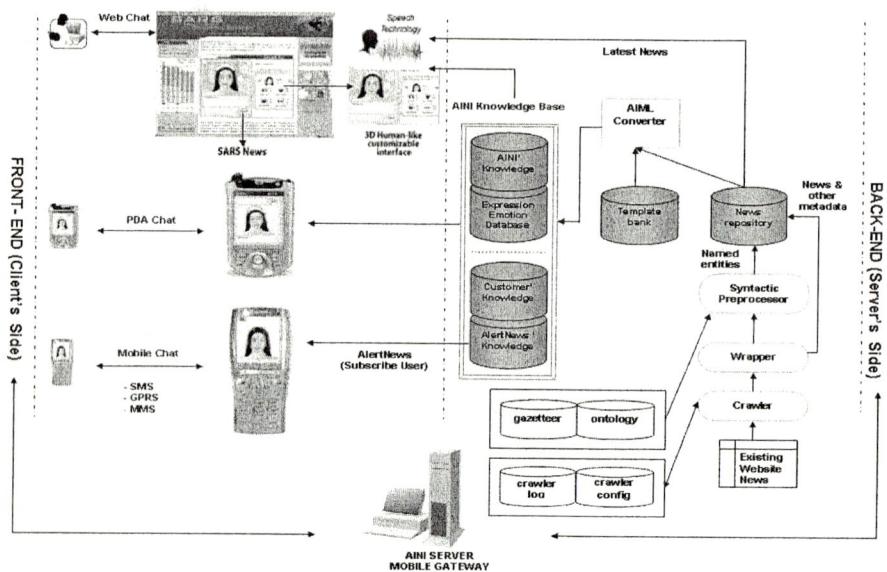

Fig. 1. Architecture Overview of CCNet

2.1 Front-End (Client's Side)

The Front-end provides the necessary interaction between the user and the system. Three different modes of communication are provided - web chat, PDA chat and mobile chat.

Web Chat
The web chat sessions allow interaction between a user and the software robot. The communication can text-based or voice-based with the animation of a 3D character. If voice is desired, Text-to-Speech technology is used to convert the text to voice using synthesizer hardware and software. This is particularly useful for someone who has difficulties or unfamiliar with the conventional keyboard. In terms of the animated character, users may customize the interface as required. They can also input the questions and receive the responses directly from the website. In addition, users may navigate through all the information on the topics or issues of their interest. If necessary, guidance may also be provided to assist the users.

AINI's Processing Engine
The main objective of AINI is to intelligently offer related information on various topics in a virtual environment where no real live agents or specialists are required to be physically involved. AINI uses natural language parsing in Artificial Intelligent

Mark-up Language (AIML) to search the predefined knowledge base as well as other data sources located in other systems via the internet. Users interact with the virtual advisor through WebGuide, WebTips and WebSearch engines. WebGuide is used to guide users through the entire portal. The WebTips engine, on the other hand, provides tips or hints to the users. The WebSearch system is an integrated search engine which can search for local sites as well as the Internet and online databases.

At the same time, the users can interact and chat with the AINI chatterbot or Virtual Agent. The chatterbot is based on natural-language processing and aimed to initiate conversations with users [1]. On the other hand, AINI also offers messaging, email and phone services to the users.

PDA Chat

Developing AINI into Personal Digital Assistance (PDA) devices is a recent approach in order to provide an alternative human and personalized interface between the computer and human. The PDA chat has the same functions as in web chats but with mobile capability. It is designed to incorporate mobile technology with natural language interface to assist interaction naturally with mobile devices. Implementation of PDA chat with the knowledge base was designed using WiFi technology.

2.2 Back-End System (Server's Side)

In this paper, the focus is on the development of a knowledge base which forms the "brain" of the CCNet portal. This knowledge base contains the domain knowledge for crisis communication based on specific discipline or topic. All the information in this knowledge base is going to be extracted from AKEA, which is explained in detail in the next section.

3 Establishment of AINI'S Knowledge Bases

In this proposed system, AINI's knowledge base consists of a common knowledge base, an expression emotion database, a customer knowledge base and an Alert-News knowledge base. From literature, it was identified that START (SynTactic Analysis using Reversible Transformations) developed by Boris Katz at MIT's Artificial Intelligence Laboratory is a natural language understanding system, and Omnibase is a virtual database that provides uniform access to heterogeneous and distributed Web sources via a wrapper-based framework [2]. A simplified version of the natural language annotation technology is employed here as the database access schemata to mediate between natural language and database queries. A detailed description of each component is provided in the following sections.

3.1 Common Knowledge Base

AIML is used to represent AINI's common knowledge base. It is an XML specification for programming chat robots created by ALICE Artificial Intelligence Foundation. A typical way of representing knowledge in an AIML file is as follows:

```
<aiml>
    <category>
            <pattern>PATTERN</pattern>
            <template>TEMPLATE</template>
    </category>
</aiml>
```

The *<aiml>* tag demonstrates that this file describes the way that knowledge is stored. The *<category>* tag indicates an AIML category and it is the basic unit of the chatterbot's knowledge. Each category has a *<pattern>* and a corresponding *<template>*. This *<pattern>* represents the question and the *<template>* represents the answer [3]. A user chats with AINI in the cyberspace and the topic may involve any topic related to crisis communication.

3.2 Expression Emotion Knowledge Base

The Expression Emotion Database, on the other hand, is used to identify and classify emotions within the context of the conversation between the user and the software robot. AINI was designed to perceive the emotion behind the human's input and it generates appropriate responses. The concept of communication between a human and the agent through AINI is depicted in Figure 2.

Human speech is passed to the natural processing unit in AINI for analysis and processing as shown in Figure 2. The natural processing unit generates proper responses by extracting the knowledge stored in the database. The emotion recognition unit is responsible for identifying the emotion found in the speech and instructs the agent to display an appropriate facial expression. For example, the agent will display a happy face to greet the user when the conversation starts; it will display a sad face when it hears something miserable; and it will show an angry face when the user says some obscene words.

An *<agplay/>* tag or "agent play" tag is created to produce an attractive expression for the character of that animated agent. Below is an example showing how the *<agplay/>* tag is used. This category is executed when the user greets the AINI by typing "HELLO". In return, the AINI will smile to the user and respond by saying "Hi there! How do you feel today?"

```
<category>
    <pattern>HELLO</pattern>
        <template>
            Hi there! How do you feel today?
                <agplay anims="greet, pleased"/>
        </template>
</category>
```

3.3 AlertNews and Customer Knowledge Base

The AlertNews knowledge base provides news and information to users who use mobile chat via SMS, MMS or GPRS technologies. There are three types of users of this system - subscribed users, non-subscribed users and the CCNet editorial. The AlertNews architecture is shown in Figure 3.

Automated Knowledge Extraction from Internet   1231

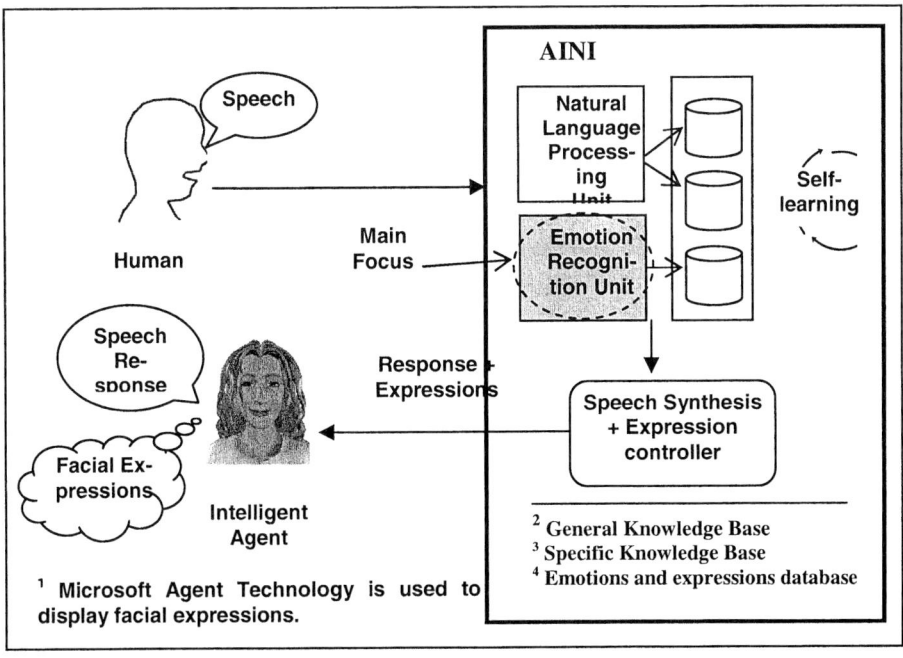

Fig. 2. Human-liked Emotion and Expression between user and AINI

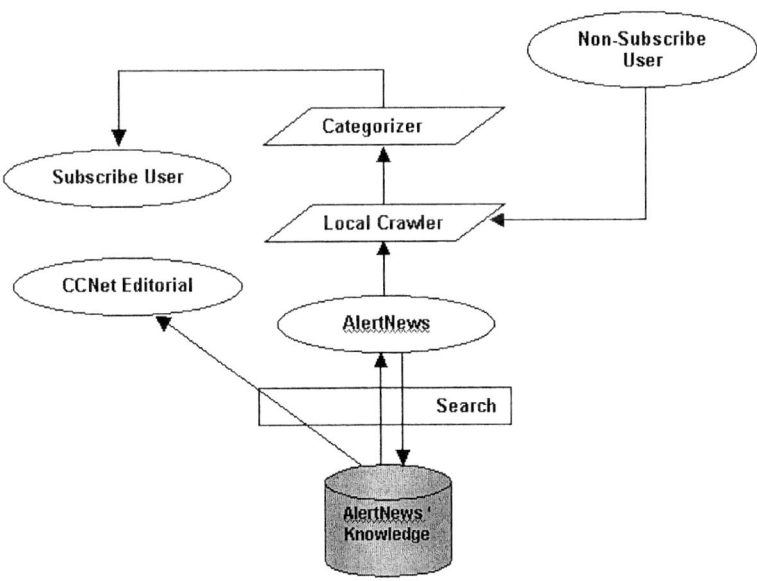

Fig. 3. AlertNews Architecture

Figure 4 shows the architecture of the CCNet Automated Knowledge extraction Agent (AKEA) focused on extracting information from the World Wide Web (WWW) for the AINI customer knowledge base. Four modules make up the agent. The modules are the Crawler, Wrapper, Named-entity Tagger and the AIML Converter. The crawler is the interface between the agent and the web. The functions of the crawler are like those used in conventional crawler-based search engines. The crawler resolves root domain names such as who.org, info.gov.hk, sars.gov.sg, etc. and follows subsequent links that are available on a page until a certain depth as defined by the user. These configurations are set in the crawler config database. For every page crawled, a copy is returned for further processing by the wrapper. The activities of the crawler are logged in the crawler log database.

The input to the syntactic preprocessor is the online news documents which may consist of several paragraphs. The functional model of the preprocessing phases required is shown in Figure 4 as part of the knowledge base construction system. Some of the stages in Figure 4 can be reorganised and even removed depending on the functions of the remaining phases. In other words, the choice of algorithm for a particular phase will determine the relevancy of other phases. This property is called *dependant optionality* where a particular phase is considered as redundant if the output it provides is already contained in the output of other phases. For example, the morphology analyzer can be put aside if the sentence parser also performs morphology analysis implicitly as part of its function.

Online news documents returned by the crawler are in the hypertext format and contain of a variety of unwanted characters. The Wrapper prepares the raw information by separating the actual news content and other meta-information from the hypertext characters or tags. This process is known as *cleaning* and the result is referred to as *cleaned news*. Once the information is processed, the key elements such as date of news, news title, news content and other relevant information are extracted and stored in the CCNet news repository.

The syntactic preprocessor performs the task of identifying the dependencies among the words. Based on the dependencies, grammatical relations (i.e. phrasal categories) like noun phrases, verb phrases and prepositional phrases are extracted using sentence parser for the English language like Link Grammar [4]. The named entities in noun phrases are assigned with tags such as disease, location and person using the weighted gazetteer approach proposed by Wong, Goh and et al. in [5]. A reference list in the Gazetteer is used by the preprocessor. These tags enable the agent to identify what type of entity the corresponding noun phrases are and in which level and node do these entities belong to in the ontology. Pronouns are also resolved whenever necessary. The named entities that have been tagged are inserted into the corresponding entry in the news repository. The syntactic preprocessor managed to identify two named entities namely meningitis and Burkina Faso. Using the gazetteer, the preprocessor will discover that meningitis is a type of disease and Burkina Faso is a country and tag them respectively using the ontology tag in the form of *named_entity[ontology tag]*.

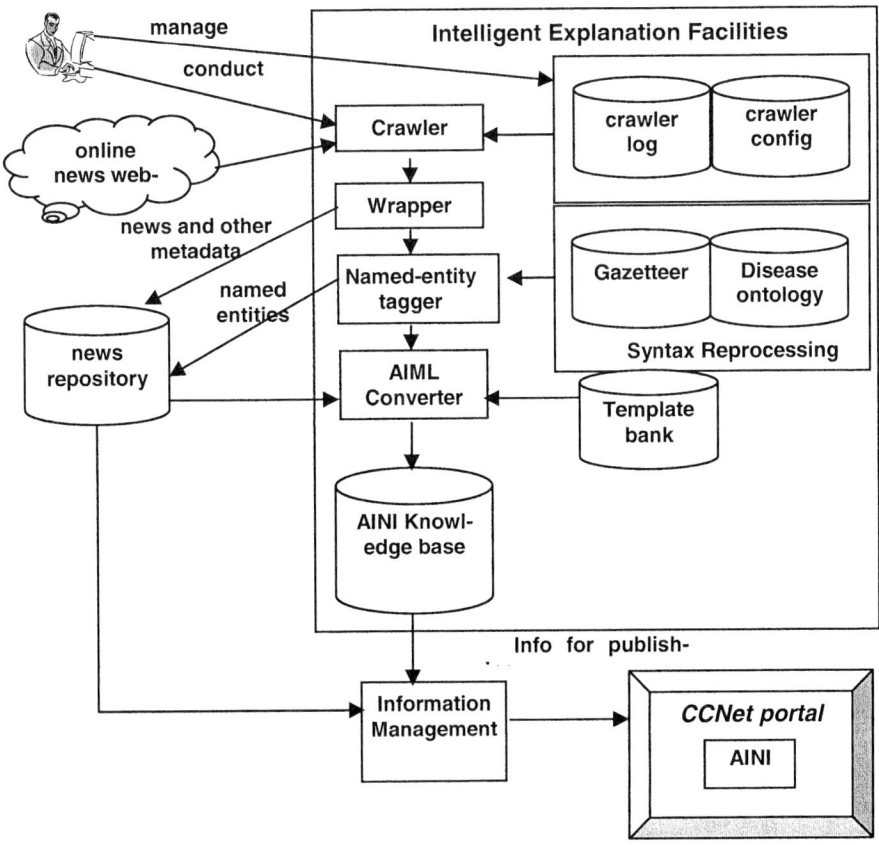

Fig. 4. CCNet Automated knowledge extraction agent architecture

In the gazetteer, each entry has additional information like *weight, ontology id* and the acceptable preceding/foregoing grammatical relations in addition to the triggering information, category and entity name. For example, a returned noun phrase *"Japanese Encephalitis disease"*, could trigger ambiguity. This could be resolved by just using the weighting mechanism without the need for any hand-crafted rules.

The information in the news repository is fed into two main components, namely the CCNet portal and the AIML converter. Information in the news repository is directly published to the CCNet portal without any further processing. The AIML converter uses the template bank to transform the news repository entries into AIML representation, which will be populated into AINI's knowledge base. The richer the template bank, the wider the scope of questions AINI will be able to handle. Substitutions will be made to the template using relevant values of each entry in the news repository. There are four fields in the template namely the *wh-token* corresponding to the ontology tag, first two lines of content, disease named entity and URL. The first and second require some processing prior to replacement.

The ontology tag associated with each named entity is resolved to obtain the corresponding wh-token. Currently, the agent is capable of handling four types of wh-token: *where*, resolved from *location* named-entities; *when*, resolved from *date* named-entities; *who*, resolved from *agent* named-entities and *what*. The *what* token is resolvable from all ontological entities with additional tokens. For example, given the named entity *Burkina Faso* and its tag *country*, we can obtain the *where* token and *what* token with the *country* tag. This is possible because the question *where does meningitis...?* is similar to asking *what country does meningitis...?*

The second processing required prior to replacement is to truncate the news content to the first two lines to be used in AINI's answers. The remaining news will be presented as part of a URL push. The AIML converter follows precedence in converting named-entities and their ontology tag into AIML representation. All questions handled by AINI are based on the concept of disease and thus, all news content will surely contain *disease* named-entities. During conversion, priority will be given to entities other than disease. Only when a news item that does not contain any other entities (i.e. there are no information about *location*, *person* or *date*) is encountered, then the converter will resolve the sole *disease* named-entity to the *what* token. Finally, these instances will be populated into AINI's knowledge base for learning and used by the AINI's chat interface in the CCNet portal for natural language question answering.

3.4 Multilevel Knowledge Base Natural Language Query

This section explains how AINI knowledge works. Firstly, AINI will search for an answer from Level 1 specific domain (Crisis Communication knowledge base) created by AKEA. If an answer is not found, it will move to Level 2 from the Frequently Asked Questions (FAQ) knowledge base which are stored in the FAQ table on MySQL database. Questions such as *what is SARS, how SARS spread, what is SARS vaccination, where SARS happen*, etc can be answered at this level. The next level is Level 3, which is metadata (News database). The search is done by identifying the keyword in the question and matching it with the content of the metadata. Since the WWW is so big that simple pattern matching techniques can often replace the need to understand both the structure and meaning of language. If an answer is still not found, AINI will proceed to Level 4 where AIML common knowledge will take place. If AINI still cannot answer the question, the last step will store the unanswered question in the database for the attention of the administrator. The answers will then be subsequently stored in the Level 2. This will enable AINI to answer the same question in the future.

4 Conclusion

The Internet has become a vital source of information and channel for effective communication during times of crisis and occurrence of global issues. This research will continue to develop and make use of the Internet to create global virtual communities by using intelligent agents and software robot. This intelligent software robot will assist the communication process by giving necessary and vital information needed by users during a crisis. It will also help in maintaining calmness and order so that the country and the communities will not be hijacked by fear and panic. It is proposed that users will have more trust in the information provided by the intelligent software robot because of its interactive features and an ongoing engagement with the users. Furthermore, the integrated Text-To-Speech Technology and 3D human-like character or avatar in the system is capable to deliver speech and interact with the user in a humanlike manner thereby generating a sense of care and comfort. The portal also provides news, advertisements, conversation logs and statistics in the system benefiting the researchers in their efforts to further enhance the system. In the anticipated forthcoming epidemics or waves of new diseases due to mutation of bacteria and viruses, the output from this research will be useful to tackle future health crises. Information on natural disasters such as tsunamis, typhoons, earthquakes and floods will also be effectively managed and disseminated by deploying CCNet System. It is believed that CCNet provides a well-engineered platform for experimentation with various Web-enabled question answering techniques by employing conversation software robot. The implementation is currently under ongoing development. Progress and preliminary results will be reported. In the future, we will endeavour to continually refine the existing technology and to develop new frameworks in order to enable an efficient Internet-based global crisis communication.

Acknowledgements

This research is funded by KUTKM Grant under contract number PJP/2003/FTMK(1) (S017) and administered by the Centre for University Industry. The project is continued with a grant supported by the Murdoch University Division of Arts Research Excellence Grant Scheme in 2005.

References

1. Goh, O.S. and Fung C.C. (2003), "Intelligent Agent Technology in E-Commerce" in Jiming Liu, Yiuming Cheung, Hujun Yin (Eds.), *Intelligent Data Engineering and Automated Learning,* LNCS, Vol. 2690, Springer-Verlag, pp. 10-17.
2. Katz B, Felshin S, Yuret D, Ibrahim A, Lin J, Marton G, McFarland AJ, Temelkuran B (2002), "Omnibase: Uniform access to heterogeneous data for question answering", Natural Language Processing and language Processing and Information Systems. Lecture in Computer Science, Vol. 2553: pp. 230-234.
3. Goh, O. S. & Teoh, K. K. (2002), "Intelligent virtual doctor system" in Proceedings of the *2nd IEE Seminar on Appropriate Medical Technology for Developing Countries,* London, United Kingdom.
4. Lafferty, J.; Sleator, D.; Temperley, D. (1993), "Grammatical trigrams: a probabilistic model of link grammar", *Probabilistic Approaches to Natural Language. Papers from the 1992 AAAI Fall Symposium*, 1993, p 89-97.
5. Wong, W., Goh, O. S., & Mokhtar Mohd Yusof. 2004. "Syntax preprocessing in cyberlaw web knowledge base construction". In M. Mohammadian (Ed.), Proceedings of the International Conference on Computational Intelligence for Modelling, Control and Automation (CIMCA 2004), Gold Coast, Australia. ISBN: 174-088-1893, pp: 174-184.

Probabilistic Principal Surface Classifier

Kuiyu Chang[1] and Joydeep Ghosh[2]

[1] School of Computer Engineering,
Nanyang Technological University, Singapore 639798, Singapore
kuiyu.chang@pmail.ntu.edu.sg,
http://www.ntu.edu.sg/home/askychang
[2] Department of Electrical and Computer Engineering,
University of Texas at Austin,
Austin Texas 78712, USA

Abstract. In this paper we propose using manifolds modeled by probabilistic principle surfaces (PPS) to characterize and classify high-D data. The PPS can be thought of as a nonlinear probabilistic generalization of principal components, as it is designed to pass through the "middle" of the data. In fact, the PPS can map a manifold of any simple topology (as long as it can be described by a set of ordered vector co-ordinates) to data in high-dimensional space. In classification problems, each class of data is represented by a PPS manifold of varying complexity. Experiments using various PPS topologies from a 1-D line to 3-D spherical shell were conducted on two toy classification datasets and three UCI Machine Learning datasets. Classification results comparing the PPS to Gaussian Mixture Models and K-nearest neighbours show the PPS classifier to be promising, especially for high-D data.

1 Introduction

Nonlinear manifolds embedded in high-dimensional space can provide a useful low-dimensional (2-D or 3-D) summary of the data that is visualizable by humans, assuming that the intrincsic data dimensionality is much lower. Principal curves and surfaces[1] can be used to compute a manifold that generalizes the property of principal components and subspaces, respectively. A discrete non-parametric approximation[2] of principal surfaces that is much simpler to compute comes in the form of Kohonen's self-organizing map (SOM)[3]. The 2-D SOM is frequently used for visualizing high-D clusters in manifold/latent space[4][5]. Moreover, the use of 3-D manifolds is not widespread.

A novel 3-D spherical (shell) manifold is proposed for modeling and visualizing high-D data. With all latent nodes on the spherical (shell) manifold equally far away from the center, it captures nicely the sparsity and peripheral property of high-D data[6]. Using the latent nodes on the spherical manifold as class reference vectors, a template-based classifier is also proposed. It is shown that the addition of the third dimension of the spherical manifold dramatically improves classification accuracy over 1-D and 2-D manifolds on two artificial classification datasets with significant class overlap. The spherical manifold classifier is also evaluated against the unconstrained Gaussian mixture model (GMM) vector quantizer, and the K-nearest neighbor (KNN) classifier on three real high-D datasets. Experimental results confirm the robustness of spherical manifolds for modeling high-D data.

2 Spherical Manifolds

2.1 Curse-of-Dimensionality

Data in very high-D space tend to lie entirely at the peripheral of a sample due to the curse-of-dimensionality[6]. To appreciate that this is indeed the case, consider data uniformly distributed within a hypercube in $\mathbb{R}^D : \mathbb{R} \in [-1, 1]$, where D denotes the dimensionality. For $D = 1$ (a line) the fraction p of data lying within the center interval $\mathbb{R}^D : \mathbb{R} \in [-0.5, 0.5]$ is 0.5, for $D = 2$ (a square) this number decreases to 0.25, and for $D = 3$ (a cube), p further decreases to 0.125. The general formula for arbitrary D is $p = 2^{-D}$, from which it can be inferred that even moderate values of D like 20 will result in only a single $(0.9537 \simeq 1)$ point out of, say, 10^6 samples to lie within the central region!

It is clear from the above example that fitting a single multivariate Gaussian distribution to high-D data is inappropriate and actually much worse than fitting a 1-D Gaussian to a 1-D uniformly distributed data, as the Gaussian density assumes the majority of data to be concentrated at the center, contrary to the peripheral property of high-D data. A Gaussian mixture model (GMM)[7] will be able to better model the high-D data by fitting a Gaussian distribution to each dense (i.e. peripheral) regions within the space. However, the uncontrained nature of the GMM makes it very sensitive to initializations; a good fit is obtained if the Gaussian centers are initialized properly and vice-versa. Consequently, a constrained mixture model incorporating some prior knowledge of the characteristics of high-D data is clearly more desirable.

The probabilistic principal surface (PPS)[8] is one such model that explicitly contrains the Gaussians to lie in a pre-defined latent topology. A PPS with 3-D spherical latent topology is introduced for approximating high-D data. The spherical manifold is comprised of nodes evenly distributed on the surface of a sphere. It possesses two attractive characteristics: (1) nodes are distributed on the peripheral and equally far away from the center, just like high-D data, and (2) it is finite but unbounded, which is intuitively suitable for estimating the boundary of high-D data. The goal is to show that the 3-D spherical manifold is capable of modeling high-D data much more accurately then 1-D and 2-D manifolds. The next section briefly describes the probabilistic principal surface model used to construct a spherical manifold.

2.2 Probabilistic Principal Surfaces

Principal surfaces (curves)[1] are nonlinear generalizations of principal subspaces (components) that formalizes the notion of a low-D manifold passing through the 'middle' of a dataset in high-D space. The probabilistic principal surface (PPS)[8], a generalization of the generative topological mapping (GTM)[9][10], is a parametric approximation of principal surfaces. The PPS manifold is comprised of M nodes $\{\mathbf{x}_m\}_{m=1}^M$ arranged typically on a uniform topological grid in latent (low-D) space $\mathbb{R}^Q$. The topology is consistently enforced via a generalized linear mapping from each latent node $\mathbf{x}_m$ in $\mathbb{R}^Q$ to it's corresponding data node $\mathbf{f}(\mathbf{x}_m)$ in data (high-D) space $\mathbb{R}^D$ (D is the data dimensionality),

$$\mathbf{f}(\mathbf{x}_m) = \mathbf{W}\phi(\mathbf{x}_m)$$

where $\mathbf{W}$ is a $D \times L$ real matrix and

$$\phi(\mathbf{x}_m) = \begin{bmatrix} \phi_1(\mathbf{x}_m) & \cdots & \phi_L(\mathbf{x}_m) \end{bmatrix}^T,$$

is the vector containing L latent basis functions $\phi_l(\mathbf{x}) : \mathbb{R}^Q \to \mathbb{R}$, $l = 1, \ldots, L$. The basis functions $\phi_l(\mathbf{x})$ are usually isotropic Gaussians with constant widths. Each data node $\mathbf{f}(\mathbf{x}_m)$ actually corresponds to the mean of a Gaussian probability distribution with noise covariance parameter,

$$\mathbf{\Sigma}_m = \frac{\alpha}{\beta} \sum_{q=1}^{Q} \mathbf{e}_q(\mathbf{x}_m) \mathbf{e}_q^T(\mathbf{x}_m)$$

$$+ \frac{(D - \alpha Q)}{\beta (D - Q)} \sum_{d=Q+1}^{D} \mathbf{e}_d(\mathbf{x}_m) \mathbf{e}_d^T(\mathbf{x}_m)$$

$$0 < \alpha < D/Q,$$

where β^{-1} is the global spherical covariance, α is the amount of clamping in the tangential direction, $\{\mathbf{e}_q(\mathbf{x}_m)\}_{q=1}^Q$ and $\{\mathbf{e}_d(\mathbf{x}_m)\}_{d=Q+1}^D$ are the set of vectors tangential and orthogonal to the manifold at $\mathbf{f}(\mathbf{x}_m)$ in data space, respectively. Figures 1 and 2 show respectively, a 1-D PPS and its noise covariance model for different values of α. Notice that the GTM is obtained for $\alpha=1$. It has been shown that the PPS with an orthogonal noise model ($\alpha<1$) yields better manifolds in terms of reconstruction error[8]. The PPS is iteratively computed using a maximum likelihood optimization procedure.

The spherical manifold can be trivially constructed using a PPS with latent nodes $\{\mathbf{x}_m\}_{m=1}^M$ arranged regularly on the surface of a sphere in $\mathbb{R}^3$. At the onset, the manifold is intialized to a hyper-sphere in dataspace via the linear transformation $\mathbf{f}(\mathbf{x}_m) = \mathbf{V}\mathbf{x}_m$ where $\mathbf{V} = \begin{bmatrix} \lambda_1 \mathbf{v}_1 & \lambda_2 \mathbf{v}_2 & \lambda_3 \mathbf{v}_3 \end{bmatrix}$ is comprised of the three largest eigenvectors $\{\mathbf{v}_q\}_{q=1}^3$ (scaled by their corresponding eigenvalues λ_q) of the data covariance matrix.

3 Spherical PPS Classifier

A template-based classifier using spherical manifolds is proposed. This is a significant improvement over the the previously proposed template-based classifier which uses principal curves (1-D manifolds)[11]. A template-based classifier models data of each class independently of all other classes; a given test sample is classified according to its similarity (distance) to the class templates (reference vectors). The K-nearest neighbor (KNN) classifier can be regarded as a template-based classifier that uses all data samples as class reference vectors, and is frequently used to obtain a rough estimate of the Bayes error. Likewise, a Gaussian mixture model (GMM) can be used to compute reference vectors for each class, known as the GMM classifier. These two related classifiers are therefore used in this paper for benchmarking the proposed classifier. The main goal is to show that the spherical manifold, with its incorporated prior knowledge of high-D data, can model the data better than the unconstrained GMM, thereby contributing to better classification performance.

4 Experiments

The spherical manifold classifier (PPS) is compared to the GMM and KNN classifiers on the 5 datasets shown in table 1. For all experiments, a 50/50 training/test random stratified partition scheme was used, and results were obtained on the test set for 10 repeated trials. An isotropic covariance model ($\alpha = 1$) was used for both the PPS and GMM classifier, unless otherwise stated. Validation was found to be unnecessary for all three types of classifiers. All real datasets were normalized to zero mean and unit variance and the same number of reference vectors (manifold nodes) were used for all models where applicable.

Table 1. N:number of samples, C:number of classes, D:number of dimensions

Dataset	D	C	N
1 gaussian	8	2	5000
2 uniform	8	2	5000
3 letter	16	26	20000
4 ocr	30	26	16280
5 satimage	36	6	6435

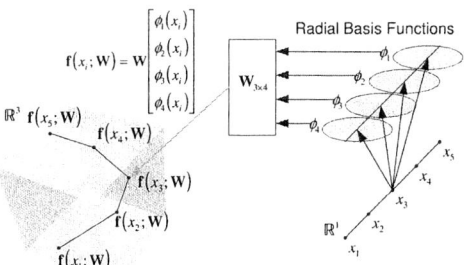

Fig. 1. A 1-D PPS in $\mathbb{R}^3$ with 5 nodes and 4 latent bases

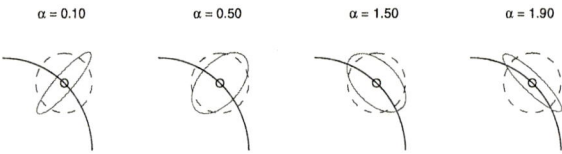

Fig. 2. Unoriented covariances $\alpha = 1$ (dashed line) and oriented covariances (solid line) for $\alpha = 0.10, 0.50, 1.50, 1.90$

4.1 Artificial Dataset

In this section, the effect of dimensionality (varied from 3 to 8) on the spherical manifold classifier is examined using two artificial datasets with highly overlapping classes. The gaussian dataset[12] is comprised of two equal-sized classes drawn from overlapping Gaussian distributions with zero mean and isotropic variances of 1 and 4, respectively. For comparison, the 1-D and 2-D PPS classifiers were also simulated. Results averaged over 10 trials are given in figure 3. It can be seen that the 1-D and 2-D PPS classifiers were the two worst performers due to their inability to model the complete overlap of the 2 classes in high-D space. While the spherical manifold (PPS-3D) classifier performed much better than the other two PPS classifiers, it was still worse than the GMM and KNN classifiers. This is expected because at lower dimensions, the Gaussian data is mostly concentrated at the core, thereby defying the peripheral assumption of the spherical manifold classifier. However, as data gets increasingly sparse at higher dimensions, both the KNN (for D>6) and GMM (for D>7) classifiers start to deteriorate, whereas the spherical manifold classifier is seen to consis-

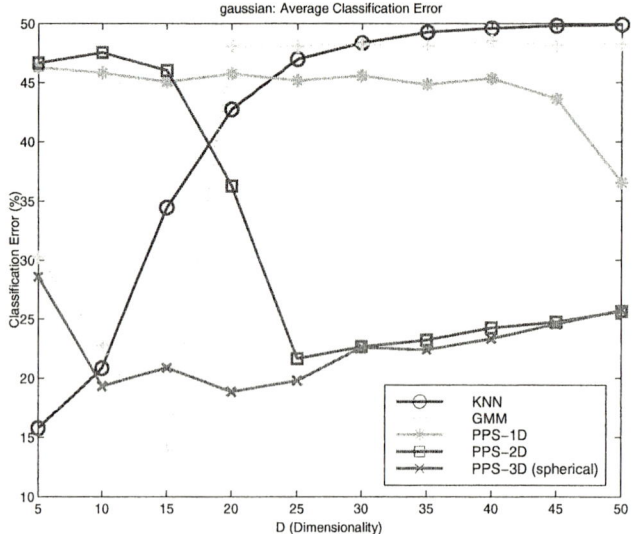

Fig. 3. Gaussian: average classification error versus dimensionality

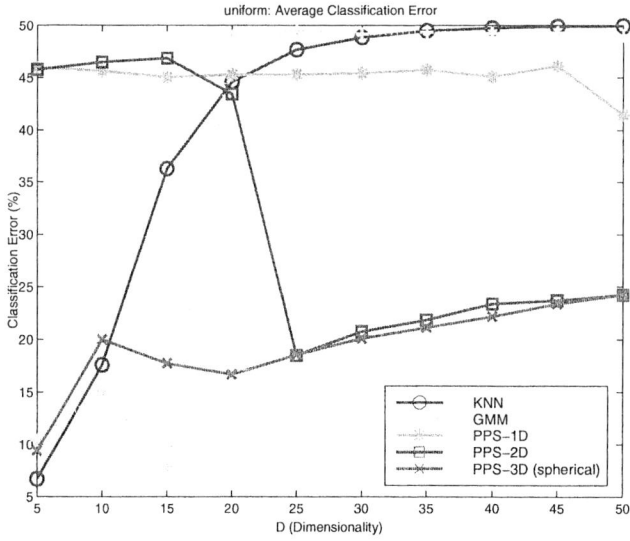

Fig. 4. uniform: average classification error versus dimensionality

tently improve with increasing dimensionality! This demonstrates the robustness of the spherical manifold classifier with respect to the curse-of-dimensionality, even where Gaussian data is concerned.

To see if the spherical manifold classifier actually performs better then GMM or KNN on high-D uniformly distributed data, a uniform dataset with features similar to the gaussian dataset was created. The first class is comprised of 2500 samples uniformly drawn from $\mathbb{R}^D : \mathbb{R} \in [-1, 1]$ and the second class contains 2500 samples drawn from $\mathbb{R}^D : \mathbb{R} \in [-2, 2]$, where $D = 8$. The results in figure 4 confirm the superiority of the spherical manifold classifier over other classifiers for high-D ($D > 6$) uniformly distributed data.

4.2 Real Dataset

In this section, the performance of the spherical manifold classifier is evaluated on three real high-D datasets–the letter (letter-recognition) dataset from the UCI machine learning database[13], the ocr (handwritten character) dataset provided by the National Institute of Science and Technology, and the remote sensing

Table 2. Letter: average classification error

Classifier	Error (%)	Std. Dev.
PPS-3D ($\alpha = 0.1$)	8.08	0.17
PPS-3D ($\alpha = 0.5$)	7.84	0.26
PPS-3D ($\alpha = 1$)	**7.82**	0.24
KNN ($k = 1$)	8.21	0.16
GMM	13.76	0.29

Table 3. ocr: average classification error

Classifier	Error (%)	Std. Dev.
PPS-3D ($\alpha = 0.1$)	10.68	0.36
PPS-3D ($\alpha = 0.5$)	**10.56**	0.34
PPS-3D ($\alpha = 1$)	10.60	0.29
KNN ($k = 5$)	11.23	0.43
GMM	16.84	1.95

Table 4. satimage: average classification error

Classifier	Error (%)	Std. Dev.
PPS-3D ($\alpha = 0.1$)	11.36	0.35
PPS-3D ($\alpha = 0.5$)	**11.03**	0.57
PPS-3D ($\alpha = 1$)	11.16	0.50
KNN ($k = 1$)	10.76	0.28
GMM	14.89	0.73

satimage dataset from the Elena database[12]. Averaged results on the three datasets are shown in tables 2 to 4. From the tables, it can be concluded that the constrained nature of the spherical manifold results in a much better set of class reference vectors compared to the GMM. Further, its classification performance was comparable to, if not occasionally better than the best KNN classifier.

5 Conclusion

From the observation that high-D data lies almost entirely at the peripheral, a 3-D spherical manifold based on probabilistic principal surfaces is proposed for modeling very high-D data. A template-based classifier using spherical manifolds as class templates is subsequently described. Experiments demonstrated the robustness of the spherical manifold classifier to the curse-of-dimensionality. In fact, the spherical manifold classifier performed better with increasing dimensionality, contrary to the KNN and GMM classifiers which deteriorates with increasing dimensionality. The spherical manifold classifier also performed significantly better than the unconstrained GMM classifier on three real datasets, confirming the usefulness of incorporating prior knowledge (of high-D data) into the manifold. In addition to giving comparable classification performance to the KNN on the real datasets, it is important to note that the spherical manifold classifier possess 2 important properties absent from the other two classifiers:

1. It defines a parametric mapping from high-D to 3-D space, which is useful for function estimation within a class, e.g object pose angles (on a viewing sphere) can be mapped to the spherical manifold[14].
2. High-D data can be visualized as projections onto the 3-D sphere, allowing discovery of possible sub-clusters within each class[15]. In fact, the PPS has been used to visualize classes of yeast gene expressions[16].

It is possible within the probabilistic formulation of the spherical manifold to use a Bayesian framework for classification (i.e. classifying a test sample to the class that gives the maximum *a posteriori* probability), thereby coming up with a rejection threshhold. However, this entails evaluating $\mathcal{O}(M)$ multivariate Gaussians, and can be computationally intensive. The PPS classifier has recently been extended to work in a committee, which was shown to improve classification rate on astronomy datasets[17]. Further studies are being done on using the spherical manifold to model data from all classes for visualization of class structure on the sphere, and also for visualizing text document vectors.

Acknowledgments

This research was supported in part by Army Research contracts DAAG55-98-1-0230 and DAAD19-99-1-0012, NSF grant ECS-9900353, and Nanyang Technological University startup grant SUG14/04.

References

1. Hastie, T., Stuetzle, W.: Principal curves. Journal of the American Statistical Association **84** (1988) 502–516
2. Mulier, F., Cherkassky, V.: Self-organization as an iterative kernel smoothing process. Neural Computation **7** (1995) 1165–1177
3. Kohonen, T.: Self-Organizing Maps. Springer, Berlin Heidelberg (1995)
4. Jain, A.K., Mao, J.: Artificial neural network for nonlinear projection of multivariate data. In: IEEE IJCNN. Volume 3., Baltimore, MD (1992) 335–340
5. Mao, J., Jain, A.K.: Artificial neural networks for feature extraction and multivariate data projection. IEEE Transactions on Neural Networks **6** (1995) 296–317
6. Friedman, J.H.: An overview of predictive learning and function approximation. In Cherkassky, V., Friedman, J., Wechsler, H., eds.: From Statistics to Neural Networks, Proc. NATO/ASI Workshop, Springer Verlag (1994) 1–61
7. Bishop, C.M.: Neural Networks for Pattern Recognition. 1st edn. Clarendon Press, Oxford. (1995)
8. Chang, K.y., Ghosh, J.: A unified model for probabilistic principal surfaces. IEEE Transactions on Pattern Analysis and Machine Intelligence **23** (2001) 22–41
9. Bishop, C.M., Svensén, M., Williams, C.K.I.: GTM: The generative topographic mapping. Neural Computation **10** (1998) 215–235
10. Bishop, C.M., Svensén, M., Williams, C.K I.: Developments of the generative topographic mapping. Neurocomputing **21** (1998) 203–224
11. Chang, K.y., Ghosh, J.: Principal curve classifier – a nonlinear approach to pattern classification. In: International Joint Conference on Neural Networks, Anchorage, Alaska, USA, IEEE (1998) 695–700
12. Aviles-Cruz, C., Guérin-Dugué, A., Voz, J.L., Cappel, D.V.: Enhanced learning for evolutive neural architecture. Technical Report Deliverable R3-B1-P, INPG, UCL, TSA (1995)
13. Blake, C., Merz, C.: UCI repository of machine learning databases (1998)

14. Chang, K.y., Ghosh, J.: Three-dimensional model-based object recognition and pose estimation using probabilistic principal surfaces. In: SPIE:Applications of Artificial Neural Networks in Image Processing V. Volume 3962., San Jose, California, USA, SPIE, SPIE (2000) 192–203
15. Staiano, A., Tagliaferri, R., Vinco, L.D.: High-d data visualization methods via probabilistic principal surfaces for data mining applications. In: International Workshop on Multimedia Databases and Image Communication, Salerno, Italy (2004)
16. Staiano, A., Vinco, L.D., Ciaramella, A., Raiconi, G., Tagliaferri, R., Longo, G., Miele, G., Amato, R., Mondo, C.D., Donalek, C., Mangano, G., Bernardo, D.D.: Probabilistic principal surfaces for yeast gene microarray data mining. In: International Conference on Data Mining. (2004) 202–208
17. Staiano, A., Tagliaferri, R., Longo, G., Benvenuti, P.: Committee of spherical probabilistic surfaces. In: International Joint Conference on Neural Networks, Budapest, Hungary (2004)

Probabilistic Based Recursive Model for Face Recognition

Siu-Yeung Cho and Jia-Jun Wong

Forensics and Security Lab, Division of Computing Systems,
School of Computer Engineering,
Nanyang Technological University,
Nanyang Avenue, Singapore 639798
assycho@ntu.edu.sg, wong0095@ntu.edu.sg

Abstract. We present a facial recognition system based on a probabilistic approach to adaptive processing of Human Face Tree Structures. Human Face Tree Structures are made up of holistic and localized Gabor Features. We propose extending the recursive neural network model by Frasconi et. al. [1] in which its learning algorithm was carried out by the conventional supervised back propagation learning through the tree structures, by making use of probabilistic estimates to acquire discrimination and obtain smooth discriminant boundaries at the structural pattern recognition. Our proposed learning framework of this probabilistic structured model is hybrid learning in locally unsupervised for parameters in mixture models and in globally supervised for weights in feed-forward models. The capabilities of the model in a facial recognition system are evaluated. The experimental results demonstrate that the proposed model significantly improved the recognition rate in terms of generalization.

1 Introduction

In most facial recognition systems, recognition of personal identity is based on geometric or statistical features derived from face images. There have been several dominant face recognition techniques such as Eigenface [2], Fisherface[3], Elastic Graph Matching [4] and Local Feature Analysis [5]. Some of the techniques employ the use of only global features while others uses only local features. Psychophysics and Neuroscientist have found that human face perception is based on both holistic and feature analysis. Fang et. al. [6] has proposed using both global and local features for their face recognition system, and the verification accuracy was much higher than those using only global or local features. Feature vectors were traditionally represented by flat vector format [7], without any feature relationship information. Wiskott et. al. [8] have used bunch graphs to represent the localized gabor jets, and using the average over similarity between pairs of corresponding jets as the similarity function.

In this paper, we proposed a method for face recognition by transforming the feature vector data into tree structure representation, which would encode the feature relationship information among the face features. Thirty-eight Localized Gabor

Features (LGF) [9] from a face, and one global Gabor features are obtained as a feature vector and transforming them into a Human Face Tree Structure (HFTS) representation.

Many researchers have explored the utilization of supervised [10] or unsupervised [11] neural network representation for classification of tree structures[1, 12]. For processing the tree structures in neural network manner, they assumed that there are underlying structures in the extracted features. A probabilistic based recursive neural network is proposed for classification of the Human Face Tree Structure (HFTS) in this paper. Probabilistic neural networks can embed discriminative information in the classification model which can be used for providing clustering analysis from the input attributes. This technique is benchmarked against Support Vector Machines (SVM) [13], K nearest neighbors (KNN) [14], Naive Bayes algorithm [15] where the flat vector files were used in the verification experiments. We have made use of the ORL Database [16] to illustrate the accuracy for the recognition system in terms of rejection rate and false acceptance rate.

2 Human Face Gabor Tree Structure Representation

2.1 Gabor Feature Extraction

The representation of the global and local features is based on Gabor wavelets transform, which are commonly used for image analysis because of their biological relevance and computational properties[17]. Gabor wavelets, which capture the properties of spatial localization, and quadrature phase relationship by its spatial frequency selectivity and orientation selectivity respectively. Gabor wavelets are known to be a good approximation to filter response profiles encountered experimentally in cortical neurons [7]. The two-dimensional Gabor wavelets $g(x,y)$ can be defined as follows [18]:

$$g(x,y) = \left(\frac{1}{2\pi\sigma_x\sigma_y}\right)\exp\left[-\frac{1}{2}\left(\frac{x^2}{\sigma_x^2}+\frac{y^2}{\sigma_y^2}\right)+2\pi j W x\right] \quad (1)$$

The mean and standard deviation of the convolution output is used as the representation for classification purpose:

$$\mu_{mn} = \iint |W_{mn}(xy)| dxdy, \text{ and } \sigma_{mn} = \sqrt{\iint (|W_{mn}(x,y)|-\mu_{mn})^2 dxdy} \quad (2)$$

2.2 Gabor Features to Human Face Tree Structure (HFTS) Transform

Four primary feature locations are located by the feature finder as suggested by FERET Evaluation Methodology [12], which will provide the coordinate location for the center of the left eye, center of the right eye, tip of the nose and the center of the lips as shown in Fig. 1a. Extending for these feature locations, extended feature identifications can be generated as follows. Each of the extended features is relative or an extension of the known features as shown in Fig. 1b.

Fig. 1. Four primary Feature Locations and 38 Extended Local Features

After extracting the localized features, human faces can be represented by a tree structure model based on the whole face acting as a root node and localized features like eyes, nose and mouth acting as its branches as shown in Fig. 2. Sub detail features from the 5 key fiducial points form the leaves of the tree structure. The branch nodes are labeled by the features number, ie. F00, F01,..., F38. The arc between the two nodes corresponds to the object relationship, and features that been extracted are attached to the corresponding nodes.

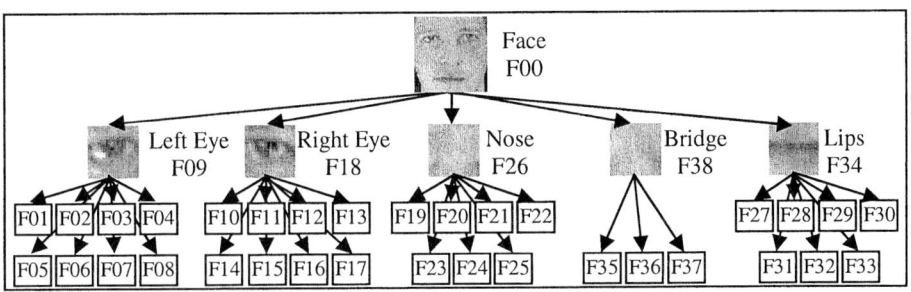

Fig. 2. Tree Structure Representation of the Human Face

3 Adaptive Processing of Human Face Tree Structures (HFTS)

3.1 Basic Idea of Adaptive Tree Processing

In this paper, the problem of devising neural network architectures and learning algorithms for the adaptive processing of human face tree structures is addressed in the context of classification of structured patterns. The encoding method by recursive neural networks is based on and modified by the research works of [19]. We consider that a structured domain and all Tree Structures are a learning set representing the task of the adaptive processing of data structures. This representation is illustrated in Fig. 3a.

Probabilistic Neural Networks (PNNs) is one of the techniques that can embed discriminative information in the classification model and are successfully used for providing clustering analysis from the input attributes, and this can be used for adaptive processing of the tree structure in Fig. 3b. Streit and Luginbuhl [20] had demonstrated that by means of the parameters of a Gaussian mixture distribution, a probabilistic neural network model can estimate the probabilistic density functions. They showed that the general homoscedastic Gaussian mixtures to approximate the optimum classifier could be implemented using a four layer feed-forward PNN using

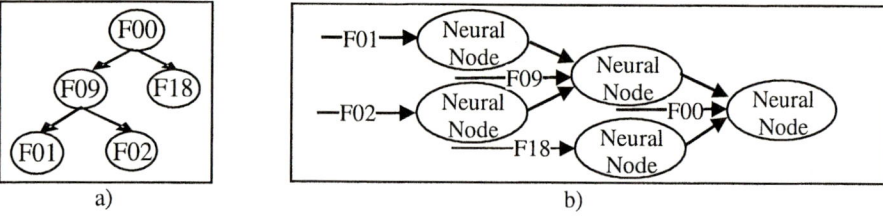

Fig. 3. Simplified/Partial Tree Structure of the Human Face and the Encoded Tree Structure Format

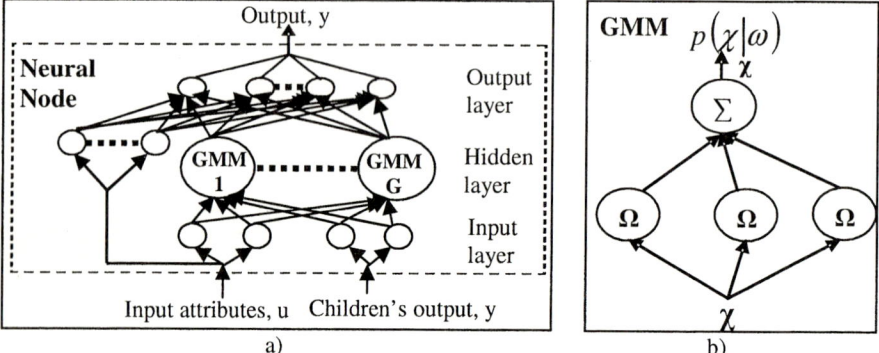

Fig. 4. Architecture of Probabilistic based recursive neural network using GMM for neural Node Representation and Structure of a Gaussian Mixture Model

general Gaussian kernel, or Parzen window. Roberts and Tarassenko [21] proposed a robust method for Gaussian Mixture Models (GMMs), using a GMM together with a decision threshold to reject unknown data during classification task.

In our study, each of the neural nodes in Fig. 3b is represented by a neural network tree classifier as illustrated in Fig. 4a. It details the architecture of the proposed tree classifier in which each neuron in the hidden layer is represented by a Gaussian Mixture model and at the output layer, each of the neurons is represented by a sigmoid activation function model. Each parameter has a specific interpretation and function in this GMM. All weights and node threshold are given explicitly by mathematical expressions involving the defining parameters of the mixture Gaussian pdf estimates and the a priori class probabilities and misclassification costs.

Suppose that a maximum branch factor of c has been predefined, each of the form q_i^{-1}, $i = 1,2,\ldots,c$, denotes the input from the ith child into the current node. This operator is similar to the shift operator used in the time series representation. Thus, the recursive network for the structural processing is formed as:

$$x = F_n(Aq^{-1}y + Bu), \tag{3}$$

$$y = F_p(Cx + Du), \tag{4}$$

where x, u, and y are the n-dimensional output vector of the n hidden layer neurons, the m-dimensional inputs to the neurons, and the p-dimensional outputs of the neurons, respectively. q^{-1} is a notation indicating the input to the node is taken from its child so that: $q^{-1}y = (q_1^{-1}y \quad q_2^{-1}y \quad \cdots \quad q_c^{-1}y)^T$. The parametric matrix A is defined as: $A = (A^1 \quad A^2 \quad \cdots \quad A^c)$, where c denotes the maximum number of children in the tree, A is a $n \times (c \times p)$ matrix such that each A^k, $k = 1,2,\ldots,c$ is a $n \times p$ matrix, which is formed by the vectors a_j^i, $j = 1,2,\ldots n$. The parameters **B**, **C**, and **D** are $(n \times m)$, $(p \times n)$ and $(p \times m)$-dimensional matrices respectively. $F_n(\cdot)$ and $F_p(\cdot)$ are n and p dimensional vectors respectively, where their elements are defined by a nonlinear function $f(\alpha) = 1/(1 + e^{-\alpha})$.

Let m be the dimension of the input attributes in the Neural Node for each node, and k be the dimension of the outputs of each node. Hence the input pattern at each GMM can be expressed as:

$$\chi = (u \quad q^{-1}y^T) = \{x_i; i = 1,2,\ldots(m + k \times c)\}, \tag{5}$$

where u and y are the m-dimensional input vector and the k-dimensional output vector respectively. The class likelihood function of structure pattern χ associated with class ω would be expressed as:

$$p(\chi|\omega) = \sum_{g=1}^{G} P(\theta_g|\omega) p(\chi|\omega, \theta_g), \tag{6}$$

where $p(\chi|\omega)$ is the class likelihood function for class ω is a mixture of G components in a Gaussian distribution. θ_g denotes the parameters of the gth mixture component and G is the total number of mixture components. $P(\theta_g|\omega)$ is the prior probability of cluster g, and is termed as the mixture coefficients of the gth component:

$$\sum_{g=1}^{G} P(\theta_g|\omega) = 1 \tag{7}$$

$p(\chi|\omega, \theta_g) \equiv \aleph(\mu_g, \Sigma_g)$ is the probability density function of the gth component, which typically is a form of Gaussian distribution with mean μ_g and covariance Σ_g, given by:

$$p(\chi|\omega, \theta_g) = \frac{1}{(2\pi)^{(m+k\times c)/2} |\Sigma_g|^{1/2}} \cdot \exp\left\{-\frac{1}{2}(\chi - \mu_g)\Sigma_g^{-1}(\chi - \mu_g)^T\right\} \tag{8}$$

Equation (4) is modified for the recursive network, and expressed as the following equation:

$$y = F_k(Wp + Vu), \tag{9}$$

where $F_k(\cdot)$ is a k-dimensional vector, and their elements are the nonlinear sigmoid activation function. $p = (p_1(\chi|\omega) \quad \cdots \quad p_r(\chi|\omega))^T$, W and V are the weighting parameters in $(k \times r)$ and $(k \times m)$ – dimensional matrices respectively.

3.2 Learning Algorithm of the Probabilistic Recursive Model

The learning scheme of the proposed probabilistic based structural model can be divided into two phases, the locally unsupervised algorithm for the GMMs and the globally structured supervised learning for recursive neural networks. Streit and Luginbuhl has shown that in the unsupervised learning phase, the Expectation-Maximization (EM) method [20] would be optimal for this type of locally unsupervised learning scheme, which requires the parameters θ to be initialized and estimated during this learning phase. There are two steps in the EM method: The first step is called the expectation (E) step and the second is called the Maximization (M) step. The E step computes the expectation of a likelihood function to obtain an auxiliary function and the M step maximizes the auxiliary function refined by the E step with respect to the parameters to be estimated. The EM algorithm can be described as follows: Using the GMM in equation (6), the goal of the EM learning is to maximize the log likelihood of input attribute set in structured pattern, $\chi^* = (\chi_1 \cdots \chi_{N_T})'$,

$$\ell(\chi^*, \theta) = \sum_{j=1}^{N_T} \sum_{g=1}^{G} \log P(\theta_g | \omega) + \log p(\chi_j | \omega, \theta_g). \tag{10}$$

where observable attributes χ^* is "incomplete" data, hence an indicator α_j^k is defined to specify which cluster the data belonged to and include it into the likelihood function as:

$$\ell(\chi^*, \theta) = \sum_{j=1}^{N_T} \sum_{g=1}^{G} \alpha_j^k \left[\log P(\theta_g | \omega) + \log p(\chi_j | \omega, \theta_g) \right], \tag{11}$$

where α_j^k is equal to one if structure pattern χ_j belongs to cluster k, else the output would be equal to zero. In E step, the expectation of the observable data likelihood in the n-th iteration would be taken as:

$$Q(\theta, \hat{\theta}(n)) = E\{\ell(\chi^*, \theta) | \chi^*, \hat{\theta}(n)\} \tag{12}$$

$$= \sum_{j=1}^{N_T} \sum_{g=1}^{G} E\{\alpha_j^k, \hat{\theta}(n)\} \left[\log P(\theta_g | \omega, \hat{\theta}(n)) + \log p(\chi_j | \omega, \theta_g, \hat{\theta}(n)) \right], \tag{13}$$

where $p(\chi_j | \omega, \theta_g, \hat{\theta}(n)) \equiv \aleph(\hat{\mu}_g(n), \hat{\Sigma}_g(n))$ and $E\{\alpha_j^k, \hat{\theta}(n)\} = P(\theta_g | \chi_j, \hat{\theta}(n))$ as the conditional posterior probabilities which can be obtained by Bayes' rule:

$$P(\theta_g | \chi_j, \hat{\theta}(n)) = \frac{P(\theta_g | \omega) p(\chi_j | \omega, \theta_g)}{\sum_{r=1}^{R} P(\theta_r | \omega) p(\chi_j | \omega, \theta_r)}, \text{ at the } n\text{-th iteration.} \tag{14}$$

In M step, the parameters of a GMM are estimated iteratively by maximizing $Q(\theta, \hat{\theta}(n))$ with respect to θ,

$$\mu_g(n+1) = \frac{\sum_{j=1}^{N_T} P(\theta_g | \chi_j, \hat{\theta}(n)) \chi_j}{\sum_{j=1}^{N_T} P(\theta_g | \chi_j, \hat{\theta}(n))}, \tag{15}$$

$$\Sigma_g(n+1) = \frac{\sum_{j=1}^{N_T} P(\theta_g | \chi_j, \hat{\theta}(n))(\chi_j - \mu_g(n+1))(\chi_j - \mu_g(n+1))'}{\sum_{j=1}^{N_T} P(\theta_g | \chi_j, \hat{\theta}(n))}, \quad (16)$$

$$P(\theta_g | \omega) = \frac{\sum_{j=1}^{N_T} P(\theta_g | \chi_j, \hat{\theta}(n))}{N_T}, \text{ at (n+1)-th iteration.} \quad (17)$$

At the next phase of supervised learning, the goal is to optimize the parameters for the entire model in the structural manner. The optimization is basically to minimize the cost function formulated by errors between the target values and the output values of the root node in the DAGs. The Levenberg-Marquardt (LM) algorithm [22] has been proven to be one of the most powerful algorithms for learning neural networks which combines the local convergence properties of Gauss-Newton method near a minimum with consistent error decrease provided by gradient descent far away from a solution. In this learning phase, LM algorithm is used to learn the parameters at the output layer of the probabilistic based structured network. The learning task could be defined as follows:

$$\min J = \min \frac{1}{2} \| d - A \cdot \phi^R \|^2, \quad (18)$$

where $d = (d_1, d_2, \ldots, d_{N_T})$ and $\phi^R = \begin{pmatrix} P_1^R, \ldots, P_{N_T}^R \\ u_1, \ldots, u_{N_T} \end{pmatrix}$ represent the matrices of inversed function of the target values and input patterns at the output layer. $d_j = F_k^{-1}(t_j)$ which is the inverse function of the target values. The matrix $A = [W \ V]$ defines the parameters at the output layer. Thus, at the (n+1)th iteration of the LM algorithm, the element a_{ij} in the matrix of A parameters is updated according to:

$$a_k^T(n+1) = a_k^T(n) + (H_j \cdot H_j^T + \alpha I)^{-1} H_j \cdot \left(\sum_{j=1}^{N_T} (d_j - A(n) \cdot \phi_j^R) \right), \quad (19)$$

where $\phi_j^R = [P_j^R \ u_j^R]^T$ is the j-th input pattern set at output layer of the root node, $d_j = F_k^{-1}(t_j)$ which is the inverse function of the target values, α is the scalar, I is a identity matrix of $(r+m) \times (r+m)$ size, and H is the Jacobian matrix which is defined as: $H_j = \phi_j^R \phi_j^{R^T} \cdot \frac{\partial q^{-1} y_k}{\partial a_k(n)}$, where $\partial q^{-1} y_k / \partial a_k(n)$ is a $(r+m) \times k$ matrix with the output gradients of the child nodes with respect to the weights.

4 Experiment and Results

The evaluation is based on the ORL Face Database [16], which comprises of 10 different images per person. A total of 40 persons are found in the database. The original images were of the size of 92 x 112 pixels as shown in Fig. 5. In order to extract the facial region properly, the images were cropped out from the original images and resized to 100x100 pixels as shown in Fig 6. The locations of the eyes, nose and center of lips are then easily detected.

Fig. 5. Original Images of 92x112 pixels of various persons in the ORL database

Fig. 6. Cropped and resized images of one of the persons in the ORL database

Table 1. Performance of HFTS model against other methods

Method	Accuracy	Verification Rate	False Accepted Rate
PCA	89.30%	80.63%	2.03%
Gabor PCA	74.89%	52.50%	2.72%
Naïve Bayes	97.92%	28.13%	1.05%
KNN	96.51%	53.13%	2.85%
SVM	97.22%	37.50%	1.91%
HFTS + PR	99.75%	100.00%	0.02%

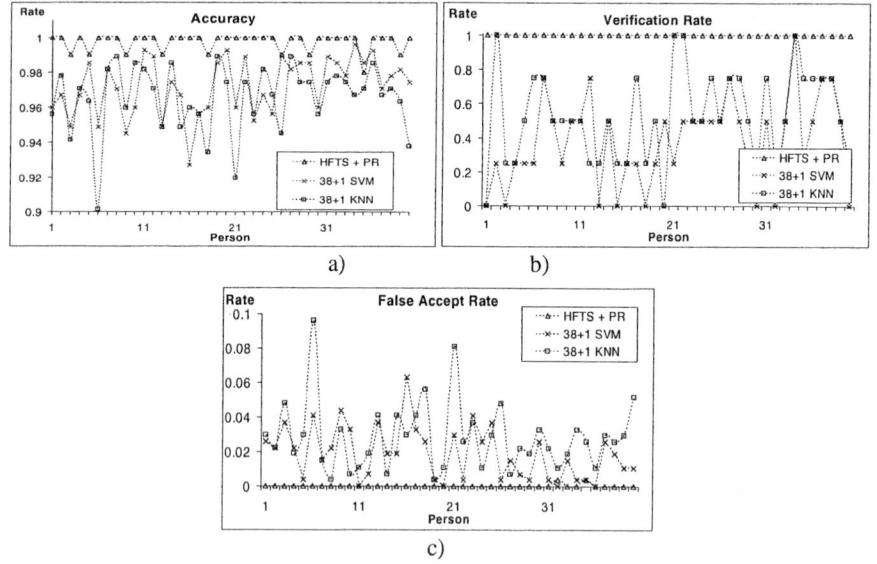

Fig. 7. Benchmark of HFTS model against other methods

Verification performance was being tested by each individual in the database enrolled with a number of images as the positive class (we used 6 out of 10 images) and a number of random images (we used 120 images) from the other individual to form the negative class. The purpose is to evaluate the performance of the systems when used as an authentication tool. The evaluation results are presented by showing the trade-off between the verification rate and false accepted rate for each of the person in the ORL database. The results illustrate that the Human Face Tree Structure representation and recognized by the proposed probabilistic structured model yields the highest accuracy against traditional representations in PCA [2], Naïve Bayes[15], KNN [14] and SVM [13]. Our proposed model can achieve the highest verification rate as well as the lowest false accepted rate as shown in Fig. 7b and Fig. 7c respectively. A low false accepted rate is critically important as it governs the amount of imposters that were successfully authenticated as the user.

In this paper, the experiment results shows that HFTS using the proposed probabilistic based model is able to produce an accuracy of 99.75%, verification rate of 100% and false accepted rate of only 0.02% on average among the 40 persons in the ORL database. Using the same Gabor features but in a flat vector form, the Naive Bayes' Rule, SVM and KNN only obtained an accuracy of about 98%, verification rate of about 53% and false accepted rate of about 1%. This clearly highlights the effectiveness of transforming the Gabor Feature Vectors to HFTS format.

5 Conclusions

This proposed approach of converting human face feature vectors to tree structure representation and using adaptive processing of tree structures method holds a strong recommendation. The probabilistic based structured model is proposed for classification of this Human Face Tree Structures patterns. The architecture represented by each node of trees is formed by a set of Gaussian Mixture Models (GMMs) at the hidden layer and a set of weighted sum of sigmoid functions at the output layers. The discriminative information can be utilized during learning in this proposed architecture, at which it is performed by an unsupervised manner. The weighting parameters in the sigmoid function model are trained by a supervised manner. Using the proposed HFTS format, new avenues are opened to solve the classification problem in a face recognition system. It also creates a possible solution towards interoperability amongst facial recognition systems. Moreover, our proposed method uses 546 features to represent in the tree structures, which is comparatively small than other feature method. This concludes that our proposed method is said to be ideal for implementing in an embedded system environment, which has limited memory and processing capabilities.

Acknowledgement

This paper was partly supported by Nanyang Technological University under an University Start-Up Grant (Ref. SUG 5/04).

References

1. P. Frasconi, M. Gori, and A. Sperduti, *A General Framework for Adaptive Processing of Data Structures,* IEEE Trans. Neural Networks, vol. 9, pp. 768-785, 1998.
2. M. Turk and A. Pentland, *Eigenfaces for Recognition,* Journal of Cognitive Neuroscience, vol. 3, pp. 71-86, 1991.

3. P.N. Belhumeur, et. al., *EigenFaces vs. FisherFaces: Recognition Using Class Specific Linear Projection,* IEEE Trans. Pattern Anal. Mach. Intell., vol. 19(7), pp. 711-20, 1996.
4. Martin Lades, et al., *Distortion Invariant Object Recognition in the Dynamic Link Architecture,* IEEE Transactions on Computers, vol. 42(2), pp. 300-311, 1993.
5. P. S. Penev and Joseph J. Atick, *Local Feature Analysis: A general statistical theory for object representation,* Network: Computation in Neural Systems, (7), pp. 477-500, 1996.
6. Y. Fang, T. Tan, and Y. Wang. *Fusion of Global and Local Features for Face Verification,* in proceedings of *International Conf. for Pattern Recognition,* pp.382-385, 2002.
7. C. Liu and Harry Wechsler, *Independent Component Analysis of Gabor Features for Face Recognition,* IEEE Transactions on neural networks, vol. 14(4), pp. 919-928, 2003.
8. Laurenz Wiskott, et al., *Face Recognition by Elastic Bunch Graph Matching,* IEEE. Trans on Pattern Analysis and Machine Intelligence, vol. 19(7), pp. 775-779, 1997.
9. S.-Y Cho and J.-J. Wong. *Robust Facial Recognition by Localised Gabor Features,* in proceedings of *Int. Workshop for Ad. Image Tech.,* 11 Jan, Jeju Island, Korea, 2005.
10. A. Sperduti and A. Starita, *Supervised neural networks for classification of structures,* IEEE Trans. Neural Networks, vol. 8, pp. 714-735, 1997.
11. Martin T. Hagan and A. C. Tsoi, *A Self-Organizing Map for Adaptive Processing of Structured Data,* IEEE Trans. Neural Networks, vol. 14(3), pp. 491-505, 2003.
12. B. Hammer and V. Sperschneider. *Neural networks can approximate mappings on structured objects,* in proceedings of *2nd Int. Conf. Comp.l Intelligence Neuroscience,* 1997.
13. J. Platt, *Fast Training of Support Vector Machines using Sequential Minimal Optimization.*, Advances in Kernel Methods - Support Vector Learning, ed. B. Scholkopf, C. Burges, and A. Smola: MIT Press, 1998.
14. D. Aha and D. Kibler, *Instance-based learning algorithms,* Mach. Learn., 6, pp37-66, 91.
15. George H. John and Pat Langley. *Estimating Continuous Distributions in Bayesian Classifiers,* in proceedings of *The Eleventh Conference on Uncertainty in Artificial Intelligence,* Morgan Kaufmann, San Mateo, pp.338-345, 1995.
16. F. Samaria and A.C. Harter. *Parameterisation of a Stochastic Model for Human Face Identification,* in *2nd IEEE Workshop Applications of Computer Vision,* 1994.
17. J.G. Daugman, *Uncertainty relation for resolution in space, spatial frequency, and orientation optimized by two-dimensional cortical filters,* J. Opt.Soc.Amer.,2(7),1160-67, 85.
18. B.S. Manjunath and W.Y. Ma, *Texture Features for Browsing and Retrieval of Image Data,* IEEE Trans. on Patt. Ana. and Machine Intellig, vol. 18(8), pp. 837-842, 1996.
19. Siu-Yeung Cho, et al., *An Improved Algorithm for learning long-term dependency problems in adaptive processing of data structures,* IEEE Trans. on NN, 14(4), pp. 781-93.
20. D. F. Streit and T. E. Luginhuhl, *Maximum likelihood training of probabilistic neural networks,* IEEE Trans. on Neural Networks, vol. 5(5), pp. 764-783, 1994.
21. S. Roberts and L. Tarassenko, *A probabilistic resource allocating network for novelty detection,* Neural Computation, vol. 6, pp. 270-284, 1994.
22. Martin T. Hagan and Mohammad B. Menhaj, *Training feedforward networks with Marquardt algorithm,* IEEE Trans. on Neural Networks, vol. 5(6), pp. 989-993, 1994.

Performance Characterization in Computer Vision: The Role of Visual Cognition Theory

Aimin Wu[1,2], De Xu[1], Xu Yang[1], and Jianhui Zheng[2]

[1] Dept. of Computer Science & Technology,
Beijing Jiaotong Univ., Beijing, China 100044
[2] Dongying Vocational College, Shandong, China 257091
wuaimin@sohu.com; xd@computer.njtu.edu.cn

Abstract. It is very difficult to evaluate the performance of computer vision algorithms at present. We argue that visual cognition theory can be used to challenge this task. Following are the reasons: (1) Human vision system is so far the best and the most general vision system; (2) The human eye and camera surely have the same mechanism from the perspective of optical imaging; (3) Computer vision problem is similar to human vision problem in theory; (4) The main task of visual cognition theory is to investigate the principles of human vision system. In this paper, we first illustrate why vision cognition theory can be used to characterize the performance of computer vision algorithms and discuss how to use it. Then from the perspective of computer science we summarize some of important assumptions of visual cognition theory. Finally, many cases are introduced, which show that our me thod can work reasonably well.

1 Introduction

The *performance* in this paper does not mean how quickly an algorithm runs, but *how well* it performs a given task. Since the early 1980s, much work has been done to challenge performance characterization in computer vision, but only a little success has been made.

Theoretical analysis. In 1986, R.M. Haralick seriously argued that computer vision lacked a completed theory to constitute an optimal solution [1]. In 1994, R. M. Haralick further argued that performance characterization in computer vision was extremely important and very difficult and proposed a general methodology to solve some basic problems about it [2,3]. In 1996, W. Forstner discussed the most disputed 10 problems to demonstrate the feasibility of computer vision algorithms evaluation, which indeed ended all objections to it [4]. In 2002, N. A. Thacker proposed a modular methodology that put the performance characterization on a sound statistics theory [5]. However, theoretical evaluation is usually too simplistic to be suitable for characterizing complicated computer vision algorithms [6], so we have to depend on empirical evaluation methods on real data.

Empirical evaluation. Though some empirical evaluation has been done in early 1970s, a large scale of works started till 1990s. Most of papers are published in several workshops and special issues. Since 1970s, *IEEE Trans. PAMI* has continually contributed to this topic and published a series of important papers. However, empirical performance characterization is far from mature, mainly because of the lack of standard free image database, the lack of a common way to get Ground Truth, and the lack of a common evaluation scheme [1-4,7]. Additionally, different experiments often get conflicting results, e.g. Bowyer's discussion about edge detection [8] and the McCane's arguments on optical flow computation [9]. So, it is very difficult to quantitatively evaluate the performance of computer vision.

We indeed agree with that the theoretical analysis and the empirical evaluation can ultimately address the complicated evaluation problem, but L. Cinque et al in [10] explicitly point out: "we realize that many difficulties in achieving such a goal may be encountered. We believe that we still have a long way to go and therefore must now principally rely on human judgment for obtaining a practical evaluation; for some specific applications we feel that this is doomed to be the only possibility." The visual cognition theory mainly investigates the principles of human vision system, such as seeing what, seeing where, how to see, so in this paper we will discuss in detail why and how to apply visual cognition theory to performance characterization of computer vision algorithms.

2 Algorithms Evaluation and Visual Cognition Theory

Methods for evaluating the performance of computer vision can be categorized into theoretical analysis and empirical evaluation. We argue that both two ways have to collaborate with visual cognition theory.

2.1 Theoretical Analysis

Three self-evident truths and two propositions will be discussed in this paragraph, which can illustrate that theoretical analysis requires visual cognition theory.

Truth 1: Assumptions in computer vision algorithms have to be made, and unsuitable assumptions must lead to poor results.

All models of computer vision algorithm are certainly not accurate description of real world in a strict sense [1-5], so some assumptions are unavoidable. For example, Gaussian Distributor is often used to model noise, though sometimes it is unsuitable for given application. Bowyer et al argue that performance of computer vision algorithm will decrease or even fall when complexity increases, so they suggest that the

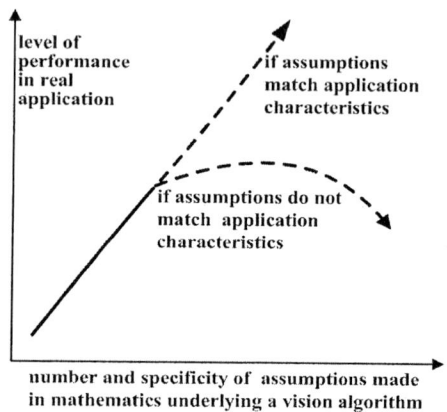

Fig. 1. Performance as a function of mathematical complexity [11]

selection and measurement of the basic assumptions must be an essential part of algorithm development [11] (see Fig.1).

Additionally, T. Poggio in [12] argues that most of computer vision issues are inverse optical problems and most of inverse problems are ill-posed. Regularization theory is a natural way to the solution for ill-posed problem. The most important criterion for ill-posed problems is the physical assumption plausibility, which means that these assumptions come from the physical world, and can constrain regularization method to get a unique solution that again has physical meanings [13, pp.75, 104]. Fig.2 shows an example

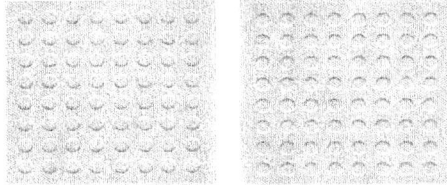

Fig. 2. An example of top-down light

of physical assumption plausibility: top-down light source assumption. The left image looks pimple, but the right one looks dimple. In fact, the left image is the result of 180-degree rotation of the right one. The law of human vision system is that dots having below shadow look pimple (the left image) and dots having upper shadow appear dimple (the right image). The law uses an assumption that the light source is always in our upper, which is indeed physical plausibility because light of sun, moon and artificial lights usually come from the above [14, pp.75-76]. Therefore this assumption is suitable for solving the issues of shape from shadow.

So it is very important to extract and validate the assumptions of algorithms, which can be used to evaluate computer vision algorithms at the theoretical level. If the assumptions used by algorithm are *unsuitable* for a given application, the results produced by this algorithm must be *poor*.

Truths 2: Each algorithm used by human vision system is the best and most general, so the assumptions used by these algorithm must be physical plausibility.

Proposition 1: To obtain optimal results for a given tasks, assumptions used by computer vision algorithm should be same as (or similar to) those employed by human vision system.

According to Marr's vision theory, each process should be investigated from three independent and loosely related levels: computational theory, representation and algorithm, and hardware implementation. From the perspective of information processing, the most critically important level is the computational theory [13,pp.10-12], whose underlying task is to find and to isolate assumptions (constraints) that are both powerful enough to define a process and generally true for the real world [13, pp.22-28]. These assumptions (constraints) are often suggested by everyday experience or by psychophysical (vision cognition theory) or even neurophysiologic findings of a quite general nature [13, pp.331].

Additionally, Computer vision problem in theory is similar to human vision problem, both of which are the process of discovering from images what is present in the world, and where it is [13, pp.1][14, pp.1-11]. The human eye and camera surely have the same mechanism from the perspective of optical imaging [14, pp.2][15, pp.1], so we can surely make use of principles of human vision to build a strong computer vision system [14, pp.19-20].

Therefore, in term of Truths 2, and above discussions, the Proposition 1 should be reasonable right.

Truths 3: One of main task of visual cognition theory is to find the assumptions used by Human Vision System.

Proposition 2: Visual cognition theory can be used to judge whether assumptions of an algorithm are suitable for given tasks, which can be further used to evaluate the algorithm.

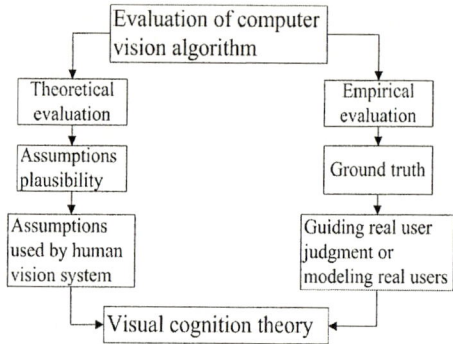

Fig. 3. The relations between evaluation of computervision and visual cognition theory

Using Truths 3, Proposition 1, and Truth 1, the Proposition 2 can be easily logical proved right. The inference procedures are shown in the left part of Fig.3.

2.2 Empirical Evaluation

It is very difficult and expensive for empirical evaluation to obtain ground truth [1-4]. All ways to do this can be classified into two classes: real user judgments, computer simulating users [7]; both of which also require vision cognition theory. The idea is shown in the right part of Fig.3.

The former, real user judgment, is better but very time-consuming because the user must give ideal results and the difference between computational results and the ideal results for any given task. There are often differences between user judgments for the same task and same inputs [7], so we have to validate them by statistical methods [6]. Luckily, it is the underlying task of visual cognition theory to investigate the difference and coherence of human vision between different human subjects [13-20]. So we argue that visual cognition theory can surely be used to guide real user judgments.

The latter, computer simulating users, is simpler, but it is difficult to model real user [6-7]. The visual cognition theory can help to model real users more exact, because its main task is to find the features of human vision system.

2.3 Evaluation Principle and Steps

Above discussions extensively illustrate that visual cognition theory can be used to evaluate computer vision algorithm both for theoretical evaluation and for empirical evaluation These ideas are paraphrased into the *Principle of Qualitative Evaluation for Computer Vision Algorithm:*

For a given task, if the assumptions used in computer vision algorithm are not consistent with assumptions of visual cognition theory (human vision system), the performance of this algorithm must be poor.

Fig.4 shows three main steps to use this principle. The step 1 extracts assumptions used by the computer vision algorithm. The difficulty is that assumptions of many algorithms are so rarely explicitly expressed that we often have to infer them. The step 2 judges whether these assumptions are consistent with assumptions of visual cognition theory. The set of assumptions of cognition theory and their applicable tasks are build offline before evaluation (see Section 3). The step 3 reports the result of evaluation, which is divided into three categories: *Good* if all assumptions match, *Fair* if some assumptions match, and *Poor* if no assumption match.

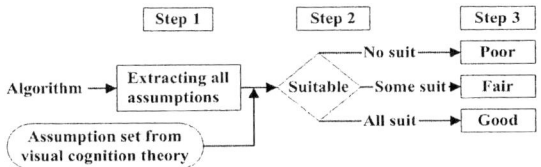

Fig. 4. Three main steps to use the Principle of Qualitative Evaluation for Computer Vision Algorithm. The set of assumptions is built offline in advance.

3 The Set of Assumptions

Most of assumptions of visual cognition theory come from [13-21], which are reorganized and reedited from the perspective of computer science.

a. Both eye and brain [14, pp.128-136][15, pp.1-13]: Human has a plenty of knowledge about physical world and how they behave, which can be used to make inferences. *Structured knowledge constraints*: If we want to design a general-purpose vision machine, we must first classify and structure knowledge about real world for it.

b. Abstract & classification principle [21, pp.1]: we use three principles of construction to understand the physical world: (1) identifying the object and its attributes, e.g. a tree and its size; (2) identifying the whole and its components, e.g. a tree and its branches; (3) identifying different classes of object, e.g. the class of trees and the class of stones.

c. Brain is a probability computer [15, pp.9-13]: Brain makes hypotheses and checks them, then makes new hypotheses and checks them again until making the best bet, during which all knowledge can be made use of. Eyes and other senses within a short time would rather provide evidence for brain to make hypotheses and to check them than give us a picture of world directly. Mechanism of inference is classified into unconscious inference and conscious inference [19, pp.1-16]. *Methodology constraint:* probability method may be better for computer vision problems.

d. See world by object not pattern [20]: Human eye receives patterns of energy (e.g. lightness, color), but we see by object not pattern. We do not generally define object by how it appears, but rather by its uses and its causal relations. Once we know what the object is, we must know its shape, size, color and so on. *Object constancy constraints:* Physical object exists continuously, uniquely, and constantly, though time is flying [13, pp.205].

e. Do we have to learn how to see? [15, pp.136-169] The inheritance only forms the basis for learning, so that we have to learn much knowledge and ability for the sake of seeing. *Computer learning constraint:* we should continuously help computer with learning by active hands-on exploration to relate the perception to conception, as do it for a baby.

f. The law of Gestalt [14, pp.113-123][17, pp.106-121]. The Grouping principle can be further summarized into five principles: (a) the principle of proximity, (b) the principle of similarity, (c) the principle of good continuation, (d) the principle of closure tendency, and (e) the principle of common fate. The Figure-ground segregation principle means that (1) in ambiguous patterns, smaller regions, symmetrical regions, vertically or horizontally oriented regions tend to be perceived as figures; (2) The enclosed region will become figure, and the enclosing one will be the ground; (3) The common borders are often assigned to the figure; (4) Generally, the ground is simpler than the figure.

g. Simultaneous contrast [13, pp.259-261] [15, pp.87-92]: Human eyes don't detect the absolute energy of brightness, lightness, color, and motion, but their difference that is directly proportional to the background energy (e.g. Weber's Laws). *Threshold constraint:* a differential value is better than absolute one. *Compensation constraint:* brightness, lightness, color, and motion should be compensated according to the background energy.

h. Constancy world [14, pp.15-52]: According to the knowledge of geometrical optical imaging, the retinal image is different from the objects' outline, and the retinal image continually varies as human moves, but the object looks the same to us, which is called Constancy. There are size constancy, color constancy, brightness constancy, lightness constancy, shape constancy, motion constancy, and so on.

i. The principle of modular design [13, pp.99-103]: Each system (e.g. vision, touch etc.) of the perception and each channel (e.g. seeing color and seeing movement of vision) of different system work independently. Sometimes, different systems and different channels may make inconsistent conclusions, which force the brain to make a final decision. *Multi-channel constraint, Information encapsulation constraint:* have been applied to Object-Oriented analysis and design by computer community [21]. Furthermore, one channel (e.g. color) of vision system may affect or even mask another channel (e.g. shape), which is called *visual masking effects.*

j. Two eyes and depth clues [14,pp.53-90] [15, pp.61-66]: Two eyes share and compare information, so they can perform feats that are impossible for the single eye, e.g. the 3-D perception from two somewhat different images. Depth perception cues include retinal disparity, convergence angle, accommodation, motion parallax and pictorial information (occlusion, perspective, shadow, and the familiar sizes of things). *Depth perception constraint: in order to yield definite depth perception, all clues must work collectively.*

k. Brightness is an experience [15, pp.84-97]: Brightness is a function not only of the intensity of light falling on a given region of retina at a certain time, but also of the intensity of light falling on other regions of retina, and of the intensity of the light that the retina has been subject to in the recent past. In the dark, the mechanisms of dark-adaptation trade eye's acuity in space and time for increase in the sensitivity *(The continuity of brightness change constraint).* The brightness can be reflected by shading and shadow, which can indicate objects' information (e.g. *Top-down light source constraint).*

l. Two seeing movement systems [14, pp.17-202] [15, pp.98-121]: One is the image/retina system that passively detects the movement. Another is the eye/head movement system that positively seeing movement. When searching for an object, the eyes move in a series of small rapid jerks *(Motion discontinuous assumption)*, but when following an object, they move smoothly *(Motion continuous assumption)*. The eyes tend to suggest that the largest object is stationary *(Motion reference frame constraint)*. Persistence and apparent movement imply *continuity, stability and uniqueness constraints*.

m. RGB is not the whole story [15, pp.121-135]: Only mixing two, not three, actual colors can give a wealth of colors. The mixture of three primary colors (e.g. RGB) can't produce some colors that we can see, such as brown, the metallic colors. Color is a sensation. It depends not only on the stimulus wavelengths and intensities, but also on the surrounding difference of intensities, and on whether the patterns are accepted as objects *(Color computational constraint)*.

n. Topological rules in visual perception [18, pp.100-158]: Local homotopy rule: we tend to accept an original image and its transformed image as identical, if the image is made a local homotopy transformation within its tolerance space. The same is true for the homeomorphism rule, homeomorphism and null-homotopy rule in cluster, the object superiority effect, and the configurable effect.

o. The whole is more than the sum of its parts [13, pp.300-327] [16][17, pp.176]: The same parts (primitive) with different relations may construct different objects. It is possible to match a number of objects with a relatively small number of templates, because it may be easier to recognize parts (primitives) with relatively simper probability methods.

p. Marr's underlying physical assumptions [13, pp.44-51]: (1) existence of smooth surface in the visible world, (2) hierarchical spatial organization of a surface with a different scale, (3) similarity of the items generated at the same scale, (4) spatial continuity generated at the same scale, (5) continuity of the loci of discontinuities, and (6) continuity of motion of an rigid object.

q. Edge perception and edge type [14, pp.49-50]: The vision system only picks up luminance difference at the edge between regions, and then assumes that the difference at the edge applies throughout a region until another edge occurs. Furthermore vision system divides the various edges into two categories: lightness edge and illumination edge. The perceptual lightness value at the edges is only determined by lightness edge.

From other psychological literatures, we can extract more assumptions such as object rigidity assumption, Gauss distribution assumption, and smooth assumption, etc.

4 Cases Study

4.1 The Problems of Optical Flow

Table 1. The discussion about optical problem. Note: (k) in the table refers to k^{th} assumption in Section 3.

Problem	Assumptions	Suitable	Result
Determining the optical flow	Flat surface	Suit (p)	Some suit Fair
	Uniform incident illumination	Ill-Suit (k, q)	
	Differentiable brightness	Suit (k)	
	Smooth optical flow	Suit (p, l)	
Recovering 3-D structure	Motion field equals to optical flow field	Ill-suit (l)	Poor

4.2 Waltz's Line Drawings [13, pp.17-18]

When all faces were planar and all edges were straight, Waltz made an exhaustive analysis of all possible local physical arrangement of these surfaces, edges, and shadows of shapes (Structured knowledge constraint and Abstract & classification principle in Section 3 a, b). Then he found an effective algorithm to interpret such actual shapes.

Fig.6 shows that some of configurations of edges are physically plausibility, and some are not. The trihedral junctions of three convex edges (a) or the three concave edges (b) are plausibility, whereas the configuration (c) is impossible. So the direction of edge E in (d) must be the same type as (a). This example shows the power of physical assumption plausibility.

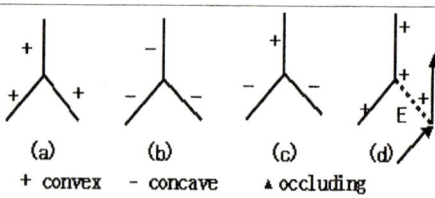

Fig. 6. The ideas behind Waltz's theory

4.3 Attention Mechanism

L. Itti et al define a set of linear "center-surround difference" operator (Simultaneous Contrast in Section 3 g) to reproduce the attention mechanism of primate visual system (visual masking effects in Section 3 i). However, it is only a bottom-up guidance of attention without using any prior knowledge. V. Navalpakkam et al proposed to use a task graph to describe the real world entities and their relationships as Top-down control [25]. A. Oliva et al used the distribution of background of scenes as knowledge constraints [26]. Both V. Navalpakkam model and A. Oliva model employed *Structured knowledge constraints* (Section 3 a), so the effect and performance of their model are better than that of pure bottom-up attention model.

Table 2. Attention models and their assumptions for object recognition or scene analysis. Note: (a) in the table refers to a[th] assumption in Section3.

Model	Assumptions	Suitable	Result
L. Itti model	Center-surround difference	Simultaneous Contrast: suit (g)	Some suit Fair
	Only bottom-up	Structured Knowledge constraints: ill-suit (a)	
	Only focus on one element	Masking effects: suit (i)	
V.Navalpakkam model	L. Itti model & Distribution of background	All suit (a, g, i)	Good
A. Oliva model	L. Itti model & Task graph (Top-down)	All suit (a, g, i)	Good

4.4 Comparison with CVIR Experiments

Many researchers have compared the performance of Content-based Visual Information Retrieval (CVIR) algorithms in an experimental way [27, 57-305], listed in Table 3. It is inherent consistency between these experimental results and the judgments of our principle.

Table 3. Comparisons between experimental results and those by our methods about CVIR algorithms. Note: NH=normal histogram; CH=cumulative histogram; EDH= edge direction histogram; Wavelet MM= wavelet Modulus Maxima; Local M= Local Motion detection; L&GM= Local motion detection after Global Motion compensation; (d) in the table refers to d^{th} assumption in Section 3.

Feature	Method name	Experimental Result	Assumptions	Suitable	Our Result
Color	NH	Poor	Color is linear	Ill-suit (m)	Poor
	CH	Fair	Color is non-linear	Suit (m)	Fair
Shape	EDH	Fair	Brightness changes in boundary	Suit (k, q)	Fair
	Wavelet MM	Good	Brightness changes in boundary, Multi-size & Multi-channel, and Gauss distribution.	All Suit (k, q, i)	Good
Color & Shape	CH	Fair	Color is nonlinear	Suit (m)	Fair
	EDH	Fair	Brightness changes in boundary	Suit (k, q)	Fair
	NH & EDH	Good	Brightness changes in boundary & Color is nonlinear	Suit (m, k, q)	Good
Motion	LocalM	Poor	Absolute motion	Ill-suit (g)	Poor
	L&GM	Fair	Relative motion	Suit (g, l)	Fair

5 Conclusion and Further Work

The preliminary study strongly suggests that vision cognition theory can be used to evaluate computer vision algorithms. In this paper, we propose the Principle of Qualitative Evaluation for computer vision algorithms. To easily use this principle, we summarize some important assumptions of psychology. Further works include: 1) to model users under the integrated framework to automatically define the ground truth; 2) to explore cognition-based methods for empirical performance characterization; 3) to find more psychological assumptions and their applicable tasks. After all, our ultimate aim is to evaluate the usefulness of a computer vision system for end users.

Acknowledgement

This work was supported by the Beijing Jiaotong University Research Project under Grant No. 2004SM013.

References

1. R. M. Haralick, "Computer Vision Theory: The Lack Thereof," *Computer Vision Graphics and Image Processing*, vol. 36, no. 2, pp. 272-286, 1986.
2. R. M. Haralick, "Performance Characterization in Computer Vision," *Computer Vision, Graphics and Image Processing: Image Understanding*, vol. 60, no. 2, pp. 245-249, 1994.

3. R. M. Haralick, "Comments on Performance Characterization Replies," *Computer Vision, Graphics, and Image Processing: Image Understanding,* vol. 60, no. 2, 264-265, 1994.
4. W. Foerstner, "10 Pros and Cons Against Performance Characterization of Vision Algorithms, " *Proc. ECCV Workshop on Performance Characteristics of Vision Algorithms,* Apr., 1996.
5. N. A. Thacker, "Using Quantitative Statistics for the Construction of Machine Vision Systems," *Keynote presentation given to Optoelectronics, Photonics and Imaging 2002,* Sept., 2002.
6. M. Heath, S. Sarkar, et al, "A Robust Visual Method for Assessing the Relative Performance of Edge Detection Algorithms," *IEEE Trans. PAMI,* vol. 19, no. 12, pp. 1338-1359, Dec. 1997.
7. H. Müller, W. Müller, et al, "Performance Evaluation in Content--Based Image Retrieval: Overview and Proposals," *Pattern Recognition Letters,* vol. 22, no. 5, pp. 593--601, 2001.
8. Min C. Shin, D. B. Goldgolf, and K. W. Bowyer, "Comparison of Edge Detector Performance Through Use in an Object Recognition Task, " *Computer Vision and Image Understanding,* vol. 84, pp. 160-178, 2001.
9. B. McCane, "On Benchmarking Optical Flow," *Computer Vision and Image Understanding,* vol. 84, pp.126–143, 2001.
10. L. Cinque, C. Guerra, and S. Levialdi, "Reply On the Paper by R.M. Haralick," *Computer Vision, Graphics, and Image Processing: Image Understanding,* vol. 60, no. 2, pp. 250-252, Sept., 1994.
11. K. W. Bowyer and P. J. Phillips, "Overview of Work in Empirical Evaluation of Computer Vision Algorithms," In *Empirical Evaluation Techniques in Computer Vision,* IEEE Computer Press, 1998.
12. T. Poggio, et al "Computational Vision and Regularization Theory," *Nature,* 317(26), pp 314-319, 1985.
13. D. Marr, *Vision,* Freeman, 1982.
14. Rock, *Perception,* Scientific American Books, Inc, 1984.
15. R. L. Gregory, *Eye and Brain,* Princeton university press, 1997.
16. Biederman, "Recognition-by-Components: A Theory of Human Image Understanding," Psychological Review, vol. 94, pp. 115-47, 1987.
17. K. Koffka, *Principle of Gestalt Psychology,* Harcourt Brace Jovanovich Company, 1935.
18. M. Zhang, *Psychology of Visual Cognition,* East China Normal University Press, 1991.
19. Rock, *The Logic of Perception,* MIT Press, 1983.
20. D. M. Sobel, et al, "Children's causal inferences from indirect evidence: Backwards blocking and Bayesian reasoning in preschoolers," *Cognitive Science,* vol. 28, pp. 303–333, 2004.
21. P. Coad and E. Yourdon, *Object-Oriented Analysis,* Yourdon Press, 1990.
22. B. K. P. Horn, et al, "Determining Optical Flow," *Artificial Intelligence,* vol.17, pp.185-203, 1981.
23. Verr, et al , "Motion Field and Optical Flow: Qualitative Properties, " *IEEE Trans. PAMI,* vol. 11, pp. 490-498, 1989.
24. L. Itti, C. Koch, and E. Neibur, "A Model of Saliency-based Visual Attention for Rapid Scene Analysis," *IEEE Trans. PAMI,* vol. 20, no. 11, 1998.
25. V. Navalpakkam and L. Itti, "A Goal Oriented Attention Guidance Model," *Lecture Notes in Computer Science,* vol. 2525, pp. 453-461, 2002.
26. Oliva, A. Torralba, M. Castelhano, and J. Henderson, "Top-down Control of Visual Attention in Object Detection," *International Conference on Image Processing,* 2003.
27. Y. J. Zhang, Content-based Visual Information Retrieval, Science Press, Beijing, 2003.

Generic Solution for Image Object Recognition Based on Vision Cognition Theory

Aimin Wu[1,2], De Xu[1], Xu Yang[1], and Jianhui Zheng[2]

[1] Dept. of Computer Science & Technology,
Beijing Jiaotong Univ., Beijing, China 100044
[2] Dongying Vocational College, Shandong, China 257091
wuaimin@sohu.com, xd@computer.njtu.edu.cn

Abstract. Human vision system can understand images quickly and accurately, but it is impossible to design a generic computer vision system to challenge this task at present. The most important reason is that computer vision community is lack of effective collaborations with visual psychologists, because current object recognition systems use only a small subset of visual cognition theory. We argue that it is possible to put forward a generic solution for image object recognition if the whole vision cognition theory of different schools and different levels can be systematically integrated into an inherent computing framework from the perspective of computer science. In this paper, we construct a generic object recognition solution, which absorbs the pith of main schools of vision cognition theory. Some examples illustrate the feasibility and validity of this solution.

Keywords: Object recognition, Generic solution, Visual cognition theory, Knowledge.

1 Introduction

Despite the fact that much success has been achieved in recognizing a relatively small set of objects in images in the past decades of research, it is currently impossible to design a generic computer algorithm to challenge this task [1]. Many causes contribute to it, but the most important one is that computer vision community is lack of effective collaboration with visual psychologists [4].

1.1 Related Work

First, successful applications focus on attention mechanism. Laurent Itti et al propose a attention model which can effectively reproduce an important performance of primate visual system, that is, while the retina potentially embraces the entire scene, attention can only focus on one or a few elements at a time, and thus facilitate their perception, their recognition, or their memorization for later recall [5]. But it only makes use of **Simultaneous Contrast Theory** of visual ccognition, which is only pure bottom-up guidance of attention.

Some more complex systems are developed on the basis of visual attention mechanism. Vidhya Navalpakkam et al. propose a goal oriented attention guide model [6]. The model uses a task graph to describe the real world entities and their relationships (top-down control). A. Oliva et al use the distribution of background of scenes as Top-down constraints to facilitate object detection in natural scenes [7]. Dirk Walther et al make a series of computer experiments to strongly demonstrate that the bottom up visual attention can effectively improve learning and recognizing performance in the presence of large mount clutters [8].

Second, some systems employ the visual inference mechanism that *perceptions are hypotheses,* a kind of inference mechanism similar to a probability computer [9, pp9-13]. Sudeep Sarkar et al [10] present an information theoretical probabilistic framework based on perceptual inference network formalism to manage special purpose visual modules trying to construct a generic solution for the problem of computer vision. Mattew Brand has built a suite of *explanation-mediated* vision systems to see, manipulate, and understand scenes in a variety of domains, including blocks, tinker toys, Lego machines, and mugs [11]. Zu Whan Kim et al present an approach for detecting and describing complex rooftops by using multiple, overlapping images of the scene [12]. More detail discussions about perception as hypothesis please refer to [13]. The same and salient characteristic of these systems is to use structured knowledge about physical world to make inference and eliminate ambiguities of images.

Third, Biederman's Recognition-by-components (RBC) theory is regarded as a promising object recognition theory and causes much attention of computer vision community [14]. Hummel et al design a Neural Network, which uses dynamic binding to represent and recognize the Geons and shape of objects [15]. Quang-Loc Nguyen propose a method to compute Geons and their connections, which employs edge characteristics and T-conjunctions to successfully recognize objects in range images [16]. More detail discusses about RBC theory please refer to [17].

Finally, Feature Integrated Theory [18] is the earliest and most frequently used by many systems, which use different visual features and their distribution (e.g. color, texture and shape) to classify and recognize objects. Jia Li et al implement automatic linguistic indexing of picture by using Wavelet coefficients of color and texture and 2D MHMMs [1]. WAN Hua-Lin et al classify image with the incorporation of the color, texture and edge histograms seamlessly [19]. Kobus Barnarda et al use a set of 40 features about size, position, color, shape and texture to translate images into text [20]. Unfortunately, there exists an enormous gap between low-level visual feature and high-level semantic information.

Each system above uses only a small subset of the visual cognition theory. The human vision system is a complicated whole, so only when the whole vision cognition theory of different schools and different levels is systematically integrated into an inherent computing framework from the perspective of computer science, it is possible to put forward a full solution of computer object recognition.

1.2 Our Approach

The logic bases of our solution include: (1) Human vision system is so far the best and the most general; (2) The human eye and camera surely have the same mechanism from the perspective of optical imaging [2, pp 19-20]; (3) Computer vision problem is similar to human vision problem in theory, both of which are the process of discovering from images what is present in the world, and where it is [3, pp1]; and (4) The main task of visual cognition theory is to investigate the principles of human vision system. So if we want to develop a vision system that can match with human vision system, we have to make full use of vision cognition theory.

The aims of the solution include:

1) Recognize many different objects accurately in an arbitrary 2D image;
2) Recognize objects in degraded image such as occlusion, deformation;
3) Recognize objects independent of viewing position;
4) Recognize objects at an appropriate level of abstraction;
5) Computational complexity is linear or sub-linear scalable;

Methods inspired by visual cognition theory include:

(a) Multiple-level objects Coding

We use Recognition-by-component (RBC) theory [14], deformed superquadrics or spheres Generalized Cylinder [21] and Feature Integrated theory [18] to code object and to produce a hierarchical **object code table**.

RBC theory is used to code the basic-level objects with specified boundaries. When shown a picture of a sparrow, most people answering quickly call it a bird not a sparrow or animal. There bird is in basic-level, and sparrow in subordinate-level, animal in superordinate-level. There are approximate 3000 basic-level terms for familiar concrete object that can be identified on the base of their shape rather than surface properties of color or texture [22]. The fundamental assumption of RBC theory is that a modest set of Geons (less than 36, 12 used in our solution), can be derived from contrasts of five readily detectable properties of edges in a two-dimensional image: curvature, collinearity, parallelism, cotermination, and symmetry. The detection of these properties is generally invariant over viewing position and image quality and consequently allows robust object perception when the image is projected from a novel viewpoint or when image is degraded. So we can roughly reach Aim (2), Aim (3). Considering relations of relative size, verticality, centering, and relative size of surfaces at joins, there are 57.6 different combinations of arrangement between the two Geons. If two Geons can be recovered from images, we can code 8294 (12*12*57.6) objects, and if three Genos, 5.73 million (12* 12* 12* 57.6*57.6) objects to be yielded, which almost be over much redundancy compared with 3000 basic-level objects. Thus we can effectively reach Aim (1).

Feature Integration Theory is used to code objects without specified boundaries or to define subordinate level objects. These surface features includes color, texture, intensity, brightness, orientation and their transformed value such as color histogram, texture co-occurrence matrix and so on. Deformed Superquadrics or spheres Generalized Cylinder suggested by [21] is quantitative method which can be used to define subordinate level objects. So the solution can roughly reach Aim (4).

The process of object recognition is mainly divided into two stages: a) to compute RBC Geons and surface features from input image; b) to simple look up the **object code table** by using Geons and features as indexes. Because the number of Geons and features is fixed, the computational complexity of the first stage is almost fixed. The process of looking-up the table is one-to-one map, so its computational complexity can be lower than linear or sub-linear. So we can reach Aim (5).

(b) Geons and feature recovering. There are two methods to **compute Geons.** One is to use Marr's prime sketch to accurately compute these Geons as Quang-Loc

Nguyen et al did [16], which needs detect edge information well. The other is to use statistical method. it is not difficult because there are only 12 Geons requiring computing. The detail computational process please see [23]. It is an easier work to compute surface features than to recognize Geons. Deformed Superquadrics or spheres Generalized Cylinder is recovered in a relatively simple numerical optimization method. The attributes of these Geons and features will be adjusted by appropriate constancy transformations controlled by depth cues and topological transformation [27]. During the process of objection recognition, attention mechanism and Gestalt Laws can be used to guide the image grouping.

(c) **Bottom-up and top-down interaction.** The whole process is the interaction of bottom-up and top-down processing, whereas top-down knowledge (such as object code table in this paper) is at the core of the whole process. Knowledge can help us resolve many of problems of noise and ambiguity. Moreover, feedback technology will be used to refresh top-down knowledge set.

2 Human Vision Cognition Framework

Gregory R.L. [9, pp 251] views human vision system as a task of information-processing, and argues that image will be translated into human internal representation during this process (see Fig.1). Fig.1 show that bottom-up signals from images are first processed unconsciously by general grammars such as the laws of Gestalt and constancy transformations (side-ways), and are then interpreted consciously by predefined knowledge (Top-down). The output about object can guide behavioral exploration. Feedback and learning of successes and failure may correct and develop the set of predefined knowledge. It is suggested that image signal processing may be affected by emotion.

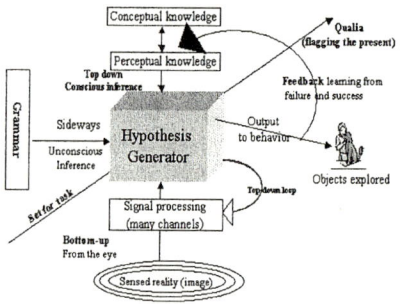

Fig. 1. the framework of image object recognition by British psychologist Richard L. Gregory

3 Our Full Solution and Main Implementation Steps

Inspired by British psychologist Richard L. Gregory, from the perspective of computer science we extract the pith of main schools of vision cognition theory to construct an object recognition solution. Fig.2 illustrates its main implementation steps. From the framework, we can easily find out that the **predefining set of knowledge is at the core of the whole process**, which not only constraints grouping and recognizing process, but also is often modified after stable output reaches. The whole process includes twelve steps and divides into four parts: initial part (step 1-2), Marr's Sketch calculation (step 3), calculation and transformation of Geons and features (step 4-8), and object recognition (step 9-12). There includes a feedback process of human-machine interaction in the later two parts.

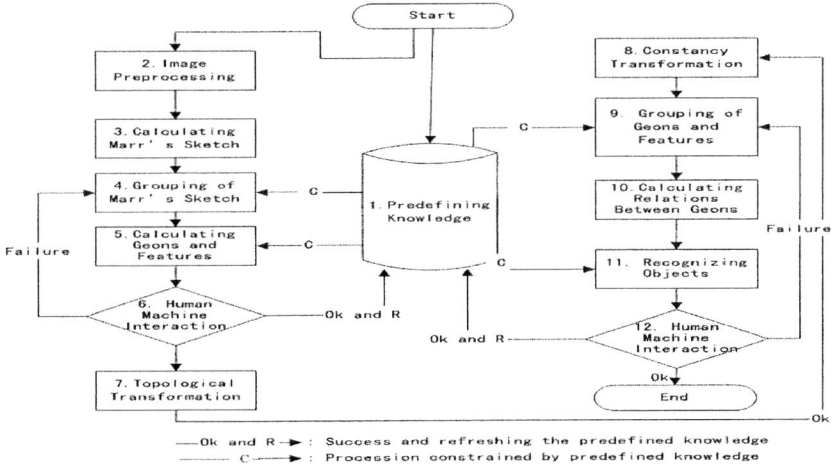

Fig. 2. Main implementation steps of image object recognition

For a given image, the main computational processes can be described as below:

(1) Predefining set of knowledge. Human has a plenty of knowledge about physical world and how they behave, which make inferences possible. So when computer vision problems are considered, don't forget the brain but only concentrate on the eye because information stored in the brain can facilitate the reasoning [9, pp 250-251]. If we want to design a general-purpose vision machine, we must first classify and structure the knowledge for it, which is also the basis of understanding human behavior. The initial set is usually created manually and it can be refreshed once when the human-machine interaction can be successfully executed (step 6, step 12 in Fig.2). After the step of human-machine interaction has been executed many times, the set of knowledge will reach a stable state and the image object recognition can be automatically completed by computer, which makes the step of human-machine interaction unnecessary. It is just as a boy can independently understand the physical world reasonably well after he has acquired a stable perception ability and knowledge by 3-6 years observation experience with adult's help.

(2) Image preprocessing. It includes image filtering, image strengthening, imaging sharpening and so on. The technologies in the domain are almost mature.

(3) Calculating Marr's Sketch. It includes Marr's prime Sketch and 2.5D sketch. The details refers to [3, pp 1-264].

(4) Grouping of Marr's Sketch. Marr's Sketch mainly consists of separate dots and lines, which does usually not completely correspond to a meaningful thing of real world. Therefore we have to further group Marr's Sketch into some meaningful perception unit. Under most of circumstances, the process can be automatically implemented controlled by the grouping principle of Gestalt laws and Pre-attention and attention mechanism [24] [25]. But when environment become very complex such as lots occlusions and local discontinuities, the predefined knowledge must be also employed to get more unambiguous units.

(5) Calculating Geons and Features. If each group of Marr's Sketch corresponds to only one Biederman's Geon, it is very easy to recognize it, because the number of Biederman's Geons is relatively small (12 used in the solution). But if there is a relation of one to more, we have to gradually adjust the grouping parameters and each time separates a single dominant Geon from all remaining Geons in the group. Therefore, all kinds of machine learning technology can be employed in this step [23]. Then the features and quantitative information of Genos will be computed for subordinate object recognition.

(6) Human-machine interaction. Benefiting from the long-term natural selection, the algorithms and mechanisms of eyes are always the best and most general. The inheritance only forms the basis for learning, so that human have to learn most knowledge and ability for the sake of seeing. What is learned by an individual can't directly be inherited by its descendants. So we have to continuously help computer with learning by active hands-on exploration to relate the perception to conceptual understanding, as do it for a baby (Gregory. R.L.1997, pp136-169). If the result of step 5 can't be confirmed by human, the process has to go back to step 4 (see Fig.2), for the wrong result mainly originate from incorrect grouping in step 4 according to psychologists' opinions. All the results of step 6 both success and failure will be regarded as a knowledge-based cases to expand or modify the predefining set of knowledge. Feedback technology sees [26].

(7) Topological transformation. The most important topological transformation is Local homotopy rule that we tend to accept an original image and its transformed image as identical, if the image is undergone a local homotopy transformation within its tolerance space. For example, a face with a mouth whether opening or closing is regarded as identical one. The process may be skipped when the image is very simple. The purpose of this step is to adjust the attributors of Geons and features, which will influence the computation of Geons' relation [27].

(8) Constancy transformation. Three of the most important transformations for our model are color constancy transformation, size constancy transformation, and shape constancy transformation, whose computational theory sees [24, pp 211-264] [2, pp 15-52]. The purpose of this step is to adjust the attributors of Geons and features, which will influence the computation of Geons' relation.

(9) Grouping of Geons and features. Each object usually includes more than one Geons. All Geons got in above steps must be grouped into proper units, each of which will construct a meaningful object. The process, principle and properties are similar to those of step 4.

(10) Calculating relations. The same Geons or features with different relations will construct different objects. So the calculation of relation is the same important for object recognition as the calculation of Geons and features. Possible relations please refers to [14].

(11) Recognizing objects. It is a relatively simple step because we only search the predefined set of knowledge by using the Geons and features and their relations as indexes to find out all possible objects in the image.

(12) Human-machine interaction. The process, principles and properties are similar to those of step 4.

Though there are only twelve steps, each step is full of extraordinary difficulties. One lies in the boundary between psychology, math and computer science. Another is how to effective define the set of structural knowledge, which can match for the memory of human brain.

4 Some Examples

Because of the huge complexity of implementation, we simulate some simple examples to illustrate the feasibility and validity of this model. The full implementation of this solution on machine will be done in further work.

4.1 Predefining Set of Knowledge

We define a small set of Biederman's Geons, some objects and their relations.

(1) The set of twelve Biederman's Geons

The set of Biederman's Geons are classified on the basis of four qualitative geometrical attributes (axis shape, cross-section edge shape, cross-section size sweeping function and cross-section symmetry). In this paper, we use the same 12 Geons as the selection of Weiwei Xing [23], which ignore the symmetry attribute of Geons because of its computational complexity. These Geons are extracted by using Support Vector Machine and Neural Network. The attributes of 12 Geons are denoted in Table 1 and their shapes are shown in Fig.3.

Table 1. is the denotation of 12 Geons

No.	axis	cross-section edge	cross-section size sweeping function	Abbr.
1	straight	straight	constant	s-s-co
2	straight	straight	increase & decrease	s-s-id
3	straight	straight	tapered	s-s-t
4	straight	curved	constant	s-c-co
5	straight	curved	increase & decrease	s-c-id
6	straight	curved	tapered	s-c-t
7	bent	straight	constant	b-s-co
8	bent	straight	increase & decrease	b-s-id
9	bent	straight	tapered	b-s-t
10	bent	curved	constant	b-c-co
11	bent	curved	increase & decrease	b-c-id
12	bent	curved	tapered	b-c-t

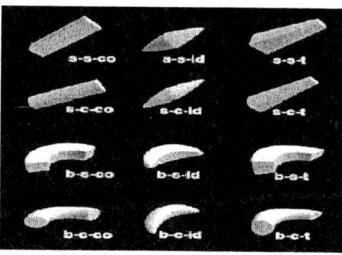

Fig. 3. shows 12 Geons with same Abbr. in Table 1

(2) The definition of objects

These definitions of below objects come from America Traditional Dictionary.

Tree is a large (usually tall) long-lasting type of plant, having a thick central wooden stem (**truck**), from which wooden **branches** grow, usually bearing **leaves**.

Door is a **movable structure** used to close off **an entrance**, typically consisting of **a panel** that swings on hinges or that slides or rotates.

Wall is an **upright structure** of masonry, wood, plaster, or other building material serving to enclose, divide, or protect an area, especially a vertical construction forming an inner partition or exterior siding of a building.

Window is an opening construct in **a wall** that functions to admit light or air to an enclosure and is often **framed** and spanned with **glass** mounted to permit opening and closing.

Roof is an exterior **vaulted surface** and its supporting structures on the top of a building.

Ground is a land surrounding or forming part of a house or another building or the solid surface of the earth.

House is a structure consisting of **walls, doors, windows, ground** and **roof,** serving as shelter or location of something, etc.

Table 2. illustrates the constitutional relations between these objects and their parts. The labels of Geons are the same as table 1 and Fig.3

Object	Part	Geons
Tree	Truck	4.s-c-co
	Branches	6.s-c-t
	Leaves	5.s-c-id
House	Roof	1.s-s-co
	Window(s)	1.s-s-co
	Wall(s)	1.s-s-co
	Door(s)	1.s-s-co
	Ground	1.s-s-co

(3) Definition of the relations between Geons

The same primitives with different relations may construct different objects, so the location relations among Geons are significant for the recognition of objects. All relations between Geons of same objects are listed in Table 3 [14]. Table 2 and Table 3 explicitly show a fact that it is the different relations between roof, window, door, wall and ground that make them different from others, though they correspond to the same Geons.

Table 3. defines all relations between Geons of same object. Symbol "/" denotes that there is not suitable relation. The Geon number is same as Table 1 and Fig.3.

Object	Part	Geon number	Size relation	Verticality relation	Centering at joins	Surface size at joins
Tree	Truck-branch	4-6	Greater	Side	Centered	Long to short
	Branches-leaf	6-5	Greater	Side	Centered	Long to short
House	Roof-wall	1-1	Greater	Above	Centered	/
	Wall-window	1-1	Greater	Below	Centered	/
	Wall-door	1-1	Greater	Above	Off center	/
	Wall-ground	1-1	Smaller	Above	Off center	/
	Door-ground	1-1	Smaller	Above	Off center	Short to long

4.2 Object Recognition from Image

In essence, the process of predefining set of knowledge is to code objects in the physical world by using different image primitives and different relations. Inversely, the object recognition is thus actually a process of looking up the predefining set of knowledge by using image primitive and their relations as indexes. Therefore image object recognition can be accurately and effectively implemented. Some examples are shown in Fig.4. Fig.4 (a) is an original input image, which will be further processed according to the flow in Section 3. Fig.4 (b) shows the result of image preprocessing such as edge extraction and image strengthening, from which we can work out Marr's Sketch such as blobs, terminations and discontinuities, edge segments, virtual lines, groups, and so on [3]. After appropriate grouping and transformation to Fig.4 (b), we can get its Biederman's Geons and relations between them, which are shown in Fig.4(c). The image in Fig.4(c) can clearly be classified into two groups by using the principle of good continuation, proximity and similarity. The left group in Fig.4(c) consists of three Geons: s-c-co (No.4), s-c-id (No.5) and s-c-t (No.6). The Geon s-c-co (No.4) is **Greater** in relatively size than Geon s-c-t (No.6) and locates to **Side** Geon s-c-t (No.6) in vertical direction. The end of Geon s-c-t (No.6) connects to the **Center** of the side of Geon s-c-co (No.4). The **Long** surface of s-c-co (No.4) joins at the **Short** surface of Geon s-c-t (No.6). So the relation between Geon 4 and Geon 6 is abbreviated into **Greater, Side, Centered and Long to short**. The relation between Geon s-c-t (No.6) and Geon s-c-id (No.5) can be got in the same way. If we can use these Geon and relations to search Table 3 and to compare with Table 2, the object of the left group in Fig.4(c) can be surely recognized as a **Tree**. Similarly, we can get object name of the right group in Fig.4(c), which is a **House**.

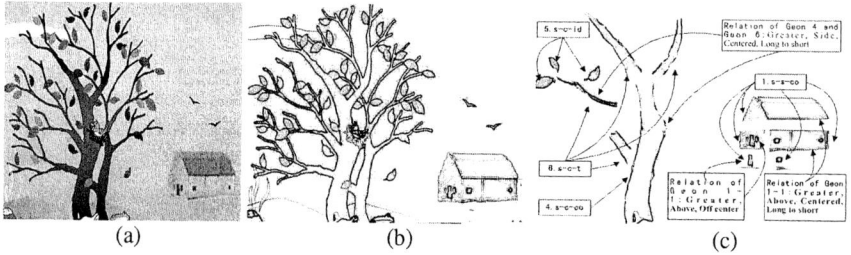

(a)   (b)   (c)

Fig. 4. shows some examples of image object recognition: (a) Original input image; (b) Result of image Preprocessing; (c) Geons and relations between Geons (the label of Geons is same as Table 1).

5 Discussions

In this paper, a general solution for image object recognition is proposed and some simple examples are given to illustrate the feasibility and validity of this solution. The intrinsic logic of this model is that computer vision problem is surely similar to human vision problem in theory, because only from the perspective of optical imaging the eye and camera have the same mechanism. So we argue that if machine can't catch up with human eye in the field of image understanding, the only reason must be that computers have not take full of use of mechanisms of human eye, in that computer has outweighed human in the aspects of information collection, data storage, computation capability and so on. Since the main tasks of psychology of visual cognition are to find human seeing what, seeing where, how to see, it is a considerably natural and reasonable selection to constitute our solution on the base of cognition science.

Though cognition science enlightens a novel, hopeful and exciting way for image object recognition, we have to overcome many difficulties. Open issues of our solution mainly include: (a) effectively defining a structural set of knowledge; (b) translating qualitative descriptions of psychology (such as the laws of Gestalt) into quantitative mathematical expression and machine implementation; (c) effectively applying technologies of human-machine interaction; and (d) recognizing Geons in the complex situations such as occlusions, and image degradation. We will continue to work in the field. After all, our ultimate aim is to make machines automatically recognize image objects like human.

Acknowledgement

This work was supported by the Beijing Jiaotong University Research Project under Grant No. 2004SM013.

References

1. Jia Li et al, Automatic Linguistic Indexing of Pictures by a Statistical Modeling Approach, IEEE Trans. on Pattern Analysis and Machine Intelligence, vol. 25, no. 10, pp 14, 2003.
2. Irvin Rock, Perception, Scientific American Books, Inc, 1984.
3. Marr. D, Vision, Freeman, 1982.
4. Wu Aimin, et al., Method for Qualitatively Evaluating CVIR Algorithms Based on Human Similarity Judgments, Proceedings of 7th ICSP, pp. 910-913, 2004.
5. Laurent Itti, et al, A Model of Saliency-based Visual Attention for Rapid Scene Analysis, IEEE Transactions PAMI, Vol. 20, No. 11, 1998.
6. Vidhya Navalpakkam and Laurent Itti, A Goal Oriented Attention Guidance Model, Lecture Notes in Computer Science, Vol. 2525, pp. 453-461, 2002.
7. Oliva, A. Torralba, M. Castelhano, and J. Henderson, Top-down Control of Visual Attention in Object Detection, in: International Conference on Image Processing, 2003.
8. Dirk Walther et al, Selective Visual Attention Enables Learning and Recognition of Multiple Objects in Cluttered Scenes, accepted by Journal of CVIU 2005.
9. Gregory. R.L. , Eye and Brain, Princeton University Press,1997.
10. S. Dickinson et al, Panel Report: The Potential of Geons for Generic 3-D Object Recognition, Image and Vision Computing, Vol. 15, No. 4, pp 277—292, 1997.
11. Matthew Brand, Physics Based on Visual Understanding, Computer Vision And Image Understanding, Vol. 65, No, 2, February, pp. 192-205, 1997.
12. Zu Whan Kim et al, An Automatic Description of Complex Building from Multiple Images, J. of Computer Vision and Image Understanding 96, pp 60–95, 2004
13. David A. Forsyth et al, Computer Vision: A modern Approach, Prentice Hall, Inc, 2003.
14. Biederman,I., Recognition-by-Components: A Theory of Human Image Understanding, Psychological Review, 94, 115-47, 1987.
15. Hummel, J. E. et al, Dynamic Binding in a Neural Network for Shape Recognition. Psychological Review, 99, 480-517, 1992.
16. Quang-Loc Nguyen et al, Representing 3-D Objects in Range Images Using Geons,J. of Computer Vision and Image Understanding, Vol. 63, No. 1, January, pp. 158–168,1996.
17. Sudeep Sarkar et al, Using Perceptual Inference Networks to Manage Vision Processes, Computer Vision and Image Understanding, Vol.62. No.1, July, pp. 27-46, 1995.

18. Treisman, A. and Gelade, G., A Feature Integration Theory of Attention, Cognitive Psychology, 12, 97-136, 1980.
19. WAN Hua-Lin and Morshed U. Chowdhury, Image Semantic Classification by Using SVM, Journal of Software, Vol.14, No.11, 1891-1899, 2003.
20. K. Barnard, Pinar Duygulub and David Forsyth, Recognition as Translating Images into Text. In Internet Imaging IV, Santa Clara, CA, USA, January, 2003.
21. Pentland, Perceptual Organization and The Representation of Natural Form, Artificial Intelligence, 28:293-331, 1986.
22. Biederman, I., Visual Object Recognition. In Visual Cognition, 2nd edition, Volume 2, MIT Press. Chapter 4, pp. 121-165, 1995.
23. Weiwei Xing, et al, Superquadric-based Geons Recognition Utilizing Support Vector Machine, Proceedings of 7th ICSP, pp. 1264-1267, 2004.
24. K. Koffka, Principle of Gestalt Psychology, Harcourt Brace Jovanovich Company, 1935.
25. Roger J. Watt and William A. Phillips, the Function of Dynamic Grouping, Trends in Cognition Science, Vol. 4, No. 12, December, 2000.
26. Rui Y and Hunag T S, Relevance Feedback Technique in Image Retrivel. In: Principles of Visual Information Retrieval. Lew M S, ed. Springer, Ch.9, 219-258, 2001.
27. Zhang Ming, Psychology of Visual Cognition, East China Normal University Press, 1999.

Cognition Theory Motivated Image Semantics and Image Language

Aimin Wu[1,2], De Xu[1], Xu Yang[1], and Jianhui Zheng[2]

[1] Dept. of Computer Science & Technology,
Beijing Jiaotong Univ., Beijing, China 100044
[2] Dongying Vocational College, Shandong, China 257091
wuaimin@sohu.com, xd@computer.njtu.edu.cn

Abstract. Much evidence from visual psychology suggests that images can be looked as a kind of language, by which image semantics can be unambiguously expressed. In this paper, we discuss the primitives and grammar of image language based on cognition theory. Hence image understanding can surely be manipulated in the same way as language analysis.

Keywords: Image semantics, Visual cognition theory, Image Language.

1 Introduction

Automatic semantics extraction of image is still a highly challenging issue. The most important cause is that computer vision community is lack of effective collaboration with visual psychologists [1]. Computer vision problem in theory is similar to the human vision problem, both of which are the process of discovering from images what is present in the world, and where it is [2, pp1][3, pp19-20]. The human vision system is the best and the most general. The visual cognition theory can guide computer to reach the performance of human vision system, because its main tasks are to find human seeing what, seeing where, how to see [4-7][9][12].

2 Definition of Image Semantics

Psychologists have given many conceptions such as Behavioral and Geographical Environment, Psychological and Physical Field, Psychological and Physical Environment, and Mental and Physical Fact, all of which are used to suggest that the same thing has different meanings [5-7]. According to these psychological conclusions, we argue that every thing in real world has three different semantics: **direct semantics, behavioral semantics, and associated semantics,** whose relations are shown in Fig.1.

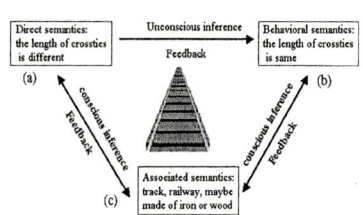

Fig. 1. Shows different semantics in a same image, (a) Direct semantics, (b) Behavioral semantics, (c) Associated semantics

The direct semantics of image refers to the values that are measured by different meters such as photometer, speedometer, retinas of animals and so on. The direct semantics directly reflects the real world.

The behavioral semantics of image refers to the values that are inferred on the basis of the direct semantics of image and may be affected by associated semanticsThe process to get behavioral semantics is called **unconscious inference** by psychologist, for only **innate knowledge** is used [4, pp 2]. The behavioral semantics of image can results in immediate and unconscious reaction to the real world.

The associated semantics refers to any information that can be induced from the direct semantics and (or) behavioral semantics. The process is constrained by predefined structured knowledge. We call such process **conscious inference** because **learned knowledge** can be employed, which is the most important difference from the unconscious inference.

3 Image Is a Language

Any semantics has to be expressed by a certain kind of language. We argue that the image itself can be looked as a kind of language, which can express image semantics.

Different nation has different language, all of which is almost equivalent, because all of them reflect the same objective world. Image can been seen as a language because it can uniquely express the objective world. In this sense, it is almost equivalent to other language. In fact, many languages are originated from the images, such as Chinese, and Japanese. Fig.2, Fig.3, Fig.4 show the evolution process of Chinese character MOUNTAIN, WATER, FIGHTE from (a) natural images to (b) ancient Chinese version, then to (c) modern Chinese version [8, pp 49,43,29].

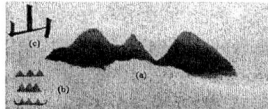
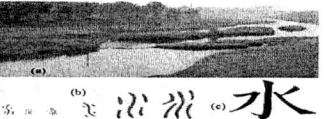
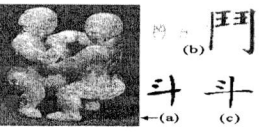

Fig. 2. Chinese "Mountain"   Fig. 3. Chinese "Water"   Fig. 4. Chinese "Fight"

4 The Primitive of Image Language

Any language can be decomposed into a relatively small set of primitives [9]. The image language can be decomposed into primitives. We choose three level primitives: Marr's prime sketch and 2.5D sketch, Generalized Cylinder, and Kobus's associated text. They are usually used to represent shape, but after necessary modification, they can also be used to represent other visual information such as color, texture, position etc.

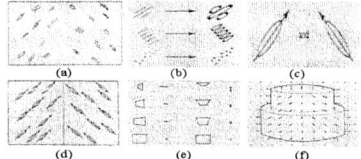

Fig. 5. Marr's primitive: (a) The original image; (b) The raw primal sketch denoted by filled circles; (c) The oriented tokens; (d) Grouping by the difference; (e)&(f) 2.5D sketch.

4.1 Marr's Prime Sketch and 2.5D Sketch [2, pp 1-264]

In general, the aim of Marr's primitives is to develop useful description of shapes and surface that form the image. These primitives have three levels from simple to complex: ***intensity value, primal sketch***, and ***2.5D sketch*** (Fig.5).

4.2 Generalized Cylinder

One is Biederman's Geons, which are some basic elements of shape that may be used for human vision. Object recognition can be achieved directly from these Geons under the constraints of predefined knowledge [4, pp 79-81][9]. Fig.6 shows Biederman's basic idea.

Another Generalized Cylinder primitive and its application see [10], which can be used to quantitatively define objects other than qualitative Biederman's Geons.

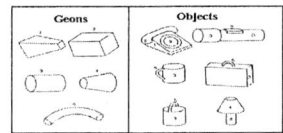

Fig. 6. Biederman's Geons —unit object shapes—fitted to some common objects

4.3 Kobus's Associated Text [11]

To eliminate image ambiguities, it is best way to represent image semantics using structural associated text from structural natural languages. Fig.7 are some examples calculated by Kobus Barnard. Once the image has been translated into structured text, they can be further analyzed, for it is relatively easier to automatically process text.

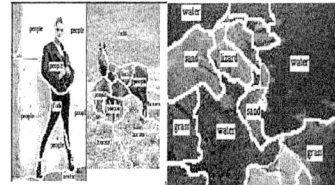

Fig. 7. Some examples of associated text from Kobus Barnard

5 The Grammar of Image Language

5.1 The Laws of Gestalt

(1) The grouping principle [3, pp 113-211][5, pp106-176]. It is used to achieve spontaneous grouping of elements. The psychologists hold that stimuli or components, which can construct the best, the simple, the stable object, will be classified into a group. The grouping principle can be further summarized into five principles: **(a)** the principle of proximity, in Fig.8-a, we tend to look these separate dots as rows and columns; **(b)** the principle of

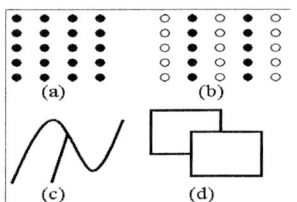

Fig. 8. Illustrates some examples of the grouping principle

similarity, in Fig.8-b, we tend to group together the spots that are similar to another, so we only see the column; **(c)** the principle of good continuation, in Fig.8-c, we tend to group together the parts of the curved, excluding the straight line; **(d)** the principle of closure tendency, in Fig.8-d, we tend to see these shapes as two rectangles, one behind the other, although we could just as well as see a rectangle and L in the same plane;

and (**e**) **the principle of common fate**: the tendency to group those units that move together in the same direction and at the same speed.

(**2**) **Figure-ground segregation principle [5, pp 177-210].** We often simple think the parts that stand out from the image as figure and those that recede into the image as background. In ambiguous patterns, **smaller** region (Fig.9-(a)), **symmetrical** regions (Fig.9-(b)), **vertically** or **horizontally** oriented regions tend to be perceived as figures. If two regions are so segregated that one encloses the other, the **enclosed** one will become figure, the enclosing the ground. The common **borders** between the figure and the ground are often assigned to figure. In generally, the ground is **simpler** than the figure.

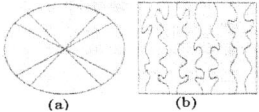

Fig. 9. Smaller regions (a) regions (b) tend to be perceived as figure

5.2 The Location Relation Among the Primitives

To visual perception, the whole is more than the sum of its parts [5, pp 176]. The same primitives with different relations may construct different objects. For example, an arc side-connected to a cylinder can yield a cup (Fig.10-a), but an arc is connected to the top of cylinder to produce a pail (Fig.10-b). Whether a component is attached to a long or short surface can also affect classification, as with the arc producing an attaché case (Fig.10-c) or a strongbox (Fig.10-d) [9].

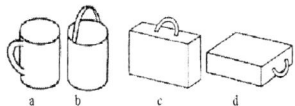

Fig. 10. Different arrangements of the same primitive can produce different objects

5.3 Constancy Transformations

The retinal image is different from the objects' outline, but the object looks the same to us, which is called **Constancy**. There mainly includes color constancy, brightness constancy, lightness constancy, motion constancy, size constancy (Fig.11-a), shape constancy, and location constancy [5, pp 211-264][3, pp 15-52]. Simultaneous contrast phenomena [2, pp. 259-261] (Fig.11-b) suggests that human eyes don't detect absolute energy of brightness, color etc, but their difference that is directly proportional to the background energy. Most of constancy can be implemented by simultaneous contrast transformation.

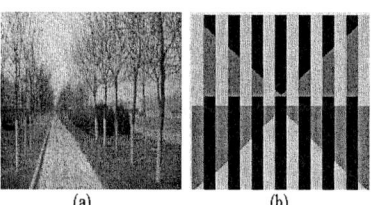

Fig. 11. (a) an example of size constancy that all trees are perceived same high in spite of different in the image; (b) simultaneous contrast: though the value of lightness are complete same, they look much different.

5.4 Topological Transformations

The most important topological transformation is Local homotopy rule that we tend to accept an original image and its transformed image as identical, if the image is undergone a local homotopy transformation within its tolerance space. For example, a face with mouth whether opening or closing is regarded as identical one. The same is true for homeomorphism rule, homeomorphism rule in cluster, null-homotopy rule in cluster, object superiority effect and so on [12, pp. 100-159].

6 Discussion and Future Works

Guided by cognition theory, we carefully discuss the definition of image semantics, image language and its grammar and primitive. Much evidence indicates that many languages are indeed originated from the images. Hence image understanding can surely be manipulated in the same way as language analysis, which is a relatively simpler and more mature field. We will continue to work in the field. In future work, semantics model and image language proposed in this paper will be used for automatic image semantics recognition.

Acknowledgement

This work was supported by Beijing Jiaotong Univ. Research Project (2004SM013).

References

1. Wu Aimin, et al, Method for Qualitatively Evaluating CVIR Algorithms Based on Human Similarity Judgments, Proceedings of 7th ICSP, pp. 910-913, 2004.
2. Marr. D. Vision, Freeman, 1982.
3. Irvin Rock, Perception, Scientific American Books, Inc, 1984.
4. Gregory R. L, Eye and Brain, Princeton University Press, 1997.
5. K. Koffka, Principle of Gestalt Psychology, Harcourt Brace Company, 1935.
6. K. Lewin, Principles of Topological Psychology, Hill Book Company Inc, 1936.
7. J.J. Gibson, The Ecological Approach to Visual perception, Houghton Mifflin, 1979.
8. C. Lindqvist, China: Empire of the Written Symbol, Shandong pictorial press, 1998.
9. Biederman, I., Recognition-by-Components: A Theory of Human Image Understanding, Psychological Review, vol. 94, 115-47,1987.
10. A. Pentland, Perceptual Organization and The Representation of Natural Form, Artificial Intelligence, 28:293-331, 1986.
11. K. Barnard et al, A Method for Comparing Content Based Image Retrieval Methods, In Internet Imaging IV, Santa Clara, CA, USA, 2003.
12. Zhang M., Psychology of Visual Cognition, East China Normal University Press, 1991.

Neuro-Fuzzy Inference System to Learn Expert Decision: Between Performance and Intelligibility

Laurence Cornez[1], Manuel Samuelides[2], and Jean-Denis Muller[3]

[1] ONERA DTIM, 2 avenue Edouard Belin, 31055 Toulouse cedex 4, France
cornez@cert.fr
[2] SUPAERO, 10 avenue Edouard BELIN, 31055 Toulouse cedex 4, France
manuel.samuelides@supaero.fr
[3] CEA DAM/DASE/LDG, BP12, 91680 Bruyres-le-Chtel, France
muller@dase.bruyeres.cea.fr

Abstract. We present a discrimation method for seismic events. One event is described by high level features. Since these variables are both quantitative and qualitative, we develop a processing line, on the crossroad of statistics ("Mixtures of Experts") and Artificial Intelligence ("Fuzzy Inference System"). It can be viewed as an original extension of Radial Basis Function Networks. The method provides an efficient trade-off between high performance and intelligibility. We propose also a graphical presentation of the model satisfying the experts' requirements for intelligibility.

1 Introduction

In the context of the CTBT ("Comprehensive Nuclear Test-Ban Treaty") the capacity to discriminate nuclear explosions from other seismic events becomes a major challenge. The CTBT provides a global verification system which will eventually include a network of 321 measurement stations worldwide with a various collection of sensors and an international data center. In the next years, the flow of data to process will be increased by an order of magnitude. Thus, there is a need for automatic methods of classifying seismic events. However, the final classification decision has to be controlled by the expert. The automatic methods will process the obvious cases and will present to the expert the more contentious cases. So the expert needs more than the final decision which is provided by the automatic system: he wants to understand the way this decision is obtained. Therefore this research was initiated by the LDG (Geophysics Laboratory) of the French Atomic Energy Agency (CEA) to select discrimination methodology with the two following joint criteria: High performance and Intelligibility.

At this stage of research, we focus the study on the seismic events recorded in France. This database gathers 13909 events which occurred between 1997 and 2003 inclusively. There are three types of events: *earthquakes*, *rock bursts* and *mine explosions*. To classify the events, we have five high level inputs that have been extracted from the seismic measurements: *magnitude* (quantitative variable ranging from 0.7 to 6.0), *latitude* (quantitative variable ranging from 42 to 51), *longitude* (quantitative variable ranging from -5 to 9), *hour* (circular variable ranging from 0 to 24) and *date*. Actually, *date* is a qualitative variable with 3 modalities: *Working day, Saturday* and *Sunday and Bank holidays*. Processing specifically this qualitative variable will contribute in this study to improve the results of the classifier. Preliminary statistical studies clearly indicate that all these features significantly contribute to the expert decision. Among the various existing discrimination methodology in the statistical literature, we selected first order Sugeno Neuro-Fuzzy inference systems for their properties:

- They are able to model intelligible and flexible decision rules
- They can be combined to build complex decision rules
- It is possible to improve them using expert knowledge through a learning procedure

This methodology allows us to build from the database a high-performance classification system which is tractable by human expert. This intelligibility issue is the key point of this work.

This paper is organized as follows. We recall first in section 2 the connection between a fuzzy inference system and an expert mixture classifier. Then we present a full data processing line which consists in different components: clustering, complementary estimation, supervised EM algorithm. They are presented successively in section 3. We show how results are improving after each step. Then, section 4 focuses on the second main point : the intelligibility of our system by human experts. We face this crucial challenge using an appropriate graphical interface. At last, we present the future orientations of this research which will focus on the interaction between the expert and the automatic system.

2 Fuzzy Inference System

Fuzzy logic became popular at the "golden age" of expert systems. Zadeh ([15]), Dubois and Prade's ([4]) pioneering works about fuzzy logic proposed a new approach to model uncertainty in inference systems. When empirical knowledge is available jointly with expert knowledge, learning is currently used to fuse them. Besides, learning process was not easy to embed into fuzzy logic system. Neuro-fuzzy systems ([8]) were designed to take avantage both of the modelling power of fuzzy inference systems and of the learning capacities of neural networks. They were used to model and to control empirical dynamical systems ([13], [14]). They were also introduced in classification problems (see for instance Frayman, Wang [5], [6]). We shall follow this approach. First we recall the mathematical model of a Sugeno rule ([12], [11]) and of a Sugeno classifier.

A rule-based classification consists in a set of rules. For instance in simple systems, each rule may be defined by its antecedent (the set of input data that check the rule) and its consequent: the classification decision that attributes a class to the elements that check the rule. A fuzzy Sugeno's rule k is defined by

- a function μ_k which is defined on the input space and which takes its values in [0,1] (the membership degree)
- a positive real ρ_k (the weight of the rule)
- a unit vector z_k in the output space (the consequent of the rule)

The membership degree of first order Sugeno's rule is the product of individual feature membership degrees (fuzzy AND). The weight of the rule is used to build the final decision by a weighted sum of overlapping rules (fuzzy agregation). The Sugeno classifier can be defined by

$$Z(x) = \frac{\sum_{k=1}^{NbRules} \rho_k \mu_k(x) z_k}{\sum_{k=1}^{NbRules} \rho_k \mu_k(x)} . \quad (1)$$

The mixture of experts classifier may be viewed as a particular case of the previous model where

- each μ_k is a probability density (with respect to a basic measure, as Lebesgue measure for quantitative variable or countable measure for qualitative one)
- the vector (ρ_k) is a stochastic vector
- each z_k is a stochastic vector

The output $Z(x)$ for a particular point of the input space is a stochastic vector: each component can be viewed as a probability of class membership. Note that a deterministic classification may be finally obtained through a max mechanism. Another mechanism can be proposed, which allows rejection when the determination of the max is not robust enough.

With a convenient normalization, it is possible to establish equivalence between Sugeno's rule and hidden variables in a stochastic framework and between Sugeno's classifier and Bayes classifier. Moreover, in that case, the classifier has a Bayes interpretation. When the membership functions are Gaussian, the classifier amounts to a classical radial basis function network. However, the processing of the qualitative variable deserves a special attention. So, we propose an original extension of this model to take into account the specific hybrid structure of our input space.

3 Procedure and Results

For estimations, we build balanced data bases (1025 events of each class) and then we separate them to make 5-fold cross-validation balanced data bases (860 events of each class in the learning data bases and 215 in the test data bases).

In this section, we expose the three steps to set the parameters of our Fuzzy Inference System of Sugeno. First, we have to estimate probability densities $(\mu_k)_{k=1\cdots NbRules}$. Among many solutions found in the scientific literature and guided by previous study made in LDG (Fabien Gravot's training [7]), we opt for Chiu's algorithm [3]. This algorithm determine cluster number ($NbClusters$) and gaussian parameters of each cluster. Each cluster found by the algorithm represente a rule used consciously or not by experts. Chiu's algorithm operate on input space spanded only by quantitative variables. Thus, to take account of our hybrid structure of input space, we add the qualitative variable by estimating modality probabilities for each cluster. At last, to improve this Fuzzy Inference System, we used the Expectation-Maximization (EM) algorithm to set parameters.

Note that our notations are :

- $X = (X_i)_{i=1...N}$ is a population sample.
- $X_i = (X_i^{QT}|X_i^{QL})$ where $X_i^{QT} = (X_{i,l}^{QT})_{l=1...(NbInputs-1)}$ is a vector of the quantitative variables and X_i^{QL} is the qualitative variable,
- $\mathbb{1}$ is the indicator function.

3.1 First Step: Clustering

Implementation. To implement this first stage, we used unsupervised learning algorithms to estimate each class localization density. Many clustering methods can be used to initialize clusters. In 1994, Yager and Filev proposed an original method based on potential computation meaning point density. Chiu [3] improve this method using data point repartition rather than grid (potential computation is faster). Alternative algorithms are fuzzy k-means ([2]), neural gas ([10])... They improve cluster localizations iteratively and the cluster number is fixed by users.

On the basis of Fabien Gravot's preliminary results, we choose Chiu's algorithm. For each class separately, we operate the algorithm of clusters research.

Algorithm parameters are:

- r_α is the radius defining a neighborhood (a positive vector). We put the value at 1.25 for each dimension of the input space.
- r_β is the radius defining a neighborhood in potential reduction ($r_\beta = 1.5\, r_\alpha$).
- $\bar{\varepsilon}$ first threshold for acceptance (its value is 0.5) and $\underline{\varepsilon}$ second threshold for acceptance (its value is 0.15). These thresholds are used to define criteria for enough potential and proximity between clusters computed.

Briefly, the Chiu's algorithm for cluster research is:

1. Compute the potential of each data point

$$P_i = \sum_{j=1}^{N} \exp\left(-4 \sum_{l=1}^{NbInputs} \frac{d(X_{i,l}, X_{j,l})^2}{r_{\alpha,l}^2}\right). \qquad (2)$$

Note that the circular variable (*hour*) is treated with the circular distance. So d can be euclidian distance or circular distance according to the input. When potential are computed, we initialize the first cluster C_1 to the point with maximum potential.

2. Look for other clusters ("while" loop)
   (a) X_k point with maximum potential (P_k)
   (b) If P_k is high enough and X_k is distant enough of other clusters then, X_k is taken as a new cluster and the potential of the other points is reduced as follows:

$$P_i = P_i - P_k \times \exp\left(-4 \sum_{l=1}^{NbInputs} \frac{d(X_{i,l}, X_{k,l})^2}{r_{\beta,l}^2}\right) \quad (3)$$

   and we return to the begining to find a new cluster.
   Else, X_k is not accepted, its potential is replaced by 0 and we return to the begining to find another cluster.
   (c) If no other cluster is found we stop.

At this stage, we initialize standard deviations (gaussian widths) by:

$$\sigma_l = \left[(r_\alpha)_l \frac{max_l - min_l}{\sqrt{8}}\right] \quad \text{where } l = 1 \cdots NbInputs, \quad (4)$$

min_l (resp. max_l) means the lth input minimum (resp. maximum). So, standard deviations are equal for each dimension of the input space for each class. This is not optimal.

Experimentation. To compute performances using (1), we initialize weights equal into the same class. For one class, the sum of weights equals to $\frac{1}{3}$ (because we have 3 classes). Output rule values z_k are deterministic : the rule conclude to one class certainly (the one associated to current cluster - we operate on each class separately-).

First, according to previous study in the LDG, we treat all the variables as quantitative variables. Later, in agreement with expert judgements and statistical theory, we operate algorithm only on the input space reduced to our quantitative variables.

Table 1 sums up performances (means and standard errors of good classification rate) computed, on 5-fold cross-validation data bases, on the complete input space (variable *date* treated as quantitative variable, see in the table "all quant.") and, on the input space without this qualitative variable (see "only quant.").

Table 1. Results after clustering: good classification rate

Method	learning data (%)	test data (%)	clusters number
Subclust (all quant.)	85.4031 ± 0.8674	84.8372 ± 1.4725	32;36;35;38;37
Subclust (only quant.)	85.6589 ± 0.8523	84.6512 ± 1.7575	25;26;28;28;29

Results in terms of good classification are similar but less clusters are found when the input space contains only the quantitative variables. So, to be more coherent and more efficient (less clusters) we keep, for the other steps, clusters obtained on the reduced input space.

3.2 Second Step: Qualitive Variable Probability Estimations

Implementation. To take into account the qualitative variable *date*, we estimate for each cluster C_k (which class is c), its probability distribution as:

$$p_k(m|X) = \frac{\sum_{X_i \in c} \mathbb{1}_{X_i^{QL}=m} A_{i,k}}{\sum_{X_i \in c} A_{i,k}} \tag{5}$$

and the corresponding model is ($NbModes$ is the modality number of *date*):

$$Z(X_i) = \sum_{k=1}^{NbClusters} \frac{z_k \, \rho_k \, A_{i,k} \, p_k(X_i^{QL})}{\sum_{k=1}^{NbClusters} \rho_k \, A_{i,k} \, p_k(X_i^{QL})} \tag{6}$$

where :

$$A_{i,k} = \frac{\exp\left(-\frac{1}{2} \sum_{l=1}^{NbInputs} \left(\frac{d(X_{i,l}^{QT}, C_{k,l})}{\sigma_{k,l}}\right)^2\right)}{\sqrt{2\pi}^{NbInputs} \prod_{l=1}^{NbInputs} \sigma_{k,l}} \tag{7}$$

and

$$p_k(X_i^{QL}) = \sum_{m=1}^{NbModes} p_k(m|X) \mathbb{1}_{X_i^{QL}=m} . \tag{8}$$

Experimentation. Table 2 sums up performances (means and standard errors of good classification rate) computed, on 5-fold cross-validation data bases, with model mixing quantitative and qualitative variables. Weights and output rule values are the same as the previous step. Good classification rates are poorly improved.

Table 2. Model mixing quantitative and qualitative variables: good classification rate

Method	learning data (%)	test data (%)
Subclust only quant.+ quali.	86.9922 ± 0.4838	86.0155 ± 0.9133

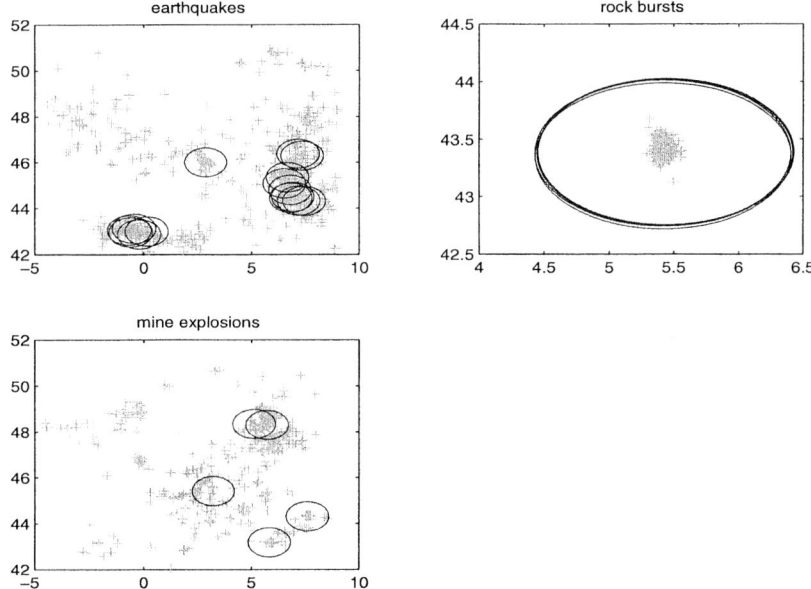

Fig. 1. Geographical cluster positions and their influence (widths=σ) for each class after clustering (best result in test among the 5-fold cross-validation data bases). The learning data points are marked by cyan crosses.

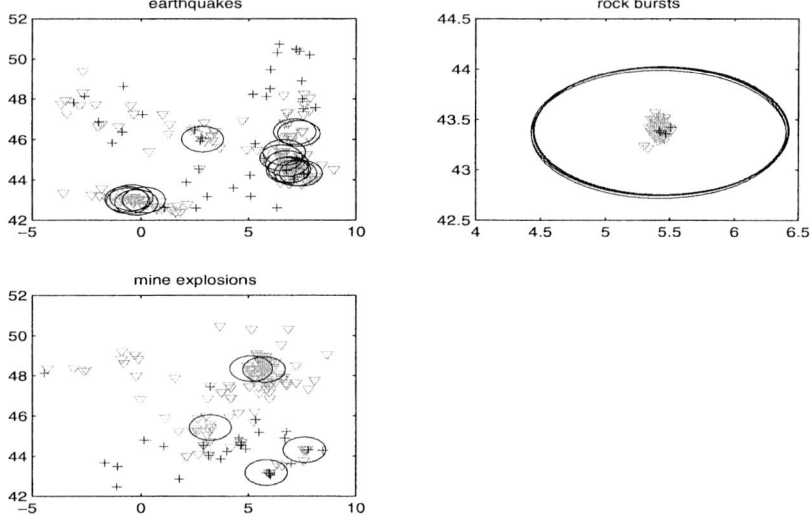

Fig. 2. Geographical positions and classification of test points (best result in test among the 5-fold cross-validation data bases). Green triangles are ill classified test points and red crosses marked well classified test points. Clusters of previous figure are recalled here.

Figure 1 presents the best model with 4 quantitative variables and the qualitative variable (best result in test among the 5 fold cross-validation data bases). We show geographical cluster localizations along longitude-latitude dimension and the learning data (cyan crosses). Clusters are marked by an ellipse which is shaped according to longitude and latitude standard deviation values computed with (4). The best model in test presents 25 clusters (15 for the first class, 5 for the second and 5 also for the third).

Figure 2 presents the test data (red crosses are ill classed points and green triangles are well classed points). The clusters (ellipses) are printed in blue here for remember.

This projection allows us to conclude that clusters have not optimal overlapping for the input space. To improve them, we are going to present EM algorithm implementation.

3.3 Third Step: Expectation-Maximization Algorithm to Improve Clusters Weight, Localizations and Widths

Implementation. To give better parameters values, we implement EM algorithm for mixture of experts. A similar case study is done in [1] and more general set-up is presented in [9]. Parameter vector, for the kth cluster is denoted θ_k. It contains cluster weight ρ_k, localizations $C_k = (C_{k,l})_{l=1..(NbInputs-1)}$ and widths $\sigma_k = (\sigma_{k,l})_{l=1..(NbInputs-1)}$. Set of parameter vectors is denoted θ. We denote $\mathcal{X}$ the distribution associated to the input space.

Briefly, EM algorithm is a "general method of finding the maximum-likelihood estimate of the parameters of an underlying distribution from a given data set when the data is incomplete or has missing values" ([1]). In fact, when we have mixture of experts, it's difficult to derivate likelihood expression because of we have to derivate the logarithm of a sum on the experts (i.e. clusters for us).

$$\log(\mathcal{L}(\theta|\mathcal{X})) = \sum_{i=1}^{N} \log \left(\sum_{k=1}^{NbClusters} \rho_k \mathrm{p}_k(X_i|\theta_k) \right) \qquad (9)$$

where $\mathrm{p}_k(X_i|\theta_k)$ is the density function parametrized by θ_k. In our case :

$$\mathrm{p}_k(X_i|\theta_k) = A_{i,k} \; \mathrm{p}_k(X_i^{QL}) \; . \qquad (10)$$

To simplify this expression, we complete the data base by a set of hidden variables Y (which distribution is denoted $\mathcal{Y}$). These hidden variables are able to associate each data point to the cluster that generated it. So y_i is the value of the hidden variable Y for the ith data point and its value ranges from 1 to $NbClusters$. Thus, the log-likelihood with this hidden variable becomes:

$$\log(\mathcal{L}(\theta|\mathcal{X},\mathcal{Y})) = \sum_{i=1}^{N} \log \left(\rho_{y_i} \mathrm{p}_{y_i}(X_i|\theta_{y_i}) \right) \; . \qquad (11)$$

This algorithm sets parameters iteratively and, after each loop, the target function (log-likelihood estimator) is improved ([9]). We operate EM algorithm on each class separately so $N = N_c$ takes number of points in the cth class.

Briefly, the implementation has two steps:

- Expectation step: computation of expected value of the complete-data log-likelihood (see [1] for details)

$$Q(\theta) = \sum_{k=1}^{NbClusters} \sum_{i=1}^{N_c} \log(\rho_k) p(k|X_i, \theta) + \sum_{k=1}^{NbClusters} \sum_{i=1}^{N_c} \log(p_k(X_i|C_k, \sigma_k)) p(k|X_i, \theta). \quad (12)$$

- Maximization step: new parameter values (optimization)

$$\rho_k^{new} = \frac{1}{N_c} \sum_{i=1}^{N_c} p(k|X_i, \theta) ,$$

$$C_{k,l}^{new} = \sum_{i=1}^{N_c} X_{i,l} \frac{p(k|X_i, \theta)}{\sum_{i=1}^{N_c} p(k|X_i, \theta)} , \quad (13)$$

$$\sigma_{k,l}^{new} = \sum_{i=1}^{N_c} d(X_{i,l}, C_{k,l}^{new})^2 \frac{p(k|X_i, \theta)}{\sum_{i=1}^{N_c} p(k|X_i, \theta)} .$$

Experimentation. Here, we stop the algorithm after 50 iterations. We keep these values to give new estimations for qualitative variable modality probabilities for each cluster. Table 3 sums up performances (means and standard errors of good classification rate) computed on 5-fold cross-validation data bases.

Table 3. Model mixing quantitative et qualitative variables after EM algorithm

Method	learning data (%)	test data (%)
Subclust only quant. + quali. + EM	93.6744 ± 0.5969	93.1163 ± 1.6575

Figure 3 presents geographical clusters position for each class by an ellipse (which is shaped according to longitude and latitude standard deviation values found after EM) in the best case (95.1938 % of well classed in test). Figure contains also cyan crosses marking learning data point. Notice that data bases giving best results after clustering are not the same that give best results after EM algorithm.

Figure 4 presents geographical position of miss classified points in test (red crosses) and the well classified (green triangles). The clusters are recalled by blue ellipses.

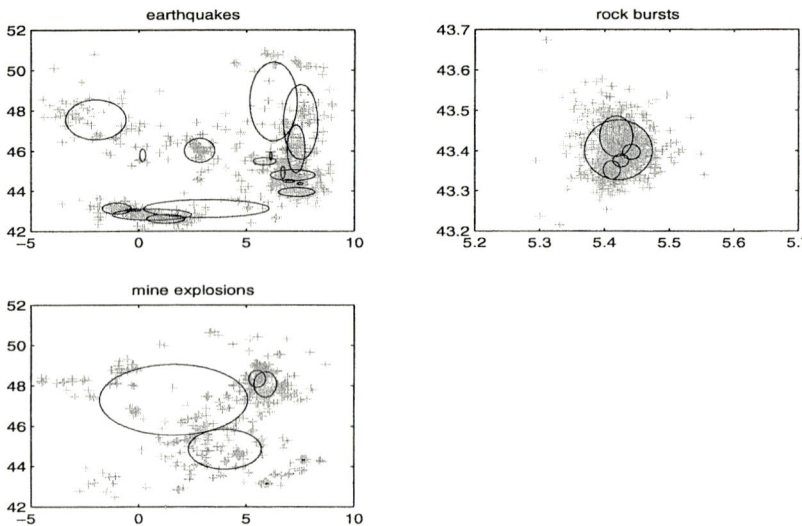

Fig. 3. Geographical clusters positions and their influence (widths=σ) for each class after EM algorithm (best result in test among the 5-fold cross-validation data bases). Learning data points are marked by cyan crosses.

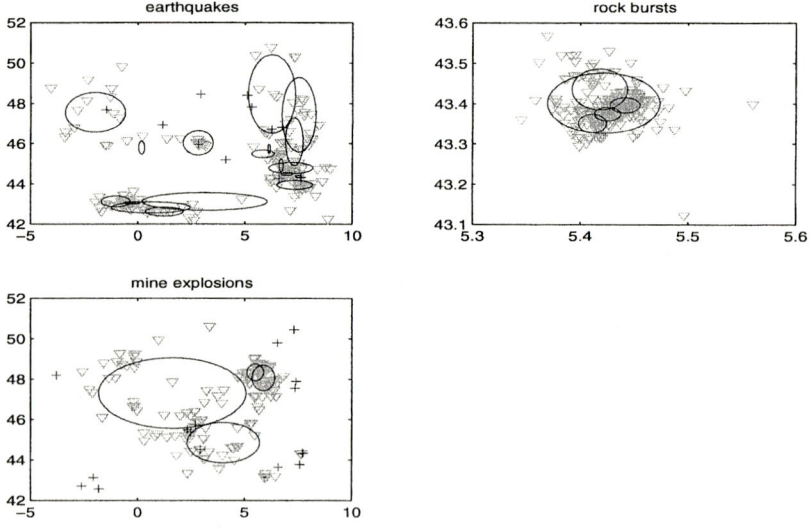

Fig. 4. Geographical positions and classification of test points (best result in test among the 5-fold cross-validation data bases). Green triangles are ill classified test points and red crosses marked well classified test points. Clusters of previous figure are recalled here.

When the cluster positioning is operated by EM, then a better overlapping is observed in the input space. Less miss classified points are found and their positions remain reasonnable.

3.4 Conclusion

With this three steps, we improve greatly good classification rates. Table 4 sums up improvement.

Table 4. Summary of the good classification rates found at each step

Method	learning data (%)	test data (%)
Subclust all quant.	85.4031 ± 0.8674	84.8372 ± 1.4725
Subclust only quant.	85.6589 ± 0.8523	84.6512 ± 1.7575
Subclust only quant. + quali.	86.9922 ± 0.4838	86.0155 ± 0.9133
Subclust only quant. + quali. + EM	93.6744 ± 0.5969	93.1163 ± 1.6575

So, we have to respect the complexity of the input space. Each input have its importance and shoud appear in the model correctly to give all its information. Evenif, Chiu's algorithm of cluster research give good classification rates, EM algorithm sems to be very efficient step to improve them.

Our model combines several methods, it gives an original processing in input space mixing both quantitative and qualitative variables. Moreover, its presents another advantage. Indeed, it gives intelligible interface to the experts of LDG, the users of our model. So, computed results and final individual classification decisions have to be understood by them. The next section develops this aspect.

4 Graphical Presentation

In this section, we present a graphical interpretation of our Sugeno's rule system ([8]). It is designed in order to be submitted to the expert. To improve legibility, we present only the 10 main rules of the system (over the 29 that the best model after EM algorithm used). Figure 5 shows rule system activation with data point Y. In fact, we want a graphical interpretation of the suitability between the data point Y and each rule. Each rule is presented along a line. The first column, presents weight, the next one, the circular gaussian membership function for the variable *hour*, the next three columns the gaussian membership functions for the variables *latitude*, *longitude* and *magnitude*, the next one presents modality probabilities for the qualitative variable *date*. In each graphic, vertical green line fixes the input value for the data point. More this line is close to the center of the gaussian function, more the data point is understanding by this rule (and the red area is large). The next three columns ("eq", "rb", "me") present output rule values which represent the membership degree for each class (at this stage of study only 0 or 1). The last one shows the activation rate associated to data point

Y. It is a graphical level giving weight multiplied by rule suitability with the data point Y (product of the area by the probability of the modality corresponding to Y).

The system issue that is associated to this data point is presented on the last three graphics (pink lines at the bottom right of figure 5). Line values are done by (1). To this input data point Y, the system concludes to an earthquake ($z1$ takes the tallest value).

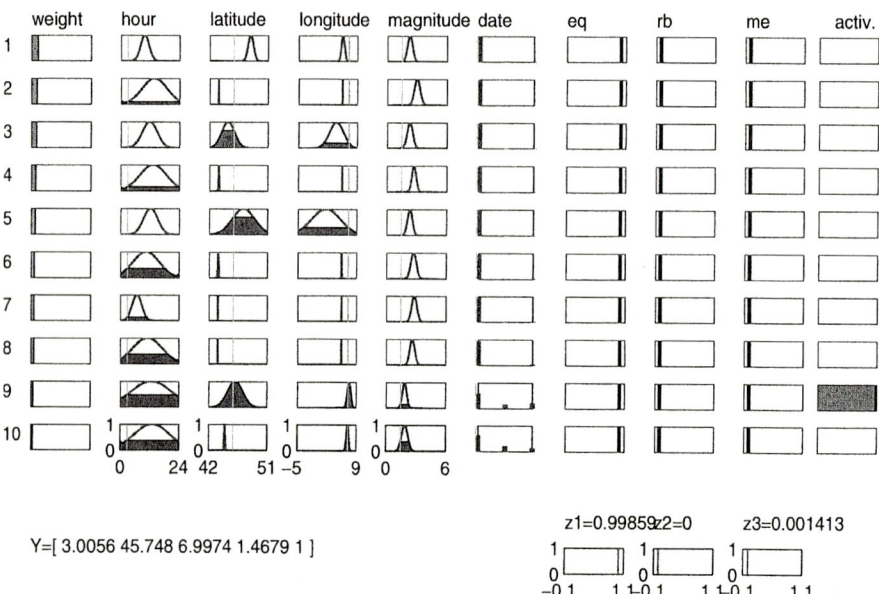

Fig. 5. Graphical interpretation of our fuzzy-inference system for a particular data point Y

So, this presentation, in agreement with expert's intuition, gives a confortable rule system interpretation for each new data point to classify. Graphically, experts understand the suitability of the event to each rule in the Fuzzy Inference System. For a given data input, the activation of each rule is visualized. Thus, it is possible for the expert to contest separately each rule. Validation of this interface by human experts is in progress.

5 Further Orientations

Next improvements will give, for each rule, stochastic output. To have a probabilistic interpretation, we want values ranging between 0 and 1 and which sum equals to 1. At this stage of research, output rule values are deterministic (the rule concludes to one class certainly). This development requires error criteria as least squares error criterion or maximum likelihood. We should also analyse more precisely cluster stability to reduce their number to the minimum by melting (but less rules means lower performance). We intend to use other statistical methods such as "Decision Trees" or "Support Vector Machines".

References

1. Bilmes, J. A Gentle Tutorial of the EM Algorithm and its applications to Parameter Estimation for Gaussian Mixture and Hidden Markov Models. International Computer Science Institute (1998)
2. Bishop, C.: Neural Networks for Pattern Recognition (1995). Oxford : Clarenton Press
3. Chiu, S., Fuzzy Model Identification Based on Cluster Estimation. Journal of Intelligent and Fuzzy Systems (1994),Volume 2, pp.267-278.
4. Dubois, D., Prade, H.: "A unifying view of comparison indices in a fuzzy set theoretic framework", in Fuzzy sets and possibility theory: recent developments, R. R. Yager (Ed.). (1994) Pergamon, NY.
5. Frayman, Y., Wang, L.P.: Data mining using dynamically constructed recurrent fuzzy neural networks. Proceedings 2nd Pacifric-Asia Conference on Knowledge Discovery and Data Mining, LNCS (1998), Volume 1394, pp.122-131.
6. Frayman, Y., Ting, K.M., Wang, L.: A Fuzzy Neural Network for Data Mining: Dealing with the Problem of Small Disjuncts. International Joint Conference on Neural Networks, IEEE (1999).
7. Gravot, F.: Rapport de stage : Etude de systmes automatiques de gnration de rgles floues . CEA-DAM/DASE/LDG.
8. Jang, J.-S. R., Sun, C.-T., Mizutani, E.: Neuro-Fuzzy and Soft Computing: A Computational Approach to Learning and Machine Intelligence (1997). Prentice Hall Upper Saddle River, NJ.
9. Jordan, M., Xu, L.: Convergence Results for the EM Approach to Mixtures of Experts Architectures (1993). Massachusetts Institute of Technology.
10. Martinetz, T.M., Berkovich, S.G., Schulten, K.J.: "Neural Gas" Network for Vector Quantization and its application to Time-Serie Prediction IEEE transactions on Systems, Man, and Cybernetics (1993), 3(1), pp. 28-44.
11. Sugeno, M., Kang, G.T.: Structure identification of fuzzy model. Fuzzy Sets and Systems (1988), 28, pp. 15-33.
12. Takagi, T., Sugeno, M.: Fuzzy identification of systems and its applications to modeling and control. IEEE transactions on Systems, Man, and Cybernetics (1985), 15, pp. 116-132.
13. Wai, R.-J., chen, P.-C.: Intelligent tracking control for robot manipulator including actuator dynamics via TSK-type fuzzy neural network. IEEE transactions on Fuzzy Systems (2004), Volume 12, pp. 552-560.
14. Wang, L.P., Frayman, Y.: A Dynamically-generated fuzzy neural network and its application to torsional vibration control of tandem cold rolling mill spindles. Engineering Applications of Artifical Intelligence (2003), Volume 15, pp. 541-550.
15. Zadeh L.A.: Outline of a new approach to the analysis of complex systems and decision processes. Journal of Intelligent and Fuzzy Systems (1973), Volume 2, pp.267-278.

Fuzzy Patterns in Multi-level of Satisfaction for MCDM Model Using Modified Smooth S-Curve MF

Pandian Vasant[1], A. Bhattacharya[2], and N.N. Barsoum[3]

[1] EEE Program Research Lecturer
Universiti Teknologi Petronas 31750 Tronoh, BSI, Perak DR Malaysia
pvasant@gmail.com
[2] Examiner of Patents & Designs, The Patent Office, Kolkata Nizam Palace,
2nd M.S.O. Building 5th, 6th & 7th Floor, Kolkata – 700 020, West Bengal, India
arijit_bhattacharya@rediffmail.com
[3] Associate Professor, School of Electrical & Electronic Engineering,
Curtin University of Technology, Sarawak, Malaysia
nader.b@curtin.edu.my

Abstract. Present research work relates to a methodology using modified smooth logistic membership function (MF) in finding out fuzzy patterns in multi-level of satisfaction (LOS) for Multiple Criteria Decision-Making (MCDM) problem. Flexibility of this MF in applying to real world problem has been validated through a detailed analysis. An example elucidating an MCDM model applied in an industrial engineering problem is considered to demonstrate the veracity of the proposed methodology. The key objective of this paper is to guide decision makers (DM) in finding out the best candidate-alternative with higher degree of satisfaction with lesser degree of vagueness under tripartite fuzzy environment. The approach presented here provides feedback to the decision maker, implementer and analyst.

1 Introduction

Existing methods to deal with fuzzy patterns of MCDM models are very cumbersome and sometimes not capable of solving many real-world problems. Present research work allows MCDM problems to take data in the forms of linguistic terms, fuzzy numbers and crisp numbers. Thus, the purpose of this paper is to propose a fuzzy methodology suitable in finding out fuzzy patterns in multi-LOS during selection of candidate-alternatives under conflicting-in-nature criteria environment.

Existing literatures on MCDM tackling fuzziness are as broad as diverse. Literatures contain several proposals on how to incorporate the inherent uncertainty as well as vagueness associated with the DM's knowledge into the model [1,2,20]. There has been a great deal of interest in the application of fuzzy sets to the representation of fuzziness and uncertainty in decision models [9,12,14,15,16,29]. Bellman and Zadeh [4] have shown fuzzy set theory's applicability to MCDM study. Yager and Basson [27]. Boucher and Gogus [6] examined certain characteristics of judgement elicitation instruments appropriate to fuzzy MCDM using a gamma function. A DM needs an MCDM assessment technique in regard of its fuzziness that can be easily used in practice [18]. By defining a DM's preference structure in fuzzy linear constraint (FLC) with soft inequality, one can operate the concerned fuzzy optimization model with S-curve MF to achieve the desired solution [25].

One form of logistic MF to overcome difficulties in using a linear MF in solving a fuzzy decision-making problem was proposed by Watada [26]. Carlsson and Korhonen [8] have illustrated the usefulness of a formulated MF. Their example was adopted to test and compare a nonlinear MF [17]. Such an attempt using the said validated non-linear MF and comparing the results was made by Vasant *et al* [24].

In the past, studies on decision-making problems were considered on the bipartite relationship of the DM and analyst [22]. This notion is now outdated. Now tripartite relationship is to be considered, as shown on Fig. 1, where the DM, the analyst and the implementer will interact in finding fuzzy satisfactory solution in any given fuzzy system. An implementer has to interact with DM to obtain an efficient and highly productive fuzzy solution with a certain degree of satisfaction. This fuzzy system will eventually be called as high productive fuzzy system.

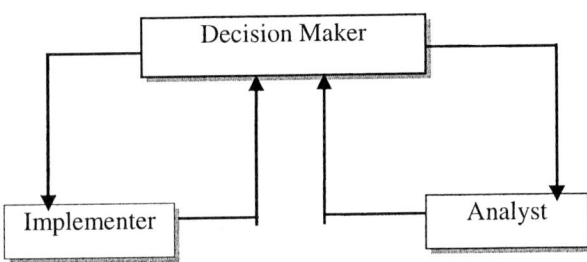

Fig. 1. The Block Diagram for the Tripartite Fuzzy System

MATLAB® fuzzy toolbox teaches 11 in-built membership functions and includes 0 and 1. In the present work 0 and 1 have been excluded and the smooth S-shaped membership function has been extensively modified accordingly.

As mentioned by Watada [26], trapezoidal MF will have some difficulties such as degeneration, i.e., some sort of deterioration of solution, while introducing in fuzzy problems. Logistic MF is found to be very useful in making decisions and implementation by DM and implementer [17,29]. This MF is used when the problems and its solutions are independent [13, 23, 28].

Vast literatures on the use of various types of MFs in finding out fuzzy patterns of MCDM methodologies force the authors to conclude with the following criticism:

(i) Data combining both ordinal and cardinal preferences are highly unreliable, unquantifiable, imperfect and sometimes contain non-obtainable information and partially ignorant facts.
(ii) Trapezoidal, triangular and even gamma functions MFs are not able to bring out fuzzy patterns in a fashion so as to delineate the degree of fuzziness inherent in MCDM model.
(iii) Designing a flexible, continuous and strictly monotonously non-increasing MF to achieve a lesser degree of fuzziness inherent in MCDM model is required.

(iv) Tripartite relationship among DM, analyst and implementer is essential, in conjunction to a more flexible MF design, to solve any industrial MCDM problem.

Among many diversified objectives of the present work, one objective is to find out fuzzy patterns of candidate-alternatives having different LOS in MCDM model. Relationships among the degree of fuzziness, LOS and the selection-indices of MCDM model guide DMs under tripartite fuzzy environment in obtaining their choice trading-off with a pre-determined allowable imprecision. Another objective of the present work is to provide a robust, quantified monitor of the level of satisfaction among DMs and to calibrate these levels of satisfaction against DMs expectations.

2 MCDM Methodology

The MCDM methodology proposed in Bhattacharya *et al* [5] deals with calculating priority weights of important attributes. Global priorities of various attributes rating are found by using AHP [1,19,20,21]. These global priority values have been used as the subjective factor measures (SFM) [5] in obtaining the LSI [5]. The candidate-alternatives are ranked according to the descending order of the LSI indices referred in equation (1).

$$LSI_i = [(\alpha \times SFM_i) + (1-\alpha) \times OFM_i] \tag{1}$$

$\forall i = 1, 2,...,n$, n being the number of criteria; α is the objective factor decision weight of the model and we call it as the level of satisfaction (LOS) of the DM.

In traditional AHP [19,21], if the inconsistency ratio (I.R.) [5,19] is greater than 10%, the values assigned to each element of the decision and pair-wise comparison matrices are said to be inconsistent. For I.R. < 10%, the level of inconsistency is acceptable [19,21]. As the very root of the judgment in constructing these matrices is the human being, some degree of inconsistency of the judgments of these matrices is deemed acceptable.

Saaty's AHP model consists of four different stages: (i) modelization, (ii) valuation, (iii) priorization and (iv) synthesis [11]. In the valuation stage of AHP fuzziness appears. Banuelas and Antony [2] raised four questions regarding the solution of classical AHP model. Among those four, three have been addressed in this work using a modified fuzzy *S*-curve membership function.

The algorithm of the MCDM model [5] in which fuzzy patterns with multi-LOS has been incorporated, considers computation of the following equations 2 to 6 in addition to equation 1. For details of the basic MCDM model readers are encouraged to refer to Bhattacharya *et al* [5].

$$Calculate \quad OFM_i = \left[OFC_i \bullet \Sigma(OFC_i^{-1})\right]^{-1} \tag{2}$$

$$Calculate \quad I.I. = \frac{\lambda_{max} - n}{n-1}, \tag{3}$$

$$Calculate \quad R.I. = \frac{[1.98 \bullet (n-2)]}{n} \tag{4}$$

$$Calculate \quad I.R. = \frac{I.I.}{R.I.} \tag{5}$$

$$SFM_i = (\text{Decision matrix PV}) \bullet PV \qquad (6)$$

$$\widetilde{LSI}_i \bigg|_{\alpha=\alpha_{SFM_i}} = LSI_L + \left(\frac{LSI_U - LSI_L}{\gamma}\right) \ln \frac{1}{C}\left(\frac{A}{\alpha_{LSI_i}} - 1\right) \qquad (7)$$

where, A = 1, C = 0.001001001, $0 < \alpha < 1$ and $3 \leq \gamma \leq 47$ and γ = a parameter to measure degree of fuzziness.

2.1 Designing a Smooth S-Curve MF

In the present work, we employ a logistic function for the non-linear MF as given by:

$$f(x) = \frac{B}{1+Ce^{\gamma x}} \qquad (8)$$

where B and C are scalar constants and γ, $0 < \gamma < \infty$ is a fuzzy parameter which measures the degree of vagueness, wherein $\gamma = 0$ indicates crisp. Fuzziness becomes highest when $\gamma \to \infty$. The logistic function, equation (8) is a monotonically non-increasing function [10]. A MF is flexible when it has vertical tangency, inflexion point and asymptotes. It can be shown that equation (8) has asymptotes at f(x) = 0 and f(x) = 1 at appropriate values of B and C [7].

It can also be shown that the said logistic function has a point of inflexion at x = x_0, such that $f''(x_0) = \infty$, $f''(x)$ being the second derivative of f(x) with respect to x. A MF of S-curve nature, in contrast to linear function, exhibits the real-life problem.

The generalized logistic MF is defined as:

$$f(x) = \begin{cases} 1 & x < x_L \\ \dfrac{B}{1+Ce^{\gamma x}} & x_L < x < x_U \\ 0 & x > x_U \end{cases} \qquad (9)$$

The S-curve MF is a particular case of the logistic function defined in equation (9). The said S-curve MF has got specific values of B, C and γ. To fit into the MCDM model [5] in order to sense its fuzzy patterns we modify and re-define the equation (9) as follows:

$$\mu(x) = \begin{cases} 1 & x < x^a \\ 0.999 & x = x^a \\ \dfrac{B}{1+Ce^{\gamma x}} & x^a < x < x^b \\ 0.001 & x = x^b \\ 0 & x > x^b \end{cases} \qquad (10)$$

In equation (10) the MF is redefined as $0.001 \leq \mu(x) \leq 0.999$. We rescale the x-axis as $x^a = 0$ and $x^b = 1$ in order to find the values of B, C and γ. The values of B, C and γ are obtained from equation (10). Since, B and γ depend on C, we require one more condition to get the values for B, C and γ. We assume that when $x_0 = \frac{x^a + x^b}{2}$, $\mu(x_0) = 0.5$. Since C has to be positive, computing equation (10) with the boundary conditions it is found that $C = 0.001001001$, $B = 1$ and $\gamma = 13.8135$.

Thus, it is evident from the preceding sections that the smooth S-curve MF can be more easily handled than other non-linear MF such as tangent hyperbola. The linear MF such as trapezoidal MF is an approximation from a logistic MF and based on many idealistic assumptions. These assumptions contradict the realistic real-world problems. Therefore, the S-curve MF is considered to have more suitability in sensing the degree of fuzziness in the fuzzy-uncertain judgemental values of a DM. The modified S-curve MF changes its shape according to the fuzzy judgemental values of a DM and therefore a DM finds it suitable in applying his/her strategy to MCDM problems using these judgemental values.

The proposed S-shaped MF is flexible due to its following characteristics:

1. $\mu(x)$ is continuous and strictly monotonously non-increasing;
2. $\mu(x)$ has lower and upper asymptotes at $\mu(x) = 0$ and $\mu(x) = 1$ as $x \to \infty$ and $x \to 0$, respectively;
3. $\mu(x)$ has inflection point at $x_0 = \frac{1}{\alpha}\ln(2 + \frac{1}{C})$ with $A = 1 + C$;

In order to benchmark the method proposed herein an illustrative example has been considered. The example has been adopted from Bhattacharya et al [5]. Their problem considers five different attributes, viz., work culture of the location, climatic condition, housing facility, transport availability, recreational facility and five different cost factor components, viz., cost of land, cost of raw material, cost of energy, cost of transportation and cost of labour. Five different sites for plant have been considered as alternatives. The approach combines both the ordinal and cardinal attributes.

3 Computing Level of Satisfaction, Degree of Fuzziness

SFM_i values, OFM_i values and LSI_i indices for five candidate-plant locations are as tabulated in Table 1. We confine our efforts assuming that differences in judgemental values are only 5%. Therefore, the upper bound and lower bound of SFM_i as well as LSI_i indices are to be computed within a range of 5% of the original value reported by Bhattacharya et al [5]. One can fuzzify the SFM_i values from the very beginning of the model by introducing modified S-curve MF in AHP and the corresponding fuzzification of LSI_i indices can also be carried out using their holistic approach.

By using equations (6) and (7) for modified S-curve MF a relationship among the LOS of the DM, the degree of vagueness and the LSI indices is found. The results are summarised in Table 2.

It may be noted that large value of γ implies less fuzziness. From Table 2 it is observed that the plot behaves as a monotonically increasing function. Fig. 2(a), (b) and (c) show three different plots depicting relation among the LOS and LSI indices for three different vagueness values. Fig. 2(a) illustrates LOS when $0.25 < LSI < 0.27$, Fig. 2(b) illustrates the same when $0.26 < LSI < 0.285$ and Fig. 2(c) depicts the results for $0.28 < LSI < 0.315$. It should always be noted that higher the fuzziness, γ, values, the lesser will be the degree of vagueness inherent in the decision. Therefore, it is understood that higher level of outcome of decision variable, LSI, for a particular LOS point, results in a lesser degree of fuzziness inherent in the said decision variable.

Table 1. SFM_i OFM_i and LSI_i indices

Candidate locations	SFM_i values	OFM_i values	LSI_i indices
P_1	0.329	0.2083	0.251
P_2	0.226	0.1112	0.153
P_3	0.189	0.2997	0.259
P_4	0.128	0.2307	0.194
P_5	0.126	0.1501	0.141

Table 2. α, γ and LSI

α	γ	LSI
0.1	3.0	0.2189
0.2	7.0	0.2312
0.3	11.0	0.2454
0.4	13.8	0.2591
0.5	17.0	0.2732
0.6	23.0	0.2882
0.7	29.0	0.3027
0.8	37.0	0.3183
0.9	41.0	0.3321
0.95	47.0	0.3465

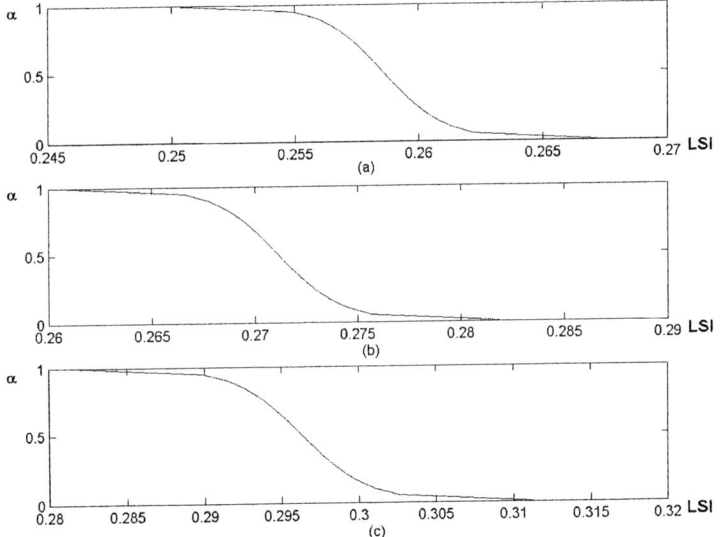

Fig. 2. Level of satisfaction and LSI indices for different fuzziness

Now, let us examine the fuzziness inherent in each plant site location decisions using equations of Table 3. Equations of Table 3 have been found using MATLAB® version 7.0. The results using these equations have been found encouraging and the corresponding results have been indicated in Figs. 3 to 8.

Fig. 8 elucidates a surface plot illustrating relationships among three parameters focusing the degree of fuzziness and fuzzy pattern of the MCDM model proposed by Bhattacharya *et al* [5]. This is a clear indication that the decision variables, as defined in equations (6) and (7), allows the MCDM model to achieve a higher LOS with a lesser degree of fuzziness.

Thus, the decision for selecting a candidate plant-location as seen from Figs. 3 to 7 is tabulated in Table 3. It is noticed from the present investigation that the present model eliciting the degree of fuzziness corroborates the MCDM model as in Bhattacharya et al [5].

Table 3. LSI equations for each plant

Candidate Plant locations	LSI	
	LSI_L	LSI_U
P_1	$LSI_L = 0.2083 + 0.1043 \bullet \alpha$	$LSI_U = 0.2083 + 0.1472 \bullet \alpha$
P_2	$LSI_L = 0.1112 + 0.1035 \bullet \alpha$	$LSI_U = 0.1112 + 0.1261 \bullet \alpha$
P_3	$LSI_L = 0.2997 - 0.1202 \bullet \alpha$	$LSI_U = 0.2997 - 0.1013 \bullet \alpha$
P_4	$LSI_L = 0.2307 - 0.1091 \bullet \alpha$	$LSI_U = 0.2307 - 0.0963 \bullet \alpha$
P_5	$LSI_L = 0.1501 - 0.0304 \bullet \alpha$	$LSI_U = 0.1501 - 0.0178 \bullet \alpha$

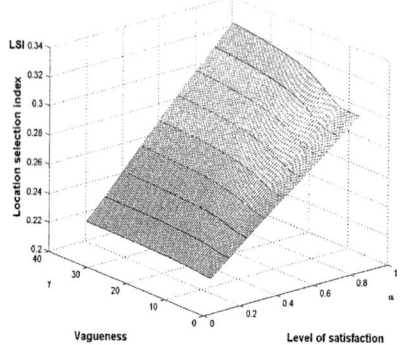

Fig. 3. Fuzzy pattern at multi-LOS for P_1 **Fig. 4.** Fuzzy pattern at multi-LOS for P_2

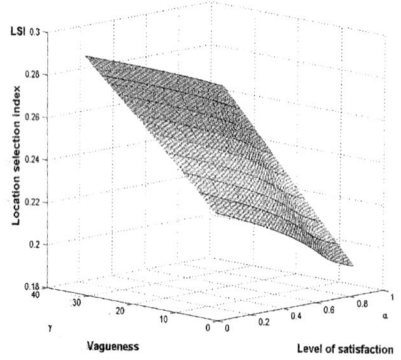
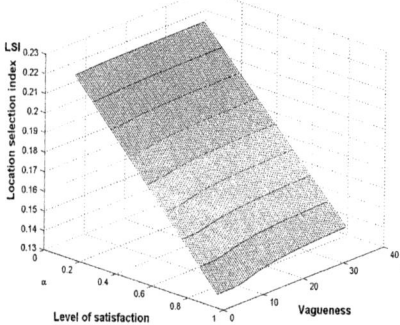

Fig. 5. Fuzzy pattern at multi-LOS for P_3 **Fig. 6.** Fuzzy pattern at multi-LOS for P_4

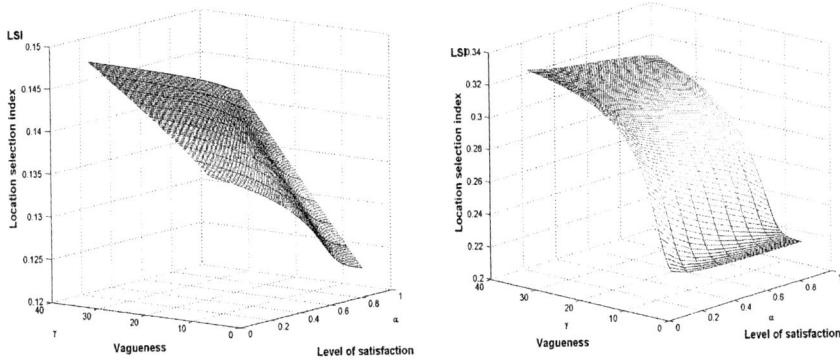

Fig. 7. Fuzzy pattern at multi-LOS for P_5 **Fig. 8.** Overall Relationship among parameters

Table 3. Final ranking based on the MCDM model at $\alpha = 0.36$ where lesser degree of fuzziness has been reflected in rank #1

Rank #	LSI_i	Plant locations
1	0.259	P_3
2	0.251	P_1
3	0.194	P_4
4	0.153	P_2
5	0.141	P_5

4 Discussion and Conclusion

Analyzing the results found from the methodology presented hereinbefore, following conclusions on the modified smooth S-curve MF in finding out fuzzy patterns with multi-LOS are drawn:

- The proposed smooth S-curve MF qualifies to be a logistic function and is said to be flexible;
- The flexibility of the proposed S-curve MF enables the analyst, in tripartite fuzzy system environment, to tackle the problem of fuzziness in various parameters of MCDM problem;
- The vagueness in the fuzzy parameters for the real-life problems decided by experts heuristically and experientially by determining the figures of MF; and
- The S-curve MF changes its shape according to the fuzzy parameter values thereby enabling DMs to apply their strategy to fuzzy problems using these parameters.

In unstructured real-world environment, there is always a chance of getting introduced fuzziness factors when an MCDM model deals with both cardinal and ordinal measures. The present methodology helps in determining the degree of fuzziness inherent in such cases. It is suggested that this methodology should be used in obtaining the degree of fuzziness and also fuzzy patterns satisfying the multi-LOS in such cases. It should always be remembered that this model is to be applied in a situation where decision alternatives are well inter-related and have both cardinal and ordinal criteria for selection. It is clear from the fuzzy patterns of the MCDM model that a fuzzy number with a low degree of fuzziness does not imply a low degree of non-specificity. Therefore, it is prudent that we become aware of the magnitude of the fuzziness introduced in a decision model when decision-making is dealt in presence of multiple and conflicting-in-nature criteria. The results delineated in various plots identify key strengths and weaknesses and prioritised areas for DM's choice.

Scope for future work is immense. There is a possibility to design self-organizing fuzzy system for the MCDM model in order to find a satisfactory solution. The risk inherent in such MCDM models can also be tackled suitably when other relevant criteria are considered in combination with the fuzzified approach. The methodology in obtaining the degree of fuzziness with multi-LOS presented here can also be extended in group-decision support systems (GDSS).

References

1. Arbel, A., Vargas, L.G.: The Analytic Hierarchy Process with Interval Judgements. Proceedings of the 9th International Conference of MCDM, Farfaix VA (1990)
2. Banuelas, R., Antony, J.: Modified Analytic Hierarchy Process to Incorporate Uncertainty and Managerial Aspects. Int. J. Prod. Res. 42 (18) (2004) 3851–3872
3. Bass, S.M., Kwakernaak, H.: Rating and Ranking of Multiple-Aspect Alternatives Using Fuzzy Sets. Automatica 13 (1) (1977) 47–58
4. Bellman, R.E., Zadeh, L.A.: Decision-making in a fuzzy environment. Management Science 17 (4) (1970) 141–164
5. Bhattacharya, A., Sarkar, B., Mukherjee, S.K.: A New Method for Plant Location Selection: A Holistic Approach. Int. J. Indus. Engg. – Theory, Applications and Practice 11 (4) (2004) 330–338
6. Boucher, T.O., Gogus, O.: Reliability, Validity and Imprecision in Fuzzy Multi-Criteria Decision Making. IEEE Trans. Sys., Man, and Cyber. – Part C: Applications and Reviews 32 (3) (2002) 1–15
7. Burzynski, D., Sanders, G.D.: Applied Calculus: Interpretation in Business, Life and Social Sciences. An International Thomson Publishing, USA (1995)
8. Carlsson, C., Korhonen, P.: A Parametric Approach to Fuzzy Linear Programming. Fuzzy Sets and Sys. 20 (1986) 17–30
9. Chen, S.J., Hwang, C.L.: Fuzzy Multiple Attribute Decision Making. Springer-Verlag, Berlin Heidelberg New York (1992)
10. Dick, T.P., Patton, C.M.: Calculus. An International Thomson Publishing, USA (1995)
11. Escobar, M.T., Moreno-Jimenez, J.M.: Reciprocal Distribution In The Analytic Hierarchy Process. European J. Oprnl. Res. 123 (2000) 154–174
12. Ghotb, F., Warren, L.: A Case Study Comparison of the Analytic Hierarchy Process and A Fuzzy Decision Methodology. Engineering Economist 40 (1995) 133–146
13. Goguen, J.A.: The Logic of Inexact Concepts. Syntheses 19 (1969) 325–373
14. Gogus, O., Boucher, T.O.: A Consistency Test for Rational Weights in Multi-Criteria Decision Analysis with Pairwise Comparisons. Fuzzy Sets and Sys. 86 (1997) 129–138
15. Lai, Y.J., Hwang, C.L.: Fuzzy Multi Objective Decision Making: Methods and Applications. Spinger-Verlag, Berlin Heidelberg New York (1994)
16. van Laarhoven, P.J.M., Pedrycz, W.: A Fuzzy Extension of Saaty's Priority Theory. Fuzzy Sets and Sys. 11 (1983) 229–241

17. Lootsma, F.A.: Fuzzy Logic for Planning and Decision Making. Kluwer Academic Publishers, London (1997)
18. Marcelloni, F., Aksit, M.: Leaving Inconsistency using Fuzzy Logic. Infor. Soft. Tech. 43 (2001) 725–741
19. Saaty, T.L.: The Analytical Hierarchy Process. McGraw-Hill, New Work (1980)
20. Saaty, T.L., Vargas, L.G.: Uncertainty and Rank Order in the Analytic Hierarchy Process. European J. Oprnl. Res. 32 (1987) 107–117
21. Saaty, T.L.: How to Make a Decision: the Analytic Hierarchy Process. European J. Oprnl. Res. 48 (1) (1990) 9–26
22. Tabucanon, M.T.: Multi Objective Programming for Industrial Engineers. In: Mathematical Programming for Industrial Engineers. Marcel Dekker, Inc., New York (1996) 487–542
23. Varela, L.R., Ribeiro, R.A.: Evaluation of Simulated Annealing to Solve Fuzzy Optimization Problems. J. Intelligent & Fuzzy Sys. 14 (2003) 59–71
24. Vasant, P., Nagarajan, R., Yaacob, S.: Fuzzy Linear Programming with Vague Objective Coefficients in an Uncertain Environment. J. Oprnl. Res. Society (Published Online) (25 August 2004) 1–7
25. Wang, H.F., Wu, K.Y.: Preference Approach to Fuzzy Linear Inequalities and Optimizations. Fuzzy Optmzn. Decision Making 4 (2005) 7–23
26. Watada, J.: Fuzzy Portfolio Selection and its Applications to Decision Making. Tatra Mountains Mathematics Publication 13 (1997) 219–248
27. Yager, R.R., Basson, D.: Decision Making with Fuzzy Sets. Decision Sciences 6 (3) (1975) 590–600
28. Zadeh, L.A.: The Concept of a Linguistic Variable and its Application to Approximate Reasoning, Part I, II, III. Information Sciences 8 (1975) 199–251, 301–357, 9 (1975) 43–80
29. Zimmermann, H.-J.: Fuzzy Sets, Decision Making and Expert Systems. Kluwer Academic Publishers, Boston (1987)
30. Zimmermann, H.-J.: Fuzzy Set Theory and its Applications. Kluwer Academic Publishers, Dordrecht (1991)

Author Index

Abraham, Ajith II-1067
Adam, Susanne I-662
Afzulpurkar, Nitin V. I-484
Ahn, Heejune II-1170
Amin, M. Ashraful I-484

Bae, Hyeon I-833
Baeg, Seung-Beom II-998
Baek, Jae-Yeon II-186
Bandara, G.E.M.D.C. II-215
Banerjee, Amit I-444
Barsoum, N.N. II-1294
Basir, Otman I-426
Batanov, Dentcho N. I-484
Batuwita, K.B.M.R. II-215
Benhabib, B. I-1217
Bhattacharya, A. II-1294
Bi, D. I-942, II-677
Bie, Rongfang II-1037
Bing, Huang I-1223
Bing, Zhou I-1151
Bloyet, Daniel I-189
Bourey, Jean-Pierre I-1025
Budiono I-1113
Byun, Doyoung I-1113
Byun, Yung-Hwan I-1081, I-1108

Cai, Hongbin I-1277
Cai, Lianhong II-600
Cai, Long-Zheng II-320
Cai, Zixing I-1217, II-921
Cansever, Galip I-981
Cao, Bing-yuan I-156, II-546
Cao, Chunling II-1022
Cao, Fei II-289
Cao, Wenliang II-339
Cao, Yijia I-79, I-882
Cao, Zhe I-285
Cao, Zhen-Fu II-596
Cao, Zhexin I-1287
Chai, Duckjin I-1175
Chai, Tianyou I-876, II-891
Chang, Chin-Chen II-551
Chang, Kuiyu II-1236

Chang, Wen-Kui II-911
Chao, Ruey-Ming I-1067
Chau, Rowena II-768
Che, Rucai I-910
Chen, Dewang II-1008
Chen, Fuzan II-420
Chen, Gang II-452
Chen, Gencai II-961
Chen, Guangsheng II-610
Chen, Guoqing I-721, II-614
Chen, Hanxiong I-584
Chen, Hongqi II-1136
Chen, Jian I-59
Chen, Jun I-969
Chen, Ling II-961
Chen, Shi-Jay I-694
Chen, Shuwei I-276
Chen, Shyi-Ming I-694
Chen, Wei II-742
Chen, Wenbin II-49
Chen, Xia I-130
Chen, Xiaoming II-778
Chen, Xiaoyun II-624
Chen, Xuerong I-672
Chen, Yan Qiu II-81, II-100
Chen, Yanmei II-275
Chen, Yanping I-189
Chen, Yen-Liang II-536
Chen, Yi II-624
Chen, Yi-Fei II-430
Chen, Yiqun II-494, II-710
Chen, Yuehui II-1067
Chen, Yun II-240
Chen, Zhe I-717
Cheng, Lishui I-505
Cheng, Wei II-408
Cheon, Seong-Pyo I-203, I-772
Cheong, Il-Ahn II-160
Chi, Chihong I-594
Chin, Kuo-Chih I-851
Cho, Dong-Sub II-561
Cho, Jinsung II-1166
Cho, Siu-Yeung II-1245
Cho, Tae-Ho II-998

Author Index

Cho, Yookun II-1154
Choi, Byung-Jae I-802
Choi, Doo-Hyun II-989
Choi, Heeyoung I-1175
Choi, Hyung-Il II-1061
Choi, Su-Il II-329
Choi, Young Chang I-1108
Chon, Tae-Soo II-186
Chu, Yayun II-230
Chun, Myung-Geun II-514, II-1132
Chun, Seok-Ju II-762
Chung, Chan-Soo II-731
Chung, Fu-lai I-1171
Chung, Henry II-677
Congfu, Xu I-1246
Constans, Jean-Marc I-189
Cornez, Laurence II-1281
Cronin, Mark T.D. II-31
Cui, Chaoyuan I-584
Cui, Wanan I-1242

Dai, Honghua II-39, II-368
Dai, Weiwei II-677
Dailey, Matthew N. I-484
Davé, Rajesh N. I-444
Deng, Hepu I-653
Deng, Ke II-362
Deng, Tingquan II-275
Deng, Yingna II-285
Deng, Zhi-Hong I-374
Dib, Marcos Vinícius Pinheiro I-1053
Ding, Jianhua II-1051
Ding, Mingli I-812
Ding, Xiying I-872, I-977
Ding, Zhan II-120
Ding, Zuquan II-869
Dong, Jinxiang II-255
Dong, Xiaoju II-1128
Dong, Yihong I-470
Dong, Zhupeng II-1128
Dou, Weibei I-189
Dou, Wen-Hua I-360
Du, Hao II-81
Du, Lei II-845, II-1184
Du, Weifeng I-1232
Du, Weiwei I-454, II-1
Duan, Hai-Xin II-774
Duan, Zhuohua II-921

Engin, Seref N. I-981
Eric, Castelli II-352
Esichaikul, Vatcharaporn I-484

Fan, Jing II-1103
Fan, Muhui II-677
Fan, Xian I-505
Fan, Xianli II-494
Fan, Xiaozhong I-571
Fan, Yushun I-26
Fan, Zhang II-865
Fan, Zhi-Ping I-130
Fang, Bin II-130
Fang, Yong II-931
Fei, Yu-lian I-609
Feng, Boqin I-580
Feng, Du II-398
Feng, Ming I-59
Feng, Zhikai I-1185
Feng, Zhilin II-255
Feng, Zhiquan II-412
Fu, Jia I-213
Fu, Tak-chung I-1171
Fu, Yuxi II-1128
Fung, Chun Che II-1226
Furuse, Kazutaka I-584

Galily, Mehdi I-900, II-976
Gang, Chen I-841
Gao, Jinwu I-304, I-321
Gao, Kai I-199, II-658
Gao, Shan II-362
Gao, Xin II-524
Gao, Yang II-698
Geng, Zhi II-362
Ghosh, Joydeep II-1236
Glass, David II-797
Goh, Ong Sing II-1226
Gong, Binsheng II-830, II-845
Gu, Wenxiang II-1118
Gu, Xingsheng I-1271
Gu, Yajun I-1287
Gu, Ying-Kui II-897
Gu, Zhimin II-110
Guo, Chonghui II-196, II-801
Guo, Chuangxin I-79, I-882
Guo, Gongde II-31, II-797
Guo, Jiankui II-1051
Guo, Jianyi I-571
Guo, Li-wei I-708

Author Index

Guo, Maozu II-861
Guo, Ping II-723
Guo, Qianjin I-743
Guo, Qingding I-872, I-977
Guo, Weiping I-792
Guo, Yaohuang I-312
Guo, Yecai I-122
Guo, Yi I-122
Guo, Zheng II-830
Gupta, Sudhir II-811

Han, Byung-Gil II-989
Han, Man-Wi II-186
Han, Qiang II-945
Hang, Xiaoshu II-39
Hao, Jingbo I-629
He, Bo II-723
He, Huacan I-31
He, Huiguang I-436
He, Lin II-1128
He, Liping II-503
He, Mingyi II-58
He, Pilian II-67
He, Ruichun I-312
He, Xiaoxian I-987
He, Xing-Jian II-727
He, Yanxiang I-865
Heo, Jin-Seok II-344
Heo, Junyoung II-1154
Higgins, Michael I-1256
Ho, Chin-Yuan II-536
Hong, Choong Seon II-1166
Hong, Dug Hun I-100
Hong, Gye Hang II-1071
Hong, Jiman II-1154, II-1158, II-1170
Hong, Kwang-Seok II-170
Hong, Won-Sin I-694
Hou, Beiping II-703
Hou, Yuexian II-67
Hu, Bo II-442
Hu, Dan II-378
Hu, Desheng II-475
Hu, Hong II-73
Hu, Huaqiang II-120
Hu, Kong-fa I-1192
Hu, Maolin I-148
Hu, Min II-742
Hu, Qinghua I-494, I-1261
Hu, Shi-qiang I-708
Hu, Wei-li I-69

Hu, Xuegang I-1309
Hu, Y.A. II-1174
Hu, Yi II-778
Hu, Yu-Shu II-1012
Hu, Yunfa II-624
Huang, Biao II-265
Huang, Chen I-26
Huang, Hailiang II-577
Huang, Hong-Zhong II-897
Huang, Jin I-59, II-945
Huang, Qian II-483
Huang, Rui II-58
Huang, Xiaochun II-21
Huang, Yanxin I-735
Huang, Yuan-sheng I-635
Huang, Zhiwei I-1063
Huawei, Guo II-398
Hui, Hong II-324
Huo, Hua I-580
Hwang, Buhyun I-1175
Hwang, Changha I-100
Hwang, Hoyon I-1092

Ikuta, Akira I-1161
Im, Younghee I-355
Inaoka, Hiroyuki I-263
Inoue, Kohei I-454, II-1

Jang, MinSeok II-249
Jang, Seok-Woo II-1061
Jeon, Kwon-Su I-1081
Jeong, Karpjoo I-1077, I-1092
Jeong, Ok-Ran II-561
Ji, Ruirui II-293
Ji, Xiaoyu I-304
Jia, Baozhu I-1011
Jia, Caiyan I-1197
Jia, Huibo I-514
Jia, Li-min I-69
Jia, Lifeng II-592
Jia, Limin I-89
Jia, Yuanhua I-118
Jian, Jiqi I-514
Jiang, Jianmin II-483
Jiang, Ping II-483
Jiang, Wei II-852, II-1184
Jiang, Y.C. II-1174
Jiang, Yunliang I-195
Jiao, Zhiping II-302
Jie, Shen I-1192

Jie, Wang I-837
Jin, Dongming I-1034, I-1044
Jin, Hanjun I-213
Jin, Hong I-861
Jin, Hui I-865
Jin, Shenyi I-1092
Jin, Weiwei I-1044
Jing, Zhong-liang I-708
Jing, Zhongliang I-672
Joo, Young Hoon I-406, I-416, I-886
Juang, Jih-Gau I-851
Jun, Byong-Hee II-1132
Jung, Chang-Gi II-989
Jung, In-Sung II-1079

Kajinami, Tomoki II-1208
Kamel, Abdelkader El I-1025
Kan, Li I-531
Kang, Bo-Yeong I-462, II-752
Kang, Dazhou I-232
Kang, Seonggoo II-1150
Kang, Yaohong I-388
Karaman, Özhan II-925
Kasabov, Nikola II-528
Keinduangjun, Jitimon II-1041
Kim, Bosoon I-1117
Kim, Byeong-Man II-752
Kim, Byung-Joo II-581
Kim, Dae-Won I-462
Kim, Deok-Eun II-994
Kim, Dong-Gyu II-170
Kim, Dong-kyoo II-205
Kim, Eun Yi I-1077
Kim, Eunkyo II-1162
Kim, Euntai I-179
Kim, Gye-Young II-1061
Kim, Ho J. II-811
Kim, Il Kon II-581
Kim, Jang-Hyun I-1015
Kim, Jee-In I-1117
Kim, Jee-in I-1077
Kim, Jeehoon II-186
Kim, Jin Y. II-329
Kim, Jong Hwa I-1141
Kim, Jonghwa I-1092
Kim, Joongheon II-1162
Kim, Jung Y. II-1158
Kim, Jung-Hyun II-170
Kim, Jungtae II-205
Kim, Kwang-Baek I-761

Kim, Kwangsik I-1092
Kim, Kyoungjung I-179
Kim, Min-Seok II-344
Kim, Min-Soo II-731
Kim, Minsoo II-160
Kim, Moon Hwan I-406
Kim, Myung Sook II-1093
Kim, Myung Won I-392
Kim, Sang-Jin I-1081
Kim, Soo-jeong I-1077
Kim, Soo-Young II-994
Kim, Sung-Ryul I-1137
Kim, Sungshin I-203, I-772, I-833
Kim, Weon-Goo II-249
Kim, Yejin I-833
Kim, Yong-Hyun I-1133
Kim, Youn-Tae I-203
Ko, Sung-Lim I-1133
Kong, Yong Hae II-1093
Koo, Hyun-jin I-1077
Kubota, Naoyuki I-1001
Kucukdemiral, Ibrahim B. I-981
Kumar, Kuldeep II-316
Kwak, Keun-Chang II-514
Kwun, Young Chel I-1

Lai, K.K. II-931
Lan, Jibin II-503
Le, Jia-Jin II-462
Lee, Changjin I-1113
Lee, Chin-Hui II-249
Lee, Choonhwa II-1162
Lee, Gunhee II-205
Lee, Ho Jae I-406
Lee, Inbok I-1137
Lee, Jae-Woo I-1081, I-1108
Lee, Jaewoo I-1092
Lee, Jang Hee II-1071
Lee, Ju-Hong II-762
Lee, Jung-Ju II-344
Lee, KwangHo II-752
Lee, Sang-Hyuk I-203
Lee, Sang-Won II-170
Lee, Sangjun II-1150
Lee, Sengtai II-186
Lee, Seok-Lyong II-762
Lee, Seoung Soo I-1141
Lee, Seungbae I-1127
Lee, Sukho II-1150
Lee, Vincent C.S. II-150

Author Index

Lee, Wonjun II-1162
Lei, Jingsheng I-388
Lei, Xusheng I-890
Leu, Fang-Yie II-911
Li, Baiheng II-1008
Li, Chang-Yun I-728
Li, Changyun II-1089
Li, Chao I-547
Li, Chuanxing II-845, II-852
Li, Gang II-368
Li, Gui I-267
Li, Guofei I-340
Li, HongXing II-378
Li, Hongyu II-49
Li, Jie II-1190
Li, Jin-tao II-689
Li, Jing II-830
Li, Luoqing II-130
Li, Ming I-619
Li, Minglu II-1027
Li, Minqiang II-420
Li, Ning II-698
Li, Peng II-1200
Li, Qing I-462, II-752
Li, Sheng-hong II-324
Li, Shu II-285
Li, Shutao II-610
Li, Stan Z. I-223
Li, Sujian II-648
Li, Xia II-830, II-836, II-845, II-852, II-869, II-1184, II-1190
Li, Xiaoli I-645
Li, Xiaolu I-837
Li, Xing II-774
Li, Xuening II-778
Li, Yanhong II-35
Li, Yanhui I-232
Li, Yinong I-822
Li, Yinzhen I-312
Li, Yu-Chiang II-551
Li, Zhijian I-1034
Li, Zhijun I-676
Lian, Yiqun II-718
Liang, Xiaobei I-140
Liao, Beishui II-1089
Liao, Lijun II-1089
Liao, Qin I-1063
Liao, Zhining II-797
Lim, Joon S. II-811
Lim, Ki Won I-1113

Lin, Bo-Shian I-851
Lin, Jie Tian Yao I-436
Lin, Zhonghua II-306
Ling, Chen I-1192
Ling, Guo I-1223
Ling, Jian II-718
Ling, Zheng I-822
Liu, Chun-Sheng II-897
Liu, Delin II-362
Liu, Diantong I-792
Liu, Fei I-969
Liu, Guoliang II-567
Liu, Haowen I-865
Liu, Jian-Wei II-462
Liu, Jing II-1031
Liu, Jun II-907
Liu, Junqiang I-580
Liu, Lanjuan II-35
Liu, Linzhong I-312
Liu, Peide I-523
Liu, Peng II-35
Liu, Ping I-728
Liu, Qihe I-1277
Liu, Qizhen II-475
Liu, Shi I-757
Liu, Wu II-774
Liu, Xiang-guan II-667
Liu, Xiao-dong I-42
Liu, Xiaodong I-53
Liu, Xiaoguang II-1031
Liu, Xin I-662
Liu, Xiyu I-523
Liu, Yan-Kui I-321
Liu, Yang I-822, II-388, II-941
Liu, Yong I-195
Liu, Yong-lin I-160
Liu, Yushu I-531
Liu, Yutian I-11
Lok, Tat-Ming II-727
Lu, Chunyan I-388
Lu, Jianjiang I-232
Lu, Ke I-436
Lu, Mingyu II-196, II-801
Lu, Naijiang II-638
Lu, Ruqian I-1197
Lu, Ruzhan II-778
Lu, Ya-dong I-922
Lu, Yinghua II-1118
Lu, Yuchang II-196, II-801
Luk, Robert I-1171

Luo, Bin II-140
Luo, Minxia I-31
Luo, Shi-hua II-667
Luo, Shuqian II-524
Luo, Ya II-723
Luo, Zongwei II-698
Lv, Sali II-830, II-836
Lyu, Michael R. II-727

Ma, and Wei I-122
Ma, Cheng I-514
Ma, Cun-bao II-466
Ma, Jing II-1027
Ma, Jixin II-140
Ma, Jun I-276
Ma, Liangyu II-339
Ma, Tianmin II-528
Ma, Weimin I-721
Ma, Yongkang II-289
Ma, Z.M. I-267
Mamun-Or-Rashid, Md. II-1166
Maoqing, Li I-1313
Mariño, Perfecto II-950
Martínez, Emilio II-950
Masuike, Hisako I-1161
Matsumura, Akio II-1208
McDonald, Mike I-782
Melo, Alba Cristina Magalhães de I-1053
Meng, Dan I-175
Meng, Xiangxu II-412
Meng, Zuqiang I-1217
Miao, Dong II-289
Miao, Zhinong I-950
Min, Fan I-1277
Modarres, Mohammad II-1012
Mok, Henry M.K. I-295
Mosavi, Morteza II-976
Muller, Jean-Denis II-1281
Mun, Jeong-Shik I-1137
Murata, Hiroshi II-1216
Murata, Tsuyoshi II-1204

Na, Eunyoung I-100
Na, Seung Y. II-329
Na, Yang I-1127
Nagar, Atulya II-821
Nam, Mi Young I-698
Neagu, Daniel II-31
Nemiroff, Robert J. II-634

Ng, Chak-man I-1171
Ng, Yiu-Kai I-557
Nguyen, Cong Phuong II-352
Ning, Yufu I-332
Niu, Ben I-987
Noh, Bong-Nam II-160
Noh, Jin Soo II-91

Oh, Hyukjun II-1170
Ohbo, Nobuo I-584
Ohta, Mitsuo I-1161
Omurlu, Vasfi E. I-981
Onoda, Takashi II-1216
Ouyang, Jian-quan II-689

Pan, De II-1051
Pan, Donghua I-537
Pan, Yingjie II-293
Park, Chang-Woo I-179
Park, Choon-sik II-205
Park, Daihee I-355
Park, Dong-Chul I-475
Park, Eung-ki II-205
Park, Geunyoung II-1154
Park, Hyejung I-100
Park, Jang-Hwan II-1132
Park, Jin Bae I-406, I-416, I-886
Park, Jin Han I-1
Park, Jin-Bae I-1015
Park, Jong Seo I-1
Park, Mignon I-179
Park, Minkyu II-1154
Park, Si Hyung I-1141
Park, Sungjun I-1117
Park, Young-Pil I-1015
Pastoriza, Vicente II-950
Pedrycz, Witold II-514
Pei, Xiaobing I-1297
Pei, Yunxia II-110
Pei, Zheng I-1232
Peng, Jin I-295
Peng, Ningsong I-370
Peng, Qin-ke I-1151
Peng, Qunsheng II-742
Peng, Wei II-120
Peng, Yonghong II-483
Pham, Thi Ngoc Yen II-352
Pham, Trung-Thanh I-1133
Piamsa-nga, Punpiti II-1041
Piao, Xuefeng II-1154

Ping, Zhao I-383
Poovorawan, Yong II-1041
Purushothaman, Sujita II-821

Qi, Gu I-1192
Qi, Jian-xun I-635
Qian, Jixin I-1266
Qian, Yuntao II-1107
Qiang, Wenyi II-567
Qiao, Yanjuan I-822
Qin, Jie I-360
Qin, Keyun I-1232
Qin, X.L. II-1174
Qin, Zhenxing I-402
Qing, Hu I-872, I-977
Qiu, Yuhui II-972
Qu, Shao-Cheng I-960
Qu, Wen-tao II-324

Rao, Shaoqi II-830, II-845, II-852, II-869, II-1184
Rao, Wei I-122
Ren, Bo II-1103
Ren, Guang I-1011
Ren, Jiangtao II-494
Ren, Yuan I-189
Rhee, Kang Hyeon II-91
Rhee, Phill Kyu I-698
Roudsari, Farzad Habibipour I-900
Ruan, Su I-189
Ryu, Joung Woo I-392
Ryu, Tae W. II-811

Sadri, Mohammadreza I-900
Samuelides, Manuel II-1281
Santamaría, Miguel II-950
Seo, Jae-Hyun II-160
Seo, Jung-taek II-205
Shamir, Lior II-634
Shan, Weiwei I-1044
Shaoqi, Rao II-865
Shen, I-Fan II-49
Shen, Jun-yi I-1151
Shi, Haoshan I-383
Shi, Jiachuan I-11
Shi, Lei II-110
Shi, Qin II-600
Shi, Wenzhong II-614
Shi, Yong-yu II-324
Shi, Yu I-20

Shi, Yuexiang I-1217
Shim, Charlie Y. II-1158
Shim, Jooyong I-100
Shin, Daejung II-329
Shin, Dongshin I-1127
Shin, Jeong-Hoon II-170
Shin, Sung Y. II-1158
Shin, Sung-Chul II-994
Shu, Tingting I-350
Shudong, Wang I-1256
Shunxiang, Wu I-1313
Sifeng, Liu I-1313
Sim, Alex T.H. II-150
Sim, Jeong Seop I-1102
Song, Chonghui I-876, II-891
Song, Jiyoung I-355
Song, Lirong I-571
Song, Qun II-528
Song, Yexin I-676
Song, Young-Chul II-989
Su, Jianbo I-890
Su, Jie II-945
Suh, Hae-Gook I-1092
Sulistijono, Indra Adji I-1001
Sun, Da-Zhi II-596
Sun, Dazhong I-1271
Sun, Jiaguang I-594
Sun, Jiantao II-196, II-801
Sun, Shiliang II-638
Sun, Xing-Min I-728
Sun, Yufang II-408
Sun, Zengqi I-910, II-567
Sun, Zhaocai II-230, II-240

Tai, Xiaoying I-470
Takama, Yasufumi II-1208
Tan, Yun-feng I-156
Tang, Bingyong I-140
Tang, Jianguo I-547
Tang, Shi-Wei I-374
Tang, Wansheng I-332, I-340
Tang, Weilong I-717
Tang, Xiang Long II-1190
Tang, Xiao-li I-1192
Tang, Yuan Yan II-130
Tanioka, Hiroki I-537
Tao, HuangFu II-723
Tawfik, Hissam II-821
Temelta, Hakan II-925
Teng, Xiaolong I-370

Thapa, Devinder   II-1079
Theera-Umpon, Nipon   II-787
Tse, Wai-Man   I-295

Um, Chang-Gun   II-989
Urahama, Kiichi   I-454, II-1

Vasant, Pandian   II-1294
Verma, Brijesh   II-316
Viswanathan, M.   I-1207

Wan Kim, Do   I-416, I-886
Wan, Wunan   II-941
Wang, Bingshu   II-339
Wang, Danli   I-861
Wang, Fei   II-1051
Wang, G.L.   I-942
Wang, Gang   II-1031
Wang, Geng   I-1266
Wang, Gi-Nam   II-1079
Wang, Haiyun   II-869
Wang, Hongan   I-861
Wang, Houfeng   II-11, II-648
Wang, Hui   I-332, I-861, II-797
Wang, J.   II-1174
Wang, Juan   I-89
Wang, Jue   I-223
Wang, Laisheng   II-980
Wang, Li-Hui   I-694
Wang, Lin   II-1142
Wang, Peng   I-232
Wang, Qi   I-812
Wang, Qian   I-676
Wang, Qianghu   II-836
Wang, Qiuju   II-852
Wang, Shi-lin   II-324
Wang, Shizhu   I-537
Wang, Shou-Yang   II-931
Wang, Shuliang   II-614
Wang, Shuqing   I-841, II-452
Wang, Tao   II-285
Wang, Tong   I-619
Wang, Xiangyang   I-370
Wang, Xiao-Feng   II-320
Wang, Xiaojing   II-941
Wang, Xiaorong   I-213
Wang, Xinya   I-1309
Wang, Xun   I-148
Wang, Yadong   II-869
Wang, Yan   I-735

Wang, Yanqiu   II-852
Wang, Yong-Ji   I-960
Wang, Yong-quan   I-160
Wang, Yongcheng   I-199, II-658
Wang, YuanZhen   I-1297
Wang, Zhe   II-592
Wang, Zhenzhen   II-836
Wang, Zhiqi   I-199, II-658
Wang, Zhongtuo   I-537
Wang, Zhongxing   II-503
Wang, Zi-cai   I-922
Wang, Ziqiang   II-388
Wei, Changhua   I-213
Wei, Lin   II-110
Wei, Zhi   II-677
Weigang, Li   I-1053
Wen, Fengxia   II-836
Wen, Weidong   I-865
Wenkang, Shi   II-398
Whang, Eun Ju   I-179
Whangbo, T.K.   I-1207
Wirtz, Kai W.   I-662
Wong, Jia-Jun   II-1245
Wu, Aimin   II-1255, II-1265, II-1276
Wu, Dianliang   II-577
Wu, Huaiyu   I-930
Wu, Jian-Ping   II-774
Wu, Jiangning   I-537, II-176
Wu, Jianping   I-118, II-1008
Wu, Ming   II-21
Wu, Wei-Dong   II-897
Wu, Wei-Zhi   I-167

Xia, Delin   II-21
Xia, Li   II-865
Xia, Yinglong   I-243
Xia, Z.Y.   II-1174
Xian-zhong, Zhou   I-1223
Xiang, Chen   I-603
Xiao, Gang   I-672
Xiao, Yegui   I-1161
Xie, Jiancang   II-1136
Xie, Jianying   II-907
Xie, Lijun   I-1277
Xie, Shengli   I-837
Xie, Wei   I-930
Xie, Xuehui   I-402
Xie, Yuan   II-907
Xin, He   I-1223
Xin, Jin   II-1037

Xin, Zhiyun I-594
Xing, James Z. II-265
Xing, Zong-yi I-69
Xiong, Feng-lan I-42
Xiong, Fenglan I-53
Xiong, Li-Rong II-1103
Xiu, Zhihong I-1011
Xu, Aidong I-743
Xu, Baochang I-717
Xu, Baowen I-232
Xu, De II-1255, II-1265, II-1276
Xu, Jia-dong II-466
Xu, Lin I-350
Xu, Lu II-306
Xu, Song I-336
Xu, Weijun I-148
Xu, Xian II-1128
Xu, Xiujuan II-592
Xu, Yang I-175, I-276, I-950
Xu, Yitian II-980
Xu, Zeshui I-110, I-684
Xu, Zhihao I-1034, I-1044
Xun, Wang I-609

Yamada, Seiji II-1200, II-1216
Yamamoto, Kenichi I-537
Yan, Li I-267
Yan, Puliu II-21
Yan, Shi-Jie II-911
Yan, Weizhen I-285
Yan, Zhang I-388
Yang, Ai-Min I-728
Yang, Aimin II-1089
Yang, Bo II-1067
Yang, Chenglei II-412
Yang, Chin-Wen I-1067
Yang, Chunyan II-891
Yang, Guangfei II-176
Yang, Guifang I-653
Yang, Hui II-891
yang, Hui I-876
Yang, Hyun-Seok I-1015
Yang, Ji Hye I-1113
Yang, Jie I-370, I-505
Yang, Ju II-1067
Yang, Mengfei I-910
Yang, Ming I-922
Yang, Shu-Qiang I-360
Yang, Shucheng I-876
Yang, Tao II-442

Yang, Wu II-723
Yang, Xiao-Ping I-1303
Yang, Xiaogang II-289
Yang, Xu II-1255, II-1265, II-1276
Yang, Y.K. I-1207
Yang, Yongjian II-1022
Yao, Liyue II-67
Yao, Min II-1107
Yao, Shengbao I-1242
Yao, Xin I-253, I-645
Yatabe, Shunsuke I-263
Ye, Bi-Cheng I-619
Ye, Bin I-79, I-882
Ye, Xiao-ling II-430
Ye, Xiuzi II-120
Ye, Yangdong I-89
Yeh, Chung-Hsing II-768
Yeh, Jieh-Shan II-551
Yerra, Rajiv I-557
Yi, Jianqiang I-792
Yi, Sangho II-1154
Yi, Zhang I-603
Yin, Jian I-59, II-494, II-710
Yin, Jianping I-629
Yin, Jianwei II-255
Yin, Minghao II-1118
Yin, Yilong II-230, II-240
Yong, Liu I-1246
Yoo, Seog-Hwan I-802
You, Xinge II-130
Yu, Daren I-494, I-1261
Yu, Dongmei I-872, I-977
Yu, Haibin I-743
Yu, Jinxia II-921
Yu, Sheng-Sheng II-320
Yu, Shiwen II-648
Yu, Shou-Jian II-462
Yu, Zhengtao I-571
Yuan, Hanning II-614
Yuan, Junpeng II-945
Yuan, Weiqi II-306
Yuan, Yueming II-1008
Yue, Chaoyuan I-1242
Yue, Wu I-603
Yun, Ling I-609
Yunhe, Pan I-1246

Zeng, Wenyi I-20
Zeng, Yurong II-1142
Zhan, Xiaosi II-230, II-240

Zhan, Yongqiang II-861
Zhang, Bao-wei I-619
Zhang, Boyun I-629
Zhang, Changshui I-243, I-253, II-638
Zhang, Chao II-466
Zhang, Chengqi I-402
Zhang, Chunkai K. II-73
Zhang, Dan II-130
Zhang, Dexian II-388
Zhang, Gexiang I-1287
Zhang, Guangmei II-836
Zhang, Hao II-324
Zhang, Hong II-972
Zhang, Honghua I-910
Zhang, Huaguang I-876
Zhang, Huaxiang I-523
Zhang, Hui II-255
Zhang, J. I-942
Zhang, Ji II-339
Zhang, Jianming I-841, II-452
Zhang, Jiashun I-336
Zhang, Jie II-845, II-852
Zhang, Jin II-624
Zhang, Jun II-677
Zhang, Junping I-223
Zhang, Ming I-374
Zhang, Peng I-1242
Zhang, Ping II-316
Zhang, Qingpu II-869
Zhang, Runtong II-880
Zhang, Shichao I-402
Zhang, Tao II-742
Zhang, Wei II-600, II-1184
Zhang, Weiguo I-148
Zhang, Wenyin I-547
Zhang, Xiao-hong I-160
Zhang, Xuefeng II-35
Zhang, Xueying II-302
Zhang, Yin II-120
Zhang, Ying-Chao II-430
Zhang, Yingchun II-567
Zhang, Yong I-69
Zhang, Yong-dong II-689
Zhang, Yongjin II-1136
Zhang, Yue II-727
Zhang, Yuhong I-1309
Zhang, Zaiqiang I-175
Zhang, Zhegen II-703
Zhang, Zhizhou II-1128

Zhao, Jianhua I-285, I-304
Zhao, Jieyu I-470
Zhao, Jizhong I-594
Zhao, Jun I-1266
Zhao, Keqin I-195
Zhao, Li II-95
Zhao, Long I-717
Zhao, Min II-667
Zhao, Ruiqing I-336, I-340, I-350
Zhao, Shu-Mei I-360
Zhao, Xiangyu I-950
Zhao, Xin I-930
Zhao, Yongqiang II-1027
Zhao, Zhefeng II-302
Zheng, Guo II-1184
Zheng, Jianhui II-1255, II-1265, II-1276
Zheng, Jie-Liang II-430
Zheng, Min II-600
Zheng, Pengjun I-782
Zheng, Qiuhua II-1107
Zheng, Quan I-1185
Zheng, Su-hua I-42
Zheng, Suhua I-53
Zheng, Wenming II-95
Zheng, Yalin I-243, I-253
Zhicheng, Liu II-865
Zhiyong, Yan I-1246
Zhou, Chang Yin II-100
Zhou, Chunguang I-735, II-592
Zhou, Jing-Li II-320
Zhou, Jun-hua I-635
Zhou, Kening II-703
Zhou, Wengang I-735
Zhou, Yuanfeng I-118
Zhu, Chengzhi I-79, I-882
Zhu, Daoli I-140
Zhu, Guohua II-830
Zhu, Hong II-285, II-293
Zhu, Hongwei I-426
Zhu, Jiaxian II-35
Zhu, Ming I-1185
Zhu, Wen II-703
Zhu, Xiaomin II-880
Zhu, Yunlong I-987
Zhuang, Ling II-39
Zhuang, Yueting I-195, II-718
Zou, Cairong II-95
Zou, Danping II-475
Zou, Xiaobing II-921

Lecture Notes in Artificial Intelligence (LNAI)

Vol. 3632: R. Nieuwenhuis (Ed.), Automated Deduction – CADE-20. XIII, 459 pages. 2005.

Vol. 3626: B. Ganter, G. Stumme, R. Wille (Eds.), Formal Concept Analysis. X, 349 pages. 2005.

Vol. 3625: S. Kramer, B. Pfahringer (Eds.), Inductive Logic Programming. XIII, 427 pages. 2005.

Vol. 3620: H. Muñoz-Avila, F. Ricci (Eds.), Case-Based Reasoning Research and Development. XV, 654 pages. 2005.

Vol. 3614: L. Wang, Y. Jin (Eds.), Fuzzy Systems and Knowledge Discovery, Part II. XLI, 1314 pages. 2005.

Vol. 3613: L. Wang, Y. Jin (Eds.), Fuzzy Systems and Knowledge Discovery, Part I. XLI, 1334 pages. 2005.

Vol. 3607: J.-D. Zucker, L. Saitta (Eds.), Abstraction, Reformulation and Approximation. XII, 376 pages. 2005.

Vol. 3596: F. Dau, M.-L. Mugnier, G. Stumme (Eds.), Conceptual Structures: Common Semantics for Sharing Knowledge. XI, 467 pages. 2005.

Vol. 3587: P. Perner, A. Imiya (Eds.), Machine Learning and Data Mining in Pattern Recognition. XVII, 695 pages. 2005.

Vol. 3584: X. Li, S. Wang, Z.Y. Dong (Eds.), Advanced Data Mining and Applications. XIX, 835 pages. 2005.

Vol. 3581: S. Miksch, J. Hunter, E. Keravnou (Eds.), Artificial Intelligence in Medicine. XVII, 547 pages. 2005.

Vol. 3577: R. Falcone, S. Barber, J. Sabater-Mir, M.P. Singh (Eds.), Trusting Agents for Trusting Electronic Societies. VIII, 235 pages. 2005.

Vol. 3575: S. Wermter, G. Palm, M. Elshaw (Eds.), Biomimetic Neural Learning for Intelligent Robots. IX, 383 pages. 2005.

Vol. 3571: L. Godo (Ed.), Symbolic and Quantitative Approaches to Reasoning with Uncertainty. XVI, 1028 pages. 2005.

Vol. 3559: P. Auer, R. Meir (Eds.), Learning Theory. XI, 692 pages. 2005.

Vol. 3558: V. Torra, Y. Narukawa, S. Miyamoto (Eds.), Modeling Decisions for Artificial Intelligence. XII, 470 pages. 2005.

Vol. 3554: A. Dey, B. Kokinov, D. Leake, R. Turner (Eds.), Modeling and Using Context. XIV, 572 pages. 2005.

Vol. 3539: K. Morik, J.-F. Boulicaut, A. Siebes (Eds.), Local Pattern Detection. XI, 233 pages. 2005.

Vol. 3538: L. Ardissono, P. Brna, A. Mitrovic (Eds.), User Modeling 2005. XVI, 533 pages. 2005.

Vol. 3533: M. Ali, F. Esposito (Eds.), Innovations in Applied Artificial Intelligence. XX, 858 pages. 2005.

Vol. 3528: P.S. Szczepaniak, J. Kacprzyk, A. Niewiadomski (Eds.), Advances in Web Intelligence. XVII, 513 pages. 2005.

Vol. 3518: T.B. Ho, D. Cheung, H. Liu (Eds.), Advances in Knowledge Discovery and Data Mining. XXI, 864 pages. 2005.

Vol. 3508: P. Bresciani, P. Giorgini, B. Henderson-Sellers, G. Low, M. Winikoff (Eds.), Agent-Oriented Information Systems II. X, 227 pages. 2005.

Vol. 3505: V. Gorodetsky, J. Liu, V. A. Skormin (Eds.), Autonomous Intelligent Systems: Agents and Data Mining. XIII, 303 pages. 2005.

Vol. 3501: B. Kégl, G. Lapalme (Eds.), Advances in Artificial Intelligence. XV, 458 pages. 2005.

Vol. 3492: P. Blache, E. Stabler, J. Busquets, R. Moot (Eds.), Logical Aspects of Computational Linguistics. X, 363 pages. 2005.

Vol. 3488: M.-S. Hacid, N.V. Murray, Z.W. Raś, S. Tsumoto (Eds.), Foundations of Intelligent Systems. XIII, 700 pages. 2005.

Vol. 3487: J. Leite, P. Torroni (Eds.), Computational Logic in Multi-Agent Systems. XII, 281 pages. 2005.

Vol. 3476: J. Leite, A. Omicini, P. Torroni, P. Yolum (Eds.), Declarative Agent Languages and Technologies II. XII, 289 pages. 2005.

Vol. 3464: S.A. Brueckner, G.D.M. Serugendo, A. Karageorgos, R. Nagpal (Eds.), Engineering Self-Organising Systems. XIII, 299 pages. 2005.

Vol. 3452: F. Baader, A. Voronkov (Eds.), Logic for Programming, Artificial Intelligence, and Reasoning. XI, 562 pages. 2005.

Vol. 3451: M.-P. Gleizes, A. Omicini, F. Zambonelli (Eds.), Engineering Societies in the Agents World V. XIII, 349 pages. 2005.

Vol. 3446: T. Ishida, L. Gasser, H. Nakashima (Eds.), Massively Multi-Agent Systems I. XI, 349 pages. 2005.

Vol. 3445: G. Chollet, A. Esposito, M. Faundez-Zanuy, M. Marinaro (Eds.), Nonlinear Speech Modeling and Applications. XIII, 433 pages. 2005.

Vol. 3438: H. Christiansen, P.R. Skadhauge, J. Villadsen (Eds.), Constraint Solving and Language Processing. VIII, 205 pages. 2005.

Vol. 3430: S. Tsumoto, T. Yamaguchi, M. Numao, H. Motoda (Eds.), Active Mining. XII, 349 pages. 2005.

Vol. 3419: B. Faltings, A. Petcu, F. Fages, F. Rossi (Eds.), Constraint Satisfaction and Constraint Logic Programming. X, 217 pages. 2005.

Vol. 3416: M. Böhlen, J. Gamper, W. Polasek, M.A. Wimmer (Eds.), E-Government: Towards Electronic Democracy. XIII, 311 pages. 2005.

Vol. 3415: P. Davidsson, B. Logan, K. Takadama (Eds.), Multi-Agent and Multi-Agent-Based Simulation. X, 265 pages. 2005.

Vol. 3403: B. Ganter, R. Godin (Eds.), Formal Concept Analysis. XI, 419 pages. 2005.

Vol. 3398: D.-K. Baik (Ed.), Systems Modeling and Simulation: Theory and Applications. XIV, 733 pages. 2005.

Vol. 3397: T.G. Kim (Ed.), Artificial Intelligence and Simulation. XV, 711 pages. 2005.

Vol. 3396: R.M. van Eijk, M.-P. Huget, F. Dignum (Eds.), Agent Communication. X, 261 pages. 2005.

Vol. 3394: D. Kudenko, D. Kazakov, E. Alonso (Eds.), Adaptive Agents and Multi-Agent Systems II. VIII, 313 pages. 2005.

Vol. 3392: D. Seipel, M. Hanus, U. Geske, O. Bartenstein (Eds.), Applications of Declarative Programming and Knowledge Management. X, 309 pages. 2005.

Vol. 3374: D. Weyns, H. V.D. Parunak, F. Michel (Eds.), Environments for Multi-Agent Systems. X, 279 pages. 2005.

Vol. 3371: M.W. Barley, N. Kasabov (Eds.), Intelligent Agents and Multi-Agent Systems. X, 329 pages. 2005.

Vol. 3369: V. R. Benjamins, P. Casanovas, J. Breuker, A. Gangemi (Eds.), Law and the Semantic Web. XII, 249 pages. 2005.

Vol. 3366: I. Rahwan, P. Moraitis, C. Reed (Eds.), Argumentation in Multi-Agent Systems. XII, 263 pages. 2005.

Vol. 3359: G. Grieser, Y. Tanaka (Eds.), Intuitive Human Interfaces for Organizing and Accessing Intellectual Assets. XIV, 257 pages. 2005.

Vol. 3346: R.H. Bordini, M. Dastani, J. Dix, A.E.F. Seghrouchni (Eds.), Programming Multi-Agent Systems. XIV, 249 pages. 2005.

Vol. 3345: Y. Cai (Ed.), Ambient Intelligence for Scientific Discovery. XII, 311 pages. 2005.

Vol. 3343: C. Freksa, M. Knauff, B. Krieg-Brückner, B. Nebel, T. Barkowsky (Eds.), Spatial Cognition IV. XIII, 519 pages. 2005.

Vol. 3339: G.I. Webb, X. Yu (Eds.), AI 2004: Advances in Artificial Intelligence. XXII, 1272 pages. 2004.

Vol. 3336: D. Karagiannis, U. Reimer (Eds.), Practical Aspects of Knowledge Management. X, 523 pages. 2004.

Vol. 3327: Y. Shi, W. Xu, Z. Chen (Eds.), Data Mining and Knowledge Management. XIII, 263 pages. 2005.

Vol. 3315: C. Lemaître, C.A. Reyes, J.A. González (Eds.), Advances in Artificial Intelligence – IBERAMIA 2004. XX, 987 pages. 2004.

Vol. 3303: J.A. López, E. Benfenati, W. Dubitzky (Eds.), Knowledge Exploration in Life Science Informatics. X, 249 pages. 2004.

Vol. 3301: G. Kern-Isberner, W. Rödder, F. Kulmann (Eds.), Conditionals, Information, and Inference. XII, 219 pages. 2005.

Vol. 3276: D. Nardi, M. Riedmiller, C. Sammut, J. Santos-Victor (Eds.), RoboCup 2004: Robot Soccer World Cup VIII. XVIII, 678 pages. 2005.

Vol. 3275: P. Perner (Ed.), Advances in Data Mining. VIII, 173 pages. 2004.

Vol. 3265: R.E. Frederking, K.B. Taylor (Eds.), Machine Translation: From Real Users to Research. XI, 392 pages. 2004.

Vol. 3264: G. Paliouras, Y. Sakakibara (Eds.), Grammatical Inference: Algorithms and Applications. XI, 291 pages. 2004.

Vol. 3259: J. Dix, J. Leite (Eds.), Computational Logic in Multi-Agent Systems. XII, 251 pages. 2004.

Vol. 3257: E. Motta, N.R. Shadbolt, A. Stutt, N. Gibbins (Eds.), Engineering Knowledge in the Age of the Semantic Web. XVII, 517 pages. 2004.

Vol. 3249: B. Buchberger, J.A. Campbell (Eds.), Artificial Intelligence and Symbolic Computation. X, 285 pages. 2004.

Vol. 3248: K.-Y. Su, J. Tsujii, J.-H. Lee, O.Y. Kwong (Eds.), Natural Language Processing – IJCNLP 2004. XVIII, 817 pages. 2005.

Vol. 3245: E. Suzuki, S. Arikawa (Eds.), Discovery Science. XIV, 430 pages. 2004.

Vol. 3244: S. Ben-David, J. Case, A. Maruoka (Eds.), Algorithmic Learning Theory. XIV, 505 pages. 2004.

Vol. 3238: S. Biundo, T. Frühwirth, G. Palm (Eds.), KI 2004: Advances in Artificial Intelligence. XI, 467 pages. 2004.

Vol. 3230: J.L. Vicedo, P. Martínez-Barco, R. Muñoz, M. Saiz Noeda (Eds.), Advances in Natural Language Processing. XII, 488 pages. 2004.

Vol. 3229: J.J. Alferes, J. Leite (Eds.), Logics in Artificial Intelligence. XIV, 744 pages. 2004.

Vol. 3228: M.G. Hinchey, J.L. Rash, W.F. Truszkowski, C.A. Rouff (Eds.), Formal Approaches to Agent-Based Systems. VIII, 290 pages. 2004.

Vol. 3215: M.G.. Negoita, R.J. Howlett, L.C. Jain (Eds.), Knowledge-Based Intelligent Information and Engineering Systems, Part III. LVII, 906 pages. 2004.

Vol. 3214: M.G.. Negoita, R.J. Howlett, L.C. Jain (Eds.), Knowledge-Based Intelligent Information and Engineering Systems, Part II. LVIII, 1302 pages. 2004.

Vol. 3213: M.G.. Negoita, R.J. Howlett, L.C. Jain (Eds.), Knowledge-Based Intelligent Information and Engineering Systems, Part I. LVIII, 1280 pages. 2004.

Vol. 3209: B. Berendt, A. Hotho, D. Mladenic, M. van Someren, M. Spiliopoulou, G. Stumme (Eds.), Web Mining: From Web to Semantic Web. IX, 201 pages. 2004.

Vol. 3206: P. Sojka, I. Kopecek, K. Pala (Eds.), Text, Speech and Dialogue. XIII, 667 pages. 2004.

Vol. 3202: J.-F. Boulicaut, F. Esposito, F. Giannotti, D. Pedreschi (Eds.), Knowledge Discovery in Databases: PKDD 2004. XIX, 560 pages. 2004.

Vol. 3201: J.-F. Boulicaut, F. Esposito, F. Giannotti, D. Pedreschi (Eds.), Machine Learning: ECML 2004. XVIII, 580 pages. 2004.

Vol. 3194: R. Camacho, R. King, A. Srinivasan (Eds.), Inductive Logic Programming. XI, 361 pages. 2004.

Vol. 3192: C. Bussler, D. Fensel (Eds.), Artificial Intelligence: Methodology, Systems, and Applications. XIII, 522 pages. 2004.